Kalman Filtering:
Theory and Application

OTHER IEEE PRESS BOOKS

Spectrum Management and Engineering, *Edited by Frederick Matos*
Digital VLSI Systems, *Edited by Mohamed T. Elmasry*
Introduction to Magnetic Recording, *Edited by Robert M. White*
Insights Into Personal Computers, *Edited by A. Gupta and H. D. Toong*
Television Technology Today, *Edited by T. S. Rzeszewski*
The Space Station: An Idea Whose Time Has Come, *By T. R. Simpson*
Marketing Technical Ideas and Products Successfully! *Edited by L. K. Moore and D. L. Plung*
The Making of a Profession: A Century of Electrical Engineering in America, *By A. M. McMahon*
Power Transistors: Device Design and Applications, *Edited by B. J. Baliga and D. Y. Chen*
VLSI: Technology Design, *Edited by O. G. Folberth and W. D. Grobman*
General and Industrial Management, *By H. Fayol; revised by I. Gray*
A Century of Honors, *An IEEE Centennial Directory*
MOS Switched-Capacitor Filters: Analysis and Design, *Edited by G. S. Moschytz*
Distributed Computing: Concepts and Implementations, *Edited by P. L. McEntire, J. G. O'Reilly, and R. E. Larson*
Engineers and Electrons, *By J. D. Ryder and D. G. Fink*
Land-Mobile Communications Engineering, *Edited by D. Bodson, G. F. McClure, and S. R. McConoughey*
Frequency Stability: Fundamentals and Measurement, *Edited by V. F. Kroupa*
Electronic Displays, *Edited by H. I. Refioglu*
Spread-Spectrum Communications, *Edited by C. E. Cook, F. W. Ellersick, L. B. Milstein, and D. L. Schilling*
Color Television, *Edited by T. Rzeszewski*
Advanced Microprocessors, *Edited by A. Gupta and H. D. Toong*
Biological Effects of Electromagnetic Radiation, *Edited by J. M. Osepchuk*
Engineering Contributions to Biophysical Electrocardiography, *Edited by T. C. Pilkington and R. Plonsey*
The World of Large Scale Systems, *Edited by J. D. Palmer and R. Saeks*
Electronic Switching: Digital Central Systems of the World, *Edited by A. E. Joel, Jr.*
A Guide for Writing Better Technical Papers, *Edited by C. Harkins and D. L. Plung*
Low-Noise Microwave Transistors and Amplifiers, *Edited by H. Fukui*
Digital MOS Integrated Circuits, *Edited by M. I. Elmasry*
Geometric Theroy of Diffraction, *Edited by R. C. Hansen*
Modern Active Filter Design, *Edited by R. Schaumann, M. A. Soderstrand, and K. B. Laker*
Adjustable Speed AC Drive Systems, *Edited by B. K. Bose*
Optical Fiber Technology, II, *Edited by C. K. Kao*
Protective Relaying for Power Systems, *Edited by S. H. Horowitz*
Analog MOS Integrated Circuits, *Edited by P. R. Gray, D. A. Hodges, and R. W. Broderson*
Interference Analysis of Communication Systems, *Edited by P. Stavroulakis*
Integrated Injection Logic, *Edited by J. E. Smith*
Sensory Aids for the Hearing Impaired, *Edited by H. Levitt, J. M. Pickett, and R. A. Houde*
Data Conversion Integrated Circuits, *Edited by D. J. Dooley*
Semiconductor Injection Lasers, *Edited by J. K. Butler*
Satellite Communications, *Edited by H. L. Van Trees*
Frequency-Response Methods in Control Systems, *Edited by A. G. J. MacFarlane*
Programs for Digital Signal Processing, *Edited by the Digital Signal Processing Committee, IEEE*
Automatic Speech & Speaker Recognition, *Edited by N. R. Dixon and T. B. Martin*
Speech Analysis, *Edited by R. W. Schafer and J. D. Markel*
The Engineer in Transition to Management, *By I. Gray*
Multidimensional Systems: Theory & Applications, *Edited by N. K. Bose*
Analog Integrated Circuits, *Edited by A. B. Grebene*
Integrated-Circuit Operational Amplifiers, *Edited by R. G. Meyer*
Modern Spectrum Analysis, *Edited by D. G. Childers*
Digital Image Processing for Remote Sensing, *Edited by R. Bernstein*
Reflector Antennas, *Edited by W. Love*
Phase-Locked Loops & Their Application, *Edited by W. C. Lindsey and M. K. Simon*
Digital Signal Computers and Processors, *Edited by A. C. Salazar*
Systems Engineering: Methodology and Applications, *Edited by A. P. Sage*
Modern Crystal and Mechanical Filters, *Edited by D. F. Sheahan and R. A. Johnson*
Electrical Noise: Fundamentals and Sources, *Edited by M. S. Gupta*
Computer Methods in Image Analysis, *Edited by J. K. Aggarwal, R. O. Duda, and A. Rosenfeld*

Kalman Filtering: Theory and Application

Edited by

Harold W. Sorenson

Professor of Engineering Sciences
Department of Applied Mechanics and Engineering Sciences
University of California, San Diego

A volume in the IEEE PRESS Selected Reprint Series, prepared under the sponsorship of the IEEE Control Systems Society.

IEEE PRESS

The Institute of Electrical and Electronics Engineers, Inc., New York

PRINTED IN THE UNITED STATES OF AMERICA

IEEE Order Number: PP03160

Library of Congress Cataloging in Publication Data
Main entry under title:

Kalman filtering.

(IEEE Press selected reprint series)
Pt. 1 consists of papers published in the 1960s; pt. 2 originally appeared as a special issue for Mar. 1983 of IEEE transactions on automatic control.
Bibliography: p.
Includes indexes.
1. Kalman filtering. I. Sorenson, H. W. (Harold Wayne), 1936- .
QA402.3.K28 1985 003 85-14253
ISBN 0-87942-191-6
ISBN 0-7803-0421-7 {pbk}

Contents

Introduction 1

Part I: Theoretical Beginnings 5

Section I-A: Historical Survey 6

Least-Squares Estimation: From Gauss to Kalman, *H. W. Sorenson* (*IEEE Spectrum*, July 1970) 7

Section I-B: Theoretical Foundations 13

A New Approach to Linear Filtering and Prediction Problems, *R. E. Kalman* (*Transactions of the ASME, Journal of Basic Engineering*, March 1960) 16

First Order Error Propagation in a Stagewise Smoothing Procedure for Satellite Observations, *P. Swerling* (*The Journal of the Astronautical Sciences*, Autumn 1959) 27

New Results in Linear Filtering and Prediction Theory, *R. E. Kalman and R. S. Bucy* (*Transactions of the ASME, Journal of Basic Engineering*, March 1961) 34

A Bayesian Approach to Problems in Stochastic Estimation and Control, *Y. C. Ho and R. C. K. Lee* (*IEEE Transactions on Automatic Control*, October 1964) 48

An Innovations Approach to Least-Squares Estimation—Part I: Linear Filtering in Additive White Noise, *T. Kailath* (*IEEE Transactions on Automatic Control*, December 1968) 55

Section I-C: General Application Considerations 65

A Statistical Optimizing Navigation Procedure for Space Flight, *R. H. Battin* (*ARS Journal*, November 1962) 67

Linear Regression Applied to System Identification for Adaptive Control Systems, *R. E. Kopp and R. J. Orford* (*AIAA Journal*, October 1963) 83

Kalman Filtering Techniques, *H. W. Sorenson* (*Advances in Control Systems Theory and Applications*, Academic Press, 1966) 90

Section I-D: Model Errors and Divergence 127

On *A Priori* Statistics in Minimum-Variance Estimation Problems, *T. T. Soong* (*Transactions of the ASME, Journal of Basic Engineering*, March 1965) 129

The Effect of Erroneous Models on the Kalman Filter Response, *H. Heffes* (*IEEE Transactions on Automatic Control*, July 1966) 133

On the A Priori Information in Sequential Estimation Problems, *T. Nishimura* (*IEEE Transactions on Automatic Control*, April 1966) 136

Divergence in the Kalman Filter, *F. H. Schlee, C. J. Standish, and N. F. Toda* (*AIAA Journal*, June 1967) 144

Divergence of the Kalman Filter, *R. J. Fitzgerald* (*IEEE Transactions on Automatic Control*, December 1971) 151

Section I-E: Divergence Control and Adaptive Filtering 163

Adaptive Filtering, *A. H. Jazwinski* (*Automatica*, July 1969) 164

On the Identification of Variances and Adaptive Kalman Filtering, *R. K. Mehra* (*IEEE Transactions on Automatic Control*, April 1970) 175

Recursive Fading Memory Filtering, *H. W. Sorenson and J. E. Sacks* (*Information Sciences*, January 1971) 185

Section I-F: Computational Considerations 195

Estimation Using Sampled Data Containing Sequentially Correlated Noise, *A. E. Bryson, Jr., and L. J. Henrikson* (*Journal of Spacecraft and Rockets*, June 1968) 197

Treatment of Bias in Recursive Filtering, *B. Friedland* (*IEEE Transactions on Automatic Control*, August 1969) 201

Discrete Square Root Filtering: A Survey of Current Techniques, *P. G. Kaminski, A. E. Bryson, Jr., and S. F. Schmidt* (*IEEE Transactions on Automatic Control*, December 1971) 210

Computational Requirements for a Discrete Kalman Filter, *J. M. Mendel* (*IEEE Transactions on Automatic Control*, December 1971) 219

Section I-G: Smoothing 230

A Survey of Data Smoothing for Linear and Nonlinear Dynamic Systems, *J. S. Meditch* (*Automatica*, March 1973) 231

Maximum Likelihood Estimates of Linear Dynamic Systems, *H. E. Rauch, F. Tung, and C. T. Striebel* (*AIAA Journal*, August 1965) 243

The Optimum Linear Smoother as a Combination of Two Optimum Linear Filters, *D. C. Fraser and J. E. Potter* (*IEEE Transactions on Automatic Control*, August 1969) 249
An Innovations Approach to Least-Squares Estimation—Part II: Linear Smoothing in Additive White Noise, *T. Kailath and P. Frost* (*IEEE Transactions on Automatic Control*, December 1968) 252

Part II: Application Papers 259
Editorial—On Applications of Kalman Filtering, *H. W. Sorenson* (*IEEE Transactions on Automatic Control*, March 1983) 261
Voyager Orbit Determination at Jupiter, *J. K. Campbell, S. P. Synnott, and G. J. Bierman* (*IEEE Transactions on Automatic Control*, March 1983) 263
Decoupled Kalman Filters for Phased Array Radar Tracking, *F. E. Daum and R. J. Fitzgerald* (*IEEE Transactions on Automatic Control*, March 1983) 276
Utilization of Modified Polar Coordinates for Bearings-Only Tracking, *V. J. Aidala and S. E. Hammel* (*IEEE Transactions on Automatic Control*, March 1983) 291
Estimation and Prediction for Maneuvering Target Trajectories, *R. F. Berg* (*IEEE Transactions on Automatic Control*, March 1983) 303
Multiconfiguration Kalman Filter Design for High-Performance GPS Navigation, *M. H. Kao and D. H. Eller* (*IEEE Transactions on Automatic Control*, March 1983) 314
Nonlinear Kalman Filtering Techniques for Terrain-Aided Navigation, *L. D. Hostetler and R. D. Andreas* (*IEEE Transactions on Automatic Control*, March 1983) 325
Application of Multiple Model Estimation to a Recursive Terrain Height Correlation System, *G. L. Mealy and W. Tang* (*IEEE Transactions on Automatic Control*, March 1983) 334
Design and Analysis of a Dynamic Positioning System Based on Kalman Filtering and Optimal Control, *S. Saelid, N. A. Jenssen, and J. G. Balchen* (*IEEE Transactions on Automatic Control*, March 1983) . . . 343
Dynamic Ship Positioning Using a Self-Tuning Kalman Filter, *P. T.-K. Fung and M. J. Grimble* (*IEEE Transactions on Automatic Control*, March 1983) 351
On the Feasibility of Real-Time Prediction of Aircraft Carrier Motion at Sea, *M. M. Sidar and B. F. Doolin* (*IEEE Transactions on Automatic Control*, March 1983) 362
An Integrated Multisensor Aircraft Track Recovery System for Remote Sensing, *W. S. Gesing and D. B. Reid* (*IEEE Transactions on Automatic Control*, March 1983) 368
Bathymetric and Oceanographic Applications of Kalman Filtering Techniques, *R. F. Brammer, R. P. Pass, and J. V. White* (*IEEE Transactions on Automatic Control*, March 1983) 376
A Kalman Filtering Approach to Natural Gamma Ray Spectroscopy in Well Logging, *G. Ruckebusch* (*IEEE Transactions on Automatic Control*, March 1983) 385
Measurement of Instantaneous Flow Rate Through Estimation of Velocity Profiles, *M. Uchiyama and K. Hakomori* (*IEEE Transactions on Automatic Control*, March 1983) 394
Estimation and Prediction of Unmeasurable Variables in the Steel Mill Soaking Pit Control System, *V. J. Lumelsky* (*IEEE Transactions on Automatic Control*, March 1983) 403
Application of Kalman Filters to the Regulation of Dead Time Processes, *W. L. Bialkowski* (*IEEE Transactions on Automatic Control*, March 1983) 416
On-Line Failure Detection in Nuclear Power Plant Instrumentation, *J. L. Tylee* (*IEEE Transactions on Automatic Control*, March 1983) 422
The Application of Kalman Filtering Estimation Techniques in Power Station Control Systems, *J. N. Wallace and R. Clarke* (*IEEE Transactions on Automatic Control*, March 1983) 432
Application of Kalman Filtering to Demographic Models, *B. G. Leibundgut, A. Rault, and F. Gendreau* (*IEEE Transactions on Automatic Control*, March 1983) 444

Author Index 452

Subject Index 453

Editor's Biography 457

Introduction

THE Kalman Filter! The term evokes many and varied responses among engineers, scientists, and managers who hear it. For some, it is a buzzword that must be used in a proposal when attempting to obtain contractual support for a new effort; for others, it is an answer in search of a problem as they already know of better ways to solve their specific problem; for some, it represents a practical set of procedures that they can use to process numerical data to obtain estimates of parameters and variables whose values are uncertain. This book is addressed to those individuals who know of the Kalman filter, accept that it may provide a computational means to solve their numerical data processing problem, and want to learn more about it.

It is probably not an overstatement to assert that the Kalman filter represents the most widely applied and demonstrably useful result to emerge from the state variable approach of "modern control theory." The application of the basic algorithm and the evolution of the body of algorithms that currently can be said to represent the "Kalman Filter" began remarkably soon after the original publication of the recursive, computational procedure proposed by Kalman and others. The earliest applications were stimulated by the "leap" into space that was beginning in the early sixties. The first proposed application of this revolutionary approach was for many (certainly this author) a NASA report published by Gerald Smith, Stanley Schmidt, and Leonard McGee in 1962 [1]. In the few years following the emergence of this report, there was a flurry of substantial activity, primarily in the aerospace industry, that provided the basis for the Kalman filter now so widely applied.

This book is divided into two, distinct parts. Part I presents the germinal papers appearing during the 1960's that provided serious consideration of the theoretical characteristics of the algorithm and the problems that are encountered in attempting to apply it to practical problems. Part II presents a broad coverage of applications that have been accomplished in the first years of the 1980's. These applications appeared in a Special Issue of the IEEE TRANSACTIONS ON AUTOMATIC CONTROL in March 1983 which is reprinted here in its entirety. The breadth of the application topics speaks clearly to the utility of the method and illustrates the distance that the methodology has come since its initial appearance and proposed applications.

The application papers in Part II are supplemented in Part I by a sampling of the theoretical papers that served to shape the development of the methodology. These papers have not been included as a reminiscence about the *good old days* of the field. Many textbooks have been published that review and synthesize the results presented in research papers that contributed to the general development. A bibliography listing many of the textbooks appears at the end of this Introduction that should provide the reader with a variety of careful and comprehensive treatments of the subject. While teaching graduate classes for several years and using several of the books as prescribed texts or references, I have been impressed by the positive response of students to the requirement that they read the early papers in the field as a means of understanding more clearly the concerns that stimulated the results that now are presented so logically in the textbooks. This pedantic observation has served as the motivation for the structure of this volume. By making the early papers conveniently available, it is hoped that the reader will be encouraged to peruse some of these papers and to apply the advantage provided to them by the passage of time to recapture the early excitement and the insights that are still valid.

The theoretical papers are organized into several subject areas. The material was organized in this manner with the intention of defining the major ingredients that must be considered in accomplishing a successful application. In a sense, the topics chosen are sufficiently independent that they can serve as a set of coordinate vectors for the consideration of the area. The volume is launched by an historical survey Section I-A that appeared in the *IEEE Spectrum*. This paper, "Least-squares estimation: from Gauss to Kalman" provides a review of the field as it had developed by the time of its appearance. Developments during the one and one-half decades since its creation suggest changes in detail but not of substance. Perhaps, this may be an indication of the rapid development and maturation of the basic methodology.

The first major group of five papers comprises Section I-B of Part I. These papers deal with the fundamental development of the algorithm and its interpretation from important and useful points-of-view. The papers in Section I-C deal with concerns that arise as one attempts to apply the algorithm to a problem. In particular, these papers discuss the manner in which the Kalman filter algorithm, derived by considering linear, dynamic systems, can be applied to nonlinear systems. From these and the other papers referenced in this section arose the *Extended Kalman Filter* which is, really, the subject of this book.

Given the algorithm and its earliest extensions to nonlinear problems, numerical simulation studies pointed to the dismaying conclusion that the estimates and associated covariances produced by the EKF did not necessarily bear a strong or discernible relationship to the actual behavior of the system and the estimation error. The term *divergence* was invented to define the problem and launched the type of investigation that any developer of an EKF for a specific problem must consider and overcome. Early papers discussing divergence and the effect of model errors that cause it are found in Section I-D.

The methods of divergence control are almost as varied as the imaginations of the people who have addressed the problem. Generally, the procedures that have been proposed are heuristic and *ad hoc*. The procedures that appear to be based on more substantial, theoretical grounds impose assumptions and conditions that permit analysis but which are seldom satisfied in practice. Some of the major approaches that were developed initially are presented in Section I-E. In addition, references containing other approaches and their variations are provided.

Putting aside concerns raised by modeling and approximation errors, the implementation of the EKF algorithm can imply a computational burden, not only in terms of its cycle-time and storage requirements, but also in terms of the effects of numerical errors and their cumulative effects as the recursion proceeds. Assessments of the computational burden and methods for reducing it by exploiting the structure of the model have received consideration. Some early analyses and results are presented in Section I-F. In addition the utilization of sound numerical procedures (e.g., square-root filters) in carrying out the calculations have consistently received attention, beginning with the earliest applications.

The Kalman filter provides an estimate of the state of the system at the current time based on all measurements of the system obtained up to and including the present time. Smoothing is defined as a more general estimation problem in which the state at some or all times in the total sampling interval is estimated from data obtained at times that both precede and succeed the time being considered. While smoothing algorithms have not seen as widespread use as the Kalman filter for a variety of reasons, this presentation cannot be complete without consideration of this more general problem. Different types of smoothing problems can be defined and each problem can be solved with mathematically equivalent but computationally different algorithms. Definitions and different solutions of smoothing problems are described in Section I-G.

The development and analysis of the Kalman filter for linear systems (Section I-B), the consideration of its application to realistic nonlinear systems (Section I-C), the effects of model errors on the filter behavior (Section I-D), methods with which the effect of model errors can be diminished to tolerable levels (Section I-E), and modifications of the basic algorithm to reduce computational burden (Section I-F) provide an outline of the major investigative directions during the evolution of the Kalman filter to its current status. The papers included in Sections I-B through I-G of this volume represent only a fraction, at best, of the published literature on the subject. A bibliography by Mendel and Gieseking [2] and the survey paper by Kailath [3] provide additional sources for information about early work. Other references are included in the papers themselves, in the books that are listed in the Bibliography at the end of this Introduction, and in the references given at the end of each section.

Part II of this book contains the application papers that appeared in the Special Issue which was identified earlier. The reader should find at least several papers that are of interest because the specific application being considered. It is informative and, hopefully, illuminating to read each paper from the perspective of the manner in which the author dealt with the general problems considered in the papers of Part I. Many clever solutions are described which, while stemming from the same concerns as stimulated the analyses presented in Part I, illustrate the variety of approaches possible in achieving a successful implementation of a Kalman filter. Alternatively, some applications leave the reader wondering how or why a generic problem was not discussed.

References

[1] G. L. Smith, S. F. Schmidt, and L. A. McGee, "Application of statistical filter theory to the optimal estimation of position and velocity on-board a circumlunar vehicle," NASA-TR-R-135, 1962.

[2] J. M. Mendel and D. L. Gieseking, "Bibliography on the linear-quadratic-Gaussian problem," *IEEE Trans. Automat. Contr.*, vol. AC-16, 1971, pp. 847–869.

[3] T. Kailath, "A view of three decades of linear filtering theory," *IEEE Trans. Inform. Theory*, vol. IT-20, 1974, pp. 146–181.

[4] B. D. O. Anderson and J. B. Moore, *Optimal Filtering*. Englewood Cliffs, NJ: Prentice-Hall, 1979.

[5] R. H. Battin, *Astronautical Guidance*. New York: McGraw-Hill, 1964.

[6] G. J. Bierman, *Factorization Methods for Discrete Sequential Estimation*. New York: Academic Press, 1977.

[7] S. M. Bozic, *Digital and Kalman Filtering*. London: Arnold Publ., 1979.

[8] R. G. Brown, *Introduction to Random Signal Analysis and Kalman Filtering*. New York: Wiley, 1983.

[9] A. E. Bryson and Y. C. Ho, *Applied Optimal Control*. Waltham, MA: Ginn, 1969.

[10] R. S. Bucy and P. D. Joseph, *Filtering for Stochastic Processes and Applications to Guidance*. New York: Interscience, 1968.

[11] M. H. A. Davis, *Linear Estimation and Stochastic Control*. New York: Halstead Press, 1977.

[12] R. Deutsch, *Estimation Theory*. Englewood Cliffs, NJ: Prentice Hall, 1965.

[13] A. Gelb, Ed., *Applied Optimal Estimation*. Cambridge, MA: MIT Press, 1974.

[14] A. H. Jazwinski, *Stochastic Processes and Filtering*. New York: Academic Press, 1970.

[15] C. L. Lawson and R. J. Hanson, *Solving Least-Squares Problems*. Englewood Cliffs, NJ: Prentice-Hall, 1974.

[16] R. C. K. Lee, *Optimal Estimation and Control*. Cambridge, MA: MIT Press, 1964.

[17] C. T. Leondes, Ed., *Theory and Applications of Kalman Filtering*. NATO AGARDograph, No. 139, 1970.

[18] C. T. Leondes, Ed., *Control and Dynamic Systems: Advances in Theory and Applications*, vols. 19, 20, 21. New York:

Academic Press, 1983 and 1984.
[19] P. B. Liebelt, *An Introduction to Optimal Estimation*. Reading, MA: Addison-Wesley, 1967.
[20] P. S. Maybeck, *Stochastic Models, Estimation, and Control*, Vol. 1. New York: Academic Press, 1979.
[21] T. P. McGarty, *Stochastic Systems and State Estimation*. New York: Wiley, 1974.
[22] J. S. Meditch, *Stochastic Optimal Estimation and Control*. New York: McGraw-Hill, 1969.
[23] N. Nahi, *Estimation Theory and Applications*. New York: Wiley, 1969.
[24] A. P. Sage and J. L. Melsa, *Estimation Theory with Applications to Communications and Control*. New York: McGraw-Hill, 1971.
[25] F. C. Schweppe, *Uncertain Dynamic Systems*. Englewood Cliffs, NJ: Prentice-Hall, 1973.
[26] H. W. Sorenson, *Parameter Estimation: Principles and Problems*. New York: Marcel Dekker Publ., 1980.

Part I
Theoretical Beginnings

Section I-A
Historical Survey

Least-squares estimation: from Gauss to Kalman

The Gaussian concept of estimation by least squares, originally stimulated by astronomical studies, has provided the basis for a number of estimation theories and techniques during the ensuing 170 years—probably none as useful in terms of today's requirements as the Kalman filter

H. W. Sorenson University of California, San Diego

This discussion is directed to least-squares estimation theory, from its inception by Gauss[1] to its modern form, as developed by Kalman.[2] To aid in furnishing the desired perspective, the contributions and insights provided by Gauss are described and related to developments that have appeared more recently (that is, in the 20th century). In the author's opinion, it is enlightening to consider just how far (or how little) we have advanced since the initial developments and to recognize the truth in the saying that we "stand on the shoulders of giants."

The earliest stimulus for the development of estimation theory was apparently provided by astronomical studies in which planet and comet motion was studied using telescopic measurement data. The motion of these bodies can be completely characterized by six parameters, and the estimation problem that was considered was that of inferring the values of these parameters from the measurement data. To solve this problem concerning the revolution of heavenly bodies, the method of least squares was invented by a "young revolutionary" of his day, Karl Friedrich Gauss. Gauss was 18 years old at the time of his first use of the least-squares method in 1795.

As happens even today (e.g., the Kalman filter), there was considerable controversy in the early 19th century regarding the actual inventor of the least-squares method. The conflict arose because Gauss did not publish his discovery in 1795. Instead, Legendre independently invented the method and published his results in 1806 in his book *Nouvelles méthodes pour la determination des orbites des comètes*. It was not until 1809, in his book *Theoria Motus Corporum Coelestium*, that Gauss published a detailed description of the least-squares method. However, in this treatise Gauss mentions Legendre's discussion of least squares and pointedly refers to his own earlier use (p. 270, *Theoria Motus*)*: "Our principle, which we have made use of since the year 1795, has lately been published by Legendre in the work *Nouvelles méthodes pour la determination des orbites des cometes, Paris, 1806*, where several other properties of this principle have been explained which, for the sake of brevity, we here omit." This reference angered Legendre who, with great indignation, wrote to Gauss and complained[3] that "Gauss, who was already so rich in discoveries, might have had the decency not to appropriate the method of least-squares." It is interesting to note that Gauss, who is now regarded as one of the "giants" of mathematics, felt that he had been eclipsed by Legendre and wrote to a friend saying,[3] "It seems to be my fate to concur in nearly all my theoretical works with Legendre. So it is in the higher arithmetic, . . . , and now again in the method of least-squares which is also used in Legendre's work and indeed right gallantly carried through." Historians have since found sufficient evidence to substantiate Gauss' claim of priority to the least-squares method, so it is Legendre rather than Gauss who was eclipsed in this instance and, indeed, in general.

* The page numbers here refer to the English translation available from Dover Publications, Inc.[1]

The method of least squares

The astronomical studies that prompted the invention of least squares were described by Gauss in *Theoria Motus*.[1] The following quotation (p. 249) not only describes the basic ingredients for Gauss' studies but captures the essential ingredients for all other data-processing studies. "If the astronomical observations and other quantities on which the computation of orbits is based were absolutely correct, the elements also, whether deduced from three or four observations, would be strictly accurate (so far indeed as the motion is supposed to take place exactly according to the laws of Kepler) and, therefore, if other observations were used, they might be confirmed but not corrected. But since all our measurements and observations are nothing more than approximations to the truth, the same must be true of all calculations resting upon them, and the highest aim of all compu-

Reprinted from *IEEE Spectrum*, vol. 7, pp. 63–68, July 1970.

tations made concerning concrete phenomena must be to approximate, as nearly as practicable, to the truth. But this can be accomplished in no other way than by a suitable combination of more observations than the number absolutely requisite for the determination of the unknown quantities. This problem can only be properly undertaken when an approximate knowledge of the orbit has been already attained, which is afterwards to be corrected so as to satisfy all the observations in the most accurate manner possible."

Let us briefly reconsider some of the ideas contained in the preceding statement and relate them to "modern" developments.

1. Gauss refers to the number of observations that are absolutely required for the determination of the unknown quantities. The problem of establishing this minimum number of observations is currently discussed in terms of the "observability of the system" and is the subject of many papers; see Refs. 4 and 5, for example.

2. Gauss notes that more observations are required than this minimum because of the errors in the measurements and observations. Thus, he notes the need for "redundant" data to eliminate the influence of measurement errors.

3. Gauss implies that the equations of motion must be exact descriptions, and therefore the problem of dynamic modeling of the system is raised.

4. Gauss requires that approximate knowledge of the orbit be available. This is currently required in virtually all practical applications of Kalman filter theory,[6] for example, and implies the use of some linearization procedure.

5. Gauss states that the parameter estimates must satisfy the observations in the most accurate manner possible. Thus, he calls for the residuals (that is, the difference between the observed values and the values predicted from the estimates) to be as small as possible.

6. Gauss refers to the inaccuracy of the observations and indicates that the errors are unknown or unknowable and thereby sets the stage for probabilistic considerations. In doing so, he anticipates most of the modern-day approaches to estimation problems.

7. Finally, Gauss refers to the "suitable combination" of the observations that will give the most accurate estimates. This is related to the definition of the structure of an estimation procedure (i.e., linear or nonlinear filtering) and to the definition of the performance criterion. These are extremely important considerations in current discussions of estimation problems.

As stated earlier, Gauss invented and used the method of least squares as his estimation technique. Let us consider Gauss' definition of the method (Ref. 1, page 260). He suggested that the most appropriate values for the unknown but desired parameters are the *most probable values*, which he defined in the following manner: ". . . *the most probable value* of the unknown quantities will be that in which the sum of the squares of the differences between the actually observed and the computed values multiplied by numbers that measure the degree of precision is a minimum." The difference between the observed and computed measurement values is generally called the *residual.*

To make the discussion more precise, consider the following statement of the estimation problem. Suppose that m measurement quantities are available at discrete instants of time $(t_1, t_2, \ldots, t_n)$ and are denoted at each time t_k as $\mathbf{z}_k$. Suppose that parameters $\mathbf{x}$ are to be determined from the data and are related according to

$$\mathbf{z}_k = H_k\mathbf{x} + \mathbf{v}_k \qquad (1)$$

where the $\mathbf{v}_k$ represent the measurement errors that occur at each observation time. As is seen in Eq. (1), the measurement data and the parameters $\mathbf{x}$ are assumed here to be linearly related, thereby making explicit the assumption that Gauss indicated was necessary in the foregoing quotation.

Denote the estimate of $\mathbf{x}$ based on the n data samples $\{\mathbf{z}_1, \mathbf{z}_2, \cdots, \mathbf{z}_n\}$ as $\hat{\mathbf{x}}_n$. Then, the residual associated with the kth measurement is

$$\mathbf{r}_k \triangleq \mathbf{z}_k - H_k\hat{\mathbf{x}}_n \qquad k = 0, 1, \cdots, n \qquad (2)$$

The least-squares method is concerned with determining the most probable value of $\mathbf{x}$ (that is, $\hat{\mathbf{x}}_n$). This most probable value is defined as the value that minimizes the sum of the squares of the residuals. Thus, choose $\mathbf{x}$ so that

$$L_n = \frac{1}{2}\sum_{k=0}^{n} [\mathbf{z}_k - H_k\mathbf{x}]^T W_k [\mathbf{z}_k - H_k\mathbf{x}] \qquad (3)$$

is minimized. The elements of the matrixes W_k are selected to indicate the degree of confidence that one can place in the individual measurements. As will be explained more fully in the discussion of the Kalman filter, W_k is equivalent to the inverse of the covariance of the measurement noise.

Gauss with his remarkable insight recognized that the simple statement of the least-squares method contains the germ of countless interesting studies. As he says in *Theoria Motus* (Ref. 1, page 269): "The subject we have just treated might give rise to several elegant analytical investigations upon which, however, we will not dwell, that we may not be too much diverted from our object. For the same reason we must reserve for another occasion the explanation of the devices by means of which the numerical calculations can be rendered more expeditious." Judging by the interest in estimation theory over the years, this statement must stand as one of the greatest understatements of all time. In passing, we note that the Kalman filter can be rightfully regarded as an efficient computational solution of the least-squares method.

Gauss did not merely hypothesize the least-squares method; it is interesting to consider his discussion of the problem of obtaining the "most probable" estimate as an introduction to more modern techniques. First, it is significant that he considered the problem from a probabilistic point of view and attempted to define the best estimate as the most probable value of the parameters. He reasoned that errors in the measurements would be independent of each other, so the joint-probability density function of the measurement residuals can be expressed as the product of the individual density functions

$$f(\mathbf{r}_0, \mathbf{r}_1, \cdots, \mathbf{r}_n) = f(\mathbf{r}_0)f(\mathbf{r}_1)\cdots f(\mathbf{r}_n) \qquad (4)$$

Next, he argued that the density $f(\mathbf{r}_k)$ would be a normal density

$$f(\mathbf{r}_k) = \frac{\sqrt{\det W}}{(2\pi)^{m/2}} \exp\left[\tfrac{1}{2}\mathbf{r}_k^T W_k \mathbf{r}_k\right] \qquad (5)$$

although he recognized that one never obtains errors of infinite magnitude, and thus Eq. (5) is not realistic.

However, he rationalized away this difficulty by stating (page 259) that: "The function just found cannot, it is true, express rigorously the probabilities of the errors for since the possible errors are in all cases confined within certain limits, the probability of errors exceeding those limits ought always to be zero while our formula always gives some value. However, this defect, which every analytical function must, from its nature, labor under, is of no importance in practice because the value of our function decreases so rapidly, when $[\mathbf{r}_k^T W_k \mathbf{r}_k]$ has acquired a considerable magnitude, that it can safely be considered as vanishing."

Gauss proceeded by noting that the maximum of the probability density function is determined by maximizing the logarithm of this function. Thus, he anticipated the *maximum likelihood method*, which was introduced by R. A. Fisher[7] in 1912 and has been thoroughly investigated up to the present time. It is interesting that Gauss rejected the maximum likelihood method[8] in favor of minimizing some function of the difference between estimate and observation, and thereby recast the least-squares method independent of probability theory. However, in maximizing the logarithm of the independent and normally distributed residuals, one is led to the least-squares problem defined by Eq. (3).

Kalman filter theory

Let us now leave the early 19th century and enter the 20th century. Consider, briefly, some of the major developments of estimation theory that preceded the introduction of the Kalman filter. As already mentioned, R. A. Fisher introduced the idea of maximum likelihood estimation and this has provided food for thought throughout the subsequent years. Kolmogorov[9] in 1941 and Wiener[10] in 1942 independently developed a linear minimum mean-square estimation technique that received considerable attention and provided the foundation for the subsequent development of Kalman filter theory.

In Wiener–Kolmogorov filter theory, Gauss' inference that linear equations must be available for the solution of the estimation problem is elevated to the status of an explicit assumption. There are, however, many conceptual differences (as one would hope after 140 years) between Gauss' problem and the problem treated by Wiener and Kolmogorov. Not the least of these is the fact that the latter considered the estimation problem when measurements are obtained continuously, as well as the discrete-time problem. To maintain the continuity of the present discussion, attention shall be restricted to the discrete formulation of Wiener–Kolmogorov (and later Kalman) filter theory.

Consider the problem of estimating a signal $\mathbf{s}_n$, possibly time-varying, from measurement data $(\mathbf{z}_0, \mathbf{z}_1, \cdots, \mathbf{z}_n)$, where the $\mathbf{s}_n$ and the $\{\mathbf{z}_i\}$ are related through knowledge of the cross-correlation functions. Assume that the estimate of $\mathbf{s}_n$, say $\hat{\mathbf{s}}_{n/n}$, is to be computed as a linear combination of the $\mathbf{z}_i$:

$$\hat{\mathbf{s}}_{n/n} = \sum_{i=0}^{n} H_{n,i}\mathbf{z}_i \tag{6}$$

The "filter gains" $H_{n,i}$ are to be chosen so that the mean-square error is minimized; that is, choose the $H_{n,i}$ in such a way that

$$M_n = E[(\mathbf{s}_n - \hat{\mathbf{s}}_{n/n})^T(\mathbf{s}_n - \hat{\mathbf{s}}_{n/n})] \tag{7}$$

is minimized. It is well known[11] that a necessary and sufficient condition for $\hat{\mathbf{s}}_{n/n}$ to minimize M_n is that the error in the estimate $(\tilde{\mathbf{s}}_{n/n} \triangleq \mathbf{s}_n - \hat{\mathbf{s}}_{n/n})$ be orthogonal to the measurement data

$$E[\tilde{\mathbf{s}}_{n/n}\mathbf{z}_i^T] = 0 \qquad i = 0, 1, \cdots, n \tag{8}$$

This is the Wiener–Hopf equation, which is frequently written as

$$E[\mathbf{s}_n\mathbf{z}_i^T] = \sum_{j=0}^{n} H_{n,j}E[\mathbf{z}_j\mathbf{z}_i^T] \qquad i = 0, 1, \cdots, n \tag{9}$$

This equation must be solved for the $H_{n,j}$ in order to obtain the gains of the optimal filter. One can rewrite this as a vector-matrix equation whose solution, theoretically speaking, is straightforward. However, the matrix inversion that is required becomes computationally impractical when n is large. Wiener and Kolmogorov assumed an infinite amount of data (that is, the lower limit of the summation is $-\infty$ rather than zero), and assumed the system to be stationary. The resulting equations were solved using spectral factorization.[9,10,12]

The problem formulated and described here is significantly different from Gauss' least-squares problem. First, no assumption is imposed that the signal is constant. Instead, the signal can be different at each n but can be described statistically by the autocorrelation and cross-correlation functions of the signal and measurement data. Second, instead of arguing that the estimate be the most probable, a probabilistic version of the least-squares method is chosen as the performance index.

It has been found that Eq. (9) is solved in a relatively straightforward manner if one introduces a "shaping filter"[13,14] to give a more explicit description of the signal. In particular, suppose that the signal and measurement processes are assumed to have the following structure. The measurements are described by

$$\begin{aligned}\mathbf{z}_i &= \mathbf{s}_i + \mathbf{v}_i \\ &= H_i\mathbf{x}_i + \mathbf{v}_i\end{aligned} \tag{10}$$

where $\mathbf{v}_i$ is a white-noise sequence (that is, $\mathbf{v}_i$ is both mutually independent and independent of $\mathbf{x}_i$). The system state vector $\mathbf{x}_i$ is assumed to be described as a dynamic system having the form

$$\mathbf{x}_{i+1} = \Phi_{i+1,i}\mathbf{x}_i + \mathbf{w}_i \tag{11}$$

where $\mathbf{w}_i$ represents a white-noise sequence. Note that if the noise $\mathbf{w}_i$ is identically zero and if $\Phi_{i+1,i}$ is the identity matrix, then the state is a constant for all i and one has returned basically to the system assumed by Gauss. With the system described by Eqs. (10) and (11), the known statistics for the initial state $\mathbf{x}_0$, and the noise sequences $\{\mathbf{w}_i\}$ and $\{\mathbf{v}_i\}$, one can proceed to the solution of Eq. (9).

Although the weighting function for the filter can be determined, a new solution must be generated for each n. It seems intuitively reasonable that estimates of $\mathbf{s}_{n+1}$ (or $\mathbf{x}_{n+1}$) could be derived, given a new measurement $\mathbf{z}_{n+1}$, from $\hat{\mathbf{s}}_{n/n}$ and $\mathbf{z}_{n+1}$ rather than from $\mathbf{z}_0, \mathbf{z}_1, \cdots, \mathbf{z}_n, \mathbf{z}_{n+1}$, since $\hat{\mathbf{s}}_{n/n}$ is based on the data $(\mathbf{z}_0, \mathbf{z}_1, \cdots, \mathbf{z}_n)$. In 1955 J. W. Follin[15] at Johns Hopkins University suggested a recursive approach based on this idea, which he carried out for a specific system. This approach had immediate appeal and essentially laid the foundation for the developments that are now referred to as the Kalman filter. It is clear (for example, see Ref. 16, p. 129) that Follin's work provided a direct stimulus for the work of Richard Bucy, which led to his subsequent collaboration with Kalman in the continuous-time version of the filter equations.[17] As frequently happens, the time was ripe for this approach, because several other people independently investigated recursive filter and prediction methods; see, for example, Refs. 18 and 19. Also, the method of stochastic approximation[20] was introduced and being studied for related problems[21] during this period.

Kalman published his first paper on discrete-time, recursive mean-square filtering in 1960.[2] It is interesting to note that, analogous to the Gauss–Legendre squabble concerning priority of the least-squares method, there is a difference of opinion concerning the originator of the Kalman filter. Peter Swerling published a RAND Corporation memorandum in 1958[18] describing a recursive procedure for orbit determination. Of further interest is the fact that orbit determination problems provided the stimulus for both Gauss' work and more modern-day developments. Swerling's method is essentially the same as Kalman's except that the equation used to update the error covariance matrix has a slightly more cumbersome form. After Kalman had published his paper and it had attained considerable fame, Swerling[22] wrote a letter to the *AIAA Journal* claiming priority for the Kalman filter equations. It appears, however, that his plea has fallen on deaf ears.

The developments beginning with Wiener's work and culminating with Kalman's reflect fundamentally the changes that have occurred in control systems theory during this period. In the "classical control theory," the emphasis was on the analysis and synthesis of systems in terms of their input–output characteristics. The basic tools used for these problems were the Laplace and Fourier transforms. The original formulation and solution of the Wiener–Kolmogorov filtering problem is consistent with this basic approach. More recent developments have stressed the "state-space" approach, in which one deals with the basic system that gives rise to the observed output. It represents in many ways a return to Gauss' approach, since he referred to the dynamic modeling problem as noted earlier. Also, the state-space approach makes use of difference and differential equations rather than the integral equations of the classical approach. Although the two approaches are mathematically equivalent, it seems to be more satisfying to work with differential equations (probably since dynamical systems are generally described in this manner).

At this point, let us summarize the Kalman filtering problem and its solution. The system that is considered is composed of two essential ingredients. First, the state is assumed to be described by

$$\mathbf{x}_{k+1} = \Phi_{k+1,k}\mathbf{x}_k + \mathbf{w}_k \qquad (11')$$

and the measurement data are related to the state by

$$\mathbf{z}_k = H_k\mathbf{x}_k + \mathbf{v}_k \qquad (10')$$

where $\{\mathbf{w}_k\}$ and $\{\mathbf{v}_k\}$ represent independent white-noise sequences. The initial state $\mathbf{x}_0$ has a mean value $\hat{\mathbf{x}}_{0/-1}$ and covariance matrix $P_{0/-1}$ and is independent of the plant and measurement noise sequences. The noise sequences have zero mean and second-order statistics described by

$$E[\mathbf{v}_k\mathbf{v}_j^T] = R_k\delta_{kj} \qquad E[\mathbf{w}_k\mathbf{w}_j^T] = Q_k\delta_{kj}$$
$$E[\mathbf{v}_k\mathbf{w}_j^T] = 0 \qquad \text{for all } k, j$$

where δ_{kj} is the Kronecker delta.

An estimate $\hat{\mathbf{x}}_{k/k}$ of the state $\mathbf{x}_k$ is to be computed from the data $\mathbf{z}_0, \mathbf{z}_1, \cdots, \mathbf{z}_k$ so as to minimize the mean-square error in the estimate. The estimate that accomplishes this is to be computed as an explicit function only of the measurement $\mathbf{z}_k$ and the previous best estimate $\hat{\mathbf{x}}_{k-1/k-1}$. This approach leads to a recursive solution that provides an estimate that is equivalent to the estimate obtained by processing all of the data simultaneously but reduces the data-handling requirements. The estimate of the signal

$$\mathbf{s}_k = H_k\mathbf{x}_k \qquad (12)$$

is given by

$$\hat{\mathbf{s}}_{k/k} = H_k\hat{\mathbf{x}}_{k/k} \qquad (13)$$

The solution of this recursive, linear, mean-square estimation problem can be determined from the orthogonality principle given in Eq. (8), as well as in a large variety of other ways, and is presented below. This system of equations has come to be known as the Kalman filter. The estimate is given as the linear combination of the estimate predicted in the absence of new data, or

$$\hat{\mathbf{x}}_{k/k-1} = \Phi_{k,k-1}\hat{\mathbf{x}}_{k-1/k-1}$$

and the residual $\mathbf{r}_k$. Thus, the mean-square estimate is

$$\hat{\mathbf{x}}_{k/k} = \Phi_{k,k-1}\hat{\mathbf{x}}_{k-1/k-1} + K_k[\mathbf{z}_k - H_k\Phi_{k,k-1}\hat{\mathbf{x}}_{k-1/k-1}] \qquad (14)$$

The gain matrix K_k can be considered as being chosen to minimize $E[(\hat{\mathbf{x}}_k - \hat{\mathbf{x}}_{k/k})^T(\mathbf{x}_k - \mathbf{x}_{k/k})]$ and is given by

$$K_k = P_{k/k-1}H_k^T(H_kP_{k/k-1}H_k^T + R_k)^{-1} \qquad (15)$$

The matrix $P_{k/k-1}$ is the covariance of the error in the predicted estimate and is given by

$$P_{k/k-1} = E[(\mathbf{x}_k - \hat{\mathbf{x}}_{k/k-1})(\mathbf{x}_k - \hat{\mathbf{x}}_{k/k-1})^T] \qquad (16)$$
$$= \Phi_{k/k-1}P_{k-1/k-1}\Phi_{k,k-1}^T + Q_{k-1} \qquad (17)$$

The $P_{k/k}$ is the covariance of the error in the estimate $\hat{\mathbf{x}}_{k/k}$.

$$P_{k/k} = E[\mathbf{x}_k - \hat{\mathbf{x}}_{k/k})(\mathbf{x}_k - \hat{\mathbf{x}}_{k/k})^T] \qquad (18)$$
$$= P_{k/k-1} - K_kH_kP_{k/k-1} \qquad (19)$$

Equations (14), (15), (17), and (19) form the system of equations comprising the Kalman filter.[2,6]

Kalman filter theory—a perspective

Let us relate elements of this problem to Gauss' earlier arguments. First, Kalman assumes that the noise is independent from one sampling time to the next. But it is clear from Eq. (5) that this is equivalent to assuming that the residual $(\mathbf{z}_k - H_k\mathbf{x}_k)$ is independent between sampling times and therefore agrees with Gauss' assumption. Next, the noise and initial state are essentially assumed by Kalman to be Gaussian. The linearity of the system causes the state and measurements to be Gaussian at each sampling time. Thus the residual is Gaussian, as Gauss assumed. Therefore, one sees that the basic assumptions of Gauss and Kalman are identical except that the latter allows the state to change from one time to the next. This difference introduces a nontrivial modification to Gauss' problem but one that can be treated within a least-squares framework if the noise $\{\mathbf{w}_k\}$ is considered as the error in the plant model at each stage. In particular, one can formulate the least-squares problem as that of choosing the estimates $\hat{\mathbf{x}}_{k/k}$ and the plant errors $\mathbf{w}_k$ to minimize the modified least-squares performance index.

$$L_n = \frac{1}{2}(\mathbf{x}_0 - \mathbf{a})^T M_0^{-1}(\mathbf{x}_0 - a) + \frac{1}{2}\sum_{i=0}^{n}(\mathbf{z}_i - H_i\mathbf{x}_i)^T R_i^{-1}(\mathbf{z}_i - H_i\mathbf{x}_i) + \frac{1}{2}\sum_{i=0}^{n-1}\mathbf{w}_i^T Q_i^{-1}\mathbf{w}_i \quad (20)$$

subject to the constraint

$$\mathbf{x}_k = \Phi_{k,k-1}\mathbf{x}_{k-1} + \mathbf{w}_{k-1} \quad (21)$$

Note that the first term essentially describes the uncertainty in the initial state. If one has no a priori information, then M_0^{-1} is identically zero and the term vanishes. Similarly, if there is no error in the plant equation, Q_i^{-1} is identically zero so this term vanishes. Then Eq. (20) is seen to reduce to Gauss' least-squares problem, as given in Eq. (3). The weighting matrices M_0^{-1}, R_i^{-1}, and Q_i^{-1} represent the matrix inverses of the a priori covariance matrices if a probabilistic interpretation is desired.

One can obtain a recursive solution to the problem of minimizing (20) by noting that

$$L_n = L_{n-1} + \frac{1}{2}(\mathbf{z}_n - H_n\mathbf{x}_n)^T R_n^{-1}(\mathbf{z}_n - H_n\mathbf{x}_n) + \frac{1}{2}\mathbf{w}_{n-1}^T Q_{n-1}^{-1}\mathbf{w}_{n-1} \quad (22)$$

and by then proceeding inductively starting with $n = 0$ to obtain recursion relations for the least-square estimate.[23] If this is done, the Kalman filter equations are obtained. It is then indicated that, for this linear problem, deterministic least-squares estimation theory and the probabilistically based mean-square estimation theory are equivalent. Further, the problem of minimizing Eq. (22) is seen to give the most probable estimate for this system.

Since the Kalman filter represents essentially a recursive solution of Gauss' original least-squares problem, it is reasonable to consider the substance of Kalman's contribution and attempt to put it into perspective. It cannot be denied that there has been a substantial contribution if for no other reason than the large number of theoretical and practical studies that it has initiated. I suggest that the contribution is significant for two basic reasons:

1. The Kalman filter equations provide an extremely convenient procedure for digital computer implementation. One can develop a computer program using the Kalman filter in a direct manner that (initially, at least) requires little understanding of the theory that led to their development. There are well-established numerical procedures for solving differential equations, so the engineer does not have to be worried about this problem. By contrast, the solution of the Wiener–Hopf equation and the implementation of the Wiener–Kolmogorov filter must be regarded as more difficult or there would have been no need for the Kalman filter. Since Gauss was very concerned with the computational aspects of least-squares applications, one can imagine that he would appreciate the computational benefits of the Kalman filter.

2. Kalman posed the problem in a general framework that has had a unifying influence on known results. Further, one can analyze the behavior of the estimates within the general framework and thereby obtain significant insights into the results obtained from computational studies. There has been a veritable "explosion" of theoretical papers that have recognizable roots in Kalman's work and thereby testify to the richness of his formulation.

Finally, a third reason for the popularity might be considered, although it is less tangible in character than the other two. It is worth noting that Kalman recognized the potential of his results, whereas others working in the area either did not or were not as successful in communicating the intrinsic worth of recursive filtering to others. One cannot overemphasize the value of recognizing and successfully communicating significant new results.

The Kalman filter, which assumes linear systems, has found its greatest application to nonlinear systems. It is generally used in these problems by assuming knowledge of an approximate solution (as Gauss proposed) and by describing the deviations from the reference by linear equations. The approximate linear model that is obtained forms the basis for the Kalman filter utilization. Commonly, such applications are accomplished with great success but, on occasion, unsatisfactory results are obtained because a phenomenon known as divergence occurs.[24,25]

Divergence is said to occur when the error covariance matrix P_k computed by the filter becomes unjustifiably small compared with the actual error in the estimate. When P_k becomes small, it causes the gain matrix to become too small and new measurement data are given too little weight. As a result, the plant model becomes more important in determining the estimate than are the data and any errors in the model can build up over a period of time and cause a significant degradation in the accuracy of the estimate. This happens most commonly when the plant is assumed to be error-free (i.e., $\mathbf{w}_k \equiv 0$ for all k). If the model were perfect and contained no random or model errors, then the vanishing of the error covariance matrix would be desirable and would represent the fact that the state could be determined precisely if sufficient redundant data were processed. However, it is naive at best to assume that any physical system can be modeled precisely, so it is necessary to account for model errors. Thus, it has become good practice[6] always to include the plant error or noise term $\mathbf{w}_k$. It should be emphasized that divergence does not occur because of any fault of the filter. If the system were actually linear, Kalman[4] showed that the filter equations are stable under very reasonable conditions. Thus, the divergence is a direct consequence of the errors introduced by the linear approximation.

To reduce approximation errors, the so-called "extended Kalman filter" is generally used in practice. In this case the nonlinear system is linearized by employing the best estimates of the state vector as the reference values used at each stage for the linearization. For example, at time t_{k-1}, the estimate $\hat{\mathbf{x}}_{k-1/k-1}$ is used as the reference in obtaining the transition matrix $\Phi_{k,k-1}$. This approximation is utilized in Eq. (17) to obtain the Kalman error covariance $P_{k/k-1}$. The estimate is given by

$$\hat{\mathbf{x}}_{k/k-1} = \mathbf{f}_k(\hat{\mathbf{x}}_{k-1/k-1}) \quad (23)$$

where $\mathbf{f}_k$ is used to denote the nonlinear plant equation.

In most cases it is obtained as the solution of an ordinary differential equation that describes the plant behavior. The processing of the data obtained at t_k is accomplished in a similar manner. The estimate $\hat{\mathbf{x}}_{k/k-1}$ serves as the reference for the determination of a linear approximation H_k to the nonlinear measurement equation. The matrix H_k is used in Eqs. (15) and (19) to determine the gain and error covariance matrices K_k and $P_{k/k}$. The filtered estimate is then given by

$$\hat{\mathbf{x}}_{k/k} = \hat{\mathbf{x}}_{k/k-1} + K_k[\mathbf{z}_k - \mathbf{h}_k(\hat{\mathbf{x}}_{k/k-1})] \tag{24}$$

where $\mathbf{h}_k$ is used to denote the measurement nonlinearity; that is, the measurement is assumed to be described by

$$\mathbf{z}_k = \mathbf{h}_k(\mathbf{x}_k) + \mathbf{v}_k \tag{25}$$

Through the use of the extended Kalman filter, one can hope to eliminate or reduce divergence. Note, however, that the $P_{k/k-1}$ and $P_{k/k}$ matrices are still linear approximations of the true error covariance matrices. Further, the nonlinear models $\mathbf{f}_k$ and $\mathbf{h}_k$ are themselves approximations of the actual physical system, so modeling errors can still exist. Thus, the extended Kalman filter does not insure the elimination of the divergence problem.

In a practical application one does not know the error in the state estimate, so there are grounds for uneasiness in using this method. Of course, the same type of problem must be considered in any least-squares application and we can return to Gauss for the means of judging the behavior of the filter. Recall that he said that the estimates should satisfy all the observations in the most accurate manner possible. Thus, one is led to further consideration of the residuals as a measure of filter performance. Kailath[26] pointed out recently that the residual sequence $(\mathbf{z}_k - H_k\hat{\mathbf{x}}_{k/k-1})$ is a white-noise sequence. Since the residual can be computed explicitly, it can be examined at each stage to verify that the residual (or innovations) sequence has the appropriate statistical characteristics. A method of controlling the divergence, based on the residuals, has been proposed[27,28] in which the plant and measurement noise covariance matrices Q_k and R_k are chosen in a manner that is appropriate to cause the residuals to have the desired properties. But this method is essentially the same as choosing the least-squares weighting matrices as a reflection of the accuracy of the measurements (or plant). Thus, the least-squares aspect continues to dominate the practical application of the method.

REFERENCES

1. Gauss, K. G., *Theory of Motion of the Heavenly Bodies.* New York: Dover, 1963.
2. Kalman, R. E., "A new approach to linear filtering and prediction problems," *J. Basic Eng.*, vol. 82D, pp. 35–45, Mar. 1960.
3. Bell, E. T., *Men of Mathematics.* New York: Simon and Schuster, 1961.
4. Kalman, R. E., "New methods in Wiener filtering theory," in *Proc. First Symp. of Engineering Application of Random Function Theory and Probability.* New York: Wiley, 1963, pp. 270–388.
5. Sorenson, H. W., "Controllability and observability of linear, stochastic, time-discrete control systems," in *Advances in Control Systems,* Vol. 6, C. T. Leondes, ed. New York: Academic Press, 1968.
6. Sorenson, H. W., "Kalman filtering techniques," in *Advances in Control Systems,* Vol. 3, C. T. Leondes, ed. New York: Academic Press, 1966.
7. Fisher, R. A., "On an absolute criterion for fitting frequency curves," *Messenger of Math.*, vol. 41, p. 155, 1912.
8. Berkson, J., "Estimation by least-squares and by maximum likelihood," *Proc. Third Berkeley Symp. on Mathematical Statistics, and Probability,* Vol. 1. Berkeley: University of California Press, 1956, pp. 1–11.
9. Kolmogorov, A. N., "Interpolation and extrapolation of stationary random sequences," translated by W. Doyle and J. Selin, Rept. RM-3090-PR, RAND Corp., Santa Monica, Calif., 1962.
10. Wiener, N., *The Extrapolation, Interpolation and Smoothing of Stationary Time Series.* New York: Wiley, 1949.
11. Sorenson, H. W., and Stubberud, A. R., "Linear estimation theory," Chapter 1 in *Theory and Application of Kalman Filtering; AGARDograph,* no. 139, 1970.
12. Yaglom, A. M., *An Introduction to the Theory of Stationary Random Functions,* translated by R. A. Silverman. Englewood Cliffs, N.J.: Prentice-Hall, 1962.
13. Bode, H. W., and Shannon C. E., "A simplified derivation of linear least-squares smoothing and prediction theory," *Proc. IRE,* vol. 38, pp. 417–425, Apr. 1950.
14. Zadeh, L. A., and Ragazzini, J. R., "An extension of Wiener's theory of prediction," *J. Appl. Phys.*, vol. 21, pp. 645–655, July 1950.
15. Carlton, A. G., and Follin, J. W., "Recent developments in fixed and adaptive filtering," *AGARDograph,* no. 21, 1956.
16. Bucy, R. S., and Joseph, P. D., *Filtering for Stochastic Processes with Applications to Guidance.* New York: Interscience, 1968.
17. Kalman, R. E., and Bucy, R. S., "New results in linear filtering and prediction theory," *J. Basic Eng.*, vol. 83D, pp. 95–108, 1961.
18. Swerling, P., "A proposed stagewise differential correction procedure for satellite tracking and prediction," Rept. P-1292, *RAND Corp.* Santa Monica, Calif., Jan. 1958; also published in *J. Astronaut. Sci.*, vol. 6, 1959.
19. Blum, M., "Recursion formulas for growing memory digital filters," *IRE Trans. Information Theory,* vol. IT-4, pp. 24–30, Mar. 1958.
20. Robbins, H., and Munro, S., "A stochastic approximation method," *Ann. Math. Stat.*, vol. 22, pp. 400–407, 1951.
21. Kiefer, J., and Wolfowitz, J., "Stochastic estimation of the maximum of a regression function," *Ann. Math. Stat.*, vol. 23, pp. 462–466, 1952.
22. Swerling, P., "Comment on 'A statistical optimizing navigation procedure for space flight,'" *AIAA J.*, vol. 1, p. 1968, Aug. 1963.
23. Sorenson, H. W., "Comparison of Kalman, Bayesian and maximum likelihood estimation techniques," Chapter 6 in *Theory and Application of Kalman Filtering; AGARDograph,* no. 139, 1970.
24. Schlee, F. H., Standish, C. J., and Toda, N. F., "Divergence in the Kalman filter," *AIAA J.*, vol. 5, pp. 1114–1120, June 1967.
25. Fitzgerald, R. J., "Error divergence in optimal filtering problems," presented at the 2nd IFAC Symposium on Automatic Control in Space, Vienna, Austria, Sept. 1967.
26. Kailath, T., "An innovations approach to least-squares estimation—Part I: linear filtering in additive white noise," *IEEE Trans. Automatic Control,* vol. AC-13, pp. 646–655, Dec. 1968.
27. Jazwinski, A. H., "Adaptive filtering," *Automatica,* vol. 5, pp. 475–485, 1969.
28. Mehra, R. K., "On the identification of variances and adaptive filtering," *Preprints 1969 Joint Automatic Control Conf.*, pp. 494–505.

Section I-B
Theoretical Foundations

THE set of recursive equations that have come to be known as the Kalman filter were presented by Kalman in the first paper contained in this section. This paper is concerned with discrete-time, linear systems and, because of the suitability of the results for digital implementation, these results provide the basis for most applications. Kalman and Bucy in the third paper of this section considered continuous-time systems and presented fundamental results regarding the behavior of the filter.

Results similar to Kalman's were published independently by other investigators. The second paper in this section, written by Peter Swerling, derives a recursive result that is equivalent to Kalman's in a sense discussed subsequently. Stratonovich in the USSR published results which are also equivalent to Kalman's. This work is listed below but will not be discussed further.

It is interesting to read and to contrast the approaches taken by Kalman and by Swerling in their papers. As a prelude to studying these papers, it may be useful to establish a common notation and to relate this notation to the two papers. The notation introduced here appears in many textbooks but is not common to all authors. The Kalman filter stems from the consideration of linear, stochastic dynamic systems described in the following manner. The *state* $\mathbf{x}$ is an n-dimensional vector which evolves according to a linear difference equation

$$\mathbf{x}_{k+1} = F_k \mathbf{x}_k + G_k \mathbf{w}_k, \qquad k = 1, 2, \cdots \tag{1.1}$$

with initial condition $\mathbf{x}_1$. The state cannot be observed directly. Instead, measurements $\mathbf{z}$ are available at discrete, sampling times and are described as

$$\mathbf{z}_k = H_k \mathbf{x}_k + \mathbf{v}_k, \qquad k = 1, 2, \cdots. \tag{1.2}$$

The vectors $\{\mathbf{w}_k\}$ and $\{\mathbf{v}_k\}$ are assumed to be white noise sequences described by their second-order statistics

$$E[\mathbf{w}_k] = 0; \; E[\mathbf{v}_k] = 0, \qquad k = 1, 2, \cdots$$
$$E[\mathbf{w}_k \mathbf{w}_j^T] = Q_k \delta_{kj}; \; E[\mathbf{v}_k \mathbf{v}_j^T] = R_k \delta_{kj}$$

where δ_{kj} is the Kronecker delta. Thus, the samples at different times are uncorrelated. Generally, it is assumed that the input noise sequence $\{\mathbf{w}_k\}$ and the output noise sequence $\{\mathbf{v}_k\}$ are uncorrelated for all sampling times but results are well-known that extend the basic results to allow for the correlation of $\mathbf{w}_k$ and $\mathbf{v}_k$. The initial condition for the state $\mathbf{x}_1$ is assumed to be random with mean value $\hat{\mathbf{x}}_{1/0}$ and covariance matrix $P_{1/0}$.

Given the model (1.1)–(1.2), the Kalman filter provides the linear, minimum mean-squared error estimator of the state $\mathbf{x}_k$ given the measurements $\{\mathbf{z}_1, \mathbf{z}_2, \cdots, \mathbf{z}_k\}$. A commonly used form of the result is the following

$$\hat{\mathbf{x}}_{k/k} = \hat{\mathbf{x}}_{k/k-1} + K_k[\mathbf{z}_k - H_k \hat{\mathbf{x}}_{k/k-1}] \tag{1.3a}$$
$$\hat{\mathbf{x}}_{k+1/k} = F_k \hat{\mathbf{x}}_{k/k}, \tag{1.3b}$$

where

$$K_k = P_{k/k-1} H_k^T [H_k P_{k/k-1} H_k^T + R_k]^{-1} \tag{1.4a}$$
$$= P_{k/k} H_k^T R_k^{-1} \tag{1.4b}$$

and

$$P_{k/k} = P_{k/k-1} - K_k H_k P_{k/k-1} \tag{1.5a}$$
$$= (I - K_k H_k) P_{k/k-1} (I - K_k H_k)^T + K_k R_k K_k^T \tag{1.5b}$$
$$P_{k+1/k} = F_k P_{k/k} F_k^T + G_k Q_k G_k^T. \tag{1.5c}$$

The vector $\hat{\mathbf{x}}_{k/k}$ is referred to as the *filtered* estimate of $\mathbf{x}_k$ and $\hat{\mathbf{x}}_{k+1/k}$ is the (one-step) *predictor* of $\mathbf{x}_{k+1}$. Equation (1.4) provides two equivalent forms of the gain matrix, assuming that R_k^{-1} exists. The covariance of the error in $\hat{\mathbf{x}}_{k/k}$ is given by $P_{k/k}$ in (1.5a) or, equivalently, in (1.5b). The covariance of the prediction error is given as (1.5c). Note that (1.5b) is valid for *any* gain and reduces to (1.5a) by substituting (1.4).

With this introduction, consider the results presented by Kalman in this first paper of this section.

The model considered below by Kalman is stated in equations (16) and (17) of the paper and are reproduced below:

$$\mathbf{x}(t+1) = \Phi(t+1; t)\mathbf{x}(t) + u(t) \tag{K-16}$$
$$\mathbf{v}(t) = M(t)\mathbf{x}(t). \tag{K-17}$$

The identification of the differences in notational conventions is made by comparing (K-16) with (1.1) and (K-17) with (1.2). The most significant difference appears in (K-17) as measurement noise does *not* appear explicitly. This apparent difference is resolved by recognizing that Kalman assumed that the state vector $\mathbf{x}$ contains among its components the measurement noise variables. For example, suppose the output noise $\mathbf{v}_k$ in (1.2) is renamed as

$$\mathbf{x}_2(k) = \mathbf{v}_k$$

and the state $\mathbf{x}_k$ in (1.1) is augmented and redefined as

$$\mathbf{x}(k) = \begin{bmatrix} \mathbf{x}_k \\ \mathbf{v}_k \end{bmatrix}.$$

Now, the model (1.1)–(1.2) can be written as

$$\mathbf{x}(k+1) = \begin{bmatrix} \mathbf{x}_{k+1} \\ \mathbf{x}_2(k+1) \end{bmatrix} = \begin{bmatrix} F_k & 0 \\ 0 & 0 \end{bmatrix} \begin{bmatrix} \mathbf{x}_k \\ \mathbf{x}_2(k) \end{bmatrix} + \begin{bmatrix} G_k \mathbf{w}_k \\ \mathbf{v}_k \end{bmatrix}$$

$$\triangleq \Phi(k+1;\, k)\mathbf{x}(k) + \mathbf{u}(k)$$

and

$$\mathbf{z}_k = [H_k \quad I] \begin{bmatrix} \mathbf{x}_k \\ \mathbf{x}_2(k) \end{bmatrix} = M(k)\mathbf{x}(k).$$

Thus, the models are seen to have the same form.

Note that Kalman does *not* require that the output noise is white. It is necessary only that it be described by a linear, stochastic difference equation (i.e., a shaping filter).

The linear, minimum mean-square estimator of $\mathbf{x}_{k+1}$ given the measurements $\mathbf{z}_1, \mathbf{z}_2, \cdots, \mathbf{z}_k$ is stated by Kalman in equations (21), (28)–(30). Note that he provides the solution in the form of a recursion for the one-step predictor $\hat{\mathbf{x}}_{k+1/k}$. In addition, the form of the filter appears to be different from the common form that appears in most textbooks (i.e., as stated above in (1.3)–(1.5)) because his model (i.e., (K-16)–(K-17)) does not correspond with the form (1.1)–(1.2) that is used commonly. Nonetheless, the results given by Kalman are equivalent to (1.3)–(1.5). Examples 1 and 2 of Kalman's paper provide useful references for understanding the notational differences between (1.1)–(1.5) and Kalman's model (K-16), (K-17) and results (K-21), (K-28)–(K-31).

The paper by Peter Swerling that appears next in the collection of this section predates Kalman's paper by approximately one year. The point-of-view is markedly different. Whereas Kalman was concerned with a general and highly theoretical result, Swerling was motivated by the need to define a practical method for smoothing satellite data. Note that the brief abstract for the paper states "A practical method of smoothing satellite data by evaluating a finite number of parameters, or elements, is presented." The early applications of the Kalman filter were motivated by problems similar to the problem considered by Swerling. Swerling discusses the features of the problem that motivated his investigation and then develops the *stagewise procedure* that we can now relate to the Kalman filter.

Kalman approached the estimation problem using the Wiener approach (i.e., as a linear, minimum mean-squared error estimation problem). Swerling took an entirely different approach by formulating a nonlinear least-squares problem (Swerling's equation (3)). Swerling did not use a state variable approach, again in contrast to Kalman, nor did he assume a linear model. His formulation initially involves constant parameters and a measurement model that is nonlinear with additive noise. To obtain his stagewise procedure, the least-squares index is written in terms of a current estimate and new measurements (i.e., equation (16)) with the weighting matrices assumed to be block diagonal (i.e., white noise).

To obtain an approximate solution of the nonlinear, least-squares problem, Swerling introduces a Taylor series expansion (equation (6)) relative to the estimate and retains only the first-order term. Thus, Swerling develops results which are comparable to the extended Kalman filter (EKF). Equations (8)–(13) provide a linearized characterization of the errors in the least-squares estimate. The matrix B in (12) is equivalent to $P_{k/k}{}^{-1}$ of the Kalman filter (recall that Swerling's dynamic model is trivial in the sense that $\mathbf{x}_{k+1} = \mathbf{x}_k$). The stagewise representation of $P_{k/k}{}^{-1}$ is stated in equation (29). The recognition that B^{-1} represents the approximation of the error covariance matrix is culminated with (44) and (45).

Swerling extends his development to allow time-variation of the parameters of the problem according to the model stated in equation (51). Note that the variation does not consider random inputs but is nonlinear and time-varying. However, in his concluding remarks on page 52, Swerling notes that the model (51) "may not be known exactly." Then, he proceeds to discuss the effect of model errors. Equations (52)–(62) define the algorithm and uses both a predictor P^- and a filter P^+.

It is clear that Swerling developed a procedure that can be regarded, essentially, as an extended Kalman filter. Because his point-of-view and notation is so different from Kalman's, the study of the two papers can serve to be both complementary and illuminating. The notation used by Swerling may seem awkward and/or clumsy to many readers who are more familiar with the state variable notation used in most descriptions of the Kalman filter. In these cases, students have found it worthwhile to redevelop Swerling's paper using matrix notation.

Kalman and Bucy, in the third paper of this section, considered the filtering problem in the context of continuous-time systems described by linear, ordinary differential equations and continuous measurement processes. The authors state on page 102 in the sentence preceding Theorem 4 that "We can now state the central theorem of the paper." In this theorem the stability of the filter is established and the limiting value of the error covariance matrix is determined for any initial condition. While the result is stated and proven for continuous-time systems, a similar result can be developed for discrete-time systems. The proof of the stability of the filter is regarded by many as the most important contribution by Kalman to the theory of linear, recursive filtering. It is this analysis that sets Kalman apart from the other developers of similar recursive solutions.

Kalman discussed the results of his paper and of his

paper with Bucy in a most remarkable paper entitled "New Methods in Wiener Filtering Theory" listed below among the entries in the References for this section. Because of its length, the paper was not included in this volume. It contains a wealth of information and useful insights. While it is not always easy to read, its study is still worthwhile for anyone interested in a better grasp of a wide variety of theoretical issues for linear estimation. Other papers dealing with filter stability are listed below as references [3] through [8].

The remaining two papers in this section provide valuable insights into the Kalman filter algorithm. Ho and Lee discuss the Bayesian approach to the problem. In other words, they cast the problem in probabilistic terms and consider the evolution of the probability density function for the state $\mathbf{x}_k$ given the set of measurements $\mathbf{Z}_k \triangleq \{\mathbf{z}_1, \mathbf{z}_2, \cdots, \mathbf{z}_k\}$. That is, how is $p(\mathbf{x}_{k+1}|\mathbf{Z}_{k+1})$ derived from $p(\mathbf{x}_k|\mathbf{Z}_k)$ and from knowledge of the prior density function for the input noise $\mathbf{v}_{k+1}$? The general result is given in their equations (23) and (24).

The general result is applied to a linear Gaussian system, equations (25)–(27), that is the same as given above as (1.1)–(1.2). Then, the Kalman filter equations are derived and shown to define the mean and covariance for the posterior density function. This result provides a useful, alternative point-of-view for the derivation of the Kalman filter equations. Additionally, it provides the basis for much of the work that has been done during the past two decades on nonlinear estimation problems.

Kailath in a series of papers beginning with the final paper contained in this section returned to Wiener and Kalman's point-of-view to establish other theoretical characteristics of the Kalman filter that are very useful and valuable. In this paper, Kailath began the examination of the properties of the *innovations sequence* generated by the Kalman filter. The innovations sequence $\{\mathbf{i}_k\}$ is defined as

$$\mathbf{i}_k \triangleq \mathbf{z}_k - H_k\hat{\mathbf{x}}_{k/k-1} \qquad k=1, 2, \cdots. \tag{1.6}$$

It is seen to represent the error in the prediction of the measurement $\mathbf{z}_k$ based on the measurements $\{\mathbf{z}_1, \mathbf{z}_2, \cdots, \mathbf{z}_{k-1}]$ since the best estimate of $\mathbf{z}_k$ given this information is

$$\hat{\mathbf{z}}_{k/k-1} = H_k\hat{\mathbf{x}}_{k/k-1}.$$

Thus, the innovation $\mathbf{i}_k$ can be regarded in a sense defined precisely by Kailath as the *new information* provided by the measurement $\mathbf{z}_k$.

The property possessed by the innovations sequence which provides the basis for both theoretical investigations and implementational testing is stated briefly. The innovations sequence, as defined by (1.6) for the system (1.1)–(1.2), is *white noise* with zero mean and covariance

$$\begin{aligned}\Omega_k &\triangleq E[\mathbf{i}_k\mathbf{i}_k^T]\\ &= H_kP_{k/k-1}H_k^T + R_k.\end{aligned} \tag{1.7}$$

This result for discrete time processes is derived in Appendix II of Kailath's paper; a similar result for continuous-time processes is also derived. Comparing and understanding the form of the covariance (i.e., (1.7) for discrete time) for the two cases provides the basis for understanding more clearly the essential differences between white noise processes and white noise sequences.

The innovations provide a very nice way of developing the filtering, prediction, and smoothing results of minimum mean-squared error estimation theory for linear systems. The properties of the innovations sequence provide a concrete basis for testing the behavior of a Kalman filter implementation for a specific application. This well-written paper provides a useful bridge between the theoretical properties of the Kalman filter and its application to practical problems.

References

[1] R. L. Stratonovich, "Application of the theory of Markoff processes in optimal signal detection," *Radio Eng. Electron. Phys.*, vol. 1, pp. 1–19, 1960.

[2] R. E. Kalman, "New methods in Wiener filtering theory," in *Proc. First Symp. Engineering Applications of Random Function Theory and Probability*. New York: Wiley, 1963, pp. 270–388.

[3] R. E. Kalman, Y. C. Ho, and K. S. Narendra, "Controllability of Linear Dynamical Systems," in *Contributions to Differential Equations*. New York: Wiley, 1962, pp. 189–213.

[4] E. Kreindler and P. E. Sarachik, "On the concepts of controllability and observability of linear systems," *IEEE Trans. Automat. Contr.*, vol. AC-9, pp. 129–136, 1964.

[5] H. W. Sorenson, "On the error behavior in linear minimum variance estimation problems," *IEEE Trans. Automat. Contr.*, vol. AC-12, pp. 557–562, 1967.

[6] ——, "Controllability and observability of linear, stochastic time-discrete control systems," in *Advances in Control Systems, Vol. 6*, C. T. Leondes, Ed. New York: Academic Press, 1968.

[7] R. A. Singer and P. A. Frost, "On the relative performance of the Kalman and Wiener filters," *IEEE Trans. Automat. Contr.*, vol. AC-14, pp. 391–394, 1969.

[8] B. D. O. Anderson, "Stability properties of Kalman–Bucy filters," *J. Franklin Inst.*, vol. 291, pp. 137–144, 1971.

A New Approach to Linear Filtering and Prediction Problems[1]

R. E. KALMAN
Research Institute for Advanced Study,[2]
Baltimore, Md.

The classical filtering and prediction problem is re-examined using the Bode-Shannon representation of random processes and the "state-transition" method of analysis of dynamic systems. New results are:

(1) The formulation and methods of solution of the problem apply without modification to stationary and nonstationary statistics and to growing-memory and infinite-memory filters.

(2) A nonlinear difference (or differential) equation is derived for the covariance matrix of the optimal estimation error. From the solution of this equation the coefficients of the difference (or differential) equation of the optimal linear filter are obtained without further calculations.

(3) The filtering problem is shown to be the dual of the noise-free regulator problem.

The new method developed here is applied to two well-known problems, confirming and extending earlier results.

The discussion is largely self-contained and proceeds from first principles; basic concepts of the theory of random processes are reviewed in the Appendix.

Introduction

An important class of theoretical and practical problems in communication and control is of a statistical nature. Such problems are: (i) Prediction of random signals; (ii) separation of random signals from random noise; (iii) detection of signals of known form (pulses, sinusoids) in the presence of random noise.

In his pioneering work, Wiener [1][3] showed that problems (i) and (ii) lead to the so-called Wiener-Hopf integral equation; he also gave a method (spectral factorization) for the solution of this integral equation in the practically important special case of stationary statistics and rational spectra.

Many extensions and generalizations followed Wiener's basic work. Zadeh and Ragazzini solved the finite-memory case [2]. Concurrently and independently of Bode and Shannon [3], they also gave a simplified method [2] of solution. Booton discussed the nonstationary Wiener-Hopf equation [4]. These results are now in standard texts [5-6]. A somewhat different approach along these main lines has been given recently by Darlington [7]. For extensions to sampled signals, see, e.g., Franklin [8], Lees [9]. Another approach based on the eigenfunctions of the Wiener-Hopf equation (which applies also to nonstationary problems whereas the preceding methods in general don't), has been pioneered by Davis [10] and applied by many others, e.g., Shinbrot [11], Blum [12], Pugachev [13], Solodovnikov [14].

In all these works, the objective is to obtain the specification of a linear dynamic system (Wiener filter) which accomplishes the prediction, separation, or detection of a random signal.[4]

[1] This research was supported in part by the U. S. Air Force Office of Scientific Research under Contract AF 49 (638)-382.

[2] 7212 Bellona Ave.

[3] Numbers in brackets designate References at end of paper.

[4] Of course, in general these tasks may be done better by nonlinear filters. At present, however, little or nothing is known about how to obtain (both theoretically and practically) these nonlinear filters.

Contributed by the Instruments and Regulators Division and presented at the Instruments and Regulators Conference, March 29–April 2, 1959, of The American Society of Mechanical Engineers.

Note: Statements and opinions advanced in papers are to be understood as individual expressions of their authors and not those of the Society. Manuscript received at ASME Headquarters, February 24, 1959. Paper No. 59—IRD-11.

Present methods for solving the Wiener problem are subject to a number of limitations which seriously curtail their practical usefulness:

(1) The optimal filter is specified by its impulse response. It is not a simple task to synthesize the filter from such data.

(2) Numerical determination of the optimal impulse response is often quite involved and poorly suited to machine computation. The situation gets rapidly worse with increasing complexity of the problem.

(3) Important generalizations (e.g., growing-memory filters, nonstationary prediction) require new derivations, frequently of considerable difficulty to the nonspecialist.

(4) The mathematics of the derivations are not transparent. Fundamental assumptions and their consequences tend to be obscured.

This paper introduces a new look at this whole assemblage of problems, sidestepping the difficulties just mentioned. The following are the highlights of the paper:

(5) *Optimal Estimates and Orthogonal Projections.* The Wiener problem is approached from the point of view of conditional distributions and expectations. In this way, basic facts of the Wiener theory are quickly obtained; the scope of the results and the fundamental assumptions appear clearly. It is seen that all statistical calculations and results are based on first and second order averages; no other statistical data are needed. Thus difficulty (4) is eliminated. This method is well known in probability theory (see pp. 75–78 and 148–155 of Doob [15] and pp. 455–464 of Loève [16]) but has not yet been used extensively in engineering.

(6) *Models for Random Processes.* Following, in particular, Bode and Shannon [3], arbitrary random signals are represented (up to second order average statistical properties) as the output of a linear dynamic system excited by independent or uncorrelated random signals ("white noise"). This is a standard trick in the engineering applications of the Wiener theory [2–7]. The approach taken here differs from the conventional one only in the way in which linear dynamic systems are described. We shall emphasize the concepts of ***state*** and ***state transition;*** in other words, linear systems will be specified by systems of first-order difference (or differential) equations. This point of view is

Reprinted with permission from *Trans. ASME, J. Basic Eng.,* Series 82D, pp. 35–45, Mar. 1960.

natural and also necessary in order to take advantage of the simplifications mentioned under (5).

(7) *Solution of the Wiener Problem.* With the state-transition method, a single derivation covers a large variety of problems: growing and infinite memory filters, stationary and nonstationary statistics, etc.; difficulty (3) disappears. Having guessed the "state" of the estimation (i.e., filtering or prediction) problem correctly, one is led to a nonlinear difference (or differential) equation for the covariance matrix of the optimal estimation error. This is vaguely analogous to the Wiener-Hopf equation. Solution of the equation for the covariance matrix starts at the time t_0 when the first observation is taken; at each later time t the solution of the equation represents the covariance of the optimal prediction error given observations in the interval (t_0, t). From the covariance matrix at time t we obtain at once, without further calculations, the coefficients (in general, time-varying) characterizing the optimal linear filter.

(8) *The Dual Problem.* The new formulation of the Wiener problem brings it into contact with the growing new theory of control systems based on the "state" point of view [17–24]. It turns out, *surprisingly*, that the Wiener problem is the *dual* of the noise-free optimal regulator problem, which has been solved previously by the author, using the state-transition method to great advantage [18, 23, 24]. The mathematical background of the two problems is identical—this has been suspected all along, but until now the analogies have never been made explicit.

(9) *Applications.* The power of the new method is most apparent in theoretical investigations and in numerical answers to complex practical problems. In the latter case, it is best to resort to machine computation. Examples of this type will be discussed later. To provide some feel for applications, two standard examples from nonstationary prediction are included; in these cases the solution of the nonlinear difference equation mentioned under (7) above can be obtained even in closed form.

For easy reference, the main results are displayed in the form of theorems. Only Theorems 3 and 4 are original. The next section and the Appendix serve mainly to review well-known material in a form suitable for the present purposes.

Notation Conventions

Throughout the paper, we shall deal mainly with *discrete* (or *sampled*) dynamic systems; in other words, signals will be observed at equally spaced points in time (*sampling instants*). By suitable choice of the time scale, the constant intervals between successive sampling instants (*sampling periods*) may be chosen as unity. Thus variables referring to time, such as t, t_0, τ, T will always be integers. The restriction to discrete dynamic systems is not at all essential (at least from the engineering point of view); by using the discreteness, however, we can keep the mathematics rigorous and yet elementary. Vectors will be denoted by small bold-face letters: $\mathbf{a}, \mathbf{b}, \ldots, \mathbf{u}, \mathbf{x}, \mathbf{y}, \ldots$ A *vector* or more precisely an *n-vector* is a set of n numbers $x_1, \ldots x_n$; the x_i are the *co-ordinates* or *components* of the vector $\mathbf{x}$.

Matrices will be denoted by capital bold-face letters: $\mathbf{A}, \mathbf{B}, \mathbf{Q}, \mathbf{\Phi}, \mathbf{\Psi}, \ldots$; they are $m \times n$ arrays of elements $a_{ij}, b_{ij}, q_{ij}, \ldots$ The *transpose* (interchanging rows and columns) of a matrix will be denoted by the prime. In manipulating formulas, it will be convenient to regard a vector as a matrix with a single column.

Using the conventional definition of matrix multiplication, we write the *scalar product* of two n-vectors $\mathbf{x}, \mathbf{y}$ as

$$\mathbf{x}'\mathbf{y} = \sum_{i=1}^{n} x_i y_i = \mathbf{y}'\mathbf{x}$$

The scalar product is clearly a scalar, i.e., not a vector, quantity. Similarly, the *quadratic form* associated with the $n \times n$ matrix $\mathbf{Q}$ is

$$\mathbf{x}'\mathbf{Q}\mathbf{x} = \sum_{i,j=1}^{n} x_i q_{ij} x_j$$

We define the expression $\mathbf{x}\mathbf{y}'$ where $\mathbf{x}$ is an m-vector and $\mathbf{y}$ is an n-vector to be the $m \times n$ matrix with elements $x_i y_j$.

We write $E(\mathbf{x}) = E\mathbf{x}$ for the expected value of the random vector $\mathbf{x}$ (see Appendix). It is usually convenient to omit the brackets after E. This does not result in confusion in simple cases since constants and the operator E commute. Thus $E\mathbf{x}\mathbf{y}'$ = matrix with elements $E(x_i y_j)$; $E\mathbf{x}E\mathbf{y}'$ = matrix with elements $E(x_i)E(y_j)$.

For ease of reference, a list of the principal symbols used is given below.

Optimal Estimates

t	time in general; present time.
t_0	time at which observations start.
$x_1(t)$, $x_2(t)$	basic random variables.
$y(t)$	observed random variable.
$x_1^*(t_1\|t)$	optimal estimate of $x_1(t_1)$ given $y(t_0), \ldots, y(t)$.
L	loss function (nonrandom function of its argument).
ϵ	estimation error (random variable).

Orthogonal Projections

$\mathcal{Y}(t)$	linear manifold generated by the random variables $y(t_0), \ldots, y(t)$.
$\bar{x}(t_1\|t)$	orthogonal projection of $x(t_1)$ on $\mathcal{Y}(t)$.
$\tilde{x}(t_1\|t)$	component of $x(t_1)$ orthogonal to $\mathcal{Y}(t)$.

Models for Random Processes

$\mathbf{\Phi}(t+1;\ t)$	transition matrix
$\mathbf{Q}(t)$	covariance of random excitation

Solution of the Wiener Problem

$\mathbf{x}(t)$	basic random variable.
$\mathbf{y}(t)$	observed random variable.
$\mathcal{Y}(t)$	linear manifold generated by $\mathbf{y}(t_0), \ldots, \mathbf{y}(t)$.
$\mathcal{Z}(t)$	linear manifold generated by $\tilde{\mathbf{y}}(t\|t-1)$.
$\mathbf{x}^*(t_1\|t)$	optimal estimate of $\mathbf{x}(t_1)$ given $\mathcal{Y}(t)$.
$\tilde{\mathbf{x}}(t_1\|t)$	error in optimal estimate of $\mathbf{x}(t_1)$ given $\mathcal{Y}(t)$.

Optimal Estimates

To have a concrete description of the type of problems to be studied, consider the following situation. We are given signal $x_1(t)$ and noise $x_2(t)$. Only the sum $y(t) = x_1(t) + x_2(t)$ can be observed. Suppose we have observed and know exactly the values of $y(t_0), \ldots, y(t)$. What can we infer from this knowledge in regard to the (unobservable) value of the signal at $t = t_1$, where t_1 may be less than, equal to, or greater than t? If $t_1 < t$, this is a *data-smoothing* (*interpolation*) problem. If $t_1 = t$, this is called *filtering*. If $t_1 > t$, we have a *prediction* problem. Since our treatment will be general enough to include these and similar problems, we shall use hereafter the collective term *estimation*.

As was pointed out by Wiener [1], the natural setting of the estimation problem belongs to the realm of probability theory and statistics. Thus signal, noise, and their sum will be random variables, and consequently they may be regarded as random processes. From the probabilistic description of the random processes we can determine the probability with which a particular sample of the signal and noise will occur. For any given set of measured values $\eta(t_0), \ldots, \eta(t)$ of the random variable $y(t)$ one can then also determine, in principle, the probability of simultaneous occurrence of various values $\xi_1(t)$ of the random variable $x_1(t_1)$. This is the conditional probability distribution function

$$Pr[x_1(t_1) \leqq \xi_1 | y(t_0) = \eta(t_0), \ldots, y(t) = \eta(t)] = F(\xi_1) \quad (1)$$

Evidently, $F(\xi_1)$ represents all the information which the measurement of the random variables $y(t_0), \ldots, y(t)$ has conveyed about the random variable $x_1(t_1)$. Any statistical estimate of the random variable $x_1(t_1)$ will be some function of this distribution and therefore a (nonrandom) function of the random variables $y(t_0), \ldots, y(t)$. This statistical estimate is denoted by $X_1(t_1|t)$, or by just $X_1(t_1)$ or X_1 when the set of observed random variables or the time at which the estimate is required are clear from context.

Suppose now that X_1 is given as a fixed function of the random variables $y(t_0), \ldots, y(t)$. Then X_1 is itself a random variable and its actual value is known whenever the actual values of $y(t_0), \ldots, y(t)$ are known. In general, the actual value of $X_1(t_1)$ will be different from the (unknown) actual value of $x_1(t_1)$. To arrive at a rational way of determining X_1, it is natural to assign a *penalty* or *loss* for incorrect estimates. Clearly, the loss should be a (i) positive, (ii) nondecreasing function of the *estimation error* $\epsilon = x_1(t_1) - X_1(t_1)$. Thus we define a *loss function* by

$$L(0) = 0$$
$$L(\epsilon_2) \geqq L(\epsilon_1) \geqq 0 \quad \text{when} \quad \epsilon_2 \geqq \epsilon_1 \geqq 0 \quad (2)$$
$$L(\epsilon) = L(-\epsilon)$$

Some common examples of loss functions are: $L(\epsilon) = a\epsilon^2, a\epsilon^4, a|\epsilon|, a[1 - \exp(-\epsilon^2)]$, etc., where a is a positive constant.

One (but by no means the only) natural way of choosing the random variable X_1 is to require that this choice should minimize the average loss or risk

$$E\{L[x_1(t_1) - X_1(t_1)]\} = E[E\{L[x(t_1) - X_1(t_1)]|y(t_0), \ldots, y(t)\}] \quad (3)$$

Since the first expectation on the right-hand side of (3) does not depend on the choice of X_1 but only on $y(t_0), \ldots, y(t)$, it is clear that minimizing (3) is equivalent to minimizing

$$E\{L[x_1(t_1) - X_1(t_1)]|y(t_0), \ldots, y(t)\} \quad (4)$$

Under just slight additional assumptions, optimal estimates can be characterized in a simple way.

Theorem 1. *Assume that L is of type* (2) *and that the conditional distribution function* $F(\xi)$ *defined by* (1) *is:*

(*A*) *symmetric about the mean* $\bar{\xi}$:

$$F(\xi - \bar{\xi}) = 1 - F(\bar{\xi} - \xi)$$

(*B*) *convex for* $\xi \leqq \bar{\xi}$:

$$F(\lambda\xi_1 + (1 - \lambda)\xi_2) \leqq \lambda F(\xi_1) + (1 - \lambda)F(\xi_2)$$

for all $\xi_1, \xi_2 \leqq \bar{\xi}$ *and* $0 \leqq \lambda \leqq 1$

Then the random variable $x_1^*(t_1|t)$ *which minimizes the average loss* (3) *is the conditional expectation*

$$x_1^*(t_1|t) = E[x_1(t_1)|y(t_0), \ldots, y(t)] \quad (5)$$

Proof: As pointed out recently by Sherman [25], this theorem follows immediately from a well-known lemma in probability theory.

Corollary. *If the random processes* $\{x_1(t)\}$, $\{x_2(t)\}$, *and* $\{y(t)\}$ *are gaussian, Theorem 1 holds.*

Proof: By Theorem 5, (*A*) (see Appendix), conditional distributions on a gaussian random process are gaussian. Hence the requirements of Theorem 1 are always satisfied.

In the control systems literature, this theorem appears sometimes in a form which is more restrictive in one way and more general in another way:

Theorem 1-a. *If* $L(\epsilon) = \epsilon^2$, *then Theorem 1 is true without assumptions* (*A*) *and* (*B*).

Proof: Expand the conditional expectation (4):

$$E[x_1^2(t_1)|y(t_0), \ldots, y(t)] - 2X_1(t_1)E[x_1(t_1)|y(t_0), \ldots, y(t)] + X_1^2(t_1)$$

and differentiate with respect to $X_1(t_1)$. This is not a completely rigorous argument; for a simple rigorous proof see Doob [15], pp. 77–78.

Remarks: (*a*) As far as the author is aware, it is not known what is the most general class of random processes $\{x_1(t)\}$, $\{x_2(t)\}$ for which the conditional distribution function satisfies the requirements of Theorem 1.

(*b*) Aside from the note of Sherman, Theorem 1 apparently has never been stated explicitly in the control systems literature. In fact, one finds many statements to the effect that loss functions of the general type (2) cannot be conveniently handled mathematically.

(*c*) In the sequel, we shall be dealing mainly with vector-valued random variables. In that case, the estimation problem is stated as: Given a vector-valued random process $\{\mathbf{x}(t)\}$ and observed random variables $\mathbf{y}(t_0), \ldots, \mathbf{y}(t)$, where $\mathbf{y}(t) = \mathbf{M}\mathbf{x}(t)$ ($\mathbf{M}$ being a singular matrix; in other words, not all co-ordinates of $\mathbf{x}(t)$ can be observed), find an estimate $\mathbf{X}(t_1)$ which minimizes the expected loss $E[L(\|\mathbf{x}(t_1) - \mathbf{X}(t_1)\|)]$, $\| \ \|$ being the norm of a vector.

Theorem 1 remains true in the vector case also, provided we require that the conditional distribution function of the n co-ordinates of the vector $\mathbf{x}(t_1)$,

$$Pr[x_1(t_1) \leqq \xi_1, \ldots, x_n(t_1) \leqq \xi_n | \mathbf{y}(t_0), \ldots, \mathbf{y}(t)] = F(\xi_1, \ldots, \xi_n)$$

be symmetric with respect to the n variables $\xi_1 - \bar{\xi}_1, \ldots, \xi_n - \bar{\xi}_n$ and convex in the region where all of these variables are negative.

Orthogonal Projections

The explicit calculation of the optimal estimate as a function of the observed variables is, in general, impossible. There is an important exception: The processes $\{x_1(t)\}$, $\{x_2(t)\}$ are gaussian.

On the other hand, if we attempt to get an optimal estimate under the restriction $L(\epsilon) = \epsilon^2$ and the additional requirement that the estimate be a linear function of the observed random variables, we get an estimate which is identical with the optimal estimate in the gaussian case, without the assumption of linearity or quadratic loss function. This shows that results obtainable by linear estimation can be bettered by nonlinear estimation only when (i) the random processes are nongaussian and even then (in view of Theorem 5, (*C*)) only (ii) by considering at least third-order probability distribution functions.

In the special cases just mentioned, the explicit solution of the estimation problem is most easily understood with the help of a geometric picture. This is the subject of the present section.

Consider the (real-valued) random variables $y(t_0), \ldots, y(t)$. The set of all linear combinations of these random variables with real coefficients

$$\sum_{i=t_0}^{t} a_i y(i) \quad (6)$$

forms a *vector space* (*linear manifold*) which we denote by $\mathcal{Y}(t)$. We regard, abstractly, any expression of the form (6) as "point" or "vector" in $\mathcal{Y}(t)$; this use of the word "vector" should not be confused, of course, with "vector-valued" random variables, etc. Since we do not want to fix the value of t (i.e., the total number of possible observations), $\mathcal{Y}(t)$ should be regarded as a finite-dimensional subspace of the space of all possible observations.

Given any two vectors u, v in $\mathcal{Y}(t)$ (i.e., random variables expressible in the form (6)), we say that u and v are *orthogonal* if $Euv = 0$. Using the Schmidt orthogonalization procedure, as described for instance by Doob [15], p. 151, or by Loève [16], p. 459, it is easy to select an *orthonormal basis* in $\mathcal{Y}(t)$. By this is meant a set of vectors $e_{t_0}, \ldots, e_t$ in $\mathcal{Y}(t)$ such that any vector in $\mathcal{Y}(t)$ can be expressed as a unique linear combination of $e_{t_0}, \ldots, e_t$ and

$$Ee_ie_j = \delta_{ij} = 1 \quad \text{if} \quad i = j \\ \qquad\quad = 0 \quad \text{if} \quad i \neq j \Bigg\} \quad (i, j = t_0, \ldots, t) \tag{7}$$

Thus any vector $\bar{x}$ in $\mathcal{Y}(t)$ is given by

$$\bar{x} = \sum_{i=t_0}^{t} a_i e_i$$

and so the coefficients a_i can be immediately determined with the aid of (7):

$$E\bar{x}e_j = E\left(\sum_{i=t_0}^{t} a_i e_i\right) e_j = \sum_{i=t_0}^{t} a_i Ee_ie_j = \sum_{i=t_0}^{t} a_i\delta_{ij} = a_j \tag{8}$$

It follows further that any random variable x (not necessarily in $\mathcal{Y}(t)$) can be uniquely decomposed into two parts: a part $\bar{x}$ in $\mathcal{Y}(t)$ and a part $\tilde{x}$ orthogonal to $\mathcal{Y}(t)$ (i.e., orthogonal to every vector in $\mathcal{Y}(t)$). In fact, we can write

$$x = \bar{x} + \tilde{x} = \sum_{i=t_0}^{t} (Exe_i)e_i + \tilde{x} \tag{9}$$

Thus $\bar{x}$ is uniquely determined by equation (9) and is obviously a vector in $\mathcal{Y}(t)$. Therefore $\tilde{x}$ is also uniquely determined; it remains to check that it is orthogonal to $\mathcal{Y}(t)$:

$$E\tilde{x}e_i = E(x - \bar{x})e_i = Exe_i - E\bar{x}e_i$$

Now the co-ordinates of $\bar{x}$ with respect to the basis $e_{t_0}, \ldots, e_t$ are given either in the form $E\bar{x}e_i$ (as in (8)) or in the form Exe_i (as in (9)). Since the co-ordinates are unique, $Exe_i = E\bar{x}e_i$ $(i = t_0, \ldots, t)$; hence $E\tilde{x}e_i = 0$ and $\tilde{x}$ is orthogonal to every base vector e_i and therefore to $\mathcal{Y}(t)$. We call $\bar{x}$ the *orthogonal projection* of x on $\mathcal{Y}(t)$.

There is another way in which the orthogonal projection can be characterized: $\bar{x}$ is that vector in $\mathcal{Y}(t)$ (i.e., that *linear* function of the random variables $y(t_0), \ldots, y(t)$) which minimizes the quadratic loss function. In fact, if $\bar{w}$ is any other vector in $\mathcal{Y}(t)$, we have

$$E(x - \bar{w})^2 = E(\tilde{x} + \bar{x} - \bar{w})^2 = E[(x - \bar{x}) + (\bar{x} - \bar{w})]^2$$

Since $\tilde{x}$ is orthogonal to every vector in $\mathcal{Y}(t)$ and in particular to $\bar{x} - \bar{w}$ we have

$$E(x - \bar{w})^2 = E(x - \bar{x})^2 + E(\bar{x} - \bar{w})^2 \geqq E(x - \bar{x})^2 \tag{10}$$

This shows that, if $\bar{w}$ also minimizes the quadratic loss, we must have $E(\bar{x} - \bar{w})^2 = 0$ which means that the random variables $\bar{x}$ and $\bar{w}$ are equal (except possibly for a set of events whose probability is zero).

These results may be summarized as follows:

Theorem 2. *Let $\{x(t)\}$, $\{y(t)\}$ random processes with zero mean (i.e., $Ex(t) = Ey(t) = 0$ for all t). We observe $y(t_0), \ldots, y(t)$.*

If either

(A) the random processes $\{x(t)\}$, $\{y(t)\}$ are gaussian; or

(B) the optimal estimate is restricted to be a linear function of the observed random variables and $L(\epsilon) = \epsilon^2$;

then

$$x^*(t_1|t) = \textit{optimal estimate of } x(t_1) \textit{ given } y(t_0), \ldots, y(t)$$
$$= \textit{orthogonal projection } \bar{x}(t_1|t) \textit{ of } x(t_1) \textit{ on } \mathcal{Y}(t). \tag{11}$$

These results are well-known though not easily accessible in the control systems literature. See Doob [15], pp. 75–78, or Pugachev [26]. It is sometimes convenient to denote the orthogonal projection by

$$\bar{x}(t_1|t) \equiv x^*(t_1|t) = \hat{E}[x(t_1)|\mathcal{Y}(t)]$$

The notation $\hat{E}$ is motivated by part (b) of the theorem: If the stochastic processes in question are gaussian, then orthogonal projection is actually identical with conditional expectation.

Proof. (A) This is a direct consequence of the remarks in connection with (10).

(B) Since $x(t)$, $y(t)$ are random variables with zero mean, it is clear from formula (9) that the orthogonal part $\tilde{x}(t_1|t)$ of $x(t_1)$ with respect to the linear manifold $\mathcal{Y}(t)$ is also a random variable with zero mean. Orthogonal random variables with zero mean are uncorrelated; if they are also gaussian then (by Theorem 5 (B)) they are independent. Thus

$$0 = E\tilde{x}(t_1|t) = E[\tilde{x}(t_1|t)|y(t_0), \ldots, y(t)]$$
$$= E[x(t_1) - \bar{x}(t_1|t)|y(t_0), \ldots, y(t)]$$
$$= E[x(t_1)|y(t_0), \ldots, y(t)] - \bar{x}(t_1|t) = 0$$

Remarks. (d) A rigorous formulation of the contents of this section as $t \to \infty$ requires some elementary notions from the theory of Hilbert space. See Doob [15] and Loève [16].

(e) The physical interpretation of Theorem 2 is largely a matter of taste. If we are not worried about the assumption of gaussianness, part (A) shows that the orthogonal projection is the optimal estimate for all reasonable loss functions. If we do worry about gaussianness, even if we are resigned to consider only linear estimates, we know that orthogonal projections are *not* the optimal estimate for many reasonable loss functions. Since in practice it is difficult to ascertain to what degree of approximation a random process of physical origin is gaussian, it is hard to decide whether Theorem 2 has very broad or very limited significance.

(f) Theorem 2 is immediately generalized for the case of vector-valued random variables. In fact, we define the linear manifold $\mathcal{Y}(t)$ generated by $\mathbf{y}(t_0), \ldots, \mathbf{y}(t)$ to be the set of all linear combinations

$$\sum_{i=t_0}^{t} \sum_{j=1}^{m} a_{ij} y_j(i)$$

of all m co-ordinates of each of the random vectors $\mathbf{y}(t_0), \ldots, \mathbf{y}(t)$. The rest of the story proceeds as before.

(g) Theorem 2 states in effect that the optimal estimate under conditions (A) or (B) is a linear combination of all previous observations. In other words, the optimal estimate can be regarded as the output of a linear filter, with the input being the actually occurring values of the observable random variables; Theorem 2 gives a way of computing the impulse response of the optimal filter. As pointed out before, knowledge of this impulse response is not a complete solution of the problem; for this reason, no explicit formulas for the calculation of the impulse response will be given.

Models for Random Processes

In dealing with physical phenomena, it is not sufficient to give an empirical description but one must have also some idea of the underlying causes. Without being able to separate in some sense causes and effects, i.e., without the assumption of causality, one can hardly hope for useful results.

It is a fairly generally accepted fact that primary macroscopic sources of random phenomena are independent gaussian processes.[5] A well-known example is the noise voltage produced in a resistor due to thermal agitation. In most cases, *observed* random phenomena are not describable by independent random variables. The statistical dependence (correlation) between random signals observed at different times is usually explained by the presence of a dynamic system between the primary random source and the observer. *Thus a random function of time may be thought of as the output of a dynamic system excited by an independent gaussian random process.*

An important property of gaussian random signals is that they remain gaussian after passing through a linear system (Theorem 5 (*A*)). Assuming independent gaussian primary random sources, if the observed random signal is also gaussian, we may assume that the dynamic system between the observer and the primary source is *linear*. This conclusion may be forced on us also because of lack of detailed knowledge of the statistical properties of the observed random signal: Given any random process with known first and second-order averages, we can find a gaussian random process with the same properties (Theorem 5 (*C*)). Thus gaussian distributions and linear dynamics are natural, mutually plausible assumptions particularly when the statistical data are scant.

How is a dynamic system (linear or nonlinear) described? The fundamental concept is the notion of the *state*. By this is meant, intuitively, some quantitative information (a set of numbers, a function, etc.) which is the least amount of data one has to know about the past behavior of the system in order to predict its future behavior. The dynamics is then described in terms of *state transitions*, i.e., one must specify how one state is transformed into another as time passes.

A *linear* dynamic system may be described in general by the vector differential equation

$$\left.\begin{aligned} d\mathbf{x}/dt &= \mathbf{F}(t)\mathbf{x} + \mathbf{D}(t)\mathbf{u}(t) \\ &\text{and} \\ \mathbf{y}(t) &= \mathbf{M}(t)\mathbf{x}(t) \end{aligned}\right\} \quad (12)$$

where $\mathbf{x}$ is an n-vector, the *state* of the system (the components x_i of $\mathbf{x}$ are called *state variables*); $\mathbf{u}(t)$ is an m-vector ($m \leq n$) representing the *inputs* to the system; $\mathbf{F}(t)$ and $\mathbf{D}(t)$ are $n \times n$, respectively, $n \times m$ matrices. If all coefficients of $\mathbf{F}(t)$, $\mathbf{D}(t)$, $\mathbf{M}(t)$ are constants, we say that the dynamic system (12) is *time-invariant* or *stationary*. Finally, $\mathbf{y}(t)$ is a p-vector denoting the outputs of the system; $\mathbf{M}(t)$ is an $n \times p$ matrix; $p \leq n$.

The physical interpretation of (12) has been discussed in detail elsewhere [18, 20, 23]. A look at the block diagram in Fig. 1 may be helpful. This is not an ordinary but a matrix block diagram (as revealed by the fat lines indicating signal flow). The integrator in Fig. 1 actually stands for n integrators such that the output of each is a state variable; $\mathbf{F}(t)$ indicates how the outputs of the integrators are fed back to the inputs of the integrators. Thus $f_{ij}(t)$ is the coefficient with which the output of the jth integrator is fed back to the input of the ith integrator. It is not hard to relate this formalism to more conventional methods of linear system analysis.

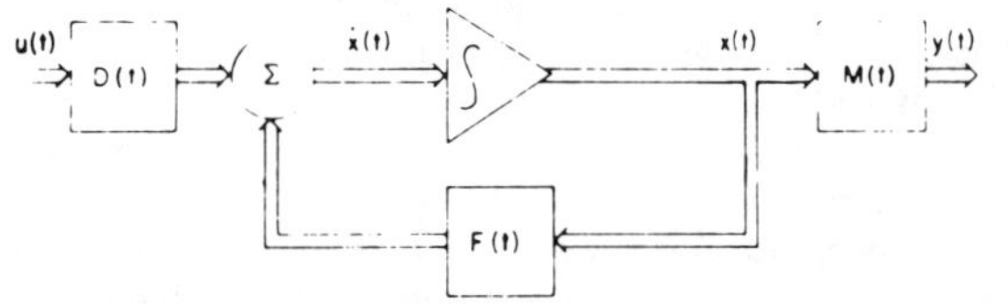

Fig. 1 Matrix block diagram of the general linear continuous-dynamic system

If we assume that the system (12) is stationary and that $\mathbf{u}(t)$ is constant during each sampling period, that is

$$\mathbf{u}(t + \tau) = \mathbf{u}(t); \; 0 \leq \tau < 1, \quad t = 0, 1, \ldots \qquad (13)$$

then (12) can be readily transformed into the more convenient discrete form

$$\mathbf{x}(t + 1) = \mathbf{\Phi}(1)\mathbf{x}(t) + \mathbf{\Delta}(1)\mathbf{u}(t); \; t = 0, 1, \ldots$$

where [18, 20]

$$\mathbf{\Phi}(1) = \exp \mathbf{F} = \sum_{i=0}^{\infty} \mathbf{F}^i/i! \quad (\mathbf{F}^0 = \text{unit matrix})$$

and

$$\mathbf{\Delta}(1) = \left(\int_0^1 \exp \mathbf{F}\tau d\tau\right) \mathbf{D}$$

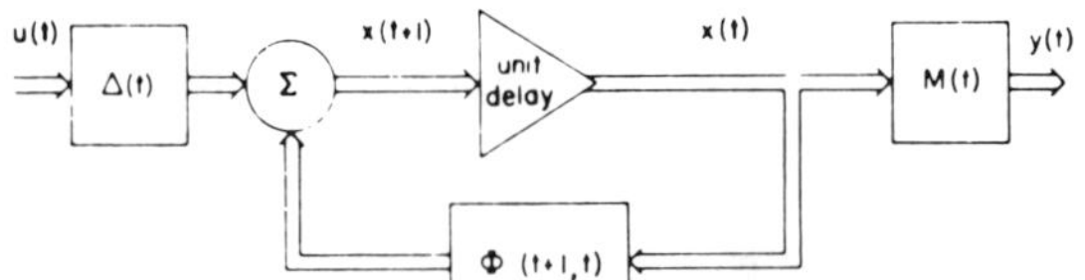

Fig. 2 Matrix block diagram of the general linear discrete-dynamic system

See Fig. 2. One could also express *exp* $\mathbf{F}\tau$ in closed form using Laplace transform methods [18, 20, 22, 24]. If $\mathbf{u}(t)$ satisfies (13) but the system (12) is nonstationary, we can write analogously

$$\left.\begin{aligned} \mathbf{x}(t + 1) &= \mathbf{\Phi}(t + 1; \; t) + \mathbf{\Delta}(t)\mathbf{u}(t) \\ \mathbf{y}(t) &= \mathbf{M}(t)\mathbf{x}(t) \end{aligned}\right\} \; t = 0, 1, \ldots \qquad (14)$$

but of course now $\mathbf{\Phi}(t + 1; \; t)$, $\mathbf{\Delta}(t)$ cannot be expressed in general in closed form. Equations of type (14) are encountered frequently also in the study of complicated sampled-data systems [22]. See Fig. 2.

$\mathbf{\Phi}(t + 1; \; t)$ is the *transition matrix* of the system (12) or (14). The notation $\mathbf{\Phi}(t_2; \; t_1)$ (t_2, t_1 = integers) indicates transition from time t_1 to time t_2. Evidently $\mathbf{\Phi}(t; \; t) = \mathbf{I}$ = unit matrix. If the system (12) is stationary then $\mathbf{\Phi}(t + 1; \; t) = \mathbf{\Phi}(t + 1 - t) = \mathbf{\Phi}(1)$ = const. Note also the product rule: $\mathbf{\Phi}(t; \; s)\mathbf{\Phi}(s; \; r) = \mathbf{\Phi}(t; \; r)$ and the inverse rule $\mathbf{\Phi}^{-1}(t; \; s) = \mathbf{\Phi}(s; \; t)$, where t, s, r are integers. In a stationary system, $\mathbf{\Phi}(t; \; \tau) = \exp \mathbf{F}(t - \tau)$.

As a result of the preceding discussion, we shall represent random phenomena by the model

$$\mathbf{x}(t + 1) = \mathbf{\Phi}(t + 1; \; t)\mathbf{x}(t) + \mathbf{u}(t) \qquad (15)$$

where $\{\mathbf{u}(t)\}$ is a vector-valued, independent, gaussian random process, with zero mean, which is completely described by (in view of Theorem 5 (*C*))

$$E\mathbf{u}(t) = \mathbf{0} \text{ for all } t;$$

$$E\mathbf{u}(t)\mathbf{u}'(s) = \mathbf{0} \quad \text{if} \quad t \neq s$$

$$E\mathbf{u}(t)\mathbf{u}'(t) = \mathbf{Q}(t).$$

Of course (Theorem 5 (*A*)), $\mathbf{x}(t)$ is then also a gaussian random process with zero mean, but it is no longer independent. In fact, if we consider (15) in the steady state (assuming it is a stable system), in other words, if we neglect the initial state $\mathbf{x}(t_0)$, then

[5] The probability distributions will be gaussian because macroscopic random effects may be thought of as the superposition of very many microscopic random effects; under very general conditions, such aggregate effects tend to be gaussian, regardless of the statistical properties of the microscopic effects. The assumption of independence in this context is motivated by the fact that microscopic phenomena tend to take place much more rapidly than macroscopic phenomena; thus primary random sources would appear to be independent on a macroscopic time scale.

$$\mathbf{x}(t) = \sum_{r=-\infty}^{t-1} \mathbf{\Phi}(t;\ r+1)\mathbf{u}(r).$$

Therefore if $t \geqq s$ we have

$$E\mathbf{x}(t)\mathbf{x}'(s) = \sum_{r=-\infty}^{s-1} \mathbf{\Phi}(t;\ r+1)\mathbf{Q}(r)\mathbf{\Phi}'(s;\ r+1).$$

Thus if we assume a linear dynamic model and know the statistical properties of the gaussian random excitation, it is easy to find the corresponding statistical properties of the gaussian random process $\{\mathbf{x}(t)\}$.

In real life, however, the situation is usually reversed. One is given the covariance matrix $E\mathbf{x}(t)\mathbf{x}'(s)$ (or rather, one attempts to estimate the matrix from limited statistical data) and the problem is to get (15) and the statistical properties of $\mathbf{u}(t)$. This is a subtle and presently largely unsolved problem in experimentation and data reduction. As in the vast majority of the engineering literature on the Wiener problem, we shall find it convenient to start with the model (15) and regard the problem of obtaining the model itself as a separate question. To be sure, the two problems *should* be optimized jointly if possible; the author is not aware, however, of any study of the *joint* optimization problem.

In summary, the following assumptions are made about random processes:

Physical random phenomena may be thought of as due to primary random sources exciting dynamic systems. The primary sources are assumed to be independent gaussian random processes with zero mean; the dynamic systems will be linear. The random processes are therefore described by models such as (15). The question of how the numbers specifying the model are obtained from experimental data will not be considered.

Solution of the Wiener Problem

Let us now define the principal problem of the paper.

Problem I. *Consider the dynamic model*

$$\mathbf{x}(t+1) = \mathbf{\Phi}(t+1;\ t)\mathbf{x}(t) + \mathbf{u}(t) \tag{16}$$

$$\mathbf{y}(t) = \mathbf{M}(t)\mathbf{x}(t) \tag{17}$$

where $\mathbf{u}(t)$ *is an independent gaussian random process of n-vectors with zero mean,* $\mathbf{x}(t)$ *is an n-vector,* $\mathbf{y}(t)$ *is a p-vector* ($p \leqq n$), $\mathbf{\Phi}(t+1;\ t)$, $\mathbf{M}(t)$ *are* $n \times n$, *resp.* $p \times n$, *matrices whose elements are nonrandom functions of time.*

Given the observed values of $\mathbf{y}(t_0), \ldots, \mathbf{y}(t)$ *find an estimate* $\mathbf{x}^*(t_1|t)$ *of* $\mathbf{x}(t_1)$ *which minimizes the expected loss.* (See Fig. 2, where $\mathbf{\Delta}(t) = \mathbf{I}$.)

This problem includes as a special case the problems of filtering, prediction, and data smoothing mentioned earlier. It includes also the problem of reconstructing all the state variables of a linear dynamic system from noisy observations of some of the state variables ($p < n!$).

From Theorem 2-*a* we know that the solution of Problem I is simply the orthogonal projection of $\mathbf{x}(t_1)$ on the linear manifold $\mathcal{Y}(t)$ generated by the observed random variables. As remarked in the Introduction, this is to be accomplished by means of a linear (not necessarily stationary!) dynamic system of the general form (14). With this in mind, we proceed as follows.

Assume that $\mathbf{y}(t_0), \ldots, \mathbf{y}(t-1)$ have been measured, i.e., that $\mathcal{Y}(t-1)$ is known. Next, at time t, the random variable $\mathbf{y}(t)$ is measured. As before let $\tilde{\mathbf{y}}(t|t-1)$ be the component of $\mathbf{y}(t)$ orthogonal to $\mathcal{Y}(t-1)$. If $\tilde{\mathbf{y}}(t|t-1) \equiv 0$, which means that the values of all components of this random vector are zero for almost every possible event, then $\mathcal{Y}(t)$ is obviously the same as $\mathcal{Y}(t-1)$ and therefore the measurement of $\mathbf{y}(t)$ does not convey any additional information. This is not likely to happen in a physically meaningful situation. In any case, $\tilde{\mathbf{y}}(t|t-1)$ generates a linear manifold (possibly 0) which we denote by $\mathcal{Z}(t)$. By definition, $\mathcal{Y}(t-1)$ and $\mathcal{Z}(t)$ taken together are the same manifold as $\mathcal{Y}(t)$, and every vector in $\mathcal{Z}(t)$ is orthogonal to every vector in $\mathcal{Y}(t-1)$.

Assuming by induction that $\mathbf{x}^*(t_1 - 1|t-1)$ is known, we can write:

$$\begin{aligned}\mathbf{x}^*(t_1|t) &= \hat{E}[\mathbf{x}(t_1)|\mathcal{Y}(t)] = \hat{E}[\mathbf{x}(t_1)|\mathcal{Y}(t-1)] + \hat{E}[\mathbf{x}(t_1)|\mathcal{Z}(t)] \\ &= \mathbf{\Phi}(t+1;\ t)\mathbf{x}^*(t_1 - 1|t-1) + \hat{E}[\mathbf{u}(t_1 - 1)|\mathcal{Y}(t-1)] \\ &\quad + \hat{E}[\mathbf{x}(t_1)|\mathcal{Z}(t)]\end{aligned} \tag{18}$$

where the last line is obtained using (16).

Let $t_1 = t + s$, where s is any integer. If $s \geqq 0$, then $\mathbf{u}(t_1 - 1)$ is independent of $\mathcal{Y}(t-1)$. This is because $\mathbf{u}(t_1 - 1) = \mathbf{u}(t + s - 1)$ is then independent of $\mathbf{u}(t-2), \mathbf{u}(t-3), \ldots$ and therefore by (16–17), independent of $\mathbf{y}(t_0), \ldots, \mathbf{y}(t-1)$, hence independent of $\mathcal{Y}(t-1)$. Since, for all t, $\mathbf{u}(t_0)$ has zero mean by assumption, it follows that $\mathbf{u}(t_1 - 1)$ ($s \geqq 0$) is orthogonal to $\mathcal{Y}(t-1)$. Thus if $s \geqq 0$, the second term on the right-hand side of (18) vanishes; if $s < 0$, considerable complications result in evaluating this term. We shall consider only the case $t_1 \geqq t$. Furthermore, it will suffice to consider in detail only the case $t_1 = t + 1$ since the other cases can be easily reduced to this one.

The last term in (18) must be a linear operation on the random variable $\tilde{\mathbf{y}}(t|t-1)$:

$$\hat{E}[\mathbf{x}(t+1)|\mathcal{Z}(t)] = \mathbf{\Delta}^*(t)\tilde{\mathbf{y}}(t|t-1) \tag{19}$$

where $\mathbf{\Delta}^*(t)$ is an $n \times p$ matrix, and the star refers to "optimal filtering."

The component of $\mathbf{y}(t)$ lying in $\mathcal{Y}(t-1)$ is $\bar{\mathbf{y}}(t|t-1) = \mathbf{M}(t)\mathbf{x}^*(t|t-1)$. Hence

$$\tilde{\mathbf{y}}(t|t-1) = \mathbf{y}(t) - \bar{\mathbf{y}}(t|t-1) = \mathbf{y}(t) - \mathbf{M}(t)\mathbf{x}^*(t|t-1) \tag{20}$$

Combining (18–20) (see Fig. 3) we obtain

$$\mathbf{x}^*(t+1|t) = \mathbf{\Phi}^*(t+1;\ t)\mathbf{x}^*(t|t-1) + \mathbf{\Delta}^*(t)\mathbf{y}(t) \tag{21}$$

where

$$\mathbf{\Phi}^*(t+1;\ t) = \mathbf{\Phi}(t+1;\ t) - \mathbf{\Delta}^*(t)\mathbf{M}(t) \tag{22}$$

Thus optimal estimation is performed by a linear dynamic system of the same form as (14). The state of the estimator is the previous estimate, the input is the last measured value of the observable random variable $\mathbf{y}(t)$, the transition matrix is given by (22). Notice that physical realization of the optimal filter requires only (*i*) the model of the random process (*ii*) the operator $\mathbf{\Delta}^*(t)$.

The estimation error is also governed by a linear dynamic system. In fact,

$$\begin{aligned}\tilde{\mathbf{x}}(t+1|t) &= \mathbf{x}(t+1) - \mathbf{x}^*(t+1|t) \\ &= \mathbf{\Phi}(t+1;\ t)\mathbf{x}(t) + \mathbf{u}(t) - \mathbf{\Phi}^*(t+1;\ t)\mathbf{x}^*(t|t-1) \\ &\quad - \mathbf{\Delta}^*(t)\mathbf{M}(t)\mathbf{x}(t)\end{aligned}$$

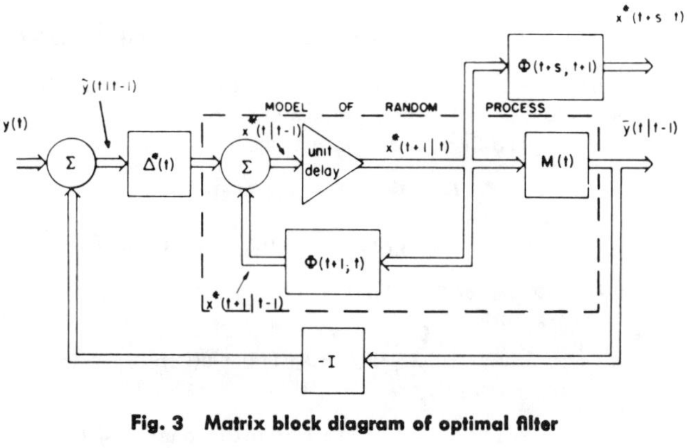

Fig. 3 Matrix block diagram of optimal filter

$$= \boldsymbol{\Phi}^*(t+1;\, t)\tilde{\mathbf{x}}(t|t-1) + \mathbf{u}(t) \tag{23}$$

Thus $\boldsymbol{\Phi}^*$ is also the transition matrix of the linear dynamic system governing the error.

From (23) we obtain at once a recursion relation for the covariance matrix $\mathbf{P}^*(t)$ of the optimal error $\tilde{\mathbf{x}}(t|t-1)$. Noting that $\mathbf{u}(t)$ is independent of $\mathbf{x}(t)$ and therefore of $\tilde{\mathbf{x}}(t|t-1)$, we get

$$\begin{aligned}
\mathbf{P}^*(t+1) &= E\tilde{\mathbf{x}}(t+1|t)\tilde{\mathbf{x}}'(t+1|t) \\
&= \boldsymbol{\Phi}^*(t+1;\, t)E\tilde{\mathbf{x}}(t|t-1)\tilde{\mathbf{x}}'(t|t-1)\boldsymbol{\Phi}^{*\prime}(t+1;\, t) + \mathbf{Q}(t) \\
&= \boldsymbol{\Phi}^*(t+1;\, t)E\tilde{\mathbf{x}}(t|t-1)\tilde{\mathbf{x}}'(t|t-1)\boldsymbol{\Phi}'(t+1;\, t) + \mathbf{Q}(t) \\
&= \boldsymbol{\Phi}^*(t+1;\, t)\mathbf{P}^*(t)\boldsymbol{\Phi}'(t+1;\, t) + \mathbf{Q}(t)
\end{aligned} \tag{24}$$

where $\mathbf{Q}(t) = E\mathbf{u}(t)\mathbf{u}'(t)$.

There remains the problem of obtaining an explicit formula for $\boldsymbol{\Delta}^*$ (and thus also for $\boldsymbol{\Phi}^*$). Since,

$$\tilde{\mathbf{x}}(t+1|\mathcal{Z}(t)) = \mathbf{x}(t+1) - \hat{E}[\mathbf{x}(t+1)|\mathcal{Z}(t)]$$

is orthogonal to $\tilde{\mathbf{y}}(t|t-1)$, it follows by (19) that

$$\begin{aligned}
0 &= E[\mathbf{x}(t+1) - \boldsymbol{\Delta}^*(t)\tilde{\mathbf{y}}(t|t-1)]\tilde{\mathbf{y}}'(t|t-1) \\
&= E\mathbf{x}(t+1)\tilde{\mathbf{y}}'(t|t-1) - \boldsymbol{\Delta}^*(t)E\tilde{\mathbf{y}}(t|t-1)\tilde{\mathbf{y}}'(t|t-1).
\end{aligned}$$

Noting that $\bar{\mathbf{x}}(t+1|t-1)$ is orthogonal to $\mathcal{Z}(t)$, the definition of $\mathbf{P}(t)$ given earlier, and (17), it follows further

$$\begin{aligned}
0 &= E\tilde{\mathbf{x}}(t+1|t-1)\tilde{\mathbf{y}}'(t|t-1) - \boldsymbol{\Delta}^*(t)\mathbf{M}(t)\mathbf{P}^*(t)\mathbf{M}'(t) \\
&= E[\boldsymbol{\Phi}(t+1;\, t)\tilde{\mathbf{x}}(t|t-1) + \mathbf{u}(t|t-1)]\tilde{\mathbf{x}}'(t|t-1)\mathbf{M}'(t) \\
&\qquad - \boldsymbol{\Delta}^*(t)\mathbf{M}(t)\mathbf{P}^*(t)\mathbf{M}'(t).
\end{aligned}$$

Finally, since $\mathbf{u}(t)$ is independent of $\mathbf{x}(t)$,

$$0 = \boldsymbol{\Phi}(t+1;\, t)\mathbf{P}^*(t)\mathbf{M}'(t) - \boldsymbol{\Delta}^*(t)\mathbf{M}(t)\mathbf{P}^*(t)\mathbf{M}'(t).$$

Now the matrix $\mathbf{M}(t)\mathbf{P}^*(t)\mathbf{M}'(t)$ will be positive definite and hence invertible whenever $\mathbf{P}^*(t)$ is positive definite, provided that none of the rows of $\mathbf{M}(t)$ are linearly dependent at any time, in other words, that none of the observed scalar random variables $y_1(t)$, . . ., $y_m(t)$ is a linear combination of the others. Under these circumstances we get finally:

$$\boldsymbol{\Delta}^*(t) = \boldsymbol{\Phi}(t+1;\, t)\mathbf{P}^*(t)\mathbf{M}'(t)[\mathbf{M}(t)\mathbf{P}^*(t)\mathbf{M}'(t)]^{-1} \tag{25}$$

Since observations start at t_0, $\tilde{\mathbf{x}}(t_0|t_0-1) = \mathbf{x}(t_0)$; to begin the iterative evaluation of $\mathbf{P}^*(t)$ by means of equation (24), we must obviously specify $\mathbf{P}^*(t_0) = E\mathbf{x}(t_0)\mathbf{x}'(t_0)$. Assuming this matrix is positive definite, equation (25) then yields $\boldsymbol{\Delta}^*(t_0)$; equation (22) $\boldsymbol{\Phi}^*(t_0+1;\, t_0)$, and equation (24) $\mathbf{P}^*(t_0+1)$, completing the cycle. If now $\mathbf{Q}(t)$ is positive definite, then all the $\mathbf{P}^*(t)$ will be positive definite and the requirements in deriving (25) will be satisfied at each step.

Now we remove the restriction that $t_1 = t+1$. Since $\mathbf{u}(t)$ is orthogonal to $\mathcal{Y}(t)$, we have

$$\mathbf{x}^*(t+1|t) = \hat{E}[\boldsymbol{\Phi}(t+1;\, t)\mathbf{x}(t) + \mathbf{u}(t)|\mathcal{Y}(t)] = \boldsymbol{\Phi}(t+1;\, t)\mathbf{x}^*(t|t)$$

Hence if $\boldsymbol{\Phi}(t+1;\, t)$ has an inverse $\boldsymbol{\Phi}(t;\, t+1)$ (which is always the case when $\boldsymbol{\Phi}$ is the transition matrix of a dynamic system describable by a differential equation) we have

$$\mathbf{x}^*(t|t) = \boldsymbol{\Phi}(t;\, t+1)\mathbf{x}^*(t+1|t)$$

If $t_1 \geq t+1$, we first observe by repeated application of (16) that

$$\begin{aligned}
\mathbf{x}(t+s) &= \boldsymbol{\Phi}(t+s;\, t+1)\mathbf{x}(t+1) \\
&\quad + \sum_{r=1}^{s-1} \boldsymbol{\Phi}(t+s;\, t+r)\mathbf{u}(t+r) \qquad (s \geq 1)
\end{aligned}$$

Since $\mathbf{u}(t+s-1), \ldots, \mathbf{u}(t+1)$ are all orthogonal to $\mathcal{Y}(t)$,

$$\begin{aligned}
\mathbf{x}^*(t+s|t) &= \hat{E}[\mathbf{x}(t+s)|\mathcal{Y}(t)] \\
&= \hat{E}[\boldsymbol{\Phi}(t+s;\, t+1)\mathbf{x}(t+1)|\mathcal{Y}(t)] \\
&= \boldsymbol{\Phi}(t+s;\, t+1)\mathbf{x}^*(t+1|t) \qquad (s \geq 1)
\end{aligned}$$

If $s < 0$, the results are similar, but $\mathbf{x}^*(t-s|t)$ will have $(1-s)(n-p)$ co-ordinates.

The results of this section may be summarized as follows:

Theorem 3. (*Solution of the Wiener Problem*)

Consider Problem I. The optimal estimate $\mathbf{x}^*(t+1|t)$ *of* $\mathbf{x}(t+1)$ *given* $\mathbf{y}(t_0), \ldots, \mathbf{y}(t)$ *is generated by the linear dynamic system*

$$\mathbf{x}^*(t+1|t) = \boldsymbol{\Phi}^*(t+1;\, t)\mathbf{x}^*(t|t-1) + \boldsymbol{\Delta}^*(t)\mathbf{y}(t) \tag{21}$$

The estimation error is given by

$$\tilde{\mathbf{x}}(t+1|t) = \boldsymbol{\Phi}^*(t+1;\, t)\tilde{\mathbf{x}}(t|t-1) + \mathbf{u}(t) \tag{23}$$

The covariance matrix of the estimation error is

$$\text{cov}\ \tilde{\mathbf{x}}(t|t-1) = E\tilde{\mathbf{x}}(t|t-1)\tilde{\mathbf{x}}'(t|t-1) = \mathbf{P}^*(t) \tag{26}$$

The expected quadratic loss is

$$\sum_{i=1}^{n} E\tilde{x}_i^2(t|t-1) = \text{trace}\ \mathbf{P}^*(t) \tag{27}$$

The matrices $\boldsymbol{\Delta}^*(t)$, $\boldsymbol{\Phi}^*(t+1;\, t)$, $\mathbf{P}^*(t)$ *are generated by the recursion relations*

$$\left.\begin{aligned}
&\boldsymbol{\Delta}^*(t) = \boldsymbol{\Phi}(t+1;\, t)\mathbf{P}^*(t)\mathbf{M}'(t)[\mathbf{M}(t)\mathbf{P}^*(t)\mathbf{M}'(t)]^{-1} && (28) \\
&\boldsymbol{\Phi}^*(t+1;\, t) = \boldsymbol{\Phi}(t+1;\, t) - \boldsymbol{\Delta}^*(t)\mathbf{M}(t) && (29) \\
&\mathbf{P}^*(t+1) = \boldsymbol{\Phi}^*(t+1;\, t)\mathbf{P}^*(t)\boldsymbol{\Phi}'(t+1;\, t) + \mathbf{Q}(t) && (30)
\end{aligned}\right\} t \geq t_0$$

In order to carry out the iterations, one must specify the covariance $\mathbf{P}^*(t_0)$ *of* $\mathbf{x}(t_0)$ *and the covariance* $\mathbf{Q}(t)$ *of* $\mathbf{u}(t)$. *Finally, for any* $s \geq 0$, *if* $\boldsymbol{\Phi}$ *is invertible*

$$\begin{aligned}
\mathbf{x}^*(t+s|t) &= \boldsymbol{\Phi}(t+s;\, t+1)\mathbf{x}^*(t+1|t) \\
&= \boldsymbol{\Phi}(t+s;\, t+1)\boldsymbol{\Phi}^*(t+1;\, t)\boldsymbol{\Phi}(t;\, t+s-1) \\
&\qquad \times \mathbf{x}^*(t+s-1|t-1) \\
&\quad + \boldsymbol{\Phi}(t+s;\, t+1)\boldsymbol{\Delta}^*(t)\mathbf{y}(t)
\end{aligned} \tag{31}$$

so that the estimate $\mathbf{x}^*(t+s|t)$ $(s \geq 0)$ *is also given by a linear dynamic system of the type* (21).

Remarks. (*h*) Eliminating $\boldsymbol{\Delta}^*$ and $\boldsymbol{\Phi}^*$ from (28–30), a nonlinear difference equation is obtained for $\mathbf{P}^*(t)$:

$$\begin{aligned}
\mathbf{P}^*(t+1) &= \boldsymbol{\Phi}(t+1;\, t)\{\mathbf{P}^*(t) - \mathbf{P}^*(t)\mathbf{M}'(t)[\mathbf{M}(t)\mathbf{P}^*(t)\mathbf{M}'(t)]^{-1} \\
&\qquad \times \mathbf{P}^*(t)\mathbf{M}(t)\}\boldsymbol{\Phi}'(t+1;\, t) + \mathbf{Q}(t) \qquad (t \geq t_0)
\end{aligned} \tag{32}$$

This equation is linear only if $\mathbf{M}(t)$ is invertible but then the problem is trivial since all components of the random vector $\mathbf{x}(t)$ are observable $\mathbf{P}^*(t+1) = \mathbf{Q}(t)$. Observe that equation (32) plays a role in the present theory analogous to that of the Wiener-Hopf equation in the conventional theory.

Once $\mathbf{P}^*(t)$ has been computed via (32) starting at $t = t_0$, the explicit specification of the optimal linear filter is immediately available from formulas (29–30). Of course, the solution of Equation (32), or of its differential-equation equivalent, is a much simpler task than solution of the Wiener-Hopf equation.

(*i*) The results stated in Theorem 3 do not resolve completely Problem I. Little has been said, for instance, about the physical significance of the assumptions needed to obtain equation (25), the convergence and stability of the nonlinear difference equation (32), the stability of the optimal filter (21), etc. This can actually be done in a completely satisfactory way, but must be left to a future paper. In this connection, the principal guide and tool turns out to be the duality theorem mentioned briefly in the next section. See [29].

(j) By letting the sampling period (equal to one so far) approach zero, the method can be used to obtain the specification of a differential equation for the optimal filter. To do this, i.e., to pass from equation (14) to equation (12), requires computing the logarithm $\mathbf{F}^*$ of the matrix $\mathbf{\Phi}^*$. But this can be done only if $\mathbf{\Phi}^*$ is nonsingular—which is easily seen *not* to be the case. This is because it is sufficient for the optimal filter to have $n - p$ state variables, rather than n, as the formalism of equation (22) would seem to imply. By appropriate modifications, therefore, equation (22) can be reduced to an equivalent set of only $n - p$ equations whose transition matrix is nonsingular. Details of this type will be covered in later publications.

(k) The dynamic system (21) is, in general, nonstationary. This is due to two things: (1) The time dependence of $\mathbf{\Phi}(t + 1; t)$ and $\mathbf{M}(t)$; (2) the fact that the estimation starts at $t = t_0$ and improves as more data are accumulated. If $\mathbf{\Phi}$, $\mathbf{M}$ are constants, it can be shown that (21) becomes a stationary dynamic system in the limit $t \to \infty$. This is the case treated by the classical Wiener theory.

(l) It is noteworthy that the derivations given are not affected by the nonstationarity of the model for $\mathbf{x}(t)$ or the finiteness of available data. In fact, as far as the author is aware, the only explicit recursion relations given before for the growing-memory filter are due to Blum [12]. However, his results are much more complicated than ours.

(m) By inspection of Fig. 3 we see that the optimal filter is a feedback system, and that the signal after the first summer is white noise since $\tilde{\mathbf{y}}(t|t - 1)$ is obviously an orthogonal random process. This corresponds to some well-known results in Wiener filtering, see, e.g., Smith [28], Chapter 6, Fig. 6-4. However, this is apparently the first *rigorous* proof that every Wiener filter is realizable by means of a feedback system. Moreover, it will be shown in another paper that such a filter is always *stable*, under very mild assumptions on the model (16-17). See [29].

The Dual Problem

Let us now consider another problem which is conceptually very different from optimal estimation, namely, the noise-free regulator problem. In the simplest cases, this is:

Problem II. *Consider the dynamic system*

$$\mathbf{x}(t + 1) = \hat{\mathbf{\Phi}}(t + 1;\ t)\mathbf{x}(t) + \hat{\mathbf{M}}(t)\mathbf{u}(t) \qquad (33)$$

where $\mathbf{x}(t)$ is an n-vector, $\mathbf{u}(t)$ is an m-vector ($m \leq n$), $\hat{\mathbf{\Phi}}$, $\hat{\mathbf{M}}$ are $n \times n$ resp. $n \times m$ matrices whose elements are nonrandom functions of time. Given any state $\mathbf{x}(t)$ at time t, we are to find a sequence $\mathbf{u}(t), \ldots, \mathbf{u}(T)$ of control vectors which minimizes the performance index

$$V[\mathbf{x}(t)] = \sum_{\tau=t}^{T+1} \mathbf{x}'(\tau)\hat{\mathbf{Q}}(\tau)\mathbf{x}(\tau)$$

where $\hat{\mathbf{Q}}(t)$ is a positive definite matrix whose elements are nonrandom functions of time. See Fig. 2, where $\mathbf{\Delta} = \hat{\mathbf{M}}$ and $\mathbf{M} = \mathbf{I}$.

Probabilistic considerations play no part in Problem II; it is implicitly assumed that every state variable can be measured exactly at each instant $t, t + 1, \ldots, T$. It is customary to call $T \geq t$ the *terminal time* (it may be infinity).

The first general solution of the noise-free regulator problem is due to the author [18]. The main result is that the optimal control vectors $\mathbf{u}^*(t)$ are nonstationary linear functions of $\mathbf{x}(t)$. After a change in notation, the formulas of the Appendix, Reference [18] (see also Reference [23]) are as follows:

$$\mathbf{u}^*(t) = -\hat{\mathbf{\Delta}}^*(t)\mathbf{x}(t) \qquad (34)$$

Under optimal control as given by (34), the "closed-loop" equations for the system are (see Fig. 4)

$$\mathbf{x}(t + 1) = \hat{\mathbf{\Phi}}^*(t + 1;\ t)\mathbf{x}(t)$$

and the minimum performance index at time t is given by

$$V^*[\mathbf{x}(t)] = \mathbf{x}'(t)\hat{\mathbf{P}}^*(t - 1)\mathbf{x}(t)$$

The matrices $\hat{\mathbf{\Delta}}^*(t)$, $\hat{\mathbf{\Phi}}^*(t + 1;\ t)$, $\hat{\mathbf{P}}^*(t)$ are determined by the recursion relations:

$$\left.\begin{aligned} &\hat{\mathbf{\Delta}}^*(t) = [\hat{\mathbf{M}}'(t)\hat{\mathbf{P}}^*(t)\hat{\mathbf{M}}(t)]^{-1}\hat{\mathbf{M}}'(t)\hat{\mathbf{P}}^*(t)\hat{\mathbf{\Phi}}(t + 1;\ t) && (35)\\ &\hat{\mathbf{\Phi}}^*(t + 1;\ t) = \hat{\mathbf{\Phi}}(t + 1;\ t) - \hat{\mathbf{M}}(t)\hat{\mathbf{\Delta}}^*(t) && (36)\\ &\hat{\mathbf{P}}^*(t - 1) = \hat{\mathbf{\Phi}}'(t + 1;\ t)\hat{\mathbf{P}}^*(t)\hat{\mathbf{\Phi}}^*(t + 1;\ t) + \hat{\mathbf{Q}}(t) && (37)\end{aligned}\right\}\ t \leq T$$

Initially we must set $\hat{\mathbf{P}}^*(T) = \hat{\mathbf{Q}}(T + 1)$.

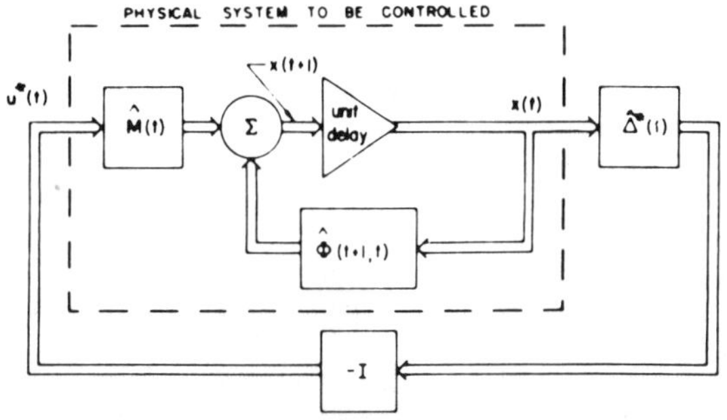

Fig. 4 Matrix block diagram of optimal controller

Comparing equations (35-37) with (28-30) and Fig. 3 with Fig. 4 we notice some interesting things which are expressed precisely by

Theorem 4. (*Duality Theorem*) *Problem I and Problem II are duals of each other in the following sense:*

Let $\tau \geq 0$. Replace every matrix $\mathbf{X}(t) = \mathbf{X}(t_0 + \tau)$ in (28-30) by $\hat{\mathbf{X}}'(t) = \hat{\mathbf{X}}'(T - \tau)$. Then one has (35-37). Conversely, replace every matrix $\hat{\mathbf{X}}(T - \tau)$ in (35-37) by $\mathbf{X}'(t_0 + \tau)$. Then one has (28-30).

Proof. Carry out the substitutions. For ease of reference, the dualities between the two problems are given in detail in Table 1.

Table 1

	Problem I	Problem II
1	$\mathbf{x}(t)$ (unobservable) state variables of random process.	$\mathbf{x}(t)$ (observable) state variables of plant to be regulated.
2	$\mathbf{y}(t)$ observed random variables.	$\mathbf{u}(t)$ control variables.
3	t_0 first observation.	T last control action.
4	$\mathbf{\Phi}(t_0 + \tau + 1;\ t_0 + \tau)$ transition matrix.	$\hat{\mathbf{\Phi}}(T - \tau + 1;\ T - \tau)$ transition matrix.
5	$\mathbf{P}^*(t_0 + \tau)$ covariance of optimized estimation error.	$\hat{\mathbf{P}}^*(T - \tau)$ matrix of quadratic form for performance index under optimal regulation.
6	$\mathbf{\Delta}^*(t_0 + \tau)$ weighting of observation for optimal estimation.	$\hat{\mathbf{\Delta}}^*(T - \tau)$ weighting of state for optimal control.
7	$\mathbf{\Phi}^*(t_0 + \tau + 1;\ t_0 + \tau)$ transition matrix for optimal estimation error.	$\hat{\mathbf{\Phi}}^*(T - \tau + 1;\ T - \tau)$ transition matrix under optimal regulation.
8	$\mathbf{M}(t_0 + \tau)$ effect of state on observation.	$\hat{\mathbf{M}}(T - \tau)$ effect of control vectors on state.
9	$\mathbf{Q}(t_0 + \tau)$ covariance of random excitation.	$\hat{\mathbf{Q}}(T - \tau)$ matrix of quadratic form defining error criterion.

Remarks. (n) The *mathematical* significance of the duality between Problem I and Problem II is that both problems reduce to the solution of the Wiener-Hopf-like equation (32).

(o) The *physical* significance of the duality is intriguing. Why are observations and control dual quantities?

Recent research [29] has shown that the essence of the Duality Theorem lies in the duality of constraints at the output (represented by the matrix $\hat{\mathbf{M}}(t)$ in Problem I) and constraints at the input (represented by the matrix $\hat{\mathbf{M}}(t)$ in Problem II).

(p) Applications of Wiener's methods to the solution of noise-free regulator problem have been known for a long time; see the recent textbook of Newton, Gould, and Kaiser [27]. However, the connections between the two problems, and in particular the duality, have apparently never been stated precisely before.

(q) The duality theorem offers a powerful tool for developing more deeply the theory (as opposed to the computation) of Wiener filters, as mentioned in Remark (i). This will be published elsewhere [29].

Applications

The power of the new approach to the Wiener problem, as expressed by Theorem 3, is most obvious when the data of the problem are given in numerical form. In that case, one simply performs the numerical computations required by (28–30). Results of such calculations, in some cases of practical engineering interest, will be published elsewhere.

When the answers are desired in closed analytic form, the iterations (28–30) may lead to very unwieldy expressions. In a few cases, $\mathbf{\Delta}^*$ and $\mathbf{\Phi}^*$ can be put into "closed form." Without discussing here how (if at all) such closed forms can be obtained, we now give two examples indicative of the type of results to be expected.

Example 1. Consider the problem mentioned under "Optimal Estimates." Let $x_1(t)$ be the signal and $x_2(t)$ the noise. We assume the model:

$$x_1(t+1) = \phi_{11}(t+1;\ t)x_1(t) + u_1(t)$$

$$x_2(t+1) = u_2(t)$$

$$y_1(t) = x_1(t) + x_2(t)$$

The specific data for which we desire a solution of the estimation problem are as follows:

1 $t_1 = t + 1;\ t_0 = 0$
2 $Ex_1^2(0) = 0$, i.e., $x_1(0) = 0$
3 $Eu_1^2(t) = a^2$, $Eu^2(t) = b^2$, $Eu_1(t)u_2(t) = 0$ (for all t)
4 $\phi_{11}(t+1;\ t) = \phi_{11} =$ const.

A simple calculation shows that the following matrices satisfy the difference equations (28–30), for all $t \geqq t_0$:

$$\mathbf{\Delta}^*(t) = \begin{bmatrix} \phi_{11}C(t) \\ 0 \end{bmatrix}$$

$$\mathbf{\Phi}^*(t+1;\ t) = \begin{bmatrix} \phi_{11}[1 - C(t)] & 0 \\ 0 & 0 \end{bmatrix}$$

$$\mathbf{P}^*(t+1) = \begin{bmatrix} a^2 + \phi_{11}^2 b^2 C(t) & 0 \\ 0 & b^2 \end{bmatrix}$$

where $$C(t+1) = 1 - \frac{b^2}{a^2 + b^2 + \phi_{11}^2 b^2 C(t)} \qquad t \geqq 0 \qquad (38)$$

Since it was assumed that $x_1(0) = 0$, neither $x_1(1)$ nor $x_2(1)$ can be predicted from the measurement of $y_1(0)$. Hence the measurement at time $t = 0$ is useless, which shows that we should set $C(0) = 0$. This fact, with the iterations (38), completely determines the function $C(t)$. The nonlinear difference equation (38) plays the role of the Wiener-Hopf equation.

If $b^2/a^2 \ll 1$, then $C(t) \cong 1$ which is essentially pure prediction. If $b^2/a^2 \gg 1$, then $C(t) \cong 0$, and we depend mainly on $x_1^*(t|t-1)$ for the estimation of $x_1^*(t+1|t)$ and assign only very small weight to the measurement $y_1(t)$; this is what one would expect when the measured data are very noisy.

In any case, $x_2^*(t|t-1) = 0$ at all times; one cannot predict independent noise! This means that ϕ^*_{12} can be set equal to zero. The optimal predictor is a first-order dynamic system. See Remark (j).

To find the stationary Wiener filter, let $t = \infty$ on both sides of (38), solve the resulting quadratic equation in $C(\infty)$, etc.

Example 2. A number of particles leave the origin at time $t_0 = 0$ with random velocities; after $t = 0$, each particle moves with a constant (unknown) velocity. Suppose that the position of one of these particles is measured, the data being contaminated by stationary, additive, correlated noise. What is the optimal estimate of the position and velocity of the particle at the time of the last measurement?

Let $x_1(t)$ be the position and $x_2(t)$ the velocity of the particle; $x_3(t)$ is the noise. The problem is then represented by the model,

$$x_1(t+1) = x_1(t) + x_2(t)$$

$$x_2(t+1) = x_2(t)$$

$$x_3(t+1) = \phi_{33}(t+1;\ t)x_3(t) + u_3(t)$$

$$y_1(t) = x_1(t) + x_3(t)$$

and the additional conditions

1 $t_1 = t;\ t_0 = 0$
2 $Ex_1^2(0) = Ex_2(0) = 0$, $Ex_2^2(0) = a^2 > 0$;
3 $Eu_3(t) = 0$, $Eu_3^2(t) = b^2$.
4 $\phi_{33}(t+1;\ t) = \phi_{33} =$ const.

According to Theorem 3, $\mathbf{x}^*(t|t)$ is calculated using the dynamic system (31).

First we solve the problem of predicting the position and velocity of the particle one step ahead. Simple considerations show that

$$\mathbf{P}^*(1) = \begin{bmatrix} a^2 & a^2 & 0 \\ a^2 & a^2 & 0 \\ 0 & 0 & b^2 \end{bmatrix} \quad \text{and} \quad \mathbf{\Delta}^*(0) = \begin{bmatrix} 0 \\ 0 \\ 1 \end{bmatrix}$$

It is then easy to check by substitution into equations (28–30) that

$$\mathbf{P}^*(t) = \frac{b^2}{C_1(t-1)} \times \begin{bmatrix} t^2 & t & -\phi_{33}t(t-1) \\ t & 1 & -\phi_{33}(t-1) \\ -\phi_{33}t(t-1) & -\phi_{33}(t-1) & \phi_{33}^2(t-1)^2 + C_1(t-1) \end{bmatrix}$$

is the correct expression for the covariance matrix of the prediction error $\tilde{\mathbf{x}}(t|t-1)$ for all $t \geqq 1$, provided that we define

$$C_1(0) = b^2/a^2$$
$$C_1(t) = C_1(t-1) + [t - \phi_{33}(t-1)]^2,\ t \geqq 1$$

It is interesting to note that the results just obtained are valid also when ϕ_{33} depends on t. This is true also in Example 1. In conventional treatments of such problems there *seems* to be an essential difference between the cases of stationary and nonstationary noise. This misleading impression created by the conventional theory is due to the very special *methods* used in solving the Wiener-Hopf equation.

Introducing the abbreviation

$$C_2(0) = 0$$
$$C_2(t) = t - \phi_{33}(t-1),\ t \geqq 1$$

and observing that

$$\text{cov}\ \tilde{\mathbf{x}}(t+1|t) = \mathbf{P}^*(t+1)$$
$$= \mathbf{\Phi}(t+1;\ t)[\text{cov}\ \tilde{\mathbf{x}}(t|t)]\mathbf{\Phi}'(t+1;\ t) + \mathbf{Q}(t)$$

the matrices occurring in equation (31) and the covariance matrix of $\tilde{\mathbf{x}}(t|t)$ are found after simple calculations. We have, for all $t \geq 0$,

$$\mathbf{\Phi}(t;\ t+1)\mathbf{\Delta}^*(t) = \frac{1}{C_1(t)}\begin{bmatrix} tC_2(t) \\ C_2(t) \\ C_1(t) - tC_2(t) \end{bmatrix}$$

$$\mathbf{\Phi}(t;\ t+1)\mathbf{\Phi}^*(t+1;\ t)\mathbf{\Phi}(t+1;\ t)$$

$$= \frac{1}{C_1(t)}\begin{bmatrix} C_1(t) - tC_2(t) & C_1(t) - tC_3(t) & -\phi_{33}tC_2(t) \\ -C_2(t) & C_1(t) - C_2(t) & -\phi_{33}C_2(t) \\ -C_1(t) + tC_2(t) & -C_1(t) + tC_2(t) & +\phi_{33}tC_2(t) \end{bmatrix}$$

and

$$\operatorname{cov} \tilde{\mathbf{x}}(t|t) = E\tilde{\mathbf{x}}(t|t)\tilde{\mathbf{x}}'(t|t) = \frac{b^2}{C_1(t)}\begin{bmatrix} t^2 & t & -t^2 \\ t & 1 & -t \\ -t^2 & -t & t^2 \end{bmatrix}$$

To gain some insight into the behavior of this system, let us examine the limiting case $t \to \infty$ of a large number of observations. Then $C_1(t)$ obeys approximately the differential equation

$$dC_1(t)/dt \cong C_2^2(t) \qquad (t \gg 1)$$

from which we find

$$C_1(t) \cong (1-\phi_{33})^2 t^3/3 + \phi_{33}(1-\phi_{33})t^2 + \phi_{33}^2 t + b^2/a^2 \qquad (t \gg 1) \quad (39)$$

Using (39), we get further

$$\mathbf{\Phi}^{-1}\mathbf{\Phi}^*\mathbf{\Phi} \cong \begin{bmatrix} 1 & 1 & 0 \\ 0 & 1 & 0 \\ -1 & -1 & 0 \end{bmatrix} \quad \text{and} \quad \mathbf{\Phi}^{-1}\mathbf{\Delta}^* \cong \begin{bmatrix} 0 \\ 0 \\ 1 \end{bmatrix} \qquad (t \gg 1)$$

Thus as the number of observations becomes large, we depend almost exclusively on $x_1^*(t|t)$ and $x_2^*(t|t)$ to estimate $x_1^*(t+1|t+1)$ and $x_2^*(t+1|t+1)$. Current observations are used almost exclusively to estimate the noise

$$x_3^*(t|t) \cong y_1(t) - x_1^*(t|t) \qquad (t \gg 1)$$

One would of course expect something like this since the problem is analogous to fitting a straight line to an increasing number of points.

As a second check on the reasonableness of the results given, observe that the case $t \gg 1$ is essentially the same as prediction based on continuous observations. Setting $\phi_{33} = 0$, we have

$$E\tilde{x}_1^2(t|t) \cong \frac{a^2b^2t^2}{b^2 + a^2t^3/3} \qquad (t \gg 1;\ \phi_{33} = 0)$$

which is identical with the result obtained by Shinbrot [11], Example 1, and Solodovnikov [14], Example 2, in their treatment of the Wiener problem in the finite-length, continuous-data case, using an approach entirely different from ours.

Conclusions

This paper formulates and solves the Wiener problem from the "state" point of view. On the one hand, this leads to a very general treatment including cases which cause difficulties when attacked by other methods. On the other hand, the Wiener problem is shown to be closely connected with other problems in the theory of control. Much remains to be done to exploit these connections.

References

1 N. Wiener, "The Extrapolation, Interpolation and Smoothing of Stationary Time Series," John Wiley & Sons, Inc., New York, N. Y., 1949.

2 L. A. Zadeh and J. R. Ragazzini, "An Extension of Wiener's Theory of Prediction," *Journal of Applied Physics*, vol. 21, 1950, pp. 645–655.

3 H. W. Bode and C. E. Shannon, "A Simplified Derivation of Linear Least-Squares Smoothing and Prediction Theory," *Proceedings IRE*, vol. 38, 1950, pp. 417–425.

4 R. C. Booton, "An Optimization Theory for Time-Varying Linear Systems With Nonstationary Statistical Inputs," *Proceedings IRE*, vol. 40, 1952, pp. 977–981.

5 J. H. Laning and R. H. Battin, "Random Processes in Automatic Control," McGraw-Hill Book Company, Inc., New York, N. Y., 1956.

6 W. B. Davenport, Jr., and W. L. Root, "An Introduction to the Theory of Random Signals and Noise," McGraw-Hill Book Company, Inc., New York, N. Y., 1958.

7 S. Darlington, "Linear Least-Squares Smoothing and Prediction, With Applications," *Bell System Tech. Journal*, vol. 37, 1958, pp. 1221–1294.

8 G. Franklin, "The Optimum Synthesis of Sampled-Data Systems," Doctoral dissertation, Dept. of Elect. Engr., Columbia University, 1955.

9 A. B. Lees, "Interpolation and Extrapolation of Sampled Data," *Trans. IRE Prof. Group on Information Theory*, IT-2, 1956, pp. 173–175.

10 R. C. Davis, "On the Theory of Prediction of Nonstationary Stochastic Processes," *Journal of Applied Physics*, vol. 23, 1952, pp. 1047–1053.

11 M. Shinbrot, "Optimization of Time-Varying Linear Systems With Nonstationary Inputs," Trans. ASME, vol. 80, 1958, pp. 457–462.

12 M. Blum, "Recursion Formulas for Growing Memory Digital Filters," *Trans. IRE Prof. Group on Information Theory*, IT-4, 1958, pp. 24–30.

13 V. S. Pugachev, "The Use of Canonical Expansions of Random Functions in Determining an Optimum Linear System," *Automatics and Remote Control* (USSR), vol. 17, 1956, pp. 489–499; translation pp. 545–556.

14 V. V. Solodovnikov and A. M. Batkov, "On the Theory of Self-Optimizing Systems (in German and Russian)," Proc. Heidelberg Conference on Automatic Control, 1956, pp. 308–323.

15 J. L. Doob, "Stochastic Processes," John Wiley & Sons, Inc., New York, N. Y., 1955.

16 M. Loève, "Probability Theory," Van Nostrand Company, Inc., New York, N. Y., 1955.

17 R. E. Bellman, I. Glicksberg, and O. A. Gross, "Some Aspects of the Mathematical Theory of Control Processes," RAND Report R-313, 1958, 244 pp.

18 R. E. Kalman and R. W. Koepcke, "Optimal Synthesis of Linear Sampling Control Systems Using Generalized Performance Indexes," Trans. ASME, vol. 80, 1958, pp. 1820–1826.

19 J. E. Bertram, "Effect of Quantization in Sampled-Feedback Systems," *Trans. AIEE*, vol. 77, II, 1958, pp. 177–182.

20 R. E. Kalman and J. E. Bertram, "General Synthesis Procedure for Computer Control of Single and Multi-Loop Linear Systems," *Trans. AIEE*, vol. 77, II, 1958, pp. 602–609.

21 C. W. Merriam, III, "A Class of Optimum Control Systems," *Journal of the Franklin Institute*, vol. 267, 1959, pp. 267–281.

22 R. E. Kalman and J. E. Bertram, "A Unified Approach to the Theory of Sampling Systems," *Journal of the Franklin Institute*, vol. 267, 1959, pp. 405–436.

23 R. E. Kalman and R. W. Koepcke, "The Role of Digital Computers in the Dynamic Optimization of Chemical Reactors," Proc. Western Joint Computer Conference, 1959, pp. 107–116.

24 R. E. Kalman, "Dynamic Optimization of Linear Control Systems, I. Theory," to appear.

25 S. Sherman, "Non-Mean-Square Error Criteria," *Trans. IRE Prof. Group on Information Theory*, IT-4, 1958, pp. 125–126.

26 V. S. Pugachev, "On a Possible General Solution of the Problem of Determining Optimum Dynamic Systems," *Automatics and Remote Control* (USSR), vol. 17, 1956, pp. 585–589.

27 G. C. Newton, Jr., L. A. Gould, and J. F. Kaiser, "Analytical Design of Linear Feedback Controls," John Wiley & Sons, Inc., New York, N. Y., 1957.

28 O. J. M. Smith, "Feedback Control Systems," McGraw-Hill Book Company, Inc., New York, N. Y., 1958.

29 R. E. Kalman, "On the General Theory of Control Systems," Proceedings First International Conference on Automatic Control, Moscow, USSR, 1960.

APPENDIX
RANDOM PROCESSES: BASIC CONCEPTS

For convenience of the reader, we review here some elementary definitions and facts about probability and random processes. Everything is presented with the utmost possible simplicity; for greater depth and breadth, consult Laning and Battin [5] or Doob [15].

A *random variable* is a function whose values depend on the outcome of a chance event. The *values* of a random variable may be any convenient mathematical entities; real or complex numbers, vectors, etc. For simplicity, we shall consider here only real-valued random variables, but this is no real restriction. Random variables will be denoted by $x, y, \ldots$ and their values by $\xi, \eta, \ldots$. Sums, products, and functions of random variables are also random variables.

A random variable x can be explicitly defined by stating the probability that x is less than or equal to some real constant ξ. This is expressed symbolically by writing

$$Pr(x \leqq \xi) = F_x(\xi); \; F_x(-\infty) = 0, \, F_x(+\infty) = 1$$

$F_x(\xi)$ is called the *probability distribution function* of the random variable x. When $F_x(\xi)$ is differentiable with respect to ξ, then $f_x(\xi) = dF_x(\xi)/d\xi$ is called the *probability density function* of x.

The *expected value* (*mathematical expectation, statistical average, ensemble average, mean*, etc., are commonly used synonyms) of any nonrandom function $g(x)$ of a random variable x is defined by

$$Eg(x) = E[g(x)] = \int_{-\infty}^{\infty} g(\xi)dF_x(\xi) = \int_{-\infty}^{\infty} g(\xi)f_x(\xi)d\xi \quad (40)$$

As indicated, it is often convenient to omit the brackets after the symbol E. A sequence of random variables (finite or infinite)

$$\{x(t)\} = \ldots, x(-1), x(0), x(1), \ldots \quad (41)$$

is called a *discrete* (or *discrete-parameter*) *random* (or *stochastic*) *process*. One particular set of observed values of the random process (41)

$$\ldots, \xi(-1), \xi(0), \xi(1), \ldots$$

is called a *realization* (or a *sample function*) of the process. Intuitively, a random process is simply a set of random variables which are indexed in such a way as to bring the notion of time into the picture.

A random process is *uncorrelated* if

$$Ex(t)x(s) = Ex(t)Ex(s) \qquad (t \neq s)$$

If, furthermore,

$$Ex(t)x(s) = 0 \qquad (t \neq s)$$

then the random process is *orthogonal*. Any uncorrelated random process can be changed into orthogonal random process by replacing $x(t)$ by $x'(t) = x(t) - Ex(t)$ since then

$$\begin{aligned} Ex'(t)x'(s) &= E[x(t) - Ex(t)]\cdot[x(s) - Ex(s)] \\ &= Ex(t)x(s) - Ex(t)Ex(s) = 0 \end{aligned}$$

It is useful to remember that, if a random process is orthogonal, then

$$E[x(t_1) + x(t_2) + \ldots]^2 = Ex^2(t_1) + Ex^2(t_2) + \ldots \; (t_1 \neq t_2 \neq \ldots)$$

If $\mathbf{x}$ is a vector-valued random variable with components $x_1, \ldots, x_n$ (which are of course random variables), the matrix

$$[E(x_i - Ex_i)(x_j - Ex_j)] = E(\mathbf{x} - E\mathbf{x})(\mathbf{x}' - E\mathbf{x}') = \text{cov } \mathbf{x} \quad (42)$$

is called the *covariance matrix* of x.

A random process may be specified explicitly by stating the probability of simultaneous occurrence of any finite number of events of the type

$$x(t_1) \leqq \xi_1, \ldots, x(t_n) \leqq \xi_n; \; (t_1 \neq \ldots \neq t_n), \text{ i.e.,}$$

$$Pr[(x(t_1) \leqq \xi_1, \ldots, x(t_n) \leqq \xi_n)] = F_{x(t_1), \ldots, x(t_n)}(\xi_1, \ldots, \xi_n) \quad (43)$$

where $F_{x(t_1), \ldots, x(t_n)}$ is called the *joint probability distribution function* of the random variables $x(t_1), \ldots, x(t_n)$. The *joint probability density function* is then

$$f_{x(t_1), \ldots, x(t_n)}(\xi_1, \ldots, \xi_n) = \partial^n F_{n(t_1), \ldots, x(t_n)}/\partial\xi_1, \ldots, \partial\xi_n$$

provided the required derivatives exist. The expected value $Eg[x(t_1), \ldots, x(t_n)]$ of any nonrandom function of n random variables is defined by an n-fold integral analogous to (40).

A random process is *independent* if for any finite $t_1 \neq \ldots \neq t_n$, (43) is equal to the product of the first-order distributions

$$Pr[x(t_1) \leqq \xi_1] \ldots Pr[x(t_n) \leqq \xi_n]$$

If a set of random variables is independent, then they are obviously also uncorrelated. The converse is not true in general. For a set of more than 2 random variables to be independent, it is not sufficient that any pair of random variables be independent.

Frequently it is of interest to consider the probability distribution of a random variable $x(t_{n+1})$ of a random process given the actual values $\xi(t_1), \ldots, \xi(t_n)$ with which the random variables $x(t_1), \ldots, x(t_n)$ have occurred. This is denoted by

$$Pr[x(t_{n+1}) \leqq \xi_{n+1} | x(t_1) = \xi_1, \ldots, x(t_n) = \xi_n]$$

$$= \frac{\int_{-\infty}^{\xi_{n+1}} f_{x(t_1), \ldots, x(t_{n+1})}(\xi_1, \ldots, \xi_{n+1})d\xi_{n+1}}{f_{x(t_1), \ldots, x(t_n)}(\xi_1, \ldots, \xi_n)} \quad (44)$$

which is called the *conditional probability distribution function* of $x(t_{n+1})$ given $x(t_1), \ldots, x(t_n)$. The *conditional expectation*

$$E\{g[x(t_{n+1})] | x(t_1), \ldots, x(t_n)\}$$

is defined analogously to (40). The conditional expectation is a random variable; it follows that

$$E[E\{g[x(t_{n+1})] | x(t_1), \ldots, x(t_n)\}] = E\{g[x(t_{n+1})]\}$$

In all cases of interest in this paper, integrals of the type (40) or (44) need never be evaluated explicitly; only the *concept* of the expected value is needed.

A random variable x is *gaussian* (or *normally distributed*) if

$$f_x(\xi) = \frac{1}{[2\pi E(x - Ex)^2]^{1/2}} \exp\left[-\frac{1}{2}\frac{(\xi - Ex)^2}{E(x - Ex)^2}\right]$$

which is the well-known bell-shaped curve. Similarly, a random vector $\mathbf{x}$ is *gaussian* if

$$f_x(\xi) = \frac{1}{(2\pi)^{n/2}(\det \mathbf{C})^{1/2}} \exp\left[-\frac{1}{2}(\xi - E\mathbf{x})'\mathbf{C}^{-1}(\xi - E\mathbf{x})\right]$$

where $\mathbf{C}^{-1}$ is the inverse of the covariance matrix (42) of $\mathbf{x}$. A *gaussian random process* is defined similarly.

The importance of gaussian random variables and processes is largely due to the following facts:

Theorem 5. (*A*) *Linear functions (and therefore conditional expectations) on a gaussian random process are gaussian random variables.*

(*B*) *Orthogonal gaussian random variables are independent.*

(*C*) *Given any random process with means* $Ex(t)$ *and covariances* $Ex(t)x(s)$, *there exists a unique gaussian random process with the same means and covariances.*

First Order Error Propagation in a Stagewise Smoothing Procedure for Satellite Observations

Peter Swerling*

Abstract

A practical method of smoothing satellite data by evaluating a finite number of parameters, or elements, is presented.

Introduction

The subject discussed is that of smoothing observational data in cases where the observations, in the absence of observational error, would all be determined by the time of observation plus a finite number of parameters, called elements. The objective is to estimate the elements.

The immediate motivation for this arises in studies of estimation of earth-satellite orbits from observational data. In this case, if, for example, the force field were known exactly, one could regard the elements as the position and velocity components at a particular instant t_0 ; if the field were a central inverse square field, the elements could alternatively be the six elements of a Keplerian elliptic orbit. (It should be mentioned, however, that satellite orbit prediction is only one of the possible applications of the results.)

In satellite tracking and prediction, it is desired to produce ephemerides—i.e., predictions of the future position as a function of time—as well as to make various other types of decisions and predictions. As new observations become available, one can improve the accuracy of these predictions. Ideally, one would like to store all previous observations of the object, and combine these in some optimum way to yield the desired predictions or decisions. The optimum method for processing the available data would be based on analysis of the error statistics for the individual observations, and on the functional dependence of future predicted quantities on the previous observations.

In satellite tracking one is dealing with a situation in which there may be a large number of observations of varying degrees of accuracy, as well as large numbers of tracked objects. Also, methods of orbit prediction (even in the absence of observation errors) are subject to various sources of error, such as

(a) uncertainty in the forces acting on the body (earth's gravitational and magnetic fields, atmospheric resistance);

(b) cumulative errors in solving the equations of motion.

Two features would be desirable in a tracking and prediction method:

(1) The data processing load per tracked object should not exceed a certain maximum, regardless of how many observations are available to be processed. On the other hand, the prediction method should not throw available observations away.

(2) The method should be adaptable to situations in which the underlying prediction functions are subject to the above-mentioned uncertainties.

The stagewise procedures described below are motivated by these considerations.

The particular methods of data smoothing to be discussed are variations of the classical method of minimizing a quadratic form in the residuals (in practise this is usually a weighted sum of squares of residuals). After defining this method, the first order dependence of the errors in the resulting element estimates on the observation errors is established.

We then go on to describe a stagewise procedure for processing the observational data, in which the element estimates at each stage are smoothed in a particular way with some additional observations. This is, in essence, a type of differential correction procedure. It is shown that the errors in the resulting element estimates, after stagewise smoothing of a given set of observations, have the same first order dependence on the observation errors as would the errors in the estimates obtained by simultaneous processing of the same total set of observations.

Some statistical properties of errors in the element estimates are derived for the case in which the observation errors are regarded as statistical variables and in which the matrix of the quadratic form to be minimized is the inverse covariance matrix of the observation errors. (For Gaussian error statistics, this would result in a maximum-likelihood method of estimation.)

The elements are at first regarded as constants, and later the treatment is extended to the case in which the elements are regarded as time-dependent.

We assume there are N observed quantities F^μ, $\mu = 1, \cdots, N$, each F^μ being a real scalar. We also assume that n real constants x_i, $i = 1, \cdots, n$ exist such that in the absence of observation error, all N observed quantities would satisfy the relations

$$F^\mu = f^\mu(x_1, \cdots, x_n, t_\mu) \qquad (1)$$

* The RAND Corporation, Santa Monica, California.

Reprinted with permission from *J. Astronautical Sci.*, vol. 6, pp. 46–52, Autumn 1959.

The vector $x = (x_1, \cdots, x_n)$ will be called the "element vector," and its components the "elements"; t_μ is the time at which the μth observation is taken.

When observation errors are present, the F^μ are given by

$$F^\mu = f^\mu(x_1, \cdots, x_n, t_\mu) + \varepsilon^\mu. \tag{2}$$

For purposes of illustration, consider the observation of a satellite following a Keplerian orbit. The elements might then be taken to be the eccentricity, semi-major axis, inclination, etc.; the quantities F^μ might be observations of such things as range, azimuth, elevation, or range rate from particular observation sites; the f^μ would be the functions describing the dependence of these observed quantities on the elements and time; and the ε^μ would be the observation errors.

The times t_μ need not be in any particular temporal order, nor are they necessarily all distinct. They are assumed to be measured without error in all cases in which the functions f^μ depend explicitly on t_μ. There is no loss of generality in this assumption, since timing errors may always be reduced to equivalent observation errors; this would of course modify the statistics of the resulting equivalent observation errors (for example, this might introduce an additional systematic component into the equivalent observation errors).

Many of the formulas below will involve the functions f^μ and their partial derivatives $\partial f^\mu/\partial x_i$. Practical application of these formulas would be possible both for cases where analytic expressions are known for the f^μ and their partial derivatives and for cases where these functions must be evaluated by numerical integration, as well as for cases where some of the functions have known analytic expressions and others must be determined by numerical methods.

Henceforth it will be supposed that $N \geqq n$. The problem to be considered is the estimation of the elements by means of smoothing, in some sense, of the observations.

A classical smoothing method is as follows: writing $P = (P_1, \cdots, P_n)$ for an estimate of the element vector, and writing $f(x, t)$ for $f(x_1, \cdots, x_n, t)$, $f(P, t)$ for $f(P_1, \cdots, P_n, t)$, and so forth, the method consists of minimizing with respect to $P_1, \cdots, P_n$ the quadratic form

$$Q = \sum_{\mu,\nu=1}^{N} \eta_{\mu\nu}[F^\mu - f^\mu(P, t_\mu)][F^\nu - f^\nu(P, t_\nu)] \tag{3}$$

where $(\eta_{\mu\nu})$ is a symmetric, positive definite matrix. Thus, the method consists in minimizing a positive definite quadratic form in the residuals. Differentiating Q with respect to P_i and setting the results equal to zero, we find that the minimizing estimates P_i must satisfy

$$\sum_{\mu,\nu=1}^{N} \frac{\partial f^\nu}{\partial x_i}(P, t_\nu)[F^\mu - f^\mu(P, t_\mu)]\eta_{\mu\nu} = 0 \qquad (i = 1, \cdots, n) \tag{4}$$

It is clear that if $\varepsilon^\mu = 0$, all μ, then $P = x$ is a solution of (4). The first question to be investigated is that of the first order propagation of errors—i.e., the first order dependence of $P - x$ on the errors ε^μ. It will be assumed that the functions f^μ are sufficiently well behaved for the following operations to be valid. We may write

$$f^\mu(P, t_\mu) = f^\mu(x, t_\mu) + \sum_{j=1}^{n} \frac{\partial f^\mu}{\partial x_j}(x, t_\mu)(P_j - x_j) + \cdots \tag{5}$$

$$\frac{\partial f^\mu}{\partial x_i}(P, t_\mu) = \frac{\partial f^\mu}{\partial x_i}(x, t_\mu) + \sum_{j=1}^{n} \frac{\partial^2 f^\mu}{\partial x_i \partial x_j}(x, t_\mu)(P_j - x_j) + \cdots \tag{6}$$

Neglecting all terms of higher order in $P - x$, and substituting (5) and (6) into (4), we obtain

$$\sum_{\mu,\nu=1}^{N} \eta_{\mu\nu}\left[\frac{\partial f^\nu}{\partial x_i}(x, t_\nu) + \sum_{j=1}^{n} \frac{\partial^2 f^\nu}{\partial x_i \partial x_j}(x, t_\nu)(P_j - x_j)\right] \times \left[F^\mu - f^\mu(x, t_\mu) - \sum_{j=1}^{n}\frac{\partial f^\mu}{\partial x_j}(x, t_\mu)(P_j - x_j)\right] = 0 \tag{7}$$

It is also clear that for sufficiently small ε^μ and $P_j - x_j$, the term involving second derivatives of f^ν may be neglected. The result may conveniently be expressed as follows: let

$$a_i^\mu(x, t_\mu) = \frac{\partial f^\mu}{\partial x_i}(x, t_\mu) \tag{8}$$

$$\rho_i(x) = \sum_{\mu,\nu=1}^{N} \eta_{\mu\nu} a_i^\nu(x, t_\nu)[F^\mu - f^\mu(x, t_\mu)] \tag{9}$$

$$\rho(x) = \{\rho_1(x), \cdots, \rho_n(x)\} \tag{10}$$

$$B_{ij}(x) = \sum_{\mu,\nu=1}^{N} \eta_{\mu\nu} a_i^\nu(x, t_\nu) a_j^\mu(x, t_\mu) \tag{11}$$

$$B(x) = \{B_{ij}(x)\} \tag{12}$$

Then, assuming $B(x)$ to be non-singular,

$$P - x = [B(x)]^{-1}\rho(x) \tag{13}$$

Eq. (13) expresses the first order dependence of $P - x$ on the errors ε^μ. This can be expressed equivalently:

$$P_i - x_i = \sum_{\mu=1}^{N} \Gamma_i^\mu(x)[F^\mu - f^\mu(x, t_\mu)] \tag{14}$$

$$\Gamma_i^\mu(x) = \sum_{j=1}^{n}\sum_{\nu=1}^{N} [B(x)]_{ij}^{-1}\eta_{\mu\nu} a_j^\nu(x, t_\nu) \tag{15}$$

A special case of this is as follows: suppose one has already an estimated element vector $p = (p_1, \cdots, p_n)$ together with K new observations. One can form a new estimate vector P by smoothing the original estimate vector p with the new observations in the following manner: P is determined by minimizing, with respect to $P_1, \cdots, P_n$, the quadratic form

$$Q = \sum_{\mu,\nu=1}^{n} \eta_{\mu\nu}(p_\mu - P_\mu)(p_\nu - P_\nu) + \sum_{\mu,\nu=n+1}^{n+K} \eta_{\mu\nu}[F^\mu - f^\mu(P, t_\mu)][F^\nu - f^\nu(P, t_\nu)] \tag{16}$$

In this case,

$$f^\mu(x, t_\mu) = x_\mu, \qquad \mu = 1, \cdots, n$$
$$a_i^\mu(x, t_\mu) = \delta_{i\mu} \qquad i = 1, \cdots, n; \quad \mu = 1, \cdots, n \tag{17}$$
$$F^\mu = p_\mu \qquad \mu = 1, \cdots, n$$

Also

$$\rho_i(x) = \sum_{j=1}^{n} \eta_{ij}(p_j - x_j) + \sum_{\mu,\nu=n+1}^{n+K} \eta_{\mu\nu}[F^\mu - f^\mu(x, t_\mu)]a_i^\nu(x, t_\nu) \qquad (i = 1, \cdots, n) \tag{18}$$

and

$$B_{ij}(x) = \eta_{ij} + \sum_{\mu,\nu=n+1}^{n+K} \eta_{\mu\nu} a_i^\nu(x, t_\nu) a_j^\mu(x, t_\mu) \qquad (i, j = 1, \cdots, n) \tag{19}$$

If we also define $r^\mu(p, t_\mu)$ for $u = n + 1, \cdots, n + K$ by

$$r^\mu(p, t_\mu) = F^\mu - f^\mu(p, t_\mu) \qquad \mu = n + 1, \cdots, n + K \tag{20}$$

then $\rho_i(x)$ becomes, to first order,

$$\rho_i(x) = \sum_{j=1}^{n} B_{ij}(x)(p_j - x_j) + \sum_{\mu,\nu=n+1}^{n+K} \eta_{\mu\nu} a_i^\nu(x, t_\nu) r^\mu(p, t_\mu) \tag{21}$$

Consequently, (13) reduces to

$$P - x = p - x + [B(x)]^{-1}\rho^*(x) \tag{22}$$

where

$$\rho_i^*(x) = \sum_{\mu,\nu=n+1}^{n+K} \eta_{\mu\nu} a_i^\nu(x, t_\nu) r^\mu(p, t_\mu) \tag{23}$$

Eq. (22) may be used as the basis for a first order differential correction to p, given the additional observations F^μ, $\mu = n + 1, \cdots, n + K$. First rewrite (22) as

$$P - p = [B(x)]^{-1}\rho^*(x) \tag{24}$$

Then, to first order, one can also write

$$P - p = [B(p)]^{-1}\rho^*(p) \tag{25}$$

Since all the quantities on the right of (25) are known, (25) gives the required first order correction to p.

A Stagewise Smoothing Procedure

Suppose the matrix $(\eta_{\mu\nu})$ can be written as a diagonal array of matrices

$$\eta = \begin{pmatrix} \eta^{(1)} & & \\ & \ddots & \\ & & \eta^{(S)} \end{pmatrix} \tag{26}$$

where $\eta^{(s)}$, $s = 1, \cdots, S$, are $N_s \times N_s$ matrices. For future notational convenience, we will regard the indices of the components $\eta_{\mu\nu}^{(s)}$ of $\eta^{(s)}$ as running over $\mu, \nu = M_{s-1} + 1, \cdots, M_s$ where

$$M_s = \sum_{r=1}^{s} N_r$$
$$M_0 = 0 \tag{27}$$
$$M_1 = N_1 \geqq n$$
$$M_S = N$$

We define a stagewise smoothing procedure as follows: Suppose initial element estimates $\{P_i^{(1)}\}$ are obtained by minimization with respect to $P^{(1)}$ of the quadratic form

$$Q^{(1)} = \sum_{\mu,\nu=1}^{N_1} \eta_{\mu\nu}^{(1)}[F^\mu - f^\mu(P^{(1)}, t_\mu)] \cdot [F^\nu - f^\nu(P^{(1)}, t_\nu)] \tag{28}$$

Now define sequences of matrices $B^{(s)}$ and $B^{(s)}(x)$ as follows:*

$$B_{ij}^{(s)} = B_{ij}^{(s-1)} + \sum_{\mu,\nu=M_{s-1}+1}^{M_s} \eta_{\mu\nu}^{(s)} a_i^\nu(P^{(s)}, t_\nu) a_j^\mu(P^{(s)}, t_\mu)$$
$$B_{ij}^{(0)} = 0 \tag{29}$$
$$B_{ij}^{(s)}(x) = B_{ij}^{(s-1)}(x) + \sum_{\mu,\nu=M_{s-1}+1}^{M_s} \eta_{\mu\nu}^{(s)} a_i^\nu(x, t_\nu) a_j^\mu(x, t_\mu)$$
$$B_{ij}^{(0)}(x) = 0$$

For $s > 1$, the sth element estimates $\{P_i^{(s)}\}$ are obtained by minimizing, with respect to $P^{(s)}$, the quadratic form

$$Q^{(s)} = \sum_{i,j=1}^{n} B_{ij}^{(s-1)}[P_i^{(s-1)} - P_i^{(s)}][P_j^{(s-1)} - P_j^{(s)}] + \sum_{\mu,\nu=M_{s-1}+1}^{M_s} \eta_{\mu\nu}^{(s)}[F^\mu - f^\mu(P^{(s)}, t_\mu)] \cdot [F^\nu - f^\nu(P^{(s)}, t^\nu)] \tag{30}$$

Let us also define $\rho_i^{(s)}(x)$, $s = 1, 2, \cdots, S$, by

$$\rho_i^{(s)}(x) = \sum_{j=1}^{n} B_{ij}^{(s-1)}(x)[P_j^{(s-1)} - x_j] + \sum_{\mu,\nu=M_{s-1}+1}^{M_s} \eta_{\mu\nu}^{(s)} a_i^\nu(x, t_\nu)[F^\mu - f^\mu(x, t_\mu)] \tag{31}$$

Then, using (13), (18), and (19), it can be shown that to first order

$$P^{(s)} - x = [B^{(s)}(x)]^{-1}\rho^{(s)}(x) \tag{32}$$

* Alternative definitions for the matrices $B^{(s)}$ are possible; see the remark at the end of this Section.

It is also not hard to show by an inductive argument that

$$P^{(S)} - x = [B^{(S)}(x)]^{-1}\rho^{(S)}(x) = [B(x)]^{-1}\rho(x) \quad (33)$$

where $B(x)$ and $\rho(x)$ are defined as in (9), (10), (11), (12). This is proved by showing that $B^{(S)}(x) = B(x)$, and $\rho^{(S)}(x) = \rho(x)$.

Thus, the first order dependence of the errors in the estimates $\{P_i^{(S)}\}$ is the same as that for the $\{P_i\}$ obtained by processing all N observations at once by minimizing Q as defined by (3). Another way of stating this is to say that, to first order, the estimate obtained by processing the N observations by the stagewise procedure just described is the same as that obtained by minimizing Q as defined by (3). (However, the range of magnitudes of $\mathcal{E}^\mu$ for which the first order expressions give good approximations to the estimation errors is not necessarily the same for the stagewise method as for the non-stagewise method.)

A stagewise smoothing procedure may be advantageous in certain situations. For example, suppose that observations are coming in at some average rate; in the stagewise procedure, it is not necessary to store all previous observations and $N \times N$ matrices $(\eta_{\mu\nu})$ with N increasing. It is at any time necessary to store only the 'current' element estimates $\{P_i^{(s)}\}$ and the matrix $B^{(s)}$, that is, $\frac{n}{2}(n + 3)$ quantities (taking into account the symmetry of $B^{(s)}$).

It can be seen that the matrices $B^{(s)}$ play the role of estimates of the matrices $B^{(s)}(x)$. Thus, the particular method of defining the sequence $\{B^{(s)}\}$ above is not the only one possible. For example, one could define

$$\tilde{B}_{ij}^{(s)} = B_{ij}^{(s)}, \qquad s = 1$$

$$= \tilde{B}_{ij}^{(s-1)} + \sum_{\mu,\nu=M_{s-1}+1}^{M_s} \eta_{\mu\nu}^{(s)} a_i^{\nu}(P^{(s-1)}, t_\nu) \quad (29')$$

$$\cdot a_j^{\mu}(P^{(s-1)}, t_\mu), \qquad s > 1$$

and define the stagewise process using the matrices $\tilde{B}^{(s)}$ instead of $B^{(s)}$. Comparing with (29), we see that the main difference is that $\tilde{B}^{(s)}$ can be computed, for $s > 1$, before one computes $P^{(s)}$.

This lends itself conveniently to first order determination, for $s > 1$, of $P^{(s)} - P^{(s-1)}$ by means of Eq's. (20), (23), and (24). One would write $p = P^{(s-1)}$; $P = P^{(s)}$; and $P^{(s)} - P^{(s-1)} = [\tilde{B}^{(s)}]^{-1}\rho^{*(s)}(P^{(s-1)})$. (The vector $\rho^{*(s)}$ would be defined by obvious modifications of (20) and (23).)

Application when the Elements are Functions of Time

Suppose the elements $x_1, \cdots, x_n$ are functions of time: $x_i = x_i(t)$. Also suppose that error-free observations are given by

$$F^\mu = g^\mu[x(t_\mu), t_\mu] \quad (34)$$

Suppose the manner in which $x(t)$ depends on its values at some t_0 is known:

$$x_i(t) = \mathfrak{F}_i[x(t_0), t_0, t] \quad (35)$$

Then we may obtain the case considered in the previous sections by defining the elements x_i in the formulas of those sections to be $x_i = x_i(t_0)$ and by defining

$$f^\mu[x, t_\mu] = g^\mu\{\mathfrak{F}_1[x, t_0, t_\mu], \cdots, \mathfrak{F}_n[x, t_0, t_\mu], t_\mu\} \quad (36)$$

(where $x = x(t_0)$).

Modification when the Functions f^μ are Imperfectly Known

Suppose the observations are given by

$$F^\mu = h^\mu[x, t_\mu] + \mathcal{E}^\mu \quad (37)$$

(where $\mathcal{E}^\mu$ are observation errors),

but that the estimated elements $\{P_i\}$ are obtained by minimizing Q as defined by (3), with the functions f^μ (which differ from h^μ).

So long as $F^\mu - f^\mu$ are sufficiently small, equations (9)–(13) still describe the dependence of $P - x$ on $F^\mu - f^\mu$. This dependence was expressed

$$P_i - x_i = \sum_{\mu=1}^{N} \Gamma_i^\mu(x)[F^\mu - f^\mu(x, t_\mu)] \quad (14)$$

where:

$$\Gamma_i^\mu(x) = \sum_{j=1}^{n}\sum_{\nu=1}^{N} [B(x)]_{ij}^{-1}\eta_{\mu\nu}a_j^\nu(x, t_\nu) \quad (15)$$

The observation errors $\mathcal{E}^\mu$ are now given by

$$\mathcal{E}^\mu = F^\mu - h^\mu(x, t_\mu). \quad (38)$$

Therefore

$$F^\mu - f^\mu(x, t_\mu) = \mathcal{E}^\mu + h^\mu(x, t_\mu) - f^\mu(x, t_\mu) \quad (39)$$

Thus, (14) becomes

$$P_i - x_i = \sum_{\mu=1}^{N} \Gamma_i^\mu(x)\mathcal{E}^\mu + \sum_{\mu=1}^{N} \Gamma_i^\mu(x)[h^\mu(x, t_\mu) - f_\mu(x, t_\mu)]. \quad (40)$$

Statistics of Propagated Errors

Eq. (13) may be used in an obvious manner to determine the means and covariance matrix of $\{P_i - x_i\}$ as functions of the means and covariance matrix of $\{\mathcal{E}^\mu\}$.

We shall deal here with a special case, namely, one in which the ensemble means and covariance matrix of $\{\mathcal{E}^\mu\}$ are known and in which the matrix $\{\eta_{\mu\nu}\}$ of the quadratic form Q has a special relation to the covariance matrix of $\{\mathcal{E}^\mu\}$.

Since the ensemble means of $\{\mathcal{E}^\mu\}$ are assumed known, it is no loss of generality to assume they are zero. In this case the covariance matrix is (denoting ensemble means by $E(\ \ )$)

$$\phi_{\mu\nu} = E(\mathcal{E}^\mu\mathcal{E}^\nu) \quad (41)$$

It will now be assumed that the matrix $(\eta_{\mu\nu})$ in (3) is

$$(\eta_{\mu\nu}) = (\phi_{\mu\nu})^{-1} \quad \text{(matrix inverse)} \tag{42}$$

If $\{\mathcal{E}^\mu\}$ were to have a Gaussian probability distribution (and if $f^\mu \equiv h^\mu$, all μ) then the resulting method of obtaining P would be the maximum likelihood method.

The covariance matrix of $\{P_i - x_i\}$ assumes a particularly simple form in this case. For generality, we will deal with the case described in the last section, in which f^μ, the functions used in the quadratic form Q, may differ from h^μ.

The means of $\{P_i - x_i\}$ are

$$E(P_i - x_i) = \sum_{\mu=1}^{N} \Gamma_i^\mu(x)[h^\mu(x, t_\mu) - f^\mu(x, t_\mu)] \tag{43}$$

The covariance matrix of $\{P_i - x_i\}$ is readily established to be

$$\begin{aligned} E\{P_i - x_i - E(P_i - x_i)\} \\ \cdot \{P_j - x_j - E(P_j - x_j)\} = [B(x)]_{ij}^{-1} \end{aligned} \tag{44}$$

This holds whether the observations are processed all together or by a stagewise procedure as described. In the latter case, of course, it must be assumed that the covariance matrix of $\{\mathcal{E}^\mu\}$ can be broken up into a diagonal array of covariance matrices corresponding to the different stages. It is, in fact, quite easy to establish that $B^{(s)}(x)$ is the inverse covariance matrix of $\{P_i^{(s)} - x_i\}$:

$$\begin{aligned} E\{P_i^{(s)} - x_i - E(P_i^{(s)} - x_i)\} \\ \cdot \{P_j^{(s)} - x_j - E(P_j^{(s)} - x_j)\} = [B^{(s)}(x)]_{ij}^{-1} \end{aligned} \tag{45}$$

This throws some further light on the stagewise method of Section II. The matrices $B^{(s-1)}$ occurring in the quadratic forms $Q^{(s)}$ are seen to be estimates of the inverse covariance matrices $B^{(s-1)}(x)$ of the element estimates $P^{(s-1)}$. Thus, if the error statistics were Gaussian, this procedure would consist at each stage of a maximum likelihood smoothing of the previous element estimates with the new observations.

The above formulas may be used to determine the rate at which the covariance matrix of the errors $P_i^{(s)} - x_i$ decreases as additional observations are processed. In fact, this information is contained in (44) and (45).

As a special case, consider the case where all observation errors are mutually uncorrelated. (If this is not true originally, it can be made true by means of linear transformations.) In this case, the matrices ϕ and η are diagonal.

If we now regard $P^{(s)}$ as the element estimate vector resulting from the processing of the first s observations, we have the inverse covariance matrix for $\{P_i^{(s)} - x_i\}$ given by

$$B_{ij}^{(s)}(x) = \sum_{\mu=1}^{s} \eta_{\mu\mu} a_i^\mu(x, t_\mu) a_j^\mu(x, t_\mu) \tag{46}$$

(This holds whether the s observations were processed all together or stagewise.)

It is also quite easy to verify that, for $s \geqq n$,

$$[B^{(s)}(x)]_{ij}^{-1} = [B^{(s-1)}(x)]_{ij}^{-1} - d_i^{(s)}(x)\, d_j^{(s)}(x) \tag{47}$$

where

$$d_i^{(s)}(x) = \frac{\sqrt{\eta_{ss}} \sum_{k=1}^{n} a_k^s(x, t_s) [B^{(s-1)}(x)]_{ik}^{-1}}{\sqrt{1 + \eta_{ss} \sum_{j,k=1}^{n} [B^{(s-1)}(x)]_{jk}^{-1} a_k^s(x, t_s) a_j^s(x, t_s)}} \tag{48}$$

We might close this section by briefly discussing the subject of systematic errors, which can be defined as errors which are highly correlated over a certain group of observations. There are several possible ways of handling systematic errors. They could be regarded as introducing off-diagonal elements into the correlation matrix $\phi_{\mu\nu}$ and the smoothing matrix $\eta_{\mu\nu}$, the magnitude of the off-diagonal elements being chosen according to the approximate magnitude of the systematic errors. Or, alternatively, a systematic error might in some cases profitably be regarded as an additional parameter to be estimated from the data i.e., as an additional element. This is one of the choices that would have to be made in actual implementation of these results.

Modified Stagewise Procedure for Time-Varying Elements

In this section the elements will be considered functions of time, $x_i = x_i(t)$. It will be assumed that the elements vary much more slowly with respect to time than do the functions describing observations.

We suppose that the matrix $(\eta_{\mu\nu})$ can be written (as in Eq. (26)) as a diagonal array of matrices $\eta^{(s)}, s = 1, 2, \cdots$; the further assumption is made that the times t_μ, $\mu = M_{s-1} + 1, \cdots, M_s$, are sufficiently close together that they may be regarded as equal insofar as variation of the elements is concerned. This will be expressed

$$\begin{aligned} x_i(t_\mu) &= x_i(T_s) \\ \mu &= M_{s-1} + 1, \cdots, M_s \end{aligned} \tag{49}$$

Henceforth, the time parameter occurring in the argument of an element or element estimate will be written T.

It will be supposed that error-free observations are given by

$$\begin{aligned} F^\mu &= g^\mu[x(T_s), t_\mu] \\ \mu &= M_{s-1} + 1, \cdots, M_s \\ s &= 1, 2, \cdots \end{aligned} \tag{50}$$

and that the variation of the elements with respect to time is given by

$$x_i(T_s) = \mathfrak{F}_i[x(T_r), T_r, T_s]. \tag{51}$$

Now consider the following stagewise procedure for processing the observations (the main reason for which is its adaptability to cases in which the prediction functions f^μ or $\mathfrak{F}_i$ are imperfectly known):

The element estimate vector $P^+(T_1)$ is obtained by minimizing with respect to $P^+(T_1)$ the quadratic form

$$Q^{(1)} = \sum_{\mu,\nu=1}^{N_1} \eta_{\mu\nu}^{(1)}\{F^\mu - g^\mu[P^+(T_1), t_\mu]\} \cdot \{F^\nu - g^\nu[P^+(T_1), t_\nu]\}. \quad (52)$$

For $T_s \leqq T < T_{s+1}$, $s = 1, 2, \cdots$,

$$P_i(T) = \mathfrak{F}_i[P^+(T_s), T_s, T] \quad (53)$$

Also,

$$P_i^-(T_{s+1}) = \mathfrak{F}_i[P^+(T_s), T_s, T_{s+1}] \quad (54)$$

The quantities $P^+(T_s)$ will be defined for $s > 1$ below. To do this, we define sequences of matrices $B^{(s)}$, $B_+^{(s)}$, $\psi^{(s)}$, $\psi_+^{(s)}$ as follows:

$$B_{+ij}^{(1)} = \sum_{\mu,\nu=1}^{N_1} \eta_{\mu\nu}^{(1)} b_i^\nu[P^+(T_1), t_\nu] b_j^\mu[P^+(T_1), t_\mu] \quad (55)$$

$$b_i^\nu[x,t_\nu] = \frac{\partial g^\nu}{\partial x_i}[x,t_\nu] \quad (56)$$

$$\psi_+^{(s)} = [B_+^{(s)}]^{-1}, \qquad s = 1, 2, \cdots \quad (57)$$

$$\psi^{(s+1)} = \sum_{k,l=1}^{n} \mathcal{A}_{ik}[P^+(T_s), T_s, T_{s+1}] \cdot \mathcal{A}_{jl}[P^+(T_s), T_s, T_{s+1}]\psi_{+kl}^{(s)} \quad (58)$$

$$\mathcal{A}_{ij}[x, T_s, T] = \frac{\partial \mathfrak{F}_i}{\partial x_j}[x, T_s, T] \quad (59)$$

$$\begin{aligned} B^{(s)} &= 0, \qquad s = 1 \\ &= [\psi^{(s)}]^{-1}, \qquad s > 1 \end{aligned} \quad (60)$$

$$B_{+ij}^{(s)} = B_{ij}^{(s)} + \sum_{\mu,\nu=M_{s-1}+1}^{M_s} \eta_{\mu\nu}^{(s)} b_i^\nu[P^+(T_s), t_\nu] b_j^\mu[P^+(T_s), t_\mu] \quad (61)$$

Then $P^+(T_s)$ is obtained for $s > 1$ by minimizing, with respect to $P^+(T_s)$, the quadratic form

$$\begin{aligned} Q^{(s)} = \sum_{i,j=1}^{n} & B_{ij}^{(s)}\{P_i^-(T_s) - P_i^+(T_s)\} \cdot \{P_j^-(T_s) - P_j^+(T_s)\} \\ &+ \sum_{\mu,\nu=M_{s-1}+1}^{M_s} \eta_{\mu\nu}^{(s)}\{F^\mu - g^\mu[P^+(T_s), t_\mu]\} \cdot \{F^\nu - g^\nu[P^+(T_s), t_\nu]\} \end{aligned} \quad (62)$$

Scrutiny of Eqs. (52)–(62) reveals that a stagewise smoothing procedure has been completely defined. Now, in order to give the first order error equations for the resulting estimates, define the following matrices:

$$B_{+ij}[T_1] = \sum_{\mu,\nu=1}^{N_1} \eta_{\mu\nu}^{(1)} b_i^\nu[x(T_1), t_\nu] b_j^\mu[x(T_1), t_\mu] \quad (63)$$

(where $b_i^\nu(x, t_\nu)$ is defined as in (56));

$$\psi_+[T_s] = \{B_+[T_s]\}^{-1} \quad (64)$$

for $T_s < T \leqq T_{s+1}$,

$$\psi_{ij}[T] = \sum_{k,l=1}^{n} \mathcal{A}_{ik}[x(T_s), T_s, T] \cdot \mathcal{A}_{jl}[x(T_s), T_s, T]\psi_{+kl}[T_s] \quad (65)$$

(where $\mathcal{A}_{ij}(x, T_s, T)$ is defined as in (59));

$$B[T] = \{\psi[T]\}^{-1}, \quad T_s < T \leqq T_{s+1} \quad (66)$$

and

$$B_{+ij}[T_s] = B_{ij}[T_s] + \sum_{\mu,\nu=M_{s-1}+1}^{M_s} \eta_{\mu\nu}^{(s)} b_i^\nu[x(T_s), t_\nu] b_j^\mu[x(T_s), t_\mu]. \quad (67)$$

Also define

$$\rho_i[T_s] = \sum_{j=1}^{n} B_{ij}[T_s][P_j^-(T_s) - x_j(T_s)] + \sum_{\mu,\nu=M_{s-1}+1}^{M_s} \eta_{\mu\nu}^{(s)} b_i^\nu[x(T_s), t_\nu]\{F^\mu - g^\mu[x(T_s), t_\mu]\}. \quad (68)$$

Then, the first order dependence of $P^+(T_s) - x(T_s)$ on the observation errors is:

$$P^+(T_s) - x(T_s) = \{B_+[T_s]\}^{-1}\rho[T_s] \quad (69)$$

Also, for $T_s < T < T_{s+1}$, the first order dependence of $P(T) - x(T)$ is

$$P_i(T) - x_i(T) = \sum_{j=1}^{n} \mathcal{A}_{ij}[x(T_s), T_s, T][P_j^+(T_s) - x_j(T_s)] \quad (70)$$

Further, suppose that $(\eta_{\mu\nu})$ represents the inverse covariance matrix of $\{F^\mu - g^\mu[x(t_\mu), t_\mu]\}$. Then, $B_+[T_s]$ is the inverse covariance matrix of $\{P_i^+(T_s) - x_i(T_s)\}$, and $B[T]$ is the inverse covariance matrix of $\{P_i(T) - x_i(T)\}$, $T_s < T < T_{s+1}$.

In the third section an alternative method was described for dealing with time-varying elements. That method involved a reduction to the case of constant elements.

Choose any time T, $T_s \leqq T < T_{s+1}$, and regard T as now fixed. Define a constant element vector x by: $x = x(T)$. Let $P^*(T)$ represent the estimate of $x = x(T)$ obtained by the method described in the third section, based on the first M_s observations.

Then, it can be verified that, to first order, $P^*(T) = P(T)$, where $P(T)$ is defined by (53). That is, the stagewise procedure described in (52)–(62) yields an estimate $P(T)$ for $x(T)$ having the same first order dependence on $F^\mu - g^\mu[x(t_\mu), t_\mu]$, $\mu = 1, \cdots, M_s$, as the estimate derived by the method of Section III.

The stagewise procedure described above actually goes through the following steps: $P^+(T_s)$ represents an estimate of $x(T_s)$, based on the first M_s observations;

for $T_s < T < T_{s+1}$, $P(T)$ is obtained by (53), that is, simply by prediction from $P^+(T_s)$ according to the functional relation by which the elements are known to vary; $P^-(T_{s+1})$ represents the estimate of $x(T_{s+1})$ just before the observations F^μ, $\mu = M_s + 1, \cdots, M_{s+1}$ are processed.

The particular method chosen above to define the matrices $B_+^{(s)}$—i.e., by (61)—is not the only one possible. One could, for example, define matrices $\tilde{B}_+^{(s)}$ for $s > 1$ by using (61) with $P^-(T_s)$ instead of $P^+(T_s)$ in the arguments of b_i^ν and b_j^μ. Then one could compute $\tilde{B}_+^{(s)}$, for $s > 1$, before computing $P^+(T_s)$. This would be convenient for purposes of using (20), (23), and (24) for a first order determination of $P^+(T_s) - P^-(T_s)$: one would put $p = P^-(T_s)$; $P = P^+(T_s)$; $P^+(T_s) - P^-(T_s) = [\tilde{B}_+^{(s)}]^{-1}\rho^{*(s)}[P^-(T)]$, with $\rho^{*(s)}$ defined by the appropriate modification of (20) and (23).

In practical applications, for example, in satellite observations where perturbation forces are imperfectly known, the functions $\mathfrak{F}_i$ may not be known exactly. Furthermore, the inaccuracy in knowledge of $\mathfrak{F}_i[x(T), T, T']$ will depend on the time difference $T' - T$—i.e., on how far ahead you are predicting the elements.

Suppose, for example, that the prediction functions used in the stagewise procedure are g^μ, $\mathfrak{F}_i$, but that the correct functions should be h^μ, $\mathfrak{F}_i^*$. Let $\mathcal{E}^\mu$ still represent the observation errors—i.e., $\mathcal{E}^\mu = F^\mu - h^\mu$—and let, for $T \geqq t_\mu$,

$$\begin{aligned} \delta^\mu(T) = {} & h^\mu\{\mathfrak{F}_1^*[x(T), T, t_\mu], \cdots, \\ & \mathfrak{F}_n^*[x(T), T, t_\mu], t_\mu\} \\ & - g^\mu\{\mathfrak{F}_1[x(T), T, t_\mu], \cdots, \\ & \mathfrak{F}_n[x(T), T, t_\mu], t_\mu\} \end{aligned} \tag{71}$$

Suppose that, in the case $g^\mu = h^\mu$, $\mathfrak{F}_i = \mathfrak{F}_i^*$, the first order error in $P(T)$, $T_s \leqq T < T_{s+1}$, could be expressed

$$P_i(T) - x_i(T) = \sum_{\mu=1}^{M_s} \Gamma_i^\mu(T)\mathcal{E}^\mu \tag{72}$$

Then, in the case $g^\mu \neq h^\mu$, $\mathfrak{F}_i \neq \mathfrak{F}_i^*$, one would have

$$P_i(T) - x_i(T) = \sum_{\mu=1}^{M_s} \Gamma_i^\mu(T)[\mathcal{E}^\mu + \delta^\mu(T)] \tag{73}$$

provided $\delta^\mu(T)$ and $\mathcal{E}^\mu$ are sufficiently small.

The errors $\delta^\mu(T)$ would, in general, grow as T increases. In such cases, it would be desirable to modify the smoothing procedure so as to give more recent observations greater weight, and continuously to diminish the weight given to past observations. There are a number of possibilities for accomplishing this; one way, for instance, would be to define the matrices $B^{(s)}$ in (60)–(62) by

$$B_{ij}^{(s)} = [\psi^{(s)}]_{ij}^{-1}\lambda_i^{(s)}\lambda_j^{(s)} \tag{60*}$$

where $\lambda_i^{(s)}$ and $\lambda_j^{(s)}$ are $\leqq 1$.

This would mean that the effective smoothing matrix for the observations would not be the matrix (η) appearing in (52)–(62); in fact, there would really be no fixed smoothing matrix. Also, the first order error equations (63)–(70) would have to be modified appropriately.

There is an analogy between this class of smoothing procedures and the process of filtering of signals. One might regard the smoothing procedures as filters with input $x(T)$ and output $P(T)$. The case where the f^μ and $\mathfrak{F}_i$ are perfectly known is equivalent to a constant signal input to the filter, in which case the filter should have infinite memory to provide maximum smoothing of observation errors. If, say, $\mathfrak{F}_i$ are not exactly known, this corresponds to the existence of an unpredictable time varying component of the filter input; in this case the filter memory must be reduced (its 'bandwidth' increased); the best 'time constant' is related to the rate of variation of the unpredictable part of $x(T)$. One could also, in effect, make the time constant different for the different elements.

New Results in Linear Filtering and Prediction Theory[1]

R. E. KALMAN
Research Institute for Advanced Study,[2] Baltimore, Maryland

R. S. BUCY
The Johns Hopkins Applied Physics Laboratory, Silver Spring, Maryland

A nonlinear differential equation of the Riccati type is derived for the covariance matrix of the optimal filtering error. The solution of this "variance equation" completely specifies the optimal filter for either finite or infinite smoothing intervals and stationary or nonstationary statistics.

The variance equation is closely related to the Hamiltonian (canonical) differential equations of the calculus of variations. Analytic solutions are available in some cases. The significance of the variance equation is illustrated by examples which duplicate, simplify, or extend earlier results in this field.

The Duality Principle relating stochastic estimation and deterministic control problems plays an important role in the proof of theoretical results. In several examples, the estimation problem and its dual are discussed side-by-side.

Properties of the variance equation are of great interest in the theory of adaptive systems. Some aspects of this are considered briefly.

1 Introduction

At present, a nonspecialist might well regard the Wiener-Kolmogorov theory of filtering and prediction [1, 2][3] as "classical'—in short, a field where the techniques are well established and only minor improvements and generalizations can be expected.

That this is not really so can be seen convincingly from recent results of Shinbrot [3], Steeg [4], Pugachev [5, 6], and Parzen [7]. Using a variety of time-domain methods, these investigators have solved some long-standing problems in *nonstationary* filtering and prediction theory. We present here a unified account of our own independent researches during the past two years (which overlap with much of the work [3–7] just mentioned), as well as numerous new results. We, too, use time-domain methods, and obtain major improvements and generalizations of the conventional Wiener theory. In particular, our methods apply without modification to multivariate problems.

The following is the historical background of this paper.

In an extension of the standard Wiener filtering problem, Follin [8] obtained relationships between time-varying gains and error variances for a given circuit configuration. Later, Hanson [9] proved that Follin's circuit configuration was actually optimal for the assumed statistics; moreover, he showed that the differential equations for the error variance (first obtained by Follin) follow rigorously from the Wiener-Hopf equation. These results were then generalized by Bucy [10], who found explicit relationships between the optimal weighting functions and the error variances; he also gave a rigorous derivation of the variance equations and those of the optimal filter for a wide class of nonstationary signal and noise statistics.

Independently of the work just mentioned, Kalman [11] gave a new approach to the standard filtering and prediction problem. The novelty consisted in combining two well-known ideas:

(i) the "state-transition" method of describing dynamical systems [12–14], and

(ii) linear filtering regarded as orthogonal projection in Hilbert space [15, pp. 150–155].

As an important by-product, this approach yielded the *Duality Principle* [11, 16] which provides a link between (stochastic) filtering theory and (deterministic) control theory. Because of the duality, results on the optimal design of linear control systems [13, 16, 17] are directly applicable to the Wiener problem. Duality plays an important role in this paper also.

When the authors became aware of each other's work, it was soon realized that the principal conclusion of both investigations was identical, in spite of the difference in methods:

Rather than to attack the Wiener-Hopf integral equation directly, it is better to convert it into a nonlinear differential equation, whose solution yields the covariance matrix of the minimum filtering error, which in turn contains all necessary information for the design of the optimal filter.

2 Summary of Results: Description

The problem considered in this paper is stated precisely in Section 4. There are two main assumptions:

(A_1) A sufficiently accurate model of the message process is given by a linear (possibly time-varying) dynamical system excited by white noise.

(A_2) Every observed signal contains an additive white noise component.

Assumption (A_2) is unnecessary when the random processes in question are sampled (discrete-time parameter); see [11]. Even in the continuous-time case, (A_2) is no real restriction since it can be removed in various ways as will be shown in a future paper. Assumption (A_1), however, is quite basic; it is analogous to but somewhat less restrictive than the assumption of rational spectra in the conventional theory.

Within these assumptions, we seek the best linear estimate of the message based on past data lying in either a finite or infinite time-interval.

The fundamental relations of our new approach consist of five equations:

---

[1] This research was partially supported by the United States Air Force under Contracts AF 49(638)-382 and AF 33(616)-6952 and by the Bureau of Naval Weapons under Contract NOrd-73861.

[2] 7212 Bellona Avenue.

[3] Numbers in brackets designate References at the end of paper.

Contributed by the Instruments and Regulators Division of The American Society of Mechanical Engineers and presented at the Joint Automatic Controls Conference, Cambridge, Mass., September 7–9, 1960. Manuscript received at ASME Headquarters, May 31, 1960. Paper No. 60—JAC-12.

Reprinted with permission from *Trans. ASME, J. Basic Eng.,* Series 83D, pp. 95–108, Mar. 1961.

(I) The differential equation governing the optimal filter, which is excited by the observed signals and generates the best linear estimate of the message.

(II) The differential equations governing the error of the best linear estimate.

(III) The time-varying gains of the optimal filter expressed in terms of the error variances.

(IV) The nonlinear differential equation governing the covariance matrix of the errors of the best linear estimate, called the *variance equation.*

(V) The formula for prediction.

The solution of the variance equation for a given finite time-interval is equivalent to the solution of the estimation or prediction problem with respect to the same time-interval. The steady-state solution of the variance equation corresponds to finding the best estimate based on all the data in the past.

As a special case, one gets the solution of the classical (stationary) Wiener problem by finding the unique equilibrium point of the variance equation. This requires solving a set of algebraic equations and constitutes a new method of designing Wiener filters. The superior effectiveness of this procedure over present methods is shown in the examples.

Some of the preceding ideas are implicit already in [10, 11]; they appear here in a fully developed form. Other more advanced problems have been investigated only very recently and provide incentives for much further research. We discuss the following further results:

(1) The variance equations are of the Riccati type which occur in the calculus of variations and are closely related to the canonical differential equations of Hamilton. This relationship gives rise to a well-known analytic formula for the solution of the Riccati equation [17, 18]. The Hamiltonian equations have also been used recently [19] in the study of optimal control systems. The two types of problems are actually duals of one another as mentioned in the Introduction. The duality is illustrated by several examples.

(2) A sufficient condition for the existence of steady-state solutions of the variance equation (i.e., the fact that the error variance does not increase indefinitely) is that the information matrix in the sense of R. A. Fisher [20] be nonsingular. This condition is considerably weaker than the usual assumption that the message process have finite variance.

(3) A sufficient condition for the optimal filter to be stable is the dual of the preceding condition.

The preceding results are established with the aid of the "state-transition" method of analysis of dynamical systems. This consists essentially of the systematic use of vector-matrix notation which results in simple and clear statements of the main results independently of the complexity of specific problems. This is the reason why multivariable filtering problems can be treated by our methods without any additional theoretical complications.

The outline of contents is as follows:

In Section 3 we review the description of dynamical systems from the state point of view. Sections 4–5 contain precise statements of the filtering problem and of the dual control problem. The examples in Section 6 illustrate the filtering problem and its dual in conventional block-diagram terminology. Section 7 contains a precise statement of all mathematical results. A reader interested mainly in applications may pass from Section 7 directly to the worked-out examples in Section 11. The rigorous derivation of the fundamental equations is given in Section 8. Section 9 outlines proofs, based on the Duality Principle, of the existence and stability of solutions of the variance equation. The theory of analytic solutions of the variance equation is discussed in Section 10. In Section 12 we examine briefly the relation of our results to adaptive filtering problems. A critical evaluation of the current status of the statistical filtering problem is presented in Section 13.

3 Preliminaries

In the main, we shall follow the notation conventions (though not the specific nomenclature) of [11], [16], and [21]. Thus τ, t, t_0 refer to the time, $\alpha, \beta, \ldots, x_1, x_2, \ldots, \phi_1, \phi_2, \ldots, a_{ij}, \ldots$ are (real) scalars; $\mathbf{a}, \mathbf{b}, \ldots, \mathbf{x}, \mathbf{y}, \ldots, \boldsymbol{\phi}, \boldsymbol{\psi}, \ldots$ are vectors, $\mathbf{A}, \mathbf{B}, \ldots, \boldsymbol{\Phi}, \boldsymbol{\Psi}, \ldots$ are matrices. The prime denotes the transposed matrix; thus $\mathbf{x}'\mathbf{y}$ is the scalar (inner) product and $\mathbf{x}\mathbf{y}'$ denotes a matrix with elements $x_i y_j$ (outer product). $\|\mathbf{x}\| = (\mathbf{x}'\mathbf{x})^{1/2}$ is the euclidean norm and $\|\mathbf{x}\|^2_{\mathbf{A}}$ (where $\mathbf{A}$ is a nonnegative definite matrix) is the quadratic form with respect to $\mathbf{A}$. The eigenvalues of a matrix $\mathbf{A}$ are written as $\lambda_i(\mathbf{A})$. The expected value (ensemble average) is denoted by $\mathcal{E}$ (usually not followed by brackets). The covariance matrix of two vector-valued random variables $\mathbf{x}(t)$, $\mathbf{y}(\tau)$ is denoted by

$$\mathcal{E}\mathbf{x}(t)\mathbf{y}'(\tau) - \mathcal{E}\mathbf{x}(t)\mathcal{E}\mathbf{y}'(\tau) \quad \text{or} \quad \operatorname{cov}[\mathbf{x}(t), \mathbf{y}(\tau)]$$

depending on what form is more convenient.

Real-valued linear functions of a vector $\mathbf{x}$ will be denoted by $\mathbf{x}^*$; the value of $\mathbf{x}^*$ at $\mathbf{x}$ is denoted by

$$[\mathbf{x}^*, \mathbf{x}] = \sum_{i=1}^{n} x^*_i x_i$$

where the x_i are the co-ordinates of $\mathbf{x}$. As is well known, $\mathbf{x}^*$ may be regarded abstractly as an element of the *dual vector space* of the $\mathbf{x}$'s; for this reason, $\mathbf{x}^*$ is called a *covector* and its co-ordinates are the x^*_i. In algebraic manipulations we regard $\mathbf{x}^*$ formally as a row vector (remembering, of course, that $\mathbf{x}^* \neq \mathbf{x}'$). Thus the inner product is $\mathbf{x}^*\mathbf{y}^{*\prime}$ and we define $\|\mathbf{x}^*\|$ by $(\mathbf{x}^*\mathbf{x}^{*\prime})^{1/2}$. Also

$$\mathcal{E}[\mathbf{x}^*, \mathbf{x}]^2 = \mathcal{E}(\mathbf{x}^*\mathbf{x})^2 = \mathcal{E}\mathbf{x}^*\mathbf{x}\mathbf{x}'\mathbf{x}^{*\prime}$$
$$= \mathbf{x}^*(\mathcal{E}\mathbf{x}\mathbf{x}')\mathbf{x}^{*\prime} = \|\mathbf{x}^*\|^2_{\mathcal{E}\mathbf{x}\mathbf{x}'}$$

To establish the terminology, we now review the essentials of the so-called *state-transition method* of analysis of dynamical systems. For more details see, for instance, [21].

A linear dynamical system governed by an ordinary differential equation can always be described in such a way that the defining equations are in the *standard form:*

$$d\mathbf{x}/dt = \mathbf{F}(t)\mathbf{x} + \mathbf{G}(t)\mathbf{u}(t) \qquad (1)$$

where $\mathbf{x}$ is an n-vector, called the *state;* the *co-ordinates* x_i of $\mathbf{x}$ are called *state variables;* $\mathbf{u}(t)$ is an m-vector, called the *control function;* $\mathbf{F}(t)$ and $\mathbf{G}(t)$ are $n \times n$ and $n \times m$ matrices, respectively, whose elements are continuous functions of the time t.

The description (1) is incomplete without specifying the *output* $\mathbf{y}(t)$ of the system; this may be taken as a p-vector whose components are linear combinations of the state variables:

$$\mathbf{y}(t) = \mathbf{H}(t)\mathbf{x}(t) \qquad (2)$$

where $\mathbf{H}(t)$ is a $p \times n$ matrix continuous in t.

The matrices $\mathbf{F}$, $\mathbf{G}$, $\mathbf{H}$ can be usually determined by inspection if the system equations are given in block diagram form. See the examples in Section 5. It should be remembered that any of these matrices may be nonsingular. $\mathbf{F}$ represents the dynamics, $\mathbf{G}$ the constraints on affecting the state of the system by inputs, and $\mathbf{H}$ the constraints on observing the state of the system from outputs. For single-input/single-output systems, $\mathbf{G}$ and $\mathbf{H}$ consist of a single column and single row, respectively.

If $\mathbf{F}$, $\mathbf{G}$, $\mathbf{H}$ are constants, (3) is a *constant* system. If $\mathbf{u}(t) = \mathbf{0}$ or, equivalently, $\mathbf{G} = \mathbf{0}$, (3) is said to be *free.*

It is well known [21–23] that the general solution of (1) may be written in the form

$$\mathbf{x}(t) = \mathbf{\Phi}(t, t_0)\mathbf{x}(t_0) + \int_{t_0}^{t} \mathbf{\Phi}(t, \tau)\mathbf{G}(\tau)\mathbf{u}(\tau)d\tau \tag{3}$$

where we call $\mathbf{\Phi}(t, t_0)$ the *transition matrix* of (1). The transition matrix is a nonsingular matrix satisfying the differential equation

$$d\mathbf{\Phi}/dt = \mathbf{F}(t)\mathbf{\Phi} \tag{4}$$

(any such matrix is a *fundamental matrix* [23, Chapter 3]), made *unique* by the additional requirement that, for all t_0,

$$\mathbf{\Phi}(t_0, t_0) = \mathbf{I} = \text{unit matrix} \tag{5}$$

The following properties are immediate by the existence and uniqueness of solutions of (1):

$$\mathbf{\Phi}^{-1}(t_1, t_0) = \mathbf{\Phi}(t_0, t_1) \quad \text{for all} \quad t_0, t_1 \tag{6}$$

$$\mathbf{\Phi}(t_2, t_0) = \mathbf{\Phi}(t_2, t_1)\mathbf{\Phi}(t_1, t_0) \quad \text{for all} \quad t_0, t_1, t_2 \tag{7}$$

If $\mathbf{F}$ = const, then the transition matrix can be represented by the well-known formula

$$\mathbf{\Phi}(t, t_0) = \exp \mathbf{F}(t - t_0) = \sum_{i=0}^{\infty} [\mathbf{F}(t - t_0)]^i/i! \tag{8}$$

which is quite convenient for numerical computations. In this special case, one can also express $\mathbf{\Phi}$ analytically in terms of the eigenvalues of $\mathbf{F}$, using either linear algebra [22] or standard transfer-function techniques [14].

In some cases, it is convenient to replace the right-hand side of (3) by a notation that focuses attention on how the state of the system "moves" in the state space as a function of time. Thus we write the left-hand side of (3) as

$$\mathbf{x}(t) \equiv \boldsymbol{\phi}(t;\ \mathbf{x}, t_0;\ \mathbf{u}) \tag{9}$$

Read: The state of the system (1) at time t, evolving from the initial state $\mathbf{x} = \mathbf{x}(t_0)$ at time t_0 under the action of a *fixed* forcing function $\mathbf{u}(t)$. For simplicity, we refer to $\boldsymbol{\phi}$ as the *motion* of the dynamical system

4 Statement of Problem

We shall be concerned with the continuous-time analog of Problem I of reference [11], which should be consulted for the physical motivation of the assumptions stated below.

(A_1) The *message* is a random process $\mathbf{x}(t)$ generated by the *model*

$$d\mathbf{x}/dt = \mathbf{F}(t)\mathbf{x} + \mathbf{G}(t)\mathbf{u}(t) \tag{10}$$

The *observed signal* is

$$\mathbf{z}(t) = \mathbf{y}(t) + \mathbf{v}(t) = \mathbf{H}(t)\mathbf{x}(t) + \mathbf{v}(t) \tag{11}$$

The functions $\mathbf{u}(t)$, $\mathbf{v}(t)$ in (10–11) are independent random processes (white noise) with identically zero means and covariance matrices

$$\begin{aligned} \text{cov}\ [\mathbf{u}(t), \mathbf{u}(\tau)] &= \mathbf{Q}(t)\cdot\delta(t - \tau) \\ \text{cov}\ [\mathbf{v}(t), \mathbf{v}(\tau)] &= \mathbf{R}(t)\cdot\delta(t - \tau) \quad \text{for all} \quad t, \tau \\ \text{cov}\ [\mathbf{u}(t), \mathbf{v}(\tau)] &= \mathbf{0} \end{aligned} \tag{12}$$

where δ is the Dirac delta function, and $\mathbf{Q}(t)$, $\mathbf{R}(t)$ are symmetric, nonnegative definite matrices continuously differentiable in t.

We introduce already here a restrictive assumption, which is needed for the ensuing theoretical developments:

(A_2) The matrix $\mathbf{R}(t)$ is positive definite for all t. Physically, this means that no component of the signal can be measured exactly.

To determine the random process $\mathbf{x}(t)$ uniquely, it is necessary to add a further assumption. This may be done in two different ways:

(A_3) The dynamical system (10) has reached "steady-state" under the action of $\mathbf{u}(t)$, in other words, $\mathbf{x}(t)$ is the random function defined by

$$\mathbf{x}(t) = \int_{-\infty}^{t} \mathbf{\Phi}(t, \tau)\mathbf{G}(\tau)\mathbf{u}(\tau)d\tau \tag{13}$$

This formula is valid if the system (10) is uniformly asymptotically stable (for precise definition, valid also in the nonconstant case, see [21]). If, in addition, it is true that $\mathbf{F}$, $\mathbf{G}$, $\mathbf{Q}$ are constant, then $\mathbf{x}(t)$ is a stationary random process—this is one of the chief assumptions of the original Wiener theory.

However, the requirement of asymptotic stability is inconvenient in some cases. For instance, it is not satisfied in Example 5, which is a useful model in some missile guidance problems. Moreover, the representation of random functions as generated by a linear dynamical system is already an appreciable restriction and one should try to avoid making any further assumptions. Hence we prefer to use:

(A_3') The measurement of $\mathbf{z}(t)$ starts at some fixed instant t_0 of time (which may be $-\infty$), at which time cov$[\mathbf{x}(t_0), \mathbf{x}(t_0)]$ is known.

Assumption (A_3) is obviously a special case of (A_3'). Moreover, since (10) is not necessarily stable, this way of proceeding makes it possible to treat also situations where the message variance grows indefinitely, which is excluded in the conventional theory.

The main object of the paper is to study the

OPTIMAL ESTIMATION PROBLEM. *Given known values of $\mathbf{z}(\tau)$ in the time-interval $t_0 \leqq \tau \leqq t$, find an estimate $\hat{\mathbf{x}}(t_1|t)$ of $\mathbf{x}(t_1)$ of the form*

$$\hat{\mathbf{x}}(t_1|t) = \int_{t_0}^{t} \mathbf{A}(t_1, \tau)\mathbf{z}(\tau)d\tau \tag{14}$$

(where $\mathbf{A}$ is an $n \times p$ matrix whose elements are continuously differentiable in both arguments) with the property that the expected squared error in estimating any linear function of the message is minimized:

$$\mathcal{E}[\mathbf{x}^*, \mathbf{x}(t_1) - \hat{\mathbf{x}}(t_1|t)]^2 = \text{minimum for all } \mathbf{x}^* \tag{15}$$

Remarks. (*a*) Obviously this problem includes as a special case the more common one in which it is desired to minimize

$$\mathcal{E}\|\mathbf{x}(t_1) - \hat{\mathbf{x}}(t_1|t)\|^2$$

(*b*) In view of (A_1), it is clear that $\mathcal{E}\mathbf{x}(t_1) = \mathcal{E}\hat{\mathbf{x}}(t_1|t) = \mathbf{0}$. Hence $[\mathbf{x}^*, \hat{\mathbf{x}}(t_1|t)]$ is the *minimum variance linear unbiased estimate* of the value of any *costate* $\mathbf{x}^*$ at $\mathbf{x}(t_1)$.

(*c*) If $\mathcal{E}\mathbf{u}(t)$ is unknown, we have a more difficult problem which will be considered in a future paper.

(*d*) It may be recalled (see, e.g., [11]) that if $\mathbf{u}$ and $\mathbf{v}$ are gaussian, then so are also $\mathbf{x}$ and $\mathbf{z}$, and therefore the best estimate will be of the type (14). Moreover, the same estimate will be best not only for the loss function (15) but also for a wide variety of other loss functions.

(*e*) The representation of white noise in the form (12) is not rigorous, because of the use of delta "functions." But since the delta function occurs only in integrals, the difficulty is easily removed as we shall show in a future paper addressed to mathematicians. All other mathematical developments given in the paper are rigorous.

The solution of the estimation problem under assumptions (A_1), (A_2), (A_3') is stated in Section 7 and proved in Section 8.

5 The Dual Problem

It will be useful to consider now the dual of the optimal estimation problem which turns out to be the optimal regulator problem in the theory of control.

First we define a dynamical system which is the *dual* (or *adjoint*) of (1). Let

$$\left.\begin{aligned} t^* &= -t \\ \mathbf{F}^*(t^*) &= \mathbf{F}'(t) \\ \mathbf{G}^*(t^*) &= \mathbf{H}'(t) \\ \mathbf{H}^*(t^*) &= \mathbf{G}'(t) \end{aligned}\right\} \tag{16}$$

Let $\mathbf{\Phi}^*(t^*, t_0^*)$ be the transition matrix of the dual dynamical system of (1):

$$d\mathbf{x}^*/dt^* = \mathbf{F}^*(t^*)\mathbf{x}^* + \mathbf{G}^*(t^*)\mathbf{u}^*(t^*) \tag{17}$$

It is easy to verify the fundamental relation

$$\mathbf{\Phi}^*(t^*, t_0^*) = \mathbf{\Phi}'(t_0, t) \tag{18}$$

With these notation conventions, we can now state the

OPTIMAL REGULATOR PROBLEM. *Consider the linear dynamical system* (17). *Find a "control law"*

$$\mathbf{u}^*(t^*) = \mathbf{k}^*(\mathbf{x}^*(t^*), t_0^*) \tag{19}$$

with the property that, for this choice of $\mathbf{u}^*(t^*)$, *the "performance index"*

$$V(\mathbf{x}^*; t^*, t_0^*; \mathbf{u}^*) = \|\phi^*(t_0^*; \mathbf{x}, t^*; \mathbf{u}^*)\|^2_{\mathbf{P}_0} + \int_{t^*}^{t_0^*} \{\|\phi^*(\tau^*; \mathbf{x}^*, t^*; \mathbf{u}^*)\|^2_{\mathbf{Q}(\tau^*)} + \|\mathbf{u}^*(\tau^*)\|^2_{\mathbf{R}(\tau^*)}\} d\tau^* \tag{20}$$

assumes its greatest lower bound.

This is a natural generalization of the well-known problem of the optimization of a regulator with integrated-squared-error type of performance index.

The mathematical theory of the optimal regulator problem has been explored in considerable detail [17]. These results can be applied directly to the optimal estimation problem because of the

DUALITY THEOREM. *The solutions of the optimal estimation problem and of the optimal regulator problem are equivalent under the duality relations* (16).

The nature of these solutions will be discussed in the sequel. Here we pause only to observe a trivial point: By (14), the solutions of the estimation problem are necessarily linear; hence the same must be true (if the duality theorem is correct) of the solutions of the optimal regulator problem, in other words, the optimal control law $\mathbf{k}^*$ must be a linear function of $\mathbf{x}^*$.

The first proof of the duality theorem appeared in [11], and consisted of comparing the end results of the solutions of the two problems. Assuming only that the solutions of both problems result in linear dynamical systems, the proof becomes much simpler and less mysterious; this argument was carried out in detail in [16].

Remark (*f*). If we generalize the optimal regulator problem to the extent of replacing the first integrand in (20) by

$$\|\mathbf{y}^*(\tau^*) - \mathbf{y}_d^*(\tau^*)\|^2_{\mathbf{Q}(\tau^*)}$$

where $\mathbf{y}_d^*(t^*) \not\equiv \mathbf{0}$ is the *desired output* (in other words, if the regulator problem is replaced by a servomechanism or follow-up problem), then we have the dual of the estimation problem with $\mathcal{E}\mathbf{u}(t) \not\equiv \mathbf{0}$.

6 Examples: Problem Statement

To illustrate the matrix formalism and the general problems stated in Sections 4–5, we present here some specific problems in the standard block-diagram terminology. The solution of these problems is given in Section 11.

Example 1. Let the model of the message process be a first-order, linear, constant dynamical system. It is not assumed that the model is stable; but if so, this is the simplest problem in the Wiener theory which was discussed first by Wiener himself [1, pp. 91–92].

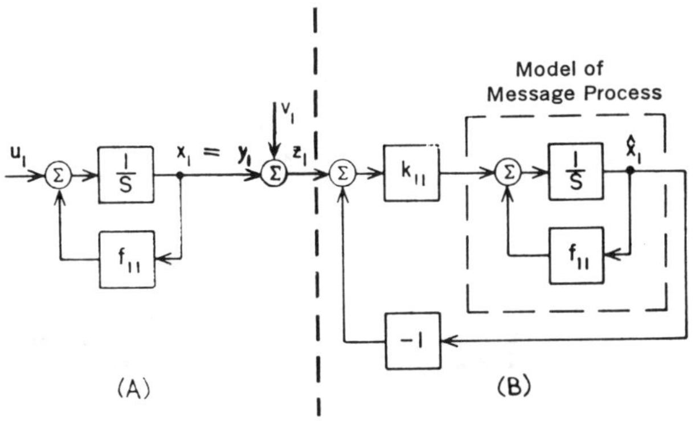

Fig. 1 Example 1: Block diagram of message process and optimal filter

The model of the message process is shown in Fig. 1(*a*). The various matrices involved are all defined by 1 × 1 and are

$$\mathbf{F}(t) = [f_{11}], \quad \mathbf{G}(t) = [1], \quad \mathbf{H}(t) = [1],$$
$$\mathbf{Q}(t) = [q_{11}], \quad \mathbf{R}(t) = [r_{11}].$$

The model is identical with its dual. Then the dual problem concerns the plant

$$dx^*_1/dt^* = f_{11}x^*_1 + u^*_1(t^*), \quad y^*_1(t) = x^*_1(t)$$

and the performance index is

$$\int_{t^*}^{t_0^*} \{q_{11}[x^*_1(\tau^*)]^2 + r_{11}[u^*_1(\tau^*)]^2\} d\tau^* \tag{21}$$

The discrete-time version of the estimation problem was treated in [11, Example 1]. The dual problem was treated by Rozonoër [19].

Example 2. The message is generated as in Example 1, but now it is assumed that two separate signals (mixed with different noise) can be observed. Hence $\mathbf{R}$ is now a 2 × 2 matrix and we assume that

$$\mathbf{H} = \begin{bmatrix} 1 \\ 1 \end{bmatrix}$$

The block diagram of the model is shown in Fig. 2(*a*).

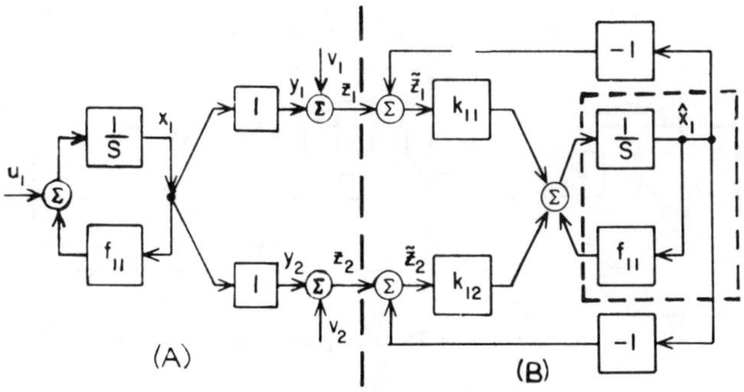

Fig. 2 Example 2: Block diagram of message process and optimal filter

Example 3. The message is generated by putting white noise through the transfer function $1/s(s+1)$. The block diagram of the model is shown in Fig. 3(*a*). The system matrices are:

$$\mathbf{F} = \begin{bmatrix} 0 & 0 \\ 1 & -1 \end{bmatrix} \quad \mathbf{G} = \begin{bmatrix} 1 \\ 0 \end{bmatrix} \quad \mathbf{H} = [0 \quad 1]$$

In the dual model, the order of the blocks $1/s$ and $1/(s+1)$ is interchanged. See Fig. 4. The performance index remains the same as (21). The dual problem was investigated by Kipiniak [24].

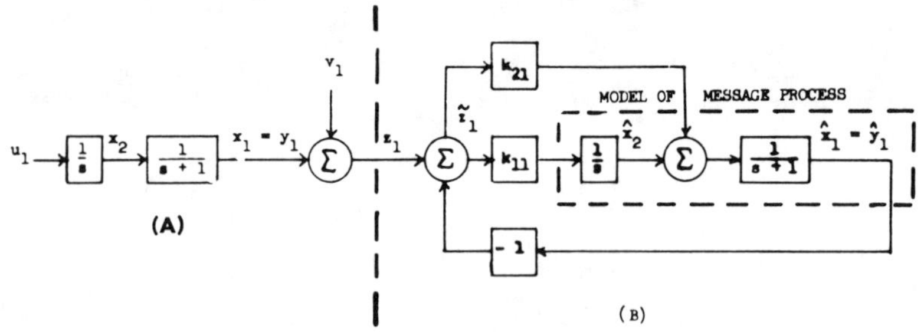

Fig. 3 Example 3: Block diagram of message process and optimal filter (x_1 and $\hat{x}_1$ should be interchanged with x_2 and $\hat{x}_2$.)

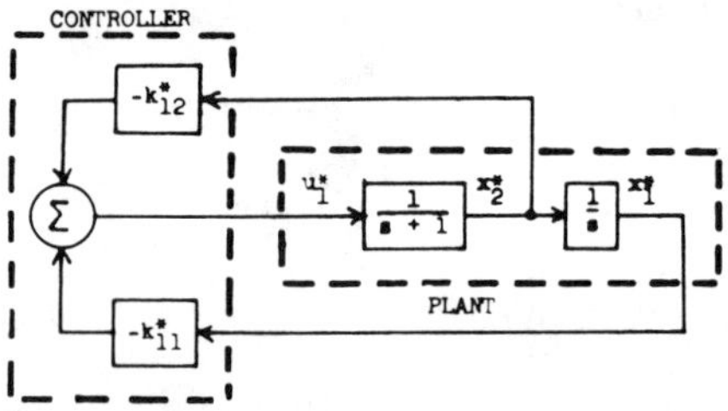

Fig. 4 Example 3: Block diagram of dual problem

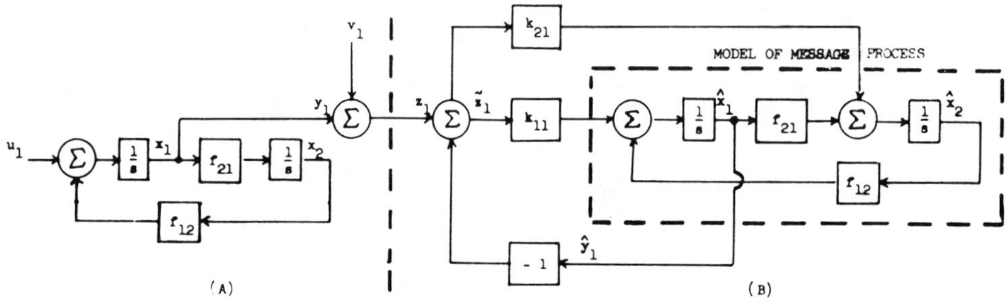

Fig. 5 Example 4: Block diagram of message process and optimal filter

Example 4. The message is generated by putting white noise through the transfer function $s/(s^2 - f_{12}f_{21})$. The block diagram of the model is shown in Fig. 5(a). The system matrices are:

$$\mathbf{F} = \begin{bmatrix} 0 & f_{12} \\ f_{21} & 0 \end{bmatrix} \quad \mathbf{G} = \begin{bmatrix} 1 \\ 0 \end{bmatrix} \quad \mathbf{H} = [1 \quad 0]$$

The transfer function of the dual model is also $s/(s^2 - f_{12}f_{21})$. However, in drawing the block diagram, the locations of the first and second state variables are interchanged, see Fig. 6. Evidently $f^*_{12} = f_{21}$ and $f^*_{21} = f_{12}$. The performance index is again given by (21).

The message model for the next two examples is the same and is defined by:

$$\mathbf{F} = \begin{bmatrix} 0 & 1 \\ 0 & 0 \end{bmatrix}$$

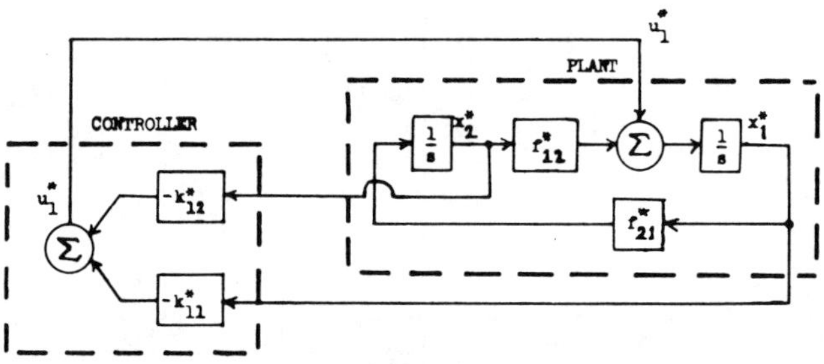

Fig. 6 Example 4: Block diagram of dual problem

The differences between the two examples lie in the nature of the "starting" assumptions and in the observed signals.

Example 5. Following Shinbrot [3], we consider the following situation. A particle leaves the origin at time $t_0 = 0$ with a fixed but unknown velocity of zero mean and known variance. The position of the particle is continually observed in the presence of additive white noise. We are to find the best estimator of position and velocity.

The verbal description of the problem implies that $p_{11}(0) = p_{12}(0) = 0$, $p_{22}(0) > 0$ and $q_{11} = 0$. Moreover, $\mathbf{G} = \mathbf{0}$, $\mathbf{H} = [1\,0]$. See Fig. 7(a).

The dual of this problem is somewhat unusual; it calls for minimizing the performance index

$$p_{22}(0)[\phi^*_2(0;\ \mathbf{x}^*,\ t^*;\ \mathbf{u}^*)]^2 + \int_{t^*}^{0} r_{11}[u^*_1(\tau^*)]^2 d\tau^* \quad (t^* < 0)$$

In words: We are given a transfer function $1/s^2$; the input u^*_1 over the time-interval $[t^*, 0]$ should be selected in such a way as to minimize the sum of (i) the square of the velocity and (ii) the control energy. In the discrete-time case, this problem was treated in [11, Example 2].

Example 6. We assume here that the transfer function $1/s^2$ is excited by white noise and that both the position x_1 and velocity x_2 can be observed in the presence of noise. Therefore (see Fig. 8a)

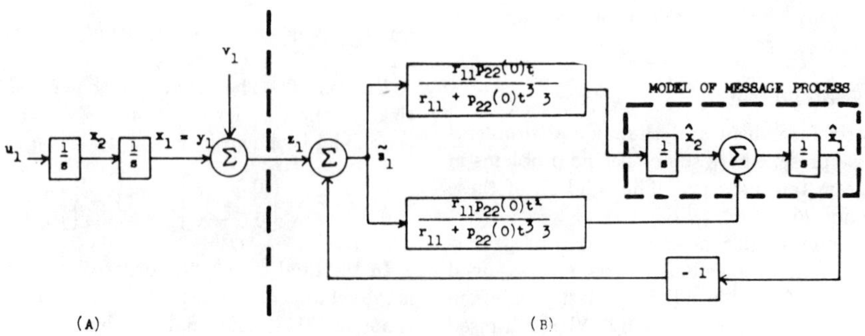

Fig. 7 Example 5: Block diagram of message process and optimal filter

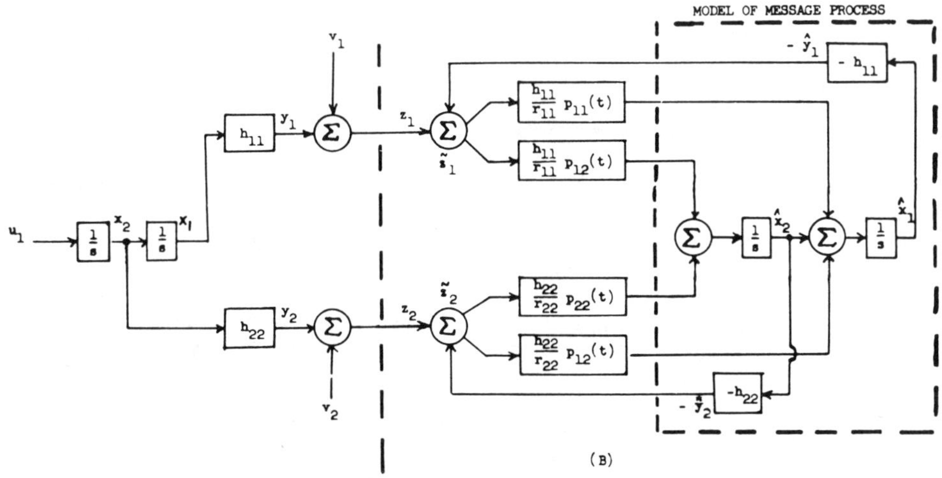

Fig. 8 Example 6: Block diagram of message process and optimal filter

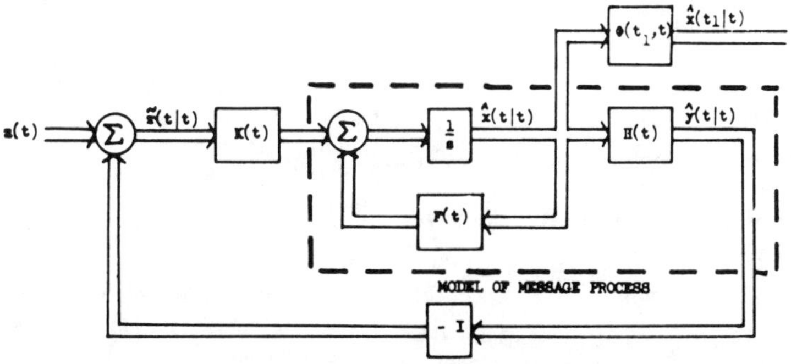

Fig. 9 General block diagram of optimal filter

$$\mathbf{G} = \begin{bmatrix} 0 \\ 1 \end{bmatrix} \qquad \mathbf{H} = \begin{bmatrix} h_{11} & 0 \\ 0 & h_{22} \end{bmatrix}$$

This problem was studied by Hanson [9] and Bucy [25, 26]. The dual problem is very similar to Examples 3 and 4.

7 Summary of Results: Mathematics

Here we present the main results of the paper in precise mathematical terms. At the present stage of our understanding of the problem, the rigorous proof of these facts is quite complicated, requiring advanced and unconventional methods; they are to be found in Sections 8–10. After reading this section, one may pass without loss of continuity to Section 11 which contains the solutions of the examples.

(1) *Canonical form of the optimal filter.* The optimal estimate $\hat{\mathbf{x}}(t|t)$ is generated by a linear dynamical system of the form

$$\begin{aligned} d\hat{\mathbf{x}}(t|t)/dt &= \mathbf{F}(t)\hat{\mathbf{x}}(t|t) + \mathbf{K}(t)\tilde{\mathbf{z}}(t|t) \\ \tilde{\mathbf{z}}(t|t) &= \mathbf{z}(t) - \mathbf{H}(t)\hat{\mathbf{x}}(t|t) \end{aligned} \qquad \text{(I)}$$

The initial state $\hat{\mathbf{x}}(t_0|t_0)$ of (I) is zero.

For optimal extrapolation, we add the relation

$$\hat{\mathbf{x}}(t_1|t) = \mathbf{\Phi}(t_1, t)\hat{\mathbf{x}}(t|t) \qquad (t_1 \geqq t) \qquad \text{(V)}$$

No similarly simple formula is known at present for interpolation ($t_1 < t$).

The block diagram of (I) and (V) is shown in Fig. 9. The variables appearing in this diagram are vectors and the "boxes" represent matrices operating on vectors. Otherwise (except for the noncommutativity of matrix multiplication) such generalized block diagrams are subject to the same rules as ordinary block diagrams. The fat lines indicating direction of signal flow serve as a reminder that we are dealing with multiple rather than single signals.

The optimal filter (I) is a feedback system. It is obtained by taking a copy of the model of the message process (omitting the constraint at the input), forming the error signal $\tilde{\mathbf{z}}(t|t)$ and feeding the error forward with a gain $\mathbf{K}(t)$. Thus the specification of the optimal filter is equivalent to the computation of the optimal time-varying gains $\mathbf{K}(t)$. This result is general and does not depend on constancy of the model.

(2) *Canonical form for the dynamical system governing the optimal error.* Let

$$\tilde{\mathbf{x}}(t|t) = \mathbf{x}(t) - \hat{\mathbf{x}}(t|t) \qquad (22)$$

Except for the way in which the excitations enter the optimal error, $\tilde{\mathbf{x}}(t|t)$ is governed by the same dynamical system as $\hat{\mathbf{x}}(t|t)$:

$$d\tilde{\mathbf{x}}(t|t)/dt = \mathbf{F}(t)\tilde{\mathbf{x}}(t|t) + \mathbf{G}(t)\mathbf{u}(t) - \mathbf{K}(t)[\mathbf{v}(t) + \mathbf{H}(t)\tilde{\mathbf{x}}(t|t)] \qquad \text{(II)}$$

See Fig. 10.

(3) *Optimal gain.* Let us introduce the abbreviation:

$$\mathbf{P}(t) = \operatorname{cov}[\tilde{\mathbf{x}}(t|t), \tilde{\mathbf{x}}(t|t)] \qquad (23)$$

Then it can be shown that

$$\mathbf{K}(t) = \mathbf{P}(t)\mathbf{H}'(t)\mathbf{R}^{-1}(t) \qquad \text{(III)}$$

(4) *Variance equation.* The only remaining unknown is $\mathbf{P}(t)$. It can be shown that $\mathbf{P}(t)$ must be a solution of the matrix differential equation

$$d\mathbf{P}/dt = \mathbf{F}(t)\mathbf{P} + \mathbf{P}\mathbf{F}'(t) - \mathbf{P}\mathbf{H}'(t)\mathbf{R}^{-1}(t)\mathbf{H}(t)\mathbf{P} + \mathbf{G}(t)\mathbf{Q}(t)\mathbf{G}'(t) \qquad \text{(IV)}$$

This is the *variance equation;* it is a system of $n(n+1)/2$[4] nonlinear differential equations of the first order, and is of the *Riccati* type well known in the calculus of variations [17, 18].

(5) *Existence of solutions of the variance equation.* Given any fixed initial time t_0 and a nonnegative definite matrix $\mathbf{P}_0$, (IV) has a unique solution

$$\mathbf{P}(t) = \mathbf{\Pi}(t;\ \mathbf{P}_0, t_0) \tag{24}$$

defined for all $|t - t_0|$ *sufficiently small,* which takes on the value $\mathbf{P}(t_0) = \mathbf{P}_0$ at $t = t_0$. This follows at once from the fact that (IV) satisfies a Lipschitz condition [21].

Since (IV) is nonlinear, we cannot of course conclude without further investigation that a solution $\mathbf{P}(t)$ exists for *all t* [21]. By taking into account the problem from which (IV) was derived, however, it can be shown that $\mathbf{P}(t)$ in (24) is defined for all $t \geqq t_0$.

These results can be summarized by the following theorem, which is the analogue of Theorem 3 of [11] and is proved in Section 8:

THEOREM 1. *Under Assumptions* (A_1), (A_2), (A_3'), *the solution of the optimal estimation problem with* $t_0 > -\infty$ *is given by relations* (I–V). *The solution* $\mathbf{P}(t)$ *of* (IV) *is uniquely determined for all* $t \geqq t_0$ *by the specification of*

$$\mathbf{P}_0 = \text{cov}[\mathbf{x}(t_0), \mathbf{x}(t_0)];$$

knowledge of $\mathbf{P}(t)$ *in turn determines the optimal gain* $\mathbf{K}(t)$. *The initial state of the optimal filter is* $\mathbf{0}$.

(6) *Variance of the estimate of a costate.* From (23) we have immediately the following formula for (15):

$$\mathcal{E}[\mathbf{x}^*, \tilde{\mathbf{x}}(t|t)]^2 = \|\mathbf{x}^*\|^2_{\mathbf{P}(t)} \tag{25}$$

(7) *Analytic solution of the variance equation.* Because of the close relationship between the Riccati equation and the calculus of variations, a closed-form solution of sorts is available for (IV). The easiest way of obtaining it is as follows [17]:

Introduce the quadratic *Hamiltonian* function

$$\mathcal{H}(\mathbf{x}, \mathbf{w}, t) = -(1/2)\|\mathbf{G}'(t)\mathbf{x}\|^2_{\mathbf{Q}(t)} - \mathbf{w}'\mathbf{F}'(t)\mathbf{x} + (1/2)\|\mathbf{H}(t)\mathbf{w}\|^2_{\mathbf{R}^{-1}(t)} \tag{26}$$

and consider the associated *canonical* differential equations

$$\left.\begin{aligned} d\mathbf{x}/dt &= \partial\mathcal{H}/\partial\mathbf{w}^5 = -\mathbf{F}'(t)\mathbf{x} + \mathbf{H}'(t)\mathbf{R}^{-1}(t)\mathbf{H}(t)\mathbf{w} \\ d\mathbf{w}/dt &= -\partial\mathcal{H}/\partial\mathbf{x} = \mathbf{G}(t)\mathbf{Q}(t)\mathbf{G}'(t)\mathbf{x} + \mathbf{F}(t)\mathbf{w} \end{aligned}\right\} \tag{27}$$

We denote the transition matrix of (27) by

$$\mathbf{\Theta}(t, t_0) = \begin{bmatrix} \mathbf{\Theta}_{11}(t, t_0) & \mathbf{\Theta}_{12}(t, t_0) \\ \mathbf{\Theta}_{21}(t, t_0) & \mathbf{\Theta}_{22}(t, t_0) \end{bmatrix} \tag{28}$$

[4] This is the number of distinct elements of the *symmetric* matrix $\mathbf{P}(t)$.

[5] The notation $\partial\mathcal{H}/\partial\mathbf{w}$ means the gradient of the scalar $\mathcal{H}$ with respect to the vector $\mathbf{w}$.

In Section 10 we shall prove

THEOREM 2. *The solution of* (IV) *for arbitrary nonnegative definite, symmetric* $\mathbf{P}_0$ *and all* $t \geqq t_0$ *can be represented by the formula*

$$\mathbf{\Pi}(t;\ \mathbf{P}_0, t_0) = [\mathbf{\Theta}_{21}(t, t_0) + \mathbf{\Theta}_{22}(t, t_0)\mathbf{P}_0]\cdot[\mathbf{\Theta}_{11}(t, t_0) + \mathbf{\Theta}_{12}(t, t_0)\mathbf{P}_0]^{-1} \tag{29}$$

Unless all matrices occurring in (27) are constant, this result simply replaces one difficult problem by another of similar difficulty, since only in the rarest cases can $\mathbf{\Theta}(t, t_0)$ be expressed in analytic form. Something has been accomplished, however, since we have shown that *the solution of nonconstant estimation problems involves precisely the same analytic difficulties as the solution of linear differential equations with variable coefficients.*

(8) *Existence of steady-state solution.* If the time-interval over which data are available is infinite, in other words, if $t_0 = -\infty$, Theorem 1 is not applicable without some further restriction.

For instance, if $\mathbf{H}(t) \equiv \mathbf{0}$, the variance of $\tilde{\mathbf{x}}$ is the same as the variance of $\mathbf{x}$; if the model (10–11) is unstable, then $\mathbf{x}(t)$ defined by (13) does not exist and the estimation problem is meaningless.

The following theorem, proved in Section 9, gives two sufficient conditions for the steady-state estimation problem to be meaningful. The first is the one assumed at the very beginning in the conventional Wiener theory. The second condition, which we introduce here for the first time, is much weaker and more "natural" than the first; moreover, it is almost a necessary condition as well.

THEOREM 3. *Denote the solutions of* (IV) *as in* (*24*). *Then the limit*

$$\lim_{t_0 \to -\infty} \mathbf{\Pi}(t;\ \mathbf{0}, t_0) = \bar{\mathbf{P}}(t) \tag{30}$$

exists for all t and is a solution of (IV) *if either*

(A_4) *the model* (10–11) *is uniformly asymptotically stable; or*

(A_4') *the model* (10–11) *is "completely observable"* [17], *that is, for all t there is some* $t_0(t) < t$ *such that the matrix*

$$\mathbf{M}(t_0, t) = \int_{t_0}^{t} \mathbf{\Phi}'(\tau, t)\mathbf{H}'(\tau)\mathbf{H}(\tau)\mathbf{\Phi}(\tau, t)d\tau \tag{31}$$

is positive definite. (*See* [21] *for the definition of uniform asymptotic stability.*)

Remarks. (*g*) $\bar{\mathbf{P}}(t)$ is the covariance matrix of the optimal error corresponding to the very special situation in which (i) an arbitrarily long record of past measurements is available, and (ii) the initial state $\mathbf{x}(t_0)$ was known exactly. When all matrices in (10–12) are constant, then so is also $\bar{\mathbf{P}}$—this is just the classical Wiener problem. In the constant case, $\bar{\mathbf{P}}$ is an equilibrium state of (IV) (i.e., for this choice of $\mathbf{P}$, the right-hand side of (IV) is zero). In general, $\bar{\mathbf{P}}(t)$ should be regarded as a moving equilibrium point of (IV), see Theorem 4 below.

(*h*) The matrix $\mathbf{M}(t_0, t)$ is well known in mathematical statistics. It is the *information matrix* in the sense of R. A. Fisher [20] corresponding to the special estimation problem when (i) $\mathbf{u}(t) \equiv \mathbf{0}$ and (ii) $\mathbf{v}(t)$ = gaussian with unit covariance matrix. In this case, the variance of any unbiased estimator $\mu(t)$ of $[\mathbf{x}^*, \mathbf{x}(t)]$ satisfies the well-known Cramér-Rao inequality [20]

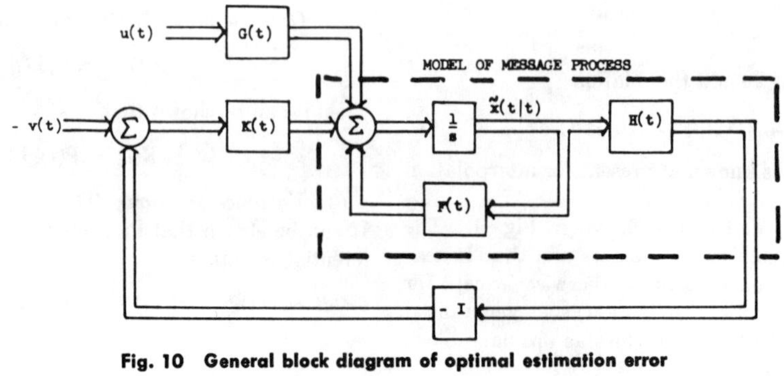

Fig. 10 General block diagram of optimal estimation error

$$\mathcal{E}[\mu(t) - \mathcal{E}\mu(t)]^2 \geqq \|\mathbf{x}^*\|^2_{\mathbf{M}^{-1}(t_0, t)} \tag{32}$$

Every costate $\mathbf{x}^*$ *has a minimum-variance unbiased estimator for which the equality sign holds in (32) if and only if* $\mathbf{M}$ *is positive definite.* This motivates the use of condition (A_4') in Theorem 3 and the term "completely observable."

(i) It can be shown [17] that in the constant case complete observability is equivalent to the easily verified condition:

$$\operatorname{rank}[\mathbf{H}', \mathbf{F}'\mathbf{H}', \ldots, (\mathbf{F}')^{n-1}\mathbf{H}'] = n \tag{33}$$

where the square brackets denote a matrix with n rows and np columns.

(9) *Stability of the optimal filter.* It should be realized now that the *optimality* of the filter (I) does not at the same time guarantee its *stability*. The reader can easily check this by constructing an example (for instance, one in which (10–11) consists of two noninteracting systems). To establish weak sufficient conditions for stability entails some rather delicate mathematical technicalities which we shall bypass and state only the best final result currently available.

First, some additional definitions.

We say that the model (10–11) is *uniformly completely observable* if there exist fixed constants, α_1, α_2, and σ such that

$$\alpha_1\|\mathbf{x}^*\|^2 \leqq \|\mathbf{x}^*\|^2_{\mathbf{M}(t-\sigma, t)} \leqq \alpha_2\|\mathbf{x}^*\|^2 \quad \text{for all} \quad \mathbf{x}^* \text{ and } t.$$

Similarly, we say that a model is *completely controllable* [*uniformly completely controllable*] if the dual model is completely observable [uniformly completely observable]. For a discussion of these motions, the reader may refer to [17]. It should be noted that the property of "uniformity" is always true for constant systems.

We can now state the central theorem of the paper:

THEOREM 4. *Assume that the model of the message process is*

(A_4'') *uniformly completely observable;*
(A_5) *uniformly completely controllable;*
(A_6) $\alpha_3 \leqq \|\mathbf{Q}(t)\| \leqq \alpha_4, \quad \alpha_5 \leqq \|\mathbf{R}(t)\| \leqq \alpha_6$ for all t;
(A_7) $\|\mathbf{F}(t)\| \leqq \alpha_7$.

Then the following is true:

(i) *The optimal filter is uniformly asymptotically stable;*

(ii) *Every solution* $\mathbf{\Pi}(t;\ \mathbf{P}_0, t_0)$ *of the variance equation* (IV) *starting at a symmetric nonnegative matrix* $\mathbf{P}_0$ *converges to* $\bar{\mathbf{P}}(t)$ *(defined in Theorem 3) as* $t \to \infty$.

Remarks. (j) A filter which is not *uniformly* asymptotically stable may have an unbounded response to a bounded input [21]; the practical usefulness of such a filter is rather limited.

(k) Property (ii) in Theorem 4 is of central importance since it shows that the variance equation is a "stable" computational method that may be expected to be rather insensitive to roundoff errors.

(l) The speed of convergence of $\mathbf{P}_0(t)$ to $\bar{\mathbf{P}}(t)$ can be estimated quite effectively using the second method of Lyapunov; see [17].

(10) *Solution of the classical Wiener problem.* Theorems 3 and 4 have the following immediate corollary:

THEOREM 5. *Assume the hypotheses of Theorems 3 and 4 are satisfied and that* $\mathbf{F}$, $\mathbf{G}$, $\mathbf{H}$, $\mathbf{Q}$, $\mathbf{R}$, *are constants.*

Then, if $t_0 = -\infty$, *the solution of the estimation problem is obtained by setting the right-hand side of* (IV) *equal to zero and solving the resulting set of quadratic algebraic equations. That solution which is nonnegative definite is equal to* $\bar{\mathbf{P}}$.

To prove this, we observe that, by the assumption of constancy, $\bar{\mathbf{P}}(t)$ is a constant. By Theorem 4, all solutions of (IV) starting at nonnegative matrices converge to $\bar{\mathbf{P}}$. Hence, if a matrix $\mathbf{P}$ is found for which the right-hand side of (IV) vanishes and if this matrix is nonnegative definite, it must be identical with $\bar{\mathbf{P}}$. Note, however, that the procedure may fail if the conditions of Theorems 3 and 4 are not satisfied. See Example 4.

(11) *Solution of the Dual Problem.* For details, consult [17]. The only facts needed here are the following: The optimal control law is given by

$$\mathbf{u}^*(t^*) = -\mathbf{K}^*(t^*)\mathbf{x}(t^*) \tag{34}$$

where $\mathbf{K}^*(t^*)$ satisfies the duality relation

$$\mathbf{K}^*(t^*) = \mathbf{K}'(t) \tag{35}$$

and is to be determined by duality from formula (III). The value of the performance index (20) may be written in the form

$$\min_{\mathbf{u}^*} V(\mathbf{x}^*;\ t^*, t_0^*, \mathbf{u}^*) = \|\mathbf{x}^*\|^2_{\mathbf{\Pi}^*(t^*;\ \mathbf{x}^*, t_0^*)}$$

where $\mathbf{\Pi}^*(t^*;\ \mathbf{x}^*, t_0^*)$ is the solution of the dual of the variance equation (IV).

It should be carefully noted that the hypotheses of Theorem 4 are invariant under duality. Hence essentially the same theory covers both the estimation and the regular problem, as stated in Section 5.

The vector-matrix block diagram for the optimal regulator is shown in Fig. 11.

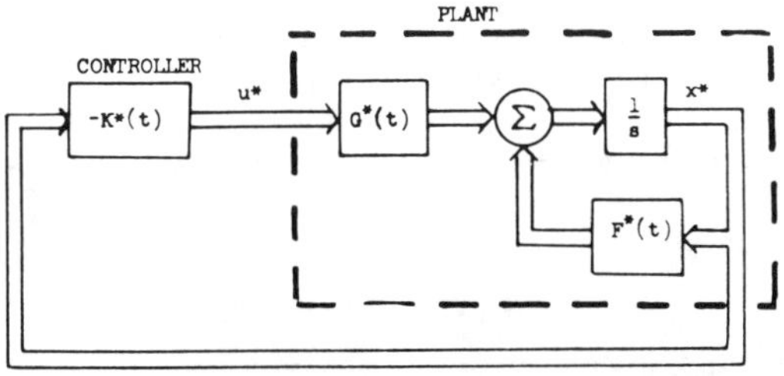

Fig. 11 General block diagram of optimal regulator

(12) *Computation of the covariance matrix for the message process.* To apply Theorem 1, it is necessary to determine cov $[\mathbf{x}(t_0), \mathbf{x}(t_0)]$. This may be specified as part of the problem statement as in Example 5. On the other hand, one might assume that the message model has reached steady state (see (A_3)), in which case from (13) and (12) we have that

$$\mathbf{S}(t) = \operatorname{cov}[\mathbf{x}(t), \mathbf{x}(t)] = \int_{-\infty}^{t} \mathbf{\Phi}(t, \tau)\mathbf{G}(\tau)\mathbf{Q}(\tau)\mathbf{G}'(\tau)\mathbf{\Phi}'(t, \tau)d\tau$$

provided the model (10) is asymptotically stable. Differentiating this expression with respect to t we obtain the following differential equation for $\mathbf{S}(t)$

$$d\mathbf{S}/dt = \mathbf{F}(t)\mathbf{S} + \mathbf{S}\mathbf{F}'(t) + \mathbf{G}(t)\mathbf{Q}(t)\mathbf{G}'(t) \tag{36}$$

This formula is analogous to the well-known lemma of Lyapunov [21] in evaluating the integrated square of a solution of a linear differential equation. In case of a constant system, (36) reduces to a system of linear algebraic equations.

8 Derivation of the Fundamental Equations

We first deduce the matrix form of the familiar Wiener-Hopf integral equation. Differentiating it with respect to time and then using (10–11), we obtain in a very simple way the fundamental equations of our theory.

Much cumbersome manipulation of integrals can be avoided by recognizing, as has been pointed out by Pugachev [27], that the Wiener-Hopf equation is a special case of a simple geometric principle: *orthogonal projection.*

Consider an abstract space $\mathfrak{X}$ such that an inner product (X, Y) is defined between any two elements X, Y of $\mathfrak{X}$. The norm is defined by $\|X\| = (X, X)^{1/2}$. Let $\mathfrak{U}$ be a subspace of $\mathfrak{X}$. We seek a vector U_0 in $\mathcal{U}$ which minimizes $\|X - U\|$ with respect to any U in $\mathcal{U}$. If such a minimizing vector exists, it may be characterized in the following way:

ORTHOGONAL PROJECTION LEMMA. $\|X - U\| \geqq \|X - U_0\|$ *for all U in* $\mathcal{U}$ (i) *if and* (ii) *only if*

$$(X - U_0, U) = 0 \text{ for all } U \text{ in } \mathcal{U} \tag{37}$$

(iii) *Moreover, if there is another vector* U_0' *satisfying* (37), *then* $\|U_0 - U_0'\| = 0$.

Proof. (i), (iii) Consider the identity

$$\|X - U\|^2 = \|X - U_0\|^2 + 2(X - U_0, U_0 - U) + \|U - U_0\|^2$$

Since $\mathcal{U}$ is a linear space, it contains $U - U_0$; hence if Condition (37) holds, the middle term vanishes and therefore $\|X - U\| \geqq \|X - U_0\|$. Property (iii) is obvious.

(ii) Suppose there is a vector U_1 such that $(X - U_0, U_1) = \alpha \neq 0$. Then

$$\|X - U_0 - \beta U_1\|^2 = \|X - U_0\|^2 + 2\alpha\beta + \beta^2\|U_1\|^2$$

For a suitable choice of β, the sum of the last two terms will be negative, contradicting the optimality of U_0. Q.E.D.

Using this lemma, it is easy to show:

WIENER-HOPF EQUATION. *A necessary and sufficient condition for* $[\mathbf{x}^*, \hat{\mathbf{x}}(t_1|t)]$ (*where* $\hat{\mathbf{x}}(t_1|t)$ *is defined by* (14)) *to be a minimum variance estimator of* $[\mathbf{x}^*, \mathbf{x}(t_1)]$ *for all* $\mathbf{x}^*$, *is that the matrix function* $\mathbf{A}(t_1, \tau)$ *satisfy the relation*

$$\operatorname{cov}[\mathbf{x}(t_1), \mathbf{z}(\sigma)] - \int_{t_0}^{t} \mathbf{A}(t_1, \tau) \operatorname{cov}[\mathbf{z}(\tau), \mathbf{z}(\sigma)]d\tau = \mathbf{0} \tag{38}$$

or equivalently,

$$\operatorname{cov}[\tilde{\mathbf{x}}(t_1|t), \mathbf{z}(\sigma)] = \mathbf{0} \tag{39}$$

for all $t_0 \leqq \sigma < t$.

COROLLARY. $\operatorname{cov}[\tilde{\mathbf{x}}(t_1|t), \hat{\mathbf{x}}(t_1|t)] = \mathbf{0}$ (40)

Proof. Let $\mathbf{x}^*$ be a fixed costate and denote by $\mathcal{X}$ the space of all scalar random variables $[\mathbf{x}^*, \mathbf{x}(t_1)]$ of zero mean and finite variance. The inner product is defined as $(X, Y) = \mathcal{E}[\mathbf{x}^*, \mathbf{x}(t_1)]\cdot[\mathbf{x}^*, \mathbf{y}(t_1)]$. The subspace $\mathcal{U}$ is the set of all scalar random variables of the type

$$U = [\mathbf{x}^*, \mathbf{u}(t_1)] = \left[\mathbf{x}^*, \int_{t_0}^{t} \mathbf{B}(t_1, \tau)\mathbf{z}(\tau)d\tau\right]$$

(where $\mathbf{B}(t_1, \tau)$ is an $n \times p$ matrix continuously differentiable in both arguments). We write U_0 for the estimate $[\mathbf{x}^*, \hat{\mathbf{x}}(t_1|t)]$.

We now apply the orthogonal projection lemma and find that condition (37) takes the form

$$\begin{aligned}(X - U_0, U) &= \mathcal{E}\,[\mathbf{x}^*, \tilde{\mathbf{x}}(t_1|t)][\mathbf{x}^*, \mathbf{u}(t_1)] \\ &= \mathbf{x}^* \operatorname{cov}[\tilde{\mathbf{x}}(t_1|t), \mathbf{u}(t_1)]\mathbf{x}^{*\prime}\end{aligned}$$

Interchanging integration and the expected value operation (permissible in view of the continuity assumptions made under (A_1), see [28]), we get

$$(X - U_0, U) = \mathbf{x}^* \left\{\int_{t_0}^{t} \operatorname{cov}[\tilde{\mathbf{x}}(t_1|t), \mathbf{z}(\sigma)]\mathbf{B}'(t_1, \sigma)d\sigma\right\} \mathbf{x}^{*\prime}$$

This expression must vanish for all $\mathbf{x}^*$. Sufficiency of (39) is obvious. To prove the necessity, we take $\mathbf{B}(t_1, \sigma) = \operatorname{cov}[\tilde{\mathbf{x}}(t_1|t), \mathbf{z}(\sigma)]$. Then $\mathbf{BB}'$ is nonnegative definite. By continuity, the integral will be positive for some $\mathbf{x}^*$ unless $\mathbf{BB}'$ and therefore also $\mathbf{B}(t_1, \sigma)$ vanishes identically for all $t_0 \leqq \sigma < t$. The Corollary follows trivially by multiplying (39) on the right by $\mathbf{A}'(t_1, \sigma)$ and integrating with respect to σ. Q.E.D.

Remark. (*m*) Equation (39) does not hold when $\sigma = t$. In fact, $\operatorname{cov}[\tilde{\mathbf{x}}(t|t), \mathbf{z}(t)] = (1/2)\,\mathbf{K}(t)\,\mathbf{R}(t)$.

For the moment we assume for simplicity that $t_1 = t$. Differentiating (38) with respect to t, and interchanging $\partial/\partial t$ and $\mathcal{E}$, we get for all $t_0 \leqq \sigma < t$,

$$\frac{\partial}{\partial t}\operatorname{cov}[\mathbf{x}(t), \mathbf{z}(\sigma)] = \mathbf{F}(t)\operatorname{cov}[\mathbf{x}(t), \mathbf{z}(\sigma)] + \mathbf{G}(t)\operatorname{cov}[\mathbf{u}(t), \mathbf{z}(\sigma)] \tag{41}$$

and

$$\begin{aligned}\frac{\partial}{\partial t}\int_{t_0}^{t} \mathbf{A}(t, \tau)\operatorname{cov}[\mathbf{z}(\tau), \mathbf{z}(\sigma)]d\tau &= \frac{\partial}{\partial t}\int_{t_0}^{t} \mathbf{A}(t, \tau)\operatorname{cov}[\mathbf{y}(\tau), \mathbf{y}(\sigma)]d\tau + \frac{\partial}{\partial t}\mathbf{A}(t, \sigma)\mathbf{R}(\sigma) \\ &= \int_{t_0}^{t} \frac{\partial}{\partial t}\mathbf{A}(t, \tau)\operatorname{cov}[\mathbf{z}(\tau), \mathbf{z}(\sigma)]d\tau + \mathbf{A}(t, t)\operatorname{cov}[\mathbf{y}(t), \mathbf{y}(\sigma)]\end{aligned} \tag{42}$$

The last term in (41) vanishes because of the independence of $\mathbf{u}(t)$ of $\mathbf{v}(\sigma)$ and $\mathbf{x}(\sigma)$ when $\sigma < t$. Further,

$$\operatorname{cov}[\mathbf{y}(t), \mathbf{y}(\sigma)] = \mathbf{H}(t)\operatorname{cov}[\mathbf{x}(t), \mathbf{z}(\sigma)] - \operatorname{cov}[\mathbf{y}(t), \mathbf{v}(\sigma)] \tag{43}$$

As before, the last term again vanishes. Combining (41-43), we get, bearing in mind also (38),

$$\int_{t_0}^{t}\left[\mathbf{F}(t)\mathbf{A}(t, \tau) - \frac{\partial}{\partial t}\mathbf{A}(t, \tau) - \mathbf{A}(t, t)\mathbf{H}(t)\mathbf{A}(t, \tau)\right]\operatorname{cov}[\mathbf{z}(\tau), \mathbf{z}(\sigma)]d\tau = \mathbf{0} \tag{44}$$

for all $t_0 \leqq \sigma < t$. This condition is certainly satisfied if the optimal operator $\mathbf{A}(t, \tau)$ is a solution of the differential equation

$$\mathbf{F}(t)\mathbf{A}(t, \tau) - \frac{\partial}{\partial t}\mathbf{A}(t, \tau) - \mathbf{A}(t, t)\mathbf{H}(t)\mathbf{A}(t, \tau) = \mathbf{0} \tag{45}$$

for all values of the parameter τ lying in the interval $t_0 \leqq \tau \leqq t$.

If $\mathbf{R}(\tau)$ *is positive definite in this interval, then condition* (45) *is necessary.* In fact, let $\mathbf{B}(t, \tau)$ denote the bracketed term in (44). If $\mathbf{A}(t, \tau)$ satisfies the Wiener-Hopf equation (38), then $\hat{\mathbf{x}}(t|t)$ given by (14) is an optimal estimate; and the same holds also for

$$\hat{\mathbf{x}}(t|t) + \int_{t_0}^{t} \mathbf{B}(t, \tau)\mathbf{z}(\tau)d\tau$$

since by (45) $\mathbf{A}(t, \tau) + \mathbf{B}(t, \tau)$ also satisfies the Wiener-Hopf equation. But by the lemma, the norm of the difference of two optimal estimates is zero. Hence

$$\mathbf{x}^*\left\{\int_{t_0}^{t}\int_{t_0}^{t} \mathbf{B}(t, \tau)\operatorname{cov}[\mathbf{z}(\tau), \mathbf{z}(\tau')]\mathbf{B}'(t, \tau')d\tau d\tau'\right\}\mathbf{x}^{*\prime} = 0 \tag{46}$$

for all $\mathbf{x}^*$. By the assumptions of Section 4, $\mathbf{y}(\tau)$ and $\mathbf{v}(\tau)$ are uncorrelated and therefore

$$\operatorname{cov}[\mathbf{z}(\tau), \mathbf{z}(\tau')] = \mathbf{R}(\tau)\delta(\tau - \tau') + \operatorname{cov}[\mathbf{y}(\tau), \mathbf{y}(\tau')]$$

Substituting this into the integral (46), the contribution of the second term on the right is nonnegative while the contribution of the first term is positive unless (45) holds (because of the positive definiteness of $\mathbf{R}(\tau)$), which concludes the proof.

Differentiating (14), with respect to t we find

$$d\hat{\mathbf{x}}(t|t)/dt = \int_{t_0}^{t} \frac{\partial}{\partial t}\mathbf{A}(t, \tau)\mathbf{z}(\tau)d\tau + \mathbf{A}(t, t)\mathbf{z}(t)$$

Using the abbreviation $\mathbf{A}(t, t) = \mathbf{K}(t)$ as well as (45) and (14), we obtain at once the differential equation of the optimal filter:

$$d\hat{\mathbf{x}}(t|t)/dt = \mathbf{F}(t)\hat{\mathbf{x}}(t|t) + \mathbf{K}(t)[\mathbf{z}(t) - \mathbf{H}(t)\hat{\mathbf{x}}(t|t)] \tag{I}$$

Combining (10) and (I), we obtain the differential equation for the error of the optimal estimate:

$$d\tilde{\mathbf{x}}(t|t)/dt = [\mathbf{F}(t) - \mathbf{K}(t)\mathbf{H}(t)]\,\tilde{\mathbf{x}}(t|t) + \mathbf{G}(t)\mathbf{u}(t) - \mathbf{K}(t)\mathbf{v}(t) \quad \text{(II)}$$

To obtain an explicit expression for $\mathbf{K}(t)$, we observe first that (39) implies that following identity in the interval $t_0 \leqq \sigma < t$:

$$\text{cov}\,[\mathbf{x}(t), \mathbf{y}(\sigma)] - \int_{t_0}^{t} \mathbf{A}(t, \tau)\,\text{cov}\,[\mathbf{y}(\tau), \mathbf{y}(\sigma)]d\tau = \mathbf{A}(t, \sigma)\mathbf{R}(\sigma) \quad (39')$$

Since both sides of (39′) are continuous functions of σ, it is clear that equality holds also for $\sigma = t$. Therefore

$$\mathbf{K}(t)\mathbf{R}(t) = \mathbf{A}(t, t)\mathbf{R}(t) = \text{cov}[\tilde{\mathbf{x}}(t|t), \mathbf{y}(t)] = \text{cov}\,[\tilde{\mathbf{x}}(t|t), \mathbf{x}(t)]\mathbf{H}'(t)$$

By (40), we have then

$$= \text{cov}\,[\tilde{\mathbf{x}}(t|t), \tilde{\mathbf{x}}(t|t)]\mathbf{H}'(t) = \mathbf{P}(t)\mathbf{H}'(t)$$

Since $\mathbf{R}(t)$ is assumed to be positive definite, it is invertible and therefore

$$\mathbf{K}(t) = \mathbf{P}(t)\mathbf{H}'(t)\mathbf{R}^{-1}(t) \quad \text{(III)}$$

We can now derive the variance equation. Let $\mathbf{\Psi}(t, \tau)$ be the common transition matrix of (I) and (II). Then

$$\mathbf{P}(t) - \mathbf{\Psi}(t, t_0)\mathbf{P}(t_0)\mathbf{\Psi}'(t, t_0) = \mathcal{E}\int_{t_0}^{t} \mathbf{\Psi}(t, \tau)[\mathbf{G}(\tau)\mathbf{u}(\tau) - \mathbf{K}(\tau)\mathbf{v}(\tau)]d\tau \times \int_{t_0}^{t} [\mathbf{u}'(\sigma)\mathbf{G}'(\sigma) - \mathbf{v}'(\sigma)\mathbf{K}'(\sigma)]\,\mathbf{\Psi}'(t, \sigma)d\sigma$$

Using the fact that $\mathbf{u}(t)$ and $\mathbf{v}(t)$ are uncorrelated white noise, the integral simplifies to

$$= \int_{t_0}^{t} \mathbf{\Psi}(t, \tau)[\mathbf{G}(\tau)\mathbf{Q}(\tau)\mathbf{G}'(\tau) + \mathbf{K}(\tau)\mathbf{R}(\tau)\mathbf{K}'(\tau)]\,\mathbf{\Psi}'(t, \tau)d\tau$$

Differentiating with respect to t and using (III), we obtain after easy calculations the variance equation

$$d\mathbf{P}/dt = \mathbf{F}(t)\mathbf{P} + \mathbf{P}\mathbf{F}'(t) - \mathbf{P}\mathbf{H}'(t)\mathbf{R}^{-1}(t)\mathbf{H}(t)\mathbf{P} + \mathbf{G}(t)\mathbf{Q}(t)\mathbf{G}'(t) \quad \text{(IV)}$$

Alternately, we could write

$$d\mathbf{P}/dt = d\,\text{cov}\,[\tilde{\mathbf{x}}, \tilde{\mathbf{x}}]/dt = \text{cov}\,[d\tilde{\mathbf{x}}/dt, \tilde{\mathbf{x}}] + \text{cov}\,[\tilde{\mathbf{x}}, d\tilde{\mathbf{x}}/dt]$$

and evaluate the right-hand side by means of (II). A typical covariance matrix to be computed is

$$\text{cov}\,[\tilde{\mathbf{x}}(t|t), \mathbf{u}(t)] = \text{cov}\left[\int_{t_0}^{t} \mathbf{\Psi}(t, \tau)[\mathbf{G}(\tau)\mathbf{u}(\tau) - \mathbf{K}(\tau)\mathbf{v}(\tau)]d\tau, \mathbf{u}(t)\right] = (1/2)\mathbf{G}(t)\mathbf{Q}(t)$$

the factor $1/2$ following from properties of the δ-function.

To complete the derivations, we note that, if $t_1 > t$, then by (3)

$$\mathbf{x}(t_1) = \mathbf{\Phi}(t_1, t)\mathbf{x}(t) + \int_{t_0}^{t} \mathbf{\Phi}(t_1, \tau)\mathbf{u}(\tau)d\tau$$

Since $\mathbf{u}(\tau)$ for $t < \tau \leqq t_1$ is independent of $\mathbf{x}(\tau)$ in the interval $t_0 \leqq \tau \leqq t$, it follows by (38) that the optimal estimator for the right-hand side above is $\mathbf{0}$. Hence

$$\hat{\mathbf{x}}(t_1|t) = \mathbf{\Phi}(t_1, t)\hat{\mathbf{x}}(t|t) \quad (t_1 \geqq t) \quad \text{(V)}$$

The same conclusion does *not* follow if $t_1 < t$ because of lack of independence between $\mathbf{x}(\tau)$ and $\mathbf{u}(\tau)$.

The only point remaining in the proof of Theorem 1 is to determine the initial conditions for (IV). From (38) it is clear that

$$\hat{\mathbf{x}}(t_0|t_0) = \mathbf{0}$$

Hence

$$\mathbf{P}_0 = \mathbf{P}(t_0) = \text{cov}[\tilde{\mathbf{x}}(t_0|t_0), \tilde{\mathbf{x}}(t_0|t_0)] = \text{cov}[\mathbf{x}(t_0), \mathbf{x}(t_0)]$$

In case of the conventional Wiener theory (see (A_3)), the last term is evaluated by means of (36).

This completes the proof of Theorem 1.

9 Outline of Proofs

Using the duality relations (16), all proofs can be reduced to those given for the regulator problem in [17].

(1) The fact that solutions of the variance equation exist for all $t \geqq t_0$ is proved in [17, Theorem (6.4)], using the fact that the variance of $\mathbf{x}(t)$ must be finite in any finite interval $[t_0, t]$.

(2) Theorem 3 is proved by showing that there exists a particular estimate of finite but not necessarily minimum variance. Under (A_4'), this is proved in [17; Theorem (6.6)]. A trivial modification of this proof goes through also with assumption (A_4).

(3) Theorem 4 is proved in [17; Theorems (6.8), (6.10), (7.2)]. The stability of the optimal filter is proved by noting that the estimation error plays the role of a Lyapunov function. The stability of the variance equation is proved by exhibiting a Lyapunov function for $\mathbf{P}$. This Lyapunov function in the simplest case is discussed briefly at the end of Example 1. While this theorem is true also in the nonconstant case, at present one must impose the somewhat restrictive conditions (A_6 – A_7).

10 Analytic Solution of the Variance Equation

Let $\mathbf{X}(t)$, $\mathbf{W}(t)$ be the (unique) matrix solution pair for (27) which satisfy the initial conditions

$$\mathbf{X}(t_0) = \mathbf{I}, \quad \mathbf{W}(t_0) = \mathbf{P}_0 \quad (47)$$

Then we have the following identity

$$\mathbf{W}(t) = \mathbf{P}(t)\mathbf{X}(t), \quad t \geqq t_0 \quad (48)$$

which is easily verified by substituting (48) with (IV) into (27). On the other hand, in view of (47–48), we see immediately from the first set of equations (27) that $\mathbf{X}(t)$ is the transition matrix of the differential equation

$$d\mathbf{x}/dt = -\mathbf{F}'(t)\mathbf{x} + \mathbf{H}'(t)\mathbf{R}^{-1}(t)\mathbf{H}(t)\mathbf{P}(t)\mathbf{x}$$

which is the adjoint of the differential equation (IV) of the optimal filter. Since the inverse of a transition matrix always exists, we can write

$$\mathbf{P}(t) = \mathbf{W}(t)\mathbf{X}^{-1}(t), \quad t \geqq t_0 \quad (49)$$

This formula may not be valid for $t < t_0$, for then $\mathbf{P}(t)$ may not exist!

Only trivial steps remain to complete the proof of Theorem 2.

11 Examples: Solution

Example 1. If $q_{11} > 0$ and $r_{11} > 0$, it is easily verified that the conditions of Theorems 3–4 are satisfied. After trivial substitutions in (III-IV) we obtain the expression for the optimal gain

$$k_{11}(t) = p_{11}(t)/r_{11} \quad (50)$$

and the variance equation

$$dp_{11}/dt = 2f_{11}p_{11} - p_{11}^2/r_{11} + q_{11} \quad (51)$$

By setting the right-hand side of (51) equal to zero, by virtue of the corollary of Theorem 4 we obtain the solution of the stationary problem (i.e., $t_0 = -\infty$, see (A_3)):

$$\bar{p}_{11} = \left[f_{11} + \sqrt{f_{11}^2 + q_{11}/r_{11}}\right] r_{11} \tag{52}$$

Since $\bar{p}_{11}$ and r_{11} are nonnegative, it is clear that only the positive sign is permissible in front of the square root.

Substituting into (50), we get the following expressions for the optimal gain

$$\bar{k}_{11} = f_{11} + \sqrt{f_{11}^2 + q_{11}/r_{11}} \tag{53}$$

and for the infinitesimal transition matrix (i.e., reciprocal time constant)

$$\bar{f}_{11} = f_{11} - \bar{k}_{11} = -\sqrt{f_{11}^2 + q_{11}/r_{11}} \tag{54}$$

of the optimal filter. We see, in accordance with Theorem 4, that the optimal filter is always stable, irrespective of the stability of the message model. Fig. 1(*b*) shows the configuration of the optimal filter.

It is easily checked that the formulas (52–54) agree with the results of the conventional Wiener theory [29].

Let us now compute the solution of the problem for a finite smoothing interval ($t_0 > -\infty$). The Hamiltonian equations (27) in this case are:

$$\left.\begin{aligned} dx_1/dt &= -f_{11}x_1 + (1/r_{11})w_1 \\ dw_1/dt &= q_{11}x_1 + f_{11}w_1 \end{aligned}\right\}$$

Let $\mathbf{T}$ be the matrix of coefficients of these equations.

To compute the transition matrix $\mathbf{\Theta}(t, t_0)$ corresponding to $\mathbf{T}$, we note first that the eigenvalues of $\mathbf{T}$ are $\pm\bar{f}_{11}$. Using this fact and constancy, it follows that

$$\mathbf{\Theta}(t, t_0) = \exp \mathbf{T}(t - t_0) = \mathbf{C}_1 \exp (t - t_0)\bar{f}_{11} + \mathbf{C}_2 \exp [-(t - t_0)\bar{f}_{11}]$$

where the constant matrices $\mathbf{C}_1$ and $\mathbf{C}_2$ are uniquely determined by the requirements

$$\mathbf{\Theta}(t_0, t_0) = \mathbf{C}_1 + \mathbf{C}_2 = \mathbf{I} = \text{unit matrix}$$

$$d\mathbf{\Theta}(t, t_0)/dt|_{t=t_0} = \mathbf{T\Theta}(t, t_0)|_{t=t_0} = \bar{f}_{11}\mathbf{C}_1 - \bar{f}_{11}\mathbf{C}_2$$

After a good deal of algebra, we obtain

$$\mathbf{\Theta}(t_0 + \tau, t_0) = \begin{bmatrix} \cosh \bar{f}_{11}\tau - \dfrac{f_{11}}{\bar{f}_{11}} \sinh \bar{f}_{11}\tau & \dfrac{1}{r_{11}\bar{f}_{11}} \sinh \bar{f}_{11}\tau \\ \dfrac{q_{11}}{\bar{f}_{11}} \sinh \bar{f}_{11}\tau & \cosh \bar{f}_{11}\tau + \dfrac{f_{11}}{\bar{f}_{11}} \sinh \bar{f}_{11}\tau \end{bmatrix} \tag{55}$$

Knowledge of $\mathbf{\Theta}(t, t_0)$ can be used to derive explicit solutions to a variety of nonstationary filtering problems.

We consider only one such problem, which was treated by Shinbrot [3, Example 2]. He assumes that $f_{11} < 0$ and that the message process has reached steady-state. From (36) we see that

$$\mathcal{E}x_1^2(t) = -q_{11}/2f_{11} \quad \text{for all } t$$

We assume that the observations of the signal start at $t = 0$. Since the estimates must be unbiased, it is clear that $\hat{x}_1(0) = 0$. Therefore

$$p_{11}(0) = \mathcal{E}\tilde{x}_1^2(0) = \mathcal{E}x_1^2(0) = -q_{11}/2f_{11}$$

substituting this into (55), we get Shinbrot's formula:

$$p_{11}(t) = q_{11}\left[\frac{(f_{11} - \bar{f}_{11})e^{\bar{f}_{11}t} - (f_{11} + \bar{f}_{11})e^{-\bar{f}_{11}t}}{-(f_{11} - \bar{f}_{11})^2e^{\bar{f}_{11}t} + (f_{11} + \bar{f}_{11})^2e^{-\bar{f}_{11}t}}\right]$$

Since $\bar{f}_{11} < 0$, we see that as $t \to \infty$, $p_{11}(t)$ converges to

$$\bar{p}_{11} = -q_{11}/(f_{11} + \bar{f}_{11}) = (f_{11} - \bar{f}_{11})r_{11}$$

which agrees with (52).

To understand better the factors affecting convergence to the steady-state, let

$$\delta p_{11}(t) = p_{11}(t) - \bar{p}_{11}$$

The differential equation for δp_{11} is

$$d\delta p_{11}/dt = 2\bar{f}_{11}\delta p_{11} - (\delta p_{11})^2/r_{11} \tag{56}$$

We now introduce a *Lyapunov function* [21] for (56)

$$V(\delta p_{11}) = (\delta p_{11}/\bar{p}_{11})^2$$

The derivative of V *along motions* of (51) is given by

$$\dot{V}(\delta p_{11}) = \frac{\partial V(\delta p_{11})}{\partial \delta p_{11}} \cdot \frac{d\delta p_{11}}{dt} = -2[p_{11}/r_{11} + q_{11}/\bar{p}_{11}]V(\delta p_{11}) \tag{57}$$

This shows clearly that the "equivalent reciprocal time constant" for the variance equation depends on two quantities: (i) the message-to-noise ratio p_{11}/r_{11} at the input of the optimal filter, (ii) the ratio of excitation to estimation error $q_{11}/\bar{p}_{11}$.

Since the message model in this example is identical with its dual, it is clear that the preceding results apply without any modification to the dual problem. In particular, the filter shown in Fig. 1(*b*) is the same as the optimal regulator for a plant with transfer function $1/(s - f_{11})$. The Hamiltonian equations (27) for the dual problem were derived by Rozonoër [19] from Pontryagin's maximum principle.

Let us conclude this example by making some observations about the nonconstant case. First, the expression for the derivative of the Lyapunov function given by (57) remains true without any modification. Second, assume $\bar{p}_{11}(t_0)$ has been evaluated somehow. Given this number, $\bar{p}_{11}(t)$ can be evaluated for $t \geq t_0$ by means of the variance equation (51); the existence of a Lyapunov function and in particular (57) shows that this computation is stable, i.e., not adversely affected by roundoff errors. Third, knowing $\bar{p}_{11}(t)$, equation (57) provides a clear picture of the transient behavior of the optimal filter, even though it might be impossible to solve (51) in closed form.

Example 2. The variance equation is

$$dp_{11}/dt = 2f_{11}p_{11} - p_{11}^2(1/r_{11} + 1/r_{22}) + q_{11}$$

If $q_{11} > 0$, $r_{11} > 0$, and $r_{22} > 0$, the conditions of Theorems 3–4 are satisfied. Therefore the minimum error variance in the steady-state is

$$\bar{p}_{11} = \frac{f_{11} + \sqrt{f_{11}^2 + q_{11}/r_{11} + q_{11}/r_{22}}}{1/r_{11} + 1/r_{22}}$$

and the optimal steady-state gains are

$$\bar{k}_{1i} = \bar{p}_{11}/r_{ii}, \qquad i = 1, 2$$

The same problem has been considered also by Westcott [30, Example]. A glance at his calculations shows that ours is the simpler and more natural approach.

Example 3. The variance equation is

$$\left.\begin{aligned} dp_{11}/dt &= -p_{12}^2/r_{11} + q_{11} \\ dp_{12}/dt &= p_{11} - p_{12} - p_{12}p_{22}/r_{11} \\ dp_{22}/dt &= 2(p_{12} - p_{22}) - p_{22}^2/r_{11} \end{aligned}\right\} \tag{58}$$

If $q_{11} > 0$, $r_{11} > 0$, the conditions of Theorems 3–4 are satisfied. Setting the right-hand side of (58) equal to zero, we get the solution of the stationary problem:

$$\bar{k}_{11} = \sqrt{q_{11}/r_{11}}$$

$$\bar{k}_{21} = -1 + \sqrt{1 + 2\sqrt{q_{11}/r_{11}}}$$

See Fig. 3(*b*).

The infinitesimal transition matrix of the optimal filter in the steady-state is:

$$\bar{\mathbf{F}} = \begin{bmatrix} 0 & -\sqrt{q_{11}/r_{11}} \\ 1 & -\sqrt{1 + 2\sqrt{q_{11}/r_{11}}} \end{bmatrix}$$

The natural frequency of the filter is $(q_{11}/r_{11})^{1/4}$ and the damping ratio is $(1/2)[2 + (r_{11}/q_{11})^{1/2}]^{1/2}$. Even for such a very simple problem, the parameters of the optimal filter are not at all obvious by inspection.

The solution of the dual problem in the steady-state (see Fig. 4) is obtained by utilizing the duality relations

$$\bar{k}^*_{11} = \bar{k}_{11}, \qquad \bar{k}^*_{12} = \bar{k}_{21}$$

The same result was obtained by Kipiniak [24], using the Euler equations of the calculus of variations.

Example 4. The variance equation is

$$\left.\begin{aligned} dp_{11}/dt &= 2f_{12}p_{12} - p_{11}^2/r_{11} + q_{11} \\ dp_{12}/dt &= f_{21}p_{11} + f_{12}p_{22} - p_{11}p_{12}/r_{11} \\ dp_{22}/dt &= 2f_{21}p_{12} - p_{12}^2/r_{11} \end{aligned}\right\} \quad (59)$$

If $f_{12} \neq 0$, $f_{21} \neq 0$, and $r_{11} > 0$, the conditions of Theorems 3–4 are satisfied. There are then two sets of possibilities for the right-hand side of (59) to vanish for nonnegative $\bar{p}_{22}$:

(A) $\bar{p}_{12} = \sqrt{q_{11}r_{11}}$, $\bar{p}_{12} = 0$, $\bar{p}_{22} = -(f_{21}/f_{12})\sqrt{q_{11}r_{11}}$

(B) $\bar{p}_{11} = \sqrt{(q_{11} + 4f_{12}f_{21}r_{11})r_{11}}$, $\bar{p}_{12} = 2f_{21}r_{11}$, $\bar{p}_{22} = (f_{21}/f_{12})\sqrt{(q_{11} + 4f_{12}f_{21}r_{11})r_{11}}$

The expression for $\bar{p}_{22}$ shows that Case (A) applies when $f_{12}f_{21}$ is negative (the model is stable but not asymptotically stable) and Case (B) applies when $f_{12}f_{21}$ is positive (the model is unstable).

The optimal filter is shown in Fig. 5(*b*). The optimal gains are given by

$$\bar{k}_{11} = \bar{p}_{11}/r_{11}, \qquad \bar{k}_{21} = \bar{p}_{12}/r_{11}$$

If $f_{12} \neq 0$ but $f_{21} = 0$, the model is completely observable but not completely controllable. Hence the steady-state variances exist but the optimal filter is not necessarily asymptotically stable since Theorem 4 is not applicable. As a matter of fact, the optimal filter in this case is partially "open loop" and it is not asymptotically stable.

If $f_{12} = 0$, then not even Theorem 3 is applicable. In this case, if $f_{21} \neq 0$, equations (59) have no equilibrium state; if $f_{21} = 0$, then equations (59) have an infinity of positive definite equilibrium states given by:

$$\bar{p}_{11} = \sqrt{q_{11}/r_{11}}, \qquad \bar{p}_{12} = 0, \qquad \bar{p}_{22} > 0$$

Thus if $f_{12} = 0$, the conclusions of Theorems 3–4 are false.

Example 5. The variance equation is

$$\left.\begin{aligned} dp_{11}/dt &= 2p_{12} - p_{11}^2/r_{11} \\ dp_{12}/dt &= p_{22} - p_{11}p_{12}/r_{11} \\ dp_{22}/dt &= -p_{12}^2/r_{11} \end{aligned}\right\}$$

We assume that $r_{11} > 0$; this assures that Theorem 3 is applicable. We then find that the steady-state error variances are all zero. The matrix of coefficients of the Hamiltonian equations (27) is:

$$\mathbf{T} = \begin{bmatrix} 0 & 0 & 1/r_{11} & 0 \\ -1 & 0 & 0 & 0 \\ 0 & 0 & 0 & 1 \\ 0 & 0 & 0 & 0 \end{bmatrix}$$

and the corresponding transition matrix is (here (4) is a *finite* series!)

$$\mathbf{\Theta}(t_0 + \tau, t_0) = \begin{bmatrix} 1 & 0 & \tau/r_{11} & \tau^2/2r_{11} \\ -\tau & 1 & -\tau^2/2r_{11} & -\tau^3/6r_{11} \\ 0 & 0 & 1 & \tau \\ 0 & 0 & 1 & 1 \end{bmatrix}$$

Using (29), we find ($t_0 = 0$):

$$\mathbf{P}(t) = \frac{r_{11}p_{22}(0)}{r_{11} + p_{22}(0)t^3/3}\begin{bmatrix} t^2 & t \\ t & 1 \end{bmatrix}$$

This formula, obtained here with little labor, is identical with the results of Shinbort [3, Example 1].

The optimal filter is shown in Fig. 7(*b*). The time-varying gains tend to 0 as $t \to \infty$; in other words, the filter pays less and less attention to the incoming signals and relies more and more on the previous estimates of x_1 and x_2.

Since the conditions of Theorem 4 are not satisfied, one might suspect that the optimal filter is *not* uniformly (and hence exponentially [21]) asymptotically stable. To check this conjecture, we calculate the transition matrix of the optimal filter. We find, for $t, \tau \geqq 0$,

$$\mathbf{\Psi}(t, \tau) = \frac{1}{\alpha(t)}\begin{bmatrix} \alpha(t) - \beta(t, \tau)t & -\alpha(t)\tau + \alpha(\tau)t + \beta(t, \tau)\tau t \\ -\beta(t, \tau) & \alpha(\tau) + \beta(t, \tau) \end{bmatrix}$$

where

$$\alpha(t) = t^3/3 + r_{11}/p_{22}(0)$$

$$\beta(t, \tau) = (t^2 - \tau^2)/2$$

Since $\psi_{11}(t\ \tau)$ does not converge to zero with $t - \tau \to \infty$, it is clear that the optimal filter is not even stable, let alone asymptotically stable.

From the transition matrix of the optimal filter, we can obtain at once its impulse response with respect to the input $z_1(t)$ and output $\hat{x}_1(t)$:

$$\psi_{11}(t, \tau)k_{11}(\tau) + \psi_{12}(t, \tau)k_{21}(\tau) = \frac{t\tau}{t^3/3 + r_{11}/p_{22}(0)}$$

This agrees with Shinbrot's result [3].

Example 6. The variance equation is:

$$\left.\begin{aligned} dp_{11}/dt &= 2p_{12} - h_{11}^2p_{11}^2/r_{11} - h_{22}^2p_{12}^2/r_{22} \\ dp_{12}/dt &= p_{22} - h_{11}^2p_{11}p_{12}/r_{11} - h_{22}^2p_{12}p_{22}/r_{22} \\ dp_{22}/dt &= -h_{11}^2p_{12}^2/r_{11} - h_{22}^2p_{22}^2/r_{22} + q_{11} \end{aligned}\right\} \quad (60)$$

If $h_{11} \neq 0$, $q_{11} > 0$, $r_{11} > 0$, $r_{22} > 0$, then the conditions of Theorems 3–4 are satisfied. Setting the right-hand side of (60) equal to zero leads to a very complicated algebraic problem. We introduce first the abbreviations:

$$\alpha = |h_{11}|\sqrt{q_{11}/r_{11}}$$

$$\beta^2 = h_{22}^2q_{11}/r_{22}$$

It follows that

$$h_{11}\bar{k}_{11} = \frac{h_{11}^2}{r_{11}}\bar{p}_{11} = \alpha\frac{\sqrt{2\alpha + \beta^2}}{\alpha + \beta^2}$$

$$h_{11}\bar{k}_{21} = \frac{h_{11}^2}{r_{11}}\bar{p}_{12} = \frac{\alpha^2}{\alpha + \beta^2}$$

$$h_{22}\bar{k}_{12} = \frac{h_{22}^2}{r_{22}}\bar{p}_{12} = \frac{\beta^2}{\alpha + \beta^2}$$

$$h_{22}\bar{k}_{21} = \frac{h_{22}^2}{r_{22}}\bar{p}_{22} = \beta^2\frac{\sqrt{2\alpha + \beta^2}}{\alpha + \beta^2}$$

It is easy to verify that the right-hand side of (60) vanishes for this set of $\bar{p}_{ij}$'s; by Theorem 5, this cannot happen for any other set. Hence the solution of the stationary Wiener problem is complete. It is interesting to note that the conventional procedure would require here the spectral factorization of a two-by-two matrix which is very much more difficult algebraically than by the present method.

The infinitesimal transition matrix of the optimal filter is given by

$$\mathbf{F}_{\text{opt}} = \begin{bmatrix} -\alpha\dfrac{\sqrt{2\alpha + \beta^2}}{\alpha + \beta^2} & \dfrac{\alpha}{\alpha + \beta^2} \\ -\dfrac{\alpha^2}{\alpha + \beta^2} & -\beta^2\dfrac{\sqrt{2\alpha + \beta^2}}{\alpha + \beta^2} \end{bmatrix}$$

The natural frequency of the optimal filter is

$$\omega = |\lambda(\mathbf{F}_{\text{opt}})| = \sqrt{\alpha}$$

and the damping ratio is

$$\zeta = |\text{Re}\,\lambda(\mathbf{F}_{\text{opt}})|/\omega = \frac{1}{\sqrt{2}}\sqrt{1 + \frac{\beta^2}{2\alpha}}$$

The quantities α and β can be regarded as signal-to-noise ratios. Since all parameters of the optimal filter depend only on these ratios, there is a possibility of building an adaptive filter once means of experimentally measuring α and β are available. An investigation of this sort was carried out by Bucy [31] in the simplified case when $h_{22} = \beta = 0$.

12 Problems Related to Adaptive Systems

The generality of our results should be of considerable usefulness in the theory of adaptive systems, which is as yet in a primitive stage of development.

An *adaptive system* is one which changes its parameters in accordance with measured changes in its environment. In the estimation problem, the changing environment is reflected in the time-dependence of **F**, **G**, **H**, **Q**, **R**. Our theory shows that such changes affect only the values of the parameters but not the structure of the optimal filter. This is what one would expect intuitively and we now have also a rigorous proof. Under ideal circumstances, the changes in the environment could be detected instantaneously and exactly. The adaptive filter would then behave as required by the fundamental equations (I–IV). In other words, our theory establishes a basis of comparison between actual and ideal adaptive behavior. It is clear therefore that *a fundamental problem in the theory of adaptive systems is the further study of properties of the variance equation* (IV).

13 Conclusions

One should clearly distinguish between two aspects of the estimation problem:

(1) *The theoretical aspect.* Here interest centers on:

(i) The general form of the solution (see Fig. 1).

(ii) Conditions which guarantee a priori the existence, physical realizability, and stability of the optimal filter.

(iii) Characterization of the general results in terms of some simple quantities, such as signal-to-noise ratio, information rate, bandwidth, etc.

An important consequence of the time-domain approach is that these considerations can be completely divorced from the assumption of stationarity which has dominated much of the thinking in the past.

(2) *The computational aspect.* The classical (more accurately, old-fashioned) view is that a mathematical problem is solved if the solution is expressed by a formula. It is not a trivial matter, however, to substitute numbers in a formula. The current literature on the Wiener problem is full of semirigorously derived formulas which turn out to be unusable for practical computation when the order of the system becomes even moderately large. The variance equation of our approach provides a practically useful and theoretically "clean" technique of numerical computation. Because of the guaranteed convergence of these equations, the computational problem can be considered solved, except for purely numerical difficulties.

Some open problems, which we intend to treat in the near future, are:

(i) Extension of the theory to include nonwhite noise. As mentioned in Section 2, this problem is already solved in the discrete-time case [11], and the only remaining difficulty is to get a convenient canonical form in the continuous-time case.

(ii) General study of the variance equations using Lyapunov functions.

(iii) Relations with the calculus of variations and information theory.

14 References

1 N. Wiener, "The Extrapolation, Interpolation, and Smoothing of Stationary Time Series," John Wiley & Sons, Inc., New York, N. Y., 1949.

2 A. M. Yaglom, "Vvedenie v Teoriya Statsionarnikh Sluchainikh Funktsii" (Introduction to the theory of stationary random Processes) (in Russian), *Ups. Fiz. Nauk.*, vol. 7, 1951; German translation edited by H. Göring, Akademie Verlag, Berlin, 1959.

3 M. Shinbrot, "Optimization of Time-Varying Linear Systems With Nonstationary Inputs," Trans. ASME, vol. 80, 1958, pp. 457–462.

4 C. W. Steeg, "A Time-Domain Synthesis for Optimum Extrapolators," *Trans. IRE*, Prof. Group on Automatic Control, Nov., 1957, pp. 32–41.

5 V. S. Pugachev, "Teoriya Sluchainikh Funktsii i Ee Primenenie k Zadacham Automaticheskogo Upravleniya" (Theory of Random Functions and Its Application to Automatic Control Problems) (in Russian), second edition, Gostekhizdat, Moscow, 1960.

6 V. S. Pugachev, "A Method for Solving the Basic Integral Equation of Statistical Theory of Optimum Systems in Finite Form," *Prikl. Math. Mekh.*, vol. 23, 1959, pp. 3–14 (English translation pp. 1-16).

7 E. Parzen, "Statistical Inference on Time Series by Hilbert-Space Methods, I," Tech. Rep. No. 23, Applied Mathematics and Statistics Laboratory, Stanford Univ., 1959.

8 A. G. Carlton and J. W. Follin, Jr., "Recent Developments in Fixed and Adaptive Filtering," Proceedings of the Second AGARD Guided Missiles Seminar (Guidance and Control) AGARDograph 21, September, 1956.

9 J. E. Hanson, "Some Notes on the Application of the Calculus of Variations to Smoothing for Finite Time, etc.," JHU/APL Internal Memorandum BBD-346, 1957.

10 R. S. Bucy, "Optimum Finite-Time Filters for a Special Nonstationary Class of Inputs," JHU/APL Internal Memorandum BBD-600, 1959.

11 R. E. Kalman, "A New Approach to Linear Filtering and Prediction Problems," Trans. ASME, Series D, Journal of Basic Engineering, vol. 82, 1960, pp. 35–45.

12 R. E. Bellman, "Adaptive Control: A Guided Tour" (to be published), Princeton University Press, Princeton, N. J., 1960.

13 R. E. Kalman and R. W. Koepcke, "The Role of Digital Computers in the Dynamic Optimization of Chemical Reactions," Proceedings of the Western Joint Computer Conference, 1959, pp. 107–116.

14 R. E. Kalman and J. E. Bertram, "A Unified Approach to the Theory of Sampling Systems," *Journal of the Franklin Institute*, vol. 267, 1959, pp. 405–436.

15 J. L. Doob, "Stochastic Processes," John Wiley & Sons, Inc., New York, N. Y., 1953.

16 R. E. Kalman, "On the General Theory of Control Systems," Proceedings of the First International Congress on Automatic Control, Moscow, USSR, 1960.

17 R. E. Kalman, "Contributions to the Theory of Optimal Control," Proceedings of the Conference on Ordinary Differential Equations, Mexico City, Mexico, 1959; Bol. Soc. Mat. Mex., 1961.

18 J. J. Levin, "On the Matrix Riccati Equation," *Trans. American Mathematical Society*, vol. 10, 1959, pp. 519–524.

19 L. I. Rozonoër, "L. S. Pontryagin's Maximum Principle in the Theory of Optimum Systems, I," *Avt. i Telemekh.*, vol. 20, 1959, pp. 1320–1324.

20 S. Kullback, "Information Theory and Statistics," John Wiley & Sons, New York, N. Y., 1959.

21 R. E. Kalman and J. E. Bertram, "Control System Analysis and Design Via the 'Second Method' of Lyapunov. I. Continuous-Time Systems," Journal of Basic Engineering, Trans. ASME, series D, vol. 82, 1960, pp. 371–393.

22 R. E. Bellman, "Introduction to Matrix Analysis," McGraw-Hill Book Company, Inc., New York, N. Y., 1960.

23 E. A. Coddington and N. Levinson, "Theory of Ordinary Differential Equations," McGraw-Hill Book Company, Inc., New York, N. Y., 1955.

24 W. Kipiniak, "Optimum Nonlinear Controllers," Report 7793-R-2, Servomechanisms Lab., M.I.T., 1958.

25 R. S. Bucy, "A Matrix Formulation of the Finite-Time Problem," JHU/APL Internal Memorandum BBD-777, 1960.

26 R. S. Bucy, "Combined Range and Speed Gate for White Noise and White Signal Acceleration," JHU/APL Internal Memorandum BBD-811, 1960.

27 V. S. Pugachev, "General Condition for the Minimum Mean Square Error in a Dynamic System," *Avt. i Telemekh.*, vol. 17, 1956, pp. 289–295, translation, pp. 307–314.

28 M. Loève, "Probability Theory," Van Nostrand and Company, New York, N. Y., 1955, Chap. 10.

29 W. B. Davenport and W. L. Root, "An Introduction to the Theory of Random Signals and Noise," McGraw-Hill Book Company, Inc., New York, N. Y., 1956.

30 J. H. Westcott, "Design of Multivariable Optimum Filters," Trans. ASME, vol. 80, 1958, pp. 463–467.

31 R. S. Bucy, "Adaptive Finite-Time Filtering," JHU/APL Internal Memorandum BBD-645, 1959.

A Bayesian Approach to Problems in Stochastic Estimation and Control

Y. C. HO, SENIOR MEMBER, IEEE AND R. C. K. LEE

***Summary*—In this paper, a general class of stochastic estimation and control problems is formulated from the Bayesian Decision-Theoretic viewpoint. A discussion as to how these problems can be solved step by step in principle and practice from this approach is presented. As a specific example, the closed form Wiener-Kalman solution for linear estimation in Gaussian noise is derived. The purpose of the paper is to show that the Bayesian approach provides; 1) a general unifying framework within which to pursue further researches in stochastic estimation and control problems, and 2) the necessary computations and difficulties that must be overcome for these problems. An example of a nonlinear, non-Gaussian estimation problem is also solved.**

Single Stage Estimation Problem

FOR THE PURPOSE of illustrating the concepts involved, the single-stage estimation problem will be discussed first. Once this is accomplished, the multistage problem can be treated straightforwardly.

Problem Statement

The following information is assumed, given;

1) A set of measurements $z_1, z_2, \cdots, z_k$ which are denoted by the vector z.
2) The physical relationship between the state of nature which is to be estimated and the measurements. This is given by

$$z = g(x, v), \tag{1}$$

where
z is the measurement vector $(k \times 1)$,
x is the state (signal) vector $(n \times 1)$, and
v is the noise vector $(q \times 1)$.

3) The joint density function $p(x, v)$. From this the respective marginal density functions, $p(x)$ and $p(v)$, are readily obtained.

It is assumed that information for 3) is available in analytical form or can be approximated by analytical distributions. Item 2) can be either in closed form or merely computable. The problem is to obtain an estimate $\hat{x}$ of x upon which to base the best measurements which will be defined later.

The Bayesian Solution

The Bayesian solution to the above problem now proceeds via the following steps:

1) Evaluate $p(z)$—This can be done analytically, at least in principle, or experimentally by Monte Carlo methods since $z = g(x, v)$ and $p(x, v)$ are given. In the latter case, it is assumed possible to fit the experimental distribution again by a member of a family of distributions.
2) At this point, two alternatives are possible, one may be superior to the other depending on the nature of the problem.
   a) Evaluate $p(x, z)$. This is possible analytically if v is of the same dimension as z and one can obtain the functional relationship $v = g^*(x, z)$ from (1). Then, using $p(x, v)$ and the theory of derived distributions, one obtains

$$p(x, z) = p(x, v = g^*(x, z))J \tag{2}$$

   where

$$J = \det\left[\frac{\partial g^*(x, z)}{\partial z}\right].$$

   b) Evaluate $p(z/x)$. This conditional density function can always be obtained either analytically, whenever possible, or experimentally from the $z = g(x, v)$ and $p(x, v)$.

Note that 2a) may be difficult to obtain in general since g^* may not exist either because of the nonlinear nature of g or the fact that z, v are of different dimensions. Nevertheless, 2b) can always be carried out. This fact will be demonstrated in the nonlinear example in the sequel.

3) Evaluate $p(x/z)$ using the following relationships:
   a) Following 2a)

$$p(x/z) = \frac{p(x, z)}{p(z)}. \tag{3}$$

   b) Following 2b), use the Bayes' rule

$$p(x/z) = \frac{p(z/x)p(x)}{p(z)}. \tag{4}$$

Depending on the class of distributions one has assumed or obtained for $p(x, v)$, $p(z)$, $p(z/x)$, this key step may be easy or difficult to carry out. Several classes of distribution which have nice properties for this purpose can be found in Raiffa and Schlaifer [1]. The density function $p(x/z)$ is known as the *a posteriori* density function of x. It is the knowledge about the state of nature *after* the measurements z. By definition, it contains all the information necessary for estimation.

Manuscript received October 31, 1963; revised July 2, 1964.
Y. C. Ho is with Harvard University, Cambridge, Mass., and is also a Consultant at the Minneapolis-Honeywell Regulator Co., Boston, Mass.
R. C. K. Lee was formerly with the Minneapolis-Honeywell Regulator Co., Minneapolis, Minn.

Reprinted from *IEEE Trans. Automat. Contr.*, vol. AC-9, pp. 333–339, Oct. 1964.

4) Depending on the criterion function for estimation, one can compute estimate $\hat{x}$ from $p(x/z)$. Some typical examples are
   a) Criterion: Maximize the Probability ($\hat{x}=x$).
      Solution: $\hat{x}=$ Mode of $p(x/z)$. (5)
      This is defined as the most probable estimate. When the *a priori* density function $p(x)$ is uniform, this estimate is identical to the classical maximum likelihood estimate.
   b) Criterion: Minimize $\int \|x-x\|^2 p(x/z)dx$.
      Solution: $\hat{x}=E(x/z)$[1] (6)
      This is the conditional mean estimate.
   c) Criterion: Minimize Maximum $|x-\hat{x}|$.
      Solution: $\hat{x}=$ Medium of $p(x/z)$. (7)
      This can be defined as the minimax estimate.

Pictorially, the three estimates are shown in Fig. 1 for a general $p(x/z)$ for a scalar case.

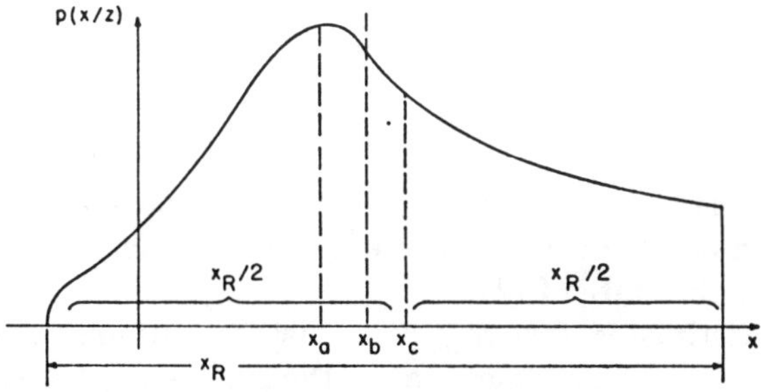

Fig. 1—Estimates based on *a posteriori* density.

Clearly, other estimates, as well as confidence intervals can be derived from $p(x/z)$ directly.

Special Case of the Wiener-Kalman Filter (single stage)

Now a special case of the above estimation problem will be considered. Let there be given;

1) A set of measurements $z=(z_1, z_2, \cdots, z_k)$.
2) The physical relationship

$$z = Hx + v. \tag{8}$$

3) The independent noise and state density functions

$$p(x, v) = p(x)p(v) \tag{9}$$

$p(x)$ be Gaussian with

$$\left.\begin{aligned} E(x) &= \bar{x} \\ \text{Cov}(x) &= P_0 \end{aligned}\right\} \tag{10}$$

$p(v)$ be Gaussian with

$$\left.\begin{aligned} E(v) &= 0 \\ \text{Cov}(v) &= R \end{aligned}\right\}. \tag{11}$$

Now, following the steps for the Bayesian solution,

1) Evaluate $p(z)$.
   Since $z=Hx+v$ and x, v are Gaussian and independent, one immediately gets

$$\left.\begin{aligned} &p(z) \text{ is Gaussian} \\ &E(z) = H\bar{x} \\ &\text{Cov}(z) = HP_0H^T + R \end{aligned}\right\}. \tag{12}$$

2a) Evaluate $p(x, z)$. Since $(\partial g^*/\partial z) =$ Identify matrix, it follows

$$\begin{aligned} p(x, z) &= p(x, v = z - Hx) \\ &= p(x)p_v(z - Hx).^2 \end{aligned} \tag{13}$$

2b) Evaluate $p(z/x)$.[3]

$$p(z/x) = \frac{p(x, z)}{p(x)} = p(v) = p_v(z - Hx). \tag{14}$$

3) Evaluate $p(x/z)$. One gets from Bayes' rule,

$$p(x/z) = \frac{p(x)p(v)}{p(z)}. \tag{15}$$

By direct substitution of (10), (11), and (12), one obtains

$$\begin{aligned} p(x/z) = &\frac{|HP_0H^T + R|^{1/2}}{(2\pi)^{n/2}|P_0|^{1/2}|R|^{1/2}} \\ &\cdot \exp\{-1/2[(x - \bar{x})^T P_0^{-1}(x - \bar{x}) \\ &+ (z - Hx)^T R^{-1}(z - Hx) \\ &- (z - H\bar{x})^T (HP_0H^T + R)^{-1}(z - H\bar{x})]\}. \end{aligned} \tag{16}$$

Now completing squares in the { }, (16) simplifies to

$$\begin{aligned} p(x/z) = &\frac{|HP_0H^T + R|^{1/2}}{(2\pi)^{n/2}|P_0|^{1/2}|R|^{1/2}} \\ &\cdot \exp\{-1/2(x - \hat{x})^T P^{-1}(x - \hat{x})\} \end{aligned} \tag{17}$$

where

$$P^{-1} = P_0^{-1} + H^T R^{-1} H \tag{18}$$

or equivalently,

$$P = P_0 - P_0H^T(HP_0H^T + R)^{-1}HP_0 \tag{19}$$

and

$$\hat{x} = \bar{x} + PH^T R^{-1}(z - H\bar{x}). \tag{20}$$

4) Now, since $p(x/z)$ is Gaussian, the most probable, conditional mean and minimax estimate all coincide and is given by $\hat{x}$.

This is the derivation of the single stage Weiner-Kalman filter [2], [3]. The pair $(P, \hat{x})$ is called a *suffi-*

[1] It is assumed that $p(x/z)$ has finite second moment.

[2] $p_v(z-Hx)$ means substituting $(z-Hx)$ for v in $p(v)$.
[3] Note: step (2b) is redundant.

cient statistic for the problem in the sense that $p(x/z) = p(x/P, \hat{x})$.

Multistage Estimation Problem

The problem formulation and the solution in this case is basically similar to the single-stage problem. The only additional complication is that now the state is changing from stage to stage according to some dynamic relationship, and that the *a posteriori* density function is to be computed recursively.

Problem Statement

It is assumed that at any stage $k+1$, the following data is given as a result of previous computation or as part of the problem statement.

1) The system equations governing the evolution of the state.

$$\left.\begin{aligned} x_{k+1} &= f(x_k, w_k) \\ z_{k+1} &= h(x_{k+1}, v_{k+1}) \end{aligned}\right\} \tag{21}$$

where

x_{k+1} is the state vector at $k+1$,

v_{k+1} is the measurement noise at $k+1$,

z_{k+1} is the additional measurement available at $k+1$,

w_k is the disturbance vector at k.

2) The complete set of measurements $Z_{k+1} \triangleq (z_1, \cdots, z_{k+1})$.

3) The density functions[4]

$p(x_k/z_1, \cdots, z_k) \triangleq p(x_k/Z_k)$

$p(w_k, v_{k+1}/x_k)$—statistics of a vector random sequence with components w_k and v_{k+1} which depends on x_k.

Now it is required to estimate x_{k+1} based on measurements $z_1, \cdots, z_{k+1}$.

The Bayesian Solution

The procedure is analogous to the single-stage case.

1) Evaluate $p(x_{k+1}/x_k)$. This can be accomplished either experimentally or analytically from knowledge of $p(w_k, v_{k+1}/x_k)$, $p(x_k/Z_k)$ and (21).

2) Evaluate $p(z_{k+1}/x_k, x_{k+1})$. This is derived from $p(w_k, v_{k+1}/x_k)$ and (21).

3) Evaluate

$$p(x_{k+1}, z_{k+1}/Z_k) = \int p(z_{k+1}/Z_k, x_{k+1})p(x_{k+1}/x_k) \cdot p(x_k/Z_k)dx_k. \tag{22}$$

From this the marginal density functions $p(x_{k+1}/Z_k)$ and $p(z_{k+1}/Z_k)$ can be directly evaluated.

4) Evaluate

$$p(x_{z+1}/Z_{k+1}) = \frac{p(x_{k+1}, z_{k+1}/Z_k)}{p(z_{k+1}/Z_k)} \tag{23}$$

from (22)

$$= \frac{\int p(z_{+1}/Z_k, x_{k+1})p(x_{k+1}/x_k)p(x_k/Z_k)dx_k}{\int\int p(z_{k+1}/Z_k, x_{k+1})p(x_{k+1}/x_k)p(x_k/Z_k)dx_{k+1}dx_k}. \tag{24}$$

Eq. (24) is a functional-integral-difference equation governing the evolution of the *a posteriori* density function of the state of (21).

5) Estimates for x_{k+1} can now be obtained from $p(x_{k+1}/Z_{k+1})$ exactly as in the single-stage case.

Special Case of the Wiener-Kalman Filter[5]

The given data at $k+1$ is specified as follows: The physical model is given by

$$\begin{aligned} x_{k+1} &= \Phi x_k + \Gamma w_k \\ z_k &= Hx_k + v_k \end{aligned} \tag{25}$$

where w and v are independent, white, Gaussian random sequences with

$$p(x_k/Z_k) \text{ is Gaussian}$$

$$\left.\begin{aligned} E(x_k/Z_k) &\triangleq \hat{x}_k \\ \text{Cov}\,(x_k/Z_k) &= P_k \end{aligned}\right\} \tag{26}$$

$$p(w_k, v_{k+1}/x_k, Z_k) = p(w_k)p(v_{k+1})$$

$$\left.\begin{aligned} E(v_k) &= E(v_{k+1}) = 0 \\ \text{Cov}\,(w_k) &= Q;\ \text{Cov}\,(v_{k+1}) = R \end{aligned}\right\}. \tag{27}$$

[4] The product of the two density functions yields $p(w_k, v_{k+1}, x_k/Z_k)$ by the Markov property of (21). It is also assumed that if $p(w\,v/x) = p(w, v)$ then w, v are white random sequences.

[5] This development of the multistage Wiener-Kalman filtering method is very similar to a paper by H. Rauch, F. Tung, and C. C. Striebel entitled "On The Maximum Likelihood Estimate for Linear Dynamic Systems" presented at the SIAM Conference on System Optimization, 1964, Monterey, Calif. The only difference between the two developments is that the Rauch-Tung-Striebel paper does not explicitly compute $p(x/z)$ but simply computes its maximum and uses it as the estimate. In the author's approach, the computation of the maximum plays a secondary role. The explicit calculation of the *a posteriori* probability is emphasized as the Bayesian viewpoint. The authors are indebted to Prof. A. E. Bryson for bringing this reference to their attention. Similar development of the Wiener-Kalman filter is also presented in NASA TR-R-135'1962 by G. L. Smith, S. Schmidt, and L. A. Megee. This was brought to the authors' attention after the publication of the JACC preprint.

Since in this case, the noise w_k, v_{k+1} is not dependent on the state, (24) simplifies to

$$p(x_{k+1}/Z_{k+1}) = \frac{p(z_{k+1}/x_{k+1})}{p(z_{k+1}/Z_k)} p(x_{k+1}/Z_k). \quad (24)'$$

Hence, the solution only involved the evaluation of the three density functions on the rhs of (24) given the data (25–27). This is carried out below. From (27), it is noted that $p(x_{k+1}/Z_k)$ is Gaussian and independent of v_{k+1}

$$\left.\begin{aligned} E(x_{k+1}/Z_k) &= \Phi\hat{x}_k \\ \text{Cov}\,(x_{k+1}/Z_k) &= \Phi P_k \Phi^T + \Gamma Q \Gamma^T \triangleq P_{k+1} \end{aligned}\right\}. \quad (28)$$

Similarly, $p(z_{k+1}/Z_k)$ is Gaussian and

$$\left.\begin{aligned} E(z_{k+1}/Z_k) &= H\Phi\hat{x}_k \\ \text{Cov}\,(z_{k+1}/Z_k) &= HP_{k+1}H^T + R \end{aligned}\right\}. \quad (29)$$

Finally $p(z_{k+1}/x_{k+1})$ is also Gaussian with

$$\left.\begin{aligned} P(z_{k+1}/x_{k+1}) &= Hx_{k+1} \\ \text{Cov}\,(z_{k+1}/x_{k+1}) &= R \end{aligned}\right\}. \quad (30)$$

Combining (28–30) using (24), one gets

$$\begin{aligned} p(x_{k+1}/Z_{k+1}) = \frac{|HM_{k+1}H^T + R|^{1/2}}{(2\pi)^{n/2}|R|^{1/2}|M_{k+1}|^{1/2}} & \cdot \exp[-1/2[(x_{k+1} - \Phi\hat{x}_k)^T M_{k+1}^{-1} \\ & \cdot (x_{k+1} - \Phi\hat{x}_k) \\ & + (z_{k+1} - Hx_{k+1})^T R^{-1}(z_{k+1} - Hx_{k+1}) \\ & - (z_{k+1} - H\Phi\hat{x}_k)^T(HM_{k+1}H^T + R)^{-1} \\ & \cdot (z_{k+1} - H\Phi\hat{x}_k)]\}. \end{aligned} \quad (31)$$

Now completing squares in { } one gets,

$$\begin{aligned} p(x_{k+1}/Z_{k+1}) &= \frac{|HM_{k+1}H^T + R|^{1/2}}{(2\pi)^{n/2}|R|^{1/2}|M_{k+1}|^{1/2}} \\ & \cdot \exp\{-1/2(x_{k+1} - \hat{x}_{k+1})^T P_{k+1}^{-1}(x_{k+1} - \hat{x}_{k+1})\} \end{aligned} \quad (32)$$

where

$$\begin{aligned} \hat{x}_{k+1} &= \Phi\hat{x}_k + M_{k+1}H^T(HM_{k+1}H^T + R)^{-1} \\ & \cdot (z_{k+1} - H\Phi\hat{x}_k) \end{aligned} \quad (33)$$

$$P_{k+1}^{-1} = M_{k+1}^{-1} + H^T R^{-1} H \quad (34)$$

or equivalently,

$$P_{k+1} = M_{k+1} - M_{k+1}H^T(HM_{k+1}H^T + R)^{-1}HM_{k+1} \quad (35)$$

and

$$M_{k+1} = \Phi P_k \Phi^T + \Gamma Q \Gamma^T. \quad (36)$$

Eqs. (33–36) are exactly the discrete Wiener-Kalman filter in the multistage case [3], [4].

A Simple Nonlinear Non-Gaussian Estimation Problem

The discussions in the above sections have been carried out in terms of continuous density functions. However, it is obvious that the same process can be applied to problems involving discrete density function and discontinuous functional relationships. It is worthwhile, at this point, to carry out one such solution for a simple *contrived* example which nevertheless illustrates the application of the basic approach.

The problems can be visualized as an abstraction of the following physical estimation problem. An infrared detector followed by a threshold device is used in a satellite to detect hot targets on the ground. However, extraneous signals, particularly reflection from clouds, obscure the measurements. The problem is to design a multistage estimation process to estimate the presence of hot targets through measurement of the output of the threshold detector.

Let s_k (target) be scalar independent Bernoulli process with,

$$p(s_k) = (1 - q)\delta(s_k) + q\delta(1 - s_k), \quad (37)^6$$

n_k (cloud noise) be a scalar Markov process with,

$$p(n_1) = (1 - a)\delta(n_1) + a\delta(1 - n_1) \quad (38)$$

$$\begin{aligned} p(n_{k+1}/n_k) &= \left(1 - a - \frac{n_k}{2}\right)\delta(n_{k+1}) \\ &\quad + \left(a + \frac{n_k}{2}\right)\delta(1 - n_{k+1}) \end{aligned} \quad (39)$$

and the scalar measurement,

$$z_k = s_k \oplus n_k \quad (40)$$

where $\oplus$ indicates the logical "OR" operation.

Essentially (37–40) indicate the fact that as the detector sweeps across the field of view, cloud reflection tends to appear in groups while targets appear in isolated dots.

Now to proceed to the Bayesian solution. First, there is,

n_1	0	0	1	1
s_1	0	1	0	1
z_1	0	1	1	1
Probability of z_1	$(1-a)(1-q)$	$q(1-a)$	$a(1-q)$	aq

$$p(z_1) = (1 - a)(1 - q)\delta(z_1) + (a + q - aq)\delta(z_1 - 1). \quad (41)$$

[6] The notation

$$\delta(x) = \begin{cases} 1 & x = 0 \\ 0 & x \neq 0 \end{cases}$$

is used here. Also, $p(x)$ is to be interpreted as mass functions.

Also,

$$p(z_1/n_1) = \delta(z_1 - 1)n_1 + (1 - q)\delta(z_1) + q\delta(z_1 - 1)](1 - n_1). \quad (42)$$

Then by direct calculation,

$$p(n_1/z_1) = \frac{p(z_1/n_1)p(n_1)}{p(z_1)} = (1 - a'(z_1))\delta(n_1) + a'(z_1)\delta(n_1 - 1), \quad (43)$$

where

$$a'(z_1) = \frac{a\delta(z_1 - 1)}{(1 - a)(1 - q)\delta(z_1) + (a + q - aq)\delta(z_1 - 1)}; \quad (44)$$

which, after straightforward but somewhat laborious manipulations, becomes

$$= \left(1 - a - \frac{a'(z_1)}{2}\right)\delta(n_2) + \left(\frac{a'(z_1)}{2} + a\right)\delta(n_2 - 1)$$

$$\triangleq (1 - a(z_1))\delta(n_2) + a(z_1)\delta(n_2 - 1).$$

Furthermore,

$$p(s_2/z_1) \triangleq p(s_2) = (1 - q)\delta(s_2) + q\delta(s_2 - 1). \quad (50)$$

Eqs. (49) and (50) now take the place of (37) and (38) and by the same process, one can get, in general,

$$p(n_k/Z_k) \overset{\Delta}{=} p(n_k/z_k, z_{k-1}, \cdots) = (1 - a'(Z_k))\delta(n_k) + a'(Z_k)\delta(n_{k-1}) \quad (51)$$

$$a'(Z_k) \overset{\Delta}{=} a'(z_k, z_{k-1}, \cdots) = \frac{a(Z_{k-1})\delta(z_k - 1)}{[1 - a(Z_{k-1})](1 - q)\delta(z_k) + (a(Z_{k-1}) + q - a(Z_{k-1})q)\delta(z_k - 1)} \quad (52)$$

$$a(Z_{k-1}) \overset{\Delta}{=} a(z_{k-1}z_{k-2}, \cdots) = a + \frac{a'(Z_{k-1})}{2} \quad (53)$$

$$p(s_k/Z_k) \overset{\Delta}{=} p(s_k/z_k, z_{k-1}, \cdots) = 1 - q'(Z_k)\delta(s_k) + q'(Z_k)\delta(s_{k-1}) \quad (54)$$

$$q'(Z_k) \overset{\Delta}{=} q'(z_k, z_{k-1}, \cdots) = \frac{q\delta(z_{k-1})}{[1 - a(Z_{k-1})](1 - q)\delta(z_k) + [a(Z_{k-1}) + q - a(Z_{k-1})q]\delta(z_k - 1)} \quad (55)$$

$$\begin{aligned} p(n_{k+1}/Z_k) &= (1 - a(Z_k))\delta(n_{k+1}) + a(Z_k)\delta(n_{k+1} - 1) \\ p(s_{k+1}/Z_k) &= p(s_{k+1}). \end{aligned} \quad (56)$$

similarly,

$$p(z_1/s_1) = \delta(z_1 - 1)s_1 + [(1 - a)\delta(z_1) + a\delta(z_1 - 1)](1 - s_1), \quad (45)$$

and

$$p(s_1/z_1) = \frac{p(z_1/s_1)p(s_1)}{p(z_1)} = (1 - q'(z_1))\delta(s_1) + q'(z_1)\delta(s_1 - 1) \quad (46)$$

where

$$q'(z_1) = \frac{q\delta(z_1 - 1)}{(1 - a)(1 - q)\delta(z_1) + (a + q - aq)\delta(z_1 - 1)}, \quad (47)$$

and a reasonable estimate is

$$\hat{s}_1 = \begin{cases} 1 & \text{if } q'(z_1) > \epsilon \text{ (Given constant)} \\ 0 & \text{if } q'(z_1) < \epsilon \end{cases} \quad (48)$$

where $\hat{s}_1 = 1$ may be interpreted as an alarm.

Now consider a second measurement z_2 has been made. One has

$$p(n_2/z_1) = \int_{-\infty}^{\infty} p(n_2/n_1)p(n_1/z_1)dn_1, \quad (49)$$

Eqs. (51–57) now represent the general recursion solution for the multistage estimation process.

As a check, two possible observed sequences for z, namely (0, 1) and (1, 1) are considered. With $a = 1/4$ and $q = 1/4$ it is found that $p(s_2/z_2, z_1) = 0.571$ and 0.337, respectively. This agrees with intuition since the sequence (1, 1) has a higher probability of being cloud reflections. On the other hand, the numbers also showed that under the circumstances, it is very difficult to detect targets with accuracy using the system contrived here.

Oftentimes, one is actually interested in $p(s_k/Z_{k+\tau})$ with $\tau > 0$ in order to obtain the so-called "smoothed" estimate for s_k. The desired density function can be computed from $p(s_k/Z_k)$ by further manipulations. However, the calculation becomes involved and will not be done here.

Relationship to General Bayesian Statistical Decision Theory

It is worthwhile to point out the relationship of the above formulation and solution of the estimation problem to and its difference from the general statistical decision problem. For simplicity, the single-stage case is considered again. In the general statistical decision problem, the input data is somewhat different. One typical form is,[7]

$p(x)$—*a priori* density of x
$\{e\}$—a set of choices of experiments from which we can derive measurements z with
$p(z/x, e)$—conditional density of z for given x and e.
$\{u\}$—a set of choices of decisions
$J(e, z, u, x)$—a criterion function which is a possible function of e, z, u and x.

The problem is then stated as the determination of e and u so that $E(J)$ is optimized. The optimal J is given by

$$J_{\text{opt}} = \underset{e}{\text{Max (Min)}} \int \Big\{ \underset{u}{\text{Max (Min)}} \cdot \Big[\int J(e, z, u, x) \cdot p(x/z, e) dx \Big] \Big\} p(z/e) dz. \quad (58)^{8}$$

Thus, the main differences between the estimation problem and the general decision problem are as follows:

1) In the estimation problem there is no choice of experiment. One always makes the same type of measurement z given by $g(x, v)$. To generalize the estimation problem, one can specify,

$$z_e = g_e(x, v); \ \{e\} = 1, 2, \cdots = \text{possible sets of measurements} \quad (59)$$

and then require that

$$\hat{x} = \underset{e}{\text{Opt}} \{ (\hat{x})_e \ e = 1, 2, \cdots \}.$$

2) In the general decision problem, the function $z = g(x, v)$ is implicit in $p(z/x, e)$. Hence step 2a) and 2b) for the estimation solution is not required. This is often a tremendous simplification.
3) In the estimation problem the criterion function J is always a simple function of x only. There is, furthermore, no choice of action (one has to make an estimate by definition). On the other hand, the general decision problem is more analogous to a combined estimation and control problem where one has a further choice of action after determining $p(x/z)$, and like a control problem, the criteria function is generally more complex.
4) It is, however, to be noted that the key step is the computation of $p(x/z)$ for both problems. The choice of action is determined only *after* the computation of $p(x/z)$. Thus, a general decision problem can be composed into two problems, namely, determination of $p(x/z)$ (estimation problem) and choice of action (control problem). In control-theoretic technology, this fact is called the Generalized Decomposition Axiom.

[7] For other equivalent forms, see Raiffa and Schlaifer [1].
[8] See [1].

As an example, consider the single stage Wiener-Kalman problem and the added requirement that,

$$J(e, z, u, x) = J(u, x) = E\|Bx + u\|^2 = \int \|Bx + u\|^2 p(x/z) dx \quad (60)$$

be a minimum. Expanding (60), one gets

$$J = E\|x\|^2_{B^TB} + 2u^T B\hat{x} + u^T u; \quad (61)$$

clearly,

$$u_{\text{opt}} = u(\hat{x}) \triangleq u(x(z)) \triangleq u(z) = -B\hat{x}(z), \quad (62)$$

which is one of the fundamental results of linear stochastic control. Thus, the control action u is only a function of the criterion J and the *a posteriori* density function $p(x/z)$. In fact, in this case only $\hat{x}$ of $p(x/z)$ is needed. We call $\hat{x}$ as the *minimal sufficient statistic for the control problem.*

In the more general multistage case, the decomposition property clearly still holds, the only difference being that $p(x_{k+1}/Z_{k+1})$ is now dependent on u_k. However, this dependence is entirely *deterministic* since, in a given situation, one always knows what u_k's are. In fact, in the Wiener-Kalman control problem, it is known that u_k is a linear function of $\hat{x}_k$ only.

Conclusion

In the above sections, the problem of estimation from the Bayesian viewpoint is discussed. It is the author's thesis that this approach offers a unifying methodology, at least conceptually, to the general problems of estimation and control.

The *a posteriori* conditional density function $p(x/z)$ is seen to be the key to the solution of the general problem. Difficulties associated with the solution of the general problem now appear more specifically as difficulties in steps leading to the computation of $p(x/z)$. From the above discussions, it is relatively obvious that these difficulties are

1) Computation of $p(z/x)$. In both the single-stage or multistage case, this problem is complicated by the nonlinear functional relationships between z and x. Except in the case when z and x are linearly related or when z and x are scalars, very little can be done in general, analytically or experimentally. As was mentioned earlier, this difficulty does not appear in the usual decision problem, since there it is assumed that $p(z/x)$ is given as part of the problem.
2) Requirement that $p(x/z)$ be in analytical form.

This is an obvious requirement if we intend to use the solution in real-time applications.

3) Requirements that $p(x)$, $p(z)$, $p(x/z)$ be conjugate distributions [1]. This is simply the requirement that $p(x)$ and $p(x/z)$ be density functions from the same family. Note that all the examples discussed in this paper possess this desirable property. This is precisely the reason that multistage computation can be done efficiently. This imposed a further restriction on the functions g, f and h.

The difficulties (1–3) listed above are formidable ones. It is not likely that they can be easily circumvented except for special classes of problems such as those discussed. However, it is worthwhile first to pinpoint these difficulties. Research toward their solution can then be effectively initiated. Finally, it is felt that the Bayesian approach offers a unified and intuitive viewpoint particularly adaptable to handling modern-day control problems where the *State* and the *Markov* assumptions play a fundamental role.

References

[1] H. Raiffa and R. Schlaifer, "Applied Statistical Decision Theory" Harvard University Press, Cambridge, Mass., chs. 1–3; 1961.

[2] A. E. Bryson and M. Frazler, "Smoothing in linear and nonlinear systems," *Proc. of Optimal System Synthesis Symp.*, Wright Field, Ohio; 1962.

[3] Y. C. Ho, "On stochastic approximation and optimal filtering method," *J. Math. Analysis and Applications*, vol. 6, pp. 152–154; February, 1963.

[4] R. E. Kalman, "New Methods and Results in Linear Prediction and Filtering Theory," Research Institute for Advanced Studies, Baltimore, Md., No. 61-1.

An Innovations Approach to Least-Squares Estimation Part I: Linear Filtering in Additive White Noise

THOMAS KAILATH, MEMBER, IEEE

Abstract—**The innovations approach to linear least-squares approximation problems is first to "whiten" the observed data by a causal and invertible operation, and then to treat the resulting simpler white-noise observations problem. This technique was successfully used by Bode and Shannon to obtain a simple derivation of the classical Wiener filtering problem for stationary processes over a semi-infinite interval. Here we shall extend the technique to handle nonstationary continuous-time processes over finite intervals. In Part I we shall apply this method to obtain a simple derivation of the Kalman-Bucy recursive filtering formulas (for both continuous-time and discrete-time processes) and also some minor generalizations thereof.**

I. Introduction

IN THE EARLY 1940's, Kolmogorov [1] and Wiener [2] first discussed problems of linear least-squares estimation for stochastic processes, but by entirely different methods. Kolmogorov [1] studied only discrete-time problems and he solved them by using a simple representation of such processes that was suggested in a 1938 doctoral dissertation by Wold [3]. This representation, which is obtained by a recursive orthonormalization procedure, is known as the Wold decomposition. [The original papers of Kolmogorov and Wold are quite readable, but a more accessible and very readable reference is the monograph by Whittle [4] (especially sec. 3.7).]

On the other hand, Wiener [2] took an almost completely nonprobabilistic approach. He mainly studied continuous-time problems and reduced them to the problem of solving a certain integral equation, the so-called Wiener–Hopf equation, that Wiener and Hopf had solved in 1931 [5] by using some of Wiener's results on harmonic analysis. Though Wiener undertook this work in response to an engineering problem (the design of antiaircraft fire-control systems), his solution was beyond the reach of his engineering colleagues, and his yellow-bound report soon came to be labeled the "Yellow Peril."

In 1950, Bode and Shannon [6] published a different derivation of Wiener's results that was, quite successfully, intended to make them more accessible to engineers. This paper was based on ideas in a classified 1944 report by Blackman, Bode, and Shannon [7]. The same approach was independently discovered by Zadeh (cf. footnote 3 in Zadeh and Ragazzini [8]).

Manuscript received January 31, 1968. This work was supported by the Applied Mathematics Division of the Air Force Office of Scientific Research under Contract AF 49(638)1517, and by the Joint Services Electronics Program at Stanford University, Stanford, Calif., under Contract Nonr 225(83).

The author is with Stanford University, Stanford, Calif.

However, it is somewhat ironic that these more engineering approaches were found later to be just the continuous-time versions of the original Wold–Kolmogorov technique, which had been developed in a purely mathematical context.

The results in [1]–[7] were all obtained for stationary processes with infinite or semi-infinite observation intervals. The paper of Zadeh and Ragazzini [8] was the first significant attempt to extend the theory. Over the last two decades, various extensions and generalizations have been obtained and many of these have been documented in textbooks, as for example, those of Doob [9], Laning and Battin [10], Pugachev [11], Lee [12], Yaglom [13], Whittle [3], Deutsch [14], Liebelt [15], Balakrishnan [16], Bryson and Ho [17], and others.

In recent years, applications in orbital mechanics and spacecraft tracking have spurred interest in recursive estimation for nonstationary processes over finite-time intervals. Such algorithms were used by Gauss in his numerical calculations of the orbit of the asteroid Ceres, but the modern interest in them is due to Swerling [18] and especially Kalman [19], [20], and Bucy [21], [22]. The great interest in recursive algorithms because of their obvious computational advantages has stimulated a great number of papers on them, providing alternate forms and derivations showing their relationship to more classical parameter estimation techniques (see, for example, the discussions and references in Deutsch [14] and Liebelt [15]). Nevertheless, it seems to us that the original derivations of Kalman [19] and Kalman and Bucy [22] still provide the most insight.

In order to obtain recursive solutions, Kalman and Bucy had to confine themselves to a special class of processes, viz., those that could be generated by passing white noise through a (possibly time-variant) "lumped" linear dynamical system, i.e., a system composed of a finite number of (possibly time-variant) R, L, C elements. (Such processes are sometimes called projections of wide-sense Markov processes, but we shall in the rest of this paper call them "lumped" processes.) They also assumed complete knowledge of this sytsem, thus sidestepping the difficult problem of spectral factorization that had been a stumbling block to the extension of Wiener's classic solution (for semi-infinite observations on a stationary process) to more general situations. In his first paper in 1959, Kalman [19] treated discrete-time processes and obtained a recursive solution by a technique that was essentially the same as Kolmogorov's. In a later paper [20], he extended these results

Reprinted from *IEEE Trans. Automat. Contr.*, vol. AC-13, pp. 646–655, Dec. 1968.

to the continuous-time case by the use of a particular limiting technique. This technique, though useful, is somewhat tedious to carry out rigorously. A careful discussion has been given by Wonham [23]. In [22], Kalman and Bucy attacked the continuous-time problem directly. However, they did not use the Wold-Kolmogorov approach because the direct continuous-time analog of Kalman's discrete-time procedure in [19] was hard to see. They therefore returned to the Wiener–Hopf integral equation and showed that (under certain assumptions on the signal and noise processes) the solution to this equation could be expressed in terms of the solution to a nonlinear Riccati differential equation. It is also worth noting that Siegert [24] had carried out essentially the same steps in a different (but mathematically isomorphic) problem.

The chief purpose of Part I is to give a derivation of the Kalman–Bucy results by the Wold–Kolmogorov method, which, for reasons that will be clear later, we shall call the innovations method. Not only does this close a gap in the preceding circle of ideas, but the insight it provides into the proof has also suggested some new results. These include some slight generalizations in the types of processes for which recursive estimation formulas can be obtained, and a very simple and general solution of the so-called smoothing (or interpolation or noncausal filtering) problem. The smoothing problem is one that has been somewhat difficult to solve by the original techniques of Kalman and Bucy, and the solutions that have been obtained are in a somewhat complicated form (see the discussions in Part II [25][1]). Our technique also enables a completely parallel method of attack for the discrete- and continuous-time problems. A new approach to linear estimation with additive colored (nonwhite) noise also follows from the present ideas (Geesey and Kailath [26]).

More strikingly, the innovations technique can also be extended to a large class of nonlinear least-squares problems, viz., those where the observation process is the sum of a non-Gaussian process and additive white Gaussian noise (cf. Kailath and Frost [27] and Frost [28]). The ideas of the present paper have also yielded some general results on the detection of general non-Gaussian signals in additive Gaussian noise (Kailath [29], discrimination between two general Gaussian processes (Kailath and Geesey [30]), and also in certain modeling problems (Kailath and Geesey [31]).

Finally, we should say a word about the level of rigor in the present work. It is difficult to work directly with white noise in a completely satisfactory and rigorous manner—one has usually, especially in the nonlinear case, to work with the integrated white noise. However, in our opinion, the key ideas can always be presented, quite simply, in the white-noise formulation. Then, after some familiarity with the appropriate mathematics has been gained, one can translate the white-noise formulation into the more rigorous (stochastic differential) framework. We shall do this in later papers. The more informal presentation here will, we hope, bring the basic ideas to a wider audience.

[1] This issue, page 655.

II. The Innovations Approach to Linear Least-Squares Estimation

The innovations approach is first to convert the observed process to a white-noise process, to be called the innovations process, by means of a *causal and causally invertible* linear transformation. The point is that the estimation problem is very easy to solve with white-noise observations. The solution to this simplified problem can then be reexpressed in terms of the original observations by means of the inverse of the original "whitening" filter.

This program, used by Bode and Shannon [6] for the stationary process problem with semi-infinite observation time, will now be carried out when the observations are made over a finite-time interval on a continuous-time (possibly nonstationary) stochastic process. Several initial sets of assumptions and several corresponding classes of problems, of varying degrees of generality, can be formulated. For simplicity, however, we shall deal largely with the following additive white-noise problem.

The given observation is a record of the form

$$y(t) = z(t) + v(t), \qquad t \in [a, b) \tag{1}$$

where

$v(\cdot)$ = a sample function of zero-mean white noise with covariance function[2]

$$\overline{v(t)v'(s)} = R(t)\delta(t-s), \qquad R(t) > 0,$$

$z(\cdot)$ = a sample function of a zero-mean "signal" process that has finite variance

$$\mathrm{tr}\,\overline{[z(t)z'(t)]} < \infty, \qquad t \in [a, b)$$

$[a, b]$ = a finite interval[3] on the real line.

We also assume that the "future" noise $v(\cdot)$ is uncorrelated from the "past" signal $z(\cdot)$, i.e.,

$$\overline{v(t)z'(s)} = 0, \qquad a \le s < t < b. \tag{2}$$

We shall be interested in the linear least-squares estimate of a related process $x(t)$. Let

$\hat{x}(t \mid b)$ = a linear function of all the data $\{y(s), a \le s < b\}$ that minimizes the mean-square error $\mathrm{tr}[z(t) - \hat{z}(t \mid b)][z(t) - \hat{z}(t \mid b)]'$. (3)

The corresponding instantaneous estimation error will be written

$$\tilde{z}(t \mid b) = z(t) - \hat{z}(t \mid b), \; \tilde{z}(t \mid t) = z(t) - \hat{z}(t \mid t). \tag{4}$$

[2] Bars will be used to denote expectations.

[3] The case of an infinite interval requires certain additional assumptions on the signal process such as stationarity, observability of models generating it, etc. Some more specific comments on this point will be made later [after (35)].

When $b=t$, the estimate is usually called the *filtered* estimate, when $b>t$ it is usually called the *smoothed* estimate, and when $b<t$ it is called the *predicted* estimate.

The major tools for the calculation of these estimates will be the following two theorems.

Theorem 1—The Projection Theorem: The best estimate $\hat{z}(t|b)$ is unique and satisfies the conditions

$$\tilde{z}(t \mid b) \triangleq z(t) - \hat{z}(t \mid b) \perp y(s), \qquad a \leq s < b \quad (5)$$

where

$$u \perp v \quad \text{means that} \quad \overline{uv'} = 0. \quad (6)$$

In words, the instantaneous error is uncorrelated with the observations.

Proof: This theorem, which was used by Kolmogorov [1], is by now fairly well known to engineers and is used in several of the textbooks cited earlier. A brief discussion of the relevant geometric picture is given in Appendix I.

Theorem 2—The Innovations Theorem: The process $\boldsymbol{\nu}(\cdot)$ defined by

$$\boldsymbol{\nu}(t) = y(t) - \hat{z}(t \mid t) = \tilde{z}(t \mid t) + v(t), \qquad a \leq t < b, \quad (7)$$

and to be called the "innovation process" of $y(\cdot)$, is a *white-noise* process with the *same covariance* as $v(\cdot)$, i.e.,

$$\overline{\boldsymbol{\nu}(t)\boldsymbol{\nu}'(s)} = \overline{v(t)v'(s)}, \qquad a \leq t, \quad s < b. \quad (8)$$

Furthermore, $y(\cdot)$ and $\boldsymbol{\nu}(\cdot)$ can be obtained from the other by causal (nonanticipative) linear operations. Therefore, $y(\cdot)$ and $\boldsymbol{\nu}(\cdot)$ are "equivalent" (i.e., they contain the same statistical information) as far as linear operations are concerned.

Proof: The proof will be deferred to Appendix II; however, a few remarks on the theorem and its significance are appropriate here.

Remark 1: The quantity $\boldsymbol{\nu}(t)=y(t)-\hat{z}(t|t)=y(t)-\hat{y}(t|t-)$ may be regarded as defining the "new information" brought by the current observation $y(t)$, being given all the past observations $y(t)$, and the old information deduced therefrom. Therefore, the name "innovation process of $y(\cdot)$" came into being. This term was first used for such processes by Wiener and has since gained wide currency. (A significant generalization, due to Frost [28] and to Kailath [29], of this theorem is that when the white noise $v(\cdot)$ is also Gaussian, but the signal $z(\cdot)$ is non-Gaussian, the innovation process $\boldsymbol{\nu}(\cdot)$ is not only white with the same covariance as $v(\cdot)$, but it is also Gaussian. Applications of this surprising result are given in [27]–[29].)

Remark 2: The fact that $\boldsymbol{\nu}(\cdot)$ is white has been noted before in the special case of lumped signal processes. In this case, the result was probably first noticed by several people (cf. [23], [32]–[34], and unpublished notes of the author and others). However, all their arguments rely, to varying degrees, on the explicit Kalman–Bucy formulas for $\hat{z}(t|t)$. Here we first obtain the result more generally [and with less computation, since we rely only on the projection properties of $\hat{z}(t|t)$], and then use it to obtain the Kalman–Bucy formulas. We note also that the equivalence of $y(\cdot)$ and $\boldsymbol{\nu}(\cdot)$ does not seem to have been explicitly pointed out before, even though, assuming knowledge of the Kalman–Bucy formulas, a proof is immediate (cf. Appendix II-D).

Remark 3: One reason the fact that $\boldsymbol{\nu}(\cdot)$ is white (for lumped processes) may have been known for a long time is that in the discrete-time solution of Kalman [19], $\boldsymbol{\nu}(\cdot)$ is shown to be white (in discrete time) with, however, a different covariance from that of the original noise. The exact formula will be given later (39).

III. Some Applications

We turn now to some applications of these two theorems. First, we present a new derivation of the Kalman–Bucy formulas for filtering of lumped signal processes in white noise. This derivation shows clearly the step at which restriction to such processes is essential to get a recursive solution, and this insight easily yields several (slight) generalizations of the Kalman–Bucy results, including some recent ones due to Kwakernaak [35], Falb [36], Balakrishnan and Lions [37] and Chang [38]. The same techniques apply to discrete-time problems as well. Our new method of proof yields, very simply, a general formula for the smoothed estimate (Part II) [26] and also, more importantly, can be generalized to the nonlinear case (Part III) [27].

A. The Kalman–Bucy Formulas for Recursive Filtering and Prediction

The Kalman–Bucy results are for lumped processes; however, we shall not begin with this assumption, but shall try to see how far we can go without any special assumptions.

We are given $\{y(s)=z(s)+v(s),\ a\leq s<t\}$ and wish to calculate the linear least-squares estimate $\hat{x}(t|t)$ of a related random variable $x(t)$.

The first step is to obtain the innovations, which, by Theorem 2, are given by

$$\boldsymbol{\nu}(t) = y(t) - \hat{z}(t/t), \qquad \overline{\boldsymbol{\nu}(t)\boldsymbol{\nu}(s)} = R(t)\delta(t-s). \quad (9)$$

Because the innovations $\boldsymbol{\nu}(\cdot)$ are equivalent to the original observations $y(\cdot)$, we can express $\hat{x}(t|t)$ as

$$\hat{x}(t \mid t) = \int_a^t g(t, s)\boldsymbol{\nu}(s)ds \quad (10)$$

where the linear filter $g(t, \cdot)$ is to be chosen so that [again using the equivalence of $y(\cdot)$ and $\boldsymbol{\nu}(\cdot)$]

$$x(t) - \hat{x}(t \mid t) \perp \boldsymbol{\nu}(s), \qquad a \leq s \leq t. \quad (11)$$

Putting together (9)–(11), we obtain

$$\overline{x(t)\boldsymbol{\nu}'(s)} = \int_a^t g(t, \sigma)\overline{\boldsymbol{\nu}(\sigma)\boldsymbol{\nu}'(s)}d\sigma \quad (12)$$

$$= g(t, s)R(s), \qquad a \leq s \leq t. \quad (13)$$

It is the last step that justifies the use of the innovation process $\nu(\cdot)$. In (10)–(12), we could equally well have used the original observations $y(\cdot)$, but now (12), instead of being trivial, becomes the Wiener–Hopf integral equation, which cannot be solved by inspection. Returning to (13), we can now write

$$\hat{x}(t \mid t) = \int_a^t \overline{x(t)\nu'(s)} R^{-1}(s)\nu(s)ds. \tag{14}$$

This is the general formula for the *linear* least-squares estimate of $x(t)$ from a white-noise process. (We may point out, in anticipation, that the *nonlinear* (nl) least-squares estimate is given by the remarkably similar formula

$$\hat{x}_{nl}(t \mid t) = \int_a^t \widehat{x(t)\nu'(s)} R^{-1}(s)\nu(s)ds \tag{15}$$

where

$$\widehat{x(t)\nu'(s)} = E[x(t)\nu'(s) \mid y(\tau), a \le \tau < s]. \tag{16}$$

This result will be derived in Part III [27].

So far we have made no special assumptions on $x(t)$. Kalman and Bucy [22] assumed that $x(t)$ satisfies the differential equation

$$\dot{x}(t) = F(t)x(t) + u(t), \qquad t \ge a, \qquad x(a) = x_a \tag{17}$$

where $u(\cdot)$ is white noise with intensity matrix $Q(\cdot)$ and uncorrelated with the observation white noise $v(\cdot)$, i.e.,

$$\overline{u(t)u'(s)} = Q(t)\delta(t-s), \qquad \overline{u(t)v'(s)} \equiv 0 \tag{18}$$[4]

and the initial value x_a is a zero-mean random variable with variance P_a and uncorrelated with $u(\cdot)$, i.e.,

$$\bar{x}_a = 0, \quad \overline{x_a x_a'} = P_a, \quad \overline{u(s)x_a'} \equiv 0, \quad a \le s < b. \tag{19}$$

To exploit this structure of $x(t)$, we can differentiate the general estimate formula (14) to obtain

$$\dot{\hat{x}}(t \mid t) = \overline{x(t)\nu'(t)}R^{-1}(t)\nu(t) + \left[\int_a^t \frac{d}{dt}\overline{x(t)\nu'(s)}R^{-1}(s)\nu(s)ds\right] \tag{20}$$

$$= \overline{x(t)\nu'(t)}R^{-1}(t)\nu(t) + \left[F(t)\int_a^t \overline{x(t)\nu'(s)}R^{-1}(s)\nu(s)ds + \int_a^t \overline{u(t)\nu'(s)}R^{-1}(s)\nu(s)ds\right]. \tag{21}$$

Now the second term in (21) is equal to $F(t)\hat{x}(t \mid t)$ [cf. (14)], and thus but for the last term, (21) would be a differential equation for $\hat{x}(t \mid t)$.

However, this last term will be zero if we assume that the white noise $u(\cdot)$ that generates the signal process $x(\cdot)$ is uncorrelated with the past observations $y(\cdot)$ [and therefore with the equivalent observations $\nu(\cdot)$]. That is, with the further assumption

$$\overline{u(t)y'(s)} \equiv 0, \qquad s < t \tag{22}$$

we shall have

$$\dot{\hat{x}}(t \mid t) = F(t)\hat{x}(t \mid t) + K(t)\nu(t), \; \nu(t) = y(t) - \hat{z}(t \mid t) \tag{23}$$

where we have defined

$$K(t) \triangleq \overline{x(t)\nu'(t)}R^{-1}(t). \tag{24}$$

A block diagram for (23) is shown in Fig. 1(a) where the box yielding $\hat{z}(t \mid t)$ will have a detailed structure similar to that for $\hat{x}(t \mid t)$ if we assume that the $z(t)$ obey a differential relation similar to (17) for $x(t)$. We can be somewhat more explicit about $\hat{z}(t \mid t)$ [and about the $K(t)$ of (24)] if we assume some specific functional relationship between $z(\cdot)$ and (past)[5] $x(\cdot)$. The simplest is, of course, the linear relationship, used by Kalman and Bucy [22],

$$z(t) = H(t)x(t) \tag{25}$$

which immediately yields (by linearity)

$$\hat{z}(t \mid t) = H(t)\hat{x}(t \mid t). \tag{26}$$

This is very useful because now [the innovations $\nu(t)$ can be obtained directly from $\hat{x}(t \mid t)$ and $\hat{y}(t)$] the estimate $\hat{x}(t \mid t)$ can be realized by the *feedback* structure of Fig. 1(b).

The (*gain*) function $K(t)$ can also be written in a simpler form under the assumption (25):

$$\begin{aligned} K(t) &= \overline{x(t)\nu'(t)}R^{-1}(t) \\ &= \overline{x(t)[\tilde{x}'(t \mid t)H'(t) + v'(t)]}R^{-1}(t) \\ &= \overline{[\hat{x}(t \mid t) + \tilde{x}(t \mid t)]\tilde{x}'(t \mid t)}H'(t)R^{-1}(t) + 0 \qquad (27) \\ &= 0 + \overline{\tilde{x}(t \mid t)\tilde{x}'(t \mid t)}H'(t)R^{-1}(t) \\ &= P(t, t)H'(t)R^{-1}(t), \qquad \text{say} \qquad (28) \end{aligned}$$[6]

where

$P(t, t)$ = the covariance function of the error in the estimate at time t.

It is easy to derive a differential equation for $P(t, t)$ by first noting from (17) and (23) that $\tilde{x}(t \mid t)$ obeys the differential equation

$$\dot{\tilde{x}}(t \mid t) = [F(t) - K(t)H(t)]\tilde{x}(t \mid t) - K(t)v(t) + u(t), \qquad \tilde{x}(a \mid a) = x_a. \tag{29}$$

Now applying a standard formula (Appendix I-B) we can show that $P(t, t)$ satisfies the (nonlinear) matrix Riccati equation

$$\dot{P}(t, t) = F(t)P(t, t) + P(t, t)F'(t) - K(t)R(t)K'(t) + Q(t), \qquad P(a, a) = P_a. \tag{30}$$

[4] The assumption $\overline{u(t)v'(s)} \equiv 0$ can be relaxed to $\overline{u(t)v'(s)} = C(t)\delta(t-s)$; cf. (31) and (32).

[5] If $z(\cdot)$ depended on future $x(\cdot)$, we could not satisfy the condition (22).

[6] Note that (28) is true for general $x(\cdot)$, not only those with the differential representation (17).

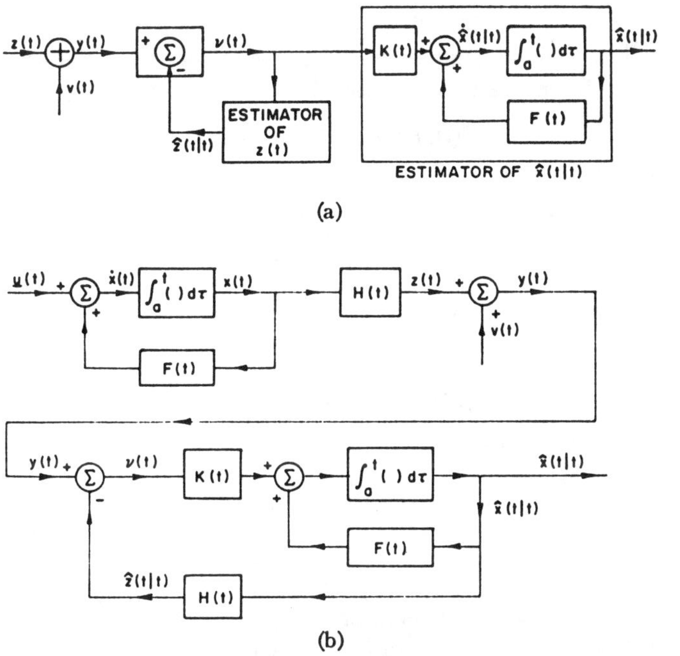

Fig. 1. (a) Filtered estimate of $x(t)$ from a related process $y(\tau)=z(\tau)+v(\tau)$, $a\leq\tau\leq t$. (b) The Kalman–Bucy filter; note that feedback of $\hat{x}(t|t)$ can be used to obtain $\nu(t)$ when $z(t)=H(t)x(t)$.

We now have obtained in formulas (23), (26), (28), and (30) the basic Kalman–Bucy formulas [for the problem defined by (1), (25), and (17)–(19)]. Our derivation is more direct than that of the original and reveals clearly the roles of the various assumptions in the Kalman-Bucy model. In Part III, we shall see the role that the corresponding assumptions play in the nonlinear problem. Our present proof also indicates some points at which the above arguments can be generalized. However, before doing this let us make a few supplementary remarks.

Correlated $u(\cdot)$ *and* $v(\cdot)$: We can, without violating the basic constraint (2) that $\overline{z(t)v'(s)}=0$, $s>t$, generalize the uncorrelatedness condition in (18) to

$$\overline{u(t)v'(s)} = C(t)\delta(t-s). \tag{31}$$

This will require minor changes in the above derivations,[7] which we shall leave for the reader's amusement. We shall only point out that finally the only change in the filter formulas [(23), (26), (28), (30)] will be that the gain function of (28) must be replaced by

$$K(t) = [P(t,t)'H'(t) + C(t)]R^{-1}(t). \tag{32}$$

The Prediction Problem: Suppose we are to estimate $x(t+\Delta)$, $\Delta>0$, given observations $\nu(\cdot)$ up to t. Then, by the innovations technique, we readily find

$$\hat{x}(t+\Delta\,|\,t) = \int_a^t \overline{x(t+\Delta)\nu'(s)}R^{-1}(s)\nu(s)ds. \tag{33}$$

[7] Notably that in (21) and (27) the terms that are zero will now be $(1/2)C(t)R^{-1}(t)$. The 1/2 arises from taking $\int_a^t\delta(t-s)\,ds=1/2$.

If $x(\cdot)$ is a lumped process described by the model (17)–(19), then it is easy to see that

$$\hat{x}(t+\Delta\,|\,t) = \Psi(t+\Delta,t)\hat{x}(t\,|\,t) \tag{34}$$

where $\Psi(t,s)$ is the fundamental (or state-transition) matrix of the differential equation (17) of the process $x(\cdot)$, i.e., $\Psi(t,s)$ is the solution of

$$\frac{d}{dt}\Psi(t,s) = F(t)\Psi(t,s), \qquad \Psi(s,s) = I. \tag{35}$$

The Steady-State Equation: In the preceding discussion, we have restricted the interval (a,t) to be finite. When the various matrices $F(\cdot)$, $H(\cdot)$, $Q(\cdot)$, and $R(\cdot)$, are time invariant, it is of interest to study the limiting behavior of the filter as the initial point a tends to $-\infty$. By a careful examination of the Riccati equation (30), Kalman and Bucy [22] have shown that when the model (17), (25) satisfies certain assumptions (stability, controllability, and observability, etc.), we can obtain a well-defined limiting solution by setting $\dot{P}=0$ in (30) and using the non-negative[8] solution of the resulting algebraic equations in the filter formulas (23) and (27). When the process $x(\cdot)$ has a rational spectral density, the above conditions are always met and the Kalman–Bucy solution reduces to the classical solution of Wiener. (The explicit equivalence has been shown by Leake [39].)

B. The Discrete-Time Problem

For discrete-time observations we will have similar results, with one rather trivial modification: the innovation process in the discrete-time case will have a different variance from that of the observation noise. Thus, let

$$\begin{aligned} y(k) &= z(k) + v(k), & k &= 0,1,2,\cdots, \\ \overline{v(k)} &= 0, & \overline{v(k)v'(l)} &= R(k)\delta_{kl} \end{aligned} \tag{36}$$

with $\{z(k)\}$ a zero-mean finite-variance signal process. The innovation process will be defined by

$$\nu(k) \triangleq y(k) - \hat{z}(k\,|\,k-1) \tag{37}$$

where

$$\begin{aligned} \hat{z}(k\,|\,k-1) = {} & \text{the linear least-squares estimate of} \\ & z(k) \text{ given } \{y(l),\ 0\leq l\leq k-1)\}. \end{aligned} \tag{38}$$

Then it is easy to calculate (cf. Appendix II) that

$$\overline{\nu(k)} = 0, \qquad \overline{\nu(k)\nu'(l)} = [P_z(k) + R(k)]\delta_{kl} \tag{39}$$

where

$$\begin{aligned} P_z(k) = {} & \text{covariance matrix of the error in the} \\ & \text{estimate } \tilde{z}(k\,|\,k-1) \\ = {} & \overline{[z(k)-\hat{z}(k\,|\,k-1)][z(k)-\hat{z}(k\,|\,k-1)]'}. \end{aligned} \tag{40}$$

Therefore, the innovation process is still (discrete-time)

[8] There are several solutions that are not non-negative definite.

white, but with a different variance. The estimation solution now proceeds essentially as in the continuous-time case; we shall rapidly outline the steps for a process $z(\cdot)$ of the form

$$\begin{aligned} z(k) &= H(k)x(k), \\ x(k+1) &= \Phi(k+1, k)x(k) + u(k), \\ \overline{u(k)u'(l)} &= Q(k)\delta_{kl}, \quad \overline{u(k)v'(l)} = C(k)\delta_{kl}. \end{aligned} \tag{41}$$

By the projection theorem, and assuming $[P_z(\cdot)+R(\cdot)]^{-1}$ exists,[9] we readily obtain the expression (42) for $\hat{x}(k+1|k)$ in terms of the $v(l)$, $l \le k$, which we can rearrange as

$$\begin{aligned} \hat{x}(k+1 \mid k) & \\ &= \sum_0^k \overline{x(k+1)v'(l)}[P_z(l) + R(l)]^{-1}v(l) \qquad (42) \\ &= \sum_0^{k-1} \overline{x(k+1)v'(l)}[P_z(l) + R(l)]^{-1}v(l) \\ &\quad + \overline{x(k+1)v'(k)}[P_z(k) + R(k)]^{-1}v(k) \qquad (43) \\ &= \Phi(k+1, k)\hat{x}(k \mid k-1) + K(k)v(k), \quad \text{say} \qquad (44) \end{aligned}$$

where we have defined

$$K(k) \triangleq \overline{x(k+1)v'(k)}[P_z(k) + R(k)]^{-1}. \tag{45}$$

Now we note that

$$\begin{aligned} &\overline{x(k)v'(k)} \\ &= \overline{[\Phi(k+1, k)x(k)+u(k)][\tilde{x}'(k \mid k-1)H'(k)+v'(k)]} \qquad (46) \\ &= \Phi(k+1, k)\overline{x(k)\tilde{x}'(k \mid k-1)}H'(k) + C(k) \\ &= \Phi(k+1, k)P(k)H'(k) + C(k) \qquad (47) \end{aligned}$$

where

$$P(k) \triangleq \overline{\tilde{x}(k \mid k-1)\tilde{x}'(k \mid k-1)}. \tag{48}$$

Therefore, using

$$P_z(k) \triangleq \overline{\tilde{z}(k \mid k-1)\tilde{z}'(k \mid k-1)} = H(k)P(k)H'(k) \tag{49}$$

we can write $K(k)$ as

$$\begin{aligned} K(k) &= \Phi(k+1, k)[P(k)H'(k) + C(k)] \\ &\quad \cdot [H(k)P(k)H'(k) + R(k)]^{-1}. \end{aligned} \tag{50}$$

Finally, with patience, we can derive a recursion relation for $P(k)$ which we quote without proof:

$$\begin{aligned} P(k+1) &= \Phi(k+1, k)A(k)\Phi'(k+1, k) + Q(k), \\ A(k) &= P(k) - K(k)[P_z(k) + R(k)]K'(k). \end{aligned} \tag{51}$$

Equations (44) and (51) define the discrete-time Kalman filter, first derived in a slightly less direct way (but still using the innovations) in Kalman [19]. Our method here is exactly parallel to the one we used in the continuous-time case.

[9] If not, we use the Moore-Penrose pseudo-inverse, but we shall not pursue this refinement here.

C. Some Generalizations

The crucial step in our derivation of the Kalman–Bucy formulas was the use of the assumptions

$$\dot{x}(t) = F(t)x(t) + u(t), \qquad \overline{u(t)v'(s)} \equiv 0$$

to write

$$\begin{aligned} \int_a^t \overline{\dot{x}(t)v'(s)}v(s)ds &= F(t)\int_a^t \overline{x(t)v'(s)}v(s)ds \\ &= F(t)\hat{x}(t \mid t). \end{aligned}$$

However, suppose we had

$$\dot{x}(t) = F(t)x(t-1) + u(t), \qquad \overline{u(t)y'(s)} \equiv 0. \tag{52}$$

Then we shall have

$$\begin{aligned} \int_a^t \overline{\dot{x}(t)v'(s)}v(s)ds &= F(t)\int_a^t \overline{x(t-1)v'(s)}v(s)ds \\ &= F(t)\hat{x}(t-1 \mid t). \end{aligned} \tag{53}$$

Kwakernaak [35] was apparently the first to point out this result. More generally, suppose

$$\begin{aligned} \dot{x}(t) &= \mathfrak{F} \circ x(\cdot) + u(t), \\ y(t) &= \mathfrak{H} \circ x(\cdot) + v(t), \qquad \overline{u(t)v'(t)} \equiv 0 \end{aligned} \tag{54}$$

where $\mathfrak{F} \circ x(\cdot)$ and $\mathfrak{H} \circ x(\cdot)$ denote some linear operation on the "past" values $\{x(s), a \le s < t\}$ of the signal process. Then

$$\begin{aligned} \int_a^t \overline{\dot{x}(t)v'(s)}v(s)ds &= \mathfrak{F} \circ \int_a^t \overline{x(\cdot)v'(s)}v(s)ds \\ &= \mathfrak{F} \circ \hat{x}(\cdot \mid t) \end{aligned} \tag{55}$$

and the obvious analogs of the Kalman–Bucy formulas (23)–(30) are again easily obtained (of course, suitable attention has to be paid to the proper topologies, etc.). Some problems of this type have been noted by Balakrishnan and Lions [36] and Falb [35] who use essentially an operator–theoretic analog of the Kalman–Bucy derivation. They given some specific examples with $\mathfrak{F}$ being a partial differential operator. For a different illustration, we note that $\mathfrak{H}$ may be a random sampling operation, a case that was recently studied in a less direct manner by Chang [37]. General representations of the form (39) often arise in describing stochastic process by evolution equations in abstract spaces and, in fact, some nonlinear processes may be made linear by such representations. We shall not explore this point further in the present elementary paper.

However, it may be of some value to point out that if $x(\cdot)$ obeys a nonlinear equation

$$\dot{x}(t) = f(x(s), s \le t, t) + u(t) \tag{56}$$

then the (linear) estimate $\hat{x}(t|t)$ obeys the equation

$$\dot{\hat{x}}(t \mid t) = \widehat{f(x(s), s \le t, t)} + \overline{x(t)v'(t)}v(t) \tag{57}$$

where $\widehat{f(x(s), s \le t, t)}$ is the best linear estimate of $f(x(\cdot), t)$ given $y(\tau)$, $a \le \tau \le t$. Such a problem was partially discussed by Chang [38].

Finally, we should make a brief comment about problems in which the additive observation noise is nonwhite. One solution is to apply a transformation that will whiten this noise and then use the Kalman–Bucy formulas. This method has been used by Bryson and Johansen [40]. However, a more powerful method is to whiten the whole observation process, the sum of the signal and the nonwhite noise; in other words, to obtain the innovations directly. This method is discussed in Geesey and Kailath [26]. It may be noted that the case of colored (finite-variance) noise plus white noise can be immediately treated by an obvious extension of Theorem 2—the observations can be whitened by subtracting out the estimates of the signal *and* the colored noise.

IV. Concluding Remarks

The main point of the innovations approach to statistical problems is that once we understand the basic probabilistic structure of the processes involved, many results can be obtained quite directly without resort to often more sophisticated (and analytical rather than probabilistic) tools like Wiener–Hopf techniques, Karhunen–Loève expansions, function space integrals, etc. In this paper we have illustrated this point for a class of nonstationary filtering problems.

In [25]–[31] applications are given for linear smoothing problems, nonlinear filtering and smoothing, covariance factorization, and detection problems.

Appendix I

A. The Projection Theorem

Formal proofs of the projection theorem are given in many textbooks. Here we shall make a few informal remarks that may aid in the understanding and application of the result. The projection theorem is probably quite familiar for linear (Hilbert) spaces of time function with inner product

$$\int_T u(t)v(t)dt \text{ or } \int_T u(t)v(t)p(t)dt, \text{ where } p(t) \geq 0. \quad (58)$$

Thus, the linear least-squares approximation to an unknown function $u(\cdot)$ in terms of a given function $v(\cdot)$ is obtained by projecting $u(\cdot)$ on $v(\cdot)$ with the given inner product (58). For our applications, we need to work with Hilbert spaces of random variables, these being values of a stochastic process $z(t)$, for different time instants $t \in [a, b]$, or linear combinations of such random variables. Now random variables are also functions not of t, but of a probability sample-space variable, say $\omega \in \Omega$. The inner product is (very heuristically) $\int_\Omega u(\omega)v(\omega)p(\omega)d\omega$ where $p(\omega)\, d\omega$ is a probability, or, as it is usually written, $\overline{uv}$. As long as we remember that ω, the probability variable, should replace time, all our intuitive notions of Hilbert function spaces (which are essentially generalizations of n-dimensional Euclidean space) carry over to random variables. The orthogonality relations of the projection theorem have a geometric setting in this space of random variables. In this context, there is often some initial confusion because the variable t is also present in the discussion of stochastic processes. However, it is essential to remember that in the Hilbert space of random variables, the elements are not functions of time but functions of ω; the variable t serves only to index some of the elements of the Hilbert space.

B. Covariance Relations for Lumped Processes

Let a random process $\mathbf{x}(t)$ be obtained as the solution of the differential equation

$$\dot{\mathbf{x}}(t) = F(t)\mathbf{x}(t) + \mathbf{u}(t), \quad \mathbf{x}(a) = \mathbf{x}_a, \quad t > a \quad (59)$$

where (the zero means are assumed for notational convenience)

$$\overline{\mathbf{u}(t)} = 0, \quad \overline{\mathbf{u}(t)\mathbf{u}'(s)} = Q(t)\delta(t-s),$$
$$\bar{\mathbf{x}}_a, \ \overline{\mathbf{u}(t)\mathbf{x}_a'} = 0, \quad t \geq a.$$

Then we can write

$$\mathbf{x}(t) = \Psi(t, a)\mathbf{x}(a) + \int_a^t \Psi(t, s)\mathbf{u}(s)ds$$

where $\Psi(t, s)$ is the state-transition matrix defined as the (unique) solution of the equation

$$\frac{d\Psi(t, s)}{dt} = F(t)\Psi(t, s), \quad \Psi(a, a) = I, \ a \leq s \leq t. \quad (60)^{10}$$

By direct computation, we obtain $\bar{\mathbf{x}}(t) \equiv 0$ and

$$\begin{aligned} R_x(t, t) &\triangleq \overline{[\mathbf{x}(t) - \bar{\mathbf{x}}(t)][\mathbf{x}(t) - \bar{\mathbf{x}}(t)]'} \\ &= \Psi(t, a)R_a\Psi'(t, a)' \\ &\quad + \int_a^t \Psi(t, s)Q(s)\Psi'(t, s)ds. \end{aligned} \quad (61)$$

Differentiating both sides of (61) with respect to t and using (60), we obtain

$$\frac{dR_x(t, t)}{dt} = F(t)R_x(t, t) + R_x(t, t)F'(t) + Q(t), \quad t \geq a \quad (62)$$
$$R_x(a, a) = R_a.$$

Furthermore, it follows by direct computation that

$$\begin{aligned} R_x(t, s) &\triangleq \overline{\mathbf{x}(t)\mathbf{x}'(s)} \\ &= \overline{\left[\Psi(t, s)\mathbf{x}(s) + \int_s^t \Psi(t, \sigma)G(\sigma)\mathbf{u}(\sigma)d\sigma\right]x'(s)} \quad (63) \\ &= \Psi(t, s)R_x(s, s) + 0 \qquad \text{for } t \geq s \\ &= R_x(t, t)\Psi'(s, t) \qquad \text{for } s \geq t \quad (64) \end{aligned}$$

where the last equation follows from the symmetry property $R_x(t, s) = R_x'(s, t)$. Equation (62), when applied to (29), yields the Riccati equation (30), as some

[10] When $F(\cdot)$ is time invariant, $\Psi(t,s) = e^{F(t-s)}$, $t \geq s$.

simple algebra will show. Equations (63) and (64) will be used for the smoothing problem in Part II. The above formulas are all well known.

Appendix II

The Innovation Process

If $y(\cdot)=z(\cdot)+v(\cdot)$, where $z(\cdot)$ is a second-order process and $v(\cdot)$ is white noise, we shall prove that the innovation process

$$\nu(t) = y(t) - \hat{z}(t \mid t), \qquad -\infty < a \leq t < b < \infty$$

is white with the same covariance as $v(\cdot)$, and that it is obtained from $y(\cdot)$ by a causal *invertible* linear operation. The first property follows easily by direct computation [and had been known for lumped process $z(\cdot)$]. The second property is more interesting and will be discussed first. (For simplicity, only the scalar case will be treated.)

A. The Relationship Between $y(\cdot)$ *and* $\nu(\cdot)$

Let $g_v(t, s)$ denote the optimum causal filter that operates on $\{y(s), s \leq t\}$ to give $\hat{z}(t\mid t)$, i.e.,

$$\hat{z}(t \mid t) = \int_a^t g_v(t, s) y(s) ds = \mathcal{G}_v y, \qquad \text{say} \tag{65}$$

where $\mathcal{G}_v$ denotes the integral operator with kernel $g_v(t, s)$.[11]

To make (65) well defined we need to assume that (cf. Doob [9], sec. 9.2)

$$\int_a^t g_v^2(t, s) ds < \infty \qquad \text{for every } t \in (a, b). \tag{66}$$

(If $g(t, \cdot)$ had delta functions in it, $\hat{z}(t\mid t)$ would have infinite variance.) From our assumption that $\int z^2(t)dt < \infty$, it can be shown that

$$\int_a^b \int_a^b g_v^2(t, s) dt ds < \infty, \tag{67}$$

a fact that will be useful presently. If we use I for the identity operator [the integral operator with kernel $\delta(t-s)$], then we can write, symbolically,

$$\nu = y - \hat{z} = y - \mathcal{G}_v y = (I - \mathcal{G}_v) y. \tag{68}$$

The problem, then, is to show that $(I-\mathcal{G}_v)$ is a causally invertible operator. The causality of $\mathcal{G}_v$ does the trick here because $\mathcal{G}_v$ is then what is called a Volterra kernel and it can be proved (see, e.g., Smithies [41], p. 34) that when $\mathcal{G}_v$ has a square-integrable kernel, then $(1-\mathcal{G}_v)^{-1}$ exists and is given by the Neumann (geometric) series

$$(1 - \mathcal{G}_v)^{-1} = 1 + \mathcal{G}_v + \mathcal{G}_v^2 + \mathcal{G}_v^3 + \cdots \tag{69}$$

where $\mathcal{G}_v^2 y = \mathcal{G}_v \mathcal{G}_v y$, and so on. The causality is obvious from (69).

In many applications, the signal process $z(\cdot)$ is continuous in the mean [which is equivalent to the continuity of the covariance function of $z(\cdot)$]. In this case, it can easily be shown that the kernel $g_v(t, s)$ is continuous in t and s (and, in this case, the arguments to establish (69) are even simpler (Riesz and Nagy [42], sec. 65).

[11] As an aside, we note that $\mathcal{G}_v$ can be regarded as an operator on L_2 (cf. Doob [9], sec. 9.2).

B. The Process $\nu(\cdot)$ *is White*

We shall establish by direct calculation that

$$\overline{\nu(t)\nu(s)} = \overline{v(t)v(s)} \quad \text{where} \quad \nu(t) = y(t) - \hat{z}(t \mid t).$$

First consider $t>s$. Then

$$\begin{aligned} \overline{\nu(t)\nu(s)} &= \overline{[\tilde{z}(t\mid t) + v(t)][\tilde{z}(s\mid s) + v(s)]} \\ &= \overline{v(t)v(s)} + \overline{v(t)\tilde{z}(s\mid s)} \\ &\quad + \overline{\tilde{z}(t\mid t)\tilde{z}(s\mid s)} + \overline{\tilde{z}(t\mid t)v(s)}. \end{aligned} \tag{70}$$

Now $\tilde{z}(s\mid s) = z(s) - \hat{z}(s\mid s)$ depends only on signal and noise up to time s. Since we have assumed that future noise is uncorrelated with past signal, the second term in (70) will be zero. Similarly, by the definition of $\tilde{z}(t\mid t)$, $\overline{\tilde{z}(t\mid t)\hat{z}(s\mid s)} = \overline{\tilde{z}(t\mid t) z(s)} - 0$ for $t>s$. Therefore, we can write (70) as

$$\begin{aligned} \overline{\nu(t)\nu(s)} &= \overline{v(t)v(s)} + \overline{\tilde{z}(t\mid t)z(s)} + \overline{\tilde{z}(t\mid t)v(s)} \\ &= \overline{v(t)v(s)} + \overline{\tilde{z}(t\mid t)[z(s) + v(s)]} \\ &= \overline{v(t)v(s)} + \overline{\tilde{z}(t\mid t)y(s)} \\ &= \overline{v(t)v(s)} + 0, \qquad t > s. \end{aligned} \tag{71}$$

A similar argument applies for $t<s$. Since $\overline{v(t)v(s)} = \delta(t-s) = 0$, $t \neq s$, we have $\overline{\nu(t)\nu(s)} = 0$, $t \neq s$. There remains only to examine the point $t=s$. Here we argue that $\overline{[\nu(t)-v(t)]^2} = \overline{\tilde{z}^2(t\mid t)} < \infty$, but $\overline{v^2(t)}$ is infinite (because $v(\cdot)$ is white), and therefore $\overline{\nu^2(t)}$ must be infinite (and we have just shown $\overline{\nu(t)\nu(s)} = 0$, $t \neq s$). This identifies $\nu(\cdot)$ as white noise. This argument seems shaky, but it is inevitable if we work with $v(\cdot)$ as an ordinary random process rather than as a generalized random process.

As long as we use the ordinary functional notation for $v(\cdot)$, all proofs, though they can be given in slightly different forms (especially with additional assumptions on $z(\cdot)$, e.g., that it is continuous in the mean or, more strongly, that it is a lumped process), must be essentially of the preceding form. A rigorous proof can be obtained by working with integrals of $v(\cdot)$ and $\nu(\cdot)$ (cf. [29]).

C. Discrete-Time Processes

Some insight is also shed on the preceding calculations by considering the discrete time

$$y(k) = z(k) + v(k), \qquad \overline{v(k)v(l)} = R(k)\delta_{kl}, \qquad \overline{v(k)z(l)} = 0, \qquad l \leq k.$$

In this case, the innovation process $\nu(\cdot)$ is defined as

$$\nu(k) = y(k) - \hat{z}(k \mid k-1) = \tilde{z}(k \mid k-1) + v(k) \tag{72}$$

and, by arguments similar to those in (71), we obtain

for $k > l$

$$\begin{aligned}\overline{\nu(k)\nu(l)} &= \overline{v(k)v(l)} + \overline{v(k)\tilde{z}(l \mid l-1)} \\ &\quad + \overline{\tilde{z}(k \mid k) - 1)\tilde{z}(l \mid l-1)} + \overline{\tilde{z}(k \mid k-1)v(l)} \\ &= \overline{v(k)v(l)} + 0 + \overline{\tilde{z}(k \mid k-1)[z(l) + v(l)]} \\ &= \overline{v(k)v(l)}.\end{aligned}$$

Similarly, we can prove the equality for $k<l$. For $k=l$, we have

$$\begin{aligned}\overline{\nu^2(k)} &= \overline{v^2(k)} + \overline{2v(k)\tilde{z}(k \mid k-1)} + \overline{\tilde{z}^2(k \mid k-1)} \\ &= \overline{v^2(k)} + \overline{\tilde{z}^2(k \mid k-1)} = R(k) + P_z(k), \qquad \text{say.}\end{aligned}$$

Therefore

$$\overline{\nu(k)\nu(l)} = [R(k) + P_z(k)]\delta_{kl} \tag{73}$$

so that $\nu(\cdot)$, like $v(\cdot)$, is white but with a different variance. The continuous-time case can be approached by a limiting procedure in which $R(k)$ becomes indefinitely large while $P_z(k)$ remains finite, so that the variances of $\nu(\cdot)$ and $v(\cdot)$ are the same.

D. A Proof of the Equivalence of $\nu(\cdot)$ and $y(\cdot)$ Using the Kalman–Bucy Formulas

We noted in our discussion of Theorem 2 (cf. Remark 2) that the equivalence of $\nu(\cdot)$ and $y(\cdot)$ was obvious if the Kalman–Bucy result was assumed. The proof is trivial. Since $\nu(t) = y(t) - \hat{z}(t \mid t)$ and $\hat{z}(t \mid t)$ can be calculated from $y(s)$, $s \le t$, $\nu(t)$ is completely determined by $y(s)$, $s \le t$. Conversely, the Kalman–Bucy formula

$$\dot{\hat{x}}(t \mid t) = F(t)\hat{x}(t \mid t) + K(t)[y(t) - H(t)\hat{x}(t \mid t)],$$
$$\hat{x}(a \mid a) = 0$$

shows that $\hat{x}(t \mid t)$ is determined if $\{\nu(s), s \le t\}$ is known, and then $y(t)$ can be obtained as $y(t) = H(t)\hat{x}(t \mid t) + \nu(t)$ since

$$\tilde{z}(t \mid t) + \nu(t) = \hat{z}(t \mid t) + \tilde{z}(t \mid t) + v(t) = z(t) + v(t).$$

Therefore, $\nu(\cdot)$ and $y(\cdot)$ can each be obtained from the other by causal operations. This argument is due to R. Geesey.

Of course, the deeper result is that this fact is true without restriction to lumped processes and can indeed be used, as we have shown, to give a simple proof of the special formulas for lumped processes.

Acknowledgment

The author thanks R. Geesey, B. Gopinath, and P. Frost, former students at Stanford University, for many stimulating, enjoyable, and instructive conversations on various aspects of the innovations concept and, in particular, for teaching him various aspects of modern control theory.

References

[1] A. N. Kolmogorov, "Interpolation and extrapolation of stationary random sequences," *Bull. Acad. Sci. USSR, Math. Ser.* vol. 5, 1941. A translation has been published by the RAND Corp., Santa Monica, Calif., as Memo. RM-3090-PR.

[2] N. Wiener, *The Extrapolation, Interpolation, and Smoothing of Stationary Time Series with Engineering Applications.* New York: Wiley, 1949. Originally issued as a classified report by M.I.T. Radiation Lab., Cambridge, Mass., February 1942.

[3] H. Wold, *A Study in the Analysis of Stationary Time Series.* Uppsala, Sweden: Almqvist & Wiksell, 1938.

[4] P. Whittle, *Prediction and Regulation.* Princeton, N. J.: Van Nostrand, 1963.

[5] N. Wiener and E. Hopf, "On a class of singular integral equations," *Proc. Prussian Acad., Math.-Phys. Ser.*, p. 696, 1931.

[6] H. W. Bode and C. E. Shannon, "A simplified derivation of linear least square smoothing and prediction theory," *Proc. IRE*, vol. 38, pp. 417–425, April 1950.

[7] R. B. Blackman, H. W. Bode, and C. E. Shannon, "Data smoothing and prediction in fire-control systems," Research and Development Board, Washington, D.C., August 1948.

[8] L. A. Zadeh and J. R. Ragazzini, "An extension of Wiener's theory of prediction," *J. Appl. Phys.*, vol. 21, pp. 645–655, July 1950.

[9] J. L. Doob, *Stochastic Processes.* New York: Wiley, 1953.

[10] H. Laning and R. Battin, *Random Processes in Automatic Control.* New York: McGraw-Hill, 1958.

[11] V. S. Pugachev, *Theory of Random Functions and Its Applications in Automatic Control.* Moscow: Goztekhizdat, 1960.

[12] Y. W. Lee, *Statistical Theory of Communication.* New York: Wiley, 1960.

[13] A. M. Yaglom, *Theory of Stationary Random Functions*, R. A. Silverman, transl. Englewood Cliffs, N. J.: Prentice-Hall, 1966.

[14] R. Deutsch, *Estimation Theory.* Englewood Cliffs, N. J.: Prentice-Hall, 1966.

[15] P. B. Liebelt, *An Introduction to Optimal Estimation Theory.* Reading, Mass.: Addison-Wesley, 1967.

[16] A. V. Balakrishnan, "Filtering and prediction theory," in *Lectures on Communication Theory*, A. V. Balakrishnan, Ed. New York: McGraw-Hill, 1968.

[17] A. E. Bryson and Y. C. Ho, *Optimal Programming, Estimation and Control.* New York: Blaisdell, 1968.

[18] P. Swerling, "First-order error propagation in a stagewise smoothing procedure for satellite observations," *J. Astronautical Sci.*, vol. 6, no. 3, pp. 46–52, Autumn 1959. See also "A proposed stagewise differential correction procedure for satellite tracking and prediction," RAND Corp., Santa Monica, Calif., Rept. P-1292, January 1958.

[19] R. E. Kalman, "A new approach to linear filtering and prediction problems," *Trans. ASME, J. Basic Engrg.*, vol. 82, pp. 34–45, March 1960.

[20] ——, "New methods in Wiener filtering theory," *Proc. 1st Symp. on Engrg. Applications of Random Function Theory and Probability*, J. L. Bogdanoff and F. Kozin, Eds. New York: Wiley, 1963.

[21] R. S. Bucy, "Optimum finite-time filters for a special nonstationary class of inputs," Johns Hopkins University, Appl. Phys. Lab., Baltimore, Md., Internal Memo. BBD-600, 1959.

[22] R. E. Kalman and R. S. Bucy, "New results in linear filtering and prediction theory," *Trans. ASME, J. Basic Engrg.*, ser. D vol. 83, pp. 95–107, December 1961.

[23] W. M. Wonham, "Lecture notes on stochastic optimal control," Div. of Appl. Math., Brown University, Providence, R.I., Rept. 67-1.

[24] A. J. F. Siegert, "A systematic approach to a class of problems in the theory of noise and other random phenomena," Pt. 2 and 3, *IRE Trans. Information Theory*, vol. IT-3, pp. 38–43, March 1957; vol. IT-4, pp. 4–14, March 1958.

[25] T. Kailath and P. Frost, "An innovations approach to least-squares estimation—Part II: Linear smoothing in additive white noise," this issue, page 655.

[26] R. Geesey and T. Kailath, "An innovations approach to least-squares estimation—Part III: Estimation in colored noise" (to be published).

[27] ——, "An innovations approach to least-squares estimation—part IV: Nonlinear filtering and smoothing in white Gaussian noise" (to be published).

[28] P. A. Frost, "Estimation in continuous-time nonlinear systems," Ph.D. dissertation, Dept. of Elec. Engrg., Stanford University, Stanford, Calif., June 1968.

[29] ——, "A general likelihood ratio formula for random signals in Gaussian noise," *IEEE Trans. Information Theory*, to appear, 1969.

[30] T. Kailath, "An RKHS approach to detection and estimation—Part III: More on gaussian detection," to be submitted to *IEEE Trans. Information Theory.*

[31] T. Kailath and R. Geesey, "Covariance factoriztion—An explication via examples," *Proc. 2nd Asilomar Conference on Circuits and Systems*, Monterey, Calif., November 1968.

[32] H. J. Kushner, *Stochastic Stability and Control.* New York:, Academic Press, 1967.

[33] L. D. Collins, "Realizable whitening filters and state-variable realizations," *Proc. IEEE (Letters)*, vol. 56, pp. 100–101, January 1968.

[34] B. D. O. Anderson and J. B. Moore, "Whitening filters: A state-space viewpoint," Dept. of Elec. Engrg., University of Newcastle, Australia, Tech. Rept. EE 6707, August 1967. Also see *Proc. JACC*, (Michigan). 1968.
[35] H. Kwakernaak, "Optimal filtering in linear systems with time delays," *IEEE Trans. Automatic Control*, vol. AC-12, pp. 169–173, April 1967.
[36] P. Falb, "Kalman-Bucy filtering in Hilbert space," *Information and Control*, vol. 11, no. 1, pp. 102–137, August–September 1967.
[37] A. V. Balakrishnan and J. L. Lions, "State estimation for infinite-dimensional systems," *J. Computer and System Sciences*, vol. 1, no. 4, pp. 391–403, December 1967.
[38] S. S. L. Chang, "Optimum filtering and control of randomly sampled systems," *IEEE Trans. Automatic Control*, vol. AC-12, pp. 537–546, October 1967.
[39] R. J. Leake, "Duality condition established in the frequency domain," *IEEE Trans. Information Theory* (*Correspondence*), vol. IT-11, p. 461, July 1965.
[40] A. E. Bryson, Jr., and D. E. Johansen, "Linear filtering for time-varying systems using measurements containing colored noise," *IEEE Trans. Automatic Control*, vol. AC-10, pp. 4–10, January 1965.
[41] F. Smithies, *Integral Equations.* London: Cambridge University Press, 1958.
[42] F. Riesz and B. S. Nagy, *Functional Analysis.* New York: Ungar, 1955.

Section I-C
General Application Considerations

THE papers in this section deal with fundamental issues that must be considered when applying a linear theory (i.e., the Kalman filter) to practical, nonlinear problems. Each paper devotes considerable attention to the linearization of nonlinear models and the treatment of the linearized equations in conjunction with the implementation of the Kalman filter. In addition, each paper discusses the use of the estimator as part of a guidance or control scheme. Problems associated with the digital implementation of the Kalman filter are discussed also. In other words, these papers provide early recognition and treatment of the various aspects that comprise the algorithm now regarded as the extended Kalman filter (EKF).

Battin provides a thorough treatment of the use of the Kalman filter for spacecraft orbit determination as part of a spacecraft guidance system. A simple derivation of the Kalman filter is presented in which Battin defines a linear estimator having the form given by equation (1.3). Battin's equation [71] describes the linear estimator that is defined and it can be compared with (1.3) to establish the notational conventions. Note that Battin deals with perturbations from reference values. The gain matrix K appearing in (1.3a) (i.e., $\mathbf{w}_n$ in Battin) is unknown and selected to minimize the mean-squared error. The covariance of the error for an arbitrary gain is stated as equation (73). This is seen to have the same form as (1.5b) for the case considered by Battin in which there are scalar measurements. The optimal gain is presented as equation (76) and the associated covariance is given in equation (81). These results correspond with (1.4a) and (1.5a), above. The development is straightforward and provides an intuitive understanding of the result. A similar development for the general model (1.3)–(1.4) is given also by Sorenson in the third paper of this section.

In Battin's paper, the basic state vector is six-dimensional, based on the position and velocity of the spacecraft, and the equations-of-motion are derived from Newton's laws. He also considers situations in which the measurement error is correlated and modeled by a random walk. Then, the measurement error is treated as a seventh state variable and estimated in concert with the position and velocity variables. We note again that this is the situation treated by Kalman in the original paper (i.e., Paper 2 of this collection). Battin devotes considerable attention to the determination of the state transition matrix. Then, the Kalman filter is used as part of the guidance system for circumlunar navigation; and detailed simulation results for an Apollo-like system are presented. The nonlinear measurement models and the associated linearizations for several devices are presented.

Kalman and Battin included correlated measurement noise in the general model by augmenting the state vector with the noise variables. The idea of state augmentation was extended in the paper by Kopp and Orford to develop an adaptive control algorithm. They assume that the dynamic model contains parameters whose values are unknown or uncertain. Then, these parameters are included in the state vector and estimated simultaneously with the basic state variables using an extended Kalman filter. It is interesting that they provide simulation results that indicate satisfactory results for the estimator. Subsequent applications of this basic approach have produced the conclusion that indicate that the approach is very sensitive to the quality of the approximations. Consequently, many examples are known in which the use of an extended Kalman filter for combined state estimation/system identification produces very untrustworthy results. Ljung published results for this problem that require a modification of the Kalman filter algorithm to ensure convergence of the parameter estimates.

In the third paper of this section, Sorenson provides a comprehensive discussion of the techniques involved in applying the Kalman filter to nonlinear systems. A review of the basic theory is provided with considerable attention given to state augmentation as a means of dealing with physical models that exhibit characteristics not satisfying the assumptions of the Kalman filter. Attention is given also to computational issues that arise during the implementation of an extended Kalman filter, and the idea of processing measurements sequentially at a sampling time is introduced. If the measurement vector at a time t_k can be partitioned into subvectors that are uncorrelated, each subvector can be processed independently of the other subvectors. When, for example, the measurement covariance matrix R_k is diagonal, then the scalar measurements can be processed individually, eliminating thereby, the need for the matrix inversion in (1.4a). It is shown that sequential processing produces results that are theoretically equivalent to processing all of the measurements simultaneously. (Computational burdens associated with nonscalar subvectors are assessed and optimal schemes for processing measurements in a sequential and/or simultaneous manner are discussed by Mendel in Section I-F of this volume.)

On page 261, Sorenson discusses the use of a simple

matrix square-root procedure, proposed by Potter and discussed in Battin's book (i.e., as listed above at the end of the Introduction). This method can be applied only in the case in which there are scalar measurements and no input noise. Potter's method was, probably, the first use of a square-root procedure to reduce the computational errors that may arise in Kalman filter implementations. Starting on page 262, the use of the input noise covariance to compensate for modeling errors is discussed. This basic idea with a large number of implementational schemes was proposed by several people. Schmidt was an active contributor and many of the approaches to the control of modeling errors were proposed first by him. A review of much of this work can be found in the papers listed below. As with Battin's paper, Sorenson directs attention to the general problem of spacecraft navigation and guidance, and results are illustrated through a specific example.

Many other papers provide interesting and useful discussions relating to the application of the Kalman filter. Several are listed below. The first substantial application of which the Editor was aware is the NASA technical report by Smith, Schmidt, and McGee. Several other reports and papers were published subsequently by Schmidt that still merit attention. Gunckel and Dusek were other early contributors to the application of the Kalman filter whose papers still merit attention.

References

[1] G. L. Smith, S. F. Schmidt, and L. A. McGee, "Application of a statistical filter-theory to the optimal estimation of position and velocity on-board a circumlunar vehicle," NASA TR-R-135, 1962.

[2] S. F. Schmidt, "Application of state-space methods to navigation problems," in *Advances in Control Systems*, Vol. 3, C. T. Leondes, Ed. New York: Academic Press, 1966, pp. 293–340.

[3] T. L. Gunckel, "Orbit determination using Kalman's method," *J. Inst. Navig.*, vol. 10, 1963, pp. 213–291.

[4] H. M. Dusek, "Theory of error compensation in astro-inertial guidance systems for low-thrust space missions," in *Progress in Aeronautics and Astronautics*. New York: Academic Press, 1964.

[5] B. E. Bona and R. J. Smay, "Optimum Reset of Ship's Inertial Navigation System," *IEEE Trans. Aerosp. Electron. Syst.*, vol. AES-2, pp. 409–414, 1966.

[6] S. F. Schmidt, "Computational techniques in Kalman filtering," in *Theory and Applications of Kalman Filtering*, C. T. Leondes, Ed. AGARDograph 139, (AD 704 306), 1970, pp. 65–86.

[7] L. Ljung, "Asymptotic behavior of the extended Kalman filter as a parameter estimation for linear systems," *IEEE Trans. Automat. Contr.*, vol. AC-24, pp. 36–50, 1979.

A Statistical Optimizing Navigation Procedure for Space Flight[1]

RICHARD H. BATTIN[2]
Massachusetts Institute of Technology
Cambridge, Mass.

In a typical self-contained space navigation system, celestial observation data are gathered and processed to produce estimated velocity corrections. T. results of this pape: provide a basis for determining the best celestial measurements and the proper times to implement velocity corrections. Fundamental to the navigation system is a procedure for processing celestial measurement data which permits incorporation of each individual measurement as it is made in order to provide an improved estimate of position and velocity. In order to "optimize" the navigation, a statistical evaluation of a number of alternative courses of action is made. The various alternatives, which form the basis of a decision process, concern the following: 1) which star and planet combination provides the "best" available observation; 2) whether the best observation gives a sufficient reduction in the predicted target error to warrant making the measurement; and 3) whether the uncertainty in the indicated velocity correction is a small enough percentage of the correction itself to justify an engine re-start and propellant expenditure. Numerical results are presented which illustrate the effectiveness of this approach to the space navigation problem.

DURING the past three years, the problems of guiding a space vehicle during the midcourse phase of its mission have been extensively explored at the Massachusetts Institute of Technology Instrumentation Laboratory. Following the specific demonstration of the technical feasibility of an unmanned photographic reconnaissance flight to the planet Mars reported by Laning, Frey, and Trageser (1),[3] the detailed navigational aspects of such a venture were developed by Laning and the present author (2). Later, a variable time of arrival navigation theory was devised (3) and contrasted with the earlier fixed time of arrival scheme. More recently, the question of optimum use of navigation data has been given considerable study. It is the solution of this problem which forms the subject of the present paper.

The general method of navigation is based on perturbation theory so that only deviations in position and velocity from a reference path are used. Data are gathered by an optical angle measuring device and processed by a spacecraft digital computer. Periodically, small changes in the spacecraft velocity are implemented by a propulsion system as directed by the computer.

Basically, three problems are considered in this paper: 1) to identify the best sources of data available to the space vehicle navigator; 2) to define the optimum linear operations for processing the data in a manner consistent with the mission objectives; and 3) to minimize both the amount of navigational data and the number of corrective maneuvers required without unduly compromising mission accuracy.

The formulation of an optimum linear estimator as a recursion operation in which the current best estimate is combined with newly acquired information to produce a still better estimate was presented by Kalman (4). The original application of Kalman's theory to space navigation was made by Schmidt (5) and his associates.

The research described in the following sections of this paper was performed without any detailed knowledge of Schmidt's activities. As a result of this independent approach, several new and interesting ideas have developed:

1 An extremely simple derivation of the optimum linear operator has been achieved using only the basic technique of least squares estimation.

2 The mathematical problem of determining the optimum plane in which to make a star-planet angular measurement has been solved.

3 A procedure for incorporating cross-correlation effects of random measurement errors in determining the optimum linear operation has been developed.

Throughout the paper, discrete information will be dealt with exclusively; observations or velocity corrections are made at specific points in time which are termed "decision points." The interval between decision points is not necessarily uniform and may be selected somewhat arbitrarily; e.g., the interval length required for accurate numerical integration of the trajectory equations was used in preparing the computational data presented in the section on application to circumlunar navigation.

Finally, a few remarks relevant to notational conventions are appropriate. Both three- and six-dimensional vectors will be dealt with generally. A column vector of any dimension is represented by a lower case boldface letter. Matrices are denoted by capital letters and can be either square or rectangular arrays. The transpose of a vector or a matrix will be denoted by a superscript T. Thus, the scalar product of two vectors **a** and **b** will be written as $\mathbf{a}^T\mathbf{b}$. In like manner, a quadratic form associated with a square matrix A will be written as $\mathbf{x}^T A \mathbf{x}$. The expected value of a random vector $\mathbf{x}$ will be indicated by a bar; thus, $\bar{\mathbf{x}}$ denotes the average value of $\mathbf{x}$.

Outline of the Navigation and Guidance Procedure

Deterministic Method

The basic process involved in determining spacecraft position by means of a celestial fix consists fundamentally of a sequence of measurements of the angles between selected pairs of celestial objects. Three independent and precise angular measurements made at a known instant of time suffice to determine uniquely the position of the vehicle. Practical constraints, however, preclude simultaneous meas-

---

Presented at the ARS Space Flight Report to the Nation, New York, October 9–15, 1961; revision received June 4, 1962.

[1] This report was prepared under the auspices of DSR Project 55-191, sponsored by NASA, under Contract NAS-9-153. The publication of this report does not constitute approval by NASA of the findings or the conclusions contained herein. It is published only for the exchange and stimulation of ideas.

[2] Assistant Director, Instrumentation Laboratory.

[3] Numbers in parentheses indicate References at end of paper.

urements without severely complicating the instrumentation. On the other hand, if the vehicle dynamics are governed by known laws and if deviations from a predetermined reference trajectory are kept sufficiently small to permit a linearization of the navigation problem, then the question of simultaneous measurements loses its significance.

Under the assumptions of a linearized theory, a single observation serves to fix the position of the spacecraft in one coordinate. For example, if A_n is the angle measured at time t_n and is defined by the lines of sight from the vehicle to a star and to a nearby celestial body, the position of the vehicle is established along a line normal to the direction toward the near body and in the plane of the measurement. It is shown in Appendix A that the deviation in position $\delta\mathbf{r}_n$ of the spacecraft from the reference position is related to the deviation in angular measurement δA_n by

$$\delta A_n = \mathbf{h}_n^T \delta\mathbf{r}_n \tag{1}$$

if the observation is made at a known instant of time t_n. The vector $\mathbf{h}_n$ depends on the geometrical configuration of the relevant celestial objects at time t_n as well as on the type of measurement made.

Because of the inherent dynamic coupling of position and velocity, the result at a later time t_{n+1} of a measurement made at time t_n does not lend itself to simple geometric interpretation. In order to provide a geometrical description, it is convenient to introduce the concept of a six-dimensional space in which the coordinates represent the components of both position and velocity deviations of the vehicle from the reference path as functions of time. Points in this space are defined by the six-dimensional deviation vector

$$\delta\mathbf{x}_n = \begin{bmatrix} \delta\mathbf{r}_n \\ \delta\mathbf{v}_n \end{bmatrix} \tag{2}$$

where $\delta\mathbf{v}_n$ is the deviation in the vector velocity of the vehicle from the reference value. The vector $\delta\mathbf{x}_n$ defines the "state" of the vehicle dynamics at time t_n. Transition from one state to another is provided by the matrix operation

$$\Phi_{n+1,n} = \Phi(t_{n+1},t_n)$$

which is frequently referred to as the "transition matrix." Indeed, the relationship between $\delta\mathbf{x}_{n+1}$ and $\delta\mathbf{x}_n$ is simply

$$\delta\mathbf{x}_{n+1} = \Phi_{n+1,n}\delta\mathbf{x}_n \tag{3}$$

as shown in the section on state transition matrix.

By means of the rectangular matrix K defined by

$$K = \begin{bmatrix} I \\ O \end{bmatrix} \tag{4}$$

Eq. [1] may be written in terms of $\delta\mathbf{x}_n$ as

$$\delta A_n = \mathbf{h}_n^T K^T \delta\mathbf{x}_n \tag{5}$$

The submatrices I and O are, respectively, the three-dimensional identity and zero matrices. Now, by combining Eqs. [3] and [5]

$$\delta A_n = \mathbf{h}_n^T K^T \Phi_{n+1,n}^{-1} \delta\mathbf{x}_{n+1} \tag{6}$$

it is clear that the effect at time t_{n+1} of an observation at time t_n is to determine the component of the six-dimensional deviation vector in the direction defined by the vector $\Phi_{n+1,n}^{T-1} K\mathbf{h}_n$. Six observations made at different times would provide a set of six equations of the form of Eq. [6]. If no two of the component directions were parallel, then the deviation vector could be obtained by inverting the six-dimensional coefficient matrix.

Statistical Parameters of the Navigation Problem

Because of the presence of instrument inaccuracies, additional observations may be used to reduce the errors associated with the simple deterministic process just described. By applying least-square techniques to the observed data, a more accurate estimate of position and velocity is frequently possible than could be obtained from the minimum number of measurements. For this purpose, it is necessary to know certain statistical information with respect to the instrument inaccuracies. In a linear least-squares estimation procedure, all statistical calculations are based on first- and second-order averages, and no additional statistical data are needed.

At this point of the discussion, it is necessary to distinguish measured values, estimated values, and true values of various quantities; e.g., $\delta\tilde{A}_n$ will be the measured value of the deviation in the angle A_n from its reference value at time t_n, δA_n the true value of the deviation, and $\delta\hat{A}_n$ the estimated value. If one writes

$$\delta\tilde{A}_n = \delta A_n + \alpha_n \tag{7}$$

then α_n will be the error in the measurement. In the subsequent analysis α_n will be regarded as a random variable with an average value $\bar{\alpha}_n$ and a variance

$$\sigma_n^2 = \overline{\alpha_n^2} - \bar{\alpha}_n^2 \tag{8}$$

The possibility of cross-correlation of measurement errors will not be excluded; i.e., in general, the average $\overline{\alpha_n\alpha_m}$ may be different from $\bar{\alpha}_n\bar{\alpha}_m$.

In the section on derivation of optimum linear estimate, an estimation procedure is developed for determining an optimal linear estimate of $\delta\mathbf{x}_n$, denoted by $\delta\hat{\mathbf{x}}_n$. As each measurement is made, the estimate $\delta\hat{\mathbf{x}}_n$ is updated by a simple recursive formula, and thereby the problem associated with inverting sixth-order matrices is avoided. An integral part of the estimation technique is the correlation matrix of the errors in the estimate. If one writes

$$\delta\hat{\mathbf{x}}_n = \delta\mathbf{x}_n + \mathbf{e}_n \tag{9}$$

then

$$\mathbf{e}_n = \begin{bmatrix} \boldsymbol{\varepsilon}_n \\ \boldsymbol{\delta}_n \end{bmatrix} \tag{10}$$

is the six-dimensional error vector and may be partitioned as shown using $\boldsymbol{\varepsilon}_n$ and $\boldsymbol{\delta}_n$ to denote, respectively, the position and velocity errors. The correlation matrix is thus defined by

$$E_n = \overline{\mathbf{e}_n\mathbf{e}_n^T} = \begin{bmatrix} \overline{\boldsymbol{\varepsilon}_n\boldsymbol{\varepsilon}_n^T} & \overline{\boldsymbol{\varepsilon}_n\boldsymbol{\delta}_n^T} \\ \overline{\boldsymbol{\delta}_n\boldsymbol{\varepsilon}_n^T} & \overline{\boldsymbol{\delta}_n\boldsymbol{\delta}_n^T} \end{bmatrix} = \begin{bmatrix} E_n^{(1)} & E_n^{(2)} \\ E_n^{(3)} & E_n^{(4)} \end{bmatrix} \tag{11}$$

For the later use in a statistical analysis of the guidance problem, the correlation matrix of the actual deviation vector will be needed. This matrix is defined by

$$X_n = \overline{\delta\mathbf{x}_n\,\delta\mathbf{x}_n^T} \tag{12}$$

and may be calculated recursively using

$$X_n = \Phi_{n,n-1}X_{n-1}\Phi_{n,n-1}^T \tag{13}$$

Initially, i.e., at injection

$$\delta\hat{\mathbf{x}}_0 = \delta\mathbf{x}_0 + \mathbf{e}_0 = 0 \tag{14}$$

so that

$$X_0 = E_0 \tag{15}$$

provides an initial value for the X_n matrix.

It is important to distinguish between a new estimate $\delta\hat{\mathbf{x}}_n$, obtained by incorporating an observation at time t_n, and an estimate simply extrapolated from a previous estimate. For the latter case, the notation $\delta\hat{\mathbf{x}}_n'$ is used where

$$\delta\hat{\mathbf{x}}_n' = \Phi_{n,n-1}\delta\hat{\mathbf{x}}_{n-1} \tag{16}$$

In like manner, an extrapolated error vector $\mathbf{e}_n'$ is defined.

The extrapolated correlation matrix is readily shown to be

$$E_n' = \Phi_{n,n-1}E_{n-1}\Phi_{n,n-1} \quad [17]$$

Note that an estimate of the deviation in the angle to be measured at time t_n may be obtained from the extrapolated estimate of $\delta\mathbf{x}_{n-1}$. Then

$$\delta\hat{A}_n' = \mathbf{h}_n^T K^T \delta\hat{\mathbf{x}}_n' \quad [18]$$

and it is this quantity, compared with the measured deviation $\delta\tilde{A}_n$, which is used in arriving at a revised estimate of $\delta\mathbf{x}_n$.

When cross-correlation of measurement errors is considered, it is convenient to use an augmented deviation vector having seven dimensions and defined as

$$\delta\mathbf{x}_n = \begin{bmatrix} \delta\mathbf{r}_n \\ \delta\mathbf{v}_n \\ \alpha_n \end{bmatrix} \quad [19]$$

Since, in this case, the error in a measurement at time t_n may be predicted on the basis of previous observations

$$\hat{\alpha}_n = \alpha_n + \beta_n \quad [20]$$

may be defined as the best estimate of the error to be expected in the measurement of A_n. The term β_n is then the error in the estimation of the measurement error. The error vector $\mathbf{e}_n$ will, of course, be seven-dimensional and expressible as

$$\mathbf{e}_n = \begin{bmatrix} \varepsilon_n \\ \delta_n \\ \beta_n \end{bmatrix} \quad [21]$$

Correspondingly, the correlation matrix becomes

$$E_n = \begin{bmatrix} \overline{\varepsilon_n\varepsilon_n^T} & \overline{\varepsilon_n\delta_n^T} & \overline{\varepsilon_n\beta_n} \\ \overline{\delta_n\varepsilon_n^T} & \overline{\delta_n\delta_n^T} & \overline{\delta_n\beta_n} \\ \overline{\beta_n\varepsilon_n^T} & \overline{\beta_n\delta_n^T} & \overline{\beta_n^2} \end{bmatrix} \quad [22]$$

It will be convenient in later work to define the correlation vector ϕ_n as the last column of the matrix E_n.

For purposes of illustration, consider the following model for correlated measurement errors. Let the error at time t_{n+1} be composed of two parts:

$$\begin{aligned} \alpha_{n+1} &= \alpha'_{n+1} + \zeta_{n+1} \\ \alpha'_{n+1} &= \alpha_n \exp[-\lambda(t_{n+1} - t_n)] \end{aligned} \quad [23]$$

where α_n and ζ_{n+1} are independent random numbers, λ is a positive constant, and $\bar{\zeta}_{n+1}$ is zero. It follows that

$$\hat{\alpha}'_{n+1} = \hat{\alpha}_n \exp[-\lambda(t_{n+1} - t_n)] \quad [24]$$

$$\beta_{n+1} = \beta_n \exp[-\lambda(t_{n+1} - t_n)] \quad [25]$$

Hence, the extrapolated error vector $\mathbf{e}'_{n+1}$ is calculated from

$$\mathbf{e}'_{n+1} = P_{n+1,n}\mathbf{e}_n \quad [26]$$

where $P_{n+1,n}$ is the augmented transition matrix

$$P_{n+1,n} = \begin{bmatrix} \Phi_{n+1,n} & 0 \\ 0 & \exp[-\lambda(t_{n+1} - t_n)] \end{bmatrix} \quad [27]$$

The augmented extrapolated correlation matrix is then computed from

$$E'_{n+1} = P_{n+1,n}E_nP^T_{n+1,n} \quad [28]$$

From the definition of the error model, Eq. [23], it is seen that the true value of the seventh component of the state vector undergoes a step change as a result of a measurement. Thus, in the cross-correlation case, one must distinguish between $\delta\mathbf{x}_n$ and $\delta\mathbf{x}_n'$. Hence, it will be convenient to define $\mathbf{e}_n''$ as $\mathbf{e}_n'$ with ζ_n subtracted from the seventh component. Then E_n'' will be the corresponding correlation matrix and ϕ_n'' the last column of E_n''.

Summary of Navigation and Guidance Equations

In the navigation and guidance theory presented here, the problem of launch guidance from Earth is ignored. It is assumed that the main propulsion stages are completed at time t_L and that the correlation matrix $E_0 = E(t_L)$ is specified initially from a statistical knowledge of injection guidance errors. The initial estimate of position and velocity deviation $\delta\hat{\mathbf{x}}_0 = \delta\hat{\mathbf{x}}(t_L)$ is zero, since, in the absence of any observation, the best unbiased estimate is that the spacecraft is on course.

The time interval from launch to arrival time t_A at the target point is considered to be subdivided into a number of smaller intervals by the sequence of times $t_1, t_2, \ldots$ called "decision points." At each decision point, one of three possible courses of action is followed: 1) a single observation is made; 2) a velocity correction is implemented; or 3) no action is taken. A revised estimate of the deviation vector $\delta\mathbf{x}(t)$ is made at each such point—the form of the revision depending, of course, on the nature of the decision. Specifically, as shown in the section on derivation of optimum linear estimate, for uncorrelated measurement errors the revised estimate at the decision time t_n is one of the following:

$$\delta\hat{\mathbf{x}}_n = \begin{cases} \delta\hat{\mathbf{x}}_n' + a_n^{-1}E_n'K\mathbf{h}_n(\delta\tilde{A}_n - \delta\hat{A}_n') & \text{measurement} \\ (I + JB_n)\,\delta\hat{\mathbf{x}}_n' & \text{correction} \\ \delta\hat{\mathbf{x}}_n' & \text{no action} \end{cases} \quad [29]$$

The scalar coefficient a_n is computed from

$$a_n = \mathbf{h}_n^T K^T E_n' K\mathbf{h}_n + \overline{\alpha_n^2} \quad [30]$$

The rectangular matrix J has six rows and three columns

$$J = \begin{bmatrix} O \\ I \end{bmatrix} \quad [31]$$

and is just the reverse of the K matrix. The matrix B_n is also rectangular, having three rows and six columns, and is partitioned as shown

$$B_n = [C_n^* \quad -I] \quad [32]$$

where C_n^* is one of the fundamental navigation matrices described in the section on vector velocity correction.

At each decision point, it is also necessary to update the correlation matrix E_n. Thus

$$E_n = \begin{cases} E_n' - a_n^{-1}(E_n'K\mathbf{h}_n)(E_n'K\mathbf{h}_n)^T & \text{measurement} \\ E_n' + J\overline{\mathbf{n}_n\mathbf{n}_n^T}J^T & \text{correction} \\ E_n' & \text{no action} \end{cases} \quad [33]$$

The vector $\mathbf{n}_n$ is the difference between the commanded velocity correction and the actual velocity change implemented at time t_n.

This collection of formulas provides the means of maintaining an up-to-date estimate of the deviation vector $\delta\mathbf{x}_n$ but, in themselves, they do not provide any clue as to what decision should be made at each point. Suggestions for reasonable decision rules are discussed in the section on numerical example and in Appendix B.

When measurement errors are correlated, the only significant change arises in the method of processing a measurement to obtain a revised estimate in the augmented deviation vector and the associated correlation matrix. Thus

$$\delta\hat{\mathbf{x}}_n = \delta\hat{\mathbf{x}}_n' + a_n^{-1}(E_n''K\mathbf{h}_n + \phi_n'') \times [\delta\tilde{A}_n - (\delta\hat{A}_n' + \hat{\alpha}_n')] \quad [34]$$

$$E_n = E_n'' - a_n^{-1}(E_n''K\mathbf{h}_n + \phi_n'')(E_n''K\mathbf{h}_n + \phi_n'')^T \quad [35]$$

where

$$a_n = \mathbf{h}_n^T K^T E_n'' K\mathbf{h}_n + 2\mathbf{h}_n^T K^T \phi_n'' + (\overline{\beta_n'^2} + \overline{\zeta_n^2}) \quad [36]$$

The remaining equations are unaltered; however, certain obvious changes are required in the definition of the matrices

J, K, and B_a in order that they be dimensionally compatible with the seven-dimensional deviation vector.

Fundamental Navigation Matrices

Basic to the solution of the navigation problem is a certain collection of matrices. The objective here is to introduce these matrices, to indicate their role in the navigation theory, and to show how they may be obtained as solutions of differential equations.

General Solution of Linearized Trajectory Equations

Let $\mathbf{r}_s(t)$ and $\mathbf{v}_s(t)$ denote the position and velocity vectors of the spacecraft in an inertial coordinate system, and let $\mathbf{g}(\mathbf{r}_s,t)$ denote the gravitational acceleration at position $\mathbf{r}_s$ and time t. Then

$$d\mathbf{r}_s/dt = \mathbf{v}_s \qquad d\mathbf{v}_s/dt = \mathbf{g}(\mathbf{r}_s,t) \qquad [37]$$

are the basic equations of motion of the spaceship except for those brief periods during which propulsion is applied.

Let the vectors $\mathbf{r}_0(t)$ and $\mathbf{v}_0(t)$ represent the position and velocity at time t associated with the prescribed reference trajectory, and define

$$\delta\mathbf{r}(t) = \mathbf{r}_s(t) - \mathbf{r}_0(t) \qquad \delta\mathbf{v}(t) = \mathbf{v}_s(t) - \mathbf{v}_0(t) \qquad [38]$$

Then, the deviations $\delta\mathbf{r}$ and $\delta\mathbf{v}$ may be approximately related by means of the linearized differential equations:

$$d(\delta\mathbf{r})/dt = \delta\mathbf{v} \qquad d(\delta\mathbf{v})/dt = G(\mathbf{r}_0,t)\,\delta\mathbf{r} \qquad [39]$$

where $G(\mathbf{r}_0,t)$ is a matrix whose elements are the partial derivatives of the components of $\mathbf{g}(\mathbf{r}_0,t)$ with respect to the components of $\mathbf{r}_0$.

A particularly useful fundamental set of solutions of Eqs. [39] may be developed in the following way. Let t_L and t_A be, respectively, the time of launch and the time of arrival at the target. Then, define the matrices $R(t)$, $R^*(t)$, $V(t)$, $V^*(t)$ as the solutions of the matrix differential equations

$$\begin{matrix} dR/dt = V & dR^*/dt = V^* \\ dV/dt = GR & dV^*/dt = GR^* \end{matrix} \qquad [40]$$

which satisfy the initial conditions

$$\begin{matrix} R(t_L) = O & R^*(t_A) = O \\ V(t_L) = I & V^*(t_A) = I \end{matrix} \qquad [41]$$

Here O and I denote, respectively, the zero and identity matrix. If one now writes

$$\delta\mathbf{r}(t) = R(t)\mathbf{c} + R^*(t)\mathbf{c}^* \qquad [42]$$

$$\delta\mathbf{v}(t) = V(t)\mathbf{c} + V^*(t)\mathbf{c}^* \qquad [43]$$

where $\mathbf{c}$ and $\mathbf{c}^*$ are arbitrary constant vectors, it follows that these expressions satisfy the perturbation differential Eqs. [39] and contain precisely the required number of unspecified constants to meet any valid set of initial or boundary conditions.

The elements of the R and V matrices represent deviations in position and velocity from the corresponding reference quantities as the result of certain specific deviations in the launch velocity from its reference value. For example, the first columns of these matrices are the vector deviations at time t due to a unit change in the first component of the velocity at time t_L. Corresponding interpretations may be ascribed to the other columns as well. A similar discussion will provide a physical meaning for the elements of R^* and V^*. For this purpose, however, it is convenient to imagine the roles of launch and target points as reversed.

Vector Velocity Correction

Associated with the position $\mathbf{r}_s$ and the time t is the vector velocity required by the spacecraft to travel in freefall from $\mathbf{r}_s(t)$ to the target point $\mathbf{r}_0(t_A)$ in the time $t_A - t$. An expression for this velocity vector is readily obtained from Eqs. [42] and [43]. The condition that the vehicle pass through the target point is met by the requirement

$$\delta\mathbf{r}(t_A) = 0 = R(t_A)\mathbf{c} + R^*(t_A)\mathbf{c}^*$$

Since $R^*(t_A) = O$, it follows that $\mathbf{c} = 0$. Eliminating $\mathbf{c}^*$ between Eqs. [42] and [43] gives for the required velocity deviation[4] at time t

$$\delta\mathbf{v}^+(t) = V^*(t)R^*(t)^{-1}\delta\mathbf{r}(t) \qquad [44]$$

Hence, the required velocity correction $\Delta\mathbf{v}^*$ is given by

$$\Delta\mathbf{v}^*(t) = C^*(t)\delta\mathbf{r}(t) - \delta\mathbf{v}^-(t) \qquad [45]$$

where the C^* matrix is defined by

$$C^*(t) = V^*(t)R^*(t)^{-1} \qquad [46]$$

The elements of the C^* matrix are deviations in vehicle velocity from the reference values, as required to place the vehicle on a trajectory to the target point, which arise from certain specific deviations in the vehicle position. The interpretation applied to the columns is made in the manner described earlier in connection with the R and V matrices.

If the spacecraft has been in a freefall status since launch, then, by employing arguments similar to those used in establishing Eq. [44], it can be shown that

$$\delta\mathbf{v}^-(t) = C(t)\,\delta\mathbf{r}(t) \qquad [47]$$

where

$$C(t) = V(t)\,R(t)^{-1} \qquad [48]$$

In this case, Eq. [45] takes the form

$$\Delta\mathbf{v}^*(t) = [C^*(t) - C(t)]\,\delta\mathbf{r}(t) \qquad [49]$$

Since $\delta\mathbf{r}(t)$ is different from zero solely as a result of an injection velocity error $\delta\mathbf{v}(t_L)$, it follows, from the definition of the R matrix, that

$$\Delta\mathbf{v}^*(t) = -\Lambda(t)\,\delta\mathbf{v}(t_L) \qquad [50]$$

Thus, the Λ matrix, defined by

$$\Lambda(t) = V(t) - C^*(t)R(t) \qquad [51]$$

relates a deviation in launch velocity to the velocity impulse required at time t. An asterisk form of the Λ matrix

$$\Lambda^*(t) = V^*(t) - C(t)R^*(t) \qquad [52]$$

will occur in the subsequent discussions.

Differential Equation Solutions

The matrices C, C^*, Λ, Λ^* may also be generated directly as solutions of differential equations. However, for C and C^*, a difficulty arises in prescribing appropriate initial conditions. From the initial values of the R and R^* matrices, it follows that $C(t_L)$ and $C^*(t_A)$ are both infinite. The singularities may be avoided by working directly with the differential equation for the inverse matrices C^{-1} and C^{*-1}.

By differentiating the identity

$$C(t)^{-1}V(t) = R(t) \qquad [53]$$

and using Eq. [40], the following equation for C^{-1} results:

$$(dC^{-1}/dt) + C^{-1}GC^{-1} = I \qquad [54]$$

Similarly, one obtains

$$(dC^{*-1}/dt) + C^{*-1}GC^{*-1} = I \qquad [55]$$

[4] The minus and plus superscripts are used to distinguish the velocity just prior to correction from the velocity immediately following the correction.

Eqs. [54] and [55] may be used to demonstrate an interesting property possessed by C and C^*. It is easy to show that the G matrix is symmetrical. It follows at once that the matrices C and C^* will be symmetrical for all values of t in the interval (t_L, t_A) if they are symmetrical for any particular time, but from Eq. [53] and a similar one involving starred matrices, one has

$$C(t_L)^{-1} = O \qquad C^*(t_A)^{-1} = O \qquad [56]$$

so that C and C^* are, indeed, symmetrical for t equal to t_L and t_A, respectively. Hence $C(t)$ and $C^*(t)$ are symmetrical for all t in the interval from launch to the target point.

In an entirely analogous manner, differential equations may be developed for Λ and Λ^*. By differentiating Eqs. [51] and [52] and using Eq. [40], one readily obtains the equations

$$(d\Lambda/dt) + C^*\Lambda = O \qquad [57]$$

$$(d\Lambda^*/dt) + C\Lambda^* = O \qquad [58]$$

with the initial conditions

$$\Lambda(t_L) = I \qquad \Lambda^*(t_A) = I \qquad [59]$$

State Transition Matrix

Let $\delta\mathbf{r}_n = \delta\mathbf{r}(t_n)$ and $\delta\mathbf{v}_n = \delta\mathbf{v}(t_n)$ be the deviations in position and velocity at time t_n, and let $R_n, V_n, \ldots$ be the corresponding values of the fundamental matrices. Then $\mathbf{c}$ and $\mathbf{c}^*$ must be obtained as solutions of

$$\delta\mathbf{r}_n = R_n\mathbf{c} + R_n^*\mathbf{c}^* \qquad [60]$$

$$\delta\mathbf{v}_n = V_n\mathbf{c} + V_n^*\mathbf{c}^* \qquad [61]$$

Multiplying Eq. [60] by R_n^{-1}, one obtains for $\mathbf{c}$

$$\mathbf{c} = R_n^{-1}(\delta\mathbf{r}_n - R_n^*\mathbf{c}^*) \qquad [62]$$

Then, by substituting this expression into Eq. [61] and using Eqs. [48] and [52], there results

$$\mathbf{c}^* = -\Lambda_n^{*-1}(C_n\delta\mathbf{r}_n - \delta\mathbf{v}_n) \qquad [63]$$

Finally, from Eq. [62]

$$\mathbf{c} = -\Lambda_n^{-1}(C_n^*\delta\mathbf{r}_n - \delta\mathbf{v}_n) \qquad [64]$$

after some simplification. Thus, with $\mathbf{c}$ and $\mathbf{c}^*$ determined, the position and velocity deviations at any other time t are given by Eqs. [42] and [43].

In terms of the six-dimensional deviation vector defined by Eq. [2], the result may be written in the form

$$\delta\mathbf{x}(t) = \begin{bmatrix} R(t) & R^*(t) \\ V(t) & V^*(t) \end{bmatrix} \begin{bmatrix} \mathbf{c} \\ \mathbf{c}^* \end{bmatrix} \qquad [65]$$

Consider now a specific value of $t = t_{n+1}$. Then substituting from Eqs. [63] and [64] into Eq. [65], a relationship between $\delta\mathbf{x}_{n+1}$ and $\delta\mathbf{x}_n$ is displayed

$$\delta\mathbf{x}_{n+1} = \Phi_{n+1,n}\delta\mathbf{x}_n \qquad [66]$$

where $\Phi_{n+1,n}$, the six-dimensional state transition matrix, is computed from

$$\Phi_{n+1,n} = \begin{bmatrix} R_{n+1} & R^*_{n+1} \\ V_{n+1} & V^*_{n+1} \end{bmatrix} \begin{bmatrix} (C_n^{*-1}\Lambda_n)^{-1} & O \\ O & (C_n^{-1}\Lambda_n^*)^{-1} \end{bmatrix} \times \begin{bmatrix} -I & C_n^{*-1} \\ -I & C_n^{-1} \end{bmatrix} \qquad [67]$$

It is not difficult to show that an alternate calculation of the transition matrix may be made directly as the solution of the sixth-order matrix differential equation

$$\frac{d\Phi(t,t_n)}{dt} = F(t)\Phi(t,t_n) \qquad [68]$$

subject to the initial condition $\Phi(t_n, t_n)$ equal to the six-dimensional identity matrix. The matrix $F(t)$ is

$$F(t) = \begin{bmatrix} O & I \\ G(t) & O \end{bmatrix} \qquad [69]$$

Finally, it has been shown (5) that the inverse of the matrix $\Phi_{n+1,n}$ is directly obtained as

$$\Phi_{n+1,n}^{-1} = \Phi_{n,n+1} = \begin{bmatrix} \Phi_1 & \Phi_2 \\ \Phi_3 & \Phi_4 \end{bmatrix}^{-1} = \begin{bmatrix} \Phi_4^T & -\Phi_2^T \\ -\Phi_3^T & \Phi_1^T \end{bmatrix} \qquad [70]$$

Derivation of Optimum Linear Estimate

Uncorrelated Measurement Errors

As noted at the beginning of this paper, the optimum linear estimate of the deviation vector may be expressed as a recursion formula. Therefore, assume that $\delta\hat{\mathbf{x}}_{n-1}$ and E_{n-1} are known and that a single measurement of the type described in Appendix A is made at time t_n. The observed deviation in the measured quantity A_n is $\delta\tilde{A}_n$, and the best estimate for δA_n, as obtained from the extrapolated estimate of $\delta\mathbf{x}_{n-1}$, is given by Eq. [18]. Then a linear estimate for the deviation vector $\delta\mathbf{x}_n$ at time t_n is expressible as a linear combination of the extrapolated estimate of $\delta\mathbf{x}_{n-1}$ and the difference between the observed and estimated deviations in the measured quantity A_n. Thus, for uncorrelated measurement errors

$$\delta\hat{x}_n \doteq \delta\hat{\mathbf{x}}_n' + \mathbf{w}_n(\delta\tilde{A}_n - \delta\hat{A}_n') \qquad [71]$$

where the vector $\mathbf{w}_n$ is a weighting factor that will be chosen so as to minimize the mean-squared error in the estimate.

For this purpose, use Eqs. [9, 7, and 5] to write

$$\begin{aligned} \mathbf{e}_n(\mathbf{w}_n) &= \delta\hat{\mathbf{x}}_n - \delta\mathbf{x}_n \\ &= \delta\hat{\mathbf{x}}_n' + \mathbf{w}_n(\delta A_n + \alpha_n - \delta\hat{A}_n') - \delta\mathbf{x}_n \\ &= (I - \mathbf{w}_n\mathbf{h}_n^T K^T)(\delta\hat{\mathbf{x}}_n' - \delta\mathbf{x}_n) + \mathbf{w}_n\,\alpha_n \qquad [72] \\ &= (I - \mathbf{w}_n\mathbf{h}_n^T K^T)\mathbf{e}_n' + \mathbf{w}_n\,\alpha_n \end{aligned}$$

where I is the six-dimensional identity matrix. Then the correlation matrix E_n defined by Eq. [11] may be expressed as a function of the weighting vector $\mathbf{w}_n$ as

$$E_n(\mathbf{w}_n) = (I - \mathbf{w}_n\mathbf{h}_n^T K^T)E_n'(I - K\mathbf{h}_n\mathbf{w}_n^T) + \mathbf{w}_n\mathbf{w}_n^T\,\overline{\alpha_n^2} \qquad [73]$$

The mean-squared errors in the estimate of position and velocity deviations $\overline{\epsilon_n^2}$ and $\overline{\delta_n^2}$ are simply the respective traces of the submatrices $E_n^{(1)}$ and $E_n^{(4)}$. If the six-dimensional weighting vector $\mathbf{w}_n$ is partitioned into two three-dimensional vectors

$$\mathbf{w}_n = \begin{bmatrix} \mathbf{w}_n^{(1)} \\ \mathbf{w}_n^{(2)} \end{bmatrix} \qquad [74]$$

then from Eq. [73] it is easy to show that $E_n^{(1)}$ is a function only of $\mathbf{w}_n^{(1)}$ and $E_n^{(4)}$ is a function only of $\mathbf{w}_n^{(2)}$. Therefore, for the purposes of the following discussion, it is legitimate to treat formally the mean-squared error in the estimate $\overline{e_n^2(\mathbf{w}_n)}$ as the trace of the six-dimensional correlation matrix $E_n(\mathbf{w}_n)$. The subvectors of the optimum weighting vector $\mathbf{w}_n$ will then each be optimum for the respective estimates of position and velocity deviations.

In order to determine the optimum weighting vector, one may apply the usual technique of the variational calculus. Let $\mathbf{w}_n$ take on a variation $\delta\mathbf{w}_n$, and obtain from Eq. [73]

$$\delta\overline{e_n^2(\mathbf{w}_n)} = 2\,tr[-\delta\mathbf{w}_n\mathbf{h}_n^T K^T E_n'(I - K\mathbf{h}_n\mathbf{w}_n^T) + \delta\mathbf{w}_n\,\mathbf{w}_n^T\overline{\alpha_n^2}] \qquad [75]$$

If $\delta\overline{e^2(\mathbf{w}_n)}$ is to vanish for all variations $\delta\mathbf{w}_n$, then it must follow that

$$a_n\mathbf{w}_n = E_n'K\mathbf{h}_n \qquad [76]$$

where the positive scalar quantity a_n is defined by Eq. [30].

It can be readily shown that the $\mathbf{w}_n$ determined from Eq. [76] actually does minimize $\overline{e_n^2(\mathbf{w}_n)}$. Suppose that the optimum $\mathbf{w}_n$ is replaced by another weighting factor $\mathbf{w}_n - \mathbf{y}_n$. Then from Eqs. [73] and [17]

$$\overline{e_n^2(\mathbf{w}_n - \mathbf{y}_n)} = tr[E_n' - 2(\mathbf{w}_n - \mathbf{y}_n)\mathbf{h}_n^TK^TE_n' + a_n(\mathbf{w}_n - \mathbf{y}_n)(\mathbf{w}_n^T - \mathbf{y}_n^T)] \quad [77]$$

and using Eq. [76]

$$\overline{e_n^2(\mathbf{w}_n - \mathbf{y}_n)} = tr[E_n' - a_n(\mathbf{w}_n - \mathbf{y}_n)(\mathbf{w}_n^T + \mathbf{y}_n^T)] \quad [78]$$

so that

$$\overline{e_n^2(\mathbf{w}_n - \mathbf{y}_n)} = \overline{e_n^2(\mathbf{w}_n)} + a_n\, tr(\mathbf{y}_n\mathbf{y}_n^T) \quad [79]$$

Thus, the mean-squared error is not decreased by perturbing $\mathbf{w}_n$ if Eq. [76] holds.

Having obtained the optimum weighting vector, the expression for the correlation matrix of the estimate errors E_n given by Eq. [73] may be written in a more convenient form. Thus, from the definition of a_n in Eq. [30], there results

$$E_n = E_n'(I - K\mathbf{h}_n\mathbf{w}_n^T) - \mathbf{w}_nh_n^TK^TE_n' + a_n\mathbf{w}_n\mathbf{w}_n^T \quad [80]$$

Substituting from Eq. [76], the final expression may be written as

$$E_n = E_n' - a_n^{-1}(E_n'K\mathbf{h}_n)(E_n'K\mathbf{h}_n)^T \quad [81]$$

Eqs. [71] and [81] then serve as recursive relations to be used in obtaining improved estimates of position and velocity deviations at each of the measurement times $t_1, t_2, \ldots$.

Correlated Measurement Errors

If the measurement errors are correlated, the derivation is only slightly altered. The linear estimate for the seven-dimensional deviation vector $\delta\mathbf{x}_n$ at time t_n is again expressible as a linear combination of the extrapolated estimate of $\delta\mathbf{x}_{n-1}$ and the difference between the observed and estimated deviations in the measured quantity A_n. However, the estimated deviation in A_n must also include the estimate of the error in the observation. Thus

$$\delta\hat{\mathbf{x}}_n = \delta\hat{\mathbf{x}}_n' + \mathbf{w}_n[\delta\tilde{A}_n - (\delta\hat{A}_n' + \hat{\alpha}_n')] \quad [82]$$

where now the weighting vector $\mathbf{w}_n$ is seven-dimensional.

The error in the estimate may be written as

$$\begin{aligned}\mathbf{e}_n &= \delta\hat{\mathbf{x}}_n - \delta\mathbf{x}_n \\ &= \delta\hat{\mathbf{x}}_n' + \mathbf{w}_n(\delta A_n - \beta_n' + \zeta_n - \delta\hat{A}_n') - \delta\mathbf{x}_n \\ &= (I - \mathbf{w}_n\mathbf{h}_n^TK^T)(\delta\hat{\mathbf{x}}_n' - \delta\mathbf{x}_n) - \mathbf{w}_n(\beta_n' - \zeta_n) \\ &= (I - \mathbf{w}_n\mathbf{h}_n^TK^T)\mathbf{e}_n'' - \mathbf{w}_n(\beta_n' - \zeta_n)\end{aligned} \quad [83]$$

The correlation matrix, expressed as a function of the weighting vector $\mathbf{w}_n$, is then

$$E_n(\mathbf{w}_n) = (I - \mathbf{w}_n\mathbf{h}_n^TK^T)E_n''(I - K\mathbf{h}_n\mathbf{w}_n^T) - (I - \mathbf{w}_n\mathbf{h}_n^TK^T)\phi_n''\mathbf{w}_n^T - \mathbf{w}_n\phi_n''^T(I - K\mathbf{h}_n\mathbf{w}_n^T) + \mathbf{w}_n\mathbf{w}_n^T(\overline{\beta_n'^2} + \overline{\zeta_n^2}) \quad [84]$$

Again, if it is required that $\delta\overline{e^2(\mathbf{w}_n)}$ vanish for all variations $\delta\mathbf{w}_n$, it is readily shown that

$$a_n\mathbf{w}_n = E_n''K\mathbf{h}_n + \phi_n'' \quad [85]$$

where a_n is defined by Eq. [36].

Correlation Between Estimate and Error

An important property of the optimum estimate, which is needed for the development of the statistical analysis procedures described in the section on statistical analysis of guidance procedure, will be derived here. The result may be stated simply as

$$\overline{\mathbf{e}_n\,\delta\hat{\mathbf{x}}_n^T} = 0 \quad [86]$$

if $\delta\hat{\mathbf{x}}_n$ is the optimum estimate; i.e., the optimum estimate and the associated error in the estimate are uncorrelated. For simplicity, only the case of uncorrelated measurement errors is considered in the proof, but the property is readily established in general.

From Eq. [75] one obtains

$$\mathbf{w}_n\overline{\alpha_n^2} - (I - \mathbf{w}_n\mathbf{h}_n^TK^T)E_n'K\mathbf{h}_n = 0 \quad [87]$$

or alternately

$$\mathbf{w}_n\overline{\alpha_n^2} - \overline{[(I - \mathbf{w}_n\mathbf{h}_n^TK^T)\mathbf{e}_n']\mathbf{e}_n'^T}K\mathbf{h}_n = 0 \quad [88]$$

Substituting for the bracketed quantity from Eq. [72] gives

$$\mathbf{w}_n\overline{\alpha_n^2} + \overline{(\mathbf{w}_n\alpha_n - \mathbf{e}_n)\mathbf{e}_n'^T}K\mathbf{h}_n = 0 \quad [89]$$

But since $\overline{\alpha_n\mathbf{e}_n'^T} = 0$, one obtains

$$\overline{(\mathbf{w}_n\alpha_n)\alpha_n} - \overline{\mathbf{e}_n\mathbf{e}_n'^T}K\mathbf{h}_n = 0 \quad [90]$$

Again substituting for $\mathbf{w}_n\alpha_n$ from Eq. [72] gives

$$\overline{[\mathbf{e}_n - (I - \mathbf{w}_n\mathbf{h}_n^TK^T)\mathbf{e}_n']\alpha_n} - \overline{\mathbf{e}_n\mathbf{e}_n'^T}K\mathbf{h}_n = 0 \quad [91]$$

or, simply

$$\overline{\mathbf{e}_n(\alpha_n - \mathbf{e}_n'^TK\mathbf{h}_n)} = 0 \quad [92]$$

Thus, $\mathbf{e}_n$ and the scalar quantity $\alpha_n - \mathbf{e}_n'^TK\mathbf{h}_n$ are uncorrelated. Hence

$$\overline{\mathbf{e}_n[\mathbf{w}_n^T(\alpha_n - \mathbf{e}_n'^TK\mathbf{h}_n)]} = 0 \quad [93]$$

or, from Eq. [72]

$$\overline{\mathbf{e}_n(\mathbf{e}_n^T - \mathbf{e}_n'^T)} = 0 \quad [94]$$

Therefore

$$\overline{\mathbf{e}_n[\delta\mathbf{x}_n^T + \mathbf{e}_n^T - (\delta\mathbf{x}_n^T + \mathbf{e}_n'^T)]} = 0 \quad [95]$$

or

$$\overline{\mathbf{e}_n\,\delta\hat{\mathbf{x}}_n^T} = \overline{\mathbf{e}_n\,\delta\hat{\mathbf{x}}_n'^T} \quad [96]$$

From this final relationship, it is easy to show that $\mathbf{e}_n$ and $\delta\hat{\mathbf{x}}_n$ are uncorrelated. For if one substitutes from Eqs. [72] and [16], it follows that

$$\begin{aligned}\overline{\mathbf{e}_n\,\delta\hat{\mathbf{x}}_n^T} &= \overline{[(I - \mathbf{w}_n\mathbf{h}_n^TK^T)\Phi_{n,n-1} \times \mathbf{e}_{n-1} + \mathbf{w}_n\alpha_n]\delta\hat{\mathbf{x}}_{n-1}^T}\Phi_{n,n-1}^T \\ &= (I - \mathbf{w}_n\mathbf{h}_n^TK^T)\Phi_{n,n-1}\overline{\mathbf{e}_{n-1}\delta\hat{\mathbf{x}}_{n-1}^T}\Phi_{n,n-1}^T\end{aligned} \quad [97]$$

Then continuing the reduction of $\overline{\mathbf{e}_{n-1}\delta\hat{\mathbf{x}}_{n-1}^T}$ gives, finally, $\overline{\mathbf{e}_n\delta\hat{\mathbf{x}}_n^T}$ related to $\overline{\mathbf{e}_0\delta\hat{\mathbf{x}}_0}$, which is zero. Thus, Eq. [86] is established, and the proof is complete.

Statistical Analysis of Guidance Procedure

From exact knowledge of the six-dimensional deviation vector $\delta\mathbf{x}_n$ at time t_n, a velocity correction may be calculated which, if implemented, will insure the vehicle's arrival at a fixed point in space at the required time. However, only the estimate $\delta\hat{\mathbf{x}}_n$ is available. From this, an estimate of the velocity correction vector $\Delta\hat{\mathbf{v}}_n$ may be determined from

$$\Delta\hat{\mathbf{v}}_n = B_n\,\delta\hat{\mathbf{x}}_n \quad [98]$$

where B_n is defined by Eq. [32]. (Refer to the discussion leading to Eq. [45].)

The need for a velocity correction arises solely from improper injection into orbit. If the first such correction is executed perfectly, then, of course, no further corrections are required. However, because of imperfect knowledge of position and velocity obtained from navigational measurements, the commanded velocity change will be in error. Furthermore, the actual velocity change experienced will differ from that commanded because of imperfect instrumentation. Therefore, subsequent corrections will be required to remove the effects produced by earlier inaccuracies.

Correlation Matrix of Velocity Correction Vector

An estimate of the required velocity correction vector $\Delta\hat{\mathbf{v}}_n$, as computed from Eq. [98], may be determined at each decision time whether or not the correction is actually implemented. The correlation matrix of the velocity correction vector may be expressed directly in terms of the extrapolated matrices E_n' and X_n'.

From Eq. [98] one has

$$\Delta\hat{\mathbf{v}}_n = B_n(\delta\mathbf{x}_n' + \mathbf{e}_n') \qquad [99]$$

so that

$$\overline{\Delta\hat{\mathbf{v}}_n\,\Delta\hat{\mathbf{v}}_n^T} = B_n(\overline{\delta\mathbf{x}_n'\delta\mathbf{x}_n'^T} + \overline{\mathbf{e}_n'\,\delta\mathbf{x}_n'^T} + \overline{\delta\mathbf{x}_n'\mathbf{e}_n'^T} + E_n')B_n^T \qquad [100]$$

On the other hand

$$\delta\hat{\mathbf{x}}_n'^T = \delta\mathbf{x}_n'^T + \mathbf{e}_n'^T \qquad [101]$$

from which

$$\overline{\mathbf{e}_n'\delta\hat{\mathbf{x}}_n'^T} = \overline{\mathbf{e}_n'\delta\mathbf{x}_n'^T} + E_n' = 0 \qquad [102]$$

according to the theorem proved in the section on correlation between estimate an error. Hence

$$\overline{\Delta\hat{\mathbf{v}}_n\,\Delta\hat{\mathbf{v}}_n^T} = B_n(X_n' - E_n')B_n^T \qquad [103]$$

The correlation matrix X_n may be calculated using Eq. [13] when no velocity correction is made. If the velocity is corrected at time t_n, the following procedure is valid.

Using Eq. [29], the following may be written:

$$\begin{aligned}\delta\mathbf{x}_n &= \delta\mathbf{x}_n' + JB_n\delta\hat{\mathbf{x}}_n' - J\mathbf{n}_n \\ &= (I + JB_n)\delta\mathbf{x}_n' + JB_n\mathbf{e}_n' - J\mathbf{n}_n\end{aligned} \qquad [104]$$

Hence

$$\overline{\delta\mathbf{x}_n\,\delta\mathbf{x}_n^T} = (I + JB_n)\,\overline{\delta\mathbf{x}_n'\,\delta\mathbf{x}_n'^T}(I + JB_n)^T + JB_nE_n'(JB_n)^T + J\overline{\mathbf{n}_n\mathbf{n}_n^T}J^T + (I + JB_n)\overline{\delta\mathbf{x}_n'\mathbf{e}_n'^T}(JB_n)^T + JB_n\,\overline{\mathbf{e}_n'\,\delta\mathbf{x}_n'^T}(I + JB_n)^T \qquad [105]$$

which may be further reduced using Eq. [102]. In summary, then

$$X_n = \begin{cases} X_n' & \text{no correction} \\ (1 + JB_n)(X_n' - E_n')(I + JB_n)^T + E_n' + J\overline{\mathbf{n}_n\mathbf{n}_n^T}J^T & \text{correction}\end{cases} \qquad [106]$$

Just as the extrapolated error vector and the associated correlation matrix are altered at an observation point, so will they change also at a correction point. Thus

$$\mathbf{e}_n = \mathbf{e}_n' + \begin{bmatrix} 0 \\ \mathbf{n}_n \end{bmatrix} \qquad [107]$$

$$E_n = E_n' + \begin{bmatrix} O & O \\ O & \overline{\mathbf{n}_n\mathbf{n}_n^T} \end{bmatrix} \qquad [108]$$

The mean-square estimate of the velocity correction is determined as the trace of the matrix $\overline{\Delta\hat{\mathbf{v}}_n\Delta\hat{\mathbf{v}}_n^T}$. As a basis for a decision theory, it is important to know something of the precision of the estimate. Clearly, a velocity correction having a large uncertainty should not be commanded if it is possible to improve substantially the estimate by future observations. The uncertainty $\mathbf{d}_n$ in the estimate $\Delta\hat{\mathbf{v}}_n$ is simply

$$\mathbf{d}_n = \Delta\hat{\mathbf{v}}_n - B_n\,\delta\mathbf{x}_n = B_n\,\mathbf{e}_n' \qquad [109]$$

Hence, the mean-square uncertainty is determined as the trace of the matrix

$$\overline{\mathbf{d}_n\mathbf{d}_n^T} = B_nE_n'B_n^T \qquad [110]$$

Uncertainty in Applied Velocity Correction

In order to complete the statistical analysis of the velocity correction, it is necessary to examine more carefully the vector uncertainty $\mathbf{n}$ in the velocity correction. The inaccuracy in establishing a commanded velocity correction $\Delta\hat{\mathbf{v}}$ is due to errors in both magnitude and orientation. In the following analysis, the two sources of error will be assumed independently random with zero means.

Consider a coordinate system in which the estimated velocity correction vector is along one of the coordinate axes. Then if M is the transformation matrix that relates the selected axis system and the original reference system, one may write

$$\Delta\hat{\mathbf{v}} = \Delta\hat{v}M\begin{bmatrix} 0 \\ 0 \\ 1 \end{bmatrix} \qquad [111]$$

Now, define a random variable κ such that

$$\Delta v = (1 + \kappa)\Delta\hat{v} \qquad [112]$$

and let γ be the random angle between $\Delta\hat{\mathbf{v}}$ and $\Delta\mathbf{v}$. It will be assumed that both κ and γ are small quantities so that powers and products are negligible compared with unity. The actual vector velocity correction is then

$$\Delta\mathbf{v} = (1 + \kappa)\Delta\hat{v}M\begin{bmatrix} \gamma\cos\beta \\ \gamma\sin\beta \\ 1 \end{bmatrix} \qquad [113]$$

where β is a polar angle defining the rotation of $\Delta\mathbf{v}$ with respect to $\Delta\hat{\mathbf{v}}$. Hence, the uncertainty vector $\mathbf{n}$ is expressible as

$$\mathbf{n} = \Delta\hat{\mathbf{v}} - \Delta\mathbf{v} = -\Delta\hat{v}M\left\{(1 + \kappa)\gamma\begin{bmatrix} \cos\beta \\ \sin\beta \\ 0 \end{bmatrix} + \kappa\begin{bmatrix} 0 \\ 0 \\ 1 \end{bmatrix}\right\} \qquad [114]$$

Assume that κ, γ, β are statistically independent random variables with zero means. Further assume that β is uniformly distributed over the interval $-\pi$ to π. Then one obtains for the correlation matrix of the velocity correction uncertainty

$$\begin{aligned}\overline{\mathbf{n}\mathbf{n}^T} &= \overline{\kappa^2}\,\overline{\Delta\hat{\mathbf{v}}\Delta\hat{\mathbf{v}}^T} + \frac{\overline{\gamma^2}}{2}\,\overline{\Delta\hat{v}^2}M\begin{bmatrix} 1 & 0 & 0 \\ 0 & 1 & 0 \\ 0 & 0 & 0 \end{bmatrix}M^T \\ &= \overline{\kappa^2}\,\overline{\Delta\hat{\mathbf{v}}\Delta\hat{\mathbf{v}}^T} + \frac{\overline{\gamma^2}}{2}(\overline{\Delta\hat{\mathbf{v}}^T\Delta\hat{\mathbf{v}}}I - \overline{\Delta\hat{\mathbf{v}}\Delta\hat{\mathbf{v}}^T})\end{aligned} \qquad [115]$$

where I is the three-dimensional identity matrix and $\overline{\kappa^2}$ and $\overline{\gamma^2}$ are the mean-squared values of κ and γ.

Miss Distance at Target

Turning now to the problem of guidance accuracy, the determination of the position deviation vector at the nominal time of arrival at the target is made by extrapolating the deviation vector from the point of the final velocity correction. Thus, if t_N is the time of the last correction and $\delta\mathbf{x}_A$ is the deviation vector at the time of arrival t_A, then

$$\delta\mathbf{x}_A = \Phi_{A,N}\,\delta\mathbf{x}_N^+ \qquad [116]$$

But from Eq. [67] and the terminal conditions for the navigation matrices, one has

$$\Phi_{A,N} = \begin{bmatrix} -R_A\Lambda_N^{-1} & O \\ -V_A\Lambda_N^{-1} & -\Lambda_N^{*-1} \end{bmatrix}\begin{bmatrix} C_N^* & -I \\ C_N & -I \end{bmatrix} \qquad [117]$$

Table 1 Typical root-mean-square injection errors

Altitude	Track	Range
10,000 ft	15,000 ft	5000 ft
15 fps	6 fps	4 fps

Hence, the position deviation vector at the target $\delta \mathbf{r}_A$ may be written as

$$\delta \mathbf{r}_A = -R_A \Lambda_N^{-1} B_N \delta \mathbf{x}_N^+ \qquad [118]$$

with a similar expression obtainable for the velocity deviation at time t_A.

The target position error may be written ultimately in terms of the error vector $\mathbf{e}_N$ according to the following self-evident steps

$$\begin{aligned} \delta \mathbf{r}_A &= -R_A \Lambda_N^{-1} B_N (\delta \mathbf{x}_N^- + J \Delta \mathbf{v}_N) \\ &= -R_A \Lambda_N^{-1} (B_N \, \delta \mathbf{x}_N^- - \Delta \mathbf{v}_N) \\ &= R_A \Lambda_N^{-1} (B_N \, \mathbf{e}_N' - \mathbf{n}_N) \\ &= R_A \Lambda_N^{-1} B_N \, \mathbf{e}_N \end{aligned} \qquad [119]$$

The mean-square position error at the target is then computed as the trace of the matrix $\overline{\delta \mathbf{r}_A \, \delta \mathbf{r}_A^T}$.

Application to Circumlunar Navigation

Decision Rules

As a necessary step in the application of the navigation and guidance scheme formulated in this paper, certain rules must be adopted concerning the course of action to be taken at each of the "decision points" described in the section on summary of navigation and guidance equations. The number and frequency of observations must be controlled in some manner—ideally by a decision rule that is realistically compatible with both the mission objectives and the capabilities of the measuring device. If an observation is to be made, a decision is required regarding the type of measurement and the celestial objects to be used. Periodic velocity corrections must be applied, and the number of impulses and times of occurrence must be decided.

Once the decision rules have been specified, it is necessary to test their effectiveness according to some measure of performance. A typical objective is to minimize the miss distance at the target. However, a reduction in miss distance usually implies an increase in either the required number of measurements or a greater expenditure of corrective propulsion or both. In the face of these conflicting objectives, compromises are clearly necessary and statistical simulation provides a means of arriving at an acceptable balance.

In the interest of minimizing the number of simulator runs, Monte Carlo techniques should be avoided if possible. Fortunately, it is unnecessary to generate the true spacecraft trajectory, as would be required for Monte Carlo simulation, in order to analyze the effects of a particular set of decision rules. The reader may readily verify that Eq. [29], which defines the estimate $\delta \hat{\mathbf{x}}_n$ and depends on actual measurement data, is never involved in any of the statistical calculations.

A specific example of a set of decision rules to be applied at each decision point is as follows:

1 The estimated mean-squared velocity correction $\overline{\Delta \hat{v}_n^2}$ and the mean-squared uncertainty $\overline{d_n^2}$ associated with the estimate are computed from Eqs. [103] and [110]. If the ratio

$$R_v = (\overline{d_n^2} / \overline{\Delta \hat{v}_n^2})^{1/2} \qquad [120]$$

is less than a specified amount $R_{v(\min)}$, a velocity correction is made at time t_n.

2 If the criterion is not met which would call for initiation of a velocity correction, the desirability of making an observation is examined. For this purpose, an abbreviated star catalog is postulated together with selected planets. Each star and planet measurement combination is analyzed to determine its effect on the reduction in position uncertainty at the target. The particular star-planet combination producing the greatest mean-square reduction is then defined as the best potential measurement.

Now let $\overline{\delta r_A^{2+}}$ and $\overline{\delta r_A^{2-}}$ be the respective mean-square position uncertainties at the target which would result with and without the best possible observation. Then, if the ratio

$$R_p = \left(\frac{\overline{\delta r_A^{2-}} - \overline{\delta r_A^{2+}}}{\overline{\delta r_A^{2-}}} \right)^{1/2} \qquad [121]$$

is greater than a specified value $R_{p(\max)}$, the best potential measurement is made at time t_n. In other words, for a measurement to be made, a significant reduction in the potential miss distance must result. If, on the other hand, the foregoing criterion is not met, no action is taken at the decision point t_n.

Numerical Example

In this section, the decision rules presented previously are applied to the circumlunar navigation problem. It was found that the velocity correction criterion worked quite well to establish the times of midcourse maneuvers with the exception of the final correction. The required velocity change increases quite rapidly as the target is approached, and the timing of this last correction is critical. After preliminary experimentation with different values of R_v, it was decided to fix a priori the correction times for the remainder of the study of the navigation problem. Cross-correlation between measurement errors was ignored, and only Earth and the moon together with the 20 brightest stars were considered for potential measurements.

The date and time of orbital injection was Julian Day 2440043.6088, with the closest point of approach some 60 miles from the lunar surface. The nominal total time of flight from injection was 126.4 hr.

The correlation matrix of injection errors E_0 was obtained from the assumed root-mean-square injection errors shown in Table 1.

The following correlation matrix was obtained by a transformation from the altitude, track, range coordinate system to a coordinate system with the x axis along the vernal equinox, z axis along the Earth polar axis, and the y axis chosen to make a right-handed coordinate system. The basic units in the E_0 matrix are miles and miles per hour:

$$E_0 = \begin{bmatrix} 0.918 & 0.063 & 0.203 & 0 & 0 & 0 \\ 0.063 & 4.580 & -1.860 & 0 & 0 & 0 \\ 0.203 & -1.860 & 7.040 & 0 & 0 & 0 \\ 0 & 0 & 0 & 7.73 & 4.65 & 2.72 \\ 0 & 0 & 0 & 4.65 & 83.80 & 36.00 \\ 0 & 0 & 0 & 2.72 & 36.00 & 36.10 \end{bmatrix}$$

At each decision point, 40 potential measurements were examined and evaluated according to the decision criterion. The minimum time between observations was required to be 15 min. For simplicity, only star elevations above an illuminated horizon of either Earth or the moon were considered. Certain practical constraints were imposed so that physically unrealizable measurements were screened out. For example, in order to keep the field of view requirements reasonable, the lines of sight to the star and to the horizon were required not to exceed 70°. Also, no measurement could be made if the line of sight to either star or planet edge was closer than 15° from the direction to the sun. Furthermore, if the illuminated face of the moon formed the background of the edge of Earth from which a star elevation was to be reckoned, that particular measurement would not be made.

The optical measuring device used for the observations was assumed to be unbiased with a random error whose variance was

$$\sigma_E^2 = (0.00005)^2 + \left(\frac{1}{r_{SE}}\right)^2 \text{ rad}$$

for Earth, and

$$\sigma_M^2 = (0.00005)^2 + \left(\frac{0.5}{r_{SM}}\right)^2 \text{ rad}$$

for the moon, where r_{SE} and r_{SM} are the distances in miles from the spacecraft to Earth and the moon, respectively. In this manner, it was possible to account for the larger uncertainty in defining the horizon that would exist when the spacecraft is close to a planet. At large distances, the rms error is approximately 0.05 mrad.

The magnitude error in applying a velocity correction was assumed to be isotropic and proportional to the commanded correction. Specifically, the relation

$$\overline{\eta_n^2} = 0.0001\ \overline{\Delta v_n^2}$$

was adopted so that the rms error would be 1% of the rms correction. The orientation error assumed was 0.01 rad.

Preliminary results of an analysis of this sample trajectory are summarized in Tables 2–10. A number of simulated guidance flights were made for which the strategy parameters R_v and R_p had various assigned values. Then, in order to evaluate the effect on the navigation data of a variation in the time of year, a set of pseudo-trajectories was generated by the simple device of rotating the direction of the sun as viewed from Earth. The trajectory was considered to be unchanged by this process, the assumption being quite adequate for the purpose of this preliminary analysis. In this manner, different illuminated portions of Earth and the moon were visible to the spacecraft, resulting, thereby, in different measurements.

In general, as R_p is increased, one requires each measurement to have a proportionately greater significance in the reduction of the potential target error, with the result that the required total number of measurements decreases. There may be a corresponding penalty, of course, in that the resulting uncertainties in position and velocity at the target can increase. The objective in preparing a measurement schedule is to arrive at an acceptable compromise.

The number of velocity corrections as well as the times of their occurrence is, of course, controlled by R_v. On the other hand, the number of measurements is not sensibly affected by variations in this parameter. As an example, in Table 2 navigation data for the Earth to moon trajectory is given for two values of the velocity correction uncertainty ratio R_v. Although the final position uncertainties are of the order of 2 miles, the deviations from the reference path are approximately 12 miles. This large difference results from the fact that measurement data were gathered after the final velocity correction so that knowledge of the orbit improved although no attempt was made to reduce the target error. It should be noted that, if one elects to eliminate the final position deviation by a velocity correction 0.1 hr before the nominal arrival time, velocity corrections of 104 and 68 mph, respectively, are required. There will, of course, be an accompanying increase in the final velocity deviations of 51 and 52 mph, respectively.

In Table 3, the navigation data for the Earth to moon trajectory are given as a function of the miss distance reduction ratio R_p for velocity corrections made at 5, 20, 52, and 61.5 hr. For the case $R_p = 0.6$, there is a noticeable decrease in the final position uncertainty compared to that for $R_p = 0.5$. This apparent anomaly arises from the fact that, for the $R_p = 0.6$ case, three observations are made after the last velocity correction, whereas, correspondingly, only two observations are made for the $R_p = 0.5$ case. Table 4 presents similar data for the moon to Earth trajectory.

In order to study the effect of variations in the illuminated portions of the planet's surfaces, one set of values for R_p and times for velocity corrections was selected and the sun direction altered in 60° steps, except for the 70° and 250° cases. These two directions were singled out because they form a line approximately perpendicular to the Earth-moon line at launch. Table 5 gives the results for the Earth to moon trajectory and further shows that the 70° and 180° cases produce significantly larger uncertainties. For the 120° case, the total velocity correction of 114 mph is somewhat higher. However, this can be improved since the times selected for velocity corrections were not optimum for all cases. Table 6 presents similar data for the moon to Earth trajectory.

In all cases, the final velocity correction just prior to arrival at perilune is significantly larger than the previous two midcourse corrections. The result is a rather large velocity deviation from the nominal value at the target point. On the return flight, this deviation causes the first velocity correction to be substantial, which accounts for the increase in fuel requirements required for the moon to Earth trip. If the objective of the flight does not include passage through a preassigned perilune position, then, obviously, the total of velocity corrections can be reduced.

Table 7 summarizes Earth to moon flight navigation data for various moon horizon uncertainties. The number of measurements remained constant (76 and 77) for the cases investigated. Total velocity corrections, final velocity deviations, and final position deviations did not increase until the uncertainty reached 5 miles. However, final position and velocity uncertainties are sensitive to moon horizon determination, as would be expected.

Table 8 presents the same data for the moon to Earth flight for various Earth horizon uncertainties. The number of measurements and total velocity correction did not vary appreciably. However, all uncertainties and deviations are sensitive to Earth horizon determination.

Table 2 Earth to moon flight navigation data as a function of velocity correction uncertainty ratio[a]

Velocity correction uncertainty ratio	Number of measurements	Times for velocity corrections, hr	Total velocity correction, mph	Final position uncertainty, miles	Final velocity uncertainty, mph	Final position deviation, miles	Final velocity deviation mph
0.2	39	7.0 18.0 61.8	107	2.5	11.1	12.5	95
0.3	40	5.5 11.5 26.0 61.4	77	1.8	4.6	12.0	39

[a] Miss distance reduction ratio = {0.1. from start to 8 hr. / 0.5 from 8 to 62.5 hr; sun line = 250°.

Table 3 Earth to moon flight navigation data as a function of miss distance reduction ratio[a]

Miss distance reduction ratio[b]	Number of measurements	Total velocity correction, mph	Final position uncertainty, miles	Final velocity uncertainty, mph	Final position deviation, miles	Final velocity deviation, mph
0.2	115	52	0.70	1.7	3.9	16
0.3	77	56	1.10	3.7	7.1	23
0.4	55	59	1.10	3.7	8.7	26
0.5	40	78	1.20	4.0	11.0	60
0.6	32	68	0.84	3.1	17.4	66

[a] Miss distance reduction ratio constant at 0.1 from 0 to 8 hr; velocity corrections at 5, 20, 52, 61.5 hr; sun line = 250°.
[b] From 8 to 62.5 hr.

Table 4 Moon to Earth flight navigation data as a function of miss distance reduction ratio[a]

Miss distance reduction ratio	Number of measurements	Total velocity correction, mph	Final position uncertainty, miles	Final velocity uncertainty, mph	Final position deviation, miles	Final velocity deviation, mph
0.2	97	82	1.5	2.8	10.0	22
0.3	44	89	1.6	3.1	12.6	28
0.4	28	99	1.8	3.4	13.9	33
0.5	12	197	2.5	4.8	15.2	94
0.6	10	211	4.3	8.0	28.0	123

[a] Velocity corrections at 64, 88, 120, 125 hr; sun line = 250°.

Table 5 Earth to moon flight navigation data for pseudo-trajectories as a function of sun direction rotation[a]

Sun direction rotation, deg	Number of measurements	Total velocity correction, mph	Final position uncertainty, miles	Final velocity uncertainty, mph	Final position deviation, miles	Final velocity deviation, mph
0	41	68	1.3	3	3	31
70	39	64	6.5	18	8	39
120	39	114	1.6	3	12	92
180	40	66	5.2	21	12	48
*250	40	78	1.2	4	11	60
300	39	88	1.2	4	4	46

[a] Miss distance reduction ratio = {0.1 from start to 8 hr. / 0.5 from 8 to 62.5 hr; velocity corrections at 5, 20, 52, 61.5 hr.

Table 6 Moon to Earth flight navigation data for pseudo-trajectories as a function of sun direction rotation[a]

Sun direction rotation, deg	Number of measurements	Total velocity correction, mph	Final position uncertainty, miles	Final velocity uncertainty, mph	Final position deviation, miles	Final velocity deviation, mph
70	16	80	3.5	6.5	24	53
120	16	227	2.9	5.5	21	66
180[b]	20	94	2.1	4.3	10	41
*250	28	99	1.8	3.4	14	33
300	14	163	2.3	3.9	31	99

[a] Miss distance reduction ratio = 0.4; velocity corrections at 64, 88, 120, 125 hr.
[b] First correction at 70 hr.

Finally, in Tables 9 and 10, a complete history of a circumlunar mission is given corresponding to the cases marked with an asterisk summarized in Tables 5 and 6.

Appendix A: Navigational Measurements

The mathematical processes are considered here in some detail for determining spacecraft position by means of both celestial observation and ground-based radar measurements. It is assumed throughout the analysis that approximations to spacecraft position and velocity are already known so that perturbation techniques may be employed.

Secondary effects arising from the finite speed of light, the finite distance of stars, etc. are ignored in this analysis. Such effects may be lumped together for a particular reference point on the trajectory as a modification to the stored data that represent reference values for the quantities to be measured at that point.

For simplicity in the present analysis, it will be assumed that the spacecraft clock is perfect so that all measurements are made at known instants of time. Methods of including clock errors in the computation are discussed thoroughly in Ref. 2.

As indicated in the section on deterministic method, each measurement establishes a component of spacecraft position along some direction in space. If Q is the quantity to be measured and δQ is the difference between the true and the reference values, then it will be shown that the relation between δQ and the deviation in spacecraft position $\delta \mathbf{r}$ is

$$\delta Q = \mathbf{h}^T \delta \mathbf{r} \quad [A1]$$

regardless of the type of measurement. Thus, the **h** vector alone will characterize the kind of measurement.

Sun-Planet Measurement

The first type of measurement to be considered is that of the angle from the sun to a planet. By passing to the limit of infinite distance from one or the other of these bodies, corresponding relations for the sun-star or planet-star type of measurement may be obtained.

Let S_0 and P_0 be, respectively, the reference positions of the spacecraft and a planet at the time of the measurement. Let **r** be the vector from the sun to S_0 and **z** the vector from S_0 to P_0. With A denoting the angle from the sun line to the planet line, one has

$$\cos A = -(\mathbf{r}\cdot\mathbf{z})/rz \qquad [A2]$$

where r and z denote magnitudes of the respective vectors **r** and **z**. Treating all changes as first-order differentials, it can be shown that

$$\delta A = \left(\frac{\mathbf{m} - (\mathbf{n}\cdot\mathbf{m})\mathbf{n}}{r \sin A} + \frac{\mathbf{n} - (\mathbf{n}\cdot\mathbf{m})\mathbf{m}}{z \sin A}\right)\cdot\delta\mathbf{r} \qquad [A3]$$

For details, the reader is referred to Ref. 2. Here **n** and **m** are, respectively, the unit vectors from S_0 toward the sun and toward P_0. The two individual vector coefficients of $\delta\mathbf{r}$ in Eq. [A3] are vectors in the plane of the measurement and normal, respectively, to the lines of sight to the sun and to the planet.

Planet Diameter Measurement

If D is the actual diameter of a planet, the apparent angular diameter A is found from

$$\sin(A/2) = D/2z \qquad [A4]$$

Again taking differentials as before, one can show that

$$\delta A = \frac{D\mathbf{m}\cdot\delta\mathbf{r}}{z^2 \cos(A/2)} \qquad [A5]$$

Star Occultations

The next type of measurement to be considered is that of noting the time at which a star is occulted by a planet. Let

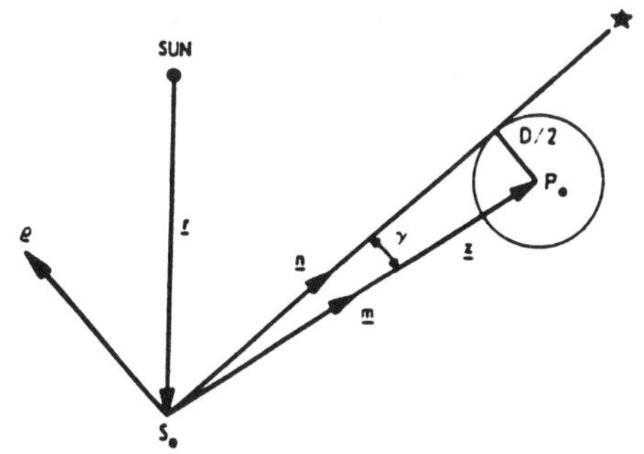

Fig. 1 Measurement of time of a star occultation

z be the vector from S_0 to P_0, **r** the vector from the sun to S_0, and **n** a unit vector in the direction of the star to be occulted. With γ denoting the angle from the star line to the planet line as shown in Fig. 1, one has, at the nominal instant of occultation

$$\mathbf{n}\cdot\mathbf{z} = z \cos\gamma \qquad [A6]$$

Treating changes as first-order differentials, one obtains

$$\begin{aligned} \mathbf{n}\cdot\delta\mathbf{z} &= \cos\gamma\, \delta z - z \sin\gamma\, \delta\gamma \\ &= \cos\gamma\, \mathbf{m}\cdot\delta\mathbf{z} - z \sin\gamma\, \delta\gamma \end{aligned} \qquad [A7]$$

where **m** is a unit vector from S_0 toward P_0.

The angle deviation $\delta\gamma$ is computed from a first-order differential of $2z \sin\gamma = D$. There results

$$\delta\gamma = -D\mathbf{m}\cdot\delta\mathbf{z}/2z^2 \cos\gamma \qquad [A8]$$

Furthermore, if $\mathbf{v}_p$ and $\mathbf{v}_s$ are the respective velocity vectors of the planet and the spacecraft, and if $\delta\tau$ is the difference between the observed and the reference occultation times, there results

$$\begin{aligned} \delta\mathbf{z} &= \mathbf{v}_p\delta\tau - (\delta\mathbf{r} + \mathbf{v}_s\,\delta\tau) \\ &= -\delta\mathbf{r} - \mathbf{v}_r\,\delta\tau \end{aligned} \qquad [A9]$$

where $\mathbf{v}_r$ is the velocity of the spacecraft relative to the planet. Then, by combining Eqs. [A7–A9] one has finally

$$\delta\tau = -\frac{\boldsymbol{\rho}\cdot\delta\mathbf{r}}{\boldsymbol{\rho}\cdot\mathbf{v}_r} \qquad [A10]$$

Table 7 Earth to moon flight navigation data as a function of moon horizon uncertainty[a]

Moon horizon uncertainty, miles	Number of measurements	Total velocity correction, mph	Final position uncertainty, miles	Final velocity uncertainty, mph	Final position deviation, miles	Final velocity deviation, mph
0.5	77	56	1.1	3.7	7.1	23
1.0	76	54	2.0	8.7	7.8	23
2.0	76	54	2.9	10.6	7.9	23
3.0	76	55	3.6	10.3	8.1	23
5.0	76	68	5.4	16.9	8.7	27

[a] Miss distance reduction ratio = {0.1 from start to 8 hr. / 0.3 from 8 to 62.5 hr; sun line = 250°; velocity corrections at 5, 20, 52, 61.5 hr.

Table 8 Moon to Earth flight navigation data as a function of Earth horizon uncertainty[a]

Earth horizon uncertainty, miles	Number of measurements	Total velocity correction, mph	Final position uncertainty, miles	Final velocity uncertainty, mph	Final position deviation, miles	Final velocity deviation, mph
1	44	89	1.6	3.1	12.6	28
2	44	88	2.6	4.8	15.3	32
3	42	89	3.8	7.1	19.1	38
5	42	91	5.8	10.7	21.6	43

[a] Miss distance reduction ratio = 0.3; sun line = 250°; velocity corrections at 64, 88, 120, 125 hr.

Table 9 Typical navigation data for Earth to moon flight[a]

Time, hr	Observation	Velocity correction, mph	Reduction in position uncertainty at target, miles	Position uncertainty at target, miles	Indicated velocity correction, mph	Uncertainty in velocity correction, mph	Position uncertainty, miles	Velocity uncertainty, mph	Position deviation, miles	Velocity deviation, mph
0.6	Moon, Antares		262	2528	0	11.9	4.8	10.9	4.9	11.0
0.9	Earth, Fomalhaut		2031	1504	1.3	12.9	4.5	7.9	7.4	11.3
1.2	Earth, Deneb		540	1404	11.0	9.2	5.3	7.9	10.4	12.0
1.5	Earth, Aldebaran		412	1342	12.6	9.2	6.5	7.6	13.9	12.7
1.8	Earth, Aldebaran		370	1290	14.0	9.2	8.2	7.6	17.6	13.5
2.2	Earth, Aldebaran		408	1224	15.5	9.4	10.6	7.6	23.1	14.6
2.6	Earth, Pollux		456	1136	16.9	9.5	12.0	7.2	28.9	15.5
3.0	Earth, Procyon		515	1013	18.3	9.2	13.2	6.8	35.2	16.4
3.4	Earth, Procyon		405	928	19.7	8.7	14.6	6.5	41.8	17.2
3.8	Earth, Pollux		426	825	20.9	8.3	15.2	6.0	48.7	17.9
4.5	Earth, Procyon		403	719	22.8	8.0	17.0	5.6	61.5	19.1
5.0		24.1								
5.5	Earth, Pollux		435	573	0	7.6	18.4	4.9	70.0	7.8
6.0	Earth, Procyon		273	504	4.6	6.4	18.3	4.4	69.3	7.6
6.5	Earth, Pollux		244	441	5.7	5.8	18.1	4.1	69.0	7.5
7.0	Moon, Antares		196	395	6.5	5.3	18.2	3.7	69.1	7.5
7.5	Earth, Pollux		186	348	7.2	4.9	17.9	3.4	69.5	7.5
8.5	Moon, Antares		187	294	7.9	4.7	18.3	3.0	71.4	7.6
9.5	Moon, Antares		158	248	8.7	4.1	18.3	2.7	74.3	7.7
10.0	Earth, Pollux		135	208	9.3	3.6	17.1	2.4	76.2	7.8
10.5	Moon, Antares		107	179	9.7	3.1	16.0	2.1	78.3	7.9
12.0	Moon, Antares		95	151	10.5	2.9	16.7	1.8	85.8	8.3
12.5	Earth, Pollux		78	130	10.8	2.5	15.9	1.7	88.7	8.4
13.5	Moon, Antares		68	111	11.3	2.3	15.6	1.5	94.9	8.6
15.0	Moon, Antares		56	96	12.0	2.0	16.0	1.3	105.2	8.9
16.0	Earth, Pollux		48	83	12.6	1.8	16.0	1.2	112.7	9.1
17.0	Moon, Antares		41	72	13.0	1.7	15.7	1.1	120.6	9.3
19.5	Moon, Antares		36	62	14.3	1.6	16.9	1.0	141.9	9.8
20.0		14.5								
22.0	Earth, Pollux		32	54	0	1.5	18.2	0.9	137.4	4.6
23.5	Moon, Antares		28	47	0.6	1.4	18.0	0.8	131.2	4.4
28.5	Moon, Antares		23	40	1.1	1.4	20.2	0.7	112.4	4.0
29.5	Earth, Pollux		20	35	1.4	1.3	20.2	0.7	109.0	3.9
37.0	Moon, Antares		17	30	1.9	1.6	23.3	0.6	86.2	3.6
40.5	Earth, Pollux		15	26	2.4	1.6	24.6	0.5	77.4	3.5
52.0		5.4								
53.5	Moon, Antares		13	22	0	3.4	28.7	0.4	54.4	5.1
57.5	Earth, Regulus		11	19	2.8	5.1	28.5	0.3	40.9	4.9
60.0	Moon, Regulus		10	17	6.8	8.7	18.2	0.7	35.5	4.8
60.5	Earth, Procyon		9	14	11.1	7.7	13.2	0.8	34.7	5.0
61.4	Moon, Regulus		8	11	21.1	10.9	11.9	1.7	33.3	6.7
61.8		33.5								
62.1	Moon, Aldebaran		10	4	0	24.6	8.4	4.4	23.6	36.2
62.4	Moon, α Crucis		4	1	59.9	22.7	1.3	2.4	14.9	51.2
62.56							1.2	3.9	11.2	60.1

[a] Miss distance reduction ratio = { 0.1 from start to 8 hr. ; 0.5 from 8 to 62.5 hr; sun line = 250°.

Table 10 Typical navigation data for moon to Earth flight[a]

Time, hr	Observation	Velocity correction mph	Reduction in position uncertainty at target, miles	Position uncertainty at target, miles	Indicated velocity correction, mph	Uncertainty in velocity correction, mph	Position uncertainty, miles	Velocity uncertainty, mph	Position deviation, miles	Velocity deviation, mph
62.6	Moon, Achernar		86	154	59.8	4.9	1.1	3.8	11.1	60.2
64.0	Moon, Fomalhaut		96	120	62.7	2.5	1.9	1.1	86.1	63.2
64.4		63.4								
65.0	Moon, Altair		105	172	0	1.4	1.7	1.1	111.4	3.4
79.0	Earth, Pollux		70	157	1.0	1.3	12.7	0.8	100.6	2.0
79.5	Moon, Fomalhaut		65	143	1.3	1.1	11.6	0.7	100.1	2.0
80.0	Earth, Aldebaran		61	130	1.4	1.0	11.2	0.7	99.6	2.0
80.5	Moon, Antares		53	118	1.4	0.9	8.3	0.5	99.1	2.0
81.0	Earth, Pollux		56	104	1.6	0.7	7.7	0.5	98.6	2.0
81.5	Earth, Pollux		44	95	1.6	0.6	7.4	0.4	98.1	2.0
83.5	Earth, Pollux		38	87	1.7	0.6	7.8	0.4	96.2	2.0
86.5	Earth, Pollux		35	79	1.8	0.6	8.5	0.4	93.5	2.0
88.5		1.9								
90.5	Earth, Pollux		32	72	0	0.6	9.4	0.4	88.7	1.7
95.5	Earth, Aldebaran		29	66	0.3	0.7	10.6	0.3	80.9	1.8
96.0	Moon, Antares		27	60	0.4	0.7	9.8	0.3	80.1	1.8
97.0	Earth, Regulus		24	55	0.5	0.6	9.5	0.3	78.5	1.8
105.0	Earth, Regulus		22	51	0.8	0.9	11.3	0.3	65.1	2.0
105.5	Moon, Antares		20	46	0.9	0.8	10.8	0.3	64.2	2.0
113.0	Earth, Regulus		19	43	1.8	1.3	12.2	0.3	50.3	2.4
114.0	Moon, Antares		17	39	2.0	1.4	11.8	0.3	48.4	2.5
120.5		5.7								
121.0	Moon, Antares		16	36	0	3.7	13.2	0.8	33.9	5.5
122.5	Moon, Fomalhaut		14	33	2.4	5.2	13.2	1.2	27.7	5.6
123.8	Earth, Procyon		14	29	5.4	8.3	11.1	1.7	23.5	6.0
124.2	Earth, Pollux		19	22	8.4	9.0	9.3	1.7	22.6	6.4
124.6	Earth, Canopus		14	17	13.1	9.1	7.9	1.7	22.1	7.0
125.0	Earth, Canopus		11	14	19.9	9.8	7.0	1.9	22.3	8.2
125.3		28.2								
125.6	Earth, Canopus		11	9	0	16.0	5.5	2.5	17.4	21.4
125.9	Earth, α Centauri		8	3	22.2	18.3	2.1	1.5	13.3	21.0
126.2	Earth, Antares		2	2	66.7	14.4	1.6	1.7	12.2	24.0
126.4							1.8	3.4	13.9	33.1

[a] Miss distance reduction ratio = 0.4; sun line = 250°.

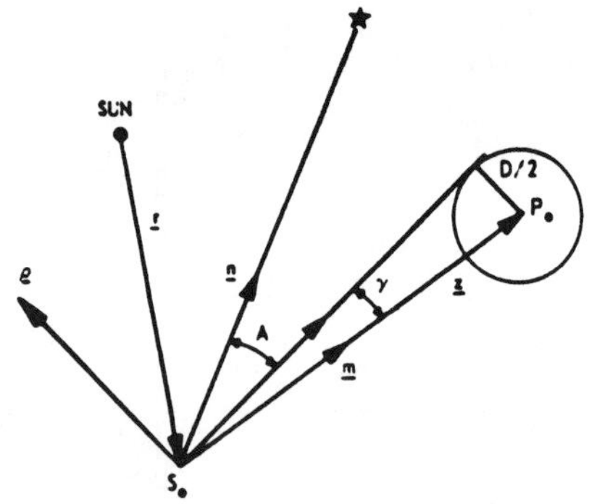

Fig. 2 Measurement of star elevation angle

where $\boldsymbol{\rho}$ is a unit vector perpendicular to $\mathbf{n}$ and lying in the plane determined by the lines of sight to the planet and the star.

Star Elevation Measurement

Consider next the measurement of the angle between the lines of sight to a star and the edge of a planet disc. From Fig. 2, one has

$$\mathbf{n}\cdot\mathbf{z} = z\cos(A+\gamma) \qquad \text{[A11]}$$

where A is the angle to be measured. Again taking total differentials and noting that $\delta\mathbf{r} = -\delta\mathbf{z}$, one obtains

$$\begin{aligned}(\boldsymbol{\rho}\cdot\delta\mathbf{r})/z &= \delta A + \delta\gamma \\ &= \delta A + D\mathbf{m}\cdot\delta\mathbf{r}/2z^2\cos\gamma \\ &= \delta A + \tan\gamma\, \mathbf{m}\cdot\delta\mathbf{r}/z\end{aligned} \qquad \text{[A12]}$$

or finally

$$\delta A = \frac{\boldsymbol{\rho}\cdot\delta\mathbf{r}}{z\cos\gamma} \qquad \text{[A13]}$$

Landmark Measurement

For the measurement of the angle between a landmark on a planet surface and a star, let $\boldsymbol{\rho}$ be a unit vector perpendicular to the line of sight to the landmark and in the plane of the measurement. Then if $\mathbf{p}$ is the vector position of the landmark relative to the center of the planet, one has

$$\delta A = \frac{\boldsymbol{\rho}\cdot\delta\mathbf{r}}{|\mathbf{z}+\mathbf{p}|} \qquad \text{[A14]}$$

Radar Range, Azimuth, and Elevation Measurements

Assume the radar site to be the origin of the coordinate system, although other origins could be used equally well. Let a Cartesian coordinate system be chosen such that the z axis is radially out from the center of Earth through the radar site; the x axis is positive in the direction from which radar azimuths are to be measured; the y axis completes the coordinate system. Then

$$\mathbf{r} = r\begin{bmatrix}\cos\beta\,\cos\theta \\ \cos\beta\,\sin\theta \\ \sin\beta\end{bmatrix} \qquad \text{[A15]}$$

may be written, where r, θ, β are, respectively, the range, azimuth, and elevation of the vehicle from the radar site. Taking differentials separately for each of the three variables gives

$$\frac{\partial\mathbf{r}}{\partial r}\delta r = \begin{bmatrix}\cos\beta\,\cos\theta \\ \cos\beta\,\sin\theta \\ \sin\beta\end{bmatrix}\delta r \qquad \text{[A16]}$$

$$\frac{\partial\mathbf{r}}{\partial\beta}\delta\beta = r\begin{bmatrix}-\sin\beta\,\cos\theta \\ -\sin\beta\,\sin\theta \\ \cos\beta\end{bmatrix}\delta\beta \qquad \text{[A17]}$$

$$\frac{\partial\mathbf{r}}{\partial\theta}\delta\theta = r\begin{bmatrix}-\cos\beta\,\sin\theta \\ \cos\beta\,\cos\theta \\ 0\end{bmatrix}\delta\theta \qquad \text{[A18]}$$

Then, by expressing each of these relations in the form of Eq. [A1], one obtains

$$\delta r = [\cos\beta\cos\theta \quad \cos\beta\sin\theta \quad \sin\beta]\delta\mathbf{r} \qquad \text{[A19]}$$

$$\delta\beta = \frac{1}{r}[-\sin\beta\cos\theta \quad -\sin\beta\sin\theta \quad \cos\beta]\delta\mathbf{r} \qquad \text{[A20]}$$

$$\delta\theta = \frac{1}{r\cos\beta}[-\sin\theta \quad \cos\theta \quad 0]\delta\mathbf{r} \qquad \text{[A21]}$$

The vector coefficients in Eqs. [A19–A21] are each unit vectors in the direction of increasing r, β, θ, respectively.

Appendix B: Optimum Selection of Navigation Measurements

In the main body of this paper, a method of processing measurement data in an optimum linear manner has been developed. The purpose of this Appendix is to treat the associated problem of selecting those measurements that are, in some sense, most effective. For example, the requirement might be to select the measurement to be made at time t_n in order to get the maximum reduction in mean-squared positional or velocity uncertainty at time t_n. Of perhaps greater significance would be the requirement of selecting the measurement that minimizes the uncertainty in any linear combination of position and velocity deviations. Specifically, one might select the measurement that minimizes the uncertainty in the required velocity correction. As a further example, one might wish to select that measurement that, if followed immediately by a velocity correction, would result in the smallest position error at the target.

Consider first the simplest case, i.e., minimizing the mean-squared positional uncertainty at time t_n. From Eq. [29], the mean-squared positional uncertainty is expressible as

$$\overline{\epsilon_n^2} = tr(E_n^{(1)\prime}) - \frac{\mathbf{h}_n^T E_n^{(1)\prime} E_n^{(1)\prime}\mathbf{h}_n}{\mathbf{h}_n^T E_n^{(1)\prime}\mathbf{h}_n + \overline{\alpha_n^2}} \qquad \text{[B1]}$$

assuming the measurement errors to be uncorrelated. In the absence of any measurement error ($\overline{\alpha_n^2} = 0$), the problem of minimizing either mean-squared error is equivalent to finding a direction for the $\mathbf{h}_n$ vector which maximizes the ratio of two quadratic forms. For the case of the mean-squared positional error, the geometrical interpretation is clear. Since the principal directions of $E_n^{(1)\prime}$ and $E_n^{(1)\prime}E_n^{(1)\prime}$ are the same, the optimal direction for $\mathbf{h}_n$ coincides with the major principal direction of $E_n^{(1)}$.

The problem of minimizing the mean-squared velocity uncertainty at time t_n by proper choice of the $\mathbf{h}_n$ vector is not as easily solved or interpreted. Again, from Eq. [29] the mean-squared velocity uncertainty may be written as

$$\overline{\delta_n^2} = tr(E_n^{(4)\prime}) - \frac{\mathbf{h}_n^T E_n^{(2)\prime} E_n^{(3)\prime}\mathbf{h}_n}{\mathbf{h}_n^T E_n^{(1)\prime}\mathbf{h}_n + \overline{\alpha_n^2}} \qquad \text{[B2]}$$

Denote by p and q the two quadratic forms

$$p = \mathbf{h}_n^T E_n^{(2)\prime} E_n^{(3)\prime}\mathbf{h}_n \qquad q = \mathbf{h}_n^T E_n^{(1)\prime}\mathbf{h}_n \qquad \text{[B3]}$$

From the theory of quadratic forms, there exists an orthogonal transformation that will reduce q to a diagonal form. Thus

$$\mathbf{h}_n = Q\mathbf{d} \qquad \text{[B4]}$$

gives

$$q = \mathbf{d}^T Q^T E_n^{(1)\prime} Q\mathbf{d} = \mu_1 d_1^2 + \mu_2 d_2^2 + \mu_3 d_3^2 \qquad \text{[B5]}$$

where μ_1, μ_2, μ_3 are the characteristic roots of the matrix

$E_n^{(1)'}$, and the columns of the Q matrix are the associated characteristic unit vectors. Since $E_n^{(1)'}$ is a positive definite matrix, the characteristic roots are positive, and a further transformation

$$\mathbf{f} = D\mathbf{d} \qquad [B6]$$

gives

$$q = \mathbf{f}^T\mathbf{f} = f_1^2 + f_2^2 + f_3^2 \qquad [B7]$$

where D is a diagonal matrix whose diagonal elements are $\mu_1^{1/2}, \mu_2^{1/2}, \mu_3^{1/2}$.

The same transformation from $\mathbf{h}_n$ to $\mathbf{f}$ applied to the quadratic form p produces

$$p = \mathbf{f}^TD^{-1}Q^TE_n^{(2)'}E_n^{(3)'}QD^{-1}\mathbf{f} \qquad [B8]$$

One final transformation applied to $\mathbf{f}$ will reduce Eq. [B8] to a diagonal form. Thus

$$\mathbf{f} = S\mathbf{m} \qquad [B9]$$

results in

$$p = \lambda_1 m_1^2 + \lambda_2 m_2^2 + \lambda_3 m_3^2 \qquad [B10]$$

where the columns of the S matrix are the characteristic unit vectors of the matrix $D^{-1}Q^TE_n^{(2)'}E_n^{(3)'}QD^{-1}$ and $\lambda_1, \lambda_2, \lambda_3$, the corresponding characteristic roots. The same transformation [B9] applied to [B7] gives

$$q = \mathbf{m}^TS^TS\mathbf{m} = m_1^2 + m_2^2 + m_3^2 \qquad [B11]$$

since S is an orthogonal matrix.

In summary, then, the transformation

$$\mathbf{h}_n = QD^{-1}S\mathbf{m} \qquad [B12]$$

produces for the ratio of the two quadratic forms

$$\frac{p}{q} = \frac{\lambda_1 m_1^2 + \lambda_2 m_2^2 + \lambda_3 m_3^2}{m_1^2 + m_2^2 + m_3^2} \qquad [B13]$$

Furthermore, if the matrix $E_n^{(2)'}$ is nonsingular, the product $E_n^{(2)'}E_n^{(3)'} = E_n^{(2)'}E_n^{(2)'T}$ is positive definite, and it would then follow that $\lambda_1, \lambda_2, \lambda_3$ are all real and positive.

The problem of maximizing the ratio p/q is now readily solved. Since no measurement error is assumed, one cannot hope to determine more than the direction for the optimum $\mathbf{h}_n$ or, equivalently, the optimum $\mathbf{m}$. Therefore, it may be assumed that $\mathbf{m}$ is a unit vector. Let

$$\lambda_k = \max(\lambda_1,\lambda_2,\lambda_3) \qquad [B14]$$

Then the optimum value of $\mathbf{m}$ is

$$m_j = \begin{cases} 1 & j = k \\ 0 & j \neq k \end{cases} \qquad [B15]$$

The same technique can be used to select that direction for $\mathbf{h}_n$ which minimizes the uncertainty in any linear combination of position and velocity deviations. Specifically, consider the selection of that measurement which minimizes the uncertainty in the velocity correction that would be required immediately following the measurement.

The correlation matrix of the velocity correction uncertainty is

$$\overline{\mathbf{d}_n\mathbf{d}_n^T} = B_nE_nB_n^T \qquad [B16]$$

and the mean-squared uncertainty may be expressed as

$$\overline{d_n^2} = tr(B_nE_n'B_n^T) - \frac{\mathbf{h}^TW\mathbf{h}_n}{\mathbf{h}_n^TE_n^{(1)'}\mathbf{h}_n + \alpha_n^2} \qquad [B17]$$

Here W is a symmetric matrix defined by

$$W = [E_n^{(1)'} \quad E_n^{(2)'}]B_n^TB_n\begin{bmatrix} E_n^{(1)'} \\ E_n^{(2)'T} \end{bmatrix} \qquad [B18]$$

so that if $[E_n^{(1)'} \quad E_n^{(2)'}]B_n^T$ is nonsingular, the matrix W will be positive definite. Under any circumstances, if the identification

$$E_n^{(2)'} \sim [E_n^{(1)'} \quad E_n^{(2)'}]B_n^T$$

is made, then the exact same procedure may be used to select the optimum direction for the $\mathbf{h}_n$ vector as was used previously to minimize the mean-squared velocity uncertainty.

In all cases of practical interest, the determination of the optimum direction for the $\mathbf{h}_n$ vector must be made subject to certain constraints. For example, one might wish to select the "best" star to be used in measuring the angle between the line of sight to the center of a planet disc and the line of sight to the star. For such a measurement, the $\mathbf{h}_n$ vector is required to be perpendicular to the line of sight to the planet. If $\mathbf{z}_n$ is the position vector of the planet from the space vehicle, then one must have

$$\mathbf{h}_n^T\mathbf{z}_n = 0 \qquad [B19]$$

Applying the transformation defined in Eq. [B12] gives

$$\mathbf{m}^TS^TD^{-1}Q^T\mathbf{z}_n = 0 \qquad [B20]$$

Let $\mathbf{p}$ be a unit vector in the direction of $S^TD^{-1}Q^T\mathbf{z}_n$. Then the problem of selecting the optimum direction for $\mathbf{h}_n$ or, equivalently, for $\mathbf{m}$ is to maximize

$$\lambda_1 m_1^2 + \lambda_2 m_2^2 + \lambda_3 m_3^2$$

subject to the conditions of constraint

$$\mathbf{m}^T\mathbf{p} = 0 \qquad \mathbf{m}^T\mathbf{m} = 1 \qquad [B21]$$

In terms of the Lagrange multipliers ρ and σ, this is equivalent to the problem of obtaining a free maximum for

$$\sum_{j=1}^{3} \lambda_j m_j^2 - 2\rho \sum_{j=1}^{3} p_j m_j - \sigma\left\{\sum_{j=1}^{3} m_j^2 - 1\right\}$$

Setting the partial derivatives with respect to each of the n_j's equal to zero, there results

$$m_j = \frac{\rho p_j}{\lambda_j - \sigma} \qquad j = 1,2,3 \qquad [B22]$$

where ρ and σ are to be determined from the requirements of Eq. [B21].

The condition that $\mathbf{m}$ be orthogonal to $\mathbf{p}$ leads to a quadratic equation for σ:

$$\sigma^2 - [p_1^2(\lambda_2 + \lambda_3) + p_2^2(\lambda_1 + \lambda_3) + p_3^2(\lambda_1 + \lambda_2)]\sigma + p_1^2\lambda_2\lambda_3 + p_2^2\lambda_1\lambda_3 + p_3^2\lambda_1\lambda_2 = 0 \qquad [B23]$$

If the λ's are ordered $\lambda_1 < \lambda_2 < \lambda_3$, then the two roots σ_1 and σ_2 will be such that $\lambda_1 < \sigma_1 < \lambda_2 < \sigma_2 < \lambda_3$. The other Lagrange multiplier ρ is determined so that $\mathbf{m}$ will be a unit vector. With the optimum vector $\mathbf{m}$ selected, the corresponding value for $\mathbf{h}_n$ is found from Eq. [B12].

It is easy to show that σ_2 provides the desired maximum, whereas σ_1 gives the minimum. From Eq. [B22], one obtains

$$\sum_{j=1}^{3} \lambda_j m_j^2 - \sigma \sum_{j=1}^{3} m_j^2 = \rho \sum_{j=1}^{3} p_j m_j \qquad [B24]$$

Using this and Eqs. [B21], it follows that

$$\sigma = \sum_{j=1}^{3} \lambda_j m_j^2 \qquad [B25]$$

Hence, σ_1 and σ_2 are the respective minimum and maximum of the original expression to be maximized.

Acknowledgments

The author is indebted to S. F. Schmidt for directing his attention to R. E. Kalman's excellent work, and to G. L.

Smith for correcting a basic mistake in the original treatment of cross-correlation errors. The author also wishes to acknowledge the extensive services of P. Philliou, who prepared the numerical data reported in the section on application to circumlunar navigation.

References

1 Laning, J. H., Jr., Frey, E. J., and Trageser, M. B., "Preliminary considerations on the instrumentation of a photographic reconnaissance of Mars," *Vistas in Astronautics*, (Pergamon Press, London, 1959), Vol. 2, pp. 63–94.

2 Battin, R. H. and Laning, J. H., Jr., "A navigation theory for round-trip reconnaissance missions to Venus and Mars," *Planetary and Space Science* (Pergamon Press, London, 1961), Vol. 7, pp. 40–56.

3 Battin, R. H., "A comparison of fixed and variable time of arrival navigation for interplanetary flight," *Proceedings of the Fifth AFBMD/STL Aerospace Symposium on Ballistic Missile and Space Technology* (Academic Press, New York, 1960), pp. 3–31.

4 Kalman, R. E., "A new approach to linear filtering and prediction problems," J. Basic Eng., Trans. Am. Soc. Mech. Engrs. **82D**, 35–45 (March 1960).

5 McLean, J. D., Schmidt, S. F., and McGee, L. A., "Optimal filtering and linear prediction applied to a space navigation system for the circumlunar mission," NASA TN D-1208 (March 1962).

Linear Regression Applied to System Identification for Adaptive Control Systems

RICHARD E. KOPP* AND RICHARD J. ORFORD†
Grumman Aircraft Engineering Corporation, Bethpage, N. Y.

The purpose of this paper is to describe a method of process identification using a linear regression technique and to indicate how this method may be applied to adaptive control systems. The unknown system parameters are considered as additional state variables. Estimates of system parameters as well as the system state are made from noise-contaminated data. Differential equations with random forcing functions describing the parameter variations are adjoined to the system of differential equations describing the process. The estimation of the error is assumed to propagate linearly about the current estimates, which are updated as new data are received.

Introduction

THE adaptive control problem has received considerable attention over the past several years. Basically, this is the problem of controlling a process (in this paper the process will be an aerospace vehicle) where imperfect or limited information is available describing the vehicle parameters that change considerably during the interval in which control is required.

This paper describes a statistical method for estimating space vehicle parameters as well as vehicle orientation from noise-contaminated data. Furthermore, it will be shown how this method might be applied to an adaptive control system. A second-order linear system, representative of an attitude rate control system for a space vehicle, will be used to illustrate the method. However, this method is not theoretically limited to second-order systems or even linear systems. The use of a linear regression technique applied to process identification was motivated by some of the recent results obtained by Kalman[1] in linear filtering. This technique appears very attractive, since no extraneous system inputs are needed. A further advantage of this method is that it is not limited to steady-state analysis as are many previously suggested schemes. A second-order system with unknown time-varying parameters is used to characterize the vehicle's attitude response. (The unknown parameters in this case would be the system damping, frequency, and gain.) The system parameters are considered as additional state variables, and differential constraints are adjoined to these variables to provide correlation between present estimates and past estimates of system parameters. A priori estimates of the vehicle parameters, that is, estimates of the behavior of the system exclusive of measurement data, are included in the adjoined differential equations describing the vehicle's parameter behavior. Estimates based on measurement data and control inputs are made for both vehicle orientation and vehicle parameters. The statistical degree of freedom is provided by introducing random variables as forcing functions in the adjoined equations for the unknown parameters.

The estimation of the vehicle parameters first will be analyzed as a sampled-data problem, linearizing the system behavior between sampling intervals. A linear regression technique is used to derive a recursive relationship for the updated estimates of vehicle parameters and orientation as a function of the last estimates and new measurement data. This approach avoids the need for large computer storage and leads directly to a filtering concept. As the sampling interval approaches zero in the limit, the recursive relationships reduce to differential equations describing time-varying nonlinear filters for the estimators of the vehicle parameters and orientation.

The basic concepts associated with this method are quite simple; the estimation of the error is assumed to propagate by a linearization of a nonlinear system (the system becomes nonlinear when one adjoins the differential equations describing the unknown parameter variations) about the current estimate of the system variables. Using these linear equations allows the estimates to be updated with new data as in Ref. 1.

The experimental results that are included in this paper are regarded as encouraging since they are quite insensitive to the assumptions made, a characteristic very desirable for engineering applications.

Problem Formulation

To illustrate how a linear regression technique can be applied to system identification for use in adaptive systems, a specific example representative of an attitude rate control system for a space vehicle will be used. It will be assumed, as is seen in Fig. 1, that the vehicle dynamics can be represented by a second-order linear system with parameters δ, ω_n, and k, which are not precisely known and which vary considerably during the interval of time when control is to be exerted. The transfer function notation, although not applicable to time-varying systems, is used in Fig. 1 merely to illustrate the problem being considered.

A realistic approach to the problem is made by assuming that the measured pitch rate $z(t)$, which will be designated as data, is the actual pitch rate $\dot{\theta}(t)$ contaminated with measurement noise $v(t)$. Furthermore, it is assumed that the actual control signal $u(t)$ is the desired control signal $u^*(t)$ plus additive noise $\delta u(t)$. Since the identification problem is of primary concern, that is, estimating the parameters $\delta(t)$, $\omega_n(t)$, and $k(t)$, as well as $\dot{\theta}(t)$ and $\ddot{\theta}(t)$, from the measured data $z(t)$, consideration of the control or "actuation problem" can be deferred until later. It is tacitly assumed that the two problems are separable, as discussed in Refs. 2 and 3.

Although the proposed method is not limited to linear systems, one can proceed to develop the method using the linear example being discussed. It is convenient to define the

Presented at the ARS 17th Annual Meeting and Space Flight Exposition, Los Angeles, Calif., November 13–18, 1962; revision received July 22, 1963. The authors wish to express their appreciation to Arthur Burns and James Hunter for their valuable assistance in the combined analog-digital simulation studies described in this paper.

* Section Head, Systems Research

† Staff Scientist; presently with Space Technology Laboratories Inc., Redondo Beach, Calif.

Reprinted with permission from *AIAA J.*, vol. 1, pp. 2300–2306, Oct. 1963 variable $x_1 \equiv d\theta/dt$. The system of differential equations describing the motion of the vehicle about the pitch axis is then written as

$$\begin{aligned} dx_1/dt &= x_2 \\ dx_2/dt &= a_1(t)x_2 + a_2(t)x_1 + a_3(t)u(t) \end{aligned} \tag{1}$$

where the unknown time-varying coefficients $a_1(t)$, $a_2(t)$, and $a_3(t)$ are defined as

$$a_1(t) \equiv -2\delta(t)\omega_n(t)$$

$$a_2(t) \equiv -\omega_n^2(t)$$

$$a_3(t) \equiv k(t)$$

Initial conditions for Eq. (1) are drawn from a set of normally distributed random variables of known statistical properties. It is assumed that estimates of $a_i(t)$, $i = 1 \ldots 3$, are equally as good as estimates of $\delta(t)$, $\omega_n(t)$, and $k(t)$ for the determination of the desired control signal $u^*(t)$.

Measured data $z(t)$ are the sum of the pitch rate $x_1(t)$ and additive noise:

$$z(t) = x_1(t) + v(t) \tag{2}$$

The noise $v(t)$ is assumed to be a normally distributed random variable with known statistical properties:

$$\begin{aligned} E[v(t)] &= 0 \\ E[v(t)v(\tau)] &= \sigma_v^2(t)\delta(t - \tau) \end{aligned} \tag{3}$$

where $\delta(t - \tau)$ is the Dirac delta, and $\sigma_v^2(t)$ is given. The case where the control signal $u(t)$ is the sum of a desired control signal $u^*(t)$ and additive noise is considered:

$$u(t) = u^*(t) + \delta u(t) \tag{4}$$

The additive noise $\delta u(t)$ is assumed to be a normally distributed random variable modulated by a function of the desired control signal:

$$\delta u(t) = S[u^*(t)]w_0(t) \tag{5}$$

where the statistical properties of $w_0(t)$ are given:

$$\begin{aligned} E[w_0(t)] &= 0 \\ E[w_0(t)w_0(\tau)] &= \sigma_{w0}^2\delta(t - \tau) \end{aligned} \tag{6}$$

Certain assumptions and approximations now will be made which we feel are reasonable, but which we will not attempt to justify. Some a priori knowledge of the expected values of $a_i(t)$ may be available which we would like to include in the analysis. These estimates, if they are available, are defined as $\bar{a}_i(t)$. It also is assumed that the coefficients $a_i(t)$ vary continuously in a random manner. With this in mind, the set of differential equations is adjoined:

$$da_i/dt = \alpha_i(t)[a_i(t) - \bar{a}_i(t)] + w_i(t) \qquad i = 1 \ldots 3 \tag{7}$$

to Eqs. (1). The differential constraints insure continuity,

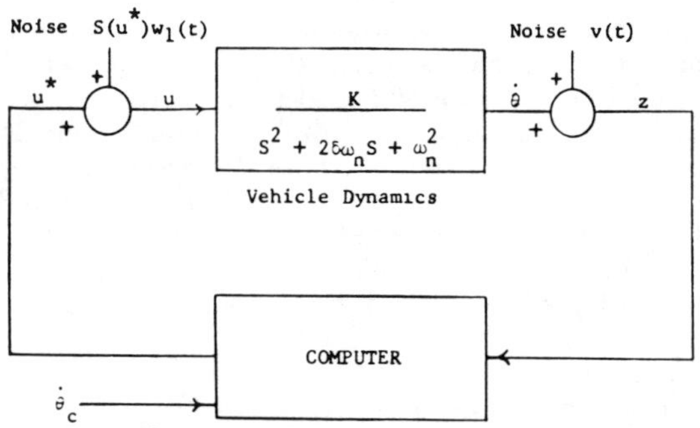

Fig. 1 Attitude rate control system.

whereas the normally distributed random functions of time $w_i(t)$ provide the needed statistical degree of freedom. Initial conditions for Eqs. (7) are drawn from a set of normally distributed random variables. The parameters $\alpha_i(t)$ are assumed given, as are the statistical properties of $w_i(t)$:

$$\begin{aligned} E[w_i(t)] &= 0 \\ E[w_i(t)w_i(\tau)] &= \sigma_{wi}^2\delta(t - \tau) \qquad i = 1 \ldots 3 \end{aligned} \tag{8}$$

One might refer to the set of Eqs. (7) as a statistical model of the vehicle parameters, but it is not the only model that could be chosen.

Sampled Data Analysis

For the sampled data analysis, time will be assumed to be divided into discrete intervals of length Δt. It further will be assumed that the control is constant over any one interval of time changing in a stepwise manner between intervals. Data are sampled at the end of each interval. For convenience, one can designate a variable at time $n\Delta t$ by $x(n)$. One also will need notation to designate the estimate of a variable at time $n\Delta t$, given all past data and desired control inputs to time $k\Delta t, n \geq k$. This will be denoted as

$$\hat{x}(n|k) \equiv \text{estimate of } x(n) \text{ given } [z(1), \ldots, z(k), u^*(1), \ldots, u^*(k)]$$

These will be the fundamental quantities, with $k = n - 1$ or n; however, one also will need estimates between sampling instants which can be designated as $\hat{x}(t|n)$:

$$\hat{x}(t|n) \equiv \text{estimate of } x(t) \text{ given } [z(1), \ldots, z(n), u^*(1), \ldots, u^*(n)] \qquad t \geq n\Delta t$$

and similarly for $a_i(t)$. Then one can define an error, e.g.,

$$\delta x(t|n) = x(t) - \hat{x}(t|n) \qquad t \geq n\Delta t \tag{9}$$

and again similarly for $\delta a_i(t)$.

One can estimate between sampling intervals by means of

$$\begin{aligned} \frac{d\hat{x}_1(t|n)}{dt} &= \hat{x}_2(t|n) \\ \frac{d\hat{x}_2(t|n)}{dt} &= \hat{a}_1(t|n)\hat{x}_2(t|n) + \hat{a}_2(t|n)\hat{x}_1(t|n) + \hat{a}_3(t|n)u^*(t) \\ \frac{d\hat{a}_2(t|n)}{dt} &= \alpha_i(t)[\hat{a}_i(t|n) - \bar{a}_i(t|n)] \end{aligned} \tag{10}$$

Substituting Eq. (9), etc., into Eqs. (1) and (7) and subtracting Eqs. (10), neglecting terms of second order, one has, using a vector matrix notation,

$$\frac{d\boldsymbol{\delta y}(t|n)}{dt} = \mathfrak{L}(t|n)\boldsymbol{\delta y}(t|n) + \boldsymbol{\Omega}(t) \tag{11}$$

where the vectors

$$\boldsymbol{\delta y}^T(t|n) \equiv [\delta x_1(t|n)\ \delta x_2(t|n)\ \delta a_1(t|n)\ \delta a_2(t|n)\ \delta a_3(t|n)]$$

and

$$\boldsymbol{\Omega}^T(t) \equiv [0\ \hat{a}_3(t|n)S(u^*)w_0(t)\ w_1(t)\ w_2(t)\ w_3(t)]$$

and the matrix

$$\mathfrak{L}(t|n) \equiv \begin{bmatrix} 0 & 1 & 0 & 0 & 0 \\ \hat{a}_2(t|n) & \hat{a}_1(t|n) & \hat{x}_2(t|n) & \hat{x}_1(t|n) & u^*(t) \\ 0 & 0 & \alpha_1(t) & 0 & 0 \\ 0 & 0 & 0 & \alpha_2(t) & 0 \\ 0 & 0 & 0 & 0 & \alpha_3(t) \end{bmatrix}$$

The superscript T denotes the transpose. The $\mathfrak{L}(t|n)$ matrix is seen to be composed of the estimates of parameters $a_i(t)$, as

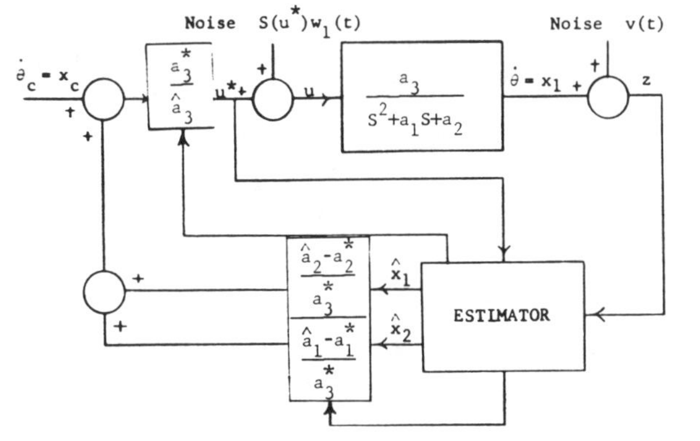

Fig. 2 Adaptive system using a linear regression method for the estimator.

well as estimates of the vehicle state $x_1(t)$ and $x_2(t)$ at time t, given data to time $n\Delta t$. Equation (11) describes the propagation of the errors between sampling intervals. At time $n\Delta t$, one can assume that the error $\delta\mathbf{y}(n|n)$ is distributed multinormally with given characteristics

$$E[\delta\mathbf{y}(n|n)] = 0$$
$$E[\delta\mathbf{y}(n|n)\delta\mathbf{y}^T(n|n)] = P(n|n) \tag{12}$$

This is certainly true initially by assumption.

For convenience, let a vector for the estimates of the state and parameters be defined as

$$\mathbf{y}(n|n) \equiv [x_1(n|n)\ x_2(n|n)\ a_1(n|n)\ a_2(n|n)\ a_3(n|n)]$$

One now has a concise formulation of the problem. We wish to estimate $\delta\mathbf{y}(n+1)$ conditioned on data at $(n+1)\Delta t$, given that $\delta\hat{\mathbf{y}}(n|n)$ is zero with known variance. The estimate of $\mathbf{y}(n+1)$ is given by

$$\hat{\mathbf{y}}(n+1|n+1) = \hat{\mathbf{y}}(n+1|n) + \delta\hat{\mathbf{y}}(n+1|n+1) \tag{13}$$

We will now proceed to discuss how one calculates $\delta\hat{\mathbf{y}}(n+1|n)$, $\delta\hat{\mathbf{y}}(n+1|n+1)$ and the variance of $\delta\mathbf{y}(n+1|n+1)$, which determines the initial conditions for the next interval of time.

One can assume that the vector $\mathbf{\Omega}(t)$ is constant in the interval $n\Delta t \leq \tau < (n+1)\Delta t$ with components selected from a multinormal distribution of known statistical properties:

$$E[\mathbf{\Omega}_n] = 0 \qquad E[\mathbf{\Omega}_n\mathbf{\Omega}_k{}^T] = Q_n\delta_{nk} \tag{14}$$

where δ_{nk} is the Kronecker delta. The definitions of a sampled random function of the type defined in Eq. (8) must be examined (see Feller,[4] p. 324). Specifically, in order to maintain consistent behavior for the output of a linear system (see Kalman[1]), where, for the continuous function,

$$E[\mathbf{\Omega}(t)\mathbf{\Omega}^T(t)] = Q(t)\delta(t-\tau) \tag{15}$$

then,

$$\lim_{\Delta t \to 0} Q_n\Delta t = Q(t) \tag{16}$$

Similarly for $v(t)$. For the specific example,

$$Q(t) \equiv \begin{bmatrix} 0 & 0 & 0 & 0 & 0 \\ 0 & a_3{}^2S^2(u)\sigma_{w0}{}^2(t) & 0 & 0 & 0 \\ 0 & 0 & \sigma_{w1}{}^2(t) & 0 & 0 \\ 0 & 0 & 0 & \sigma_{w2}{}^2(t) & 0 \\ 0 & 0 & 0 & 0 & \sigma_{w3}{}^2(t) \end{bmatrix}$$

By use of the transition matrix,[5] the solution of Eq. (11) is written:

$$\delta\mathbf{y}(n+1) = \Phi(n+1,n)\delta\mathbf{y}(n) + \Gamma(n+1,n)\mathbf{\Omega}(n) \tag{17}$$

To obtain $\Phi(n+1,n)$, $\Gamma(n+1,n)$ in Eq. (17), one first must solve Eqs. (10), from which one also obtains $\mathbf{y}(n+1|n)$. An estimate of $\delta\mathbf{y}(n+1)$, given data to $(n+1)\Delta t$, is made by linear regression, i.e.,

$$\delta\hat{\mathbf{y}}(n+1|n+1) = \Psi(n+1)\tilde{z}(n+1) \tag{18}$$

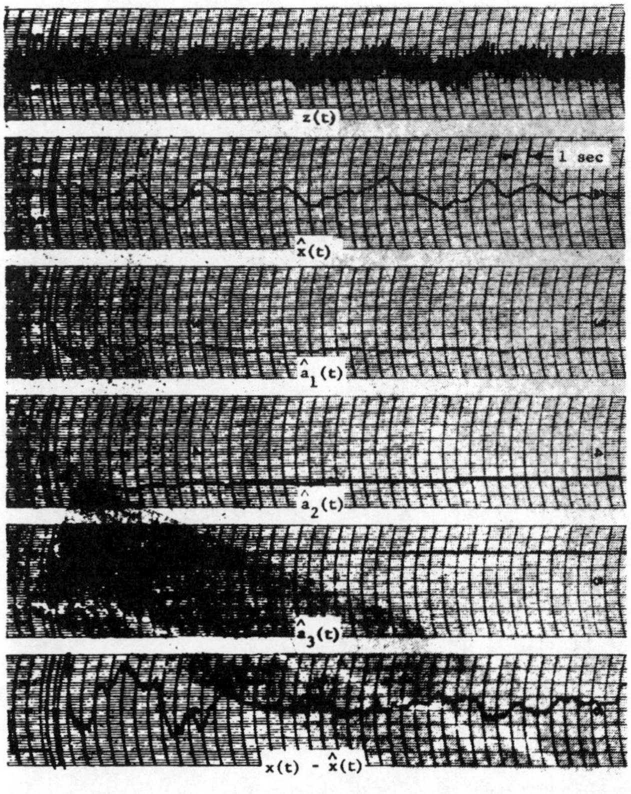

Fig. 3 Estimation of damping coefficient $a_1(t)$, natural frequency $a_2(t)$, and gain $a_3(t)$ for a random input [$a_1(t) = -1$, $a_2(t) = -1$, $a_3(t) = 1$, $\hat{a}_1(0) = 0$, $\hat{a}_2(0) = 0$, $\hat{a}_3(0) = 0$].

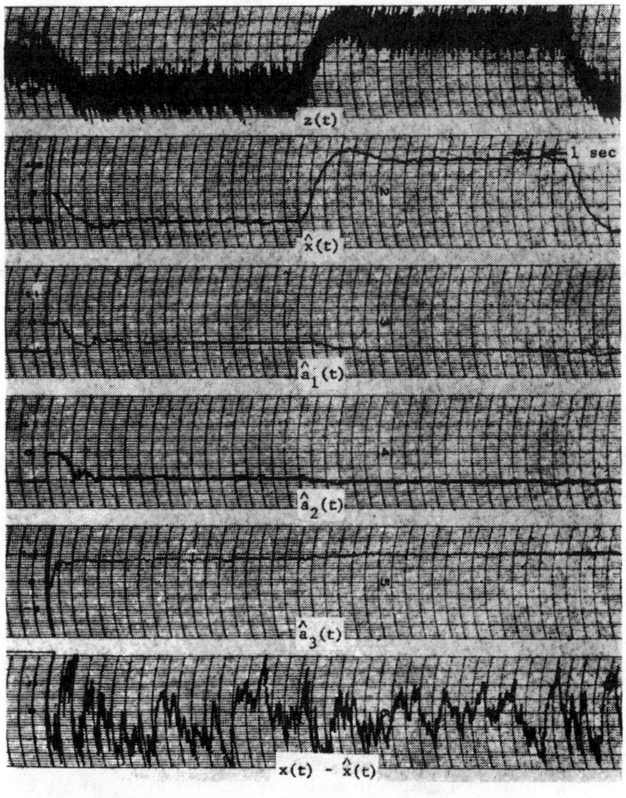

Fig. 4 Estimation of damping coefficient $a_1(t)$, natural frequency $a_2(t)$, and gain $a_3(t)$ for a step input [$a_1(t) = -1$, $a_2(t) = -1$, $a_3(t) = 1$, $\hat{a}_1(0) = 0$, $\hat{a}_2(0) = 0$, $a_3(0) = 0$].

where $\tilde{z}(n+1)$ are new data, defined as the difference between the actual data received and the estimate of the data conditioned on the previous sampling instant:

$$\begin{aligned}\tilde{z}(n+1) &= z(n+1) - \hat{z}(n+1|n) = \\ &\qquad x_1(n+1) + v(n+1) - \hat{x}_1(n+1|n) \\ &= \delta x(n+1|n) + v(n+1)\end{aligned} \tag{19}$$

To write Eq. (19) in terms of the y vector and preserve dimensionality, the row vector is introduced:

$$\mathbf{M} \equiv [1 \quad 0 \quad 0 \quad 0 \quad 0]$$

then

$$\tilde{z}(n+1) = z(n+1) - \mathbf{M}\hat{\mathbf{y}}(n+1|n) \tag{20a}$$

$$= \mathbf{M}\delta\mathbf{y}(n+1|n) + v(n+1) \tag{20b}$$

The column vector $\Psi(n+1)$ is determined by minimizing

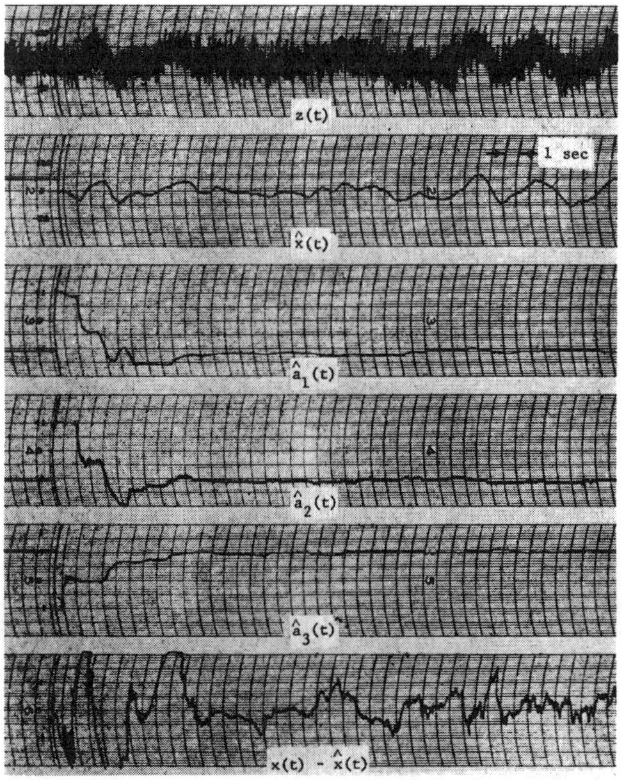

Fig. 5 Estimation of damping coefficient $a_1(t)$, natural frequency $a_2(t)$, and gain $a_3(t)$ with initial estimates of opposite sign [$a_1(t) = -1$, $a_2(t) = 1$, $a_3(t) = 1$, $\hat{a}_1(0) = 1$, $\hat{a}_2(0) = 1$, $\hat{a}_3(0) = -1$].

the diagonal elements of the covariance matrix $P(n+1|n+1)$ of the error $\delta\mathbf{y}(n+1|n+1)$:

$$\Psi(n+1) = P(n+1|n)\mathbf{M}^T[\mathbf{M}P(n+1|n)\mathbf{M}^T + \sigma_v^2(n+1)]^{-1} \tag{21}$$

where

$$P(n+1|n) = \Phi(n+1,n)P(n|n)\Phi(n+1,n) + \Gamma(n+1,n)Q(n)\Gamma^T(n+1,n) \tag{22}$$

and

$$P(n+1|n+1) = [I - \Psi(n+1)\mathbf{M}]P(n+1|n) \tag{23}$$

At $n = 0$, which is the start of the process, initial values of $\hat{\mathbf{y}}(0)$ and $P(0|0)$ are given. That is, for the specific example, initial estimates of $x_1(0)$, $x_2(0)$, and $a_i(0)$ are given, along with the variances of their respective errors. From Eqs. (21–23), $P(1|0)$, $P(1|1)$, and $\Psi(1)$ are calculated. When data are received at the first sampling interval, $\delta\hat{x}_1(1|1)$,

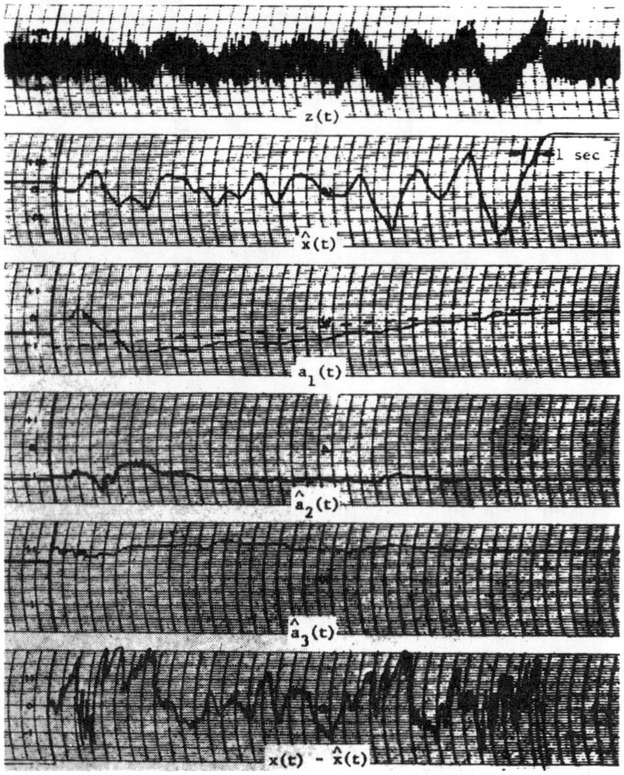

Fig. 6 Estimation of a time-varying damping coefficient $a_1(t)$ [$a_1(t)$ is time varying, $a_2(t) = -1$, $a_3(t) = 1$].

$\delta\hat{x}_2(1|1)$, and $\delta a_i(1|1)$ are calculated from Eq. (18), and the estimates of the respective variables are determined from Eq. (13). These values form initial conditions for the next interval, and the calculation proceeds. The important relationships are summarized as

$$P(n+1|n) = \Phi(n+1,n)P(n|n)\Phi^T(n+1,n) + \Gamma(n+1,n)Q(n)\Gamma^T(n+1,n)$$

$$\Psi(n+1) = P(n+1|n)\mathbf{M}^T[\mathbf{M}P(n+1|n)\mathbf{M}^T + \sigma_v^2(n+1)]^{-1}$$

$$\hat{\mathbf{y}}(n+1|n) \qquad \text{[obtained from Eqs. (10)]} \tag{24}$$

$$\delta\hat{y}(n+1|n+1) = \Psi(n+1)[z(n+1) - \mathbf{M}\hat{\mathbf{y}}(n+1|n)]$$

$$\hat{\mathbf{y}}(n+1|n+1) = \hat{\mathbf{y}}(n+1\,|\,n) + \delta\hat{\mathbf{y}}(n+1\,|\,n+1)$$

$$P(n+1|n+1) = [I - \Psi(n+1)\mathbf{M}]P(n+1|n)$$

Continuous Analysis

To analyze the continuous case, that is, where one receives data continuously, one allows the sampling interval Δt to approach zero in the limit. A state vector $x(t)$ is defined as

$$\mathbf{x}^T(t) \equiv [x_1(t)x_2(t)]$$

and the estimate of $\mathbf{x}$ at $(n+1)\Delta t$, given data to $n\Delta t$ in terms of the transition matrix for the vehicle $\Phi^x(n+1,n)$, is written as

$$\hat{\mathbf{x}}(n+1|n) = \Phi^x(n+1,n)\hat{\mathbf{x}}(n|n) + \Gamma^x(n+1,n)u^*(n) \tag{25}$$

The estimate of x at $(n+1)\Delta t$, given data to $(n+1)\Delta t$, is therefore

$$\hat{\mathbf{x}}(n+1|n+1) = \Phi^x(n+1,n)\hat{\mathbf{x}}(n|n) + \Gamma^x(n+1,n)u^*(n|n) + \delta\hat{\mathbf{x}}(n+1|n+1) \tag{26}$$

Subtracting $\hat{\mathbf{x}}(n|n)$ from each side of Eq. (26), dividing by

Δt, and passing to the limit gives

$$\frac{d\hat{\mathbf{x}}(t)}{dt} = \hat{A}(t)\hat{\mathbf{x}}(t) + \hat{B}(t)u^*(t) + \lim_{\Delta t \to 0} \boldsymbol{\delta}\hat{\mathbf{x}} \frac{(t + \Delta t | t + \Delta t)}{\Delta t} \quad (27)$$

where

$$\hat{A}(t) \equiv \begin{bmatrix} 1 & 0 \\ \hat{a}_1(t) & \hat{a}_2(t) \end{bmatrix} \qquad \hat{B}(t) \equiv \begin{bmatrix} 0 \\ \hat{a}_3(t) \end{bmatrix}$$

Here it is understood that $\hat{\mathbf{x}}(t)$ designates the estimates of $\mathbf{x}(t)$, given data to t, and likewise for $\hat{A}(t)$ and $\hat{B}(t)$. Similarly, a vector $\boldsymbol{\xi}(t)$ is defined as

$$\boldsymbol{\xi}(t) \equiv [a_1(t)\ a_2(t)\ a_3(t)]$$

and the following is obtained:

$$\frac{d\hat{\boldsymbol{\xi}}(t)}{dt} = C(t)[\hat{\boldsymbol{\xi}}(t) - \bar{\boldsymbol{\xi}}(t)] + \lim_{\Delta t \to 0} \boldsymbol{\delta}\hat{\xi} \frac{(t + \Delta t | t + \Delta t)}{\Delta t} \quad (28)$$

where

$$C(t) \equiv \begin{bmatrix} \alpha_1(t) & 0 & 0 \\ 0 & \alpha_2(t) & 0 \\ 0 & 0 & \alpha_3(t) \end{bmatrix}$$

Carrying through this limiting process, one sees that the $P(n + 1 | n)$ and $P(n + 1 | n + 1)$ matrices become the same and will be designated by $P(t)$. Equations (27) and (28) for the updated estimates reduce to

$$\begin{aligned} d\hat{\mathbf{x}}(t)/dt &= \hat{A}(t)\hat{\mathbf{x}}(t) + \hat{B}(t)u^*(t) + P_1(t)M_1{}^T\sigma_v{}^{-2}(t)\mathbf{z}(t) \\ d\hat{\boldsymbol{\xi}}(t)/dt &= C(t)[\hat{\boldsymbol{\xi}}(t) - \bar{\boldsymbol{\xi}}(t)] + P_2(t)M_2{}^T\sigma_v{}^{-2}(t)\mathbf{z}(t) \end{aligned} \quad (29)$$

where $P_1(t)$ and $P_2(t)$ are partitions of $P(t)$:

$$P_1(t) \equiv \mathrm{cov}\,\delta\mathbf{x}(t)\ \boldsymbol{\delta}\mathbf{x}^T(t)$$

$$P_2(t) \equiv \mathrm{cov}\,\delta\boldsymbol{\xi}(t)\ \boldsymbol{\delta}\mathbf{x}^T(t)$$

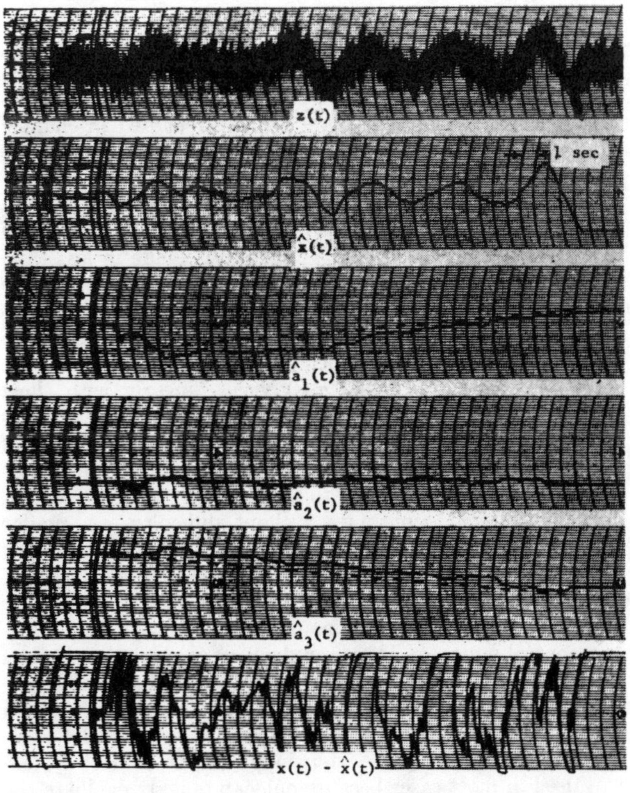

Fig. 7 Estimation of a time-varying damping coefficient $a_1(t)$ and a time-varying gain $a_3(t)$ [$a_1(t)$ is time varying, $a_2(t) = -1$, $a_3(t)$ is time varying].

$$P(t) \equiv \begin{bmatrix} P_1(t) \\ \hline P_2(t) \end{bmatrix}$$

Again $\mathbf{M}_1$ and $\mathbf{M}_2$ are introduced to preserve dimensionality:

$$\mathbf{M}_1 \equiv [1\ 0] \qquad \mathbf{M}_2 \equiv [1\ 0\ 0]$$

The covariance matrix $P(t)$ obeys the matrix equation

$$\dot{P}(t) = \mathfrak{L}(t)P(t) + P(t)\mathfrak{L}^T(t) - P(t)\mathbf{M}^T(t)\sigma_v{}^{-2}(t)\mathbf{M}(t)P(t) + Q(t) \quad (30)$$

Determination of Control Signal

A discussion of how the desired control signal $u^*(t)$ is to be generated has been avoided deliberately until now, because the primary purpose of this paper is to develop a method for measuring (estimating) the vehicle system parameters. However, if this technique is to be applied to adaptive control systems, some discussion is in order. As was pointed out earlier, the separation of the estimation of system parameters and control signal generation is justifiably open to criticism. It would be desirable to formulate the problem such that $u^*(t)$ is determined directly by minimizing the conditional expectation of a system performance measure, such as an integral square criterion. However, upon analysis one soon becomes discouraged with this approach. Therefore, a compromise between what one desires and what one can obtain seems in order. The separation of system parameter estimation and control signal generation certainly has intuitive appeal and motivation. It has been shown[2] for linear systems with known parameters that, for an integral square criterion, the filtering of data can be analyzed separately from the control signal generation, the control law being the same as for the deterministic case with the state variables being replaced with their conditional expectations. With these thoughts in mind, one can proceed to develop a very simple (actually an open-loop) control law to obtain a desired dynamic behavior of the process. Additional feedback would no doubt be required in many applications that would depend on the specific system requirements.

Let it be assumed that $u^*(t)$ is a linear combination of the command attitude rate $\dot{\theta}_c \equiv x_c$, $x_1(t)$, and $\dot{x}_1(t)$, i.e.,

$$u^*(t) = \eta_1 x_c(t) + \eta_2 x_1(t) + \eta_3 \dot{x}_1(t) \quad (31)$$

It is specified further that the overall system is to respond like a specified model that is described by the differential equations

$$\ddot{x} = a_1{}^*\dot{x} + a_2{}^*x + a_3{}^*x_c \quad (32)$$

Under these conditions and assumptions,

$$\eta_1 = \frac{a_3{}^*}{a_3} \qquad \eta_2 = \frac{a_2 - a_2{}^*}{a_3} \qquad \eta_3 = \frac{a_1 - a_1{}^*}{a_3} \quad (33)$$

A typical block diagram of the complete system is shown in Fig. 2.

Simulation

The system discussed was simulated using a combined analog-digital hybrid computing system. This provided what was felt to be a realistic simulation of the problem. The process to be controlled as well as the desired response were simulated on an analog computer, and Eqs. (29) and (30) for the state and parameter estimaters were solved simultaneously on a digital computer using the most elementary integration techniques. The sampling interval was of the order of seven msec. Although the data were sampled, a continuous analysis was used because of the small sampling interval compared to the dynamics of the system.

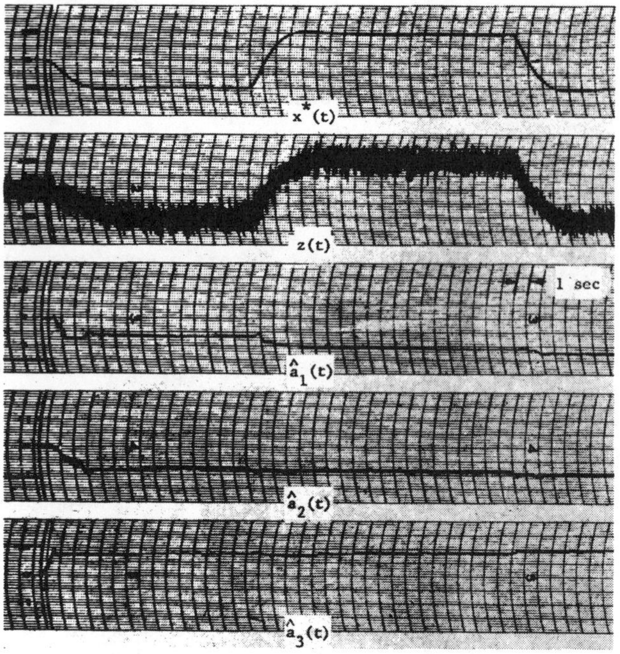

Fig. 8 Control system with desired response 0.7 critical damping, initial estimates of all parameters zero [$a_1^*(t)$ = -1.4, $a_2^*(t) = -1$, $a_3^*(t) = 1$, $a_1(t) = -1.4$, $a_2(t) = -1$, $a_3(t) = 1$, $\hat{a}_1(0) = 0$, $\hat{a}_2(0) = 0$, $\hat{a}_3(0) = 0$].

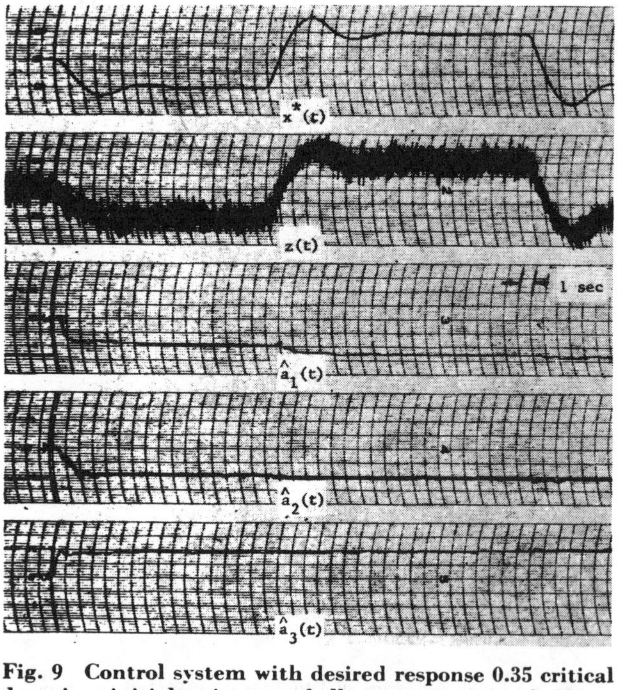

Fig. 9 Control system with desired response 0.35 critical damping, initial estimates of all parameters zero [$a_1^*(t)$ = -0.7, $a_2^*(t) = -1$, $a_3^*(t) = 1$, $a_1(t) = -1.4$, $a_2(t) = -1$, $a_3(t) = 1$, $\hat{a}_1(0) = 0$, $\hat{a}_2(0) = 0$, $\hat{a}_3(0) = 0$].

The control signal $u^*(t)$ also was calculated on the digital computer using Eqs. (31) and (33).

Before simulating the entire control system, many open-loop time histories were run off to illustrate process identification for varied conditions. For all conditions, the variance σ_r^2 of the additive noise to the data was assumed to be 25. (The equations have been nondimensionalized.) Perfect information was assumed to be available for the control input $u^*(t)$ and, therefore, $\sigma_{u_c}^2 = 0$.

The values of $\bar{a}_1(t)$, $\bar{a}_2(t)$, and $\bar{a}_3(t)$ were assumed zero, as were α_1, α_2, and α_3 [no a priori information available as to how the parameters $a_1(t)$, $a_2(t)$, and $a_3(t)$ might vary].

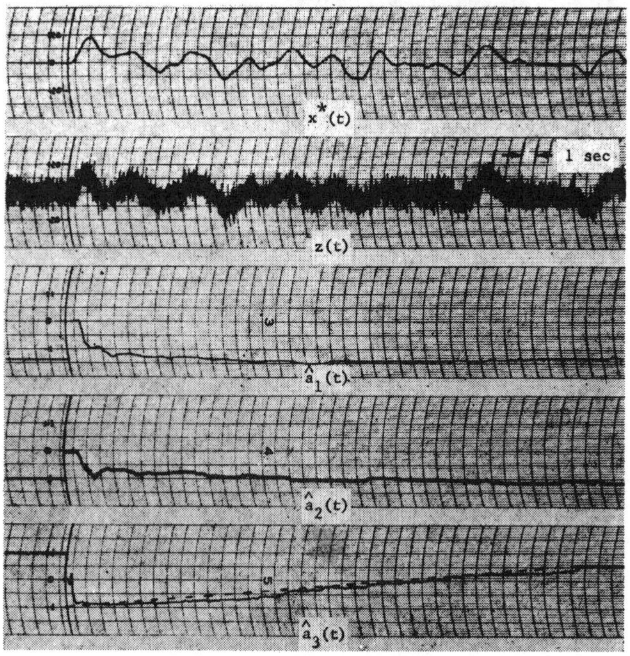

Fig. 10 Control system with desired response 0.7 critical damping, $a_3(t)$ time varying, random input [$a_1^*(t) = -1.4$, $a_2^*(t) = -1$, $a_3^* = 1$, $a_1(t) = -1$, $a_2(t) = -1$, $a_3(t)$ is time varying].

The variances $\sigma_{w_1}^2$, $\sigma_{w_2}^2$, and $\sigma_{w_3}^2$ were all given values of 0.1. Initial estimates of the diagonal elements of the covariance matrix P were all equal to 25 and the off-diagonal initial estimates 0. This, it is felt, might be a representative situation, with the possible exception of letting $\sigma_{w1} = 0$. This condition was brought about by the immediate lack of enough random function generators as well as a consideration of not investigating the effect of too many conditions at one time.

In Fig. 3 time histories are shown for the condition where $a_1(t)$, $a_2(t)$, and $a_3(t)$ are constants with respective values of -1, -1, and $+1$. The initial estimates of $\hat{a}_1(t)$, $\hat{a}_2(t)$, and $\hat{a}_3(t)$ are zero. The input $u^*(t)$ to the process is generated by a random noise generator with a bandwidth of approximately 2 cps. The noisy data $z(t)$ are recorded in channel 1, from which estimates of $a_1(t)$, $a_2(t)$, $a_3(t)$, $\hat{x}(t)$, and $dx(t)/dt$ are made. The estimates $\hat{x}(t)$, $\hat{a}_1(t)$, $\hat{a}_2(t)$, and $\hat{a}_3(t)$ are recorded in channels 3, 4, and 5, respectively, and the error in the estimate of $\hat{x}(t)$ is recorded in channel 6. It is observed that a good estimate of all parameters is made in less than one period of the natural frequency of the system. A similar situation is shown in Fig. 4 for a step input. Figure 5 illustrates an extreme condition when the initial estimates are completely reversed from their actual values. In Fig. 6 is shown the condition when $a_1(t)$ varies linearly with time. The broken line is the actual value of $a_1(t)$. Initial estimates of $a_2(t)$ and $a_3(t)$ are correct, whereas the initial estimate of $a_1(t)$ is zero. It should be observed that, although the process goes from a stable condition to an unstable condition, parameter identification continues without difficulty. A similar situation is shown in Fig. 7, where both $a_1(t)$ and $a_3(t)$ vary linearly with time, illustrating that the method is capable of recognizing the phenomenon of control reversal.

The remaining runs were made with simulation of the entire control system. The desired response was taken, unless otherwise specified, as that of a system with 0.7 critical damping. For this condition, the parameters $a_1^*(t)$, $a_2^*(t)$, and $a_3^*(t)$ which generate the desired response are -1.4, -1, and $+1$, respectively. A step input to the control system was used in most cases because one can quickly evaluate the systems response. Figure 8 illustrates the condition when $a_1(t)$, $a_2(t)$, and $a_3(t)$ are constants with the desired values of -1.4, -1, and $+1$, with initial estimates of the parameters all zero. The desired response $x^*(t)$ is recorded in channel 1, and the output $x(t)$ with additive noise which collectively represents the data $z(t)$ is recorded in channel 2. The estimates $\hat{a}_1(t)$, $\hat{a}_2(t)$, and $\hat{a}_3(t)$ are recorded in channels 3, 4, and 5. The estimation of $\hat{a}_1(t)$ takes considerably longer than that of $\hat{a}_2(t)$ and $\hat{a}_3(t)$, because, once the steady-state response to the step input (square wave input) is reached, no further information about the relative damping can be obtained. In Fig. 9 the desired response has been changed to 0.35 critical damping [$a_1^*(t) = 0.7$, $a_2^*(t) = -1$, $a_3^*(t) = +1$] with $a_1(t)$, $a_2(t)$, and $a_3(t)$ equal to -1.4, -1, and $+1$. Initial estimates $a_1(t)$, $a_2(t)$, and $a_3(t)$ are all zero. Figure 10 shows the condition when $a_3(t)$ varies linearly with time, with all initial estimates zero and a random input. It should be noted that the system can recognize the cope with control reversal.

The particular runs shown here represent only a few of the many runs made and were chosen to be representative of different conditions under which such a system might operate. A first-order system, with variable time constant and gain that was actually studied prior to the second-order system illustrated throughout this paper, was also simulated, with equally good results.

Conclusions

The primary purpose of this paper was to develop a statistical means of identifying the parameters of a linear system from noisy data and to show how the method might be applied to an adaptive control system. Although many approximations and assumptions were made, excellent results have been demonstrated by experiment.

If additional measurement data are available (x_2 in example discussed), **M** and Ψ become rectangular matrices, v becomes a column vector, and the variance σ_v^2 becomes a covariance matrix. The vector matrix equations remain the same.

The reader might feel that this particular demonstration is indeed trivial compared to a more general formulation that has been postulated. Certainly this is a valid concern, to which the following comment is offered. Observe that Eq. (1) is a nonlinear equation in the variables x_1, x_2, a_1, a_2, and a_3. One seeks to estimate these variables given data **z**, albeit postulating certain not unreasonable behavior of the elements a_1, a_2, and a_3. From this point of view, the reader is referred to the works of Battin[6] and others[7] which successfully use a similar technique of sequential linear regression to estimate the state of a six-dimensional nonlinear system (space vehicle trajectory determination). Actually, it was that application of Kalman's techniques which led the authors to consider the identification problem from the point of view expressed in this paper. The excellent results obtained by Battin et al. would seem to justify the authors' expectation of successful application of the proposed identification scheme to higher-order systems.

References

[1] Kalman, R. E., "New methods and results in linear prediction and filtering," Symposium on Engineering Applications of Random Function Theory and Probability, Purdue Univ. (November 1960); also Rias Rept. 61-1 (1961).

[2] Joseph, P. D. and Tou, J. T., "On linear control theory," AIEE Summer General Meeting, Ithaca, N. Y. (June 1961).

[3] Orford, R. J., "Optimization of stochastic final value control systems subject to constraints," Ph.D. Dissertation, Polytech. Inst. Brooklyn (June 1962).

[4] Feller, W., *An Introduction to Probability Theory and Its Applications* (John Wiley and Sons Inc., New York, 1957), p. 324.

[5] Coddington, E. A. and Levinson, N., *Theory of Ordinary Differential Equations* (McGraw-Hill Book Co., Inc., New York, 1955), Chap. I.

[6] Battin, R. H., "A statistical optimizing navigation procedure for space flight," ARS Preprint 2297-61 (October 1961).

[7] McLean, J. D., Schmidt, S. F., and McGee, L. A., "Optimum filtering and linear prediction applied to a midcourse navigation system for the circumlunar mission," NASA TN D-1208 (March 1962).

Kalman Filtering Techniques

H. W. SORENSON

Space Systems Group
AC Electronics Division, General Motors Corporation
El Segundo, California

I. Introduction

In 1809 the great German mathematician K. F. Gauss, while discussing in his classic treatise "Theoria Motus" (*1*) the problem of determining the orbital elements of a celestial body from available measurement data, made the following statement:

If the astronomical observations and other quantities, on which the computations of orbits is based, were absolutely correct, the elements also, whether deduced from three or four observations, would be strictly accurate (so far indeed as the motion is supposed to take place exactly according to the laws of Kepler), and, therefore, if other observations were used, they might be confirmed but not corrected. But since all our measurements and observations are nothing more than approximations to the truth, the same must be true of all calculations resting upon them, and the highest aim of all computations made concerning concrete phenomena must approximate, as nearly as practicable, to the truth. But this can be accomplished in no other way than by a suitable combination of more observations than the number absolutely requisite for the determination of the unknown quantities. This problem can only be properly undertaken when an approximate knowledge of the orbit has been already attained, which is afterwards to be corrected so as to satisfy all the observations in the most accurate manner possible.

The technique that Gauss suggested for obtaining the approximations (or estimates) of the unknown quantities has come to be known as the *method of least squares*. Note that he called for approximate knowledge of the orbit in order to determine the least-squares estimates. These estimates were then used to correct the reference orbit.

A modern version of Gauss' least-squares technique is developed and discussed in the paragraphs that follow. There are two major differences between the classical and modern techniques. The mathematical theory of probability did not exist at the start of the nineteenth century, so Gauss' considerations were necessarily of a deterministic nature. The additional information that accrues through the introduction of probabilistic and statistical considerations allows a more accurate theory to be devised, although the estimation procedure becomes more complex. The complexity leads to the second significant difference. The advent of modern digital-computer technology permits more sophisticated mathematical models to be used in scientific investigations. Investigators of Gauss' time were restricted to techniques that permitted closed-form solutions or involved computations that could be accomplished in a reasonable time using pencil and paper. Analytic solutions are no longer as important, because numerical solutions can be obtained for a wide class of problems.

There are three main sections to this chapter. Section II contains some of the more important theoretical considerations relating to the

Reprinted with permission from *Advances in Control Systems Theory and Applications,* vol. 3, Edited by C. T. Leondes, 1966, pp. 219–292.

linear estimation theory as developed by Kalman (*2–4*) (and hereafter referred to as the Kalman filter) and others (*5, 6, 8, 9*). This estimation theory can be formulated in terms of either a time-discrete (*2, 3*) or time-continuous (*2, 4*) model. The major portion of the presentation is concerned with the time-discrete model, since it seems to be the most natural version for implementation on a digital computer. Further, it more accurately describes the common physical situation in which measurement data are obtained at discrete instants of time. The filter equations for the time-continuous model are discussed briefly in Section II, G.

The derivation of the Kalman filter is accomplished in a manner that relies more upon physical intuition than upon mathematical sophistication. Some familiarity with matrix algebra and basic probabilistic concepts (e.g., probability density functions, mathematical expectation) is assumed. State vector and state space notions are used throughout. The development is carried out by first considering an unforced dynamical system. Then the modifications that are required when a more complex dynamical model is introduced are discussed.

The time-discrete Kalman filter is composed of a group of matrix recursion relations. The simplicity of these relations make them particularly amenable to implementation on a digital computer. To study a particular system which requires state vector estimation, the system is generally simulated and the filter equations are exercised. In accomplishing the simulation many difficulties are uncovered. Considerations of this nature are discussed in Section III.

The present-day extension of Gauss' problem of determining the orbits of celestial bodies is that of determining the orbital characteristics of a manmade spacecraft. To apply the methodology of the Kalman filter, it is necessary to assume the existence of an approximating orbit (as Gauss did) or, in the parlance of the day, a reference trajectory. This reference trajectory is used to reduce the nonlinear equations which describe the motion of the spacecraft to a set of linear equations. The application of these techniques to space navigation is discussed in a general manner in Section IV. The discussion is designed to provide insight into the procedure that can be used for other problems involving nonlinear differential and/or algebraic equations. The specific problem of estimating the state of a spacecraft in a nearly circular orbit is examined to illustrate these procedures.

The concluding section provides a brief summary of the basic results presented in each subsection of Sections II–IV. By referring to this summary, the reader should be able to determine the subsections of the main body of the text that are of particular interest to him. In addition to the summary, this section briefly discusses other topics relating to estimation theory that have been dealt with in the literature.

II. Kalman Filter Theory

A. The Mathematical Model and Problem Statement

Consider a dynamical system whose state is described by a linear, vector difference equation:

$$\mathbf{x}_k = \Phi_{k,k-1}\mathbf{x}_{k-1} + \Gamma_{k,k-1}\mathbf{u}_{k-1} + \mathbf{f}_{k-1} + \mathbf{w}_{k-1} \qquad \text{(I)}$$

The independent variable t can assume the values $t_0 \leqslant t_1 \leqslant \cdots t_N$, where the t_i are not necessarily equidistant. The state of the system at t_k is given by the n-dimensional vector $\mathbf{x}_k$. The p-dimensional control vector for the interval $[t_{k-1}, t_k)$ is $\mathbf{u}_{k-1}$. The $\mathbf{f}_{k-1}$ is a vector forcing function that is a known function of time. The $\mathbf{w}_{k-1}$ is a vector random sequence with known statistics

$$E[\mathbf{w}_k] = \mathbf{0} \qquad \text{for all} \quad k$$

$$E[\mathbf{w}_k \mathbf{w}_j^T] = Q_k \delta_{kj}$$

where δ_{kj} is the Kronecker delta. The matrix Q_k is assumed to be nonnegative-definite, so it is possible that $\mathbf{w}_k \equiv \mathbf{0}$.

In many applications (I) is derived from the linear perturbation equations relating to a dynamical system, so the matrix $\Phi_{k,k-1}$ may be assumed to be a state transition matrix with the following properties:

$$\Phi_{k,k} = I \qquad \text{for all} \quad k$$

$$\Phi_{k,j}\Phi_{j,i} = \Phi_{k,i}$$

These properties imply (*27*) that

$$\Phi_{k,j}^{-1} = \Phi_{j,k}$$

This assumption is not required in much of the discussion that follows, but $\Phi_{k,k-1}$ shall be referred to as the state transition matrix anyway. The $\Gamma_{k,k-1}$ is a known $(n \times p)$ matrix that relates the control inputs to the state vector.

The initial state $\mathbf{x}_0$ is considered to be a vector random variable with the known statistics

$$E[\mathbf{x}_0] = \mathbf{0} \qquad E[\mathbf{x}_0 \mathbf{x}_0^T] = M_0$$

and

$$E[\mathbf{w}_k \mathbf{x}_0^T] = O \qquad \text{for all} \quad k$$

Suppose that at each time t_k there are available m measurements $\mathbf{z}_k$ that are linearly related to the state and which are corrupted by additive noise.

$$\mathbf{z}_k = H_k \mathbf{x}_k + \mathbf{v}_k \tag{II}$$

H_k is a known $(m \times n)$-dimensional observation matrix. The vector $\mathbf{v}_k$ is an additive, random sequence with known statistics.

$$E[\mathbf{v}_k] = \mathbf{0} \qquad \text{for all} \quad k$$

$$E[\mathbf{v}_k \mathbf{v}_j^T] = R_k \delta_{kj}$$

The matrix R_k is assumed to be nonnegative-definite unless otherwise stated.

Further, assume that the random processes $\mathbf{v}_k$ and $\mathbf{w}_k$ are uncorrelated. *These processes will be called white noise sequences.*

$$E[\mathbf{v}_k \mathbf{w}_j^T] = O \qquad \text{for all} \quad k, j$$

and

$$E[\mathbf{v}_k \mathbf{x}_0^T] = O \qquad \text{for all} \quad k$$

The mathematical model described above provides the basis for all succeeding discussion. In this discussion we shall deal with a problem of considerable importance in engineering practice.

Recursive Linear Estimation Problem. Given the preceding model, determine an estimate $\hat{\mathbf{x}}_k$ of the state at t_k that is a *linear* combination of an estimate at t_{k-1} and the measurement data $\mathbf{z}_k$. The estimate must be "best" in the sense that the expected value of the sum of the squares of the error in the estimate is a minimum. That is, the $\hat{\mathbf{x}}_k$ is to be chosen so that

$$E[(\hat{\mathbf{x}}_k - \mathbf{x}_k)^T (\hat{\mathbf{x}}_k - \mathbf{x}_k)] = \text{minimum}$$

B. Simplified Derivation for an Unforced Dynamical System

Before dealing with the system described by (I) let us consider a somewhat simpler model. Suppose the dynamics are described by the homogeneous, linear difference equation

$$\mathbf{x}_k = \Phi_{k,k-1} \mathbf{x}_{k-1} \tag{Iu}$$

The Kalman filter equations shall be derived now for the model described by (Iu) and (II).

First, the form of a linear estimation equation can be hypothesized from the physical characteristics of the system. The state evolves according to (Iu) so, given an estimate $\hat{\mathbf{x}}_{k-1}$ at t_{k-1}, it is reasonable to predict the estimate at t_k as

$$\hat{\mathbf{x}}_k' = \Phi_{k,k-1} \hat{\mathbf{x}}_{k-1}$$

when no other information is available. A measurement at t_k can be used to modify this estimate. Based on $\hat{\mathbf{x}}_k'$ and (II), one would expect the measurement value at t_k to be $H_k \hat{\mathbf{x}}_k'$. An error in the estimate is reflected by an error in this expected measurement value.

$$\mathbf{e}_k = \mathbf{z}_k - H_k \Phi_{k,k-1} \hat{\mathbf{x}}_{k-1}$$

According to the problem statement, the estimate is to be a linear function of the new measurements. Define an unknown matrix K_k such that the estimate $\hat{\mathbf{x}}_k$ is given by

$$\hat{\mathbf{x}}_k = \Phi_{k,k-1} \hat{\mathbf{x}}_{k-1} + K_k [\mathbf{z}_k - H_k \Phi_{k,k-1} \hat{\mathbf{x}}_{k-1}] \tag{IIIu}$$

The matrix K_k shall be determined so that $E[(\hat{\mathbf{x}}_k - \mathbf{x}_k)^T (\hat{\mathbf{x}}_k - \mathbf{x}_k)]$ is minimized. It shall be referred to as the *weighting* or *gain matrix.*

Let

$$\tilde{\mathbf{x}}_k \overset{\text{Df}}{=} \hat{\mathbf{x}}_k - \mathbf{x}_k$$

so

$$E[(\hat{\mathbf{x}}_k - \mathbf{x}_k)^T (\hat{\mathbf{x}}_k - \mathbf{x}_k)] = E[\tilde{\mathbf{x}}_k^T \tilde{\mathbf{x}}_k]$$

It is obvious that this can be rewritten as

$$E[\tilde{\mathbf{x}}_k^T \tilde{\mathbf{x}}_k] = \text{trace } E[\tilde{\mathbf{x}}_k \tilde{\mathbf{x}}_k^T]$$

where the trace is defined as the sum of the diagonal elements of a matrix. Define the matrix P_k as

$$P_k \overset{\text{Df}}{=} E[\tilde{\mathbf{x}}_k \tilde{\mathbf{x}}_k^T]$$

Now, form $\tilde{\mathbf{x}}_k$

$$\begin{aligned}
\tilde{\mathbf{x}}_k &= [\Phi_{k,k-1} \hat{\mathbf{x}}_{k-1} + K_k(\mathbf{z}_k - H_k \Phi_{k,k-1} \hat{\mathbf{x}}_{k-1})] - \Phi_{k,k-1} \mathbf{x}_{k-1} \\
&= \Phi_{k,k-1} \tilde{\mathbf{x}}_{k-1} - K_k H_k \Phi_{k,k-1} \hat{\mathbf{x}}_{k-1} + K_k (H_k \mathbf{x}_k + \mathbf{v}_k) \\
&= (I - K_k H_k) \Phi_{k,k-1} \tilde{\mathbf{x}}_{k-1} + K_k \mathbf{v}_k
\end{aligned}$$

The matrix P_k can now be formed.

$$P_k = E\{[(I - K_kH_k)\Phi_{k,k-1}\tilde{\mathbf{x}}_{k-1} + K_k\mathbf{v}_k][(I - K_kH_k)\Phi_{k,k-1}\tilde{\mathbf{x}}_{k-1} + K_k\mathbf{v}_k]^T\}$$
$$= (I - K_kH_k)\Phi_{k,k-1}E[\tilde{\mathbf{x}}_{k-1}\tilde{\mathbf{x}}_{k-1}^T]\Phi_{k,k-1}^T(I - H_k^TK_k^T)$$
$$+ K_kE[\mathbf{v}_k\tilde{\mathbf{x}}_{k-1}^T]\Phi_{k,k-1}^T(I - H_k^TK_k^T)$$
$$+ (I - K_kH_k)\Phi_{k,k-1}E[\tilde{\mathbf{x}}_{k-1}\mathbf{v}_k^T]K_k^T + K_kE[\mathbf{v}_k\mathbf{v}_k^T]K_k^T$$

By definition

$$E[\tilde{\mathbf{x}}_{k-1}\tilde{\mathbf{x}}_{k-1}^T] = P_{k-1}$$

and

$$E[\mathbf{v}_k\mathbf{v}_k^T] = R_k$$

Also, it follows that

$$E[\mathbf{v}_k\tilde{\mathbf{x}}_{k-1}^T] = O = E[\tilde{\mathbf{x}}_{k-1}\mathbf{v}_k^T]$$

since

$$E[\mathbf{v}_k\mathbf{v}_{k-1}^T] = O$$

and

$$E[\mathbf{v}_k\mathbf{x}_0^T] = O$$

Thus P_k can be rewritten

$$P_k = (I - K_kH_k)P_k'(I - K_kH_k)^T + K_kR_kK_k^T \tag{2.0}$$
$$P_k' \overset{\text{Df}}{=} \Phi_{k,k-1}P_{k-1}\Phi_{k,k-1}^T \tag{IVu}$$

Expanding (2.0) gives

$$P_k = P_k' - K_kH_kP_k' - P_k'H_k^TK_k^T + K_k(H_kP_k'H_k^T + R_k)K_k^T \tag{2.1}$$

The matrix P_k' does not depend upon K_k, so it is unaffected by the selection of K_k. The matrix $(H_kP_k'H_k^T + R_k)$ is symmetric and non-negative-definite, so it can be written as the product of a matrix S_k and its transpose (i.e., a matrix square root).

$$S_kS_k^T \overset{\text{Df}}{=} H_kP_k'H_k^T + R_k \tag{2.2}$$

Observe that the last three terms of (2.1) have the form of a quadratic matrix polynomial in terms of the unknown K_k. Introduce (2.2) into (2.1) and hypothesize the existence of a matrix A_k such that

$$P_k = P_k' + (K_kS_k - A_k)(K_kS_k - A_k)^T - A_kA_k^T \tag{2.3}$$

This procedure is the matrix equivalent of completing the square of a quadratic polynomial. Assuming that $S_kS_k^T$ is positive-definite, it follows directly that

$$A_k = P_k'H_k^T(S_k^{-1})^T \tag{2.4}$$

When the S_k is singular, the pseudo inverse (*2*) $S_k^\dagger$ is used instead of S_k^{-1}. Only the product term in (2.3) involves the gain matrix K_k. The product of a matrix and its transpose is nonnegative-definite, so the trace of P_k is minimized by choosing

$$K_kS_k = P_k'H_k^T(S_k^{-1})^T$$

Thus the optimal gain matrix is

$$K_k = P_k'H_k^T[H_kP_k'H_k^T + R_k]^{-1} \tag{V}$$

Inserting (V) and (2.4) in (2.3) shows that the matrix P_k reduces to

$$P_k = P_k' - K_kH_kP_k' \tag{VI}$$

Equations (IIIu), (IVu), (V), and (VI) constitute the Kalman filter for the model of (Iu) and (II).

C. Extension to Deterministic and Random Forcing Functions

Extend the dynamical model of (Iu) to include deterministic and random forcing terms.

$$\mathbf{x}_k = \Phi_{k,k-1}\mathbf{x}_{k-1} + \mathbf{f}_{k-1} + \mathbf{w}_{k-1} \tag{IF}$$

These terms have been defined in Section II, A. We note that the control term $\Gamma_{k,k-1}\mathbf{u}_{k-1}$ of (I) has not yet been included.

It is possible to hypothesize the estimate of $\mathbf{x}_k$ based only on the estimate at t_{k-1} (i.e., no measurements at $\mathbf{t}_k$). The noise vector $\mathbf{w}_k$ is independent of the state at time t_k and has mean zero so one would not expect it to effect the estimate at t_k. Also, $\mathbf{f}_{k-1}$ is a known vector function that acts on the interval $[t_{k-1}, t_k)$ so the estimate can be taken as

$$\hat{\mathbf{x}}_k' = \Phi_{k,k-1}\hat{\mathbf{x}}_{k-1} + \mathbf{f}_{k-1}$$

It is a straightforward matter to establish this estimate on a more rigorous basis. First, an important lemma shall be introduced and proved. The reader should note that the result provided by this lemma is true for *all* mean-square estimates, including nonlinear estimates. We assume that the necessary probability densities exist.

LEMMA. *Suppose that a vector random variable* $\mathbf{x}$ *is to be estimated from the known variables* $\mathbf{z}_1, \mathbf{z}_2, \ldots, \mathbf{z}_q$. *The* $\mathbf{x}$ *and the* $\mathbf{z}_j$ *have the joint probability density* $p(\mathbf{x}, \mathbf{z}_1, \mathbf{z}_2, \ldots, \mathbf{z}_q)$. *The estimate* $\hat{\mathbf{x}}$ *is to be chosen so that*

$$E[(\hat{\mathbf{x}} - \mathbf{x})^T(\hat{\mathbf{x}} - \mathbf{x})] = \textit{minimum}$$

Then the minimum mean-square estimate $\hat{\mathbf{x}}$ *of* $\mathbf{x}$ *given* $\mathbf{z}_1, \mathbf{z}_2, \ldots, \mathbf{z}_q$ *is*

$$\hat{\mathbf{x}} = E[\mathbf{x} \mid \mathbf{z}_1, \mathbf{z}_2, \ldots, \mathbf{z}_q]$$

Proof. Write $E[(\hat{\mathbf{x}} - \mathbf{x})^T(\hat{\mathbf{x}} - \mathbf{x})]$ in terms of the density function

$$E[(\hat{\mathbf{x}} - \mathbf{x})^T(\hat{\mathbf{x}} - \mathbf{x})] = \iint \cdots \int (\hat{\mathbf{x}} - \mathbf{x})^T(\hat{\mathbf{x}} - \mathbf{x}) p(\mathbf{x}, \mathbf{z}_1, \ldots, \mathbf{z}_q)\, d\mathbf{x}\, d\mathbf{z}_1 \cdots d\mathbf{z}_q \tag{2.5}$$

In this equation each integral sign represents a multiple integral since $\mathbf{x}, \mathbf{z}_1, \ldots, \mathbf{z}_q$ are vectors.

The density function can be written

$$p(\mathbf{x}, \mathbf{z}_1, \ldots, \mathbf{z}_q) = p(\mathbf{x} \mid \mathbf{z}_1, \ldots, \mathbf{z}_q) p(\mathbf{z}_1, \ldots, \mathbf{z}_q)$$

so (2.5) is equivalent to

$$E[(\hat{\mathbf{x}} - \mathbf{x})^T(\hat{\mathbf{x}} - \mathbf{x})] = \int \cdots \int \left\{ \int (\hat{\mathbf{x}} - \mathbf{x})^T(\hat{\mathbf{x}} - \mathbf{x}) p(\mathbf{x} \mid \mathbf{z}_1, \ldots, \mathbf{z}_q)\, d\mathbf{x} \right\} \times p(\mathbf{z}_1, \ldots, \mathbf{z}_q)\, d\mathbf{z}_1 \cdots d\mathbf{z}_q$$

Consider the integral in brackets. Since $\hat{\mathbf{x}}$ depends only upon the $\mathbf{z}_i$, the integral can be written

$$\begin{aligned}
\int (\hat{\mathbf{x}} - \mathbf{x})^T(\hat{\mathbf{x}} - \mathbf{x}) p(\mathbf{x} \mid \mathbf{z}_1, \ldots, \mathbf{z}_q)\, d\mathbf{x} &\\
= \hat{\mathbf{x}}^T\hat{\mathbf{x}} - \{E[\mathbf{x} \mid \mathbf{z}_1, \ldots, \mathbf{z}_q]\}^T\hat{\mathbf{x}} &\\
- \hat{\mathbf{x}}^T E[\mathbf{x} \mid \mathbf{z}_1, \ldots, \mathbf{z}_q] + E[\mathbf{x}^T\mathbf{x} \mid \mathbf{z}_1, \ldots, \mathbf{z}_q] &\\
= (\hat{\mathbf{x}}^T - \{E[\mathbf{x} \mid \mathbf{z}_1, \ldots, \mathbf{z}_q]\}^T)(\hat{\mathbf{x}} - E[\mathbf{x} \mid \mathbf{z}_1, \ldots, \mathbf{z}_q]) &\\
- \{E[\mathbf{x} \mid \mathbf{z}_1, \ldots, \mathbf{z}_q]\}^T E[\mathbf{x} \mid \mathbf{z}_1, \ldots, \mathbf{z}_q] + E[\mathbf{x}^T\mathbf{x} \mid \mathbf{z}_1, \ldots, \mathbf{z}_q] & \qquad (2.6)
\end{aligned}$$

By definition, this quantity is positive, so to minimize $E[(\hat{\mathbf{x}} - \mathbf{x})^T(\hat{\mathbf{x}} - \mathbf{x})]$ it is sufficient to choose $\hat{\mathbf{x}}$ to minimize the integrand (2.6). Only the first term involves $\hat{\mathbf{x}}$ and the smallest value it can assume is zero. Thus the minimizing estimate is given by

$$\hat{\mathbf{x}} = E[\mathbf{x} \mid \mathbf{z}_1, \ldots, \mathbf{z}_q] \qquad \text{Q.E.D.}$$

This lemma can be applied to the specific problem under consideration. What is the best estimate of $\mathbf{x}_k$ given measurement $\mathbf{z}_1, \mathbf{z}_2, \ldots, \mathbf{z}_{k-1}$? We know that the answer to this question is provided by the lemma.

$$\begin{aligned}
\hat{\mathbf{x}}_k' &= E[\mathbf{x}_k \mid \mathbf{z}_1, \ldots, \mathbf{z}_{k-1}]\\
&= E[(\Phi_{k,k-1}\mathbf{x}_{k-1} + \mathbf{f}_{k-1} + \mathbf{w}_{k-1}) \mid \mathbf{z}_1, \ldots, \mathbf{z}_{k-1}]\\
&= \Phi_{k,k-1}E[\mathbf{x}_{k-1} \mid \mathbf{z}_1, \ldots, \mathbf{z}_{k-1}] + E[\mathbf{f}_{k-1} \mid \mathbf{z}_1, \ldots, \mathbf{z}_{k-1}]\\
&\quad + E[\mathbf{w}_{k-1} \mid \mathbf{z}_1, \ldots, \mathbf{z}_{k-1}]
\end{aligned}$$

The first expectation is by definition

$$E[\mathbf{x}_{k-1} \mid \mathbf{z}_1, \ldots, \mathbf{z}_{k-1}] = \hat{\mathbf{x}}_{k-1}$$

The $\mathbf{f}_{k-1}$ is not a random vector, so

$$E[\mathbf{f}_{k-1} \mid \mathbf{z}_1, \ldots, \mathbf{z}_{k-1}] = \mathbf{f}_{k-1}$$

By hypothesis, the noise $\mathbf{w}_{k-1}$ is independent of the state at all times earlier than t_k and it also independent of the measurement noise. Thus

$$E[\mathbf{w}_{k-1} \mid \mathbf{z}_1, \ldots, \mathbf{z}_{k-1}] = E[\mathbf{w}_{k-1}] = \mathbf{0}$$

The best estimate of $\mathbf{x}_k$ based on the estimate at t_{k-1} is

$$\hat{\mathbf{x}}_k' = \Phi_{k,k-1}\hat{\mathbf{x}}_{k-1} + \mathbf{f}_{k-1} \tag{2.7}$$

This provides corroboration for the intuitive form of the estimate.

Assume that a measurement at t_k is made available. Proceeding as in Section II, B, the new estimate is taken to have the form

$$\hat{\mathbf{x}}_k = \hat{\mathbf{x}}_k' + K_k[\mathbf{z}_k - H_k\hat{\mathbf{x}}_k'] \tag{IIIF}$$

where K_k is the unknown gain matrix. Let us form $\tilde{\mathbf{x}}_k$.

$$\begin{aligned}
\tilde{\mathbf{x}}_k &= \hat{\mathbf{x}}_k - \mathbf{x}_k\\
&= (\Phi_{k,k-1}\hat{\mathbf{x}}_{k-1} + \mathbf{f}_{k-1} + K_k[\mathbf{z}_k - H_k\hat{\mathbf{x}}_k']) - \Phi_{k,k-1}\mathbf{x}_{k-1} - \mathbf{f}_{k-1} - \mathbf{w}_{k-1}\\
&= \Phi_{k,k-1}\tilde{\mathbf{x}}_{k-1} - K_kH_k\Phi_{k,k-1}\tilde{\mathbf{x}}_{k-1} + K_kH_k\mathbf{w}_{k-1} - \mathbf{w}_{k-1} + K_k\mathbf{v}_k\\
&= (I - K_kH_k)(\Phi_{k,k-1}\tilde{\mathbf{x}}_{k-1} - \mathbf{w}_{k-1}) + K_k\mathbf{v}_k
\end{aligned} \tag{2.8}$$

The matrix P_k can be formed.

$$\begin{aligned}
P_k = E\{&[(I - K_kH_k)(\Phi_{k,k-1}\tilde{\mathbf{x}}_{k-1} - \mathbf{w}_{k-1}) + K_k\mathbf{v}_k]\\
&\times [(I - K_kH_k)(\Phi_{k,k-1}\tilde{\mathbf{x}}_{k-1} - \mathbf{w}_{k-1}) + K_k\mathbf{v}_k]^T\}
\end{aligned}$$

It follows immediately that

$$E\{[\Phi_{k,k-1}\tilde{\mathbf{x}}_{k-1} - \mathbf{w}_{k-1}][\Phi_{k,k-1}\tilde{\mathbf{x}}_{k-1} - \mathbf{w}_{k-1}]^T\} = \Phi_{k,k-1}P_{k-1}\Phi_{k,k-1}^T + Q_{k-1}$$

because of the statistical properties of $\mathbf{w}_{k-1}$.

Let P_k' be defined as

$$P_k' \overset{\mathrm{Df}}{=} \Phi_{k,k-1}P_{k-1}\Phi_{k,k-1}^T + Q_{k-1} \tag{IV}$$

Thus P_k can be written

$$P_k = (I - K_kH_k)P_k'(I - K_kH_k)^T + K_kR_kK_k^T$$

But this equation is identical in form to (2.0), so the optimal gain matrix K_k is given by (V). The P_k is represented by (VI) in this case.

Equations (IIIF), (IV), (V), and (VI) constitute the Kalman filter for the model given by (IF) and (II). It should be noted that the only effect of random noise ($\mathbf{w}_{k-1}$) acting upon the dynamical system is to increase the uncertainty of the estimate by entering additively through (IV). On the other hand, the deterministic term $\mathbf{f}_k$ appears only in the estimation equation (IIIF).

D. Other Considerations

1. Initial Conditions and Unbiased Estimates

The recursion relations developed in the preceding section are valid for all k. Thus far, the question of *starting* the estimation procedure has been avoided. If the first measurement is obtained at t_1, questions immediately arise concerning the values of $\hat{\mathbf{x}}_0$ and P_0 that must be introduced in (IIIF) and (IV).

The initial state is not known precisely; instead it is described by its mean and covariance. In the absence of any measurement data at t_0, the initial estimate will be chosen as

$$\hat{\mathbf{x}}_0 = E[\mathbf{x}_0]$$

with covariance

$$P_0 = E[\mathbf{x}_0\mathbf{x}_0^T] = M_0$$

This choice for the initial estimate, besides being intuitively satisfying, has the advantage that it causes the estimation scheme to be unbiased for all t_k. This fact shall be verified.

Definition. *An estimate* $\hat{\mathbf{x}}$ *of a vector random variable* $\mathbf{x}$ *is said to be unbiased if and only if*

$$E[\hat{\mathbf{x}}] = E[\mathbf{x}]$$

The estimate $\hat{\mathbf{x}}_k$ is a random variable, so it is possible, formally, to form its expected value using (IIIF).

$$E[\hat{\mathbf{x}}_k] = \Phi_{k,k-1}E[\hat{\mathbf{x}}_{k-1}] + \mathbf{f}_{k-1} + K_k\{E[\mathbf{z}_k] - H_kE[\hat{\mathbf{x}}_k']\} \tag{2.9}$$

where

$$E[\mathbf{z}_k] = H_kE[\mathbf{x}_k] = H_k\Phi_{k,k-1}E[\mathbf{x}_{k-1}] + H_k\mathbf{f}_{k-1}$$

$$E[\hat{\mathbf{x}}_k'] = \Phi_{k,k-1}E[\hat{\mathbf{x}}_{k-1}] + \mathbf{f}_{k-1}$$

since

$$E[\mathbf{w}_{k-1}] = \mathbf{0} = E[\mathbf{v}_k]$$

Equation (2.9) can be simplified to the following:

$$E[\hat{\mathbf{x}}_k] = \Phi_{k,k-1}E[\hat{\mathbf{x}}_{k-1}] + \mathbf{f}_{k-1} + K_kH_k\Phi_{k,k-1}\{E[\mathbf{x}_{k-1}] - E[\hat{\mathbf{x}}_{k-1}]\} \tag{2.10}$$

Suppose $k = 1$.

$$E[\hat{\mathbf{x}}_1] = \Phi_{1,0}E[\hat{\mathbf{x}}_0] + \mathbf{f}_0 + K_1H_1\Phi_{1,0}\{E[\mathbf{x}_0] - E[\hat{\mathbf{x}}_0]\} \tag{2.11}$$

Since $\hat{\mathbf{x}}_0$ must be specified,

$$E[\hat{\mathbf{x}}_0] = \hat{\mathbf{x}}_0$$

Suppose $\hat{\mathbf{x}}_0$ is chosen to be equal to $E[\mathbf{x}_0]$. Then (2.11) becomes

$$E[\hat{\mathbf{x}}_1] = \Phi_{1,0}E[\mathbf{x}_0] + \mathbf{f}_0$$

Consider the expected value of the actual state.

$$E[\mathbf{x}_1] = \Phi_{1,0}E[\mathbf{x}_0] + \mathbf{f}_0 + E[\mathbf{w}_0]$$
$$= \Phi_{1,0}E[\mathbf{x}_0] + \mathbf{f}_0$$

Thus,

$$E[\mathbf{x}_1] = E[\hat{\mathbf{x}}_1]$$

It follows by induction that

$$E[\hat{\mathbf{x}}_k] = E[\mathbf{x}_k]$$

so (IIIF) represents an unbiased estimator. Observe that the $E[\mathbf{x}_0]$ has not been required to be equal to zero.

The fact that the estimator is unbiased is convenient, because it causes P_k and the covariance of the state

$$M_k = E[\mathbf{x}_k \mathbf{x}_k^T] \tag{2.12}$$

to be distributed about the same point. This aspect will be discussed further in Section III, C in terms of the error ellipsoids.

2. Weighted Least-Square Estimators

Consider for a moment an apparently more general problem. Instead of choosing the gain matrix to minimize

$$E[\tilde{\mathbf{x}}_k^T \tilde{\mathbf{x}}_k]$$

we shall introduce an arbitrary $(n \times n)$ positive-definite, symmetric matrix W_k. Then, choose K_k to minimize the weighted mean-square error

$$E[\tilde{\mathbf{x}}_k^T W_k \tilde{\mathbf{x}}_k]$$

Note that this can be rewritten

$$\begin{aligned} E[\tilde{\mathbf{x}}_k^T W_k \tilde{\mathbf{x}}_k] &= \text{trace}\,\{E[\tilde{\mathbf{x}}_k \tilde{\mathbf{x}}_k^T] W_k\} \\ &= \text{trace}\,\{P_k W_k\} \end{aligned}$$

The minimum value is seen to be attained by choosing K_k as given by (V). Thus, the weighting matrix will be omitted from our considerations in the remainder of this presentation.

This result has an important application that should be mentioned. Suppose that a vector $\mathbf{y}_k$ is to be estimated where it is known that

$$\mathbf{y}_k = A_k \mathbf{x}_k$$

A_k is a known $q \times n$ matrix. Then

$$E[\tilde{\mathbf{y}}_k^T \tilde{\mathbf{y}}_k] = E[(\hat{\mathbf{y}}_k - A_k \mathbf{x}_k)^T (\hat{\mathbf{y}}_k - A_k \mathbf{x}_k)]$$

Since A_k is known, it follows that

$$\hat{\mathbf{y}}_k = A_k \hat{\mathbf{x}}_k$$

where $\hat{\mathbf{x}}_k$ is chosen to minimize

$$\begin{aligned} E[\tilde{\mathbf{y}}_k^T \tilde{\mathbf{y}}_k] &= E[(\hat{\mathbf{x}}_k - \mathbf{x}_k)^T A_k^T A_k (\hat{\mathbf{x}}_k - \mathbf{x}_k)] \\ &= \text{trace}\,\{P_k A_k^T A_k\} \end{aligned}$$

Therefore, $\hat{\mathbf{y}}_k$ is computed directly from the mean-square estimate of $\mathbf{x}_k$. This result verifies our use of $H_k \hat{\mathbf{x}}_k'$ as the estimate of $\mathbf{z}_k$ in Sections II, B and II, C.

3. Orthogonal Projections

Equations (V) and (VI) can be derived in a manner that appeals to the geometric ideas associated with vector spaces. Suppose that the estimate of the state at t_k is given as the linear combination of $\hat{\mathbf{x}}_{k-1}$ and the measurements $\mathbf{z}_k$ [i.e., (IIIF)]. Introduce an inner product (*11*) as

$$\langle \mathbf{x}, \mathbf{y} \rangle \overset{\text{Df}}{=} E[\mathbf{x}^T \mathbf{y}] \tag{2.13}$$

It is easily verified that this satisfies the conditions (*10*) for an inner product. The gain matrix shall be selected so that the inner product $\langle \tilde{\mathbf{x}}_k, \tilde{\mathbf{x}}_k \rangle$ is minimized. But,

$$\begin{aligned} \langle \tilde{\mathbf{x}}_k, \tilde{\mathbf{x}}_k \rangle &= \langle \tilde{\mathbf{x}}_k, \hat{\mathbf{x}}_k - \mathbf{x}_k \rangle \\ &= \langle \tilde{\mathbf{x}}_k, \hat{\mathbf{x}}_k \rangle - \langle \tilde{\mathbf{x}}_k, \mathbf{x}_k \rangle \end{aligned}$$

Since $\hat{\mathbf{x}}_k$ is to be formed as a linear combination of the $\mathbf{z}_k$ and $\hat{\mathbf{x}}_{k-1}$ (where $\hat{\mathbf{x}}_{k-1}$ actually is determined implicitly as the linear combination of $\mathbf{z}_1, \mathbf{z}_2, \ldots, \mathbf{z}_{k-1}$), the estimate must lie in the hyperplane formed by these vectors. If $\mathbf{x}_k$ does not lie in this plane, it is clear that the error $\tilde{\mathbf{x}}_k$ can never be made to be identically zero. Furthermore, the smallest magnitude of the error $|\tilde{\mathbf{x}}_k|$ occurs when the estimate is equal to the orthogonal projection of the $\mathbf{x}_k$ onto the hyperplane defined by the $\{\mathbf{z}_j\}$. Thus the inner product $\langle \tilde{\mathbf{x}}_k, \tilde{\mathbf{x}}_k \rangle$ is minimized by choosing the K_k so that the error in the estimate and the estimate are orthogonal.

$$\langle \tilde{\mathbf{x}}_k, \hat{\mathbf{x}}_k \rangle = 0 \tag{2.14}$$

In this case, the minimum error is equal to

$$\min \langle \tilde{\mathbf{x}}_k, \tilde{\mathbf{x}}_k \rangle = -\langle \tilde{\mathbf{x}}_k, \mathbf{x}_k \rangle \tag{2.15}$$

To verify the preceding arguments, we shall demonstrate that the K_k and P_k are again described by (V) and (VI).

First, the K_k shall be derived to satisfy (2.14). From (2.8) and (IIIF) we have that

$$\begin{aligned} \langle \tilde{\mathbf{x}}_k, \hat{\mathbf{x}}_k \rangle = \langle (I - K_k H_k)(\Phi_{k,k-1} \tilde{\mathbf{x}}_{k-1} - \mathbf{w}_{k-1}) + K_k \mathbf{v}_k, \Phi_{k,k-1} \hat{\mathbf{x}}_{k-1} \\ + \mathbf{f}_{k-1} + K_k [\mathbf{z}_k - H_k (\Phi_{k,k-1} \hat{\mathbf{x}}_{k-1} + \mathbf{f}_{k-1})] \rangle \end{aligned} \tag{2.16}$$

By hypothesis, we require that

$$\langle \tilde{\mathbf{x}}_{k-1}, \hat{\mathbf{x}}_{k-1} \rangle = 0$$

From the assumptions that have been imposed on the noise processes, it follows that

$$\langle \mathbf{w}_{k-1}, \hat{\mathbf{x}}_{k-1} \rangle = 0 = \langle \mathbf{v}_k, \hat{\mathbf{x}}_{k-1} \rangle$$

and

$$\langle \mathbf{w}_{k-1}, \mathbf{f}_{k-1} \rangle = 0 = \langle \mathbf{v}_k, \mathbf{f}_{k-1} \rangle$$

Following Section II, D, 1, suppose that the estimate is unbiased. Then

$$\langle \tilde{\mathbf{x}}_{k-1}, \mathbf{f}_{k-1} \rangle = 0$$

Using the properties of the inner product and the preceding statements, (2.16) can be rewritten

$$\langle \tilde{\mathbf{x}}_k, \hat{\mathbf{x}}_k \rangle = \langle (I - K_k H_k)(\Phi_{k,k-1}\tilde{\mathbf{x}}_{k-1} - \mathbf{w}_{k-1}) + K_k \mathbf{v}_k, K_k \mathbf{z}_k \rangle$$

But $\mathbf{z}_k$ is given by (II), so

$$\langle \tilde{\mathbf{x}}_k, \hat{\mathbf{x}}_k \rangle = \langle (I - K_k H_k)\Phi_{k,k-1}\tilde{\mathbf{x}}_{k-1}, K_k H_k \mathbf{x}_k \rangle - \langle (I - K_k H_k)\mathbf{w}_{k-1}, K_k H_k \mathbf{x}_k \rangle + \langle K_k \mathbf{v}_k, K_k \mathbf{v}_k \rangle$$

The state $\mathbf{x}_k$ is given by (IF), so this reduces to

$$\langle \tilde{\mathbf{x}}_k, \hat{\mathbf{x}}_k \rangle = \langle (I - K_k H_k)\Phi_{k,k-1}\tilde{\mathbf{x}}_{k-1}, K_k H_k \Phi_{k,k-1}\mathbf{x}_{k-1} \rangle - \langle (I - K_k H_k)\mathbf{w}_{k-1}, K_k H_k \mathbf{w}_{k-1} \rangle + \langle K_k \mathbf{v}_k, K_k \mathbf{v}_k \rangle$$

Since

$$\langle \tilde{\mathbf{x}}_k, \hat{\mathbf{x}}_k \rangle = E[\tilde{\mathbf{x}}_k^T \hat{\mathbf{x}}_k] = \text{trace } E[\tilde{\mathbf{x}}_k \hat{\mathbf{x}}_k^T]$$

and since

$$\mathbf{x}_{k-1} = \hat{\mathbf{x}}_{k-1} - \tilde{\mathbf{x}}_{k-1}$$

$$\begin{aligned}\langle \tilde{\mathbf{x}}_k, \hat{\mathbf{x}}_k \rangle &= -\text{trace}\{(I - K_k H_k)\Phi_{k,k-1}P_{k-1}\Phi_{k,k-1}^T H_k^T K_k^T \\ &\quad + (I - K_k H_k)Q_{k-1}H_k^T K_k^T - K_k R_k K_k^T\} \\ &= -\text{trace}\{P_k' H_k^T K_k^T - K_k(H_k P_k' H_k^T + R_k)K_k^T\}\end{aligned}$$

where P_k' is defined by (IV).

If K_k is selected to be (V), it is clear that

$$\langle \tilde{\mathbf{x}}_k, \hat{\mathbf{x}}_k \rangle = 0$$

It follows in a straightforward manner that

$$\begin{aligned}-\langle \tilde{\mathbf{x}}_k, \mathbf{x}_k \rangle &= -\text{trace } E[\tilde{\mathbf{x}}_k \mathbf{x}_k^T] \\ &= \text{trace}[P_k' - K_k H_k P_k']\end{aligned}$$

so

$$\langle \tilde{\mathbf{x}}_k, \tilde{\mathbf{x}}_k \rangle = \text{trace } P_k$$

The proof shall be omitted.

The results demonstrate that the error in the estimate is orthogonal to the estimate. The idea of orthogonal projections can be used for time-continuous systems. It provides a particularly simple derivation of the Wiener-Hopf equation for continuous, linear estimation problems (*11*). Also, Balakrishnan (*12*) uses this concept to develop nonlinear estimators.

4. Relation to Deterministic Least Squares

In the classical least-squares estimation procedure, an approximate relationship between the measurements and the state is assumed. For the linear estimation problem one would suppose that

$$\mathbf{z} \approx H\mathbf{x}$$

where H is a known $(m \times n)$ matrix with rank n. Because of errors in the relationship and in the measurements, one can solve this equation for $\mathbf{x}$ only in some approximate manner. In the least-squares sense the $\hat{\mathbf{x}}$ is chosen so that

$$(\mathbf{z} - H\mathbf{x})^T W (\mathbf{z} - H\mathbf{x})$$

is minimized. The W is an arbitrary positive-definite weighting matrix. Elementary techniques lead to the conclusion that the least-squares estimate is

$$\hat{\mathbf{x}} = (H^T W^{-1} H)^{-1} H^T W^{-1} \mathbf{z} \tag{2.17}$$

We want to indicate the interesting relation that exists between this estimate and the estimate given by the Kalman filter (*13*). To do this, let $k = 1$ and associate the measurements $\mathbf{z}_1$ with the $\mathbf{z}$ of (2.17), $\mathbf{x}_1$ with $\mathbf{x}$, and H_1 with H.

Implicit in the estimate of (2.17) is the assumption that $\mathbf{x}$ is constant. This is equivalent to requiring that the state transition matrix is the identity matrix

$$\Phi_{1,0} = I$$

so that

$$\mathbf{x}_1 = \mathbf{x}_0$$

For this dynamical model, (IIIu), (IVu), (V), and (VI) are applicable. First, note that

$$P_1' = P_0$$

It is proved in Section III, D that K_k and $(P_k')^{-1}$ can be written

$$P_k^{-1} = (P_k')^{-1} + H_k^T R_k^{-1} H_k \tag{VII}$$

and

$$K_k = P_k H_k^T R_k^{-1} \tag{VIII}$$

Using these relations, we obtain

$$P_1^{-1} = P_0^{-1} + H_1^T R_1^{-1} H_1 \qquad K_1 = P_1 H_1^T R_1^{-1}$$

If *no information* is available about the initial state, the P_0^{-1} matrix can be assumed to vanish (i.e., the covariance matrix of the initial state has infinite norm. This will sometimes be written as

$$P_0 = \infty I)$$

Make this assumption. Then

$$P_1^{-1} = H_1^T R_1^{-1} H_1$$

or

$$P_1 = (H_1^T R_1^{-1} H_1)^{-1}$$

The estimate is given by (IIIu). Substituting the preceding results gives

$$\begin{aligned} \hat{\mathbf{x}}_1 &= \hat{\mathbf{x}}_0 + K_1(\mathbf{z}_1 - H_1\hat{\mathbf{x}}_0) \\ &= \hat{\mathbf{x}}_0 + (H_1^T R_1^{-1} H_1)^{-1} H_1^T R_1^{-1}(\mathbf{z}_1 - H_1\hat{\mathbf{x}}_0) \\ &= (H_1^T R_1^{-1} H_1)^{-1} H_1^T R_1^{-1}\mathbf{z}_1 \end{aligned} \tag{2.18}$$

The similarity between (2.17) and (2.18) is striking. If the weighting matrix is taken to be equal to the covariance of the measurement noise, the two equations are identical. An important case occurs when the weighting matrix is set equal to a scalar matrix. In the deterministic sense, this is equivalent to causing each measurement to have the same influence. In a probabilistic sense, this is equivalent to assuming that the noise in each measurement is independent and identically distributed.

E. Optimal Control and the Separation Theorem

Let us extend our considerations to the dynamical model described by (I). The estimation procedure that has been developed in the preceding sections is optimal in the sense that the mean-square error is minimized. When the control term $\Gamma_{k,k-1}\mathbf{u}_{k-1}$ is added, it is natural to ask that the composite system of control and estimation be jointly optimum in some well-defined sense. For linear systems with quadratic performance indices this system optimization can be accomplished in a straightforward manner through the use of the so-called "separation theorem" (*14–16*).

The control shall be determined for the system of (I) under the constraint that it minimizes the expected value of a quadratic performance index V_N.

$$E[V_N] = E\left\{\sum_{i=1}^{N}[\mathbf{x}_i^T W_i^X \mathbf{x}_i + \mathbf{u}_{i-1}^T W_{i-1}^U \mathbf{u}_{i-1}]\right\} \tag{2.19}$$

The terminal time for the process is t_N. The matrices W_i^X and W_i^U are arbitrary, symmetric, positive-definite weighting matrices.

The noise processes $\mathbf{w}_k$ and $\mathbf{v}_k$ have been described by their first- and second-order moments. If these moments are assumed to completely describe the processes, then these processes must be gaussian. Let us explicitly assume that this is the case. Under these conditions, the following theorem can be shown to be valid.

Separation Theorem. *Consider a dynamical system described by* (I). *The only information about* $\mathbf{x}_k$ *is given by measurements described by* (II). *The random processes* $\mathbf{w}_k$ *and* $\mathbf{v}_k$ *are gaussian with known statistics. Then, the stochastic control policy that minimizes* (2.19) *is given by*

$$\mathbf{u}_k = \Lambda_k \hat{\mathbf{x}}_k \tag{IX}$$

where $\hat{\mathbf{x}}_k$ *is the optimal linear estimate obtained by treating* $\Gamma_{k,k-1}\mathbf{u}_{k-1}$ *as a known forcing function. The control matrix* Λ_k *is obtained as the solution of the deterministic control problem.*

The proof of this theorem shall be omitted. The first proofs were obtained independently by Joseph (*17*) and by Gunckel (*18*) at approximately the same time.

When the conditions of the theorem are satisfied, the estimate of the state is provided by the following relation:

$$\hat{\mathbf{x}}_k = \hat{\mathbf{x}}_k' + K_k[\mathbf{z}_k - H_k\hat{\mathbf{x}}_k']$$

where

$$\hat{\mathbf{x}}_k{}' = [\Phi_{k,k-1} + \Gamma_{k,k-1}\Lambda_{k-1}]\hat{\mathbf{x}}_{k-1} + \mathbf{f}_{k-1}$$

Let

$$\Omega_{k,k-1} = \Phi_{k,k-1} + \Gamma_{k,k-1}\Lambda_{k-1}$$

Then the estimate can be written

$$\hat{\mathbf{x}}_k = \Omega_{k,k-1}\hat{\mathbf{x}}_{k-1} + \mathbf{f}_{k-1} + K_k[\mathbf{z}_k - H_k(\Omega_{k,k-1}\hat{\mathbf{x}}_{k-1} + \mathbf{f}_{k-1})] \qquad \text{(III)}$$

Thus (III), (IV), (V), and (VI) provide the filter equations for the model (I) and (II) when the control is chosen to minimize (2.19). Note that (III) has the same form as (IIIF) except that $\Phi_{k,k-1}$ is replaced by $\Omega_{k,k-1}$. We shall generally refer to (IIIF) in the subsequent discussion.

This estimate can be rewritten in a slightly different form (*14*) by using a result from Section III, D. This alternative form is analogous to the structure that is generally stated for the time-continuous case. Equation (III) can be rearranged as

$$\hat{\mathbf{x}}_k = \Phi_{k,k-1}\hat{\mathbf{x}}_{k-1} + \mathbf{f}_{k-1} + K_k[\mathbf{z}_k - H_k(\Phi_{k,k-1}\hat{\mathbf{x}}_{k-1} + \mathbf{f}_{k-1})]$$
$$+ (I - K_kH_k)\Gamma_{k,k-1}\mathbf{u}_{k-1}$$

But from (VII) and (VIII) of Section III, D,

$$(I - K_kH_k) = P_k(P_k{}')^{-1}$$

so

$$\hat{\mathbf{x}}_k = \Phi_{k,k-1}\hat{\mathbf{x}}_{k-1} + \mathbf{f}_{k-1} + K_k[\mathbf{z}_k - H_k(\Phi_{k,k-1}\hat{\mathbf{x}}_{k-1} + \mathbf{f}_{k-1})]$$
$$+ P_k(P_k{}')^{-1}\Gamma_{k,k-1}\mathbf{u}_{k-1} \qquad \text{(III')}$$

F. Linear Estimation for Correlated Sequences

1. A More General Probabilistic Model

Consider the model of (I) and (II) with one modification. Replace the white noise sequences $\mathbf{w}_k$ and $\mathbf{v}_k$ with sequences $\boldsymbol{\omega}_k$ and $\boldsymbol{\nu}_k$ that are *not* necessarily independent between sampling times. Then we have

$$\mathbf{x}_k = \Phi_{k,k-1}\mathbf{x}_{k-1} + \Gamma_{k,k-1}\mathbf{u}_{k-1} + \mathbf{f}_{k-1} + \boldsymbol{\omega}_{k-1} \qquad \text{(I')}$$

and

$$\mathbf{z}_k = H_k\mathbf{x}_k + \boldsymbol{\nu}_k \qquad \text{(II')}$$

Assume that $\boldsymbol{\omega}_k$ and $\boldsymbol{\nu}_k$ have mean zero with covariance matrices

$$E[\boldsymbol{\omega}_k\boldsymbol{\omega}_j{}^T] = D_{kj} \qquad E[\boldsymbol{\nu}_k\boldsymbol{\nu}_j{}^T] = N_{kj} \qquad E[\boldsymbol{\omega}_k\boldsymbol{\nu}_j{}^T] = O \qquad \text{for all } k, j$$

Because of the correlation between sampling times, the Kalman filter equations that have been developed in the preceding sections cannot be applied without introducing modifications somewhere. Specifically, this problem shall be circumvented through the introduction of *shaping filters* (*2, 19*). For this discussion the following definitions of a shaping filter shall be employed although more general definitions are possible.

Definition. *Consider a random sequence* $\boldsymbol{\omega}_k$ *that has mean zero and covariance matrix* D_{kj}. *A linear dynamical system whose output has covariance* D_{kj} *when the input is a white noise sequence* $\mathbf{w}_k$ *shall be called a "shaping filter".*

This definition implies that the correlated process $\boldsymbol{\omega}_k$ can be described by a vector difference equation.

$$\boldsymbol{\omega}_k = \Phi^D_{k,k-1}\boldsymbol{\omega}_{k-1} + \mathbf{w}_{k-1} \qquad (2.20)$$

where $\mathbf{w}_{k-1}$ represents the white noise sequence. When $\boldsymbol{\omega}_k$ represents a gaussian sequence, the output of the shaping filter corresponds in distribution with $\boldsymbol{\omega}_k$ so we shall write $\boldsymbol{\omega}_k$ for the output. The matrix $\Phi^D_{k,k-1}$ must be a transition matrix for (2.20) to define a linear, dynamical system.

Suppose the processes $\boldsymbol{\omega}_{k-1}$ and $\boldsymbol{\nu}_k$ of (I') and (II') are generated by shaping filters

$$\boldsymbol{\omega}_k = \Phi^D_{k,k-1}\boldsymbol{\omega}_{k-1} + \mathbf{w}_{k-1} \qquad (2.21)$$

$$\boldsymbol{\nu}_k = \Phi^N_{k,k-1}\boldsymbol{\nu}_{k-1} + \mathbf{s}_{k-1} \qquad (2.22)$$

Let us treat $\boldsymbol{\omega}_k$ and $\boldsymbol{\nu}_k$ as state variables and adjoin (2.21) and (2.22) to (I') and (II'). Combining the result into partitioned matrices, we obtain

$$\begin{pmatrix}\mathbf{x}_k\\ \boldsymbol{\omega}_k\\ \boldsymbol{\nu}_k\end{pmatrix} = \begin{pmatrix}\Phi_{k,k-1} & I & O\\ O & \Phi^D_{k,k-1} & O\\ O & O & \Phi^N_{k,k-1}\end{pmatrix}\begin{pmatrix}\mathbf{x}_{k-1}\\ \boldsymbol{\omega}_{k-1}\\ \boldsymbol{\nu}_{k-1}\end{pmatrix} + \begin{pmatrix}\Gamma_{k,k-1}\\ O\\ O\end{pmatrix}\mathbf{u}_{k-1} + \begin{pmatrix}\mathbf{f}_{k-1}\\ \mathbf{0}\\ \mathbf{0}\end{pmatrix} + \begin{pmatrix}\mathbf{0}\\ \mathbf{w}_{k-1}\\ \mathbf{s}_{k-1}\end{pmatrix} \qquad (2.23)$$

and

$$\mathbf{z}_k = [H_k \quad O \quad I]\begin{pmatrix}\mathbf{x}_k\\ \boldsymbol{\omega}_k\\ \boldsymbol{\nu}_k\end{pmatrix} \qquad (2.24)$$

The system described by (2.23) and (2.24) corresponds with the model used for the Kalman filter after the obvious definitions are introduced. Note that there is no white noise sequence in (2.24), so the R_k associated with (II) is the zero matrix. Thus, whenever the mathematical model includes correlated processes which are of such a nature as to permit the derivation of the appropriate shaping filter, the system can be reduced to the form of (I) and (II). Note that this procedure results in an estimation of the random processes $\boldsymbol{\omega}_k$ and $\boldsymbol{\nu}_k$.

2. Shaping Filters for Wide-Sense Markov Sequences

It is not known how to derive a shaping filter for an arbitrary random sequence. If the sequence is stationary and has a rational spectral density, a shaping filter can be devised by frequency-domain techniques (*20*). Then, the transfer function of the filter is determined by factoring the power spectral density of the process into the product of a lower triangular matrix and its conjugate transpose. This can always be done. Since this technique involves frequency rather than time-domain techniques, we shall not discuss it any further.

An important class of nonstationary random sequences are *wide-sense Markov sequences* or, equivalently, *sequentially correlated sequences* (*9*).

Definition. *A random sequence* $\boldsymbol{\omega}_k$ *with zero mean is said to be "wide-sense Markov" if its covariance matrix* D_{kj} *satisfies the relation*

$$D_{kj}D_{jj}^{-1}D_{ji} = D_{ki} \qquad (2.25)$$

where

$$t_k \geqslant t_j \geqslant t_i$$

It is possible to derive a shaping filter for this type of sequence.

Assume that the wide-sense Markov sequence $\boldsymbol{\omega}_k$ can be generated by a linear difference equation

$$\boldsymbol{\omega}_k = \Phi^D_{k,k-1}\boldsymbol{\omega}_{k-1} + \mathbf{w}_{k-1} \qquad (2.26)$$

For (2.26) to represent a shaping filter, $\Phi^D_{k,k-1}$ must be determined and must be a transition matrix. First, form

$$\begin{aligned} E[\boldsymbol{\omega}_{k+1}\boldsymbol{\omega}_k{}^T] &= D_{k+1,k} \\ &= E[(\Phi^D_{k+1,k}\boldsymbol{\omega}_k + \mathbf{w}_k)\boldsymbol{\omega}_k{}^T] \\ &= \Phi^D_{k+1,k}E[\boldsymbol{\omega}_k\boldsymbol{\omega}_k{}^T] + E[\mathbf{w}_k\boldsymbol{\omega}_k{}^T] \end{aligned}$$

But $\mathbf{w}_k$ is by assumption a white noise sequence, so

$$E[\mathbf{w}_k\boldsymbol{\omega}_k{}^T] = O$$

Therefore,

$$D_{k+1,k} = \Phi^D_{k+1,k}D_{k,k}$$

Assuming that $D_{k,k}$ is positive-definite,

$$\Phi^D_{k+1,k} = D_{k+1,k}D_{k,k}^{-1} \qquad (2.27)$$

We must verify that this matrix is a transition matrix. Obviously,

$$\Phi^D_{k,k} = I \qquad \Phi^D_{k+1,k}\Phi^D_{k,k-1} = D_{k+1,k}D_{k,k}^{-1}D_{k,k-1}D_{k-1,k-1}^{-1}$$

Since $\boldsymbol{\omega}_k$ is wide-sense Markov, this equation reduces to

$$\Phi^D_{k+1,k}\Phi^D_{k,k-1} = D_{k+1,k-1}D_{k-1,k-1}^{-1} = \Phi^D_{k+1,k-1}$$

Thus, $\Phi^D_{k+1,k}$ satisfies the requirement for a transition matrix stated in Section II, A. To complete the derivation, the matrix

$$Q_k = E[\mathbf{w}_k\mathbf{w}_k{}^T]$$

must be determined. From (2.26),

$$\begin{aligned} E[\mathbf{w}_{k-1}\mathbf{w}_{k-1}^T] &= E\{(\boldsymbol{\omega}_k - \Phi^D_{k,k-1}\boldsymbol{\omega}_{k-1})(\boldsymbol{\omega}_k - \Phi^D_{k,k-1}\boldsymbol{\omega}_{k-1})^T\} \\ &= D_{k,k} - D_{k,k-1}D_{k-1,k-1}D_{k-1,k} \end{aligned} \qquad (2.28)$$

It can be shown that this matrix is nonnegative-definite.

As an example of a wide-sense Markov sequence, consider a scalar sequence ω_k with an exponential covariance function

$$D_{k,j} = e^{-|t_k - t_j|} \qquad t_k \geqslant t_j$$

Then

$$\begin{aligned} D_{k,j}D_{j,j}^{-1}D_{j,i} &= e^{-|t_k-t_j|}e^{-|t_j-t_i|} \\ &= e^{-|t_k-t_i|} \end{aligned}$$

since

$$t_k \geqslant t_j \geqslant t_i$$

For this sequence the transition matrix of the shaping filter is

$$\begin{aligned} \Phi^D_{k,k-1} &= D_{k,k-1}D_{k-1,k-1}^{-1} \\ &= e^{-|t_k-t_{k-1}|} \end{aligned}$$

The covariance of the white noise sequence is

$$E[\mathbf{w}_k^2] = 1 - e^{-|t_{k+1}-t_k|}e^{-|t_{k+1}-t_k|}$$
$$= 1 - e^{-2|t_{k+1}-t_k|}$$

3. Parameter Estimation

It often happens in physical situations that parameters appear in the dynamic model that are known imprecisely and which can be considered to be constant during the interval of interest. The measurement data may also contain constant random bias errors. When these parameters enter the system linearly (see Section IV, B for discussion concerning this assumption), (I) and (II) are modified.

$$\mathbf{x}_k = \Phi_{k,k-1}\mathbf{x}_{k-1} + \Gamma_{k,k-1}\mathbf{u}_{k-1} + Z_{k,k-1}\boldsymbol{\alpha}_{k-1} + \mathbf{f}_{k-1} + \mathbf{w}_{k-1} \qquad (\text{I}'')$$

and

$$\mathbf{z}_k = H_k\mathbf{x}_k + \mathbf{b}_k + \mathbf{v}_k \qquad (\text{II}'')$$

Since the parameters $\boldsymbol{\alpha}_{k-1}$ and $\mathbf{b}_k$ are assumed to be constant, we know

$$\boldsymbol{\alpha}_k = \boldsymbol{\alpha}_{k-1} \qquad (2.29)$$

and

$$\mathbf{b}_k = \mathbf{b}_{k-1} \qquad (2.30)$$

Further,

$$E[\boldsymbol{\alpha}_k\boldsymbol{\alpha}_j^T] = A_0 \qquad \text{for all} \quad k, j$$
$$E[\mathbf{b}_k\mathbf{b}_j^T] = B_0 \qquad \text{for all} \quad k, j$$

The matrices A_0 and B_0 must be known. Thus (I″) and (II″) can be rewritten by considering $\boldsymbol{\alpha}_k$ and $\mathbf{b}_k$ as additional components of the state vector. Then, in terms of partitioned matrices, we have

$$\begin{pmatrix}\mathbf{x}_k\\ \boldsymbol{\alpha}_k\\ \mathbf{b}_k\end{pmatrix} = \begin{pmatrix}\Phi_{k,k-1} & Z_{k,k-1} & O\\ O & I & O\\ O & O & I\end{pmatrix}\begin{pmatrix}\mathbf{x}_{k-1}\\ \boldsymbol{\alpha}_{k-1}\\ \mathbf{b}_{k-1}\end{pmatrix} + \begin{pmatrix}\Gamma_{k,k-1}\\ O\\ O\end{pmatrix}\mathbf{u}_{k-1} + \begin{pmatrix}\mathbf{f}_{k-1}\\ \mathbf{0}\\ \mathbf{0}\end{pmatrix} + \begin{pmatrix}\boldsymbol{\omega}_{k-1}\\ \mathbf{0}\\ \mathbf{0}\end{pmatrix} \qquad (2.31)$$

and

$$\mathbf{z}_k = [H_k \quad O \quad I]\begin{pmatrix}\mathbf{x}_k\\ \boldsymbol{\alpha}_k\\ \mathbf{b}_k\end{pmatrix} + \mathbf{v}_k \qquad (2.32)$$

Equations (2.31) and (2.32) have the form of (I) and (II). The Kalman filter will provide estimates of the parameters as well as the covariance of the error in the estimates. This technique is utilized in (*21*) to estimate the universal gravitational constant and other constants that appear in the equations of motion of a spacecraft.

G. Linear Estimation for Time-Continuous Systems

The filter equations (III) to (VI) for the time-discrete models (I) and (II) can be used to derive the Kalman filter for time-continuous dynamical systems and measurement processes. This is accomplished with a semi-rigorous limiting argument employed by Kalman (*2*). The general structure of the resulting filter equations is quite similar although differential rather than difference equations are involved, and white noise processes replace the white noise sequences defined in Section II, A. Consider a dynamical system described by a linear, vector differential equation

$$\frac{d\mathbf{x}}{dt} = F(t)\mathbf{x} + G(t)\mathbf{w}(t) \qquad (\text{Ic})$$

Let $\mathbf{w}(t)$ be a gaussian white noise process. A p-dimensional vector random process is called a gaussian white noise process if it is a gaussian process with moments prescribed as

$$E[\mathbf{w}(t)] = \mathbf{0} \qquad \text{for all} \quad t$$
$$E[\mathbf{w}(t)\mathbf{w}^T(\tau)] = Q(t)\,\delta(t-\tau) \qquad \text{for all} \quad t, \tau$$

The $Q(t)$ is a symmetric, nonnegative-definite $(p \times p)$ matrix and $\delta(t-\tau)$ represents the Dirac delta function (*10*). Note that the elements of the covariance matrix are infinite-valued but that

$$\int_0^T E[\mathbf{w}(t)\mathbf{w}^T(\tau)]\,d\tau = \int_0^T Q(\tau)\,\delta(t-\tau)\,d\tau$$
$$= Q(t) \qquad \text{for} \quad 0 < t < T \qquad (2.33)$$

The separation principle discussed in Section II, E can be extended to time-continuous systems (*16*) so control terms and known forcing functions have been omitted from (Ic). They can be included and will enter the filter equations in the trivial manner demonstrated above for the time-discrete model.

The measurement model is assumed to be

$$\mathbf{z}(t) = H(t)\mathbf{x}(t) + \mathbf{v}(t) \qquad (\text{IIc})$$

where $\mathbf{v}(t)$ is a gaussian white noise process with

$$E[\mathbf{v}(t)] = \mathbf{0} \qquad \text{for all} \quad t$$
$$E[\mathbf{v}(t)\mathbf{v}^T(\tau)] = R(t)\,\delta(t-\tau) \qquad \text{for all} \quad t, \tau$$

[see *(29)* and *(36)* when $\mathbf{v}(t)$ is not white noise.] The Kalman filter for the system (Ic), (IIc) is obtained from the time-discrete model by causing the sampling interval to become infinitesimal. However, fundamental differences exist between white noise processes and the white noise sequences of (I) and (II). These differences must be accounted for prior to introducing the limiting arguments.

It is the property delineated by (2.33) that must be approximated by the random sequences in (I) and (II). The covariance of the random sequence $\mathbf{v}_k$ has been defined as

$$E[\mathbf{v}_k\mathbf{v}_j^T] = R_k\,\delta_{kj} \qquad \text{for all} \quad k, j$$

If the time interval Δt between adjacent sampling times is permitted to become arbitrarily small, the noise will contain no power. That is, analogous to (2.33) for $0 \leqslant t_k \leqslant T$,

$$\lim_{\substack{n\to\infty \\ \Delta t\to 0}} \sum_{j=1}^{n} R_k\,\delta_{kj}\,\Delta t = O$$

When the sampling interval is made arbitrarily small and the elements of the covariance matrix at any time t_k remain finite, the sequence is essentially reduced to a deterministic sequence whose values are zero. In other words, there is no noise in the measurements so the estimation problem is uninteresting.

To circumvent this difficulty, introduce the constraint that

$$E[\mathbf{v}_k\mathbf{v}_j^T]\,\Delta t = R_k\,\delta_{kj} \qquad \text{for all} \quad k, j \tag{2.34}$$

for any sampling interval Δt and a prescribed matrix R_k. With this restriction it is apparent that

$$\lim_{\substack{n\to\infty \\ \Delta t\to 0}} \sum_{j=1}^{n} E[\mathbf{v}_k\mathbf{v}_j^T]\,\Delta t = R_k$$

The constraint (2.34) is equivalent to requiring that the noise sequence $\{\mathbf{w}_k\}$ and $\{\mathbf{v}_k\}$ of (I) and (II) be replaced by

$$\left\{\frac{\mathbf{w}_k}{(\Delta t)^{1/2}}\right\} \quad \text{and} \quad \left\{\frac{\mathbf{v}_k}{(\Delta t)^{1/2}}\right\}$$

where $\mathbf{w}_k$ and $\mathbf{v}_k$ continue to represent the sequences of Section II, A.

Consider the dynamical system

$$\mathbf{x}_k = \Phi_{k,k-1}\mathbf{x}_{k-1} + \Delta_{k,k-1}\frac{\mathbf{w}_{k-1}}{(\Delta t)^{1/2}} \tag{2.35}$$

and the measurement process

$$\mathbf{z}_k = H_k\mathbf{x}_k + \frac{\mathbf{v}_k}{(\Delta t)^{1/2}} \tag{2.36}$$

The sampling interval is assumed to be constant with value Δt. Suppose that $\Phi_{k,k-1}$ is the state transition matrix associated with the system (Ic). Then, it can be expanded as

$$\Phi_{k,k-1} = I + F(t_{k-1})\,\Delta t + o(\Delta t) \tag{2.37}$$

where $o(\Delta t)$ denotes terms of greater than first order in Δt (i.e., $o(\Delta t)/\Delta t$ vanishes as $\Delta t \to 0$). Furthermore, the matrix $\Delta_{k,k-1}$ can be written

$$\Delta_{k,k-1} = G(t_{k-1})\,\Delta t + o(\Delta t) \tag{2.38}$$

See Section IV, A, for further discussion of the derivation of difference equations from the differential equation model.

Using these relations in (2.35), one obtains after rearranging terms,

$$\frac{\mathbf{x}_k - \mathbf{x}_{k-1}}{\Delta t} = F(t_{k-1})\mathbf{x}_{k-1} + G(t_{k-1})\frac{\mathbf{w}_{k-1}}{(\Delta t)^{1/2}} \tag{2.39}$$

Now, define the processes $\mathbf{v}(t)$ and $\mathbf{w}(t)$ such that

$$\mathbf{v}(t) \overset{\text{Df}}{=} \frac{\mathbf{v}_k}{(\Delta t)^{1/2}} \qquad \text{for} \quad t_{k-1} \leqslant t < t_k$$
$$\mathbf{w}(t) \overset{\text{Df}}{=} \frac{\mathbf{w}_k}{(\Delta t)^{1/2}} \qquad \text{for} \quad t_{k-1} \leqslant t < t_k$$

and let

$$\lim_{\Delta t\to 0} \frac{\delta_{kj}}{\Delta t} = \delta(t-\tau)$$

Letting $\Delta t \to 0$ in (2.39), the differential equation (Ic) is obtained. Similarly (2.36) is seen to become equivalent to (IIc).

We are now in a position to derive the Kalman filter equations for (Ic) and (IIc). From (IV) the extrapolated error covariance matrix for the modified noise sequence is

$$P_k' = \Phi_{k,k-1}P_{k-1}\Phi_{k,k-1}^T + \Delta_{k,k-1}\frac{Q_{k-1}}{\Delta t}\Delta_{k,k-1}^T$$

Substituting (2.37) and (2.38), this equation becomes

$$P_k' = P_{k-1} + F(t_{k-1})P_{k-1}\Delta t + P_{k-1}F^T(t_{k-1})\Delta t + G(t_{k-1})Q_{k-1}G^T(t_{k-1})\Delta t + o(\Delta t) \tag{2.40}$$

Using (VI) and rearranging terms, we obtain

$$\frac{P_k' - P'_{k-1}}{\Delta t} = F(t_{k-1})P'_{k-1} + P'_{k-1}F^T(t_{k-1}) - \frac{1}{\Delta t}K_{k-1}H_{k-1}P'_{k-1}$$
$$- F(t_{k-1})K_{k-1}H_{k-1}P'_{k-1} - K_{k-1}H_{k-1}P'_{k-1}F^T(t_{k-1})$$
$$+ G(t_{k-1})Q_{k-1}G^T(t_{k-1}) + \frac{o(\Delta t)}{\Delta t} \tag{2.41}$$

Let us examine $(1/\Delta t)K_{k-1}$.

$$\frac{1}{\Delta t}K_{k-1} = \frac{1}{\Delta t}P'_{k-1}H^T_{k-1}\left[H_{k-1}P_{k-1}H^T_{k-1} + \frac{R_{k-1}}{\Delta t}\right]^{-1}$$
$$= P'_{k-1}H^T_{k-1}[R_{k-1} + H_{k-1}P'_{k-1}H^T_{k-1}\,\Delta t]^{-1} \tag{2.42}$$

Assume that R_{k-1} is positive-definite.

Let $\Delta t \to 0$ in (2.41), remembering (2.42). Then

$$\lim_{\Delta t\to 0}\frac{P_k' - P'_{k-1}}{\Delta t} \stackrel{\text{Df}}{=} \frac{dP}{dt}$$

so

$$\frac{dP}{dt} = F(t)P + PF^T(t) - PH^T(t)R^{-1}(t)H(t)P + G(t)Q(t)G^T(t) \tag{Vc}$$

This equation has the form of a matrix Ricatti equation and we shall discuss it below. Observe from (2.42) that

$$\lim_{\Delta t\to 0}\frac{1}{\Delta t}K_{k-1} = P(t)H^T(t)R^{-1}(t)$$

Denote this as

$$K(t) \stackrel{\text{Df}}{=} P(t)H^T(t)R^{-1}(t) \tag{IVc}$$

The estimate described by (III) becomes

$$\frac{\hat{\mathbf{x}}_k - \hat{\mathbf{x}}_{k-1}}{\Delta t} = F(t_{k-1})\hat{\mathbf{x}}_{k-1} + \frac{K_k}{\Delta t}[\mathbf{z}_k - H_k\hat{\mathbf{x}}_{k-1} - H_kF(t_{k-1})\hat{\mathbf{x}}_{k-1}\,\Delta t]$$

upon introducing (2.37) and rearranging terms.

Let

$$\lim_{\Delta t\to 0}\frac{\hat{\mathbf{x}}_k - \hat{\mathbf{x}}_{k-1}}{\Delta t} \stackrel{\text{Df}}{=} \frac{d\hat{\mathbf{x}}}{dt}$$

Then

$$\frac{d\hat{\mathbf{x}}}{dt} = F(t)\hat{\mathbf{x}} + K(t)[\mathbf{z}(t) - H(t)\hat{\mathbf{x}}] \tag{IIIc}$$

Equations (IIIc) through (Vc) comprise the Kalman filter for the time-continuous system (Ic) and (IIc).

Let us consider the variance equation (Vc) in somewhat greater detail. As has already been mentioned, (Vc) is a matrix Ricatti equation. From the theory of these equations (*2, 58*) it is known that the solution $P(t)$ of (Vc) for given initial conditions P_0 can be found in terms of the solution of an equivalent set of linear differential equations. This set of $2n$ linear equations is given by

$$\frac{d\mathbf{r}}{dt} = -F^T(t)\mathbf{r} + H^T(t)R^{-1}(t)H(t)\mathbf{s} \tag{2.43}$$

$$\frac{d\mathbf{s}}{dt} = G(t)Q(t)G^T(t)\mathbf{r} + F(t)\mathbf{s} \tag{2.44}$$

The **r** and **s** are arbitrary symbols and are not important to the discussion. The transition matrix associated with this homogeneous system is denoted as $\theta(t, t_0)$, where in partitioned form

$$\theta(t, t_0) \stackrel{\text{Df}}{=} \begin{bmatrix}\theta_1(t, t_0) & \theta_2(t, t_0)\\ \theta_3(t, t_0) & \theta_4(t, t_0)\end{bmatrix}$$

The $\theta_i(t, t_0)$ are $(n \times n)$ matrices. The solution of the variance equation (Vc) can be written in terms of the $\theta_i(t, t_0)$ and the initial conditions P_0.

$$P(t) = [\theta_3(t, t_0) + \theta_4(t, t_0)P_0][\theta_1(t, t_0) + \theta_2(t, t_0)P_0]^{-1} \tag{2.45}$$

The solution exists whenever the matrix inverse $[\theta_1(t, t_0) + \theta_2(t, t_0)P_0]^{-1}$ exists.

Let us develop a special case of this equation that is interesting when the plant is unforced. Suppose that there is no noise in the plant, so that

$$Q(t) = O$$

Then (2.44) reduces to

$$\frac{d\mathbf{s}}{dt} = F(t)\mathbf{s}$$

But the solution of this equation produces the state transition matrix $\Phi(t, t_0)$, so

$$\theta_3(t, t_0) = O \qquad \theta_4(t, t_0) = \Phi(t, t_0)$$

The expression (2.43) becomes

$$\frac{d\mathbf{r}}{dt} = -F^T(t)\mathbf{r} + H^T(t)R^{-1}(t)H(t)\Phi(t, t_0)\mathbf{s}_0$$

If the adjoint matrix is denoted as $\Psi(t, t_0)$, then (*27*) it follows that

$$\theta_1(t, t_0) = [\Phi^{-1}(t, t_0)]^T \equiv \Psi(t, t_0)$$

and

$$\theta_2(t, t_0) = \Phi^T(t_0, t) \int_{t_0}^{t} \Phi^T(\tau, t_0)H^T(\tau)R^{-1}(\tau)H(\tau)\Phi(\tau, t_0)\, d\tau$$

Substituting the $\theta_i(t, t_0)$ in (2.45) yields

$$P(t) = \Phi(t, t_0)P_0 \left[I + \int_{t_0}^{t} \Phi^T(\tau, t_0)H^T(\tau)R^{-1}(\tau)H(\tau)\Phi(\tau, t_0)\, d\tau\, P_0\right]^{-1} \Phi^T(t, t_0) \tag{2.46}$$

III. Computational Considerations

A. Basic Computational Logic

The basic equations of the Kalman filter consist of four matrix recursion relations. These equations are logically simple to mechanize on a digital computer and this fact accounts for a great deal of the appeal of the Kalman filter. Let us summarize the quantities that must be provided for the digital mechanization. They fall naturally into two types: first, the physical model must be defined, and, second, the statistics of the random processes must be determined. In many cases the statistics represent parameters that are varied during a computer simulation study to determine hardware requirements. The minimum sampling interval is determined by the statistics and is another parameter that plays an important role in simulation studies.

Matrices of the Physical System. At each sampling time t_k, it is necessary to specify in some manner the following matrices: $\Phi_{k.k-1}$, H_k, $\mathbf{f}_{k-1}$, $\Gamma_{k.k-1}$, Λ_{k-1}.

Statistics. To start the computations at t_0, the following statistical quantities are required:

$$\hat{\mathbf{x}}_0 = E[\mathbf{x}_0]$$
$$P_0 = M_0$$

At all sampling times, $(t_1, t_2, \ldots, t_k, \ldots)$ the noise covariance matrices Q_{k-1} and R_k must be defined. The covariance matrices P_0, R_k, Q_{k-1}, and the sampling intervals (t_k, t_{k-1}) are of special interest because they represent parameters that can be manipulated during the computer study of a particular system.

When the Kalman filter is used to estimate the state of an actual physical process, the measurement data $\mathbf{z}_k$ is provided as the output of specific measuring devices. Assuming that the first observation occurs at t_1, the sequence of operations that is performed at each sampling time can be described by the following steps:

(1) At t_0, initialize P_0, $\hat{\mathbf{x}}_0$, Q_0. Let $t = t_1$.

(2) Form the *projected estimate* of the covariance of the estimate error.

$$P_1' = \Phi_{1.0}P_0\Phi_{1.0}^T + Q_0$$

(3) Compute the gain matrix.

$$K_1 = P_1'H_1^T[H_1P_1'H_1^T + R_1]^{-1}$$

(4) Form the estimate of the state at t_1 from the measurement $\mathbf{z}_1$.

$$\hat{\mathbf{x}}_1 = \Omega_{1.0}\hat{\mathbf{x}}_0 + \mathbf{f}_0 + K_1[\mathbf{z}_1 - H_k(\Omega_{1.0}\hat{\mathbf{x}}_0 + \mathbf{f}_0)]$$

where

$$\Omega_{1.0} = \Phi_{1.0} + \Gamma_{1.0}\Lambda_0$$

(5) Compute the covariance of the error in the estimate

$$P_1 = P_1' - K_1H_1P_1'$$

(6) Update time to t_2 and return to step (2) with all indices incremented by 1.

When the Kalman filter is used as part of a simulation of a physical system, it becomes necessary to generate the measurement data $\mathbf{z}_k$. To accomplish this, the actual state $\mathbf{x}_k$ must be computed. Since the initial conditions are random, it is necessary to select $\mathbf{x}_0$ from a statistical ensemble with the statistics

$$E[\mathbf{x}_0] = \mathbf{0} \qquad E[\mathbf{x}_0\mathbf{x}_0^T] = M_0$$

The covariance matrix M_0 is prescribed. The selection can be performed through the use of a noise generator (*22, 56*). More will be said on this subject in Section III, B. In addition, the random process $\mathbf{w}_k$ must be generated according to the statistics

$$E[\mathbf{w}_k] = \mathbf{0} \quad \text{and} \quad E[\mathbf{w}_k \mathbf{w}_j^T] = Q_k \delta_{kj}$$

When these operations are accomplished, the actual state of the system can be computed.

The measurement process contains the additive random process $\mathbf{v}_k$, which also must be determined with a noise generator. Then the measurement data $\mathbf{z}_k$ can be computed.

The calculation of the $\hat{\mathbf{x}}_k$ in the Kalman filter "feeds back" into the simulation of the actual system when there is control, since

$$\mathbf{u}_k = \Lambda_k \hat{\mathbf{x}}_k$$

B. Generation of Correlated Vector Random Variables

It has been noted in Section III, A, that in order to simulate a physical system it is necessary to generate the vector random variables $\mathbf{x}_0$, $\mathbf{w}_{k-1}$, and $\mathbf{v}_k$ ($k = 1, 2, \ldots, N$) from a known distribution. We shall assume the distribution is gaussian, since only the first- and second-order moments are prescribed. The procedure for generating independent *scalar* random variables is well defined. Thus, if the covariance matrices M_0, Q_{k-1}, and R_k are diagonal, the desired vectors can be computed. The discussion of the actual procedure is outside the scope of this discussion, so the interested reader is referred to (*22, 56*).

When the components of the vectors are correlated, an additional computation must be introduced before the random number generator can be utilized. Since covariance matrices are nonnegative-definite, they can be written as the product of a lower triangular matrix T and its transpose. In this case we seek a lower triangular matrix T such that a correlated random vector $\mathbf{v}$ can be written

$$\mathbf{v} = T\boldsymbol{\nu} \tag{3.1}$$

where

$$E[\boldsymbol{\nu}\boldsymbol{\nu}^T] = D$$

The matrix D is diagonal.

The proof that the matrix T can be found is given by describing the computation of T (*23*). We assume that the mean value of $\mathbf{v}$ is zero. Let

$$\mathbf{v} = \begin{pmatrix} v_1 \\ \vdots \\ v_n \end{pmatrix} \quad \text{and} \quad \boldsymbol{\nu} = \begin{pmatrix} \nu_1 \\ \vdots \\ \nu_n \end{pmatrix}$$

(1) Let $\nu_1 = v_1$.

(2) Let $\nu_2 = v_2 - \alpha_{21}\nu_1$. Compute the coefficient α_{21} so that ν_1 and ν_2 are uncorrelated.

$$\begin{aligned} E[\nu_1\nu_2] &= 0 \\ &= E[\nu_1(v_2 - \alpha_{21}\nu_1)] \\ &= E[\nu_1 v_2] - \alpha_{21}E[\nu_1^2] \end{aligned}$$

Thus

$$\alpha_{21} = \frac{E[\nu_1 v_2]}{E[\nu_1^2]}$$

(3) Let $\nu_3 = v_3 - \alpha_{31}\nu_1 - \alpha_{32}\nu_2$. Select the coefficients α_{31} and α_{32} so that ν_1, ν_2, and ν_3 are uncorrelated. Thus

$$E[\nu_1\nu_3] = 0 \quad \text{and} \quad E[\nu_2\nu_3] = 0$$

$$\begin{aligned} E[\nu_1\nu_3] &= E[\nu_1 v_3] - \alpha_{31}E[\nu_1^2] - \alpha_{32}E[\nu_1\nu_2] \\ &= E[\nu_1 v_3] - \alpha_{31}E[\nu_1^2] \end{aligned}$$

Therefore,

$$\alpha_{31} = \frac{E[\nu_1 v_3]}{E[\nu_1^2]}$$

Similarly, it follows that

$$\alpha_{32} = \frac{E[v_3\nu_2]}{E[\nu_2^2]}$$

(4) Proceeding inductively, we see that in general

$$\nu_k = v_k - \alpha_{k1}\nu_1 - \alpha_{k2}\nu_2 - \cdots - \alpha_{k,k-1}\nu_{k-1} \tag{3.2}$$

The coefficients are computed from

$$\alpha_{kj} = \frac{E[v_k\nu_j]}{E[\nu_j^2]} \qquad j = 1, 2, \ldots, k-1 \tag{3.3}$$

Note that if $E[\nu_j^2] = 0$ for any j, then the corresponding coefficient α_{kj} can be set equal to zero.

A general relation for $E[\nu_k^2]$ can be found from (3.2):

$$E[\nu_k^2] = E\left\{\left[v_k - \sum_{i=1}^{k-1}\alpha_{ki}\nu_i\right]\left[v_k - \sum_{j=1}^{k-1}\alpha_{kj}\nu_j\right]\right\}$$

After some manipulation, this becomes

$$E[\nu_k^2] = E[v_k^2] - \sum_{i=1}^{k-1} \alpha_{ki}^2 E[\nu_i^2] \tag{3.4}$$

It is obvious that the α_{kj} provide the desired transformation matrix of (3.1):

$$T = \begin{pmatrix} 1 & 0 & \cdots & & 0 \\ \alpha_{21} & 1 & \cdots & & \vdots \\ \vdots & \vdots & & 1 & 0 \\ \alpha_{n1} & \alpha_{n2} & \cdots & \alpha_{n,n-1} & 1 \end{pmatrix} \tag{3.5}$$

since from (3.2)

$$v_k = \alpha_{k1}\nu_1 + \alpha_{k2}\nu_2 + \cdots + \alpha_{k,k-1}\nu_{k-1} + \nu_k \qquad k = 1, 2, \ldots, n$$

Also, we see that $\boldsymbol{\nu}$ has a diagonal covariance matrix with elements given by (3.4).

Let T^Q_{k-1} and $T_k{}^R$ be the transformation matrices for $\mathbf{w}_{k-1}$ and $\mathbf{v}_k$, and let $\boldsymbol{\mu}_{k-1}$ and $\boldsymbol{\eta}_k$ be the associated uncorrelated random vectors. Then, (I) and (II) can be rewritten as

$$\mathbf{x}_k = \Phi_{k,k-1}\mathbf{x}_{k-1} + \Gamma_{k,k-1}\mathbf{u}_{k-1} + \mathbf{f}_{k-1} + T^Q_{k-1}\boldsymbol{\mu}_{k-1} \tag{I$'$}$$

$$\mathbf{z}_k = H_k\mathbf{x}_k + T_k{}^R\boldsymbol{\eta}_k \tag{II$'$}$$

The components of $\boldsymbol{\mu}_{k-1}$ and $\boldsymbol{\eta}_k$ can now be generated as independent random variables with prescribed statistics. The filter equations are unaffected by this transformation, although one could rewrite them in terms of the transformations T^Q_{k-1} and $T_k{}^R$ and $\boldsymbol{\mu}_{k-1}$ and $\boldsymbol{\eta}_k$. Note that the procedure that has been described represents another variant of the shaping filter concept. In fact, the shaping filter that was derived in Section II, F, in general, has a nondiagonal covariance matrix, so that white noise sequence would actually have to be replaced by the above procedure to simulate the shaping filter.

C. The Error Ellipsoids

The P_k matrix represents the covariance of the error in the estimate $\tilde{\mathbf{x}}_k$ since, as was shown in Section II, D, 1, the estimate is unbiased (i.e., if the estimate were biased, P_k would represent the second-moment matrix rather than the covariance matrix). As such, it provides significant information about the accuracy of the estimate. If the physical model is accurately described by (I) and (II), the P_k can be used to describe the manner in which the estimate converges (or diverges) to the true state. Kalman (*2*) derived conditions (i.e., complete controllability and complete observability) under which the P_k matrix will be positive-definite. Those results are either difficult to apply or do not apply in many cases, so the behavior of the filter is more conveniently studied by examining the output of computer programs.

Examination of the P_k matrix directly (i.e., element by element) is not a satisfactory approach, since it involves n^2 elements. To introduce some rationale into the considerations, the concept of the error ellipsoid (*24, 25*) is proposed.

DEFINITION. *Suppose the n-dimensional vector random variable* $\mathbf{x}$ *has a multivariate gaussian distribution with a mean value of zero and covariance* $E[\mathbf{x}\mathbf{x}^T] = P$. *The "error ellipsoids" are defined as n-dimensional surfaces of constant probability density.*

Since $\mathbf{x}$ has a multivariate gaussian distribution, it has a probability density function given by

$$p(\mathbf{x}) = [(2\pi)^{n/2} \mid \det P \mid^{1/2}]^{-1} \exp -\tfrac{1}{2}\mathbf{x}^T P^{-1}\mathbf{x} \tag{3.6}$$

This equation shows that the surface of constant probability density is described as

$$\mathbf{x}^T P^{-1}\mathbf{x} = c^2 \tag{3.7}$$

where c^2 is an arbitrary constant. Since P is nonnegative-definite, the surface is an ellipsoid; hence, the name *error ellipsoid*. In fact, P must be positive-definite for the ellipsoid to be n-dimensional. If P is nonnegative-definite, the surface collapses into an m-dimensional ($m < n$) ellipsoid. The m is equal to the rank of P. The matrix P is symmetric and its elements are real. Under these conditions (*10*) there exists an orthogonal matrix S such that P can be transformed to a diagonal matrix D by the relation

$$D = S^T P S$$

The diagonal elements of D are the eigenvalues of P and the column vectors of S are the orthonormal eigenvectors of P. Since P is nonnegative-definite, the eigenvalues of P are nonnegative. They are *all* positive when P is positive-definite. The number of nonzero eigenvalues is equal to the rank of P. Let

$$\mathbf{x} = S^{-1}\mathbf{y} = S^T\mathbf{y}$$

and suppose that P is positive-definite. The inverse of S exists and is equal to S^T since S is orthogonal. Thus,

$$\begin{aligned} \mathbf{x}^T P^{-1} \mathbf{x} &= \mathbf{y}^T S P^{-1} S^T \mathbf{y} \\ &= \mathbf{y}^T D^{-1} \mathbf{y} \\ &= \sum_{i=1}^{n} \frac{y_i^2}{\lambda_i} = c^2 \end{aligned}$$

Divide this relation by c^2,

$$\sum_{i=1}^{n} \frac{y_i^2}{c^2 \lambda_i} = 1 \tag{3.8}$$

Equation (3.8) is the normal form of an n-dimensional ellipsoid. The n-principal semiaxes of the ellipsoid are

$$c(\lambda_i)^{1/2} \qquad i = 1, 2, \ldots, n$$

Since the columns $\mathbf{e}_i$ of the matrix S are orthonormal eigenvectors of P, it follows immediately that the $\mathbf{e}_i$ define the directions of the axes of the ellipsoid. When an eigenvalue is zero, the corresponding eigenvector indicates the direction normal to the subspace which contains the $(n - 1)$ dimensional ellipsoid.

The term *error ellipsoids* often refers to the specific case when c is set equal to one. We shall adhere to this convention. The significance of the ellipsoids stems from the fact that they have a simple probabilistic interpretation. For a given value of c, it is possible to integrate the probability density over the surface of the ellipsoid to obtain the probability that a particular sample point will lie within the ellipsoid. For the important case when $n = 3$, the probability that a sample point will be within the ellipsoid for $c = 1, 2$, or 3 is, respectively, 0.2, 0.74, or 0.94 (*25*).

In (I) and (II), the noise processes $\mathbf{w}_k$ and $\mathbf{v}_k$ can be assumed to be gaussian. Since the state $\mathbf{x}_k$ is a linear function of $\mathbf{w}_{k-1}$, it is also gaussian. The same conclusion can be reached about the estimate $\hat{\mathbf{x}}_k$ since it is a linear combination of $\mathbf{x}_k$ and $\mathbf{v}_k$; so the error in the estimate must be a gaussian variable with zero mean and covariance P_k. Thus the error ellipsoid can be used to characterize the concentration of the estimate about the true value of the state. When the magnitude of an axis of the ellipsoid decreases, the conclusion is that the error in the estimate is decreasing in that direction.

It often happens that the components of the state vector represent entirely different types of variables. For instance, in the space-navigation problem considered in Section IV, the first three components represent position, the next three represent velocity, and different systems parameters appear as the remaining components. In a situation of this nature, the ellipsoid for the total matrix is a conglomerate mess; so it is more reasonable to examine submatrices relating state variables of the same character.

D. An Alternative Form of the Kalman Filter

The equations that describe the gain matrix K_k and the error covariance matrix P_k can be written in a different form; a form that is interesting because it permits us to eliminate P_0 from consideration during a parametric study and is useful in certain other theoretical considerations (e.g., see Section III, E). First, we shall state and prove an interesting matrix inversion lemma (*13*).

Matrix Inversion Lemma. *Suppose* $(n \times n)$ *matrices B and R are positive-definite. Let H be any, possibly rectangular, matrix. Let A be an $n \times n$ matrix related to B, R, and H according to*

$$A = B - BH^T[HBH^T + R]^{-1}HB \tag{3.9}$$

Then, A^{-1} is given by

$$A^{-1} = B^{-1} + H^T R^{-1} H \tag{3.10}$$

Proof. The proof follows by direct multiplication.

$$\begin{aligned} AA^{-1} &= [B - BH^T(HBH^T + R)^{-1}HB][B^{-1} + H^T R^{-1} H] \\ &= I - BH^T[(HBH^T + R)^{-1} - R^{-1} + (HBH^T + R)^{-1}HBH^T R^{-1}]H \\ &= I - BH^T[(HBH^T + R)^{-1}(I + HBH^T R^{-1}) - R^{-1}]H \\ &= I \end{aligned}$$

Q.E.D.

This result can be applied immediately to (VI). Introduce the following correspondences to relate quantities in (VI) and (3.9). Let

$$A \sim P_k, \quad B \sim P_k', \quad H \sim H_k, \quad R \sim R_k$$

Assume that R_k and P_k' are positive-definite and apply (3.10).

$$P_k^{-1} = (P_k')^{-1} + H_k^T R_k^{-1} H_k \tag{VII}$$

where

$$(P_k')^{-1} = [\Phi_{k,k-1}P_{k-1}\Phi_{k,k-1}^T + Q_{k-1}]^{-1}$$

Now, rewrite the gain matrix. From (V),

$$\begin{aligned} K_k &= P_k'H_k^T[H_kP_k'H_k^T + R_k]^{-1} \\ &= (P_kP_k^{-1})P_k'H_k^TR_k^{-1}[H_kP_k'H_k^TR_k^{-1} + I]^{-1} \end{aligned}$$

Using (VII), we obtain

$$\begin{aligned} K_k &= P_k[I + H_k^TR_k^{-1}H_kP_k']H_k^TR_k^{-1}[H_kP_k'H_k^TR_k^{-1} + I]^{-1} \\ &= P_kH_k^TR_k^{-1} \end{aligned} \quad \text{(VIII)}$$

Equations (II), (IV), (VII), and (VIII) provide an alternative form of the Kalman filter when P_k' and R_k are positive-definite. This version was derived by Swerling (5) in 1958 with $Q_{k-1} \equiv O$.

This formulation has been introduced because it is useful for certain theoretical manipulations, and for certain parametric studies. As mentioned in Section III, A, the matrices P_0, R_k, and Q_{k-1} provide the parameters that can be manipulated during a study of a particular dynamical system and measurement process. If the principal concern of the study is the effect of measurement errors on the behavior of the estimate, it may be advantageous to eliminate as many other parameters as possible. In order to be conservative, the P_0 should be assigned the most pessimistic values possible. This would correspond to a total lack of information about the initial state. That is, when the initial state is *completely* unknown,

$$P_0 = \infty I$$

But if this value is introduced in (V), the gain matrix cannot be computed immediately because of the manner in which P_0 enters. However, (VII) and (VIII) avoid the difficulty. Then, for t_1, if the rank of H_1 is equal to n,

$$P_1 = [H_1^TR_1^{-1}H_1]^{-1}$$

and

$$K_1 = P_1H_1^TR_1^{-1}$$

If H_1 has rank less than n, it is not possible to let $P_0^{-1} = O$, for then P_1 does not exist. In this case P_0 is set equal to $(1/\epsilon)I$, where ϵ is a small number.

For later sampling times (i.e., t_2, t_3, ...) it is generally advisable to return to the use of (V) and (VI) rather than to continue to utilize (VII) and (VIII). The preference arises because the dimension of P_k is generally larger than the dimension of R_k and it is always preferable to invert the matrix of smallest dimension.

E. Sequential Processing

Suppose that at each sampling time t_k, q statistically independent sources provide measurement data. The m-dimensional measurement vector can be represented as

$$\mathbf{z}_k \overset{\text{Df}}{=} \begin{pmatrix} \mathbf{z}_k^1 \\ \mathbf{z}_k^2 \\ \vdots \\ \mathbf{z}_k^q \end{pmatrix} \overset{\text{Df}}{=} \begin{pmatrix} H_k^1 \\ H_k^2 \\ \vdots \\ H_k^q \end{pmatrix} \mathbf{x}_k + \begin{pmatrix} \mathbf{v}_k^1 \\ \mathbf{v}_k^2 \\ \vdots \\ \mathbf{v}_k^q \end{pmatrix} \quad (3.11)$$

Since the sources are statistically independent,

$$E[\mathbf{v}_k^i(\mathbf{v}_k^j)^T] = R_k^i\delta_{ij}$$

The R_k^i have the dimension $(m^i \times m^i)$, where

$$\sum_{i=1}^{q} m^i = m$$

In (V) the inverse of the matrix $[H_kP_k'H_k^T + R_k]$ must be computed. If there are several data sources, the dimension m of this matrix could be quite large. The inversion on a digital computer of a matrix of large dimension is undesirable for several reasons—the amount of storage cells that must be used, the time that is consumed in obtaining the inverse, and the accuracy of the end result. Thus, if the inversion can be circumvented, it is advisable to do so. In this case an alternative is available in which the largest matrix that has to be inverted has dimension equal to max (m^i). This procedure will be designated by the term *sequential processing*. Another policy that can be used involves the alternative formulation described in Section III, D. The disadvantages inherent in this scheme are the same as those discussed in Section III, D. It shall be called *simultaneous processing*.

In sequential processing, each set of data are treated separately. Specifically, the data $\mathbf{z}_k^1$ are used to obtain an estimate $\hat{\mathbf{x}}_k$ and covariance P_k with

$$\mathbf{z}_k = \mathbf{z}_k^1$$

When these calculations are completed, $\mathbf{z}_k^2$ is processed to obtain new values for $\hat{\mathbf{x}}_k$ and P_k. Each set of data are processed in this manner until the final set $\mathbf{z}_k^q$ has been included. Then, time is advanced to t_{k+1} and the cycle is repeated. The simultaneous processing technique involves the direct application of the alternative formulation.

These two schemes shall be applied for an arbitrary time t_k and the form of the final estimates will be compared to prove their equivalence.

1. Simultaneous Processing

When the data are processed simultaneously and (VIII) is used, the gain matrix is given in partitioned form by

$$K_k = P_k H_k^T R_k^{-1} = P_k[(H_k^1)^T (H_k^2)^T \cdots (H_k^q)^T] \begin{pmatrix} R_k^1 & & O \\ & \ddots & \\ O & & R_k^q \end{pmatrix}^{-1}$$

$$= [P_k(H_k^1)^T(R_k^1)^{-1} \; P_k(H_k^2)^T(R_k^2)^{-1} \cdots P_k(H_k^q)^T(R_k^q)^{-1}] \tag{3.12}$$

K_k has the dimension $(n \times m)$ and the submatrices $P_k(H_k^i)^T(R_k^i)^{-1}$ have the dimension $(n \times m^i)$.

The covariance of the error in the estimate is given by (VIII). Consider $H_k^T R_k^{-1} H_k$ in partitioned form.

$$H_k^T R_k^{-1} H_k = [(H_k^1)^T (H_k^2)^T \cdots (H_k^q)^T] \begin{bmatrix} (R_k^1)^{-1} & & O \\ & \ddots & \\ O & & (R_k^q)^{-1} \end{bmatrix} \begin{bmatrix} H_k^1 \\ H_k^2 \\ \vdots \\ H_k^q \end{bmatrix}$$

$$H_k^T R_k^{-1} H_k = \sum_{i=1}^{q} (H_k^i)^T (R_k^i)^{-1} H_k^i$$

so

$$P_k^{-1} = (P_k')^{-1} + \sum_{i=1}^{q} (H_k^i)^T(R_k^i)^{-1}(H_k^i) \tag{3.13}$$

Again we observe that the $(n \times n)$ matrix P_k and the q matrices R_k^i must be inverted.

The estimate in this technique is obtained from (III) by introducing (3.12).

$$\hat{\mathbf{x}}_k = \hat{\mathbf{x}}_k' + [P_k(H_k^1)^T(R_k^1)^{-1} \cdots P_k(H_k^q)^T(R_k^q)^{-1}] \begin{bmatrix} \mathbf{z}_k^1 - H_k^1\hat{\mathbf{x}}_k' \\ \vdots \\ \mathbf{z}_k^q - H_k^q\hat{\mathbf{x}}_k' \end{bmatrix} \tag{3.14}$$

where

$$\hat{\mathbf{x}}_k' = \Phi_{k,k-1}\hat{\mathbf{x}}_{k-1} + \mathbf{f}_{k-1}$$

2. Sequential Processing

In this policy the data $\mathbf{z}_k^j$ is processed without consideration of the other available data. Denote the gain, estimate, and covariance obtained from this data with a superscript j. Then, for $j = 1$,

$$(P_k^1)^{-1} = (P_k')^{-1} + (H_k^1)^T(R_k^1)^{-1}H_k^1 \tag{3.15}$$

where

$$P_k' = \Phi_{k,k-1}P_{k-1}\Phi_{k,k-1}^T + Q_{k-1} \qquad K_k^1 = P_k^1(H_k^1)^T(R_k^1)^{-1} \tag{3.16}$$

and

$$\hat{\mathbf{x}}_k^1 = \hat{\mathbf{x}}_k' + K_k^1[\mathbf{z}_k^1 - H_k^1\hat{\mathbf{x}}_k'] \tag{3.17}$$

Using $\hat{\mathbf{x}}_k^1$, K_k^1, and P_k^1, process $\mathbf{z}_k^2$. Since $\mathbf{z}_k^2$ relates to the same time as $\mathbf{z}_k^1$, there is no need to extrapolate.

The estimate $\hat{\mathbf{x}}_k^2$ is given by

$$\hat{\mathbf{x}}_k^2 = \hat{\mathbf{x}}_k^1 + K_k^2[\mathbf{z}_k^2 - H_k^2\hat{\mathbf{x}}_k^1] \tag{3.18}$$

Substitute (3.17) in (3.18):

$$\hat{\mathbf{x}}_k^2 = \hat{\mathbf{x}}_k' + K_k^1[\mathbf{z}_k^1 - H_k^1\hat{\mathbf{x}}_k'] + K_k^2\{\mathbf{z}_k^2 - H_k^2[\hat{\mathbf{x}}_k' + K_k^1(\mathbf{z}_k^1 - H_k^1\hat{\mathbf{x}}_k')]\} \tag{3.19}$$

In partitioned form, this becomes

$$\hat{\mathbf{x}}_k^2 = \hat{\mathbf{x}}_k' + [(I - K_k^2H_k^2)K_k^1 \; K_k^2] \begin{bmatrix} \mathbf{z}_k^1 - H_k^1\hat{\mathbf{x}}_k' \\ \mathbf{z}_k^2 - H_k^2\hat{\mathbf{x}}_k' \end{bmatrix}$$

The covariance matrix is

$$\begin{aligned} (P_k^2)^{-1} &= (P_k^1)^{-1} + (H_k^2)^T(R_k^2)^{-1}H_k^2 \\ (P_k^2)^{-1} &= (P_k')^{-1} + (H_k^1)^T(R_k^1)^{-1}H_k^1 + (H_k^2)^T(R_k^2)^{-1}H_k^2 \end{aligned} \tag{3.20}$$

The P_k^j is independent of the estimate, so (3.20) generalizes immediately. After the q sets of data have been processed,

$$(P_k^q)^{-1} \overset{\text{Df}}{=} (P_k)^{-1} = (P_k')^{-1} + \sum_{i=1}^{q} (H_k^i)^T(R_k^i)^{-1}H_k^i$$

But this agrees precisely with (3.13) of the simultaneous processing policy. Therefore, the covariance matrix P_k is not affected by the differences in the processing techniques. Since K_k was chosen to minimize the trace of P_k, the estimate should also be unaffected. We shall complete the algebraic proof. The gain K_k^2 of (3.19) is

$$K_k^2 = P_k^2(H_k^2)^T(R_k^2)^{-1}$$

Proceed to include $\mathbf{z}_k^3$. In this case,

$$\hat{\mathbf{x}}_k^3 = \hat{\mathbf{x}}_k^2 + K_k^3[\mathbf{z}_k^3 - H_k^3\hat{\mathbf{x}}_k^2] \tag{3.21}$$

$$\hat{\mathbf{x}}_k^3 = \hat{\mathbf{x}}_k' + [(I - K_k^3H_k^3)(I - K_k^2H_k^2)K_k^1 \; (I - K_k^3H_k^3)K_k^2 \; K_k^3] \begin{bmatrix} \mathbf{z}_k^1 - H_k^1\hat{\mathbf{x}}_k' \\ \mathbf{z}_k^2 - H_k^2\hat{\mathbf{x}}_k' \\ \mathbf{z}_k^3 - H_k^3\hat{\mathbf{x}}_k' \end{bmatrix}$$

The last equation was obtained by introducing (3.19) and rearranging terms. The extension through $\mathbf{z}_k^q$ is straightforward. To avoid the algebra, we shall demonstrate that the gain matrix in (3.21) is identical with (3.12). When this is accomplished, it is obvious that (3.14) and (3.21) are the same for $q = 3$.

The gain matrix K_k^3 is

$$K_k^3 = P_k(H_k^3)^T(R_k^3)^{-1}$$

since

$$P_k^3 = P_k$$

This shows that the third submatrix is the same in (3.21) and (3.12).

The P_k can be written as

$$P_k = P_k^2 - K_k^3H_k^3P_k^2$$

so

$$P_kH_k^{2T}R_k^{2^{-1}} = (I - K_k^3H_k^3)K_k^2$$

Furthermore, it follows that

$$P_kH_k^1R_k^{1^{-1}} = (I - K_k^3H_k^3)(I - K_k^2H_k^2)K_k^1$$

Therefore, the final gain matrix is identical in each approach, so sequential and simultaneous processing are equivalent.

The following computational procedure can be used when there are q sets of statistically independent observation data:

$$\mathbf{z}_k^1, \mathbf{z}_k^2, \ldots, \mathbf{z}_k^q$$

(1) Form P_k', P_k^1, and K_k^1, and $\hat{\mathbf{x}}_k^1$ using (III), (IV), (V), and (VI), and the first set of data

$$\mathbf{z}_k^1 \qquad \text{(i.e., treat } \mathbf{z}_k^1 \text{ as } \mathbf{z}_k)$$

(2) Based on the P_k^1 and $\hat{\mathbf{x}}_k^1$, process $\mathbf{z}_k^2$ using the relations

$$K_k^2 = P_k^1H_k^{2^T}[H_k^2P_k^1H_k^{2^T} + R_k^2]^{-1}$$

$$\hat{\mathbf{x}}_k^2 = \hat{\mathbf{x}}_k^1 + K_k^2[\mathbf{z}_k^2 - H_k^2\hat{\mathbf{x}}_k^1]$$

$$P_k^2 = P_k^1 - K_k^2H_k^2P_k^1$$

(3) Repeat step (2) with the next set of data and with all indices incremented by 1 in the equations. Continue this policy until all the data have been processed. Then the estimate and the covariance matrix are

$$\hat{\mathbf{x}}_k^q = \hat{\mathbf{x}}_k \qquad \text{and} \qquad P_k^q = P_k$$

F. Special Case: Unforced Dynamics

In this section we shall consider the particular case when the dynamical system contains no forcing functions.

$$\mathbf{x}_k = \Phi_{k,k-1}\mathbf{x}_{k-1} \tag{IU}$$

This case requires special consideration because the P_k matrix tends to become singular if the filter converges. In fact, the P_k matrix tends to vanish completely. This feature has the disadvantage that the filter becomes extremely sensitive to computational inaccuracies. In fact, these inaccuracies can cause the P_k to lose its property of nonnegative definiteness. Numerical results have been observed in which the diagonal elements of P_k, which represent variances, become negative. In some instances the P_k matrix is used in other parts of a system as a criterion for determining different courses of action. When the validity of the P_k is made suspect because of computational inaccuracies, the over-all system performance must suffer.

An additional disadvantage caused by the vanishing of the P_k arises because the gain matrix K_k also vanishes. When this happens, new measurement data have no effect upon the estimate. In essence the estimation equation (IIIU) reduces to

$$\hat{\mathbf{x}}_k = \Phi_{k,k-1}\hat{\mathbf{x}}_{k-1}$$

Unless the estimate and the true state agree exactly [an extremely unlikely situation in actual physical problems where (IU) is an approximation of the true dynamical system], the ultimate result is that the estimate diverges from the correct value. Numerical results and additional discussion regarding this unsavory circumstance can be found in (*26*).

There are many different ways in which one can attempt to combat this problem. Certainly, great pains can and should be taken in the construction of programs to avoid unnecessary loss of accuracy. Double-precision operations can be used in particularly sensitive computations. Computational inaccuracies will always occur though; thus, improving the precision of the operations serves to delay the time at which the inaccuracies begin to have significant effects. In the remainder of this section, two additional alternatives will be proposed. In the first, a mathematical technique through which the nonnegative definiteness of the P_k is ensured is presented. The second technique is heuristic. Mainly, it represents an argument questioning the validity of (IU) in any simulation.

One immediately available method of preserving the nonnegative definiteness of the P_k is to utilize (2.0) rather than (VI). Because of the symmetry of (2.0) this equation is less sensitive to computational inaccuracy. The use of (2.0) also provides the covariance of the error that occurs when a suboptimal gain is used to compute the estimate.

1. Factoring the P_k Matrix

The matrix P_k is symmetric and nonnegative definite, so it can be written as the product of a nonnegative-definite matrix Π_k and its transpose (i.e., a matrix square root).

$$P_k = \Pi_k \Pi_k^T \tag{3.22}$$

or

$$\Pi_k \Pi_k^T = P_k' - K_k H_k P_k'$$

Let

$$\Pi_k' = \Phi_{k,k-1} \Pi_{k-1} \tag{3.23}$$

so that we obtain

$$\Pi_k \Pi_k^T = \Pi_k'[I - \Pi_k'^T H_k^T (H_k P_k' H_k^T + R_k)^{-1} H_k \Pi_k'] \Pi_k'^T \tag{3.24}$$

A recursion relation for Π_k can be developed by factoring the matrix in the square brackets. This can always be done theoretically. We shall derive an analytical relationship for the case of a single observation (*7*).

When there is only one observation, the term $(H_k P_k' H_k^T + R_k)$ is scalar. Define r_k as

$$r_k = H_k \Pi_k' \Pi_k'^T H_k^T + R_k \tag{3.25}$$

Equation (3.24) becomes

$$\Pi_k \Pi_k^T = \Pi_k' \left[I - \frac{1}{r_k} \Pi_k'^T H_k^T H_k \Pi_k' \right] \Pi_k'^T \tag{3.26}$$

The bracketed term must be factored.

Suppose that

$$I - \frac{1}{r_k} \Pi_k'^T H_k^T H_k \Pi_k' = (I - a_k \Pi_k'^T H_k^T H_k \Pi_k')(I - a_k \Pi_k'^T H_k^T H_k \Pi_k')^T \tag{3.27}$$

where a_k is an, as yet, unknown scalar. Expanding and recombining terms, we find that

$$I - \frac{1}{r_k} \Pi_k'^T H_k^T H_k \Pi_k' = I - a_k(2 - a_k H_k \Pi_k' \Pi_k'^T H_k^T) \Pi_k'^T H_k^T H_k \Pi_k'$$

To obtain this result it is necessary to recognize that $H_k \Pi_k' \Pi_k'^T H_k^T$ is a scalar.

For equality to actually hold, the a_k must be selected such that

$$a_k(2 - a_k H_k \Pi_k' \Pi_k'^T H_k^T) = \frac{1}{r_k}$$

But this is a quadratic equation in a_k.

$$(H_k \Pi_k' \Pi_k'^T H_k^T) a_k^2 - 2a_k + \frac{1}{r_k} = 0$$

The roots of this scalar equation are

$$a_k = \frac{1 + (R_k/r_k)^{1/2}}{H_k \Pi_k' \Pi_k'^T H_k^T}$$

and

$$a_k = \frac{1 - (R_k/r_k)^{1/2}}{H_k \Pi_k' \Pi_k'^T H_k^T} \tag{3.28}$$

In order to be definite, (3.28) shall be assumed to define a_k. Using this a_k, we see from (3.26) and (3.27) that Π_k is generated by the recursion relation

$$\Pi_k = \Pi_k'(I - a_k \Pi_k'^T H_k^T H_k \Pi_k') \tag{3.29}$$

Using (3.29), a computational procedure different from that of Section III, A, becomes available. In place of steps (1) to (6) of that section, the following procedure is suggested. Note that the filter corresponding to the dynamical model (IU) is used. Only one observation is obtained at each time.

(1) At t_0, initialize P_0 and $\hat{\mathbf{x}}_0$. If necessary, factor P_0 to get Π_0.

(2) Form the *projected estimate* of Π_1'.

$$\Pi_1' = \Phi_{k,k-1}\Pi_0'$$

(3) Compute the gain matrix

$$K_1 = \frac{1}{r_1}\Pi_1'\Pi_1'^T H_1^T$$

where

$$r_1 = (H_1\Pi_1')(H_1\Pi_1')^T + R_1$$

(4) Form the estimate of the state at t_1 from the measurement z_1.

$$\hat{\mathbf{x}}_1 = \Phi_{1,0}\hat{\mathbf{x}}_0 + (z_1 - H_k\hat{\mathbf{x}}_0)K_1$$

(5) Compute the factored covariance matrix Π_1 from (3.29) with the assistance of (3.28).

$$\Pi_1 = \Pi_1'(I - a_1(H_1\Pi_1')^T(H_1\Pi_1'))$$

(6) Update time and return to step (2) with all indices incremented by 1.

The advantage of this technique is that P_k' is generated as the product of a matrix and its transpose. The *computational* realization of this multiplication must yield a matrix that is at least nonnegative-definite. Thus the problem of the loss of sign-definiteness of P_k is removed. The method has the disadvantage that it is restricted to scalar measurements. Of course, if there are several uncorrelated measurements at each sampling time, the sequential processing technique of Section III, E, can be used. This permits an extension of this method from scalar to vector measurements when R_k is diagonal.

2. A White Noise Approximation

If the dynamical system for a particular problem were described precisely by (IU) and if the computations contained no inaccuracies, then the fact that the P_k becomes singular would not represent a source of difficulty. It would only describe the situation as it evolves. However, neither hypothesis is true, in general, so it can be argued that (IU) does not represent any system of engineering significance. As a first step toward a more accurate model, the computer-generated noise that acts on the dynamical system can be assumed to be equivalent to additive white noise. Then, the state evolves according to

$$\mathbf{x}_k = \Phi_{k,k-1}\mathbf{x}_{k-1} + \mathbf{w}_{k-1} \qquad \text{(IF}')$$

The addition of $\mathbf{w}_{k-1}$ makes the system controllable in the sense of (*2*) and prevents P_k from vanishing. The introduction of this term can be considered to be equivalent to assuming the computer-generated errors are analogous to thermal noise in a resistor. The $\mathbf{w}_{k-1}$ describes a process that truly acts on the system of equations during a simulation, so does *not* have to be formed using a noise generator as described in Section III, A.

The covariance Q_{k-1} of $\mathbf{w}_{k-1}$ can be used as a matrix of parameters to counteract the effects of computational noise and deficient mathematical model. It is not suggested that this term can be used to compensate for major deficiences in the model or inordinately careless programming. However, numerical studies indicate that the addition of Q_{k-1} can eliminate the problem of the loss of nonnegative definiteness of P_k without causing the validity of $\hat{\mathbf{x}}_k$ to deteriorate, significantly. The filter for this model is described by (IIIF), (IV), (V), and (VI) with $\mathbf{f}_{k-1} \equiv 0$.

IV. Application of the Kalman Filter to Nonlinear Problems

A. General Discussion

The discussion of the preceding sections has been based upon the mathematical model described in Section II, A. The physical system has been assumed to be described by a *first-order* system of *linear difference equations*. The *output* of the system is provided by quantities that are *linearly related* to the *state variables*. This model is not immediately applicable in most engineering problems of significance.

The dynamical system for a particular problem is frequently found to be described by a system of *n*th-*order nonlinear differential equations*. The fact that *n*th-order derivatives appear does not offer any theoretical difficulty because any *n*th-order differential equation can be written as a system of *n* first-order equations. Let us assume that this transformation has been accomplished. The variables that are described by the first-order equations will be referred to as the *state variables* (*27*).

The output of the system (i.e., the measured quantities) commonly appears in terms of quantities that have a *nonlinear algebraic relation* with the state variables. Also, measurements are sometimes available which relate to forcing functions that appear in the dynamical systems.

The nonlinearity of the dynamical and output equations must be removed. To accomplish this, the following fundamental assumption must be introduced.

FUNDAMENTAL ASSUMPTION. *A nominal solution of the nonlinear differential equations must exist. This solution must provide a "good" approximation to the actual behavior of the system. The approximation is "good" if the difference between the nominal and actual solutions can be described by a system of linear differential equations. These equations shall be called "linear perturbation equations."*

Suppose that the state $\mathbf{X}$ of a dynamical system evolves according to the vector differential equation

$$\dot{\mathbf{X}} = \mathbf{f}(\mathbf{X}, \mathbf{U}) \tag{4.1}$$

$\mathbf{X}$ is n-dimensional. The p-dimensional vector $\mathbf{U}$ is included to represent the system forcing functions.

Define nominal initial conditions for (4.1),

$$\mathbf{X}^*(t_0) = \mathbf{X}_0^*$$

and let $\mathbf{U}^*(t)$ be a nominal set of forcing functions. The solution that results from these conditions shall be the nominal solution $\mathbf{X}^*$. Equation (4.1) can be expanded in a Taylor series about the nominal values. Neglect all terms except those of first order. Then we obtain

$$\delta\dot{\mathbf{X}} \overset{\mathrm{Df}}{=} (\dot{\mathbf{X}} - \dot{\mathbf{X}}^*) = \frac{\partial \mathbf{f}}{\partial \mathbf{X}^*}\,\delta\mathbf{X} + \frac{\partial \mathbf{f}}{\partial \mathbf{U}^*}\,\delta\mathbf{U} \tag{4.2}$$

where

$$\frac{\partial \mathbf{f}}{\partial \mathbf{X}^*} \overset{\mathrm{Df}}{=} \left.\begin{pmatrix} \dfrac{\partial f_1}{\partial X_1} & \cdots & \dfrac{\partial f_1}{\partial X_n} \\ \vdots & & \vdots \\ \dfrac{\partial f_n}{\partial X_1} & \cdots & \dfrac{\partial f_n}{\partial X_n} \end{pmatrix}\right|_{\text{evaluated with nominal values}}$$

$$\frac{\partial \mathbf{f}}{\partial \mathbf{U}^*} \overset{\mathrm{Df}}{=} \left.\begin{pmatrix} \dfrac{\partial f_1}{\partial U_1} & \cdots & \dfrac{\partial f_1}{\partial U_p} \\ \vdots & & \vdots \\ \dfrac{\partial f_n}{\partial U_1} & \cdots & \dfrac{\partial f_n}{\partial U_p} \end{pmatrix}\right|_{\text{evaluated with nominal values}}$$

Let us redefine the terms in (4.2) according to the following schedule:

$$\mathbf{x} = \delta\mathbf{X} \qquad \mathbf{u} = \delta\mathbf{U}$$

$$F(t) = \frac{\partial \mathbf{f}}{\partial \mathbf{X}^*} \qquad G(t) = \frac{\partial \mathbf{f}}{\partial \mathbf{U}^*}$$

$$\dot{\mathbf{x}} = F(t)\mathbf{x} + G(t)\mathbf{u} \tag{4.3}$$

Equation (4.3) represents the linear perturbation equation. This equation must describe the motion of the actual system relative to the nominal. We shall not discuss how to verify that this requirement is satisfied for a specific case. We only emphasize the importance of the fact that the requirement must be satisfied.

The linear differential equation (4.3) is reduced to a linear difference equation through the use of the following lemma.

LEMMA (27). *The solution of a linear nonhomogeneous differential equation of the form* (4.3) *with the initial conditions*

$$\mathbf{x}(t_0) = \mathbf{x}_0$$

is

$$x(t) = \Phi(t, t_0)\mathbf{x}_0 + \int_{t_0}^{t} \Phi(t, \tau)G(\tau)\mathbf{u}(\tau)\, d\tau \tag{4.4}$$

The matrix $\Phi(t, t_0)$ ***is the "state transition matrix" and is the solution of the matrix differential equation***

$$\frac{d\Phi(t, t_0)}{dt} = F(t)\Phi(t, t_0) \qquad \Phi(t_0, t_0) = I$$

[properties of $\Phi(t, t_0)$ are discussed in Section II, A].

This lemma can be verified by substituting (4.4) in (4.3). The details shall be omitted.

Let $t = t_1$ and suppose that $\mathbf{u}(\tau)$ is constant for $t_0 \leqslant \tau < t_1$. Then (4.4) can be rewritten as the difference equation

$$\mathbf{x}(t_1) = \Phi(t_1, t_0)\mathbf{x}(t_0) + \Gamma(t_1, t_0)\mathbf{u}(t_0)$$

where

$$\Gamma(t_1, t_0) \overset{\mathrm{Df}}{=} \int_{t_0}^{t_1} \Phi(t_1, \tau)G(\tau)\, d\tau$$

No generality is lost if k and $k - 1$ are substituted for the subscripts 1 and 0. We obtain

$$\mathbf{x}(t_k) = \Phi(t_k, t_{k-1})\mathbf{x}(t_{k-1}) + \Gamma(t_k, t_{k-1})\mathbf{u}(t_{k-1})$$

or, in the notation of Section II, A,

$$\mathbf{x}_k = \Phi_{k,k-1}\mathbf{x}_{k-1} + \Gamma_{k,k-1}\mathbf{u}_{k-1} \tag{4.5}$$

The similarity between this equation and (I) are obvious. The main difference is the absence of the random noise process $\mathbf{w}_{k-1}$. The $\mathbf{w}_{k-1}$ enters because of physical considerations relating to the actual problem. Certainly, if the initial conditions

$$\mathbf{x}(t_0) = \mathbf{x}_0$$

and the time history of the control vector $\mathbf{u}_k$ were known precisely, there would be no estimation problem. Equation (4.5) would provide an exact description of the behavior of the system. Thus the noise vector $\mathbf{w}_k$ and the statistics of the initial conditions P_0 are added to improve the model.

The observation equations must also be linearized. Suppose the measurements are related to the state by

$$\mathbf{Y} = \mathbf{h}(\mathbf{X})$$

The linearized equations are

$$\mathbf{y}(t) = H(t)\mathbf{x}(t) \tag{4.6}$$

where

$$\mathbf{y}(t) \overset{\text{Df}}{=} \mathbf{Y}(t) - \mathbf{Y}^*(t)$$

and

$$H(t) \overset{\text{Df}}{=} \left.\begin{pmatrix} \dfrac{\partial h_1}{\partial X_1} & \cdots & \dfrac{\partial h_1}{\partial X_n} \\ \vdots & & \vdots \\ \dfrac{\partial h_m}{\partial X_1} & \cdots & \dfrac{\partial h_m}{\partial X_n} \end{pmatrix}\right|_{\text{evaluated with nominal values}}$$

The measurements will not be precise, so the errors $\mathbf{v}_k$ are assumed to be additive and uncorrelated between sampling times; correlated noise processes are introduced using the results of Section II, F. Equation (4.6) then assumes the form of (II).

This completes the general discussion. In the succeeding paragraphs the space-navigation problem will be discussed within the context of these remarks.

B. Navigation of a Space Vehicle

In this discussion we shall restrict ourselves to the general form of the relevant equations and thereby avoid unnecessary and confusing details.

The motion of a spacecraft is described by a system of nonlinear differential equations.

$$\ddot{\mathbf{r}} = \mathbf{g}(\mathbf{r}, \mathbf{A}^G) + \mathbf{f}(\mathbf{r}, \dot{\mathbf{r}}, \mathbf{U}, \mathbf{A}^F) \tag{4.7}$$

where $\ddot{\mathbf{r}}$, $\dot{\mathbf{r}}$, and $\mathbf{r}$ represent the acceleration, velocity, and position, respectively, of the spacecraft relative to some inertially fixed cartesian coordinate system. The term $\mathbf{g}(\mathbf{r}, \mathbf{A}^G)$ contains the gravitational acceleration. Atmospheric and propulsive effects are described by $\mathbf{f}(\mathbf{r}, \dot{\mathbf{r}}, \mathbf{U}, \mathbf{A}^F)$. The control term $\mathbf{U}$ is p-dimensional. The vectors $\mathbf{A}^G$ and $\mathbf{A}^F$ have been included in order to define variables that appear in a system and whose values are not known precisely. For example, $\mathbf{A}^F$ might represent the thrust magnitude and specific impulse of the propulsion system, or $\mathbf{A}^G$ the universal gravitational constant. The uncertainty might itself be described by a system of differential equations, but we shall assume that $\mathbf{A}^G$ and $\mathbf{A}^F$ are unknown constants in the subsequent discussion.

The first item that should be observed about (4.7) is that it is a second-order vector differential equation. Let us reduce it to a system of first-order equations by introducing the following definitions:

$$\mathbf{X}^p \overset{\text{Df}}{=} \mathbf{r} \qquad \mathbf{X}^v \overset{\text{Df}}{=} \dot{\mathbf{r}}$$

Equation (4.7) can now be written as a system of six first-order equations

$$\dot{\mathbf{X}} = \begin{bmatrix} \dot{\mathbf{X}}^p \\ \dot{\mathbf{X}}^v \end{bmatrix} = \begin{bmatrix} \mathbf{X}^v \\ \mathbf{g}(\mathbf{X}^p, \mathbf{A}^G) + \mathbf{f}(\mathbf{X}^p, \mathbf{X}^v, \mathbf{U}, \mathbf{A}^F) \end{bmatrix} \tag{4.8}$$

Two different types of measurement data are sometimes available.

a. State-Related Measurements. Most measurements bear some algebraic relation to the state variables. Radar systems measure slant range and various angles. These quantities are functions of the position vector $\mathbf{X}^p$. Range-rate data are also provided. These data involve the position and velocity vectors. Horizon sensors and space sextants provide position-related data.

b. Acceleration Measurements. When the vehicle is subjected to aerodynamic or propulsive forces, the accelerations resulting from these forces can be determined using an inertial measurement unit (IMU).

The output of the IMU frequently is the integral of the acceleration rather than the nongravitational acceleration itself, so the procedure for putting these measurements into the form of (II) can be more involved. Discussion of IMU data will be deferred to Section IV, C.

Suppose the state-related measurements are related to the state by

$$\mathbf{Y}(t) = \mathbf{h}(\mathbf{X}^p, \mathbf{X}^v) \tag{4.9}$$

Equations (4.8) and (4.9) must be linearized. It is usually possible to define a nominal trajectory that will provide a good approximation to the path that is actually followed by the spacecraft. As stated in Section IV, A, the existence of a *good nominal trajectory* is fundamental to all considerations relating to the application of the Kalman filter to this problem.

Let $\mathbf{X}^*(t)$, $\mathbf{U}^*(t)$, $\mathbf{A}^{G*}$, and $\mathbf{A}^{F*}$ define the nominal trajectory. Applying the results of Section IV, A, the linear perturbation equation is

$$\dot{\mathbf{x}} = F_1(t)\mathbf{x} + F_2(t)\mathbf{x} + E_1(t)\boldsymbol{\alpha}^G + E_2(t)\boldsymbol{\alpha}^F + G(t)\mathbf{u} \tag{4.10}$$

where

$$F_1(t) \overset{\text{Df}}{=} \left.\begin{pmatrix} O & I \\ \dfrac{\partial \mathbf{g}}{\partial \mathbf{X}^p} & O \end{pmatrix}\right|_{\text{evaluated with nominal values}}$$

$$F_2(t) \overset{\text{Df}}{=} \left.\begin{pmatrix} O & O \\ \dfrac{\partial \mathbf{f}}{\partial \mathbf{X}^p} & \dfrac{\partial \mathbf{f}}{\partial \mathbf{X}^v} \end{pmatrix}\right|_{\text{evaluated with nominal values}}$$

$$G(t) \overset{\text{Df}}{=} \left.\begin{pmatrix} O \\ \dfrac{\partial \mathbf{f}}{\partial \mathbf{U}} \end{pmatrix}\right|_{\text{evaluated with nominal values}}$$

$$E_1(t) \overset{\text{Df}}{=} \left.\begin{pmatrix} O \\ \dfrac{\partial \mathbf{g}}{\partial \mathbf{A}^G} \end{pmatrix}\right|_{\text{evaluated with nominal values}}$$

$$E_2(t) \overset{\text{Df}}{=} \left.\begin{pmatrix} O \\ \dfrac{\partial \mathbf{f}}{\partial \mathbf{A}^F} \end{pmatrix}\right|_{\text{evaluated with nominal values}}$$

and $\mathbf{x} \overset{\text{Df}}{=} \mathbf{X} - \mathbf{X}^*$; $\mathbf{u} \overset{\text{Df}}{=} \mathbf{U} - \mathbf{U}^*$; $\boldsymbol{\alpha}^G \overset{\text{Df}}{=} \mathbf{A}^G - \mathbf{A}^{G*}$, $\boldsymbol{\alpha}^F \overset{\text{Df}}{=} \mathbf{A}^F - \mathbf{A}^{F*}$; $\mathbf{x}$ is the *perturbation* state vector. It will usually be referred to simply as the "state vector." Equation (4.10) reduces to a linear difference equation by using (4.4).

$$\mathbf{x}_k = \Phi_{k,k-1}\mathbf{x}_{k-1} + \Gamma_{k,k-1}\mathbf{u}_{k-1} + A_{k,k-1}\boldsymbol{\alpha}_{k-1}^G + B_{k,k-1}\boldsymbol{\alpha}_{k-1}^F \tag{4.11}$$

The $\mathbf{A}^G$ and $\mathbf{A}^F$ have been defined to be composed of constant parameters so

$$\boldsymbol{\alpha}_k^G = \boldsymbol{\alpha}_{k-1}^G \qquad \boldsymbol{\alpha}_k^F = \boldsymbol{\alpha}_{k-1}^F$$

This model is still incomplete. The term $\mathbf{u}_{k-1}$ represents the control that is commanded by guidance considerations. The control action that is actually experienced by the vehicle will be slightly different because of inaccuracies in the spacecraft control system. These errors are of a random nature so a statistical model for these errors must be developed. For this discussion, we shall assume that this model leads to the introduction of an additive, white noise sequence $\mathbf{w}_{k-1}$ with known statistics. This assumption results in a dynamical system of the form of (I). More general random processes can be treated by the methods of Section II, F, 1. Additional errors in the commanded control can be introduced because of errors in the guidance computations (see Section III, F, 2 also). Assuming the noise enters the dynamical system (4.10) as $E_3(t)\mathbf{w}(t)$, the stochastic version of (4.11) is seen to be

$$\mathbf{x}_k = \Phi_{k,k-1}\mathbf{x}_{k-1} + A_{k,k-1}\boldsymbol{\alpha}_{k-1}^G + B_{k,k-1}\boldsymbol{\alpha}_{k-1}^F + \Gamma_{k,k-1}\mathbf{u}_{k-1} + C_{k,k-1}\mathbf{w}_{k-1} \tag{4.12}$$

In this equation, $\Phi_{k,k-1}$ is computed as the solution of the matrix differential equation

$$\frac{d\Phi(t, t_0)}{dt} = [F_1(t) + F_2(t)]\Phi(t, t_0); \quad \Phi(t_0, t_0) = I$$

The other system matrices are defined as

$$A_{k,k-1} \overset{\text{Df}}{=} \int_{t_{k-1}}^{t_k} \Phi(t_k, \tau)E_1(\tau)\,d\tau \qquad C_{k,k-1} \overset{\text{Df}}{=} \int_{t_{k-1}}^{t_k} \Phi(t_k, \tau)E_3(\tau)\,d\tau$$

$$B_{k,k-1} \overset{\text{Df}}{=} \int_{t_{k-1}}^{t_k} \Phi(t_k, \tau)E_2(\tau)\,d\tau \qquad \Gamma_{k,k-1} \overset{\text{Df}}{=} \int_{t_{k-1}}^{t_k} \Phi(t_k, \tau)G(\tau)\,d\tau$$

Equation (4.9) must also be linearized. It assumes the form

$$\mathbf{z}_k = H_k\mathbf{x}_k + \mathbf{b}_k + \mathbf{v}_k \tag{4.13}$$

where

$$H_k = \frac{\partial \mathbf{h}}{\partial \mathbf{X}}\Big/_{\text{evaluated with nominal values}}$$

The measurements will contain errors. The $\mathbf{b}_k$ term has been included to describe constant bias errors in the measurements. The $\mathbf{v}_k$ is a white noise sequence. These two types of random errors are the antithesis of one another.

As discussed in Section II, F, 3, the $\boldsymbol{\alpha}_k^G$, $\boldsymbol{\alpha}_k^F$, and $\mathbf{b}_k$ can be treated as new state variables. Then (4.12) and (4.13) coincide with (I) and (II) of Section II, A, after the obvious redefinitions are introduced. The space-navigation problem has been put into the form that is required for the application of the Kalman filter. To summarize, we state the system in terms of the position and velocity variables, the dynamical parameters, and the measurement bias errors.

$$\begin{bmatrix} \mathbf{x}_k \\ \boldsymbol{\alpha}_k^G \\ \boldsymbol{\alpha}_k^F \\ \mathbf{b}_k \end{bmatrix} = \begin{bmatrix} \Phi_{k,k-1} & A_{k,k-1} & B_{k,k-1} & O \\ O & I & O & O \\ O & O & I & O \\ O & O & O & I \end{bmatrix} \begin{bmatrix} \mathbf{x}_{k-1} \\ \boldsymbol{\alpha}_{k-1}^G \\ \boldsymbol{\alpha}_{k-1}^F \\ \mathbf{b}_{k-1} \end{bmatrix} + \begin{bmatrix} \Gamma_{k,k-1} \\ O \\ O \\ O \end{bmatrix} \mathbf{u}_{k-1} + \begin{bmatrix} C_{k,k-1} \\ O \\ O \\ O \end{bmatrix} \mathbf{w}_{k-1} \tag{Is}$$

$$\mathbf{z}_k = [H_k \quad O \quad O \quad I] \begin{bmatrix} \mathbf{x}_k \\ \boldsymbol{\alpha}_k^G \\ \boldsymbol{\alpha}_k^F \\ \mathbf{b}_k \end{bmatrix} + \mathbf{v}_k \tag{IIs}$$

The covariance matrix for the initial state is

$$E\left\{ \begin{bmatrix} \mathbf{x}_0 \\ \boldsymbol{\alpha}_0^G \\ \boldsymbol{\alpha}_0^F \\ \mathbf{b}_0 \end{bmatrix} [\mathbf{x}_0^T \boldsymbol{\alpha}_0^{G^T} \boldsymbol{\alpha}_0^{F^T} \mathbf{b}_0^T] \right\} = \begin{bmatrix} M_0 & O & O & O \\ O & E[\boldsymbol{\alpha}_0^G \boldsymbol{\alpha}_0^{G^T}] & O & O \\ O & O & E[\boldsymbol{\alpha}_0^F \boldsymbol{\alpha}_0^{F^T}] & O \\ O & O & O & E[\mathbf{b}_0 \mathbf{b}_0^T] \end{bmatrix}$$

where the $\mathbf{x}_0$, $\boldsymbol{\alpha}_0^G$, $\boldsymbol{\alpha}_0^F$, and $\mathbf{b}_0$ are assumed to be pairwise-uncorrelated.

C. Incorporation of IMU Data

Only state-related measurements have been considered in the preceding section. Suppose now that IMU data are available in the form of integrals of the nongravitational accelerations (i.e., assume integrating accelerometers) acting upon the spacecraft. (In the following the term ***acceleration measurement*** shall refer to the difference between actual and nominal values of nongravitational acceleration.) Naturally, it is desirable to process these data in order to improve the estimate of the state, so it is necessary to describe the IMU outputs with an equation having the form of (II).

To arrive at the desired relationship, we observe from (4.10) that the acceleration $\mathbf{a}(t)$ measured by the IMU is

$$\mathbf{a}(t) = F_2(t)\mathbf{x}(t) + E_2(t)\boldsymbol{\alpha}^F + G(t)\mathbf{u}(t) + E_3(t)\mathbf{w}(t) \tag{4.14}$$

The noise is assumed to have the nature of acceleration disturbances, so it is measured by the IMU. The $\mathbf{a}(t)$ is a (6×1) vector, but the first three components are identically zero. The output of a set of integrating accelerometers at t_k is

$$\begin{aligned} \mathbf{s}(t_k) &\overset{\text{Df}}{=} \int_{t_0}^{t_k} \mathbf{a}(t)\,dt \\ &= \int_{t_0}^{t_k} [F_2(t)\mathbf{x}(t) + E_2(t)\boldsymbol{\alpha}^F + G(t)\mathbf{u}(t) \\ &\qquad + E_3(t)\mathbf{w}(t)]\,dt \end{aligned} \tag{4.15}$$

The state $\mathbf{x}(t)$ appearing in (4.15) can be expressed in terms of $\mathbf{x}(t_k)$, $\mathbf{u}(\tau)$, and $\mathbf{w}(\tau)$ $(t \leqslant \tau < t_k)$.

$$\begin{aligned} \mathbf{x}(t) = \Phi(t, t_k)\mathbf{x}(t_k) + \int_{t_k}^{t} \Phi(t,\tau)[E_1(\tau)\boldsymbol{\alpha}^G + E_2(\tau)\boldsymbol{\alpha}^F + G(\tau)\mathbf{u}(\tau) \\ + E_3(\tau)\mathbf{w}(\tau)]\,d\tau \end{aligned} \tag{4.16}$$

Inserting (4.16) in (4.15) produces

$$\mathbf{s}_k = {}_aJ_k\mathbf{x}_k + {}_1J_k\boldsymbol{\alpha}^G + {}_2J_k\boldsymbol{\alpha}^F + \boldsymbol{\sigma}_k + \boldsymbol{\eta}_k \tag{4.17}$$

where

$${}_aJ_k \overset{\text{Df}}{=} \int_{t_0}^{t_k} F_2(t)\Phi(t, t_k)\,dt$$

$${}_1J_k \overset{\text{Df}}{=} \int_{t_0}^{t_k} F_2(t) \int_{t_k}^{t} \Phi(t,\tau)E_1(\tau)\,d\tau\,dt$$

$${}_2J_k \overset{\text{Df}}{=} \int_{t_0}^{t_k} [F_2(t) \int_{t_k}^{t} \Phi(t,\tau)E_2(\tau)\,d\tau + E_2(t)]\,dt$$

$$\boldsymbol{\sigma}_k \overset{\text{Df}}{=} \int_{t_0}^{t_k} [F_2(t) \int_{t_k}^{t} \Phi(t,\tau)G(\tau)\mathbf{u}(\tau)\,d\tau + G(t)\mathbf{u}(t)]\,dt$$

$$\boldsymbol{\eta}_k \overset{\text{Df}}{=} \int_{t_0}^{t_k} [F_2(t) \int_{t_k}^{t} \Phi(t,\tau)E_3(\tau)\mathbf{w}(\tau)\,d\tau + E_3(t)\mathbf{w}(t)]\,dt$$

The term $\boldsymbol{\sigma}_k$ depends upon the control $\mathbf{u}(t)$ which is known for all $t < t_k$. Thus $\boldsymbol{\sigma}_k$ can be considered as a known quantity and the measurement $\mathbf{s}_k$ modified to be $(\mathbf{s}_k - \boldsymbol{\sigma}_k)$.

Equation (4.17) has the desired form except for the $\boldsymbol{\eta}_k$. Let us examine this term in more detail. The plant noise $\mathbf{w}(t)$ has been assumed to be piecewise constant in Section IV, B, so

$$\mathbf{w}(t) = \mathbf{w}_k, \qquad t_k \leqslant t < t_{k+1}$$

When this fact is utilized in the defining relation for $\boldsymbol{\eta}_k$, the $\boldsymbol{\eta}_k$ can be rewritten after considerable algebraic manipulation as

$$\boldsymbol{\eta}_k = \sum_{i=0}^{k-1} \left[C_i + \int_{t_i}^{t_{i+1}} E_3(t)\, dt \right] \mathbf{w}_i \tag{4.18}$$

where

$$C_i \overset{\mathrm{Df}}{=} \int_{t_0}^{t_i} F_2(t) \int_{t_{i+1}}^{t_i} \Phi(t,\tau) E_3(\tau)\, d\tau\, dt + \int_{t_i}^{t_{i+1}} F_2(t) \int_{t_{i+1}}^{t} \Phi(t,\tau) E_3(\tau)\, d\tau\, dt$$

From (4.18) we see that $\boldsymbol{\eta}_k$ can be described by a recursion relation

$$\boldsymbol{\eta}_k = \boldsymbol{\eta}_{k-1} + \beta_{k-1} \mathbf{w}_{k-1} \tag{4.19}$$

where

$$\beta_{k-1} \overset{\mathrm{Df}}{=} C_{k-1} + \int_{t_{k-1}}^{t_k} E_3(t)\, dt$$

But (4.19) can be interpreted as a shaping filter and $\boldsymbol{\eta}_k$ can be considered as additional state variables. When this redefinition is performed, (4.17) is observed to have the desired form.

The control term $\boldsymbol{\sigma}_k$ can be changed in a manner analogous to that used for $\boldsymbol{\eta}_k$ by recalling that

$$\mathbf{u}(t) = \mathbf{u}_k \qquad \text{for} \quad t_k \leqslant t < t_{k+1}$$

Then it follows that

$$\boldsymbol{\sigma}_k = \boldsymbol{\sigma}_{k-1} + \gamma_{k-1} \mathbf{u}_{k-1} \tag{4.20}$$

where

$$\gamma_{k-1} = A_{k-1} + \int_{t_{k-1}}^{t_k} G(t)\, dt$$

and

$$A_{k-1} = \int_{t_0}^{t_{k-1}} F_2(t) \int_{t_k}^{t_{k-1}} \Phi(t,\tau) G(\tau)\, d\tau\, dt + \int_{t_{k-1}}^{t_k} F_2(t) \int_{t_k}^{t} \Phi(t,\tau) G(\tau)\, d\tau\, dt$$

The matrices ${}_aJ_k$, ${}_1J_k$, ${}_2J_k$, A_{k-1}, and C_{k-1} can be rewritten as recursion relations. This form is more convenient for computational purposes.

$${}_aJ_k = {}_aJ_{k-1} \Phi_{k-1,k} + \int_{t_{k-1}}^{t_k} F_2(t) \Phi(t, t_k)\, dt$$

$${}_1J_k = {}_1J_{k-1} - {}_aJ_{k-1} \Phi_{k-1,k} A_{k,k-1} + \int_{t_{k-1}}^{t_k} F_2(t) \int_{t_k}^{t} \Phi(t,\tau) E_1(\tau)\, d\tau\, dt$$

$${}_2J_k = {}_2J_{k-1} - {}_aJ_{k-1} \Phi_{k-1,k} B_{k,k-1} + \int_{t_{k-1}}^{t_k} F_2(t) \int_{t_k}^{t} \Phi(t,\tau) E_2(\tau)\, d\tau\, dt + \int_{t_{k-1}}^{t_k} E_2(t)\, dt$$

$$A_{k-1} = -{}_aJ_{k-1} \Phi_{k-1,k} \Gamma_{k,k-1} + \int_{t_{k-1}}^{t_k} F_2(t) \int_{t_k}^{t} \Phi(t,\tau) G(\tau)\, d\tau\, dt$$

$$C_{k-1} = -{}_aJ_{k-1} \Phi_{k-1,k} C_{k,k-1} + \int_{t_{k-1}}^{t_k} F_2(t) \int_{t_k}^{t} \Phi(t,\tau) E_3(\tau)\, d\tau\, dt$$

No consideration has yet been given to measurement noise in the IMU data. Generally, the IMU errors are assumed to be constant random variables (*59*) during a particular phase of flight and are linearly related to the acceleration measurements. Let the IMU error sources be denoted by $\boldsymbol{\epsilon}$ and let the linear transformation be

$$\int_{t_0}^{t_k} B(t)\, dt$$

Assuming some white noise effects $\boldsymbol{\nu}_k$, the linear measurement model for the IMU data becomes

$$\boldsymbol{\zeta}_k - \boldsymbol{\sigma}_k = \mathbf{s}_k + \int_{t_0}^{t_k} B(t)\, dt\, \boldsymbol{\epsilon} + \boldsymbol{\nu}_k \tag{4.21}$$

When both state-related and acceleration-related data are available, the results of this and the preceding section combine to give the augmented state vector ${}_A\mathbf{x}_k$.

$${}_A\mathbf{x}_k \overset{\mathrm{Df}}{=} \begin{bmatrix} \mathbf{x}_k \\ \boldsymbol{\alpha}_k{}^G \\ \boldsymbol{\alpha}_k{}^F \\ \mathbf{b}_k \\ \boldsymbol{\eta}_k \\ \boldsymbol{\epsilon}_k \end{bmatrix}$$

The dynamical model is

$$\begin{bmatrix} \mathbf{x}_k \\ \boldsymbol{\alpha}_k^G \\ \boldsymbol{\alpha}_k^F \\ \mathbf{b}_k \\ \boldsymbol{\eta}_k \\ \boldsymbol{\epsilon}_k \end{bmatrix} = \begin{bmatrix} \Phi_{k,k-1} & A_{k,k-1} & B_{k,k-1} & O & O & O \\ O & I & O & O & O & O \\ O & O & I & O & O & O \\ O & O & O & I & O & O \\ O & O & O & O & I & O \\ O & O & O & O & O & I \end{bmatrix} \begin{bmatrix} \mathbf{x}_{k-1} \\ \boldsymbol{\alpha}_{k-1}^G \\ \boldsymbol{\alpha}_{k-1}^F \\ \mathbf{b}_{k-1} \\ \boldsymbol{\eta}_{k-1} \\ \boldsymbol{\epsilon}_{k-1} \end{bmatrix} + \begin{bmatrix} \Gamma_{k,k-1} \\ O \\ O \\ O \\ O \\ O \end{bmatrix} \mathbf{u}_{k-1} + \begin{bmatrix} C_{k,k-1} \\ O \\ O \\ O \\ \beta_{k-1} \\ O \end{bmatrix} \mathbf{w}_{k-1}$$

or, with the obvious definitions,

$$_A\mathbf{x}_k = {}_A\Phi_{k,k-1}\, {}_A\mathbf{x}_{k-1} + {}_A\Gamma_{k,k-1}\mathbf{u}_{k-1} + {}_AC_{k,k-1}\mathbf{w}_{k-1} \tag{4.22}$$

The measurement model is

$$_A\mathbf{z}_k \overset{\mathrm{Df}}{=} \begin{bmatrix} \mathbf{z}_k \\ \boldsymbol{\zeta}_k - \boldsymbol{\sigma}_k \end{bmatrix}$$

$$\begin{bmatrix} \mathbf{z}_k \\ \boldsymbol{\zeta}_k - \boldsymbol{\sigma}_k \end{bmatrix} = \begin{bmatrix} H_k & O & O & I & O & O \\ {}_aJ_k & {}_1J_k & {}_2J_k & O & I & \int_{t_0}^{t_k} B(t)\, dt \end{bmatrix} \begin{bmatrix} \mathbf{x}_k \\ \boldsymbol{\alpha}_k^G \\ \boldsymbol{\alpha}_k^F \\ \mathbf{b}_k \\ \boldsymbol{\eta}_k \\ \boldsymbol{\epsilon}_k \end{bmatrix} + \begin{bmatrix} \mathbf{v}_k \\ \boldsymbol{\nu}_k \end{bmatrix}$$

or

$$_A\mathbf{z}_k = {}_AH_k\, {}_A\mathbf{x}_k + {}_A\mathbf{v}_k \tag{4.23}$$

The IMU data can be incorporated into the model in a way that is radically different from the technique just discussed. Suppose that the output of the IMU is described by

$$\boldsymbol{\rho}(t) = \mathbf{a}(t) + B(t)\boldsymbol{\epsilon} \tag{4.24}$$

where $\mathbf{a}(t)$ is defined by (4.14) and $B(t)\boldsymbol{\epsilon}$ is found in (4.21).

Using (4.14), the perturbation equations (4.10) become

$$\dot{\mathbf{x}} = F_1(t)\mathbf{x} + E_1(t)\boldsymbol{\alpha}^G + \mathbf{a}(t)$$

or, using (4.24),

$$\dot{\mathbf{x}} = F_1(t)\mathbf{x} + E_1(t)\boldsymbol{\alpha}^G + \boldsymbol{\rho}(t) - B(t)\boldsymbol{\epsilon} \tag{4.25}$$

The $\boldsymbol{\rho}(t)$ represents the difference between the measured and nominal accelerations [the first three components of the (6×1) vector are identically zero] so is a known function of time. Equation (4.25) can be reduced to a difference equation in which the transition matrix is based upon $F_1(t)$ rather than $[F_1(t) + F_2(t)]$ as in (4.12). The observation equation (4.13) is unchanged when this method of including the IMU data is employed. The $\boldsymbol{\epsilon}$ are treated as new state variables.

When this formulation is utilized, it is interesting to observe that an aiding instrument (i.e., a state-related measurement device) must always be included in order that z_k exist. Without the z_k, no filtering can occur because the gain matrix K_k is identically zero. Furthermore, note that the control $u(t)$ no longer appears explicitly in the model and that the IMU output must be assumed to be the nongravitational acceleration rather than its integral.

D. Example

To illustrate the procedure for reducing a nonlinear system to the form required for the application of the Kalman filter, let us consider the problem of determining the state of a spacecraft moving in a nearly circular orbit about the earth. The necessary system matrices, $\Phi_{k,k-1}$ and H_k, shall be derived but no attempt to evaluate the filter equations shall be made. This problem is discussed in *(30)*.

Suppose that the position and velocity of a spacecraft in a nearly circular orbit is to be estimated from the angular measurements of a horizon sensor. This instrument shall be assumed to measure two angles: (1) the angle γ between the line of sight to the edge of the planet and the local vertical, and (2) the angle α between the local vertical and a known reference direction.

To simplify the equations the earth shall be assumed to be spherical with radius R_e and have a spherical potential U described by

$$U = \frac{\mu}{R}$$

The μ is a constant equal to the product of the mass of the earth and the universal gravitational constant. Let R be the distance from the center of the earth to the spacecraft. Furthermore, the motion shall be restricted to be two-dimensional.

A coordinate system must be defined and, as is frequently true, its selection is worthy of considerable contemplation. It is known that the numerical accuracy obtained in the evaluation of the Kalman filter equations can be affected significantly by the choice of coordinate system. Let us couch this problem in terms of three different coordinate systems and examine the system matrices that result for their relative simplicity.

A possible choice of coordinate system would certainly be a nonrotating cartesian system. In this frame the equations of motion are known to be

$$\ddot{\mathbf{R}} = -\frac{\mu}{R^3}\mathbf{R}$$

Since the motion of the spacecraft is nearly circular, let us assume that the reference trajectory is a circular orbit with radius R_0. This statement completely defines the nominal trajectory. Using the knowledge of the nominal orbit, let us change the independent variable from time to a variable redefined by

$$\tau = \omega t$$

where ω is the rate of rotation of an object moving in the nominal orbit

$$\omega^2 = \frac{\mu}{R_0{}^3}$$

Also, consider the nondimensional variable

$$\boldsymbol{\xi} = \frac{\mathbf{R}}{R_0}$$

The equations of motion now become

$$\frac{d^2\boldsymbol{\xi}}{d\tau^2} = \frac{1}{r^3}\boldsymbol{\xi} \tag{4.26}$$

where

$$r^2 = \boldsymbol{\xi}^T\boldsymbol{\xi}$$

The linear perturbation equations must be formed from (4.26) using the nominal trajectory stated above. For this case the perturbation equations (which are omitted for brevity) can be solved in closed form to give the following state transition matrix:

$$\Phi(\tau, 0) = \begin{bmatrix} \Phi_{11} & \Phi_{12} & \Phi_{13} & \Phi_{14} \\ \Phi_{21} & \Phi_{22} & \Phi_{23} & \Phi_{24} \\ \Phi_{31} & \Phi_{32} & \Phi_{33} & \Phi_{34} \\ \Phi_{41} & \Phi_{42} & \Phi_{43} & \Phi_{44} \end{bmatrix} \tag{4.27}$$

The Φ_{ij} are defined as

$$\Phi_{11} = -\tfrac{3}{2} + 2\cos\tau + \tfrac{1}{2}\cos 2\tau + 3\tau\sin\tau$$

$$\Phi_{12} = \sin\tau - \tfrac{1}{2}\sin 2\tau$$

$$\Phi_{13} = 2\sin\tau - \tfrac{1}{2}\sin 2\tau$$

$$\Phi_{14} = -3 + 2\cos\tau + \cos 2\tau + 3\tau\sin\tau$$

$$\Phi_{21} = 2\sin\tau + \tfrac{1}{2}\sin 2\tau - 3\tau\cos\tau$$

$$\Phi_{22} = \tfrac{3}{2} - \cos\tau + \tfrac{1}{2}\cos 2\tau$$

$$\Phi_{23} = \tfrac{3}{2} - 2\cos\tau + \tfrac{1}{2}\cos 2\tau$$

$$\Phi_{24} = 2\sin\tau + \sin 2\tau - 3\tau\cos\tau$$

$$\Phi_{31} = \sin\tau - \sin 2\tau + 3\tau\cos\tau$$

$$\Phi_{32} = \cos\tau - \cos 2\tau$$

$$\Phi_{33} = 2\cos\tau - \cos 2\tau$$

$$\Phi_{34} = \sin\tau - 2\sin 2\tau + 3\tau\cos\tau$$

$$\Phi_{41} = -\cos\tau + \cos 2\tau + 3\tau\sin\tau$$

$$\Phi_{42} = \sin\tau - \sin 2\tau$$

$$\Phi_{43} = 2\sin\tau - \sin 2\tau$$

$$\Phi_{44} = -\cos\tau + 2\cos 2\tau + 3\tau\sin\tau$$

The angles measured by the horizon sensor are given by

$$\gamma = \sin^{-1}\frac{\xi_e}{r} \qquad \text{where} \qquad \xi_e = \frac{R_e}{R_0} \tag{4.28}$$

and

$$\alpha = -\sin^{-1}\frac{\xi_2}{r} \tag{4.29}$$

The $\boldsymbol{\xi}$ has been defined as

$$\boldsymbol{\xi} \overset{\mathrm{Df}}{=} \begin{bmatrix} \xi_1 \\ \xi_2 \end{bmatrix}$$

and ξ_1 is assumed to be the reference axis for the measurement α.

Equations (4.28) and (4.29) must be linearized to obtain the observation matrix H.

$$H \overset{\mathrm{Df}}{=} \begin{bmatrix} \dfrac{\partial\alpha}{\partial\xi_1} & \dfrac{\partial\alpha}{\partial\xi_2} & \dfrac{\partial\alpha}{\partial\dot{\xi}_1} & \dfrac{\partial\alpha}{\partial\dot{\xi}_2} \\ \dfrac{\partial\gamma}{\partial\xi_1} & \dfrac{\partial\gamma}{\partial\xi_2} & \dfrac{\partial\gamma}{\partial\dot{\xi}_1} & \dfrac{\partial\gamma}{\partial\dot{\xi}_2} \end{bmatrix}$$

$$= \begin{bmatrix} \dfrac{\xi_2}{r^2} & -\dfrac{\xi_1}{r^2} & 0 & 0 \\ \dfrac{-\xi_1\xi_e}{r^2(r^2-\xi_e^2)^{1/2}} & \dfrac{-\xi_2\xi_e}{r^2(r^2-\xi_e^2)^{1/2}} & 0 & 0 \end{bmatrix} \tag{4.30}$$

In the nonrotating coordinate system it is clear from (4.30) that the observation matrix H is a function of τ since ξ_1 and ξ_2 vary with τ.

Instead of the nonrotating system, suppose that a rectangular coordinate system is introduced that rotates at the rate ω of the nominal orbit. In this system with the dimensionless variables τ, ξ_1, ξ_2 the Lagrangian L is

$$L = \tfrac{1}{2}(\dot{\xi}_1 - \xi_2)^2 + \tfrac{1}{2}(\dot{\xi}_2 + \xi_1)^2 + \frac{1}{r} \tag{4.31}$$

The equations of motion are formed from

$$\frac{d}{d\tau}\left(\frac{\partial L}{\partial\dot{\xi}_i}\right) - \left(\frac{\partial L}{\partial\xi_i}\right) = 0 \qquad i = 1, 2$$

and are

$$\frac{d\xi_1^2}{d\tau^2} - 2\frac{d\xi_2}{d\tau} - \xi_1 + \frac{\xi_1}{r^3} = 0 \qquad \frac{d\xi_2^2}{d\tau^2} + 2\frac{d\xi_1}{d\tau} - \xi_2 + \frac{\xi_2}{r^3} = 0 \tag{4.32}$$

Define the nominal position as

$$\xi_1^* = 1 \qquad \xi_2^* = 0$$

so

$$r^* = 1$$

The linearized equations obtained directly from (4.32) are found to be

$$\delta\ddot{\xi}_1 - 2\delta\dot{\xi}_2 - 3\delta\xi_1 = 0 \qquad \delta\ddot{\xi}_2 + 2\delta\dot{\xi}_1 = 0$$

Writing these in state vector notation, the linear perturbation equations are

$$\begin{bmatrix} \dot{x}_1 \\ \dot{x}_2 \\ \dot{x}_3 \\ \dot{x}_4 \end{bmatrix} = \begin{bmatrix} 0 & 0 & 1 & 0 \\ 0 & 0 & 0 & 1 \\ 3 & 0 & 0 & 2 \\ 0 & 0 & -2 & 0 \end{bmatrix} \begin{bmatrix} x_1 \\ x_2 \\ x_3 \\ x_4 \end{bmatrix} \tag{4.33}$$

or

$$\dot{\mathbf{x}} = F\mathbf{x}$$

This stationary system can be solved in a straightforward manner to give the state transition matrix

$$\Phi(\tau, 0) = \begin{bmatrix} 4 - 3\cos\tau & 0 & \sin\tau & 2(1-\cos\tau) \\ 6(\sin\tau - \tau) & 1 & 2(\cos\tau - 1) & 4\sin\tau - 3\tau \\ 3\sin\tau & 0 & \cos\tau & 2\sin\tau \\ 6(\cos\tau - 1) & 0 & -2\sin\tau & 4\cos\tau - 3 \end{bmatrix} \tag{4.34}$$

Observe that (4.34) is appreciably simpler than (4.27).

The measurements are still described by (4.28) and (4.29). However, since the coordinate system rotates at the rate ω, the H is seen to be

$$H = \begin{bmatrix} 0 & -1 & 0 & 0 \\ k & 0 & 0 & 0 \end{bmatrix} \tag{4.35}$$

where

$$k \overset{\mathrm{Df}}{=} \frac{-\xi_e}{(1-\xi_e^2)^{1/2}}$$

Thus, the H is a constant matrix which is a considerable improvement over (4.30) when the nonrotating system is used.

The rotating coordinate system is a natural choice for this problem. In a recent paper, Tschauner and Hempel (*60*) describe another and less intuitively obvious system in which the transition matrix assumes an even less complicated form than in (4.34). In fact, their results apply also to elliptic orbits and have the property that the transition matrix has the same form for both circular and elliptic orbits. For circular orbits the transformation is constant, so the observation matrix remains constant in the new system. A time-varying transformation is used for elliptic orbits, however, so the H matrix is no longer time-invariant in the new system.

For the circular case consider the transformation of the state $\mathbf{x}$ from the rotating system to a new system defined by

$$\begin{bmatrix} \zeta_1 \\ \zeta_2 \\ \zeta_3 \\ \zeta_4 \end{bmatrix} = \begin{bmatrix} 2 & 0 & 0 & 1 \\ 0 & -\frac{1}{3} & \frac{2}{3} & 0 \\ \frac{3}{2} & 0 & 0 & 1 \\ 0 & 0 & -\frac{1}{2} & 0 \end{bmatrix} \begin{bmatrix} x_1 \\ x_2 \\ x_3 \\ x_4 \end{bmatrix} \tag{4.36}$$

or

$$\zeta_k = T\mathbf{x}_k$$

It follows from the relation

$$\psi(\tau, 0) = T\Phi(\tau, t_0)T^{-1}$$

that the transition matrix $\psi(\tau, 0)$ for the ζ_k is

$$\psi(\tau, 0) = \begin{pmatrix} 1 & 0 & 0 & 0 \\ \tau & 1 & 0 & 0 \\ 0 & 0 & \cos\tau & \sin\tau \\ 0 & 0 & -\sin\tau & \cos\tau \end{pmatrix} \tag{4.37}$$

In this system the observation matrix becomes, using

$$H_\zeta = HT^{-1} \qquad H_\zeta = \begin{bmatrix} 0 & 3 & 0 & 4 \\ 2k & 0 & -2k & 0 \end{bmatrix} \tag{4.38}$$

The $\psi(\tau, 0)$ and H_ζ exhibit a frequently occurring circumstance when different coordinate systems are compared. The $\psi(\tau, 0)$ has a simpler form than $\Phi(\tau, 0)$, but H_ζ contains more nonzero terms than H.

The $\Phi(\tau, 0)$ and H provide the system matrices for the Kalman filter for this problem. It remains to describe the statistical properties of the initial state and the horizon sensor errors to complete the formulation of the problem. For a detailed solution of the problem, see *(30)*.

V. Summary and Concluding Remarks

A. Summary of Important Results

In this section we intend to summarize the significant results of each section and to present the important equations.

Section II, A. The basic dynamical and observational model is defined. Equations (I) and (II) describe the dynamical model and the observation process.

$$\mathbf{x}_k = \Phi_{k,k-1}\mathbf{x}_{k-1} + \Gamma_{k,k-1}\mathbf{u}_{k-1} + \mathbf{f}_{k-1} + \mathbf{w}_{k-1} \tag{I}$$

$$\mathbf{z}_k = H_k\mathbf{x}_k + \mathbf{v}_k \tag{II}$$

The noise sequences $\mathbf{w}_{k-1}$ and $\mathbf{v}_k$ are assumed to be mutually uncorrelated and uncorrelated between sampling times. They are referred to as white noise sequences.

$$E[\mathbf{w}_k] = E[\mathbf{v}_k] = \mathbf{0} \qquad \text{for all} \quad k\ (k = 0, 1, 2, \ldots, N)$$
$$E[\mathbf{w}_k\mathbf{w}_j^T] = Q_k\delta_{kj}$$
$$E[\mathbf{v}_k\mathbf{v}_j^T] = R_k\delta_{kj}$$
$$E[\mathbf{w}_k\mathbf{v}_j^T] = O$$

The $\mathbf{f}_k$ is assumed to be deterministic.

The linear estimation problem is stated as that of choosing the linear estimate $\hat{\mathbf{x}}_k$ that minimizes the mean-square error

$$E[(\hat{\mathbf{x}}_k - \mathbf{x}_k)^T(\hat{\mathbf{x}}_k - \mathbf{x}_k)] = \text{trace } E[\hat{\mathbf{x}}_k - \mathbf{x}_k)(\hat{\mathbf{x}}_k - \mathbf{x}_k)^T] \overset{\text{Df}}{=} \text{trace } P_k$$

Section II, B. The dynamical model is reduced to an unforced system $\mathbf{x}_k = \Phi_{k,k-1}\mathbf{x}_{k-1}$, and the filter equations are derived. The estimate is hypothesized to be given by

$$\hat{\mathbf{x}}_k = \Phi_{k,k-1}\hat{\mathbf{x}}_{k-1} + K_k[\mathbf{z}_k - H_k\Phi_{k,k-1}\hat{\mathbf{x}}_{k-1}] \tag{IIIu}$$

and the gain is shown to be

$$K_k = P_k'H_k^T[H_kP_k'H_k^T + R_k]^{-1} \tag{V}$$

The P_k' in this case is

$$P_k' = \Phi_{k,k-1}P_{k-1}\Phi_{k,k-1}^T \tag{IVu}$$

and

$$P_k \overset{\text{Df}}{=} E[\hat{\mathbf{x}}_k - \mathbf{x}_k)(\mathbf{x}_k - \mathbf{x}_k)^T] = P_k' - K_kH_kP_k' \tag{VI}$$

Section II, C. The dynamical model is extended to include random and deterministic forcing terms

$$\mathbf{x}_k = \Phi_{k,k-1}\mathbf{x}_{k-1} + \mathbf{f}_{k-1} + \mathbf{w}_{k-1} \tag{IF}$$

A lemma is proved that shows that the estimate $\hat{\mathbf{x}}$ that minimizes the mean-square error given the measurement data $\mathbf{z}$ is

$$\hat{\mathbf{x}} = E[\mathbf{x}/\mathbf{z}]$$

This is true for *all* mean-square estimates, whether linear or nonlinear. For this model (IF) the estimate is shown to be

$$\hat{\mathbf{x}}_k = \Phi_{k,k-1}\hat{\mathbf{x}}_{k-1} + \mathbf{f}_{k-1} + K_k[\mathbf{z}_k - H_k(\Phi_{k,k-1}\mathbf{x}_{k-1} + \mathbf{f}_{k-1})] \quad \text{(IIIF)}$$

and

$$P_k' = \Phi_{k,k-1}P_{k-1}\Phi^T_{k,k-1} + Q_{k-1} \quad \text{(IV)}$$

The K_k and P_k are still defined by (V) and (VI). Note that the deterministic term $\mathbf{f}_{k-1}$ only enters into the estimation equation itself, whereas the noise $\mathbf{w}_{k-1}$ introduces uncertainty into the variance of estimate by appearing only in (IV).

Section II, D. The estimate $\mathbf{x}_k$ is shown to be unbiased if the initial estimate is selected as

$$\hat{\mathbf{x}}_0 = E[\mathbf{x}_0]$$

The initial covariance matrix P_0 is set equal to the uncertainty M_0 in the state at t_0. In part 2 of this section, the equations of the filter are shown to be unaffected by the introduction of a weighting matrix in the least-square estimate. That is, the $\hat{\mathbf{x}}_k$ that minimizes $E[(\hat{\mathbf{x}}_k - \mathbf{x}_k)^T W_k(\hat{\mathbf{x}}_k - \mathbf{x}_k)]$ is identical to those derived with $W_k \equiv I$.

The error in the estimate $\tilde{\mathbf{x}}_k = \hat{\mathbf{x}}_k - \mathbf{x}_k$ is shown to be orthogonal to the estimate $\hat{\mathbf{x}}_k$ (i.e., $E[\tilde{\mathbf{x}}_k{}^T\hat{\mathbf{x}}_k] = 0$). This fact indicates that the best linear estimate of $\mathbf{x}_k$ is given by the orthogonal projection of $\mathbf{x}_k$ onto the linear subspace formed by the measurement data $\mathbf{z}_k$.

In the final part of this section the relationship between the classical deterministic least-squares estimate and the Kalman filter estimates is discussed.

Section II, E. The dynamical model is extended to include the control term $\Gamma_{k,k-1}\mathbf{u}_{k-1}$. The separation theorem is stated without proof. This result permits the joint problems of optimal control and optimal estimation to be solved separately. In this case the estimate is given by

$$\hat{\mathbf{x}}_k = \Omega_{k,k-1}\hat{\mathbf{x}}_{k-1} + \mathbf{f}_{k-1} + K_k[\mathbf{z}_k - H_k(\Omega_{k,k-1}\hat{\mathbf{x}}_{k-1} + \mathbf{f}_{k-1})] \quad \text{(III)}$$

where

$$\Omega_{k,k-1} \stackrel{\text{Df}}{=} \Phi_{k,k-1} + \Gamma_{k,k-1}\Lambda_{k-1}$$

The Λ_{k-1} is the deterministic optimal control matrix.

An alternative formulation is presented that draws upon a result presented in Section III, D.

$$\hat{\mathbf{x}}_k = \Phi_{k,k-1}\hat{\mathbf{x}}_{k-1} + K_k[\mathbf{z}_k - H_k\Phi_{k,k-1}\mathbf{x}_{k-1}] + P_kP_k'^{-1}\mathbf{u}_{k-1} \quad \text{(III')}$$

where

$$\mathbf{u}_{k-1} = \Lambda_{k-1}\hat{\mathbf{x}}_{k-1}$$

The remaining filter equations are given by (IV), (V), and (VI), except that $\Phi_{k,k-1}$ is replaced by $\Omega_{k,k-1}$.

Section II, F. A more general statistical model is introduced. The noise processes are permitted to be correlated between sampling times. The concept of a shaping filter is introduced and this model is reduced to that of Section II, A through the augmentation of the state vector.

The class of wide-sense Markov sequences is considered. The shaping filter for these sequences $\boldsymbol{\omega}_k$ is shown to be described by

$$\boldsymbol{\omega}_k = \Phi^D_{k,k-1}\boldsymbol{\omega}_{k-1} + \mathbf{w}_{k-1} \quad (2.20)$$

where

$$\Phi^D_{k,k-1} = D_{k,k-1}D^{-1}_{k-1,k-1}$$

and

$$E[\boldsymbol{\omega}_{k+1}\boldsymbol{\omega}_k{}^T] = D_{k+1,k}$$

The $\mathbf{w}_{k-1}$ is a white noise sequence with covariance

$$E[\mathbf{w}_{k-1}\mathbf{w}^T_{k-1}] = D_{k,k} - D_{k,k-1}D^{-1}_{k-1,k-1}D_{k-1,k}$$

The topic of parameter estimation was considered in part 3.

Section II, G. The time-discrete form of the filter is modified so that the white noise sequences will exhibit the properties of white noise processes for infinitesimal sampling intervals. Then the time-continuous Kalman filter is derived from (III) to (VI) by allowing the sampling interval to become arbitrarily small. In this case the state evolves according to

$$\frac{d\mathbf{x}}{dt} = F(t)\mathbf{x} + G(t)\mathbf{w} \quad \text{(Ic)}$$

and the measurements are described by

$$\mathbf{z}(t) = H(t)\mathbf{x}(t) + \mathbf{v}(t) \quad \text{(IIc)}$$

where $\mathbf{w}(t)$ and $\mathbf{v}(t)$ are white noise processes. The filter is given by

$$\frac{d\hat{\mathbf{x}}}{dt} = F(t)\hat{\mathbf{x}} + K(t)[\mathbf{z}(t) - H(t)\hat{\mathbf{x}}] \quad \text{(IIIc)}$$

$$K(t) = P(t)H^T(t)R^{-1}(t) \quad \text{(IVc)}$$

$$\frac{dP}{dt} = F(t)P + PF^T(t) + PH^T(t)R^{-1}(t)H(t)P + G(t)QG^T(t) \quad \text{(Vc)}$$

The solution of the matrix Ricatti equation is discussed.

Section III, A. The basic cycle of computations that must be followed to obtain the estimate at each t_k is stated. The flow within the Kalman filter equation is described and then the external details that must be supplied to simulate a system within which the filter is exercised is discussed.

Section III, B. Random noise must be generated to simulate a system for which the estimation process is to be utilized. When independent random variables are generated, well-known techniques can be used. However, when the variables are correlated (i.e., R_k and/or Q_{k-1} are nondiagonal) a transformation must be introduced to obtain independent variables. The computational algorithm is described in this section.

Section III, C. The error ellipsoids are introduced as a measure of the filter response.

Section III, D. It is possible through the use of an interesting matrix inversion lemma to obtain an alternative form for the equations describing the covariance matrix P_k and the gain matrix K_k. P_k^{-1} rather than P_k is generated.

$$P_k^{-1} = (P_k')^{-1} + H_k^T R_k^{-1} H_k \qquad \text{(VII)}$$

and

$$K_k = P_k H_k^T R_k^{-1} \qquad \text{(VIII)}$$

Section III, E. When there are a large number, say m, of measured quantities, the matrix R_k will be $m \times m$. Since R_k (or $H_k P_k' H_k^T + R_k$) must be inverted, a computational problem of considerable significance can be engendered. This can be avoided through the use of a sequential processing technique when most of the measurements are uncorrelated with one another. The procedure is described in detail and its equivalence with direct evaluation of the equations is proved.

Section III, F. The evaluation of the filter equation can be extremely sensitive to computational inaccuracies, particularly when the dynamical system is unforced, and can result in the loss of the sign-definiteness properties of the covariance matrix P_k. This effect causes the validity of estimates of the state (and, when required, the guidance estimates) to deteriorate. To counteract these errors, a formulation in terms of the matrix square root of P_k is derived. This formulation is presented for the special case of a single observation. Alternatively, the possibility of modeling the computer-generated errors as forcing terms on the dynamical system is discussed.

Section IV, A. The technique by which a dynamical system described by nth-order nonlinear vector differential equations is reduced to a system described by first-order linear vector difference equations is discussed. The key step in this procedure is the assumption of the existence of a nominal solution that approximates the actual response of the system. Then the linear differential equations are rewritten in the linear difference equation form required by this formulation of the Kalman filter.

Section IV, B. The general procedure presented in the preceding section is applied to the problem that arises in the navigation and guidance of space vehicles. The equations of motion are written in a general form that indicates the functional relationships, but they have not been specialized to a specific problem (e.g., a central body or n-bodies). The measurement data are distinguished as being either state-related (e.g., radar measurements) or acceleration-related (e.g., accelerometer outputs). Only state-related measurements are considered in detail in this section.

Section IV, C. The formulation for acceleration-related measurements is developed. In this approach the accelerometer outputs are written in the form of (II) by a suitable redefinition of the state vector. Then the acceleration-related measurements are treated in the same manner as the state-related measurements.

Section IV, D. The problem of estimating the position and velocity of a spacecraft in a nearly circular orbit using horizon sensor measurements is discussed. The form of the state transition matrix $\Phi(t_k, t_0)$ and the observation matrix $H(t_k)$ in three different coordinate systems are compared.

B. Other Aspects of the Estimation Problem

Many fundamental aspects of the linear estimation problem have been discussed in the preceding paragraphs. By no stretch of the imagination can the topic be assumed to have been exhausted. To conclude the presentation the remainder of this section shall be devoted to many additional subjects that have been treated in the voluminous literature. The reader is directed to the indicated references for more thorough presentations.

Correlation between $\mathbf{v}_k$ *and* $\mathbf{w}_k$. The white noise sequences $\mathbf{v}_k$ and $\mathbf{w}_{k-1}$ have been assumed to be uncorrelated throughout the discussion. This restriction is not necessary. Kalman (*2*) considered this more general model. The equations that result are considerably more complicated.

Random Sampling Times, Random Multiplicative Parameters, etc. The measurement data have been assumed to occur at prespecified times. It can happen that the time to which a measurement is related is random. Also, the transition matrix and/or the observation matrix might contain parameters that are random. These cases have been considered by several investigators including Tou (*15*), Rauch (*33*), Gunckel (*34*), and others.

Effect of Inaccurate Mathematical Model and/or Statistics. The validity of the estimates provided by this estimation technique require an accurate mathematical description of the dynamical and observational processes. The omission of terms that actually enter these processes can result in extremely poor estimates. If the statistics (i.e., P_0, R_k, Q_k) are not accurate descriptions of the second-order moments of the noise processes (*7, 35*), the covariance matrix P_k might not be a realistic measure of the accuracy of the estimate. The effects of P_0 generally vanish as t_k increases, so the instrument statistics provide the more significant parameters.

Fixed-Memory Filters (Moving-Window Estimates) (*37*). In the Kalman filter each measurement is considered, loosely speaking, equally in the computations. As time advances, the "memory" upon which the estimates are based grows. In some cases (e.g., when the model is inaccurate), it might become desirable to eliminate measurement data that were obtained more than $N\,\Delta T$ seconds before. Then the estimate would be based only upon N data samples. This can be accomplished by a recursive filter that represents a slight modification of the Kalman filter equations.

Filter Response. The matrix P_k is used as a measure of the accuracy of the estimates and of the general response of the estimation procedure. The structure of the mathematical model (i.e., $\Phi_{k,k-1}$, H_k, $\Gamma_{k,k-1}$, etc.) can sometimes be used to determine the general nature of the response that can be expected (*2*).

Monte Carlo Simulation. The estimates of the state are random variables. To determine the validity of the estimates and the significance of the P_k, it is often necessary to perform a Monte Carlo simulation of the physical system under consideration (*38–40, 56*).

Suboptimal Filtering. The dimension of the matrices that are involved sometimes become so large as to render their manipulation virtually impossible in a particular digital computer. Also, when the state vector is large, it is often impracticable to compute estimates for every component and it becomes necessary to consider only a subset of the components. To reduce the dimensionality of the over-all system and still retain a reasonable model, methods which are not optimal in the strict sense have been devised (*41, 57*).

Non-Wide-Sense Markov Sequences. The covariance matrix of a sequence might not satisfy the conditions of a wide-sense Markov sequence. Stubberud (*42*) has suggested a technique whereby this covariance matrix can be approximated by a matrix that has the desired characteristics.

Nonlinear Estimators. The linear estimation techniques that have been discussed are subsumed into the large class of nonlinear mean-square estimators. The lemma of Section II, B, showed that any minimum mean-square estimate, whether nonlinear or linear, is given by the conditional expectation of the quantity to be estimated $\mathbf{x}$ given the measurement data $\mathbf{z}$. The Bayesian estimation scheme discussed by Ho and Lee (*43*) reduces this problem to that of determining the conditional density function $p(\mathbf{x} \mid \mathbf{z})$. Kushner (*44*) and others (*45*) have shown that $p(\mathbf{x} \mid \mathbf{z})$ can be described by a partial differential equation when the processes are Markovian (*31*). Another approach to nonlinear estimation in terms of polynomial estimators has been proposed by Balakrishnan (*12*). Wiener (*46*) devised a similar technique. Cox (*32*) has considered nonmean-square estimators.

Linear Smoothing Problem. In many applications (e.g., postflight analysis), it is desirable to estimate the state $\mathbf{x}_j$ using measurement data that relates to sampling times t_k, where $t_k > t_j$, as well as for times earlier than t_j. This smoothing problem has been solved for time-discrete systems by Rauch *et al.* (*28*) and for time-continuous systems by Bryson and Frazier (*61*).

Computational Results. Many computational results have been published that relate to the space navigation problem. Schmidt (*47, 48*), in conjunction with Smith, was possibly the first person to apply the Kalman filter. Since then, Smith (*21, 48, 49*) has published the results of several studies. Other results have been discussed by Mendelson (*26*), White *et al.* (*50*), Meditch *et al.* (*30*), Friedlander (*51*), Gunckel (*52*), Blackman (*53*), Claus (*54*), and Dusek (*55*).

The list of topics and references to papers could be continued interminably. However, we shall truncate the discussion at this point.

List of Symbols

- t_k time at kth sampling instant ($k = 0, 1, 2, \ldots, N$)
- $\mathbf{b}_k$ bias errors in the state-related measurement data
- $\mathbf{f}_k$ deterministic vector forcing function in the linear dynamical system at t_k
- $\mathbf{u}_k$ vector control function in linear, dynamical system at t_k
- $\mathbf{v}_k$ white noise sequence corrupting measurement data
- $\mathbf{w}_k$ white noise sequence corrupting linear dynamical system at t_k
- $\mathbf{x}_k$ the state of the linear dynamical system at t_k
- $\hat{\mathbf{x}}_k$ the linear estimate of the state $\mathbf{x}_k$ at t_k using the data $\mathbf{z}_k$
- $\hat{\mathbf{x}}_k'$ the predicted linear estimate of the state $\mathbf{x}_k$ at t_k before $\mathbf{z}_k$ is processed
- $\tilde{\mathbf{x}}_k$ the error in the estimate at t_k: $\tilde{\mathbf{x}}_k = \hat{\mathbf{x}}_k - \mathbf{x}_k$
- $\mathbf{z}_k$ measurement data at t_k linearly related to $\mathbf{x}_k$
- $\boldsymbol{\alpha}_k$ constant random parameters in the dynamical system
- $\boldsymbol{\epsilon}_k$ constant random errors in inertial measuring unit (IMU)
- H_k observation matrix at t_k: $\mathbf{z}_k = H_k\mathbf{x}_k + \mathbf{v}_k$
- K_k optimal gain matrix for Kalman filter at t_k
- M_k covariance of state at t_k: $M_k = E[\mathbf{x}_k\mathbf{x}_k^T]$
- P_k covariance of the error in the estimate of state at t_k: $P_k = E[\tilde{\mathbf{x}}_k\tilde{\mathbf{x}}_k^T]$
- P_k' covariance of error in the predicted estimate of state at t_k: $P_k' = E[\tilde{\mathbf{x}}_k'\tilde{\mathbf{x}}_k'^T]$
- Q_k covariance of $\mathbf{w}_k$: $Q_k = E[\mathbf{w}_k\mathbf{w}_k^T]$
- R_k covariance of $\mathbf{v}_k$: $R_k = E[\mathbf{v}_k\mathbf{v}_k^T]$
- $\Gamma_{k,k-1}$ matrix that relates $\mathbf{u}_k$ to $\mathbf{x}_k$ in dynamical system
- Λ_k guidance matrix: $\mathbf{u}_k = \Lambda_k\hat{\mathbf{x}}_k$
- Π_k the factored P_k matrix: $P_k = \Pi_k\Pi_k^T$
- $\Phi_{k,k-1}$ the state transition matrix relating $\mathbf{x}_{k-1}$ to $\mathbf{x}_k$
- $\Omega_{k,k-1}$ $\Omega_{k,k-1} \stackrel{\mathrm{Df}}{=} \Phi_{k,k-1} + \Gamma_{k,k-1}\Lambda_k$
- $\mathbf{0}$ vector whose elements are identically zero
- I identity matrix
- O zero matrix
- $\mathbf{(\ )}$ the boldface quantity ( ) is a vector
- $(\ )^T$ the matrix transpose of ( )
- $(\ )^{-1}$ the matrix inverse of ( )
- $(\ )^*$ the nominal value of ( )
- $E(\ )$ the expected value of ( )
- $p(\ )$ the probability density function of ( )
- $(\ )_k$ the quantity ( ) at the time t_k
- δ_{kj} the Kronecker delta: $\delta_{kj} = \begin{cases} 1 & \text{if } k = j \\ 0 & \text{if } k \neq j \end{cases}$
- $\langle \mathbf{x}, \mathbf{y} \rangle$ inner product of $\mathbf{x}$ and $\mathbf{y}$
- $(\ ) \stackrel{\mathrm{Df}}{=} [\ ]$ the quantity ( ) is defined to be equivalent to [ ]

References

1. K. F. Gauss, "Theory of the Motion of the Heavenly Bodies Moving about the Sun in Conic Sections." Dover, New York, 1963.

2. R. E. Kalman, Fundamental study of adaptive control systems. *Tech. Rept. ASD-TR-61*, Vol. 1 (NASA N62-15355), see Appendix I.

3. R. E. Kalman, A new approach to linear filtering and prediction problems. *J. Basic Eng.* **82D** (1960).

4. R. E. Kalman, New results in linear filtering and prediction theory. *J. Basic Eng.* **83D** (1961).

5. P. Swerling, First order error propagation in a stage-wise smoothing procedure for satellite observations. *J. Astronautical Sci. 6*, 46–52 (1959).

6. R. H. Battin, A statistical optimizing navigation procedure for space flight. *ARS (Am. Rocket Soc.) J.* **32**, 1681–1696 (1962).

7. R. H. Battin, "Astronautical Guidance." McGraw-Hill, New York, 1964.

8. A. G. Carlton, Linear estimation in stochastic processes. The Johns-Hopkins University, Applied Physics Laboratory, *Bumblebee Series Rept. 311*, Baltimore, Maryland, 1962.

9. C. G. Pfeiffer, Sequential estimation of correlated stochastic variables. *Jet Propulsion Lab. Tech. Rept.* 32-445, July 1963.

10. B. Friedman, "Principles and Techniques of Applied Mathematics." Wiley, New York, 1956.

11. A. M. Yaglom, "An Introduction to the Theory of Stationary Random Functions," translated by R. A. Silverman. Prentice-Hall, Englewood Cliffs, New Jersey, 1962.

12. A. V. Balakrishnan, A general theory of non-linear estimation problems in control systems. *J. Math. Anal. Appl.* **8** (**1**) (1964).

13. Y. C. Ho, The method of least squares and optimal filtering theory. *Rand Corp. Mem. RM-3329-PR*, 1962.

14. R. C. K. Lee, "Optimal Estimation, Identification, and Control," *MIT Res. Monograph 28*. MIT Press, Cambridge, Massachusetts, 1964.

15. J. T. Tou, "Optimum Design of Digital Control Systems." Academic Press, New York, 1963.

16. W. M. Wonham, Stochastic problems in optimal control, *RIAS Tech. Rept. 63-14*, 1963.

17. P. Joseph and J. T. Tou, On linear control theory. *AIEE Trans. (Applications and Industry)* **80**, 193-196 (1961).

18. T. L. Gunckel and G. F. Franklin, A general solution for linear sampled data control. *J. Basic Eng.* **85D**, 197–201 (1963).

19. E. Stear, "Synthesis of Shaping Filters for Non-Stationary Stochastic Processes and Their Uses." *Univ. Calif. Rept. 61-50*, Los Angeles, California, 1961.

20. I. Matyash, and Ya. Shilkanikh, A generation of random processes from their given spectral density matrices. *Automation and Remote Control*, **21**, 1 (1960).

21. G. L. Smith, Secondary errors and off-design conditions in optimal estimates of space vehicle trajectories. *NASA TN D-2129*, 1964.

22. G. Marsaglia, A method for producing random variables in a computer. *Boeing Sci. Res. Lab. Math. Note 342*, 1964.

23. V. S. Pugachev, "Theory of Random Functions and Its Application to Control Problems," p. 78. Addison-Wesley, Reading, Massachusetts, 1965.

24. H. Cramer, "Mathematical Methods of Statistics," Princeton Univ. Press, Princeton, New Jersey, 1961.

25. I. I. Shapiro, "The Prediction of Ballistic Missile Trajectories from Radar Observations." McGraw-Hill, New York, 1958.

26. J. Mendelson, W. O'Dwyer, and A. Kaerchner, A numerical comparison of two orbit determination methods, *in* "Transactions of the Ninth Symposium on Ballistic Missiles," Vol. 1, DDC, 1964.

27. L. A. Zadeh, and C. A. Desoer, "Linear System Theory." McGraw-Hill, New York, 1963.

28. H. E. Rauch, F. Tung, and C. T. Striebel, Maximum likelihood estimates of linear dynamic systems. *AIAA J.* **3**(**8**) (1965).

29. A. E. Bryson and D. E. Johansen, Linear filtering for time-varying systems using measurements containing colored noise. *Sylvania Res. Rept. 385*, 1964.

30. J. S. Meditch, J. L. Lemay, and J. P. Janus, Analysis of a horizon scanner autonomous orbital navigation system, *Aerospace Corp. Rept. TDR-469 (5540-10)-6*, 1965.
31. J. L. Doob, "Stochastic Processes," Wiley, New York, 1959.
32. H. Cox, Estimation of state variables via dynamic programming, *Preprints of JACC*, pp. 376-381 (1964).
33. H. E. Rauch, Linear estimation of sampled stochastic processes with random parameters. *Stanford Electronics Lab. Tech. Rept. 2108-1*, Stanford, California, April, 1962.
34. T. L. Gunckel, Optimum design of sampled-data systems with random parameters. *Stanford Electronics Lab. Tech. Rept. 2102-2* (ASTIA. AD 255857). Stanford, California, 1961.
35. T. T. Soong, and C. G. Pfeiffer, On a priori statistics in orbit determination. *Jet Propulsion Lab. Space Programs Summary 37-21.*
36. C. G. Pfeiffer, Continuous estimation of sequentially correlated random variables. *Jet Propulsion Lab. Tech. Rept. 32-445*, 1963.
37. R. C. K. Lee, The moving window approach to problems of estimation and identification. *Aerospace Corp. Tech. Mem. ATM-65(5540-10)-18*, 1965.
38. H. A. Meyer (ed.), "Symposium on Monte Carlo Methods." Wiley, New York, 1956.
39. G. Itzelberger, Ueber die Grundlagen der Monte-Carlo-Methoden, *Wissenschaftliche Gesellschaft fuer Luft- und Raumfahrt Rept. 6*, pp. 14–26. Cologne, 1963.
40. F. Hammelrath, Allgemeine Einfuehrung in die Monte-Carlo-Methoden, *Wissenschaftliche Gesellschaft fuer Luft- und Raumfahrt Rept. 6*, 5-13, Cologne, 1963.
41. J. S. Meditch, Suboptimal linear filtering for continuous dynamic process. *Aerospace Corp. Rept. TOR-469(5107-35)-2*, 1964.
42. A. R. Stubberud, On the generation of random noise for a digital simulation. *Aerospace Corp. Rept.*, ATM 65(5107-40)-21, 1965.
43. Y. C. Ho, and R. C. K. Lee, A Bayesian approach to problems in stochastic estimation and control *Preprints of JACC*, pp. 382-387 (1964).
44. H. J. Kushner, On the differential equations satisfied by conditional probability densities of markov processes, with applications. *SIAM J. Control* **2**(1) (1964).
45. R. L. Stratonovich, Conditional Markov Processes, *Theory of Probability and Its Application* (transl. from Russian), Vol. 5. 1960.
46. N. Wiener, "Nonlinear Problems in Random Theory." *An MIT Res. Monograph*, MIT Press, Cambridge, Massachusetts, 1958.
47. S. F. Schmidt, State space techniques applied to the design of a space navigation system." *Preprints of JACC*, session II, paper 3, pp. 1-10 (1962).
48. G. L. Smith, S. F. Schmidt, and L. A. McGee, Application of statistical filter-theory to the optimal estimation of position and velocity on board a circumlunar vehicle. *NASA TR-R-135*, 1962.
49. G. L. Smith, and E. V. Harper, Midcourse guidance using radar tracking and on-board estimation data. *NASA TN D-2238*, 1964.
50. J. S. White, G. P. Callas, and L. S. Cicolani, Application of statistical filter to the interplanetary navigation and guidance problem. *NASA TN D-2697*, 1965.
51. A. L. Friedlander, A midcourse guidance procedure for electrically propelled interplanetary spacecraft. M.S. Thesis, Case Institute of Technology, Cleveland, Ohio, 1963.
52. T. L. Gunckel, Orbit determination using Kalman's method. *J. Inst. Navigation* **10** (3), 213-291 (1963).
53. R. B. Blackman, Methods of orbit refinement. *Bell System Tech. J.* **43(3)** (1964).
54. A. J. Claus, Orbit determination in the presence of systematic errors, *in* "Progress in Astronautics and Aeronautics," Vol. 14. Academic Press, New York, 1964.
55. H. M. Dusek, Theory of error compensation in astro-inertial guidance systems for low-thrust space missions, *in* "Progress in Astronautics and Aeronautics," Vol. 13. Academic Press, New York, 1964.
56. H. Kahn, Application of Monte Carlo. *Rand Corp. Mem. RM-1237*, 1956.
57. E. E. Pentecost, Synthesis of computationally efficient sequential linear estimators. Univ. Calif. Ph.D. Dissertation, Los Angeles, California, June 1965.
58. W. T. Reid, A matrix differential equation of Ricatti type. *Am. J. Math.* **68**, 237–246 (1946).
59. M. Fernandez, and G. R. Macomber, "Inertial Guidance Engineering." Prentice-Hall, Englewood Cliffs, New Jersey, 1962.
60. J. Tschauner, and P. Hempel, Rendezvous zu einem in Elliptischer Bahn Umlaufenden Ziel. *Astronaut. Acta* **11**(**2**), 104-109 (1965).
61. A. E. Bryson, and M. Frazier, Smoothing for linear and non-linear dynamic systems. *Wright-Patterson Air Force Base, Ohio, Aeronautical Systems Div. TDR-63-119*, pp. 353–364, 1962.

Section I-D
Model Errors and Divergence

IT is apparent that any mathematical model of a system is an approximation to reality. The results obtained from using the Kalman filter in a computer simulation, based on models of the system, often provide very optimistic results. The error variances provided by the Riccati equation generally diminish to values that indicate that system objectives can be achieved. To temper these results with a dash of realism, an integral part of any simulation study is the consideration of the effects of modeling errors on the actual behavior of the estimates.

Divergence is said to occur when the covariance matrix produced by the Kalman filter $P_{k/k}$ no longer provides a reasonable characterization of the error in estimate. The theoretical study of the effects of model errors on the Kalman filter may have been launched in the journal literature with a paper by Soong, the first paper collected in this section. In this paper, Soong considers the effects of errors in the prior statistics provided to the filter for a single-stage operation. For the analysis, he defines three *covariance* matrices for the problem: the *calculated* covariance produced by evaluating the Kalman filter using prescribed, but incorrect, prior statistics; the *actual* covariance of the filter error; and the *true* covariance that could be obtained were the filter to use correct values for the prior statistics. The *only* errors considered in the paper appear in the initial state covariance $P_{1/0}$. Explicit relations for the differences between the calculated and the actual or true covariances are derived.

Soong's paper was soon followed by a paper by Nishimura, a colleague at JPL, who considered the multistage problem. He established inequalities for quadratic forms based on the three types of covariance matrices defined by Soong. Again, Nishimura considered errors only in the prior covariance for the initial state. This work was followed closely by the publication by Heffes of the third paper printed below. Heffes extended the considerations to include errors in the covariance matrices defined for the input noise and the measurement noise.

The papers by Soong, Nishimura, and Heffes only consider errors in the prior covariance or in the input and measurement noise covariances. These errors, while producing incorrect covariances from the Kalman filter, can be compensated for by choosing filter covariances that are sufficiently large. Thus, they do not have to cause divergence in the sense that the Kalman covariance represents an optimistic approximation of the actual error covariance. Effects that can cause a serious divergence problem were discussed increasingly and the term "divergence" may have first appeared in journal publications with the appearance of the paper by Schlee, Standish, and Toda.

Schlee, Standish and Toda consider modeling errors in which incorrect values for parameters are used for the filter model. They also consider the effects of computer round-off errors. Recursions are derived with which the state error covariances, state and parameter cross-covariances, and parameter covariances can be computed in a digital simulation of a system. Then, simulation results for an orbit determination problem, including two dynamic model parameter errors. In this paper, the authors suggest ways of compensating for modeling errors and present results that demonstrate the success of their approach.

In 1971, Fitzgerald published the final paper included in this section. A thorough analysis of the divergence problem that subsumes the analyses of the earlier papers is presented. Fitzgerald examines the asymptotic behavior of the errors and incorporates filter stability properties based on the observability and controllability of the system. He distinguishes between *true* divergence (i.e., errors may become unbounded) and *apparent* divergence (i.e., finite filter errors). The results of the analysis are supplemented by a scalar example which provides both analytical and numerical results. Finally, he discusses *numerical* divergence with brief mention of square-root formulations that overcome or at least reduce these difficulties. In some measure, this paper can be regarded as a summary of the considerations and progress in the development of the Kalman filtering technique during the decade following its first appearance.

The rate of appearance of references dealing with the problems addressed in this section increased as the Kalman filter saw broadening application. Because the problems must be addressed with each implementation, work continues to this day, and the applications presented in Part II of this book consider, in varying degrees, the effects of modeling errors on their algorithm. A representative list of useful papers preceding Fitzgerald's paper are given in the following references.

References

[1] S. L. Fagin, "A unified approach to the error analysis of augmented dynamically exact inertial navigation systems," *IEEE Trans. Aerosp. Navig. Elect.*, vol. ANE-11, pp. 234–248.

[2] ——, "Recursive linear regression theory, optimal filter theory, and error analysis of optimal systems," *IEEE Int. Conv. Record*, vol. 12, pp. 216–240.

[3] T. Nishimura, "Error bounds of continuous Kalman filters and the application to orbit determination problems," *IEEE Trans. Automat. Contr.*, vol. AC-12, pp. 268–275, 1967.

[4] R. E. Griffin and A. P. Sage, "Large and small scale sensitivity analysis of optimum estimation algorithms," *IEEE Trans. Automat. Contr.*, vol. AC-13, pp. 320–329, 1968.

[5] C. F. Price, "An analysis of the divergence problem in the Kalman filter," *IEEE Trans. Automat. Contr.*, vol. AC-13, pp. 699–702,

[6] W. F. Denham and S. Pines, "Sequential estimation when measurement function nonlinearity is comparable to measurement error," *AIAA J.*, vol. 4, pp. 1071–1076, 1966.

[7] H. Kwaakernaak, "Sensitivity analysis of discrete Kalman filters," *Int. J. Contr.*, vol. 12, pp. 657–699, 1970.

[8] R. E. Griffin and A. P. Sage, "Sensitivity analysis of discrete filtering and smoothing algorithm," *AIAA J.*, vol. 10, pp. 1890–1897, 1969.

[9] J. A. D'Appolito and C. A. Hutchinson, "Low sensitivity filters for state estimation in the presence of large parameter uncertainties," *IEEE Trans. Automat. Contr.*, vol. AC-14, pp. 310–312, 1969.

T. T. SOONG[2]
Senior Research Engineer,
Jet Propulsion Laboratory,
California Institute of Technology,
Pasadena, Calif.

On *A Priori* Statistics in Minimum-Variance Estimation Problems[1]

Simple general formulas are derived for investigating the effect of errors in a priori statistics on the minimum-variance estimates of linear regression parameters from observations obscured by noise. These formulas permit a direct evaluation of the covariance matrix of the errors of a posteriori estimates, showing the sensitivity to errors in a priori weighting matrix. A simple example illustrates that, for slight variations in the assumed a priori statistics, the calculated a posteriori error standard deviations of the estimates can deviate substantially from the correct values.

Introduction

THE present work is concerned with the minimum-variance estimation of linear regression parameters from a set of observations contaminated with noise. In the application of the theory of optimal estimation to practical problems in filtering and prediction, orbit determination, optimal control and other related areas, the optimality of the estimation process depends upon the completeness of the mathematical model representing dynamics of the system involved, and our *a priori* knowledge of all uncertainties entering the system and its environment. In reality, the increasing complexity of dynamical systems makes the task of specifying accurately the mathematical model extremely difficult, as well as impractical. One of the major causes of imperfection in the model can be attributed to the errors introduced, either unintentionally or deliberately, to the *a priori* statistics employed in the estimation process about the uncertainties present in the system and the environment within which the system operates.

For example, an important application of the theory of linear estimation arises in the estimation of deep-space trajectories from noisy radar tracking or on-board optical observations [1, 2].[3] While the estimation of orbital parameters (six position and velocity components at a specified epoch, for instance) is of our primary concern, the uncertainties in the nonorbital parameters (physical constants, biases in observations, tracking station coordinates, and so on) also must be considered for the estimates to be optimum. In assigning *a priori* statistics to the uncertainties in this complex set of parameters, errors are inevitable as uncertainties in certain parameters are either deliberately ignored because of computational limitations in handling vectors and matrices of large dimensions, or unintentionally ignored as a result of our unawareness in their presence. Furthermore, the assigned *a priori* data for the parameters considered are based on our limited information on hand which may deviate considerably from the "true" information. Hence, it is of practical importance to examine the effect of errors in *a priori* data on the optimality of the calculated results and, more important, on the deviation between the calculated and actual *a posteriori* statistics of the estimated parameters.

[1] This paper presents the results of one phase of research carried out at the Jet Propulsion Laboratory, California Institute of Technology, under Contract No NAS 7-100, sponsored by the National Aeronautics and Space Administration.

[2] Now Assistant Professor, Division of Interdisciplinary Studies and Research, State University of New York at Buffalo, Buffalo, N. Y.

[3] Numbers in brackets designate References at end of paper.

Contributed by the Automatic Control Division of THE AMERICAN SOCIETY OF MECHANICAL ENGINEERS and presented at the Joint Automatic Control Conference, Stanford, Calif., June 24–26, 1964. Manuscript received at ASME Headquarters, September 6, 1963.

Our purpose in this paper is to present formulas for investigating this problem area for a linear dynamic model described in the next section. The formulas presented are completely general and quite simple to manipulate. The application of these formulas to various practical situations is illustrated by an example.

Basic Equations

The basic concept of a linear, unbiased, minimum-variance estimate is introduced for the analysis to follow. We state without proof a version of the well-known Gauss-Markov theorem [3].

Theorem. Given two column random vectors x and y (not necessarily of the same dimension) with zero means, then the linear, unbiased, minimum-variance estimate, y^*, of y given x is

$$y^* = \langle yx^T \rangle \langle xx^T \rangle^{-1} x \tag{1}$$

where x^T denotes the transpose of the vector x, and the angular brackets $\langle \ \rangle$ denote the expected value operator. Note that the estimate y^* is unbiased ($\langle y^* \rangle = 0$).

The estimate y^* given by equation (1) is the minimum-variance estimate of y in the sense that the covariance matrix of the error of the estimate is a minimum among the error covariance matrices of all other linear unbiased estimates of y. Equivalently, given any other linear unbiased estimate, y', of y, the quadratic form associated with

$$\langle (y^* - y)(y^* - y)^T \rangle - \langle (y' - y)(y' - y)^T \rangle$$

is nonnegative definite.

The covariance matrix $\langle (y^* - y)(y^* - y)^T \rangle$, denoted by $\langle \epsilon\epsilon^T \rangle$, is given by

$$\langle \epsilon\epsilon^T \rangle = \langle yy^T \rangle - \langle yx^T \rangle \langle xx^T \rangle^{-1} \langle xy^T \rangle \tag{2}$$

We are concerned here with the application of the theorem just stated to the dynamic model described next:

Let a set of observations be denoted by a u-vector ϕ, and q denotes a set of p-parameters ($p \leqslant u$) to be estimated. Consider a physical process described by

$$\delta\phi = A\delta q + n \tag{3}$$

where δq is the *unknown* variation of q from its *a priori* estimate q_0, $\delta\phi$ the observed variation of ϕ from its reference value, A is a $u \times p$ matrix of regression coefficients, and n denotes a u-vector of zero-mean random noise associated with the observations. The elements of the matrix A are assumed known and nonrandom functions of time.

While equation (3) assumes linear mapping characteristics between noise-free observables and parameters to be estimated, the equation is also fundamental as linear approximation to problems such as orbit determination where observables are in general nonlinear functions of the state variables and parameters [1]. The solution to nonlinear regression can be obtained by iteratively solving equation (3).

Reprinted with permission from *Trans. ASME, J. Basic Eng.*, Series 87D, pp. 109–112, Mar. 1965.

If the *a priori* covariance matrix about the δq is denoted by $\Lambda_0\ (=\langle\delta q\delta q^T\rangle)$, the noise moment matrix by $\lambda(=\langle nn^T\rangle)$, and the covariance matrix of the error of the estimate by $\Lambda(=\langle\epsilon\epsilon^T\rangle)$, equations (1) and (2) then lead to

$$\delta q^* = [\Lambda_0^{-1} + A^T\lambda^{-1}A]^{-1}A^T\lambda^{-1}\delta\phi \tag{4}$$

$$\Lambda = [\Lambda_0^{-1} + A^T\lambda^{-1}A]^{-1} \tag{5}$$

It is noted in passing that the δq^* given by equation (4) is also the maximum likelihood estimator for δq if the noise is Gaussian; it is the same as the least-square estimate when n is white noise.

Equations (4) and (5) provide means of calculating the minimum-variance estimate δq^* and the *a posteriori* error covariance matrix Λ provided, among other considerations,[4] that the *a priori* covariance matrix employed, Λ_0, is correct. Since the matrix Λ_0 is inevitably in error, the estimate δq^* as calculated from equation (4) is not truly optimum. In particular, equation (5) does not give correct *a posteriori* error covariance matrix for δq^* when incorrect *a priori* data are used in the estimation procedure. Hence, it is of practical value to consider the following:

(i) Calculated $\Lambda(\Lambda_c)$. Ignoring errors in *a priori* data employed, Λ_c is calculated from equation (5) assuming employed *a priori* is the true *a priori* information.

(ii) Actual $\Lambda(\Lambda_a)$. The *a priori* data employed are used in making the estimate from equation (4). Λ_a is the actual *a posteriori* error covariance matrix of the estimate; i.e.,

$$\Lambda_a = \langle(\delta q^* - \delta q)(\delta q^* - \delta q)^T\rangle$$

(iii) True $\Lambda(\Lambda_t)$. The true *a posteriori* covariance matrix associated with the minimum-variance estimate using true *a priori* statistics.

Principal Results

Let the employed *a priori* covariance matrix of δq be denoted by Λ_0'. For convenience in carrying out matrix manipulation, we shall relate Λ_0' to Λ_0, the true *a priori* covariance matrix, by the equation

$$\Lambda_0'^{-1} = \Lambda_0^{-1} + E \tag{6}$$

where E is a square $p \times p$ matrix, not necessarily nonsingular. The only physical restriction on E is that, due to nonnegative definite nature of covariance matrices, E has a "lower bound" equal to $-\Lambda_0^{-1}$.

Since the inverses $\Lambda_0'^{-1}$ and Λ_0^{-1} always exist in practice, the use of the matrix E to describe the errors in Λ_0 constitutes no restriction. In fact, if one desires to express Λ_0' by

$$\Lambda_0' = \Lambda_0 + F$$

instead of equation (6), it can be verified easily that the matrix E can be calculated from F by

$$E = -\Lambda_0^{-1}F(\Lambda_0 + F)^{-1} = -\Lambda_0^{-1}F\Lambda_0'^{-1} \tag{7}$$

Returning to definition (6) it is clear from equation (5) that

$$\Lambda_c = (E + K)^{-1} \tag{8}$$

and

$$\Lambda_t = K^{-1} \tag{9}$$

where

$$K = \Lambda_0^{-1} + A^T\lambda^{-1}A$$

The form of Λ_a is, of course, our main concern here. In particular, we are interested in the differences $\Lambda_a - \Lambda_c$ and $\Lambda_a - \Lambda_t$.

The derivation of Λ_a is given in the Appendix. After a great deal of manipulation of matrices we obtain the results in the following simple forms:

$$\Lambda_a - \Lambda_c = (E + K)^{-1}(I + E\Lambda_0)E(E + K)^{-1} = \Lambda_c(I + E\Lambda_0)E\Lambda_c \tag{10}$$

$$\Lambda_a - \Lambda_t = (E + K)^{-1}E(\Lambda_0 - K^{-1})E(E + K)^{-1} = \Lambda_cE(\Lambda_0 - \Lambda_t)E\Lambda_c \tag{11}$$

also, from equations (10) and (11)

$$\Lambda_t - \Lambda_c = \Lambda_tE\Lambda_c = \Lambda_c(I - E\Lambda_c)^{-1}E\Lambda_c \tag{12}$$

where I is the unit matrix.

Based on the linear dynamical model, equations (10), (11), and (12) are completely general and quite simple to manipulate. Since E is arbitrary, these formulas can be applied to various practical situations in a straightforward and simple fashion.

Before carrying the analysis any further, some comments concerning the general and limiting properties of the formulas are in order. Some of them are seen to conform with intuition.

(i) $\Lambda_a \geqslant \Lambda_t$. This is true by definition; since the estimate associated with Λ_t is the true minimum-variance estimate. This conclusion also can be deduced from equation (11) which indicates that $\Lambda_a - \Lambda_t$ is nonnegative definite for all E.

(ii) $\Lambda_t \geqslant \Lambda_c$ if E is nonnegative definite; $\Lambda_t \leqslant \Lambda_c$ if E is nonpositive definite. The statement follows directly from equation (12) and is intuitively obvious (it is recalled that $E_{\min} = -\Lambda_0^{-1}$). E being nonnegative definite implies that the *a priori* data employed are optimistic for *all* components of δq. The *a priori* employed is conservative when E is nonpositive definite. No general statements can be made, of course, if E is of an indefinite form.

(iii) $\Lambda_a \geqslant \Lambda_c$ if E is nonnegative definite; $\Lambda_a \leqslant \Lambda_c$ if E is nonpositive definite. This follows from equation (10). It is also noteworthy that, as seen from equation (10), $\Lambda_a = \Lambda_c$ not only when $E = 0$ but also when $E = -\Lambda_0^{-1}(\Lambda_0' \to \infty)$. This indicates that in the event when no *a priori* data are given in the determination of δq^*, $\Lambda_a = \Lambda_c = (A^T\lambda^{-1}A)^{-1}$ irrespective of the true *a priori* information.

(iv) As $E \to \infty$, equations (10), (11), and (12) reduce to

$$\Lambda_a = \Lambda_0$$
$$\Lambda_c = 0$$
$$\Lambda_t = K^{-1}$$

a well-known result when δq is assumed perfectly known.

A Numerical Example

In this section the formulas (10), (11), and (12) are discussed on a quantitative basis. A simple example is given here so that the problem can be studied parametrically, and some essential features can be deduced from the results.

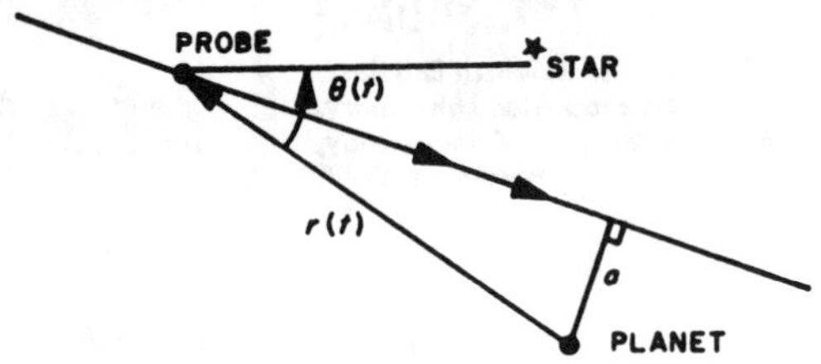

Fig. 1 Geometry of probe trajectory

Consider the motion of a deep-space probe approaching a target planet as shown in Fig. 1. The motion is assumed linear with known velocity. Our problem is to determine the miss distance a, from noisy cone angle data $\theta(t)$ measured between the center of the target planet and a star whose line of sight is at a fixed constant angle from the probe trajectory. The only uncertainties considered are those in the miss distance, a, and a constant bias, b,

[4] The degradation of the optimal estimation process may also be caused by, for example, the errors in the noise moment matrix. This has been studied extensively in the literature (e.g., see [1], p. 87).

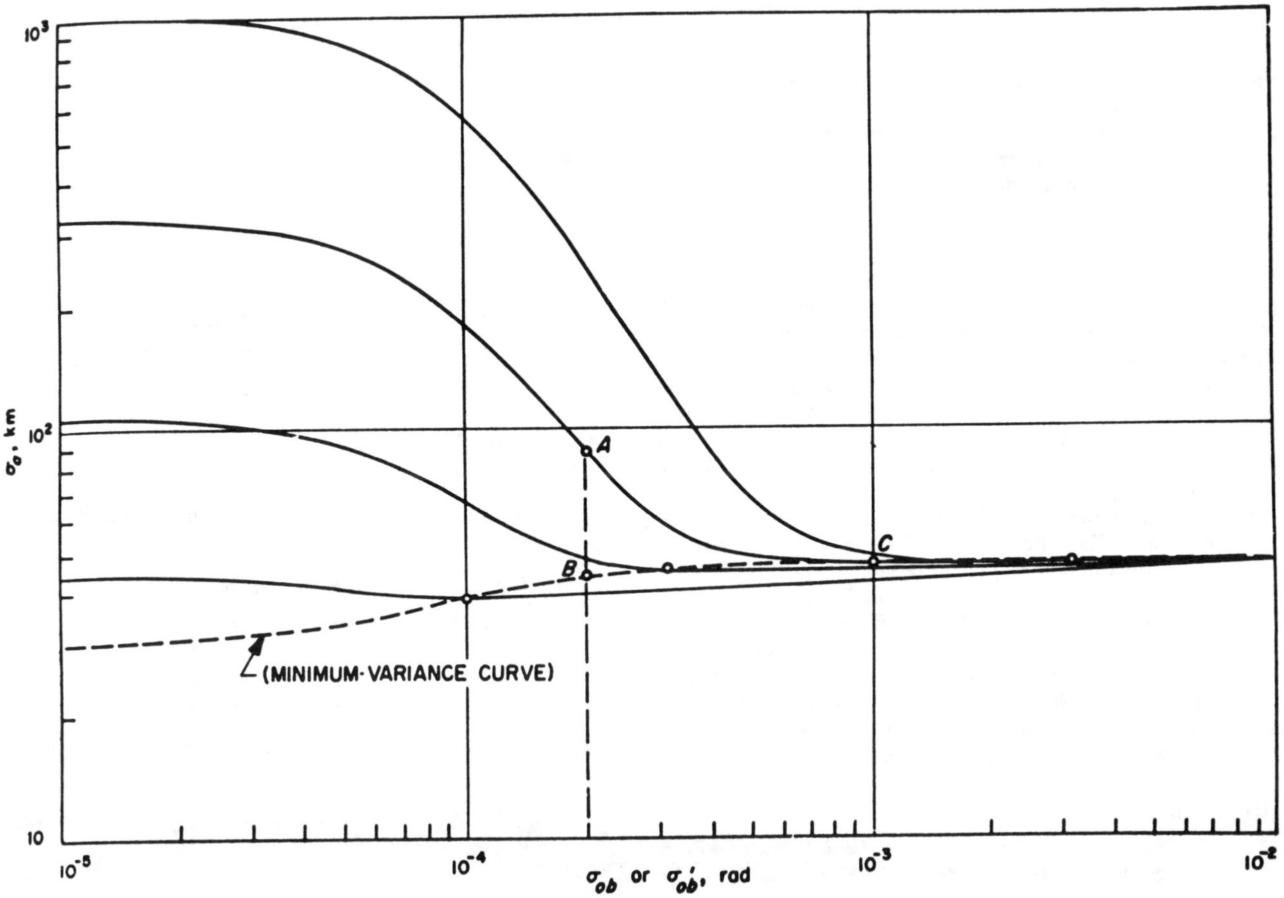

Fig. 2 Comparison of σ_{ac}, σ_{aa}, and σ_{at}

in cone-angle measurements. Finally, only two angle measurements are assumed taken at planetocentric distances r_1 and r_2 whose values are much larger than a.

The regression equation for this simple case is

$$\delta\theta = A\delta q + n \cong \begin{bmatrix} \dfrac{1}{r_1} & 1 \\ \dfrac{1}{r_2} & 1 \end{bmatrix} \begin{bmatrix} \delta a \\ \delta b \end{bmatrix} + \begin{bmatrix} n_1 \\ n_2 \end{bmatrix} \tag{13}$$

where we assume that

$$\lambda = \langle nn^T \rangle = \sigma^2 I$$

$$\Lambda_0 = \begin{bmatrix} \sigma_{0a}^2 & 0 \\ 0 & \sigma_{0b}^2 \end{bmatrix}$$

and

$$\Lambda_0' = \begin{bmatrix} \sigma_{0a}'^2 & 0 \\ 0 & \sigma_{0b}'^2 \end{bmatrix}$$

We shall further set $\sigma_{0a} = \sigma_{0a}'$ and examine only σ_a, the *a posteriori* standard deviation of the error in δa^*, when the assigned *a priori* standard deviation for the bias is different from the true value ($\sigma_{0b}' \neq \sigma_{0b}$). The calculated $\sigma_a(\sigma_{ac})$, true $\sigma_a(\sigma_{at})$, and actual $\sigma_a(\sigma_{aa})$ for a wide range of values of σ_{0b} and σ_{0b}' are computed in a straightforward manner from equations (10), (11), and (12) and results are presented graphically in Fig. 2. The following numerical values are used in the computation: $\sigma^2 = 10^{-8}$ rad², $\sigma_{0a}^2 = \sigma_{0a}'^2 = 10^7$ km², $r_1 = 3 \times 10^6$ km, and $r_2 = 3 \times 10^5$ km.

The values of σ_{at} or σ_{ac} are presented by the dashed curve, depending upon whether the abscissa is interpreted as σ_{0b} or σ_{0b}'. The solid curves represent values of σ_{aa}. The ordinate of a point on a solid curve gives the value of σ_{aa} when σ_{0b}' is equal to the abscissa of that point and σ_{0b} is the abscissa of the point of intersection of the solid curve and the dashed curve. To be more precise, let us consider three points A, B, and C, in Fig. 2. The ordinates of A, B, and C give, respectively, the values of σ_{aa}, σ_{ac}, and σ_{at} when the assigned *a priori* standard deviation for the bias is 2×10^{-4} radians where the true *a priori* is 1×10^{-3} radians. These solid curves clearly indicate that the actual values of σ_a can be significantly larger than the calculated ones when unduely optimistic *a priori* data are assumed for the bias.

It is seen in Fig. 2 that all solid curves converge to the dashed curve as σ_{0b}' becomes large, as observed in the preceding section. On the other hand, the difference between σ_{ac} and σ_{aa} becomes large as σ_{0b}' becomes smaller than σ_{0b}. It is of interest to calculate this difference in the limiting case when σ_{0b}' becomes zero (equivalently, in the case when we ignore the uncertainty in the bias). We find from equations (6) and (10) that, when $\sigma_{0b}' \to 0$

$$E = \lim_{\Delta\to\infty} \begin{bmatrix} 0 & 0 \\ 0 & \Delta \end{bmatrix}$$

and

$$\Lambda_a - \Lambda_c = \Lambda_c(I + E\Lambda_0)E\Lambda_c$$

$$= \sigma_{0b}^2 \begin{bmatrix} \dfrac{r_1^2 r_2^2 (r_1 + r_2)^2}{(r_1^2 + r_2^2)^2} & -\dfrac{r_1 r_2 (r_1 + r_2)}{r_1^2 + r_2^2} \\ -\dfrac{r_1 r_2 (r_1 + r_2)}{r_1^2 + r_2^2} & 1 \end{bmatrix} \tag{14}$$

Equation (14) becomes

$$\Lambda_a - \Lambda_c \cong \sigma_{0b}^2 \begin{bmatrix} r_2^2 & -r_2 \\ -r_2 & 1 \end{bmatrix} \tag{15}$$

if $r_2 \ll r_1$.

The physical interpretation of equation (15) is clear. The error in the variance σ_a^2 by ignoring the uncertainty in the bias is the additional variance of the arc due to the true *a priori* uncertainty in the angle bias, with radius r_2 if $r_2 \ll r_1$. Incidentally, equation (15) indicates that the error in σ_b^2 is σ_{0b}^2. This is obvious since the calculated σ_b^2 is zero and the actual σ_b^2 becomes σ_{0b}^2 when no uncertainty is assumed for the bias.

Conclusion

Based on a linear dynamical model defined by equation (3), general formulas have been presented for the study of the effect of errors in *a priori* information on the minimum-variance estimation procedures. These formulas permit a simple and direct evaluation of the calculated result regarding its optimality, and its sensitivity and accuracy in describing the actual *a posteriori* statistics of the estimated parameters. It is noteworthy that equations (10), (11), and (12) are functions of only Λ_0', Λ_0, and Λ_e.

In addition to the classical approach to minimum-variance estimation as outlined here, it should be pointed out that an equivalent estimation procedure has been developed by Kalman [4] for linear processes excited by uncorrelated error sources. The technique has been extended by Pfeiffer [5] to "sequentially correlated" processes. In applying this approach to practical problems, accurate *a priori* information is essential since the most recent estimate is "weighted" against new observational data at each step in the iterative estimation procedure. The formulas presented in this paper have an important application in this case.

References

1 I. I. Shapiro, *The Prediction of Ballistic Missile Trajectories From Radar Observations*, McGraw-Hill Book Company, Inc., New York, N. Y., 1958.

2 R. E. Carr and R. H. Hudson, "Tracking and Orbit-Determination Program of the Jet Propulsion Laboratory," TR 32-7, Jet Propulsion Laboratory, Pasadena, Calif., 1960.

3 H. Scheffe, *The Analysis of Variance*, John Wiley & Sons, Inc., New York, N. Y., 1959, p. 14.

4 R. E. Kalman, "A New Approach to Linear Filtering and Prediction Theory," JOURNAL OF BASIC ENGINEERING, TRANS. ASME, Series D, vol. 82, 1960, pp. 35–50.

5 C. G. Pfeiffer, "Sequential Estimation of Correlated Stochastic Variables," TR 32-544, Jet Propulsion Laboratory, Pasadena, Calif., 1963.

APPENDIX

Derivation of Λ_a

It is seen from equation (4) that

$$\delta q^* = [\Lambda_0'^{-1} + A^T\lambda^{-1}A]^{-1}A^T\lambda^{-1}\delta\phi \qquad (16)$$

where

$$\delta\phi = A\delta q + n$$
$$\langle nn^T\rangle = \lambda \qquad (17)$$
$$\langle \delta q\delta q^T\rangle = \Lambda_0$$

By definition

$$\Lambda_a = \langle(\delta q^* - \delta q)(\delta q^* - \delta q)^T\rangle$$
$$= \langle\delta q^*\delta q^{*T}\rangle - \langle\delta q^*\delta q^T\rangle - \langle\delta q\delta q^{*T}\rangle + \langle\delta q\delta q^T\rangle \qquad (18)$$

where, from equations (16) and (17)

$$\langle\delta q^*\delta q^{*T}\rangle = (\Lambda_0'^{-1} + A^T\lambda^{-1}A)^{-1}A^T\lambda^{-1}(A\Lambda_0A^T + \lambda)\lambda^{-1}A[\Lambda_0'^{-1} + A^T\lambda^{-1}A]^{-1}$$
$$\langle\delta q^*\delta q^T\rangle = [\Lambda_0'^{-1} + A^T\lambda^{-1}A]^{-1}A^T\lambda^{-1}A\Lambda_0 \qquad (19)$$
$$\langle\delta q\delta q^{*T}\rangle = \Lambda_0A^T\lambda^{-1}A[\Lambda_0'^{-1} + A^T\lambda^{-1}A]^{-1}$$
$$\langle\delta q\delta q^T\rangle = \Lambda_0$$

In view of equation (7) and with the aid of a matrix inversion formula we may write

$$[\Lambda_0'^{-1} + A^T\lambda^{-1}A]^{-1} = [E + \Lambda_0^{-1} + A^T\lambda^{-1}A]^{-1}$$
$$= K^{-1} - K^{-1}[I + EK^{-1}]^{-1}EK^{-1} \qquad (20)$$

where

$$K = \Lambda_0^{-1} + A^T\lambda^{-1}A$$

The substitution of equations (19) and (20) into (18) hence gives, upon simplifying,

$$\Lambda_a = K^{-1} + \Delta_{at} = \Lambda_t + \Delta_{at} \qquad (21)$$

with

$$\Delta_{at} = (E + K)^{-1}E\Lambda_0A^T\lambda^{-1}A(E + K)^{-1}EK^{-1} \qquad (22)$$

which can be further simplified by means of the matrix identity[4]

$$(A + B)^{-1}AB^{-1} = B^{-1}A(A + B)^{-1} \qquad (23)$$

provided the indicated inverses exist. The final form of Δ_{at} can now be written in the form

$$\Delta_{at} = (E + K)^{-1}E(\Lambda_0 - K^{-1})E(E + K)^{-1} \qquad (24)$$

which is equation (11). Equations (10) and (12) follow immediately.

[4] Proof of equation (23). Consider $(A + B)^{-1}(AB^{-1}A + A)(A + B)^{-1} = C$. We see that

$$C = (A + B)^{-1}AB^{-1}(A + B)(A + B)^{-1} = (A + B)^{-1}AB^{-1}$$

Also

$$C = (A + B)^{-1}(A + B)B^{-1}A(A + B)^{-1}$$
$$= B^{-1}A(A + B)^{-1}. \quad \text{QED}$$

The Effect of Erroneous Models on the Kalman Filter Response

H. HEFFES, MEMBER, IEEE

***Abstract*—The optimal filtering equations, as derived by Kalman [1], [2], require the specification of a number of models for a given application. This paper concerns itself with the effect of errors in the assumed models on the filter response. The types of errors considered are those in the covariance of the initial state vector, the covariance of the stochastic inputs to the system, and the covariance of the uncorrelated measurement noise.**

Presented here is a derivation of a recursive equation for the actual covariance matrix of the estimation error when the filter design is based upon erroneous models. The derived equation can also be used to obtain the covariance matrix of the estimation error when the optimal filter gains are approximated by simple functions of time to be used in a real-time filtering application.

A numerical example illustrates the use of the derived equations.

Introduction

This paper is concerned with the determination of the performance of the suboptimal filter obtained when a Kalman [1] filter design is based upon erroneous noise models and a priori statistics. The performance is specified by the covariance matrix of the suboptimal estimation error. A recursive equation is derived for this matrix which makes possible a quantitative evaluation of the statistical quality of the estimate obtained.

System Model

The system model used is as follows:

$$x(t_{k+1}) = \Phi(t_{k+1}; t_k)x(t_k) + u(t_k) \tag{1}$$

$$y(t_k) = M(t_k)x(t_k) + v(t_k) \tag{2}$$

where

1) $x(t_k)$ is the n vector of states at time t_k

$$E\{x(t_0)\} = 0 \quad \text{and} \quad E\{x(t_0)x'(t_0)\} = P^*(t_0)$$

2) $\Phi(t_{k+1}; t_k)$ is the transition matrix $(n \times n)$ of the system
3) $u(t_k)$ is an n vector of stochastic inputs such that

$$E\{u(t_k)\} = 0 \quad \text{and} \quad E\{u(t_k)u'(t_j)\} = \delta_{k,j}Q(t_k)$$

4) $y(t_k)$ is an m vector of measurements taken at time t_k
5) $M(t_k)$ is an $m \times n$ matrix (Measurement Sensitivity Matrix)
6) $v(t_k)$ is an m vector of additive measurement noise with the following properties:

a) $E\{v(t_k)\} = 0$
b) $E\{v(t_k)v'(t_j)\} = \delta_{k,j}R(t_k)$
c) $E\{u(t_k)v'(t_j)\} = 0.$

Optimal Filter

The optimal filter can be represented as a linear dynamic system by the following system of equations [2]:

$$x^*(t_{k+1} \mid t_k) = \psi^*(t_{k+1}; t_k)x^*(t_k \mid t_{k-1}) + K^*(t_k)y(t_k) \tag{3}$$

$$\psi^*(t_{k+1}; t_k) = \Phi(t_{k+1}; t_k) - K^*(t_k)M(t_k) \tag{4}$$

$$K^*(t_k) = \Phi(t_{k+1}; t_k)P^*(t_k)M'(t_k)[M(t_k)P^*(t_k)M'(t_k) + R(t_k)]^{-1} \tag{5}$$

$$P^*(t_{k+1}) = \psi^*(t_{k+1}; t_k)P^*(t_k)\Phi'(t_{k+1}; t_k) + Q(t_k) \tag{6}$$

where

1) $x^*(t_{k+1}|t_k)$ is the optimal estimate of the state vector $x(t_{k+1})$, having observed $y(t_0), y(t_1), \cdots, y(t_k)$
2) $\psi^*(t_{k+1}; t_k)$ is the transition matrix $(n \times n)$ of the filter
3) $K^*(t_k)$ is an $n \times m$ matrix of optimal gains
4) $P^*(t_{k+1})$ is the covariance matrix of the estimation error.

Manuscript received December 16, 1965; revised April 11, 1966.
The author is with Bell Telephone Laboratories, Inc., Whippany, N. J.

Analytic Results

Now assume that the actual gains employed in the filter design deviate from the optimal gains, i.e., that

$$K(t_k) = K^*(t_k) + \delta K^*(t_k) \tag{7}$$

and

$$\psi(t_{k+1}; t_k) = \psi^*(t_{k+1}; t_k) - \delta K^*(t_k)M(t_k). \tag{8}$$

In order to determine the effect of these errors we ask the following question. How close do we come to the optimum by using the following filter equation

$$\hat{x}(t_{k+1} \mid t_k) = \psi(t_{k+1}; t_k)\hat{x}(t_k \mid t_{k-1}) + K(t_k)y(t_k) \tag{9}$$

where $K(t_k)$ and $\psi(t_{k+1}; t_k)$ are given by (7) and (8), respectively?

The answer is given by the covariance matrix of the estimation error obtained when (9) is used instead of (3). This matrix, denoted by $P_a(t)$, indicates the statistical quality of the suboptimal estimate. We now proceed to derive a recursive equation for the suboptimal covariance matrix $P_a(t)$.

Consider the error in the suboptimal estimate:

$$x_e(t_{k+1} \mid t_k) = x(t_{k+1}) - \hat{x}(t_{k+1} \mid t_k).$$

Using both the system model (1) and (9) we have

$$x_e(t_{k+1} \mid t_k) = \Phi(t_{k+1}; t_k)x(t_k) + u(t_k) - \psi(t_{k+1}; t_k)\hat{x}(t_k \mid t_{k-1}) - K(t_k)y(t_k). \tag{10}$$

Using (2) we obtain

$$x_e(t_{k+1} \mid t_k) = [\Phi(t_{k+1}; t_k) - K(t_k)M(t_k)]x(t_k) - \psi(t_{k+1}; t_k)\hat{x}(t_k \mid t_{k-1}) + u(t_k) - K(t_k)v(t_k).$$

Using (4), (7), and (8) we obtain

$$x_e(t_{k+1} \mid t_k) = \psi(t_{k+1})x_e(t_k \mid t_{k-1}) + u(t_k) - K(t_k)v(t_k).$$

The covariance matrix of the error is given by

$$P_a(t_{k+1}) = E\{x_e(t_{k+1} \mid t_k)x_e'(t_{k+1} \mid t_k)\}.$$

Now $x_e(t_k|t_{k-1})$ is a linear combination of the vectors

$$u(t_0), u(t_1), \cdots, u(t_{k-1}), v(t_0), v(t_1), \cdots, v(t_{k-1}).$$

Since $u(t_k)$ and $v(t_k)$ are orthogonal to these vectors and to each other, the orthogonality of $x_e(t_k|t_{k-1})$, $u(t_k)$, and $K(t_k)v(t_k)$ follows. Using these facts we have

$$P_a(t_{k+1}) = \psi(t_{k+1}; t_k)P_a(t_k)\psi'(t_{k+1}; t_k) + Q(t_k) + K(t_k)R(t_k)K'(t_k) \tag{11}$$

with $P_a(t_0) = P^*(t_0)$. Equation (11) represents the desired recursion.

In order to discuss the sources of error in the optimal gains we introduce the following:

1) $\hat{P}(t_0)$ is the covariance of the initial state used in the filter design rather than the correct $P^*(t_0)$
2) $\hat{Q}(t)$ is the covariance of the stochastic input used in the filter design rather than the correct $Q(t)$
3) $\hat{R}(t)$ is the covariance of the uncorrelated measurement noise used in the filter design rather than the correct $R(t)$
4) let $P_c(t)$ represent the matrix computed from (4), (5), and (6), assuming that the employed models, $\hat{P}(t_0)$, $\hat{Q}(t)$ and $\hat{R}(t)$ are correct.

The design based upon the incorrect models will yield the following:

$$P_c(t_{k+1}) = \psi(t_{k+1}; t_k)P_c(t_k)\Phi'(t_{k+1}; t_k) + Q(t_k) \tag{12}$$

with

$$P_c(t_0) = \hat{P}(t_0) \tag{12a}$$

and

$$\psi(t_{k+1}; t_k) = \Phi(t_{k+1}; t_k) - K(t_k)M(t_k) \tag{13}$$

$$K(t_k) = \Phi(t_{k+1}; t_k)P_c(t_k)M'(t_k)[M(t_k)P_c(t_k)M'(t_k) + \hat{R}(t_k)]^{-1}. \tag{14}$$

Reprinted from *IEEE Trans. Automat. Contr.*, vol. AC-11, pp. 541–543, July 1966.

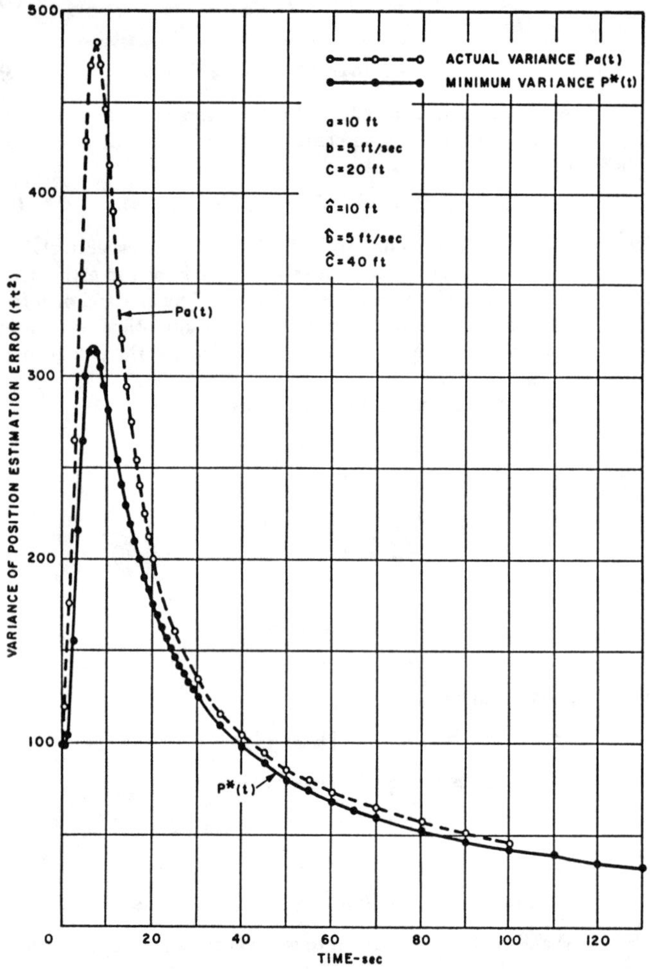

Fig. 1. Comparison between minimum and actual variance of position estimation error.

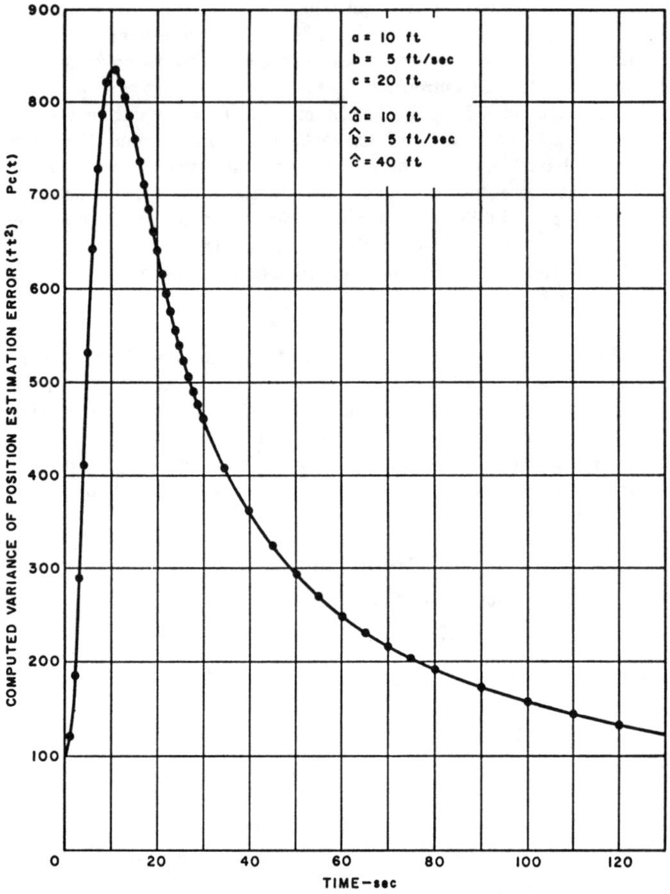

Fig. 2. Computed variance of position estimation error.

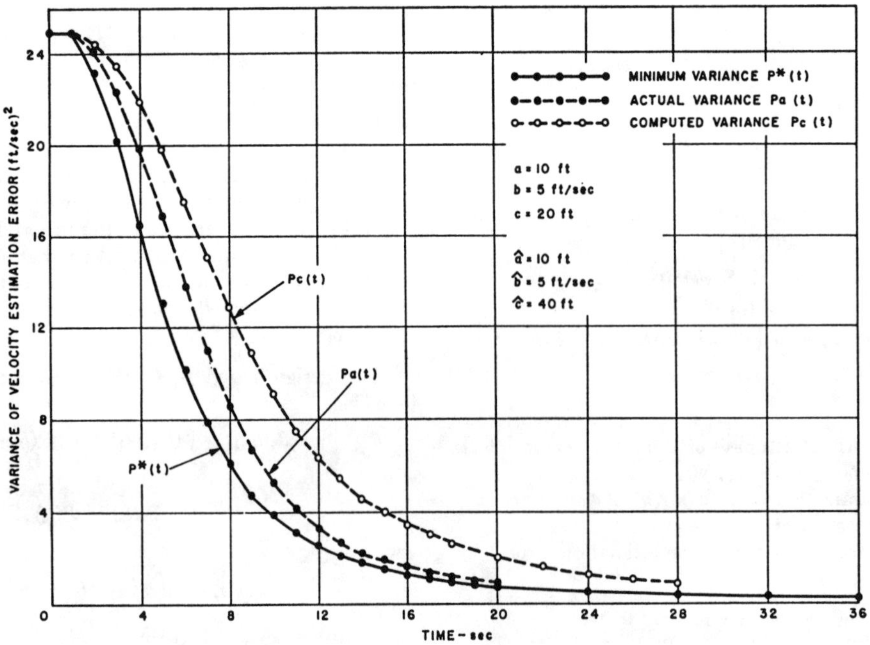

Fig. 3. Comparison of minimum, actual, and computed variance of velocity estimation error.

When the transition matrix $\psi(t_{k+1}; t_k)$ obtained from (12), (13), and (14) is substituted into (11), the actual covariance of the estimation error is obtained.

Numerical Example

To illustrate the use of the derived equations we consider a constant velocity system

$$\begin{bmatrix} x_1(t+1) \\ x_2(t+1) \\ x_3(t+1) \end{bmatrix} = \begin{bmatrix} 1 & 1 & 0 \\ 0 & 1 & 0 \\ 0 & 0 & \frac{1}{2} \end{bmatrix} \begin{bmatrix} x_1(t) \\ x_2(t) \\ x_3(t) \end{bmatrix} + \begin{bmatrix} 0 \\ 0 \\ u(t) \end{bmatrix}$$

whose position measurement is corrupted by correlated, additive, measurement noise $x_3(t)$, i.e.,

$$M(t) = [1 \quad 0 \quad 1].$$

The state variables x_1, x_2, and x_3 represent position, velocity, and correlated measurement noise, respectively. Let

$$P^*(0) = \begin{bmatrix} a^2 & 0 & 0 \\ 0 & b^2 & 0 \\ 0 & 0 & c^2 \end{bmatrix} = \begin{bmatrix} 100 & 0 & 0 \\ 0 & 25 & 0 \\ 0 & 0 & 400 \end{bmatrix}$$

and

$$\hat{P}(0) = \begin{bmatrix} \hat{a}^2 & 0 & 0 \\ 0 & \hat{b}^2 & 0 \\ 0 & 0 & \hat{c}^2 \end{bmatrix} = \begin{bmatrix} 100 & 0 & 0 \\ 0 & 25 & 0 \\ 0 & 0 & 1600 \end{bmatrix}.$$

The correlated noise $x_3(t)$ is generated by

$$x_3(t+1) = \tfrac{1}{2}x_3(t) + u(t).$$

Now $x_3(0)$ can be written

$$x_3(0) = \sum_{k=-\infty}^{-1} \left(\frac{1}{2}\right)^{-k-1} u(k).$$

By means of the properties of the independent noise it is found that

$$c^2 = E\{x_3^2(0)\} = \frac{4}{3} E\{u^2\}.$$

Thus

$$Q(t) = \begin{bmatrix} 0 & 0 & 0 \\ 0 & 0 & 0 \\ 0 & 0 & 300 \end{bmatrix}$$

and

$$\hat{Q}(t) = \begin{bmatrix} 0 & 0 & 0 \\ 0 & 0 & 0 \\ 0 & 0 & 1200 \end{bmatrix}.$$

The effect of the use of the erroneous model on the filter response is shown in Figs. 1, 2, and 3.

Figure 1 is a comparison between the minimum and the actual variance of the position estimation error. The incorrect design is seen to increase the peak variance, which can be of importance for a short filtering interval. For smoothing times greater than 20 seconds the estimation errors are comparable.

Figure 2 shows the computed variance of the position estimation error, which is seen to differ significantly from the actual estimation error.

Figure 3 shows a comparison of the minimum, actual, and computed variances of the velocity estimation error.

Conclusion

A general tool for error analysis in filter design has been devised which makes possible the quantitative evaluation of the effect of errors in the assumed models on the filter response. In a recent paper, T. Nishimura [3] has considered the effect of incorrect initial covariance matrix upon the filter response. The results obtained here agree with those obtained by Nishimura but are extended to the case of incorrect noise models. The use of the derived equation for erroneous noise models serves as an important extension of his results. The numerical example illustrates the use of the derived equation for erroneous noise models.

The important cases of an incorrect plant description and incorrect measurement sensitivity matrix have not been considered here. They are important sources of error, especially in nonlinear systems and in systems where the measurement matrix is obtained from geometry, and are accordingly the subject of present research [4].

Note Added in Proof: The author has recently become aware of Fagin's work which considered the above problem as well as that considered in [3].

References

[1] R. E. Kalman, "A new approach to linear filtering and prediction problems," *J. Basic Engrg., Trans. ASME*, pp. 35–45, March 1960.
[2] ——, "New methods in Weiner filtering theory," *Proc. 1st Symp. on Engrg. Applications of Random Function Theory and Probability.* New York: Wiley, 1963, pp. 270–388.
[3] T. Nishimura, "On the a priori information in sequential estimation problems," *1965 Proc. NEC*, vol. 21, pp. 511–516.
[4] S. L. Fagin, "Recursive linear regression theory, optimal filter theory, and error analyses of optimal systems," *1964 IEEE Conv. Rec.*, pp. 216–240.

On the a priori Information in Sequential Estimation Problems

T. NISHIMURA, MEMBER, IEEE

***Abstract*—In this paper, the effect of errors in the a priori information is studied when the sequential estimations are carried out on the states of linear systems disturbed by white noise. Four theorems are derived to describe the mutual relations among the three covariance matrices, namely the optimum, calculated, and actual covariance matrices, where the last two are based on the incorrect a priori information. By finding the upper bound for the variance of the actual estimate, performance of the Kalman filter is prescribed and the knowledge is utilized for design of the combined system of analog and digital filters. A phase-locked loop receiver is used as an example of analog filter and the considerable improvement on the estimation process is deduced by the theory and it is confirmed by the experimental simulation on the digital computer.**

I. Introduction

THE OPTIMAL FILTER which has been introduced by Kalman[1,2] into the field of system theory yields the minimum variance estimate of states of linear systems, which are contaminated by white Gaussian noises, when a set of sequential observations are carried out. The basic feature of this filter is that the estimate of states is up-dated by a sequence of observations so as to minimize its variance, or equivalently, if noise is Gaussian, to maximize the conditional probability density of the current states after having a set of observations.

The original theory assumes an a priori information on the initial states and its variances. However, there are cases when this a priori information may contain certain errors or some information may not be available at the beginning of the estimating process.

Soong[3] examined the effect of errors in the a priori data on the a posteriori variance of estimates for the single-stage case, i.e., for the nonsequential case in which estimation is carried out after a sequence of observation has been completed, and derived the deviations of the calculated and actual variances from the true minimum variance. Thus no real-time estimation is performed in this case. Also the state being estimated in this model is a constant state vector so that the state transition matrix is an identity matrix and no system noise is assumed there.

The same problem is studied in this paper for the multistage case, namely for the case when sequential observations and estimations are performed on the state in real-time. Also a general transition form of states is assumed subject to excitation of the system noise. Then there is an investigation of how the error in the a priori information propagates through the sequential estimation process and how it affects the final result of estima-

Manuscript received June 17, 1965; revised January 18, 1966. This paper represents one phase of research carried out at the Jet Propulsion Laboratory, California Institute of Technology, under NASA Contract NAS 7-100.

The author is with the Jet Propulsion Laboratory, California Institute of Technology, Pasadena, Calif.

[1] R. E. Kalman, "A new approach to linear filtering and prediction problems," *Trans. ASME, ser. D, J. Basic Engrg.*, vol. 82, pp. 35–45, March 1960.

[2] R. E. Kalman and R. S. Bucy, "New results in linear filtering and prediction theory," *Trans. ASME, ser. D, J. Basic Engrg.*, vol. 83, pp. 95–107, March 1961.

[3] T. T. Soong, "On a priori statistics in minimum variance estimation problems," *Trans. ASME, ser. D, J. Basic Engrg.*, vol. 87, pp. 109–112, March 1965.

Reprinted from *IEEE Trans. Automat. Contr.*, vol. AC-11, pp. 197–204, Apr. 1966.

tion by finding the mutual relationships among the optimum, calculated, and actual covariances of the estimate. This study is the major objective of the paper and is presented in the first half.

In the second half, an application of one of the theorems derived in the first part which puts an upper-bound on the covariance of the estimation error will be demonstrated using a phase-locked loop system as an example. It is intended to improve the performance of the existing analog filter, a phase-locked loop receiver in this case, by the application of the optimal filter, which can be synthesized in the form of a program on the digital computer, thus assuming a form of a digital filter. By means of this tandem configuration of analog and digital filters, shown in Fig. 1, detection of states as well as signals can be speeded up considerably without increasing the covariance of the estimation error. Without the digital filter, speeding-up of the estimation process normally results in an increase of the estimation error. Thus the practical importance of the extended results on the effect of errors in the a priori information will be demonstrated in this specific example.

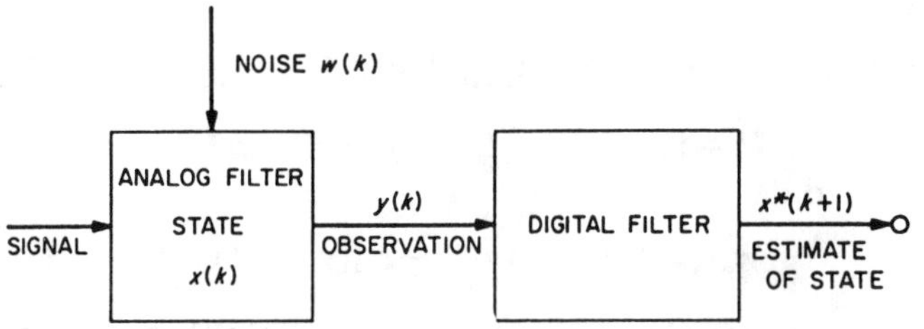

Fig. 1.

II. Fundamental Equations

The system treated is a discrete-time linear system where the discrete interval T and its state vector representation are described by the following difference equations:

$$x(k+1) = \Phi(k)x(k) + G(k)w(k). \tag{1}$$

The observation is

$$y(k) = M(k)x(k) \tag{2}$$

where

x: a column vector of dimension n_x, the state vector of the system

w: d column vector of dimension n_w, white, Gaussian noise inputs to the system

y: a column vector of dimension n_y, the observation of the system output

Φ: an $n_x \times n_x$ matrix, the state transition matrix

G: an $n_x \times n_w$ matrix, the noise coefficient matrix

M: an $n_y \times n_x$ matrix, the observation matrix.

It is assumed that the observation noise, if there is any, is treated as a state or states.[1]

The problem is now to find an optimum estimate of $x(k+1)$ based on the observation $y(0), y(1), \cdots, y(k)$. One solution to this problem employing a Bayesian approach is given by Ho.[4] In this solution, the conditional density function $p[x(k+1) \mid y(0), y(1), \cdots, y(k)]$ is derived by means of the Bayes' rule, observing that both x and y are Gaussian when w is Gaussian. Then the conditional expectation of a quadratic loss function

$$\langle (x - x^*)^T(x - x^*) \rangle$$

is derived and minimized, thus yielding the following equations for the optimum estimator x^* of the state vector x. Here, $\langle\ \rangle$ denotes the expected-value operator. Let Q be the covariance matrix of the noise w (assuming its mean is zero),

$$\langle w \rangle = 0 \tag{3}$$

$$\langle w \cdot w^T \rangle = Q, \tag{4}$$

and let $P(k)$ be a square matrix of dimension n_x which represents a covariance matrix of the estimator:

$$P(k) = \langle [x(k) - x^*(k)][x(k) - x^*(k)]^T \rangle. \tag{5}$$

Then given a priori information on initial conditions $x^*(0)$ and $P(0)$, the optimal estimator $x^*(k)$ of the state vector $x(k)$ is recursively defined by,[1,4]

$$x^*(k+1) = \Phi(k)x^*(k) + K(k)[y(k) - M(k)x^*(k)] \tag{6}$$

$$K(k) = \Phi(k)P(k)M^T(k)[M(k)P(k)M^T(k)]^{-1} \tag{7}$$

and

$$P(k+1) = [\Phi(k) - K(k)M(k)]P(k)[\Phi(k) - K(k)M(k)]^T + G(k)QG^T(k). \tag{8}$$

The digital filter of Fig. 1 is represented by (6) to (8) and its matrix block diagram is drawn in Fig. 2.

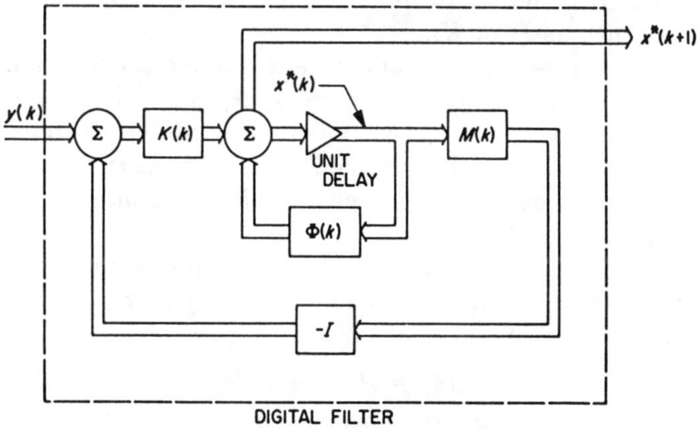

Fig. 2.

[4] Y. C. Ho and R. C. K. Lee, "A Bayesian approach to problems in stochastic estimation and control," *IEEE Trans. on Automatic Control*, vol. AC-9, pp. 333–339, October 1964.

Combining (7) and (8),

$$P(k+1) = \Phi(k)\{P(k) - P(k)M^T(k)[M(k)P(k)M^T(k)]^{-1} \cdot M(k)P(k)\}\Phi^T(k) + G(k)QG^T(k). \quad (9)$$

These recurrence relations yield the optimum estimate of the state vectors sequentially, provided the a priori information on the initial conditions $x^*(0)$ and $P(0)$ is available. However, if such a priori information is incorrect, the subsequent $x^*(k)$ and $P(k)$ will be inevitably in error. Therefore, an analysis becomes necessary on the behavior of the error caused by the incorrect a priori information, since it is highly probable that in practical circumstances the a priori information is not sufficiently correct, or that it is not available at all.

As mentioned in the Introduction, Soong[3] considered this problem for the single stage case (that is, the case when estimation is performed after a sequence of observations has been completed).

This paper extends the analysis to the multistage case, that is, the case when sequential estimation is carried out in real-time following sequential observations on the states.

III. Derivation of Error Matrices

Adopting Soong's notations,[3] three different kinds of covariance matrices are introduced, namely P_o, P_c, and P_a:

- P_o: optimum covariance matrix when correct a priori information is employed.
- P_c: calculated covariance matrix derived by (9), when incorrect information is employed for the initial condition.
- P_a: actual covariance matrix derived by (5), when the estimate is computed by using the calculated covariance matrix with the incorrect a priori data.

In the subsequent analysis, a subscript o is attached to the optimum P and K described by (7) and (8) in order to differentiate them from other similar quantities such as P_c or K_c.

Picking up an arbitrary instant k, the error function at the next instant, i.e., the $(k+1)$th instant, is derived as a function of the error function at the kth instant. Then, by means of induction, a useful property of the error function shall be proved over all the instants, starting from the 0th instant.

Since there are three kinds of covariance matrices, three error functions are defined as their mutual differences. They are:

$$E_{ca}(k) = P_c(k) - P_a(k) \quad (10)$$

$$E_{co}(k) = P_c(k) - P_o(k) \quad (11)$$

$$E_{ao}(k) = P_a(k) - P_o(k). \quad (12)$$

It is known that $P_o(k)$, which is an optimum covariance matrix defined by (5), is non-negative definite for all $k \geq 0$.[1] Further, it is assumed that MP_oM^T and MP_cM^T are invertible for the sake of simplicity throughout this paper.

A. Calculation of E_{ca}

It is noticed that the calculated covariance matrix P_c is derived by recurrence formulas (7) and (8) starting from the given initial condition $P_c(0)$. On the other hand, the actual covariance matrix is derived from (5) by definition, in which the estimate x^* is computed by (6) using the calculated K and P in it.

In order to simplify the description, the time (k) shall be dropped from all the quantities concerned except when it is more advantageous to indicate it explicitly in the subsequent analysis. Hence, from (7) and (8)

$$P_c(k+1) = (\Phi - K_cM)P_c(\Phi - K_cM)^T + GQG^T \quad (13)$$

where

$$K_c = \Phi P_cM^T[MP_cM^T]^{-1}. \quad (14)$$

On the other hand, $P_a(k+1)$ is described as follows: denoting the actual, nonoptimum estimate as x_a^*, when it is derived using K_c and P_c in (7) and (8),

$$P_a(k+1) = \langle[x_a^*(k+1) - x(k+1)][x_a^*(k+1) - x(k+1)]^T\rangle \quad (15)$$

where $x\ (k+1)$ is given by (1), while $x_a^*(k+1)$ is described as

$$x_a^*(k+1) = \Phi x_a^* + K_c(y - Mx_a^*). \quad (16)$$

Substitute (1) and (16) into (15). Note that $x(k)$ and $x^*(k)$ are independent from $w(k)$ because w is white noise. Then (15) can be reduced to the following simple form:

$$P_a(k+1) = (\Phi - K_cM)P_a(\Phi - K_cM)^T + GQG^T. \quad (17)$$

The difference between (13) and (17) yields the desired recurrence relationship for E_{ca}.

$$\begin{aligned} E_{ca}(k+1) &= (\Phi - K_cM)(P_c(k) - P_a(k))(\Phi - K_cM)^T \\ &= (\Phi - K_cM)E_{ca}(k)(\Phi - K_cM)^T. \end{aligned} \quad (18)$$

B. Calculation of E_{co}

Since $P_c(k+1)$ and $P_o(k+1)$ are given by (13) and (8), respectively, their difference is

$$\begin{aligned} E_{co}(k+1) &= P_c(k+1) - P_o(k+1) \\ &= (\Phi - K_cM)P_c(\Phi - K_cM)^T \\ &\quad - (\Phi - K_oM)P_o(\Phi - K_oM)^T \\ &= (\Phi - K_cM)E_{co}(\Phi - K_cM)^T \\ &\quad + (\Phi - K_oM)[E_{co}M^T\overline{P}_c^{-1}\overline{P}_o\overline{P}_c^{-1}ME_{co} \\ &\quad - P_cM^T\overline{P}_c^{-1}ME_{co} - E_{co}M^T\overline{P}_c^{-1}MP_c] \\ &\quad \cdot(\Phi - K_oM)^T \end{aligned} \quad (19)$$

where

$$K_c = K_o + (\Phi - K_oM)E_{co}M^T\overline{P}_c^{-1}. \quad (20)$$

The bar above indicates the multiplication of M to the left and M^T to the right of the underlying quantity, e.g.,

$$\overline{P}_c = MP_cM^T. \quad (21)$$

When (19) is expanded and arranged together, it can be finally reduced to the following concise form:

$$\begin{aligned} &E_{co}(k+1) \\ &= (\Phi - K_oM)[E_{co}(k) - E_{co}(k)M^T\overline{P_c(k)}^{-1}ME_{co}(k)]^{-1} \\ &\cdot(\Phi - K_oM)^T. \quad (22)\end{aligned}$$

C. Calculation of E_{ao}

$E_{ao}(k+1)$ is obtained either by taking the difference between $P_a(k+1)$ and $P_o(k+1)$ of (17) and (8),

$$\begin{aligned} E_{ao}(k+1) &= P_a(k+1) - P_o(k+1) \\ &= (\Phi - K_cM)P_a(\Phi - K_cM)^T \\ &\quad - (\Phi - K_oM)P_o(\Phi - K_oM)^T \quad (23)\end{aligned}$$

or by subtracting E_{ca} from E_{co},

$$E_{ao}(k+1) = E_{co}(k+1) - E_{ca}(k+1) \quad (24)$$

where $E_{co}(k+1)$ and $E_{ca}(k+1)$ are given by (22) and (18), respectively.

Either method leads to the following result which is represented as a sum of two symmetric matrices after certain matrix operations.

$$\begin{aligned} &E_{ao}(k+1) = (\Phi - K_cM)E_{ao}(k)(\Phi - K_cM)^T \\ &+ (\Phi - K_oM)[E_{co}(k)M^T\overline{P_c}(k)^{-1}\overline{P_o}(k)\overline{P_c}(k)^{-1}ME_{co}(k)] \\ &\cdot(\Phi - K_oM)^T. \quad (25)\end{aligned}$$

IV. Evaluation of Results

It may be observed that $E_{ca}(k+1)$, $E_{co}(k+1)$ and $E_{ao}(k+1)$ derived in Section III are all symmetric, square matrices. For the purpose of evaluating these recurrence formulas, the following three lemmas concerning real matrices are useful.

Lemma 1a: Given an arbitrary matrix B, $(-)B \cdot B^T$ is (nonpositive) non-negative definite.

Lemma 1b: Conversely, if H is a (nonpositive) non-negative definite symmetric matrix, it can be factored into the form $H = (-)B \cdot B^T$.

Lemma 2: If H_1 and H_2 are symmetric and non-negative (nonpositive) definite matrices, then $H_1 + H_2$ is non-negative (nonpositive) definite.

Lemma 3: If H is a symmetric non-negative (nonpositive) definite matrix, then every diagonal component of H is non-negative (nonpositive).

Then the following four theorems are presented and proved as the principal results of this paper.

Theorem 1

If $E_{ca}(k)$ is non-negative (nonpositive) definite, then $E_{ca}(k+1)$ is non-negative (nonpositive) definite.

Proof: If $E_{ca}(k)$ is non-negative definite, by Lemma 1b, it can be factored as

$$E_{ca}(k) = B \cdot B^T. \quad (26)$$

Then by letting

$$C = (\Phi - K_cM)B \quad (27)$$

and referring to (18), $E_{ca}(k+1)$ can be factored as

$$E_{ca}(k+1) = C \cdot C^T. \quad (28)$$

Thus it becomes clear that $E_{ca}(k+1)$ is non-negative definite according to Lemma 1a. Extension to the case when $E_{ca}(k)$ is nonpositive definite is obvious.

Theorem 2

If $E_{co}(k)$ is non-negative (nonpositive) definite, then $E_{co}(k+1)$ is non-negative (nonpositive) definite.

Proof: When $E_{co}(k)$ is positive definite, the content of the parenthesis in the middle of (22) can be rewritten as a sum of two symmetric matrices as follows, assuming that $ME_{co}(k)M^T$ is also positive definite and hence invertible.

$$\begin{aligned} &E_{co}(k) - E_{co}(k)M^T\overline{P_c}(k)^{-1}ME_{co}(k) \\ &= E_{co}(k)M^T[\overline{E_{co}}(k) + \overline{E_{co}}(k)\overline{P_o}(k)^{-1}\overline{E_{co}}(k)]^{-1}ME_{co}(k) \\ &\quad + [(I - E_{co}(k)M^T\overline{E_{co}}(k)^{-1}M] \\ &\quad \cdot E_{co}(k)[I - E_{co}(k)M^T\overline{E_{co}}(k)^{-1}M]^T. \quad (29)\end{aligned}$$

Since the inverse of a positive definite matrix is also positive definite, the positive definiteness of the quantity in the right hand side of the above equation can be proved with the help of Lemma 2 and using the result in the proof of Theorem 1. Therefore, the non-negative definiteness of $E_{co}(k+1)$ in (22) is claimed, referring again to the proof in Theorem 1.

When $E_{co}(k)$ is non-negative definite, $E_{co}(k) + \epsilon I$ ($\epsilon > 0$) is positive definite. Substituting this sum for $E_{co}(k)$ in (29), the non-negative definiteness of the left-hand side of (29) can be proved as a limiting case as $\epsilon \to 0$. Thus the statement at the beginning is verified. When $E_{co}(k)$ is nonpositive definite, the left-hand side of (29) is clearly nonpositive definite. Thus the nonpositive definiteness of $E_{co}(k+1)$ can be proved by following a similar process to (26)–(28).

Theorem 3

If $E_{ao}(k)$ is non-negative definite then $E_{ao}(k+1)$ is non-negative definite.

Proof: In (25), the second term is always non-negative definite because the covariance matrix P_o is non-negative definite, so that a proof similar to the one in

Theorem 1 can be applied. Also, the first term is non-negative definite when $E_{ao}(k)$ is non-negative definite. Therefore, the above theorem is verified, referring again to Lemma 2.

Finally, an important conclusion is drawn by applying induction to these results.

Theorem 4

The error matrices $E_{ca}(k)$ and $E_{co}(k)$ are non-negative (nonpositive) definite for every $k \geq 1$ whenever their respective initial conditions $E_{ca}(0)$ and $E_{co}(0)$ are non-negative (nonpositive) definite, and $E_{ao}(k)$ is always non-negative definite[5] because $E_{ao}(0)$ is non-negative definite as deduced from the definition of $P_o(0)$

Consequently, the following corollary is stated referring to Lemma 3.

Corollary 1

All the diagonal components of $E_{ca}(k)$, $E_{co}(k)$ are non-negative (nonpositive) (for all $k \geq 1$) when $E_{ca}(0)$ and $E_{co}(0)$ are, respectively, non-negative (nonpositive) definite. The diagonal components of $E_{ao}(k)$ are non-negative for all $k \geq 0$ without any condition.

This result will be very useful in practical applications. Especially, the non-negativeness of $e_{ca_{ii}}(k)$, which represents a diagonal component of $E_{ca}(k)$, implies

$$p_{c_{ii}}(k) \geq p_{a_{ii}}(k) \geq 0 \tag{30}$$

where $p_{c_{ii}}(k)$ and $p_{a_{ii}}(k)$ denote diagonal components of $P_c(k)$ and $P_a(k)$, respectively. It is clear that $p_{a_{ii}}(k)$ is non-negative because the covariance matrix $P_a(k)$ is non-negative definite.

Therefore, the actual variance of estimates (which is nonaccessible because an accurate information on $P_a(0)$ is not available) based on an incorrect a priori information is bounded by the calculated variance (which is accessible) if the initial error matrix $E_{ca}(0)$ is so selected as to be non-negative definite.

V. Application to the Phase-Locked Loop System

For the purpose of demonstrating an application of the theory developed so far to the tandem configuration of analog and digital filter, a phase-locked loop which is extensively used for ranging in the deep space, is taken as an example of such an analog filter.

The phase-locked loop receiver is designed to detect a narrow-band signal in the presence of wide-band noise. This receiver continuously tracks the phase of the incoming signal by correlating it with the local signal which is generated from the voltage controlled oscillator (VCO). The VCO operates on a low-frequency output of the correlation filter, which is proportional to the sine of the phase difference between the incoming and local signals, and yields a sinusoid having a phase which is linearly related to the integral of the input.

When a block diagram in terms of phases of the incoming and local signals is drawn, the VCO is represented by an integrator and the combined gain of the loop filter and the VCO is by a. As long as the phase difference remains small, the nonlinear operation of multiplication may be linearized to subtraction in the feedback configuration of Fig. 3, together with an amplifier of gain A after the subtractor where A^2 is the power of the incoming signal. It is assumed that the incoming noise is white and Gaussian with density N_o (one-sided). Further details of the derivation leading to Fig. 3 are left to the references.[6,7]

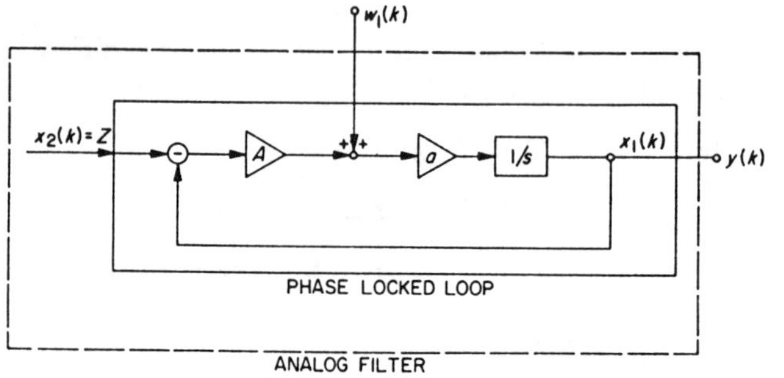

Fig. 3.

When the input to the loop is a phase step, it can be considered as a state of an expanded system constructed by adding a constant voltage to the original system at its input. This input state is described as x_2 and the output of the phase-locked loop as x_1. Then the state vector at $t=(k+1)T$ is described as a linear function of the state at $t=kT$.

$$x_1(k+1) = e^{-bT}x_1(k) + (1 - e^{-bT})x_2(k) + \frac{(1 - e^{-bT})}{A} w_1(k) \tag{31}$$

$$x_2(k+1) = x_2(k) \tag{32}$$

where the equivalent loop gain is given by

$$b = aA = \frac{1}{\tau} \tag{33}$$

and the τ above is the time-constant of the loop. Equations (31) and (32) constitute the analog filter in Fig. 1.

The observation y is identical to the phase output x_1.

$$y(k) = x_1(k). \tag{34}$$

Then the matrices Φ, G and M in (1) and (2) can be easily derived from (31)–(34).

[5] This is also clear from the definition that $P_o(k)$ is the minimum variance associated with the optimum estimate.

[6] R. Jaffe and E. Rechtin, "Design and performance of phase-lock circuits capable of near-optimum performance over a wide range of input signals and noise levels," *IRE Trans. on Information Theory*, vol. IT-1, pp. 66–76, March 1955.

[7] A. J. Viterbi, "Phase-locked loop dynamics in the presence of noise by Fokker-Planck techniques," *Proc. IEEE*, vol. 51, pp. 1737–1753, December 1963.

The direct substitution of these matrices into (6) yields the optimum estimator as

$$x^*(k+1) = \Phi \hat{x}^*(k) \tag{35}$$

where $\hat{x}_1^*(k)$ is given by

$$\hat{x}_1^*(k) = y(k) \tag{36}$$

$$\hat{x}_2^*(k) = x_2^*(k) + \frac{p_{12}(k)}{p_{11}(k)} [y(k) - x_1^*(k)]. \tag{37}$$

The recurrence formulas for the P matrix are

$$p_{11}(k+1) = \frac{p_{11}(k)p_{22}(k) - p_{12}^2(k)}{p_{11}(k)} \phi_{12}^2 + \sigma_u^2 \tag{38}$$

$$p_{12}(k+1) = \frac{p_{11}(k)p_{22}(k) - p_{12}^2(k)}{p_{11}(k)} \phi_{12} = p_{21}(k+1) \tag{39}$$

$$p_{22}(k+1) = \frac{p_{11}(k)p_{22}(k) - p_{12}^2(k)}{p_{11}(k)} \tag{40}$$

where

$$\phi_{12} = 1 - e^{-bT} \tag{41}$$

and

$$\sigma_u^2 = \frac{(1 - e^{-bT})^2 N_0}{2A^2T} \cdot \tag{42}$$

Equations (35) to (40) specify the optimal digital filter shown in Figs. 1 and 2.

The principal objective of this digital filter is, of course, to find the optimum estimate of the state x_2 at every instant of time, by suppressing the effect of noise disturbance as much as possible.

Assuming that the system is at rest for $t<0$, and that the a priori information on $x^*(0)$ and $P(0)$ is correct, the derivation of the optimum estimate $x^*(k)$ and the covariance matrix $P(k)$ is straight forward with the help of (6)–(8).[8]

However, the a priori information may be incorrect, especially on $x_2^*(0)$ and $p_{22}(0)$ (if they are correct, there is no need to proceed on the estimation). Therefore, the estimation process of (6) to (9) is inevitably started by assigning a set of initial conditions which are not the optimum ones. Fortunately, the discussion in Section IV assures that the actual variance of the estimate is bounded by the calculated variance if $E_{ca}(0)$ is selected to be non-negative definite. Hence, it becomes possible to study the real-time behavior of the digital filter using the P_c matrix as an upper bound for the actual covariance matrix.

[8] T. Nishimura, "On the a priori information in multi-stage estimation problems," Jet Propulsion Laboratory, Calif. Inst. Tech., Pasadena, Space Program Summary 37-32, vol. IV, April 1965.

The behavior of P_c as a function of time is obtained as a solution of three nonlinear difference equations described by (38) to (40) starting from the assumed a priori information $P_c(0)$. Certain manipulations of (38) to (40) yield a nonlinear difference equation of $p_{c_{22}}$ alone which plays the most important role among the components of P_c matrix in this analysis.

$$\phi_{12}^2 p_{c_{22}}(k+1) p_{c_{22}}(k) + \sigma_u^2 [p_{c_{22}}(k+1) - p_{c_{22}}(k)] = 0. \tag{43}$$

When the discrete interval T is very short, the above difference equation can be replaced by the corresponding differential equation. Further, it is assumed that $T(dp_{c_{22}}/dt)$ is much smaller than $p_{c_{22}}$ when T is a sufficiently short interval. In this case, the above difference equation is reduced to the following Riccati type of differential equation

$$\frac{dp_{c_{22}}}{dt} = -\frac{\alpha}{2\tau} p_{c_{22}}^2 \tag{44}$$

where

$$\alpha = \frac{2\tau\phi_{12}^2}{\sigma_u^2 T} = \frac{A^2}{N_0 B_L} \cdot \tag{45}$$

Thus α is the signal-to-noise ratio in terms of the loop bandwidth B_L, where

$$B_L = \frac{b}{4} = \frac{1}{4\tau} \cdot \tag{46}$$

The solution of this differential equation is[9]

$$p_{c_{22}}(t) = \frac{p_{c_{22}}(0)}{\alpha p_{c_{22}}(0)(t/2\tau) + 1} \cdot \tag{47}$$

We observe that $p_{c_{22}}(t)$ is a monotone decreasing function of t and it behaves as $2\tau/\alpha t$ for large t. Also $p_{a_{22}}$ will behave in a similar manner because it is bounded by $p_{c_{22}}$ by means of (30).

The observation time $t_o(\lambda)$ is defined as the time required for the variance of the estimate to become smaller than a specified bound λ. This is immediately computed from (47) as

$$t_o(\lambda) = \frac{2\tau}{\alpha\lambda}\left(1 - \frac{\lambda}{p_{c_{22}}(0)}\right). \tag{48}$$

Since $p_{c_{22}}(0)$ usually assumes much larger value than λ, $t_o(\lambda)$ is simply reduced to

$$t_o(\lambda) \approx \frac{2\tau}{\alpha\lambda} = \frac{1}{\lambda} \frac{N_0}{2A^2} \cdot \tag{49}$$

[9] Equation (43) can be solved directly as a nonlinear difference equation. The solution is identical to (47) if t is replaced by nT there.

Therefore, the observation time $t_o(\lambda)$ is inversely proportional to λ and its coefficient is determined by the signal-to-noise ratio of the input. It is neither dependent on the system gain b nor on the estimation period T. This is a noteworthy result and it is valid for $1/B_{w_1} \ll T \ll 1/b$ (B_{w_1} is the bandwidth of noise).

Now a comparison of this result is made against the case without the digital filter. The steady state variance λ_o of the phase-locked loop having the configuration of Fig. 3 is given by

$$\lambda_o = \frac{b}{4A^2} N_o = \frac{1}{\alpha} \cdot \tag{50}$$

On the other hand, 4τ ($\tau = 1/b$ is the time-constant of the loop) may be sufficient for the transient to vanish, so that it is selected as the observation time

$$t_o'(\lambda_o) = \frac{4}{b} = \frac{1}{\lambda_o} \frac{N_o}{A^2} \cdot \tag{51}$$

Compared to (49)

$$t_o(\lambda_o) = \tfrac{1}{2} t_o'(\lambda_o). \tag{52}$$

Thus the observation time is reduced at least, to half, by the application of the digital filter for $\lambda = \lambda_o$. Since $t_o(\lambda)$ is the decaying time for $p_{c_{22}}$, the decaying time of the actual variance $p_{a_{22}}$, which is bounded by $p_{c_{22}}$, must be shorter than that in most cases, as is observed in the experimental simulation at the end of this paper.

Moreover, the variance of the estimate with the digital filter is monotonically decreasing with time, thus assuring the continuous improvement of the estimate, while the variance remains constant after $t_o'(\lambda_o)$ for the system without the digital filter and no further improvement is expected on the estimate beyond that point. This is also observed on the experimental result.

Another view of the advantage gained by this filter can be seen in the frequency domain. The bandwidth $B_c(t)$ of the combined system is a function of t; viz.

$$B_c(t) = B_L \cdot \frac{2\tau}{t} \qquad \text{for } t \gg 4\tau \frac{\lambda_o}{p_{c_{22}}(0)} \cdot \tag{53}$$

Thus, the combined system behaves as a time-varying filter whose bandwidth decreases inversely proportional to time for sufficiently large t, and at time $t = 4\tau$, B_c is already one half of B_L.

VI. Experimental Simulation

The entire system including a model of the phase-locked loop, the digital filter, and signal and noise generators are simulated on the digital computer. Several experiments have been conducted varying the initial settings of $P_c(0)$ and $x^*(0)$ matrices.

The simplest form of the non-negative definite $E_{ca}(0)$ matrix is a diagonal one having

$$e_{ca_{11}}(0) \geq 0 \tag{54}$$

$$e_{ca_{22}}(0) > 0. \tag{55}$$

Based on the above relations, diagonal components of $P_c(0)$ matrix are selected to satisfy the following condition:

$$p_{c_{11}}(0) \geq \sigma_u^2 \tag{56}$$

$$p_{c_{22}}(0) > p_{a_{22}}(0) = Z^2 \tag{57}$$

where Z is the amplitude of the actual input.

The initial estimates of states are set at zero.

Although Z^2 is an unknown quantity, it is assumed that a certain upper bound max (Z^2) is known. This is a reasonable assumption because normally a small deviation (Z in this case) of the actual range from the ephemeris data is tracked by the phase-locked loop, so that this scheme is only valid for a limited Z. Then $p_{c_{22}}(0)$ is determined at some value no smaller than max (Z^2).

$$p_{c_{22}}(0) \geq \max(Z^2). \tag{58}$$

Since the effect of the initial value $p_{c_{22}}(0)$ vanishes very quickly, as observed from the solution of the differential equation in (47), the choice is rather arbitrary as long as it satisfies the above condition.

The rest of the components are assigned as

$$p_{c_{12}}(0) = p_{c_{21}}(0) = 0. \tag{59}$$

In Fig. 4(a), $x_1(k)$ [$=y(k)$], $x_2(k)$($=Z$), and $x_{2a}^*(k)$ (nonoptimum estimate) are plotted. The chained line indicates the range of the calculated variance $p_{c_{22}}(k)$. The experimental results prove that the ensemble of $x_{2a}^*(k)$ remains inside this range verifying the conclusion of Theorem 1. In this particular example, it is observed that $x_{2a}^*(k)$ enters into the range of $\lambda_o = 0.0025$, the steady-state variance of the phase-locked loop, of this example, at $t = 1.7\tau$, and remains within that range afterwards, further approaching to the actual input valve. This proves the previous conclusion that $t_o(\lambda_o)$ is no larger than 2τ.

Fig. 4(b) and (c) are the cases when the loop gain b is changed to some other values. Observe that $t_o(\lambda)$ maintains almost the same value irrespective of the variation of the loop gain and hence the time-constant of the loop. This is also expected from the previous analysis.

VII. Conclusion

The effect of errors in the a priori information has been investigated for the sequential estimation problems, and the four theorems which specify the mutual relations among the three covariance matrices P_o, P_c and P_a have been derived. With the aid of these theorems, it becomes possible to prescribe the behavior of the nonoptimum estimate $x_a^*(k)$ by confining its variance $p_{a_{ii}}(k)$ (nonaccessible) within the calculated (accessible) variance $p_{c_{ii}}(k)$.

Having the information on $p_{c_{ii}}(k)$ which is supplied

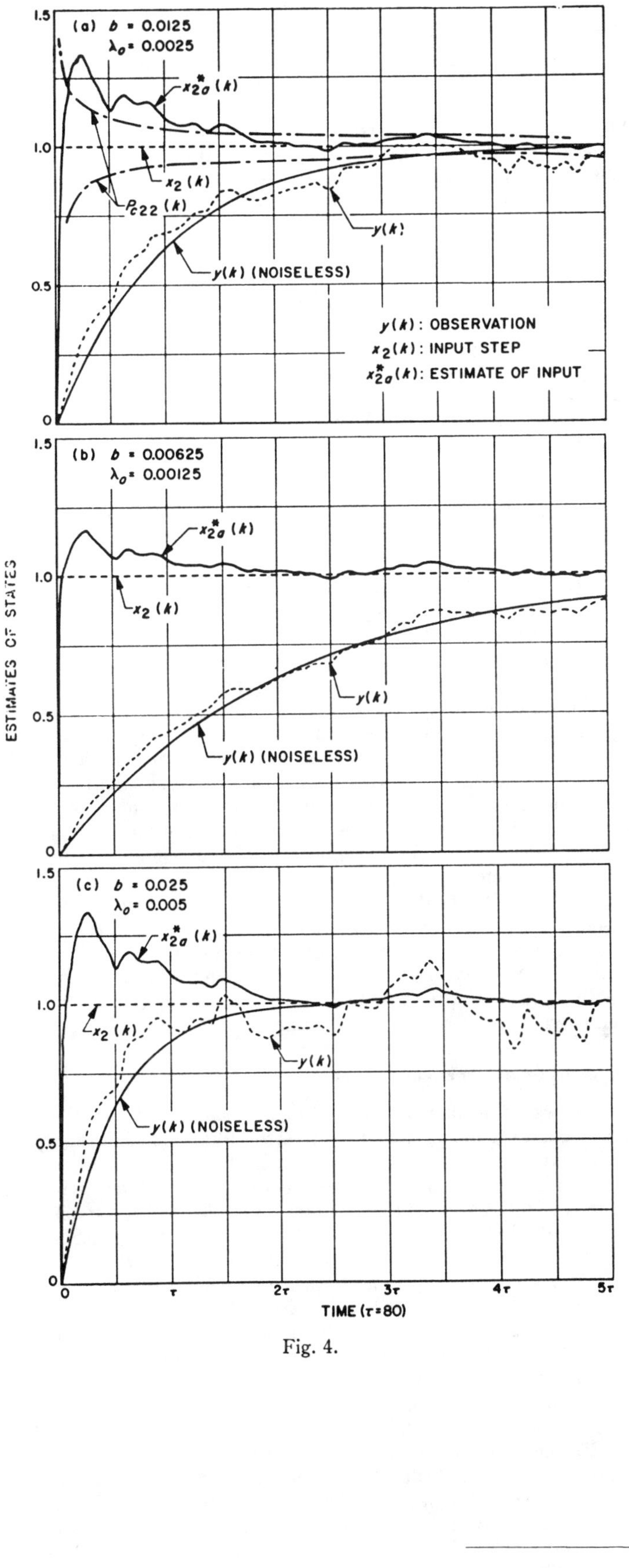

Fig. 4.

as solutions of Riccati-type differential equations, the overall performance of the combined analog and digital filters can be controlled in certain optimum ways.

According to Kalman,[10] the optimal filter is uniformly and asymptotically stable if the system is uniformly completely observable and uniformly completely controllable. Hence it can be expected that the initial uncertainty will decay sooner or later if the system satisfied the above condition. However, the result of this paper will answer the question of how soon such initial uncertainty will decay when the observability and controllability condition is satisfied. Also, it will specify a bound for the additional inaccuracy due to the initial uncertainty. Furthermore, the results here also apply to the case when the above condition is not satisfied, because no such condition is required in the derivation of the theorems. Therefore, it becomes possible to prescribe the behavior of the filter even when it is not uniformly, asymptotically stable, or when its stability is not known for certain.

The phase-locked loop receiver is picked up as an example of analog filter, for which a suitable digital filter is designed. Both the theoretical and experimental (simulation) results prove that the effective noise bandwidth of the combined filters can be reduced to, at least, half at $t=4\tau$ by the application of the digital filter for the phase step input.

The technique developed in this paper is useful for the estimation problem of deterministic signals in the presence of stochastic noise, while neither the Wiener theory nor the conventional design technique of feedback systems is suitable for such problems when they are applied separately.

Although the signal (phase) input has been treated as a state in this analysis, it is possible to describe the input contribution[11] in an explicit form in (1). Especially when the input to the phase-locked loop takes a form of frequency step or some other forms, such explicit description of the input contribution may simplify the treatment of the problem, though it is immaterial for the step input.

[10] R. E. Kalman, "New methods and results in linear prediction and estimation theory," Research Institute for Advanced Study, Baltimore, Md., Tech. Rept. 61-1, 1961.

[11] H. Cox, "Estimation of state variables via dynamic programming," Preprints of *1964 Joint Automatic Control Conf., at Stanford University*, pp. 376–381, June 1964.

Divergence in the Kalman Filter

F. H. Schlee,* C. J. Standish,† and N. F. Toda‡
International Business Machines Corporation, Endicott, N. Y.

Under certain conditions, the orbit estimated by a Kalman filter has errors that are much greater than predicted by theory. This phenomenon is called divergence, and renders the operation of the Kalman filter unsatisfactory. This paper investigates the control of divergence in a Kalman filter used for autonomous navigation in a low earth orbit. The system studied utilizes stellar-referenced angle sightings to a sequence of known terrestrial landmarks. A Kalman filter is used to compute differential corrections to spacecraft position, spacecraft velocity, and landmark location. A variety of filter modifications for the control of divergence was investigated. These included the Schmidt-Pines analytical modification and an "empirical" modification based upon Pines' machine noise treatment. Several simplified approximations to the theoretically optimum analytical modifications were also investigated. The principal numerical results are presented in graphs of the magnitude of the error in estimated position and velocity vs time for sixteen orbits. These graphs compare actual position and velocity errors with the theoretical estimates furnished by the trace of the position and velocity covariance matrices. Numerical results indicate that a properly modified filter achieves a steady-state operating level.

Introduction

THE performance of the Kalman filter under actual operating conditions can be seriously degraded from the theoretical performance indicated by the state covariance matrix. The Kalman filter theoretically produces an increasingly accurate estimate as additional observation data are processed. The magnitude of estimation errors as measured by the determinant of the estimation error covariance matrix is a monotonically decreasing function of the number of observations. However, it has been noted that under actual operating conditions, error levels in the Kalman filter are significantly higher than predicted by theory. Errors can, in fact, increase continuously although additional data are being processed. The possibility of such unstable or divergent behavior was first suggested by Kalman.[1] It was later noted by Pines[2] and Knoll[3] and others in application of the Kalman filter to space navigation and orbit determination.

Autonomous Navigation Filter

A precise determination of the orbit of a space vehicle can be made by employing observational data from on board a spacecraft and a differential correction technique. The data-processing technique, called a filter, assumes a linear relationship between the deviations in the measurements and the corresponding deviations in the orbital elements. These deviations, called observations and states, respectively, are the differences between the measured or estimated values and the reference or nominal values. A large number of observations provides an overdetermined set of linear equations in the unknown states. The filter "fits" an orbit to the observational data by employing some criterion such as least squares, maximum likelihood, or minimum variance.

Presented at the AIAA/JACC Guidance and Control Conference, Seattle, Wash., August 15–17, 1966 (no preprint number; published in bound volume of preprints of the meeting); submitted August 30, 1966. The authors wish to thank D. Moore and J. Ducey, who wrote the simulation programs used in this investigation. [8.02, 8.04, 8.07]

* Advisory Engineer, Federal Systems Division, Space Systems Center.

† Senior Mathematician, Federal Systems Division, Space Systems Center; now at Electronics Systems Center, Owego, N. Y.

‡ Senior Engineer, Federal Systems Division, Space Systems Center. Member AIAA.

The traditional filtering methods store the observational data obtained over a long arc, and then process the whole "batch of data" at once to determine a state vector or, equivalently, the orbital elements. Batch processing requires the numerical computation of the inverse of a 6 by 6 or higher-order matrix.

The Kalman filter employs the same linearity assumptions and minimum variance criterion as the batch filter, but processes each piece of data as it is obtained. Because the estimation process is expressed as a set of recursion relations, the Kalman filter is referred to as a recursive process. This recursive estimation process computes a new set of orbital elements after each measurement, thus affording the opportunity to change or rectify the reference trajectory after each measurement. Frequent rectification of the reference orbit reduces the effect of nonlinearities. Except for the residual effects of nonlinearities, a recursive filter is equivalent to an iterative batch process, since both employ the same linearity assumptions and are optimal under the same criteria. In fact, the recursive formulation of the Kalman filter can be derived from the nonrecursive minimum variance filter by employing a matrix inversion lemma.[8]

Causes of Divergence

One cause of filter divergence is discrepancies between the mathematical model used to derive the filter equations and the actual conditions under which the filter must operate. Examples of such discrepancies are the neglect of terms in the gravitational potential or an inaccurate knowledge of the constants in the potential. Another source of divergence is errors in the ballistic coefficient or in the air density model used in computing drag accelerations. Such errors are assumed to be bias errors. Since they affect the dynamical equations of motion, and to differentiate them from observation biases,[4] these errors are termed dynamical biases.

Figure 1 shows the divergence caused by a 25% error in the drag acceleration. A different source of divergence is the round-off errors inherent in the implementation of the filter equations on a finite word length digital computer. Figure 2 shows the effect of computational errors induced by single precision arithmetic (IBM 7090) on the Kalman filter.

One manifestation of machine-caused errors occurs in the computation of the state covariance matrix. After the Kalman filter has been operating for some time, this matrix ceases to be positive definite and symmetric. Filter weighting coefficients computed using this matrix are then wrong and, consequently, orbit estimates are incorrect. J. E. Potter and R. Battin[9] derived a variation of the Kalman filter in which the covariance matrix remains at least symmetric and non-negative. This technique eliminates some but not all of the effects of computational errors and at the same time requires a somewhat more complex filter algorithm. In batch processing filters, round-off errors in the filter become evident as serious computational errors in the matrix inversion operation.

Control of Divergence

Several approaches have been suggested for preventing filter divergence. One approach argues that divergence occurs when the filter assigns too small a weight to the last measured data. Thus the current data make only a small correction in the estimate, so small in fact, that errors actually grow because of the natural interaction of position and velocity errors. An obvious "fix" is to more or less arbitrarily increase the weighting of current data. One such fix involves increasing the state covariance matrix while holding the orbital period uncertainty constant. The frequency and amount of the increase must be determined empirically, and may be employed in either recursive or nonrecursive filters.

Schmidt[5] and Pines[2] have suggested analytical modifications to the filter equations to account for dynamical biases without increasing the number of states to be estimated by the filter. Additionally, Pines[2] has developed a simple machine noise modification based on an assumed model of the errors caused by round-off in the digital computer. Other techniques, such as filter reset to keep the diagonal elements of the state covariance matrix above a specified threshold, were investigated by Holick et al.[6]

Ditto[7] experienced divergence in the nonrecursive Gemini-Bayes differential corrector, where numerical problems appear in the inversion of ill-conditioned matrices as well as in the neglect of dynamical biases. The occurrence of filter divergence in a nonrecursive filter is not surprising in view of the equivalence of the Kalman filter and the nonrecursive minimum variance filter.[8] The filter modifications employed by Ditto bear a similarity to those of Schmidt and Pines.

Illustrative Example

To provide the reader with some intuitive understanding of the divergence problem, we give a simple analytical example, where the Kalman filter estimate of a trajectory diverges from the true trajectory as the number of observations increases. This phenomenon is brought about in the present case by the omission of a component of the state vector describing the system. This situation often occurs in practice, either deliberately in order to decrease the complexity of representation of the system or inadvertently because the existence of certain states is unknown. The example considered here is estimation of the altitude X of a vehicle from altimeter data. A Kalman filter is first designed under the assumption that X is a constant, i.e., the vehicle is at some constant altitude. The filter is then shown to diverge when the vehicle is actually climbing (or falling) at a constant rate. A minor modification to the Kalman filter is then shown to eliminate this divergence.

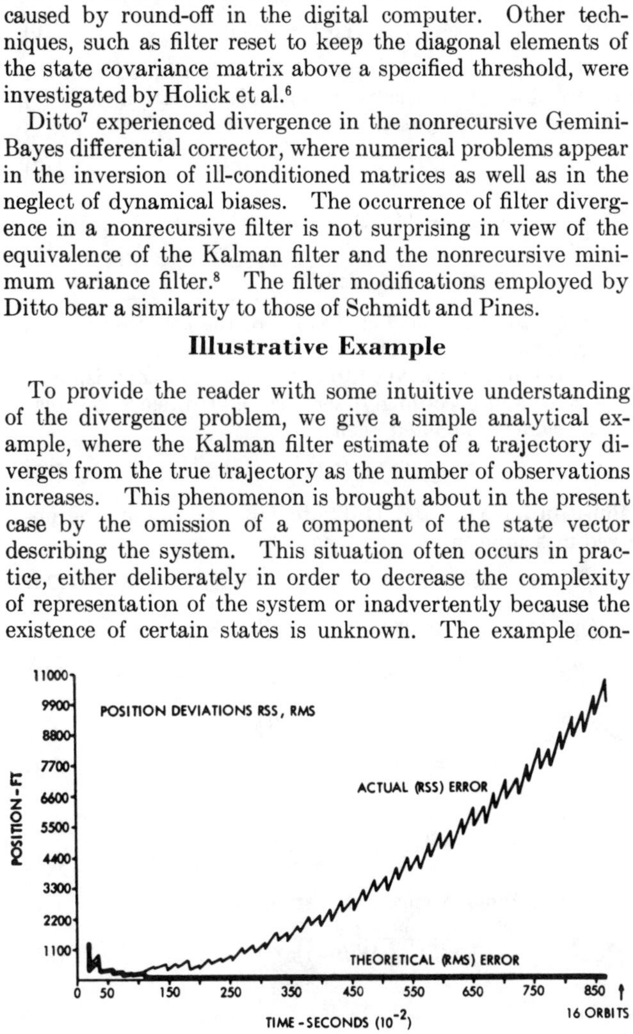

Fig. 1. Kalman filter divergence caused by 25% error in drag acceleration.

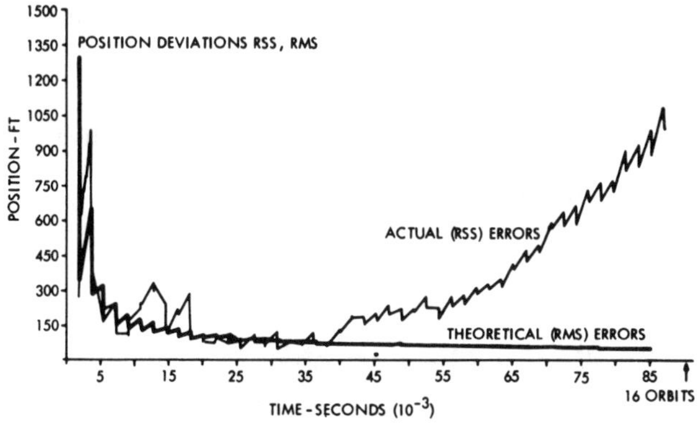

Fig. 2 Divergence of the Kalman filter caused by computional errors resulting from single precision arithmetic (IBM 7090).

To design the Kalman filter, we assume that altitude measurements are corrupted by zero mean independent random noise errors $n(k)$ possessing common standard deviation σ. In addition, the noise errors are assumed bounded, $|n(k)| \leq B$.

The Kalman filter estimate of altitude is given by

$$\left.\begin{aligned} X(k+) &= X(k-) + W(k)[\tilde{y}(k) - y_c(k)] \\ W(k) &= P(k-)/[P(k-) + \sigma^2] \\ P(k-) &= P(k-1+) \qquad X(k-) = X(k-1+) \\ P(k+) &= P(k-) - P(k-)[P(k-) + \sigma^2]^{-1}P(k-) \\ \tilde{y}(k) &= X(k) + n(k) \qquad y_c(k) = X(k-) \end{aligned}\right\} \quad (1)$$

where σ = standard deviation of altimeter noise error $n(k)$ $P(0-)$ = variance of the a priori altitude estimate; and the the symbols $-$ and $+$ appearing in the preceding equations denote estimates immediately prior to and immediately subsequent to an observation.

Suppose now that the vehicle, instead of remaining at a constant altitude, is actually climbing at a constant rate and its trajectory is described by

$$X(t) = X(0) + Vt \quad (2)$$

Suppose further that altitude measurements are taken once per second. An easy calculation yields for the Kalman filter altitude estimate after N observations:

$$X(N+) = X(0) + \frac{VN}{2} + \frac{1}{N+1}\sum_{k=1}^{N} n(k)$$

Subtracting the true trajectory

$$X(N) = X(0) + VN$$

one finds for the error $\epsilon(N)$

$$\epsilon(N) = X(N) - X(N+) = \frac{VN}{2} + \frac{1}{N+1}\sum_{k=1}^{N} n(k) \quad (3)$$

and since noise errors are bounded,

$$|\epsilon(N)| \geq (VN/2) - B$$

Hence the deviation of the Kalman filter estimate from the true trajectory increases with increasing N.

As we have seen, the divergence of the filter was brought about by a constant altitude rate, a phenomenon not provided for in the filter construction. One solution to divergence in this case is to design a filter to estimate both vehicle altitude and altitude rate. Although such a two-state filter would prevent divergence, it would do so at the cost of a considerably increased number of computations. An alternative is to modify the weighting sequence $W(k)$ of the one-state filter to include the effects of a climbing vehicle.

To implement this alternative, we assume that the true trajectory is given by

$$X^*(k+) = X^*(k-) + W(k)^*[\tilde{y}(k) - y_c^*(k)] \quad (4)$$

$$W(k)^* = P^*(k-1)/[P^*(k-1) + \sigma^2] \quad (5)$$

$$P^*(k+) = [1 - W^*(k)]\, P^*(k-) \quad (6)$$

$$P^*(k-) = P^*(k-1+) + \alpha \quad (7)$$

where α is the a priori variance of the rate of change of vehicle altitude.

Denoting the filter error $X(N) - X^*(N+)$ by $\epsilon^*(N)$, one finds from (4)

$$\epsilon^*(N) = \epsilon^*(N-1)[1 - W(N)^*] - V[1 - W(N)^*] + W(N)^* n(N) \quad (8)$$

Combine (5–7) to write

$$P(k+1-) = \frac{\sigma^2 P(k-)}{P(k-) + \sigma^2} + \alpha$$

It is easily seen that $P(k-)$ tends to a limit L as $k \to \infty$ given by

$$L = [\alpha + (\sigma^2 + 4\alpha\sigma^2)^{1/2}]/2$$

and hence $[1 - W(k)^*]$ tends to $L^* = \sigma^2(L + \sigma^2)^{-1}$. Since $|L^*| < 1$,

$$|1 - W^*(k)| < L_0 < 1$$

and for all k sufficiently large, the error recursion relation (8) becomes

$$|\epsilon^*(k)| \leqq L_0|\epsilon^*(k-1)| + L_0(|V| + B)$$

Thus, the $\epsilon^*(k)$ are majorized by a suitable solution of the difference equation

$$z(k) = L_0 z(k-1) + L_0(|V| + B)$$

Since with $|L_0| < 1$ all the solutions $Z(k)$ are bounded, then the $|\epsilon^*(k)|$ are bounded and the modified filter does not diverge.

Derivation of the Modified Equations

In this section, we present the equation for the modified Kalman filter employed in the simulation studies. The development follows the work of Pines[2] and Schmidt.[5]

Let $\mathbf{x}(k)$, $\mathbf{y}(k)$, $\mathbf{u}(k)$ denote true values of the orbital elements, measurements, and unestimated parameters, respectively, and let $\delta\mathbf{x}(k)$, $\delta\mathbf{y}(k)$, $\delta\mathbf{u}(k)$ denote deviations in the preceding quantities at time t_k. These deviations will be called states, observations, and biases, respectively. Let $\mathbf{x}$ ref (k), $\mathbf{y}$ ref (k), $\mathbf{u}$ ref (k) denote reference or nominal values. Then

$$\delta\mathbf{x}(k) = \mathbf{x}\text{ ref }(k) - \mathbf{x}(k)$$

$$\delta\mathbf{y}(k) = \mathbf{y}\text{ ref }(k) - \mathbf{y}(k) \qquad \delta\mathbf{u}(k) = \mathbf{u}\text{ ref }(k) - \mathbf{u}(k)$$

The true dynamic situation represented by a linearization about the reference trajectory is governed by

$$\begin{aligned}\delta\mathbf{x}(k+1) &= \Phi(k+1, k)\delta\mathbf{x}(k) + U(k+1, k)\delta\mathbf{u}(k) \\ \delta\mathbf{u}(k+1) &= \psi(k+1, R)\delta\mathbf{u}(k)\end{aligned} \quad (9)$$

The observables are

$$\delta\mathbf{y}(k) = H(k)\,\delta\mathbf{x}(k) + F(k)\,\delta\mathbf{u}(k) \quad (10)$$

The estimated states are denoted by

$$\begin{aligned}\delta\hat{\mathbf{x}}(k+) &= \mathbf{x}\text{ ref }(k) - \hat{\mathbf{x}}(k+) \\ \delta\hat{\mathbf{x}}(k-) &= \mathbf{x}\text{ ref }(k) - \hat{\mathbf{x}}(k-)\end{aligned} \quad (11)$$

where $k-$ and $k+$ denote estimates immediately before and immediately after the processing of the kth observation.

The measurement deviation $\delta\tilde{\mathbf{y}}(k)$ is given by

$$\delta\tilde{\mathbf{y}}(k) = H(k)\delta\mathbf{x}(k) + F(k)\delta\mathbf{u}(k) + \mathbf{n}(k) \quad (12)$$

where $\mathbf{n}(k)$ is the noise in the kth observation.

We now seek to find an estimate $\delta\hat{\mathbf{x}}(k+)$ of $\delta\mathbf{x}(k)$ of the form

$$\delta\hat{\mathbf{x}}(k+) = \delta\hat{\mathbf{x}}(k-) + W(k)[\delta\tilde{\mathbf{y}}(k) - \delta\mathbf{y}_c(k)] \quad (13)$$

where $\delta\mathbf{y}_c(k)$ is the computed estimate of the kth measurement based on the previous observations and $W(k)$ is chosen to minimize the covariance matrix

$$P(k+) = \overline{[\delta\mathbf{x}(k) - \delta\hat{\mathbf{x}}(k+)][\delta\mathbf{x}(k) - \delta\hat{\mathbf{x}}(k+)]^T} \quad (14)$$

where $(\overline{\ })$ denotes ensemble average and $(\ )^T$ denotes the matrix transpose.

$$\delta\mathbf{x}(k) - \delta\hat{\mathbf{x}}(k+) = \delta\mathbf{x}(k) - \delta\hat{\mathbf{x}}(k-) - W(k)\delta q(k) \quad (15)$$

where

$$\delta q(k) = \delta\tilde{\mathbf{y}}(k) - \delta\mathbf{y}_c(k) \quad (16)$$

Now it is clear that $P(k+)$ given by (14) is a function of the filter weighting coefficients $W(k)$. Minimization of $P(k)$ with respect to $W(k)$ yields

$$W(k) = A(k)[Y(k)]^{-1} \quad (17)$$

where

$$\begin{aligned}A(k) &= \overline{\{\delta\mathbf{x}(k) - \delta\hat{\mathbf{x}}(k-)\}\delta q^T(k)} \\ Y(k) &= \overline{\delta q(k)\,\delta q^T(k)}\end{aligned} \quad (18)$$

Now $\delta\mathbf{y}_c(k)$ is the predicted value of the kth observation, computed on the basis of observations taken prior to time t_k. If the initial estimate $\delta\hat{\mathbf{u}}(0)$ of the bias parameter is taken to be zero, and accordingly, zero henceforth, since the bias estimates are not updated, then

$$\delta\mathbf{y}_c(k) = H(k)\,\delta\hat{\mathbf{x}}(k-) \quad (19)$$

Substituting (12) and (19) into (16), we obtain the observed measurement deviation $\delta q(k)$ as

$$\delta\mathbf{q}(k) = H(k)\{\delta\mathbf{x}(k) - \delta\hat{\mathbf{x}}(k-)\} + F(k)\,\delta\mathbf{u}(k) + \mathbf{n}(k) \quad (20)$$

We now introduce the covariance matrices

$$\overline{[\delta\mathbf{x}(k) - \delta\hat{\mathbf{x}}(k-)][\delta\mathbf{x}(k) - \delta\hat{\mathbf{x}}(k-)]^T} = P(k-) \quad (21)$$

$$\overline{\mathbf{n}(k)\mathbf{n}^T(k)} = R(k) \quad (22)$$

$$\overline{\delta\mathbf{u}(k)\ \delta\mathbf{u}^T(k)} = D(k) \quad (23)$$

and the cross-correlation matrix

$$\overline{[\delta\mathbf{x}(k) - \delta\hat{\mathbf{x}}(k-)][\delta\mathbf{u}(k)]^T} = C(k-) \quad (24)$$

Following Pines, we make the assumptions that

$$\overline{[\delta\mathbf{x}(k) - \delta\hat{\mathbf{x}}(k-)]\mathbf{n}^T(k)} = 0 \quad (25)$$

$$\overline{\delta\mathbf{u}(k)\mathbf{n}^T(k)} = 0 \quad (26)$$

i.e., errors in the orbit estimate based on $k-1$ observations and errors in the bias states are uncorrelated with the noise in the kth observation. Subject to these assumptions, one easily expresses $A(k)$ and $Y(k)$ as

$$A(k) = P(k-)H^T(k) + C(k-)F^T(k) \quad (27)$$

$$Y(k) = H(k)\,P(k-)H^T(k) + R(k) + F(k)D(k)F^T(k) + H(k)\,C(k-)F^T(k) + F(k)C^T(k-)H^T(k) \quad (28)$$

One observes that if bias errors $\delta\mathbf{u}(k)$ were neglected then

$$W(k) = P(k-)H(k)^T[H(k)\,P(k-)H^T(k) + Q(k)]^{-1} \quad (29)$$

which is the conventional Kalman filter gain matrix.

Updating and Propagation of *P*(*k*), *D*(*k*), and *C*(*k*)

To complete the filter construction, one needs a procedure for updating the matrices $P(k-)$, $C(k-)$ after an observation has been taken, and a procedure for propagating the updated matrices forward to the time of the next observation.

We address the updating problem first. Let

$$P(k+) = \overline{[\delta\mathbf{x}(k) - \delta\hat{\mathbf{x}}(k+)][\delta\mathbf{x}(k) - \delta\hat{\mathbf{x}}(k+)]^T}$$

denote the covariance matrix of the errors in the state estimates immediately after processing the kth observation. Combining (15–17), one obtains

$$\delta\mathbf{x}(k) - \delta\hat{\mathbf{x}}(k+) = \delta\mathbf{x}(k) - \delta\hat{\mathbf{x}}(k-) - A(k)Y^{-1}(k)\delta q(k)$$

From this one easily obtains

$$\begin{aligned} P(k+) &= P(k-) - W(k)A^T(k) \\ &= P(k-) - W(k)[H(k)P(k-) + F(k)C^T(k-)] \end{aligned} \quad (30)$$

by employing (17) and (27).

The updated C matrix is obtained by employing (15, 20, 21, and 24) to yield

$$C(k+) = C(k-) - W(k)[H(k)C(k-) + F(k)\,D(k)] \quad (31)$$

The updated matrices $P(k+)$ and $C(k+)$ must now be propagated forward by the transition matrix to obtain $P(k+1-)$, $C(k+1-)$. The propagation of P is accomplished as follows. The true value of the state at t_{k+1} are related to the state and bias at t_k by Eq. (9). The state vector estimates $\delta\hat{\mathbf{x}}(k+1-)$ based on k observations is

$$\delta\hat{\mathbf{x}}(k+1-) = \Phi(k+1, k)\,\delta\hat{\mathbf{x}}(k+)$$

The computer actually produces a "rounded off" version of the foregoing, given by

$$\delta\hat{\mathbf{x}}(k+1-) = \Phi(k+1, k)\,\delta\hat{\mathbf{x}}(k+) + \mathbf{n}_x(k+1) \quad (32)$$

Assuming the "round-off noise" $\mathbf{n}_x(k+1)$ is uncorrelated with the state estimate $\delta\hat{\mathbf{x}}(k+)$ and bias, one finds

$$\begin{aligned} P(k+1-) = {} & \Phi(k+1, k)P(k+)\Phi^T(k+1, k) + \\ & \Phi(k+1, k)C(k+)U^T(k+1, k) + \\ & U(k+1, k)C^T(k+)\Phi^T(k+1, k) + \\ & U(k+1)D(k)U^T(k+1, k) + Q_x(k+1) \end{aligned} \quad (33)$$

where

$$Q_x(k) = \overline{\mathbf{n}_x(k)\mathbf{n}_x{}^T(k)} \quad (34)$$

also

$$C(k+1) = \Phi(k+1, k)C(k)\psi^T(k+1, k) + U(k+1, k)D(k)\psi^T(k+1, k) \quad (35)$$

$$D(k+1) = \psi(k+1, k)D(k)\psi^T(k+1, k) \quad (36)$$

The modified filter equations previously discussed are applied to three types of error sources: round-off errors due to the finite word length of a digital computer, errors in the gravity constant, and errors in the drag acceleration.

Round-Off Error

Each computation performed in a digital computer will be in error because of the finite word length limitations. For instance, a computer with an eight-digit word length will store a number as a sequence of eight digits: $z_1, z_2, \ldots z_8$ and an exponent p. Thus, the computer stores z as

$$\tilde{z} = \cdot z_1\, z_2 \ldots z_8 10^p$$

The difference between z and $\tilde{z}$ is the round-off error $n_z = z - \tilde{z}$. This error has the same sign as z. Since the computer can include only the first eight digits, the round-off error is of the order of magnitude of 10^{p-8}. Thus, an estimate of the round-off error is

$$n_z = z10^{p-8}/|z| \quad (37)$$

Since, to order-of-magnitude accuracy, $|z| = 10^p$, Eq. (37) reduces to $n_z = z10^{-8}$. This is the round-off error model suggested by Pines[2] and utilized in the present study.

Level I Modification

A so-called Level I modification to the Kalman filter was used to compensate for round-off errors. This modification is based on assuming that all effects of round-off errors can be expressed as an error in (32) and that

$$\mathbf{n}_x(k+1) = \begin{bmatrix} x_1(k+1) \\ x_2(k+1) \\ x_3(k+1) \\ x_4(k+1) \\ x_5(k+1) \\ x_6(k+1) \end{bmatrix} (10^{-\alpha})$$

where $x_1(k+1) \ldots x_6(k+1)$ are the components of the six-dimensional computed state vector at time $k+1$, and α is a parameter that was varied during the study. (Investigations were made between 5.0 and 8.0.)

The covariance matrix $Q_x(k+1)$ given by (22) is then

$$Q_x(k+1) = 10^{-2\alpha} \begin{bmatrix} x_1(k+1)^2 & & & 0 \\ & \ddots & & \\ & & \ddots & \\ 0 & & & x_6(k+1)^2 \end{bmatrix} \quad (38)$$

Assumptions

This matrix is diagonal because round-off errors are assumed to be independent. Round-off errors are assumed to have no effect upon the observation $\delta\tilde{\mathbf{y}}(k+1)$. Thus, the matrix $F(k+1)$ is zero (null). Given $F = 0$, the filter equations (27) and (28) reduce to (29), the unmodified Kalman filter. Round-off errors at time $k+1$ are assumed to be independent of the estimated state vector $\hat{\mathbf{x}}(k)$. Thus, the matrix $C(k-)$ given by (24) is zero (null). Given $F = 0$ and $C = 0$, Eqs. (30) and (33) reduce to

$$P(k+) = P(k-) - W(k)[H(k)P(k-)] \quad (39)$$

$$P(k+1-) = \Phi(k+1, k)P(k+)\Phi^T(k+1, k) + Q_x(k+1) \quad (40)$$

The only difference between the ordinary unmodified Kalman filter and the machine noise modified filter is the added term $Q_x(k+1)$ in Eq. (40), the covariance matrix propagation. Note the similarity between (40) and (7), the analogous relation for the illustrative example. Modifications in both cases are in the form of additions to the filter covariance matrix P.

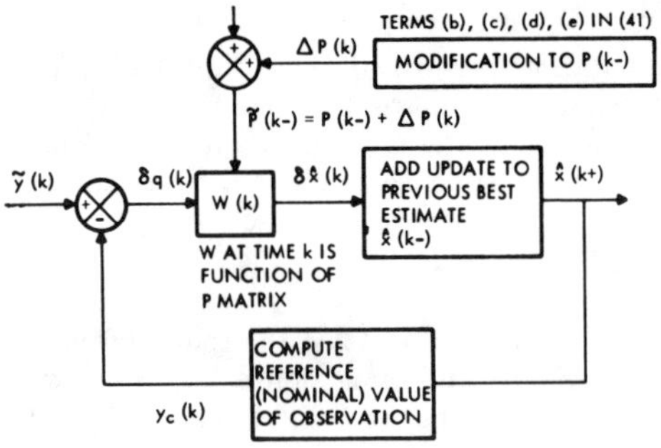

Fig. 3 Modification to the Kalman filter.

Dynamical Bias Errors

Bias errors may be divided into two categories: dynamical biases—errors in parameters which enter into the equations of motion of the vehicle; and instrument biases—systematic errors in the observational data (for example, errors caused by misalignment of instruments or miscalibration). We shall limit discussion to the dynamical biases and in particular, errors in the gravity and drag computation.

Form of Modified Filter

The sensitivity matrix F in the modified filter equation is a null matrix because dynamical biases do not affect observational data. With this simplification, the modified filter equations again reduce to (29).

The update equations (30) and (31) become:

$$P(k+) = P(k-) - W(k)\, H(k)\, P(k-)$$

$$C(K+) = C(k-) - W(k)\, H(k)\, P(k-)$$

Propagation equations (33, 35, and 36) become:

$$\begin{aligned} P(k+1-) = {} & \Phi(k+1, k)P(k+)\Phi^T(k+1, k) && \text{a)} \\ & + \Phi(k+1, k)C(k+)U^T(k+1, k) && \text{b)} \\ & + U(k+1, k)C(k+)\Phi^T(k+1, k) && \text{c)} \\ & + U(k+1, k)DU^T(k+1, k) && \text{d)} \\ & + Q_x(k+1) && \text{e)} \end{aligned} \quad (41)$$

$$C(k+1-) = \Phi(k+1, k)\, C(k+) + U(k+1, k)D \quad (42)$$

$$D(k+1) = D(\text{a constant for all } k) \quad (43)$$

In this study, the gravity bias was assumed to be a (constant) error in the gravitational constant; the drag bias was assumed to be a (constant) error in the ballistic drag coefficient. Since the biases do not change with time, $\psi(k+1, k)$ in (35) and (36) is the identity matrix and (42) and (43) follow.

Note that the filter equation (29) and the $P(k+)$ update equation are identical to the unmodified Kalman filter. Only the propagation equation (41) differs, and the differences are in the form of additive modification terms. Term a in (41) is the term that appears in the unmodified Kalman filter, terms b and c are modifications due to correlation between the bias state and the estimation error, term d is the modification due to bias uncertainty (as represented by the bias error covariance matrix D, term e is the effect of computational noise. Figure 3 shows the modified Kalman filter.

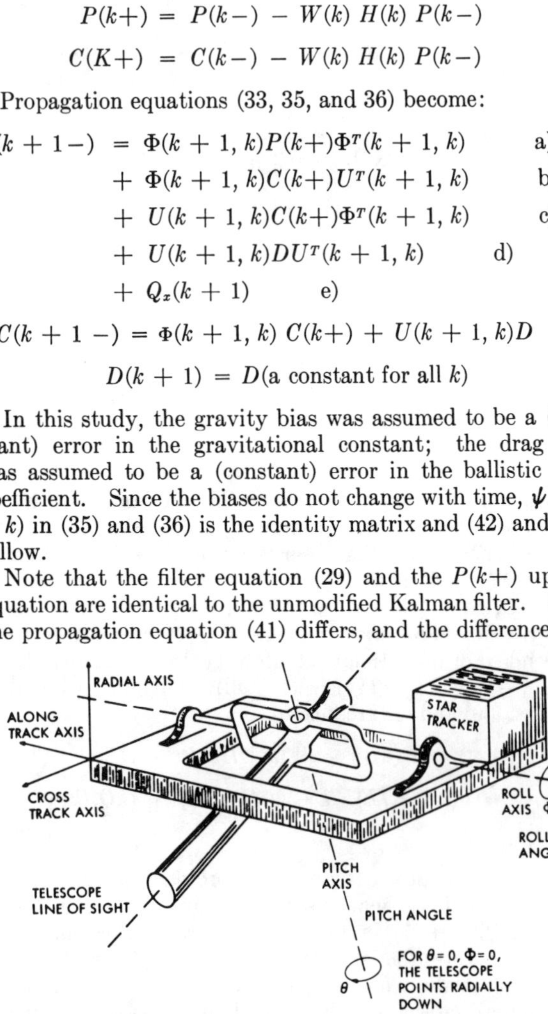

Fig. 4 Telescope gimbal layout.

Simplified Versions of the Modified Filter

Simulation studies were performed using three levels of approximation to the modified filter equations:

Level I: the simplest approximation to the modified filter assumed that $C(k)$ and D in (41) are null matrices with the result:

$$P(k+1-) = \Phi(k+1, k)\, P(k+)\Phi^T(k+1, k) + Q_x(k+1)$$

The matrix $Q_x(k+1)$ is of the form given by (38)

Level II: Only the cross-correlation matrix $C(k)$ in (41) is assumed null. Thus,

$$P(k+1-) = \Phi(k+1, k)\, P(k+)\Phi^T(k+1, k) + U(k+1, k)DU^T(k+1, k) + Q_x(k+1)$$

Level III: All terms in (41) are included in the filter propagation.

There are considerable differences in the computational complexity of the three levels of filter modification. Level I is clearly only a trivial increase in the complexity of the unmodified filter. Level II, however, requires a significantly greater number of computations, primarily in the computation of the bias state transition matrix $U(k+1, 1)$.

The Level III modified filter is more complex than Level II because of the additional terms b and c in (41) and the addition of the update and propagation relations for matrix $C(k)$. Some idea of the relative complexity of the three filter schemes may be derived from the execution times of the computer simulations: a Level I filter simulation executed in 3.00 mins, a comparable Level II filter simulation executed in 6.60 min, and a Level III simulation, 9.18 min.

Simulation Studies

Simulation studies investigated the divergence of the Kalman filter used for autonomous navigation in a low earth orbit. Observations consist of pitch and roll line-of-sight angles derived from optical tracking of known terrestrial landmarks. Figure 4 defines telescope pitch and roll angles. Inertial attitude reference is maintained by an automatic star-tracker system. A Kalman filter is used to compute differential corrections to nine states: three components of spacecraft position, three components of spacecraft velocity, and three components of landmark location.

Divergence was investigated by simulating the entire navigation process including computation of the true orbit and the estimated orbit on an IBM 7090 computer. Investigations considered both a six-state filter, which estimated

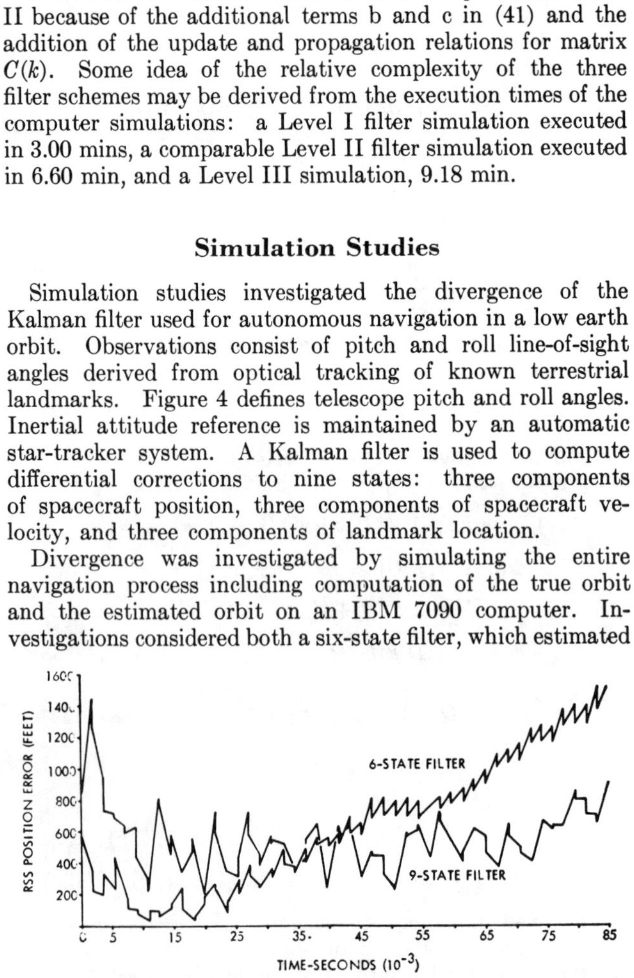

Fig. 5 Effect of filter state dimension upon divergence.

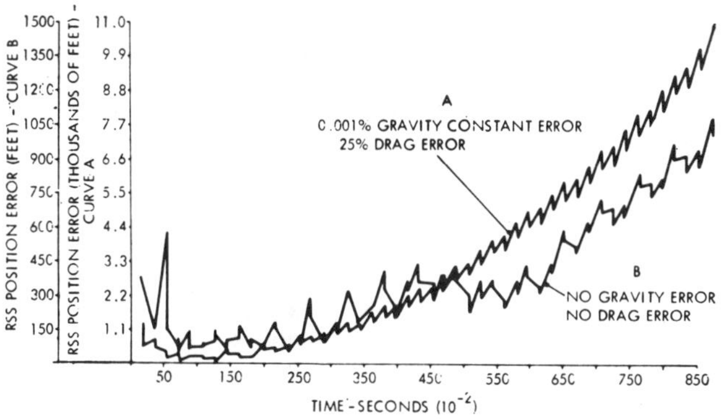

Fig. 6 Effect of dynamic bias errors upon divergence.

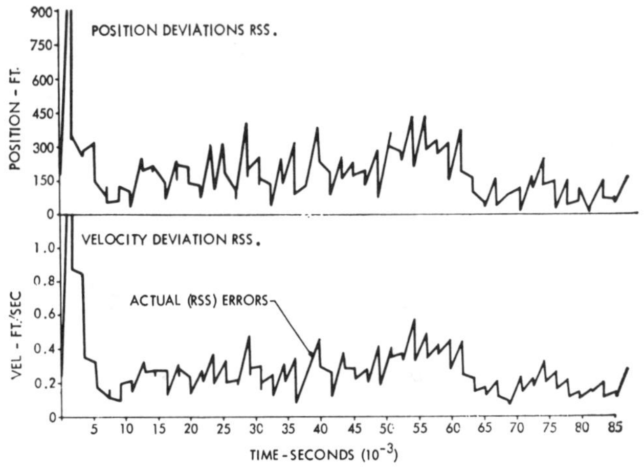

Fig. 8 Level I modified filter.

only spacecraft states, and a nine-state filter. The effect of adding altimeter measurements was also investigated. Simulation studies utilized a nominal 160-naut-mile circular orbit. Three landmark tracking exercises were performed each orbit. Landmarks were assumed to be first acquired at 30° (forward) pitch. Out-of-plane orientation of the landmark at acquisition was random but limited to ±30° roll angles. Level I, II, and III modified filters, in addition to the unmodified Kalman filter, were studied under a variety of conditions reflecting effects of gravity error, drag error, data rate, pointing accuracy, and landmark location errors.

Simulations showed that an unmodified Kalman filter always diverged after a sufficiently long period of operation. A nine-state variable filter simulation was less divergent than an equivalent six-state filter. Figure 5 shows that the nine-state variable filter diverges later and more slowly than the six-state variable filter. The divergence is caused by round-off in the computations.

The rss position error plotted in Fig. 5 is the length of the vector difference between true spacecraft position and the position estimated by the Kalman filter. Rss position error is thus a measure of the actual navigation error. Rms position error is the square root of the trace of the position covariance matrix. As such, rms is an ensemble statistic that represents the average or expected position error. If there were no machine computation errors and no uncompensated bias errors, the rms position error would be a measure of the actual position errors. Rms errors are also referred to as theoretical errors whereas rss are referred to as actual errors.

Presence of errors in the gravity model and drag parameters used to compute spacecraft ephemeris increased filter divergence. Figure 6 shows the divergence caused by an error of 0.001% in the gravity constant and a 25% error in ballistic drag coefficient. Addition of another measurement such as altimeter data increases the divergence. Simulations show that the rss error increased from 11,000 to 16,000 ft after 16 orbits for the gravity and drag error case shown in Fig. 6. An increase in the assumed (a priori) data accuracy likewise aggravates divergence. Simulations that assumed no gravity and no drag error showed a position error of 1085 ft at the end of 16 orbits when the one-sigma telescope pointing error was assumed to be 40 sec, and 1500 ft when the telescope pointing error was assumed to be 10 sec. Increased data rate increases filter divergence. Figure 7 compares a run with eight sightings (eight pairs of pitch and roll angle measurements are made) on each landmark with a run in which only three sightings are made on each landmark.

These results all support the intuitive notion that filter divergence occurs when filter weights become too small. Addition of an altimeter, increased data rate, and increased data accuracy all reduce the filters estimation error covariance matrix and thereby decrease the filter weights. On the other hand, including an additional error source, landmark uncertainty, as in the nine-state filter, increases the filter covariance matrix for sufficiently large landmark uncertainty and would be expected to reduce divergence. The landmark uncertainty used in Fig. 5 is 600 ft (in each axis).

Simulations of the modified filter have shown that divergence can be avoided and that a form of steady-state behavior is achieved. Figure 8 shows that rss error for a Level I modified filter. Figure 9 shows a Level III modified filter. A series of simulations was performed to determine the effects of data rate and magnitude of the α parameter in (38) upon performance of a Level I modified filter. A histogram-like plot was used to quantitatively evaluate these results. These

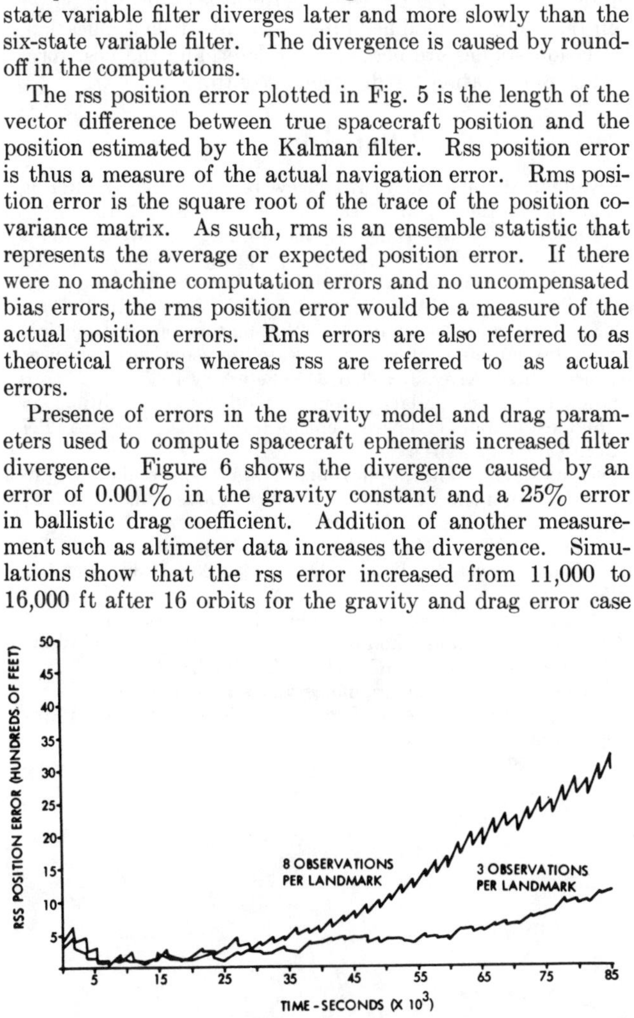

Fig. 7 Effect of data rate upon divergence.

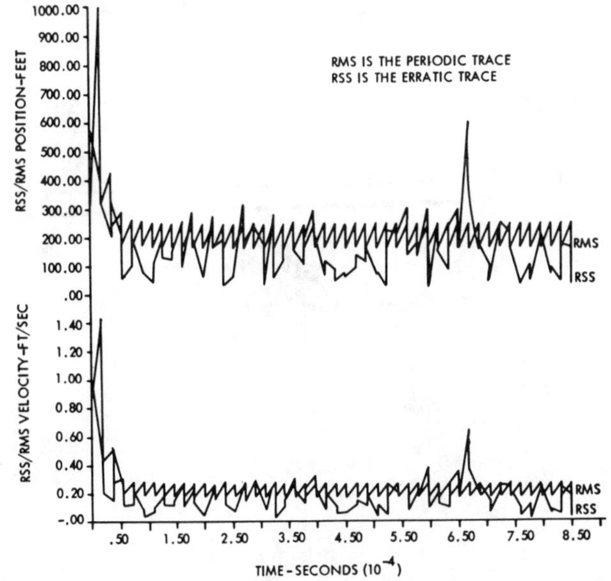

Fig. 9 Level III modified filter.

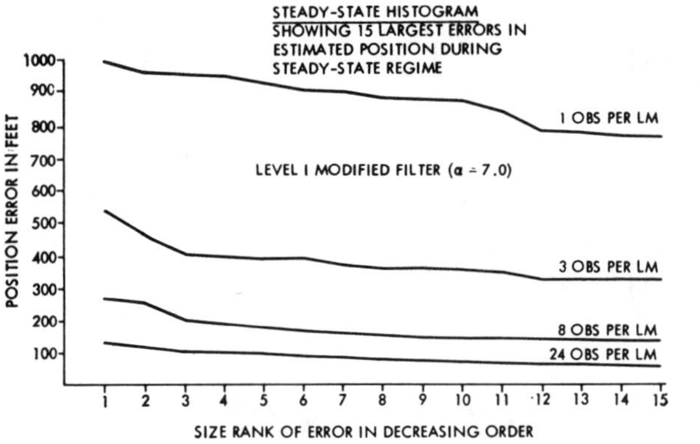

Fig. 10 Effect of varying number of observations per landmark.

plots contain the 15 largest rss position error excursions that occur after steady state is achieved. Figure 10 shows the histograms for a Level I modified filter for various values of the data rate. Figure 11 shows a similar plot for various values of the α modification parameter. A comparison between Level I, Level II, and Level III filters is made in Fig. 12.

Conclusions

Simulation studies have shown that divergence depends upon the a priori instrument accuracy, data rate, and number of observations processed; divergence is more severe for sensors with increased a priori accuracy, for increased data rates, and for landmark tracking systems augmented by a radar altimeter. The addition of three landmark bias states, with moderate standard deviations, retards but does not prevent divergence. Although dynamical biases could be included in the state to be estimated, long-duration Kalman filter operation will still require modifications to control divergence caused by round-off errors.

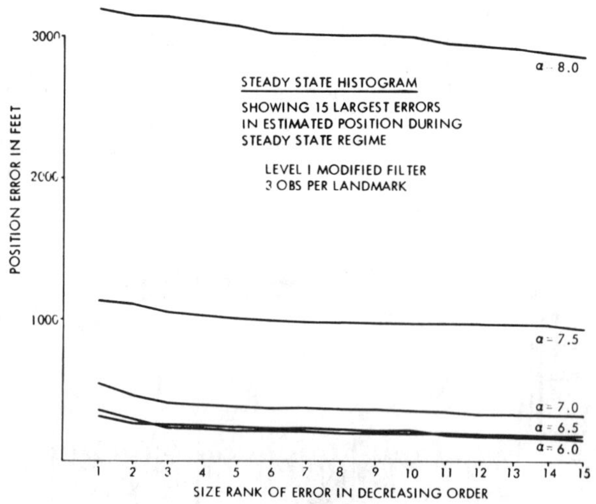

Fig. 11 Effect of varying a parameter.

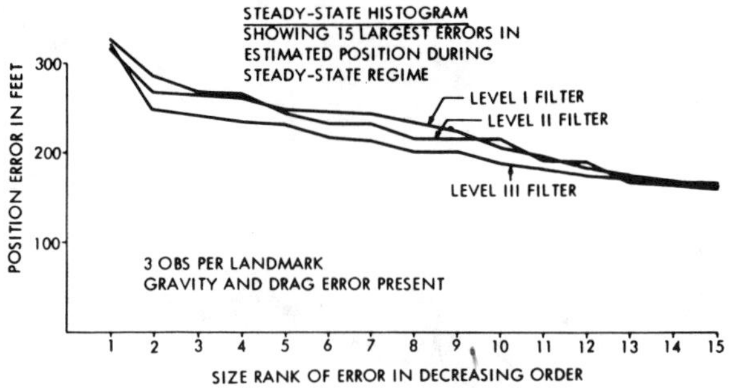

Fig. 12 Comparison of three level filter modifications.

The suppression of divergence by modifications to account for both gravity and drag dynamic biases is demonstrated. The Level I modification is shown to control filter divergence caused by finite word length (single precision IBM 7090) computations, and with properly chosen parameters, is as effective as the Level II and Level III modifications in controlling divergence due to dynamic biases in the drag and gravity models.

Level III modifications would be optimal (since they are derived as a constrained optimal filter) if computer round-off errors were not present. Since round-off errors are always present in a finite word length machine, the Level III modification is not optimal. Thus it is not surprising to find in Fig. 12 that the simple Level I modification performs as well as the complex Level III modifications. It is entirely conceivable that, for certain combinations of dynamic bias error and data rate, data accuracy and computer word length, the Level I filter could perform significantly better than the Level III modified filter. In any given application, simulation studies are necessary to determine the best form of filter modification for divergence control.

References

[1] Kalman, R. and Bucy, R., "New results in linear filtering and prediction theory," J. Basic Eng., 95–108 (March 1961).

[2] Pines, S., Wolf, H., Baile, A., and Mohan, J., "Modifications of the Goddard minimum variance program for the processing of real data," Analytical Mechanics Associates, Uniondale, N.Y., Tech. Rept., NASA Contract 5-2535 (October 16, 1964).

[3] Knoll, A. and Edelstein, M., "Estimation of local vertical and orbital parameters for an earth satellite using horizon sensor measurements," AIAA J. **3**, 338–345 (February 1965).

[4] Smith, G., "Secondary errors and off-design conditions in optimal estimation of space vehicle trajectories," NASA TN D-2129 (January 1964).

[5] Schmidt, S., "The application of state space methods to navigation problems," Philco Western Development Labs., Palo Alto, Calif., Rept. TR4 (October 1963).

[6] Sward, D. V. and Holick, A., "Computer precision study II," IBM Federal Systems Div., IBM Rept. 65-504-30 (October 1965).

[7] Ditto, F., private communication (March 1965).

[8] Ho, Y. C., "The method of least squares and optimal filtering," Rand Corp., Memo RM-3329-PR (October 1962).

[9] Potter, J. E., "Space guidance analysis memo 40," Massachusetts Institute of Technology Instrumentation Lab., Cambridge, Mass. (April 3, 1963).

Divergence of the Kalman Filter

ROBERT J. FITZGERALD, MEMBER, IEEE

Abstract—The Kalman estimation technique is examined from the point of view of the asymptotic behavior of the errors in the estimates. It is shown that, under certain conditions, the mean-square errors may become unbounded with time, and that this divergence may or may not be correctable by increasing the intensity of process noise assumed in the filtering model. General results are derived for multidimensional systems, and both "true" and "apparent" divergence are demonstrated by a simple scalar system. Divergence due to numerical inaccuracies is considered, and an example problem in orbital navigation is used to demonstrate divergence and its elimination.

I. Introduction

ALTHOUGH the Kalman filtering technique has gained wide acceptance in recent years, and has proved to be extremely useful in a wide variety of applications, it has become apparent that insufficient care in modeling the system can easily lead to entirely unacceptable results. In particular, the usefulness of the filter may be completely nullified by the phenomenon known as "divergence" [1], [2]: after an extended period of operation of the filter, the errors in the estimates eventually diverge to values entirely out of proportion to the rms values predicted by the equations of the filtering procedure.

The explanation most often offered for this phenomenon is that the calculated covariance matrix becomes unrealistically small, so that undue confidence is placed in the estimates and subsequent measurements are effectively ignored. This effect can be due to a variety of causes; in general, any inaccuracy in the model used or in the computational operations performed offers the possibility of causing the calculated covariance matrix to become overly optimistic. In orbital navigation problems, where the divergence phenomenon is often observed, these error sources may include inaccuracies in the density and gravity

Manuscript received July 29, 1971. Paper recommended by D. G. Luenberger, Associate Guest Editor.

The author is with the Missile Systems Division, Raytheon Company, Bedford, Mass. 01730.

Reprinted from *IEEE Trans. Automat. Contr.,* vol. AC-16, pp. 736–747, Dec. 1971.

models used in the filter equations, nonlinearities, biases which are not compensated for, lack of knowledge of statistical models for filter derivation, and roundoff and truncation errors in the computations.

The purpose of this paper is to investigate the mechanisms by which such error sources degrade the performance of the filter, and hopefully to provide some clues to techniques which may be used to lessen such degradations. We approach the problem by considering both "apparent" divergence (finite degradations due to modeling inaccuracies) and "true" divergence, in which the errors may actually become unbounded. Sections II and III present the appropriate equations for analysis of suboptimal filters, and some considerations of the steady-state behavior of the calculated covariance matrix. In Section IV we consider the causes of true divergence, and in Section V apparent divergence (including the effects of ignored biases). A simple scalar system is used for parametric studies where appropriate. The related phenomenon of numerical divergence is considered in Section VI, and a numerical example is presented in Section VII.

II. Effects of Model Errors

In this section we examine the behavior of a multidimensional Kalman filter in the presence of errors in the values of some of the system parameters assumed by the filter.

We assume a dynamic system and measurement device described by the equations

$$\dot{x} = Fx + v \tag{1}$$

$$m = Hx + w \tag{2}$$

where the "process noise" $v(t)$ and the "measurement noise" $w(t)$ are white-noise vectors with correlation matrices

$$\langle v(t)v^T(\tau)\rangle = Q\delta(t-\tau) \tag{3}$$

$$\langle w(t)w^T(\tau)\rangle = R\delta(t-\tau). \tag{4}$$

In many cases the linear equations (1) and (2) will represent a linearization of the true nonlinear relationships around some nominal solution. The applicability of the linear filter therefore requires that the errors in the estimates be "sufficiently" small. The Kalman filter is described by

$$\dot{\hat{x}} = F_c\hat{x} + K(m - H_c\hat{x}) \tag{5}$$

$$\dot{P}_c = F_cP_c + P_cF_c^T - P_cH_c^TR_c^{-1}H_cP_c + Q_c \tag{6}$$

where

$$K = P_cH_c^TR_c^{-1} \tag{7}$$

and we have used the subscript c to indicate values calculated by the filter or parameter values used in the calculations. The vector $\hat{x}$ is the estimate of x, and the matrix P_c is the "calculated" covariance matrix of the errors e in the estimates. If the estimation error e is defined by

$$e = \hat{x} - x,$$

then (1) and (5) yield a differential equation for e in the form

$$\begin{aligned}\dot{e} &= (F_c - F)x + F_ce + K(m - H_c\hat{x}) - v\\ &= G\dot{e} + Cx + Kw - v\end{aligned} \tag{8}$$

where

$$G = F_c - KH_c \tag{9}$$

and

$$\begin{aligned}C &= (F_c - F) - K(H_c - H)\\ &= \Delta F - K\Delta H.\end{aligned} \tag{10}$$

We may now derive the equation for propagation of the "actual" covariance matrix

$$P_a = \langle ee^T\rangle \tag{11}$$

from the relation

$$\dot{P}_a = \langle\dot{e}e^T\rangle + \langle e\dot{e}^T\rangle. \tag{12}$$

Letting

$$S = \langle ex^T\rangle \tag{13}$$

we have

$$\langle\dot{e}e^T\rangle = GP_a + CS^T + K\langle we^T\rangle - \langle ve^T\rangle \tag{14}$$

and since

$$\langle we^T\rangle = \langle w\hat{x}^T\rangle = \tfrac{1}{2}RK^T \tag{15}$$

and

$$\langle ve^T\rangle = -\langle vx^T\rangle = -\tfrac{1}{2}Q \tag{16}$$

we may write (12) as

$$\dot{P}_a = GP_a + P_aG^T + CS^T + SC^T + KRK^T + Q. \tag{17}$$

Solution of (17) requires the simultaneous determination of $S(t)$, whose differential equation can be shown, by a similar procedure, to be

$$\dot{S} = SF^T + GS + CX - Q \tag{18}$$

where the covariance matrix of the system state

$$X = \langle xx^T\rangle \tag{19}$$

is the solution of the differential equation

$$\dot{X} = FX + XF^T + Q. \tag{20}$$

The behavior of the filtering system, in the presence of the assumed parameter errors, can now be observed by simultaneous propagation of (6), (17), (18), and (20). Analysis of the behavior of the solutions is difficult in the general time-varying case, but conclusions drawn from consideration of the time-invariant case have direct bearing on the more general problem. Accordingly, in what follows we shall consider only the constant-coefficient case.

III. Asymptotic Behavior of the Calculated Covariance Matrix

Some results directly applicable to this problem have been obtained by Potter [3], who has examined the properties of (6) in the case of constant-coefficient systems which, besides exhibiting positive semidefiniteness of R_c and Q_c, obey the following "regularity" conditions.

1) No eigenvector of F_c whose eigenvalue has a nonnegative real part is a null vector of $R_c^{-1}H_c$ (i.e., every unstable mode of the system represented by F_c must be observable).

2) No eigenvector of F_c^T whose eigenvalue has a nonnegative real part is a null vector of Q_c (i.e., every unstable mode of F_c^T must be excited by the white-noise input).

Potter's main result is that, if the (constant-coefficient) system $(F_c, R_c^{-1}H_c, Q_c)$ is regular, then (6) has a unique positive semidefinite critical point (a solution of $\dot{P}_c = 0$), which is asymptotically stable and to which all other solutions converge exponentially fast.

The uniqueness of the critical point may not hold if one of the regularity conditions is violated. In the scalar case, for example, if Q_c (a scalar) vanishes and $F_c > 0$ in (6), the system is irregular. The equation then has two nonnegative critical points, only one of which is stable. If F_c also vanishes, then the solution $P_c = 0$ is approached asymptotically, but not exponentially fast.

It may also be pointed out that the steady-state solution of (6) is not unique, but rather depends on the initial conditions, if there exists a vector $\boldsymbol{e}$ such that $F_c\boldsymbol{e} = R_c^{-1}H_c\boldsymbol{e} = \boldsymbol{0}$ (which violates the first regularity condition), i.e., if the system model has an unobservable mode corresponding to a zero eigenvalue. For, in this case, it is apparent that the right-hand side of (6) is unchanged by the addition of any multiple of $\boldsymbol{e}\boldsymbol{e}^T$ to P_c.

At this point it would be well to point out some facts concerning the stability properties of a matrix differential equation of the form

$$\dot{Y} = AY + YB^T + Z \tag{21}$$

where Y and Z are $(m \times n)$, A is $(m \times m)$, and B is $(n \times n)$. If the columns of Y are placed end to end to form an mn-vector $\boldsymbol{y}$, and similarly with Z to form $\boldsymbol{z}$, the equation may be rewritten in the form

$$\dot{\boldsymbol{y}} = M\boldsymbol{y} + \boldsymbol{z} \tag{22}$$

in which the $(mn \times mn)$ matrix M is given by [4]

$$M = A \times I_n + I_m \times B \tag{23}$$

where I_n and I_m are the $(n \times n)$ and $(m \times m)$ identity matrices, and the matrix products are to be interpreted as Kronecker products.

If $\lambda_i (i = 1, 2, \ldots, m)$ and $\mu_j (j = 1, 2, \ldots, n)$ are the eigenvalues of A and B, respectively, then the mn eigenvalues $\eta_{i,j}$ of M are given by

$$\eta_{i,j} = \lambda_i + \mu_j, \qquad i = 1, 2, \ldots, m; j = 1, 2, \ldots, n. \tag{24}$$

The stability properties of (21) therefore depend on the pairwise sums of the eigenvalues of A and B.

The eigenvector of M corresponding to the eigenvalue $\eta_{i,j} = \lambda_i + \mu_j$ is the mn-vector

$$\boldsymbol{m}_{i,j} = \begin{bmatrix} a_{i1}\boldsymbol{b}_j \\ a_{i2}\boldsymbol{b}_j \\ \cdot \\ \cdot \\ \cdot \\ a_{im}\boldsymbol{b}_j \end{bmatrix} \tag{25}$$

where $\boldsymbol{a}_i$ and $\boldsymbol{b}_j$ are the corresponding eigenvectors of A and B, and a_{ik} is the kth element of $\boldsymbol{a}_i$.

Returning to (6), several conclusions can be demonstrated concerning the properties of the asymptotic values of $P_c(t)$. Let P_{cs} be a positive semidefinite critical point[1] of (6), and let $\boldsymbol{e}$ be an eigenvector of F_c^T corresponding to the eigenvalue $(\lambda + i\mu)$. Then from (6) we have

$$2\lambda \boldsymbol{e}^* P_{cs}\boldsymbol{e} = \boldsymbol{e}^* P_{cs}H_c^T R_c^{-1} H_c P_{cs}\boldsymbol{e} - \boldsymbol{e}^* Q_c \boldsymbol{e}. \tag{26}$$

The following conclusions can be drawn concerning the properties of P_{cs}.

1) If $\boldsymbol{e}$ is "stable" $(\lambda < 0)$ and "unforced" $(Q_c\boldsymbol{e} = \boldsymbol{0})$, then it must be a null vector of P_{cs} (i.e., $P_{cs}\boldsymbol{e} = \boldsymbol{0}$).

2) If $\boldsymbol{e}$ is forced $(Q_c\boldsymbol{e} \neq \boldsymbol{0})$, whether stable or not, it cannot be a null vector of P_{cs}.

3) If $\boldsymbol{e}$ is "unstable" $(\lambda > 0)$ and P_{cs} is a stable critical point, then $\boldsymbol{e}$ cannot be a null vector of P_{cs}.

The first two statements follow directly from the fact that all three quadratic forms in (26) are at least positive semidefinite. In the case where $\boldsymbol{e}$ is unstable, we make use of the fact that the perturbation of $P_c(t)$ from P_{cs}, $\Delta P_c = P_c - P_{cs}$, satisfies

$$\Delta\dot{P}_c = G_s\Delta P_c + \Delta P_c G_s^T - \Delta P_c H_c^T R_c^{-1} H_c \Delta P_c \tag{27}$$

and linearizing around $\Delta P = 0$,

$$\delta\dot{P}_c = G_s\delta P_c + \delta P_c G_s^T \tag{28}$$

so that the stability of the steady state depends on the eigenvalues of G_s. If $P_{cs}\boldsymbol{e} = \boldsymbol{0}$, then $G_s^T\boldsymbol{e} = F_c^T\boldsymbol{e}$ and $\boldsymbol{e}$ is an eigenvector of G_s^T with the same eigenvalue. Hence if $\lambda > 0$, P_{cs} cannot be a stable steady state.

The significance of null vectors of P_c lies in the fact that they indicate certain linear combinations of the error variables which are believed to vanish, i.e., certain functions of the state variables which are thought to be perfectly known; for if $P\boldsymbol{e} = \boldsymbol{0}$, then $\boldsymbol{e}^T P\boldsymbol{e} = \langle[\boldsymbol{e}^T(\hat{\boldsymbol{x}} - \boldsymbol{x})]^2\rangle$ also vanishes.

The cases where $\lambda = 0$ are more difficult to analyze. The first conclusion above is still valid in modified form

[1] The subscript s denotes "steady state."

(i.e., if $Q_c e = 0$ and a critical point P_{cs} exists, then $R_c^{-1}H_cP_{cs}e = 0$). In cases where μ does not vanish, however, there arises the possibility of undamped oscillations in $P_c(t)$, in which case (26) is not necessarily valid.

One result which can be demonstrated in any case is the following. If $P_c(t)$ (in general, time varying) possesses a zero eigenvalue, it will retain this zero eigenvalue as long as the corresponding (time-varying) eigenvector is a null vector of Q_c. To show this, we note that if $P_c v = \eta v$, then by differentiation

$$v^*\dot{P}_c v + v^*P_c\dot{v} = \dot{\eta}v^*v + \eta v^*\dot{v} \tag{29}$$

and substituting ηv^* for v^*P_c, we obtain

$$\dot{\eta} = \frac{v^*\dot{P}_c v}{v^*v} = v^*\dot{P}_c v \tag{30}$$

if we assume that v is normalized. But in the case $\eta = 0$ we have $P_c v = 0$, so that using (6) we find

$$\dot{\eta} = v^*\dot{P}_c v = v^*Q_c v \tag{31}$$

so that η will remain zero as long as $Q_c v = 0$.

A particular case worthy of mention is that of $Q_c = 0$. Such a model is often used in space navigation problems, where a vehicle in "free fall" is assumed to be entirely free of disturbances. This case may not be covered by Potter's results since it violates the second regularity condition if F_c is unstable. If $Q_c = 0$, then $P_c = 0$ is a critical point of (6), i.e., causes $\dot{P}_c$ to vanish. However, in this case, since $K_s = 0$ so that (28) becomes

$$\delta\dot{P}_c = F_c\delta P_c + \delta P_c F_c^T$$

we see that this particular critical point is stable or unstable, accordingly, as F_c is stable or unstable. Hence in the unstable case the stable steady-state covariance matrix, if such a solution exists, cannot be the null matrix.

As has been pointed out in [3], it may be shown, from the definition of G, that no stable steady state exists for $P_c(t)$ if any eigenvector of F_c (not F_c^T), corresponding to an eigenvalue with positive real part, is a null vector of $R_c^{-1}H_c$, i.e., if the system has an unstable mode which is not observable.

If it is assumed that the solution of (6) approaches a stable steady state, the investigation of the divergence phenomenon reduces to an examination of the asymptotic behavior of the solution $P_a(t)$ of (17). The stability properties of this equation do not, however, completely resolve the problem for practical purposes. If some of the mean-square errors approach stable, but very large, steady-state values, the filter may be as useless, in a practical sense, as if the errors were truly unbounded. We are therefore forced to consider two different phenomena.

1) True, or "mathematical," divergence, in which the mean-square errors can actually be shown to approach infinity with increasing time.

2) Apparent, or practical, divergence, in which a steady state is reached but the associated errors are too large to allow the estimates to be useful. Obviously this case can only be demonstrated by parametric studies.

We shall examine the first of the above effects, that of true divergence, before proceeding to an investigation of apparent divergence.

In what follows, we shall largely ignore the difficult case of the oscillatory steady-state covariance matrix (which arises when an undamped oscillatory mode of the system is unobservable), so that the term "steady-state" will refer to a critical point ($\dot{P} = 0$) of the differential equation. In the large class of problems where $P_c(t)$ asymptotically approaches a constant matrix, it is generally sufficient to consider P_c to be constant in examining the asymptotic behavior of (17).

IV. True Divergence

The situation we are interested in investigating is that in which some of the elements of $P_a(t)$ increase without limit while the solution $P_c(t)$ of (6) approaches a stable steady state. In view of (28), this implies that G_s has no eigenvalues with positive real parts, and no multiple eigenvalues whose real parts vanish. This assumption will be retained throughout the remainder of the paper. Furthermore, in accordance with the above comment about oscillatory steady states, we shall generally assume that G_s has no pure imaginary eigenvalues. Hence in (17), if G_s has no zero eigenvalues the asymptotic behavior of $P_a(t)$ depends only on the properties of C and the behavior of $S(t)$.

Since the presence or absence of zero eigenvalues of G_s is of crucial importance, we present the following theorem.

Theorem 1: G_s has a zero eigenvalue if and only if F_c^T and Q_c have a common null vector.

Proof: If F_c^T has a zero eigenvalue with eigenvector e, so that $F_c^Te = 0$, and if $Q_ce = 0$ also, then any critical point P_{cs} of (6) must satisfy

$$R_c^{-1}H_cP_{cs}e = K_s^Te = 0$$

which causes e to be a null vector of G_s^T. This is proven by letting $\dot{P}_c = 0$ in (6), premultiplying by e^* and postmultiplying by e. To prove necessity, we note that from (9)

$$e^*GP_ce = e^*F_cP_ce - e^*KR_cK^Te \tag{32}$$

and since the last term is nonnegative, if e is a null vector of G^T the scalar $e^*F_cP_ce$ must be nonnegative. Combining the above equation with (6) we find in the steady state ($\dot{P}_c = 0$)

$$e^*F_cP_{cs}e + e^*Q_ce = 0 \tag{33}$$

and since both terms are nonnegative they must both vanish. Thus $Q_ce = 0$ and (32) shows that $K_s^Te = 0$. It follows from (9) that $F_c^Te = 0$ also. (End of proof.)

The existence of a zero eigenvalue of G_s can be shown to imply a lack of complete controllability of the system represented by the assumed parameters F_c and Q_c. It is apparent from Theorem 1 that this condition can always be prevented by proper choice of Q_c; in particular, positive definiteness of Q_c will always ensure that all eigenvalues of G_s are nonzero, since in this case Q_c has no nontrivial null vectors.

We may now discuss two separate cases, depending on whether G_s has a zero eigenvalue or not.

Case 1: G_s does not have a zero eigenvalue.

a) If $C = \mathbf{0}$ (i.e., the filter uses precise values for F and H), then it is apparent from (17) that $P_a(t)$ will exhibit a stable asymptotic behavior, and no (true) divergence can occur. This conclusion holds regardless of any errors present in Q_c and/or R_c. In other words, no matter how small $P_c(t)$ is allowed to become, this cannot cause $P_a(t)$ to become unbounded. The idea of "over-optimism" of the filter as a cause of divergence is therefore contradicted in this case.

b) If $C \neq \mathbf{0}$, it provides a coupling from (20) to (18) to (17). The behavior of $P_a(t)$ is then dependent on that of $S(t)$, as determined by (18). Because of the stability properties of G_s, it can be seen from (18) and (20) that if F is a stable matrix, then $S(t)$ is stable and will not cause divergence of $P_a(t)$. More precisely, if the divergence is to be caused by the divergence of the CX term in (18), it is necessary that F [in (20)] have an eigenvalue $\lambda + i\mu$ for which $\lambda > 0$ or $\lambda = \mu = 0$, or a multiple eigenvalue with $\lambda = 0$ and $\mu \neq 0$. If the divergence is to be caused by divergence of S without divergence of the product CX, there must be an eigenvalue of F with $\lambda > 0$ (since (18) must exhibit an unstable or vanishing eigenvalue and, by assumption, all the eigenvalues of G_s have negative real parts).

Thus we conclude that if G_s has no zero eigenvalues, divergence requires that $C \neq \mathbf{0}$ and that F be unstable (i.e., have an eigenvalue which vanishes or has a positive real part, or a multiple imaginary eigenvalue).

Case 2: G_s has a zero eigenvalue.

If G_s possesses a zero eigenvalue, it is quite possible for elements of $P_a(t)$ to increase without limit due to integration of the constant forcing terms in (17). If $G_s^T \boldsymbol{e} = \mathbf{0}$ we can show from (17) (assuming $C = \mathbf{0}$) that

$$\frac{d}{dt}(\boldsymbol{e}^* P_a \boldsymbol{e}) = \boldsymbol{e}^* Q \boldsymbol{e} \tag{34}$$

so that true divergence occurs in $P_a(t)$ if $Q\boldsymbol{e} \neq 0$. In simpler terms, if the matrix F_c^T possesses a zero eigenvalue, and the corresponding eigenvector is excited by the process noise but the filter assumes it is not, then divergence will occur. Since an effective process noise is always present due to numerical inaccuracies in propagating the estimates, it can probably be assumed that divergence will occur eventually if G_s has a zero eigenvalue.

We note that the term KRK^T of (17) has disappeared in (34) due to the vanishing of $K_s^T\boldsymbol{e}$ as shown in the proof of Theorem 1. One might therefore wonder whether this term is, in fact, capable of causing divergence when G has a zero eigenvalue, $Q_c = Q$ and $C = \mathbf{0}$. Such an occurrence could properly be interpreted as divergence due to errors in R_c, for if $R_c = R$ we know that $P_a = P_c$, and we have already assumed stability of P_c.

Before investigating this question, we must point out a possible property of the $n \times n$ matrix G_s. Since G_s is not necessarily symmetric, if it has repeated eigenvalues it may not be diagonalizable by a similarity transformation. This is equivalent to saying that its Jordan normal form is not diagonal, or that G_s possesses less than n linearly independent eigenvectors. By ruling out such cases, we can prove the following result.

Theorem 2: If the Jordan form of G_s is diagonal, errors in R_c cannot cause divergence.

We have already seen the truth of this theorem when G_s has no zero eigenvalues. (See Case 1.) The proof in the opposite case is deferred to the Appendix, since it is more easily accomplished after the simpler proof of Theorem 3 has been developed.

When C does not vanish, the divergence conditions for Case 2 are similar to those for Case 1, with one exception, namely that divergence of S without divergence of CX does not necessarily require an eigenvalue of F with $\lambda > 0$, but may be caused also by a zero eigenvalue or a multiple imaginary one.

It should be emphasized that only some of the possible sources of error have been considered here, namely errors in the assumed values of the matrices F, H, Q, and R. Other inaccuracies, such as those which arise from the numerical operations performed in the filtering procedure, may not be adequately described by errors of this sort. An important source of error, which is not considered in the model used here, arises from the necessity of approximating the system by a mathematical model with a reduced number of state variables. This is especially common when correlated errors must be dealt with in a reasonably simple fashion, or when biases must be ignored. (The effects of biases will be discussed below.) One such case has been investigated in [2], in which the ignored states represent errors in *a priori* knowledge of the position coordinates of a landmark used for navigation.

The Effects of Biases

We consider now the way in which the divergence phenomenon is affected by the presence of (constant) biases $\boldsymbol{b}$ and $\boldsymbol{d}$, both with zero mean and with

$$\langle \boldsymbol{b}\boldsymbol{b}^T \rangle = B$$
$$\langle \boldsymbol{d}\boldsymbol{d}^T \rangle = D \tag{35}$$

which affect the system in such a way that (1) and (2) become

$$\dot{\boldsymbol{x}} = F\boldsymbol{x} + \boldsymbol{v} + \boldsymbol{b} \tag{36}$$

$$\boldsymbol{m} = H\boldsymbol{x} + \boldsymbol{w} + \boldsymbol{d}. \tag{37}$$

It is assumed that $\boldsymbol{v}$, $\boldsymbol{w}$, $\boldsymbol{b}$, and $\boldsymbol{d}$ are not correlated with each other, and that the presence of the biases is ignored in the filtering computations, so that (5) and (6) are still employed.

Although $\boldsymbol{b}$ and $\boldsymbol{d}$ are assumed to be constant, the results derived here can be applied, in a qualitative manner, to cases where they represent imperfectly known time-varying forcing functions.

The effects of ignored biases are also indicative of the effects of *nonlinearities* in the dynamics or the measurements. (Recall that a nonlinear element generally introduces a nonzero mean into a zero-mean process.) Most second-order nonlinear filters [5] differ from the "extended" (linearized) Kalman filter mainly in the presence of bias-like terms (generally of constant sign) in the dynamics and measurement portions of the estimation equations. In fact, Jazwinski's "modified" second-order filter [5, p. 339] differs *only* in this way.

When (1) and (2) are replaced by (36) and (37), (8) is replaced by

$$\dot{e} = Ge + Cx + K(w + d) - (v + b) \tag{38}$$

and (17) by

$$\begin{aligned}\dot{P}_a = {} & GP_a + P_aG^T + CS^T + SC^T + KRK^T \\ & + Q - \langle be^T\rangle - \langle eb^T\rangle + K\langle de^T\rangle \\ & + \langle ed^T\rangle K^T\end{aligned} \tag{39}$$

where

$$\frac{d}{dt}\langle eb^T\rangle = G\langle eb^T\rangle + C\langle xb^T\rangle - B \tag{40}$$

$$\frac{d}{dt}\langle ed^T\rangle = G\langle ed^T\rangle + KD \tag{41}$$

and

$$\frac{d}{dt}\langle xb^T\rangle = F\langle xb^T\rangle + B. \tag{42}$$

The matrix $S(t)$ is now the solution of

$$\dot{S} = SF^T + GS + CX - Q - \langle bx^T\rangle + \langle eb^T\rangle \tag{43}$$

and $X(t)$ is given by

$$\dot{X} = FX + XF^T + Q + \langle xb^T\rangle + \langle bx^T\rangle. \tag{44}$$

Once again we look for the possibility of divergence of $P_a(t)$ while $P_c(t)$ behaves in a stable fashion. We will again consider the two cases discussed previously.

Case 1: G_s does not have a zero eigenvalue.

In this case, the solution of (41) is stable, hence $\langle ed^T\rangle$ does not cause divergence in (39). Divergence can thus be caused by the biases only through one, or both, of the terms CS^T and $\langle eb^T\rangle$ in (39).

a) If $C = 0$, the divergence of (39) must be caused by divergence of $\langle eb^T\rangle$, which cannot take place because of the stability of (40).

b) If $C \neq 0$, the divergence may be caused by divergence of either S or $\langle eb^T\rangle$. Equation (40) shows that the latter occurrence requires divergence of $\langle xb^T\rangle$, and through (42)–(44) we are led to the conclusion that F must be unstable for either S or $\langle eb^T\rangle$ to diverge. More precisely, if divergence is to be caused by X through (43), or by $\langle xb^T\rangle$ of (42), acting through (40), (43), or (44)–(43) to (39), then F must have an eigenvalue with $\lambda > 0$ or $\lambda = \mu = 0$, or a multiple imaginary eigenvalue. If the divergence is to be due to S without divergence of $\langle xb^T\rangle$ or X, then $\lambda > 0$ is necessary in (43).

Thus in Case 1 we have the same necessary conditions as we had in the absence of biases, namely $C \neq 0$ and F unstable.

Case 2: G_s has a zero eigenvalue.

If G_s does possess a zero eigenvalue, the process bias (represented by B) can cause $\langle eb^T\rangle$ to diverge in (40), causing in turn a divergence of P_a. It would appear that D could cause a similar effect, but this will be shown to be impossible for a large class of problems, as follows.

Theorem 3: If G_s has a diagonal Jordan form, divergence cannot be caused by a measurement bias.

Proof: Consider, for simplicity, a single column of the matrix (41), which may be written (after P_c has reached a steady state)

$$\dot{z} = G_s z + K_s c \tag{45}$$

where z is a column of $\langle ed^T\rangle$ and c is a column of D. Let the eigenvalues of G_s be $\lambda_i + i\mu_i$ $(i = 1, 2, \cdots, n)$, corresponding to the eigenvectors v_i, and denote the corresponding left eigenvectors (eigenvectors of G_s^T) by w_i. Then, because G_s is assumed to have n linearly independent eigenvectors, the vector z may be written (as may any n-vector)

$$z = \sum_{j=1}^{n} a_j v_j \tag{46}$$

so that (45) becomes

$$\sum_{j=1}^{n} \dot{a}_j v_j = \sum_{j=1}^{n} a_j(\lambda_j + i\mu_j)v_j + K_s c \tag{47}$$

and

$$\sum_{j=1}^{n} \dot{a}_j w_i^* v_j = \sum_{j=1}^{n} a_j(\lambda_j + i\mu_j)w_i^* v_j + w_i^* K_s c. \tag{48}$$

Now because of the biorthogonality property of the eigenvectors [4], we have (assuming appropriate orthogonalization in the case of repeated eigenvalues)

$$\dot{a}_i = (\lambda_i + i\mu_i)a_i + w_i^* K_s c. \tag{49}$$

Thus if $\lambda_i < 0$, or if $\lambda_i = 0$ and $\mu_i \neq 0$, a_i will remain finite due to the stability of (49). If $\lambda_i = \mu_i = 0$, then it follows from the proof of Theorem 1 that $K_s^T w_i = 0$, so that the forcing term vanishes in (49), again assuring stable behavior of a_i. We have already seen that $\lambda \leq 0$; thus we conclude that each column z of $\langle ed^T\rangle$ behaves in a stable manner. (End of proof.)

A Scalar System

Many of the effects considered here can be conveniently demonstrated if we restrict ourselves to the scalar (one state-variable) system described by the equations

$$\begin{aligned}\dot{x} &= fx + v \\ m &= hx + w\end{aligned} \tag{50}$$

with

$$\langle v(t)v(\tau)\rangle = q\delta(t-\tau)$$

$$\langle w(t)w(\tau)\rangle = r\delta(t-\tau).$$

The four essential equations, (6), (17), (18), and (20), now become

$$\dot{p}_c = 2f_c p_c - \frac{h_c^2}{r_c} p_c^2 + q_c$$

$$\dot{p}_a = 2gp_a + 2cs + \left(\frac{h_c p_c}{r_c}\right)^2 r + q$$

$$\dot{s} = (f+g)s + c\langle x^2\rangle - q$$

$$\frac{d}{dt}\langle x^2\rangle = 2f\langle x^2\rangle + q. \tag{51}$$

The number of quantities to be varied in a parametric study of these equations can be reduced by introduction of the dimensionless parameters

$$\begin{aligned}\tau &= t|f|\\ \pi_c &= p_c|f|/q\\ \pi_a &= p_a|f|/q\\ \sigma &= s|f|/q\\ \eta &= \langle x^2\rangle|f|/q\\ r_f &= f_c/f\\ r_h &= h_c/h\\ r_q &= q_c/q\\ r_r &= r_c/r\end{aligned} \tag{52}$$

and

$$N = f\sqrt{r/qh^2}. \tag{53}$$

The dimensionless parameter N can be considered as a sort of "noise-to-signal" ratio having the same sign as f. With the above substitutions, and using the prime to denote differentiation with respect to τ, the differential equations (51) become

$$\pi_c' = \pm 2r_f\pi_c - U\pi_c^2 + r_q \tag{54}$$

$$\pi_a' = 2\gamma\pi_a + 2\beta\sigma + (U/r_r)\pi_c^2 + 1 \tag{55}$$

$$\sigma' = (\pm 1 + \gamma)\sigma + \beta\eta - 1 \tag{56}$$

$$\eta' = \pm 2\eta + 1 \tag{57}$$

where we have used

$$U = \frac{1}{N^2}\frac{r_h^2}{r_r} \tag{58}$$

$$\gamma = \pm r_f - U\pi_c \tag{59}$$

$$\beta = \pm(r_f - 1) - U\pi_c\left(1 - \frac{1}{r_h}\right) \tag{60}$$

and the minus sign in "$\pm$" is to be used when f is negative (i.e., when the system is stable).

Computational experience with (51) has shown that their divergence properties are not noticeably affected by the initial condition $p_c(0)$. It is therefore reasonable to set p_c to its steady-state value at $t = 0$ in order to investigate the phenomenon. This eliminates the only nonlinear equation in the set (51), and allows the solutions to be written very simply in closed form. The solutions are easily shown to be (for the unstable case $f > 0$)

$$\eta(\tau) = \left(\eta_0 + \frac{1}{2}\right)e^{2\tau} - \frac{1}{2} \tag{61}$$

$$\sigma(\tau) = k_1 e^{(1+\gamma)\tau} + \frac{\beta}{1-\gamma}\left(\eta_0 + \frac{1}{2}\right)e^{2\tau} + \frac{2+\beta}{2(1+\gamma)} \tag{62}$$

$$\pi_a(\tau) = k_2 e^{2\gamma\tau} + \frac{2\beta k_1}{1-\gamma}e^{(1+\gamma)\tau} + \left(\frac{\beta}{1-\gamma}\right)^2\left(\eta_0 + \frac{1}{2}\right)e^{2\tau} - k_3 \tag{63}$$

where

$$k_1 = \sigma_0 - \frac{\beta\eta_0}{1-\gamma} - \frac{1-\gamma+\beta}{1-\gamma^2} \tag{64}$$

$$k_2 = \pi_{a_0} - \frac{2\beta k_1}{1-\gamma} - \left(\frac{\beta}{1-\gamma}\right)^2\left(\eta_0 + \frac{1}{2}\right) + k_3 \tag{65}$$

and

$$k_3 = \frac{1}{2\gamma}\left[\frac{\beta(2+\beta)}{(1+\gamma)} + \frac{U}{r_r}\pi_{cs}^2 + 1\right]. \tag{66}$$

The steady-state value π_{cs} of π_c can be found from the vanishing of π_c' in (54). The acceptable (positive) solution is

$$\pi_{cs} = \frac{1}{U}\left[r_f + \sqrt{r_f^2 + Ur_q}\right] \tag{67}$$

which together with (59) gives

$$\gamma = -\sqrt{r_f^2 + Ur_q}. \tag{68}$$

The constant γ, which corresponds to the matrix G of (9), is therefore negative. Since r_f and r_q will often be close to unity, the quantity $(1 + \gamma)$ will also be negative in many cases. In any case, even if positive it will always be smaller than unity, so that the important term in (63), as far as divergence is concerned, is the term in $e^{2\tau}$, namely,

$$\pi_{ad}(\tau) = \left(\frac{\beta}{1-\gamma}\right)^2\left(\eta_0 + \frac{1}{2}\right)e^{2\tau} \tag{69}$$

where γ is given by (68) and β can be found from (60) and (67) as

$$\beta = \left(\frac{r_f}{r_h} - 1\right) + \left(\frac{1}{r_h} - 1\right)\sqrt{r_f^2 + Ur_q}. \tag{70}$$

The constant β corresponds to the matrix C of (10), and when $\beta = 0$ the quantity $\pi_a(\tau)$ of (63) approaches a constant value.

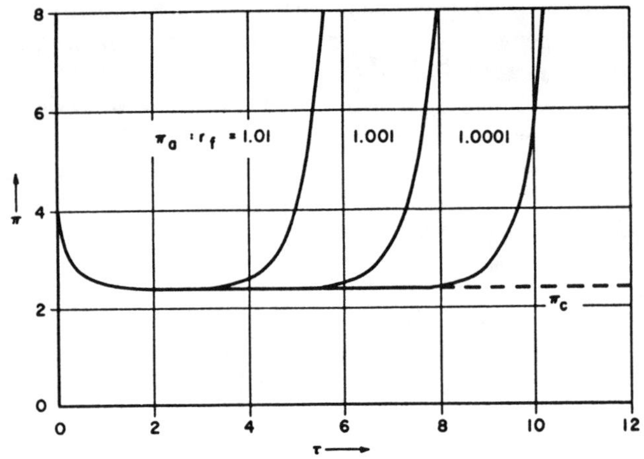

Fig. 1. True divergence due to errors in $f_c(N = +1)$.

Fig. 1 illustrates the behavior of $\pi_a(\tau)$ for errors of 1, 0.1, and 0.01 percent in f_c. The case considered is that of $N = 1$ (unstable system), $r_h = r_q = r_r = 1$, with initial conditions $\pi_c(0) = \pi_a(0) = \eta(0) = 4$, $\sigma(0) = -4$. The broken line indicates the stable behavior of $\pi_c(\tau)$ in all three cases. This is an example of divergence occurring because $C \neq 0(r_f \neq 1)$ and F is unstable ($f > 0$). It is apparent from (70) that a similar effect could be caused by errors in $h_c(r_1 \neq 1)$.

V. Apparent Divergence

Perhaps of even greater importance in practical situations is the phenomenon we have chosen to call "apparent" divergence, in which the errors, though not truly unbounded with increasing time, do reach intolerably high levels, equivalent, in a practical sense, to true divergence. In a real situation, in fact, it may be impossible to distinguish the two for reasonable periods of operation of the filter.

Since this phenomenon can be present in varying degrees of severity, depending on the magnitudes of the errors which cause it, it must be demonstrated by means of parametric studies. We therefore confine ourselves largely to a consideration of the scalar problem discussed earlier. We shall assume that the filter uses the correct values for the parameters f and h, and investigate the effects of errors in q_c and r_c, and of the presence of ignored biases of the two types introduced previously.

When the scalar equivalents of (39) to (44) are solved under these conditions, we may derive an expression for the (finite) steady-state actual variance p_{as} (or for π_{as}). It can be shown, after considerable manipulation, that

$$\pi_{as} = |N| \left\{ N + \frac{(N^2 + 1) + (N^2 + (r_q/r_r))}{2\sqrt{N^2 + (r_q/r_r)}} + \frac{B_p}{(N^2 + (r_q/r_r))} + \frac{B_m}{(N^2 + (r_q/r_r))} \cdot [N + \sqrt{N^2 + (r_q/r_r)}]^2 \right\} \tag{71}$$

where

$$B_p = \frac{\langle b^2 \rangle}{q} \sqrt{\frac{r}{qh^2}} \tag{72}$$

and

$$B_m = \frac{\langle d^2 \rangle}{r} \sqrt{\frac{r}{qh^2}}. \tag{73}$$

For the case of an optimal filter with no biases present, π_{as} will be called π_{onb} (optimal, no bias), and is given by

$$\pi_{\text{onb}} = |N| [N + \sqrt{N^2 + 1}]. \tag{74}$$

We shall examine the behavior of the ratio p_{as}/p_{onb} ($= \pi_{as}/\pi_{\text{onb}}$), which, in the special case $f = 0$ ($N = 0$), can be written

$$\left.\frac{p_{as}}{p_{\text{onb}}}\right|_{N=0} = \frac{1}{2}\left[\sqrt{\frac{r_q}{r_r}} + \sqrt{\frac{r_r}{r_q}}\right] + \frac{r_r}{r_q} B_p + B_m. \tag{75}$$

It is apparent from (71) that the value of π_{as} depends only on the ratio of r_q and r_r, and not on either individually. Thus in the absence of biases, if the filter uses values of q_c and r_c in the same ratio as q and r, it will achieve a steady-state performance equivalent to the optimal. (This does not, of course, guarantee optimal results before the steady state is reached.) The same effect is apparent in the multidimensional case as well, for if Q_c and R_c are both in error by a given multiplicative factor, the steady-state solution P_{cs} of (6) is scaled by the same factor, so that the gain matrix K remains correct and the filter is optimal in the steady state. (In fact, if $P_c(0)$ is scaled by the same factor, the gains are correct at *all* times.)

The form of (75) puts in evidence the possibility of true divergence in the case $f = 0$, when r_q is allowed to vanish (the zero eigenvalue case discussed previously). (True divergence would also occur if $r_r = 0$, but the assumption $r_c = 0$ can be assumed never to be made.)

The dependence of the ratio p_{as}/p_{onb} on the "noise-to-signal" ratio N, is depicted in Fig. 2 for various values of r_q/r_r, with no biases present ($B_p = B_m = 0$). For all values $r_q/r_r < 1$, the ratio p_{as}/p_{onb} peaks in the vicinity of $N = 0$, which may explain the observation [2] that divergence is aggravated by more frequent or more accurate measurements. The peak is of infinite magnitude when $r_q = 0$ (i.e., when the process noise is completely ignored).

In the multidimensional case with correct F_c and H_c, some conclusions can be drawn about the effects of changes in the *true* values of R and Q when the *assumed* values R_c and Q_c are fixed (so that the gains K remain fixed). In this case (17) may be written

$$\dot{P}_a = (F - KH)P_a + P_a(F - KH)^T + KRK^T + Q. \tag{76}$$

Since this equation is linear, the steady-state solution P_{as} is the sum of the solutions due to R and Q, separately, and these solutions are proportional to R and Q, respectively. If separate solutions to the equation are generated to find the contributions to P_{as} due to each noise matrix (or even each element of each matrix), one can predict the effects on

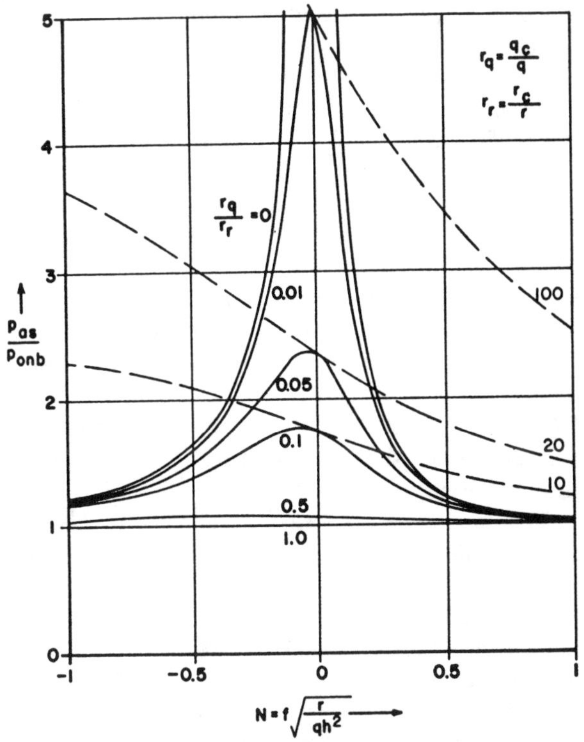

Fig. 2. Dependence of steady-state mean-square error on noise-to-signal ratio N.

P_{as} of changes in these quantities. In general, it can be concluded that any change in P_{as} will be less than proportional to the change in R or Q causing it, since only a portion of P_{as} is due to that particular noise source.

The Effects of Biases

We now consider the significance of (71) when a process or measurement bias is present. For a process bias intensity such that $B_p = 1$, Fig. 3 illustrates the dependence of p_{as} on the ratio r_q/r_r, for a positive and negative N (i.e., an unstable and a stable system). In both cases, of course, the presence of the bias degrades the performance of the filter (increases p_{as}), but we note that in both cases the minimum of p_{as}, with respect to r_q/r_r, occurs at a value of r_q/r_r greater than unity. This verifies that it is possible to partially compensate for the effects of process bias, by increasing the intensity of process noise assumed by the filter (q_c), even beyond a value known to be the correct value of q. A similar steady-state effect would be achieved by decreasing r_c, but this would probably produce inferior transient performance of the filter.

In the case of a measurement bias, a different situation is observed. In Fig. 4 (with $B_m = 1$) we see that, for an unstable system, the minimum p_{as} again occurs at some value $r_q/r_r > 1$. For a stable system, however, the opposite is the case, and r_q/r_r should be *reduced* to minimize p_{as}. One might suspect that, in the interests of good transient behavior, this should be accomplished by increasing r_c rather than by decreasing q_c. Limited studies have shown, however, that this is not necessarily always true.

When $N = 0$ ($f = 0$), the effects of biases are as shown in Fig. 5. The value of r_q/r_r corresponding to the mini-

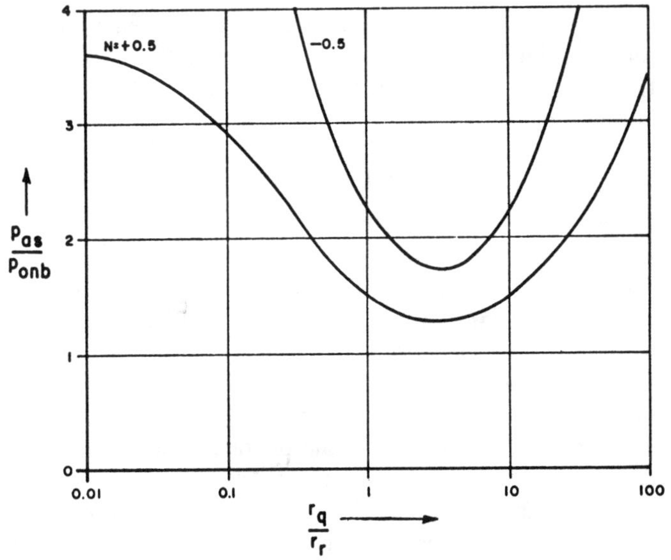

Fig. 3. Effects of process bias ($B_p = 1$).

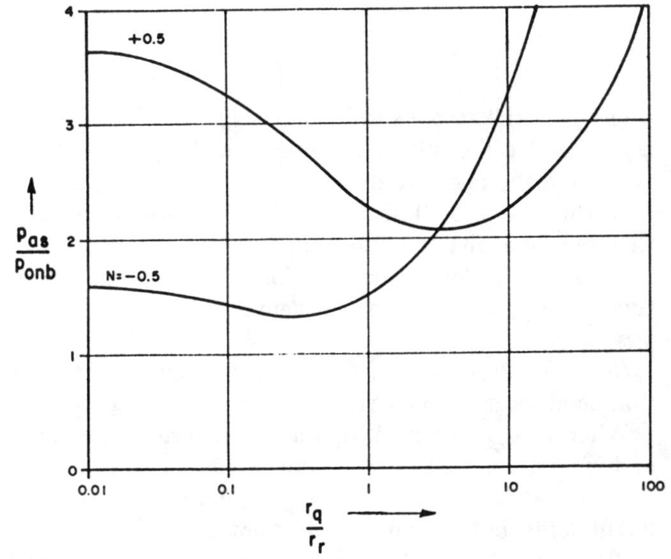

Fig. 4. Effects of measurement bias ($B_m = 1$).

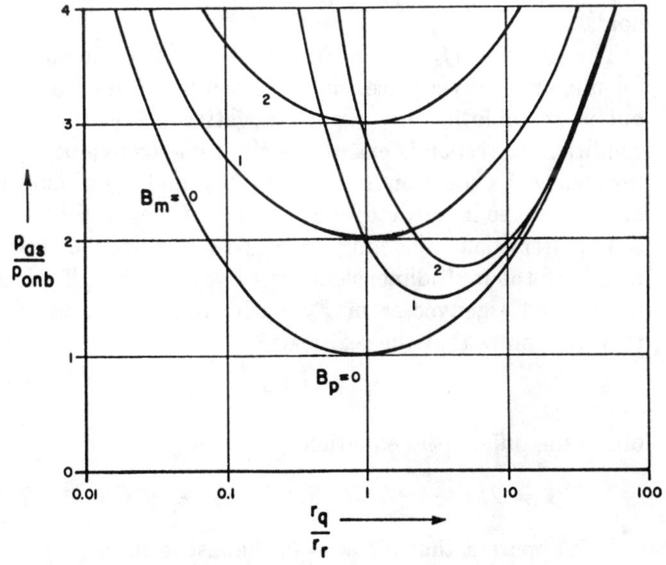

Fig. 5. Effects of process and measurement biases when $f = 0$.

mum of p_{as} increases with increasing process bias, but is totally unaffected by measurement bias. In the latter case, therefore, the bias cannot be compensated for by manipulation of r_c or q_c.

VI. Numerical Divergence

One effect somewhat unrelated to those discussed above, but which may have equally serious consequences, can arise through numerical inaccuracies which cause the calculated covariance matrix P_c to lose its positive semidefinite character. Recently a square-root technique has been developed to overcome these difficulties [6], [8], [9], but is rather inconvenient to apply. Since such an occurrence can be disastrous to the operation of the filter, a consideration of its causes is of interest.

The phenomenon is easiest to investigate in the scalar case, in which the calculated variance p_c is generated by the first equation in (51). Examination of this equation shows that it has two critical points (solutions of $\dot{p}_c = 0$), $p_1 \geq 0$ and $p_2 \leq 0$, given by

$$p_1, p_2 = \frac{f_c r_c}{h_c^2}\left[1 \pm \sqrt{1 + \frac{h_c^2 q_c}{f_c^2 r_c}}\right] \tag{77}$$

where p_1 is the stable steady-state solution. Furthermore, if $p_c > p_2$ it will converge towards p_1, and if $p_c < p_2$ it will diverge in the negative direction.

In the case $q_c > 0$ both critical points are nonzero, so that the following conditions hold.

Condition 1: The correct solution tends towards a positive value $p_1 > 0$, so that accidental negative values are less likely.

Condition 2: Since $p_2 < 0$ a region of safety exists, such that small negative values of p_c will still converge to p_1.

When q_c is allowed to vanish the critical points are 0 and $2f_c r_c/h_c^2$, so that the following situations can exist.

1) If $f_c < 0$, then $p_1 = 0$ and $p_2 < 0$, so that Condition 2 still holds, but Condition 1 does not.

2) If $f_c > 0$, then $p_1 > 0$ and $p_2 = 0$, so that Condition 1 holds but Condition 2 does not.

3) If $f_c = 0$, then $p_1 = p_2 = 0$ and neither condition holds.

The last case ($f_c = q_c = 0$) is therefore the critical one for numerical divergence, since the solution tends to zero but is unstable for any slightly negative value. But this condition corresponds exactly to the zero-eigenvalue case (existence of a common null vector of F_c^T and Q_c) which we have seen is so important for divergence of the multidimensional filter due to modeling errors. To generalize the above results to the multidimensional case, we note that, if $\boldsymbol{e}$ is a normalized eigenvector of F_c^T with eigenvalue $\lambda + i\mu$, then (6) shows that the real scalar

$$z = \boldsymbol{e}^* P_c \boldsymbol{e} \tag{78}$$

obeys the differential equation

$$\dot{z} = 2\lambda z - \boldsymbol{e}^*(P_c H_c^T R_c^{-1} H_c P_c)\boldsymbol{e} + \boldsymbol{e}^* Q_c \boldsymbol{e}. \tag{79}$$

It is apparent that if $Q_c\boldsymbol{e} \neq \boldsymbol{0}$, the last term in (79) will tend to prevent a negative divergence of z (negative $\dot{z}$). If $Q_c\boldsymbol{e} = \boldsymbol{0}$, however, our previous results on steady-state behavior of P_c show the following.

1) If $\lambda < 0$, then $P_c\boldsymbol{e}$ (and hence z) tends to zero, but the first term in (79) provides a stabilizing effect for small negative z. Thus Condition 2 holds for z but Condition 1 does not.

2) If $\lambda > 0$, the term $2\lambda z$ is destabilizing, but $P_c\boldsymbol{e}$ (and hence z) does not tend to zero. Thus Condition 1 holds but Condition 2 does not.

3) If $\lambda = 0$, only the second term in (79) exists, and this term tends to zero since, as stated previously, $R_c^{-1}H_c P_{cs}\boldsymbol{e} = \boldsymbol{0}$. The term is nonpositive, however, so that if z becomes slightly negative in such a way that $R_c^{-1}H_c P_c\boldsymbol{e} \neq \boldsymbol{0}$, then $\dot{z} < 0$ and z will diverge negatively. This implies that P_c has an eigenvalue which is diverging in the negative direction.

The phenomenon of numerical divergence involves the property of "finite escape time" of the matrix Riccati equation, which is normally impossible in filtering problems as long as P_c remains positive semidefinite [3]. When the phenomenon occurs, the errant eigenvalue should be expected to diverge to $-\infty$, then converge again from $+\infty$ to finite values. Of course, we would not expect to observe this in solving (6) by computer. In the case of discrete measurements, however, a corresponding phenomenon *can* be observed, since infinite numbers are never involved. Such a case has been reported, for example, in [6]. The effect is easily seen in the scalar case, where the updating equation for p_c is

$$p_c = \frac{r_c p_c'}{p_c' h_c^2 + r_c} \tag{80}$$

where the prime indicates the value before the measurement. If p_c' is positive (as it should be), then p_c is also positive but smaller in magnitude than p_c'. If $0 > p_c' > -r_c/h_c^2$, then $p_c < p_c'$, i.e., the updating causes p_c to diverge negatively. But when p_c becomes sufficiently negative, so that $p_c' < -r_c/h_c^2$, then the denominator in (80) is negative and p_c will revert to a *positive* value after the updating.

In the multivariable case, an explanation is given in [6] by noting that the updating operation is equivalent to an addition, to P_c^{-1}, of a positive semidefinite matrix $H_c^T R_c^{-1} H_c$, and by showing that this operation causes nonnegative changes in the eigenvalues of P_c^{-1}. (Although the proof of this in [6] is lengthy, it can be shown quite simply by making appropriate use of (30)). When an eigenvalue of P_c^{-1} increases through zero, its reciprocal (the corresponding eigenvalue of P_c) exhibits the divergent-convergent behavior described above.

VII. Numerical Example

We present here some numerical results obtained by computer simulation of the problem of navigating an earth satellite by means of horizon-sensor measurements. The vehicle is in a circular equatorial orbit at an altitude of 200 nmi, and has a ballistic coefficient $C_D A/2m = 1.6$ ft²/slug. At alternating intervals of 40° and 115° of traversed geocentric angle, the horizon sensor observes the horizon at four points, bisects the included angles to find the direction of the vertical, and compares this direction with an accurate inertial reference. The measurement errors are assumed to arise solely from random variations in the altitude of the apparent horizon. These horizon variations are generated as a correlated random function on the sphere of the earth [7], with an rms value of 1 km, exponentially correlated with correlation distance equivalent to 40° of geocentric angle. The filtering techniques employed are described in [7].

We show only the most significant component of the error in estimation of vehicle position, namely the horizontally forward component. Fig. 6 demonstrates a divergence caused by the fact that the filtering computer completely ignores the effects of atmospheric drag. Such an error is properly regarded, from the point of view of the linearized equations, as constituting a process bias, as represented by the vector $\boldsymbol{b}$ in (36). The figure also shows (heavy line) the time history of the corresponding rms value calculated by the filtering computer.

In Fig. 7 the divergence has been eliminated by allowing the filter to assume the presence of some process noise. This was done by altering the matrix Q_c of (6), adding three equal constants to the diagonal elements corresponding to the three velocity variables in the state vector. These additions were introduced at approximately $t = 200$ min, and the appropriate magnitude was found by numerical experimentation.

The most probable explanation for the observed behavior in this case is as follows. The orbital system exhibits a zero eigenvalue, as evidenced by the fact that an initial error in true anomaly will not increase or decrease in the absence of measurements. Furthermore, in Fig. 6 no process noise was assumed, so that we have a situation equivalent to one in which the matrix G exhibits a zero eigenvalue in the steady state. The process bias is therefore capable of causing an early divergence (even though a later one might occur anyway due to numerical errors). The insertion of a nonzero Q_c eliminates the zero eigenvalue from G_s and prevents the divergence.

Additional examples of divergence and its elimination can be found in [2].

VIII. Conclusions

The principal results concerning the causes of true divergence are summarized in Table I, where $J(G_s)$ denotes the Jordan normal form of G_s. The most important conclusion is the increased likelihood of divergence when G_s has a zero eigenvalue, a situation which may always be prevented by a proper choice of the process-noise matrix Q_c. We have also seen that manipulation of Q_c can partially compensate for filter performance deterioration due to biases.

In general, when G_s does not have a zero eigenvalue, divergence may take place (for the causes considered here) only if F is unstable. Such cases require special filtering schemes such as those of the limited memory type [5].

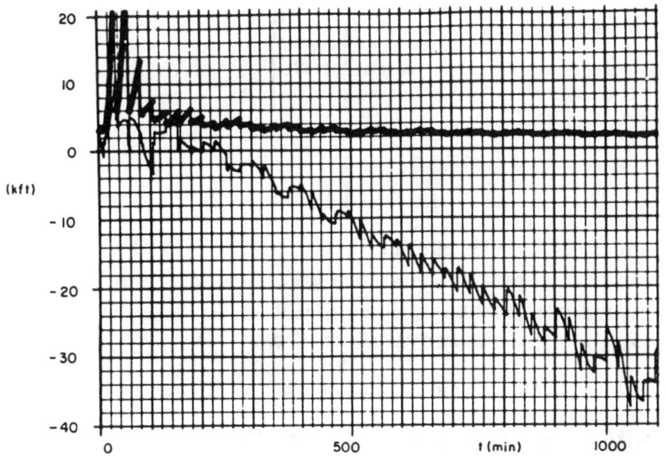

Fig. 6. Divergence due to ignoring drag in satellite navigation (forward component of position error).

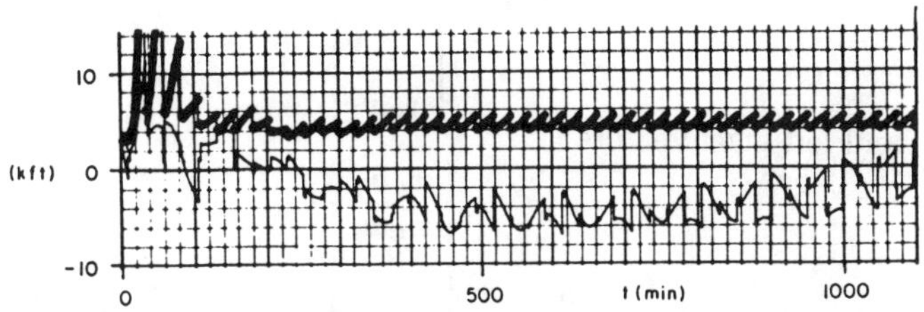

Fig. 7. Elimination of divergence by increasing assumed process noise.

TABLE I
Causes of Divergence

Cause of Divergence	Case 1 (No zero eigenvalue of G_s)	Case 2 (G_s has a zero eigenvalue)
ΔF, ΔH	Requires F unstable	Requires F unstable
ΔQ	No	Possible (certain if $Qe \neq 0$)
ΔR	No	Not if $J(G_s)$ diagonal
Process bias	Requires $C \neq 0$ and F unstable	Possible
Measurement bias	No	Not if $J(G_s)$ diagonal
Numerical	Unlikely	Possible

Appendix

The proof of Theorem 2 closely parallels that of Theorem 3, if we make use of the properties outlined in (21)–(24). We are interested in the possibility of divergence of the linear equation (17) [or of (39)] due to the presence of the term KRK^T. For this purpose it suffices to consider the equation

$$\dot{P}_a = G_s P_a + P_a G_s^T + K_s R K_s^T. \tag{81}$$

Using the approach of (22), we can write the vector equation

$$\dot{y} = My + z \tag{82}$$

where M is the $n^2 \times n^2$ matrix $G_s \times I + I \times G_s$, z is of the form

$$z = \begin{bmatrix} K_s s_1 \\ K_s s_2 \\ \cdot \\ \cdot \\ \cdot \\ K_s s_n \end{bmatrix} \tag{83}$$

and s_i is the ith column of the constant matrix RK_s^T. If the n (complex) eigenvalues of G_s are $\lambda_i + i\mu_i$, with corresponding right eigenvectors v_i and left eigenvectors w_i, then the n^2 eigenvalues of M are

$$\lambda_{i,j} + i\mu_{i,j} = (\lambda_i + \lambda_j) + i(\mu_i + \mu_j) \tag{84}$$

with corresponding right eigenvectors

$$v_{i,j} = \begin{bmatrix} v_{i1}v_j \\ v_{i2}v_j \\ \cdot \\ \cdot \\ \cdot \\ v_{in}v_j \end{bmatrix} \tag{85}$$

and left eigenvectors

$$w_{i,j} = \begin{bmatrix} w_{i1}w_j \\ w_{i2}w_j \\ \cdot \\ \cdot \\ \cdot \\ w_{in}w_j \end{bmatrix} \tag{86}$$

Under our assumption that the eigenvectors of G_s are all linearly independent, it follows easily that those of M are also, so that the vector y can be expressed as

$$y = \sum_{i,j=1}^{n} a_{i,j}v_{i,j} \tag{87}$$

and (82) becomes

$$\sum_{i,j=1}^{n} \dot{a}_{i,j}v_{i,j} = \sum_{i,j=1}^{n} a_{i,j}(\lambda_{i,j} + i\mu_{i,j})v_{i,j} + z \tag{88}$$

and again making use of biorthogonality,

$$\dot{a}_{i,j} = (\lambda_{i,j} + i\mu_{i,j})a_{i,j} + w_{i,j}{}^*z. \tag{89}$$

As in the proof of Theorem 3, the only possibility for divergence (because of the assumed stability of G_s) occurs when $\lambda_{i,j} = \mu_{i,j} = 0$. But (since we have assumed that G_s has no pure imaginary eigenvalues) this is only possible if $\lambda_i = \lambda_j = \mu_i = \mu_j = 0$, which causes w_i and w_j to be null vectors of K_s^T. Thus it follows, from (83) and (86), that the forcing term in (89) vanishes, and divergence does not occur. (It is interesting to note the effect of relaxing the requirement that G_s have no pure imaginary eigenvalues. For a pair of imaginary eigenvalues $i\mu_i = i\mu$, $i\mu_j = -i\mu$, we have $\lambda_{i,j} = \mu_{i,j} = 0$, but the forcing term in (89) need not vanish, so that the theorem no longer holds. Similarly, some of our previous conclusions would be modified if F and G_s possessed a common pair of imaginary eigenvalues, since this could result in divergence of (18) without actual instability of F.)

References

[1] T. L. Gunckel, "Orbit determination using Kalman's method," *Navigation*, vol. 10, pp. 273–291, autumn 1963.

[2] F. H. Schlee, C. J. Standish, and N. F. Toda, "Divergence in the Kalman filter," *AIAA J.*, vol. 5, pp. 1114–1120, June 1967.

[3] J. E. Potter, "A matrix equation arising in statistical filter theory," NASA Contractor Rep., NASA CR-270, Aug. 1965.

[4] R. Bellman, *Introduction to Matrix Analysis*. New York: McGraw-Hill, 1960.

[5] A. H. Jazwinski, *Stochastic Processes and Filtering Theory*. New York: Academic Press, 1970.

[6] J. F. Bellantoni and K. W. Dodge, "A square root formulation of the Kalman-Schmidt filter," *AIAA J.*, vol. 5, pp. 1309–1314, July 1967.

[7] R. J. Fitzgerald, "Filtering horizon-sensor measurements for orbital navigation," *J. Spacecr. Rockets*, vol. 4, pp. 428–435, Apr. 1967.

[8] A. Andrews, "A square root formulation of the Kalman covariance equations," *AIAA J.*, vol. 6, pp. 1165–1166, June 1968.

[9] P. Dyer and S. McReynolds, "Extension of square-root filtering to include process noise," *J. Optimiz. Theory Appl.*, vol. 3, pp. 444–458, June 1969.

Section I-E
Divergence Control and Adaptive Filtering

THE divergence problem, as discussed in the preceding section, was well-known to practitioners. A variety of algorithmic modifications were invented in an attempt to compensate for the model errors that caused the misbehavior of the filter. Some of these methods have been described in the preceding sections. In the first paper contained in this section, Jazwinski reviews many of the approaches that had been invented and, then, proposes a very simple scheme for estimating the input noise from the sample averages for the innovations sequence (or residuals as he refers to them). It is interesting to note that no mention is made of the property that the innovations is a white noise sequence when the filter is optimal. His moment-matching technique uses only the property that the innovations have zero mean with a known covariance. Simulation results are presented that indicate that divergence can be controlled with this simple, adaptive filter. It is worth noting that Jazwinski gives a thorough discussion of divergence control techniques in his book, listed above in the introductory section.

In the second paper of this section, Mehra develops an algorithm for estimating the measurement noise covariance R and the input noise covariance Q. He observes that, possibly, only a portion of the elements of Q can be estimated. Alternatively, he proposes a procedure for estimating the optimal gain directly without attempting to estimate Q. Mehra's adaptive filter is based on statistical tests of the innovations sequence in which the whiteness of the innovations plays a key role. For this development, a constant coefficient (i.e., F, G, H are constant), linear system is considered, and it is assumed that, except for R and Q, the system matrices are known. Furthermore, the Kalman filter is assumed to be operating in steady-state so that the gain matrix K is constant. Although the model used by Mehra is restricted, the utility of the algorithm is demonstrated with an example. The paper stimulated the appearance of several other papers that deal with the problem of estimating the noise covariances and/or the filter gains.

Divergence is a problem because the Kalman error covariance matrix indicates that the estimates are more accurate than is actually the case. Then, decisions and actions based on the supposed quality of the estimates can lead to unsatisfactory results. Thus, the methods proposed to control the problem, beginning with the work of Stanley Schmidt and continuing in subsequent developments, have the effect of increasing the size of the covariance matrix and, concomitantly, the magitude of the filter gains. This can be accomplished in a variety of ways, including the use of the adaptive filters discussed above. Most methods, however, do not attempt to identify parameters of the model explicitly. Instead, changes are made directly to the error covariance matrix P or to the gain matrix K. In the final paper of this section, Sorenson and Sacks develop an algorithm for modifying the covariance and the gain that causes the filter to have a "fading memory" in which the fading parameter is chosen to insure that the innovations sequence is zero mean and white with appropriate covariance values. The method developed extends the ideas of exponential data weighting that are used in least-squares estimation for constant parameters, first proposed for use with the Kalman filter by Fagin.

The methods for divergence control are *ad hoc* procedures for masking the effects of modeling errors. Depending upon the significance of the model imperfections, the divergence control methods may produce satisfactory responses from an otherwise divergent filter. Their utility is difficult to establish since the characterization of the error models is, generally, not available. Consequently, the practice followed for Kalman filter implementations is to always include an input covariance matrix Q and choose the elements of Q, often a single parameter by assuming $Q=qI$, to "tune" the filter response.

References

[1] B. D. O. Anderson, "Exponential data weighting in the Kalman–Bucy filter," *Inform. Sci.*, vol. 5, pp. 217–230, 1973.

[2] B. Carew and P. R. Belanger, "Identification of optimal filter steady-state gain for systems with unknown noise covariance," *IEEE Trans. Automat. Contr.*, vol. AC-18, pp. 582–587, 1973.

[3] S. L. Fagin, "A unified approach to the error analysis of augmented dynamically exact intertial navigation systems," *IEEE Trans. Aerosp. Navig. Elect.*, vol. ANE-11, pp. 234–248, 1964.

[4] ——, "Recursive linear regression theory, optimal filter theory, and error analysis of optimal systems," *IEEE Int. Conv. Record*, vol. 12, pp. 216–240, 1964.

[5] N. E. Nahi and B. M. Schaefer, "Decision-directed adaptive recursive estimators: Divergence prevention," *IEEE Trans. Automat. Contr.*, vol. AC-17, pp. 61–67, 1972.

[6] A. H. Jazwinski, "Limited memory optimal filtering," *IEEE Trans. Automat. Contr.*, vol. AC-13, pp. 558–563, 1968.

[7] T. S. Tarn and J. Zaborsky, "A practical nondiverging filter," *AIAA J.*, vol. 9, pp. 1127–1133, 1970.

[8] R. K. Mehra, "On-line identification of linear dynamic systems with applications to Kalman filtering," *IEEE Trans. Automat. Contr.*, vol. AC-16, pp. 12–21, 1971.

[9] ——, "Approaches to adaptive filtering," *IEEE Trans. Automat. Contr.*, vol. AC-17, pp. 693–698, 1972.

Adaptive Filtering*

Filtrage adaptatif

Adaptive Filterung

Адаптативное фильтрование

A. H. JAZWINSKI†

The linear Kalman filter is known to diverge when the dynamical model is in error. Divergence is prevented by covering model errors with noise and adaptively estimating the noise level.

Summary—Applications of the Kalman filter in orbit determination problems have sometimes encountered a difficulty which has been referred to as divergence. The phenomenon is a growth in the residuals; the state and its estimate diverge. This problem can often be traced to insufficient accuracy in modeling the dynamics used in the filter. Although more accurate modeling is an obvious solution, it is often an impractical, and sometimes an impossible, one. Model errors are here approximated by a white, Gaussian noise input, and its covariance (Q) is determined so as to produce consistency between residuals and their statistics. In this way, realtime feedback is provided from the residuals to the filter gain. Onset of divergence produces an increase in the filter gain and the adaptive filter is able to continue tracking. This scheme has a probabilistic interpretation. Under certain conditions the estimate of Q produces the most probable finite sequence of residuals.

INTRODUCTION

THE LINEAR filtering theory [1, 2] assumes that system dynamics are completely known and are precisely modeled in the filter. Clearly, this is never true in practice, and furthermore, finite arithmetic precludes the exact computation of the filter state. The modeling and computational errors which are invariably present may not present particular difficulties when the noise inputs to the system are large. When these are small, however; when model errors, such as dynamic biases, exist; and when the filter operates over long time intervals, over much data, its operation is sometimes rendered totally unacceptable.

This is often the case in the determination of space-vehicle trajectories [3, 4] via the extended‡ Kalman filter. The observed phenomenon is a "divergence" of the errors in the estimates to values totally inconsistent with the rms values predicted by theory. The covariance matrix becomes unrealistically small and optimistic; the filter gain thus becomes small, and subsequent measurements are ignored. The state and its estimate then diverge, due to model errors in the filter.

Analyses of error "divergence" may be found in [4, 5]. Some techniques found useful in controlling divergence are outlined in [3–5]. These range from arbitrary incrementation of the covariance matrix, to keep its elements above an *a priori* lower bound; to a computational error noise model [6]; experimentally determined system noise input levels [4]; and modeling additional state variables (biases) and including their uncertainties in the filter, with or without actually estimating such biases [6, 7]. This last technique requires state augmentation and may not be practical from the computational point of view.

More recently, SCHMIDT [8] proposed two new schemes. One computes an estimate which is a linear combination of the estimate given all prior data with the estimate given no prior data. Past information, the data, is thus degraded. The other scheme imposes *a priori* lower bounds on certain projections of the covariance matrix. JAZWINSKI [9] recently developed a limited memory optimal filter which computes the best (minimum variance) estimate based on a "moving window" of most

* The original version of this paper was presented at the IFAC Symposium on Multivariable Control Systems which was held in Düsseldorf, Germany during October 1968; received 15 November 1968, and in revised form 10 March 1969. It was recommended for publication in revised form by associate editor A. Sage. Research supported by the NASA Goddard Space Flight Center under Contract NAS 5-11048.

† Manager, Guidance and Control, Analytical Mechanics Associates, Inc. (a subsidiary of Scientific Resources Corporation), 9430 Lanham Severn Road, Seabrook, Maryland, 20801, U.S.A.

‡ This is explained later in the text; also see Ref. [5] for a complete treatment of filtering theory, including applications of the linear theory to non-linear problems, the extended Kalman filter, etc., filter divergence, error sensitivity, and model error compensation techniques. (Also see the next paper in this issue—Editor.)

Reprinted with permission from *Automatica,* vol. 5, pp. 475–485, July 1969.

recent data. Thus, data beyond the accuracy or predictability (time) range of the model is discarded.

The approach to the problem of divergence taken in this paper is to "cover" modeling errors with noise and adaptively estimate the noise variance. This is an extension of work previously reported [10]. In "real-world" applications of the Kalman filter, or, for that matter, of batch processing least-squares type methods as well, the only quantities available to the engineer in judging filter performance are the residuals and their predicted statistics. If the residuals are sufficiently small and consistent with their predicted statistics, then the filter is deemed to be operating satisfactorily. It should be noted that a common fallacy in batch processing schemes is to judge the scheme's efficacy solely on the basis of the size of the residuals, and not on their statistical consistency. Rather than analyzing residuals after the fact to determine filter performance, the adaptive filter provides feedback from residuals, in real-time, in terms of system noise input levels. These degrade the estimation error covariance matrix, increase the filter gains, and thus open the filter to incoming data. DENNIS [11] reports some related ideas.

More precisely, the approach is this. A requirement is imposed that certain residuals be consistent with their statistics. This leads to an algorithm which produces estimates of the system noise covariance matrix Q. Under certain conditions [10, 12] these estimates produce the most probable sequences of residuals. Short sequences of residuals are used in the estimation of Q; the estimator never learns Q. This is clearly desirable, since Q is a fiction designed to account for system model errors which are nonstationary and generally of rather low fequency. It is for this reason that the work of [13] and [14] is not particularly applicable to the problem of divergence. These authors apply Bayesian estimation theory to 'the estimation of noise variances or other unknown parameters. This essentially means that, given sufficient data, these parameters can be learned arbitrarily well.

The present technique is related to the work of SCHMIDT [8]. That relationship is pursued in [10]. These adaptive techniques can also be applied to the estimation of the measurement noise variance [10, 12].

ANALYSIS

Problem definitions

The following linear dynamical system model is assumed

$$\mathbf{x}_{k+1}=\Phi_{k+1,k}\mathbf{x}_k+G_k\mathbf{w}_k\,;\qquad k=0,1,\ldots;$$
$$x_0\sim N(\hat{x}_0, P_0)$$
$$y_k=M_kx_k+v_k, \tag{1}$$

where $\mathbf{x}_k$ is the n-vector* state, Φ is the $n\times n$ state-transition matrix, G is $n\times r$, y_k is a scalar† observation, and M is $1\times n$. $\{\mathbf{w}_k\}$ is an r-vector, zero mean, white Gaussian sequence with

$$\varepsilon\{w_k w_l^T\}=Q\delta_{kl}\,; \tag{2}‡$$

and $\{v_k\}$ is a scalar, zero-mean, white Gaussian sequence with

$$\varepsilon\{v_k v_l\}=R_k\delta_{kl},\qquad R_k>0 \text{ scalar}. \tag{3}$$

$\{w_k\}$ and $\{v_k\}$ are assumed uncorrelated. As stated in the Introduction, the noise input G_kw_k is provided to "cover" errors made in modeling the dynamics.

Assuming that the statistical parameters Q and R_k are known, the well-known Kalman filter [1] for this model is

$$\hat{\mathbf{x}}_{k+}=\hat{\mathbf{x}}_{k-}+P_{k-}M_k^T(M_kP_{k-}M_k^T+R_k)^{-1}(y_k-M_k\hat{\mathbf{x}}_{k-}),$$
$$P_{k+}=P_{k-}-P_{k-}M_k^T(M_kP_{k-}M_k^T+R_k)^{-1}M_kP_{k-}, \tag{4}$$

$$\hat{\mathbf{x}}_{(k+1)-}=\Phi_{k+1,k}\hat{\mathbf{x}}_{k+},$$
$$P_{(k+1)-}=\Phi_{k+1,k}P_{k+}\Phi_{k+1,k}^T+G_kQG_k^T \tag{5}$$

with initial conditions $\hat{x}_0$ and P_0. $-(+)$ denotes the time instant prior to (after) an observation.

Let $\vartheta_k=(\ldots, y_{k-1}, y_k)$. Then

$$\hat{\mathbf{x}}_{k-}\underline{\Delta}\,\varepsilon\{\mathbf{x}_k|\vartheta_{k-1}\}=\hat{\mathbf{x}}_{k|k-1},$$
$$\hat{\mathbf{x}}_{k+}\underline{\Delta}\,\varepsilon\{\mathbf{x}_k|\vartheta_k\}=\hat{\mathbf{x}}_{k|k},$$
$$P_{k-}\underline{\Delta}\,\varepsilon\{(\mathbf{x}_k-\hat{\mathbf{x}}_{k-})(\mathbf{x}_k-\hat{\mathbf{x}}_{k-})^T|\vartheta_{k-1}\}=P_{k|k-1},$$
$$P_{k+}\underline{\Delta}\,\varepsilon\{(\mathbf{x}_k-\hat{\mathbf{x}}_{k+})(\mathbf{x}_k-\hat{\mathbf{x}}_{k+})^T|\vartheta_k\}=P_{k|k}. \tag{6}$$

Define the following (*predicted*) residuals

$$r_{k+l}\underline{\Delta}\,y_{k+l}-\varepsilon\{y_{k+l}|\vartheta_k\},\qquad l>0. \tag{7}$$

These residuals are zero-mean, Gaussian. It is straightforward to compute

$$r_{k+l}=M_{k+l}\Phi_{k+l,k}(\mathbf{x}_k-\hat{\mathbf{x}}_{k|k})+M_{k+l}\sum_{i=1}^{l}\Phi_{k+l,k+i}G_{k+i-1}w_{k+i-1}+v_{k+l} \tag{8}$$

---

* A vector is a column vector: superscript T is transposition.

† This presents no loss of generality.

‡ $\varepsilon\{\cdot\}$ is the expectation operator.

and

$$\varepsilon\{r_{k+l}r_{k+m}\} = M_{k+l}\Phi_{k+l,k}P_{k|k}\Phi^T_{k+m,k}M^T_{k+m} + M_{k+l}\sum_{i=1}^{l}\Phi_{k+l,k+i}G_{k+i-1}QG^T_{k+i-1} \times \Phi^T_{k+m,k+i}M^T_{k+m} + R_{k+l}\delta_{lm}, \quad m \geq l. \tag{9}$$

Estimation of Q

Suppose that the filter is at time $k+$, so that $\hat{x}_{k+}$ and P_{k+} are available, and (5) is about to be used to compute $(k+1)-$ values. Fix N. Determine the Q to be used in (5) (call it $\hat{Q}_{k,N}$) by the requirement that

$$r^2_{k+l} = \varepsilon\{r^2_{k+l}\}, \quad l = 1, \ldots, N. \tag{10}$$

This will produce consistency between the residuals and their statistics. Note that the resulting filter will have a time lag (N), since $k+N$ observations have to occur before prediction to time $k+1$ can be made.

Uncorrelated and identically distributed noise inputs. Suppose $Q=qI$, where I is the identity matrix. Let $N=1$ (one residual). Then with the aid of (9), (10) becomes

$$qM_{k+1}G_kG_k^TM^T_{k+1} = r^2_{k+1} - \varepsilon\{r^2_{k+1}|q\equiv 0\} \tag{11}$$

where, in slight abuse of notation

$$\varepsilon\{r^2_{k+1}|q\equiv 0\} = M_{k+1}\Phi_{k+1,k}P_{k+}\Phi^T_{k+1,k}M^T_{k+1} + R_{k+1}. \tag{12}$$

Now if

$$r^2_{k+1} < \varepsilon\{(r^2_{k+1}|q\equiv 0\}$$

no noise input is required. This leads to the estimator

$$\hat{q}_{k,1} = \begin{cases} \dfrac{r^2_{k+1} - \varepsilon\{r^2_{k+1}|q\equiv 0\}}{M_{k+1}G_kG_k^TM^T_{k+1}}, & \text{if positive} \\ 0, & \text{otherwise.} \end{cases} \tag{13}$$

It is shown in [10] that $\hat{q}_{k,1}$ maximizes the probability density function

$$p(r_{k+1})$$

with respect to $q \geq 0$. The restriction $\hat{q}_{k,1} \geq 0$ is consistent with the notion of a variance. It is, of course, assumed that $M_{k+1}G_kG_k^TM^T_{k+1} > 0$; that M_{k+1} is not orthogonal to G_k, an observability condition.

It is seen that, except for the proper scaling factor, $\hat{q}_{k,1}$ is the excess, if any, of the residual squared over the expected value of that residual squared under the assumption of no input noise.

Since

$$\varepsilon[r^2_{k+1} - \varepsilon(r^2_{k+1}|q\equiv 0)] = qM_{k+1}G_kG_k^TM^T_{k+1},$$

this estimator has a positive bias.

Filter (4, 5), with $\hat{q}_{k,1}I$ replacing Q in (5), is adaptive in the following sense. As long as residuals remain within their 1 sigma limits, the noise input is zero. This is as it should be since residuals are small and consistent with their statistics; the filter is operating properly. When residuals become large *relative* to their 1 sigma values, the filter is diverging, $P_{(k+1)-}$ is increased, the filter gains are thus increased, and the filter is "open" to incoming observations.

The estimator of Q described by (13) has several limitations. Because of the restrictions placed on the noise input ($Q=qI$), the noise coefficient matrix G_k must be modeled,* meaning that the distribution of the noise among the components of the state has to be specified *a priori*. This may be a difficult task. Furthermore, the estimate is based on one residual and therefore is not statistically significant. This difficulty is overcome by smoothing, to be discussed later. Thirdly, this estimator will respond to measurement noise (v_k) if

$$R_{k+1} \gtrsim M_{k+1}G_kQG_k^TM^T_{k+1}$$

and the best performance, in terms of absolute size of residuals, will not be realized.

The last difficulty is overcome by the following simple device. Instead of one residual, the *sample mean* of N predicted residuals is used

$$m_r \triangleq \frac{1}{N}\sum_{l=1}^{N} r_{k+l}/R^{\frac{1}{2}}_{k+l}. \tag{14}$$

It is easy to compute

$$\varepsilon\{m_r\} = 0, \tag{15}$$

$$\varepsilon\{m_r^2\} = S_N\Phi_{k+1,k}P_{k+}\Phi^T_{k+1,k}S_N^T + qS + 1/N, \tag{16}$$

* See [5, Ch. 8] for several approaches to modeling G_k.

where

$$S_N=\frac{1}{N}\sum_{l=1}^{N}(1/R_{k+l}^{\frac{1}{2}})M_{k+l}\Phi_{k+l,k+1},$$

$$S_{N-1}=\frac{1}{N}\sum_{l=2}^{N}(1/R_{k+l}^{\frac{1}{2}})M_{k+l}\Phi_{k+l,k+2},$$

$$\cdot$$

$$\cdot$$

$$\cdot$$

$$S_1=\frac{1}{N}(1/R_{k+N}^{\frac{1}{2}})M_{k+N},$$

$$S=S_NG_kG_k^TS_N^T+S_{N-1}G_{k+1}G_{k+1}^TS_{N-1}^T+\dots+S_1G_{k+N-1}G_{k+N-1}^TS_1^T. \quad (17)$$

Thus the variance of the measurement noise level in the sample mean goes to zero as $N\to\infty$, the familiar square root of N law.*

The same considerations as above, applied to m_r instead of r_{k+1}, now leads to the estimator

$$\hat{q}_{k,N}=\begin{cases}\dfrac{m_r^2-\varepsilon\{m_r^2|q\equiv 0\}}{S}, & \text{if positive}\\ 0, & \text{otherwise}\end{cases} \quad (18)$$

where

$$\varepsilon\{m_r^2|q\equiv 0\}=S_N\Phi_{k+1,k}P_{k+}\Phi_{k+1,k}^TS_N^T+1/N. \quad (19)$$

The structure of (18) is the same as that of (13). $\hat{q}_{k,N}$ maximizes

$$p(m_r)$$

with respect to $q\geq 0$.

Note that $\hat{q}_{k,N}$ need not be re-computed at every step k. It might be re-computed every l steps, $l\leq N$.

Independent noise inputs. Return now to the general case of N residuals (10), and assume that Q is diagonal. This provides independent noise inputs to each component of the state with the added assumption that $G_k\equiv I$ (Q is $n\times n$). Here, no modeling of the noise coefficients is required, which represents a more completely adaptive situation.

For this case (9) becomes

$$\varepsilon\{r_{k+l}^2\}=\varepsilon\{r_{k+l}^2|Q\equiv 0\}+\sum_{j=1}^{n}a_{lj}q_{jj} \quad (20)$$

where

$$\varepsilon\{r_{k+l}^2|Q\equiv 0\}=M_{k+l}\Phi_{k+l,k}P_{k+}\Phi_{k+l,k}^TM_{k+l}^T+R_{k+l}, \quad (21)$$

$$a_{lj}=\sum_{i=1}^{l}(M_{k+l}\Phi_{k+l,k+i})_j^2, \quad (22)$$

and q_{jj} are the diagonal elements of Q. Introducing the vectors and matrix

$$\varepsilon^T=[r_{k+1}^2-\varepsilon\{r_{k+1}^2|Q\equiv 0\},\dots,r_{k+N}^2-\varepsilon\{r_{k+N}^2|Q\equiv 0\}],$$

$$(\operatorname{diag}Q)^T=[q_{11},\dots,q_{nn}],\quad A=[a_{lj}](N\times n),$$

the imposed consistency requirement (10) becomes

$$A(\operatorname{diag}Q)=\varepsilon. \quad (23)$$

The relationship between (23) and the maximization of

$$p(r_{k+1},\dots,r_{k+N})$$

is pursued in [12].

Now (23) is a generalization of (11). Various special cases, depending on N, are discussed in [10, 12]. They are handled all together here by writing the solution of (23) as

$$\operatorname{diag}\bar{Q}_{k,N}=(A^TA)^{\#}A^T\epsilon \quad (24)$$

where $B^{\#}$ is the pseudo-inverse of B. This leads to the estimator

$$\operatorname{diag}\hat{Q}_{k,N}=\begin{cases}0, & \epsilon\leq 0\\ \operatorname{diag}\bar{Q}_{k,N}, & \text{otherwise},\end{cases} \quad (25)^*$$

subject to the rule that if $(\hat{q}_{jj})_{k,N}<0$, set $(\hat{q}_{jj})_{k,N}=0$. This last restriction is made *a posteriori* to preserve the properties of a variance.

Note that it is possible to define N sample means of residuals [such as (14)] and impose a consistency condition, such as (10), on these sample means. This leads to an estimator identical in structure to (25), but these details will not be pursued here. Also, note that diag $\hat{Q}_{k,N}$ need not

* Recall that the measurement noise is assumed white.

* A vector is "non-positive" ($\varepsilon\leqslant 0$) if $\varepsilon_i\leqslant 0$ for all i.

be computed at every step k, but only every l steps, $l \leq N$.

Filter stabilization. In order to enhance the statistical significance of the estimates (18), a simple smoothing† operation is performed. Defining

$$\bar{q}_{k,N} = \frac{m_r^2 - \varepsilon\{m_r^2 | q \equiv 0\}}{S} \tag{26}$$

the smoothed estimate is

$$\hat{q}_{k,N}^L = \begin{cases} \frac{1}{L} \sum_{i=k-L+1}^{k} \bar{q}_{i,N}, & \text{if positive} \\ 0, & \text{otherwise.} \end{cases} \tag{27}$$

Similarly, smoothing (25),

$$\operatorname{diag} \hat{Q}_{k,N}^L = \frac{1}{L} \sum_{i=k-L+1}^{k} \operatorname{diag} \bar{Q}_{i,N} \quad (28$$

subject to the rule that if $(\hat{q}_{jj}^L)_{k,N} < 0$, set $(\hat{q}_{jj}^L)_{k,N} = 0$.

Filter stabilization might also be achieved by making the *ad hoc* modification to (24)

$$\operatorname{diag} \bar{Q}_{k,N} = (C_o^{-1} + A^T A)^{\#}(A^T \varepsilon + C_o^{-1} \operatorname{diag} Q^o). \tag{29}$$

$\operatorname{diag} Q^o$ is an *a priori* estimate of $\operatorname{diag} \hat{Q}_{k,N}$ (perhaps $\operatorname{diag} \hat{Q}_{k-1,N}$). C_o (diagonal) is a measure of the uncertainty in that prior estimate. Equation (29) might be computed by the equivalent recursion

$$\operatorname{diag} Q^i = C_{i-1} A_i^T (A_i C_{i-1} A_i^T + 1)^{-1} (\varepsilon_i - A_i \operatorname{diag} Q^{i-1}) + \operatorname{diag} Q^{i-1} \tag{30}$$

$$C_i = C_{i-1} - C_{i-1} A_i^T (A_i C_{i-1} A_i^T + 1)^{-1} A_i C_{i-1},$$

$$i = 1, \ldots, N$$

where A_i, is the ith row of A and ε_i is the ith element of ε. Then

$$\operatorname{diag} \bar{Q}_{k,N} = \operatorname{diag} Q^N. \tag{31}$$

Recursion (30) involves the inversion of scalars, no matrix inversion, but it has nothing to offer over using a pseudo-inverse routine in (29) [15].

† For $N > 1$, smoothing involves using the same measurements over again. The residuals, however, are different since the state estimate is itself updated at each measurement.

Theoretical performance

All the estimates of Q given above are random variables since they depend on the residuals. When these estimators are installed in the filter equations (4, 5), the difference equations for the covariance matrix P in the filter become random difference equations. These equations are also non-linear. As a result, it is impossible to average them in closed form to obtain a measure of the adaptive filter performance. In essence, then, little can be said about the performance of the adaptive filter *a priori*. Its efficacy must be determined by numerical simulations.

Clearly, from the computational point of view, it is desirable to use only a "few" predicted residuals in the estimation of Q. To use more than a few residuals presents no theoretical difficulties, however. In computing predicted residuals and their mean square values, a noise input level Q is postulated after each measurement time. Thus, in prediction, the error covariance matrix P is sequentially adjusted, by adding Q, so that it is consistent with the predicted residuals. It is recalled, however, that the adaptive filter has a time lag N. The fewer residuals used, the shorter the filter lag.

SIMULATIONS

Simulation model

The dynamical system chosen for numerical simulation is the "rectilinear" orbit problem with dynamics

$$\ddot{x} = -\frac{\mu}{x^2} (x \text{ scalar}). \tag{32}$$

μ is the Gravitational Constant times earth mass, 19·9094165 er^3/hr^2 (er--earth radii). In first order form

$$\dot{x}_1 = x_2$$

$$\dot{x}_2 = -\frac{\mu}{x_1^2}, \tag{33}$$

where x_1 is position and x_2 velocity. Observations consist of positional, range, data

$$y_k = (x_1)_k + v_k, R_k = 1{\cdot}0 \times 10^{-7} \text{ er}^2. \tag{34}$$

The Kalman filter is applied to this problem by assuming that the estimate approximately satisfies (33) between observations. A closed form, although implicit, solution of (33) is available. This solution is re-linearized about the most current estimate for the purpose of computing the state transition matrix, for the propagation of the covariance matrix only. Thus the problem is completely discretized, and no numerical integration is required. To summarize, between observations the estimate evolves according to (33) but via a closed form solution, and the covariance matrix according to

$$P_{k+1|k}=\Phi_{k+1,k}P_{k|k}\Phi^T_{k+1,k}, \tag{35}$$

where $\Phi_{k+1,\,k}$ is computed as described above. This is precisely the well-known extended Kalman filter [5, Ch. 8].

the RSS/RMS ratio, where

$$\mathrm{RSS}(k)=\left\{\frac{1}{k}\sum_{i=1}^{k}[(x_i)_i-(\hat{x}_{i+})_i]^2\right\}^{\frac{1}{2}},$$

$$\mathrm{RMS}(k)=[(P_{k+})_{ii}]^{\frac{1}{2}},$$

where again $i=1$, 2 for position and velocity, respectively.

The adaptive filter with estimator (27) was simulated on a noisy trajectory to determine how well the (input) noise levels can be recovered. The data, or observation, rate is 10/hr. The results

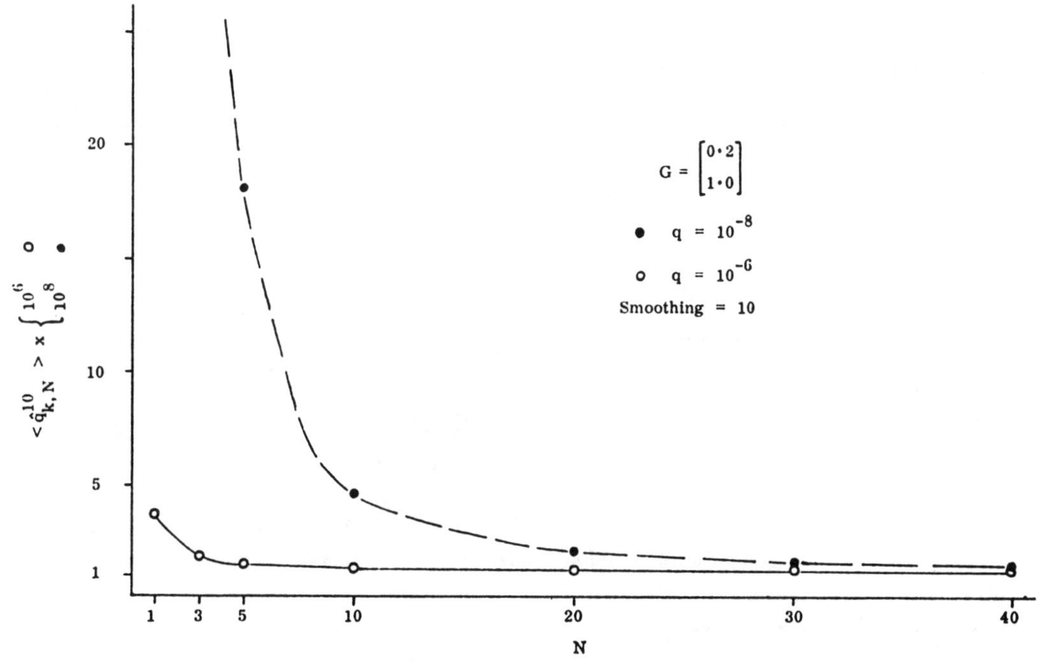

FIG. 1. Average noise variance estimates.

Discussion

Several trajectories of (32) were used in simulations. The one presented here is a rectilinear "ellipse" roughly described by the following points

Time (hr)	Position (er)	Velocity (er/hr)
0	8·0	2·0
3	13·0	1·4
44	33·0	0 (apogee)
50	32·0	−0·1 (termination)

Adaptive filter performance is evaluated on the basis of the estimation errors

$$(x_k)_i-(\hat{x}_{k+})_i,$$

$i=1$, 2 for position and velocity respectively, and on the basis of statistical consistency measured by

for two different noise levels and smoothing (L) of 10 are presented in Fig. 1. Plotted are the (time) averages of the estimates (averages over k), denoted by $<\cdot>$, as a function of N, the number of residuals in the sample mean. It is seen that the recovery is good for $N=20$, and improves with N. The behavior in the two cases plotted is substantially different for small N. In case $q=10^{-6}$ the system noise level is comparable to the measurement noise and good results are obtained for N small. Measurement noise dominates the system noise for $q=10^{-8}$, and a larger number of residuals are required (N larger) in order to effectively average it out. The efficacy of that simple device, the sample mean, is demonstrated.

Subsequent simulations of the adaptive filter employ estimator (28). The adaptive filter is first simulated on a noisy trajectory, with independent noise inputs to position and velocity. In this situation it is compared with the optimal filter, which knows and uses the input noise levels, and with what we shall call the Kalman filter, which uses a system noise level of zero.* The adaptive filter is then simulated on a trajectory in which a bias in the value of μ is introduced. It is here compared with the Kalman filter which knows nothing of this bias. Of course the adaptive filter is also ignorant of the bias.

The results for the noise simulations are given in Figs. 2–5. The data rate here is 100/hr; noise levels are indicated in the figures. Smoothing (L)

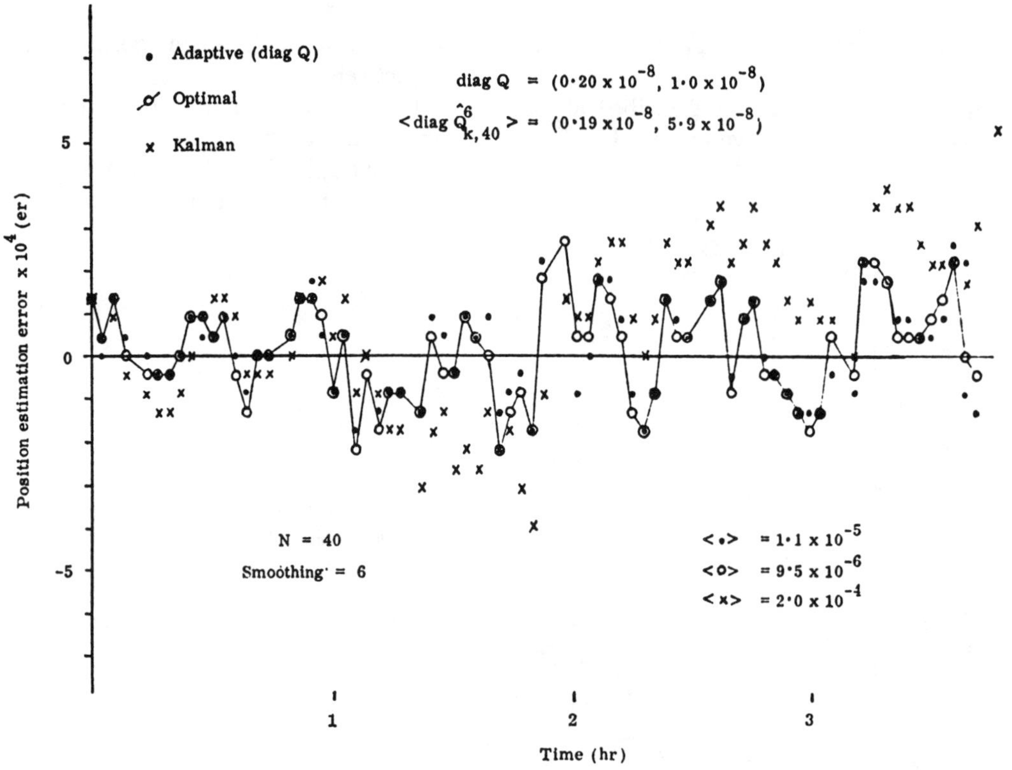

FIG. 2. Position estimation error, noise case.

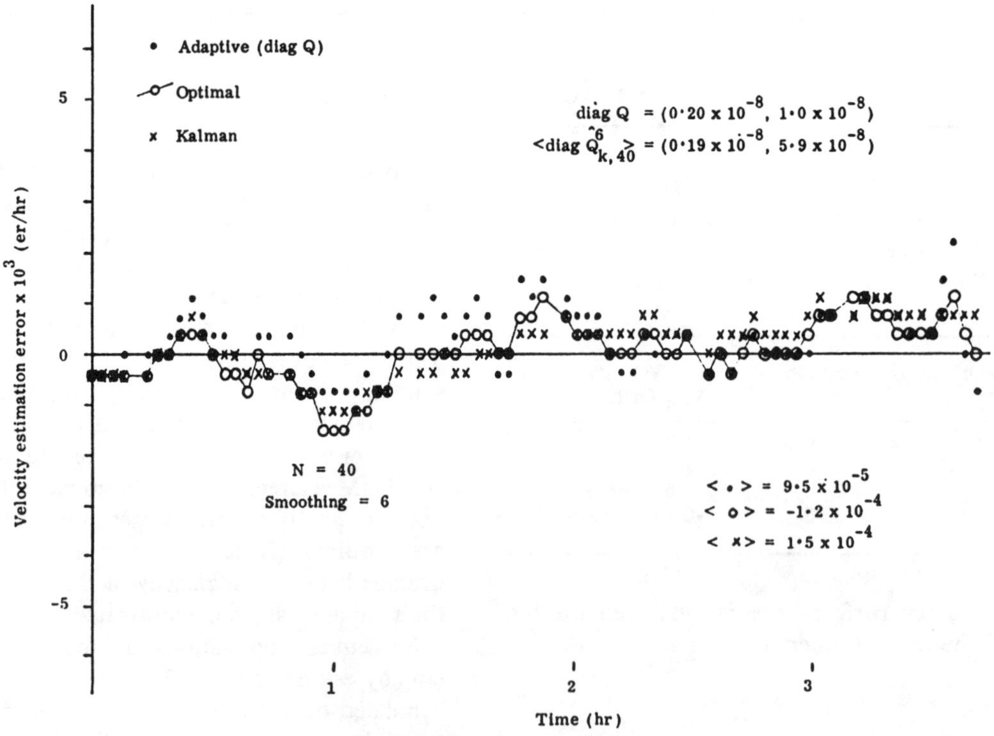

FIG. 3. Velocity estimation error, noise case.

* All the filters simulated are of course based on the extended Kalman filter.

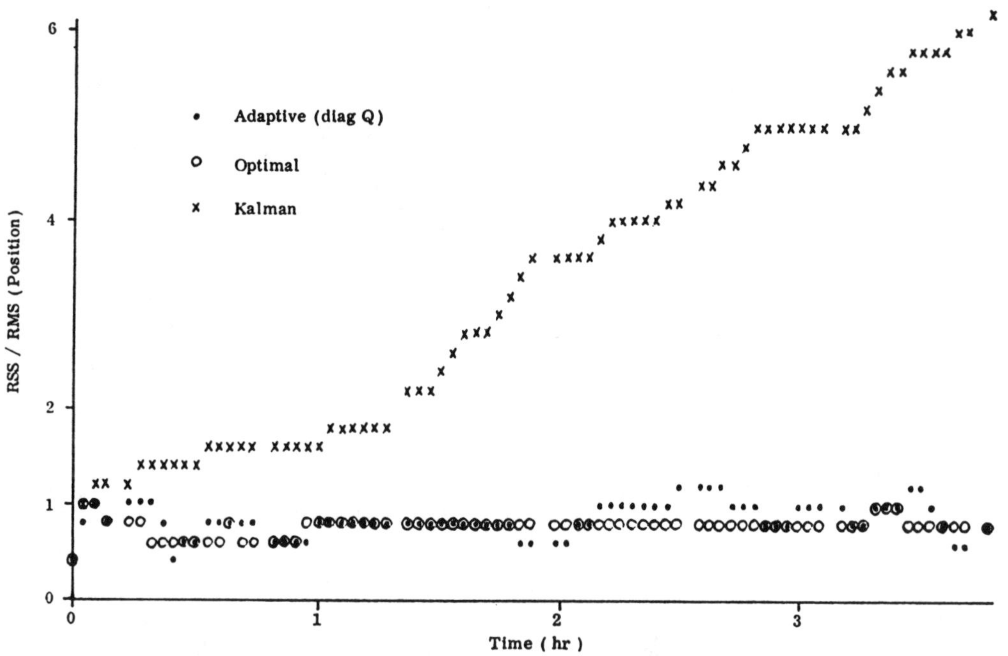

FIG. 4. Position RSS/RMS, noise case.

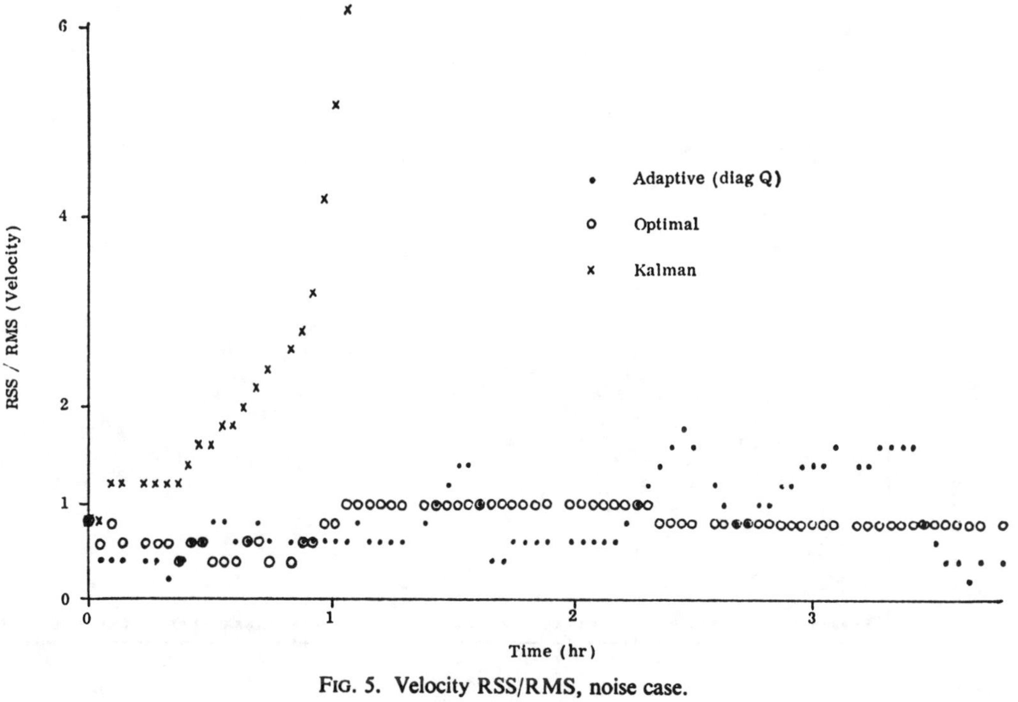

FIG. 5. Velocity RSS/RMS, noise case.

is 6 and $N=40$. We see that the adaptive filter performance is comparable to that of the optimal filter. The Kalman estimates are biased and the Kalman filter diverges, as evidenced by its RSS/RMS ratios.

Bias simulations are presented in Figs. 6–9. The bias in μ is 0·0075, which is about 50 standard deviations. A large bias is used so that long-term effects can be observed in short time. The data rate here is 10/hr. It is seen that the Kalman filter diverges; an example of the divergence problem discussed in the Introduction. The adaptive filter is however tracking the orbit. The errors are substantially random and about as small as the μ-bias permits. Statistical consistency is achieved since RSS/RMS$\sim$1. It is seen that a relatively small $N(10)$ and $L(6)$ are required.

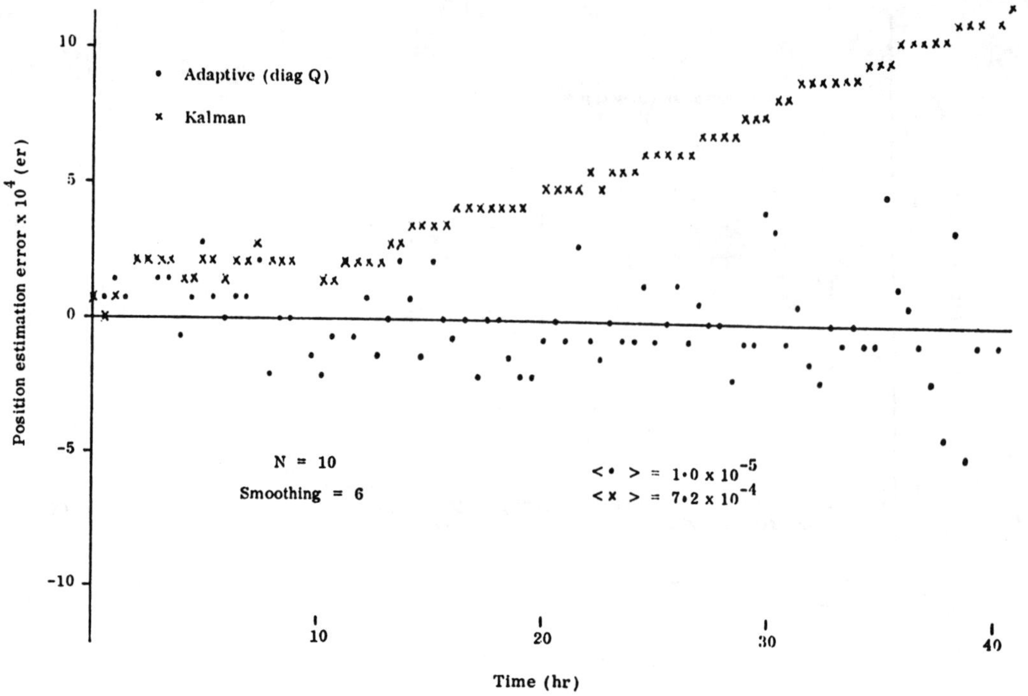

FIG. 6. Position estimation error, (50 σ)μ—bias case.

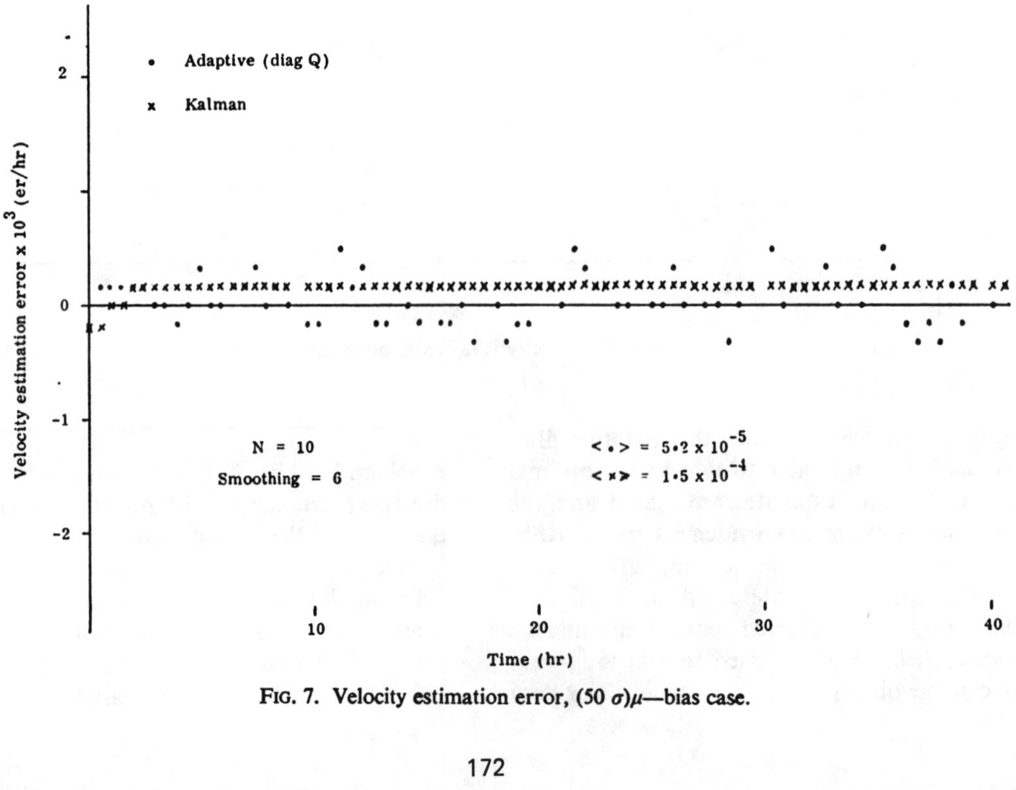

FIG. 7. Velocity estimation error, (50 σ)μ—bias case.

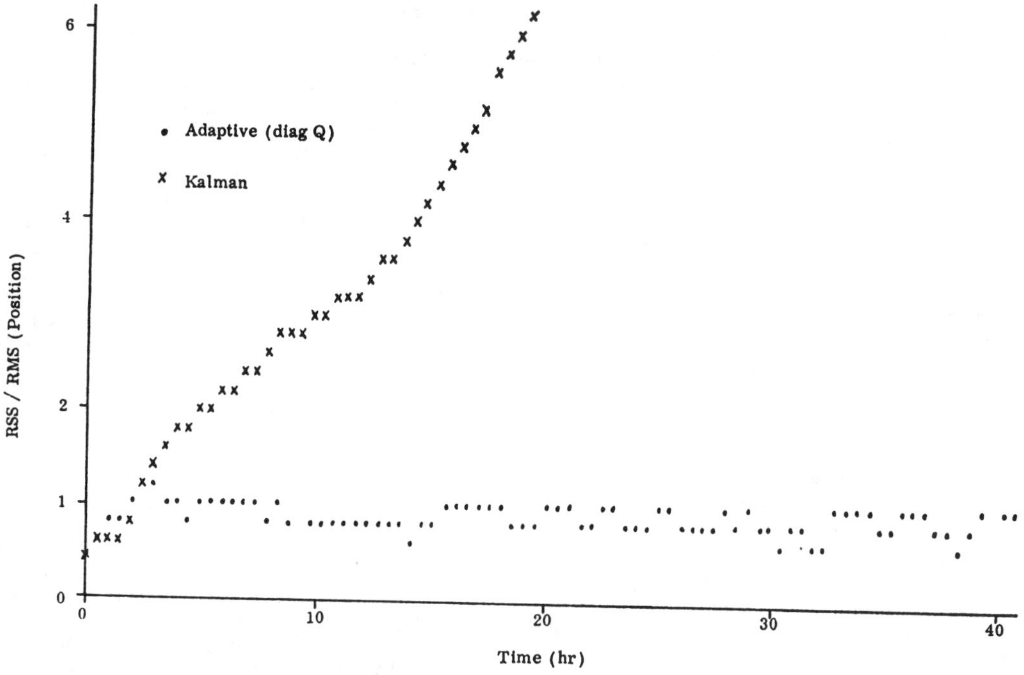

FIG. 8. Position RSS/RMS, (50 σ)μ—bias case.

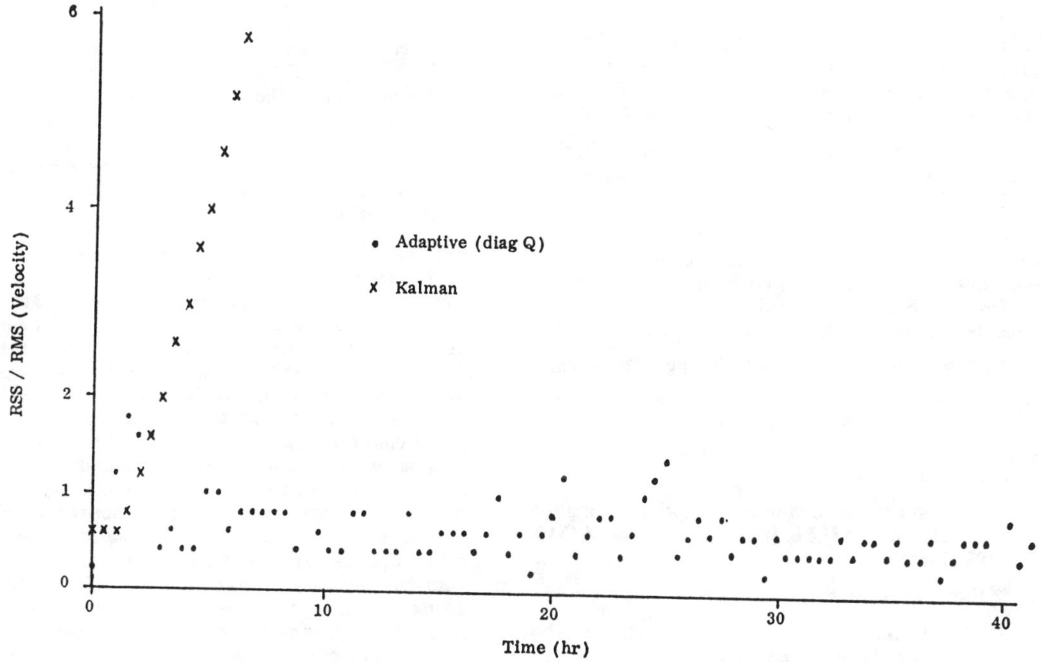

FIG. 9. Velocity RSS/RMS, (50 σ)μ—bias case.

CONCLUSIONS

The simulations indicate that the adaptive filter concepts presented in this paper are useful in preventing the divergence phenomenon often observed in applying the Kalman filter to orbit determination. This conclusion is further substantiated by numerical results reported elsewhere [16].

It is of course possible to determine a Q to use by experimentation, or from insight into a particular problem, or both. Various methods which have been developed as means of preventing divergence are described in [5, Ch. 8]. The limited memory filter [9], in particular, gives another rational approach to the problem of divergence.

Acknowledgements—The author wishes to express his appreciation for the active interest shown by Mr. R. K. Squires of Goddard Space Flight Center and Dr. H. Wolf of Analytical Mechanics Associates, Inc. in this problem and its solution. The basic idea of providing some sort of feedback from the residuals to the filter gains is due to Mr. Squires. The author is also indebted to Mrs. A. Bailie for generating the data for this paper, and to Mr. N. Levine for the computer program development.

REFERENCES

[1] R. E. KALMAN: A new approach to linear filtering and prediction problems. *Trans. ASME, J. Basic Eng.* **82**, 35–44 (1960).

[2] R. E. KALMAN and R. S. BUCY: New results in linear filtering and prediction theory. *Trans. ASME, J. Basic Eng.* **83**, 95–108 (1961).

[3] F. H. SCHLEE, C. J. STANDISH and N. F. TODA: Divergence in the Kalman filter. *AIAA J.* **5**, 1114–1120 (1967).

[4] R. J. FITZGERALD: Error divergence in optimal filtering problems. Second IFAC Symposium on Automatic Control in Space, Vienna, Austria (1967).

[5] A. H. JAZWINSKI: *Stochastic Processes and Filtering Theory*. Academic Press, New York, to appear Fall 1969.

[6] S. PINES, H. WOLF, A. BAILIE and J. MOHAN: Modifications to the Goddard minimum variance program for the processing of real data. Analytical Mechanics Associates, Inc., Tech. Rept. Contract NAS5-2535 (1964).

[7] S. F. SCHMIDT: Application of state-space methods to navigation problems. In *Advance in Control Systems*, Vol. 3 (Edited by C. T. LEONDES). Academic Press, New York (1966).

[8] S. F. SCHMIDT: Estimation of state with acceptable accuracy constraints. Analytical Mechanics Associates, Inc., Interim Report 67–4 (1967).

[9] A. H. JAZWINSKI: Limited memory optimal filtering. Proc. 1968 Joint Automatic Control Conference, Ann Arbor, Michigan, June (1968). Also in *IEEE Trans. Aut. Control* **AC-13**, No. 5, 558–563, October (1968).

[10] A. H. JAZWINSKI and A. E. BAILIE: Adaptive filtering. Analytical Mechanics Associates, Inc., Interim Report 67–6 (1967).

[11] A. R. DENNIS: Functional updating and adaptive noise variance determination in recursive-type trajectory estimators. The Special Projects Branch Astrodynamics conference, NASA Goddard Space Flight Center, Greenbelt, Maryland, May (1967).

[12] A. H. JAZWINSKI: On adaptive filtering. To appear.

[13] G. L. SMITH: Sequential estimation of observation error variances in a trajectory estimation problem. Paper No. 67–89, AIAA 5th Aerospace Sciences Mtg., New York, January (1967).

[14] D. G. LAINIOTIS: A non-linear adaptive estimation recursive algorithm. *IEEE Trans. Aut. Control* **AC-13**, 197–198 (1968).

[15] T. S. ENGLAR: Unpublished notes.

[16] A. H. JAZWINSKI: Adaptive filtering in satellite orbit estimation. Conference Record of Second Asilomar Conference on Circuits and Systems, pp. 153–158, Pacific Grove, California, October (1968).

Résumé—Les applications du filtre de Kalman aux problèmes de détermination de l'orbite ont parfois rencontré une difficulté mentionnée sous le nom de divergent. Le phénomène est celui de l'augmentation des résidus; l'état et son estimation divergent. Ce problème peut être souvent expliqué par une précision insuffisante de la simulation de la dynamique utilisée dans le filtre. Quoique une simulation plus précise constitue une solution évidente celle-ci est souvent peu pratique et quelquefois impossible. L'article donne une méthode d'approximation des erreurs de simulation par un signal d'entrée de bruit blanc gaussien et sa covariance (Q) est déterminée de manière à produire un rapprochement entre les résiduset leur statistiques. De cette manière, une réaction en temps réél est fournie par les résidus au gain du filtre. Une apparition de la divergence produit une augmentation du gain du filtre et le filtre adaptatif est capable de continuer la recherche. Ce schéma possède une interpretation probabiliste. Dans certaines conditions l'estimation de Q a produit la plus probable séquence finie de résidus.

Zusammenfassung—Anwendungen des Kalman-Filters bei Orbitbestimmungsproblemen stiessen manchmal auf eine Schwierigkeit, die als Divergenz bezeichnet wurde. Das Phänomen ist eine Zunahme der Reste; der Zustand und seine Bestimmung laufen auseinander. Das Problem lässt sich oft auf die ungenügende Genauigkeit der Modellbildung der in dem Filter angewendeten Dynamik zurückführen. Wenn auch die genauere Modellbildung eine auf der Hand liegende Lösung darstellt, ist sie oft unpraktisch und manchmal unmöglich. Modellfehler werden hier durch die Eingabe eines weissen Gauss'schen Rauschens angenähert, und dessen Kovarianz (Q) wird bestimmt, um die Konsistenz zwischen Resten und ihrer Statistik herzustellen. Auf diese Weise wird ein Echtzeit-Feedback von den Resten zur Filterverstärkung geliefert. Das Einsetzen der Divergenz erzeugt eine Filterverstärkung, und das Adaptivfilter vermag das Tracking fortzusetzen. Es handelt sich hier um eine Wahrscheinlichkeitsdeutung. Unter bestimmten Bedingungen liefert die Bestimmung von Q die wahrscheinlichste endliche Folge von Resten.

Резюме—Применения фильтра Кальмана к проблемам опроделения орбиты иногда встречали затруднение известное под именем расхождения. Это явление соответствует увеличению остатков; состояние и его оценка расходятся. Эта проблема может быть часто сведена к недостаточной точности моделирования используемого в фильтре. Хотя более точное моделирование составляет очевидное решение, оно часто не практично а иногда и невозможно. Статья дает метод аппроксимации ошибок моделирования входным сигналом гауссовского белого шума и его ковариантность Q определена таким путем что она производит сближение остатков с их статистиками. Таким образом, обратная связь в реальном времени устанавливается от остатков к коэффициенту усиления фильтра. Появление расхождения производит увеличение коэффициента фильтра и адаптативный фильтр находится в состоянии продолжать поиск. Эта схема имеет вероятностное истолкование. В некоторых условиях оценка Q привела к наиболее вероятной конечной серии остатков.

On the Identification of Variances and Adaptive Kalman Filtering

RAMAN K. MEHRA, MEMBER, IEEE

Abstract—A Kalman filter requires an exact knowledge of the process noise covariance matrix Q and the measurement noise covariance matrix R. Here we consider the case in which the true values of Q and R are unknown. The system is assumed to be constant, and the random inputs are stationary. First, a correlation test is given which checks whether a particular Kalman filter is working optimally or not. If the filter is suboptimal, a technique is given to obtain asymptotically normal, unbiased, and consistent estimates of Q and R. This technique works only for the case in which the form of Q is known and the number of unknown elements in Q is less than $n \times r$ where n is the dimension of the state vector and r is the dimension of the measurement vector. For other cases, the optimal steady-state gain K_{op} is obtained directly by an iterative procedure without identifying Q. As a corollary, it is shown that the steady-state optimal Kalman filter gain K_{op} depends only on $n \times r$ linear functionals of Q. The results are first derived for discrete systems. They are then extended to continuous systems. A numerical example is given to show the usefulness of the approach.

I. Introduction

THE OPTIMUM filtering results of Kalman and Bucy [1], [2] for linear dynamic systems require an exact knowledge of the process noise covariance matrix Q and the measurement noise covariance matrix R. In a number of practical situations, Q and R are either unknown or are known only approximately. Heffes [3] and Nishimura [4] have considered the effect of errors in Q and R on the performance of the optimal filter. Several other investigators [5]–[9] have proposed on-line schemes to identify Q and R. Most of these schemes do well in identifying R but run into difficulties in identifying Q. Moreover, their extension to continuous cases is not clear. A different approach has been taken in this paper. It is assumed that the system under consideration is time invariant, completely controllable, and observable [2]. Both the system and the filter (optimal or suboptimal) are assumed to have reached steady-state conditions. First, a correlation test is performed on the filter to check whether it is working optimally or not. The test is based on the innovation property of an optimal filter [10]. If the filter is suboptimal, the auto-correlation function of the innovation process is used to obtain asymptotically unbiased and consistent estimates of Q and R. The method has the limitation that the number of unknown elements in Q must be less than $n \times r$ where n is the dimension of the state vector and r is the dimension of the measurement vector. It is shown that in spite of this limitation, the optimal steady-state filter gain can be obtained by an iterative procedure. As a corollary, it is shown that the Kalman filter gain depends only on $n \times r$ linear relationships between the elements of Q.

A numerical example is included to illustrate the application of the results derived in the paper. The extension of the results to the continuous case is straightforward and is given in the last section.

Manuscript received August 2, 1968; revised May 16, 1969. This paper was presented at the 1969 Joint Automatic Control Conference, Boulder, Colo.

The author was with the Analytic Sciences Corporation, Reading, Mass. 01867. He is now with Systems Control, Inc., Palo Alto, Calif. 94306.

II. Statement of the Problem

System

Consider a multivariable linear discrete system

$$x_{i+1} = \Phi x_i + \Gamma u_i \tag{1}$$

$$z_i = Hx_i + v_i \tag{2}$$

where x_i is $n \times 1$ state vector, Φ is $n \times n$ nonsingular transition matrix, Γ is $n \times q$ constant input matrix, z_i is $r \times 1$ measurement vector, and H is $r \times n$ constant output matrix.

The sequences $u_i(q \times 1)$ and $v_i(r \times 1)$ are uncorrelated Gaussian white noise sequences with means and covariances as follows:

$$E\{u_i\} = 0; \quad E\{u_i u_j^T\} = Q\delta_{ij}$$

$$E\{v_i\} = 0; \quad E\{v_i v_j^T\} = R\delta_{ij}$$

$$E\{u_i v_j^T\} = 0, \quad \text{for all } i,j$$

where $E\{\cdot\}$ denotes the expectation, and δ_{ij} denotes the Kronecker delta function.

Q and R are bounded positive definite matrices ($Q > 0$, $R > 0$). Initial state x_0 is normally distributed with zero mean and covariance P_0.

The system is assumed to be completely observable and controllable, i.e.,

$$\text{rank}\,[H^T, (H\Phi)^T, \cdots, (H\Phi^{n-1})^T] = n$$

$$\text{rank}\,[\Gamma, \Phi\Gamma, \cdots, \Phi^{n-1}\Gamma] = n.$$

Filter

Let Q_0 and R_0 be the initial estimates of Q and R ($Q_0 > 0$, $R_0 > 0$). Using these estimates, let the steady-

Reprinted from *IEEE Trans. Automat. Contr.*, vol. AC-15, pp. 175–184, Apr. 1970.

state Kalman filter gain be K_0 ($n \times r$ matrix)[1]

$$K_0 = M_0H^T(HM_0H^T + R_0)^{-1} \tag{3}$$

$$M_0 = \Phi[M_0 - M_0H^T(HM_0H^T + R_0)^{-1}HM_0]\Phi^T + \Gamma Q_0\Gamma^T. \tag{4}$$

M_0 may be recognized as the steady-state solution to the covariance equations of Kalman [1].

The filtering equations are

$$\hat{x}_{i+1/i} = \Phi\hat{x}_{i/i} \tag{5}$$

$$\hat{x}_{i/i} = \hat{x}_{i/i-1} + K_0(z_i - H\hat{x}_{i/i-1}) \tag{6}$$

where $\hat{x}_{i+1/i}$ is the estimate of x_{i+1} based on all the measurements up to i, i.e., $\{z_0,\cdots,z_i\}$.

In an optimal Kalman filter (i.e., when $Q_0 = Q$ and $R_0 = R$), M_0 is the covariance of the error in estimating the state. But in a suboptimal case, the covariance of the error (M_1) is given by the following equation [3]:

$$M_1 = \Phi[M_1 - K_0HM_1 - M_1H^TK_0^T + K_0(HM_1H^T + R)K_0^T]\Phi^T + \Gamma Q\Gamma^T \tag{7}$$

where $M_1 = E\{(x_i - \hat{x}_{i/i-1})(x_i - \hat{x}_{i/i-1})^T\}$.

Problem

The true values of Q and R are unknown. It is required to

1) check whether the Kalman filter constructed using some estimates of Q and R is close to optimal or not (hypothesis testing),
2) obtain unbiased and consistent estimates of Q and R (statistical estimation), and
3) adapt the Kalman filter at regular intervals using all the previous information (adaptive filtering).

To solve these problems, we make use of the innovation property of an optimal filter [10].

III. The Innovation Property of an Optimal Filter[2]

Statement

For an optimal filter, the sequence $\nu_i = (z_i - H\hat{x}_{i/i-1})$, known as the innovation sequence, is a Gaussian white noise sequence.

Proof: A direct proof is obtained using the orthogonality principle of linear estimation [10].[3] Let $e_i = x_i - \hat{x}_{i/i-1}$ denote the error in estimating the state.

Then

$$\nu_i = He_i + v_i$$

$$E\{\nu_i\nu_j^T\} = E\{(He_i + v_i)(He_j + v_j)^T\}. \tag{8}$$

For $i > j$, v_i is independent of e_j and v_j:

$$E\{\nu_i\nu_j^T\} = E\{He_i(He_j + v_j)^T\} = E\{He_i(z_j - H\hat{x}_{j/j-1})^T\}.$$

The orthogonality principle states that e_i is orthogonal to $\{z_k, k < i\}$. Since $\hat{x}_{j/j-1}$ depends only on $\{z_k, k < j\}$, we conclude that

$$E\{\nu_i\nu_j^T\} = 0, \quad \text{for } i > j.$$

Similarly, $E\{\nu_i\nu_j^T\} = 0$, for $i < j$.

For $i = j$, $E\{\nu_i\nu_i^T\} = HMH^T + R$. Further, since ν_i is a linear sum of Gaussian random variables, it is also Gaussian. Hence ν_i is a Gaussian white noise sequence.

Heuristically, the innovation ν_i represents the new information brought by z_i. Kailath [10] shows that ν_i and z_i contain the same statistical information and are equivalent as far as linear operations are concerned. Schweppe [12] shows that ν_i can be obtained from z_i by a Gram–Schmidt orthogonalization (or a whitening) procedure.

In this paper, we use the innovation sequence to check the optimality of a Kalman filter and to estimate Q and R. With this in mind, we investigate the effect of suboptimality on the innovation sequence.

IV. Innovation Sequence for a Suboptimal Filter

Let K denote the steady-state filter gain. We will show that under steady state, the innovation sequence ν_i is a stationary Gaussian sequence:

$$\nu_i = z_i - H\hat{x}_{i/i-1} = He_i + v_i$$

$$E\{\nu_i\nu_{i-k}^T\} = HE\{e_ie_{i-k}\}H^T + HE\{e_iv_{i-k}^T\}, \quad \text{for } k > 0.$$

A recursive relationship can be obtained for e_i by using (1), (2), (5), and (6):

$$e_i = \Phi(I - KH)e_{i-1} - \Phi Kv_{i-1} + \Gamma u_{i-1}. \tag{9}$$

Carrying (9) k steps back,

$$e_i = [\Phi(I - KH)]^ke_{i-k} - \sum_{j=1}^{k}[\Phi(I - KH)]^{j-1}\Phi Kv_{i-j} + \sum_{j=1}^{k}[\Phi(I - KH)]^{j-1}\Gamma u_{i-j}. \tag{10}$$

Postmultiplying (10) by e_{i-k}^T and taking expectations,

$$E\{e_ie_{i-k}^T\} = [\Phi(I - KH)]^kM$$

where M is the steady-state error covariance matrix. An expression for M is obtained directly from (9) or from (7):

$$M = \Phi(I - KH)M(I - KH)^T\Phi^T + \Phi KRK^T\Phi^T + \Gamma Q\Gamma^T. \tag{11}$$

Postmultiplying (10) by v_{i-k}^T and taking expectations,

$$E\{e_iv_{i-k}^T\} = -[\Phi(I - KH)]^{k-1}\Phi KR.$$

[1] The conditions of complete controllability and observability together with the positive definiteness of Q_0 and R_0 ensure the asymptotic global stability of the Kalman filter. See Deyst and Price [11].

[2] For a detailed discussion, see Kailath [10].

[3] An alternate proof will be given in Section IV.

Therefore,

$$E\{\nu_i\nu_{i-k}^T\} = H[\Phi(I-KH)]^{k-1} \cdot\Phi[MH^T - K(HMH^T+R)], \quad k>0.$$

When $k=0$, $E\{\nu_i\nu_i^T\} = HMH^T+R$.

It is seen that the autocorrelation function of ν_i does not depend on i. Therefore, ν_i is a stationary Gaussian random sequence (Gaussian because of linearity) and we can define

$$C_k \equiv E\{\nu_i\nu_{i-k}^T\}.$$

Then

$$C_k = HMH^T + R, \qquad k=0 \quad (12)$$

$$= H[\Phi(I-KH)]^{k-1}\Phi[MH^T - KC_0], \quad k>0. \quad (13)$$

Furthermore,

$$C_{-k} = C_k^T.$$

Notice that the optimal choice of K, viz. $K = MH^T(HMH^T+R)^{-1}$ makes C_k vanish for all $k \neq 0$ (the innovation property).

V. A Test of Optimality for a Kalman Filter

From the discussion of the preceding two sections, it is clear that a necessary and sufficient condition for the optimality of a Kalman filter is that the innovation sequence ν_i be white. This condition can be tested statistically by a number of different methods [13], [16]–[19]. Here we consider a particular method given in Jenkins and Watts [13].

In this method, we obtain an estimate of C_k, denoted as $\hat{C}_k$, by using the ergodic property of a stationary random sequence

$$\hat{C}_k = (1/N)\sum_{i=k}^{N} \nu_i\nu_{i-k}^T \quad (14)$$

where N is the number of sample points.

The estimates $\hat{C}_k$ are biased for finite sample sizes:

$$E\{\hat{C}_k\} = (1-k/N)C_k. \quad (15)$$

In case an unbiased estimate is desired, we divide by $(N-k)$ instead of N in (14). However, it is shown in [13] that the estimate of (14) is preferable since it gives less mean-square error than the corresponding unbiased estimate.

An expression for the covariance of $\hat{C}_k$ can be derived by straightforward manipulation, but the general results are rather involved. We quote here approximate results for large N given in Bartlett [14]:

$$\operatorname{cov}([\hat{C}_k]_{ij},[\hat{C}_l]_{pq}) \approx (1/N)\sum_{t=-\infty}^{\infty}([C_t]_{ip}[C_{t+l-k}]_{jq} + [C_{t+l}]_{iq}[C_{t-k}]_{jp}) \quad (16)$$

where $[\hat{C}_k]_{ij}$ denotes the element in the ith row and the jth column of the matrix $\hat{C}_k$ and cov (a,b) denotes the covariance of a and b; viz.

$$\operatorname{cov}(a,b) \equiv E\{[a-E(a)][b-E(b)]\}.$$

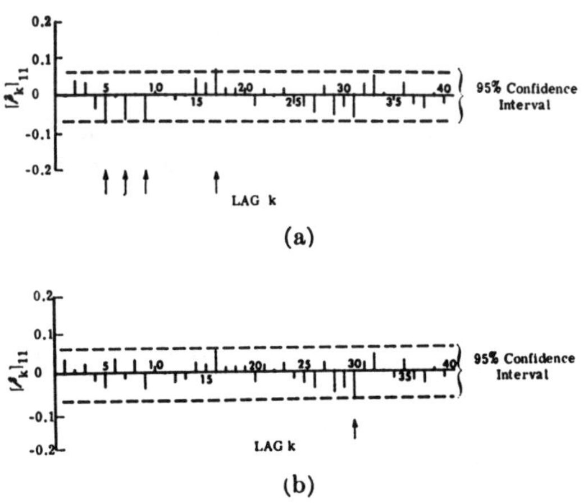

Fig. 1. Normalized autocorrelation function of innovation process. (a) Suboptimal filter. (b) Optimal filter. (Arrows indicate points for which 95 percent confidence limits do not enclose zero.)

It is seen from (13) that $C_k \to 0$ for large k. It can be shown[4] that the infinite series in (16) has a finite sum so that the covariance of $\hat{C}_k$ is proportional to $1/N$. Thus, the estimates $\hat{C}_k$ are asymptotically unbiased and consistent. Moreover, since all the eigenvalues of $\Phi(I-KH)$ lie inside the unit circle, ν_i belongs to the class of linear processes [14] for which Parzen [15] has shown that $\hat{C}_k$ are asymptotically normal.

For the white noise case, (16) is greatly simplified by putting $C_k = 0$, for all $k \neq 0$:

$$\operatorname{cov}([\hat{C}_k]_{ij},[\hat{C}_l]_{pq}) = 0, \quad k\neq l$$
$$= (1/N)[C_0]_{ip}[C_0]_{jq}, \quad k=l>0$$
$$= (1/N)[C_0]_{ip}[C_0]_{jq} + [C_0]_{iq}[C_0]_{jp}, \quad k=l=0. \quad (17)$$

Estimates of the normalized autocorrelation coefficients ρ_k are obtained by dividing the elements of $\hat{C}_k$ by the appropriate elements of $\hat{C}_0$, e.g.,

$$[\hat{\rho}_k]_{ij} = \frac{[\hat{C}_k]_{ij}}{\{[\hat{C}_0]_{ii}[\hat{C}_0]_{jj}\}^{1/2}}. \quad (18)$$

Of particular interest here are the diagonal elements of $\hat{\rho}_k$ for the case of white noise. Using (17), we can show that

$$\operatorname{var}[\hat{\rho}_k]_{ii} = 1/N + 0(1/N^2). \quad (19)$$

Further, $[\hat{\rho}_k]_{ii}$ like $[\hat{C}_k]_{ii}$ are asymptotically normal [15]. Therefore, the 95 percent confidence limits for $[\hat{\rho}_k]_{ii}$, $k>0$ are $\pm(1.96/N^{1/2})$, or equivalently the 95 percent confidence limits for $[\hat{C}_k]_{ii}$ are $\pm(1.96/N^{1/2})[\hat{C}_0]_{ii}$.

Test

Look at a set of values for $[\hat{\rho}_k]_{ii}$, $k>0$ and check the number of times they lie outside the band $\pm(1.96/N^{1/2})$. If this number is less than 5 percent of the total, the sequence ν_i is white. (Examples of a nonwhite and a white

[4] The proof is essentially similar to the one for proving the stability of a Kalman filter [11].

sequence are shown in Fig. 1. See the example in Section IX.)

This test is based on the assumption of large N. If N is small, other tests proposed by Anderson [17] and Hannan [16], etc., may be used. Jenkins and Watts [13] also give a frequency domain test which is useful if there are slow periodic components in the time series.

VI. Estimation of Q and R

If the test of Section V reveals that the filter is suboptimal, the next step will be to obtain better estimates of Q and R. This can be done using C_k computed earlier. The method proceeds in three steps.

1) Obtain an estimate of MH^T using (13). Rewriting (13) explicitly,

$$C_1 = H\Phi MH^T - H\Phi KC_0$$

$$C_2 = H\Phi^2 MH^T - H\Phi KC_1 - H\Phi^2 KC_0$$

$$\vdots$$

$$C_n = H\Phi^n MH^T - H\Phi KC_{n-1} - \cdots - H\Phi^n KC_0.$$

Therefore

$$MH^T = B^* \begin{bmatrix} C_1 + H\Phi KC_0 \\ C_2 + H\Phi KC_1 + H\Phi^2 KC_0 \\ \vdots \\ C_n + H\Phi KC_{n-1} + \cdots + H\Phi^n KC_0 \end{bmatrix} \quad (20)$$

where B^* is the pseudo-inverse of matrix B [1] defined as

$$B \equiv \begin{bmatrix} H \\ H\Phi \\ \vdots \\ H\Phi^{n-1} \end{bmatrix} \cdot \Phi.$$

Notice that B is the product of the observability matrix and the nonsingular transition matrix Φ. Therefore

$$\text{rank}\,(B) = n$$

and

$$B^* = (B^TB)^{-1}B^T.$$

Denoting[5] by $\widehat{MH}^T$ the estimate of MH^T and using (20), we can write

$$\widehat{MH}^T = B^* \begin{bmatrix} \hat{C}_1 + H\Phi K\hat{C}_0 \\ \hat{C}_2 + H\Phi K\hat{C}_1 + H\Phi^2 K\hat{C}_0 \\ \vdots \\ \hat{C}_n + H\Phi K\hat{C}_{n-1} + \cdots + H\Phi^n K\hat{C}_0 \end{bmatrix}. \quad (21)$$

An alternate form for MH^T can be obtained directly from (13):

$$\widehat{MH}^T = K\hat{C}_0 + A^* \begin{bmatrix} \hat{C}_1 \\ \vdots \\ \hat{C}_n \end{bmatrix} \quad (22)$$

[5] The symbol $\hat{A}\hat{B}$ always implies $\widehat{AB}$ as a single symbol.

where

$$A = \begin{bmatrix} H\Phi \\ H\Phi(I - KH)\Phi \\ H[\Phi(I - KH)]^{n-1}\Phi \end{bmatrix}.$$

In numerical computation, it has been found preferable to use (22) since matrix A is better conditioned than matrix B. (This is an experimental observation.)

2) Obtain an estimate of R using (12):

$$\hat{R} = \hat{C}_0 - H(\widehat{MH}^T). \quad (23)$$

3) Obtain an estimate of Q using (11).

This step gets complicated due to the fact that only the estimate of MH^T instead of M is available. Consequently, only $n \times r$ linear relationships between the unknown elements of Q are available. If the number of unknowns in Q is $n \times r$ or less, a solution can be obtainedl But if the number of unknowns in Q is greater than $n \times r$. a unique solution cannot be obtained. However, it wil, be shown in the next section that a unique solution for the optimal gain K_{op} can still be obtained.

Restricting ourselves to the case in which the number of unknowns in Q is $n \times r$ or less, we can solve for the unknown elements of Q by rewriting (11) as follows:

$$M = \Phi M\Phi^T + \Omega + \Gamma Q\Gamma^T \quad (24)$$

where

$$\Omega = \Phi[-KHM - MH^TK^T + KC_0K^T]\Phi^T.$$

Substituting back for M on the right-hand side of (24),

$$M = \Phi^2 M(\Phi^2)^T + \Phi\Omega\Phi^T + \Omega + \Phi\Gamma Q\Gamma^T\Phi^T + \Gamma Q\Gamma^T. \quad (25)$$

Repeating the same procedure n times and separating the terms involving Q on the left-hand side of the equation, we obtain

$$\sum_{j=0}^{k-1} \Phi^j\Gamma Q\Gamma^T(\Phi^j)^T = M - \Phi^k M(\Phi^k)^T - \sum_{j=0}^{k-1} \Phi^j\Omega(\Phi^j)^T, \quad \text{for } k = 1,\cdots,n. \quad (26)$$

Premultiplying both sides of (26) by H and postmultiplying by $(\Phi^{-k})^TH^T$, we obtain

$$\sum_{j=0}^{k-1} H\Phi^j\Gamma Q\Gamma^T(\Phi^{j-k})^TH^T = HM(\Phi^{-k})^TH^T - H\Phi^k MH^T - \sum_{j=0}^{k-1} H\Phi^j\Omega(\Phi^{j-k})^TH^T, \quad k = 1,\cdots,n. \quad (27)$$

The right-hand side of (27) is completely determined from MH^T and C_0. Substituting their estimated values,

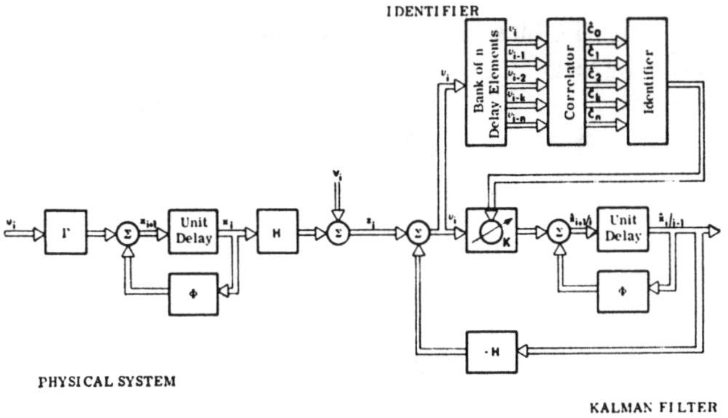

Fig. 2. Identification scheme.

we obtain

$$\sum_{j=0}^{k-1} H\Phi^{j}\Gamma\hat{Q}\Gamma^{T}(\Phi^{j-k})^{T}H^{T} = \hat{H}\hat{M}(\Phi^{-k})^{T}H^{T} - H\Phi^{k}\hat{M}\hat{H}^{T} - \sum_{j=0}^{k-1} H\Phi^{j}\hat{\Omega}(\Phi^{j-k})^{T}H^{T}, \quad k = 1,\cdots,n \quad (28)$$

where

$$\hat{\Omega} = \Phi[-K\hat{H}\hat{M} - \hat{M}\hat{H}^{T}K^{T} + K\hat{C}_0K^{T}]\Phi^{T}. \quad (29)$$

The set of equations (28) is not linearly independent. In any particular case, one has to choose a linearly independent subset of these equations. The procedure will be illustrated by an example in Section IX.

The preceding identification scheme is shown schematically in Fig. 2.

VII. Direct Estimation of the Optimal Gain

If the number of unknowns in Q is more than $n \times r$, or the structure of Q is unknown, the method of the previous section for estimating Q does not work. However, it is still possible to estimate the optimal gain K_{op} by an iterative procedure.

Following the notation of Section II, let K_0 denote the initial gain of the Kalman filter. Let M_1 be the error covariance matrix corresponding to K_0. Then M_1 satisfies the following equation [cf., (7)]:

$$M_1 = \Phi[M_1 - K_0HM_1 - M_1H^TK_0^T + K_0(HM_1H^T + R)K_0^T]\Phi^T + \Gamma Q\Gamma^T. \quad (30)$$

Define

$$K_1 \equiv M_1H^T(HM_1H^T + R)^{-1}.$$

Let the error covariance matrix corresponding to gain K_1 be called M_2. Then

$$M_2 = \Phi[M_2 - K_1HM_2 - M_2H^TK_1^T + K_1(HM_2H^T + R)K_1^T]\Phi^T + \Gamma Q\Gamma^T. \quad (31)$$

Subtracting (30) from (31) and simplifying

$$(M_2 - M_1) = \Phi(I - K_1H)(M_2 - M_1)(I - K_1H)^T\Phi^T - \Phi(K_1 - K_0)(HM_1H^T + R)(K_1 - K_0)^T\Phi^T. \quad (32)$$

The solution to (32) can be written as an infinite sum. Then, using observability and controllability conditions, it can be shown that[6]

$$M_2 - M_1 < 0 \quad \text{or} \quad M_2 < M_1.$$

Similarly, define $K_2 \equiv M_2H^T(HM_2H^T + R)^{-1}$ and M_3 as the corresponding error covariance matrix. Then, by a similar argument,

$$M_3 < M_2 < M_1.$$

The above sequence of monotonically decreasing matrices must converge since it is bounded from below ($M > 0$). Hence, the sequence $K_0, K_1, K_2, \cdots$ must converge to K_{op}.

Based on the preceding property of K, we now construct the following scheme for estimating K_{op}.

1) Obtain an estimate of K_1, denoted as $\hat{K}_1$ from (22):

$$\hat{K}_1 = K_0 + A^* \begin{bmatrix} \hat{C}_1 \\ \hat{C}_2 \\ \vdots \\ \hat{C}_n \end{bmatrix} \hat{C}_0^{-1}. \quad (33)$$

Also, obtain estimates of M_1H^T and R from (22) and (23).

2) Define $\delta M_1 = M_2 - M_1$. Obtain $\delta\hat{M}_1$, an estimate of δM_1, using (32):

$$\delta\hat{M}_1 = \Phi(I - \hat{K}_1H)\delta\hat{M}_1(I - \hat{K}_1H)^T\Phi^T - \Phi(\hat{K}_1 - K_0)\hat{C}_0(\hat{K}_1 - K_0)^T\Phi^T. \quad (34)$$

$\delta\hat{M}_1$ can be calculated recursively in the same manner as M_0 is calculated for a Kalman filter. For convergence, it is sufficient that $\Phi(I - \hat{K}_1H)$ be stable, i.e., all eigenvalues be inside the unit circle.

3) Obtain $\hat{M}_2\hat{H}^T$ and $\hat{K}_2$ as follows:

$$\hat{M}_2\hat{H}^T = \hat{M}_1\hat{H}^T + \delta\hat{M}_1H^T \quad (35)$$

$$\hat{K}_2 = \hat{M}_2\hat{H}^T(H\hat{M}_2\hat{H}^T + \hat{R})^{-1}. \quad (36)$$

4) Repeat steps 2) and 3) until $\|\delta\hat{M}_i\|$ or $\|\hat{K}_i - \hat{K}_{i-1}\|$ become small compared to $\|\hat{M}_i\|$ or $\|K_i\|$ where $\|\cdot\|$ denotes a suitable matrix norm. An alternative way to get K_2 would be to filter data z again using K_1 and then use (33).

This procedure for obtaining K_{op} reveals an interesting relationship between K_{op} and Q. It is seen that the equa-

[6] The proof is similar to the one by Kalman [1] for showing the positive definiteness of M in (11).

tion for $(M_2 - M_1)$ does not involve Q. We need Q only to calculate M_1H^T. This leads us to the following corollary.

Corollary: It is sufficient to know $n \times r$ linear functions of Q in order to obtain the optimal gain of a Kalman filter.

Proof: Consider (30) which can be written as

$$M_1 = \Phi(I - K_0H)M_1(I - K_0H)^T\Phi^T + \Phi K_0RK_0^T\Phi^T + \Gamma Q\Gamma^T. \quad (37)$$

Writing the solution as an infinite series,

$$M_1H^T = \sum_{j=0}^{\infty} [\Phi(I - K_0H)]^j(\Phi K_0RK_0^T\Phi^T + \Gamma Q\Gamma^T) \cdot[(I - K_0H)^T\Phi^T]^jH^T. \quad (38)$$

M_1H^T depends on $n \times r$ linear functions of Q; viz.,

$$\sum_{j=0}^{\infty} [\Phi(I - K_0H)]^j\Gamma Q\Gamma^T[(I - K_0H)^T\Phi^T]^jH^T.$$

If these linear functions are given, we do not need to know Q itself to obtain M_1H^T. Furthermore, since the equation for $(M_2 - M_1)$ does not involve Q explicitly, the optimal gain K_{op} can be obtained by knowing the preceding $n \times r$ linear functions of Q only.

Notice that a complete knowledge of Q is required to obtain the covariance matrix M of a Kalman filter. If one is interested only in K_{op}, the preceding corollary shows that a complete knowledge of Q is not essential. Since our iterative scheme tries to identify K_{op} by whitening the residuals ν_i, it fails to identify the complete Q matrix if the unknowns in Q are more than $n \times r$.

VIII. Statistical Properties of the Estimates

It was shown in Section V that the estimates $\hat{C}_k$ are asymptotically normal, unbiased, and consistent. Since $\widehat{MH^T}$, $\hat{R}$, and $\hat{Q}$ are linearly related to $\hat{C}_k$, it is easy to show that they are also asymptotically normal, unbiased, and consistent.

The general expressions for the mean and covariance of the estimates are rather involved. We, therefore, specialize to the case of a scalar measurement.

Using (22), (23), and (28),

$$E[\widehat{MH^T}] = KC_0 + A^* \begin{bmatrix} C_1 \\ \vdots \\ C_n \end{bmatrix} - \frac{A^*}{N} \begin{bmatrix} C_1 \\ 2C_2 \\ \vdots \\ nC_n \end{bmatrix}$$

or

$$E[\widehat{MH^T}] = MH^T - (1/N)A^* \begin{bmatrix} C_1 \\ 2C_2 \\ \vdots \\ nC_n \end{bmatrix}. \quad (39)$$

For $N \gg n$, the bias in $\widehat{MH^T}$ is negligible. The covariance of $\widehat{MH^T}$ for large N is

$$\text{cov}(\widehat{MH^T}) \approx K \text{ var}(\hat{C}_0)K^T + A^* \text{cov}\left(\begin{bmatrix} \hat{C}_1 \\ \vdots \\ \hat{C}_n \end{bmatrix}\right) A^{*T} + K \text{cov}\left(\hat{C}_0, \begin{bmatrix} \hat{C}_1 \\ \vdots \\ \hat{C}_n \end{bmatrix}\right) A^{*T} + A^* \text{cov}\left(\begin{bmatrix} \hat{C}_1 \\ \vdots \\ \hat{C}_n \end{bmatrix}, C_0\right) K^T. \quad (40)$$

Expressions for

$$\text{cov}\left(\begin{bmatrix} \hat{C}_1 \\ \vdots \\ \hat{C}_n \end{bmatrix}\right), \quad \text{etc.}$$

can be obtained from (16). It can be seen that $\text{cov}(\widehat{MH^T})$ decreases as $1/N$ for large sample sizes. Similarly,

$$E(\hat{R}) = R + (1/N)HA^* \begin{bmatrix} C_1 \\ 2C_2 \\ \vdots \\ nC_n \end{bmatrix} \quad (41)$$

$$\text{var}(\hat{R}) = \text{var}(\hat{C}_0) + H \text{cov}(\widehat{MH^T})H^T - \text{cov}(\hat{C}_0, \widehat{MH^T})H^T - H \text{cov}(\widehat{MH^T}, \hat{C}_0). \quad (42)$$

The expressions for $E[\hat{Q}]$ and $\text{var}([\hat{Q}]_{ij})$ can be obtained similarly.

The usefulness of the preceding expressions is limited by the fact that they depend on the actual values of Q and R which are unknown. If the values of Q and R are known to lie within a certain range, one might use these expressions to plot curves of $\text{var}(\hat{R})$ and $\text{var}([\hat{Q}]_{ij})$ versus N for different values of Q and R. The dependence on Q and R may be removed by considering the covariance of $\hat{K}$:

$$\text{cov}(\hat{K}) = \text{cov}(\widehat{MH^T}\hat{C}_0^{-1}) \approx A^* \text{cov}\left(\begin{bmatrix} \hat{\rho}_1 \\ \vdots \\ \hat{\rho}_n \end{bmatrix}\right) A^{*T}. \quad (43)$$

It can be shown [14], [15] that $\hat{\rho}_k$ are asymptotically normal with mean ρ_k and covariance

$$\text{cov}(\hat{\rho}_k, \hat{\rho}_l) \approx (1/N) \sum_{j=-\infty}^{\infty} \rho_j\rho_{j+l-k}.$$

A satisfactory estimate of $\text{cov}(\hat{\rho}_k, \hat{\rho}_l)$ is provided by

$$(1/2N) \sum_{j=-(N-1)}^{N-1} \hat{\rho}_j\hat{\rho}_{j+l-k}$$

which can be used in (43) to calculate $\text{cov}(\hat{K})$. For the special case of an optimal filter, (43) reduces to

$$\text{cov}(\hat{K}) \approx (1/N)A^*A^{*T} = (1/N)(A^TA)^{-1}. \quad (44)$$

Equation (44) gives us a simple expression for the minimum variance in estimating K. It can be used in deciding upon the minimum sample size N.

We now consider the asymptotic convergence (N large) of the iterative scheme of Section VII. Equation (34) shows that $E[\delta\hat{M}_1]$ depends on the second- and higher order moments of $\hat{K}_1$ which for a normal process are finite and tend to zero asymptotically. Therefore, $\lim_{n\to\infty} E[\delta\hat{M}_1] = \delta M_1$.

Similarly, the covariance of $\delta\hat{M}_1$ asymptotically tends to zero. Thus, $\delta\hat{M}_1$ tends to δM_1 with probability one. Extending the same argument, $\hat{K}_2 \to K_2$, $\hat{K}_3 \to K_3$, $\cdots$, $\hat{K}_{op} \to K_{op}$ with probability one.

IX. A Numerical Example from Inertial Navigation

The results of Sections V and VI are applied to a damped Schuler loop forced by an exponentially correlated stationary random input. Two measurements are made on the system, both of which are corrupted by exponentially correlated as well as white noise type errors. The state of the system is augmented to include all the correlated random inputs so that the augmented state vector x is 5×1, the random input vector u is 3×1, and the measurement noise vector v is 2×1. The system is discretized using a time step of 0.1 and the resultant system matrices are

$$\Phi = \begin{bmatrix} 0.75 & -1.74 & -0.3 & 0 & -0.15 \\ 0.09 & 0.91 & -0.0015 & 0 & -0.008 \\ 0 & 0 & 0.95 & 0 & 0 \\ 0 & 0 & 0 & 0.55 & 0 \\ 0 & 0 & 0 & 0 & 0.905 \end{bmatrix}$$

$$\Gamma = \begin{bmatrix} 0 & 0 & 0 \\ 0 & 0 & 0 \\ 24.64 & 0 & 0 \\ 0 & 0.835 & 0 \\ 0 & 0 & 1.83 \end{bmatrix}, \quad H = \begin{bmatrix} 1 & 0 & 0 & 0 & 1 \\ 0 & 1 & 0 & 1 & 0 \end{bmatrix}$$

$$Q = \begin{bmatrix} q_1 & 0 & 0 \\ 0 & q_2 & 0 \\ 0 & 0 & q_3 \end{bmatrix}, \qquad R = \begin{bmatrix} r_1 & 0 \\ 0 & r_2 \end{bmatrix}.$$

The actual values of q_1, q_2, q_3, r_1, and r_2 are unity, but they are assumed unknown. It is required to identify these values using measurements $\{z_i, i = 1,N\}$.

The starting values of Q and R are taken as

$$Q_0 = \begin{bmatrix} 0.25 & 0 & 0 \\ 0 & 0.5 & 0 \\ 0 & 0 & 0.75 \end{bmatrix}, \quad R_0 = \begin{bmatrix} 0.4 & 0 \\ 0 & 0.6 \end{bmatrix}.$$

Using these values, the innovation sequence $\nu_i = (z_i - H\hat{x}_{i/i-1})$ is generated from (3) to (6). The estimates $\hat{C}_0, \hat{C}_1, \cdots, \hat{C}_k$ of the autocorrelation are calculated using (14). For a typical sample of 950 points, Fig. 1(a) shows a plot of the first diagonal element of $\hat{\rho}_k$ for $k = 0{,}40$. The 95 percent confidence limits are ± 0.0636 and four points lie outside this band (i.e., 10 percent of the total). Therefore, we reject the hypothesis that ν_i is white. The same conclusion is reached by looking at the second diagonal element of $\hat{\rho}_k$.

We now proceed to the identification of Q and R. Since the number of unknowns in Q is less than $n \times r = 10$, we can identify Q completely. The set of equations (28) gives us a large number of linear equations for $\hat{q}_1$, $\hat{q}_2$, and $\hat{q}_3$. However, the most important of these occur along the diagonal for $k = 1$ and $k = 5$.

For $k = 1$ the left-hand side of (28) is

$$\begin{bmatrix} 4.37\hat{q}_3, & -0.0326\hat{q}_3 \\ 0, & 1.27\hat{q}_2 \end{bmatrix}.$$

For $k = 5$ the left-hand side of (28) is

$$\begin{bmatrix} -8.38\hat{q}_1 + 22.3\hat{q}_3, & 1.22\hat{q}_1 - 1.47\hat{q}_2 \\ -1.25\hat{q}_1 - 0.87\hat{q}_3, & 0.141\hat{q}_1 + 20\hat{q}_2 + 0.023\hat{q}_3 \end{bmatrix}.$$

The diagonal elements of the first equation are used to calculate $\hat{q}_3$ and $\hat{q}_2$. The first diagonal element of the second equation is then used to calculate $\hat{q}_1$.

It is possible to use a few other equations and to make a least-squares fit for $\hat{q}_1$, $\hat{q}_2$, and $\hat{q}_3$. This, however, does not alter the results significantly in the present example.

The results obtained by using the identification scheme repeatedly on the same batch of data are shown in Table I. It is seen that most of the identification is done during the first iteration. Further iterations do not increase the likelihood function[7] much, even though the changes in Q and R are significant. A check case using true values of Q and R is also shown in Table I. It is seen that the value of the likelihood function in the check case is very close to that in the first iteration. This indicates that the estimates obtained are quite close to the maximum likelihood estimates. It was further noticed that even if different starting values are used for Q and R, the identification scheme converges to the same values.

[7] The likelihood function $L(Q,R)$ has been given by Schweppe [12]:

$$L(Q,R) = -(1/N) \sum_{i=1}^{N} \nu_i^T (HMH^T + R)^{-1} \nu_i - \ln |HMH^T + R|.$$

TABLE I
ESTIMATES OF Q AND R BASED ON A SET OF 950 POINTS

Number of Iterations	$\hat{q}_1$	$\hat{q}_2$	$\hat{q}_3$	$\hat{r}_1$	$\hat{r}_2$	Likelihood Function $L(\hat{Q},\hat{R})$	Percentage of Points Lying Outside the 95 Percent Confidence Limits: First Measurement (percent)	Percentage of Points Lying Outside the 95 Percent Confidence Limits: Second Measurement (percent)	Estimate of Actual Mean-Square Error*	Calculated Mean-Square Error†
0	0.25	0.5	0.75	0.4	0.6	−5.17	10	10	2915	902
1	0.731	1.31	0.867	1.444	0.776	−4.676	2.5	5	2755	2390
2	0.87	1.39	0.797	1.537	0.767	−4.673	2.5	5	2720	2725
3	0.91	1.40	0.776	1.565	0.765	−4.672	2.5	5	2714	2814
4	0.92	1.41	0.77	1.573	0.7646	−4.671	2.5	5	2712	2840
Check case	1.0	1.0	1.0	1.0	1.0	−4.669	2.5	5	2720	2900

* Estimate of mean-square error is

$$(1/N) \sum_{i=1}^{N} (x_i - \hat{x}_{i|i-1})^T (x_i - \hat{x}_{i|i-1})$$

where x_i is obtained by actual simulation.
† Calculated mean-square error is tr (M_e) where M_e is obtained from the variance equation using $\hat{Q}$ and $\hat{R}$ [cf. (4)].

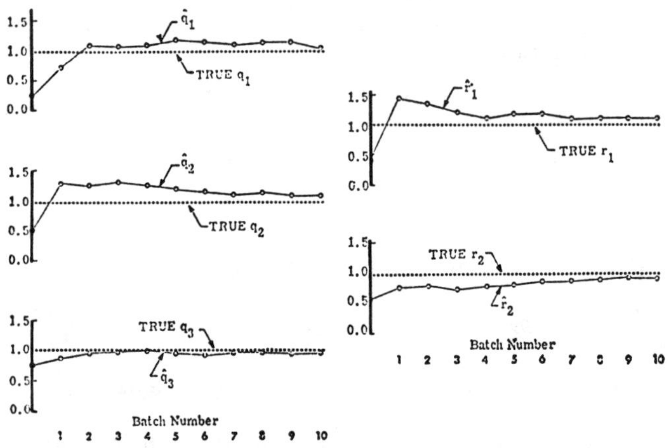

Fig. 3. On-line identification of Q and R.

We now check the optimality of the filter after identification. Fig. 1(b) shows a plot of the first diagonal element of $\hat{\rho}_k$, for $k = 0,40$. It is seen that only one point lies outside the band of 95 percent confidence limits (2.5 percent of the total). This supports the hypothesis that ν_i is white.

The asymptotic convergence of Q and R towards their actual values is shown in Fig. 3. The estimates of Q and R are updated after every batch of N points ($N = 950$). In the absence of any knowledge about the variances of the estimates, a simple averaging of all the previous values is performed. This is equivalent to the following stochastic approximation scheme [5]:

$$\hat{Q}_{k+1} = \hat{Q}_k + [1/(k+1)](\hat{Q}_{k+1,k} - \hat{Q}_k) \tag{45}$$

where k denotes the batch number,

- $\hat{Q}_k$ the estimate of Q after k batches
- $\hat{Q}_{k+1,k}$ the estimate of Q based on the $(k+1)$th batch
- $\hat{Q}_{k+1}$ the estimate of Q after $(k+1)$ batches.

Similarly,

$$\hat{R}_{k+1} = \hat{R}_k + [1/(k+1)](\hat{R}_{k+1,k} - \hat{R}_k). \tag{46}$$

X. CONTINUOUS SYSTEM

The results of the previous sections can be easily extended to continuous systems. We simply state the results below.[8]

System

$$\dot{x} = Fx + Gu \tag{47}$$

$$z = Hx + v. \tag{48}$$

Filter

$$\dot{\hat{x}} = F\hat{x} + K_0(z - H\hat{x}) \tag{49}$$

where

$$K_0 = P_0 H^T R_0^{-1} \tag{50}$$

and

$$FP_0 + P_0F^T + GQ_0G^T - P_0H^TR_0^{-1}HP_0 = 0. \tag{51}$$

The error covariance P_1 is given as

$$(F - K_0H)P_1 + P_1(F - K_0H)^T + GQG^T + K_0RK_0^T = 0. \tag{52}$$

Innovation Process

$$\nu = z - H\hat{x}$$

$$= He + v \tag{53}$$

where $e = (x - \hat{x})$. For an optimal filter, ν is white with the same covariance as v [10]. For a suboptimal filter,

$$\dot{e} = (F - K_0H)e + Gu - K_0v \tag{54}$$

[8] These results have not been applied to a practical problem so far.

and the autocorrelation function $C(\tau)$ of ν is given as

$$C(\tau) = E\{\nu(t)\nu^T(t-\tau)\}$$
$$= HE\{e(t)e^T(t-\tau)\}H^T$$
$$+ HE\{e(t)v^T(t-\tau)\} + R\delta(\tau), \quad \tau > 0$$
$$= He^{F'\tau}[P_1H^T - K_0R] + R\delta(\tau), \qquad F' = F - K_0H. \tag{55}$$

Let $S(w)$ denote the Fourier transform of $C(\tau)$

$$S(w) = H(iw - F')^{-1}(P_1H^T - K_0R) + (HP_1 - RK_0^T) \cdot(-iw - F'^T)^{-1}H^T + R. \tag{56}$$

Test of Optimality and the Estimation of Q and R

We may use either the estimates of $C(\tau)$ or of $S(w)$ to test the optimality of the Kalman filter and to identify Q and R. These estimates are obtained by using methods given in [13].

P_1H^T and R may be obtained from the set of equations (55) or (56) by using methods very similar to the discrete case. If the number of unknowns in Q is $n \times r$ or less, Q can be obtained using (52). We obtain expressions for

$$\sum_{j=0}^{k-1} (-1)^j HF^jGQG^TF^{k-j}H^T, \quad \text{for } k = 0,1,\cdots$$

[the set of equations analogous to (28)].

If the number of unknowns in Q is more than $n \times r$, K_{op} is obtained directly without identifying Q. The procedure is as follows. Define

$$K_1 = P_1H^TR^{-1}. \tag{57}$$

Let P_2 be the error covariance corresponding to K_1. Then it can be shown that

$$(F - K_1H)(P_2 - P_1) + (P_2 - P_1)(F - K_1H)^T - (K_1 - K_0)R(K_1 - K_0)^T = 0. \tag{58}$$

Therefore

$$P_2 < P_1.$$

Similarly, define $K_2 = P_2H^TR^{-1}$ and let P_3 be the error covariance for K_2. Then

$$P_3 < P_2 < P_1.$$

In this way, P is decreased at each step and the sequence $K_0,K_1,K_2\cdots$ converges to K_{op}.

Equation (58) is now used to obtain $\hat{K}_{op}$, an estimate of K_{op}. After obtaining $\hat{K}_1 = \hat{P}_1\hat{H}^T\hat{R}^{-1}$, we substitute it in (58) to get an estimate of $\delta P_1 = P_2 - P_1$:

$$(F - \hat{K}_1H)\hat{\delta P}_1 + \hat{\delta P}_1(F - \hat{K}_1H)^T - (\hat{K}_1 - K_0)\hat{R}(\hat{K}_1 - K_0)^T = 0. \tag{59}$$

Then

$$\hat{P}_2\hat{H}^T = \hat{P}_1\hat{H}^T + \hat{\delta P}_1H^T \tag{60}$$

and

$$\hat{K}_2 = \hat{P}_2\hat{H}^T\hat{R}^{-1} \tag{61}$$

and so on until the relative changes in $\hat{K}$ become small.

We omit the proof of the asymptotic convergence of these estimates since they are essentially similar to the discrete case. All the estimates obtained are asymptotically unbiased and consistent.

XI. Summary and Conclusions

The problem of optimal filtering for a linear time-invariant system with unknown Q (process noise covariance matrix) and R (measurement noise covariance matrix) is considered. Based on the innovation property of an optimal filter, a statistical test is given to check whether a particular filter is working optimally or not. In case the filter is suboptimal, an identification scheme is given to obtain asymptotically unbiased and consistent estimates of Q and R. For the case in which the form of Q is unknown or the number of unknowns in Q is more than $n \times r$ (n is the dimension of the state vector and r is the dimension of the measurement vector), the preceding scheme fails and an alternate scheme is given to obtain an estimate of the optimal gain directly without identifying Q. A numerical example is given to illustrate the results and to show the usefulness of the approach. The results are first derived for a discrete system. They are then extended to continuous systems.

Nomenclature

Matrices

Φ,Γ,H,F,G,B,A	System matrices
Q,R,M,P,M_1,P_1	covariance matrices
Q_0,R_0,M_0,P_0	initial values of Q,R,M,P
$K,K_0,K_1,K_{op},\cdots$	Kalman filter gains
$\hat{Q},\hat{R},\hat{M},\hat{K},\cdots$	estimated values of Q,R,M,K (caret over any quantity denotes an estimate)
$C_k,\hat{C}_k,C(\tau)$	autocorrelation function
$\rho_k,\hat{\rho}_k$	normalized autocorrelation function
$S(w)$	power spectral density
$\delta M,\delta P,\hat{\delta M},\hat{\delta P}$	increment matrices.[5]

Vectors

$x_i,\hat{x}_{i/i-1}$	Actual and estimated states
u_i,v_i	white noise sequences
z_i	measurements
ν_i	innovations
e_i	error in state estimation.

Scalars

n,q,r	Dimension variables
N	sample size
$[C_k]_{ij}$	element in the ith row and the jth column of the matrix C_k
$\delta_{ij},\delta(\tau)$	Kronecker delta and the delta function
$L(Q,R)$	likelihood function.

Operations

$E\{\cdot\}$	Expected value operator
cov $(\cdot,\cdot)$	covariance operator
var $(\cdot)$	variance operator
$(\ )^T$	transpose of a matrix
$(\ )^\#$	pseudo-inverse of a matrix
$(\ )^{-1}$	inverse of a matrix
$\|\|\cdot\|\|$	norm of a matrix.

Acknowledgment

The author wishes to thank all his colleagues at The Analytic Sciences Corporation for their help during the course of this research, and C. L. Bradley in particular for many stimulating discussions and ideas.

References

[1] R. E. Kalman, "New methods and results in linear prediction and filtering theory," *Proc. Symp. on Engineering Applications of Random Function Theory and Probability*. New York: Wiley, 1961.

[2] R. E. Kalman and R. S. Bucy, "New results in linear filtering and prediction theory," *Trans. ASME, J. Basic Engrg.*, ser. D, vol. 83, pp. 95–108, March 1961.

[3] H. Heffes, "The effects of erroneous models on the Kalman filter response," *IEEE Trans. Automatic Control* (Short Papers), vol. AC-11, pp. 541–543, July 1966.

[4] T. Nishimura, "Error bounds of continuous Kalman filters and the application to orbit determination problems," *IEEE Trans. Automatic Control*, vol. AC-12, pp. 268–275, June 1967.

[5] R. C. K. Lee, "Optimal estimation, identification and control," Massachusetts Institute of Technology, Cambridge, Res. Mono. 28, 1964.

[6] D. T. McGill, "Optimal adaptive estimation of sampled stochastic process," Stanford Electronics Labs., Stanford, Calif., Tech. Rept. SEL-63-143(TR 6302-3), December 1963.

[7] J. S. Shellenberger, "A multivariance learning technique for improved dynamic system performance," *1967 Proc. NEC*, vol. 23, pp. 146–151.

[8] G. L. Smith, "Sequential estimation of observation error variances in a trajectory estimation problem," *AIAA J.*, vol. 5, pp. 1964–1970, November 1967.

[9] R. L. Kashyap, "Maximum likelihood identification of stochastic linear systems," Purdue University, Lafayette, Ind., Tech. Rept. TR-EE 68-28, August 1968.

[10] T. Kailath, "An innovations approach to least-squares estimation, pt. I: linear filtering in additive white noise," *IEEE Trans. Automatic Control*, vol. AC-13, pp. 646–655, December 1968.

[11] J. J. Deyst, Jr., and C. F. Price, "Conditions for asymptotic stability of the discrete minimum variance linear estimator," *IEEE Trans. Automatic Control* (Short Papers), vol. AC-13, pp. 702–705, December 1968.

[12] F. C. Schweppe, "Evaluation of likelihood functions for Gaussian signals," *IEEE Trans. Information Theory*, vol. IT-11, pp. 61–70, January 1965.

[13] G. M. Jenkins and D. G. Watts, *Spectral Analysis and its Applications*. San Francisco: Holden Day Publ, 1968.

[14] M. S. Bartlett, *An Introduction to Stochastic Processes*. London: Cambridge University Press, 1962.

[15] E. Parzen, "An approach to time series analysis," *Ann. Math. Statist.*, vol. 32, no. 4, December 1961.

[16] E. J. Hannan, *Time Series Analysis*. New York: Wiley, 1960.

[17] R. L. Anderson, "Distribution of the serial correlation coefficient," *Ann. Math. Statist.*, vol. 13, 1942.

[18] G. S. Watson and J. Durbin, "Exact tests of serial correlation using noncircular statistics," *Ann. Math. Statist.*, vol. 22, pp. 446–451, 1951.

[19] C. W. J. Granger, "A quick test for serial correlation suitable for use with non-stationary time series," *Am. Statist. Assoc. J.*, pp. 728–736, September 1963.

Recursive Fading Memory Filtering

H. W. SORENSON AND J. E. SACKS
Department of Aerospace and Mechanical Engineering Sciences
University of California at San Diego
La Jolla, California

Communicated by John M. Richardson

ABSTRACT

A recursive, fading memory filter for time-continuous and time-discrete systems is presented as a means for overcoming the destructive influence of model errors in Kalman filter applications that lead to the occurrence of divergence. This fading memory filter is shown to be uniformly asymptotically stable under basically the same conditions as the Kalman filter and bounds on the error covariance matrix of the filter are given. An adaptive procedure for implementing the procedure is discussed in terms of a scalar example. The ease of implementation of this filter and the highly satisfactory nature of numerical results indicate the efficacy of the procedure as a desirable recursive data processing method.

1. INTRODUCTION

The linear Kalman–Bucy filter [1] has seen considerable application to engineering problems involving both linear and nonlinear systems. The generally successful application of this procedure has been marred by the not uncommon appearance of the so-called divergence phenomenon. Divergence is said to occur when the actual error in the estimate of the state becomes inconsistent with the error covariance predicted by the filter equations and essentially represents a breakdown in the data processing method.

Divergence is caused by errors in the model assumed for the filter; most commonly, errors in the model of the plant constitute the dominant source of difficulty. Since the filter requires a linear model, errors can result either from the basic description of the system (possibly nonlinear) or from the approximations required to obtain a linear system. The filter uses the plant model to relate data obtained at sampling times $t_0, t_1, \ldots, t_{k-1}$ to the state at the "current" time t_k. As the time interval increases, the model errors generally become larger thereby destroying the validity of these older data as a source of information about the current state.

Many methods [e.g., 2–5] have been devised to combat the divergence problem in Kalman filter applications and each can be considered as a means of diminishing or eliminating the influence of past data on the estimate of the current state. While this goal is sometimes achieved in rather indirect ways, a procedure which explicitly discounts past data is discussed in this paper. The basic idea is not new but it does not appear that it has received the attention that would appear to be warranted in the recursive filtering context. It was introduced in this context by Fagin [6] for discrete systems with no plant noise and is generalized here to include both time-continuous and time-discrete systems with plant noise. Fagin referred to it as a method for the exponential age-weighting of old data. More recently, Morrison [7] designated the procedure as "fading memory filtering."

Fading memory filtering is an outgrowth of considerations relating to deterministic least-squares. In this problem, the discounting of past data is accomplished through the choice of the least-squares weighting matrices. If all data are to be treated equally, then the weighting matrices are all the same. If the data obtained at earlier times are to have a smaller influence than more recent data, then these data are discounted by assigning smaller values to the associated weighting matrices. One can show that (e.g., see [8]) the Kalman filter equations can be obtained as a solution to the unbiased, linear, mean-square filtering problem or as the solution of a deterministic least-squares problem. The two solutions yield the result that the least-square weighting matrices play the same role as the *a priori* covariance matrices of the noise processes of the filtering problem. Thus, the discounting of data in the filtering context is achieved by the appropriate selection of the noise covariances. This approach is taken in the following discussion to obtain the fading memory filter equations for time-continuous and time-discrete systems. The behavior of the fading memory filter is then discussed by examining the behavior of the error covariance matrix of the resulting filter. Also the fading memory filter is applied in an adaptive manner to a scalar system in order to emphasize the character of the resulting filtering procedure.

2. FADING MEMORY FILTERING

For this discussion it will be assumed that the basic system can be described approximately by the following system. The n-dimensional state $\mathbf{x}$ is represented for some finite interval by

$$\dot{\mathbf{x}}(t) = F(t)\mathbf{x}(t) + \mathbf{w}(t) \tag{2.1}$$

and is observed imperfectly at each time through measurement quantities

$$\mathbf{z}(t) = H(t)\mathbf{x}(t) + \mathbf{v}(t). \tag{2.2}$$

Reprinted with permission from *Inform. Sci.*, vol. 3, pp. 101–119, Jan. 1971.

The initial state $\mathbf{x}(t_0)$ is assumed to have mean value $\mathbf{a}$ and covariance matrix M_0. The plant and measurement noise processes, $\mathbf{w}(t)$ and $\mathbf{v}(t)$, are zero mean white noise with covariance matrices

$$E[\mathbf{w}(t)\,\mathbf{w}^T(\tau)] = Q(t)\,\delta(t-\tau),$$

$$E[\mathbf{v}(t)\,\mathbf{v}^T(\tau)] = R(t)\,\delta(t-\tau).$$

The noise processes are assumed to be mutually independent and independent of $\mathbf{x}(t_0)$ for all t.

It is frequently true that the behavior of the state can be represented adequately by (2.1) over some finite interval of time. However, the use of (2.1) for more extending periods cannot be justified with the consequence in many filtering problems that the resulting model errors cause the occurrence of divergence. The filter bases its estimates $\hat{\mathbf{x}}(t/t)$ on all data $\mathbf{z}(s)$, $t_0 \leqslant s < t$, and is being misled because the early data (e.g., $\mathbf{z}(\tau)$, $\tau \ll t$) is no longer accurately related to $\mathbf{x}(t)$ in the manner implied by (2.1) and (2.2). Thus, divergence control is essentially the problem of reducing the influence of these data on the determination of $\hat{\mathbf{x}}(t/t)$. Since the errors resulting from the use of (2.1) would frequently be expected to accumulate gradually rather than to manifest themselves suddenly, the data should themselves be discounted gradually. It is proposed that this be accomplished in the manner described in the following paragraphs.

2.1. *Continuous-Time Fading Memory Filtering*

As has already been mentioned, the discounting of data can be accomplished through the choice of the noise covariance matrices. With this in mind define a filter model in the following manner. At the current time T let the actual state $\mathbf{x}(t)$, $t \leqslant T$, be modeled by $\mathbf{x}_T(t)$, where

$$\dot{\mathbf{x}}_T(t) = F(t)\,\mathbf{x}_T(t) + \mathbf{w}_T(t) \tag{2.3}$$

and the data be given by

$$\mathbf{z}(t) = H(t)\,\mathbf{x}_T(t) + \mathbf{v}_T(t). \tag{2.4}$$

Note that $\mathbf{z}(t)$ appears on the left-hand side of both (2.2) and (2.4). This is done to emphasize that the data $\mathbf{z}(t)$ in a given application are known realizations so are not affected by a change in the model. However, to compensate for model inaccuracies that result from the continued utilization of (2.1)–(2.2), the fading memory filter assumes the data are more adequately represented by (2.4) in which the measurement noise $\mathbf{v}(t)$ is replaced by noise $\mathbf{v}_T(t)$ with a larger covariance.

The initial state $\mathbf{x}_T(t_0)$ is again assumed to have mean value $\mathbf{a}$ but the covariance is modified to be

$$E[(\mathbf{x}_T(t_0) - \mathbf{a})(\mathbf{x}_T(t_0) - \mathbf{a})^T] = M_0 \exp \int_{t_0}^{T} c(\tau)\,d\tau \overset{\Delta}{=} M_T(t_0).$$

The noise processes, $\mathbf{w}_T(t)$ and $\mathbf{v}_T(t)$, are again assumed to be zero mean white noise but the covariances are taken to be

$$E[\mathbf{w}_T(t)\,\mathbf{w}_T{}^T(\tau)] = Q(t)\,\delta(t-\tau)\exp\int_t^T c(\tau)\,d\tau \overset{\Delta}{=} Q_T(t)\,\delta(t-\tau)$$

$$E[\mathbf{v}_T(t)\,\mathbf{v}_T{}^T(\tau)] = R(t)\,\delta(t-\tau)\exp\int_t^T c(\tau)\,d\tau \overset{\Delta}{=} R_T(t)\,\delta(t-\tau)$$

The factor $\exp\int_t^T c(\tau)\,d\tau$ is included to accomplish an exponential discounting of the past data where $c(t) \geqslant 0$ for all $t_0 \leqslant t \leqslant T$. Thus, the covariance of the noise increases as the length of the time interval $(T-t)$ increases and causes the data at t to be given a lesser weight in obtaining the estimate of $\mathbf{x}(T)$.

The problem of continuous-time fading memory filtering can now be stated. Given an observation process $\mathbf{z}(t)$, determine a recursive estimate $\hat{\mathbf{x}}(t/t)$ which satisfies the condition that for each fixed $t = T$, $\hat{\mathbf{x}}(T/T)$ is the unbiased, linear mean-square estimate of $\mathbf{x}_T(T)$. The state $\mathbf{x}_T(T)$ and the measurement data $\mathbf{z}(t)$, $t_0 \leqslant t < T$, are described by (2.3) and (2.4).

The solution of the problem stated above is provided by the following theorem:

THEOREM 2.1. *Let* $\hat{\mathbf{x}}(t/t)$, $P_c(t/t)$, $K_c(t)$ *satisfy the following system of equations.*

$$\frac{d}{dt}\hat{\mathbf{x}}(t/t) = F(t)\,\hat{\mathbf{x}}(t/t) + K_c(t)\,[\mathbf{z}(t) - H(t)\,\hat{\mathbf{x}}(t/t)], \tag{2.5}$$

$$\frac{d}{dt}P_c(t/t) = F_c(t)\,P_c(t/t) + P_c(t/t)\,F_c{}^T(t) - P_c(t/t)\,H^T(t)\,R^{-1}(t)\,H(t)\,P_c(t/t) + Q(t), \tag{2.6}$$

where

$$F_c(t) \overset{\Delta}{=} F(t) + \frac{c(t)}{2}I$$

and

$$K_c(t) = P_c(t/t)\,H^T(t)\,R^{-1}(t), \qquad P_c(t_0/t_0) = M_0, \qquad \hat{\mathbf{x}}(t_0/t_0) = \mathbf{a}. \tag{2.7}$$

Then $\hat{\mathbf{x}}(T/T)$ *is the unbiased linear mean-square estimate of* $\mathbf{x}_T(T)$.

Proof. The proof is obtained by considering the Kalman filter equations for the system (2.3)–(2.4) for the fixed time T. It is known that the Kalman filter provides the unbiased, linear, mean-square estimate and is described by

$$\frac{d}{dt}\hat{\mathbf{x}}_T(t/t) = F(t)\,\hat{\mathbf{x}}_T(t/t) + K_T(t)\,[\mathbf{z}(t) - H(t)\,\hat{\mathbf{x}}_T(t/t)], \tag{2.8}$$

$$\frac{d}{dt}P_T(t/t) = F(t)\,P_T(t/t) + P_T(t/t)\,F^T(t) - P_T(t/t)\,H^T(t)\,R_T^{-1}(t)\,H(t)\,P_T(t/t) + Q_T(t), \tag{2.9}$$

$$K_T(t) = P_T(t/t)\,H^T(t)\,R^{-1}(t), \tag{2.10}$$

with initial conditions

$$\hat{\mathbf{x}}_T(t_0/t_0) = \mathbf{a}, \qquad P_T(t_0/t_0) = M_T(t_0).$$

But

$$P_T(t/t) = P_c(t/t)\exp\int_t^T c(\tau)\,d\tau, \tag{2.11}$$

since substitution of the right-hand side of (2.11) into (2.9) reduces directly to (2.6), and therefore

$$K_T(t) = K_c(t). \tag{2.12}$$

Consequently, $\hat{\mathbf{x}}(t/t) = \hat{\mathbf{x}}_T(t/t)$ so $\hat{\mathbf{x}}(T/T)$ is the unbiased, linear, mean square estimate of $\mathbf{x}_T(T)$ and the proof is complete.

Observe in (2.6) that the exponential weighting $\exp\int_t^T c(\tau)d\tau$ has had the effect of modifying the plant model from $F(t)$ to $F_c(t)$. Except for this change, the Riccati equation is identical with the equations for the ordinary, unfaded, Kalman filter. Note also that (2.6) can be rewritten as

$$\frac{d}{dt}P_c(t/t) = F(t)\,P_c(t/t) + P_c(t/t)\,F^T(t) - P_c(t/t)\,H^T(t)\,R^{-1}(t)\,H(t)\,P_c(t/t) + Q(t) + c(t)\,P_c(t/t). \tag{2.13}$$

From (2.13), one can see that the data aging has the effect of modifying the plant noise covariance matrix $Q(t)$.

2.2. *Discrete-Time Fading Memory Filtering*

The results of the preceding section can be extended without difficulty to time-discrete systems. That is, suppose the state behavior is described by a linear difference equation instead of (2.1) and that the measurements (2.2) occur only at discrete instants:

$$\mathbf{x}_k = \Phi_{k,k-1}\,\mathbf{x}_{k-1} + \mathbf{w}_{k-1}, \tag{2.14}$$

$$\mathbf{z}_k = H_k\,\mathbf{x}_k + \mathbf{v}_k. \tag{2.15}$$

The sequences $\mathbf{w}_k$ and $\mathbf{v}_k$ are zero mean white noise with covariance matrices Q_k and R_k.

Suppose that the filter model is given by

$$\mathbf{x}_k^N = \phi_{k,k-1}\,\mathbf{x}_{k-1}^N + \mathbf{w}_{k-1}, \tag{2.16}$$

$$\mathbf{z}_k = H_k\,\mathbf{x}_k^N + \mathbf{v}_k^N, \tag{2.17}$$

where the superscript N is viewed as current time.

The initial state and the noise covariances will now be assumed to be[1]

$$E[(\mathbf{x}_0^N - \mathbf{a})(\mathbf{x}_0^N - \mathbf{a})^T] = M_0 \exp\sum_{i=0}^{N-1} c_i,$$

$$E[\mathbf{w}_{k-1}^N\,\mathbf{w}_{k-1}^{N^T}] = Q_{k-1}\exp\sum_{i=k}^{N-1} c_i,$$

$$E[\mathbf{v}_k^N\,\mathbf{v}^{N^T}] = R_k \exp\sum_{i=k}^{N-1} c_i,$$

where $c_i \geqslant 0$ for all i.

The time-discrete fading memory filtering problem can now be stated. Given a measurement sequence $(\mathbf{z}_0, \mathbf{z}_1, \ldots, \mathbf{z}_N)$, determine a recursive estimate $\hat{\mathbf{x}}_{n/n}$ such that for each fixed $n = N$, $\hat{\mathbf{x}}_{N/N}$ is the unbiased, linear, mean-square estimate of $\mathbf{x}_N^N$.

The solution of this problem is given by the following theorem:

THEOREM 2.2. *Let* $\hat{\mathbf{x}}_{n/n}$, $P^c_{n/n-1}$, $P^c_{n/n}$, *and* K_n^c *satisfy the following system:*

$$\hat{\mathbf{x}}_{n/n} = \phi_{n,n-1}\,\hat{\mathbf{x}}_{n-1/n-1} + K_n^c[\mathbf{z}_n - H_n\,\Phi_{n,n-1}\,\hat{\mathbf{x}}_{n-1/n-1}], \tag{2.18}$$

$$P^c_{n/n-1} = \Phi_{n,n-1}\,P^c_{n-1/n-1}\,\Phi^T_{n,n-1}(\exp c_{n-1}) + Q_{n-1}, \tag{2.19}$$

$$K_n^c = P^c_{n/n-1}\,H_n^T(H_n\,P^c_{n/n-1}\,H_n^T + R_n)^{-1}, \tag{2.20}$$

$$P^c_{n/n} = P^c_{n/n-1} - K_n^c\,H_n\,P^c_{n/n-1}, \tag{2.21}$$

with initial conditions

$$\hat{\mathbf{x}}_{0/-1} = \mathbf{a}, \qquad P_{0/-1} = M_0.$$

Then, $\hat{\mathbf{x}}_{N/N}$ *is the unbiased, linear, mean-square estimate of* $\mathbf{x}_N^N$.

The proof follows that given for time-continuous system and is omitted.

Note that the past data are discounted by multiplying $P_{n-1/n-1}$ by the number $\exp c_{n-1} \geqslant 1$ before updating the estimate to the nth sampling time. Also, it can be shown by a formal limiting argument that (2.18)–(2.21) yield

[1] A vacuous sum is to be interpreted as 0.

the time-continuous fading memory filter (2.5)–(2.7) when the length of the sampling interval is allowed to vanish. It should also be observed that the aging factor does not have to have the exponential form introduced above. Instead, any constant $\beta_{n-1} \geq 1$ can be used in place of $(\exp c_{n-1})$ in (2.19).

3. ASYMPTOTIC BEHAVIOR OF THE FADING MEMORY FILTER

The time-continuous system (2.1)–(2.2) is considered in this section. The filter equations for this system are given by

$$\frac{d}{dt}\hat{\mathbf{x}}(t/t) = F(t)\hat{\mathbf{x}}(t/t) + K(t)[\mathbf{z}(t) - H(t)\hat{\mathbf{x}}(t/t)], \tag{3.1}$$

$$\frac{d}{dt}P(t/t) = F(t)P(t/t) + P(t/t)F^T(t) - P(t/t)H^T(t)R^{-1}(t)H(t)P(t/t) + Q(t), \tag{3.2}$$

$$K(t) = P(t/t)H^T(t)R^{-1}(t), \tag{3.3}$$

with initial conditions

$$\hat{\mathbf{x}}(t_0/t_0) = \mathbf{a}, \qquad P(t_0/t_0) = M_0.$$

In the discussion that follows, model (2.1)–(2.2) will be employed for $t \geq t_0$; and it is tacitly assumed that this model is a valid representation of the actual system over at least time intervals of length σ for which the system is uniformly completely observable and controllable [9].

3.1. *Stability of the Fading Memory Filter*

In 1961 Kalman [1,9] introduced the concepts of uniformly completely controllable and uniformly completely observable for the system (2.1)–(2.2) (hereafter denoted as UCCO). He then proved for UCCO systems that the Kalman–Bucy filter is uniformly asymptotically stable. Furthermore, he showed that there exists an equilibrium solution $\bar{P}(t/t)$ for (3.2) which is uniformly bounded from above and to which all other solutions of (3.2) with positive semidefinite initial conditions M_0 converge exponentially fast. The following theorem extends this result to the fading memory filter.

THEOREM 3.1. *Let the system* (2.1)–(2.2) *be* UCCO *and let* $0 \leq c(t) \leq b < \infty$ *for all t. Then, the fading memory filter* (2.5)–(2.7) *is uniformly asymptotically stable. Also, there exists an equilibrium solution* $\bar{P}_c(t/t)$ *to the Riccati equation* (2.6) *which is uniformly bounded from above and to which all other solutions of* (2.6) *with positive semidefinite initial conditions* M_0 *converge exponentially fast.*

Proof. As seen in (2.6), the exponential aging factor has had the effect in the Riccati equation of modifying the linear system (2.1) from a system with coefficient matrix $F(t)$ to a system with coefficient matrix $F_c(t)$. Now, if $\Phi(t, t_0)$ is the transition matrix for the system

$$\dot{\mathbf{x}} = F(t)\mathbf{x} \tag{3.4}$$

and if $\psi(t, t_0)$ is the transition matrix for the modified system

$$\dot{\mathbf{x}} = \left[F(t) + \frac{c(t)}{2}I\right]\mathbf{x}, \tag{3.5}$$

then Φ and ψ are related by

$$\psi(t, t_0) = \Phi(t, t_0)\exp\int_{t_0}^{t}\frac{c(\tau)}{2}d\tau. \tag{3.6}$$

If $\sigma > 0$ is the interval of observability of (2.1)–(2.2) (see [9]), then the boundedness of $c(t)$ implies that there exists a constant $k > 1$ such that

$$1 \leq \exp\int_{t}^{t+\sigma} c(\tau)\,d\tau \leq k,$$

where k is independent of t. From this, it follows that the system (2.1)–(2.2) is UCCO if and only if the system

$$\dot{\mathbf{x}} = \left[F(t) + \frac{c(t)}{2}I\right]\mathbf{x} + \mathbf{w}, \tag{3.7}$$

$$\mathbf{z} = H(t)\mathbf{x} + \mathbf{v} \tag{3.8}$$

is UCCO. But then Kalman's results imply that the equilibrium solution $\bar{P}_c(t/t)$ exists which has the properties asserted in the theorem.

Applying Kalman's result again, it follows that the matrix

$$A_c(t) \stackrel{\Delta}{=} F_c(t) - P_c(t/t)H^T(t)R^{-1}(t)H(t)$$

describes a system that is uniformly asymptotically stable with transition matrix denoted as $\theta(t, t_0)$. But the fading memory estimate is given by

$$\frac{d}{dt}\hat{\mathbf{x}}(t/t) = [F(t) - P_c(t/t)H^T(t)R^{-1}(t)H(t)]\hat{\mathbf{x}}(t/t) + K_c(t)\mathbf{z}(t) \tag{3.9}$$

and the coefficient matrix

$$A(t) \stackrel{\Delta}{=} F(t) - P_c(t/t)H^T(t)R^{-1}(t)H(t)$$

is related to $A_c(t)$ according to

$$A(t) = A_c(t) - \frac{c(t)}{2}I.$$

This implies that the fading memory filter has transition matrix

$$\theta(t, t_0) \exp\left(-\tfrac{1}{2} \int_{t_0}^{t} c(\tau)\, d\tau\right).$$

Since

$$0 \leqslant \exp\left(-\int_{t_0}^{t} c(\tau)\, d\tau\right) \leqslant 1 \qquad \text{for all } t,$$

the fading memory filter must be uniformly asymptotically stable and the proof is complete.

3.2. *Bounds for the Error Covariance Matrix*

It is possible to obtain bounds that demonstrate that the fading memory filter gives an error covariance matrix $P_c(t/t)$ that is larger than that of the unfaded Kalman–Bucy filter and thereby prevents the filter gains from becoming unjustifiably small. The too rapid diminishing of the gains has been frequently noted as the cause of the divergence problem.

Suppose that the filter model is assumed to be defined by (2.1)–(2.2). Then, the Kalman–Bucy filter would provide an error covariance $P(t/t)$ which is the solution of (3.2). Now, if the gain $K_c(t)$ were used with the system (2.1), it would represent a suboptimal gain and would give rise to an actual error covariance matrix $P_a(t/t)$. The $P_c(t/t)$ would represent only a matrix of numbers under the assumption that (2.1)–(2.2) describes the true system. With these preliminaries, the following theorem results:

THEOREM 3.2.[2] $P(t/t) \leqslant P_a(t/t) \leqslant P_c(t/t) \leqslant P(t/t) \exp \int_{t_0}^{t} c(\tau)\, d\tau$.

Proof. (i) $P(t/t) \leqslant P_a(t/t)$ follows immediately from the optimality of the Kalman–Bucy filter.

(ii) Consider

$$P_a(t/t) \leqslant P_c(t/t).$$

If $K_c(t)$ is a suboptimal gain, it is well known that the error covariance associated with an arbitrary gain is described by

$$\frac{d}{dt} P_a(t/t) = [F(t) - K_c(t) H(t)] P_a(t/t) + P_a(t/t) [F(t) - K_c(t) H(t)]^T + K_c(t) R(t) K_c^T(t) + Q(t), \tag{3.10}$$

where

$$P_a(t_0/t_0) = M_0.$$

[2] The notation that a matrix $A \leqslant B$ is used to signify that the matrix $B - A$ is positive semidefinite.

The $P_c(t/t)$ is the solution of (2.6) which can be rewritten as

$$\frac{d}{dt} P_c(t/t) = [F_c(t) - K_c(t) H(t)] P_c(t/t) + P_c(t/t) [F_c(t) - K_c(t) H(t)]^T + K_c(t) R(t) K_c^T(t) + Q(t), \tag{3.11}$$

where

$$P_c(t_0/t_0) = M_0.$$

Using the notation defined above for the transition matrix of the linear system with coefficient matrix $[F_c - K_c H]$ and letting

$$B(t) \stackrel{\Delta}{=} K_c(t) R(t) K_c^T(t) + Q(t) \tag{3.12}$$

it follows that

$$P_a(t/t) = \theta(t, t_0) M_0 \theta^T(t, t_0) \exp - \int_{t_0}^{t} c(\tau)\, d\tau + \int_{t_0}^{t} \left[\exp - \int_{t_0}^{\tau} c(\sigma)\, d\sigma\right] \theta(t, \tau) B(\tau) \theta^T(t, \tau)\, d\tau \tag{3.13}$$

and

$$P_c(t/t) = \theta(t, t_0) M_0 \theta^T(t, t_0) + \int_{t_0}^{t} \theta(t, \tau) B(\tau) \theta^T(t, \tau)\, d\tau \tag{3.14}$$

Since $c(t) \geqslant 0$, the conclusion follows directly.

(iii) The proof of the inequality

$$P_c(t/t) \leqslant P(t/t) \exp \int_{t_0}^{t} c(\tau)\, d\tau$$

is accomplished by observing that $P(t/t) \exp \int_{t_0}^{t} c(\tau)\, d\tau$ satisfies the Riccati equation

$$\dot{S} = F_c(t) S + S F_c^T(t) - S H^T(t) \left[R^{-1}(t) \exp - \int_{t_0}^{t} c(\tau)\, d\tau\right] H(t) S + Q(t) \exp \int_{t_0}^{t} c(\tau)\, d\tau.$$

Since

$$R(t) \exp \int_{t_0}^{t} c(\tau)\, d\tau \geqslant R(t)$$

and

$$Q(t)\exp\int_{t_0}^{t} c(\tau)\,d\tau \geqslant Q(t),$$

one can apply the following result (for a proof, see Nishimura [10]).

LEMMA. *If S_α, $\alpha = 1, 2$, are given by the differential equations*

$$\dot{S}_\alpha = FS_\alpha + S_\alpha F^T - S_\alpha H^T R^{-1} HS_\alpha + Q_\alpha \tag{3.12}$$

where Q_α, R_α, $S_\alpha(t_0)$ are symmetric and such that

$$Q_2(t) \geqslant Q_1(t) \geqslant 0; \qquad R_2(t) \geqslant R_1(t) > 0, \qquad \textit{for all } t$$

and

$$S_2(t_0) \geqslant S_1(t_0),$$

then

$$S_2(t) \geqslant S_1(t) \qquad \textit{for all } t.$$

This completes the *proof of the theorem.*

If the system is UCCO, then it follows from Theorems 3.1 and 3.2 that $P_c(t/t)$ and $P_a(t/t)$ are uniformly bounded from above for all t. As a result, the inequality $P_c(t/t) \leqslant P(t/t)\exp\int_{t_0}^{t} c(\tau)d\tau$ appears to be unduly conservative. However, in two limiting cases:

(1) $Q = 0$, $R^{-1} = 0$,
(2) $c(t) = \delta(t-\lambda)b$ with b and λ constants, and $t \leqslant \lambda$,

the inequality becomes an equality.

4. A SCALAR APPLICATION OF THE FADING MEMORY FILTER

Consider a scalar system which is described by

$$\dot{x}(t) = Fx(t), \tag{4.1}$$

$$z(t) = x(t) + v(t). \tag{4.2}$$

The system is stationary and the plant noise has been omitted in order to simplify the subsequent discussion. Also, it is assumed that the measurement noise variance is $R\,\delta(t-\tau)$ where R is constant for all t. The notation introduced in Section 2.1 is adhered to here. Note that the system is asymptotically stable if $F < 0$ and is unstable if $F > 0$.

The error variance of the Kalman filter for this system (see (3.1)–(3.2)) is described by

$$\frac{d}{dt}P(t/t) = 2FP(t/t) - P^2(t/t)/R, \tag{4.3}$$

and it is easily shown that the solution $P(t/t)$ is

$$P(t/t) = P_0\exp 2Ft/[1 + P_0(\exp 2Ft - 1)/2FR]. \tag{4.4}$$

Note that, if $F < 0$, then

$$\lim_{t\to\infty} P(t/t) = 0, \tag{4.5}$$

which implies that the error in the estimate of the state tends to vanish. However, if $F > 0$, then

$$\lim_{t\to\infty} P(t/t) = 2FR, \tag{4.6}$$

which shows that, even when there is no plant noise, the error variance does not vanish. These results are specific examples of the general behavior of the error covariance as discussed in [11].

The filter gain for this system is given by

$$K(t) = P(t/t)/R > 0 \tag{4.7}$$

and it is clear that $K(t)$ either tends to vanish (when $F < 0$) or approaches $2F$ (when $F > 0$).

It is interesting to consider the effect of the filter on the system. Based on (3.1) the estimate $\hat{x}(t/t)$ for the system (4.1)–(4.2) is given by

$$\frac{d}{dt}\hat{x}(t/t) = [F - K(t)]\,\hat{x}(t/t) + K(t)\,z(t). \tag{4.8}$$

The transition matrix for the filter is

$$\Phi(t,0) = \exp\left[Ft - \int_0^t k(\tau)d\tau\right] \tag{4.9}$$

Thus, the plant with system coefficient F has been replaced by a time varying system with coefficient $F^*(t) = F - K(t)$. Note that $K(t) > 0$ for all t and if in addition $F > 0$ then $\int_0^t K(\tau)d\tau \to \infty$ as $t \to \infty$. Thus, F^* eventually become and remains negative regardless of the value of F. Therefore, even if the system (4.1) is unstable, the filter (4.8) is asymptotically stable.

To demonstrate the effect of the fading memory filter, consider a modification to (4.1). In particular suppose that at $t = 0$ the coefficient F of the plant is equal to minus two and suppose that the filter uses this value. Then, suppose at some time, say $t = \frac{1}{2}$, the system changes so that F is equal to one but the filter model continues to use the original value. Thus, at $t = \frac{1}{2}$, the filter has error variance and gain given by

$$P(\tfrac{1}{2}/\tfrac{1}{2}) = P_0\exp(-2)/[1 + P_0(1 - \exp(-2))/4R], \tag{4.10}$$

$$K(\tfrac{1}{2}) = K(0)\exp(-2)/[1 + K(0)(1 - \exp(-2))/4] < K(0)\exp(-2).$$

Thus the gain and error variance have been reduced by more than exp(2) and implies that $\hat{x}(\frac{1}{2}/\frac{1}{2})$ is a very accurate estimate of $x(\frac{1}{2})$.

For $t > \frac{1}{2}$, the actual plant is changed so that

$$\begin{aligned} z(t) &= x(t) + v(t) \\ &= e^{(t-\frac{1}{2})} x(\tfrac{1}{2}) + v(t). \end{aligned} \tag{4.11}$$

But the estimate of the measurement is given essentially by

$$\begin{aligned} \hat{z}(t/t) &= \hat{x}(t/t) \\ &= e^{-2(t-\frac{1}{2})} \hat{x}(\tfrac{1}{2}/\tfrac{1}{2}) \end{aligned} \tag{4.12}$$

so that

$$z(t) - \hat{z}(t/t) = (c_1 e^t - c_2 e^{-2t}) \hat{x}(\tfrac{1}{2}/\tfrac{1}{2}) + v(t), \tag{4.13}$$

where $x(\frac{1}{2})$ and $\hat{x}(\frac{1}{2}/\frac{1}{2})$ have been assumed equal and c_1 and c_2 are unequal positive constants. But the residual is supposed to be a zero mean white noise process [12] and the term $(c_1 e^t - c_2 e^{-2t}) \hat{x}(\frac{1}{2}/\frac{1}{2})$ certainly violates this property. Thus, the onset of divergence is indicated by the residual, and the need for a modification of the data processing procedure is apparent. The modification is accomplished through the introduction of the fading memory filtering.

The behavior of the fading memory filter is best demonstrated by numerical example and it is convenient to consider discrete measurement data for ease of computation. Suppose that the data are available every tenth of a second. Then, the discrete model is described by

$$x_k = \exp(F/10)\, x_{k-1}, \tag{4.14}$$

$$z_k = x_k + v_k, \tag{4.15}$$

and the filter equations are

$$\hat{x}_{k/k} = \exp(F/10)\, \hat{x}_{k-1/k-1} + K_k[z_k - \exp(F/10)\, \hat{x}_{k-1/k-1}], \tag{4.16}$$

$$K_k = P_{k/k-1}/(P_{k/k-1} + R), \tag{4.17}$$

$$P_{k/k-1} = \exp(2F/10)\, P_{k-1/k-1}, \tag{4.18}$$

$$P_{k/k} = P_{k/k-1} - K_k P_{k/k-1}. \tag{4.19}$$

Note that

$$\begin{aligned} P_{k/k} &= P_{k/k-1} - P_{k/k-1}^2/(P_{k/k-1} + R) \\ &= P_{k/k-1} R/(P_{k/k-1} + R). \end{aligned}$$

Thus, it follows from the above equations that

$$0 < K_k < 1, \qquad \text{for all } k,$$

$$0 < P_{k/k} < R, \qquad \text{for all } k.$$

One also sees that, when K_k equals one,

$$\hat{x}_{k/k} = z_k.$$

From (2.19) it is seen that the fading memory filter for this system is described by (4.16), (4.17), and (4.19) and the $P_{k/k-1}$ is given by

$$P_{k/k-1} = \exp(2F/10 + c_{k-1})\, P_{k-1/k-1}, \tag{4.20}$$

where $c_{k-1} \geqslant 0$ for all k. In the following paragraphs, a means of selecting the c_k based on the residual

$$r_k \overset{\Delta}{=} z_k - \hat{x}_{k/k} \tag{4.21}$$

is discussed. Although the application here is quite simple, the basic ideas and procedures can be generalized to the system (2.14)–(2.15) without difficulty.

The residual is supposed to be a zero mean white noise process with variance

$$E[r_k^2] = P_{k/k} + R. \tag{4.22}$$

To test the data relative to this hypothesis, one can form the sample mean and sample correlations of the residuals. The sample mean is

$$\bar{r}_N \overset{\Delta}{=} \frac{1}{N+1} \sum_{i=0}^{N} r_i. \tag{4.23}$$

The sample correlations are computed according to

$$s_{m,N} = \frac{1}{N} \sum_{i=m}^{N} r_i r_{i-m}, \qquad m = 1, 2, \ldots, N. \tag{4.24}$$

It is easily shown that

$$E[\bar{r}_N] = 0 = E[s_{m,N}] \tag{4.25}$$

and

$$E[\bar{r}_N^2] = \frac{1}{N^2} \sum_{i=0}^{N-1} E[r_i^2], \tag{4.26a}$$

$$E[s_{m,N}^2] = \frac{1}{N^2} \sum_{i=m}^{N} E[r_i^2]\, E[r_{i-m}^2], \tag{4.26b}$$

where $E[r_i^2]$ is given by (4.22). Under ordinary circumstances (i.e., the filter model is valid), the error variance becomes significantly smaller than the measurement noise variance so that (4.26) can be approximated by neglecting $P_{i/i}$. Then, one has

$$E[\bar{r}_N^2] = \frac{R}{N+1}, \quad E[s_{m,N}^2] = \frac{R^2(N-m)}{N^2}, \tag{4.27}$$

and for $N \gg m$

$$E[s_{m,N}^2] = R^2/N.$$

Assuming the distribution of $\bar{r}_N$ and $s_{m,N}$ are gaussian, then one has the probabilities

$$P\left[-2/\sqrt{N} < \frac{\bar{r}_N}{\sqrt{R}} < 2/\sqrt{N}\right] = 0.95,$$

$$P\left[-2/\sqrt{N} < \frac{s_{m,N}}{R} < 2/\sqrt{N}\right] = 0.95.$$

It has been shown (e.g., see [13]) that the gaussian assumption for $s_{m,N}$ is a good approximation when N is much larger than m. The sample mean is gaussian for all N if the measurement noise is gaussian.

On the basis of the preceding discussion, consider the following adaptive procedure for choosing the fading parameters c_k.

Assuming that the procedure stated below has been used for all $j < k$, consider the kth sampling time:

1. Form $\hat{x}_{k/k}$ using (4.16)–(4.19) and compute the residual r_k using (4.21).
2. Form the sample mean according to

$$\bar{r}_k = \frac{k-1}{k}\bar{r}_{k-1} + \frac{r_k}{k} \tag{4.28}$$

and the sample correlation

$$s_{1,k} = \frac{k-1}{k}s_{1,k-1} + \frac{r_k r_{k-1}}{k} \tag{4.29}$$

The correlations $s_{m,k}(m = 2,\ldots,k-1)$ could also be formed if so desired.

3. Compare $|\bar{r}_k/\sqrt{R}|$ and $|s_{1,k}/R|$ with $2/\sqrt{k}$. If they are both less, then set $k = k+1$ and return to Step 1. If either or both are greater than $2/\sqrt{k}$ then the residual is inconsistent with the zero mean white noise assumption and the fading memory filter must be used.
4. Choose c_{k-1} so that either (a) $|r_k/\sqrt{R}|$ and $|s_{1,k}/R|$ become consistent or (b) the residual becomes very small.

One or the other of these alternatives will be satisfied. If the divergence problem is severe, it will generally be the latter one. In this case the fading parameter will be very large (thereby discounting the influence of all past data) and the gain will become very close to its maximum value of one.

The choice of c_{k-1} can be accomplished by solving (4.20), (4.17), and (4.16) iteratively until either of the two conditions are satisfied. Then, set $k = k+1$ and return to Step 1.

This procedure was applied to the system with the following choices for the parameter values.

$$a = 0, \qquad M_0 = 0.5, \qquad R = 0.0625.$$

Three cases were considered in which the plant coefficient F was varied:

CASE 1:
$$F = \begin{cases} -2, & \text{for } k \leqslant 5, \\ 1, & \text{for } 5 < k \leqslant 20, \\ -2, & \text{for } 20 < k. \end{cases}$$

CASE 2:
$$F = \begin{cases} -2, & \text{for } k \leqslant 5, \\ 1, & \text{for } 5 < k \leqslant 20, \\ 0, & \text{for } 20 < k. \end{cases}$$

CASE 3:
$$F = \begin{cases} -2, & \text{for } k \leqslant 5, \\ 1, & \text{for } 5 < k. \end{cases}$$

In the first case the filter model is incorrect for 15 samples following the initial filter convergence. In Case 2 the filter model is incorrect for all samples after the first five but the plant itself does not remain unstable. In the final case the plant becomes and remains unstable. Certainly, the divergence problem becomes increasingly difficult in going from Case 1 to Case 3.

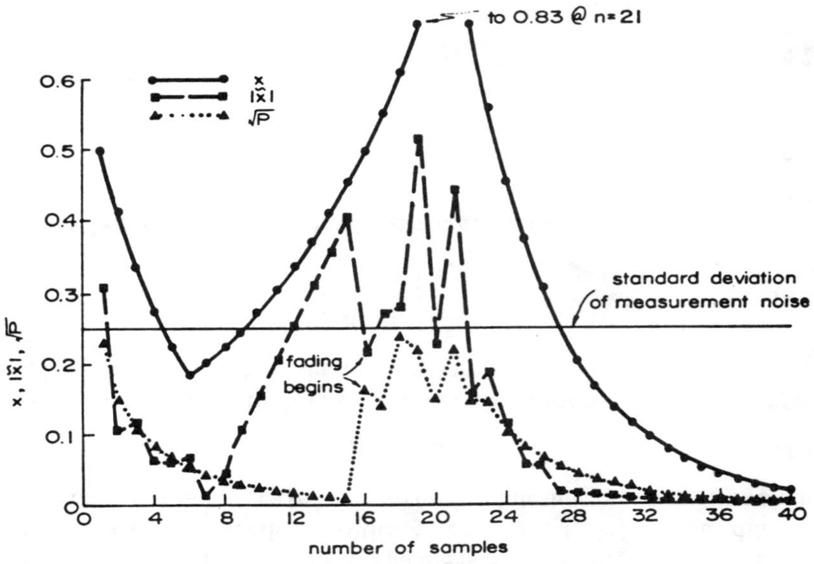

FIGURE 1. Exponentially stable plant.

The results of this study are depicted in Figures 1 through 4. The state, the absolute value of the estimation error, and the standard deviation of the error as predicted by the filter are shown for Cases 1 to 3 in Figures 1 through 3, respectively. The gains for all three cases are shown in Figure 4.

It is apparent that the error in the plant is not detected immediately. This

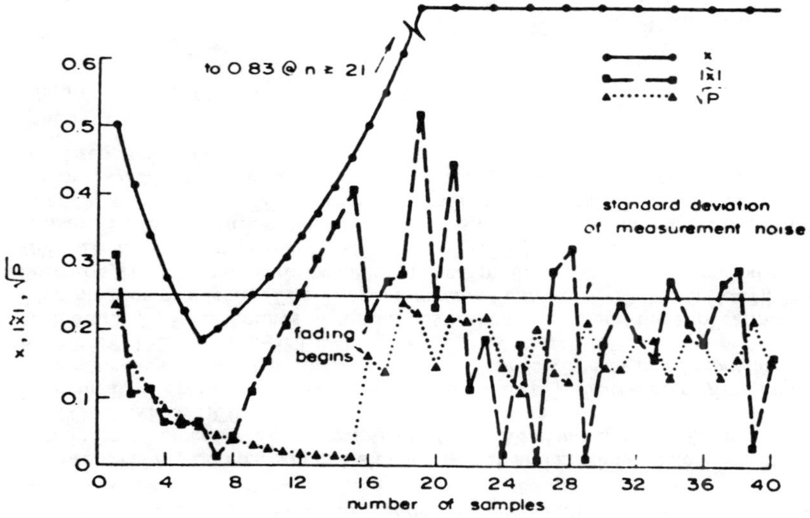

FIGURE 2. Stable plant.

is caused primarily because the magnitude of the error is small relative to the measurement noise standard deviation for several stages following the changes of F from -2 to 1. However, the error is consistent with the error standard deviation throughout the filtering period of 40 samples. The plots of the filter gains indicate, as expected, that very little filtering of the data is accomplished in Case 3 (i.e., $\sqrt{P}$ is very close to its maximum value of $\sqrt{R}$) because of the

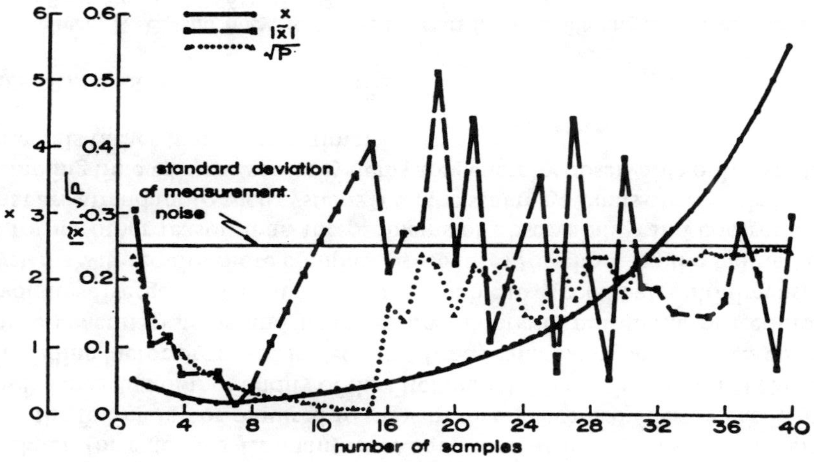

FIGURE 3. Unstable plant.

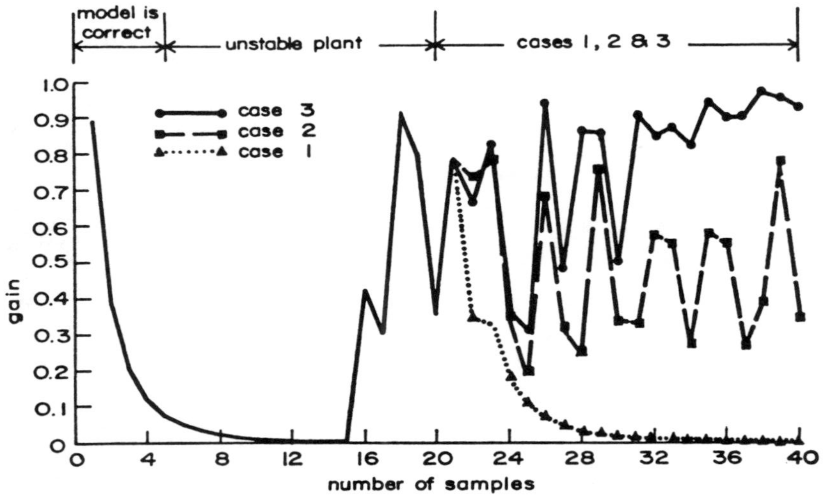

FIGURE 4. Filter gains for all three cases.

complete inadequacy of the dynamic model. Nonetheless, the improved data processing provided by the fading memory filter prevents the occurrence of divergence which is all that could be hoped for in such a circumstance.

Although not shown, the behavior of $\bar{r}_k$ and $s_{1,k}$ requires comment. It was found in these and in other cases that the consistency of the sample mean $\bar{r}_k$ was generally the more difficult constraint to satisfy. However, the $s_{1,k}$ was more sensitive and exhibited more rapid change than $\bar{r}_k$ when the plant errors first began to exhibit themselves and when the fading memory filter began to exert control. For the cases shown here, testing only the consistency of the sample mean provided the same results as testing both $\bar{r}_k$ and $s_{1,k}$

5. CONCLUSIONS

The fading memory filter is presented as a straightforward and easily interpreted way for discounting old data in order to eliminate the influence of filter modeling errors and the subsequent appearances of the divergence phenomenon. It provides an explicit means for weighting past data by a prescribed factor and thereby to reduce or essentially eliminate their influence on the determination of the estimate of the current state.

When the unfaded linear model used to describe a system is uniformly completely observable and controllable, then it follows that the fading memory filter is uniformly asymptotically stable and bounds for the error covariance are given. Thus, the fading memory filter can be expected to be well behaved in the face of nonnegligible errors in the system model used by the filter.

Since the fading memory filter requires a trivial modification to the standard Kalman (or extended Kalman) filter equations, it is very easy to implement and the fading factor required to prevent divergence provides an easily interpreted measure of the validity of the filter model. The basic method of choosing the fading factor proposed in Section 4 for a scalar example can be extended to more general systems and provides an essentially automatic procedure for avoiding divergence. Of course, it is reasonable to expect that multidimensional systems will require more complex procedures for choosing the fading factor if for no other reason than the problems of observability and controllability that are attendant to such systems (i.e., the scalar system is observable at every sampling time whereas vector systems are generally observable only over finite intervals rather than at each time).

REFERENCES

1 Kalman, R. E., and Bucy, R. S., New results in linear filtering and prediction theory, *J. Basic Eng.* **83D** (1961), 95–108.

2 Jazwinski, A. H., Adaptive filtering, *Automatica* 5 (1969), 475–485.

3 Jazwinski, A. H., Limited memory optimal filtering, *IEEE Trans. Automatic Control*, **AC-13**, No. 5 (1968), 558–563.

4 Cosaert, R., and Gottzein, E., *A Decoupled Shifted Memory Filter Method for Radio Tracking of Space Vehicles*, ELDO Technical Memo F5, March, 1967 (also presented at XVIII International Astronautical Congress, Belgrade, Yugoslavia, September, 1967).

5 Schmidt, S. F., *Estimation of State with Acceptable Accuracy Constraints*, Analytical Mechanics Associates, Inc., Interim Report 67–4, 1967.

6 Fagin, S. F., Recursive linear regression theory, optimal filter theory, and error analysis of optimal systems, 1964 *IEEE Convention Record*, 216–240.

7 Morrison, N., *Introduction to Sequential Smoothing and Prediction*, McGraw-Hill New York, 1969.

8 Sorenson, H. W., Comparison of Kalman Bayesian and maximum likelihood estimation techniques, Chapter 6 in *Theory and Application of Kalman Filtering*, AGARDograph 13, NATO-AGARD, 1970.

9 Kalman, R. E., New methods in Wiener filtering theory, *Proceedings of the First Symposium on Engineering Application of Random Function Theory and Probability*, Wiley, New York, 1963, pp. 270–388.

10 Nishimura, T., Error bounds of continuous Kalman filters and the application to orbit determination problems, *IEEE Trans. Automatic Control*, **AC-12**, No. 3 (1967), 268–275.

11 Sorenson, H. W., On the error behavior in linear minimum variance estimation problems, *IEEE Trans. Automatic Control*, **AC-12**, No. 5 (1967), 557–562.

12 Kailath, T., An innovations approach to least-squares estimation. Part I: Linear filtering in additive white noise, *IEEE Trans. Automatic Control*, **AC-13**, No. 6 (1968), 646–665.

13 Jenkins, G. M., and Watts, D. G., *Spectra Analysis and Its Applications*, Holden-Day, San Francisco, 1968.

Received June 15, 1970

Section I-F
Computational Considerations

EVEN with a good model of the process and its associated statistics, computational errors can produce results that are incorrect and misleading. The problem of reducing the computational burden, both in terms of storage and arithmetic operations, has drawn and continues to draw attention. The papers by Bryson and Henrikson and by Friedland that appear below are concerned with general situations in which the burden can be reduced. Even more attention has been given in the archival literature of methods for implementing the estimator using sound numerical practice. The paper by Kaminski, Bryson, and Schmidt presents a survey of methods whose use had been proposed by 1970. Further extensions and alternatives have been explored thoroughly by Bierman and others since this paper. The final paper of this section by Mendel provides a useful reference for computational requirements associated with different forms of the algorithms that can be implemented.

Correlated measurement noise can be incorporated into the Kalman filter by augmenting the state vector with the noise variables. Kalman's original formulation used a model having this form. Battin proposed a similar treatment. Bryson and Henrickson observed that the state augmentation can be avoided by using a form of measurement differencing that leads to a redefinition of the measurements processed by the filter. Since the Kalman filter generally require, $O(n^3)$ operations, the elimination of colored measurement noise variables from the state can produce a substantial reduction in the computational burden. Bryson and Henrikson consider a constant coefficient system, define the appropriate differencing of the measurements in their equation (7), and then derive the revised form for the filtering equations that appears as equation (9). Then, they extend the result to time-varying systems and present the result in equation (20).

Many system models include bias variables, generally assumed to be constant, in the state vector defined for the system. Because of the simple dynamical model for constant biases, one might expect that some simplification of the computations can be achieved by making use of these characteristics. Friedland developed an algorithm for treating biases in the Kalman filter that is described in the second paper included below. References are given to papers that confirm the computational reductions that are possible with this algorithm and to further extensions of his basic ideas. Friedland considers both continuous-time and discrete-time systems. The result for discrete-time systems is developed in Section III of the paper and leads to a parallel computation of bias-free estimates of the state and of the bias. The algorithm is described in Fig. 2.

Kaminski, Bryson, and Schmidt provide a survey of numerical techniques for solving systems of linear equations as they can be applied to the Kalman filter equations. The paper traces the developments from the square-root algorithm proposed by Potter to the new (i.e., at the time) implementations based on Householder or modified Gram–Schmidt procedures. The paper considers both the covariance and information matrix forms of the filter equations. It also provides an analysis of the computational burden imposed by the use of a square-root algorithm. With its breadth of coverage, this paper still provides a useful introduction to the algorithms that have been introduced subsequently, some of which are included as references at the end of this section.

A thorough analysis of the storage and operational requirements for various implementations of the Kalman filter is provided by the paper by Mendel. Particular emphasis is placed on the relative merits of sequential and simultaneous processing. Optimal ways for accomplishing sequential processing are discussed. An independent analysis of this implementation is provided in the paper by Singer and Sea which is listed among the references at the end of this section.

References

[1] A. E. Bryson and D. E. Johansen, "Linear filtering for time-varying systems using measurements containing colored noise," *IEEE Trans. Automat. Contr.*, vol. AC-10, pp. 4–10, 1965.

[2] R. S. Bucy, "Optimal filtering for correlated noise," *J. Math Appl.*, 20, pp. 1, 1967.

[3] J. F. Bellantoni and K. W. Dodge, "A square-root formulation of the Kalman–Schmidt filter," *AIAA J.*, vol. 5, pp. 1309–1314, 1967.

[4] J. E. Potter and D. C. Fraser, "A formula for updating the determinant of the covariance matrix," *AIAA J.*, vol. 5, pp. 1352–1354, 1967.

[5] E. B. Stear and A. R. Stubberud, "Optimal filtering for Gauss–Markov noise," *Int. J. Contr.*, vol. 8, pp. 123, 1968.

[6] A. Andrews, "A square-root formulation for the Kalman covariance equation," *AIAA J.*, vol. 6, pp. 1165–1166, 1968.

[7] P. Dyer and S. McReynolds, "Extension of Square-Root Filtering to Include Process Noise," *J. Opt. Theory Appl.*, vol. 3, pp. 444–458, 1969.

[8] K. W. Simon and A. R. Stubberud, "Reduced order Kalman filter," *Int. J. Contr.*, vol. 10, pp. 501–509, 1969.

[9] D. R. Vaughn, "A nonrecursive algebraic solution for the discrete Riccati equation," *IEEE Trans. Automat. Contr.*, vol. AC-15, pp. 597–599, 1970.

[10] R. A. Singer and R. G. Sea, "Increasing the computational efficiency of discrete Kalman filters," *IEEE Trans. Automat. Contr.*, vol. AC-16, pp. 254–257, 1971.
[11] I. Gura and A. B. Bierman, "On computational efficiency of linear filtering algorithms," *Automatica*, vol. 7, pp. 299–314, 1971.
[12] G. J. Bierman, "A comparison of discrete linear filtering algorithms," *IEEE Trans. Aerosp. Elect. Syst.*, vol. AES-9, pp. 28–37, 1973.
[13] N. A. Carlson, "Fast triangular formulation of the square-root filter," *AIAA J.*, vol. 11, pp. 1259–1265, 1973.
[14] G. J. Bierman, "Sequential square root filtering and smoothing of discrete linear systems," *Automatica*, vol. 10, pp. 147–158, 1974.
[15] M. Morf, G. S. Sidhu, and T. Kailath, "Some new algorithms for recursive estimation in constant, linear, discrete-time systems," *IEEE Trans. Automat. Contr.*, vol. AC-19, pp. 315–323, 1974.
[16] V. S. Samant and H. W. Sorenson, "On reducing computational burden in the Kalman filter," *Automatica*, vol. 10, pp. 61–68, 1974.
[17] M. Morf and T. Kailath, "Square-root algorithms for least-squares estimation," *IEEE Trans. Automat. Contr.*, vol. AC-20, pp. 487–497, 1975.
[18] G. J. Bierman and C. L. Thornton, "Numerical comparison of Kalman filter algorithms: Orbit determination case study," *Automatica*, vol. 13, pp. 23–35, 1977.

Estimation Using Sampled Data Containing Sequentially Correlated Noise

A. E. BRYSON JR.* AND L. J. HENRIKSON†
Harvard University, Cambridge, Mass.

This paper presents improved filtering, prediction, and smoothing procedures for multistage linear dynamic systems when the measured quantities are linear combinations of the state variables with additive sequentially correlated noise. The "augmented state" procedure suggested by Kalman may lead to ill-conditioned computations in constructing the data processing filter. The design procedure described here eliminates these ill-conditioned computations and reduces the dimension of the filter required. The results include explicit relations for prediction, filtering, and smoothing procedures and the associated covariance matrices.

I. Introduction

THE problem considered is that of estimating the state variables of a multistage linear dynamic system based on measurements of linear combinations of the state variables containing additive sequentially correlated noise.‡ A design procedure for the data processing estimation filters is developed which eliminates the ill-conditioned computations of the augmented state approach, and which is of a lower dimension than the augmented state filters. It was suggested by the work of Bryson and Johansen[1] on the related problem for continuous linear dynamic systems. Considering the measurement vector as a set of constraints among the augmented state variables, a measurement differencing scheme is used to reduce the dimension of the estimation problem. The estimation theory of Kalman[2] is then applied to this reduced problem.

II. The Problem

For simplicity of presentation, a constant coefficient system will be studied. Results for more general systems with time-varying coefficients are presented at the end of the paper. With this restriction, a fairly general system of the type we are considering is described by

$$\text{state:} \qquad x_{i+1} = \Phi x_i + w_i \qquad x:(n \times 1)$$

$$\text{measurement:} \qquad z_i = Hx_i + \epsilon_i \qquad z:(m \times 1) \qquad (1)$$

$$\text{measurement noise:} \qquad \epsilon_{i+1} = \Psi \epsilon_i + u_i$$

Here w_i and u_i are independent gaussian purely random vector sequences ("white noise") with zero means and covariances Q and $\bar{Q}$, respectively, and $HQH^T + \bar{Q}$ is assumed nonsingular. The more general case where the dimension of ϵ_i is greater than the dimension of z_i (i.e., $z_i = Hx_i + G\epsilon_i$), where there is cross-coupling between x_i and ϵ_i, and where some measurements contain purely random noise, is treated in Ref. 3.

The problem is to obtain the maximum likelihood estimate of x_i from the measurements up to and including z_k. If $k < i$, the estimate is called a prediction; if $k = i$, the estimate is called filtering; and if $k > i$, the estimate is called smoothing. (See the Appendix for a basic estimation problem and solution.)

III. Augmented State Approach

The method of optimal filtering developed by Kalman[2] would be applied to the problem as follows. The state is first augmented to include ϵ_i:

Presented as Paper 67-541 at the AIAA Guidance, Control, and Flight Dynamics Conference, Huntsville, Ala., August 14–16, 1967; submitted September 13, 1967; revision received February 7, 1968. This research was supported by the Office of Naval Research Contract NONR-1866(16) and NASA Grant NGR-22-007-068.

* Professor, Division of Engineering and Applied Physics. Associate Fellow AIAA.

† PhD. Candidate, Division of Engineering and Applied Physics; also Engineer Dynamic Research Corp., Stoneham, Mass.

‡ Other names for sequentially correlated noise are "colored noise," "correlated noise," and noise with serial correlation.

Reprinted with permission from *J. Spacecr. Rockets,* vol. 5, pp. 662–665, June 1968.

$$x_i^a \triangleq \begin{bmatrix} x_i \\ \epsilon_i \end{bmatrix} \quad \text{and} \quad H^a \triangleq [H \vdots I]$$
$$\Phi^a = \begin{bmatrix} \Phi & 0 \\ 0 & \Psi \end{bmatrix} \qquad Q^a = \begin{bmatrix} Q & 0 \\ 0 & \bar{Q} \end{bmatrix} \tag{2}$$

The system description is then

$$\text{state:} \qquad x_{i+1}^a = \begin{bmatrix} \Phi & 0 \\ 0 & \Psi \end{bmatrix} x_i^a + \begin{bmatrix} w_i \\ u_i \end{bmatrix} \tag{3}$$
$$\text{measurement:} \qquad z_i = H^a x_i^a$$

For this augmented system the measurements are "perfect," i.e., contain no noise. Calling P_i^a the covariance of the best estimate of x_i^a after measurement z_i, and M_i^a the covariance of the best estimate of x_i^a before measurement z_i (see Appendix), the relation between P_i^a and M_i^a for this case can be written as[4]

$$P_i^a = M_i^a - M_i^a H^{aT}(H^a M_i^a H^{aT})^{-1} H^a M_i^a$$
$$M_{i+1}^a = \Phi^a P_i^a \Phi^{aT} + Q^a \tag{4}$$

Now P_i^a must be singular, since linear combinations of the components of x_i^a are known perfectly. In fact, it follows easily from (4) that

$$H^a P_i^a H^{aT} = 0 \tag{5}$$

Thus, if Φ^a is near unity§ and Q^a is small, the covariance updating may become ill-conditioned (i.e., $M_{i+1}^a \to P_i^a$).

Another way to look at the estimation problem is to observe that the measurements represent m linear constraints among the augmented state variables. Thus, although the augmented state vector is of dimension $n + m$, there are only n linearly independent variables in the estimation problem. This implies that the $(n + m)$ by $(n + m)$ matrix P_i^a is singular (of rank $\leq n$), and also points to the fact that the estimation filter need only be of dimension n, rather than of dimension $n + m$ as it is for this augmented state filter.

IV. Measurement Differencing Approach

In this section, we develop estimation filters of dimension n for the system (1). In particular, since ϵ_i is often not of interest, we design estimation filters dealing only with the original state vector x_i. Using the state transition relations for x_i and ϵ_i, z_{i+1} can be expressed in terms of x_i, ϵ_i, and the purely random vectors w_i and u_i. Having done this, it is possible to use the constraint relations (i.e., the measurements) to eliminate ϵ_i. In effect, we determine a linear combination of z_{i+1} and z_i (two measurement vectors in sequence) which does not contain ϵ_i. From (1) the proper linear combination is easily seen to be

$$\zeta_i \triangleq z_{i+1} - \Psi z_i = (H\Phi - \Psi H)x_i + Hw_i + u_i \tag{6}$$

The "measurement" ζ_i now contains only the purely random sequence $u_i + Hw_i$ instead of the sequentially correlated sequence ϵ_i. Since ζ_i is based on z_{i+1}, it will prove convenient to state the problem as

$$\text{state:} \qquad x_i = \Phi x_{i-1} + w_{i-1} \qquad w_{i-1}:(0,Q)$$
$$\text{measurement:} \quad \zeta_{i-1} = H^r x_{i-1} + u_{i-1} + Hw_{i-1}$$
$$u_{i-1}:(0,\bar{Q}) \tag{7}$$
$$\text{with } H^r = H\Phi - \Psi H \qquad u_{i-1} \text{ and } w_{i-1}$$
$$(\zeta_{i-1} = z_i - \Psi z_{i-1}) \qquad \text{independent}$$

Note that the process noise w_{i-1} and the measurement noise $u_{i-1} + Hw_{i-1}$ are correlated. The formal problem (7) can be solved with the basic solutions given in the Appendix. However, the fact that ζ_{i-1} is calculated from z_i requires further consideration.

V. Filtering Solution

At first glance, it would appear that the problem in the form (7) is immediately solved by application of the basic estimation results in the Appendix. Thus, based on ζ_{i-1}, one would obtain an "estimate" $\hat{x}_{i-1}$ of x_{i-1} and a "prediction" $\bar{x}_i$ of x_i. However, ζ_{i-1} is based on z_i; this means that the prediction of x_i based on ζ_{i-1} is, in fact, the best estimate of x_i based on z_i. Another way of stating this is that the mean of x_i conditioned on ζ_{i-1} is the mean of x_i conditioned on z_i, which is the desired optimal estimate of x_i.

By stating the problem in the form (7), the dimension of the problem has been reduced from $n + m$ in form (3) to n, and the basic estimation solution of the Appendix formally applies. However, the formal "filtering" and prediction solutions are actually "single-stage smoothing" and filtering solutions, respectively. In order to distinguish between the formal and the actual estimates, the following notation will be adopted for the actual estimates: $\hat{x}_{i/k}$ = optimal estimate of x_i given measurements up to and including z_k. $P_{i/k}$ = covariance of $\hat{x}_{i/k} = E\{(x_i - \hat{x}_{i/k})(x_i - \hat{x}_{i/k})^T\}$. The $\bar{x}_i$, $\hat{x}_i$ notation will be reserved for the formal application of the basic solutions in the Appendix to the problem in the form (7), and results in the following equivalences:

$$\text{actual} \qquad \text{formal}$$
$$\hat{x}_{i/i} = \bar{x}_i \tag{8}$$
$$\hat{x}_{i-1/i} = \hat{x}_{i-1}$$

where $\bar{x}_i$ and $\hat{x}_{i-1}$ are the formal prediction and estimate based on ζ_{i-1}. Note that a single-stage smoothing estimate is obtained automatically if it is desired.

Using the equivalences in (8), the filtering solution can be written by the formal application of Eq. (A2) to the problem in the form (7) as

$$\hat{x}_{i-1/i} = \hat{x}_{i-1/i-1} + K_{i-1}(\zeta_{i-1} - H^r\hat{x}_{i-1/i-1})$$
$$\hat{x}_{i/i} = \Phi\hat{x}_{i-1/i} + D(\zeta_{i-1} - H^r\hat{x}_{i-1/i})$$

where

$$D = SR^{-1} \qquad R = \bar{Q} + HQH^T \qquad S = QH^T$$
$$H^r = H\Phi - \Psi H \qquad \zeta_{i-1} = z_i - \Psi z_{i-1}$$
$$K_{i-1} = M_{i-1}H^{rT}(H^r M_{i-1} H^{rT} + R)^{-1}$$
$$P_{i-1} = (I - K_{i-1}H^r)M_{i-1}(I - K_{i-1}H^r)^T + K_{i-1}RK^T_{i-1} \tag{9}$$
$$M_i = (\Phi - DH^r)P_{i-1}(\Phi - DH^r)^T + Q - DRD^T$$
$$P_{i/i} = M_i \qquad P_{i-1/i} = P_{i-1}$$

If the single-stage smoothing estimate $\hat{x}_{i-1/i}$ is not explicitly desired, the filter can be written as

$$\hat{x}_{i/i} = \Phi\hat{x}_{i-1/i-1} + [D + (\Phi - DH^r)K_{i-1}] \times (\zeta_{i-1} - H^r\hat{x}_{i-1/i-1}) \tag{10}$$

After the first measurement there is not yet sufficient information to calculate ζ_1, so the augmented state

$$x_1^a = \begin{bmatrix} x_1 \\ \epsilon_1 \end{bmatrix}$$

approach must be used to obtain the best estimate of x_1^a and the associated covariance based on the "perfect measurement" $z_1 = H^a x_i^a$ and single-stage estimation theory.¶ After

§ For example, this would be the case in sampling a continuous system at instants of time close together relative to the time constants of the system.

¶ Note the analogy with the continuous linear dynamic system problem[1] where a "starting procedure" is also required. In fact, the present problem helps one to understand that requirement.

the second measurement, ζ_1 can be calculated, and the filter (10) can be used with the estimate of x_1 from $\hat{x}_1^a$ and its covariance as the a priori starting statistics.

Since $z_i = Hx_i + \epsilon_i$, an estimate of ϵ_i can be obtained any time after the first measurement by

$$\hat{\epsilon}_{i/i} = z_i - H\hat{x}_{i/i} \qquad \text{cov}\{\epsilon_{i/i}\} = HP_{i/i}H^T \tag{11}$$

The reduced filter (10) requires the storage of one measurement vector (z_{i-1} is needed in addition to z_i to calculate ζ_{i-1}), but it has two distinct advantages over the augmented state filter: 1) The dimension is n instead of $n + m$. 2) The potentially ill-conditioned inversion $(H^a M_i^a H^{aT})^{-1}$ is eliminated.

Prediction Solution

The prediction of x_{i+1} given the estimate $\hat{x}_{i/i}$ follows immediately from the state equation

$$x_{i+1} = \Phi x_i + w_i \tag{12}$$

and from single-stage estimation theory as

$$\hat{x}_{i+1/i} = \Phi\hat{x}_{i/i} \qquad P_{i+1/i} = \Phi P_{i/i}\Phi^T + Q \tag{13}$$

Similarly, if prediction of ϵ_{i+1} given $\hat{\epsilon}_{i/i}$ is desired, use of

$$\epsilon_{i+1} = \Psi\epsilon_i + u_i \tag{14}$$

and (12) yields

$$\hat{\epsilon}_{i+1} = \Psi\hat{\epsilon}_i \qquad \text{cov}\{\hat{\epsilon}_{i+1/i}\} = \Psi HP_{i/i}H^T\Psi^T + \bar{Q} \tag{15}$$

Smoothing Solution

The smoothing solution also follows from the formulation (7) and the basic solution, Eq. (A2), again noting that one must be careful of the nomenclature. When smoothing backwards from the Nth stage, the formal solution smooths backwards from ζ_{N-1}. In other words, the smoothing estimate at the $(N - 1)$st stage has already been obtained from the filtering estimate (9). In terms of the formal smoothed estimate, $\hat{x}(i/\zeta_{N-1})$, obtained from applying Eq. (A2) to the formulation (7), the actual estimate is

$$\hat{x}_{i/N} = \hat{x}(i/\zeta_{N-1}) \tag{16}$$

With these observations, the smoothing solution for x_i is

$$\hat{x}_{i/N} = \hat{x}_{i/i+1} - C_i(\hat{x}_{i+1/i+1} - \hat{x}_{i+1/N}) \qquad \hat{x}_{N-1/N} \text{ given}$$

$$P_{i/N} = P_{i/i+1} - C_i(P_{i+1/i+1} - P_{i+1/N})C_i^T \qquad P_{N-1/N} \text{ given} \tag{17}$$

$$\text{where} \qquad C_i = P_{i/i+1}(\Phi - DH^r)^T P^{-1}_{i+1/i+1}$$

If a smoothed estimate of ϵ_i is desired, it is given directly from the constraint relations by

$$\hat{\epsilon}_{i/N} = z_i - H\hat{x}_{i/N} \qquad \text{cov}\{\hat{\epsilon}_{i/N}\} = HP_{i/N}H^T \tag{18}$$

Generalization to Systems with Time-Varying Coefficients

A fairly general system of this kind can be described by

$$\begin{aligned}
&\text{state:} && x_{i+1} = \Phi_i x_i + w_i \qquad w_i:(0,Q_i)\\
& && x:(n \times 1)\\
& && z:(m \times 1)\\
&\text{measurement:} && z_i = H_i x_i + \epsilon_i \qquad u_i:(0,\bar{Q}_i)\\
& && \epsilon:(m \times 1)\\
&\text{measurement noise:} && \epsilon_{i+1} = \Psi_i\epsilon_i + u_i\\
& && w_i \text{ and } u_i \text{ independent}
\end{aligned} \tag{19}$$

As mentioned in Sec. II, further generalizations can be found in Ref. 3. The technique of determining filtering, prediction, and smoothing solutions for (19) is the same as that used previously.

In (1), it was assumed that R was nonsingular. In (19), the corresponding quantity R_i is the covariance of $u_i + H_{i+1}w_i$. If R_i is nonsingular, the elimination of the ill-conditioning in constructing the filters is guaranteed. However, even if R_i is singular, there will be cases where there is no ill-conditioning and the reduction in dimension of the data processing filters is desirable. For this reason the basic solution, Eq. (A3), is used to obtain the solution to (19) as it is valid independent of the rank of R_i. (If R_i is singular, a further reduction in the dimension of the filters is possible; see Ref. 3.) In fact, given the reduced problem (7), any set of filtering equations may be used, but the results, which are based on ζ_{i-1}, must be "interpreted" in terms of z_i.

Using Eq. (A3), the estimation filters for (19) are

filtering:

$$\hat{x}_{i-1/i} = \hat{x}_{i-1/i-1} + K_{i-1}(\zeta_{i-1} - H^r_{i-1}\hat{x}_{i-1/i-1})$$

$$\hat{x}_{i/i} = \Phi_{i-1}\hat{x}_{i-1/i} + S_{i-1}(H^r_{i-1}M_{i-1}H^r_{i-1}{}^T + R_{i-1})^{-1}(\zeta_{i-1} - H^r_{i-1}\hat{x}_{i-1/i-1})$$

prediction:

$$\hat{x}_{i+1/i} = \Phi_i\hat{x}_{i/i}$$

smoothing:

$$\hat{x}_{i/N} = \hat{x}_{i/i+1} - C_i(\hat{x}_{i+1/i+1} - \hat{x}_{i+1/N}) \qquad \hat{x}_{N-1/N} \text{ given}$$

where

$$S_{i-1} = Q_{i-1}H_i^T \qquad R_{i-1} = \bar{Q}_{i-1} + H_iQ_{i-1}H_i^T$$

$$H_{i-1}{}^r = H_i\Phi_{i-1} - \Psi_{i-1}H_{i-1}$$

$$\zeta_{i-1} = z_i - \Psi_{i-1}z_{i-1} \tag{20}$$

$$K_{i-1} = M_{i-1}H^r_{i-1}{}^T(H^r_{i-1}M_{i-1}H^r_{i-1}{}^T + R_{i-1})^{-1}$$

$$C_i = (P_{i/i+1}\Phi_i^T - K_iS_i^T)P^{-1}_{i+1/i+1}$$

$$P_{i-1/i} = P_{i-1} = (I - K_{i-1}H^r_{i-1})M_{i-1}(I - K_{i-1}H^r_{i-1})^T + K_{i-1}R_{i-1}K^T_{i-1}$$

$$P_{i/i} = M_i = \Phi_{i-1}P_{i-1}\Phi^T_{i-1} + Q_{i-1} - S_{i-1}(H^r_{i-1}M_{i-1}H^r_{i-1}{}^T + R_{i-1})^{-1}S^T_{i-1} - \Phi_{i-1}K_{i-1}S^T_{i-1} - S_{i-1}K^T_{i-1}\Phi^T_{i-1}$$

$$P_{i+1/i} = \Phi_iP_{i/i}\Phi_i^T + Q_i$$

$$P_{i/N} = P_{i/i+1} - C_i(P_{i+1/i+1} - P_{i+1/N})C_i^T \qquad P_{N-1/N} \text{ given}$$

Note that the starting procedure using the augmented state must be used as described in Sec. IV.

Summary and Conclusions

This paper has considered the estimation problem for multistage linear dynamic systems based on measurements of linear combinations of the state variables with additive sequentially correlated noise. By using a weighted first difference of the present and previous measurements, a filter, predictor, and smoother have been developed of lower dimension than those obtained from the augmented state approach. Further, the potential ill-conditioning of the augmented state approach is eliminated (R_i nonsingular) or reduced (R_i singular).

The results include explicit relations for prediction, filtering, and smoothing procedures and the associated covariances. These are summarized in Eq. (20). These improved methods should be useful in orbit determination, guidance, control, navigation, and flight testing.

Appendix: Basic Estimation Solution

The results of basic estimation theory are summarized here for use in this paper (see Refs. 1, 3, and 4). The general problem may be stated as

$$\text{state:} \quad x_{i+1} = \Phi_i x_i + w_i \qquad w_i:(\bar{w}_i, Q_i)$$
$$x:(n \times 1)$$
$$\text{measurement:} \quad z_i = H_i x_i + v_i \qquad v_i:(0, R_i) \quad \text{(A1)}$$
$$z:(m \times 1)$$
$$E\{w_i v_j^T\} = S_i \delta_{ij}$$

where w_i and v_i are gaussian purely random vector sequences. δ_{ij} is the Kronecker delta function.

For the estimation solution, the following definitions are used: $\bar{x}_i$ = estimate of x_i using measurements up to z_{i-1} (single-stage prediction); $\hat{x}_i$ = estimate of x_i using measurements up to z_i (filtering); $\hat{x}(i/z_N)$ = estimate of x_i using measurements up to z_N ($N > i$) (smoothing); M_i = covariance of $\bar{x}_i = E\{(x_i - \bar{x}_i)(x_i - \bar{x}_i)^T\}$; P_i = covariance of $\hat{x}_i = E\{(x_i - \hat{x}_i)(x_i - \hat{x}_i)^T\}$; and $P(i/N)$ = covariance of $\hat{x}(i/z_N) = E\{[x_i - \hat{x}(i/z_N)][x_i - \hat{x}(i/z_N)]^T\}$.

With these definitions the estimation solution for the problem (A1) will be given in two separate forms:

Form 1 (R_i nonsingular):

$$\hat{x}_i = \bar{x}_i + K_i(z_i - H_i\bar{x}_i)$$
$$\bar{x}_{i+1} = \Phi_i\hat{x}_i + D_i(z_i - H_i\hat{x}_i) + \bar{w}_i$$
$$\hat{x}(i/z_N) = \hat{x}_i - C_i(\bar{x}_{i+1} - \hat{x}_{i+1/N}) \qquad \text{(A2)}$$
$$\hat{x}(N/Z_N) = \hat{x}_N$$

where

$$K_i = M_i H_i^T (H_i M_i H_i^T + R_i)^{-1}$$
$$C_i = P_i(\Phi_i - D_i H_i)^T M_{i+1}^{-1}$$
$$D_i = S_i R_i^{-1} \qquad D_i R_i D_i^T = S_i R_i^{-1} S_i^T$$
$$P_i = (I - K_i H_i) M_i (I - K_i H_i)^T + K_i R_i K_i^T$$
$$M_{i+1} = (\Phi_i - D_i H_i) P_i (\Phi_i - D_i H_i)^T + Q_i - D_i R_i D_i^T$$
$$P(i/N) = P_i - C_i(M_{i+1} - P_{i+1/N})C_i^T \qquad P_{N/N} = P_N$$

Form 2 (R_i singular):

$$\hat{x}_i = \bar{x}_i + K_i(z_i - H_i\bar{x}_i)$$
$$\bar{x}_{i+1} = \Phi\hat{x}_i + S_i(H_i M_i H_i^T + R_i)^{-1}(z_i - H_i\bar{x}_i) + \bar{w}_i \qquad \text{(A3)}$$
$$\hat{x}(i/z_N) = \hat{x}_i - C_i(\bar{x}_{i+1} - \hat{x}_{i+1/N}) \qquad \hat{x}(N/Z_N) = \hat{x}_N$$

where

$$K_i = M_i H_i^T (H_i M_i H_i^T + R_i)^{-1}$$
$$C_i = (P_i \Phi_i^T - K_i S_i^T) M_{i+1}^{-1}$$
$$P_i = (I - K_i H_i) M_i (I - K_i H_i)^T + K_i R_i K_i^T$$
$$M_{i+1} = \Phi_i P_i \Phi_i^T + Q_i - S_i(H_i M_i H_i^T + R_i)^{-1} S_i^T - \Phi_i K_i S_i^T - S_i K_i^T \Phi_i^T$$
$$P(i/N) = P_i - C_i(M_{i+1} - P_{i+1/N})C_i^T \qquad P_{N/N} = P_N$$

It is assumed that the a priori statistics M_1 and $\bar{x}_1$ are given.

References

[1] Bryson, A. E., Jr. and Johansen, D. E., "Linear Filtering for Time-Varying Systems Using Measurements Containing Colored Noise," *IEEE Transactions*, Vol. AC-10, Jan. 1965, pp. 4–10.

[2] Kalman, R. E., "New Methods in Wiener Filtering Theory," *Proceedings of the First Symposium on Engineering Applications of Random Function Theory and Probability*, edited by J. L. Bogdanoff and F. Kozin, Wiley, 1963, pp. 270–388.

[3] Henrikson, L. J., Ph.D. thesis, Div. of Engineering and Applied Physics, Harvard University, Cambridge, Mass.

[4] Bryson, A. E., Jr. and Ho, Y. C., *Optimization, Estimation, and Control*, Blaisdell (to be published).

Treatment of Bias in Recursive Filtering

BERNARD FRIEDLAND, MEMBER, IEEE

Abstract—The problem of estimating the state x of a linear process in the presence of a constant but unknown bias vector b is considered. This bias vector influences the dynamics and/or the observations. It is shown that the optimum estimate $\hat{x}$ of the state can be expressed as

$$\hat{x} = \tilde{x} + V_x \hat{b} \tag{1}$$

where $\tilde{x}$ is the bias-free estimate, computed as if no bias were present, $\hat{b}$ is the optimum estimate of the bias, and V_x is a matrix which can be interpreted as the ratio of the covariance of $\tilde{x}$ and $\hat{b}$ to the variance of $\hat{b}$. Moreover, $\hat{b}$ can be computed in terms of the residuals in the bias-free estimate, and the matrix V_x depends only on matrices which arise in the computation of the bias-free estimates. As a result, the computation of the optimum estimate $\hat{x}$ is effectively decoupled from the estimate of the bias $\hat{b}$, except for the final addition indicated by (1).

I. Introduction

IN THE APPLICATION of Kalman–Bucy recursive filtering techniques [1], [2] an accurate model of the process dynamics and observations is required. In many instances this model contains parameters which may deviate by constant but unknown amounts from their nominal values. Use of the nominal values of the parameters in the filter design may lead to unacceptably large errors [3] in the estimate provided by the filter. To reduce the estimation errors which could arise from such incorrect modeling, it is a common practice to augment the state vector of the original problem by adding additional components to represent the uncertain parameters, which are conveniently designated as bias terms. The filter then estimates the bias terms as well as those of the original problem. This method is reasonably effective when the number of bias terms is small relative to the state variables of the original problem; the dimension of the state vector is then not significantly increased by adjoining the state variables which are used to represent the bias terms. When the number of bias terms is comparable to the number of state variables of the original problem, however, the new state vector is substantially larger in dimension than that of the original problem, and the computations required by the filtering algorithm may become excessive: accuracy as well as computational speed may be severely compromised by the use of this technique.

The present paper was motivated by the need for a method whereby the numerical inaccuracies introduced by computations with large vectors and matrices can be avoided. The results presented here are based on expressing the solution of the variance equation of the problem with

Manuscript received October 23, 1968.
The author is with the Research Center, Kearfott Division, Singer-General Precision, Inc., Little Falls, N. J.

Reprinted from *IEEE Trans. Automat. Contr.*, vol. AC-14, pp. 359–367, Aug. 1969.

bias present in terms of the solution of the variance equation for bias-free estimation and other matrices which depend only on the bias-free computations. By this technique the estimation of the bias is essentially decoupled from the computation of the bias-free estimate of the state.

Section II deals with the continuous-time case and Section III deals with the discrete-time case. Although the continuous-time case is of lesser practical importance because most actual implementations are by digital means, the formulas developed for continuous-time processes are simpler and hence more readily interpreted. Thus the development of the continuous-time case serves as an introduction to the more practical but also more complicated discrete-time case.

II. Continuous-Time Filtering

Problem Statement

The following notation is used:

- x original or physical state vector (n components)
- b bias vector (r components)
- y observation vector (s components)
- ξ process noise vector, with $E[\xi(t)\xi'(\tau)] = Q\delta(t-\tau)$
- η observation noise vector, with $E[\eta(t)\eta'(\tau)] = R\delta(t-\tau)$.

For simplicity, ξ and η are assumed to be independent.

The dynamic equations can be expressed as follows

$$\dot{x} = Ax + Bb + \xi \tag{2a}$$

$$\dot{b} = 0 \tag{2b}$$

and the observation equation is expressed as

$$y = Hx + Cb + \eta. \tag{3}$$

The matrices B and C determine how the components of the bias vector b enter into the dynamics and observations, respectively, and represent the general case. For the case in which only the observations are biased, $B = 0$; similarly, if only the dynamics are biased, $C = 0$.

By adjoining b to x we obtain a new state vector

$$z = \begin{bmatrix} x \\ \hline b \end{bmatrix}$$

of $n + r$ components. Equations (2a), (2b), and (3) can then be written as

$$\dot{z} = Fz + G\xi \tag{4}$$

$$y = Lz + \eta \tag{5}$$

where

$$F = \overset{\quad n \;\; r}{\begin{bmatrix} A & B \\ 0 & 0 \end{bmatrix}} \begin{matrix} \updownarrow n \\ \updownarrow r \end{matrix}, \qquad G = \overset{n}{\begin{bmatrix} I \\ 0 \end{bmatrix}} \begin{matrix} \updownarrow n \\ \updownarrow r \end{matrix}$$

$$L = \overset{\; n \;\; r}{[H \mid C]} \updownarrow s.$$

Application of the well-known [2] Kalman–Bucy recursive filtering theory to the process governed by (4) and (5) results in the following equation for the optimum estimate $\hat{z}$, where

$$\hat{z} = \begin{bmatrix} \hat{x} \\ \hline \hat{b} \end{bmatrix}$$

$$\dot{\hat{z}} = F\hat{z} + PL'R^{-1}(y - L\hat{z}) \tag{6}$$

where P is the a posteriori covariance matrix which is the solution to the variance equation

$$\dot{P} = FP + PF' - PL'R^{-1}LP + GQG'. \tag{7}$$

The covariance matrix P is partitioned as follows

$$P = \overset{\quad n \;\; r}{\begin{bmatrix} P_x & P_{xb} \\ P_{xb}' & P_b \end{bmatrix}} \begin{matrix} \updownarrow n \\ \updownarrow r \end{matrix}. \tag{8}$$

Note that

- P_x autocovariance of estimate of original state x
- P_b autocovariance of bias b
- P_{xb} cross covariance of x and b.

In terms of the submatrices of (8), the variance equation (7) takes the form of the following three equations

$$\begin{aligned} \dot{P}_x &= AP_x + P_xA' + BP_{xb}' + P_{xb}B' \\ &\quad - (P_xH' + P_{xb}C')R^{-1}(HP_x + CP_{xb}') + Q \end{aligned} \tag{9a}$$

$$\begin{aligned} \dot{P}_{xb} &= AP_{xb} + BP_b - (P_xH' + P_{xb}C')R^{-1}(HP_{xb} + CP_b) \\ &= [A - (P_xH' + P_{xb}C')R^{-1}H]P_{xb} \\ &\quad + [B - (P_xH' + P_{xb}C')R^{-1}C]P_b \end{aligned} \tag{9b}$$

$$\dot{P}_b = -(P_{xb}'H' + P_bC')R^{-1}(HP_{xb} + CP_b). \tag{9c}$$

Transformation of Variance Equation

It is noted that (9b) and (9c) together are homogeneous in P_{xb} and P_b. Hence, if

$$P_{xb}(0) = 0$$

$$P_b(0) = 0$$

then

$$\begin{aligned} P_{xb} &\equiv 0 \\ P_b &\equiv 0 \end{aligned} \tag{10}$$

for $t > 0$, and hence P_x satisfies

$$\dot{P}_x = AP_x + P_xA' - P_xH'R^{-1}HP_x + Q. \tag{11}$$

The interpretation of (10) and (11) is that if the bias b is perfectly known ($P_b(0) = 0$) at $t = 0$, then by virtue of (2b), it is perfectly known thereafter, and the estimation problem reduces to that in which there is no bias; the variance equation (11) is the same as would result if b were known to be zero (with probability one). The covariance matrix for the $(n + r)$ dimensional problem in this case is denoted by

$$\tilde{P} = \begin{bmatrix} \tilde{P}_x & 0 \\ 0 & 0 \end{bmatrix} \tag{12}$$

where $\tilde{P}_x$ is the solution to (11) with $\tilde{P}_x(0)$ given. It is noted that $\tilde{P}$ is the solution to (7) for the initial condition

$$P(0) = \begin{bmatrix} \tilde{P}_x(0) & 0 \\ 0 & 0 \end{bmatrix}. \tag{13}$$

If the bias is not perfectly known, however, (13) is not the correct initial condition for the problem to be solved. The question is how much do P_x, P_{xb}, and P_b change as a result of changing the initial condition so that

$$P(0) = \begin{bmatrix} \tilde{P}_x(0) & P_{xb}(0) \\ P_{xb}'(0) & P_b(0) \end{bmatrix}$$

where

$$P_b(0) \neq 0$$

and $P_{xb}(0)$ may or may not be zero. To answer this question we make use of the fact that if $\tilde{P}$ is a solution to (7) then any other solution can be expressed as follows [4]

$$P = \tilde{P} + VMV' \tag{14}$$

where

$$\dot{V} = (F - \tilde{P}L'R^{-1}L)V \tag{15}$$

$$\dot{M} = -MV'L'R^{-1}LVM. \tag{16}$$

In (14)–(16) M is an $r \times r$ symmetric matrix, and V can be partitioned as follows

$$V = \overset{\overset{r}{\leftrightarrow}}{\begin{bmatrix} V_x \\ \hline V_b \end{bmatrix}} \begin{matrix} \updownarrow n \\ \updownarrow r \end{matrix}. \tag{17}$$

Then, on using the definition of F (below (5)) and $\tilde{P}$ given by (12), it is found that (15) can be expressed by

$$\dot{V}_x = (A - \tilde{P}_x H'R^{-1}H)V_x + (B - \tilde{P}_x H'R^{-1}C)V_b \tag{18}$$

$$\dot{V}_b = 0 \quad \text{i.e., } V_b = \text{constant}. \tag{19}$$

Similarly (16) becomes

$$\dot{M} = -M(V_x'H' + V_b'C')R^{-1}(HV_x + CV_b)M. \tag{20}$$

It is noted that (18) and (20) are somewhat similar to (9b) and (9c), respectively. There is an important difference, however. In (9) the equations for P_x, P_{xb}, and P_b are all coupled and hence must be solved together, whereas (18) and (20) depend on $\tilde{P}_x$, the covariance matrix for the problem without bias. Consequently, it is possible to solve for $\tilde{P}_x$ independent of P_{xb} and P_b, then find V_x and V_b, and finally compute the actual values of the desired covariance matrix using (14), which in partitioned form is

$$\begin{aligned} P_x &= \tilde{P}_x + V_x M V_x' \\ P_{xb} &= V_x M V_b' \\ P_b &= V_b M V_b'. \end{aligned} \tag{21}$$

The initial conditions $V_x(0)$, $V_b(0)$ on (18)–(20) must be selected so that

$$\begin{aligned} 0 = P_x(0) - \tilde{P}_x(0) &= V_x(0)M(0)V_x'(0) \\ P_{xb}(0) &= V_x(0)M(0)V_b'(0) \\ P_b(0) &= V_b(0)M(0)V_b'(0). \end{aligned} \tag{22}$$

These initial conditions are not unique. In the important special case in which $P_{xb}(0) = 0$, i.e., there is no a priori correlation between the state and the bias, a convenient choice of initial conditions is

$$\begin{aligned} M(0) &= P_b(0) \\ V_x(0) &= 0 \\ V_b(0) &= I. \end{aligned} \tag{23}$$

In this case

$$\begin{aligned} P_{xb} &= V_x M \\ P_b &= M. \end{aligned}$$

Computational Considerations

By the introduction of the transformation $P = \tilde{P} + VMV'$, the variance equation for the combined state $z = \{x,b\}$, given in partitioned form (9a)–(9c) has been replaced by the three equations (11) (with $\tilde{P}_x$ in place of P_x), (18) and (20). The total number of these equations are equal in both sets of equations. The latter set of equations has decided computational advantages, however. In the original set (9a)–(9c) are mutually coupled, hence there is no way of avoiding the integration of $(n + r)(n + r + 1)/2$ simultaneous nonlinear equations. On the other hand, (18) depends only on the solution of the variance equation (11) for $\tilde{P}_x$ and (20), in turn, depends only on the solution of (18). Since the numerical integration error increases rapidly with the number of simultaneous equations integrated, the solution for $\tilde{P}_x$ from (11) can be expected to be much more accurate than the solution for P_x, P_{xb}, and P_b obtained from (9a)–(9c). Moreover, (18) is a linear equation having a theoretically stable fundamental matrix [5]. If the errors in computation of $\tilde{P}_x$ are not large enough to upset the stability of (18) this differential equation for V_x can reasonably be expected to be well behaved, and to converge rapidly to its asymptotic solution. Finally, since $M(0)$ can be selected to be positive definite (see (23) for example) $\dot{M}$ is negative semidefinite, and hence (20) is stable. In fact, letting

$$W = M^{-1}$$

(20) can be expressed as

$$\dot{W} = (V_x'H' + V_b'C')R^{-1}(HV_x + CV_b) \tag{24}$$

so that

$$W(t) = W(0) + \int_0^t (V_x'H' + V_b'C')\, R^{-1}\, (HV_x + CV_b)\, ds. \tag{25}$$

Since the integrand in (25) is positive semidefinite, the integral is nondecreasing, and $W(t)$ is always larger (by use of any suitable norm) than $W(0)$. Hence $M(t) = W^{-1}(t)$ remains positive definite for all finite t and is always smaller than $M(0)$. In the limit as $t \to \infty$ $M(t) \to 0$ or a singular, i.e., positive semidefinite, matrix. The matrix cannot be nonsingular because $M(t)$ tends to a limit $\dot{M} \to 0$, and hence the right-hand side of (20) must be zero, which can happen only for

$$M(V_x'H' + V_b'C') = 0.$$

The results of this analysis can also be employed to obtain a better form of the estimation equation (6), which in component form is

$$\dot{\hat{x}} = A\hat{x} + B\hat{b} + (P_xH' + P_{xb}C')R^{-1}(y - H\hat{x} - C\hat{b}) \tag{26}$$

$$\dot{\hat{b}} = (P_{xb}'H' + P_bC')R^{-1}(y - H\hat{x} - C\hat{b}). \tag{27}$$

Using $P_b = M$, $P_{xb} = V_xM$ for the special case $P_{xb}(0) = 0$, these become

$$\dot{\hat{x}} = A\hat{x} + B\hat{b} + [\tilde{P}_xH + V_xM(V_x'H' + C)]R^{-1}(y - H\hat{x} - C\hat{b}) \tag{28}$$

$$\dot{\hat{b}} = M(V_x'H' + C')R^{-1}(y - H\hat{x} - C\hat{b}). \tag{29}$$

Since the estimate $\hat{b}$ of the bias is a forcing term in (26) or (28) it is necessary to integrate (26) and (27) or (28) and (29) simultaneously, and there is a possibility of significant error because of the high dimension $(n + r)$ of the combined system. To reduce the computational error we introduce the transformation

$$\hat{x} = \tilde{x} + S\hat{b} \tag{30}$$

where $\tilde{x}$ is the estimate of x which would be obtained in the absence of bias, i.e., $\tilde{x}$ satisfies

$$\dot{\tilde{x}} = A\tilde{x} + \tilde{P}_xHR^{-1}(y - H\tilde{x}). \tag{31}$$

In order for (30) to hold we must have

$$= \dot{\tilde{x}} + S\dot{\hat{b}} + \dot{S}\hat{b}. \tag{32}$$

Substitution of (28), (29), and (31) into (32) results in the following

$$[-\dot{S} + (A - \tilde{P}_xH'R^{-1}H)S + B - \tilde{P}_xH'R^{-1}C]\hat{b} = 0$$

which is satisfied by

$$\dot{S} = (A - \tilde{P}_xH'R^{-1}H)S + B - \tilde{P}_xH'R^{-1}C.$$

This is the same as the differential equation (18) for V_x. Since $\hat{x}(0) = \tilde{x}(0)$, when $P_{xb}(0) = 0$, $S(0) = 0 = V_x(0)$. Hence $S = V_x$ and (30) can be written

$$\hat{x} = \tilde{x} + V_x\hat{b} \tag{33}$$

and (29) can be written

$$\dot{\hat{b}} = -M(V_x'H' + C')R^{-1}(HV_x + C)\hat{b} + M(V_x'H' + C')R^{-1}(y - H\tilde{x}). \tag{34}$$

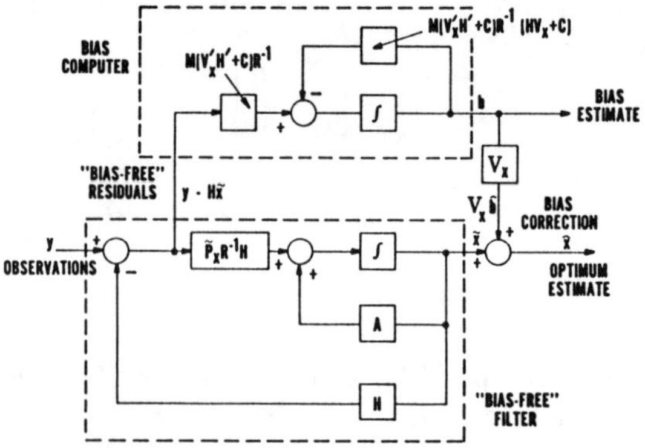

Fig. 1. Schematic representation of computation of estimates in presence of bias.

Thus it is possible to compute x by first computing the bias-free estimate $\tilde{x}$ then correcting the latter by the quantity $V_x\hat{b}$. The bias estimate $\hat{b}$, by virtue of (34), can be computed in terms of the bias-free estimate $\tilde{x}$, or, more properly, the bias-free residuals $y - H\tilde{x}$. Hence the computation may be done by two uncoupled computers as shown in Fig. 1.

III. Discrete-Time Filtering

Problem Statement

The following notation is employed, analogous to that of the continuous-time process

- x_k original or physical process state (at kth observation instant) n components
- b_k bias vector (r components)
- ξ_k process noise vector, with $E[\xi_k\xi_l'] = Q_k\delta_{kl}$
- η_k observation noise vector, with $E[\eta_k\eta_l] = R_k\delta_{kl}$.

Again ξ_k is assumed independent of η_l for all k and l.

The dynamic equations can be written as

$$x_k = A_{k-1}x_{k-1} + B_{k-1}b_{k-1} + \xi_{k-1} \tag{35a}$$

and

$$b_{k+1} = b_k. \tag{35b}$$

The second equation denotes the assumption that $b_k =$ constant, as required for a bias term, and B_k denotes how the bias vector $b_k = b$ enters into the dynamics.

The observation equation is

$$y_k = H_kx_k + C_kb_k + \eta_k. \tag{36}$$

Upon defining the new state

$$z_k = \begin{bmatrix} x_k \\ \hdashline b_k \end{bmatrix} \begin{matrix} \updownarrow n \\ \updownarrow r \end{matrix}$$

the dynamic equations and observation equations can be expressed as

$$z_{k+1} = F_kz_k + G\xi_k \tag{37}$$

$$y_k = L_kz_k + \eta_k \tag{38}$$

where

$$F_k = \begin{bmatrix} A_k & B_k \\ 0 & I \end{bmatrix} \begin{matrix} \updownarrow n \\ \updownarrow r \end{matrix} \qquad G = \begin{bmatrix} I \\ 0 \end{bmatrix} \begin{matrix} \updownarrow n \\ \updownarrow r \end{matrix}$$

$$L_k = [H_k \mid C_k] \updownarrow s.$$

The application of the well-known recursive filtering formulas [1] to the optimum estimation of z_k given the sequence of observations $\{y_1, \cdots, y_k\}$ results in the following estimation equation[1]

$$\hat{z}_k = F_{k-1}\hat{z}_{k-1} + K(k)(y_k - L_kF_{k-1}\hat{z}_{k-1}) \tag{39}$$

where $K(k)$ is the so-called gain matrix which is usually expressed as

$$K(k) = P(k)L_k'[L_kP(k)L_k' + R_k]^{-1} \tag{40}$$

where $P(k)$ is the covariance matrix of the estimate of x_k given the observations $\{y_1, \cdots, y_{k-1}\}$ but not y_k. This matrix is sometimes called the a priori covariance matrix, and is recursively computed by means of the variance equation

$$P(k+1) = F_k[I - K(k)L_k]P(k)F_k' + GQ_{k+1}G'. \tag{41}$$

In the normal sequence of computations, another matrix

$$T(k) = [I - K(k)L_k]P(k) \tag{42}$$

is often introduced. This matrix is the covariance of the estimate of z_k given $\{y_1, \cdots, y_k\}$. Because the current observation y_k is included, $T(k)$ is often called the a posteriori covariance matrix. In terms of $T(k)$, (41) can be expressed as

$$P(k+1) = F_kT(k)F_k' + GQ_{k+1}G' \tag{43}$$

and the gain matrix can be written

$$K(k) = T(k)L_k'R_k^{-1} \tag{44}$$

when R_k is nonsingular, as we shall assume here.

To verify (44), write the right-hand side as

$$\begin{aligned}
&[I - PL'(LPL' + R)^{-1}L]PL'R^{-1} \\
&= PL'R^{-1} - PL'(LPL' + R)^{-1}LPL'R^{-1} \\
&= PL'[(LPL' + R)^{-1}(LPL' + R) \\
&\quad - (LPL' + R)^{-1}\,LPL']R^{-1} \\
&= PL'(LPL' + R)^{-1}RR^{-1} = PL'(LPL' + R)^{-1}
\end{aligned}$$

which was to be shown.

For subsequent calculations it will be convenient to use (44) as well as (40) to express the gain matrix.

Transformation of Variance Equation

As in the continuous case, it is possible to transform the variance equation (41) with the aid of a transformation of the form of (14). The result, proved in the Appendix, is

$$P(k) = \tilde{P}(k) + U(k)M(k)U'(k) \tag{45}$$

where $\tilde{P}(k)$ is any solution to (41) with $\tilde{P}(0) \neq P(0)$ and where

$$U(k+1) = F_k(I - \tilde{P}(k)L_k'(L_k\tilde{P}(k)L_k' + R_k)^{-1}L_k)U(k) \tag{46}$$

$$M(k+1) = M(k) - M(k)U'(k)L_k'[L_k\tilde{P}(k)L_k' + R_k + L_kU(k)M(k)U'(k)L_k']^{-1}L_kU(k)M(k). \tag{47}$$

Note that (46) and (47) are analogous to (15) and (16), respectively.

In terms of this transformation (45) can be written

$$\begin{aligned}
P(k+1) &= \tilde{P}(k+1) + U(k+1)M(k+1)U'(k+1) \\
&= \tilde{P}(k+1) + F_k[I - \tilde{K}(k)L_k]U(k) \\
&\quad \cdot M(k+1)U'(k)[I - \tilde{K}(k)L_k]'F_k'
\end{aligned} \tag{48}$$

where

$$\tilde{K}(k) = \tilde{P}(k)L_k'(L_k\tilde{P}(k)L_k' + R_k)^{-1}.$$

Now, from (43),

$$P(k+1) = F_kT(k)F_k' + GQ_{k+1}G'$$

$$\tilde{P}(k+1) = F_k\tilde{T}(k)F_k' + GQ_{k+1}G'.$$

Hence (48) can be expressed as

$$T(k) = \tilde{T}(k) + V(k)M(k+1)V'(k) \tag{49}$$

where

$$V(k) = [I - \tilde{K}(k)L_k]U(k) \tag{50}$$

and

$$U(k+1) = F_kV(k). \tag{51}$$

Equation (49) is the analog of (45) for the a posteriori covariance matrix. Note, however, that $M(k+1)$, not $M(k)$, appears in this equation. As a result it is necessary to calculate $M(k+1)$ before $T(k)$ can be calculated. Equations (50) and (51) are interpreted as resulting in a two-stage calculation of $U(k+1)$ from $U(k)$, just as $P(k+1)$ is calculated from $P(k)$ in two stages using (42) and (43).

In the present application $P(k)$ is partitioned as in the continuous-time case

$$P(k) = \begin{bmatrix} P_x(k) & P_{xb}(k) \\ P_{xb}'(k) & P_b(k) \end{bmatrix} \begin{matrix} \updownarrow n \\ \updownarrow r \end{matrix}. \tag{52}$$

The matrices $\tilde{P}(k)$, $T(k)$, and $\tilde{T}(k)$ are similarly partitioned. It is readily established, however, that if $\tilde{P}_b(0) = 0$, $\tilde{P}_{xb}(0) = 0$ then

$$\tilde{P}(k) = \begin{bmatrix} \tilde{P}_x(k) & 0 \\ 0 & 0 \end{bmatrix}, \quad \text{for all } k. \tag{53}$$

[1] For certain matrices, e.g., $P(k)$, $K(k)$, the index is written as an argument instead of a subscript. Subscripts will be used to denote submatrices in the sequel.

In this case

$$\tilde{P}(k)L_k' = \left[\begin{array}{c}\tilde{P}_x(k)H_k' \\ \hline 0\end{array}\right]$$

and

$$\tilde{K}(k) = \left[\begin{array}{c}\tilde{P}_x(k)H_k'(H_k\tilde{P}_x(k)H_k' + R_k)^{-1} \\ \hline 0\end{array}\right] = \left[\begin{array}{c}\tilde{K}_x(k) \\ \hline 0\end{array}\right] \tag{54}$$

where $\tilde{K}_x(k)$ is the gain matrix for filtering in the absence of bias. Hence

$$I - \tilde{K}(k)L_k = \left[\begin{array}{c|c}I - \tilde{K}_k(k)H_k & -\tilde{K}_x(k)C_k \\ \hline 0 & I\end{array}\right]. \tag{55}$$

The vectors $U(k)$ and $V(k)$ are also partitioned as follows

$$U(k) = \overset{\leftrightarrow r}{\left[\begin{array}{c}U_x(k) \\ \hline U_b(k)\end{array}\right]}\begin{array}{l}\updownarrow n \\ \updownarrow r\end{array} \quad V(k) = \overset{\leftrightarrow r}{\left[\begin{array}{c}V_x(k) \\ \hline V_b(k)\end{array}\right]}\begin{array}{l}\updownarrow n \\ \updownarrow r\end{array}. \tag{56}$$

In terms of these partitioned matrices (50) and (51) become

$$V_x(k) = (I - \tilde{K}_x(k)H_k)U_x(k) - \tilde{K}_x(k)C_kU_b(k) \tag{57}$$
$$V_b(k) = U_b(k)$$
$$U_x(k+1) = A_kV_x(k) + B_kV_b(k) \tag{58}$$
$$U_b(k+1) = V_b(k).$$

The second equations in (57) and (58) imply that

$$U_b(k) = V_b(k) = U_b(0) = \text{constant}, \quad \text{for all } k. \tag{59}$$

Finally (45) becomes

$$P_x(k) = \tilde{P}_x(k) + U_x(k)M(k)U_x'(k)$$
$$P_{xb}(k) = U_x(k)M(k)U_b'(k) \tag{60}$$
$$P_b(k) = U_b(k)M(k)U_b'(k)$$

and (49) becomes

$$T_x(k) = \tilde{T}_x(k) + V_x(k)M(k+1)V_x'(k)$$
$$T_{xb}(k) = V_x(k)M(k+1)V_b'(k) \tag{61}$$
$$T_b(k) = V_b(k)M(k+1)V_b'(k).$$

Assuming x and b are independent at the beginning ($k = 0$), then

$$P_{xb}(0) = 0.$$

In this case we can make the identification

$$U_b(0) = I$$

and, hence, using (59), we can write (57)–(61) as follows

$$V_x(k) = U_x(k) - \tilde{K}_x(k)S(k) \quad (U_x(0) = 0) \tag{62}$$

where

$$S(k) = H_kU_x(k) + C_k \tag{63}$$
$$U_x(k+1) = A_kV_x(k) + B_k \tag{64}$$
$$P_x(k) = \tilde{P}_x(k) + U_x(k)M(k)U_x'(k) \tag{65}$$
$$P_{xb}(k) = U_x(k)M(k) \tag{66}$$
$$P_b(k) = M(k) \tag{67}$$
$$T_x(k) = \tilde{T}_x(k) + V_x(k)M(k+1)V_x'(k) \tag{68}$$
$$T_{xb}(k) = V_x(k)M(k+1) \tag{69}$$
$$T_b(k) = M(k+1) \tag{70}$$

and (47) becomes

$$M(k+1) = M(k) - M(k)S'(k)[H_k\tilde{P}_x(k)H_k' + R_k + S(k)M(k)S'(k)]^{-1}S(k)M(k). \tag{71}$$

The gains $K_x(k)$ and $K_b(k)$, using (44), are given by

$$K_x(k) = \tilde{K}_x(k) + V_x(k)K_b(k) \tag{72}$$

where

$$K_b(k) = M(k+1)[V_x'(k)H_k' + C_K']R_k^{-1}. \tag{73}$$

Computational Considerations

The calculations required are embodied in (62)–(73). Using these equations the covariance matrices $\tilde{P}_x$ and $\tilde{T}_x$ for the bias-free estimates are computed in the normal manner, as is the bias-free gain $\tilde{K}_x$ using the formulas

$$\tilde{K}_x(k) = \tilde{P}_x(k)H_k'[H_k\tilde{P}_x(k)H_k' + R_k]^{-1}$$
$$\tilde{T}_x(k) = (I - \tilde{K}_xH_k)\tilde{P}_x(k) \tag{74}$$
$$\tilde{P}_x(k+1) = A_k\tilde{T}_x(k)A_k' + Q_{k+1}.$$

The gain $\tilde{K}_x(k)$ thus computed and the matrix $H_k\tilde{P}_x(k)H_k' + R_k$ are stored for the computation of the auxiliary quantities in (62)–(73). In addition the auxiliary matrices $U_x(k)$ and $V_x(k)$ are recursively computed and stored as follows. With $U_x(k)$ stored, $S(k)$ is computed from (63), and then $V_x(k)$ is computed from (62). Then $M(k+1)$ is computed from (71), $K_b(k)$ is computed from (73), and $K_x(k)$ is computed from (72); finally $V_x(k)$ is updated to $U_x(k+1)$ using (64).

It is noted that the bias-free computation is independent of the computation of the correction terms $K_b(k)$ and $V_x(k)K_b(k)$. Since these correction terms are often small, the precision with which the calculations are made need not be as high as that used for the calculation of the bias-free terms.

The use of these results also leads to a transformation of the filtering equations to permit the computation of a bias-free estimate independent of the bias estimation, and then addition of a correction due to the bias. First, consider the estimation equations in the usual form, i.e., the partitioned form of (39)

$$\hat{x}_k = A_{k-1}\hat{x}_{k-1} + B_{k-1}\hat{b}_{k-1} + K_x(k)\,[y_k - H_k(A_{k-1}\hat{x}_{k-1} + B_{k-1}\hat{b}_{k-1}) - C_k\hat{b}_{k-1}] \tag{75}$$

$$\hat{b}_k = \hat{b}_{k-1} + K_b(k)[y_k - H_k(A_{k-1}\hat{x}_{k-1} + B_{k-1}\hat{b}_{k-1}) - C_k\hat{b}_{k-1}] \tag{76}$$

where $K_x(k)$ and $K_b(k)$ are given by (72) and (73), respectively. Let $\tilde{x}_k$ be the bias-free estimate, i.e.,

$$\tilde{x}_k = A_{k-1}\tilde{x}_{k-1} + \tilde{K}_x(k)[y_k - H_k A_{k-1}\tilde{x}_{k-1}] \quad (77)$$

and $\tilde{K}_x(k)$ is the bias-free gain, given by (54). The result we will obtain is expressed by the following formula

$$\hat{x}_x = \tilde{x}_x + V_x(k)\hat{b}_k. \quad (78)$$

This equation is analogous to (33) of the continuous-time case.

To prove (78) we write the bracketed term in (75) and (76), using (78), as

$$\begin{aligned} y_k &- H_k(A_{k-1}\hat{x}_{k-1} + B_{k-1}\hat{b}_{k-1}) - C_k\hat{b}_{k-1} \\ &= y_k - H_k A_{k-1}\tilde{x}_{k-1} \\ &\quad - [H_k(A_{k-1}V_x(k-1) + B_{k-1}) + C_k]\hat{b}_{k-1} \\ &= \tilde{r}_k - S(k)\hat{b}_{k-1} \end{aligned} \quad (79)$$

where $\tilde{r}_k = y_k - H_k A_{k-1}\tilde{x}_{k-1}$ = the residual of bias-free estimation. The last line of (79) is obtained using (63) and (64). Substitution of (78) and (79) into (75) and (76) results in

$$\begin{aligned} \hat{x}_k = A_{k-1}\tilde{x}_{k-1} &+ A_{k-1}V_x(k-1)\hat{b}_{k-1} \\ &+ B_{k-1}\hat{b}_{k-1} + K_x(k)[\tilde{r}_k - S(k)\hat{b}_{k-1}] \end{aligned} \quad (80)$$

$$\hat{b}_k = \hat{b}_{k-1} + K_b(k)[\tilde{r}_k - S(k)\hat{b}_{k-1}]. \quad (81)$$

Equation (81) gives the expression for the estimate $\hat{b}_k$ of the bias in terms of the bias-free residual $\tilde{r}_k$. This can be expressed as

$$\hat{b}_k = [I - K_b(k)S(k)]\hat{b}_{k-1} + K_b(k)\tilde{r}_k. \quad (82)$$

To complete the proof, we show that (80) is satisfied with $\hat{x}_k$ given by (78). We must show that

$$\begin{aligned} \tilde{x}_k &+ V_x(k)[I - K_b(k)S(k)]\hat{b}_{k-1} + V_x(k)K_b(k)\tilde{r}_k \\ &= A_{k-1}\tilde{x}_{k-1} + [A_{k-1}V_x(k-1) + B_{k-1} \\ &\quad - K_x(k)S(k)]\hat{b}_{k-1} + K_x(k)\tilde{r}_k \end{aligned}$$

is satisfied for all k, $\tilde{r}_k$, and $\hat{b}_k$. This in turn requires that

$$K_x(k) - V_x(k)K_b(k) = \tilde{K}_x(k) \quad (83)$$

to agree with (77), and

$$\begin{aligned} V_x(k)[I - K_b(k)S(k)] = A_{k-1}V_x(k-1) &+ B_{k-1} \\ &- K_x(k)S(k). \end{aligned} \quad (84)$$

Equation (83) is satisfied by virtue of (72), and (84) then becomes

$$V_x(k) = A_{k-1}V_x(k-1) + B_{k-1} - \tilde{K}_x(k)S(k). \quad (85)$$

But $A_{k-1}V_x(k-1) + B_{k-1} = U_x(k)$ by (64), so that (85) becomes

$$V_x(k) = U_x(k) - \tilde{K}_x(k)S(k)$$

which is (62). Hence the desired result, (78), is proved.

Although all the required calculations can be performed by a single computer, this result makes it possible to implement the optimum filter by the use of two separate computers. One computer computes the bias-free estimate $\tilde{x}_k$ and the corresponding bias-free residuals $\tilde{r}_k$. The latter are used as inputs to the second computer to obtain the optimum estimates $\hat{b}_k$ of the bias vector. The computations (62)–(73) are also performed in the second computer. The output of this calculation is $V_x(k)\hat{b}_k$ which is added to $\tilde{x}_k$ to obtain the desired estimate $\hat{x}_k$. A flow chart depicting this calculation is given in Fig. 2.

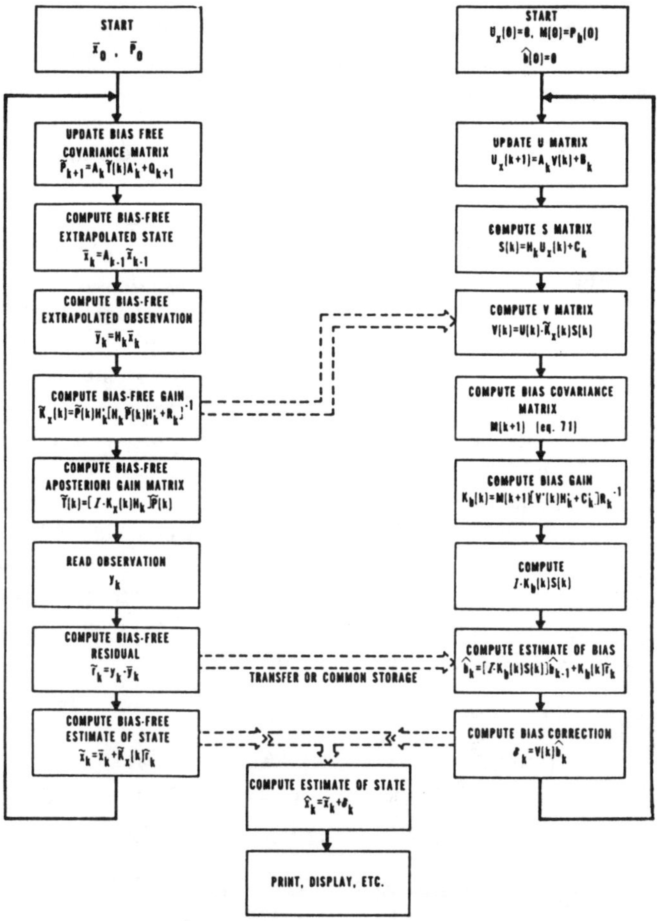

Fig. 2. Flow chart illustrating parallel computation of bias-free estimate and estimate of bias.

IV. Conclusions and Applications

By rearranging the equations of continuous-time and discrete-time Kalman filtering, computational methods have been devised in which the estimate $\tilde{x}$ of the state is computed as if there were no bias present; this estimate is then corrected to account for the bias. The computation of the corrections depend only on the variance of the bias-free estimates. In effect, a single calculation involving large matrices is reduced to a sequence of calculations involving smaller matrices. Common experience with matrix calculations would indicate that such a reduction could mitigate the computational difficulties associated with large matrices.

The decoupling of the bias estimation from the state estimation also makes the effect of the bias on the estimate of x more readily apparent. If, for example, preliminary simulation studies reveal that the difference between

$\tilde{x}$ and $\hat{x}$ is small (relative to $\hat{x}$), then the computation of $\hat{b}$ and V_x can be made approximate, possibly by use of lower precision or by computing $\hat{b}$ and V_x at less frequent intervals than used for computing $\tilde{x}$.

There are many applications in which a large number of bias variables are present. In aided-inertial navigation, for example, models have been developed in which the bias terms which are used in an accurate error model may number as high as 50. In orbit determination, bias terms may include uncertainty (survey error) in station locations, atmospheric density, and the coefficients of the earth's geopotential. The inclusion of all bias terms of this nature has not been possible in previous applications because of the numerical problems which have arisen.

Although these results have been derived for an assumption of constant bias, they can readily be extended to time-varying bias models, i.e., $\dot{b} = Zb$ for continuous time processes, or $b_{n+1} = Z_n b_n$ for discrete-time processes. The extension to randomly varying bias terms, e.g., $\dot{b} = Zb + \delta$, where δ is Gaussian white noise, however, appears to be a more difficult problem.

Appendix

Transformation of the Variance Equation in Discrete-Time Processes

Consider the variance equation

$$P_{k+1} = F_k(P_k - P_kH_k'(H_kP_kH_k' + R_k)^{-1}H_kP_k)F_k' + \Gamma_k, \quad k = 0, 1, 2, \cdots. \tag{86}$$

Let $\tilde{P}_k$ be any solution to (86) with

$$\tilde{P}_0 \neq P_0$$

i.e.,

$$\tilde{P}_{k+1} = F_k(\tilde{P}_k - \tilde{P}_kH_k'(H_k\tilde{P}_kH_k' + R_k)^{-1}H_k\tilde{P}_k)F_k' + \Gamma_k, \quad n = 0, 1, 2, \cdots. \tag{87}$$

We wish to show that

$$P_k = \tilde{P}_k + U_kM_kU_k' \tag{88}$$

where

$$U_{k+1} = F_k(I - \tilde{P}_kH_k'(H_k\tilde{P}_kH_k' + R_k)^{-1}H_k)U_k \tag{89}$$

$$M_{k+1} = M_k - M_kU_k'H_k'(H_k\tilde{P}_kH_k' + R_K + H_kU_kM_kU_k'H_k')^{-1}H_kU_kM_k. \tag{90}$$

For notational simplicity we shall designate all quantities with the index $k + 1$ by a subscript $+$ and all quantities with index k by an unsubscripted symbol, e.g.,

P_{k+1} is written P_+

P_k is written P.

Let

$$Z = P - \tilde{P} = UMU' \tag{91}$$

then

$$Z_+ = P_+ - \tilde{P}_+ = U_+M_+U_+'. \tag{92}$$

Using (86) and (87) the left-hand side of (92) is

$$Z_+ = F\{P - PH'(HPH' + R)^{-1}HP - \tilde{P} + \tilde{P}H'(H\tilde{P}H' + R)^{-1}H\tilde{P}\}F' \tag{93}$$

while the right-hand side of (92), using (89) and (90), is

$$Z_+ = F\{[I - \tilde{P}H'(H\tilde{P}H' + R)^{-1}H]UM_+U'\,[I - H'(H\tilde{P}H' + R)^{-1}H\tilde{P}]\}F'. \tag{94}$$

Hence (88) is established if we can verify that the terms in braces $\{\ \}$ in (93) and (94) are identical. Using (90)

$$\begin{aligned} UM_+U' &= U[M - MU'H'(A^{-1} + HZH')^{-1}HUM]U' \\ &= Z - ZH'(A^{-1} + HZH')^{-1}HZ \end{aligned}$$

where

$$A = (H\tilde{P}H' + R)^{-1}. \tag{95}$$

Hence the identity to be established is

$$\begin{aligned} Z - (\tilde{P} + Z)H'(A^{-1} + HZH')^{-1}H(\tilde{P} + Z) + \tilde{P}H'AH\tilde{P} &= (I - \tilde{P}H'AH)[Z - ZH'(A^{-1} \\ &\quad + HZH')^{-1}HZ](I - H'AH\tilde{P}) \end{aligned} \tag{96}$$

or

$$Z - (\tilde{P} + Z)H'QH(\tilde{P} + Z) + \tilde{P}H'AH\tilde{P} = (I - \tilde{P}H'AH)(Z - ZH'QHZ)(I - H'AH\tilde{P}) \tag{97}$$

where

$$Q = (A^{-1} + HZH')^{-1}.$$

Expansion of the left-hand side (L.H.S.) of (97) gives

$$\text{L.H.S.} = Z - \tilde{P}H'QH\tilde{P} - ZH'QH\tilde{P} - \tilde{P}H'QHZ - ZH'QHZ + \tilde{P}H'AH\tilde{P}$$

and expansion of the right-hand side (R.H.S.) gives

$$\begin{aligned} \text{R.H.S.} = Z &- ZH'QHZ - \tilde{P}H'AHZ + \tilde{P}H'AHZH'QHZ \\ &- ZH'AH\tilde{P} + ZH'QHZH'AH\tilde{P} \\ &+ \tilde{P}H'AHZH'AH\tilde{P} - \tilde{P}H'AHZH'QHZH'AH\tilde{P}. \end{aligned}$$

Upon canceling $Z - ZH'QHZ$ on both sides and then bringing all terms to one side, we obtain from (97)

$$\begin{aligned} &\tilde{P}H'[A - Q - AHZH'A + AHZH'QHZH'A]H\tilde{P} \\ &+ ZH'[-Q + A - QHZH'A]H\tilde{P} \\ &+ \tilde{P}H'\,[-Q + A - AHZH'Q]HZ = 0. \end{aligned}$$

The third term in brackets is zero because

$$\begin{aligned} &A(I - HZH'Q) - Q \\ &= A(I - HZH'(A^{-1} + HZH')^{-1}] - Q \\ &= A[HZH' + A^{-1} - HZH'](A^{-1} + HZH')^{-1} - Q \\ &= AA^{-1}Q - Q = 0. \end{aligned}$$

Similarly the second term in brackets is zero. Also since, as a result

$$A - Q = QHZH'A$$

the first term in brackets is thus

$$QHZH'A - AHZH'A + AHZH'QHZH'A$$
$$= (Q - A)HZH'A + AHZH'QHZH'A$$
$$= (Q - A + AHZH'Q)HZH'A = 0.$$

Hence the required identity is proved.

References

[1] R. E. Kalman, "A new approach to linear filtering and prediction problems," *Trans. ASME, J. Basic Engrg.*, ser. D, vol. 82, pp. 35–45, March 1960.

[2] R. E. Kalman and R. S. Bucy, "New results in linear filtering and prediction theory," *Trans. ASME, J. Basic Engrg.*, ser. D, vol. 83, pp. 95–108, March 1961.

[3] J. R. Huddle and D. A. Wismer, "Degradation of linear filter performance due to modeling error," *IEEE Trans. Automatic Control* (Short Papers), vol. AC-13, pp. 421–423, August 1968.

[4] B. Friedland, "On solutions of the Riccati equation in optimization problems," *IEEE Trans. Automatic Control* (Short Papers), vol. AC-12, pp. 303–304, June 1967.

[5] R. E. Kalman, "New methods in Wiener filtering theory," in *Proc. 1st Symp. on Engineering Applications of Random Function Theory and Probability*, J. L. Bogdanoff and F. Kozin, Eds. New York: Wiley, 1963, pp. 270–388.

Discrete Square Root Filtering: A Survey of Current Techniques

PAUL G. KAMINSKI, MEMBER, IEEE, ARTHUR E. BRYSON, JR., AND
STANLEY F. SCHMIDT, MEMBER, IEEE

Abstract—The conventional Kalman approach to discrete filtering involves propagation of a state estimate and an error covariance matrix from stage to stage. Alternate recursive relationships have been developed to propagate a state estimate and a square root error covariance instead. Although equivalent algebraically to the conventional approach, the square root filters exhibit improved numerical characteristics, particularly in ill-conditioned problems.

In this paper, current techniques in square root filtering are surveyed and related by applying a duality association. Four efficient square root implementations are suggested, and compared with three common conventional implementations in terms of computational complexity and precision. The square root computational burden should not exceed the conventional by more than 50 percent in most practical problems. An examination of numerical conditioning predicts that the square root approach can yield twice the effective precision of the conventional filter in ill-conditioned problems. This prediction is verified in two examples.

The excellent numerical characteristics and reasonable computation requirements of the square root approach make it a viable alternative to the conventional filter in many applications, particularly when computer word length is limited, or the estimation problem is badly conditioned.

I. Introduction

In A significant class of filtering problems, propagation of the error covariance matrix by means of the Kalman filter equations [1] results in a matrix which is not positive semidefinite—a theoretical impossibility. This may occur when 1) the covariance matrix is rapidly reduced by processing very accurate measurements, 2) a linear combination of state vector components is known with great precision, while other combinations are essentially unobservable. The source of trouble in both cases is numerical computation of ill-conditioned quantities in finite word length.

Manuscript received July 12, 1971. Paper recommended by D. G. Luenberger, Associate Guest Editor. This research was sponsored in part by Research Grant NASA-NgL-05-020-007, a National Science Foundation graduate fellowship, and the U.S. Air Force. The views expressed herein are the authors' and do not necessarily reflect those of Air University, the U.S. Air Force, or the Department of Defense.

P. G. Kaminski is with the U.S. Air Force Space and Missile Systems Organization, El Segundo, Calif.

A. E. Bryson, Jr., is with the Department of Aeronautics and Astronautics, Stanford University, Stanford, Calif.

S. F. Schmidt is with Analytical Mechanics Associates, Inc., Mountain View, Calif.

To circumvent this difficulty, Potter [2] gave a method for propagating the error covariance matrix in a square root form in the absence of process noise. This method is completely successful in maintaining the positive semidefinite nature of the error covariance, and it can provide twice the effective precision of the conventional filter implementation in ill-conditioned problems. The outstanding numerical characteristics and relative simplicity of this Potter square root approach led to its implementation in the Apollo navigation filters [3].

Extensions of the Potter square root approach, and the development of an information square root filter have provided several recursive square root solutions to the discrete filtering problem. The purpose of this paper is to review and relate these square root filters, summarize the most promising computational approaches, and compare these with the conventional approach in terms of computational complexity and precision.

II. Conventional Filtering

The following notation is used to describe the discrete time problem:

$$\text{process: } \boldsymbol{x}(k+1) = \boldsymbol{\Phi}(k)\boldsymbol{x}(k) + \boldsymbol{B}(k)\boldsymbol{\xi}(k) \tag{1}$$

$$\text{observation: } \boldsymbol{z}(k) = \boldsymbol{C}(k)\boldsymbol{x}(k) + \boldsymbol{\theta}(k) \tag{2}$$

Reprinted from *IEEE Trans. Automat. Contr.*, vol. AC-16, pp. 727–735, Dec. 1971.

where $\xi(k) \in R_m$ and $\theta(k) \in R_r$ are independent purely random Gaussian sequences with zero mean and covariance $\Xi(k)$ and $\Theta(k)$, respectively.

The conventional filter equations for this process [1] can be summarized in the following covariance form.

Covariance Filter

Time Update:

$$\hat{x}(k+1) = \Phi(k)\hat{x}_+(k) \tag{3}$$

$$\Sigma(k+1) = \Phi(k)\Sigma_+(k)\Phi^T(k) + B(k)\Xi(k)B^T(k). \tag{4}$$

Measurement Update:

$$\hat{x}_+(k) = \hat{x}(k) + K(k)[z(k) - C(k)\hat{x}(k)] \tag{5}$$

$$\Sigma_+(k) = [I - K(k)C(k)]\Sigma(k) \tag{6}$$

$$K(k) = \Sigma(k)C^T(k)[\Theta(k) + C(k)\Sigma(k)C^T(k)]^{-1}. \tag{7}$$

The conventional filter may also be implemented [4] to propagate Σ^{-1}, the information matrix, accentuating the recursive least squares nature of filtering.

Information Filter[1]

Time Update:

$$d(k+1) = [I - L(k)B^T(k)]\Phi^{-T}(k)d_+(k) \tag{8}$$

$$\Sigma^{-1}(k+1) = [I - L(k)B^T(k)]F(k) \tag{9}$$

$$F(k) = \Phi^{-T}(k)\Sigma_+^{-1}(k)\Phi^{-1}(k) \tag{10}$$

$$L = FB[\Xi^{-1} + B^TFB]^{-1}. \tag{11}$$

Measurement Update:

$$d_+(k) = d(k) + C^T(k)\theta^{-1}(k)z(k) \tag{12}$$

$$\Sigma_+^{-1}(k) = \Sigma^{-1}(k) + C^T(k)\theta^{-1}(k)C(k) \tag{13}$$

where

$$d(k) \triangleq \Sigma^{-1}(k)\hat{X}(k) \tag{14}$$

$$d_+(k) \triangleq \Sigma_+^{-1}(k)\hat{X}_+(k). \tag{15}$$

Although the covariance and information implementations of the filter are algebraically equivalent, the numerical properties of the two forms may differ substantially. The suitability of one form or the other to a particular problem depends upon the nature of the *a priori* information, the available computational facilities, and the relative size of m and r, the dimensions of ξ and z. The covariance filter is the more popular form, probably due to its relative computational simplicity when $r \ll n$. However, the information filter has been successfully applied when either very poor or no *a priori* information leads to starting difficulties in the covariance filter. There is no problem in initializing the information filter in the absence of *a priori* information, as $\Sigma^{-1}(0)$ and $d(0)$ are set to zero, then $\Sigma^{-1}(k)$ and $d(k)$ are propagated from measurement to measurement until Σ^{-1} can be inverted (it can always be pseudo-inverted) to find $\hat{x}$.

[1] The notation $(\ )^{-T}$ implies $[(\ )^{-1}]^T$, which is equivalent to $[(\ )^T]^{-1}$.

TABLE I

DUAL ASSOCIATIONS

Time Update	Measurement Update
$\Sigma(k+1)$	$\Sigma_+^{-1}(k)$
$\Phi(k)\Sigma_+(k)\Phi^T(k)$	$\Sigma^{-1}(k)$
$B(k)$	$C^T(k)$
$\Xi(k)$	$\Theta^{-1}(k)$

Note the similarity between the covariance time update equation (4) and the information measurement update equation (13). The same similarity of form is observed in (6) and (9), indicating that a time update of Σ and a measurement update of Σ^{-1} are equivalent using the duality associations in Table I.

Note that a time update of $\hat{x}_+$ and a measurement update of d are not quite equivalent, as (3) and (12) have slightly different forms. Thus the dual associations in Table I are useful only in relating covariance and information matrix updates.

III. Matrix Square Roots

The basic idea of matrix square roots can be traced back to Cholesky and Banachiewicz [5], who gave independent but equivalent algorithms for factoring any positive semidefinite symmetric (PSDS) matrix into the product of a lower triangular matrix (zeros above the diagonal) and its transpose. This Cholesky decomposition algorithm, given in Appendix I, was developed to solve the equation $Ax = b$ (with A = PSDS matrix), and is still recognized as the best numerical approach to the problem. More recently, Schmidt [6] independently derived an equivalent decomposition algorithm using the covariance measurement update (6) in a factored form.

The covariance square roots S, S_+, U, and V are defined by

$$\Sigma \triangleq SS^T \tag{16}$$

$$\Sigma_+ \triangleq S_+S_+^T \tag{17}$$

$$\Xi \triangleq UU^T \tag{18}$$

$$\Theta \triangleq VV^T. \tag{19}$$

The covariance square roots are not uniquely determined by these relationships. This lack of uniqueness is not generally bothersome, as a unique square root may be defined by Cholesky decomposition. In this paper, a Cholesky decomposition of Σ is denoted by writing $\Sigma = \Sigma^{1/2}\Sigma^{T/2}$, with the superscript 1/2 reserved for a lower triangular Cholesky square root.

The basic idea of the square root filter is replacement of the covariance matrix by the square root covariance, then replacing (4) and (6) by equivalent relationships for propagating S. This approach is motivated by two considerations: 1) the product SS^T can never be indefinite, even in the presence of roundoff errors, while roundoff errors sometimes cause the computed value of Σ to be indefinite; 2) the numerical conditioning of S is generally much better than that of Σ.

A quantitative illustration of this improvement can be obtained by examining the condition number [7] $K(A)$ defined by

$$K(A) = \sigma_1/\sigma_n \tag{20}$$

where σ_1^2 is the maximum eigenvalue of A^TA and σ_n^2 is the minimum eigenvalue of A^TA.

This condition number, often used to analyze the effect of perturbations in linear equations, is relevant to analysis of numerical operations with the covariance matrix. When computing in base 10 arithmetic with p significant digits, numerical difficulties may be expected as $K(\Sigma)$ approaches 10^p. The advantage of the square root approach becomes obvious by relating $K(\Sigma)$ and $K(S)$.

$$K(\Sigma) = K(SS^T) = [K(S)]^2. \tag{21}$$

Thus the condition number of S is the square root of $K(\Sigma)$, and while numerical operations with Σ may encounter difficulties when $K(\Sigma) = 10^p$, the square root filter should function until $K(\Sigma) = 10^{2p}$. This effective double precision of the square root filter in ill-conditioned problems is illustrated explicitly by the example problems in Section VI.

IV. Development of Square Root Filtering

Covariance Square Root Filter

Potter [2] gave a square root filter implementation for space navigation applications in which Ξ was essentially zero. Using (3) and (4) with $\Xi = 0$, the time update is obtained with

$$\hat{x}(k+1) = \Phi(k)\hat{x}_+(k) \tag{22}$$

$$S(k+1) = \Phi(k)S_+(k). \tag{23}$$

To obtain the measurement update, Potter limited his attention to scalar measurements, for which (6) may be written as

$$\Sigma_+ = S_+S_+^T = \Sigma - KC\Sigma = S[I - aFF^T]S^T \tag{24}$$

with

$$F = S^TC^T \tag{25}$$

$$1/a = F^TF + \Theta. \tag{26}$$

Potter showed that, by proper choice of the constant γ, the bracketed term in (24) can be factored as

$$[I - aFF^T] = [I - a\gamma FF^T][I - a\gamma FF^T]. \tag{27}$$

Multiplication in (27) and solving the resultant quadratic for γ yields

$$\gamma = 1/(1 \pm \sqrt{a\Theta}). \tag{28}$$

Using (27) in (24) gives the square root covariance update

$$S_+ = S - a\gamma SFF^T \tag{29}$$

and $\hat{x}_+$ is computed using (5) with

$$K = aSF. \tag{30}$$

Bellantoni and Dodge [8] extended Potter's results to handle a vector measurement. This extension requires an eigenvalue decomposition of an $n \times n$ matrix with r nonzero eigenvalues. The technique is effective, but becomes inefficient when $r \ll n$, as is often the case. Subsequently Andrews [9] showed that Potter's algorithm can always be applied r times to process an r-vector measurement after first diagonalizing Θ. Andrews also gave a new procedure for handling a vector measurement without diagonalizing Θ. This procedure is summarized in (31)–(34), and may be verified by multiplication of (33) by its transpose and comparing the result with (6).

$$F = S^TC^T \tag{31}$$

$$G = [\Theta + F^TF]^{1/2} \tag{32}$$

$$S_+ = S - SFG^{-T}(G + V)^{-1}F^T \tag{33}$$

$$\hat{x}_+ = \hat{x} + SFG^{-T}G^{-1}(z - C\hat{x}). \tag{34}$$

This procedure is more efficient than the Bellantoni and Dodge extension when $r \ll n$, as it requires only a Cholesky decomposition of an $r \times r$ matrix, plus inversion of two triangular $r \times r$ matrices.

To make the Potter algorithm applicable to a wider class of problems, it was extended to include process noise. Andrews [9] gave a differential equation for propagating S between measurements, including the effect of process noise. These results include an algorithm which keeps S in triangular form by addition of a constructed $n \times n$ skew symmetric matrix. The triangular form of S facilitates the computation of S^{-1} which is required to evaluate $\dot{S}$.

Schmidt [6] showed that process noise can be handled by constructing an $(n+m) \times (n+m)$ orthogonal matrix T such that

$$T\underbrace{\begin{bmatrix} S_+^T(k) \\ \hdashline U^T(k)B^T(k) \end{bmatrix}}_{n}\begin{matrix} \}n \\ \}m \end{matrix} = \begin{bmatrix} S^T(k+1) \\ \hdashline 0 \end{bmatrix}. \tag{35}$$

To verify (35), multiply each side on the left by its transpose and compare the result with (4). Choosing the appropriate T may be interpreted as "completing the square" in a matrix sense. Schmidt derived an algorithm for constructing T which is equivalent to the modified Gram–Schmidt algorithm [10] given in Appendix II.

Information Square Root Filter

Golub [11] and Businger and Golub [12] showed that an efficient matrix triangularization procedure[2] developed by Householder [7] for matrix inversion could be applied to obtain a square root solution to the least squares problem. The numerical superiority of the square root approach has been validated experimentally by Jordan [13], and

[2] See Appendix II for a summary of this procedure.

this technique is generally regarded as the best numerical approach for solving least squares problems.

Hanson and Lawson [14] extended the work of Golub and Businger to rank deficient systems, and adapted the Householder algorithm to sequential least squares. Using the dynamic programming formulation of sequential estimation given by Cox [15], Dyer and McReynolds [16] applied the Householder algorithm to obtain both measurement and time updates of the information square root.

To briefly outline these results, consider the cost function $J(k)$ which is relevant to a measurement update [15].

$$J(k) = \|x(k) - \hat{x}(k)\|^2_{\Sigma^{-1}(k)} + \|z(k) - C(k)x(k)\|^2_{\Theta^{-1}(k)}. \tag{36}$$

To find the minimizing value of $x(k)$, denoted by $\hat{x}(k)$, (36) is first placed in a more convenient form using the square root covariances, and the properties of the norm:

$$J(k) = \left\| \begin{matrix} n\{ \\ r\{ \end{matrix} \underbrace{\begin{bmatrix} S^{-1}(k) \\ \hdashline V^{-1}(k)C(k) \end{bmatrix}}_{n} x(k) - \begin{bmatrix} b(k) \\ \hdashline V^{-1}(k)z(k) \end{bmatrix} \right\|^2 \tag{37}$$

where

$$b(k) \triangleq S^{-1}(k)\hat{x}(k). \tag{38}$$

Note that $b(k)$ is analogous to $d(k)$ in (14). The Householder algorithm is applied to construct an orthogonal T such that

$$T\begin{bmatrix} S^{-1}(k) \\ \hdashline V^{-1}(k)C(k) \end{bmatrix} = \begin{bmatrix} S_+^{-1}(k) \\ \hdashline 0 \end{bmatrix}. \tag{39}$$

Using b_+ and e as defined in (40),

$$T\begin{bmatrix} b(k) \\ \hdashline V^{-1}(k)z(k) \end{bmatrix} = \begin{bmatrix} b_+(k) \\ \hdashline e(k) \end{bmatrix} \tag{40}$$

(37) can be written

$$J(k) = \|S_+^{-1}(k)x(k) - b_+(k)\|^2 + \|e(k)\|^2. \tag{41}$$

Thus $e(k)$ is the residual after processing the measurement and $\hat{x}_+(k)$ is the solution to

$$S_+^{-1}\hat{x}_+(k) = b_+(k). \tag{42}$$

Using the same approach, Dyer and McReynolds showed that $\bar{J}(k+1)$, the cost at the $(k+1)$th stage before incorporating the $(k+1)$th measurement, can be written as

$$\bar{J}(k+1) = \left\| \begin{matrix} m\{ \\ n\{ \end{matrix} \begin{bmatrix} \underbrace{\begin{matrix} U^{-1}(k) \\ \hdashline S_+^{-1}(k)\Phi^{-1}(k)B(k) \end{matrix}}_{m} & \vdots & \underbrace{\begin{matrix} 0 \\ \hdashline S_+^{-1}(k)\Phi^{-1}(k) \end{matrix}}_{n} \end{bmatrix} \begin{bmatrix} \xi(k) \\ \hdashline x(k+1) \end{bmatrix} - \begin{bmatrix} 0 \\ \hdashline b_+(k) \end{bmatrix} \right\|^2 + \|e(k)\|^2. \tag{43}$$

Thus if an $(n+m) \times (n+m)$ orthogonal matrix T is constructed such that

$$T\begin{bmatrix} U^{-1}(k) & \vdots & 0 \\ \hdashline S_+^{-1}(k)\Phi^{-1}(k)B(k) & \vdots & S_+^{-1}(k)\Phi^{-1}(k) \end{bmatrix} = \begin{bmatrix} F(k+1) & \vdots & G(k+1) \\ \hdashline 0 & \vdots & S^{-1}(k+1) \end{bmatrix} \tag{44}$$

and

$$\begin{bmatrix} a(k+1) \\ \hdashline b(k+1) \end{bmatrix} \triangleq T\begin{bmatrix} 0 \\ \hdashline b_+(k) \end{bmatrix} \tag{45}$$

(43) may be written

$$\bar{J}(k+1) = \|F(k+1)\xi(k) + G(k+1)x(k+1) - a(k+1)\|^2 + \|S^{-1}(k+1)x(k+1) - b(k+1)\|^2 + \|e(k)\|^2. \tag{46}$$

The first norm in (46) is zeroed by choosing

$$\xi(k) = F^{-1}(k+1)[a(k+1) - G(k+1)x(k+1)]. \tag{47}$$

Equation (47) yields the smoothed estimate for $\xi(k)$ when $x(k+1)$ is replaced by its smoothed estimate. The second norm in (46) is zeroed with

$$S^{-1}(k+1)\hat{x}(k+1) = b(k+1). \tag{48}$$

Dyer and McReynolds showed that $S^{-1}(k+1)$ can also be obtained using Potter's algorithm instead of (44). This idea relates to the duality associations given in Table I.

V. Suggested Square Root Filter Implementations

The square root update procedures described in Section IV may be tied together by using the duality relationships given in Table I, but expressed now in terms of the square root covariances given in Table II.

The associations in Table II may be interpreted in the following way: given a procedure to compute $S(k+1)$, that same procedure may be used to compute $S_+^{-T}(k)$ if $\Phi(k)S(k)$ is replaced by $S^{-T}(k)$, $B(k)$ by $C^T(k)$, etc. For example, (35) which computes $S^T(k+1)$ is the dual form of (39) which computes $S_+^{-1}(k)$. This duality association implies that Potter's algorithm may be used to compute $S^{-1}(k+1)$, as was noted by Dyer and McReynolds [16], and it also implies that the procedure given by Dyer and McReynolds to compute $S^{-1}(k+1)$ can be used to form a new measurement update procedure.

When performing measurement updates, it is *always* possible to process an r-vector measurement as r scalar measurements. When Θ is diagonal, the components of z can be treated as independent measurements and processed one at a time. When Θ is not diagonal, use the following procedure [6], [16]. Compute the Cholesky decomposition

$$\Theta = \Theta^{1/2}\Theta^{T/2} \tag{49}$$

and solve for z^* and C^* using back substitution ($\Theta^{1/2}$ is lower triangular), where

TABLE II
SQUARE ROOT DUAL ASSOCIATIONS

Time Update	Measurement Update
$S(k+1)$	$S_+^{-T}(k)$
$\Phi(k)S(k)$	$S^{-T}(k)$
$B(k)$	$C^T(k)$
$U(k)$	$V^{-T}(k)$
$\Xi(k)$	$\Theta^{-1}(k)$

$$\Theta^{1/2}z^* = z \tag{50}$$

$$\Theta^{1/2}C^* = C. \tag{51}$$

This transformation yields a new process noise Θ^* with unit covariance. Thus the components of z^* may be processed one at a time. Note that the triangular form of $\Theta^{1/2}$ is such that z_i^*, the ith component of z^*, is simply a linear combination of the first i components of z. This transformation procedure appears more attractive than the vector measurement extensions of either Bellantoni and Dodge [8], or Andrews [9], as it is more efficient and appears to be less sensitive to badly conditioned measurements.

The following square root filter implementations represent a combination of the techniques discussed in Section IV, with some minor extension. Note that scalar updates are suggested in many implementations, as this improves both the numerical characteristics and the efficiency of the procedure. Vector updates are suggested in (52)–(53) and (70)–(71) to improve efficiency without degrading numerical characteristics.

Covariance Square Root Filter

Time Update: Use a single vector update.

$$\hat{x}(k+1) = \Phi(k)\hat{x}_+(k) \tag{52}$$

$$\begin{matrix} n\{ \\ m\{ \end{matrix}\underbrace{\begin{bmatrix} S^T(k+1) \\ \hdashline 0 \end{bmatrix}}_{n} = T\begin{bmatrix} S_+^T(k)\Phi^T(k) \\ \hdashline U^T(k)B^T(k) \end{bmatrix} \tag{53}$$

(see Appendix II for construction of T).

Measurement Update: Use r scalar updates.

Implementation I:

$$\hat{x}_+(k) = \hat{x}(k) + K[z(k) - C(k)\hat{x}(k)] \tag{54}$$

$$S_+(k) = S(k) - \gamma K F^T \tag{55}$$

$$K = aS(k)F \tag{56}$$

$$F = S^T(k)C^T(k) \tag{57}$$

$$1/a = F^TF + \Theta(k) \tag{58}$$

$$\gamma = 1/(1 + \sqrt{a\Theta(k)}). \tag{59}$$

Implementation II:

$$\hat{x}_+(k) = \hat{x}(k) + (G^T/F)[z(k) - C(k)\hat{x}(k)] \tag{60}$$

$$\begin{bmatrix} F & G \\ \hdashline 0 & S_+^T(k) \end{bmatrix}\begin{matrix} \}1 \\ \}n \end{matrix} = T\begin{bmatrix} V(k) & 0 \\ \hdashline S^T(k)C^T(k) & S^T(k) \end{bmatrix} \tag{61}$$

(column widths: 1, n)

See Appendix II for construction of T. Note that T can be constructed by passing through (83)–(89) only one time.

Information Square Root Filter

Time Update: Use m scalar updates.[3]

Implementation I:

$$b(k+1) = b_+(k) - a\gamma FF^Tb_+(k) \tag{62}$$

$$S^{-1}(k+1) = S_+^{-1}(k)\Phi^{-1}(k) - \gamma FL \tag{63}$$

$$L = aF^TS_+^{-1}(k)\Phi^{-1}(k) \tag{64}$$

$$F = S_+^{-1}(k)\Phi^{-1}(k)B(k) \tag{65}$$

$$1/a = F^TF + 1/\Xi(k) \tag{66}$$

$$\gamma = 1/(1 + \sqrt{a/\Xi(k)}). \tag{67}$$

Implementation II[4]*:*

$$\begin{matrix} 1\{ \\ n\{ \end{matrix}\begin{bmatrix} F & G \\ \hdashline 0 & S^{-1}(k+1) \end{bmatrix} = T\begin{bmatrix} 1/U(k) & 0 \\ \hdashline S_+^{-1}(k)\Phi^{-1}(k)B(k) & S_+^{-1}(k)\Phi^{-1}(k) \end{bmatrix} \tag{68}$$

(column widths: 1, n)

$$\begin{matrix} 1\{ \\ n\{ \end{matrix}\begin{bmatrix} a(k+1) \\ \hdashline b(k+1) \end{bmatrix} = T\begin{bmatrix} 0 \\ \hdashline b_+(k) \end{bmatrix}. \tag{69}$$

Measurement Update: Use a single vector update.

$$\begin{matrix} n\{ \\ r\{ \end{matrix}\underbrace{\begin{bmatrix} S_+^{-1}(k) \\ \hdashline 0 \end{bmatrix}}_{n} = T\begin{bmatrix} S^{-1}(k) \\ \hdashline V^{-1}(k)C(k) \end{bmatrix} \tag{70}$$

$$\begin{matrix} n\{ \\ r\{ \end{matrix}\begin{bmatrix} b_+(k) \\ \hdashline e(k) \end{bmatrix} = T\begin{bmatrix} b(k) \\ \hdashline V^{-1}(k)z(k) \end{bmatrix}. \tag{71}$$

Computational Complexity

The minimum number of operations required for one time update and one measurement update have been estimated for several filter implementations and summarized in Table III. In computing these estimates, all implementations are assumed to take advantage of matrix symmetry and zeros as they appear in general forms. In ordering computations to minimize the number of operations, r and m are assumed to be less than n, and both Ξ and Θ are treated as diagonal matrices. For square root procedures requiring a transformation to a triangular form, the computational requirements of the Householder algo-

[3] After the first of m updates, Φ is replaced by I.
[4] See comments in Covariance Implementation II.

TABLE III

NUMBER OF OPERATIONS FOR ONE TIME UPDATE AND ONE MEASUREMENT UPDATE

Filter Implementation	Square Roots	Divides	Multiplies (divide entry below by 6)	Adds (divide entry below by 6)
Square Root Covariance I	$n + r$	$n + 2r$	$10n^3 + 3n^2(6r + 2m + 2) + n(24r + 6m + 8) + 12r$	$10n^3 + 3n^2(6r + 2m - 1) + n(6r + 5) + 6r$
Square Root Covariance II	$n + r$	$n + 2r$	$10n^3 + 3n^2(6r + 2m + 2) + n(30r + 6m + 8) + 12r$	$10n^3 + 3n^2(6r + 2m - 1) + n(6r + 5) + 12r$
Conventional Covariance	0	r	$9n^3 + 3n^2(3r + m + 3) + n(27r + 9m)$	$9n^3 + 3n^2(3r + m - 1) + n(15r + 3m - 6)$
Joseph Covariance	r	$2r - 1$	$18n^3 + 3n^2(5r + m + 4) + n(9r^2 + 24r + 9m) + 3r^3 + 9r^2 - 6r$	$18n^3 + 3n^2(5r + m - 10) + n(9r^2 + 6r + 3m) + 3r^3 - 6r^2 + 3r$
Square Root Information I	$n + m$	$2(n + m)$	$7n^3 + 3n^2(2r + 6m + 6) + n(12r + 24m + 11) + 6m$	$7n^3 + 3n^2(2r + 6m + 2) + n(12r + 6m + 5) + 6m$
Square Root Information II	$n + m$	$2n + m$	$7n^3 + 3n^2(2r + 6m + 6) + n(12r + 30m + 11)$	$7n^3 + 3n^2(2r + 6m + 2) + n(12r + 6m + 5) + 6m$
Conventional Information	n	$2n + m - 1$	$10n^3 + 3n^2(r + 3m + 6) + n(15r + 21m - 10)$	$10n^3 + 3n^2(r + 3m + 2) + n(9r + 9m - 16)$

rithm are used. In all information implementations, assume that $\hat{x}_+$ is computed by solving

$$\boldsymbol{\Sigma}^{-1}\hat{x}_+ = d_+ \tag{72}$$

or its square root equivalent after performing the measurement update.

The summarized results in Table III show that the square root and conventional implementations have about the same complexity when both r and m are much less than n. Where both r and m approach n, the square root burden slightly exceeds $1^1/_2$ times the conventional. Note that the Joseph implementation, which computes $\boldsymbol{\Sigma}_+$ using the improved computational form

$$\boldsymbol{\Sigma}_+ = (I - KC)\boldsymbol{\Sigma}(I - KC)^T + K\Theta K^T \tag{73}$$

requires about twice the computation of the conventional covariance filter for any r and m less than n. Implementations I and II of the square root filters have essentially the same computation requirements, and the choice of implementation should be influenced primarily by the compatibility of time and measurement update procedures for the particular problem.

To put these results in a computation time perspective, consider the case with $n = 10$, $m = 5$, $r = 1$. If subscript operations and storage transfer times are neglected, the run time of the various filter implementations can be compared using the following single precision instruction times which are approximate values for the IBM 360 Model 67-1.

add: 2.7 μs   multiply: 4.1 μs
divide: 6.6 μs   square root: 60 μs.

The approximate run times given in Table IV indicate that the square root implementations require less than 30 percent more computation time than the conventional implementation in either the covariance or information filters. When filtering housekeeping chores, such as computing $\boldsymbol{\Phi}$, C, and manipulating input and output data are considered, this 30 percent time difference may become negligible.

TABLE IV

APPROXIMATE RUN TIME FOR ONE TIME AND ONE MEASUREMENT UPDATE

Filter Implementation	Time (ms)
Square Root Covariance I	18
Square Root Covariance II	18
Conventional Covariance	14
Joseph Covariance	24
Square Root Information I	23
Square Root Information II	23
Conventional Information	20

VI. Example Problems

To illustrate and compare the performance of the square root and conventional filter implementations, two simple example problems have been constructed.

Problem 1: Initialization ill-conditioning.

Given:

$$\boldsymbol{\Sigma} = \begin{bmatrix} 1 & 0 \\ 0 & 1 \end{bmatrix} \qquad C = [1, 0] \qquad \boldsymbol{\Theta} = e^2$$

where $e \ll 1$, and to simulate roundoff assume $1 + e \neq 1$ but $1 + e^2 \stackrel{r}{=} 1$.

Find: $\boldsymbol{\Sigma}_+$.

This problem illustrates the initialization problems that result when $C\boldsymbol{\Sigma}C^T + \boldsymbol{\Theta}$ is rounded to $C\boldsymbol{\Sigma}C^T$. The exact answer, and the rounded answer using $1 + e^2 \stackrel{r}{=} 1$ in all calculations, are given for various filter implementations in Table V. Note that all but the conventional covariance implementation give a nonsingular and nearly exact answer. Although the difference between the zero in the diagonal element of the conventional covariance result

TABLE V
COMPARISON OF ROUNDED SOLUTIONS TO PROBLEM 1

Filter Implementation	Rounded Solution
Exact Answer	$\Sigma_+ = \begin{bmatrix} \frac{e^2}{1+e^2} & 0 \\ 0 & 1 \end{bmatrix}$
Conventional Covariance	$\Sigma_+ \stackrel{r}{=} \begin{bmatrix} 0 & 0 \\ 0 & 1 \end{bmatrix}$
Joseph Covariance	$\Sigma_+ \stackrel{r}{=} \begin{bmatrix} e^2 & 0 \\ 0 & 1 \end{bmatrix}$
Square Root Covariance I or II	$S_+ \stackrel{r}{=} \begin{bmatrix} e & 0 \\ 0 & 1 \end{bmatrix}$
Conventional Information	$\Sigma_+^{-1} \stackrel{r}{=} \begin{bmatrix} 1/e^2 & 0 \\ 0 & 1 \end{bmatrix}$
Square Root Covariance I or II	$S_+^{-1} \stackrel{r}{=} \begin{bmatrix} 1/e & 0 \\ 0 & 1 \end{bmatrix}$

TABLE VI
COMPARISON OF ROUNDED SOLUTIONS TO PROBLEM 2

Filter Implementation	Rounded Solution
Exact Answer	$\Sigma_+ = \frac{1}{2+e^2} \begin{bmatrix} 1+e^2 & -1 \\ -1 & 1+e^2 \end{bmatrix}$
Conventional Covariance	$\Sigma_+ \stackrel{r}{=} \frac{1}{2} \begin{bmatrix} 1 & -1 \\ -1 & 1 \end{bmatrix}$
Joseph Covariance	$\Sigma_+ \stackrel{r}{=} \frac{1}{2} \begin{bmatrix} 1 & -1 \\ -1 & 1 \end{bmatrix}$
Square Root Covariance I or II	$S_+ \stackrel{r}{=} \frac{1}{2} \begin{bmatrix} 1+e/\sqrt{2} & e/\sqrt{2}-1 \\ e/\sqrt{2}-1 & 1+e/\sqrt{2} \end{bmatrix}$
Conventional Information	$\Sigma_+^{-1} \stackrel{r}{=} 1/e^2 \begin{bmatrix} 1 & 1 \\ 1 & 1 \end{bmatrix}$
Square Root Information I or II	$S_+^{-1} \stackrel{r}{=} 1/e \begin{bmatrix} -1 & -1 \\ 0 & e/\sqrt{2} \end{bmatrix}$

and e^2 may seem unimportant, note that the gain K computed in the conventional covariance filter will be zero, while the correct gain for a second measurement of the same type should be

$$K \cong \frac{1}{2}\begin{bmatrix} 1 \\ 0 \end{bmatrix}.$$

Thus, divergence in the conventional covariance filter might be anticipated until Σ_+ becomes nonsingular due to addition of process noise.

Problem 2: Direction ill-conditioning.

Given: same as Problem 1, but $C = [1, 1]$.

Find: Σ_+.

The exact solution and the rounded solutions are summarized in Table VI. With this more general type of ill-conditioning, only the square root implementations give nonsingular results. Just as in Problem 1, a singular Σ_+ will lead to a zero gain if a second measurement of the same type is processed, while the square root filters will compute a gain for the second measurement which is nearly exact.

If S_+ is used to compute Σ_+, the result is

$$\Sigma_+ = S_+S_+^T = \frac{1}{2}\begin{bmatrix} 1+\frac{e^2}{2} & -1+\frac{e^2}{2} \\ -1+\frac{e^2}{2} & 1+\frac{e^2}{2} \end{bmatrix} \stackrel{r}{=} \frac{1}{2}\begin{bmatrix} 1 & -1 \\ -1 & 1 \end{bmatrix}.$$

Thus even though S_+ is nonsingular, the associated value of Σ_+ would be rounded to a singular matrix in storage. This is a clear illustration of the improvement in numerical conditioning provided by the square root approach.

Although these example problems are trivial, they provide an accurate representation of the numerical characteristics associated with specific filter implementations. The direction ill-conditioning illustrated in Problem 2 is generally the more serious and insidious problem, as it can occur not only when one accurate measurement is processed, but also after processing, in a relatively short time compared to system time constants, many less accurate measurements taken in approximately the same state space direction. For a confirmation of these characteristics in actual numerical computation see Bellatoni and Dodge [8], Dyer and McReynolds [16], and Kaminski [17].

VII. SUMMARY AND CONCLUSIONS

Current techniques in discrete square-root filtering have been surveyed, and a measurement-time update duality association was applied to relate these techniques and obtain a new measurement update procedure. Four square-root filter implementations were suggested, two of the covariance and two of the information type. The computational complexity of these four square root and three conventional filters was compared, indicating that the square root computational burden should not exceed the conventional by more than 50 percent in most practical problems.

An investigation of numerical conditioning predicted that a single precision square root filter would provide the accuracy equivalent of a double precision conventional filter in ill-conditioned problems. This prediction was verified by filter performance comparison in two ill-conditioned example problems.

The combination of excellent numerical characteristics and only moderate computational complexity should make fixed point square root filtering a viable alternative to floating point conventional filtering in many aerospace-borne computers. Double precision square root filtering can provide a substantial improvement in large-scale data reduction problems which tend to be ill conditioned. The square root approach can extend the effective numerical observability in these problems, providing meaningful solutions beyond the point where double precision conventional algorithms fail. The basic square root filter procedures may also be applied [17] to obtain numerically improved solutions to the smoothing problem.

APPENDIX I

CHOLESKY DECOMPOSITION

Any symmetric positive semidefinite $n \times n$ matrix P may be written in the factored form:

$$\begin{bmatrix} P_{11} & P_{12} & \cdots & P_{1n} \\ P_{12} & P_{22} & & \\ \cdot & & & \cdot \\ \cdot & & & \cdot \\ \cdot & & & \cdot \\ P_{1n} & & \cdots & P_{nn} \end{bmatrix} = \begin{bmatrix} S_{11} & 0 & \cdots & 0 \\ S_{21} & S_{22} & \cdots & 0 \\ \cdot & & \cdot & \cdot \\ \cdot & & & \cdot \\ \cdot & & & 0 \\ S_{n1} & & \cdots & S_{nn} \end{bmatrix} \cdot \begin{bmatrix} S_{11} & S_{21} & \cdots & S_{n1} \\ 0 & S_{22} & & \\ \cdot & \cdot & \cdot & \cdot \\ \cdot & \cdot & \cdot & \cdot \\ \cdot & \cdot & \cdot & \cdot \\ 0 & 0 & \cdots & S_{nn} \end{bmatrix}. \quad (74)$$

Using the correspondence with the scalar case, S is often called the square root, or Cholesky square root, of P, as Cholesky (and Banachiewicz) [5] gave the following recursive algorithm for computing S.

For $i = 1, n$,

$$S_{ii} = \sqrt{P_{ii} - \sum_{j=1}^{i-1} S_{ij}^2} \quad (75)$$

$$S_{ji} = \begin{cases} 0, & j < i \\ \dfrac{1}{S_{ii}}\left(P_{ji} - \sum_{k=1}^{i-1} S_{jk}S_{ik}\right), & j = i+1, n. \end{cases} \quad (76)$$

Example:

$$P = \begin{bmatrix} 1 & 2 & 3 \\ 2 & 8 & 2 \\ 3 & 2 & 14 \end{bmatrix}$$

$$S = \begin{bmatrix} 1 & 0 & 0 \\ 2 & \sqrt{8-4} & 0 \\ 3 & (2-6/2) & \sqrt{14-9-4} \end{bmatrix} = \begin{bmatrix} 1 & 0 & 0 \\ 2 & 2 & 0 \\ 3 & -2 & 1 \end{bmatrix}.$$

APPENDIX II

TRIANGULARIZATION ALGORITHMS

The matrix triangularization required in some square root update algorithms can be accomplished by at least four different algorithms. The two most promising algorithms, Householder transformation and modified Gram-Schmidt orthogonalization (MGS), are presented here.

The triangularization problem can be stated in the following general form.

Given:

$$Ax = b$$

where $A \in R_{(n+r)}xR_n$, $x \in R_n$, $b \in R_{(n+r)}$.

Find: an orthogonal transformation T such that

$$TA = \overbrace{\begin{bmatrix} W \\ \hline 0 \end{bmatrix}}^{n} \begin{matrix} \}n \\ \}r \end{matrix}$$

where W is upper triangular, (or equivalently, find W, and $b' = Tb$ directly).

Householder Algorithm

This algorithm was derived by Householder [7] and applied to the problem as stated here by Golub [11].

Definitions: Let

$$T \triangleq T^{(n)}T^{(n-1)} \cdots T^{(1)}$$

where $T^{(k)}$ is an orthogonal $(n+r) \times (n+r)$ matrix defined by

$$T^{(k)} \triangleq I - \beta_k u^{(k)} u^{(k)^T} \quad (77)$$

with

$$\beta_k \triangleq \frac{2}{u^{(k)^T}u^{(k)}}. \quad (78)$$

Also let $A^{(1)} \triangleq A$, $b^{(1)} \triangleq b$, and $A_j^{(k)} \triangleq j$th column of $A^{(k)}$, $A_{ij}^{(k)} \triangleq$ element in ith row and jth column of $A^{(k)}$.

Basic Algorithm: The algorithm proceeds in a recursive fashion for $k = 1, n$ using

$$A^{(k+1)} = T^{(k)}A^{(k)} \quad (79)$$

$$b^{(k+1)} = T^{(k)}b^{(k)}. \quad (80)$$

At stage k, the first $(k-1)$ columns of $A^{(k)}$ are zero below the diagonal, and $u(k)$ is chosen such that the subdiagonal elements of $A_k^{(k+1)}$ will be zero. Thus when $k = n$, we have

$$A^{(n+1)} = \begin{bmatrix} W \\ \hline 0 \end{bmatrix} \begin{matrix} \}n \\ \}r \end{matrix} \quad (81)$$

$$b^{(n+1)} = b'. \quad (82)$$

Computational Algorithm: The following computational algorithm carries out the Householder algorithm without ever storing or computing T explicitly.

For $k = 1, n$

$$\sigma_k = \sqrt{\sum_{i=k}^{n+r} [A_{ik}^{(k)}]^2} \cdot \operatorname{sgn}(A_{kk}^{(k)}). \quad (83)$$

(This simple algorithm does not use pivoting. See [12] for a pivoting algorithm.)

$$\beta_k = \frac{1}{\sigma_k(\sigma_k + A_{kk}^{(k)})} \quad (84)$$

$$u_i^{(k)} = \begin{cases} 0, & i < k \\ \sigma_k + A_{kk}^{(k)}, & i = k \\ A_{ik}^{(k)}, & i > k \end{cases} \quad (85)$$

$$y_j^{(k)} = \begin{cases} 0, & j < k \\ 1, & j = k \\ \beta_k u^{(k)^T} A_j^{(k)}, & j > k \end{cases} \quad (86)$$

$$z^{(k)} = \beta_k u^{(k)^T} b^{(k)} \quad (87)$$

$$A^{(k+1)} = A^{(k)} - u^{(k)} y^{(k)^T} \quad (88)$$

$$b^{(k+1)} = b^{(k)} - u^{(k)} z^{(k)}. \quad (89)$$

If $\sigma_k = 0$ at any stage in the algorithm, then the rank of

A is less than n. A Householder triangularization of an $(n + r) \times n$ matrix requires the following number of computations:

$$n \sqrt{\ }, n \div, 1/6[4n^3 + 6n^2(r + 1) + 2n] \times, 1/6[4n^3 + 6rn^2 + 8n] +.$$

Modified Gram–Schmidt (MGS) Algorithm [10]

This algorithm is a numerically improved adaptation of the classical Gram–Schmidt orthogonalization procedure. When computations are made exactly (no roundoff) the result is equivalent to the classical Gram–Schmidt result. However, when roundoff errors occur, Björck [10] has shown that the MGS procedure is much more accurate.

The algorithm can be derived from the classical Gram–Schmidt orthogonalizing procedure, and is essentially the classical Gram–Schmidt procedure in reverse order [17].

Using the same notation used to describe the Householder algorithm, the MGS algorithm can also be stated in a form which computes W and b' directly.

MGS Algorithm: For $k = 1, n$

$$\sigma_k = \sqrt{A_k^{(k)T} A_k^{(k)}} \tag{90}$$

$$W_{kj} = \begin{cases} 0, & j = 1, k - 1 \\ \sigma_k, & j = k \\ \dfrac{1}{\sigma_k} A_k^{(k)T} A_j^{(k)}, & j = k + 1, n \end{cases} \tag{91}$$

$$b_k' = \frac{1}{\sigma_k} A_k^{(k)T} b^{(k)} \tag{92}$$

$$A_j^{(k+1)} = A_j^{(k)} - \frac{W_{kj}}{\sigma_k} A_k^{(k)}, \qquad j = k + 1, n \tag{93}$$

$$b^{(k+1)} = b^{(k)} - \frac{b_k'}{\sigma_k} A_k^{(k)}. \tag{94}$$

If $\sigma_k = 0$ at any stage in the algorithm, then the rank of A is less than n. An MGS triangularization of an $(n + r) \times n$ matrix requires the following computations: $n\sqrt{\ }$, $n \div$, $1/6[6n^3 + 6n^2(r + 1) - 6n] \times$, $1/6[6n^3 + 3n^2(2r - 1) - 3n] +$. Thus the MGS algorithm requires approximately $1/3n^3$ more multiplications and additions than the Householder algorithm. However, the MGS algorithm has been shown [13] to become slightly more accurate as the residual increases. For r approaching n, the MGS procedure might be preferred over the Householder.

Acknowledgment

The authors express their appreciation to S. McReynolds for many helpful comments and suggestions for improvement which were incorporated into this paper.

References

[1] R. E. Kalman, "A new approach to linear filtering and prediction problems," *Trans. ASME*, vol. 82D, pp. 35–50, 1960.
[2] R. H. Battin, *Astronautical Guidance*. New York: McGraw-Hill, 1964, pp. 338–339.
[3] R. H. Battin and G. M. Levine, "Application of Kalman filtering techniques in the Apollo program," in *Theory and Applications of Kalman Filtering*, NATO Advisory Group for Aerospace Research and Development, AGARDograph 139, Feb. 1970.
[4] D. C. Fraser, "A new technique for the optimal smoothing of data," M.I.T. Instrumentation Lab., Rep. T-474, Jan. 1967, p. 114.
[5] V. N. Fadeeva, *Computational Methods of Linear Algebra*. New York: Dover, 1959, pp. 81–84.
[6] S. F. Schmidt, "Computational techniques in Kalman filtering in *Theory and Applications of Kalman Filtering*, NATO Advisory Group for Aerospace Research and Development, AGARDograph 139, Feb. 1970.
[7] A. S. Householder, *The Theory of Matrices in Numerical Analysis*. Waltham, Mass.: Blaisdell, 1964, ch. 5.
[8] J. F. Bellantoni and K. W. Dodge, "A square root formulation of the Kalman-Schmidt filter," *AIAA J.*, vol. 5, pp. 1309–1314, July 1967.
[9] A. Andrews, "A square root formulation of the Kalman covariance equations," *AIAA J.*, vol. 6, pp. 1165–1166, June 1968.
[10] A. Björck, "Solving linear least squares problems by Gram-Schmidt orthogonalization," *BIT*, vol. 7, pp. 1–21, 1967.
[11] G. H. Golub, "Numerical methods for solving linear least squares problems," *Numer. Math.*, vol. 7, pp. 206–216, 1965.
[12] P. Businger and G. H. Golub, "Linear least squares solution by Householder transformations," *Numer. Math.*, vol. 7, pp. 269–276, 1965.
[13] T. Jordan, "Experiments on error growth associated with some linear least square procedures," *Math. Comput.*, vol. 20, pp. 325–328, 1966.
[14] R. J. Hanson and C. L. Lawson, "Extensions and applications of the Householder algorithm for solving linear least squares problems," *Math. Comput.*, vol. 23, pp. 787–812, Oct. 1969.
[15] H. C. Cox, "Estimation of state variables via dynamic programming," in *Proc. 1964 Joint Automatic Control Conf.*, pp. 376–381.
[16] P. Dyer and S. McReynolds, "Extension of square-root filtering to include process noise," *J. Optimiz. Theory Appl.*, vol. 3, no. 6, pp. 444–459, 1969.
[17] P. G. Kaminski, "Square root filtering and smoothing for discrete processes," Ph.D. dissertation, Dep. Aeronaut. Astronaut., Stanford Univ., Stanford, Calif., Sept. 1971.

Computational Requirements for a Discrete Kalman Filter

JERRY M. MENDEL, MEMBER, IEEE

Abstract—How practical is a Kalman filter? One answer to this question is provided by the computational requirements for the filter. Computational requirements—computational time per cycle (iteration) and required storage—determine minimum sampling rates and computer memory size. These requirements are provided in this paper as functions of the dimensions of the important system matrices for a discrete Kalman filter.

Two types of measurement processing are discussed: simultaneous and sequential. It is shown that it is often better to process statistically independent measurements in more than one batch and then to use sequential processing than to process them together via simultaneous processing.

I. Introduction

KALMAN filtering is a popular method of estimating states from noisy measurements. Kalman filter stability, convergence, and sensitivity have been thoroughly examined in the literature; however, the explicit computational requirements of Kalman filtering have not received as much attention. By computational requirements, we mean computation time per iteration and storage. Obviously, in a practical application of a Kalman filter, it is important to know both of these quantities in advance. Computation time per iteration establishes meaningful data sampling rates and storage requirements establish memory size.

In this paper, computational requirements are obtained for a discrete Kalman filter. The discrete Kalman filter, which is easily programmed on a digital computer, provides minimum variance estimates for the following dynamic system:

$$x(k+1) = \Phi(k+1, k)x(k) + \Psi(k+1, k)u(k) + \Gamma(k+1, k)\xi(k) \quad (1)$$

$$z(k+1) = C(k+1)x(k+1) + \theta(k+1). \quad (2)$$

Computational requirements are obtained as explicit functions of the dimensions of the system's state, measurement, and disturbance vectors for the situation in which all matrices that appear in the equations for the Kalman filter are assumed to be full. (The Kalman filter equations are listed in the column entitled "Defining Equation" of Table IV.)

Manuscript received June 8, 1971. Paper recommended by S. K. Mitter, Associate Guest Editor. The work described here was conducted by the McDonnell Douglas Astronautics Company for NASA's Langley Research Center under Contract NAS1-9566.

The author is with the McDonnell Douglas Astronautics Company, Huntington Beach, Calif. 92647.

The following baseline digital computer configuration was assumed.

1) A word length of 28–32 bits. Based on this assumption, single-precision arithmetic was used in all programs.
2) Eight hardware index registers (R_i, $i = 1, 2, \cdots, 8$).
3) The instruction list given in Table I.

In Table I, the assumption has been made that it is possible to express execution time for each instruction as some multiple of a unit time. Unit time (in μs) is a function of the specific computer actually used. Note also that multiplication and division execution times are highly dependent upon the specific computer implementation. Representative execution times are given in Table II.

Additional assumptions made during this study are as follows.

1) Indexed or relative addressed operands contribute no additional execution time.
2) Indirect addressed operands contribute one additional unit of execution time for each level of indirect addressing.
3) The measurement vector $z(k)$ is required by the flight controller and is therefore available to the Kalman filter computer at no computing expense.
4) $\Phi(k+1, k)$, $\Gamma(k+1, k)$, and $\Psi(k+1, k)$ are computed outside of the Kalman filter computer and are therefore available to it at no computing expense.

II. Basic Operations

The basic operations required for programming the Kalman filter equations are matrix addition, matrix subtraction, matrix multiplication, matrix transpose multiplication, matrix inversion, and multiplication of a vector by a scalar.

Computer language subroutines were written for each of these operations, using the instructions in Table I.[1] From these subroutines it was possible to determine their memory requirements and total computing times. Results are tabulated in Table III. The matrix inversion routine uses the standard Gauss–Jordan algorithm. It also includes the evaluation of a determinant which, if zero, indicates that the matrix is singular. Observe that computing times are given as functions of the dimensions of the matrices

[1] These subroutines were written by J. J. Zapalac of the Advance Electronics Department, McDonnell Douglas Astronautics Company, Huntington Beach, Calif.

Reprinted from *IEEE Trans. Automat. Contr.*, vol. AC-16, pp. 748–758, Dec. 1971.

TABLE I
INSTRUCTION LIST

Instruction	Execution Time (in unit times)
Add to A	2
Sub from A	2
Load A or B	2
Store A or B	2
Mul A with memory	MUL
Div A by memory	DIV
Mark place and transfer	2
Return branch	2
Transfer on zero or minus	1
Transfer unconditional	1
Transfer on plus	1
Increment A	1
Increment index register R_i	1
Decrement index register R_i	1
Add immediate to index register R_i	1
Sub immediate from index register R_i	1
Increment B register	1
Skip on index register zero ($R_i = 0$)	1

TABLE II
REPRESENTATIVE EXECUTION TIMES

Add Times (μs)	Multiply Times (μs)	Divide Times (μs)	Remarks
4–9	20–40	20–60	Typical for 1967–1968 aerospace computers [1]
2	6	11.4	Minimum for 1967–1968 aerospace computers [1]
1	8	10	Fictitious 1971 computer [1]

TABLE III
ROUTINE CHARACTERISTICS

Name	Function	Program Size (Words)	Total Computing Time (in Unit Times)	Logic Time (in Unit Times)
Matrix addition[a]	$C_{MN} = A_{MN} + B_{MN}$	18	$\text{MUL} + 27 + 7MN$	$\text{MUL} + 27 + 5MN$
Matrix subtraction	$C_{MN} = A_{MN} - B_{MN}$	18	$\text{MUL} + 27 + 7MN$	$\text{MUL} + 27 + 5MN$
Matrix multiply[b]	$C_{ML} = A_{MN}B_{NL}$	36	$10 + 8MNL + 19ML + 16M + \text{MUL}(MNL)$	$10 + 6MNL + 21ML + 16M$
Matrix transpose multiply	$C_{MN} = A_{ML}(B_{NL}')$	36	$10 + 8MNL + 19ML + 16M + \text{MUL}(MNL)$	$10 + 6MNL + 21ML + 16M$
Matrix inversion[c]	$A_{NN} \rightarrow A_{NN}^{-1}$	232	$10 + 7.5N^4 + 43N^3 + 139.5N^2 + 92N + \text{DIV}(2N^2 + N) + \text{MUL}(N^3 + 0.5N^2 + 2.5N)$	$10 + 7.5N^4 + 41N^3 + 139.5N^2 + 92N + \text{DIV}(2N^2 + N) + \text{MUL}(0.5N^2 + 2.5N)$
Scalar–vector product	$C_{N1} = \mu A_{N1}$	8	$8N + \text{MUL}(N)$	$8N$

[a] C_{MN} requires MN additions, each requiring two units of computing time (Table I).
[b] C_{ML} requires MNL multiplications, each requiring MUL units of computing time and $ML(N - 1)$ additions, each requiring two units of computing time.
[c] A_{NN}^{-1} requires N^3 multiplications, each requiring MUL units of computing time and N^3 additions, each requiring two units of computing time [2].

utilized by the routine, as well as execution times for the operations of multiplication and division.

A computer implementation of mathematical operations requires additional steps for properly controlling and sequencing the operations, as is evident from the large numbers of logic operations required for each routine. The data in the last column of Table III were obtained by subtracting from the total computing time the product of the number of operations a human performs when he goes through the matrix operations listed in the first column (such as add, subtract, or multiply; see footnotes to Table III, for example) and the respective number of execution times.

III. COMPUTATION TIME PER CYCLE

The defining equations for the discrete Kalman filter, which are listed in Table IV (all of the filter's variables being defined in Table V), can be programmed as shown in the column of Table IV labeled "Computations," using the instruction list in Table I and the subroutines listed in Table III. The total numbers of multiplications, additions, and logic times for each computation are listed in Table

TABLE IV
DISCRETE KALMAN FILTER COMPUTATION TIME REQUIREMENTS

Variable	Defining Equation[b]	Computations	Number of Multiplications	Number of Additions	Logic Time
$\hat{x}(k+1\|k)$	$\Phi(k+1,k)\hat{x}(k\|k) + \Psi(k+1,k)u(k)$	$\Phi\hat{x}$	n^2	$n^2 - n$	$10 + 6n^2 + 37n$
		Ψu	n	0	$8n$
		$\Phi\hat{x} + \Psi u$	0	n	$\mathrm{MUL} + 27 + 5n$
$\boldsymbol{\Sigma}(k+1\|k)$	$\Phi(k+1,k)\boldsymbol{\Sigma}(k\|k)\Phi'(k+1,k) + \Gamma(k+1,k)\boldsymbol{\Xi}(k)\Gamma'(k+1,k)$	$\boldsymbol{\Sigma}\Phi'$	n^3	$n^3 - n^2$	$10 + 6n^3 + 21n^2 + 16n$
		$\Phi(\boldsymbol{\Sigma}\Phi')$	n^3	$n^3 - n^2$	$10 + 6n^3 + 21n^2 + 16n$
		$\boldsymbol{\Xi}\Gamma'$	ns^2	$ns^2 - ns$	$10 + 6ns^2 + 21ns + 16s$
		$\Gamma(\boldsymbol{\Xi}\Gamma')$	n^2s	$n^2s - n^2$	$10 + 6n^2s + 21n^2 + 16n$
		$\Phi(\boldsymbol{\Sigma}\Phi') + \Gamma(\boldsymbol{\Xi}\Gamma')$	0	n^2	$\mathrm{MUL} + 27 + 5n^2$
$\boldsymbol{K}(k+1)$	$\boldsymbol{\Sigma}(k+1\|k)C'(k+1) \cdot [C(k+1)\boldsymbol{\Sigma}(k+1\|k)C'(k+1) + \Theta(k+1)]^{-1}$	$\boldsymbol{\Sigma}C'$	n^2r	$n^2r - nr$	$10 + 6n^2r + 21nr + 16n$
		$C(\boldsymbol{\Sigma}C')$	nr^2	$nr^2 - r^2$	$10 + 6nr^2 + 21r^2 + 16r$
		$C(\boldsymbol{\Sigma}C') + \Theta$	0	r^2	$\mathrm{MUL} + 27 + 5r^2$
		$[C(\boldsymbol{\Sigma}C') + \Theta]^{-1}$	r^3	r^3	$10 + 7.5r^4 + 41r^3 + 139.5r^2 + 92r + \mathrm{DIV}(r + 2r^2) + \mathrm{MUL}(2.5r + 0.5r^2)$
		$\boldsymbol{\Sigma}C'[C(\boldsymbol{\Sigma}C') + \Theta]^{-1}$	nr^2	$nr^2 - nr$	$10 + 6nr^2 + 21nr + 16n$
$\hat{x}(k+1\|k+1)$	$\hat{x}(k+1\|k) + \boldsymbol{K}(k+1) \cdot [z(k+1) - C(k+1)\hat{x}(k+1\|k)]$	$C\hat{x}$	nr	$nr - r$	$10 + 6nr + 37r$
		$z - (C\hat{x})$	0	r	$\mathrm{MUL} + 27 + 5r$
		$\boldsymbol{K}[z - (C\hat{x})]$	nr	$nr - n$	$10 + 6nr + 37n$
		$\hat{x} + \boldsymbol{K}[z - (C\hat{x})]$	0	n	$\mathrm{MUL} + 27 + 5n$
[a]$\boldsymbol{\Sigma}(k+1\|k+1)$	$[I - \boldsymbol{K}(k+1)C(k+1)]\boldsymbol{\Sigma}(k+1\|k)$	$\boldsymbol{K}C$	n^2r	$n^2r - n^2$	$10 + 6n^2r + 21n^2 + 16n$
		$I - (\boldsymbol{K}C)$	0	n^2	$\mathrm{MUL} + 27 + 5n^2$
		$[I - (\boldsymbol{K}C)]\boldsymbol{\Sigma}$	n^3	$n^3 - n^2$	$10 + 6n^3 + 21n^2 + 16n$

[a] $\boldsymbol{\Sigma}(k+1|k+1)$ is symmetric and positive definite. If it is difficult to maintain these properties using the defining equation, the following alternate, symmetric form can be used:

$$\boldsymbol{\Sigma}(k+1|k+1) = [I - \boldsymbol{K}(k+1)C(k+1)]\boldsymbol{\Sigma}(k+1|k)[I - \boldsymbol{K}(k+1)C(k+1)]' + \boldsymbol{K}(k+1)\Theta(k+1)\boldsymbol{K}'(k+1)$$

Computational time will increase. For example, the total numbers of multiplications and additions required to obtain $\boldsymbol{\Sigma}(k+1|k+1)$ increase by $n^3 + n^2r + nr$ and $n^3 + n^2r - n^2$, respectively.

[b] All variables which appear in these equations are defined in Table V.

TABLE V
FILTER VARIABLES AND STORAGE REQUIREMENTS

Variable	Definition	Dimension	Storage Requirement
$\hat{x}(k\|k)$	State estimate at t_k given $z(k)$	$n \times 1$	n
$\boldsymbol{\Sigma}(k\|k)$	Covariance matrix of the error in $\hat{x}(k\|k)$	$n \times n$	n^2
$\Phi(k+1,k)$	State transition matrix (from t_k to t_{k+1})	$n \times n$	n^2
$\Gamma(k+1,k)$	System disturbance distribution matrix	$n \times s$	ns
$\boldsymbol{\Xi}(k)$	System disturbance covariance matrix	$s \times s$	s^2
$\hat{x}(k+1\|k)$	State estimate at t_{k+1} given z_k	$n \times 1$	n
$\boldsymbol{\Sigma}(k+1\|k)$	Covariance matrix of the error in $\hat{x}(k+1\|k)$	$n \times n$	n^2
$C(k+1)$	Measurement matrix	$r \times n$	nr
$\Theta(k+1)$	Measurement noise covariance matrix	$r \times r$	r^2
$\boldsymbol{K}(k+1)$	Filter (Kalman) gain matrix at t_{k+1}	$n \times r$	nr
$z(k+1)$	Measurement (observation) at t_{k+1}	$r \times 1$	r
$\Psi(k+1,k)$	Control distribution vector	$n \times 1$	n
$u(k)$	Control at t_k	Scalar[a]	1

[a] The extension of the results of this paper to the vector control situation is trivial, since the control appears only in the prediction equation $\hat{x}(k+1|k)$.

IV. These results were obtained by direct application of the results in Table III to each computation.

Total computation time requirements are summarized in Table VI. These results are obtained directly from those in Table IV. In Table VI $M(n, r, s) \equiv$ total number of multiplications for a discrete Kalman filter, $A(n, r, s) \equiv$ total number of additions for a discrete Kalman filter, and $L(n, r, s, \mathrm{MUL}, \mathrm{DIV}) \equiv$ total logic time for a discrete Kalman filter.

We define total computational time as $C(n, r, s, \mathrm{MUL}, \mathrm{DIV})$, where

$$C(n, r, s, \mathrm{MUL}, \mathrm{DIV}) = [M(n, r, s) \cdot \mathrm{MUL} + 2A(n, r, s) + L(n, r, s, \mathrm{MUL}, \mathrm{DIV})] \cdot T_u \quad (3)$$

TABLE VI
TOTAL COMPUTATION TIME REQUIREMENTS

Variable	Number of Multiplications.	Number of Additions	Logic Time
$\hat{\mathbf{x}}(k+1\|k)$	n^2+n	n^2	$6n^2+50n+37+\mathrm{MUL}$
$\mathbf{\Sigma}(k+1\|k)$	$2n^3+n^2s+ns^2$	$2n^3+n^2s+ns^2-2n^2-sn$	$12n^3+68n^2+48n+67+6ns^2+21ns+6n^2s+16s+\mathrm{MUL}$
$\mathbf{K}(k+1)$	$n^2r+2nr^2+r^3$	$n^2r+2nr^2-2nr+r^3$	$67+6n^2r+42nr+32n+12nr^2+7.5r^4+41r^3+165.5m^2+108r+\mathrm{DIV}(r+2r^2)+\mathrm{MUL}(0.5r^2+2.5r+1)$
$\hat{\mathbf{x}}(k+1\|k+1)$	$2nr$	$2nr$	$74+12nr+42n+42r+2\mathrm{MUL}$
$\mathbf{\Sigma}(k+1\|k+1)$	n^3+n^2r	$n^3+n^2r-n^2$	$47+6n^2r+47n^2+32n+6n^3+\mathrm{MUL}$
Totals	$M(n,r,s)=M(n,r,0)+M^*(n,0,s)$	$A(n,r,s)=A(n,r,0)+A^*(n,0,s)$	$L(n,r,s,\mathrm{MUL},\mathrm{DIV})=L^*(n,r,0,0,0)+L^*(n,0,s,0,0)+L^*(0,r,0,\mathrm{MUL},0)+L^*(0,r,0,0,\mathrm{DIV})+292$
	$M(n,r,0)=3n^3+2n^2r+2nr^2+r^3+n^2+2nr+n$	$A(n,r,0)=3n^3+2n^2r+2nr^2+r^3-2n^2$	$L^*(n,r,0,0,0)=18n^3+121n^2+12n^2r+204n+54nr+12nr^2+7.5r^4+41r^3+165.5r^2+150r$
	$M^*(n,0,s)=n^2s+ns^2$	$A^*(n,0,s)=n^2s+ns^2-ns$	$L^*(n,0,s,0,0)=6n^2s+21ns+6ns^2+16s$
			$L^*(0,r,0,\mathrm{MUL},0)=\mathrm{MUL}(0.5r^2+2.5r+6)$
			$L^*(0,r,0,0,\mathrm{DIV})=\mathrm{DIV}(2r^2+r)$

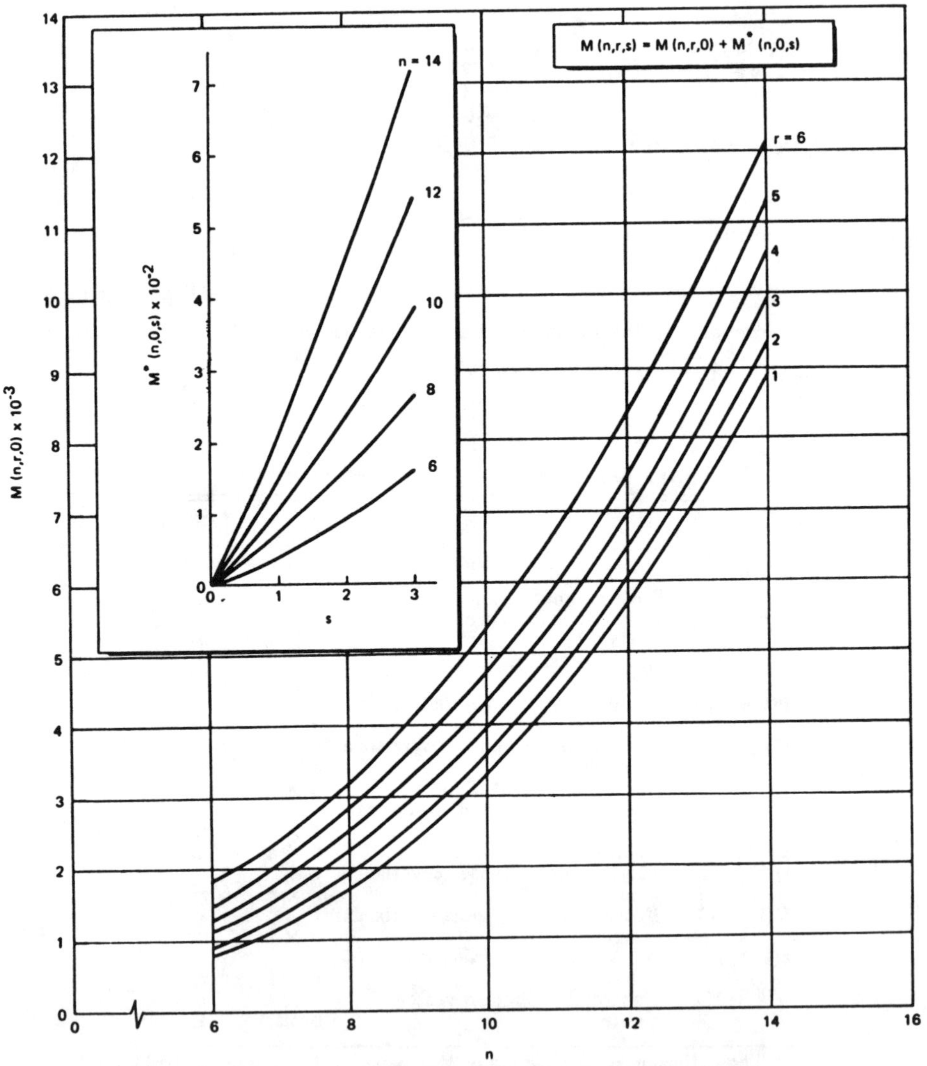

Fig. 1. Number of multiplications $M(n, r, s)$ for a discrete Kalman filter.

and T_u denotes the basic unit time (e.g., 1 μs/unit time).

Plots from which $M(n, r, s)$, $A(n, r, s)$, and $L(n, r, s, \mathrm{MUL}, \mathrm{DIV})$ can be obtained, for representative values of n, r, and s ($n = 6, 8, 10, 12, 14$, $r = 1, 2, 3, 4, 5, 6$, $s = 0, 1, 2, 3$) are given in Figs. 1, 2, and 3, respectively. These plots were made by observing that $M(n, r, s)$, $A(n, r, s)$, and $L(n, r, s, \mathrm{MUL}, \mathrm{DIV})$ can be decomposed, as shown in Table VI. Note, from these figures, that effects of s (disturbances) are second order, compared to those of r (measurements) and n (order). Note also that effects of MUL and DIV are third order. Observe also that $M(n, r, s)$ and $A(n, r, s)$ are more sensitive to n than r. Logic time is more sensitive to n than to r for small values of r. As r increases, however, $L^*(n, r, 0, 0, 0)$ is much more sensitive to r be-

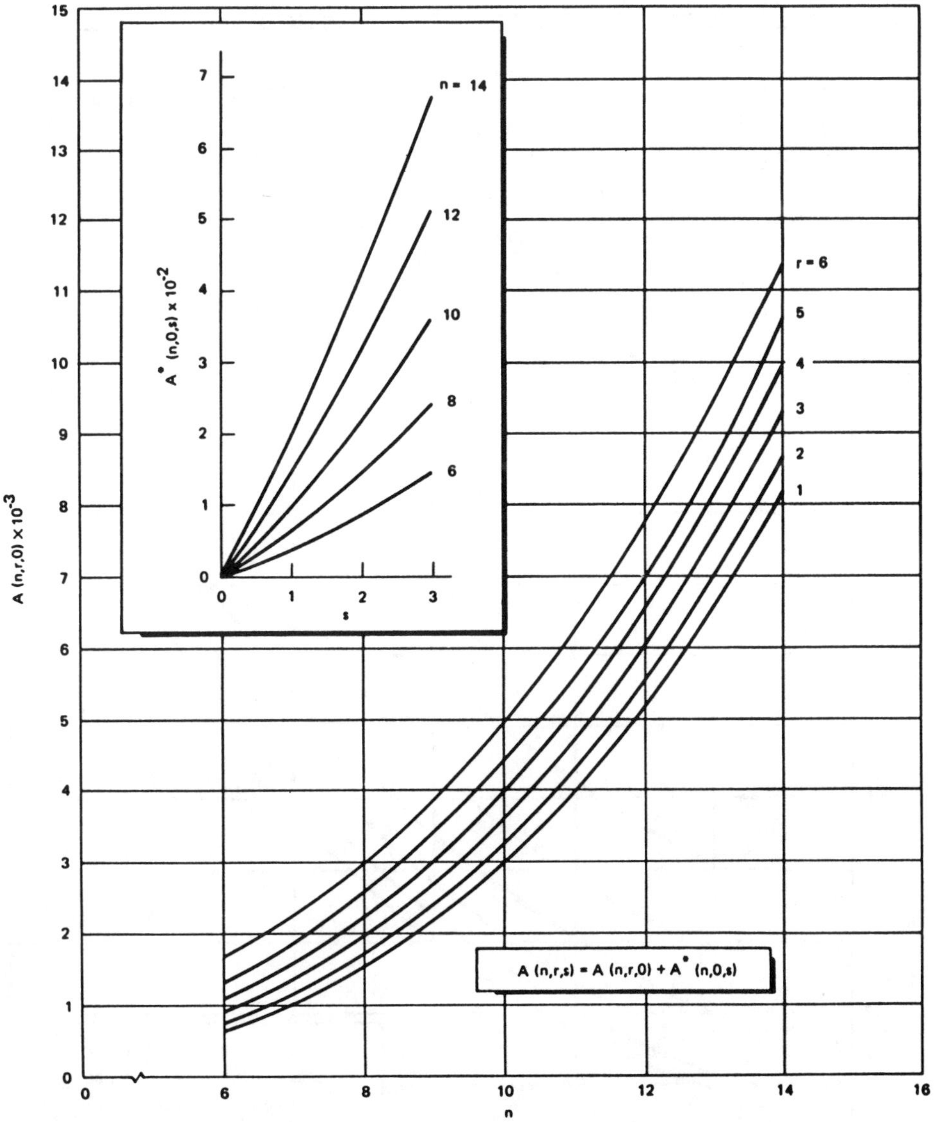

Fig. 2. Number of additions $A(n, r, s)$ for a discrete Kalman filter.

cause of the term $7.5r^4$ (see Table VI). These observations suggest two courses: 1) reduce n, if possible and 2) process measurements in small batches. Course 2 is investigated in Section V of this paper.

In order to illustrate the application of some of this section's results, consider the case in which $n = 14$, $r = 2$, and $s = 0$, MUL = 6, DIV = 12, and $T_u = 1$ μs. It is easy to compute, from the equations in Table VI, that

$$C(14, 2, 0, 6, 12) = [9400(6) + 2(8700) + 76\,523]*10^{-6} = 0.152323 \text{ s.}$$

If the order of the system's model is reduced to $n = 12$, then

$$C(12, 2, 0, 6, 12) = [6050(6) + 2(5550) + 52\,156]*10^{-6} = 0.099556 \text{ s.}$$

Hence, reducing the system's order from 14 to 12 decreases the computation time approximately 35 percent.

IV. Storage

Total memory storage requirements are obtained from Tables V and III. It has been assumed that separate memory is available for storing all of the items listed in Table V. No doubt, memory requirements can be reduced by an efficient allocation of a fixed amount of memory. In addition to the memory required to store the items in Table V, the six subroutines in Table III require 348 words of memory; and the main program that was written for the Kalman filter equations requires an additional 140 words. Finally, $2n^2$ words of working storage were assumed. The total storage requirements, denoted $S(n, r, s)$, are

$$S(n, r, s) = S^*(n, r, 0) + S^*(n, 0, s) + 489 \qquad (4)$$

where

$$S^*(n, r, 0) = 5n^2 + 3n + 2nr + r^2 + r \qquad (5)$$

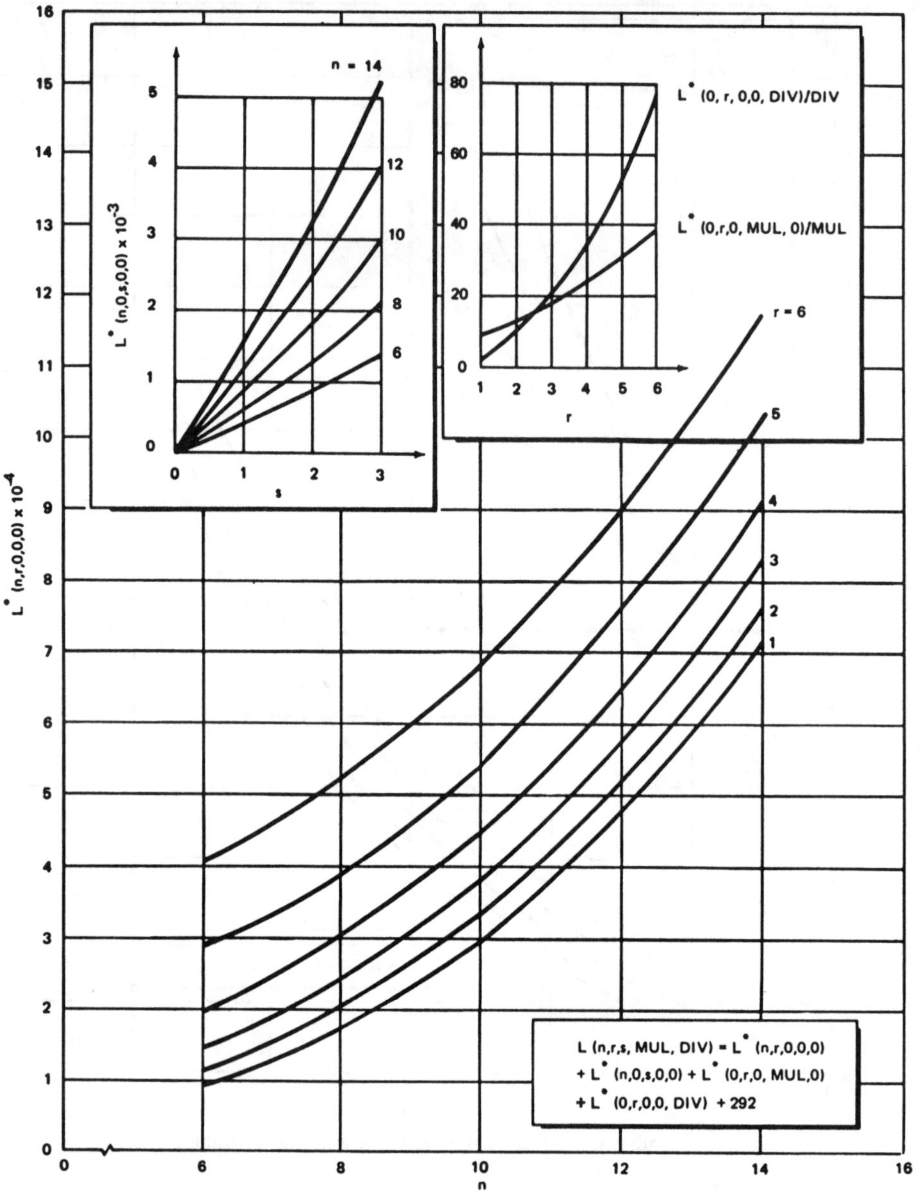

Fig. 3. Logic time in unit times $L(n, r, s, \text{MUL}, \text{DIV})$ for a discrete Kalman filter.

and

$$S^*(n, 0, s) = s^2 + ns. \tag{6}$$

$S^*(n, r, 0)$ and $S^*(n, 0, s)$ are plotted in Fig. 4. Observe, for example, that $S(14, 2, 0) = 1559$ words, whereas $S(12, 2, 0) = 1284$ words, i.e., reducing the system's order from 14 to 12 decreases memory requirements by approximately 17 percent. Observe also that memory is most sensitive to system order n.

V. Sequential Processing

Suppose[2] that at each sampling time t_k, q statistically independent sources provide measurement data. The r-dimensional measurement vector can be represented as

$$z(k) \triangleq \begin{pmatrix} z_1(k) \\ z_2(k) \\ \cdot \\ \cdot \\ \cdot \\ z_q(k) \end{pmatrix} = \begin{pmatrix} C_1(k) \\ C_2(k) \\ \cdot \\ \cdot \\ \cdot \\ C_q(k) \end{pmatrix} x(k) + \begin{pmatrix} \theta_1(k) \\ \theta_2(k) \\ \cdot \\ \cdot \\ \cdot \\ \theta_q(k) \end{pmatrix}. \tag{7}$$

Since the sources are statistically independent,

[2] Except for some changes in notation, the discussions in this section have been excerpted directly from [3, Sect. III-E] with the kind permission of Prof. Sorenson.

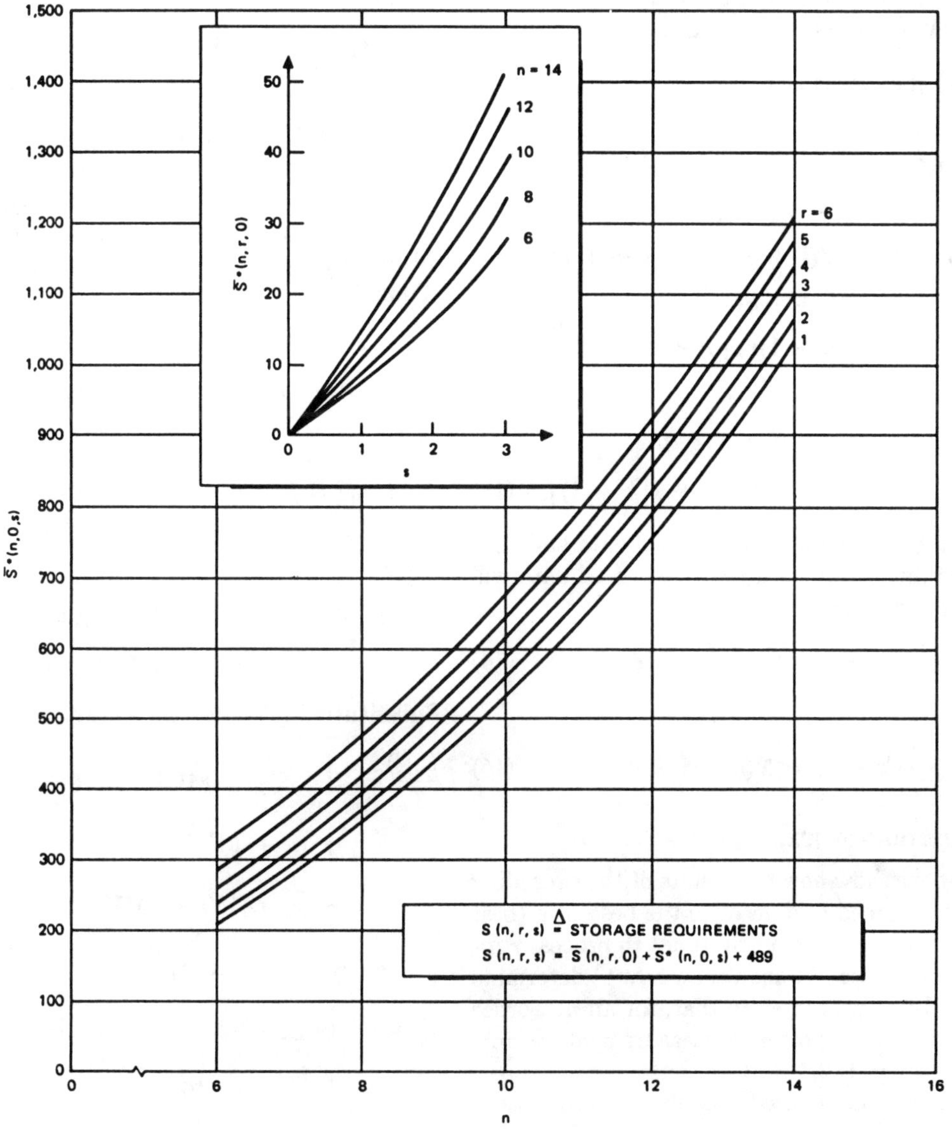

Fig. 4. Storage requirements, in cells, for a discrete Kalman filter.

$$E\{\theta_i(k)\theta_j'(k)\} = \Theta_i(k)\delta_{ij}. \tag{8}$$

The $\Theta_i(k)$ have the dimension $(r_i \times r_i)$, where

$$\sum_{i=1}^{q} r_i = r. \tag{9}$$

In the equation for $K(k)$ (third row of Table IV), the inverse of the matrix $[C(k)\Sigma'(k|k-1)C'(k) + \Theta(k)]$ must be computed. If there are several data sources, the dimension r of this matrix could be quite large. The inversion on a digital computer of a matrix of large dimension is undesirable because of the number of storage cells that must be used, the time that is consumed in obtaining the inverse, and the accuracy of the end result. Thus, if the inversion can be circumvented, it is advisable to do so. In this case, an alternative is available in which the largest matrix that must be inverted has dimension equal to max (r_i). This procedure will be designated by the term "sequential processing." Another policy that can be used involves the formulation described in Table IV. The disadvantage inherent in this scheme is the inversion of a large matrix. It shall be called "simultaneous processing."

In sequential processing, each set of data is treated separately. Specifically, the data $z_1(k)$ are used to obtain an estimate $\hat{x}(k|k)$ and covariance $\Sigma(k|k)$ with

$$z(k) \triangle z_1(k). \tag{10}$$

When these calculations are completed, $z_2(k)$ is processed to obtain new values for $\hat{x}(k|k)$ and $\Sigma(k|k)$. Each set of data is processed in this manner until the final set $z_q(k)$ has been included. Then, time is advanced to t_{k+1} and the cycle is repeated. The simultaneous processing technique involves the direct application of the defining equations in Table IV. Sorenson [3] proves the equivalence of sequential and simultaneous processing and gives the following computational procedure that can be used when there are q sets of statistically independent observation data: $z_1(k)$, $z_2(k)$, $\cdots$, $z_q(k)$.

Step 1: Form $\hat{x}(k+1|k)$, $\Sigma(k+1|k)$, $K_1(k+1)$, $\hat{x}_1(k$

$+ 1|k + 1)$, and $\Sigma_1(k + 1|k + 1)$ using the defining equations in Table IV and the first set of data:

$$z_1(k + 1) \text{ [i.e., treat } z_1(k + 1) \text{ as } z(k + 1)].$$

Step 2: Based on the $\Sigma_1(k + 1|k + 1)$ and $\hat{x}_1(k + 1|k + 1)$, process $z_2(k + 1)$ using the relations[3] (for $j = 2$):

$$K_j(k + 1) = \Sigma_{j-1}(k + 1|k + 1)C_j'(k + 1)[C_j(k + 1)\Sigma_{j-1} \cdot (k + 1|k + 1)C_j'(k + 1) + \Theta_j(k + 1)]^{-1} \quad (11)$$

$$\hat{x}_j(k + 1|k + 1) = \hat{x}_{j-1}(k + 1|k + 1) + K_j(k + 1) \cdot [z_j(k + 1) - C_j(k + 1)\hat{x}_{j-1}(k + 1|k + 1)] \quad (12)$$

and

$$\Sigma_j(k + 1|k + 1) = [I - K_j(k + 1)C_j(k + 1)] \cdot \Sigma_{j-1}(k + 1|k + 1)]. \quad (13)$$

Step 3: Repeat Step 2 with the next set of data and with $j = 3$. Continue this policy until $j = q$, when all data will have been processed. Then

$$\hat{x}_q(k + 1|k + 1) = \hat{x}(k + 1|k + 1) \quad (14)$$

and

$$\Sigma_q(k + 1|k + 1) = \Sigma(k + 1|k + 1). \quad (15)$$

VI. Optimal Sequential Processing

In this section we show how the results of the preceding sections can be combined to determine the best way (best in the sense of minimum computing time) to process r independent measurements. We shall show how to determine $r_1, r_2, \cdots, r_l$, and l so we can state that, for an nth-order system, the best way to process r measurements is in l batches, of size r_i, $i = 1, 2, \cdots, l$.

Sequential processing is used because it provides a means for processing l batches of measurements. Letting r_1 denote the number of measurements in the first batch, we observe that the r_1 measurements are processed completely in Step 1 of the three-step sequential processing computational procedure (Section V) and that the remaining $r - r_1$ measurements, which are now distributed in the $l - 1$ batches ($r_2, r_3, \cdots$, and r_l), are processed in Steps 2 and 3 of this procedure. Focusing our attention on the computation of $\hat{x}_1(k + 1|k + 1)$ from the r_1 measurements, we observe from Table VI that the first two computations are independent of $r = r_1$; hence, for the purposes of the present study, we need not include the computation time associated with computing $\hat{x}_1(k + 1|k)$ and $\Sigma_1(k + 1|k)$. We shall only be interested in the computation time associated with (11), (12), and (13) for $j = 1, 2, \cdots$, and l. Denoting this computation time as $C_I(l)$, it is easily shown that

$$C_I(l) = \sum_{j=1}^{l} C_I(n, r_j, \text{MUL}, \text{DIV}) \quad (16)$$

[3] Observe that, relative to the second batch of data, $\Sigma_1(k + 1|k + 1)$ acts like the *a priori* covariance matrix.

where

$$C_I(n, r_j, \text{MUL}, \text{DIV}) = [M_I(n, r_j)\cdot\text{MUL} + 2A_I(n, r_j) + L_I(n, r_j, \text{MUL}, \text{DIV})]\cdot T_u \quad (17)$$

and where $M_I(n, r_j)$, $A_I(n, r_j)$, and $L_I(n, r_j, \text{MUL}, \text{DIV})$ are obtained from the rows in Table VI labeled $K(k + 1)$, $\hat{x}(k + 1|k + 1)$, and $\Sigma(k + 1|k + 1)$, by replacing r by r_j and summing the three columns labeled "Number of Multiplications," "Number of Additions," and "Logic Time:"

$$M_I(n, r_j) = n^3 + r_j^3 + 2nr_j + 2nr_j^2 + 2n^2r_j \quad (18)$$

$$A_I(n, r_j) = n^3 - n^2 + r_j^3 + 2nr_j^2 + 2n^2r_j \quad (19)$$

and

$$L_I(n, r_j, \text{MUL}, \text{DIV}) = 188 + 106n + 47n^2 + 6n^3 + 12n^2r_j + 54nr_j + 12nr_j^2 + 165.5r_j^2 + 150r_j + 41r_j^3 + 7.5r_j^4 + \text{MUL}(4 + 2.5r_j + 0.5r_j^2) + \text{DIV}(r_j + 2r_j^2). \quad (20)$$

Substituting (17)–(20) into (16), it is easily shown that

$$\frac{C_I(l)}{T_u} = l[(188 + 4\ \text{MUL}) + 106n + 45n^2 + (8 + \text{MUL})n^3] + \sum_{j=1}^{l} r_j[150 + \text{DIV} + \text{MUL}\ (2.5 + 2n + 2n^2) + 54n + 16n^2] + \sum_{j=1}^{l} r_j^2[2\ \text{DIV} + \text{MUL}(2n + 0.5) + 165.5 + 16n] + \sum_{j=1}^{l} r_j^3[43 + \text{MUL}] + 7.5 \sum_{j=1}^{l} r_j^4. \quad (21)$$

For the purposes of this paper, r_1, r_2, $\cdots$, and r_l, which must be computed before $C_I(l)/T_u$ can be evaluated, are obtained as the solution to the following problem in combinatorial analysis:

$$r_1 + r_2 + \cdots + r_l = r \quad (22a)$$

$$r_j = 0, 1, \cdots, \text{or at most } r, \text{ for } j = 1, 2, \cdots, l \quad (22b)$$

$$\text{the collection of nonzero } r_j \text{ precedes zero } r_j \quad (22c)$$

$$\text{for fixed } r_1, \text{ only one combination of } r_2, \cdots, r_l \text{ allowed (no permutations of } r_2, \cdots, r_l \text{ allowed).} \quad (22d)$$

Equation (22c) guarantees that

$$l = \text{total number of nonzero } r_j. \quad (23)$$

An example that illustrates the solution to these equations for $r = 4$ is given in Table VII.

TABLE VII

SOLUTIONS TO (22A)–(22C) WHEN $r = 4$[a]

r_1	r_2	r_3	r_4
1	1	1	1
1	1	2	0
1	3	0	0
2	1	1	0
2	2	0	0
3	1	0	0
4	0	0	0

[a] Equation (22d) excludes the possibility of the two solutions (1, 1, 2, 0) and (1, 2, 1, 0). Both of these solutions would give exactly the same computation time in (17); hence, only one is needed.

At this point, we could proceed to illustrate some solutions to (17); however, we observe from Table VII that $r_j = 1$ is a frequently occurring value for the r_j. When $r_j = 1$, computational time can be further reduced by evaluating $[C(\Sigma C') + \Theta]^{-1}$ in the expression for $K(k+1)$ by division rather than by using the matrix inverse subroutine (as was done in Table IV). This changes the fourth row of $K(k+1)$ in Table IV to DIV/MUL multiplications, zero additions, and zero logic time. Now, setting $r = 1$ in the rows of Table IV for $K(k+1)$, subject to its changed fourth row, $\hat{x}(k+1|k+1)$ and $\Sigma(k+1|k+1)$, we find that

$$M_I(n, 1) = n^3 + 2n^2 + 4n + \text{DIV/MUL} \tag{24}$$

$$A_I(n, 1) = n^3 + n^2 + 2n \tag{25}$$

and

$$L(n, 1, \text{MUL}, \text{DIV}) = 266 + 4\text{MUL} + 176n + 59n^2 + 6n^3. \tag{26}$$

Hence

$$C_I(n, 1, \text{MUL}, \text{DIV}) = [266 + \text{DIV} + \text{MUL}(4 + 4n\ 2n^2 + n^3) + 180n + 61n^2 + 8n^3]. \tag{27}$$

Finally, letting

$$l_1 \equiv \text{number of unity } r_j \text{ from the solution to (22a)–(22d)} \tag{28}$$

we combine (21), (27), and (28) and rewrite $C_I(l)/T_u$ as

$$\begin{aligned}\frac{C_I(l)}{T_u} = {} & (l - l_1)[(188 + 4\,\text{MUL}) + 106n + 45n^2 \\ & + (8 + \text{MUL})n^3] \\ & + \Sigma^* r_j[150 + \text{DIV} + \text{MUL}(2.5 + 2n + 2n^2) \\ & + 54n + 16n^2] \\ & + \Sigma^* r_j^2[2\,\text{DIV} + \text{MUL}(2n + 0.5) \\ & + 165.5 + 16n] \\ & + \Sigma^* r_j^3[43 + \text{MUL}] + 7.5\,\Sigma^* r_j^4 \\ & + l_1[266 + \text{DIV} + \text{MUL}(4 + 4n + 2n^2 + n^3) \\ & + 180n + 61n^2 + 8n^3]\end{aligned} \tag{29}$$

where

$$\Sigma^* \equiv \sum_{j=1}^{l} \quad r_j \neq 1. \tag{30}$$

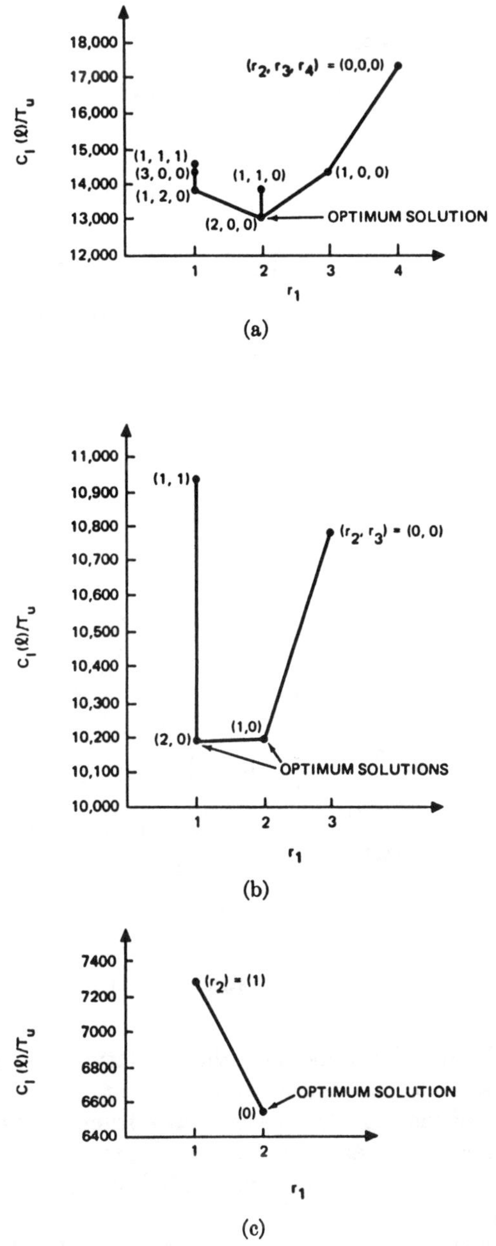

Fig. 5. Cost curves for $n = 4$ and (a) $r = 4$, (b) $r = 3$, and (c) $r = 2$.

Example 1: Fig. 5 depicts $C_I(l)/T_u$ versus r_1 for a fourth-order system ($n = 4$), with either 4, 3, or 2 measurements. These results were obtained by a direct solution of (22a)–(22d) and then by direct computation of $C_I(l)/T_u$ in (29).

Observe from (22a) and (22b) that, for r measurements, there can be at most r batches; hence, $(r_2, r_3, \cdots, r_r)$, which appears next to each test point in Fig. 5, denotes the way in which $(r - r_1)$ measurements are allocated into batches 2, 3, $\cdots$, and r.

The plots in Fig. 5 demonstrate the existence of local minima as well as global minima. If r_1 is fixed *a priori* (i.e., perhaps only $r - r_1$ measurements are statistically independent), it is clear that there is an optimum way to batch the remaining $r - r_1$ measurements. If, on the other hand, r_1 is not fixed *a priori*, then it is clear that there is an optimum way to batch the r measurements. Note, in Fig. 5(b), that, regarding computing time, $(r_1, r_2, r_3) = (1, 2, 0)$ and

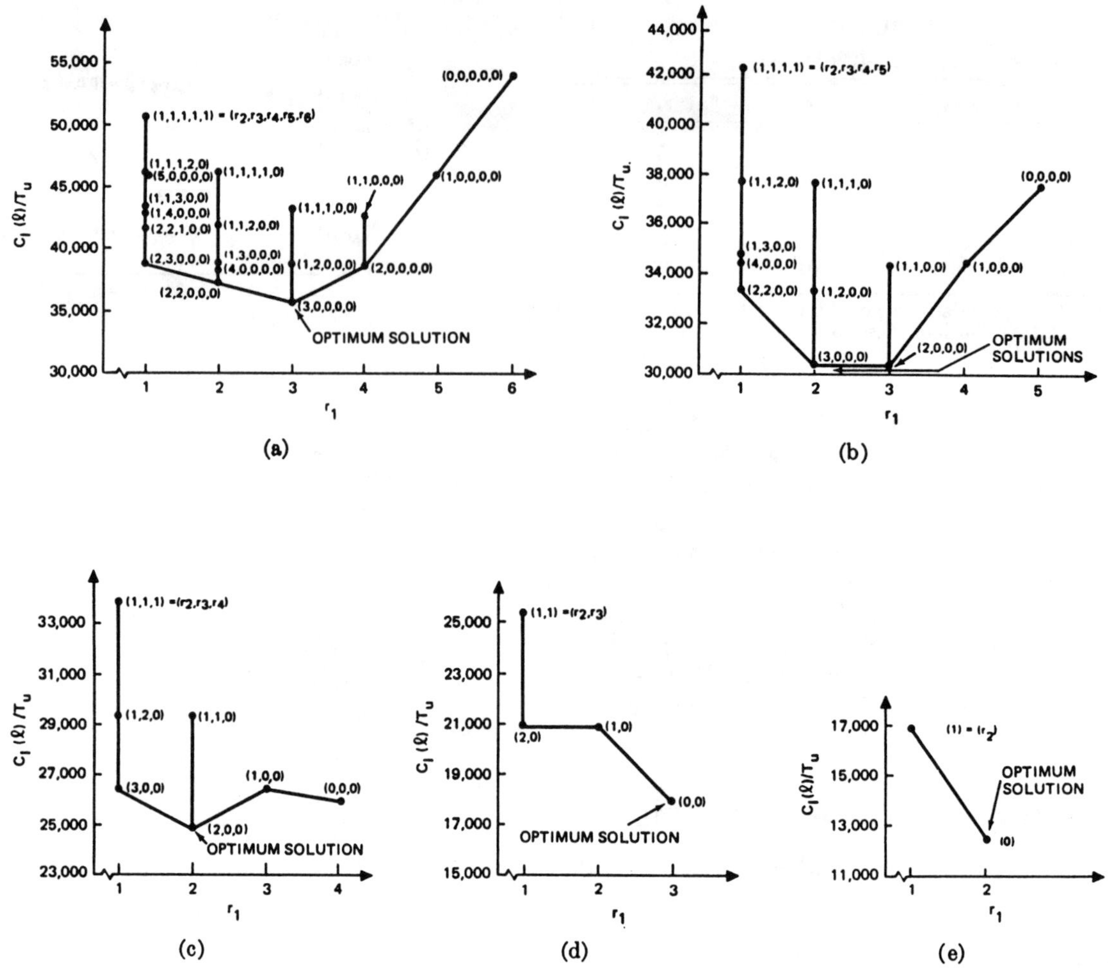

Fig. 6. Cost curves for $n = 6$ and (a) $r = 6$, (b) $r = 5$, (c) $r = 4$, (d) $r = 3$, and (e) $r = 2$.

(2, 1, 0) are identical. Since these solutions are merely permutations of each other, either one can be used.

The following conclusions are drawn from Fig. 5(a)–(c).

1) Processing four measurements in two batches of two measurements is 25 percent more efficient than processing them in one batch.

2) Processing three measurements in two batches, one containing one measurement and the other containing two measurements, is 5.5 percent more efficient than processing them in one batch.

3) Processing two measurements in one batch is 11.5 percent more efficient than processing them in two batches of one measurement.

Example 2: Fig. 6 depicts $C_I(l)/T_u$ versus r_1 for a sixth-order system ($n = 6$), with either 6, 5, 4, 3, or 2 measurements. The following conclusions are drawn from Fig. 6(a)–(e).

1) Processing six measurements in two batches of three measurements is 34 percent more efficient than processing them in one batch.

2) Processing five measurements in two batches, one containing three measurements and the other containing two measurements, is 19 percent more efficient than processing them in one batch.

3) Processing four measurements in two batches of two measurements is 4.5 percent more efficient than processing them in one batch.

4) Processing three measurements in one batch is 16.6 percent more efficient then the next best other way.

5) Processing two measurements in one batch is 36 percent more efficient than processing them in two batches of one measurement.

Example 3: Here we compare the results from Examples 1 and 2 and conjecture the following.

1) For large values of r (relative to n), a unique minimum of $C_I(l)/T_u$ occurs for some value of r_1 *between* 1 and r [Figs. 5(a) and 6(a)].

2) For $r = n$, any other way of processing the measurements is better than processing them in one batch [Figs. 5(a) and 6(a)].

3) For $r < n$, processing the measurements in one batch is not as bad as some other ways [Figs. 5(b), 6(b), and 6(c)].

4) For values of $r \leq n/2$, the best way to process the measurements is in one batch [Figs. 5(c), 6(d), and 6(e)].

5) Conclusions about optimal processing depend not only on r, but also on n. For example [Figs. 5(b) and 6(d)], given $r = 3$, the best way to process these measurements for $n = 4$ is in two batches; whereas for $n = 6$, the best way to process the three measurements is in one batch.

VII. Conclusions

Computational requirements—computing time per iteration and storage—have been obtained for a discrete Kalman filter. These requirements were obtained as explicit functions of the system's order, the number of measurements, and the number of plant disturbances. They were obtained for the situation in which all of the system's matrices were assumed to be full.

Computing time has been viewed as a linear combination of the times associated with total numbers of multiplications and additions and logic time. Quite often, logic time is not included in computation studies and statements; however, it has been demonstrated herein that logic time is at least as important, if not more important, than multiplication time.

Total computing time for a discrete Kalman filter can be greatly reduced by making explicit use of the structure of the system's matrices (they often contain many zeros). This reduction will be obtained at the expense of greater computer storage requirements. In the design of a Kalman filter for a 14th-order model of a flexible space station, a 50 percent saving in computing time is realized when system-structure information is used [4]; however, computer storage requirements are increased by 15 percent. In light of the large savings in computing time that can be realized by using information about the structure of system matrices, one can view the results from this paper as an upper bound on computing time.

This paper has also shown that rather large savings in computing time can be realized by grouping independent measurements optimally into batches and then using sequential processing on the batches. The optimal way to process r measurements is strongly dependent on the relative sizes of r and n. From examples, a general rule of thumb appears to be that, if $r \leq n/2$, process the r measurements in one batch; if $r > n/2$, process the measurements in l batches, each with $r_1, r_2, \cdots$, and r_l measurements, as described in Section VI. As $r \rightarrow n$, large savings in computing time are realized when optimal sequential processing is employed. The economic advantages of this method of data processing in this situation are obvious and it is reasonable to expect that they apply to ground-based simulations as well as to airborne computations.

References

[1] D. O. Baechler, "Trends in aerospace digital computer design," *IEEE Comput. Group News*, pp. 18–23, Jan. 1969.
[2] Gradde, Ramo, and Woolridge, Eds., *Handbook of Automation, Computation, and Control*, vol. I. New York: Wiley, 1958.
[3] H. W. Sorenson, "Kalman filtering techniques," in *Advances in Control Systems*, vol. 3, C. T. Leondes, Ed. New York: Academic Press, 1966.
[4] J. M. Mendel, "Feasibility and design study of adaptive control of flexible, highly variable spacecraft," NASA-CR-111781, Aug. 1970.

Section I-G
Smoothing

THE Kalman filter provides the linear minimum mean-squared estimator of the state $\boldsymbol{x}_k$ given the measurements $\boldsymbol{z}_1$ through $\boldsymbol{z}_k$. Many applications require the best estimator of the state $\boldsymbol{x}_j$ at some time t_j where $j<k$. This is referred to as the *smoothing* problem. The Kalman filter has the advantage that it is entirely recursive and can be used for real-time implementations. It is particularly well-suited to this type of application because there is no need to store measurements obtained at times earlier than the current time. While it is possible to obtain recursive solutions for the smoothing problem, these solutions require the storage of filtered estimates and covariances at all times in the interval. This requirement limits their suitability for real-time applications. Consequently, smoothing algorithms have seen relatively little application when compared with their close relative, the Kalman filter. A major exception appears in the work on seismic deconvolution by Mendel and his students. Representative papers are listed in the bibliography for this section.

Four papers which discuss the linear smoothing problem are contained in this section. Meditch introduced a classification of smoothing into fixed-interval, fixed-point, and fixed-lag problems. In *fixed-interval* smoothing, the state $\boldsymbol{x}_j$ is to be estimated for all j from measurements obtained in an interval (t_1, t_N) where the final time t_N and the number of samples N are fixed in value. *Fixed-point smoothing* arises when the time t_j does not change, while the last time t_N increases. In some cases, the state $\boldsymbol{x}_{N-j}$ at a *fixed-lag* j is estimated as N increases. This classification scheme provides a useful perspective for understanding the types of requirements that might arise in a specific application. In the first paper of this section, Meditch surveys the literature and results of linear and nonlinear smoothing and provides an extensive bibliography to the literature as it existed at that time.

The papers by Rauch, Tung, and Striebel and by Fraser and Potter provide different algorithms for solving the fixed-interval smoothing problem. A third algorithm was proposed by Bryson and Frazier in [1]. Comparing the algorithms, one finds very different requirements and insights into the structure of the solution. The Rauch–Tung–Striebel algorithm, given in eqs. (3.28)–(3.33), uses the filter estimates obtained for each sampling time in the entire interval. It proceeds "backwards" from the terminal time to the initial time using a recursion in which the filter estimates serve as "measurements" in updating the smoothed estimate. The Fraser–Potter algorithm uses a "forward filter" (i.e., eqs. (3)–(4)) and a "backward filter" (i.e., eqs. (5)–(6)). The smoothed estimator is formed as the convex combination of the two filter estimators (i.e., eqs. (1)1–(2)). The two algorithms have a very different appearance but are equivalent solutions of the same problem, as is shown by Fraser–Potter (i.e., eq. (16)). While the discussion considers continuous time systems, Table I states the equivalent results for discrete-time systems. Consideration of each algorithm provides the reader with an understanding of the increased computational burden inherent in their implementation.

The derivation of solutions to the smoothing problem can be unified by using the innovations approach as presented by Ljung and Kailath in [9]. The property that the innovations sequence generated by the Kalman filter is white can be used to obtain a relatively straightforward solution of the Wiener–Hopf equation. The solution is given in Section II of this paper. General results for each type of smoothing problem are derived in Section III. These results can be converted to any of the proposed smoothing algorithms by algebraic manipulations. The innovations approach provides a unifying perspective on the smoothing problem that does not appear to be possible in any other way. Ljung and Kailath's paper provides a useful complement to Meditch's survey article.

References

[1] A. E. Bryson and M. Frazier, "Smoothing for linear and nonlinear dynamic systems," Tech. Rep. ASD-TDR-63-119, Wright–Patterson Air Force Base, OH, pp. 353–364, 1962.

[2] H. E. Rauch, "Optimal estimation of satellite trajectories including random fluctuations in drag," *AIAA J.*, vol. 3, pp. 717–722, 1965.

[3] D. Q. Mayne, "A solution of the smoothing problem for linear dynamic systems," *Automatica*, vol. 4, pp. 73–92, 1966.

[4] J. S. Meditch, "On optimal smoothing theory," *J. Info. Contr.*, vol. 10, pp. 598–615, 1967.

[5] ——, "On optimal fixed-point smoothing," *Int. J. Contr.*, vol. 6, pp. 189–199, 1967.

[6] C. N. Kelley and B. D. O. Anderson, "On the stability of fixed-lag smoothing algorithms," *J. Franklin Inst.*, vol. 291, pp. 271–281, 1971.

[7] J. M. Mendel, "Minimum-variance deconvolution," *IEEE Trans. Geosci. Remote Sensing*, vol. GE-19, pp. 161–171, 1981.

[8] J. J. Kormylo and J. M. Mendel, "Maximum-likelihood seismic deconvolution," *IEEE Trans. Geosci. Remote Sensing*, vol. GE-21, pp. 72–82, 1983.

[9] L. Ljung and T. Kailath, "A Unified approach to smoothing formulas," *Automatica*, vol. 12, pp. 147–157, 1976.

A Survey of Data Smoothing for Linear and Nonlinear Dynamic Systems*

Une Étude du Nivellement de Données pour les Systèmes Dynamiques Linéaires et non-Linéaires

Ein Überblick über Datenglättung für lineare und nichtlineare dynamische Systeme

Обзор сглаживания данных для линейных и нелинеиных динамических систем

J. S. MEDITCH†

The field of data smoothing for noisy linear and nonlinear dynamic systems has reached sufficient maturity to warrant a detailed survey of its development.

Summary—A survey of the field of data smoothing for lumped-parameter, linear and nonlinear, dynamic systems is presented. The survey beings with the work of Kepler and Gauss, proceeds through that of Kolmogorov and Wiener, and concludes with the studies of numerous researchers during the past 10–12 years. The purpose of the survey is to place in perspective the development of the field of data smoothing relative to the broader area of estimation theory of which it is a part.

1. INTRODUCTION

THE field of data smoothing, as defined here, deals with the problem of obtaining an approximation to a state or parameter, vector at a time point which is intermediate or prior to a span of imperfect measurements that are related to the vector. This field is part of the broader field of estimation theory, an area whose origin dates back at least to the time of KEPLER [1], if not to earlier astronomers, and their attempts to "fit" orbits to celestial observations. However, the analytical tool that was needed to cope with these orbit determination problems did not appear until GAUSS [2, 3] presented his least-squares method.

It is the purpose of this paper to present a survey of the field of data smoothing, as an area within estimation theory, beginning with the work of Kepler and Gauss, through that of KOLMOGOROV [4] and WIENER [5], and concluding with the concerted efforts of numerous individuals throughout the past 10-12 years. The latter work, of course, was stimulated in large measure by the filter theory results of SWERLING [6], KALMAN [7, 8] and KALMAN and BUCY [9].

In order to keep the paper of reasonable length, attention will be focused primarily on finite-dimensional problems. Further, since smoothing for infinite-dimensional processes is a relatively new area of endeavor, a survey of it is perhaps best left until sufficient time to gain perspective has elapsed. However, linear and nonlinear problems in both discrete and continuous-time will be treated here.

To begin, the assumed form of the measurement process is taken to be

$$z(i)=y(i)+v(i)$$

in the discrete-time case, and

$$z(t)=y(t)+v(t)$$

in the continuous-time case. Here, z, y and v are m-vectors, respectively, the measurement, message, and measurement error vectors; $i=0, 1, \ldots$ is the discrete-time index denoting the sequence of fixed time points $t_o<t_1< \ldots$; and t denotes continuous time with $t \geqslant t_o$, t_o fixed. Specific forms for y are $H(i)x(i)$ and $H(t)x(t)$ in the linear case where H is an $m\times n$ matrix, $H(i)$ bounded, $H(t)$ continuous, and x is the state vector which is to be estimated. In the nonlinear case, one has $h(x(i), i)$ and $h(x(t), t)$, both continuous in x, the former bounded in i, and the latter continuous in t. For the present, no particular structure is attached to the processes $\{x(i), i=0, 1, \ldots\}$, $\{x(t), t\geqslant t_o\}$, $\{v(i), i=0, 1, \ldots\}$, or $\{v(t), t\geqslant t_o\}$.

Any estimate of $x(k)$, $k=0, 1, \ldots,$ which is based on the sequence of measurements

$$\{z(0), \ldots, z(j)\},$$

* This work was supported by the Air Force Office of Scientific Research, Air Force Systems Command, USAF, under Grant No. AFOSR-71-2116.

† University of California, Irvine, California 92664.

Reprinted with permission from *Automatica,* vol. 9, pp. 151–162, Mar. 1973.

$j \geq 0$, is denoted $\hat{x}(k|j)$. Similarly, in continuous-time, one has, for any t for which x is defined, $\hat{x}(t|s)$ as an estimate of $x(t)$ derived from the measurement set $\{z(\tau), t_0 \leq \tau \leq s\}$. These estimates are classified according to the value of k or t relative to that of j or s, respectively, in the following manner now in common usage [7]:

1. Prediction, $k > j$ $(t > s)$
2. Filtering, $k = j$ $(t = s)$
3. Smoothing, $k < j$ $(t < s)$.

Further classification of smoothed estimates as introduced in [10] has proven useful and will be adhered to in the sequel:

1. Fixed-interval, $k = 0, 1, \ldots, N-1$; N fixed $j = N$
2. Fixed-point, k fixed $j = k+1, k+2, \ldots$
3. Fixed-lag, $k = 0, 1, \ldots$ $j = k + M$, M = fixed positive integer Δlag

and similarly in the continuous-time case.

Usually, an estimate, whether predicted, filtered, or smoothed, is determined so that it is "best" or optimal with regard to some estimation accuracy functional. The specific functionals which have been used by the various authors will be noted in the sequel.

The presentation here will generally follow the growth of the field historically. However, where relevant, works will be discussed out of chronological order to point out their relationship to each other and to the field's development.

2. KEPLER, GAUSS, KOLMOGOROV AND WIENER

It is generally agreed that estimation theory has its analytical foundation in GAUSS' least-squares method [2, 3]. Gauss developed the technique in 1795 to establish and predict the orbit of Ceres. Prior to this, geometric ideas had been utilized extensively in orbit determination, generally with great frustration. Kepler, for example, began his study of the orbit of Mars [1] in 1601 having at his disposal 12 observations, 10 from Brahe and two of his own. He began by attempting to fit a circle to these data, a task that he abandoned in 1603 in favor of the elliptic orbit. While his method was awkward, it was workable; but it appears [1] that Kepler was very much a victim of his own numerical errors. It thus remained for Gauss to point the way and present a procedure that has proven extremely useful in engineering and the sciences.

Following Gauss' work, relatively little appeared on estimation theory until the 1930's. By that time, the theory of stochastic processes had matured to the point where estimation problems could be reexamined and reformulated from a new viewpoint. There followed the definitive works of KOLMOGOROV [4] and WIENER [5]. Both researchers adopted the linear least-squares estimation viewpoint and considered stationary random processes; Kolmogorov treated the discrete-time problem, while Wiener dealt with the continuous-time one. Their methods of attack were substantially different, although both apparently first approached the prediction problem: "In both cases the object of study is the optimum prediction" [5, p. 59].

Kolmogorov's approach made use of a fundamental decomposition theorem on regular processes due to WOLD [11] (see also DOOB [12, Ch. 12, Theorems 4.1, 4.2 and 4.3]). Wiener described the message and measurement error processes in terms of their correlation functions, or equivalently their power spectral densities, and obtained the now familiar time–domain, Wiener–Hopf, integral equation. The equation could then be solved via the spectral factorization technique of harmonic analysis [5]. In fact, it is both interesting and curious that WIENER and HOPF had solved this integral equation much earlier [13, 14].

It appears that Wiener's work was motivated by a number of physical problems, while Kolmogorov's research stemmed more from mathematical considerations. Wiener had in mind applications in weather forecasting, flood control, servomechanisms, economics, and even the commodities market [5, pp. 1-24]. His work in prediction stemmed from the need in the late 1930's and early 1940's to cope with fire control problems. However, Wiener had for some time been intensely interested in communication problems, and it was undoubtedly this interest that led him to the study of filtering and fixed-lag smoothing—the latter he referred to as "lag filtering".

Typical of the type of problem that can be handled via the Wiener theory is the following [15-17]. In the continuous-time case, let $y(t)$ and $v(t)$ be scalar-valued, stationary, random processes which are independent of each other and have power spectral densities

$$\Phi_{yy}(s) = \frac{a^2}{-s^2 + b^2}$$

and

$$\Phi_{vv}(s) = c^2$$

where a^2, b^2 and c^2 are positive, real numbers and s is complex frequency.

Let it be desired to reproduce the message $y(t)$ with a fixed lag T from the measurement data $\{z(s), -\infty < s \leq t\}$. Then it can be shown that the system which minimizes the mean-square error $E\{y(t)-y(t-T|t)]^2\}$, where E denotes the expected value, has the transfer function

$$W_o(s)=(\alpha-b)\frac{e^{-\alpha T}(s+b)-(\alpha+b)e^{-sT}}{s^2-\alpha^2}$$

where

$$\alpha=\sqrt{b^2+\frac{a^2}{c^2}}.$$

The corresponding impulse response is

$$w_o(t)=\begin{cases} 0 & \text{for } t<0 \\ \frac{\alpha-b}{\alpha}e^{-\alpha T}(\alpha\cosh\alpha t+b\sinh\alpha t) & \text{for } 0\leqslant t\leqslant T \\ \frac{\alpha-b}{\alpha}(\alpha\cosh\alpha T+b\sinh\alpha T)e^{-\alpha t} & \text{for } t\geqslant T \end{cases}$$

and has the form shown in Fig. 1.

The corresponding mean-square error is

$$c^2(\alpha-b)\left[1-\frac{\alpha-b}{2\alpha}(1-e^{-2\alpha T})\right]$$

which clearly indicates the improvement in accuracy that is possible with fixed-lag smoothing, $T>0$, relative to filtering, $T=0$.

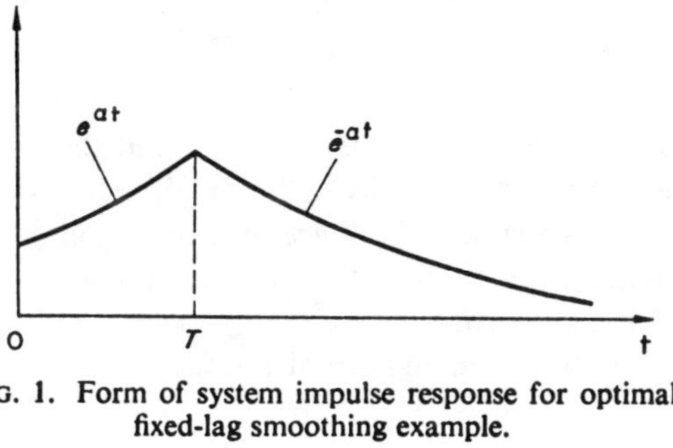

FIG. 1. Form of system impulse response for optimal fixed-lag smoothing example.

From its transfer function, it is clear that the smoother is infinite-dimensional which poses an obvious problem in terms of physical implementation. WIENER [5, p. 94] suggests approximation with the following argument: "If we thus fix on the reasonable delay for a filter, there is still another error dependent on the fact that the theoretically optimum design is not, in fact, realizable with a finite network of resistances, capacities, and inductances. This is the error implicit in our approximation to $e^{i\alpha\omega}$ (Wiener's notation for the time delay). Again, such an error may be estimated, and there is no point in reducing this part of the error to a level substantially below that of the error implicit in the delay already chosen. It is this last error which determines the number of meshes and ports appropriate to the complete filter network. Once this final error is decided upon, the rational voltage ratio characteristic of the network is determined, and its design is a matter of known technique."

For the present example, it is easy to show that

$$\int_0^\infty |w_o(t)|dt<\infty$$

so that the smoother is bounded-input/bounded-output stable. This conclusion is not immediately evident from the transfer function although one can readily show that there is a pole-zero cancellation at $s=\alpha$.*

In [5] WIENER examines the prediction problem for stationary random sequences, but does not address the filtering and smoothing problems for this class of processes. In Appendix B of [5], and also in [19], LEVINSON deals with minimum mean-square error prediction, filtering, and fixed-lag smoothing for these processes subject to a constraint on the estimator. Specifically, he requires that the estimate be of the form

$$\sum_{j=0}^{M} w(j)z(k-j)$$

and seeks the weighting sequence $w(j)$, $j=0, 1, \ldots, M$ "for some suitably chosen M" such that

$$I=\lim_{N\to\infty}\frac{1}{2N+1}\sum_{k=-N}^{N}e^2(k)$$

is minimized where

$$e(k)=y(k+s)-\sum_{j=0}^{M}w(j)z(k-j).$$

Here, s is an integer which is positive, zero, or negative in the case of prediction, filtering, or fixed-lag smoothing, respectively.

The minimization of I with respect to the $w(j)$ leads to $M+1$ linear algebraic equations in as many unknowns.

Among other estimation problems, Kolmogorov treated the following problem of linear interpolation for stationary random sequences (see YAGLOM [20, Ch. 8])†. For some fixed k and positive n, let

* In the state-space formulation of fixed-lag smoothing, this corresponds to the presence of an uncontrollable block which, when removed, renders the smoother stable without affecting its input–output properties [18]. This point is discussed further in the sequel.

† The original version of this book appeared in 1952 as a lengthy review paper in the Russian journal *Uspekhi Mathematicheskikh Nauk.*, Vol. 7, No. 5.

$\{y(i), i=k-n, \ldots, k-1, k+1, \ldots, k+n\}$denote a sequences of measurements which constitute a realization of a stationary random sequence. From these data, it is desired to "reconstruct" the missing value $y(k)$. It should be noted that one has "error-free" measurements here.

In this formulation, a linear estimate

$$\hat{y}(k)=\sum_{\substack{i=k-n\\ i\neq k}}^{k+n} a(i)y(i)$$

is assumed and the $a(i)$ are determined to minimize the mean-square error

$$\sigma_n^2=E\{[y(k)-\hat{y}(k)]^2\}$$

under the assumption that the autocorrelation function of the random sequence is known. This obviously reduces to solving a system of linear algebraic equations as above.

It can be shown that

$$\sigma^2=\lim_{n\to\infty}\sigma_n^2$$

exists. This limit is termed the mean-square interpolation error.

3. THE PAST TWENTY-FIVE YEARS

Numerous interpretations, extensions and generalizations of the Wiener–Kolmogorov work were made during the late 1940's and early to middle 1950's. Both time and frequency domain methods were employed, but the latter strongly dominated the work since it was the prevailing systems analysis method at that time. Also, attention was focused primarily on the prediction and filtering problems.

BODE and SHANNON [21] and ZADEH and RAGAZZINI [22] made Wiener's work more accessible to engineers with the latter including also an extension to the finite memory case. These authors independently introduced the "shaping filter" approach which considerably expedited solution. The approach allows one to model a random process as the output of a dynamic system excited by white noise. It has considerable engineering appeal and has proven to be a powerful modeling tool. A somewhat different technique, algebraic in nature, was presented by BATTIN [23; 15, Sec. 7.5]. Extensions to discrete-time problems were made by FRANKLIN [24] and LEES [25] among others.

At about this time, interest began to develop in coping with estimation problems for nonstationary random processes, principally via time–domain methods. There ensued the early work of BOOTON [26] and DAVIS [27] in 1952; that of BENDAT [28] in 1954; BENDAT [29], CARLTON and FOLLIN [30], PUGACHEV [31] and SOLODOVNIKOV and BATKOV [32] in 1956; STEEG [33] and HANSON [34] in 1957; BLUM [35], SHINBROT [36], and SWERLING [6] in 1958; BUCY [37] in 1959; and PARZEN [38–40] in 1959 and 1960. From the literature, it appears that CARLTON and FOLLIN [30] pioneered in adopting the viewpoint of sequential calculation of optimal, unbiased, linear estimates.

However, the work which has made the greatest impact is that of KALMAN [7, 8] on both the discrete and continuous-time problems, and KALMAN and BUCY [9] on the continuous-time one. They treated the prediction and filtering problems in detail, but only alluded to the smoothing problem. Nonetheless, they had laid the foundation for solution of the latter. Further, they dealt extensively with the question of stability in filtering. Their work has found numerous applications, particularly in the aerospace field, and is documented in a number of textbooks, e.g. [41–50].

KALMAN and BUCY [7-9] stressed the use of state space methods in the time domain to develop sequential, and, therefore, computationally efficient algorithms. This approach was subsequently first applied to the smoothing problem by CARLTON [51], RAUCH [52, 53], BRYSON and FRAZIER [54], RAUCH *et al.* [55] and WEAVER [56].

In addition to the problems of prediction and filtering, Carlton treated the problem of fixed-point smoothing for the following class of wide-sense, vector, Markov processes:

$$x_k=\Phi_{k-1}x_{k-1}+w_{k-1}$$

where $k=1, 2, \ldots$ is the discrete-time index, x is the state vector, Φ_{k-1} is a matrix, and the $\{w_{k-1}\}$ are zero mean, mutually orthogonal, vector-valued, random variables which are also orthogonal to x_o. It is assumed further that second moments of x_k and w_{k-1} are finite.

The measurements are of the form

$$z_k=H_kx_k$$

where H_k is a matrix. The matrix $E(z_kz_k')$, where E denotes expected value and prime the transpose, is assumed to be nonsingular.

Carlton's approach is best summarized in his own words: "... ; best linear estimates $\hat{x}(\tau)$ are by definition linear estimates minimizing the variance of estimate, $E[x(\tau)-x(\tau)][\hat{x}(\tau)-x(\tau)]'$". [51, p. 1], and "The basis of the analysis is the fact that best linear estimates and observations on which they are based are orthogonal to resulting errors of estimate; this and the linear structure of $\{x(\tau)\}$ permit orthogonalization of successive observations by subtracting from each observation its best linear prediction" [50, p. 2].*

Using orthogonality arguments, similar to those employed by KALMAN [7], Carlton developed the following sequential algorithm for optimal, linear, fixed-point smoothing:

$$\hat{x}_{k|j}=\hat{x}_{k|j-1}+M_{k|j-1}\Phi'_{j-1}H'_j(H_jP_{j|j-1}H'_j)^{-1}\tilde{z}_{j|j-1}$$

$$\tilde{z}_{j|j-1}=z_j-H_j\hat{x}_{j|j-1}$$

$$\hat{x}_{j|j-1}=\Phi_{j-1}\hat{x}_{j-1|j-1}$$

where k is fixed, $j=k+1,\ k+2,\ \ldots,$ and prime denotes the transpose. In these expressions, $\hat{x}_{k|j}$ is the optimal estimate of x_k based on measurements up to and including the one at time j, $\tilde{z}_{j|j-1}$ is the measurement residual, and

$$M_{k|j}=E[(x_k-\hat{x}_{k|j})(x_j-\hat{x}_{j|j})']$$

$$\underline{\Delta}E(\tilde{x}_{k|j}\tilde{x}'_{j|j})$$

where E denotes the expected value, and $\tilde{x}_{k|j}$ and $\tilde{x}_{j|j}$ are, respectively, the fixed-point smoothing and filtering errors.

The matrix $M_{k|j}$ is generated recursively by the relation

$$M_{k|j}=M_{k|j-1}\Phi'_{j-1}[I-H'_j(H_jP_{j|j-1}H_j)^{-1}]H_jP_{j|j-1}$$

subject to the initial condition $M_{k|k}=P_{k|k}$. Here, $P_{k|k}$ is the covariance matrix of the filtering error $\tilde{x}_{k|k}=x_k-\hat{x}_{k|k}$, $P_{j|j-1}$ is the covariance matrix of the prediction error $\hat{x}_{j|j-1}=x_j-\hat{x}_{j-1}$, and I is the identity matrix.

Finally, the covariance matrix of the fixed point smoothing error $\tilde{x}_k|_j$ is determined by the expression.

$$P_{k|j}=P_{k|j-1}-M_{k|j-1}\Phi'_{j-1}H'_j(H_jP_{j|j-1}H'_j)^{-1}H_j\Phi_{j-1}M'_{k|j-1}$$

for $j+k+1,\ k+2,\ \ldots,$ where $P_{k|k}$, the filtering error covariance matrix at time k, is the initial condition.

* The difference referred to here has traditionally been termed the "measurement residual". Kalman was quite explicit in pointing out and exploiting a particularly important property of the random process defined by the measurement residuals [7, p. 42]: "..., and that the signal after the first summer is white noise since $y(t|t-1)$ is obviously an orthogonal random process. This corresponds to some well-known results in Wiener filtering, see, e.g. Smith 28], Chapter 6 Figs. 6–4."

RAUCH [52, 53] and RAUCH *et al.* [55] developed sequential algorithms for discrete-time, optimal, fixed-interval, fixed-point, and fixed-lag smoothing where optimality is in the sense of marginal maximum likelihood. The model they adopted was of the form

$$x_{k+1}=\Phi_kx_k+w_k$$

$$z_k=H_kx_k+v_k$$

where $k=0, 1, \ldots$ is the time index, Φ_k and H_k are matrices, $\{w_k\}$ and $\{v_k\}$ are gaussian, white sequences, independent of each other, with zero means and known covariances. The initial state x_o was assumed to be gaussian with given mean, covariance, and to be independent of $\{w_k\}$ and $\{_kv\}$.

The fixed-interval algorithm is of some interest here. It is characterized by the set of relations

$$\hat{x}_{k|N}=\hat{x}_{k|k}+A_k(\hat{x}_{k+1|N}-\hat{x}_{k+1|k})$$

$$P_{k|N}=P_{k|k}-A_k(P_{k+1|k}-P_{k+1|N})A'_k$$

$$A_k=P_{k|k}\Phi'_kP^{-1}_{k+1|k}$$

where $k=N-1,\ N-2,\ \ldots,$ for N fixed and $[\ ]^{-1}$ denotes the matrix inverse. Here, it is noted that the algroithm is sequential in reverse time. Further, $P_{k+1|k}$ is the covariance matrix of the prediction error $\tilde{x}_{k+1|k}=x_{k+1}-\hat{x}_{k+1|k}$ and $P_{k|N}$ the covariance matrix of the fixed-interval smoothing error $\hat{x}_{k|N}=x_k-\hat{x}_{k|N}$. The required inputs to this algorithm, as stored values from optimal filtering and prediction, are evident. However, the principal disadvantage lies in the required inversion of $P_{k+1|k}$ at each step. Not only is this computationally inefficient, but numerical problems arise when $P_{k+1|k}$ is ill-conditioned as it often is in practice.

Earlier, LEE [57] had solved the fixed-interval smoothing problem for discrete-time, linear systems utilizing the Bayesian estimation viewpoint.

In [56], WEAVER used the same viewpoint to obtain the sequential relations for fixed-point smoothing for discrete-time, linear systems.

BRYSON and FRAZIER [54] were the first to treat the fixed-interval smoothing problem for continuous-time, linear and nonlinear, dynamic systems. They adopted the maximum likelihood estimation viewpoint and reformulated the problem as one of minimizing an integral quadratic performance functional subject to a dynamic constraint, viz., the system dynamics. This reduced the problem to a Bolza problem in the calculus of variations. For linear processes, the result was a linear, two-point boundary-value problem which they solved in closed-form via the sweep method. In the nonlinear case, they proposed solution via iterative procedures utilizing again the sweep method.

The linear, discrete-time, Bryson–Frazier algorithm has proven to be of some interest in potential applications. It has the advantage over the Rauch–Tung–Striebel formulation of avoiding the matrix inversion discussed earlier.

The requisite process model is the same as that given above. The algorithm consists of a "forward sweep" of the measurement data utilizing the Kalman filter over the fixed-interval $k=0, 1, \ldots, N$, and a "reverse sweep" governed by the "reverse-time" pair of equations

$$\hat{x}_{k|N}=\hat{x}_{k|k}-P_{k|k}\Phi_k'\lambda_k$$

and

$$\lambda_{k-1}=(I-K_kH_k)'(\Phi_k'\lambda_k-H_k'R_k^{-1}\tilde{z}_{k|k-1})$$

for $k=N-1, N-2, \ldots, 0$ where $\lambda_N=0$, K_k is the Kalman filter gain matrix, and R_k is the covariance matrix of the measurement noise v_k. The new variable here is the vector λ which is of the same dimension as the state vector. For purposes of calculation, it is an auxiliary variable, but it does have the following mathematical interpretation:

$$\lambda_k=P_{k+1|k}^{-1}(\hat{x}_{k+1|N}-\hat{x}_{k+1|k})$$

which makes clear the manner in which the present algorithm avoids the problem of inverting the prediction error covariance matrix.

In this formulation, one also obtains the optimal estimate of the system noise, viz.,

$$\hat{w}_{k|N}=-Q_k\lambda_k$$

where Q_k is the covariance matrix of the system noise w_k. Further details on the algorithm are given in [47].

In [52], RAUCH *et al.* applied a formal limiting procedure of KALMAN's [8] to obtain an algorithm for fixed-interval smoothing for continuous-time linear systems. Their result is equivalent, as they showed, to that of Bryson and Frazier, but is in a simpler form.

The work of the above authors constituted the major step in the development of data smoothing for dynamic systems in the spirit of Kalman and Bucy. What followed were alternate derivations and interpretations, extensions to nonlinear systems by means of various methods, and applications.

In 1964, Cox [58] published his work on estimation for discrete-time nonlinear systems wherein he obtained the discrete analog of the two-point, boundary-value problem obtained by BRYSON and FRAZIER [54]. Cox' system model, subsequently adopted by numerous authors, was

$$x(k+1)=f(x(k), k)+G(k)w(k)$$

$$z(k)=h(x(k), k)+v(k)$$

with $\{w(k)\}$ and $\{v(k)\}$ independent, gaussian, white sequences with known means and covariances. The initial state $x(0)$ was assumed to be gaussian with known mean and covariance, and to be independent of $\{w(k)\}$ and $\{v(k)\}$.

Utilizing the method of orthogonal projection as KALMAN [7] had done in the case of prediction and filtering, MEDITCH [59] presented alternate derivations of RAUCH *et al.* [52, 53, 55] algorithms for discrete-time, fixed-interval and fixed-point linear smoothing, and pointed out how the same could be done for the fixed-lag case.

At about the same time, MAYNE [60] studied the problem of linear, fixed-interval smoothing for continuous-time processes, giving the smoothed estimate as a linear combination of two filtered estimates. The two filtered estimates were the outputs of forward-time and reverse-time optimal filters. This approach was also examined after by FRASER [61] and MEHRA [62] whose results were in a different, but equivalent form.

An alternate formulation of this approach, suitable for computational purposes, has been given by FRASER and POTTER [63].

In [64] and [10], MEDITCH applied Kalman's formal limiting procedure to Rauch's results to derive algorithms for continuous-time, fixed-point, and fixed-lag, linear smoothing, respectively.

KWAKERNAAK [65] studied the problem of optimal linear filtering for continuous-time, linear systems possessing time delays. As a by-product, he was able to obtain a general formula for continuous-time, linear smoothing. It appears that the same result had been obtained earlier by FRASER [61] and more recently by FROST [66], all three authors having utilized different methods.

Extensions to fixed-interval, linear smoothing in the presence of time-correlated measurement noise were given by BRYSON and HENRIKSON [67] in the discrete-time case, and MEHRA and BRYSON [68] in the continuous-time problem. In a subsequent publication [69], the latter authors have given an application of their results to the analysis of inertial navigation system performance.

MCLANE [70] obtained linear algorithms for all three smoothing classifications for continuous-time, linear systems with state-dependent noise in both the dynamics and measurements. Since the message process is non-gaussian, by virtue of the state-dependent noise, the linear algorithms, which were obtained by solving the Wiener–Hopf equation, are suboptimal. They are, however, superior to the algorithms which would follow from a linearization of the message process about the filtered estimate.

As in the case of filtering, considerable attention has been given to the problem of smoothing for continuous-time, nonlinear processes, i.e. nonlinear dynamic systems with white noise excitation wherein the measurements are nonlinear functions of the state in the presence of additive white noise. Formally, the system description is

$$\dot{x}(t)=f(x(t),\ t)+G(x(t),\ t)w(t)$$

$$z(t)=h(x(t),\ t)+v(t)$$

where $\{w(t)\}$ and $\{v(t)\}$ are white noise processes.

While closed-form relations have been obtained, principally, but not exclusively, in the form of a modified Fokker–Planck equation, all have thus far been in a form unsuitable for either on-line or off-line computation. Hence, the emphasis has been on approximations. As already noted, in the work of BRYSON and FRAZIER [54] on the continuous-time, fixed-interval problem, use of the sweep method was proposed for coping with the resulting two-point, boundary-value problem. COX [58] proposed the same in the discrete-time case.

Adopting the least-squares/invariant imbedding approach that DETCHMENDY and SRIDHAR [71] used on the continuous-time, nonlinear filtering problem, KAGIWADA *et al.* [72] derived partial differential equations for sequential, nonlinear smoothing.

A unified approach to all three cases of smoothing for both discrete-time and continuous-time nonlinear systems has been given by SAGE and EWING [73, 74] and SAGE [75]; see also Ch. 9 of [49]. The method involves the maximum likelihood viewpoint and invariant imbedding. The continuous-time algorithms follow from the discrete-time ones via KALMAN's formal limiting argument [8].

Utilizing the marginal maximum likelihood estimation approach of RAUCH *et al.* [55], MEDITCH [76–79] has presented both sequential and iterative-sequential (Newton–Raphson) algorithms for discrete and continuous-time, fixed-interval smoothing. The method is second-order in both dynamic and measurement nonlinearity.

In a study of filtering, prediction, and smoothing, FROST, GEESEY and KAILATH [66, 80–86] have employed the "innovations approach". The method is due to the early work of KOLMOGOROV [4] and WOLD [11], and later that of WIENER and MASANI [87, 88]. The approach had been utilized in the BODE–SHANNON [21] and ZADEH–RAGAZZINI [22] work on constant-coefficient, linear systems.

The essence of the method is as follows. Referring to the second equation in the introduction, if $\{v(t)\}$ is a zero mean, white noise and $\hat{y}(t|s)$ is the least-squares estimate of $y(t)$ given the measurement data $\{z(\tau),\ \tau<s\}$, then

$$\tilde{y}(t|s)=y(t)-\hat{y}(t|s),$$

which is the estimation error, is orthogonal to (uncorrelated with) $z(t)$, $t<s$. Further, if $\hat{z}(t|t)$ denotes the least-squares (filtered) estimate of $z(t)$, then

$$\nu(t)=z(t)-\hat{z}(t|t)$$

which is termed the "innovations process", $t<s$, is a white noise process whose covariance is identically that of $\{v(t),\ t<s\}$. As a consequence, $\{z(t),\ t<s\{$ and $\{\nu(t),\ t<s\}$ are "equivalent" processes in the sense that they possess the same statistical information.* These results had been noted earlier by numerous authors (see, for example, [7, 51, 89, 90] and the footnote in Section 3 of this paper).

One new result that has been proved [66, 92, 93] is that if $\{v(t)\}$ is gaussian, then $\{\nu(t)\}$ is also gaussian even if $\{y(t)\}$ is not.

Utilizing the above results, previous algorithms for linear filtering and smoothing have been re-derived. In the case of smoothing, general relationships are given and the specific algorithms developed therefrom are in a somewhat simpler form than the earlier ones.

General relations for nonlinear smoothing are also given [66, 85], but are subject to the following: "Assuming that equivalence holds, the rest of our derivation below is valid for processes more general than in (1)–(4). However, until a completely rigorous proof of the equivalence is obtained, the nonlinear estimation formulas for these more general processes can only be regarded as sub-optimum estimates; more precisely, as nonlinear least-squares estimates based on a possibly information-reducing prewhitening operation on the given observations" [85, p. 218].

Other work on nonlinear smoothing has also been presented by LEONDES *et al.* [94] and LEE [95]. In [94], the continuous-time, fixed-interval problem is examined for systems described by nonlinear, vector, differential equations with additive, gaussian white noise inputs. Exact functional differential equations for the smoothing density function and the smoothed expectation of an arbitrary function of the system's state are derived. Since solution of these equations is not possible except in extremely simple cases, approximations are developed for sequential, minimum-variance smoothing. An iterative technique is also suggested for cases where the nonlinearities may be severe.

LEE [95] presents a general solution to the continuous-time, fixed-point smoothing problem for the class of systems

$$dx=f(x)dt+F(x)dz$$

$$dy=g(x)dt+dw$$

* Equivalence has been proven under rather weak conditions for gaussian $\{y(t)\}$, but a completely rigorous proof is lacking—see [81], [85, p. 218], [91, 92].

where $\{z(t)\}$ and $\{w(t)\}$ are Wiener processes, and are uncorrelated. He utilizes techniques similar to KUSHNER's [96] to develop the, exact, partial differential equation for the evolution of the fixed-point smoothing density function. Solution is presented for the finite state case and an interesting example involving the processing of binary video telemetry data is given.

In the case of continuous-time, nonlinear, fixed-interval smoothing ANDERSON [97] has derived an equation for the probability density of the system state, conditioned on the measurement data over the interval. His approach makes novel use of the modified Fokker–Planck equation for the filtering probability density function.

LAINIOTIS [98] has obtained the *a posteriori* probability density functions for nonlinear prediction and smoothing utilizing the "partition theorem". For situations where the model is linear, the noise processes are gaussian, but the probability distribution of the initial state is non-gaussian, he has obtained explicit results for the optimal estimates and their corresponding error covariance.

FROST [100] and SNYDER [101] have initiated research on nonlinear smoothing that is along paths differing somewhat from those discussed above. Frost has examined the problem of estimating certain parameters of conditionally independent increment processes and the associated two-hypothesis detection problem. He has found cases wherein the estimation or hypothesis discrimination can be executed exactly with an arbitrarily short measurement period. Such problems are termed "singular".

Snyder develops a recursive relation for the evolution of the *a posteriori* statistics of a markov process that modulates the intensity of an observed, self-exciting, doubly-stochastic, point process, e.g. a doubly-stochastic renewal process. From this result, he develops a non-recursive equation for determining the smoothed estimate of a gaussian process that modulates the intensity.

In a different vein, KELLY and ANDERSON [99] have shown that the algorithms for both discrete and continuous-time, linear, fixed-lag smoothing given in [10, 53, 80, 82] are unstable when considered with the filtered estimate and the innovations process as their external inputs. However, with the optimal filter present to generate the required filtered estimate, recent work [18] has shown that the "total system" when viewed with the measurement data as the input and the fixed-lag smoothed estimate as the output is stable. The apparent culprit is an uncontrollable block in the smoother state equations which can be removed without affecting the input-output characteristics. In the case of stationary processes, this amounts in both the discrete and continuous-time situations to pole-zero cancellation of the type already noted in the scalar example of Section 2. For vector-valued processes, this pole-zero cancellation has been explicitly demonstrated in [18] for discrete-time problems and in [102] for continuous-time problems.

For nonstationary processes, two synthesis procedures leading to stable fixed-lag smoothers have been given for the discrete-time case. The first [18] has already been noted. The second [103] is quite novel and appears to be the most practical approach currently available for fixed-lag smoothing. In this reference, it is shown that the optimal, fixed-lag, smoothed estimate is the optimal filtered estimate of the state delayed by the amount of lag. This leads basically to the result that synthesis requires nothing more than a higher-dimension Kalman filter. Specifically, the dimension is nN where n is the number of elements in the state vector and N the number of units of lag. Many simplifications are also given in [103] including reduced-order algorithms. The approach taken is based in part on the work of ZACHRISSON [104] on continuous-time smoothing and WILLMAN [105] on discrete-time smoothing.

In the continuous-time case, the fixed-lag smoothing problem is an infinite-dimensional one and no straightforward synthesis procedure exists at this time.

Finally, it is generally known that, in applications, considerable success has been achieved in prediction and filtering for nonlinear systems using the "linearized Kalman filter". However, because of imperfect modeling the "divergence phenomenon" [48, 106] arises. It is characterized by the fact that the calculated filtered estimate of the state and its associated error covariance diverge from the "true" state and error covariance, respectively, as a result of model errors. To combat this difficulty, numerous schemes based on "discounting" past measurement data have been proposed and studied; see, for example [48, 106–108]. The same approach has recently been applied by SACKS and SORENSON [109] to the fixed-interval, smoothing problem for both the discrete and continuous-time cases.

4. CONCLUSION

A survey of the field of data smoothing, dating from early studies in astronomy to the current time, has been presented. The intent has been to place the field's development in perspective relative to the broader area of estimation theory of which it is a part. In this connection, it is noted that while applications of optimal filter theory have been numerous especially in aerospace navigation and guidance during the past 10 or so years, the same has not been generally true for smoothing theory.

However, some very worthwhile studies which indicate tremendous promise for applications have been carried out and a few of these are noted here.

The potential for applications of smoothing in communications systems was noted very early by WIENER [5] and LEE [17] among others.

More recently, RAUCH *et al.* [55] treated an orbit determination problem involving estimation of intrack position for a satellite in a nominal circular orbit. In addition to position sensor error, they included the effect of drag and modeled it as a constant plus a stochastic component. Assuming 25 equally spaced measurements of intrack position over a single orbit, they showed that for roughly 60 per cent of the orbit, smoothing position error variance was factors of 5–8 times less than that obtained via filtering. A more complete study of the problem can be found in [110].

BRYSON and MEHRA [69] examined filtering and smoothing of data from a ship inertial navigation system wherein the LORAN and EM log data contained time-correlated measurement errors. Their model also included the effects of accelerometer bias and gyro drift. The most significant result in their study was that for continuous velocity data and a position fix every $1\frac{1}{2}$ hr over a 3 hr data span, the RMS error in estimating gyro drift using smoothing was roughly half of the RMS error using filtering.

LEE's [95] use of nonlinear fixed-point smoothing of binary video telemetry data has already been noted. Other communication system application of smoothing are also given in SAGE and MELSA [49].

A quite recent study by NASH *et al.* [111] has examined the use of smoothing in the testing and evaluation of inertial navigation systems. This has been done at both the component and system level, and includes (1) gyro testing, (2) system testing under laboratory conditions, (3) identification of component failure during a mission, and (4) postmission analysis.

For the future, the most important theoretical problems facing the area of smoothing are the same as those confronting the areas of prediction and filtering—the intimate relationship among these three areas being rather obvious. The question of equivalence, touched on in Section 3, is theoretically of extreme importance as are the issues of stability and convergence for nonlinear problems. Indeed, the entire area of estimation for noisy, nonlinear dynamic systems is fragmented. Various authors adopting various estimation viewpoints have developed numerous algorithms for prediction, filtering, and smoothing. All of these algorithms appear to "work" only on some specific problem(s). Thus, perhaps the most significant problem of all is that of development of a unified theory in the nonlinear case.

The lack of widespread applications of smoothing, at least if one is to judge from the open literature, is puzzling. While some positive results have been noted along the lines discussed above, no one appears to have reported any unfavorable experience. Thus, it is not known if any researchers have applied smoothing algorithms only to find that the improvement in estimation accuracy over filtering is insufficient to warrant the additional computation that is required. The availability of useful error bounds that provide a direct filtering vs smoothing comparison would clearly be most helpful here. Some results in this direction have been obtained [112].

REFERENCES

[1] O. GINGERICH: The computer versus Kepler. *Am. Scientist* **52**, 218–226 (1964).

[2] K. F. GAUSS: *Theory of the Motion of Heavenly Bodies.* Dover, New York (1963).

[3] H. W. SORENSON: Least-squares estimation: From Gauss to Kalman. *IEEE Spectrum* **7**, 63–68 (1970).

[4] A. N. KOLMOGOROV: Interpolation und Extrapolation von Stationaren Zufalligen Folgen. *Bull. Acad. Sci. U.S.S.R., Ser. Math.* **5**, 3–14 (1941).

[5] N. WIENER: *Extrapolation, Interpolation, and Smoothing of Stationary Time Series.* Wiley, New York (1949).

[6] P. SWERLING: A proposed stagewise differential correction procedure for satellite tracking and prediction. *J. Astronautic. Sci.* **6**, 46–52 (1959).

[7] R. E. KALMAN: A new approach to linear filtering and prediction problems. *Trans. ASME, Ser. D., J. Basic Engng* **82**, 35–45 (1960).

[8] R. E. KALMAN: New Methods and Results in Linear Prediction and Filtering Theory. Tech. Rept. No. 61-1, RIAS, Martin Co., Baltimore, Md. (1961).

[9] R. E. KALMAN and R. S. BUCY: New results in linear filtering and prediction theory. *Trans. ASME, Ser. D., J. Basic Engng* **83**, 95–108 (1961).

[10] J. S. MEDITCH: On optimal linear smoothing theory. *J. Inform. Control* **10**, 598–615 (1967).

[11] H. WOLD: *A Study in the Analysis of Stationary Time Series.* Almquist and Wiksell, Uppsala, Sweden (1938).

[12] J. L. DOOB: *Stochastic Processes.* Wiley, New York (1953).

[13] N. WIENER: Generalized harmonic analysis. *Acta Mathematica* **55**, 117–258 (1930).

[14] N. WIENER and E. HOPF: On a class of singular integral equations. *Proc. Prussian Acad., Math. Phys. Ser.* 696–718 (1931).

[15] J. H. LANING and R. H. BATTIN: *Random Processes in Automatic Control.* McGraw-Hill, New York (1956).

[16] G. C. NEWTON, JR., L. A. GOULD and J. F. KAISER: *Analytical Design of Linear Feedback Controls.* Wiley, New York (1957).

[17] Y. W. LEE: *Statistical Theory of Communication* Wiley, New York (1960).

[18] S. CHIRARATTANANON and B. D. O. ANDERSON: The fixed-lag smoother as a stable finite-dimensional linear system. *Automatica* **7**, 657–669 (1971).

[19] N. LEVINSON: The Wiener RMS error criterion in filter design and prediction. *J. Math. Phys.* **25**, 261–278 (1947).

[20] A. M. YAGLOM: *An Introduction to the Theory of Stationary Random Functions.* Prentice-Hall, Englewood Cliffs, N.J. (1962).

[21] H. W. BODE and C. E. SHANNON: A simplified derivation of linear least-squares smoothing and prediction. *Proc. IRE* **38**, 417–425 (1950).

[22] L. A. ZADEH and J. R. RAGAZZINI: An extension of Wiener's theory of prediction. *J. appl. Phys.* **21**, 645–655 (1950).

[23] R. H. BATTIN: A Simplified Approach to the Wiener Filter Theory. Rept. R-38, M.I.T., Instrumentation Laboratory, Cambridge, Mass. (1952).

[24] G. F. FRANKLIN: The Optimum Synthesis of Sampled-Data Systems. Sc.D. dissertation, Department of Electrical Engineering, Columbia University, New York, N.Y. (1955).

[25] A. B. LEES: Interpolation and extrapolation of sampled data. *Trans. IRE, Prof. Group on Info. Th.* **IT-2,** 173–175 (1956).

[26] R. C. BOOTON: An optimization theory for time-varying linear systems with nonstationary statistical inputs. *Proc. IRE* **40,** 977–981 (1952).

[27] R. C. DAVIS: On the theory of prediction of non-stationary stochastic processes. *J. appl. Phys.* **23,** 1047–1053 (1952).

[28] J. S. BENDAT: Optimum Time-Variable Filtering for Nonstationary Random Processes. Tech. Rept. NAI-54-771, Northrop Aircraft, Inc., Hawthorne, California (1954).

[29] J. S. BENDAT: A general theory of linear prediction and filtering. *J. Soc. Ind. Appl. Math.* **4,** 131–151 (1956).

[30] A. G. CARLTON and J. W. FOLLIN, JR.: Recent developments in fixed and adaptive filtering. *Proc. Second AGARD Guided Missiles Seminar,* AGARD ograph 21 (1956).

[31] V. S. PUGACHEV: General condition for the minimum mean square error in a dynamic system. *Aut. i Telemekh.* **17,** 289–295 (1956). (English translation, pp. 307–314.)

[32] V. V. SOLODOVNIKOV and A. M. BATKOV: On the Theory of Self-Optimizing Systems. Proceedings of the Heidelberg Conference on Automatic Control, pp. 308–323 (1956).

[33] C. W. STEEG: A time-domain synthesis for optimum extrapolators. *Trans. IRE Prof. Group Aut. Control* **PGAC-3,** 32–41 (1957).

[34] J. E. HANSON: Some Notes on the Application of the Calculus of Variations to Smoothing for Finite Time. JHU/APL Internal Memo. BBD-346, Applied Physics Lab., Johns Hopkins University, Silver Springs, Md. (1957).

[35] M. BLUM: Recursion formulas for growing memory digital filters. *Trans. IRE, Prof. Group on Info. Theory* **IT-4,** 24–30 (1958).

[36] M. SHINBROT: Optimization of time-varying linear systems with nonstationary inputs. *Trans. ASME* **80,** 457–462 (1958).

[37] R. S. BUCY: Optimum Finite-Time Filters for a Special Nonstationary Class of Inputs. JHU/APL Internal Memo. BBD-600, Applied Physics Lab., Johns Hopkins University, Silver Springs, Md. (1959).

[38] E. PARZEN: Statistical Inference on Time Series by Hilbert-Space Methods. Tech. Rept. 23, Applied Mathematics and Statistics Laboratory, Stanford University, Stanford, California (1959).

[39] E. PARZEN: A new approach to the synthesis of optimal smoothing and prediction systems. Tech. Rept. 34, Applied Mathematics and Statistics Laboratory, Stanford University, Stanford, California (1960).

[40] E. PARZEN: A survey of time-series analysis. Tech. Rept. 37, Applied Mathematics and Statistics Laboratory University, Stanford, California (1960).

[41] R. DEUTSCH: *Estimation Theory.* Prentice-Hall, Englewood Cliffs, N.J. (1966).

[42] P. B. LIEBELT: *An Introduction to Optimal Estimation Theory.* Addison-Wesley, Reading, Mass. (1967).

[43] M. AOKI: *Optimization of Stochastic Systems.* Academic Press, New York (1967).

[44] A. P. SAGE: *Optimum Systems Control.* Prentice-Hall, Englewood Cliffs, N.J. (1968).

[45] R. S. BUCY and P. D. JOSEPH: *Filtering for Stochastic Processes with Applications to Guidance.* Interscience, New York (1968).

[46] J. S. MEDITCH: *Stochastic Optimal Linear Estimation and Control.* McGraw-Hill, New York (1969).

[47] A. E. BRYSON, JR. and Y. C. HO: *Applied Optimal Control.* Blaisdell, Waltham, Mass. (1969).

[48] A. H. JAZWINSKI: *Stochastic Processes and Filtering Theory.* Academic Press, New York (1970).

[49] A. P. SAGE and J. L. MELSA: *Estimation Theory with Applications to Communications and Control.* McGraw-Hill, New York (1971).

[50] H. KUSHNER: *Introduction to Stochastic Control.* Holt, Rinehart and Winston, New York (1971).

[51] A. G. CARLTON: Linear Estimation in Stochastic Processes. Bumblebee Series Rept. No. 311, Applied Physics Laboratory, Johns Hopkins University, Silver Springs, Md. (1962).

[52] H. E. RAUCH: Linear Estimation of Sampled Stochastic Processes with Random Parameters. Tech. Rept. 2108-1, Stanford Electronics Laboratory, Stanford University, Stanford, California (1962).

[53] H. E. RAUCH: Solutions to the linear smoothing problem. *IEEE Trans. Aut. Control* **AC-8,** 371–372 (1963).

[54] A. E. BRYSON, JR. and M. FRAZIER: Smoothing for Linear and Nonlinear Dynamic Systems. TDR 63-119, Aero. Sys. Div., pp. 353–364, Wright-Patterson Air Force Base, Ohio (1963).

[55] H. E. RAUCH, F. TUNG and C. T. STRIEBEL: Maximum likelihood estimates of linear dynamic systems. *AIAA J.* **3,** 1445–1450 (1965).

[56] C. S. WEAVER: Estimating the output of a linear discrete system with Gaussian inputs. *IEEE Trans. Aut. Control* **AC-8,** 372–374 (1963).

[57] R. C. K. LEE: *Optimal Estimation, Identification and Control.* M.I.T. Press, Cambridge, Mass. (1964).

[58] H. COX: On the estimation of state variables and parameters for noisy dynamic systems. *IEEE Trans. Aut. Control* **AC-9,** 5–12 (1964).

[59] J. S. MEDITCH: Orthogonal projection and discrete optimal linear smoothing. *SIAM J. Control* **5,** 74–89 (1967).

[60] D. Q. MAYNE: A solution to the smoothing problem for linear dynamic systems. *Automatica* **4,** 73–92 (1966).

[61] D. C. FRASER: A New Technique for the Optimal Smoothing of Data. Sc.D. dissertation, Dept. of Aero. and Astro., M.I.T., Cambridge, Mass. (1967).

[62] R. K. MEHRA: Ph.D. dissertation, Division of Engineering and Applied Physics, Harvard University, Cambridge, Mass. (1968).

[63] D. C. FRASER and J. E. POTTER: The optimum linear smoother as a combination of two optimum linear filters. *IEEE Trans. Aut. Control* **AC-14,** 387–390 (1969).

[64] J. S. MEDITCH: Optimal Fixed-Point Continuous Linear Smoothing. Proceedings of the 1967 Joint Automatic Control Conference, Philadelphia, Pa., pp. 249–257 (1967).

[65] H. KWAKERNAAK: Optimal filtering in linear systems with time delays. *IEEE Trans. Aut. Control* **AC-12,** 169–173 (1967).

[66] P. A. FROST: Nonlinear Estimation in Continuous Time Systems. Tech. Rept. No. 6304-4, Systems Theory Laboratory, Stanford University, Stanford, California (1968).

[67] A. E. BRYSON, JR. and L. J. HENRIKSON: Estimation Using Sampled-Data Containing Sequentially Correlated Noise. Tech. Rept. No. 533, Division of Engineering and Applied Physics, Harvard University, Cambridge, Mass. (1967).

[68] R. K. MEHRA and A. E. BRYSON, JR.: Smoothing for Time-Varying Systems Using Measurements Containing Colored Noise. Tech. Rept. No. 1, Division of Engineering and Applied Physics, Harvard University, Cambridge, Mass. (1967).

[69] R. K. MEHRA and A. E. BRYSON, JR.: Linear smoothing using measurements containing correlated noise with an application to inertial navigation. *IEEE Trans. Aut. Control* **AC-13,** 496–503 (1968).

[70] P. J. McLane: Ph.D. Thesis, Department of Electrical Engineering, University of Toronto, Toronto, Ontario, Canada (1968).
[71] D. M. Detchmendy and R. Sridhar: Sequential estimation of states and parameters in noisy nonlinear dynamical systems. *Trans. ASME, Ser. D, J. Basic Engrs* **88**, 362–368 (1966).
[72] H. H. Kagiwada, R. E. Kalaba, A. Schumitzky and R. Sridhar: Invariant Imbedding and Sequential Interpolating Filters for Nonlinear Processes, Memo. RM-5507-PR, Rand Corp., Santa Monica, California (1967).
[73] A. P. Sage and W. S. Ewing: On Smoothing Algorithms for Nonlinear State and Parameter Estimation. Proceedings of the Second Hawaii International Conference on System Science, pp. 373–376 (1969).
[74] A. P. Sage and W. S. Ewing: On filtering and smoothing algorithms for nonlinear state estimation. *Int. J. Control* **11**, 1–18 (1970).
[75] A. P. Sage: Maximum *a posteriori* filtering and smoothing algorithms. *Int. J. Control* **11**, 171–183 (1970).
[76] J. S. Meditch: A Successive Approximation Procedure for Nonlinear Data Smoothing. Proceedings of the Symposium on Information Processing, pp. 555–568, Purdue University, Lafayette, Ind. (1969).
[77] J. S. Meditch: Sequential Estimation for Discrete-Time Nonlinear Systems. Proceedings of the Seventh Annual Allerton Conference on Circuit and System Th., pp. 293–302, University of Illinois, Urbana, Ill. (1969).
[78] J. S. Meditch: Formal algorithms for continuous-time nonlinear filtering and smoothing. *Int. J. Control* **11**, 1061–1068 (1970).
[79] J. S. Meditch: Newton's method in discrete-time nonlinear data smoothing. *Computer J.* **13**, 387–391 (1970).
[80] T. Kailath: A Wold-Kolmogorov Approach to Linear Least-Squares Estimation, Part I: The Filtering Problem, Part II: The Smoothing Problem, TR 7050-13, Stanford Electronics Laboratory, Stanford University, Stanford, California (1968).
[81] T. Kailath: An innovations approach to least-squares estimation—Part I: Linear filtering in additive white noise. *IEEE Trans. Aut. Control* **AC-13**, 646–654 (1968).
[82] T. Kailath and P. A. Frost: An innovations approach to least-squares estimation—Part II: Linear smoothing in additive white noise. *IEEE Trans. Aut. Control* **AC-13**, 655–660 (1968).
[83] T. Kailath: The innovations approach to detection and estimation theory. *Proc. IEEE* **58**, 680–695 (1970).
[84] P. A. Frost: The Innovations Process and Its Application to Nonlinear Estimation and Detection of Signals in Additive White Noise. Proceedings of the UMR-Mervin J. Kelly Comm. Conference (1970).
[85] P. A. Frost and T. Kailath: An innovations approach to least-squares estimation—Part III: Nonlinear estimation in white Gaussian noise. *IEEE Trans. Aut. Control* **AC-16**, 217–226 (1971).
[86] T. Kailath and R. Geesey: An innovations approach to least-squares estimation—Part IV: Recursive Estimation given the covariance function. *IEEE Trans. Aut. Control* **AC-16**, pp. 720–727 (1971).
[87] P. Masani and N. Wiener: Nonlinear Prediction. Proceedings of the 4th Berkeley Symposium on Mathematics Statistics and Problems, Vol. 2, pp. 403–419 (1961).
[88] P. Masani: Wiener's contributions to generalized harmonic analysis, prediction theory and filtering theory. *Bull. Am. Math. Soc.* **72**, part II, 73–125 (1966).
[89] H. J. Kushner: *Stochastic Stability and Control.* Academic Press, New York (1967).
[90] B. D. O. Anderson and J. B. Moore: State Estimation via the Whitening Filter. Proceedings of the 1968 Joint Automatic Control Conference, pp. 123–129, Ann Arbor, Michigan (1968).
[91] M. Hitsuda: Representation of Gaussian processes equivalent to Wiener processes. *Osaka J. Math.* **5**, 299–312 (1968).
[92] T. Kailath: Likelihood ratios for Gaussian processes. *IEEE Trans. Inform. Th.* **IT-16**, 276–288 (1970).
[93] A. N. Shiryaev: On stochastic equations in the theory of conditional Markov processes. *Theory Prob. appl.* (*U.S.S.R.*) **11**, 179–184 (1946).
[94] C. T. Leondes, J. B. Peller and E. B. Stear: Nonlinear smoothing theory. *IEEE Trans. Sys. Sci. Cyb.* **SSC-6**, 63–71 (1970).
[95] G. M. Lee: Nonlinear interpolation. *IEEE Trans. Inform. Theory* **IT-17**, 45–49 (1971).
[96] H. J. Kushner: On the differential equations satisfied by conditional probability densities of Markov processes with applications. *SIAM J. Control* **2**, 106–119 (1964).
[97] B. D. O. Anderson: Fixed internal smoothing for nonlinear continuous time systems. *J. Inform. Control* (to appear).
[98] D. G. Lainiotis: Optimal non-linear estimation. *Int. J. Control* **14**, 1137–1148 (1971).
[99] C. N. Kelly and B. D. O. Anderson: On the stability of fixed-lag smoothing algorithms. *J. Franklin Inst.* **291**, 271–281 (1971).
[100] P. A. Frost: Some Singular Estimation and Detection Problems for Conditionally Independent Increment Processes. Proceedings of the Ninth Annual Allerton Conference on Circuit and System Theory, pp. 232–241, University of Illinois, Urbana, Ill. (1971).
[101] D. L. Snyder: Non-Recursive Smoothing for Gaussian Modulated Point Processes. Proceedings of the Ninth Annual Allerton Conference on Circuit and System Theory, University of Illinois, Urbana, Ill. (1971).
[102] J. S. Meditch: Stability in Continuous-time Fixed Lag Smoothing. Proceedings 1972 Joint Automatic Control Conference, Stanford, California (1972).
[103] J. B. Moore: Discrete-time fixed-lag smoothing algorithms. *Automatica* (this issue).
[104] L. E. Zachrisson: On optimal smoothing of continuous-time Kalman processes. *Inform. Sciences* **1**, 143–172 (1969).
[105] W. W. Willman: On the linear smoothing problem. *IEEE Trans. Aut. Control* **AC-14**, 116–117 (1969).
[106] F. H. Schlee, C. J. Standish and N. F. Toda: Divergence in the Kalman filter. *AIAA J.* **5**, 1114–1120 (1967).
[107] T. J. Tarn and J. Zaborszky: A practical non-diverging filter. *AIAA J.* **8**, 1127–1133 (1970).
[108] J. S. Meditch and F. C. Johnson: Review and Critique of Some Procedures and Results in Nonlinear Estimation. Proceedings Symposium on Nonlinear Estimation Theory and Its Application, pp. 91–107, San Diego, California (1970).
[109] J. E. Sacks and H. W. Sorenson: Recursive fading memory smoothing. *Inform. Sciences* (to appear).
[110] H. E. Rauch: Optimal estimation of satellite trajectories including random fluctuations in drag. *AIAA J.* **3**, 717–722 (1965).
[111] R. A. Nash, Jr., J. F. Kasper, Jr., B. S. Crawford and S. A. Levine: Application of optimal smoothing to the testing and evaluation of inertial navigation systems and components. *IEEE Trans. Aut. Control* **AC-16**, 806–816 (1971).
[112] B. D. O. Anderson and S. Chirarattananon: Smoothing as an improvement on filtering: A universal bound. *Electronics Let.* **7** (1971).

Résumé—On présente une étude du domaine du nivellement de données pour les systèmes dynamiques linéaires et non-linéaires à paramètres agglomérés. L'étude débute avec les travaux de Kepler et de Gauss, passant par ceux de Kolmogorov et de Wiener et conclut avec les recherches de nombreux savants des dernières dix à douze années. Le but de l'étude est de mettre en perspective les développements du domaine de nivellement de données relativement au domaine plus étendu de la théorie d'estimation dont elle fait partie.

Zusammenfassung—Gegeben wird eine Übersicht über das Gebiet der Datenglättung für lineare und nichtlineare dynamische Systeme. Die Übersicht beginnt mit den Arbeiten von Kepler und Gauss, geht dann zu denen von Kolmogorov und Wiener über und schließt mit Studien zahlreicher Forscher in den letzten 10-12 Jahren ab. Der Zweck der Übersicht leigt darin, die Entwicklung des Gebietes der Datenglättung in den Blickpunkt zu rücken und zwar relativ zu dem breiteren Bereich der Schätztheorie, von der sie ein Teil ist.

Резюме—В этой работе дается обзор сглаживания области данных для динамических систем: сосредоточенных параметров, линейных и нелинейных. Обзор начинается с работ Кеплера и Гаусса, идет через работы Холмогорова и Винера и заканчивается изучением работ многих исследователей за последние 10–12 лет. Цель обзора: приведение к перспективе развития сглаживания области данных в отношении к более широкой области теории оценки, частью которой является сглаживание.

Maximum Likelihood Estimates of Linear Dynamic Systems

H. E. RAUCH,* F. TUNG,* AND C. T. STRIEBEL*
Lockheed Missiles and Space Company, Palo Alto, Calif.

This paper considers the problem of estimating the states of linear dynamic systems in the presence of additive Gaussian noise. Difference equations relating the estimates for the problems of filtering and smoothing are derived as well as a similar set of equations relating the covariance of the errors. The derivation is based on the method of maximum likelihood and depends primarily on the simple manipulation of the probability density functions. The solutions are in a form easily mechanized on a digital computer. A numerical example is included to show the advantage of smoothing in reducing the errors in estimation. In the Appendix the results for discrete systems are formally extended to continuous systems.

1. Introduction

THE pioneer work of Wiener[1] on the problem of linear smoothing, filtering and prediction has received considerable attention over the past few years in fields such as space science, statistical communication theory, and many others that often require the estimates of certain variables that are not directly measurable. Many papers have appeared since then giving different solutions to this problem. A summary of these solutions can be found in a paper by Parzen[2] who gives a general treatment of the problem from the point of view of reproducing kernel Hilbert Space. The most widely used solution in practice in linear filtering and prediction is probably the one derived by Kalman[3] using the method of projections. The primary advantage of Kalman's solution is that the equations that specify the optimum filter are in the form of difference equations, so that they can be mechanized easily on the present-day digital computer. However, Kalman does not consider the important problem of smoothing. (The filtering and prediction solution allows one to estimate current and future values of the variables of interest, whereas the smoothing solution permits one to estimate past values.) The purpose of this paper is to provide a solution of the linear smoothing problem based on the principle of the maximum likelihood, and a derivation of the filtering problem based on the same principle. It is shown that the equations describing the smoothing solution also can be easily implemented on a digital computer and a numerical example is presented to show the advantage of smoothing in reducing the errors in estimation.

Solutions of the smoothing problem in different forms have been obtained recently by Rauch[4] for discrete systems and by Bryson and Frazier[5] for continuous systems. The elegant proof and the tools used by Bryson and Frazier are based on the calculus of variations and the method of maximum likelihood. Our derivation differs from their work in that the method used here depends primarily on the simple manipulation of the probability density functions and hence leads immediately to recursion equations. Our results are also different. The derivation leads directly to a smoothing solution that uses processed data instead of the original measurements.

An early version of this paper was published as a company report.[6] During the period in which the paper was being revised for publication, Cox[7] had also presented some similar results using a slightly different approach.

Received December 18, 1964; also presented at the Joint AIAA-IMS-SIAM-ONR Symposium on Control and System Optimization, Monterey, Calif., January 27–29, 1964 (no preprint number); revision received May 13, 1965.

* Research Scientist.

2. Statement of the Problem

2.1 Dynamic System

a) Given†:

$$x_{k+1} = \Phi(k+1, k)x_k + w_k \tag{2.1}$$

$$y_k = M_k x_k + v_k \tag{2.2}$$

where

x_k	=	state vector $(n \times 1)$
y_k	=	output vector $(r \times 1)$, $r \leqq n$
w_k	=	Gaussian random disturbance $(n \times 1)$
v_k	=	Gaussian random disturbance $(r \times 1)$
$\Phi(k+1, k)$	=	transition matrix $(n \times n)$
M_k	=	output matrix $(r \times n)$

and w_k and v_k are independent Gaussian vectors with zero mean and covariances

$$\text{cov}(w_j, w_k) = Q_k \delta_{jk} \tag{2.3}$$

$$\text{cov}(v_j, v_k) = R_k \delta_{jk} \tag{2.4}$$

$$\text{cov}(w_j, v_k) = 0 \tag{2.5}$$

where δ_{jk} is the Kronecker delta, and we assume that R_k is positive definite.

b) Initial condition x_0 is a Gaussian vector with the a priori information

$$E(x_0) = \bar{x}_0 \qquad \text{cov}(x_0) = \bar{P}_0 \tag{2.6}$$

c) Observations: $y_0, y_1, \ldots, y_N$ $(N = 0, 1, \ldots)$.

The problem is to find an estimate of x_k from the observations $y_0, \ldots, y_N$. Such an estimate will be denoted by $\hat{x}_{k/N} = \hat{x}_{k/N}(y_0, \ldots, y_N)$. It is commonly called the problem of 1) filtering if $k = N$, 2) prediction with filtering if $k \geqq N$, and 3) smoothing if $k \leqq N$.

2.2. Estimation Criteria

Three possible estimation criteria will be presented in this section. For the linear Gaussian case defined in Sec. 2.1 these three criteria result in the same estimate. The distinction is made here in order to see how this problem can be extended to the nonlinear case and how it compares with other work in this field.

The standard procedure is to specify a loss function

$$l(x_0, \hat{x}_{0/N};\ x_1, \hat{x}_{1/N};\ \ldots;\ x_K, \hat{x}_{K/N}) \tag{2.7}$$

† If the original problem is described by nonlinear equations, the linear system can be obtained from equations governing small deviations from a reference path.

Reprinted with permission from *AIAA J.*, vol. 3, pp. 1445–1450, Aug. 1965.

and then to find the functions $\hat{x}_{k/N}$ for $k = 0, \ldots, K$ which minimize the expected loss. In order to do this, the distribution of interest is the joint distribution of $x_0, \ldots, x_K$ conditioned on $y_0, \ldots, y_N$:

$$p(x_0, \ldots, x_K/y_0, \ldots, y_N)$$

If the loss function (2.7) is zero near $x_k = \hat{x}_{k/N}$ for $k = 0, \ldots, K$ and very large otherwise, the optimum estimating procedure is the maximum likelihood, and the estimate will be called the joint maximum likelihood estimate. It is obtained by solving the simultaneous equations

$$(\partial/\partial x_k)p(x_0, \ldots, x_K/y_0, \ldots, y_N) = 0 \qquad k = 0, \ldots, K \tag{2.8}$$

If the loss function (2.7) has the special form

$$\sum_{k=0}^{K} l_k(x_k, \hat{x}_{k/N})$$

or equivalently, if $K + 1$ distinct estimation problems with losses $l_k(x_k, \hat{x}_{k/N})$ are considered, the distribution of interest is the marginal distribution of x_k conditioned on $y_0, \ldots, y_N$

$$p(x_k/y_0, \ldots, y_N)$$

The distribution can be obtained from (2.7) by integrating out the variables x_j for $j \neq k$. If $l_k(x_k, \hat{x}_{k/N})$ is zero near $x_k = \hat{x}_{k/N}$ and very large otherwise, the optimum estimate is the marginal maximum likelihood estimate obtained as a solution to the single equation

$$(\partial/\partial x_k)p(x_k/y_0, \ldots, y_N) = 0 \tag{2.9}$$

The marginal maximum likelihood estimate (MLE) is the estimate that will be derived in this paper. The estimate used by Bryson and Frazier[4] is the joint maximum likelihood estimate; so that, although the estimates they obtain in the linear case are the same as the MLE to be derived here, it should be expected that in the nonlinear cases they would not necessarily agree.

Another estimation criterion that is often appropriate is the conditional mean given by $\hat{x}_{k/N} = \int x_k p(x_k/y_0, \ldots, y_N) dx_k$. The conditional mean has the advantage that it is the same for the joint and marginal distributions, and that it minimizes a large class of loss functions.

3. Solutions

3.1. Filtering and Prediction

We shall first consider the case of estimating x_k given all the data up to t_k, i.e., $y_0, \ldots, y_k$. The estimate will be denoted $\hat{x}_{k/k}$, whereas the data $y_0, \ldots, y_k$ will be denoted by Y_k. From the discussion in the previous section, we know that $\hat{x}_{k/k}$ is the solution of x_k which maximizes the conditional probability density function $p(x_k/Y_k)$. This is the same as maximizing the log of the density given by

$$L(x_k, Y_k) = \log p(x_k/Y_k) = \log p(x_k, Y_k) - \log p(Y_k) \tag{3.1}$$

Using the concept of conditional probabilities and the fact that the v_k are independent random vectors, we see

$$\begin{aligned} p(x_k, Y_k) &= p(y_k/x_k, Y_{k-1})p(x_k, Y_{k-1}) \\ &= p(y_k/x_k)\, p(x_k/Y_{k-1})\, p(Y_{k-1}) \end{aligned} \tag{3.2}$$

Let $\hat{x}_{k-1/k-1}$ and $\hat{x}_{k/k-1}$ be the estimates of x_{k-1} and x_k given Y_{k-1}, respectively, and let $\tilde{x}_{k-1/k-1}$ and $\tilde{x}_{k/k-1}$ be the errors in these estimates. Define

$$\operatorname{cov}(\tilde{x}_{k-1/k-1}) = P_{k-1/k-1} \tag{3.3}$$

and

$$\operatorname{cov}(\tilde{x}_{k/k-1}) = P_{k/k-1} \tag{3.4}$$

Since all the random disturbances v_k are statistically independent, it follows that

$$\hat{x}_{k/k-1} = \Phi(k, k-1)\hat{x}_{k-1/k-1} \tag{3.5}$$

and

$$P_{k/k-1} = \Phi(k, k-1)P_{k-1/k-1}\Phi'(k, k-1) + Q_{k-1} \tag{3.6}$$

This is, in fact, the solution of the prediction problem. Using (2.1–2.4) and the assumption that the random disturbances are normally distributed, we see that the conditional random vector x_k given Y_{k-1} has a mean

$$E(x_k/Y_{k-1}) = \hat{x}_{k/k-1} \tag{3.7}$$

and a covariance

$$\operatorname{cov}(x_k/Y_{k-1}) = P_{k/k-1} \tag{3.8}$$

whereas the conditional vector y_k given x_k has a mean

$$E(y_k/x_k) = M_k x_k \tag{3.9}$$

and a covariance

$$\operatorname{cov}(y_k/x_k) = R_k \tag{3.10}$$

Substituting (3.7) to (3.10) into (3.2) and using the fact that all the vectors are normally distributed, we find‡

$$\begin{aligned} p(x_k, Y_k) = {} & (2\pi)^{-r/2}(\det R_k)^{-1/2} \times \\ & \exp(-1/2\|y_k - M_k x_k\|^2_{R_k^{-1}})\cdot(2\pi)^{-n/2}(\det P_{k/k-1})^{-1/2} \times \\ & \exp(-1/2\|x_k - \hat{x}_{k/k-1}\|^2_{(P_{k/k-1})^{-1}})\cdot p(Y_{k-1}) \end{aligned} \tag{3.11}$$

Substitution of (3.11) into (3.1) shows that the terms in L which depend on x_k can be written as

$$J = \|y_k - M_k x_k\|^2_{R_k^{-1}} + \|x_k - \hat{x}_{k/k-1}\|^2_{(P_{k/k-1})^{-1}} \tag{3.12}$$

Setting the gradient of J to zero, we find

$$\hat{x}_{k/k} = (M_k'R_k^{-1}M_k + P_{k/k-1}^{-1})(M_k'R_k^{-1}y_k + P_{k/k-1}^{-1}\hat{x}_{k/k-1}) \tag{3.13}$$

which is essentially the solution of the filtering problem. Equation (3.13) may be put into a more convenient form by using a well-known matrix inversion lemma.§ This lemma, for instance, has beeen used by Ho[8] to show the relations between the stochastic approximation method and the optimal filter theory.

Lemma: If $S_{k+1}^{-1} = S_k^{-1} + M_k'R_k^{-1}M_k$ where S_k and R_k are symmetric and positive definite, then S_{k+1} exists and is given by $S_{k+1} = S_k - S_kM_k'(M_kS_kM_k' + R_k)^{-1}M_kS_k$. The proof is by direct substitution. By making use of this lemma, it is seen that (3.13) also can be written as

$$\begin{aligned} \hat{x}_{k/k} &= \hat{x}_{k/k-1} + B_k(y_k - M_k\hat{x}_{k/k-1}) \\ &= \Phi(k, k-1)\hat{x}_{k-1/k-1} + \\ &\qquad B_k[y_k - M_k\Phi(k, k-1)\hat{x}_{k-1/k-1}] \end{aligned} \tag{3.14}$$

where

$$B_k = P_{k/k-1}M_k'(M_kP_{k/k-1}M_k' + R_k)^{-1} \tag{3.15}$$

Remark: The computation of $\hat{x}_{k/k}$ by way of (3.13) requires the inversion of a $n \times n$ matrix $(M_k'R_k^{-1}M_k + P_{k/k-1}^{-1})$ whereas Eqs. (3.14) and (3.15) require only the inversion of the matrix $(M_kP_{k/k-1}M_k' + R_k)$ which is $r \times r$ ($r \leqq n$). Hence, the representation given by (3.14) and (3.15) appears to be more desirable for the purpose of computation.

Substituting (2.1) and (2.2) into (3.14) shows that the estimation error satisfies the recursive equation

$$\tilde{x}_{k/k} = (I - B_kM_k)[\Phi(k, k-1)\tilde{x}_{k-1/k-1} + w_{k-1}] - B_kv_k \tag{3.16}$$

---

‡ $\|a\|^2_R = a'Ra$.

§ The authors wish to acknowledge Y. C. Ho of Harvard for pointing out this identity.

where I is the identity matrix. Since $\tilde{x}_{k-1/k-1}$, v_k, and w_{k-1} are statistically independent, it follows that

$$P_{k/k} = \text{cov}(\tilde{x}_{k/k}) = (I - B_kM_k)P_{k/k-1} \quad (3.17)$$

where use is made of (3.15). Equations (3.14–3.17) are the same as those derived originally by Kalman.[2] To start the recursive equation, we need $\hat{x}_{0/-1}$ and $P_{0/-1}$. From the a priori information about x_0, we see

$$\hat{x}_{0/-1} = \bar{x}_0 \quad (3.18)$$

and

$$P_{0/-1} = \bar{P}_0 \quad (3.19)$$

This completes the solution of the filtering problem. The solution of the prediction problem has already been obtained. For any $N \geqq k$,

$$\hat{x}_{N/k} = \Phi(N, k)\hat{x}_{k/k} \quad (3.20)$$

3.2 Smoothing

From the principle of the MLE, we know that the estimate of x_k given Y_N, denoted by $\hat{x}_{k/N}$, is that value of x_k which maximizes the function

$$L(x_k, Y_N) = \log p(x_k/Y_N) \quad (3.21)$$

Similarly, $\hat{x}_{k/N}$ and $\hat{x}_{k+1/N}$ are the values of x_k and x_{k+1} which maximize

$$L(x_k, x_{k+1}, Y_N) = \log p(x_k, x_{k+1}/Y_N) \quad (3.22)$$

Let us now inspect the joint probability density function $p(x_k, x_{k+1}, Y_N)$. Using the concept of conditional probabilities, we see

$$p(x_k, x_{k+1}, Y_N) = p(x_k, x_{k+1}, y_{k+1}, \ldots, y_N/Y_k)p(Y_k) \quad (3.23)$$

Now

$$\begin{aligned} p(x_k, x_{k+1}, y_{k+1}, \ldots, y_N/Y_k) &= p(x_{k+1}, y_{k+1}, \ldots, y_N/x_k, Y_k)\, p(x_k/Y_k) \\ &= p(x_{k+1}, y_{k+1}, \ldots, y_N/x_k)\, p(x_k/Y_k)\P \\ &= p(y_{k+1}, \ldots, y_N/x_{k+1}, x_k)\, p(x_{k+1}/x_k)\, p(x_k/Y_k) \\ &= p(y_{k+1}, \ldots, y_N/x_{k+1})\, p(x_{k+1}/x_k)\, p(x_k/Y_k) \quad (3.24) \end{aligned}$$

Substituting (3.24) into (3.23) shows that

$$p(x_k, x_{k+1}, Y_N) = p(x_{k+1}/x_k)\, p(x_k/Y_k)\, p(y_{k+1}, \ldots, y_N/x_{k+1}) \cdot p(Y_k) \quad (3.25)$$

Let us assume that $\hat{x}_{k/k}$ has already been obtained. Substituting (3.25) into (3.22) and using the same reasoning as that given in the previous section, we see

$$\max_{x_k, x_{k+1}} L(x_k, x_{k+1}, Y_N) = \max_{x_k, x_{k+1}} \{-||x_{k+1} - \Phi(k+1, k)x_k||^2_{Q_k^{-1}} - ||x_k - \hat{x}_{k/k}||^2_{P_{k/k}^{-1}}\} + \text{terms which do not involve } x_k \quad (3.26)$$

It follows immediately that $\hat{x}_{k/N}$ is the solution that minimizes the expression

$$J = ||\hat{x}_{k+1/N} - \Phi(k+1, k)x_k||^2_{Q_k^{-1}} + ||x_k - \hat{x}_{k/k}||^2_{P_{k/k}^{-1}} \quad (3.27)$$

Setting the gradient of J to zero and using the matrix inversion lemma, we find

$$\hat{x}_{k/N} = \hat{x}_{k/k} + C_k[\hat{x}_{k+1/N} - \Phi(k+1, k)\hat{x}_{k/k}] \quad (3.28)$$

where

$$\begin{aligned} C_k &= P_{k/k}\Phi'(k+1, k) \times [\Phi(k+1, k)P_{k/k}\Phi'(k+1, k) + Q_k]^{-1} \\ &= P_{k/k}\Phi'(k+1, k)P_{k+1/k}^{-1} \quad (3.29) \end{aligned}$$

This is the solution of the smoothing problem. It is in the form of a backward recursive equation that relates the MLE of x_k given Y_N in terms of the MLE of x_{k+1} given Y_N and the MLE of x_k given Y_k. Hence, the smoothing can be obtained from the filtering solution by computing backwards using (3.29).

Subtracting x_k from both sides of (3.28) and rearranging the terms, we find

$$\tilde{x}_{k/N} + C_k\hat{x}_{k+1/N} = \tilde{x}_{k/k} + C_k\,\Phi(k+1, k)\hat{x}_{k/k} \quad (3.30)$$

Using the facts that

$$E(\tilde{x}_{k/N}\hat{x}_{k+1/N}') = E(\tilde{x}_{k/k}\,\hat{x}_{k/k}') = 0^{**}$$

$$\text{cov}(\hat{x}_{k+1/N}) = \text{cov}(x_{k+1}) - P_{k+1/N}$$

$$\text{cov}(\hat{x}_{k/k}) = \text{cov}(x_k) - P_{k/k}$$

and

$$\text{cov}(x_{k+1}) = \Phi(k+1, k)\,\text{cov}(x_k)\,\Phi'(k+1, k) + Q_k$$

we see from (3.30) that $P_{k/N}$ satisfies the recursive equation

$$P_{k/N} = P_{k/k} + C_k(P_{k+1/N} - P_{k+1/k})C_k' \quad (3.31)$$

The computation is initiated by specifying $P_{N/N}$. This essentially completes the solution for the smoothing problem. It should be noted that the estimates $\hat{x}_{k/k}$ $(k \leqq N)$ are assumed to have been obtained in the process of computing $\hat{x}_{N/N}$ and hence can be made available by storing them in the memory. The covariance $P_{k/k}$ also may be stored. However, it can be easily computed. We will now give a formula for computing $P_{k/k}$ from $P_{k+1/k+1}$ and hence eliminate the storage problem for $P_{k/k}(k = 0, \ldots, N)$.

Substituting (3.15) into (3.17) shows

$$P_{k/k-1} = (P_{k/k}^{-1} - M_k'R_k^{-1}M_k)^{-1} \quad (3.32)$$

which can be written as

$$P_{k/k-1} = P_{k/k} - P_{k/k}M_k'(M_kP_{k/k}M_k' - R_k)^{-1}M_kP_{k/k} \quad (3.33)$$

after applying the matrix inversion lemma. From $P_{k/k-1}$, $P_{k-1/k-1}$ can be computed by using (3.6) which can be written as

$$P_{k-1/k-1} = \Phi^{-1}(k-1, k)(P_{k/k-1} - Q_{k-1})\Phi'^{-1}(k-1, k) \quad (3.34)$$

The terminal condition for (3.33) is again $P_{N/N}$. It is of interest to note from (3.33) that the computation for $P_{k/k}$ requires only the inversion of a $r \times r$ matrix.

Remark:

1) Another formulation of the smoothing problem which relates $\hat{x}_{k/N}$ to $\hat{x}_{k+1/N}$ and all the data $y_j (j \geq k+1)$ and hence requires the storage of the data can be obtained by noting that $\hat{x}_{i/N}(i = 0, 1, \ldots, N)$ is the solution which maximizes the function

$$L(x_0, x_1, \ldots, x_N, Y_N) = \log p(x_0, x_1, \ldots, x_N/Y_N) \quad (3.35)$$

Now

$$\begin{aligned} p(x_0, x_1, \ldots, x_N, Y_N) &= p(Y_N/x_0, x_1, \ldots, x_N)\, p(x_0, x_1, \ldots, x_N) \\ &= p(Y_N/x_0, x_1, \ldots, x_N)\, p(x_N/x_{N-1}) \\ &\quad p(x_{N-1}/x_{N-2}) \ldots p(x_1/x_0)\, p(x_0) \quad (3.36) \end{aligned}$$

where use is made of the fact that x is a Markov process,

$$p(x_k/x_{k-1}, \ldots, x_0) = p(x_k/x_{k-1}) \quad (3.37)$$

Substituting (3.36) into (3.35) shows that maximizing L is

¶ This is because $x_{k+1}, y_{k+1}, \ldots, y_N$ given x_k is independent of y_i, $i \leqq k$, and $p(a/bc) = p(a/b)$ if a/b is independent of c.

** This can be verified after somewhat lengthy manipulation of Eqs. (3.16) and (3.30) using the properties of $\tilde{x}_{k/k}$ and $\tilde{x}_{k/N}$.

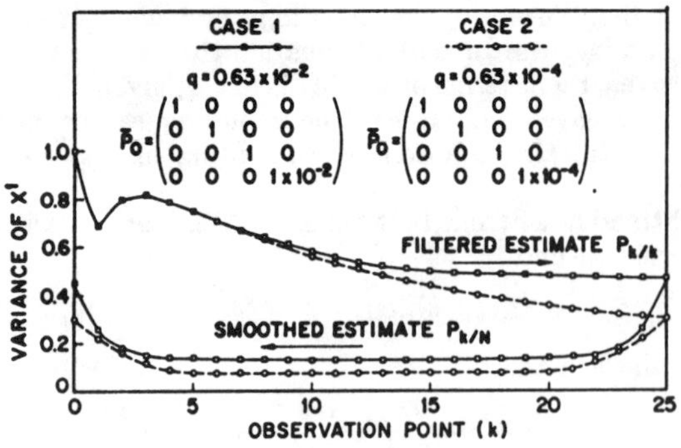

Fig. 1 Variance history for two levels of random disturbances (q).

equivalent to

$$\min_{x_0, \ldots x_N} \left\{ \sum_{i=0}^{N} ||y_i - M_i x_i||^2_{R_i^{-1}} + \sum_{i=0}^{N} ||x_i - \Phi(i, i-1)x_{i-1}||^2_{Q_{i-1}^{-1}} \right\} \quad (3.38)$$

with the initial condition

$$\Phi(0, -1)x_{-1} = \bar{x}_0 \quad (3.39)$$

and

$$Q_{-1} = \bar{P}_0 \quad (3.40)$$

This is the equivalent discrete formulation of the continuous smoothing problem recently given by Bryson and Frazier.[5] The scalar version of (3.38) may also be found in a book by Bellman.[9]

To show the equivalence of our solution with the results of Bryson and Frazier and to obtain the solution of (3.38) in terms of the observations y_k, we define a new variable

$$w_k = P_{k+1/k}^{-1}[\hat{x}_{k+1/N} - \Phi(k+1, k)\hat{x}_{k/k}] \quad (3.41)$$

It follows that

$$w_N = 0 \quad (3.42)$$

Substituting (3.41) into (3.28) and using (3.14, 3.17, and 3.32), we obtain, after many algebraic manipulation, a set of $2n$ difference equations

$$\hat{x}_{k+1/N} = \Phi(k+1, k)\hat{x}_{k/N} + Q_k w_k$$

$$w_k = \Phi'(k, k+1)M_k'R_k^{-1}M_k\,\hat{x}_{k/N} + \Phi'(k, k+1)w_{k-1} - \Phi'(k, k+1)M_k R_k^{-1} y_k \quad (3.43)$$

Notice that if $\hat{x}_{N/N}$ is given, then the set of equations given by (3.43) may be computed backwards from the index N. Otherwise, it involves the solution of a two point boundary value problem.

2) It has been shown[10] that by simple manipulations of the results derived in this paper, namely Eqs. (3.28) and (3.31), the smoothing solution can be written in still another form that directly relates the smoothed estimate at a particular time to the new observations as they are received. This form is preferable for the class of problems where one is only interested in the smoothing solution for the state at a particular time.

3) The problem of interpolation is concerned with estimating the state between measurement points. If it is desired to estimate the state x_k at a point where no measurement was taken, the equations presented here for the smoothing solution can be used by assuming that a measurement is taken at that point with covariance R_k very large and with B_k equal to zero.

4. Numerical Example

Consider the dynamical system given by

$$x_{k+1} = \begin{pmatrix} 1 & 1 & 0.5 & 0.5 \\ 0 & 1 & 1 & 1 \\ 0 & 0 & 1 & 0 \\ 0 & 0 & 0 & 0.606 \end{pmatrix} x_k + w_k$$

$$y_k = (1, 0, 0, 0)x_k + v_k \quad (4.1)$$

where x_k is the (4×1) state vector composed of four state variables (x^1, x^2, x^3, and x^4), and y_k is the (1×1) output vector that is a noisy measurement of the state variable x^1. The disturbances w_k and v_k are independent Gaussian vectors with zero mean and covariances

$$\text{cov}(w_k) = \begin{pmatrix} 0 & 0 & 0 & 0 \\ 0 & 0 & 0 & 0 \\ 0 & 0 & 0 & 0 \\ 0 & 0 & 0 & q \end{pmatrix}$$

$$\text{cov}(v_k) = 1 \quad (4.2)$$

The initial condition x_0 is a Gaussian vector with a priori information such that the covariance of x_0 is given by $\bar{P}_0$.

The entire dynamic system can be considered as a linearized version of the in-track motion of a satellite traveling in a circular orbit. The satellite motion is affected by both constant and stochastic drag.[11] The state variables x^1, x^2, and x^3 can be considered as angular position, velocity, and (constant) acceleration, respectively. The state variable x^4 is a stochastic component of acceleration generated by a first-order Gauss-Markov process.

Three cases will be considered:

Case 1:

$$q = 0.63 \times 10^{-2}$$

$$\bar{P}_0 = \begin{pmatrix} 1 & 0 & 0 & 0 \\ 0 & 1 & 0 & 0 \\ 0 & 0 & 1 & 0 \\ 0 & 0 & 0 & 1 \times 10^{-2} \end{pmatrix}$$

Table 1 Diagonal elements of the covariance

Observation point (k)	Filtered estimate ($P_{k/k}$)			
	cov(x^1)	cov(x^2)	cov(x^3)	cov(x^4)
0	1.00	1.00	1.00	0.0100
1	0.69	1.31	0.92	0.0100
2	0.80	1.31	0.54	0.0100
3	0.82	0.96	0.26	0.0100
4	0.79	0.68	0.13	0.0100
5	0.75	0.49	0.07	0.0100
10	0.58	0.15	0.008	0.00995
15	0.50	0.10	0.004	0.0099
20	0.48	0.093	0.0026	0.0099
25	0.47	0.089	0.0020	0.0099
Observation point (k)	**Smoothed estimate ($P_{k/N}$)**			
	cov(x^1)	cov(x^2)	cov(x^3)	cov(x^4)
25	0.47	0.089	0.0020	0.0099
24	0.26	0.058	0.0020	0.0096
23	0.18	0.036	0.0020	0.0091
22	0.15	0.023	0.0020	0.0085
21	0.15	0.017	0.0020	0.0080
20	0.15	0.015	0.0020	0.0078
15	0.135	0.014	0.0020	0.0078
10	0.135	0.014	0.0020	0.0078
5	0.14	0.015	0.0020	0.0078
1	0.26	0.053	0.0020	0.0089
0	0.45	0.082	0.0020	0.0094

Case 2:

$$q = 0.63 \times 10^{-4}$$

$$\bar{P}_0 = \begin{pmatrix} 1 & 0 & 0 & 0 \\ 0 & 1 & 0 & 0 \\ 0 & 0 & 1 & 0 \\ 0 & 0 & 0 & 1 \times 10^{-4} \end{pmatrix}$$

Case 3:

$$q = 0.63 \times 10^{-2}$$

$$P_0 = \begin{pmatrix} 100 & 0 & 0 & 0 \\ 0 & 100 & 0 & 0 \\ 0 & 0 & 100 & 0 \\ 0 & 0 & 0 & 1 \times 10^{-2} \end{pmatrix}$$

In each case 25 measurements are taken starting with y_1. The diagonal elements of the covariance of the estimates of the state for case 1 are presented in Table 1 for both the filtered and smoothed estimate. Notice how smoothing the estimate decreases the errors. In Fig. 1 the variance of the filtered and smoothed estimates of the state variable x^1 are plotted for both case 1 and case 2. Reducing the variance of the random disturbance reduces the variance of the estimates. In Fig. 2 the variance of the estimates of x^1 are plotted for both case 1 and case 3. Notice how the effect of initial conditions (the a priori information about the state) rapidly dies out.

5. Conclusions

The solution to the discrete version of the filtering and smoothing problem has been derived using the principal of maximum likelihood and simple manipulation of the probability density function. The filtered estimate is calculated forward point by point as a linear combination of the previous filtered estimate and the current observation. The smoothing solution starts with the filtered estimate at the last point and calculates backward point by point determining the smoothed estimate as a linear combination of the filtered estimate at that point and the smoothed estimate at the previous point. A numerical example has been presented to illustrate the advantage of smoothing in reducing the error in the estimate.

Appendix: Extension to the Continuous Case

The MLE of the states with continuous observations can be obtained formally from the MLE of the discrete system. The difference equations in the previous section become differential equations in the limit as the time between observations approaches zero. No rigorous proof of the limiting process is attempted here. A discussion of the conditions under which it is valid can be found elsewhere.[3]

Let us assume that the discrete indices k and $k + 1$ in all the variables have been replaced by t and $t + q$, and let the disturbances w_k be replaced by $qu(t)$. A Taylor series expansion in q is made of the transition matrix $\Phi(t + q, t)$ so that Eqs. (2.1) and (2.2) can be written as

$$x(t + q) = \Phi(t + q, t)x(t) + qu(t)$$

$$= [I + qF(t) + 0(q^2)]x(t) + qu(t) \tag{A1}$$

and

$$y(t) = M(t)x(t) + v(t) \tag{A2}$$

where $0(q^2)$ represents the terms of the order of q^2. The covariances Q_k and R_k are replaced by $qQ(t)$ and $R(t)/q$, respectively, so that††

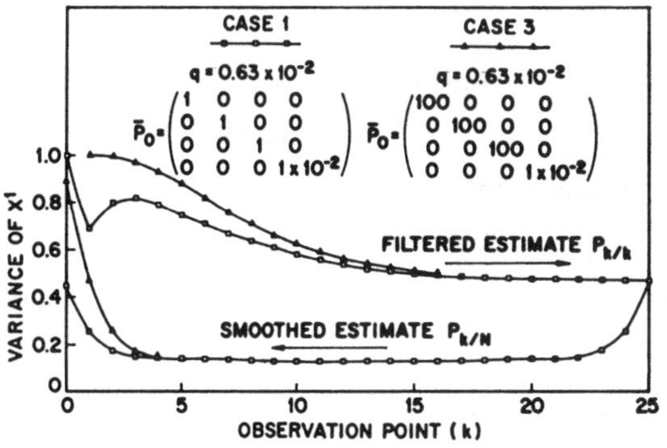

Fig. 2 Variance history for two sets of initial conditions ($\bar{P}_0$).

$$\text{cov}[u(t)] = Q(t)/q \tag{A3}$$

and

$$\text{cov}[v(t)] = R(t)/q$$

In the limit as q approaches zero, we find that (A1) and (A2) become

$$[dx(t)]/dt = F(t)x(t) + u(t) \tag{A4}$$

and

$$y(t) = M(t)x(t) + v(t) \tag{A5}$$

where $u(t)$ and $v(t)$ are white noises such that

$$\text{cov}[u(t), u(s)] = Q(t)\,\delta(t - s) \tag{A6}$$

$$\text{cov}[v(t), v(s)] = R(t)\,\delta(t - s) \tag{A7}$$

$\delta(t - s)$ being the Dirac delta function.

The same limiting process will now be applied to the solutions of the MLE derived for the discrete system in the previous section. For the purpose of clarification, the following notation will be used: $\hat{x}_t(t)$ = estimate of $x(t)$ using the data over the interval $(0, t)$, $\hat{x}_T(t)$ = estimate of $x(t)$ using the data over the interval $(0, T)$, $P_t(t) = \text{cov}[x(t) - \hat{x}_t(t)]$, and $P_T(t) = \text{cov}[x(t) - \hat{x}_T(t)]$.

Filtering Solution

Applying the limiting process to Eqs. (3.14) and (3.15) and the corresponding covariances given by (3.6) and (3.17), we find that the filtering solution for the continuous case can be written as

$$[d\hat{x}_t(t)]/dt = F(t)\hat{x}_t(t) + P_t(t)M'(t)R^{-1}(t)[y(t) - M(t)\hat{x}_t(t)] \tag{A8}$$

†† The replacement in (A3) keeps the statistical properties of the random disturbances nearly the same as can be shown by the following explanation. Divide the interval between k and $k + 1$ (which is of length T_k) into n equally spaced intervals with an observation made at each interval. The time between observations is $q = T_k/n$. Assume, for the moment, that there are no dynamics between k and $k + 1$. Because the errors in the observations are Gaussian, the accuracy obtained from n observations, each with covariance nR_k, would be the same as the accuracy obtained from one observation with covariance R_k. Therefore, if $v(t)$ is the noise on the observation at time t, $\text{cov}[v(t)] = nR_k = R_kT_k/q = R(t)/q$. Furthermore, the sum of n identically distributed independent Gaussian random inputs with covariance Q_k/n would have the same distribution as one random input with covariance Q_k. Therefore, if $qu(t)$ is the random input at time t, $\text{cov}[u(t)] = q^{-2}\,Q_k/n = q^{-1}\,Q_k/T_k = Q(t)/q$.

and

$$[dP_t(t)]/dt = F(t)P_t(t) + P_t(t)F'(t) - P_t(t)M'(t)R^{-1}(t)\,M(t)P_t(t) + Q(t) \quad \text{(A9)}$$

with the initial conditions

$$\hat{x}_0(0) = \bar{x}_0 \text{ and } P_0(0) = \bar{P}_0 \quad \text{(A10)}$$

Equations (A8) and (A9) are the same as those given by Kalman.[3]

Smoothing Solution

In a similar manner, the continuous version of the MLE for the smoothing problem given by Eqs. (3.28, 3.29, and 3.31) can be written as

$$[d\hat{x}_T(t)]/dt = F(t)\hat{x}_T(t) + Q(t)P_t^{-1}(t)[\hat{x}_T(t) - \hat{x}_t(t)] \quad \text{(A11)}$$

and

$$[dP_T(t)]/dt = [F(t) + Q(t)P_t^{-1}(t)]P_T(t) + P_T(t)[F(t) + Q(t)P_t^{-1}(t)]' - Q(t) \quad \text{(A12)}$$

with the terminal condition $\hat{x}_T(T)$ and $P_T(T)$.

To show the equivalence of our solution with the results of Bryson and Frazier, we define a new variable

$$\omega(t) = P_t^{-1}(t)[\hat{x}_T(t) - \hat{x}_t(t)] \quad \text{(A13)}$$

It follows that

$$\omega(T) = 0 \quad \text{(A14)}$$

Substituting (A13) into (A11) and using (A8), (A7), as well as (A11), we obtain a set of $2n$ differential equations

$$[d\hat{x}_T(t)]/dt = F(t)\hat{x}_T(t) + Q(t)\omega(t) \quad \text{(A15)}$$

$$[d\omega(t)]/dt = M'(t)R^{-1}(t)M(t)\hat{x}_T(t) - F'(t)\omega(t) - M'(t)R^{-1}(t)y(t) \quad \text{(A16)}$$

which are precisely those derived by Bryson and Frazier. Hence, we have given a physical interpretation of the Lagrange multipliers $\omega(t)$ used in their derivation. Moreover, it can be readily shown that

$$\text{cov}[\tilde{x}_T(t), \omega(t)] = P_T^{-1}(t)P_t(t) - I \quad \text{(A17)}$$

and

$$\text{cov}[\omega(t)] = P_t^{-1}(t)P_T(t)P_t^{-1}(t) - P_t^{-1}(t) \quad \text{(A18)}$$

where

$$\tilde{x}_T(t) = x(t) - \hat{x}_T(t)$$

References

[1] Wiener, N., *The Extrapolation, Interpolation and Smoothing of Stationary Time Series* (John Wiley & Sons, Inc., New York, 1949).

[2] Parzen, E., "An approach to time series analysis," Ann. Math. Statist. **32**, 951–989 (1961).

[3] Kalman, R. E., "New methods and results in linear prediction and filtering theory," Research Institute for Advanced Studies Rept. 61-1, Martin Co., Baltimore, Md. (1960).

[4] Rauch, H. E., "Linear estimation of sampled stochastic processes with random parameters," TR 2108-1, Stanford Electronics Lab., Stanford Univ. (April 1962).

[5] Bryson, A. E. and Frazier, M., "Smoothing for linear and nonlinear dynamic systems," Wright-Patterson Air Force Base, Ohio, Aeronautical Systems Division TDR-63-119, pp. 353–364 (September 1962).

[6] Rauch, H. E., Tung, F., and Striebel, C. T., "On the maximum likelihood estimates for linear dynamic systems," Lockheed Missiles and Space Co., Palo Alto, Calif., TR 6-90-63-62 (June 1963).

[7] Cox, H., "On the estimation of state variables and parameters for noisy dynamic systems," *IEEE Transactions on A.C.* (Institute of Electrical and Electronic Engineering, New York, January 1964), Vol. AC-9, pp. 5–12.

[8] Ho, Y. C., "On the stochastic approximation method and the optimal filtering theory," *Math. Analysis and Applications* (February 1963), Vol. 6, pp. 152–155.

[9] Bellman, R., *Introduction to Matrix Analysis* (McGraw-Hill Book Co., New York, 1950), pp. 154–155.

[10] Rauch, H. E., *Solutions to the Linear Smoothing Problem*, *IEEE Transactions on A.C.* (Institute of Electrical and Electronic Engineering, New York, October 1963), Vol. AC-8, pp. 371–372.

[11] Rauch, H. E., "Optimum estimation of satellite trajectories including random fluctuations in drag," AIAA J. **3**, 717–722 (April 1965).

The Optimum Linear Smoother as a Combination of Two Optimum Linear Filters

DONALD C. FRASER AND JAMES E. POTTER

Abstract—**A solution to the optimum linear smoothing problem is presented in which the smoother is interpreted as a combination of two optimum linear filters. This result is obtained from the well-known equation for the maximum likelihood combination of two independent estimates and equivalence to previous formulations is demonstrated. Forms of the solution which are convenient for practical computation are developed.**

Introduction

The optimum linear filtering and prediction methods introduced by Kalman [1] and others which have received much attention in recent years provide an estimate of current and future values of the quantities of interest. Optimum linear smoothing, which has received much less attention, provides an estimate of past values of the desired quantities. As such, optimum linear smoothing is a generally useful technique which may be employed in any data reduction situation where a linear or linearized model of the process dynamics is reasonable.

The solutions to the fixed data length smoothing problem [2]–[9] have taken three computational forms. Two of these, typified by the work of Rauch, Tung, and Striebel [2] and Bryson and Frazier [3] give the smoother estimate as a correction to the Kalman filter estimate for the same point. The third types of solutions, which have been documented by Fraser [4] and Mayne [5], do not have the appearance of a correction to the Kalman filter estimate. It is demonstrated here that this third formulation results from the fact that the optimum linear smoother may be interpreted as the optimum combination of two optimum linear filter estimates. Results presented in [4] indicate that this third formulation offers some numerical advantages in the computation of the smoothed covariance matrix in those cases where the smoother gives the most improvement over filtering. This is supported by an analysis similar to that used by Joseph [10] for the filter.

Interpreting the optimum linear smoother as a combination of two optimum linear filters, one of which works forward over the data and the other of which runs backward over the interval, is also appealing for physical reasons. The filter which is processed from the beginning of the data interval to some time t within the interval yields the best[1] estimate of the state at time t based upon all the information from the beginning of the interval to time t. If, however, the filter equations are processed from the end of the data interval backward to time t they will yield the best estimate of the state at time t based upon all the measurements from time t to the end of the data interval. Together these two filters utilize precisely all the available information. If both the measurement and driving noises have white distributions in time these two estimates are uncorrelated. One would thus expect that by properly combining these two optimum estimates he could obtain the optimum smoother estimate which is based upon all the available information.

Only the case where both the measurement and driving noises exist and have white distributions in time is considered here. The correlated noise case has been treated by Mehra and Bryson [11], and the technique presented here can be adapted to those situations where there are perfect measurements by using Deyst's filter reduction method [12]. Due to space limitations only the continuous smoother equations are developed here. Corresponding results for the discrete case are given in Table I.

Manuscript received April 16, 1968; revised January 23, 1969.
The authors are with the Department of Aeronautics and Astronautics, Massachusetts Institute of Technology, Cambridge, Mass.

[1] For linear systems with Gaussian noises this is best in the maximum likelihood, minimum variance, and least squares senses.

Basic Equations

The well-known formulas [13] for the optimum combination of two independent estimates, $\hat{x}_1$ (covariance P_1) and $\hat{x}_2$ (covariance P_2), of x are

$$\hat{x} = (P_1^{-1} + P_2^{-1})^{-1}(P_1^{-1}\hat{x}_1 + P_2^{-1}\hat{x}_2) \tag{1}$$

$$P = \operatorname{cov}(\hat{x} - x) = (P_1^{-1} + P_2^{-1})^{-1} \tag{2}$$

where the statistics have been assumed to be Gaussian with zero mean. This is both the minimum variance and maximum likelihood estimate of x. It is also unbiased.

Interpretation as Two Filters

If $\hat{x}_1(t)$ and $\hat{x}_2(t)$ are the estimates at time t obtained from two optimum linear filters, one of which runs from the beginning of the data interval forward to time t and the other of which works backward to time t from the end of the data interval, $\hat{x}(t)$ is indeed the optimum smoother estimate for time t. The following are equations which describe the continuous form of these two filters.

Forward Filter

$$\dot{\hat{x}}_1 = F\hat{x}_1 + \bar{u} + P_1H^TR^{-1}(z - H\hat{x}_1) \tag{3}$$

$$\dot{P}_1 = FP_1 + P_1F^T + Q - P_1H^TR^{-1}HP_1 \tag{4}$$

$\hat{x}_1(0)$, $P_1(0)$ obtained from a priori knowledge.

Backward Filter

$$\dot{\hat{x}}_2 = F\hat{x}_2 + \bar{u} - P_2H^TR^{-1}(z - H\hat{x}_2) \tag{5}$$

$$\dot{P}_2 = FP_2 + P_2F^T - Q + P_2H^TR^{-1}HP_2 \tag{6}$$

$$\lim_{t \to T} P_2^{-1}(t) = 0 \tag{7a}$$

$$\lim_{t \to T} [P_2^{-1}(t)\hat{x}_2(t)] = 0 \tag{7b}$$

where the new symbols are defined by the basic system equations for state and measurement

$$\dot{x} = Fx + u, \quad z = Hx + \eta. \tag{8a}$$

Gaussian Statistics

$$E(u) = \bar{u}, \quad E\{[u(t) - \bar{u}(t)][u(s) - \bar{u}(s)]^T\} = Q(t)\delta(t - s) \tag{8b}$$

$$E(\eta) = 0, \quad E\{\eta(t)\eta(s)^T\} = R(t)\delta(t - s) \tag{8c}$$

$$T = \text{terminal time of the data interval.}$$

The infinite terminal covariance matrix and absence of a terminal state estimate for the backward filter reflect the fact that if no information obtained after the end of the data interval is used there is complete uncertainty about the state estimate at that time due to the complete absence of information. The estimate $\hat{x}_2(t)$ generally will not have a limiting value at time T because the large value of $P_2(t)$ multiplying the measurement noise in (5) causes $\hat{x}_2(t)$ to have large amplitude fluctuations as $t \to T$. It follows by the transformation of (21) that specifying $\lim_{t\to T}[P_2^{-1}(t)\hat{x}_2(t)]$ is sufficient to determine the solution of (5). Physically (7b) means that the limit of the gradient of the likelihood function of the state based upon the backward filter information goes to zero as $t \to T$. This reflects the fact that there is no information about the distribution of x at time T.

Reprinted from *IEEE Trans. Automat. Contr.*, vol. AC-14, pp. 387–390, Aug. 1969.

TABLE I
EQUATION SUMMARY

Continuous Version	Discrete Equivalent
State Equations	
$x = Fx + u$	$x_{k+1} = \Phi_k x_k + u_k$
$z = Hx + \eta$	$z_k = H_k x_k + \eta_k$
$E(u) = \bar{u}$	$E(u_k) = \bar{u}_k$
$\text{cov}\,(u - \bar{u}) = Q$	$\text{cov}\,(u_k - \bar{u}_k) = Q_k$
$\text{cov}\,(\eta) = R$	$\text{cov}\,(\eta_k) = R_k$
Basic Smoother Forms	
$P = (P_1^{-1} + P_2^{-1})^{-1}$	$P_{k/N} = (P_{1k}^{-1} + P_{2k}'^{-1})^{-1}$
$\hat{x} = P(P_1^{-1}\hat{x}_1 + P_2^{-1}\hat{x}_2)$	$\hat{x}_{k/N} = P_{k/N}(P_{1k}^{-1}\hat{x}_{1k} + P_{2k}'^{-1}\hat{x}_{2k}')$
Smoother Computational Forms	
$P = (I - WU)P_1(I - WU)^T + WUW^T$	$P_{k/N} = (I - W_kU_k')P_{1k}(I - W_kU_k')^T + W_kU_k'W_k^T$
$W = P_1(I + P_1U)^{-1T}$	$W_k = P_{1k}(I + P_{1k}U_k')^{-1T}$
$\hat{x} = (I + P_1U)^{-1}\hat{x}_1 + Pw$	$\hat{x}_{k/N} = (I + P_{1k}U_k')^{-1}\hat{x}_k + P_{k/N}w_k'$
Backward Filter Computational Forms	
$\dot{U} = -UF - F^TU + UQU - H^TR^{-1}H$	$J_k = U_k(U_k + Q_{k-1}^{-1})^{-1}$
	$U_{k-1}' = \Phi_{k-1}^T[(I - J_k)U_k(I - J_k)^T + Q_{k-1}]\Phi_{k-1}$
	$U_k = U_k' + H_k^TR_k^{-1}H_k$
$w = -(F - QU)^Tw + U\bar{u} - H^TR^{-1}z$	$w_{k-1}' = \Phi_k^T(I - J_k)(w_k - U_k\bar{u}_{k-1})$
	$w_k = w_k' + H_k^TR_k^{-1}z_k$
Terminal Conditions	
$U(T) = 0$	$U_N' = 0$
$w(T) = 0$	$w_N' = 0$

1) Subscript 1 denotes forward filter quantity.
2) Subscript 2 denotes backward filter quantity.
3) $U = P_2^{-1}$, $U_K = P_{2K}^{-1}$, $U_K' = P_{2K}'^{-1}$.
4) Prime denotes measurement has not been incorporated, and lack of prime means measurement has been incorporated (filter variables only).

That this combination does indeed give the optimum smoother estimate can be demonstrated by deriving one of the more familiar smoother forms from (1) and (2). This will now be done first for the smoother covariance matrix $P(t)$, and then for the smoothed state estimate $\hat{x}(t)$.

Differential equations for the inverses of P_1 and P_2 may be obtained from (4) and (6). The result is

$$\frac{d}{dt}(P_1^{-1}) = -P_1^{-1}F - F^TP_1^{-1} - P_1^{-1}QP_1^{-1} + H^TR^{-1}H \tag{9}$$

$$\frac{d}{dt}(P_2^{-1}) = -P_2^{-1}F - F^TP_2^{-1} + P_2^{-1}QP_2^{-1} - H^TR^{-1}H. \tag{10}$$

Inverting and differentiating (2) yields, after substitution of (9) and (10)

$$\frac{d}{dt}(P^{-1}) = -(P_1^{-1} + P_2^{-1})F - F^T(P_1^{-1} + P_2^{-1}) - P_1^{-1}QP_1^{-1} + P_2^{-1}QP_2^{-1}. \tag{11}$$

Using (2) to substitute for P_2^{-1} and $(P_1^{-1} + P_2^{-1})$ in (11), collecting terms, and inverting the result gives

$$\dot{P} = (F + QP_1^{-1})P + P(F + QP_1^{-1})^T - Q \tag{12}$$

which is the form of the smoother covariance matrix equation obtained in [2]. Since this is a linear differential equation, equivalence at all times is guaranteed if P can be shown to be the smoother covariance matrix at any time. Using the boundary conditions for the backward filter specified by (7) results in

$$P(T) = P_1(T) \tag{13}$$

which is the familiar terminal condition for the smoother covariance matrix, hence (2) is indeed an expression for the smoother covariance matrix when P_1 and P_2 are the covariance matrices associated with the forward and backward filters described above.

Differentiating (1) we obtain

$$\dot{\hat{x}} = \dot{P}[P_1^{-1}\hat{x}_1 + P_2^{-1}\hat{x}_2] + P\left[\left(\frac{d}{dt}P_1^{-1}\right)\hat{x}_1 + P_1^{-1}\dot{\hat{x}}_1 + \left(\frac{d}{dt}P_2^{-1}\right)\hat{x}_2 + P_2^{-1}\dot{\hat{x}}_2\right]. \tag{14}$$

Substitution of (1), (3), (5), (9), (10), and (12) into (14) yields, after collecting and cancelling terms

$$\dot{\hat{x}} = F\hat{x} + QP_1^{-1}(\hat{x} - \hat{x}_1) + P(P_1^{-1} + P_2^{-1})\bar{u} + (PP_1^{-1}QP_2^{-1} - QP_2^{-1} + PP_2^{-1}QP_2^{-1})\hat{x}_2. \tag{15}$$

Using (2) in the third and fourth terms on the right side of (15) one obtains

$$\dot{\hat{x}} = F\hat{x} + \bar{u} + QP_1^{-1}(\hat{x} - \hat{x}_1) \tag{16}$$

which is the same linear differential equation as that which describes the smoothed state in [2]. The boundary condition on $\hat{x}(T)$ is also the same as that of [2] since by (1) and (7)

$$\hat{x}(T) = \hat{x}_1(T) \tag{17}$$

which is the familiar terminal condition for the smoothed state estimate. Hence, (1) is an expression for the smoother state estimate when the two estimates $\hat{x}_1$ and $\hat{x}_2$ come from the forward and backward filters described above.

A More Convenient Form

Equation (2) is not convenient for the calculation of the smoother covariance matrix for two reasons. 1) $P_2(t)$ is not finite at the terminal time. 2) Three inversions of matrices which are of the dimension of the state are necessary. The problem with the infinite $P_2(t)$ can be avoided by using its inverse. An expression can then be derived which does not involve any infinite quantities and which requires only one inversion of a matrix of the dimension of the state. To begin this derivation, rearrange (2) and premultiply the right side by $(I + P_1P_2^{-1} - P_1P_2^{-1})$. The result is

$$P = P_1 - P_1P_2^{-1}(I + P_1P_2^{-1})^{-1}P_1. \tag{18}$$

This result may then be expanded into a symmetric form which is the sum of two positive definite matrices

$$P = (I - WP_2^{-1})P_1(I - WP_2^{-1})^T + WP_2^{-1}W^T \tag{19}$$

where

$$W = P_1(I + P_1P_2^{-1})^{-1T}. \tag{20}$$

This result may be verified by expanding (19) and using (20) to obtain (18).

Equations (19) and (20) are equivalent to (2). These expanded equations are free of infinite terms, require fewer matrix inversions than (2), avoid forming the result as a difference of two positive definite matrices, and are of a symmetric form.

Calculation of the backward filter state estimate using (5) involves the backward filter covariance matrix $P_2(t)$ and the terminal condition $\hat{x}_2(T)$. Both of these are undesirable. The infinite value of $P_2(T)$ and the boundary condition to specify $\hat{x}_2(T)$ are difficult to apply. These problems may be avoided by introducing the new variable

$$w = P_2^{-1}\hat{x}_2 \tag{21}$$

which has the terminal value

$$w(T) = 0 \tag{22}$$

according to (7).

Differentiating (21) and substituting (5), (10), and (21) in the result yields

$$w = -(F - QP_2^{-1})^T w + P_2^{-1}\bar{u} - H^TR^{-1}z. \tag{23}$$

Equation (23) must be integrated backward from the terminal condition specified by (22). The resulting value of w can then be used in the calculation of the smoothed state estimate via the equation

$$\hat{x} = P(P_1^{-1}\hat{x}_1 + w). \tag{24}$$

Note that (10) and (23) together, with the signs appropriately changed, provide a convenient method of treating the unbiased forward filter.

Equation (24) involves P_1^{-1}, a matrix inversion not yet required. This can be avoided by replacing it with the matrix inversion needed in the computation of the smoother covariance matrix P according to (19) and (20). By expanding (24) and using (2), one can obtain the desired result

$$\hat{x} = (I + P_1P_2^{-1})^{-1}\hat{x}_1 + Pw. \tag{25}$$

Table I summarizes these results and gives the equivalent results for the discrete case.

Summary

It has been demonstrated that interpreting the optimum linear smoother as a combination of two optimum linear filters leads to a new and useful solution to the smoothing problem. One of these filters is the conventional forward Kalman filter which runs forward over the data to the point of interest. The other works backward over the data to the same point. Forms of this solution which are convenient for practical computer computation have been presented.

References

[1] R. E. Kalman, "A new approach to linear filtering and prediction theory," *Trans. ASME, J. Basic Engrg.*, ser. D, vol. 83, pp. 95–108, March 1961.

[2] H. E. Rauch, F. Tung, and C. T. Striebel, "Maximum likelihood estimates of linear dynamic systems," *AIAA J.*, vol. 3, no. 8, pp. 1445–1450, August 1965.

[3] A. E. Bryson and M. Frazier, "Smoothing for linear and nonlinear dynamic systems," Aeronautical Sys. Div., Wright-Patterson AFB, Ohio, Tech. Rept. TDR-63-119, pp. 353–364, September 1962.

[4] D. C. Fraser, "A new technique for the optimal smoothing of data," Sc.D. thesis, Massachusetts Institute of Technology, Cambridge, Mass., January 1967.

[5] D. Q. Mayne, "A solution to the smoothing problem for linear dynamic systems," *Automatica*, vol. 4, pp. 73–92, December 1966.

[6] H. Cox, "On the estimation of state variables and parameters for noisy dynamic systems," *IEEE Trans. Automatic Control*, vol. AC-9, pp. 5–12, January 1964.

[7] J. S. Meditch, "Orthogonal projection and discrete optimal linear smoothing," *SIAM J. Control*, vol. 5, pp. 74–89, February 1967.

[8] T. Kailath, "A Wold-Kolmogorov approach to linear least-squares estimation, pt. II: the smoothing problem," Stanford University, Stanford, Calif., Tech. Rept. TR 7050-13, January 1968.

[9] R. K. Mehra, "Studies in smoothing and conjugate gradient methods applied to optimal control problems," Ph.D. thesis, Harvard University, Cambridge, Mass., December 1967.

[10] P. D. Joseph, "Automatic rendezvous, pt. II: on board navigation for rendezvous missions," course notes for "Space Control Systems—Attitude, Rendezvous, and Docking," Engineering Extension Course, University of California, Los Angeles, Calif., 1964.

[11] R. K. Mehra and A. E. Bryson, "Smoothing for linear time varying systems using measurements containing colored noise," Preprints, 1968 JACC (Ann Arbor, Mich.).

[12] J. J. Deyst, "Optimum continuous estimation of nonstationary random variables," S.M. thesis, Dept. of Aeron. and Astron., Massachusetts Institute of Technology, Cambridge, Mass., January 1964.

[13] P. B. Liebelt, *An Introduction to Optimal Estimation.* Reading, Mass.: Addison-Wesley, 1967.

An Innovations Approach to Least-Squares Estimation Part II: Linear Smoothing in Additive White Noise

THOMAS KAILATH, MEMBER, IEEE, AND PAUL FROST, MEMBER, IEEE

***Abstract*—The innovations method of Part I is used to obtain, in a simple way, a general formula for the smoothed (or noncausal) estimation of a second-order process in white noise. The smoothing solution is shown to be completely determined by the results for the (causal) filtering problem. When the signal is a lumped process, differential equations for the smoothed estimate can easily be derived from the general formula. In several cases, both the derivations and the forms of the solution are significantly simpler than those given in the literature.**

I. Introduction

IN THIS PAPER, we apply the innovations technique of Part I[1] to solve the smoothing problem. We shall show that the smoothing solution is completely determined in a simple way by the optimum causal filter and its adjoint. This result is valid for a general second-order (finite-variance) signal process in white noise, without restriction to lumped signal processes. For lumped processes, recursive solutions are available for the filtered estimate (Part I), and from these similar solutions can easily be found for the smoothed estimate. In the literature, recursive solutions to the smoothing problem have generally been obtained rather laboriously and often in less convenient form than ours (cf. Section III).

The problem we shall begin with is the following. We are given observations

$$y(t) = H(t)x(t) + v(t), \qquad a \leq t < b \tag{1}$$

where

$$\overline{v(t)} = 0, \quad \overline{v(t)v'(s)} = R(t)\delta(t-s), \quad R(t) > 0 \tag{2}$$

$$\overline{x(t)} = 0, \quad \overline{x(t)x'(t)} < \infty, \quad \overline{x(t)v'(s)} \equiv 0, \quad s > t. \tag{3}$$

It is required to find the linear least-squares smoothed estimate

$$x(t \mid b) = \text{the linear function of the data } \{y(s), a \leq s < b\} \text{ that minimizes } \overline{[x(t) - \hat{x}(t \mid b)]'[x(t) - \hat{x}(t \mid b)]}. \tag{4}$$

Manuscript received January 31, 1968. This work was supported by the Applied Mathematics Division of the Air Force Office of Scientific Research under Contract AF 49(638)1517, by the Air Force Avionics Laboratory under Contract F 33615-67-C-1245, and by the Joint Services Electronics Program at Stanford University, Stanford, Calif., under Contract Nonr 225(83).

The authors are with Stanford University, Stanford, Calif.

[1] This issue, page 646.

Reprinted from *IEEE Trans. Automat. Contr.*, vol. AC-13, pp. 655–660, Dec. 1968.

We shall show, and the proof is trivial by the innovations method, that the smoothed estimate $\hat{x}(t|b)$ can be expressed

$$\hat{x}(t\mid b) = \int_a^b \overline{x(t)\nu'(s)}R^{-1}(s)\nu(s)ds \qquad (5)$$

where

$$\nu(s) = y(s) - H(s)\hat{x}(s\mid s) \triangleq \text{the innovation process.}$$

By breaking up the range of integration, $\hat{x}(t|b)$ can be written as the sum of the filtered estimate $\hat{x}(t|t)$ and a correction term,

$$\hat{x}(t\mid b) = \hat{x}(t\mid t) + \int_t^b P(t,s)H'(s)R^{-1}(s)\nu(s)ds \qquad (6)$$

where

$P(t,s) \triangleq \overline{\tilde{x}(t\mid t)\tilde{x}'(s\mid s)}$ = the covariance function of the error in the filtered estimate; viz., $\tilde{x}(\cdot) = x(\cdot) - \hat{x}(\cdot\mid\cdot)$.

We shall show that the covariances of the smoothed and filtered errors are related by a fairly simple equation that exhibits the reduction in estimation error due to the additional observations from t to b:

$$\overline{\tilde{x}(t\mid b)\tilde{x}'(t\mid b)} = \sum(t\mid b) = P(t\mid t) - \int_t^b P(t,s)H'(s)R^{-1}(s)H(s)P(s,t)ds. \qquad (7)$$

Adjoint Filter Interpretation

When the signal and noise process $z(\cdot)$ and $v(\cdot)$ are *completely uncorrelated*, i.e.,

$$\overline{z(t)v'(s)} = 0 \qquad \text{for all } a \le t, s \le b, \qquad (8)^2$$

we can express $\hat{x}(t|b)$ in a form that has a useful interpretation:

$$\hat{x}(t\mid b) = \int_a^b G_{y_m}(t,s)y_m(s)ds + \int_a^b G_{y_m}'(s,t)\nu_m(s)ds \qquad (9)$$

where

$$G_{y_m}(t,s) \triangleq \begin{cases} P(t,s), & t \ge s \\ 0, & t < s \end{cases}$$

and

$$\nu_m(s) \triangleq H'(s)R^{-1}(s)\nu(s) \triangleq \text{modified innovation process}$$

$$y_m(s) \triangleq H'(s)R^{-1}(s)y(s) \triangleq \text{modified observation process.}$$

The first term on the right-hand side of (9) will be shown to be a representation of the filtered estimate, i.e.,

$$\hat{x}(t\mid t) = \int_a^t G_{y_m}(t,s)y_m(s)ds = \int_a^t P(t,s)H'(s)R^{-1}(s)y(s)ds. \qquad (10)$$

Thus, $G_{y_m}(t,s)$ is the impulse response of the optimum causal filter with input $y_m(\cdot)$, and $G_{y_m}'(s,t)$ may be regarded as the impulse response of a filter that is *adjoint* to the optimum causal filter. It is clear from the definition of $G_{y_m}(\cdot,\cdot)$ that the adjoint filter is a completely noncausal filter (see, e.g., Zadeh and Desoer [1, sec. 8.3] for a discussion of adjoint filters). Equation (9) states that the smoothed estimate $\hat{x}(t|b)$ is obtained by adding a term that is acquired by passing the modified innovation $\nu_m(\cdot)$ through the adjoint of the optimum causal filter with $y_m(\cdot)$ [not $\nu_m(\cdot)$] as its input. The smoothing formula (9) was first obtained in a different way (see Kailath [2]) by using certain resolvent identities of Siegert [24′].[3] For *lumped* processes the formula (6) was obtained independently by Fraser [3], Kwakernaak [35′] and Frost, who all used different methods. The fact that the result holds for general second-order processes and the present simple proof, based on the innovations theorem of Part I, are due to Kailath. In the lumped case, other formulas for $\hat{x}(t|b)$ have been obtained by us and other authors, but as we shall show in Section III, these formulas are most easily deduced from the general formula (6).

The Discrete-Time Case

In the discrete-time case, we have the analog of the general formula (6) and (7)

$$\hat{x}(k\mid b) = \hat{x}(k\mid k) + \sum_{k+1}^{b} \overline{x(k)\nu(l)}[P_z(l) + R(l)]^{-1}\nu(l) \qquad (11)$$

where

$$\nu(l) = y(l) - H(l)\hat{x}(l\mid l-1),$$
$$P_z(l) = \overline{\tilde{z}(l\mid l-1)\tilde{z}'(l\mid l-1)}.$$

But, somewhat surprisingly, there is no analog of the adjoint interpretation (9). The reason is that the adjoint interpretation really depends upon the resolvent identity used in [2] and, unfortunately, this identity does not hold in discrete time.

II. Derivation of General Smoothing Formulas

We start (as for $\hat{x}(t|t)$ in Part I) by noting that because of the equivalence of the observations $y(\cdot)$ and the innovations $\nu(\cdot)$, we can write $\hat{x}(t|b)$ as a linear

[2] For a lumped signal process, this condition can always be arranged by a special change of variables (cf. Kalman [20′]).[3]

[3] Primes are used to denote references and equations that were given in Part I.

combination of the $\{\nu(s), a \le s < b\}$,

$$\hat{x}(t|b) = \int_a^b G_\nu(t, s)\nu(s)ds$$

where $G_\nu(\cdot, \cdot)$ is such that

$$\tilde{x}(t|b) = x(t) - \hat{x}(t|b) \perp \nu(s), \quad a \le s < b. \tag{12}$$

From these, we obtain

$$\begin{aligned}\overline{x(t)\nu'(s)} &= \int_a^b G_\nu(t, \sigma)\overline{\nu(\sigma)\nu'(s)}d\sigma \\ &= G_\nu(t, s)R(s), \qquad a \le t, \quad s < b.\end{aligned} \tag{13}$$

With this explicit expression for G_ν, we can rewrite $\hat{x}(t|b)$ as

$$\begin{aligned}\hat{x}(t|b) &= \int_a^b \overline{x(t)\nu'(s)}R^{-1}(s)\nu(s)ds \\ &= \int_a^t \overline{x(t)\nu'(s)}R^{-1}(s)\nu(s)ds \\ &\quad + \int_t^b \overline{x(t)\nu'(s)}R^{-1}(s)\nu(s)ds.\end{aligned} \tag{14}$$

By setting $b=t$ in the above equations, it is obvious that the first integral in (14) is just the filtered estimate $\hat{x}(t|t)$ [cf. (14′)].

To complete the proof of (6), we note that (the orthogonality property (12) is repeatedly used below)

$$\begin{aligned}&G_\nu(t, s)R(s) \\ &= \overline{x(t)\nu'(s)} = \overline{x(t)[\tilde{x}'(s|s)H'(s) + v'(s)]} \\ &= \overline{x(t)\tilde{x}'(s|s)}H'(s) = \overline{[\hat{x}(t|t) + \tilde{x}(t|t)]\tilde{x}'(s|s)}\,H'(s) \\ &= \overline{\tilde{x}(t|t)\tilde{x}'(s|s)}H'(s) \qquad \text{for } s > t \\ &= P(t, s)H'(s) \qquad \text{for } s > t.\end{aligned} \tag{15}$$

To obtain (7), we note that using (6) we can write

$$\tilde{x}(t|t) = \tilde{x}(t|b) + \int_t^b P(t, s)H'(s)R^{-1}(s)\nu(s)ds. \tag{16}$$

From (12), $\tilde{x}(t|b)$ is orthogonal to $\nu(s)$ for $a \le s < b$. Hence (7) follows immediately.

Finally, with the assumption (8), viz., $\overline{x(t)v'(s)} \equiv 0$, we shall prove (10) from which (9) easily follows. We note [by arguments similar to those for (15)] that

$$\begin{aligned}&P(t, s)H'(s) \\ &\triangleq \overline{\tilde{x}(t|t)\tilde{x}'(s|s)}H'(s) = \overline{\tilde{x}(t|t)[y(s) - H(s)\hat{x}(s|s) - v(s)]'} \\ &= -\overline{\tilde{x}(t|t)v'(s)}, \qquad \text{for } s < t \\ &= \overline{\hat{x}(t|t)v'(s)} = \int_a^t G_{y_m}(t, \sigma)H'(\sigma)R^{-1}(\sigma)\overline{y(\sigma)v'(s)}d\sigma \\ &= G_{y_m}(t, s)H'(s), \qquad s < t.\end{aligned} \tag{17}$$

Equation (10) follows easily from (17). This completes our derivation of the main smoothing formulas.

III. Recursive Smoothing Formulas for Lumped Processes

We shall now assume that the signal $x(\cdot)$ is a lumped process described by

$$\dot{x}(t) = F(t)x(t) + u(t) \tag{18}$$

where

$$\overline{u(t)u'(s)} = Q(t)\delta(t - s), \quad \overline{u(t)v'(s)} = C(t)\delta(t - s). \tag{19}$$

Some Earlier Results

Judging by the literature, recursive solutions for the smoothing problem have been thought to be harder to obtain than recursive-filtering solutions. We have already mentioned the work of Kwakernaak [35′] and Fraser [3].

The first recursive solutions were given by Bryson and Frazier [4] (see also Cox [5]) who obtained an estimate that minimized a certain deterministic quadratic form.

Similar solutions and some alternative forms were obtained (by discrete-time Bayes-rule manipulations) by Rauch [6], and Rauch, Tung, and Striebel [7] for discrete-time processes. Lee [8] derived these results using the calculus of variations. Rauch *et al.* extended (all but one of) the forms to the continuous-time case by the formal limiting procedure of Kalman [20′]. Meditch [9] has very recently done the same for the one that had been omitted by Rauch *et al.* In a somewhat earlier paper, Meditch [10] derived smoothing solutions for discrete-time processes by using projection arguments. A very direct projection method was given in an early report by Carlton [13].

We may also mention some recent theses on the smoothing problem. Fraser [3] studied a different form for the smoothed estimate, expressing it as the combination of two filtered estimates obtained by the use of forward and backward Kalman–Bucy filters. He showed the equivalence to earlier results and discussed the computational aspects of the problem in several numerical examples. The two-filter solution was first given by Mayne [14] and has been further studied by Mehra [15]. Henrikson [16] gives a solution for the discrete-time problem with colored observation noise and Mehra [15] does the same in continuous time, both essentially combining the smoothing proofs of Bryson and Frazier [4] and the colored noise techniques of Bryson and Johansen [40′]. Baggeroer [17] gives a proof similar to that of Kwakernaak [35′] based on the Wiener-Hopf equation. A recent thesis by Lindquist [18] also obtains results similar to those of Kwakernaak [35′], but by an ingenious discretization procedure.

In this paper, we shall derive all these earlier results very directly from the recursive Kalman–Bucy formulas for $\hat{x}(t|t)$ and $P(t, s)$; of course, such deductions could also have been made by Kwakernaak and Fraser, who independently obtained (6) for lumped processes, but in a much less direct way. In fact, Fraser [3] made a partial beginning in this direction.

The filtered estimate for the model (18)–(19) is

$$\dot{\hat{x}}(t|t) = F(t)\hat{x}(t|t) + K(t)\nu(t) = [F(t) - K(t)H(t)]\,\hat{x}(t|t) + K(t)y(t), \quad t \geq a \tag{20}$$

$$\nu(t) = y(t) - H(t)\hat{x}(t|t), \qquad K(t) = [P(t,t)H'(t) + C(t)]R^{-1}(t) \tag{21}$$

$$\dot{P}(t,t) = F(t)P(t,t) + P(t,t)F'(t) - K(t)R(t)K'(t) + Q(t). \tag{22}$$

We also note that we can write (cf. Appendix II of Part I)

$$P(t,s) \triangleq \tilde{x}(t|t)\tilde{x}'(s|s) = P(t,t)\Phi'(s,t) \qquad \text{for } s \geq t \tag{23}$$

where $\Phi(t, s)$ is the fundamental (state-transition) matrix of the error equation

$$\dot{\tilde{x}}(t|t) = [F(t) - K(t)H(t)]\tilde{x}(t|t) + u(t) - K(t)v(t) \tag{24}$$

i.e., $\Phi(t, s)$ obeys the equation

$$\frac{d\Phi}{dt}(t,s) = [F(t) - K(t)H(t)]\Phi(t,s), \quad \Phi(s,s) = I. \tag{25}$$

For later use, we note here the (well-known) formula[4]

$$\frac{d\Phi}{dt}(s,t) = -\Phi(s,t)[F(t) - K(t)H(t)], \qquad \Phi(s,s) = I. \tag{26}$$

Now returning to the basic smoothing formula (6), we can write, using (23),

$$\hat{x}(t|b) = \hat{x}(t|t) + \int_t^b P(t,s)H'(s)R^{-1}(s)\nu(s)ds, \qquad t \geq a \tag{27}$$

$$= \hat{x}(t|t) + P(t,t)\int_t^b \Phi'(s,t)H'(s)R^{-1}(s)\nu(s)ds$$

$$= \hat{x}(t|t) + P(t,t)\lambda(t), \quad \text{say} \tag{28}$$

where

$$\lambda(t) = \int_t^b \Phi'(s,t)\nu_m(s)ds \tag{29}$$

and $\nu_m(s)$ is defined by (9). The function $\lambda(t)$ is closely related to the adjoint system associated with (24), as we shall soon see. It will be useful to distinguish (as perhaps first done by Meditch [9]) three classes of smoothing problems, defined by the nature of the observation interval (a, b).

[4] This can be obtained by differentiation of the identity $\Phi(t,s)\Phi(s,t) = I$ and use of (25).

Fixed-Interval Smoothing: a and b Fixed

From (29),

$$\dot{\lambda}(t) = -\Phi'(t,t)\nu_m(t) + \int_t^b \frac{d}{dt}\Phi'(s,t)\nu_m(s)ds \tag{30}$$

which, by use of (26) and (27), can be written

$$\dot{\lambda}(t) = -[F'(t) - H'(t)K'(t)]\lambda(t) - \nu_m(t). \tag{31}$$

Equation (31) for $\lambda(\cdot)$ (with $\nu(\cdot)$ set equal to zero) is (cf. Zadeh and Desoer [1, ch. 6) the differential equation of the system that is *adjoint* to the systems of (20) or (24) (with driving terms $y(\cdot)$, $u(\cdot)$, $v(\cdot)$ set equal to zero). Therefore, the $\{\lambda(t)\}$ are often called the *adjoint* variables.

In any case, the formulas (20), (22), (28), and (31) define the smoothed estimate $\hat{x}(t|b)$ by a set of algebraic and differential equations. This set was the form in which the first recursive smoothing solutions were obtained (in a much less transparent manner) by Bryson and Frazier [4].

We shall now derive an alternative representation of $\hat{x}(t|t)$ which was first obtained by Rauch *et al.* [7], also in a more complicated way. We merely note that by differentiation of (28) we will have

$$\dot{\hat{x}}(t|b) = \dot{\hat{x}}(t|t) + \dot{P}(t,t)\lambda(t) + P(t,t)\dot{\lambda}(t) \tag{32}$$

which by use of (20), (22), (29), (31), and some simple algebra can be reduced to

$$\dot{\hat{x}}(t|b) = F(t)\hat{x}(t|b) + Q(t)\lambda(t) \tag{33}$$

and, by one more use of (28), this becomes

$$\dot{\hat{x}}(t|b) = F(t)\hat{x}(t|b) + Q(t)P^{-1}(t,t)[\hat{x}(t|b) - \hat{x}(t|t)]. \tag{34a}$$

By differentiating (7), it easily is shown that the smoothing error covariance can be represented as

$$\dot{\Sigma}(t|b) = (F(t) + Q(t)P^{-1}(t,t))\,\Sigma(t|b) + \Sigma(t|b)(F(t) + Q(t)P^{-1}(t|t))' - Q. \tag{34b}$$

The formulas (34a) and (34b) of Rauch *et al.* are noteworthy because they point out that knowledge of the filtering solution is sufficient for determining the smoothing solution. That is, once $\hat{x}(t|t)$ and $P(t, t)$ are determined for $a \leq t \leq b$, there is no further need to retain the original observations $\{y(t)\}$. A more detailed discussion of this sufficiency property of $\hat{x}(t|t)$ is given in Frost [28′, sec. II-K]. Practical implementation or (34a) and (34b) would dictate obtaining $P^{-1}(t, t)$ as the solution of

$$\dot{P}^{-1}(t,t) = -P^{-1}(t,t)F(t) - F'(t)P^{-1}(t,t) - P^{-1}(t,t)QP^{-1}(t,t) + H'(t)R^{-1}(t)H(t) \tag{35}$$

since inverting $P(t, t)$ for each t is not feasible.

There are various other representations for the solution to the fixed-interval smoothing problem. In particular, we note the two-filter solution introduced by Mayne [14] and Fraser [3]. For a fixed-time interval problem, the notion of a time direction is rather artificial and it is clear that recursive solutions can be defined for a time index running from $t=b$ to $t=a$. Thus, if we set $\tau=b-t$, the model equation (18) becomes

$$\frac{d}{d\tau}x(b-\tau) = -F(b-\tau)x(b-\tau) - u(b-\tau)$$

and the backward time filter equations are

$$\frac{d}{d\tau}\hat{x}_B(\tau) = -E(b-\tau)\hat{x}_B(\tau) + P_B(\tau,\tau)H'(b-\tau) \cdot R^{-1}(b-\tau)[y(b-\tau) - H(b-\tau)\hat{x}_B(\tau)]$$

$$\dot{P}_B(\tau,\tau) = -F(b-\tau)P_B(\tau,\tau) - P_B(\tau,\tau)F'(b-\tau) - P_B(\tau,\tau)H'(b-\tau)R^{-1}(b-\tau)H(b-\tau) \cdot P_B(\tau,\tau) + Q(b-\tau).$$

The backward filter equations expressed in terms of t are then

$$\dot{\hat{x}}_B(b-t) = F(t)\hat{x}_B(t) - P_B(t,t)H'(t)R^{-1}(t)[y(t) - H(t)\hat{x}_B(t)] \tag{36a}$$

$$\dot{P}_B(b-t, b-t) = -F(t)P_B(b-t,b-t) - P_B(b-t,b-t)F'(t) - P_B(t-b,t-b)H(t)R^{-1}(t)H(t) \cdot P_B(t-b,t-b) + Q(t). \tag{36b}$$

A simple calculation shows that the smoothed error covariance $\Sigma(t|b)$ can be expressed by

$$\Sigma^{-1}(t|b) = P^{-1}(t,t) + P_B^{-1}(b-t,b-t) \tag{37a}$$

provided we choose $P_B^{-1}(0,0)=0$.[5] Moreover, the smoothed estimate $\hat{x}(t|b)$ satisfies the relation[6]

$$\hat{x}(t|b) = \Sigma(t|b)[P^{-1}(t,t)\hat{x}(t|t) + P_B^{-1}(b-t,b-t)\hat{x}_B(b-t,b-t)]. \tag{37b}$$

It is useful to be able to express the solution to a smoothing problem in a number of different forms because of the various computational advantages that may accrue to each form. Such computational aspects have been partially considered by Fraser [3] and Mehra [16].

Fixed-Point Smoothing: t is Fixed, Say $t=a$, While b Increases

Our basic smoothing formula (6) yields, with $t=a$,

$$\hat{x}(a|b) = \hat{x}(a|a) + \int_a^b P(a,s)\nu_m(s)ds \tag{38}$$

where we recall that $P(t,s)$ is the covariance function of the error in the filtered estimate $\hat{x}(t|t)$, $a \le t < b$. Moreover, our formula holds for second-order processes $x(\cdot)$. For lumped processes, Rauch *et al.* [7] and Meditch [11] have given different appearing but equivalent formulas. Thus, Meditch has[7]

$$\hat{x}(a|b) = \hat{x}(a|a) + \int_a^b B(s,a)P(s,s)\nu_m(s)ds \tag{39}$$

where $B(s,a)$ is the solution of the matrix differential equation

$$\frac{d}{ds}P(s,a) = -B(s,a)[F(s) + Q(s)P^{-1}(s,s)], \quad B(a,a) = I. \tag{40}$$

The two solutions (38) and (39) are equivalent because the matrices $P(a,s)$ and $B(s,a)P(s,s)$ are equal. This can be shown [suggested by Rauch (private communication)] by simple differentiation because both matrices have the same initial value $[P(a,a)]$ and obey the same differential equations.

Fixed-Lag Smoothing: $b=t+\Delta$, $\Delta=A$ Fixed Positive Constant

We proceed as for fixed-interval smoothing, the only difference being that we have an additional term in the expression (31) for $\lambda(t)$. Because

$$\lambda(t) = \int_t^{t+\Delta} \Phi'(s,t)\nu_m(s)ds, \tag{41}$$

we now have

$$\dot{\lambda}(t) = -[F'(t) - H'(t)K'(t)]\lambda(t) - \nu_m(t) + \Phi'(t+\Delta,t)\cdot\nu_m(t+\Delta). \tag{42}$$

Substituting into (32), we have after some algebra

$$\dot{\hat{x}}(t|t+\Delta) = F(t)\hat{x}(t|t+\Delta) + Q(t)P^{-1}(t,t) \cdot (\hat{x}(t|t+\Delta) - \hat{x}(t|t)) + P(t,t)\Phi(t+\Delta,t)H'(t+\Delta) \cdot R^{-1}(t+\Delta)\nu(t+\Delta). \tag{43a}$$

The fixed-lag smoothing error covariance matrix $\Sigma_L(t)$ is given by

$$\dot{\Sigma}_L(t) = [F(t) + Q(t)P^{-1}(t,t)]\Sigma_L(t) + \Sigma_L(t)[F(t) + Q(t)P^{-1}(t,t)]' - (Q(t) + J(t)) \tag{43b}$$

where

$$J(t) = \Phi'(t+L,t)H'(t+L)R^{-1}(t+L)H(t+L)\Phi(t+L,t).$$

[5] Equation (37a) is trivially correct for $t=b$; equality for all t is simply verified by showing that the derivative of the right-hand side of (37a) is given by (34b).

[6] Equation (37b) is verified by showing that the derivative of the right-hand side is given by (34a).

[7] Very recently in another paper [12], Meditch has obtained the form (38) but in a less direct way (similar to that of Kwakernaak [35']).

We observe that the fixed-lag smoothing equations are essentially the same as those for a fixed interval [cf. (34a) and (34b)] except for an additional forcing term.

For this problem Meditch [9] obtains, by a tedious limiting argument on a discrete-time formula of Rauch [6], the formula

$$\begin{aligned}\dot{\hat{x}}(t \mid t + \Delta) &= F(t)\hat{x}(t \mid t + \Delta) + Q(t)\lambda(t) + C(t + \Delta, t) \\ &\quad \cdot P(t + \Delta, t + \Delta) \cdot \nu_m(t + \Delta) \qquad (44)\end{aligned}$$

where the matrix $C(t+\Delta, t)$ is the solution of the differential equation

$$\begin{aligned}\dot{C}(t + \Delta, t) &= [F(t) + Q(t)P^{-1}(t, t)]C(t + \Delta, t) \\ &\quad - C(t + \Delta, t)[F(t + \Delta) \\ &\quad + Q(t + \Delta)P^{-1}(t + \Delta, t + \Delta)] \qquad (45)\end{aligned}$$

with initial condition $C(a+\Delta, a) = B(a+\Delta, a)$ where $B(\cdot, \cdot)$ was defined in (40). The equivalence of (43a) and (44)–(45) can be proved in the same way as for (38) and (39), viz., it is easy to check that $P(t, t)\Phi'(t+\Delta, t)$ and $C(t+\Delta, t)P(t+\Delta, t+\Delta)$ have the same initial values and obey the same differential equations.

IV. Concluding Remarks

This paper gives a direct and simple derivation, by the innovations method of Part I, of all previously known smoothing formulas for lumped processes in additive white noise. It also gives a new general formula, (6), for arbitrary second-order signal processes and points out an interesting adjoint filter interpretation of the general smoothing solution.

References

[1] L. A. Zadeh and C. Desoer, *Linear System Theory—A State-Space Approach.* New York: McGraw-Hill, 1963.
[2] T. Kailath, "Application of a resolvent identity to a problem in linear smoothing," to appear in *J. SIAM on Control*, 1969.
[3] D. C. Fraser, "A new technique for the optimal smoothing of data," Sc.D. dissertation, Dept. of Aeronautical Engrg., Massachusetts Institute of Technology, Cambridge, Mass., January 1967.
[4] A. E. Bryson and M. Frazier, "Smoothing for linear and nonlinear dynamic systems," Aeronautical Systems Div., Wright-Patterson AFB, Ohio, Tech. Rept. ASD-TDR-63-119, February 1963.
[5] H. Cox, "On the estimation of state variables and parameters for noisy dynamic systems," *IEEE Trans. Automatic Control*, vol. AC-9, pp. 5–12, January 1964. Also, see Sc.D. dissertation, Dept. of Elec. Engrg., Massachusetts Institute of Technology, Cambridge, Mass., June 1963.
[6] H. E. Rauch, "Solutions to the linear smoothing problem," *IEEE Trans. Automatic Control* (*Short Papers*), vol. AC-8, pp. 371–372, October 1963. Also, see Ph.D. dissertation, Dept. of Elec. Engrg., Stanford University, Stanford, Calif., 1962.
[7] H. E. Rauch, F. Tung, and C. T. Striebel, "Maximum likelihood estimates of linear dynamic systems," *AIAA J.*, vol. 3, no. 8, pp. 1445–1450, August 1965.
[8] R. C. K. Lee, *Optimal Estimation, Identification and Control.* Cambridge, Mass.: M.I.T. Press, 1964.
[9] J. S. Meditch, "On optimal linear smoothing theory," *J. Information and Control*, vol. 10, pp. 598–615, 1967.
[10] ——, "Orthogonal projection and discrete optimal linear smoothing," *J. SIAM on Control*, vol. 5, pp. 74–89, February 1967.
[11] ——, "Optimal fixed-point continuous linear smoothing," *Proc. 1967 Joint Automatic Control Conf.* (Philadelphia, Pa.).
[12] ——, "On optimal fixed-point linear smoothing," *Internat'l J. Control*, vol. 6, pp. 189–199, 1967.
[13] A. G. Carlton, "Linear estimation in stochastic processes," Johns Hopkins University, Appl. Phys. Lab., Baltimore, Md., Internal Rept., 1962.
[14] D. Q. Mayne, "A solution of the smoothing problem for linear dynamic systems," in *Automatica*, vol. 4. New York: Pergamon, 1966, pp. 73–92.
[15] R. K. Mehra, Ph.D. dissertation, Dept. of Appl. Phys. and Engrg., Harvard University, Cambridge, Mass., February 1968.
[16] J. Henrikson, Ph.D. dissertation, Dept. of Appl. Phys. and Engrg., Harvard University, Cambridge, Mass., September 1967.
[17] A. E. Baggeroer, Ph.D. dissertation, Dept. of Elec. Engrg., Massachusetts Institute of Technology, Cambridge, Mass., February 1968.
[18] A. Lindquist, "On optimal stochastic control with smoothed information," Royal Inst. Tech., Stockholm, Sweden, IOS Rept. R-25, May 1968.

Part II
Application Papers

THIS part is reprinted from the "Special Issue on Applications of Kalman Filtering," IEEE TRANSACTIONS ON AUTOMATIC CONTROL, vol. AC-28, pp. 254–434, March 1983. This special issue was edited by Harold W. Sorenson. The first paper of this part is an edited version of the editorial from that March issue giving a summary of the contents of Part II.

Editorial
On Applications of Kalman Filtering

The idea for a Special Issue, based on applications of the Kalman filter, was suggested by M. Grimble to A. Harvey. At the time, Harvey was Chairman of the Technical Committee on Applications for the CSS. He discussed the possibility with me and a detailed proposal was made to the IDC which was approved in December 1980. The Editorial Board was formed with Associate Editors J. Balchen, P. Belanger, G. Blankenship, B. Friedland, M. Grimble, J. LeMay, J. Mendel, H. Titus, and K. Wall. I want to thank the members of the Editorial Board for their contributions. Without their conscientious and thoughtful efforts, this issue would not exist. Their recommendations were based on the evaluations provided by many reviewers whose objective and constructive comments are greatly appreciated. Finally, we need to express our gratitude to the authors who contributed their ideas and manuscripts to the Special Issue. We regret that every contribution could not be included.

A basic objective defined for the issue was to obtain and present a set of applications that is as wide ranging as possible. By seeking breadth, it was felt that conceptual similarities between disparate problems could be identified and approaches to these problems could be contrasted. We feel that this objective has been achieved. As described below, the applications range from spacecraft orbit determination to the demographics of cattle production. The commonality of concerns that are addressed in these papers is striking. In reading these papers, one is impressed by the cleverness exhibited in achieving successful implementations. But, also, the results may suggest intriguing questions not addressed by the authors. Certainly, other approaches to specific problems may suggest themselves. As many of the systems that are described have been implemented operationally or have been subjected to extensive experimental verification, the "solutions" have the weight of experience and success.

The Kalman filter can be used for a variety of end-purposes. Its basic function is to provide estimates of the current state of the system. But it also serves as the basis for predicting future values of prescribed variables or for improving estimates of variables at earlier times. In many of the papers in this issue, the performance of the Kalman filter is assessed and evaluated, and the estimates serve as the end product for the application. In other papers the filter serves as a component of a system such as a controller or a detector. The performance of the larger system assumes paramount importance and the behavior of the filter is assessed against the broader context.

The papers for the Special Issue have been grouped according to general types of applications (as labeled by the Guest Editor). The breadth of applications is gauged, easily, by reviewing the types. Furthermore, papers have been arranged, in a pseudochronological order. The earliest applications of the Kalman filter dealt with satellite orbit determination. The first paper in this issue, by Campbell *et al.*, describes the use of the filter for orbit determination for the Voyager spacecraft during its Jupiter fly by. The issue concludes with a type of application that is of a more recent vintage and, certainly, very different from orbit determination. Leibundgut *et al.* describe the use of the Kalman filter for predicting cattle populations in France.

The papers, included between those dealing with Voyager orbit determination and cattle demographics, have been categorized as tracking, navigation, ship motion, remote sensing, geophysical exploration, industrial processes, and power systems. Orbit determination, tracking, and navigation problems represent, probably, the first major applications of the Kalman filter and the first seven papers in this issue describe mature and sophisticated examples of these applications. The specific problems are different, but the papers share many common aspects and concerns. For example, dynamic and error models generally have similar bases and origins.

The problems of dynamic modeling and of state variable descriptions tend to be less well-defined in the types of applications other than tracking and navigation. In many instances, a dynamic model for the system can be defined, conceptually, but the resulting description is so complex that it is not very useful. The development of acceptable models upon which the Kalman filter can be based becomes a fundamental concern. This aspect of successful filter implementation is a dominating theme throughout the issue.

The papers on ship motion prediction and position control have many features in common with the tracking and navigation problems. But the modeling of ship/wave interactions requires the development of simple, but effective, models using signal processing or identification techniques. In addition, these papers are concerned with the Kalman filter performance primarily in the context of a control system. For the remote sensing papers, the Kalman filter appears as part of a signal processor that is used for offline analysis. These papers, as well as the geophysical exploration paper, are not concerned with real-time processing. Nonetheless, simple models are used to accomplish the processing, and the performance is based on the achievement of the general goals of the system.

Estimation and control problems abound in industrial

Reprinted, with modifications, from *IEEE Trans. Automat. Contr.*, vol. AC-28, pp. 254–255, Mar. 1983.

processes and power systems. These applications impose the necessity for real-time operation, generally with limited computational capability, using poorly defined models of the process. Simplicity and adaptability are basic concerns in these applications.

It is interesting in reading this issue to identify common problems and to compare the manner in which their solution has been approached and evaluated. As mentioned above, modeling problems are a recurrent topic throughout the issue. Modeling concerns that are addressed include the definition of variables to be included in the state vector, coordinate systems to be used, linearization methods, observability and conditioning of the model, the effects of neglected variables or model errors, etc. Given the basic model, the Kalman filter can be implemented in a variety of ways. In fact, the model and the filter algorithm are interrelated and several papers describe the feedback process that leads to the form of the filter that is implemented.

Requirements of the system impose significant constraints on the design of the filter algorithm. The need for real-time operation can force the development of an implementation that is very different from that which would occur for offline analysis. In many applications, limited computational capability demands simple models and algorithms, and these considerations can exaggerate the divergence control problem. Model errors or uncertainties can dictate the use of adaptive filters or system identification procedures. Since the introduction of the Kalman filter and its initial applications in the early 1960's, many theoretical papers have been stimulated by problems encountered in applying the Kalman filter to practical problems. The insights gained from these analyses manifest themselves throughout the presentations contained in this issue.

HAROLD W. SORENSON
Guest Editor

Voyager Orbit Determination at Jupiter

JAMES K. CAMPBELL, STEPHEN P. SYNNOTT, AND GERALD J. BIERMAN, SENIOR MEMBER, IEEE

***Abstract*—This paper summarizes the Voyager 1 and Voyager 2 orbit determination activity extending from encounter minus 60 days to the Jupiter encounter, and includes quantitative results and conclusions derived from mission experience. The major topics covered include an identification and quantification of the major orbit determination error sources and a review of salient orbit determination results from encounter, with emphasis on the Jupiter approach phase orbit determination. Special attention is paid to the use of combined spacecraft-based optical observations and Earth-based radiometric observations to achieve accurate orbit determination during the Jupiter encounter approach phase.**

I. Introduction

On March 5, 1979 after a journey of 546 days and slightly more than 1 billion km, the Voyager 1 spacecraft passed within 0.3 Jupiter radii of the innermost Galilean satellite Io. Four months later, on July 8, Voyager 2 flew by the third Galilean satellite Ganymede at 0.8 Jupiter radii, and then by Jupiter the next day, at 10 Jupiter radii. Fig. 1 shows the Voyager heliocentric trajectory, and Figs. 2 and 3 show the near Jupiter trajectory for each mission. These Voyager encounters with Jupiter proved to be both spectacular and historic, with each mission returning voluminous data from 11 scientific instruments, including some 15 000 high resolution pictures of Jupiter and five of its satellites. After passing Jupiter, both spacecraft flew on to Saturn. Voyager 1 had encounters with several Saturn satellites, including a close flyby of the massive satellite Titan, on November 12, 1980. Voyager 2 encountered Saturn and its satellites on August 26, 1981; this spacecraft has undertaken the long journey to Uranus and will encounter that planet in January 1986, providing the first closeup view of that planet and its satellites. If the spacecraft remains healthy, it will then embark on its fourth interplanetary cruise, arriving at the planet Neptune in 1989. The orbit determination discussed in this paper is confined to the Jupiter encounter phase.

The spacecraft flight path had to be accurately controlled to achieve its scientific objectives. Since the final trajectory correction maneuver and its associated execution errors were small, the Jupiter delivery accuracies were determined by the orbit estimation accuracies available at the time of maneuver specification. From postencounter reconstruction of each flyby trajectory it has been determined that the final orbit control was within the accuracy predicted by the filter/smoother covariance analyses. The accurate instrument sequences required to obtain the near encounter science data were highly dependent upon accurate postcontrol knowledge of the spacecraft trajectory and satellite orbits. The results from the reconstructed orbits indicate that the near encounter spacecraft orbits were predicted to within 50 km and that spacecraft-satellite pointing was predicted to within 3 mrad, even for the close (20 000 km) Voyager 1 Io flyby.

This paper briefly describes the navigational measurement system and error source modeling used to produce accurate Jupiter-relative orbit determination. We will focus

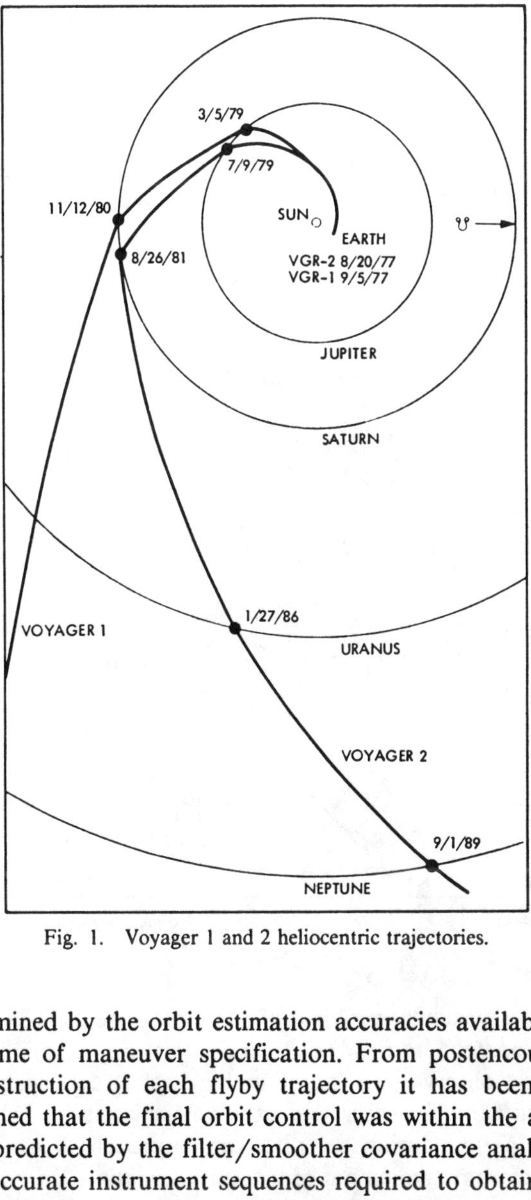

Fig. 1. Voyager 1 and 2 heliocentric trajectories.

Manuscript received March 5, 1982; revised September 7, 1982 and September 20, 1982. This work was supported by NASA under Contract NAS7-100 and Factorized Estimation Applications, Inc.

J. K. Campbell and S. P. Synnott are with the Jet Propulsion Laboratory, California Institute of Technology, Pasadena, CA 91109.

G. J. Bierman was with the Navigation Systems Section, Jet Propulsion Laboratory, California Institute of Technology, Pasadena, CA 91109. He is now with Factorized Estimation Applications, Inc., Canoga Park, CA 91307.

Reprinted from *IEEE Trans. Automat. Contr.*, vol. AC-28, pp. 256–268, Mar. 1983.

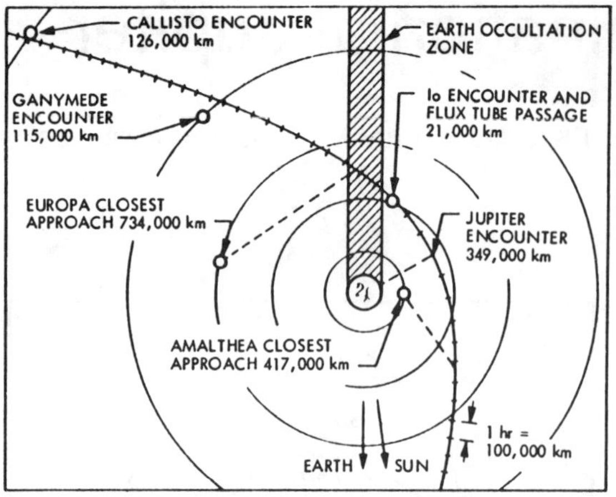

Fig. 2. Voyager 1 Jupiter flyby geometry.

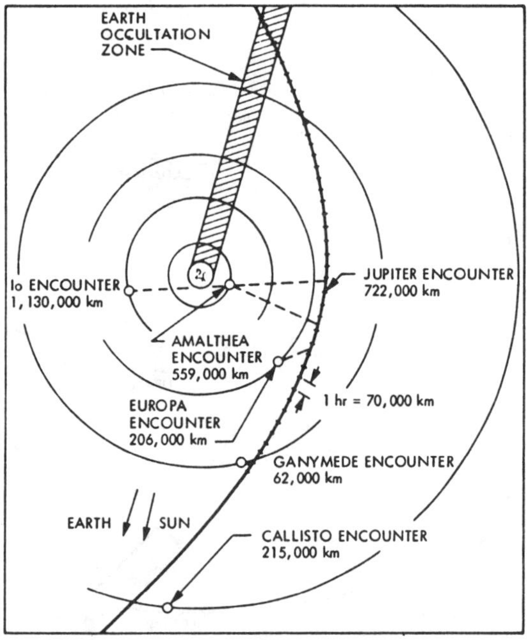

Fig. 3. Voyager 2 Jupiter flyby geometry.

on the particular problem of combining the orbit information content from two distinct sources; Earth-based radiometric observations (range and Doppler) and spacecraft-based optical observations to achieve accurate orbit determination during the planetary approach. Because this paper is an application of modern estimation, the reader is assumed to be familiar with the estimation concepts that are employed. The algorithmic formulations based on matrix factorization that are used to compute the orbit determination estimates, estimate error sensitivities, and estimate error covariances (filter/smoother and consider filter/smoother) are described (in great detail) in [1], [2], [6], and [7]. The latter part of Section II contains a discussion of the merits of the SRIF/SRIS and $U-D$ covariance factorized estimation algorithms that are used throughout this application.

The research reported here is extracted in large part from the Voyager navigation team report [8]. Interested readers are urged to consult this reference and also [10], which documents the Voyager Saturn encounter orbit determination.

II. Navigation Filter Design Rationale

There were two principal reasons for using a sequential stochastic filtering algorithm to process the Voyager tracking data. The first reason has to do with modeling of nongravitational accelerations and the second has to do with modeling of the optical data to account for pointing errors. Let us first focus on the nongravitational force problem.

The Voyager spacecraft, shown in Fig. 4, are three-axis stabilized vehicles, and remain in an Earth-pointed orientation relative to the Sun and a star, usually Canopus, for long periods of time. Notable features of the spacecraft are the large antenna dish, the science scan platform, and the groups of thrusters. The thrusters are unbalanced, since they do not fire in pairs, and are separated from the center of mass on opposite moment arms. A consequence of this configuration is that each time that a thruster is fired to either maintain or change the spacecraft attitude, there is a net translational velocity imparted to the spacecraft. Motion of the scan platform puts torque on the spacecraft, requiring thruster firings to maintain alignment. The effective center of solar pressure does not coincide with the spacecraft center of mass, and this can cause one-sided thruster firings.

A design flaw of the spacecraft is that the exhaust plumes from the positive and negative pitch attitude thrusters, which have velocity components along the radiometric measurement direction, strike the spacecraft structure. This fact was known before launch, but it was believed that the effect would be negligible. The conclusion of a postlaunch study was that, in fact, the plume impingement effect is significant.

Spacecraft outgassing as a byproduct of attitude control was considered to be basically a dynamic stochastic process which imparted ΔV velocity impulses to the spacecraft. The nature of these pulses was such that over a daily period the net translational effect on the spacecraft was zero. However, the Doppler tracking data were significantly corrupted by this attitude control pulsing. It has been known for some time [9] that estimate accuracy severely degrades when such disturbances are not accounted for in the filter model, and it was demonstrated early in the flight that the OD estimates produced from radiometric data without taking into account a dynamic stochastic process propagated poorly and gave inaccurate orbit predictions. Doppler residuals could be more accurately predicted from a previous fit which assumed a stochastic process than from a fit to the same data where stochastic effects were ignored.

These spacecraft generated forces are commonly termed spacecraft nongravitational forces. For Voyager, small attitude control impulses were averaged over a daily period and treated as piecewise constant stochastic accelerations

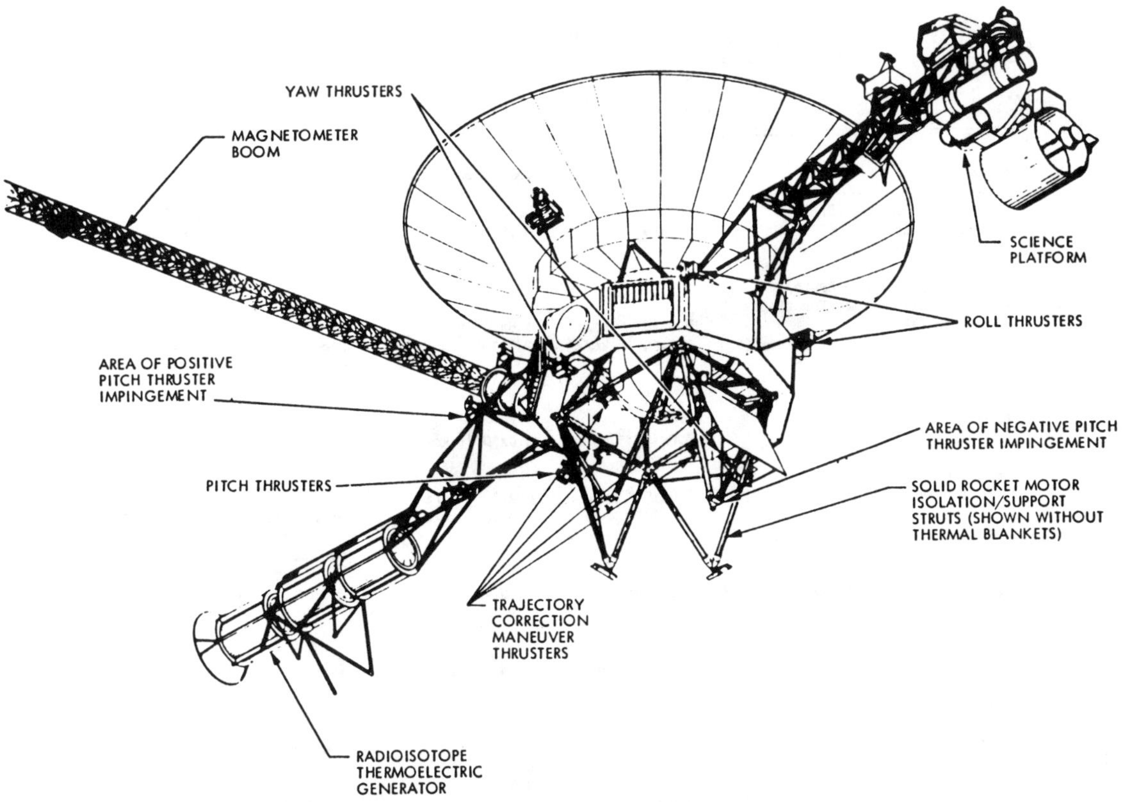

Fig. 4. Voyager spacecraft configuration.

along the respective spacecraft axes. Short term larger magnitude velocity impulses resulting from spacecraft turns were explicitly modeled as velocity impulses and were included in the list of parameters to be solved for (see Table I). Because of observability considerations, the stochastic effect was only significant in the axis which was aligned with the Doppler measurement direction, i.e., the Earth-spacecraft direction. Being able to eliminate unnecessary filter model state variables was doubly important because storage limitations imposed by the JPL Voyager ODP allows at most 70 parameters to be included in the filter model and, as a perusal of Table I shows, there are a considerable number of items to be estimated.

Let us turn attention now to the second reason for utilizing a sequential stochastic model, which has to do with the processing of optical measurements. These approach measurements were required to allow accurate orbit control at the final approach maneuver time. Doppler measurements which are sufficiently sensitive to Jupiter's gravity to allow determination of accurate planet-relative orbits occur too late for the information to be used for the final control maneuvers.

The optical data consisted of about 95 pictures processed for Voyager 1, and 113 pictures processed for Voyager 2. Galilean satellites (130 Voyager 1 satellite images, 143 Voyager 2 satellite images) were imaged against a star background with the narrow-angle imaging science instrument, which has a resolution of 10 μrad. The exposure time of the frames was 960 ms, sufficiently long to ensure the detection of dim stars up to a 9.5 effective visual magnitude. The optical data for orbit estimate solutions that are in agreement with the chosen Voyager 1 and 2 reconstruction solutions were fit to 0.25 pixels rms and essentially zero mean. No remaining systematic trends were detected *a posteriori*, indicating that there were no pixel level biases or center-finding errors inherent in the image extraction process.

The major navigational error source to be accounted for after extracting the star locations and satellite center locations is associated with inertial camera pointing. These errors were accounted for on a frame-by-frame basis, i.e., camera pointing errors were modeled as (frame dependent) white noise, and known inertial positions of the background stars in each frame were utilized. This type of stochastic model structure invites the use of a Kalman filter. Processing of these images produced accurate Jupiter-relative spacecraft orbit estimates that were used to construct the final control maneuvers. In addition, accurate satellite ephemerides were also determined [15] and these were used to define science instrument pointing.

Since this paper is a Kalman filter application, it is important to discuss the estimation processing techniques that were employed. It is a standard practice at JPL to implement navigation filters using square-root factorization techniques; even unweighted least-squares parameter fits are generally implemented using orthogonal transformation triangularization. Our experience with such things is that there are so many modeling and mission-related problems to worry about that we prefer to minimize numerical error effects as even a possible cause of poor

TABLE I
67 STATE VOYAGER NAVIGATION MODEL WITH ESTIMATION RESULTS BASED ON ENCOUNTER DATA FROM FEBRUARY 9–MARCH 18, 1979

State	A Priori Sigma	A Posteriori (smoothed sigmas)	Remarks
Cartesian positions (3)	500.0 km	88.0 km	Smoothed values are RMS over the 108-step time arc.
Cartesian velocities (3)	.5 m/s	0.06 m/s	
Line-of-sight (s/c - Earth) non-gravitational accelerations			
bias (1)	5.0×10^{-12} km/s^2	1.5×10^{-12} km/s^2	
piecewise constant random (1)	5.0×10^{-12} km/s^2	2.5×10^{-12} km/s^2	Smoothed value is RMS over the 108-step time arc.
Cross-track accelerations			
biases (2)	5.0×10^{-12} km/s^2	5.0×10^{-12} km/s^2	No improvement (unobservable).
Maneuver ΔV's (12)			
line-of-sight components (12)	1.0×10^{-4} km/s	0.2×10^{-6} km/s	
cross-track components (6)	1.0×10^{-4} km/s	1.0×10^{-4} km/s	Only 6 cross-track (unobservable) ΔV components included.
Satellite positional parameters (19)	400.0 km	50.0 km	RMS of the 19 parameters.
Satellite masses (4)	50.0 km^3/s^2	5.0 km^3/s^2	
Jupiter mass (1)	600.0 km^3/s^2	40.0 km^3/s^2	
Jupiter Ephemeris (3)	400.0 km	200.0 km	
Station Location errors:			3 tracking stations.
spin-axes (3)	1.5 m	0.4 m	
Longitudes (3)	3.0 m	0.5 m	
Range Biases (3)	100.0 km	1.0 km	Range measurement biases, one for each station.
Camera pointing angles (3)			Independent from frame to frame (white)
clock, cone	0.3 deg	1.0×10^{-4}	Smoothed values RMS over 87 frames.
twist	1.0 deg	0.2 deg	

TABLE II
MEASUREMENT MODEL WITH *A POSTERIORI* FIT RESULTS

Data Type	A Priori Uncertainty	Smoothed Estimate RMS residuals	# Data Points
Radiometric:			
Doppler (60 sec. sample average)	1.0 mm/s	0.4 mm/s	2400
Range	1.0 km	0.1 km	27
Optical	0.5 pixels	0.25 pixels	139 stars 101 satellite images

filter performance. Our concerns are illustrated in [11] which summarizes some of the peculiar (and wrong) numerical results that were obtained in premission simulations of the Voyager Jupiter approach using conventional Kalman filter covariance formulations.

The Earth-based radiometric data are processed using a batch sequential, square-root information filter/smoother [1], [2], and [7], and the spacecraft-based optical data are processed with a $U-D$ covariance factorized Kalman filter [2], [6]. The radiometric observation set is much larger than the set of optical observations; for the encounter application reported here the radiometric data consisted of 2400 Doppler and 27 range measurements and the optical data set consisted of 139 stars and 101 satellite images. The data types and associated statistics are summarized in Table II. As is pointed out in [2] the SRIF is computationally more efficient for larger data sets than is the Kalman filter. It is of interest to note that, despite this, when the other ancillary aspects of the problem are included (such as integration of the variational equations and computation of the measurement observables and differential correction estimate partial derivatives), the difference in computational cost (as will be shown in the following paragraphs) turns out to be relatively unimportant.

To give an idea of the computational burden that is involved, consider a typical radiometric SRIF/SRIS solution with 67 state variables (Table I). This model contains only 4 process noise states (line-of-sight acceleration and 3 camera pointing errors); there are 3500 data points and 132 time propagation steps. The problem run on a UNIVAC 1110, in double precision, used 275 CPU s for filtering; smoothed solutions and covariance computation used 265 CPU s. The entire run scenario including trajectory variational equation integration, observable partials generation, solution mapping, and generation of smoothed residuals used 4320 CPU s. Thus, the filter and smoother portion each involved little more than 6 percent of the CPU time. Two points of note, in regard to these sample run times are as follows.

1) We generally iterate and generate several filter/smoother solution sets for a given nominal trajectory and file of observation partials. Smoothed covariances use the lion's share of the smoother computation and these need only be computed for the last iteration.

2) The program inputs were not configured to minimize filter/smoother CPU requirements.

For this dynamic state configuration optimally arranged smooth code should require less than 25 percent of the CPU time required for the filter phase. The point we are aiming at is that, based on our experience in [11], a conventional Kalman filter covariance formulation would execute in essentially the same amount of CPU time (± 15 percent), and evidently the filter CPU cost is small compared with total run CPU time requirements.

As pointed out in [1] the pseudoepoch state formulation model is an evolutionary outgrowth of the least-squares initial condition estimator [3]. If x_j^c represents current time position and velocity differential correction estimates, then the pseudoepoch state x_j^o is defined by the equation

$$x_j^c = \Phi_{xx}(t_j, t_0)x_j + \Phi_{xy}(t_j, t_0)y + \Phi_{xp}(t_j, t_{j-1})p_{j-1}$$

where the components of y are bias parameters and the components of vector p_{j-1} are piecewise constant stochastic model parameters. The transition matrices $\Phi_{xx}(t_j, t_0)$, $\Phi_{xy}(t_j, t_0)$ and

$$\Phi_{xp}(t_j, t_{j-1}) = \Phi_{xx}^{-1}(t_j, t_0)\Phi_{xp}(t_j, t_0) - \Phi_{xx}^{-1}(t_{j-1}, t_0)\Phi_{xp}(t_{j-1}, t_0)$$

are obtained by integrating variational differential equations for

$$\left[\Phi_{xx}(t, t_0), \Phi_{xy}(t, t_0), \Phi_{xp}(t, t_0)\right]$$

from an epoch time t_0. In Table I the filter state vector x, y, and p components are defined.

Orbit determination problems with radiometric data are ill-conditioned. This is due, in the main, to poor observability, and large dynamic ranges of the variables that are involved (viz. acceleration errors $\sim 10^{-12}$ km/s^2, range values $\sim 10^8$ km, etc). The observability problem is aggravated by the inclusion of large numbers of parameters (such as ephemerides) that are very weakly coupled to the spacecraft observables. The two features of the SRIF/SRIS algorithmic formulation that are most important for this application are as follows.

1) *Numerical reliability:* The SRIF/SRIS algorithms are the most (numerically) accurate and reliable formulation of the Kalman filter/smoother known. Because of the complexity and difficulty inherent in the formulation of the deep space navigation problem it is most important that the estimates and covariances be computed correctly.

2) *Computational efficiency:* The algorithms are formulated so as to exploit the structure of the orbit determination problem. In particular, there are many measurements to be processed per time propagation step, the dynamical model involves only a small number of stochastic variables, there are a preponderance of bias parameters, and the position-velocity states are cast in a pseudoepoch state formulation. Early tests demonstrated that the SRIF, for this structure, was more efficient than a conventional (and less reliable) Kalman filter mechanization.

Using the SRIF it is relatively easy to take the processed results and generate estimates that correspond to models with a smaller number of bias parameters. This feature is especially useful for confirming that parameters thought to be of little significance turn out, in fact, to have little effect on the key estimate state vector components.

The $U-D$ covariance factorization was chosen to mechanize the optical navigation filter for many of the same reasons (numerical reliability and computational efficiency), except in this case the measurement set per time propagation is small, and for such problems the $U-D$ formulation is more efficient than the SRIF. In [11] the $U-D$ formulation and the Kalman filter both conventional (optimal) and stabilized (Joseph/suboptimal) forms are compared for a Jupiter approach simulation with radiometric data. In the tests reported there the covariance mechanized filters performed very poorly; they gave results that ranged from inaccurate, but which might be thought correct (20–50 percent errors) to impossible (negative variances and estimates that were absurd). It happens that optical navigation data are not nearly as ill-conditioned as are the radiometric data, and in fact, the early optical navigation studies successfully carried out in [5] used a conventional Kalman filter mechanization. The decision to use a $U-D$ factorization in place of the conventional mechanization was based on the following facts:

1) Comparisons (operation counts, actual CPU, and storage requirements) show that *optimally coded $U-D$* and conventional covariance mechanizations are nearly indistinguishable in terms of storage and computational requirements.

2) $U-D$ factor mechanization has accuracy that is comparable with the SRIF. On the other hand, one cannot be certain when the covariance mechanization will degrade or fail (e.g., when the *a priori* uncertainties are too large, the measurement uncertainties are too small, the data geometry is near linearly dependent, etc., one can expect stability and accuracy problems).

It is the conviction of (one of) the authors (who is believed by the others!) that covariance mechanized Kalman filters should *never* be computer implemented. Further, it is believed that if the Kalman filter applications community had more experience with efficiently and reliably mechanized factorization alternatives, there would be few instances where a covariance mechanized Kalman filter would find application. We note in closing this factoriza-

TABLE III
SUMMARY OF JUPITER APPROACH ORBIT CONTROL

Voyager 1 TCM	Execution Time	ΔV, m/sec Designed	ΔV, m/sec Achieved	$\Delta\bar{B}$, Km ΔB·R	$\Delta\bar{B}$, Km B·T	ΔT
3	E - 35 days	4.146	4.208	-1725.	+13125.	$+0^h\ 14^m$
4	E - 12.5 days	0.586	0.594	+710.	-100.	$-0^h\ 0^m\ 14^s$
Voyager 2 TCM						
3	E - 45 days	1.442	1.384	+3330.	+5040.	$-0^h\ 4^m\ 42^s$
4	E - 12 days	0.576	0.574	-895.	+210.	$+0^h\ 0^m\ 03^s$

tion algorithm discussion that the SRIF and $U-D$ algorithms that were used in this application have been refined and generalized, and are commercially available in the form of portable Fortran subroutines [12].

III. OD PERFORMANCE

This section will summarize the near-encounter OD results for both spacecraft. These results are 1) the orbit determination that was done to deliver the Voyager spacecraft to their required final target points met the required accuracies; and 2) that the final postdelivery orbit determination needed to accurately point science instruments in the near and postencounter phases also exceeded specification accuracies. A limitation of the Voyager orbit determination program is that one can include at most 70 filter states, and this includes the sum of both estimated and consider filter states. For completeness we remind the reader that consider parameters, cf. [1] and [2], play no role in the estimation except to provide a deweighting to the confidence one would otherwise put on the estimates. As noted earlier, the size limitation does not allow us to include all the known error sources in the filter model. Table I lists one of the several filter models that were used, and quantifies the results obtained. Observe that although there are 12 maneuvers only six cross-track maneuver correction terms are included, and these terms have a negligible effect on estimator performance. This conclusion is based both on estimate comparisons with and without such terms included and on the incremental change in the computed covariance due to the addition of consider parameter sensitivity effects. The discussion that follows is an elaboration of the results summarized in Table I.

Jupiter Approach Solutions

For approximately the last 60 days of each approach, both radio and optical data were continuously acquired to allow orbit estimates to be revised every few days. Optical data in the time period beginning 60 days from encounter (E) to $E-30$ days were acquired at the rate of approximately one picture per day, and this rate increased to about four or five pictures per day in the last few days before encounter. Radio tracking (coherent Doppler and range) was virtually continuous for the last 60 days of approach. For each encounter, there were several critical events for which orbit estimate deliveries to certain elements of the project were required. The events consisted of two trajectory correction maneuvers, at about $E-40$ days and $E-12$ days, and several updates to the onboard instrument pointing, a critical element in the near encounter science return. The detailed summary of the delivery events for each spacecraft is shown in Table III.

Based on a preliminary data acquisition schedule, covariance analysis estimates were computed preflight of the orbit determination errors which would result at these various delivery times. The primary purpose of this section is to compare the near real-time filter and the current best reconstructed smoothed orbit estimates, and also to compare both of these to the expected covariance performance computed preflight. It will be shown that all orbit estimates but one fell well within the expected preflight capability ($1-\sigma$) even though those early statistics were derived with unrealistic, very optimistic, assumptions about the "small" nongravitational forces caused by spacecraft attitude changes.

More specifically, as discussed in Section II, the spacecraft's attitude control system did not employ coupled thrusters, and hence there were many small velocity impulses imparted to the spacecraft. These velocity impulses are impossible to model individually, and affect (to a sensible level) the meter level ranging and submillimeter/second Doppler observables if they are imparted along the Earth-line. In addition, whenever the spacecraft orientation was changed to allow an engineering calibration, or a science scan of the Jovian system, velocity impulses of tens of millimeters per second were imparted. These larger impulses were not accounted for in the preflight analysis, and the more continuous smaller impulses resulted effectively in accelerations of the order of $5-10\times10^{-12}$ km/s^2 along all three spacecraft axes. The larger impulsive velocity changes have two effects. When they occur within the data arc, an estimate of their components becomes confused with the estimates of the dynamically important parameters, and the result is an incorrect orbit; when they occur past the end of the data arc they cause incorrect mapping to the encounter time, and the result is an incorrect prediction of encounter conditions. Based on covariance analyses, it did not appear possible to accurately predict the velocity components at the maneuver event times from *a priori* information.

Data Weighting

In the vicinity of a planetary encounter, the radio and optical data are geometrically complementary. The Earth-spacecraft line-of-sight range and velocity measurements indirectly provide the spacecraft-planet distance through observable dynamic effects, and the optical data effectively measure the instantaneous spacecraft-satellite cross line-of-sight position relative to a particular satellite. The radio and optical data were planned to allow determination of both the satellite orbits and the spacecraft trajectory relative to the planet.

In the estimation process using both radio and optical observations, the solutions at certain stages were found to be sensitive to the relative weighting of the two data types. Based on the image extraction analysis, the optical data were thought to measure the star and satellite center locations to 0.5 pixels or less in a random measurement noise sense. In addition, if there were any systematic optical image extraction errors, they were thought to be small (< 0.5 pixels) and to behave as biases in the 2-D picture space, or as long term slowly varying functions, i.e., nearly biases. Using these arguments the optical data usually were assumed to have 0.5 pixel measurement noise; 0.5 pixel optical biases were used as consider parameters to compute the expected uncertainty in the orbit estimates. Incidentally, these two consider parameters, together with the 67 estimated parameters listed in Table I, essentially exhausts the 70 parameter ODP storage limitation.

After calibrating for troposphere, ionosphere, and space plasma effects, the measurement errors for radio data are equivalent to only a few meters in range and 0.3–0.5 mm/s for a 60 s count time for Doppler. However, because of the known level of unmodelable nongravitational accelerations, range was more usually weighted at 100 m, and Doppler at 1 mm/s, for a 60 s count time. These values, together with the postfit rms residual results are displayed in Table II. For the radio data, the most likely sources of error are transmission media calibration errors and Earth station location errors, both of which are usually combined into an assumed systematic error with a diurnal signature. Data noise for Voyager 2 was larger because its view periods occurred during local daylight hours, when transmission media effects were larger. The Voyager 2 encounter also coincides with a more active solar period.

Reconstruction of Encounter Orbits

Since the encounters, it has been possible to do a detailed analysis of the larger nongravitational impulsive events for Voyager 1, using data from approximately 60 days after encounter. In addition to errors associated with the two-approach TCM's, there occurred 10 preencounter events whose effects were directly observable along the Earth-line and which therefore were estimated as impulsive

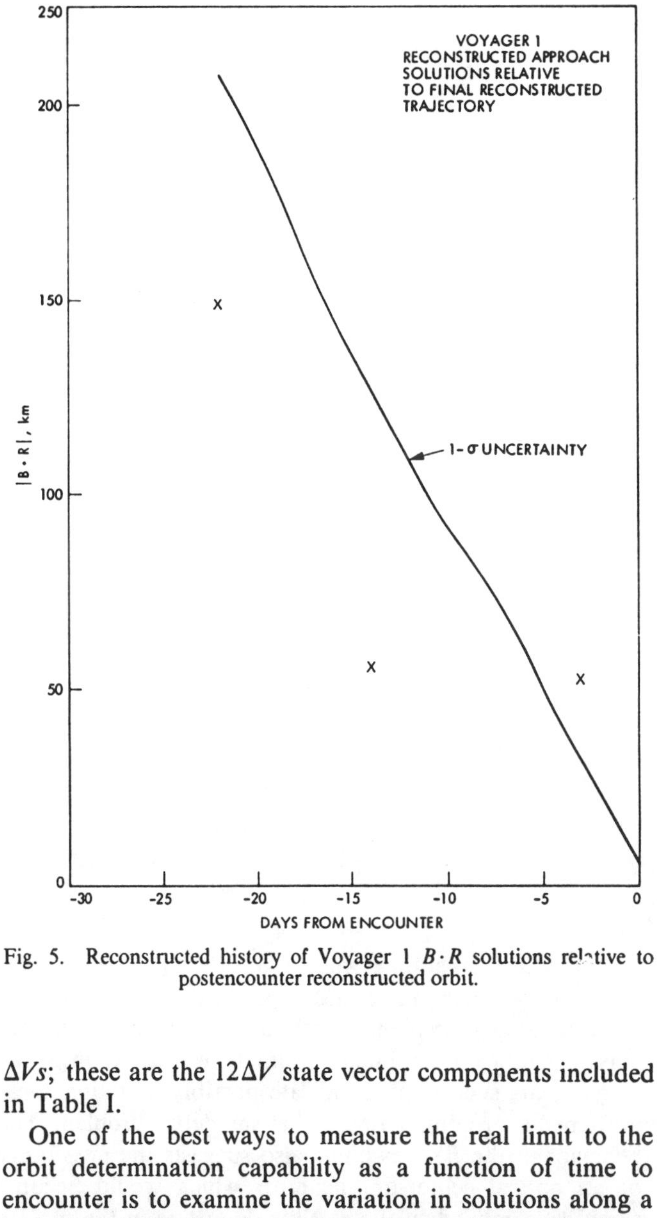

Fig. 5. Reconstructed history of Voyager 1 $B \cdot R$ solutions relative to postencounter reconstructed orbit.

ΔVs; these are the $12\Delta V$ state vector components included in Table I.

One of the best ways to measure the real limit to the orbit determination capability as a function of time to encounter is to examine the variation in solutions along a trajectory which has already passed through several estimation iterations for the magnitude of the impulsive components. With as detailed a treatment of the larger impulsive events as is possible, and with stochastic and constant acceleration components solved for in the sequential filter to remove any excess nongravitational effects, the relative time history *B*-plane[1] solutions for the last 25 days before Voyager 1 encounter were computed and are displayed in Figs. 5 and 6 along with the $1-\sigma$ uncertainties in these solutions. These uncertainties were calculated using no assumed systematic errors in either radio or optical data, such as optical biases, and therefore represent lower limits to the error that could be expected. In general, the solutions fall within even this optimistic uncertainty level. The 55 km $\Delta B \cdot R$ solution at $E-3.5$ days is the only anomaly

[1]Planetary targeting is usually expressed in terms of the *B-plane*, a plane passing through the center of a target planet and perpendicular to the incoming approach hyperbola asymptote of a spacecraft. "$B \cdot T$" is the intersection of the *B*-plane with the ecliptic and "$B \cdot R$" is a vector in the *B*-plane that is perpendicular to $B \cdot T$ and making a right-handed system R, S, T, where S is the incoming asymptote.

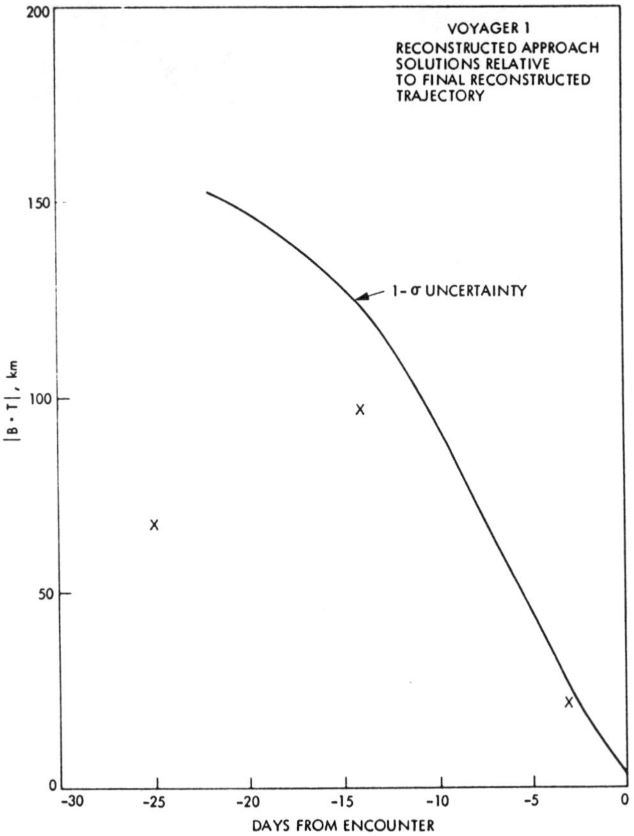

Fig. 6. Reconstructed history of Voyager 1 $B \cdot T$ solutions relative to postencounter reconstructed orbit.

and it occurs at a time when significant ($>1-\sigma$) transients are occurring in several important dynamic parameter estimate components, such as, for example, planet mass. Significant parameter estimate variations in static components such as this indicate possibly a station location, media, or nongravitational mismodeling problem. The behavior in the $B \cdot R$ estimate also suggests the possibility of an optical center-finding error which would become significant only when the satellite diameters in the picture become very large (approximately 100 pixels). However, an analysis of the center-finding at this time period for all the different satellites indicates that this is an unlikely possibility. A center finding error would also imply that earlier satellite residuals would show strong systematic effects. This was not, however, observed when the entire arc of optical residuals was plotted from the current best reconstruction run. (The orientation of the Voyager camera in space was such that to within about 20°, pixels represent a measurement in the ecliptic or trajectory plane, and lines represent measurements that are normal to this plane.)

A slight slope in the smoothed estimate line residuals seems to indicate that there is some remaining out-of-trajectory-plane mismodeling, but the error at encounter caused by this slope should be no more than 10–15 km. This same error behavior was observed to occur in the near-real time approach solutions. The general results and conclusions stated here were derived mostly from the Voyager 1 orbit reconstruction experience; they are applicable directly, however, to Voyager 2.

Real-Time Solutions

In this subsection we compare the near-real time trajectory estimates for both spacecraft that were delivered to the Voyager project at certain stages of each approach, to the final reconstruction solutions, and compare the accuracies of these deliveries to the attainable accuracies.

The Voyager 1 approach solutions relative to reconstruction are shown in Figs. 7 and 8. The lump, or flattening, in the Voyager 1 uncertainty curves reflects the fact that the data arc for this and subsequent solutions was shortened to include data only within 30 days of encounter. In general, near-real time orbit determination accuracies for both spacecraft fell well within the $1-\sigma$ uncertainties, which are taken from the computer run that was made in near-real time. The errors (with smoothed estimates regarded as truth) also generally fell well within the preflight covariance predictions. The only point on approach at which the near-real time solutions were not substantially below the $1-\sigma$ uncertainty level occurred on Voyager 1 at the final delivery for the last approach trajectory correction maneuver TCM4. After considerable study it became apparent that the relative weighting of the radio and optical data at this epoch was responsible for this anomalous estimate.

The variability of the Voyager 1 *B*-plane solutions leading up to the final TCM4 delivery at $E-15$ days are shown in Figs. 9 and 10 as functions of time, relative data weightings, and estimated parameter list. The correct (best postfit) answers are indicated in each figure. The solutions plotted by the small circles in Fig. 9 were generated by solving for the state vector whose components are listed in Table I; the estimate list includes spacecraft state, stochastic and constant nongravitational accelerations, several impulsive ΔVs, planet mass, and planet and satellite ephemerides, as well as the three camera orientation angles, which were always estimated. The solutions of Fig. 10 also included Earth station locations in the estimate list. In both figures there are significant sensitivities for all three curves to both data arc length and relative radio/optical weightings. Comparison of the results of Figs. 9 and 10 indicates that there is a large sensitivity (of the order of several hundred kilometers for the radio only case) to station location, or station location-like, errors. This type of sensitivity typically is an indication of mismodeling of the radio observables.

The orbit determination performance for the last approach TCM and the pointing update for Voyager 2 are summarized in Fig. 11 in which the *B*-plane conditions predicted using data to $E-17$ days and $E-9.5$ days and their uncertainty ellipses are compared to the final reconstruction orbit estimate. Because of the Voyager 1 experience and because of the difficulty encountered with a long arc of Voyager 2 radio data, many different arc lengths and combinations of data and parameter estimate lists were used for the second spacecraft. By thus experimenting we were able to eliminate the larger effect of systematic mismodeling and arrived at solutions that fell well within all of the uncertainty ellipses.

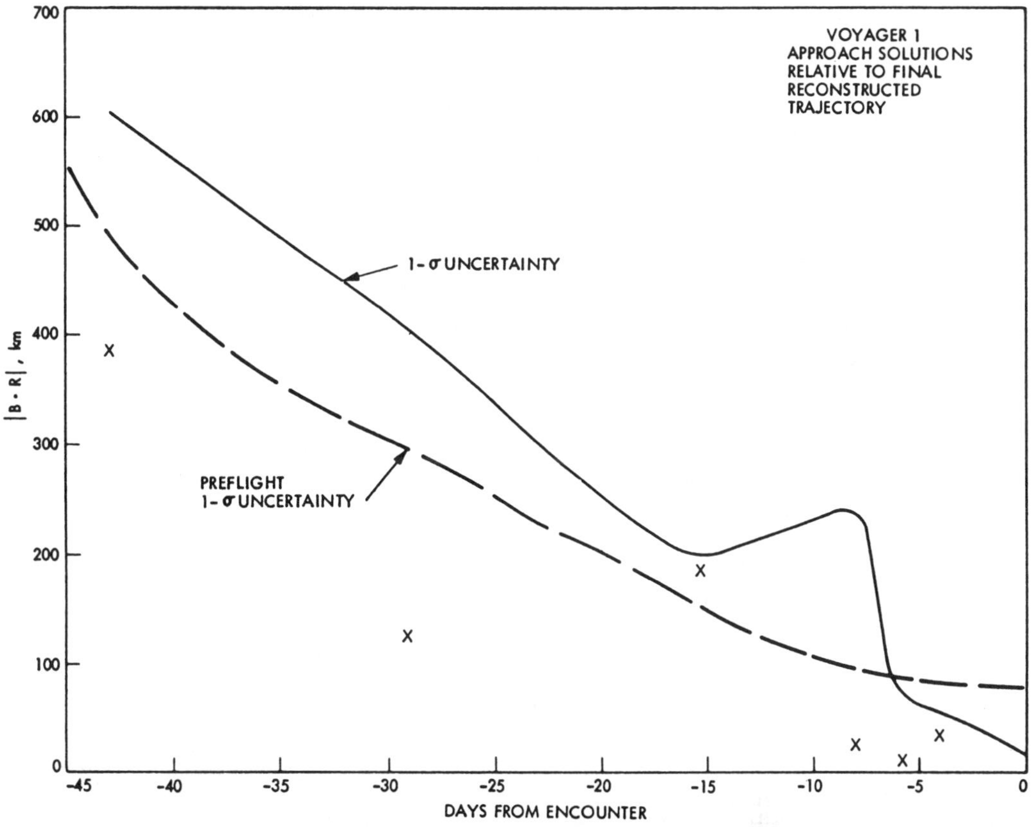

Fig. 7. History of Voyager 1 real-time $B \cdot R$ solutions relative to postencounter reconstructed orbit.

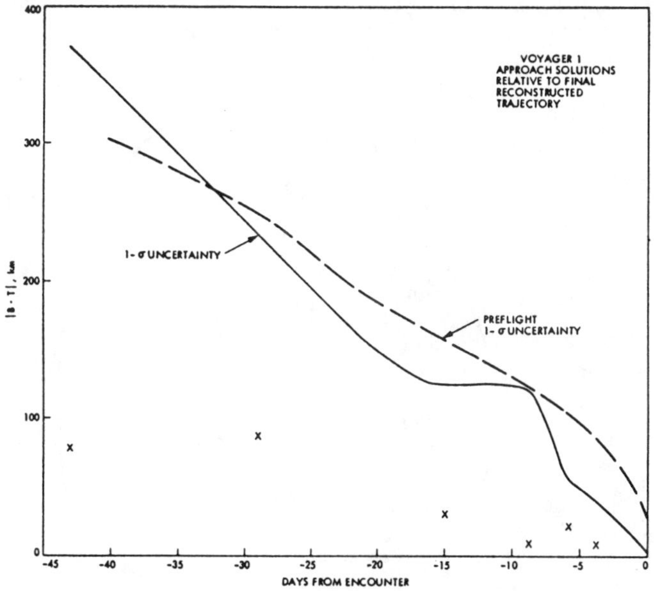

Fig. 8. History of Voyager 1 real-time $B \cdot T$ solutions relative to postencounter reconstructed orbit.

IV. Summary of OD Performance

Tables IV and V indicate the OD delivery performance for each spacecraft. The last columns of each table show that the delivery OD performance for each spacecraft easily met the established accuracy criteria. In the case of Voyager 2, the delivery requirements were less stringent, and the achieved delivery was nearly equivalent to 0.5 pixel, Jupiter relative. Figs. 12 and 13 give a brief indication of the overall capability and performance of the Voyager radiometric data. Fig. 12 shows, in geocentric angular coordinates, the shift in Jupiter position coordinates obtained from a set of orbit estimates based on radiometric data arcs extending from $E-60$ to $E+30$ days. This data arc allows for an accurate Jupiter-relative orbit estimate and some sensitivity to Jupiter ephemeris error. One solution estimates only ephemeris error; the second solution estimates both ephemeris and station location errors, and essentially trades off these two error sources to give the same net geocentric angular offset as the ephemeris-only solution. This plot indicates the general inability of Earth-based radiometric data to fully distinguish station location error from ephemeris error, and sets a lower bound of about 0.25 μrad to the Voyager radiometric performance. Fig. 13 plots in geocentric angular coordinates the radio-based OD that was done for the final Voyager 1 delivery, and compares this real-time result with the best reconstructed orbit at the delivery point, obtained using a combined radio plus optical data set. This plot shows that the radiometric OD for Voyager 1 performed to about the 0.5 μrad level.

V. Satellite State Estimates

A natural fallout of using the optical and radio data in the spacecraft trajectory determination process is the improvement in the knowledge of the orbits of the four Galilean satellites. The optical data are directly sensitive to satellite position changes. Near the satellite close ap-

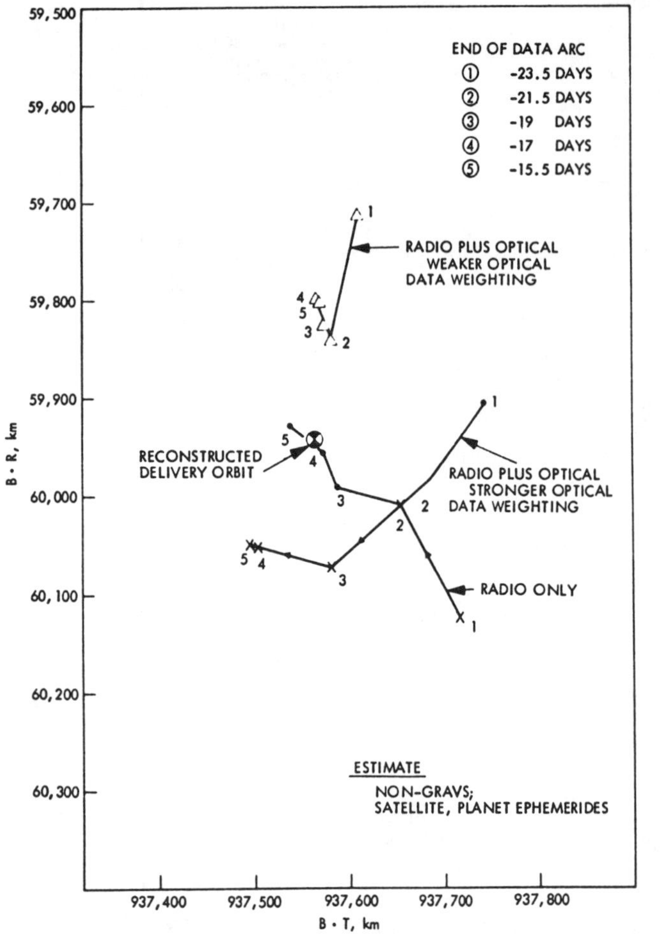

Fig. 9. Variation in Voyager 1 real-time *B*-plane solutions; station locations not estimated.

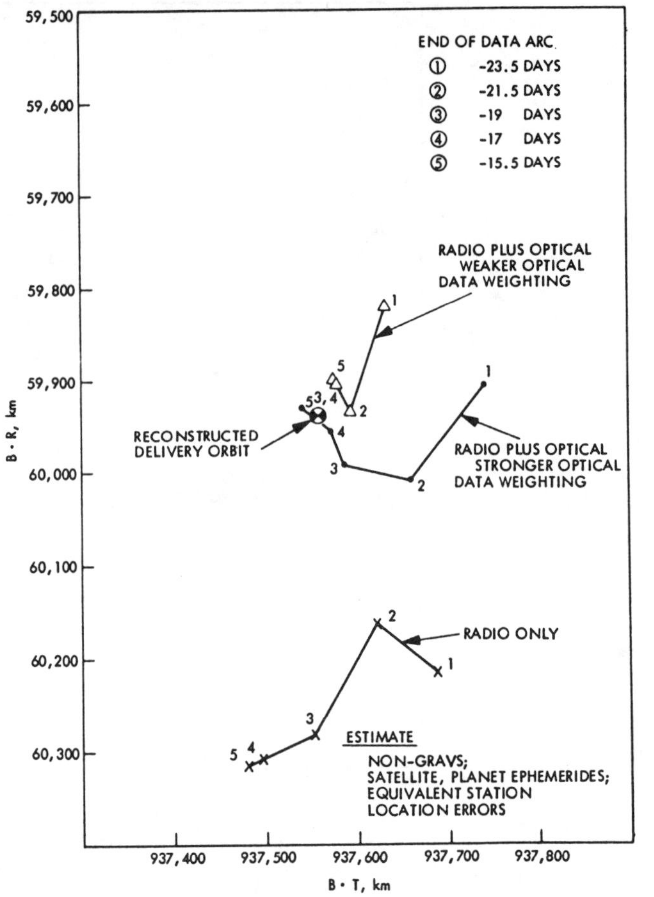

Fig. 10. Variation in Voyager 1 real-time *B*-plane solutions; station locations estimated.

proaches, the radio data are sensitive normal to the line-of-sight to the dynamic effects on spacecraft motion due to satellite mass or position errors in certain directions. The combination of both data types from both Voyager encounters results in a Galilean satellite ephemeris significantly improved over that available to Voyager from Earth-based observations [15]. A brief summary of the orbit estimates based on the Voyager 1 data is briefly discussed here.

A dynamical theory for the Galilean satellites developed by Lieske [13], [14] was used to compute both residuals and partials for the optical data and to develop the ephemerides which were the starting point for the Voyager analyses. The adjustable Lieske parameters (included as satellite positional errors in Table I) consist of three constants for each satellite which are fractional errors in the mean motion, the eccentricity and the sine of the inclination and three angular parameters which are essentially the initial longitude, the argument of periapse, and the nodal position. The changes to the estimated Lieske parameters, and the *a priori* and *a posteriori* errors for these parameters are shown in Table VI. In addition, the masses of the satellites, and Jupiter, and the right ascension and declination of the north pole of Jupiter are also adjustable.

In the analysis described here the arguments of periapse were not estimated because their effect on the data was

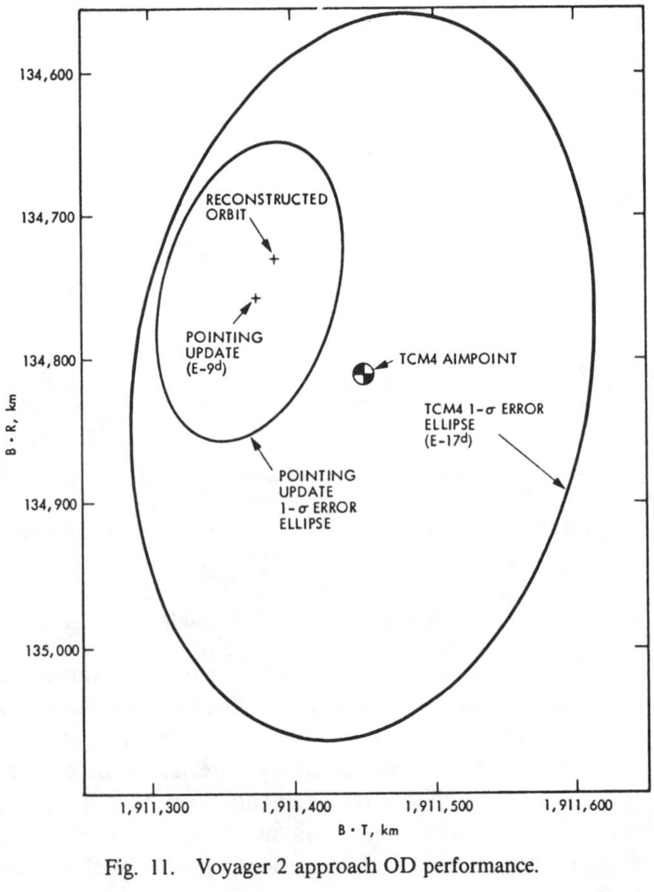

Fig. 11. Voyager 2 approach OD performance.

TABLE IV
SUMMARY OF VOYAGER 1 NAVIGATION DELIVERY PERFORMANCE

Control Parameter	Rationale	Target Value	Achieved Value	Allowable Error	Actual Error
Geocentric occulation entry	Timing of Jupiter limb scan at occulation	March 5, 1979 15:45:20.7	March 5, 1979 15:45:22.1	30 sec	+1.4 sec
Distance off center-line of flux tube model, km	Guarantee sampling of current in Io flux tube	-7.4 km	+101 km	1000 km	108 km
Io closest approach time	Io mosaicking sequence control	March 5, 1979 15:13:18.9	March 5, 1979 15:13:20.7	20 sec	+1.8 sec

Final delivery maneuver was executed 12.5 days prior to Jupiter closest approach.

TABLE V
SUMMARY OF VOYAGER 2 NAVIGATION DELIVERY PERFORMANCE

Control Parameter	Rationale	Target Value	Achieved Value	Expected 1-σ Performance	Actual Error
Jupiter B-plane coordinates	Minimize magnitude of post-Jupiter TCM5	B.R = 134,804 km B.T = 1,911,454 km	134,730 km 1,911,387 km	220 km 161 km	-74 km -67 km
Jupiter closest approach time	Pointing control of Earthline TCM5	7/9/79 $22^h29^m0^s$ GMT	7/9/79 $22^h29^m1.6^s$ GMT	22 sec	+1.6 sec

Final delivery maneuver was executed 12 days prior to Jupiter closest approach.

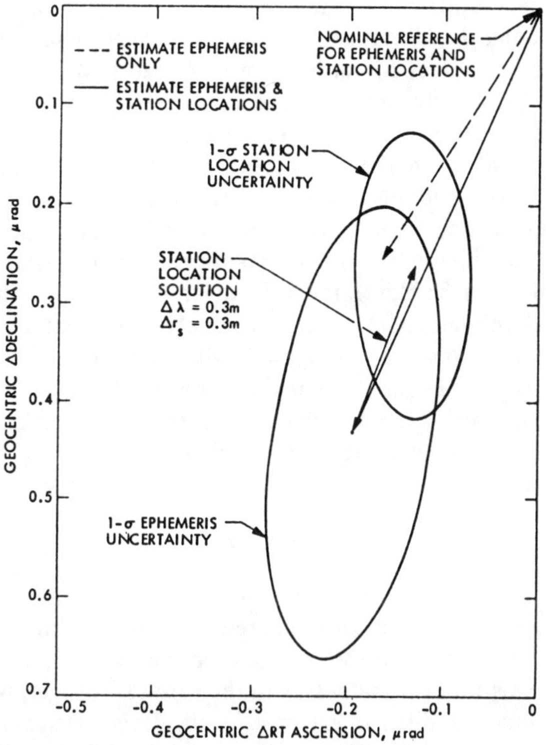

Fig. 12. Indication of Voyager 1 radiometric capability.

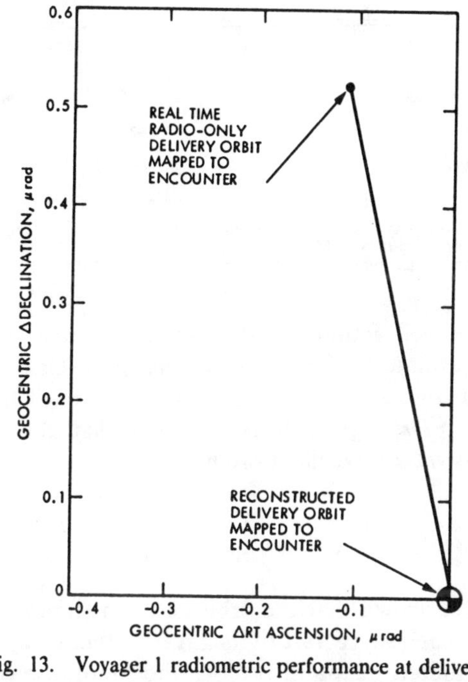

Fig. 13. Voyager 1 radiometric performance at delivery.

known to be small (less than about 15 km). Mean motions, whose values were extremely well-determined from many years of Earth-based data were also not included.

A comparison of the formal standard errors in Cartesian coordinates determined from the Earth-based and Voyager 1 data (using the Earth-based estimates as *a priori*) is shown in Table VII in which the entries represent the RSS of the sigmas and the changes in the three components of position of each satellite at the epoch of Voyager 1 Jupiter closest approach. The Voyager 1 *a posteriori* errors are very similar for all the satellites with the slight differences explainable by the variability of optical data distribution and quantity, and by the ratio of the satellite periods which affects the "average quality" of the data around the observed orbits. The Cartesian changes at the encounter

TABLE VI
CHANGES TO ESTIMATED LIESKE PARAMETERS

		Lieske Parameter	Δ	a posteriori σ	Earth-based a priori σ
eccentricity		5EPS 16	-.175	.213	.42
		5EPS 17	-.0022	.179	.26
		5EPS 18	-.013	.0071	.019
		5EPS 19	.0024	.0013	.0036
sine i		5EPS 21	-.134	.174	.41
		5EPS 22	-.0082	.0066	.025
		5EPS 23	.0032	.020	.050
		5EPS 24	-.0114	.018	.11
long		5BET 01	-.0014	.0028	.017
		5BET 02	-.0045	.0011	.004
		5BET 04	-.0054	.0005	.005
node		5BET 11	7.5	5.7	18.0
		5BET 12	.512	.575	1.05
		5BET 13	.407	1.16	2.5
		5BET 14	-.551	.817	2.4
pole	RA	ZACPL 5	.0064	.0084	.05
	DEC	ZDEPL 5	-.0003	.0049	.026
Satellite, Jupiter masses		501 GM	29.4	4.6	56
		503 GM	3.6	3.5	76
		504 GM	7.5	1.4	47
		GM 5	657.5	24.7	1200

TABLE VII
COMPARISON OF CARTESIAN FORMAL STANDARD ERRORS AND CORRECTIONS

	Earth Based a priori σ	Voyager a posteriori σ	Voyager Δ (RSS)
Io	87	28	35
Europa	137	28	43
Ganymede	161	34	80
Callisto	448	35	180

Computed from Voyager 1 Encounter Data.
Numbers shown are RSS of x, y, z components.
Epoch of Corrections = March 5, 1979 12:05 GMT

epoch are well within the Earth-based *a priori*, but may vary somewhat at other points of the orbit. Only changes to the longitudes of Europa and Callisto were of the order of an *a priori* sigma. It is apparent that the *a priori* ephemeris was an excellent product.

VI. CONCLUSIONS

We have shown that the orbit determination performed to deliver each spacecraft to its target at Jupiter was within the predicted capability of the radiometric and optical-based navigation system for Voyager, and in fact improved for Voyager 2 as a result of the Voyager 1 experience. The postdelivery OD knowledge solutions done to refine scan platform pointing were well within the stated requirement, even though the knowledge epochs were earlier than anticipated prelaunch.

Numerous tests were made throughout the mission to test estimate consistency and accuracy of the SRIF/SRIS and *U–D* algorithm implementations. The navigation estimation software performed flawlessly. Indeed, the estimation software performed so well that most of the time it was taken for granted by the navigation team, and that is the ultimate compliment.

As a result of having actually used satellite images to perform Jupiter and satellite-relative navigation, it has been determined that several prelaunch hypotheses regarding error sources for optical measurements were not correct. Specifically, before the encounter experience it was assumed that center-finding errors for large satellite images would scale with the size of the image. From a detailed analysis of hundreds of images it was found that the center finding errors were more likely to either decrease with increasing image size, or to remain essentially constant. As indicated in Table II the rms noise associated with the optical measurement postfit residuals were found to be about 0.25 pixel as compared with the 1.0 pixel error assumed prelaunch. It is to be expected that the postfit residuals should have a smaller rms than the data noise σ because it is known that the postfit residual $z_j - Hx_{j:n}$ has variance

$$\sigma^2 - HP_{j:n}H^T.$$

In addition, we have learned that the combination of radiometric with optical measurements must be done carefully, with regard to the information content of each data type. This was especially true in the case of Voyager, which represented a distinct extreme in the level of dynamical corruption of the radiometric signal by small spacecraft-generated velocity pulses, that did not affect the corresponding optical data.

ACKNOWLEDGMENT

J. E. Reidel provided much of the postencounter computer support for this paper.

References

[1] G. J. Bierman, "Sequential least-squares using orthogonal transformations," Jet Propulsion Lab., Pasadena, CA, Aug. 1, 1975, Tech. Memorandum 33-735.

[2] G. J. Bierman, *Factorization Methods For Discrete Sequential Estimation*. New York: Academic, 1977.

[3] T. D. Moyer, "Mathematical formulation of the double-precision orbit determination program (DPODP)," Jet Propulsion Lab., Pasadena, CA, May 15, 1971, Tech. Rep. 32-1527.

[4] J. Ellis, "Voyager software requirements document for the orbit determination program (ODP) Jupiter encounter version," Jet Propulsion Lab., Pasadena, CA, May 10, 1977, Rep. 618-772.

[5] N. Jerath, "Interplanetary approach optical navigation with applications," Jet Propulsion Lab., Pasadena, CA, June 1, 1978, Pub. 78-40.

[6] S. P. Synnott, "Voyager software requirements document," Jet Propulsion Lab., Pasadena, CA, July 10, 1977, Pub. 618-729.

[7] G. J. Bierman, "Square-root information filtering and smoothing for precision orbit determination," *Mathematical Programming Study 18, Algorithms and Theory in Filtering and Control*, 1982, pp. 61–75.

[8] J. K. Campbell, S. P. Synnott, J. E. Reidel, S. Mandel, L. A. Morabito, and G. C. Rinker, "Voyager 1 and Voyager 2 Jupiter encounter orbit determination," presented at the AIAA 18th Aerospace Sci. Meet., Pasadena, CA, Jan. 14–16, 1980, paper AIAA 80 0241.

[9] C. S. Christensen, "Performance of the square-root information filter for navigation of the Mariner 10 spacecraft," Jet Propulsion Lab., Pasadena, CA, Jan. 1976, Tech. Memorandum 33-757.

[10] J. K. Cambell, R. A. Jacobson, J. E. Reidel, S. P. Synnott, and A. H. Taylor, "Voyager I and Voyager II Saturn encounters," presented at the 20th Aerospace Sci. Meet., Jan. 1982, paper AIAA 8-82-0419.

[11] G. J. Bierman and C. L. Thornton, "Numerical comparison of Kalman filter algorithms; Orbit determination case study," *Automatica*, vol. 13, pp. 23–35, 1977.

[12] G. J. Bierman and K. H. Bierman, "Estimation subroutine library, directory and preliminary user's guide," Factorized Estimation Applications, Inc., Sept. 1982.

[13] J. H. Lieske, "Improved ephemeris of the Galilean satellites," *Astrophysics*, vol. 82, pp. 340–348, 1980.

[14] J. H. Lieske, "Theory of motion of the Galilean satellites," *Astron. and Astrophys.*, vol. 56, pp. 333–352, 1977.

[15] S. P. Synnott and J. K. Campbell, "Orbits and masses of the Jovian system from Voyager data: Preliminary," presented at the IAU-Colloquium 57, Institute for Astronomy, Univ. Hawaii, May 1980.

Decoupled Kalman Filters for Phased Array Radar Tracking

FREDERICK E. DAUM AND ROBERT J. FITZGERALD

Abstract—Kalman filters have been used in numerous phased array radars to track satellites, reentry vehicles, and missiles. This paper considers the design of these filters to reduce computational requirements, ill-conditioning, and the effects of nonlinearities. Several special coordinate systems used to represent the Kalman filter error covariance matrix are described. These covariance coordinates facilitate the approximate decoupling required for practical filter design. A tutorial discussion and analysis of ill-conditioning in Kalman filters is used to motivate these design considerations. This analysis also explains several well-known phenomena reported in the literature. In addition, a discussion of nonlinearities and methods to mitigate their ill effects is included.

I. Introduction

KALMAN filters have been applied in many phased array radars to track satellites, missiles, reentry vehicles, and other objects. Table I summarizes the major characteristics of both the radar and the particular Kalman filter in these applications. All of these radars have rather large antennas (up to 60 m), they all use pulse compression and monopulse angle measurement techniques, the antenna beamwidths are on the order of 1°, and most of them are capable of tracking objects at slant ranges of several thousand miles, with range measurement accuracies on the order of 1–10 m. Further details on these particular radars and phased arrays, pulse compression, and monopulse techniques in general can be found in [1]–[3]. This paper will discuss some aspects of the design of Kalman tracking filters for such radars. Although the discussions are applicable to radar tracking in general, they reflect the experiences of the authors with the several large, long-range phased array radars designed and built at the Raytheon Company, which constitute the majority of the systems in Table I. These are exemplified by the radars shown in Fig. 1(a)–(d).

Our presentation begins with a tutorial on ill-conditioning (Section II), which also includes an analysis that explains two rather common types of filter divergence in applications of this type. Although this analysis is elementary, it provides considerable insight into these phenomena. This is followed in Section III by a discussion of the special coordinate systems used to represent the Kalman filter error covariance matrices. These so-called "covariance coordinates" include range-velocity Cartesian coordinates (RVCC), radar principal Cartesian coordinates (RPCC), and track-oriented Cartesian coordinates (TOCC). Two of these coordinate systems (RPCC and RVCC) were initially developed by Brown and others at IBM [7], although numerous refinements and modifications have been made for the applications in Table I. Section IV considers some computational aspects of these filters, and we conclude in Sections V and VI with a brief review of certain nonlinear aspects of Kalman filtering in applications such as those in Table I. Although the uncertain origin of radar observations [26]–[28] is an important problem in systems such as those in Table I, there is insufficient space to discuss this topic here.

The basic reason for introducing special coordinate systems for representing the Kalman filter error covariance matrix (Section III) is to allow a decoupling of this matrix without suffering a significant loss of state vector estimation accuracy. Decoupling the error covariance matrix has three benefits: 1) reduction in computational requirements, 2) reduction of ill-conditioning (Section II), and 3) mitigation of the ill effects of certain nonlinearities (Sections V and VI). As will be illustrated in subsequent sections, there is often no noticeable loss of accuracy, and in certain cases, there is a substantial improvement in accuracy. It should be noted, however, that for some applications, even partial decoupling may result in significantly suboptimal performance. (Some examples of performance degradation due to decoupling are given in [37].)

The term "decoupling," as used in this context, refers to the approximation of certain covariance matrix elements as zero. For example, the "fully coupled" covariance matrix

$$P = \begin{bmatrix} P_{11} & P_{12} & P_{13} \\ P_{12}^T & P_{22} & P_{23} \\ P_{13}^T & P_{23}^T & P_{33} \end{bmatrix} \tag{1.1}$$

where the P_{ij} are nonzero submatrices, might be reduced to a "partially coupled" matrix by approximating P_{13} and P_{23} by zero matrices. It may be further reduced to a "decoupled" matrix by also letting P_{12} be zero.

The decoupling defined above and the special coordinate systems described in Section III all refer to the error

Manuscript received March 17, 1982; revised August 26, 1982.
F. E. Daum is with the Equipment Division, Raytheon Company, Wayland, MA 01778.
R. J. Fitzgerald is with the Missile Systems Division, Raytheon Company, Bedford, MA 01730.

Reprinted from *IEEE Trans. Automat. Contr.*, vol. AC-28, pp. 269–283, Mar. 1983.

TABLE I
LARGE PHASED ARRAY RADARS

Radar System	Manufacturer of Radar	Operating Frequency	Kalman Filter Characteristics		
			Dimension of State Vector	Degree of Coupling of Covariance Matrix	Coordinate System For Covariance Matrix
AN/FPS-85	Bendix	UHF	7	Full*	Standard * Cartesian
MSR	Bell Labs/Raytheon	S-Band	6 or 7	Partial	RVCC
PAR	Bell Labs/GE	UHF	6	Decoupled	RPCC
Site Defense	McDonnell-Douglas/GE		7	Partial	RVCC
Cobra Dane	Raytheon	L-Band	9	Decoupled	RPCC
PAVE PAWS (Sites I & II)	Raytheon	UHF	6	Partial	RVCC/CR
Cobra Judy	Raytheon				

*Modification to RVCC under consideration.

covariance matrix rather than the state vector itself. The prediction of the state vector from one time to the next is generally performed in a Cartesian coordinate system fixed with respect to the radar. The differential equations used to predict the state vector include gravity, Coriolis and centrifugal accelerations due to earth rotation, and also (if needed for a particular application or flight regime) drag, lift, and thrust acceleration. There is no "decoupling" of the equations of motion used to represent or predict the state vector.

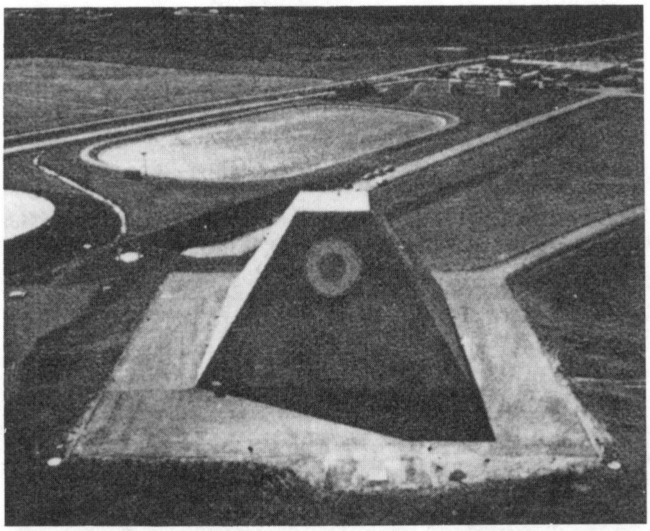

(a)

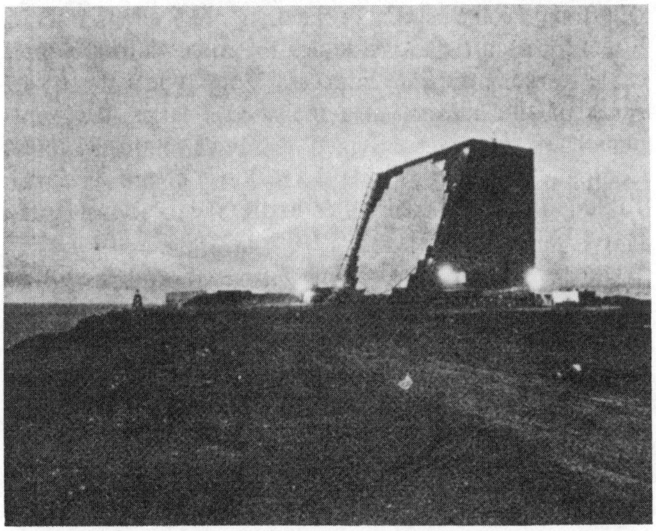

(b)

Fig. 1. (a) Missile site radar (MSR) S-band phased array radar, designed for tracking of reentry vehicles and Spartan missiles in Safeguard ballistic missile defense system. (b) Cobra Dane (AN/FPS-108) L-band phased array radar, designed for collection of data (exoatmospheric) on flight tests of foreign missile systems (also satellite tracking and ICBM early warning).

II. Ill-Conditioning in Kalman Filters

Numerical analysts have been keenly aware of the phenomenon known as "ill-conditioning" in connection with matrix inversion for many years (see [10]–[12]). Least squares problems generally give rise to particularly ill-conditioned matrix inversion problems, owing to the redundancy of the data being processed. Considering that a Kalman filter is simply a recursive solution to a certain weighted least squares problem, it is not surprising that Kalman filters tend to be ill-conditioned. An awareness of ill-conditioning of Kalman filters was achieved soon after the first nontrivial applications [13] and [14]. A particularly fine exposition of this topic is given in [4] and [15]; the following material is intended to complement and illuminate the discussion in these two references. Also, it is interesting to compare Kalman's comments in [5], concerning ill-conditioning in optimal control problems, to our analysis.

The first problem to overcome is defining the notoriously elusive term "ill-conditioning." We will do this in the context of solving simultaneous linear equations, which will be adequate for our purposes. Roughly speaking, we say that the problem of solving the linear equation $Ax = b$ for the vector x is "ill-conditioned" if "small" fractional errors in the $n \times n$ matrix A or the vector b result in "large" fractional errors in x. The fractional errors in x, A, and b are $\|\Delta x\|/\|x\|$, $\|\Delta A\|/\|A\|$, and $\|\Delta b\|/\|b\|$, respectively, in which Δx, ΔA, and Δb denote the errors in x, A, and b, $\|x\|$ is the norm of x, and $\|A\|$ is the corresponding matrix norm induced by the vector norm (see [12, pp.

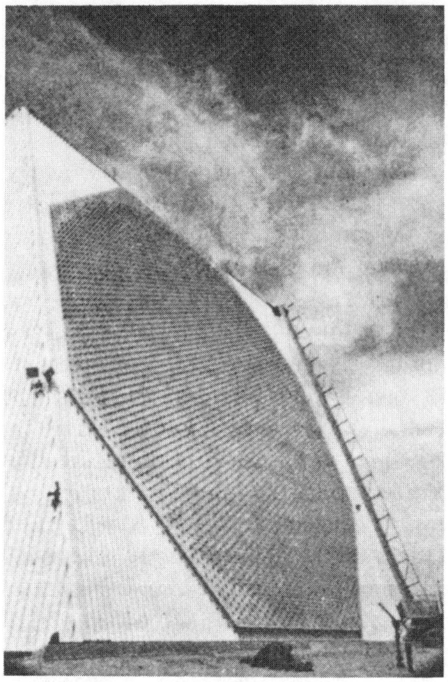

(c)

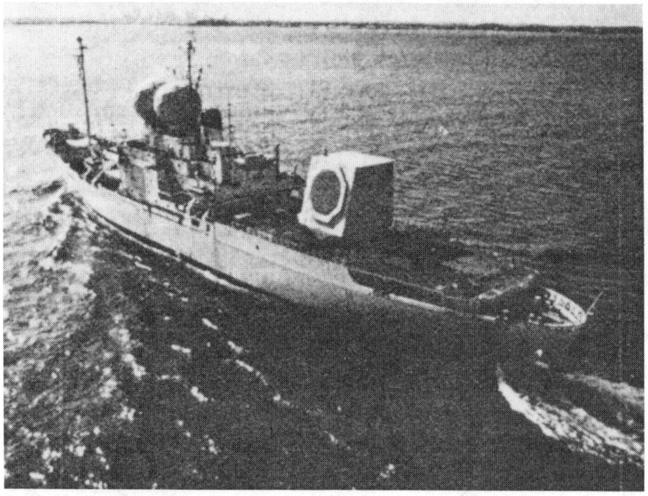

(d)

Fig. 1. (*Continued.*) (c) PAVE PAWS (AN/FPS-115) solid-state UHF phased array radar, designed for early warning of submarine-launched ballistic missiles (also satellite tracking). (d) Cobra Judy (AN/SPQ-11) shipboard phased array radar, designed for collection of data on tests of foreign missile systems (also satellite tracking).

763–771]). Usually we think of the errors in A and $\boldsymbol{b}$ arising from the finite word length in digital computers, although these errors can arise from other sources (e.g., modeling the equations of motion and linearization); scaling and bias errors in an analog computer implementation would qualify as well. Various vector norms are possible, but for this discussion, we will consider only the standard Euclidean norm. The corresponding matrix norm is $\|A\| = \sqrt{\lambda_{\max}(A^T A)}$ in which $\lambda_{\max}(\cdot)$ denotes the largest eigenvalue of $(\cdot)$. At this point, the term "ill-conditioning" is still rather vague, owing to the lack of a rule to determine when to call $\|\Delta x\|/\|x\|$ "large." If fact, the threshold of "largeness" depends on the specific application and the particular norms involved.

The reason that we only need to consider the ill-conditioning of the linear equation $A\boldsymbol{x} = \boldsymbol{b}$ is that the Kalman filter corrector equation has precisely this form. In particular, one standard form of the state vector corrector equation

$$\hat{x} = \bar{x} + MH^T(HMH^T + R)^{-1}(z - H\bar{x}) \tag{2.1}$$

is mathematically equivalent to

$$(M^{-1} + H^T R^{-1} H)\hat{x} = M^{-1}\bar{x} + H^T R^{-1} z \tag{2.2}$$

which is of the form $A\boldsymbol{x} = \boldsymbol{b}$ if we make the following identifications:

$$A = M^{-1} + H^T R^{-1} H \tag{2.3}$$

$$\boldsymbol{b} = M^{-1}\bar{x} + H^T R^{-1} z. \tag{2.4}$$

In the above, $\hat{x}$ and $\bar{x}$ denote the corrected and predicted state vectors, respectively; R is the $m \times m$ measurement error covariance matrix; M is the $n \times n$ error covariance matrix of $\bar{x}$; z is the measurement vector ($z = Hx + v$) where x is the true state vector; and v is the measurement noise vector. With this formulation, the vector $\boldsymbol{b}$ itself is the sum of two terms, each of which involves matrix inversions. Consequently, we must consider the conditioning of the three matrices A, M, and R in order to determine the overall conditioning of the Kalman filter corrector equation. Our focus here is different from other treatments of Kalman filter ill-conditioning which concentrate on numerical errors in the nonlinear covariance matrix equations [9], [13], [22].

Having defined the concept of ill-conditioning, we are now at a point to define a precise mathematical bound [12, p. 809]

$$\frac{\|\Delta x\|}{\|x\|} \leqslant \frac{c(A)}{1 - \|\Delta A\|\|A^{-1}\|}\left[\frac{\|\Delta b\|}{\|b\|} + \frac{\|\Delta A\|}{\|A\|}\right] \tag{2.5}$$

in which $c(A) = \|A\|\|A^{-1}\|$ is called the "condition number." The bound (2.5) is valid under the assumption that $\|\Delta A\|\|A^{-1}\| < 1$, and the norm is such that $\|I\| = 1$ where I is the identity matrix. For obvious reasons, in most interesting practical applications, $\|\Delta A\|\|A^{-1}\|$ is much less than unity, and therefore the condition number is the key parameter. If A is a real symmetric matrix, then the condition number is

$$c(A) = \frac{\lambda_{\max}(A)}{\lambda_{\min}(A)} \tag{2.6}$$

for the Euclidean vector norm considered here. If $c(A)$ is large, then the problem $A\boldsymbol{x} = \boldsymbol{b}$ may be ill-conditioned, but because (2.5) is only an upper bound, $A\boldsymbol{x} = \boldsymbol{b}$ might be well-conditioned. In fact, one of the main goals of this paper (as well as [9]–[17]) is to show how to make $A\boldsymbol{x} = \boldsymbol{b}$ well-conditioned despite enormous values of $c(A)$. The failure to recognize this point is very common (e.g., [5, pp. 27–28]). The bound (2.5) can be extremely pessimistic, depending on the exact form of A. For example, if A is the matrix

$$A = \begin{bmatrix} 10^{100} & 0 \\ 0 & 1 \end{bmatrix}, \tag{2.7}$$

then $c(A) = 10^{100}$, but the problem $Ax = b$ is not ill-conditioned at all. This and other defects in the classical definition of condition number have led to alternative definitions (e.g., [7], [15], and [18]), but we will be content with (2.6). Also, with some care, one can define a condition number for the solution of nonlinear problems [19]; however, as indicated earlier, we do not need such a generalization to analyze the Kalman filter corrector equation.

Using (2.5) and (2.6), we can now analyze the (potential) ill-conditioning of Kalman filters. In particular, assuming that M is positive definite, and using the convexity of $\lambda_{\max}(\cdot)$ as a function of $(\cdot)$ and the concavity of $\lambda_{\min}(\cdot)$ [8, p. 72], it is easy to show that

$$c(A) \leqslant \frac{\lambda_{\max}(M^{-1}) + \lambda_{\max}(H^T R^{-1} H)}{\lambda_{\min}(M^{-1}) + \lambda_{\min}(H^T R^{-1} H)} \tag{2.8}$$

in which A is given by (2.3). In some applications, the ill-conditioning is so severe that M eventually fails to be positive definite (e.g., [6]), in which case (2.8) is invalid. Moreover, if M is not positive definite, then the filter is likely to be unstable, and the error in $\hat{x}$ will be unbounded. In any case, (2.8) is a valid bound before M loses positive definiteness; it follows from (2.8) that

$$c(A) \leqslant \frac{1/\lambda_{\min}(M) + \lambda_{\max}(H^T R^{-1} H)}{1/\lambda_{\max}(M) + \lambda_{\min}(H^T R^{-1} H)}. \tag{2.9}$$

For the majority of practical applications, including all of those listed in Table I, the measurement matrix H has rank less than n, and therefore $\lambda_{\min}(H^T R^{-1} H) = 0$. It follows from (2.9) and (2.6) that

$$c(A) \leqslant c(M) + \lambda_{\max}(M)\lambda_{\max}(H^T R^{-1} H). \tag{2.10}$$

If H has rank n, then (2.10) is still valid, but is more pessimistic. Combining (2.10) and (2.5), and assuming that $\|\Delta A\| \|A^{-1}\|$ is much less than unity, yields the main result of this (approximate) analysis:

$$\frac{\|\Delta \hat{x}\|}{\|\hat{x}\|} \leqslant \left[c(M) + \lambda_{\max}(M)\lambda_{\max}(H^T R^{-1} H)\right] \cdot \left[\frac{\|\Delta A\|}{\|A\|} + \frac{\|\Delta b\|}{\|b\|}\right] \tag{2.11}$$

in which A and b are given by (2.3) and (2.4), respectively. Referring to (2.4), the errors in b arise principally from errors in $\bar{x}$, M, and H, and to a lesser extent from errors in z and R. Note that radar measurement noise should not be considered a source of error in z because z is defined as $Hx + v$, in which v is the noise-induced error. A further application of (2.5) to (2.4) will result in a bound on $\|\Delta b\|/\|b\|$ which explicitly displays its dependence on $c(R)$ and $c(M)$. We shall not pursue this analysis here because (2.11) is sufficient for our immediate purposes; it is interesting to note, however, that such an analysis can produce a bound that is quadratic in $c(M)$. It should be emphasized that (2.2) is not the form of the Kalman filter corrector equation used in most practical applications because its computation generally requires many more operations than the form given in (2.1). Nevertheless, the error bound derived here (2.11) is obviously still valid. The only question is: how tight a bound is (2.11) if (2.1) is used rather than (2.2)? Some insight into this question is derived from the following comments. In particular, [21, pp. 470–479] and [1, p. 371] indicate that, based on practical experience, the computation of A^{-1} via (2.3) is much better conditioned than the mathematically equivalent form $A^{-1} = M - MH^T(HMH^T + R)^{-1}HM$ deduced from Schur's matrix inversion formula. This is true in spite of the fact that the former method requires, in general, many more operations than Schur's alternate method. This suggests that, in general, (2.11) is a tight bound, except for the basic defects in (2.6) noted earlier. This discussion also illustrates the error in the popular notion that fewer operations imply smaller error. (For example, see [5, p. 40].) Other counterexamples to this idea include iterative improvement [12, pp. 831–833] and the so-called Joseph form of the covariance update equation [9, pp. 7, 84]. Moreover, these examples show that ill-conditioning is not related to the number of operations in a simple way, but rather a much more subtle and powerful mechanism is at work.

We will now use (2.11) to interpret a number of results reported in the literature and/or experienced in some of the applications in Table I. First, errors in A and b will be propagated strongly in the initial phase of track when $\lambda_{\max}(M)$ may be large; this is especially so immediately after an exceptionally accurate radar measurement because $\lambda_{\max}(H^T R^{-1} H)$ would be large in this case. An example of this well-known effect is reported in [6, p. 32]. Second, a result often reported in the literature (e.g., [13], [16], [41]) is the divergence of the Kalman filter after an extended period of tracking. In this case, $\lambda_{\max}(M)$ should be relatively small, and therefore the source of difficulty evidently lies in the term involving $c(M)$. The condition number of M can become enormous as the ostensible accuracy of the state vector estimate (as measured by M) improves. To gain some insight into this, consider the simple Kalman filter defined by $H = [1,0]$, $R =$ constant, with no *a priori* data and no process noise modeled, and the state transition matrix

$$\Phi = \begin{bmatrix} 1 & T \\ 0 & 1 \end{bmatrix} \tag{2.12}$$

in which T, the time between measurements, is a constant. For this example, it is well known [20] that

$$M = \begin{bmatrix} 2(2k-1) & 6 \\ 6 & 12/(k-1) \end{bmatrix} \frac{R}{k(k+1)} \tag{2.13}$$

in which k is the number of measurements and $T = 1$ is assumed without loss of generality. It can be shown, using (2.13), that $c(M) = \lambda_{\max}(M)/\lambda_{\min}(M)$ is unbounded for increasing k. Moreover, both $\lambda_{\max}(M)$ and $\lambda_{\min}(M)$ approach zero for this example. Therefore, contrary to the untutored intuition, as the Kalman filter thinks the estimate of $\hat{x}$ becomes more and more accurate, the error

TABLE II
METHODS TO REDUCE ILL-CONDITIONING

Method	Comment
1. Approximate decoupling of covariance matrix	See Section III.
2. Diagonalization of covariance matrix	e.g. Gram-Schmidt technique used in adaptive arrays (Ref. 23).
3. Covariance matrix factorization	See Refs. 6, 9 and 22.
4. Lower bound main diagonal elements of covariance matrix	Encourages positive definite covariance matrix, but does not guarantee it.
5. Introduce or increase process noise	Makes plant model controllable.
6. Tikhonov regularization (also called ridge analysis)	Implicit in Kalman-Wiener filter formulation; see Ref. 11.
7. Extra precision arithmetic (e.g. double precision)	For real-time applications such as those in Table 1, this would impose a significant memory and throughput burden.
8. Iterative improvement	See Refs. 10 and 12; implicit in some iterated least squares methods.
9. Joseph's stabilized form of covariance update equation	See pp. 25 and 31 of Ref. 6 for caveats on this method.
10. Updating the covariance matrix with a sequence of scalar measurements	See Refs. 16 and 17; avoids non-scalar matrix inversion.
11. Preferred order of processing scalar measurements	In Ref. 24 this technique is shown to reduce errors due to nonlinearities, but it should also reduce ill-conditioning for certain applications.
12. Scaling or equilibration of covariance matrix	See Refs. 10 and 12.
13. Rearrangement of filter equations to preserve precision	See pp. 91-96 of Ref. 9.
14. H matrix of rank n	Usually cannot be achieved in practical applications.

propagation given in (2.11) becomes arbitrarily large. This is precisely the phenomenon described in [13] and [16] among others. Finally, it is useful to note that M is (theoretically) positive definite for all finite k; moreover, the pair (H, Φ) defines an observable system for $k \geqslant 2$ [20], although the plant model is obviously uncontrollable. The Kalman filter in this example also happens to be asymptotically unstable, having both poles on the unit circle. (Recall that Kalman's theorem relating controllability and observability to filter stability [45] allows, but does not imply, asymptotic instability of the filter in this case.)

As a result of this phenomenon, it has become standard engineering practice to model some process noise in the Kalman filter design (to make the plant model controllable) and/or explicitly bound the main diagonal elements of M from below. With these safeguards, $\lambda_{\min}(M)$ cannot become arbitrarily close to zero, and $c(M)$ is bounded from above. Nevertheless, it is well known that these ad hoc devices are insufficient to prevent ill-conditioning in many applications. The situation is particularly severe in radars such as those listed in Table I, in which $\lambda_{\max}(M)$ is approximately the variance of position estimation error normal to the radar line of sight, whereas $\lambda_{\min}(M)$ is typically the variance of range-rate estimation error. Under these circumstances, $c(M)$ increases quadratically in track time, and can easily attain values of 10^7 or more [15] before the process noise or lower bounds on the main diagonal elements of M take effect. In view of this, further steps to reduce ill-conditioning are usually required. Table II is a list of the most common methods to mitigate ill-conditioning. These techniques can be used singly or in various combinations. For example, in one of the applications listed in Table I, we have used Methods 1, 4, 5, 6, 9, and 13 in one Kalman filter. Also, it may be useful to note that Methods 1 and 3 are compatible. The method of decoupling the covariance matrix (Section III) has the advantage over all the others of dramatically reducing ill-conditioning, as well as significantly decreasing the number of operations required in real time. Table III quantifies these considerations for a typical application in Table I. Table III also indicates the problems associated with too much decoupling. The results in Table III are based on well-tuned process noise models for each filter with 60-bit floating-point arithmetic used for all computations. Also, the track rates, average signal-to-noise ratio, radar cross-section fluctuation statistics, duration of track, missile trajectories, radar model, etc., were identical for all three filters. Table IV gives some further details of the comparison between the fully coupled and partially coupled filters for three different missile trajectories.

In order to understand why approximate decoupling (Method 1) reduces ill-conditioning, one should think of this method as a step in the direction of diagonalization of the covariance matrix (Method 2 in Table II). In fact,

TABLE III
COMPARISON OF THREE SIX-STATE KALMAN FILTERS

Degree of Coupling	Full	Partial	Decoupled
Coordinate system for covariance representation	Standard Cartesian	RVCC	RPCC
Normalized one-sigma impact prediction errors for various missile trajectories	1.0 to 2.0	0.6 to 1.0	0.6 to 4.0
Number of operations per measurement	3650	1750	840

TABLE IV
COMPARISON OF FULLY COUPLED AND PARTIALLY COUPLED SIX-STATE KALMAN FILTERS

Local Flight Path Angle of Missile	Angle Off Array Boresight	Trajectory Orientation Relative to Radar	Normalized One-Sigma Impact Prediction Errors	
			Fully Coupled Filter	Partially Coupled Filter
60°	45°	Crossing	1.4	1.0
30°	60°	Radial	2.0	1.0
30°	0°	Radial	1.0	0.6

Method 1 is a type of approximate block diagonalization. It is obvious that a diagonal matrix is not ill-conditioned at all. Diagonalization is a very old and well-known technique in least squares problems. The use of orthogonal polynomials rather than the standard basis $(1, x, x^2, \cdots)$ to perform least squares curve fitting is an application of this idea, which is associated with such names as Legendre, Laguerre, Hermite, etc. In the context of dynamical systems, it is known that the eigenvalues of the transition matrix (Φ) of a linear system are least sensitive to numerical errors if Φ is realized in diagonal form [34].

Finally, it should be noted that some progress has been made towards formulating the precise connection between decoupling and ill-conditioning in [44, ch. II].

As an illustration of this idea applied to covariance matrices, consider the simple example

$$P_1 = \begin{bmatrix} 1 & 1-\epsilon \\ 1-\epsilon & 1 \end{bmatrix} \tag{2.14}$$

where $0 < \epsilon \ll 1$. P_1 has positive eigenvalues $\lambda_1 = 2 - \epsilon$ and $\lambda_2 = \epsilon$. In a computer of finite word length (using floating-point or well-scaled fixed-point arithmetic), roundoff errors will tend to produce fractional errors of a particular size in the variables. It is evident from (2.14) that a small fractional error in the off-diagonal element will be equivalent to a large fractional error (or even a sign error) in ϵ, and hence in the smaller eigenvalue λ_2. Thus, the rounded-off matrix may have eigenvalues greatly in error, and may even lose its positive definiteness. In contrast, if principal coordinates are utilized so that the matrix has the diagonal form

$$P_2 = \begin{bmatrix} 2-\epsilon & 0 \\ 0 & \epsilon \end{bmatrix}, \tag{2.15}$$

then fractional errors of moderate magnitude will not seriously alter the eigenvalues or the essential properties of the matrix.

The same ideas can be presented graphically by construction of a Mohr circle [42] to represent the covariance matrix of (2.14) and (2.15). The circle is shown in Fig. 2. In general, the circle representing the covariance matrix P of a random vector in the xy plane is drawn on a diameter defined by the two points $X(P_{11}, -P_{12})$ and $Y(P_{22}, +P_{12})$. Then if the coordinate axes are rotated through an angle θ, a rotation of the diameter through an angle 2θ, in the same sense, yields a new diameter whose end points represent the elements of the transformed covariance matrix. In the present case, the matrix P_1 is represented by the points $X_1(1, -1+\epsilon)$ and $Y_1(1, 1-\epsilon)$. A rotation through 45° (90° on the Mohr circle) brings us to the principal-axis representation of (2.15), represented by the points $X_2(2-\epsilon, 0)$ and $Y_2(\epsilon, 0)$. It is evident from the figure that small fractional errors in the coordinates of X_1 and Y_1 may produce large errors in the smaller eigenvalue (the abscissa of Y_2), even to the extent of destroying positive definiteness (in which case the abscissa becomes negative). No such problem exists when the matrix is represented as in (2.15) (i.e., represented directly by the coordinates of X_2 and Y_2).

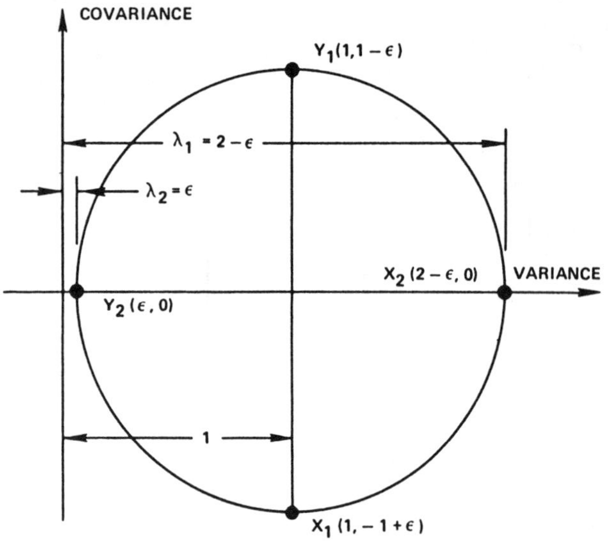

Fig. 2. Mohr circle representation of the covariance matrix of (2.14) and (2.15).

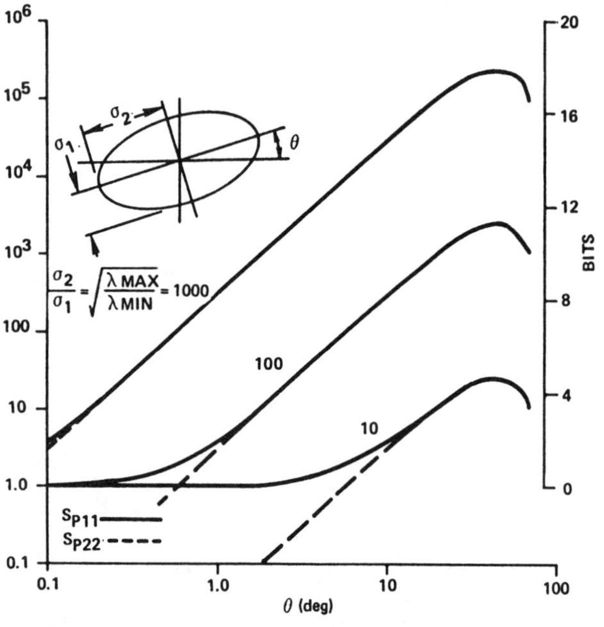

Fig. 3. Eigenvalue sensitivities for a 2×2 covariance matrix.

The above concepts can be quantified by computing the sensitivities (partial derivatives) of the smallest eigenvalue

$$\lambda_{\min} = \frac{P_{11} + P_{22}}{2} - \sqrt{\left(\frac{P_{11} - P_{22}}{2}\right)^2 + P_{12}^2} \quad (2.16)$$

with respect to the individual covariance-matrix elements $\partial \lambda_{\min} / \partial P_{ij}$. In order to deal with relative errors, we may normalize these derivatives to yield sensitivity functions of the form

$$S_{P_{ij}} = \frac{P_{ij}}{\lambda_{\min}} \frac{\partial \lambda_{\min}}{\partial P_{ij}}. \quad (2.17)$$

For 2×2 covariance matrices with high flattening ratio σ_2/σ_1, Fig. 3 shows the extreme sensitivities which can result from even small misorientation angles θ between the coordinate axes and the principal axes. (The right-hand scale, expressed in powers of two, can be interpreted as the number of bits of precision lost in $\lambda_{\min}$ for a 1-bit error in P_{ij}.)

It is interesting to note that if P is expressed in the factorized form $P = SS^T$ (S triangular), all of the sensitivities $S_{S_{ij}}$ remain small, regardless of θ. Such a statement cannot be made, however, when we employ the factorization $P = UDU^T$ (D diagonal, U triangular and unitary). Furthermore, scaling of the covariance matrix (so that all of its diagonal elements are equal) does not change any of these conclusions.

Although the applications discussed here all involve computers of considerable word length, short word-length implementations (for example, 16 bits) are common. The result is larger fractional errors due to roundoff, and hence an increase in the importance of ill-conditioning, as demonstrated by the above examples.

III. Decoupled Tracking

In using covariance coordinates for tracking filter decoupling, the basic objective is to perform the covariance computations in coordinates which are, as nearly as possible, principal coordinates of the error covariances. If the covariance matrix can be visualized as representing an error ellipsoid (which is easily done, for example, if it represents only position errors in three-dimensional space), then the appropriate coordinate axes are the principal axes of this ellipsoid.

Tracking problems are usually more amenable to this technique than are estimation problems in general because it is often possible to predict, quite readily, which coordinate axes will produce the desired result. For example, when the target dynamics exhibit no significant directional properties, one might immediately surmise that the principal axes of the measurement error ellipsoid would constitute an appropriate axis system in which to perform the covariance computations. Provided that these axes do not change their orientation at an appreciable rate, the resulting estimation errors in each axis will be largely uncoupled from those in the other axes. If, for example, the covariance matrix is a 6×6 matrix representing estimation errors in position and velocity in three Cartesian coordinate directions, it can be expected that the only significant off-diagonal elements will be those which represent cross correlations between position and velocity errors in the same axis. If the other (small) elements are neglected, the resulting matrix can be represented as three distinct uncoupled 2×2 matrices. If we use the rough rule of thumb that the computational cost of covariance propagation is proportional to n^3 (where n is the number of states), we find that the computation has been reduced by a factor of $6^3/(3 \times 2^3) = 9$.

Even when Kalman filters (i.e., filters which use covariance propagation for gain determination) are not utilized,

TABLE V
APPROACHES TO DECOUPLED TRACKING

Coordinate System	Number of States	Applications	Examples
RPCC (Radar Principal Cartesian Coords)	3,3,3 or 2,2,2 or 3,2,2	Meas. errors (ME) highly directional; LOS rotation slow relative to filter memory	Aircraft or missiles in the atmosphere
TOCC (Track-Oriented Cartesian Coords)	3,3,3 or 2,2,2 or 2,3,3	Dynamic errors (DE) highly directional (or ME directional along velocity); LOS rate slow	Air traffic control
RVCC (Range-Velocity Cartesian Coords)	6,3 or 4,2 or 5,2	ME and DE both highly directional; LOS rate slow	Reentry vehicles
RVCC/CR (RVCC with Covariance Rotation)	6,3 or 4,2	LOS rate significant relative to filter memory (perhaps because of long memory due to small DE)	Exoatmospheric missiles, satellites, space vehicles

the concept of covariance coordinates facilitates decoupling of the tracking problem, and indeed makes possible the successful application of simple precomputed-gain filters, such as those of the so-called "$\alpha\beta\gamma$" or "*ghk*" type [29]. In this sense, the concept, in its simplest form, is implicit in many of the approaches commonly used today for decoupled tracking.

In more complex situations, such as those involving directional target dynamics or a rapidly rotating line of sight, the concept must be modified appropriately. This leads to a family of related decoupling techniques which have repeatedly proven their effectiveness in various applications over the past decade. These concepts were originally developed for the Safeguard system where they were utilized for tracking Sprint missiles and reentry vehicles [7], [30]; they have since been applied to a variety of other systems, and the results of some of these applications are discussed here.

Four classes of decoupled tracking schemes are described briefly below. Table V provides a summary of the four approaches, and outlines the conditions under which each may be appropriate.

RPCC Filters

Radar principal Cartesian coordinates (RPCC's) are defined as the directions of the principal axes of the error ellipsoid of the radar position measurements. Such an ellipsoid is depicted in Fig. 4. In the case of a phased array radar, it is customary to define a set of face (F) Cartesian coordinates with the z_F axis normal to the array face and the x_F and y_F axes in the face. Then the target position vector in face coordinates is given by

$$X_F = \begin{bmatrix} x_F \\ y_F \\ z_F \end{bmatrix} = \begin{bmatrix} ru \\ rv \\ rw \end{bmatrix} \tag{3.1}$$

where r is the range and u, v, and w are the direction cosines of the range vector or line of sight (LOS), related by $u^2+v^2+w^2=1$ or $w=\sqrt{1-u^2-v^2}$.

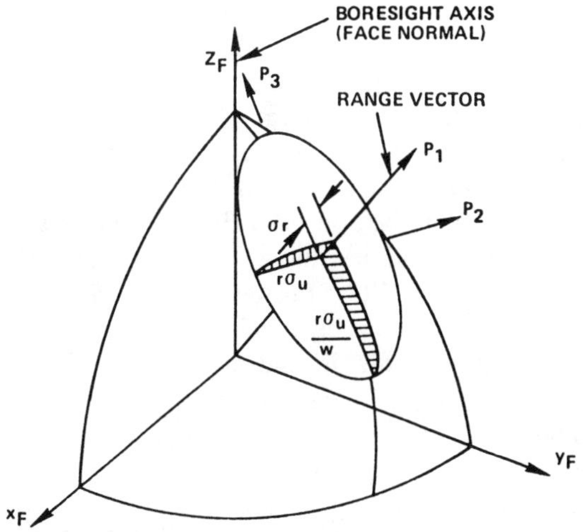

Fig. 4. Measurement error ellipsoid and RPCC coordinates for a phased array radar.

The quantities normally measured by the radar are r, u, and v, and their measurement errors are characterized by a diagonal covariance matrix $R_M = \mathrm{diag}(\sigma_r^2, \sigma_u^2, \sigma_v^2)$. When $\sigma_v = \sigma_u$ (as is generally the case for a circular or square array), it can be shown that the RPCC (or "P") axes are oriented as shown. The P_1 axis lies along the range vector, P_2 is parallel to the radar face plane, and P_3 lies in the plane of P_1 and the face normal. In this coordinate system, the measurement error covariance matrix is

$$R_P = [PF][FM]R_M[FM]^T[PF]^T \tag{3.2}$$

where $[PF]$ is the transformation matrix from F to P coordinates,

$$[PF] = \begin{bmatrix} u & v & w \\ -v/a & u/a & 0 \\ -uw/a & -vw/a & a \end{bmatrix} \tag{3.3}$$

with $a=\sqrt{1-w^2}=\sqrt{u^2+v^2}$, and $[FM]$ is the linearized transformation matrix from measurement (ruv or M) coordinates to F, which may be found by differentiation of (3.1):

$$[FM] = \begin{bmatrix} u & r & 0 \\ v & 0 & r \\ w & -ru/w & -rv/w \end{bmatrix}. \quad (3.4)$$

It is readily verified that (3.2) results in a diagonal measurement-error covariance matrix in the Cartesian P frame, given by

$$R_P = \text{diag}\left(\sigma_r^2, r^2\sigma_u^2, r^2\sigma_u^2/w^2\right). \quad (3.5)$$

If rotations of the line of sight (LOS) can be neglected, and if the dynamic error effects due to unpredictable target motions can be assumed to be approximately the same in all directions (spherically isotropic), then the principal directions for the estimation errors can be assumed to remain along the RPCC axes, and hence covariance computations can be decoupled in these axes.

Such an approach can often be applied successfully to the tracking of airborne targets such as aircraft and missiles, and of sea targets such as ships and submarines. In propagating the three independent covariance matrices, rotation of the line of sight is ignored completely. Significant LOS rates will result in the true error principal directions lagging behind the RPCC system; resulting is a degree of cross coupling and a degradation of performance of the decoupled filter. Furthermore, LOS rotation produces a triangulation effect which can significantly reduce the estimation errors in the cross-range direction; this effect is not usually taken into consideration in the decoupled propagation of the covariance matrices, resulting in an additional loss of performance relative to that of a fully optimum approach.

For these reasons, the fully decoupled RPCC tracker yields near-optimum performance only when the LOS rate is small. In this context, the significance of the LOS rate must be judged in relation to the memory lengths of the filters (i.e., the significant parameter is the LOS rotation per memory length); the governing filter in this regard is the one with the largest measurement errors, and hence the narrowest bandwidth (longest memory), which is usually the filter operating in the cross-range direction rather than the range direction. Furthermore, since filter memory is longer when dynamic errors are small, even small LOS rates can cause performance degradation in cases where the filter's dynamic equations accurately model the target behavior, as is often the case in exoatmospheric tracking.

The above implies that considerable LOS rates can be tolerated as long as the tracking filters have wide bandwidths (i.e., large gains). This has been shown to be the case in the Safeguard application [30] where RPCC filters have been used successfully to track fast-moving Sprint missiles at short ranges.

In many applications (including those described here), the radar measurement error ellipsoid is flattened to an extreme degree, especially at long range. Flattening ratios (ratios of maximum to minimum rms error) of 1000 or more are not uncommon. (By way of comparison, the corresponding ratio of diameter to thickness of an ordinary LP phonograph record is generally in the vicinity of 200.) With such extreme flattening, small changes in LOS angle, if properly accounted for in a fully coupled Kalman filter, can yield drastic reductions in the estimation errors in the cross-range directions. Thus, extreme flattening and significant LOS rotation rates conspire together, in a decoupled filter, to degrade performance from the theoretical optimum.

In contrast to these high flattening ratios, the differences in cross-range rms errors (the two largest axes of the error ellipsoid) are often not very significant. For a circular phased array like those discussed here, their ratio (Fig. 4) is equal to $w = z_F/r$, the cosine of the off-boresight angle of the beam, and generally lies between unity and approximately 1/2. This difference can often be ignored, and an average value can be used in both cross-range axes so that the same covariance matrix and filter gains can be used in both axes. Since much of the computational burden is normally in the covariance-matrix propagation, this can lead to significant further savings. By the same token, such an assumed circularization of the error ellipsoid means that often any other Cartesian coordinate system, rotated from the RPCC set around the range vector, can be utilized just as well as the original RPCC coordinates. The important thing is that the range direction itself, with its much smaller measurement errors, be retained as one of the coordinate axes.

For application to dish antennas rather than phased arrays, identical considerations apply for filter decoupling. In this case, the z_F axis in Fig. 4 can be interpreted as the antenna's azimuth axis, and the rms measurement errors may be defined somewhat differently. (In particular, a fan-beam radar may have quite different accuracies in the P_2 and P_3 directions.)

Of course, it is apparent that similar principles would apply if the range errors were much *larger* than the cross-range errors (as they often are in CW radars), in which case the error ellipsoid would be shaped like a cigar rather than a pancake.

TOCC Filters

When the directionality properties of the dynamic errors are much more important than those of the measurement errors, so that the latter may be approximated as isotropic, it may be appropriate to align the filtering coordinate axes with the principal dynamic-error directions. In air traffic control applications where target accelerations are almost entirely normal to the direction of flight, this has led to the use of the TOCC (track-oriented Cartesian coordinates) system for tracking filter decoupling [31]. A similar approach could be appropriate for tracking a reentry vehicle (RV); in this case, variations in the ballistic coefficient (or our lack of knowledge of its magnitude) and atmospheric uncertainties and inhomogeneities result in error propagation predominantly in the direction of the velocity vector.

The appropriateness of the TOCC approach may also result from measurement-error directionality in cases where target-induced measurement errors such as radar glint predominate. Such may be the case, for example, when a homing missile or torpedo nears an elongated target such as an airplane fuselage, a waking RV, or a submarine.

In the TOCC case, of course, it is the target's turn rate, rather than LOS rotation, which is likely to degrade the performance of the decoupled filter.

RVCC Filters

In cases where both the measurement errors and dynamic errors are directional in character, in general no single coordinate system will result in complete decoupling. The most common such case is when the measurement accuracy is predominantly in the range direction $\boldsymbol{r}$, while the target velocity direction $\boldsymbol{v}$ is the unique one for dynamic error propagation; this is usually the case, for example, in reentry vehicle tracking [32]. In this case, the plane defined by the $\boldsymbol{r}$ and $\boldsymbol{v}$ vectors is the only one which constitutes a common plane of symmetry for the measurement and dynamic errors. It now becomes appropriate to *partially* decouple the covariance computations in the RVCC (range-velocity Cartesian coordinates) system. This "V" coordinate system is usually defined with V_1 along the range vector $\boldsymbol{r}$, V_2 normal to V_1 in the $\boldsymbol{r}$-$\boldsymbol{v}$ plane, and V_3 normal to that plane. The unit vectors along the RVCC axes are thus given by

$$\begin{aligned} \boldsymbol{u}_{V1} &= \boldsymbol{u}(\boldsymbol{r}) \\ \boldsymbol{u}_{V2} &= \boldsymbol{u}_{V3} \times \boldsymbol{u}_{V1} \\ \boldsymbol{u}_{V3} &= \boldsymbol{u}(\boldsymbol{r} \times \boldsymbol{v}) \end{aligned} \tag{3.6}$$

and these three vectors constitute the rows of the transformation matrix $[VF]$, from face to RVCC.

The covariance computations for the V_3 axis can now be performed independently, while a second coupled covariance matrix is utilized in the $\boldsymbol{r}$-$\boldsymbol{v}$ (or $V_1 - V_2$) plane. Thus, for example, a nine-state filter can be decoupled into a three-state and a six-state filter. For RV tracking in the atmosphere, a two-state (position and velocity) filter is often used in the V_3 direction, and a five-state filter (with some function of ballistic coefficient as the fifth state) in the $\boldsymbol{r}$-$\boldsymbol{v}$ plane [7].

The availability of a Doppler (range rate) measurement also makes such a partial filter coupling desirable. Consider, for example, a six-state (position and velocity) tracking filter for which the state corrections are to be computed in the RPCC frame of Fig. 4. Then the appropriate Kalman filter H matrix (the partial derivative matrix of the measurement with respect to the states) contains derivatives of $\dot{r}$ with respect to elements of the state vector

$$\boldsymbol{x} = [r_1 \; r_2 \; r_3 \; v_1 \; v_2 \; v_3]^T. \tag{3.7}$$

Since

$$\dot{r} = \frac{\boldsymbol{r} \cdot \boldsymbol{v}}{\sqrt{\boldsymbol{r} \cdot \boldsymbol{r}}}, \tag{3.8}$$

we have, by differentiation,

$$H = \frac{1}{r^2}\left[(rv_1 - \dot{r}r_1) \; (rv_2 - \dot{r}r_2) \; (rv_3 - \dot{r}r_3) \; rr_1 \; rr_2 \; rr_3\right]. \tag{3.9}$$

Now in RPCC, $r_1 = r$, $r_2 = r_3 = 0$, and $v_1 = \dot{r}$. Hence,

$$H = \left[0 \;\; \frac{v_2}{r} \;\; \frac{v_3}{r} \;\; 1 \; 0 \; 0\right]. \tag{3.10}$$

Thus, the $\dot{r}$ measurement exhibits sensitivities to cross-range position errors in the P_2 and P_3 directions, as well as to velocity errors in the P_1 direction. In general, therefore, all three axes are coupled together, and a fully coupled filter is called for. If we utilize RVCC coordinates, however, then $v_3 = 0$ by definition, and the third axis remains uncoupled from the other two.

Even when a Doppler measurement is not used, a similar situation occurs when a linear-FM (chirp) pulse is utilized because of the range-Doppler coupling effect [33]. The "range" quantity measured is

$$m = r + T_c \dot{r} \tag{3.11}$$

where

$$T_c = T_p \frac{f_0}{f_2 - f_1}. \tag{3.12}$$

T_c is the range-Doppler coupling constant, T_p is the pulse length, f_0 is the center frequency of the pulse, and the frequency sweep within the pulse is from f_1 to f_2. Thus, $(f_2 - f_1)$ is the swept bandwidth, and may be positive (upsweep) or negative (downsweep). For conventional radars, the coupling constant T_c is usually on the order of a few milliseconds or less, but for sensitive long-range phased arrays such as those discussed here, extreme pulse lengths may result in much larger values for T_c (up to 36 s for PAVE PAWS).

RVCC/CR Filters (with Covariance Rotation)

In the RPCC and RVCC filters described above, it was tacitly assumed that LOS rotation is ignored in propagating the Kalman filter covariance matrices. The covariance coordinate frame is nonrotating between measurements, but is redefined (with a slightly different orientation) at each measurement time. The filter ignores the effects of these reorientations on the covariance matrices, and simply assumes that the matrices predicted in the old frame are valid in the new. As discussed above, this leads to a loss of performance due to covariance-matrix misorientation and disregard of triangulation effects when the LOS rate is significant.

Since both of these degradation effects are connected with covariance rotations in the plane of LOS rotation, they can be virtually eliminated by utilizing an RVCC filter to preserve covariance coupling in that plane, and performing a small rotational transformation on the in-plane covariance matrix to account for intermeasurement LOS rotation. The transformation takes the form

$$P' = LPL^T \tag{3.13}$$

where P is the in-plane covariance matrix and (for the case of a 4×4 P matrix for two positions and two velocities, for example)

$$L = \left[\begin{array}{cc|cc} \cos\epsilon & \sin\epsilon & 0 & 0 \\ -\sin\epsilon & \cos\epsilon & 0 & 0 \\ \hline 0 & 0 & \cos\epsilon & \sin\epsilon \\ 0 & 0 & -\sin\epsilon & \cos\epsilon \end{array}\right]. \tag{3.14}$$

The LOS rotation angle ϵ is computed as

$$\epsilon = \frac{\Delta t}{r}(\boldsymbol{u}_{V2} \cdot \boldsymbol{v}) \tag{3.15}$$

where $\boldsymbol{u}_{V2}$ is a unit vector in the direction of the second (in-plane cross-range) axis of the RVCC frame.

In practice, computation can be saved by combining the rotation operation with covariance prediction in the form $(\Phi L)P(\Phi L)^T$ where Φ is the transition matrix. Small-angle approximations may also be employed.

With the added operation of covariance rotation, the RVCC filter becomes capable of tracking, in near-optimum fashion, in the presence of rapid LOS rotation and extreme directionality in measurements and/or dynamics. The principal remaining obstacle to complete optimality is the possibility of rapid rotation of the $\boldsymbol{r} - \boldsymbol{v}$ plane around the line of sight, such as might occur in the case of a violently maneuvering target. In such cases, it may be necessary to maintain full coupling of the covariance matrix; however, covariance coordinates may still be employed to advantage to reduce ill-conditioning effects. Such an implementation might include the propagation of a fully coupled covariance matrix in RPCC, with covariance rotation around *both* cross-range axes; the resulting algorithm is more costly computationally than a standard fully coupled filter, but the alleviation of ill-conditioning may make this a desirable approach.

Performance Comparison

A typical performance comparison is presented in Fig. 5 in the form of rms position errors from a 25-run Monte Carlo simulation of a reentry vehicle tracking problem. The RV is tracked from 41 km altitude on a flight path inclined at 35°, and passes within 15 km of the radar at an altitude of 6.7 km. Measurements are taken every 0.1 s, with typical accuracies in range and angle; the ballistic coefficient β has a correlated random component added to its nominal value.

The standard of comparison is provided by a seven-state fully coupled filter operating in earth-fixed Cartesian coordinates. A decoupled filter in *ruv* coordinates (with the β state in the range filter) operates well at small aspect angles (high altitude), but shows considerable degradation when the crossing geometry becomes severe. The difficulty is largely eliminated by the RVCC filter, even without covariance rotation. (With covariance rotation, the degradation due to decoupling could probably be expected to be negligible.)

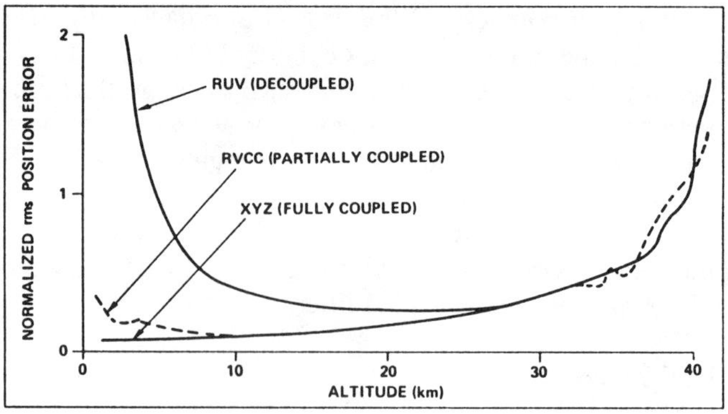

Fig. 5. Performance comparison for three RV tracking filters.

IV. Implementation Alternatives

The implementation of these decoupled filters is exemplified by the flowchart in Fig. 6, which represents an RVCC filter with covariance rotation. In this particular configuration, the measurement residuals are computed in *ruv*(M) coordinates and transformed to RVCC where they are multiplied by the gain matrix

$$K_{VV} = P_V H_{VV}^T R_V^{-1} \tag{4.1}$$

to yield state corrections in RVCC, which are then transformed to F before being applied to the state estimates.

The state estimates are propagated in radar face coordinates. Although it is possible to utilize *ruv* coordinates for this portion of the algorithm, Cartesian coordinates are usually preferable because of the less complex differential equations involved and the ease with which these equations can be integrated using simple algorithms [25].

The configuration shown here, in which the gains transform residuals to state corrections in the same decoupled coordinate frame, results in H matrices of a particularly simple form. For a six-state filter (three positions, three velocities) we have, assuming simple position measurements,

$$H_{VV} = \left[\begin{array}{c|c} I & 0 \\ 3\times3 & 3\times3 \end{array}\right]. \tag{4.2}$$

In addition, especially for RPCC filters, this type of configuration sometimes makes possible the use of simple filters with constant or precomputed gains, or highly simplified gain algorithms without covariance propagation.

A variety of other configurations is possible. For example, the residuals may be left in M coordinates and the corrections still generated in V, in which case the gain matrix is given by

$$K_{VM} = P_V H_{MV}^T R_M^{-1}. \tag{4.3}$$

In general, the R matrix in the gain equation must be in the

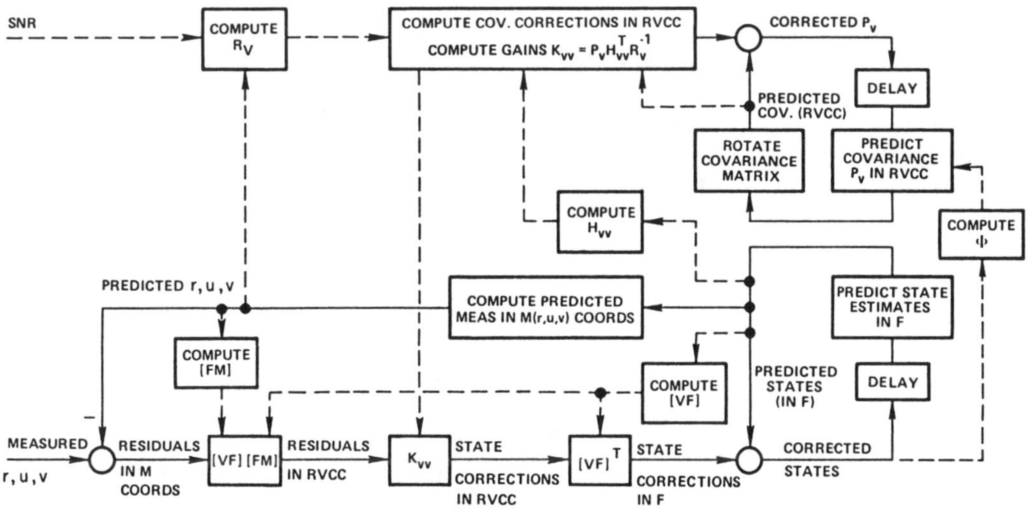

Fig. 6. RVCC decoupled tracker (with covariance rotation).

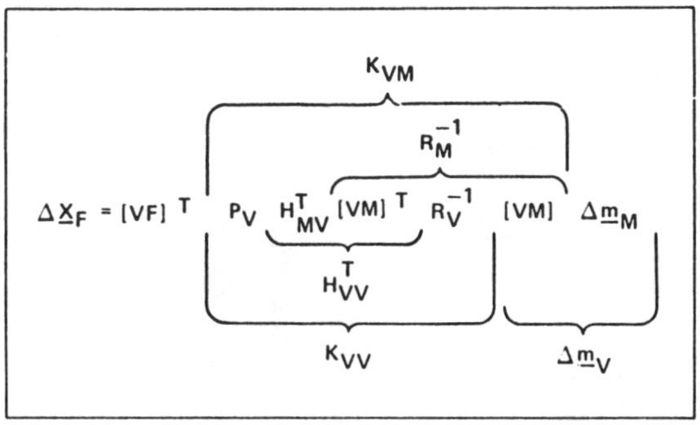

Fig. 7. Alternative interpretations of the state-correction operation.

same coordinates as the residuals to which the gains are to be applied, and the P matrix is in the coordinates in which the state corrections are to be generated. In the example of (4.3), the transformation between V and M is now absorbed into the H matrix, which is related to H_{VV} by

$$H_{MV} = [MV] H_{VV}. \tag{4.4}$$

The relationship between the two approaches is depicted in Fig. 7, in which $\Delta \boldsymbol{m}$ represents the residual vector and $\Delta \boldsymbol{x}$ the state-correction vector. (In Figs. 6 and 7, the matrix $[VF]^T$, which transforms the state corrections from V to F, must be regarded as an expanded version of the 3×3 matrix generated by (3.6), enlarged to the dimension of the state vector.)

Another possible modification is to transform the measurements directly to F coordinates and peiform the state differencing (form the residuals) in F. This procedure alters the effects of measurement nonlinearities, although not necessarily in a beneficial way. An approach which does appear to be beneficial in dealing with nonlinearities is that of [24]: sequential (rather than simultaneous) measurement processing is used, with the range measurement being processed last. (The state estimates after angle measurement processing are used to redefine the filtering coordinates, and in computing the range residual.)

V. Nonlinearities

All of the Kalman filters listed in Table I are based on linearizing the differential equations of motion and the measurement equations to compute the Φ and H matrices. Errors in this linearization process lead to degraded tracking performance to some extent. Moreover, the basic Kalman filter theory assumes Gaussian state estimation error distributions, an assumption that is only approximate for the applications of interest here. Even if the measurement noise and the process noise were both Gaussian, the resulting errors in $\hat{x}$ would be non-Gaussian, owing to the nonlinearities noted above. These nonlinearities can lead to biases, instability, and/or filter divergence. Many methods have been developed to mitigate these effects, a number of which are summarized in Table VI. Several of these techniques were used in early tracking filter simulations for some of the applications in Table I with very poor results. In particular, Methods 1 and 2 in Table VI produced tracking errors in position and velocity that were larger than without these techniques. The poor performance of the second-order filter may be related to the theoretical errors noted in [38], and some insight into the failure of single-stage iteration (i.e., relinearization using an improved state vector estimate) is given in [43]. Although Method 3 in Table VI has worked well in one radar application not listed in Table I, this batch technique is reported to be inferior (as measured by the radius of convergence) to a standard Kalman filter in certain other applications [40]. Concerning Method 4 of Table VI, [32] and [37] indicate possible advantages of *ruv* coordinates over Cartesian coordinates because of the linearity of the measurements in *ruv*. However, [15] and [25] indicate that such is not necessarily the case. In [25], a minor modification to the initialization procedure for the Cartesian filter eliminated large bias-like errors, and resulted in performance virtually identical to that of an *ruv* filter. We have not applied Methods 5, 6, and 7 to radar tracking problems, but they appear to be promising.

TABLE VI
METHODS TO MITIGATE NONLINEARITIES

Method	Comments
1. Second Order Filter	See Ref. 37 and 38; may produce results worse than the first order filter.
2. Single-Stage Iteration	See Refs. 37 and 43; may produce results worse than no iteration.
3. Batch Least Squares	See Refs. 39 and 40; may be worse than the standard Kalman filter.
4. Covariance matrix in ruv coordinates	See Refs. 37 and 32; also Sec. V and Refs. 15 and 25.
5. Add gradient of Kalman filter gain with respect to unknown parameter to algorithm	See Ref. 35.
6. Preferred order of scalar measurement processing	See Ref. 24 and Sec. IV.
7. Virtual decoupling of covariance matrix	See Ref. 36.
8. Quasi-decoupling of covariance	See Sec. VI.

VI. FALSE OBSERVABILITY

A phenomenon that has not been widely noted in the literature, but which is common to most of the applications in Table I, is called "false cross-range observability." Briefly, this effect is due to an overoptimism on the part of the (linearized) Kalman filter with respect to its ability to derive cross-range estimates from accurate measurements in the range direction. The effect is particularly bothersome early in track, when the inevitable noisiness of the cross-range position estimates gives the illusion of a changing measurement direction and an associated triangulation effect. In a fully coupled filter, this false observability is present in both cross-range directions. If the covariance matrix is completely decoupled, as in an RPCC filter, this effect is removed; however, the filter is no longer capable of taking advantage of the true cross-range observability (in the range-velocity plane) provided by the line of sight rotation. The partially decoupled RVCC/CR filter benefits from the true observability, while eliminating the false observability in the out-of-plane direction; the benefits are obtained, however, only if the velocity direction is well known so that the RVCC system can be properly oriented, and this is generally not true early in track.

The identification of this problem is due to K. Brown of IBM. The effect is particularly acute in radars such as in Table I, owing to the large ratio of angle to range measurement error (both expressed in rectilinear coordinates). Thus, the basic cause of this phenomenon is the same as ill-conditioning, but it is demonstrably a distinct nonlinear effect. It turns out, however, that a special type of decoupling mitigates this effect also. The details of decoupling the covariance matrix to avoid false cross-range observability without suffering a significant loss in the benefit of covariance coupling were developed by Daum and Brown. A key ingredient in this algorithm is the so-called Joseph form of

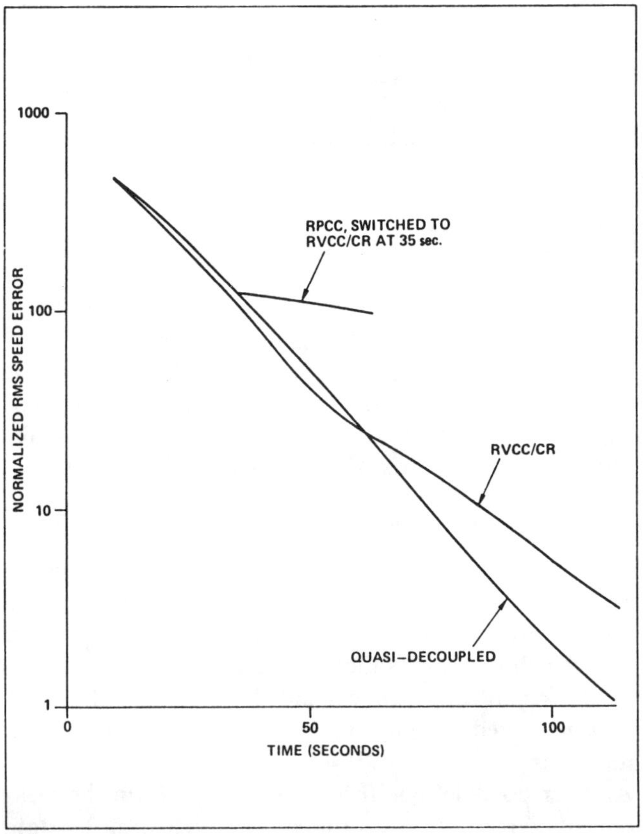

Fig. 8. Use of quasi-decoupling to overcome effects of false observability.

the error covariance update equations, for reasons that have nothing to do with ill-conditioning. In particular, the Kalman filter gain matrix is computed using a totally decoupled covariance matrix early in track; however, the error covariance matrix itself is propagated using the partially coupled form with Joseph's form of the update equation. This is the etymology of the term "quasi-decoupled"; in one sense, the filter is totally decoupled, but in another sense, it is not. The Joseph form of the update equation is used here because it correctly propagates the error covariance matrix (except for nonlinear effects) for any Kalman filter gain matrix, whether or not the gain is optimal. In the present case, the decoupled filter in early track ignores the degradation of range estimation which results from the poor angle rate estimates via the centrifugal ($r\dot{\theta}^2$) component of $\ddot{r}$. It is this degradation which is kept track of by the Joseph algorithm in order to avoid overoptimism in range. It is essential that the algorithm be told that the early gains are suboptimal due to the decoupling, and that the Joseph form be used to maintain a realistic covariance matrix; otherwise, performance can deteriorate markedly.

Quasi-decoupling improves performance by deferring the time at which covariance coupling affects the Kalman filter gain. At this time, the cross-range rate estimates deduced from radar angular measurements alone are sufficiently accurate to suppress false observability. Intuitively, quasi-decoupling makes the Kalman filter less greedy. In some sense, this method is an ad hoc approximation to Ljung's algorithm (Method 5). Fig. 8 presents results from a Monte Carlo comparison (25 runs) of three six-state RVCC tracking filters, and shows the improvement in speed estimation when quasi-decoupling is employed. Also presented is the performance achieved when the filter is simply changed from RPCC to RVCC/CR (at 35 s), without utilizing the Joseph form for covariance propagation.

VII. Summary

The use of covariance coordinates of various kinds for decoupling of Kalman trackers yields the threefold advantage of reduced computational cost, alleviation of ill-conditioning, and mitigation of nonlinear effects. In comparison to covariance-matrix factorization techniques (item 3 of Table II), these decoupling approaches are attractive because of their computational efficiency. They are particularly appropriate for tracking problems because their success depends on statistical symmetries, and these are generally readily predictable in such problems. The relative appropriateness of the various decoupling approaches in any particular application depends on the directionality properties of the measurement errors and dynamic errors. Significant line-of-sight rotation rates may require modifications to the algorithms, including partial coupling and covariance-matrix rotation operations.

References

[1] E. Brookner, *Radar Technology*. Dedham, MA: Artech, 1977.

[2] K. J. Stein, "Cobra Judy phased array radar tested," *Aviat. Week Space Technol.*, vol. 115, pp. 70–73, Aug. 1981.

[3] E. Filer and J. Hartt, "Cobra Dane wideband pulse compression system," presented at IEEE EASCON, Washington, DC, Sept. 1976.

[4] G. W. Johnson, "Choice of coordinates and computational difficulty," *IEEE Trans. Automat. Contr.*, vol, AC-19, pp. 77–78, Feb. 1974.

[5] R. E. Kalman, "Toward a theory of difficulty of computation in optimal control," in *Proc. IBM Scientific Comput. Symp. on Contr. Theory and Appl.*, 1966.

[6] G. J. Bierman and C. L. Thornton, "Numerical comparison of Kalman filter algorithms," *Automatica*, vol. 13, pp. 23–35, 1977.

[7] K. R. Brown, A. O. Cohen, E. F. Harrold, and G. W. Johnson, "Covariance coordinates—A key to efficient radar tracking," presented at IEEE EASCON, 1977.

[8] E. F. Beckenbach and R. Bellman, *Inequalities*. Berlin, Germany: Springer-Verlag, 2nd rev. printing, 1965.

[9] G. J. Bierman, *Factorization Methods for Discrete Sequential Estimation*. New York: Academic, 1977.

[10] J. H. Wilkinson, *Rounding Errors in Algebraic Processes*. Englewood Cliffs, NJ: Prentice-Hall, 1963.

[11] A. N. Tikhonov and V. Y. Arsenin, *Solutions of Ill-Posed Problems*. Washington, DC: Winston, 1977.

[12] D. M. Young and R. T. Gregory, *A Survey of Numerical Mathematics, Vol. II*, Reading, MA: Addison-Wesley, 1973.

[13] F. H. Schlee, C. J. Standish, and N. F. Toda, "Divergence in the Kalman filter," *AIAA J.*, vol. 5, pp. 1114–1120, June 1967.

[14] S. F. Schmidt, J. D. Weinberg, and J. S. Lukesh, "Application of Kalman filtering to the C-5 guidance and control system," ch. 13 in *Theory and Applications of Kalman Filtering*, C. T. Leondes, Ed. AGARD 139, 1970, pp. 312–313.

[15] G. W. Johnson, "Controllability, observability, and computational difficulty," in *24th Annu. South-West IEEE Conf. Rec.*, 1972.

[16] R. B. Blackman, "Methods of orbit refinement," *Bell Syst. Tech. J.*, vol. 43, May 1964.

[17] R. H. Battin, "A statistical optimizing navigation procedure for space flight," *ARS J.*, vol. 32, pp. 1681–1696, Nov. 1962.

[18] B. M. Irons, "Roundoff criteria in direct stiffness solutions," *AIAA J.*, vol. 6, pp. 1308–1312, 1968.

[19] W. C. Rheinbolt, "On measures of ill-conditioning for nonlinear equations," *Math. Comput.*, vol. 30, pp. 104–111, Jan. 1976.

[20] H. W. Sorenson, "On the error behavior in linear minimum variance estimation problems," *IEEE Trans. Automat. Contr.*, vol. AC-12, pp. 557–562, Oct. 1967.

[21] N. Morrison, *Introduction to Sequential Smoothing and Prediction*. New York: McGraw-Hill, 1969.

[22] P. G. Kaminski, A. E. Bryson, and S. F. Schmidt, "Discrete square root filtering: A survey of current techniques," *IEEE Trans. Automat. Contr.*, vol. AC-16, pp. 727–735, Dec. 1971.

[23] R. A. Monzingo and T. W. Miller, *Introduction to Adaptive Arrays*. New York: Wiley, 1980.

[24] D. M. Leskiw and K. S. Miller, "Nonlinear estimation with radar observations," *IEEE Trans. Aerosp. Electron. Syst.*, vol. AES-18, pp. 192–200, Mar. 1982.

[25] R. J. Fitzgerald, "On reentry vehicle tracking in various coordinate systems," *IEEE Trans. Automat. Contr.*, vol. AC-19, pp. 581–582, Oct. 1974.

[26] M. Athans, R. H. Whiting, and M. Gruber, "A suboptimal estimation algorithm with probabilistic editing for false measurements with applications to target tracking with wake phenomena," *IEEE Trans. Automat. Contr.*, vol. AC-22, pp. 372–384, June 1977.

[27] R. A. Singer, R. G. Sea, and K. Housewright, "Derivation and evaluation of improved tracking filters for use in dense multitarget environments," *IEEE Trans. Inform. Theory*, vol. IT-20, pp. 423–432, July 1974.

[28] Y. Bar-Shalom, "Tracking methods in a multitarget environment," *IEEE Trans. Automat. Contr.*, vol. AC-23, pp. 618–626, Aug. 1978.

[29] R. J. Fitzgerald, "Simple tracking filters: Steady-state filtering and smoothing performance," *IEEE Trans. Aerosp. Electron. Syst.*, vol. AES-16, pp. 860–864, Nov. 1980. See also corrections, vol. AES-17, p. 305, Mar. 1981.

[30] K. R. Brown, A. O. Cohen, E. F. Harrold, and G. W. Johnson, "Safeguard track filter design, Part I," IBM Fed. Syst. Div., IBM Rep. 75-0103-M19, Dec. 1976.

[31] M. Wold, G. Kelly, B. Birkholz, and L. Cady, "ARTS-III augmented tracking study," Fed. Aviation Admin. Rep. FAA-RD-73-27, AD-758 886, June 1972.

[32] R. K. Mehra, "A comparison of several nonlinear filters for reentry vehicle tracking," *IEEE Trans. Automat. Contr.*, vol. AC-16, pp. 307–319, Aug. 1971.

[33] R. J. Fitzgerald, "Effects of range-Doppler coupling on chirp radar tracking accuracy," *IEEE Trans. Aerosp. Electron. Syst.*, vol. AES-10, pp. 528–532, July 1974.

[34] P. E. Mantey, "Eigenvalue sensitivity and state-variable selection," *IEEE Trans. Automat. Contr.*, vol. AC-13, pp. 263–269, June 1968.

[35] L. Ljung, "Asymptotic behavior of the extended Kalman filter as a parameter estimator for linear systems," *IEEE Trans. Automat. Contr.*, vol. AC-24, pp. 36–50, Feb. 1979.

[36] K. Brown, D. Bedford, and D. Sweeney, "The virtually coupled sufficient statistics filter," IBM Int. Rep., Jan. 18, 1980.

[37] R. P. Wishner, R. E. Larson, and M. Athans, "Status of radar tracking algorithms," presented at the Symp. on Nonlinear Estimation Theory and Appl., 1970.
[38] R. Henriksen, "The truncated second-order nonlinear filter revisited," *IEEE Trans. Automat. Contr.*, vol. AC-27, pp. 247–251, Feb. 1982.
[39] B. T. Fang, "A nonlinear counterexample for batch and extended sequential estimation algorithms," *IEEE Trans. Automat. Contr.*, vol. AC-21, pp. 138–139, Feb. 1976.
[40] B. E. Schutz, J. D. McMillan, and B. D. Tapley, "Comparison of statistical orbit determination methods," *AIAA J.*, Nov. 1974.
[41] R. J. Fitzgerald, "Divergence of the Kalman filter," *IEEE Trans. Automat. Contr.*, vol. AC-16, pp. 736–747, Dec. 1971.
[42] —, "Graphical transformations of 2×2 covariance matrices," *IEEE Trans. Automat. Contr.*, vol. AC-13, pp. 751–753, Dec. 1968.
[43] M. L. Andrade Netto, L. Gimeno, and M. J. Mendes, "On the optimal and suboptimal nonlinear filtering problem for discrete-time systems," *IEEE Trans. Automat. Contr.*, vol. AC-23, pp. 1062–1067, Dec. 1978.
[44] P. J. Courtois, *Decomposability*. New York: Academic, 1977.
[45] R. E. Kalman, "New methods in Wiener filtering theory," in *Proc. Symp. Eng. Appl. of Random Function Theory and Probability*, F. Kozin and J. L. Bogdanoff, Eds. New York: Wiley, 1963.

Utilization of Modified Polar Coordinates for Bearings-Only Tracking

VINCENT J. AIDALA, MEMBER, IEEE, AND SHERRY E. HAMMEL, MEMBER, IEEE

Abstract—**Previous studies have shown that the Cartesian coordinate extended Kalman filter exhibits unstable behavior characteristics when utilized for bearings-only target motion analysis (TMA). In contrast, formulating the TMA estimation problem in modified polar (MP) coordinates leads to an extended Kalman filter which is both stable and asymptotically unbiased. Exact state equations for the MP filter are derived without imposing *any* restrictions on own-ship motion; thus, prediction accuracy inherent in the traditional Cartesian formulation is completely preserved. In addition, these equations reveal that MP coordinates are well-suited for bearings-only TMA because they automatically decouple observable and unobservable components of the estimated state vector. Such decoupling is shown to prevent covariance matrix ill-conditioning, which is the primary cause of filter instability. Further investigation also confirms that the MP state estimates are asymptotically unbiased. Realistic simulation data are presented to support these findings and to compare algorithm performance with respect to the Cramer–Rao lower bound (ideal) as well as the Cartesian and pseudolinear filters.**

Manuscript received March 12, 1982; revised August 27, 1982. This work was supported by the Naval Sea Systems Command under Code 63-R, Program Element 62633N, Project F33341/SF33323602, and the Naval Material Command under Code 08T1, Program Element 61152N, Project ZR00001/ZR0000101.

The authors are with the U.S. Naval Underwater Systems Center, Newport, RI 02840.

Reprinted from *IEEE Trans. Automat. Contr.*, vol. AC-28, pp. 283–294, Mar. 1983.

INTRODUCTION

PASSIVE localization and tracking problems arise in a variety of important practical applications [1]–[4]. In the ocean environment, two-dimensional bearings-only target motion analysis (TMA) is perhaps familiar [4]–[8]. Here, a single moving observer (own-ship) monitors noisy sonar bearings from a radiating acoustic source (target) assumed to be traveling with constant velocity, and subsequently processes these measurements to obtain estimates of source position and velocity. The geometric configuration is depicted in Fig. 1, where own-ship and target are presumed to lie in the same horizontal plane.

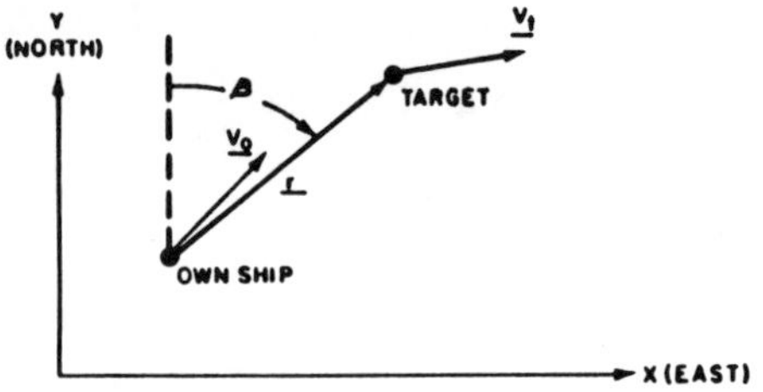

Fig. 1.

Unfortunately, this particular estimation problem is not amenable to simple solution. Intrinsic system nonlinearities preclude the rigorous application of conventional linear analysis. When pseudolinear formulations [9], [10] are employed, the resulting algorithms exhibit biased estimation properties [11], [12]. Moreover, since bearing measurements are extracted from only one sensor, the process remains unobservable until own-ship executes a maneuver [5], [13]. It is this prerequisite maneuver which distinguishes bearings-only TMA from more conventional localization and tracking procedures (e.g., classical triangulation ranging, etc.) and introduces added complexity to the problem.

Despite the aforementioned difficulties, numerous techniques have been devised for bearings-only TMA [6]. One method of solution which has received considerable attention in recent years is the extended Kalman filter [14]. As is well known, utilization of this filter requires explicit mathematical models for both the measurement process and the state dynamics. When addressing these modeling requirements, it is important to recognize that the pertinent analytical equations often acquire entirely dissimilar properties when expressed in different coordinate systems (e.g., linear equations are transformed into nonlinear equations and vice versa). Accordingly, the final filter configuration for any specific problem will ultimately depend upon which reference frame is employed during problem formulation. It is not surprising then, to find that Cartesian coordinates are used extensively, if not exclusively, to formulate TMA estimation problems in the context of an extended Kalman filter. Indeed, this reference frame permits a simple linear representation of the state dynamics; all system nonlinearities are embedded in a single scalar measurement equation [10]. Such a modeling structure is especially appealing for practical applications because it minimizes filter computational requirements.

While the question of choosing "optimal" tracking coordinates has been previously addressed in the literature [3], [6], [15], [16], until recently, sufficient evidence was not available to legitimately infer that non-Cartesian filters may possess significantly different, and perhaps better, performance characteristics than their Cartesian counterparts. Furthermore, the underlying cause of these differences was never clearly identified or well understood. Now, however, new theoretical and experimental findings have been published [10], [17]–[19] which conclusively demonstrate that the Cartesian filter is unstable for single sensor bearings-only TMA. Specifically, these results show that the unique interaction and feedback of estimation errors within this filter render it highly susceptible to premature covariance collapse and solution divergence.

One method which has been successfully utilized to eliminate Cartesian filter instability involves replacing the measured bearings with pseudolinear measurement residuals [9], [10]. While this procedure is particularly simple to implement, it has not gained widespread acceptance within the TMA community because the resulting algorithm (pseudolinear filter) generates biased estimates whenever noisy measurements are processed [12]. Instead, research efforts have focused upon the analysis and development of alternative estimation schemes which are both stable *and* asymptotically unbiased.

In this paper, a candidate TMA algorithm with the desired attributes is rigorously analyzed and subsequently evaluated under realistic operating conditions. The pertinent equations of state and measurement are formulated in modified polar (MP) coordinates [20], [21], while the algorithm itself is configured as an extended Kalman filter [14]. This coordinate system is shown to be well-suited for bearings-only TMA because it automatically decouples observable and unobservable components of the estimated state vector. Such decoupling prevents covariance matrix ill-conditioning, which is the primary cause of filter instability. Further investigation also confirms that the resulting state estimates are asymptotically unbiased, as required.

The MP state vector is comprised of the following four components: bearing, bearing rate, range rate divided by range, and the reciprocal of range. In theory, the first three can be determined from single-sensor bearing data without an own-ship maneuver; the fourth component, however, should remain unobservable until this maneuver requirement is satisfied. These theoretical observability properties are implicitly preserved in the MP filter formulation. In essence, the state estimates are constrained to behave as predicted by theory, even in the presence of measurement errors. Under similar conditions, standard Cartesian filters often experience covariance matrix ill-conditioning which precipitates false observability.

Exact state equations for the MP filter are rigorously derived without imposing *any* restrictions on own-ship motion; thus, prediction accuracy inherent in the traditional Cartesian formulation is completely preserved. These equations also reveal that the choice of reciprocal range as the fourth state is optimal, at least from the viewpoint of minimizing system nonlinearities. In addition, the aforementioned expressions can be readily generalized to satisfy the nonlinear differential equations of *arbitrary* particle motion. To the authors' knowledge, these exact solutions have not been previously documented in the literature.

Realistic tactical scenarios are simulated to illustrate and evaluate MP filter performance. Employing an "idealized" filter, optimal performance (i.e., the Cramer–Rao lower bound [22]) is first determined, and subsequently used as an evaluation basis. Simulation data depicting relative performance are then presented. Finally, the MP filter is compared to the standard Cartesian and pseudolinear filters, both of which are utilized in operational systems. These results clearly demonstrate the stability and overall efficacy of the MP filter.

II. Development of the Filter Equations

As noted earlier, formulating the TMA estimation problem in Cartesian coordinates leads to a linear representation of the state dynamics and a nonlinear scalar measurement relation,[1] viz.

$$x(t) = A_x(t,t_o)x(t_o) - w_o(t,t_o) \tag{1a}$$

$$\tilde{\beta}(t) = h_x[x(t)] + n(t) \tag{1b}$$

where $x(t)$ defines the Cartesian state vector of unknown target motion parameters. To facilitate the ensuing analysis, it is implicitly assumed that $w_o(t,t_o)$ is deterministic and $n(t)$ is zero-mean Gaussian white noise with variance $\sigma^2(t)$. For completeness, initial estimates of the state vector and its associated error covariance matrix are also presumed to be specified.

Ostensibly, it is difficult to obtain an equivalent representation of equation set (1) in MP coordinates. This is especially true if conventional modeling techniques are employed. Significant analytical complications arise because the transformed equations of motion are highly nonlinear and not readily amenable to direct integration. These encumbrances can be avoided, however, by recognizing that the desired representation is also derivable entirely by algebraic manipulation (see Appendix B for details). Briefly, if $y(t)$ denotes the MP state vector, then $x(t)$ and $y(t)$ will be related at all instants of time by the nonlinear one-to-one transformations

$$x(t) = f_x[y(t)] \tag{2a}$$

$$y(t) = f_y[x(t)]. \tag{2b}$$

Letting $t = t_o$ in (2a), and substituting the result into (1a) to eliminate $x(t_o)$, yields

$$x(t) = A_x(t,t_o)\, f_x[y(t_o)] - w_o(t,t_o). \tag{3}$$

[1]Mathematical details of the Cartesian and modified polar coordinate modeling processes are presented in Appendix A and B, respectively, along with a precise description of the various quantities.

Substitution of (3) into (2b) subsequently leads to the relation

$$y(t) = f[y(t_o); t, t_o] \tag{4}$$

where

$$f[y(t_o); t, t_o] \triangleq f_y[A_x(t,t_o)\, f_x[y(t_o)] - w_o(t,t_o)]. \tag{5}$$

In a similar manner, the measurement relation can be formally expressed in MP coordinates by substituting (2a) into (1b), viz.

$$\tilde{\beta}(t) = h_y[y(t)] + n(t) \tag{6}$$

where

$$h_y[y(t)] \triangleq h_x[f_x[y(t)]] = [0,0,1,0]\,y(t). \tag{7}$$

Equations (4) and (6) are the exact MP analogs of (1a) and (1b), respectively. As might be expected, the state equations now exhibit nonlinearities and are considerably more complicated than their Cartesian counterparts. The derivation outlined here, however, remains comparatively simple; all difficulties associated with integrating the transformed equations of motion have been expeditiously circumvented without sacrificing mathematical rigor. Moreover, since (4) is valid for *arbitrary* own-ship motion, prediction accuracy inherent in the traditional Cartesian formulation is completely preserved.

Although the preceding results are expressed in continuous form, discrete time equations of state and measurement may be readily deduced by assigning appropriate values to t and t_o. In particular, if $t = kT$ and $t_o = (k-1)T$ where $k = 1,2,3,\cdots$ and T = constant sampling period, straightforward application of the extended Kalman filter [14] to (4) and (6) will yield the MP estimation algorithm given below:

$y(0/0)$ = initial estimate of the MP state vector

$P(0/0)$ = initial estimate of the MP state vector error covariance matrix

$$y(k/k-1) = f[y(k-1/k-1); kT, (k-1)T] \tag{8a}$$

$$A_y(k,k-1) = \frac{\partial f[y(k-1/k-1); kT, (k-1)T]}{\partial y(k-1/k-1)} \tag{8b}$$

$$P(k/k-1) = A_y(k,k-1)P(k-1/k-1)A_y'(k,k-1) \tag{8c}$$

$$H = [0,0,1,0] \tag{8d}$$

$$G(k) = P(k/k-1)H'[HP(k/k-1)H' + \sigma^2(k)]^{-1} \tag{8e}$$

$$y(k/k) = y(k/k-1) + G(k)[\tilde{\beta}(k) - Hy(k/k-1)] \tag{8f}$$

$$P(k/k) = [I - G(k)H]P(k/k-1) \qquad k = 1,2,3,\cdots \tag{8g}$$

where the "prime" symbol (′) denotes matrix transposition. Here, $y(i/j)$ and $P(i/j)$ for $i, j = 1,2,3,\cdots$ denote estimates of the true state vector $y(i)$ and its associated error covariance matrix $P(i)$, respectively, based upon j data measurements.

While equation set (8) is well-suited for numerical work, it is not readily amenable to analysis. In contrast, nonrecursive representations of this filter are often more tractable, despite their computational inefficiency. One such representation is obtained by reformulating the original dynamic estimation problem for $y(t)$ as a static estimation problem for $y(t_o)$. Note that since $w_o(t, t_o)$ is deterministic, (4) allows the unknown state vector to be uniquely specified at any *arbitrary* time t in terms of its value at some *fixed* time t_o. The required measurement relation then follows by substituting (4) into (6). It is unnecessary to explicitly specify state equations because $y(t_o)$ is time invariant. Subsequent application of the extended Kalman filter to the discretized expressions obtained by letting $t = kT$ and $t_o = 0$ yields a recursive algorithm which may be algebraically recast into the nonrecursive form shown below:

$y(0/0)$ = initial estimate of the MP state vector

$P(0/0)$ = initial estimate of the MP state vector error covariance matrix

$$\begin{aligned} y(0/k) &= P(0/k)P^{-1}(0/0)\,y(0/0) \\ &\quad + P(0/k)\sum_{j=1}^{k} M'(j)\sigma^{-2}(j)M(j)\,y(0/j-1) \\ &\quad + P(0/k)\sum_{j=1}^{k} M'(j)\sigma^{-2}(j) \\ &\quad \cdot\left[\tilde{\beta}(j) - Hy(j/j-1)\right] \end{aligned} \tag{9a}$$

$$P(0/k) = \left[P^{-1}(0/0) + \sum_{j=1}^{k} M'(j)\sigma^{-2}(j)M(j)\right]^{-1} \tag{9b}$$

where H is given by (8d), and

$$M(j) = HA_y(j,0) \tag{10a}$$

$$A_y(j,0) = \frac{\partial f\left[y(0/j-1); jT, 0\right]}{\partial y(0/j-1)} \tag{10b}$$

$$y(j/j-1) = f\left[y(0/j-1); jT, 0\right]. \tag{10c}$$

Finally, discrete estimates of the current state vector and its associated error covariance matrix can be obtained via the relations

$$y(k/k) = f\left[y(0/k); kT, 0\right] \tag{11a}$$

$$P(k/k) = A_y(k,0)P(0/k)A'_y(k,0). \tag{11b}$$

We remark that the static and dynamic estimation algorithms described here possess similar mathematical properties and exhibit virtually identical behavior characteristics.

III. Behavior Characteristics of the Filter

In practical applications, $y(k/k)$ and $P(k/k)$ are of special interest since they statistically characterize the *current* state vector. For analysis purposes, however, it is more convenient to work directly with $y(0/k)$ and $P(0/k)$ which do not vary explicitly with time. By utilizing this approach, extraneous time variations are automatically surpressed. There is no loss of generality either because all four quantities are related by equation set (11). Pertinent behavior characteristics of $y(k/k)$ and $P(k/k)$ can always be determined *a posteriori* from knowledge of the behavior of $y(0/k)$ and $P(0/k)$. Consequently, an examination of equation set (9) will reveal the underlying mechanism governing MP filter stability.

To begin the discussion, consider the closed form expression for $P(0/k)$ given by (9b). Utilizing conventional matrix multiplication rules, it can be shown that

$$\sum_{j=1}^{k} M'(j)\sigma^{-2}(j)M(j) = C'(k)D^{-1}(k)C(k) \tag{12}$$

where $D(k)$ is a $k \times k$ diagonal matrix given by

$$D(k) = \operatorname{diag}\left[\sigma^2(1), \sigma^2(2), \cdots, \sigma^2(k)\right] \tag{13}$$

and $C(k)$ is a $k \times 4$ matrix which may be written in the partitioned form

$$C(k) = \left[M'(1)|M'(2)|\cdots|M'(k)\right]'. \tag{14}$$

Important structural characteristics of $P(0/k)$ can now be discerned by analyzing the ancillary matrix $C(k)$. Combining equation set (10) with (B10) of Appendix B yields

$$M(j) = \left[M_{11}(j) \;\Big|\; \frac{\partial y_3(j/j-1)}{\partial y_4(0/j-1)}\right] \tag{15}$$

where

$$M_{11}(j) = \left[\frac{\partial y_3(j/j-1)}{\partial y_1(0/j-1)}, \frac{\partial y_3(j/j-1)}{\partial y_2(0/j-1)}, \frac{\partial y_3(j/j-1)}{\partial y_3(0/j-1)}\right] \tag{16}$$

$$\begin{aligned} y_3(j/j-1) &= f_3\left[y(0/j-1); jT, 0\right] \\ &= y_3(0/j-1) \\ &\quad + \tan^{-1}\left[S_3(jT,0)/S_4(jT,0)\right] \end{aligned} \tag{17}$$

and

$$\begin{aligned} S_3(jT,0) &= jTy_1(0/j-1) - y_4(0/j-1) \\ &\quad \cdot\left[w_{o3}(jT,0)\cos y_3(0/j-1)\right. \\ &\quad \left. - w_{o4}(jT,0)\sin y_3(0/j-1)\right] \end{aligned} \tag{18a}$$

$$\begin{aligned} S_4(jT,0) &= 1 + jTy_2(0/j-1) - y_4(0/j-1) \\ &\quad \cdot\left[w_{o3}(jT,0)\sin y_3(0/j-1)\right. \\ &\quad \left. + w_{o4}(jT,0)\cos y_3(0/j-1)\right]. \end{aligned} \tag{18b}$$

The quantities $w_{o3}(jT,0)$ and $w_{o4}(jT,0)$ that appear in equation set (18) depend upon own-ship acceleration and can be deduced from (A9) of Appendix A by letting $t = jT$ and $t_o = 0$. It is especially important to recognize that $w_{o3}(jT,0) = 0$ and $w_{o4}(jT,0) = 0$ prior to the first own-ship maneuver. In addition, the fourth element of $M(j)$ will also vanish under these circumstances, as can be seen by performing the indicated differentiation, viz.

$$\frac{\partial y_3(j/j-1)}{\partial y_4(0/j-1)} = \frac{w_{o4}(jT,0)\sin y_3(j/j-1) - w_{o3}(jT,0)\cos y_3(j/j-1)}{\sqrt{S_3^2(jT,0)+S_4^2(jT,0)}}. \tag{19}$$

These important null characteristics may be exploited further by repartitioning the matrix $C(k)$ as follows:

$$C(k) = [C_{11}(k) | C_{12}(k)] \tag{20}$$

where $C_{11}(k)$ is a $k \times 3$ matrix given by

$$C_{11}(k) = [M'_{11}(1) | M'_{11}(2) | \cdots | M'_{11}(k)]' \tag{21a}$$

and $C_{12}(k)$ is a $k \times 1$ column matrix of the form

$$C_{12}(k) = \left[\frac{\partial y_3(1/0)}{\partial y_4(0/0)}, \frac{\partial y_3(2/1)}{\partial y_4(0/1)}, \cdots, \frac{\partial y_3(k/k-1)}{\partial y_4(0/k-1)}\right]'. \tag{21b}$$

Next, assume that $P(0/0)$ is chosen so as to satisfy the relation

$$P(0/0) = \left[\begin{array}{c|c} \Gamma_o & 0 \\ \hline 0 & \sigma_o^2 \end{array}\right] \tag{22}$$

where Γ_o is a 3×3 positive definite symmetric matrix and σ_o is a nonzero scalar. Note that this covariance matrix structure is physically realistic and implies that errors associated with observable and unobservable components of the initial state vector are uncorrelated. Combining (9b) with (12), (20), and (22) subsequently yields

$$P(0/k) = \left[\begin{array}{c|c} \Gamma_o^{-1} + C'_{11}(k)D^{-1}(k)C_{11}(k) & C'_{11}(k)D^{-1}(k)C_{12}(k) \\ \hline C'_{12}(k)D^{-1}(k)C_{11}(k) & \sigma_o^{-2} + C'_{12}(k)D^{-1}(k)C_{12}(k) \end{array}\right]^{-1}. \tag{23}$$

Recall that, in theory, the first three components of $y(0/k)$ can be determined without an own-ship maneuver, whereas the fourth component remains unobservable until this maneuver requirement is satisfied. Formulating the TMA estimation problem in MP coordinates leads to a natural decoupling of observable and unobservable states. The resulting structure of $P(0/k)$ depicted by (23) illustrates this decoupling property and provides the key to MP filter stability. From (19) and (21b) it is evident that $C_{12}(k)$ reduces to a null matrix for unaccelerated own-ship motion. In this case, the off-diagonal terms in (23) vanish and the lower diagonal term reduces to σ_o^{-2}. Straightforward inversion [23] of the resulting simplified matrix leads to

$$P(0/k) = \left[\begin{array}{c|c} [\Gamma_o^{-1} + C'_{11}(k)D^{-1}(k)C_{11}(k)]^{-1} & 0 \\ \hline 0 & \sigma_o^2 \end{array}\right]. \tag{24}$$

Examination of (24) reveals that the variance associated with $y_4(0/k)$ remains unchanged (i.e., equal to σ_o^2) prior to the first own-ship maneuver. The automatic decoupling that accrues from utilizing MP coordinates thus prevents covariance matrix ill-conditioning and premature collapse. As a result, $P(0/k)$ will behave exactly as predicted by theory, even in the presence of measurement errors.

The implications of component decoupling on $y(0/k)$ are also readily exposed by substituting (15), (19), (22), and (24) into (9a). Assuming no own-ship acceleration, and partitioning the state vector into observable and unobservable components, leads to the expression

$$y(0/k) = \left[\begin{array}{c} y_o(0/k) \\ \hline y_4(0/k) \end{array}\right] = \begin{bmatrix} y_1(0/k) \\ y_2(0/k) \\ y_3(0/k) \\ y_4(0/k) \end{bmatrix} \tag{25}$$

where

$$y_o(0/k) = [\Gamma_o^{-1} + C'_{11}(k)D^{-1}(k)C_{11}(k)]^{-1} \cdot \left[\Gamma_o^{-1} y_o(0/0) + \sum_{j=1}^{k} M'_{11}(j)\sigma^{-2}(j)M_{11}(j)y_o(0/j-1) + \sum_{j=1}^{k} M'_{11}(j)\sigma^{-2}(j)[\tilde{\beta}(j) - \beta(j/j-1)]\right] \tag{26a}$$

$$y_4(0/k) = y_4(0/0) \tag{26b}$$

and $\beta(j/j-1)$ is given by the formula

$$\beta(j/j-1) = y_3(0/j-1) + \tan^{-1}\left[\frac{jTy_1(0/j-1)}{1 + jTy_2(0/j-1)}\right] \tag{27}$$

which follows from (17) for $w_{o3}(jT,0) \equiv 0$ and $w_{o4}(jT,0) \equiv 0$.

Note that any measurements processed before the first own-ship maneuver will affect only $y_o(0/k)$, while $y_4(0/k)$

remains unaltered. This result is not surprising since data acquired under such conditions cannot realistically yield information about the unobservable component. In fact, for unaccelerated own-ship motion, (27) clearly shows that bearings extracted from a single sensor are independent of the fourth state. Hence, $y(0/k)$ exhibits theoretically consistent behavior characteristics analogous to $P(0/k)$.

Unlike its MP counterpart, the Cartesian filter does not provide a natural decoupling of observable and unobservable states. Consequently, there is no internal mechanism to effectively prevent covariance matrix ill-conditioning and premature collapse. The unimpeded feedback and interaction of estimation errors resulting from this deficiency often precipitate false observability. Under such conditions, the filter erroneously attempts to estimate all four state components, even though only three are theoretically observable. Since fundamental uniqueness requirements are violated, the corresponding state estimates will necessarily exhibit unstable behavior characteristics. In essence, Cartesian filter instability is intrinsically related to its development.

IV. Simulation Results

To illustrate MP filter performance characteristics, the estimation algorithm described by equation set (8) was programmed on a digital computer and subsequently tested with simulated data. Representative results for two typical TMA scenarios are summarized in Figs. 2–4 (additional data depicting MP, Cartesian, and pseudolinear filter performance may also be found in [7], [10]–[12], [17], [20], [21]). Fig. 2 describes the target and own-ship trajectories for both scenarios, which differ only in their respective values of initial range. The solution errors plotted in Fig. 3 correspond to an initial range of 2700 yds, while those in Fig. 4 correspond to an initial range of 27 000 yds. The initial bearing is 0°, and the target maintains a steady course of 0° with a constant speed of 20 knots. Own-ship also maintains a constant speed of 28.28 knots, but periodically executes 90° course changes as follows:

$$\text{from } 45° \text{ to } 315° \text{ at } t=(4+17k)\text{min} \quad [k=0,1,2,3,4]$$

$$\text{from } 315° \text{ to } 45° \text{ at } t=(12.5+17k)\text{min} \quad [k=0,1,2,3,4].$$

Own-ship turning rate is constrained to 3°/s; thus, each maneuver requires 0.5 min to complete.

To realistically simulate measurement errors and account for the effects of data preprocessing, all raw bearings were corrupted by additive zero-mean Gaussian noise and then sequentially time-averaged in blocks of twenty 1 s samples. Three different noise levels were used.[2] The graphs labeled $(A-a)$–$(A-d)$ depict TMA solution errors obtained with

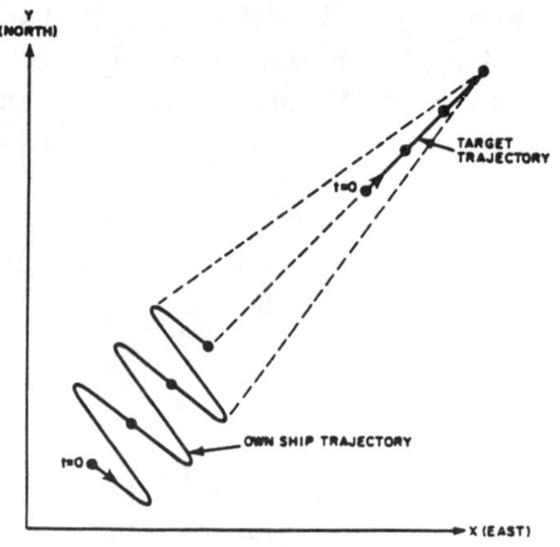

Fig. 2.

a rms noise level of 2°; those labeled $(B-a)$–$(B-d)$ and $(C-a)$–$(C-d)$ correspond to rms noise levels of 4° and 6°, respectively. Estimation errors associated with four different TMA filters are plotted simultaneously on each graph labeled as follows:

A—the "idealized" MP filter, viz. an extended Kalman filter formulated in MP coordinates and linearized about the *true* state vector. This filter provides a measure of optimal performance since the error covariance matrix coincides with the Cramer–Rao lower bound [22],

B—the MP filter defined by equation set (8),

C—the pseudolinear filter described in [10], [12],

D—the Cartesian filter described in [10].

Finally, all results have been ensemble averaged over 25 Monte Carlo runs.

Some comments on filter initialization are perhaps in order here. Under actual operating conditions, it is extremely difficult, and indeed rare, to obtain reliable initial estimates of the state vector and associated covariance matrix. Consequently, existing procedures for accomplishing this task are necessarily ad hoc and somewhat arbitrary. While no claim of optimality is implied, the MP filter and its "idealized" counterpart were found to perform quite satisfactorily when initialized according to the scheme

$$y(0/0)=[0,0,\tilde{\beta}(0),10^{-4}]'$$
$$P(0/0)=\text{diag}[10^{-4},10^{-4},10^{-4},10^{-8}].$$

The pseudolinear and Cartesian filters were subsequently initialized in accordance with commonly prescribed procedures [10], viz.

$$x(0/0)=[0,0,0,0]'$$
$$P(0/k)=\text{diag}[1,1,1,1]$$

for the pseudolinear filter, and

$$x(0/0)=[0,0,10^4\cdot\sin\tilde{\beta}(0),10^4\cdot\cos\tilde{\beta}(0)]'$$
$$P(0/0)=\text{diag}[15^2,15^2,10^8,10^8]$$

for the Cartesian filter.

Examination of Figs. 3 and 4 reveals that the MP filter performs exactly as predicted by theory. Indeed, for both

[2] Note that data preprocessing reduces the rms random noise level by a factor of $1/\sqrt{20}$.

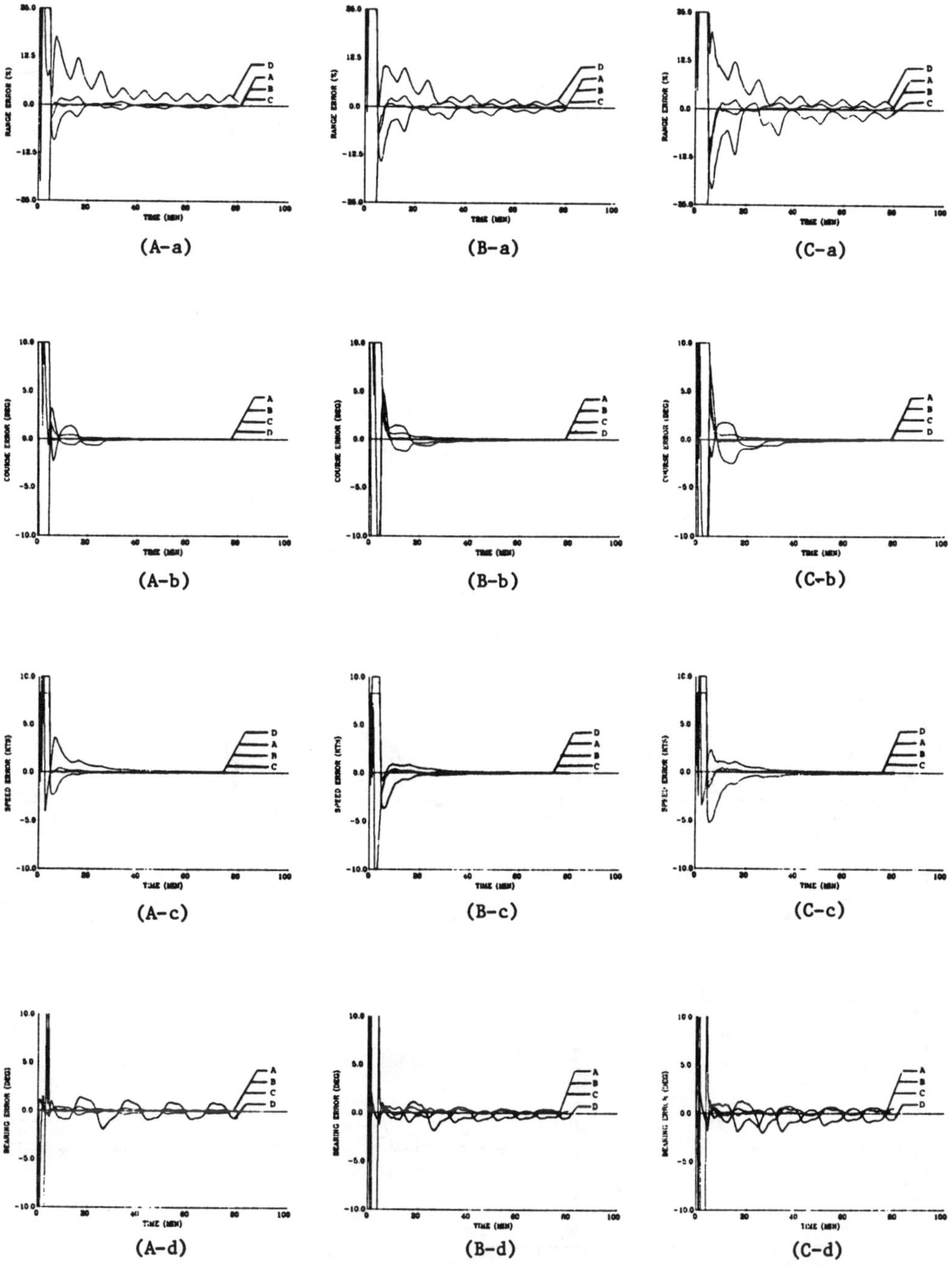

Fig. 3.

short and long range scenarios, the state estimates begin converging to their true values immediately after own-ship executes a maneuver. There is also no evidence of instability; subsequent own-ship maneuvers simply enhance convergence so that the final estimates become asymptotically unbiased. Overall efficacy of the MP filter can also be discerned by noting how rapidly it approaches "idealized" filter performance.

Behavior characteristics exhibited by the pseudolinear filter were also in agreement with previously documented theoretical and experimental findings [12]. Specifically, this filter generated biased range estimates which are readily apparent in the long range scenario data (see Fig. 4). However, the pertinent bias errors are known to be geometrically dependent and become negligibly small for TMA scenarios characterized by high bearing rates and/or low measurement noise levels. This fact is substantiated by the simulation results presented in Fig. 3. Under such conditions, the pseudolinear filter possesses excellent tracking capabilities, comparable even to the MP filter.

In contrast with the other filters tested, Cartesian filter performance was generally poor, and erratic at best. For the short range scenario, this filter converged very slowly, despite relatively favorable tracking conditions. Indeed, five own-ship maneuvers were required to obtain steady-state range estimation errors of less than 5 percent. Abnormal behavior was also manifest in the long range scenario. Here, the Cartesian filter converges to the wrong

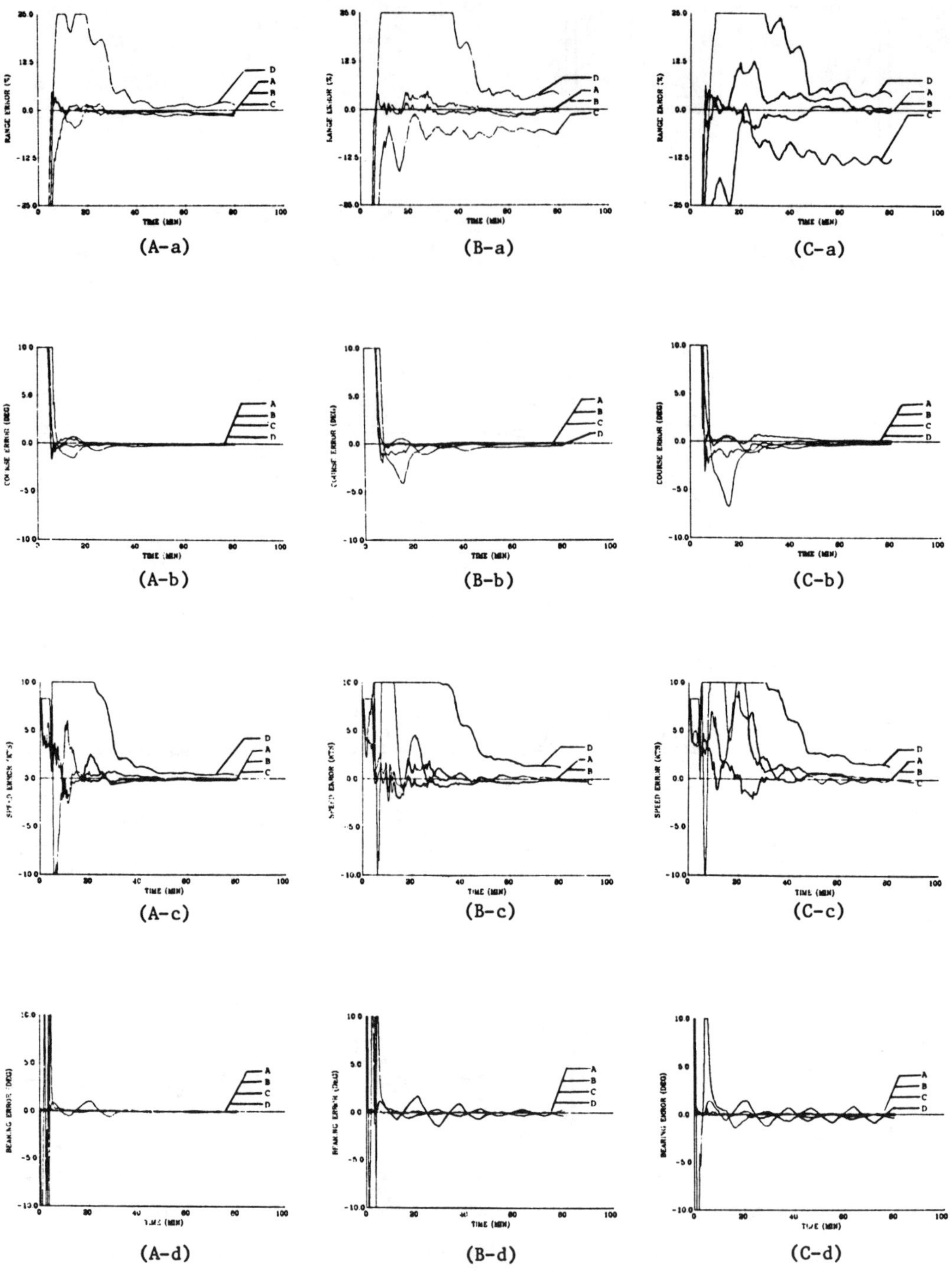

Fig. 4.

solution after erratic transient response. The unstable estimation characteristics of this filter are clearly evident.

V. Summary and Conclusions

We have attempted to elucidate the advantages of utilizing MP coordinates for bearings-only tracking via an extended Kalman filter. Exact state and measurement relations were rigorously derived for the general case of arbitrary vehicle motion. Subsequent analysis revealed that observable and unobservable components of the MP state vector are automatically decoupled prior to the first ownship maneuver. It was further shown that the covariance matrix structure accurately reflects this decoupling property. As a result, estimates generated by the MP filter are constrained to behave exactly as predicted by theory, even in the presence of measurement errors.

A realistic description of MP filter performance is provided by the data shown in Figs. 3 and 4. While only a small representative sampling of results is presented here, additional experiments have also been conducted elsewhere [24], [25] using both real and simulated measurements. A wide variety of geometries and noise levels were examined.

In all these tests, the MP filter behaved in a completely predictable manner. No evidence of instability was observed, and the state estimates were always asymptotically unbiased. Under actual operating conditions, it is important to recognize that filter performance will still be affected by such factors as own-ship tactics, environmental disturbances, measurement errors, etc.. However, estimation difficulties associated with bearings-only TMA unobservability have been eliminated.

The foregoing theoretical and experimental findings demonstrate that filter performance is intrinsically related to its development. Since other reference frames that permit a natural decoupling of observable and unobservable components also lead to more complicated equations of state and measurement, it is concluded that MP coordinates are ideally suited for bearings-only TMA.

Appendix A
Cartesian Coordinate Formulation of the Bearings-Only TMA Problem

Derivation of the State Equations

Consider the geometry depicted in Fig. 1, with target and own-ship confined to the same horizontal plane.[3] The Cartesian state vector for this two-dimensional configuration is defined by

$$x(t) \triangleq \begin{bmatrix} x_1(t) \\ x_2(t) \\ x_3(t) \\ x_4(t) \end{bmatrix} = \begin{bmatrix} v_x(t) \\ v_y(t) \\ r_x(t) \\ r_y(t) \end{bmatrix} \tag{A1}$$

where

$$\begin{bmatrix} v_x(t) \\ v_y(t) \end{bmatrix} = \begin{bmatrix} v_{tx}(t) - v_{ox}(t) \\ v_{ty}(t) - v_{oy}(t) \end{bmatrix} \tag{A2a}$$

and

$$\begin{bmatrix} r_x(t) \\ r_y(t) \end{bmatrix} = \begin{bmatrix} r_{tx}(t) - r_{ox}(t) \\ r_{ty}(t) - r_{oy}(t) \end{bmatrix} \tag{A2b}$$

denote *relative* target velocity and position, respectively. A mathematical model of the system dynamics may now be specified in general form via the linear differential equations of motion

$$\begin{bmatrix} \dot{x}_1(t) \\ \dot{x}_2(t) \\ \dot{x}_3(t) \\ \dot{x}_4(t) \end{bmatrix} = \begin{bmatrix} a_x(t) \\ a_y(t) \\ x_1(t) \\ x_2(t) \end{bmatrix} \tag{A3}$$

[3]The bearings-only TMA problem is typically formulated under this assumption.

where

$$\begin{bmatrix} a_x(t) \\ a_y(t) \end{bmatrix} = \begin{bmatrix} a_{tx}(t) - a_{ox}(t) \\ a_{ty}(t) - a_{oy}(t) \end{bmatrix} \tag{A4}$$

depicts *relative* acceleration. Integrating (A3) directly and expressing the result in matrix notation subsequently yields

$$x(t) = A_x(t, t_o)x(t_o) + w(t, t_o) \tag{A5}$$

where

$$A_x(t, t_o) \triangleq \begin{bmatrix} 1 & 0 & 0 & 0 \\ 0 & 0 & 0 & 0 \\ (t - t_o) & 0 & 1 & 0 \\ 0 & (t - t_o) & 0 & 1 \end{bmatrix} \tag{A6}$$

$$w(t, t_o) \triangleq \begin{bmatrix} w_1(t, t_o) \\ w_2(t, t_o) \\ w_3(t, t_o) \\ w_4(t, t_o) \end{bmatrix} = \begin{bmatrix} \int_{t_o}^{t} a_x(\lambda)\, d\lambda \\ \int_{t_o}^{t} a_y(\lambda)\, d\lambda \\ \int_{t_o}^{t} (t - \lambda) a_x(\lambda)\, d\lambda \\ \int_{t_o}^{t} (t - \lambda) a_y(\lambda)\, d\lambda \end{bmatrix} \tag{A7}$$

and t_o denotes any arbitrary fixed value of time.

Although (A5) remains valid for unconstrained vehicle motion, solution uniqueness requirements necessitate that the bearings-only TMA estimation problem be formulated under more restrictive assumptions [5], e.g., constant target velocity. In this case $a_{tx}(t) \equiv a_{ty}(t) \equiv 0$ and $w(t, t_o)$ reduces to a deterministic input vector which depends only upon the characteristics of own-ship acceleration. Specifically,

$$w(t, t_o) = -w_o(t, t_o) \quad \text{iff} \quad \begin{bmatrix} a_{tx}(t) \\ a_{ty}(t) \end{bmatrix} \equiv \mathbf{0} \tag{A8}$$

$$w_o(t, t_o) \triangleq \begin{bmatrix} w_{o1}(t, t_o) \\ w_{o2}(t, t_o) \\ w_{o3}(t, t_o) \\ w_{o4}(t, t_o) \end{bmatrix} = \begin{bmatrix} \int_{t_o}^{t} a_{ox}(\lambda)\, d\lambda \\ \int_{t_o}^{t} a_{oy}(\lambda)\, d\lambda \\ \int_{t_o}^{t} (t - \lambda) a_{ox}(\lambda)\, d\lambda \\ \int_{t_o}^{t} (t - \lambda) a_{oy}(\lambda)\, d\lambda \end{bmatrix}. \tag{A9}$$

The Cartesian state equations then take the familiar form

$$x(t) = A_x(t, t_o)x(t_o) - w_o(t, t_o). \tag{A10}$$

Derivation of the Measurement Equation

As the name implies, bearings-only TMA is distinguished by the fact that measured data consist entirely of passive sonar bearings extracted from a single sensor. Consequently, the measurement process is described by a time-varying scalar equation of the form (see Fig. 1)

$$\tilde{\beta}(t) = h_x[x(t)] + \eta(t) \tag{A11}$$

where

$$h_x[x(t)] = \tan^{-1}[x_3(t)/x_4(t)] \tag{A12}$$

and $\tilde{\beta}(t)$ represents the measured target bearing corrupted by additive measurement noise $\eta(t)$. If the pertinent acoustic sensor is properly aligned and calibrated, it may be realistically assumed that $\eta(t)$ is zero-mean Gaussian white noise with variance $\sigma^2(t)$, i.e.,

$$E[\eta(t)] = 0 \tag{A13a}$$

$$E[\eta(t)\eta(t+\lambda)] = \begin{cases} \sigma^2(t) & \lambda = 0 \\ 0 & \lambda \neq 0 \end{cases} \tag{A13b}$$

where $E[\cdot]$ denotes the statistical expectation operator.

Appendix B
Modified Polar Coordinate Formulation of the Bearings-Only TMA Problem

Derivation of the State Equations

Again referring to the geometric configuration depicted in Fig. 1, the MP state vector is defined by

$$y(t) \triangleq \begin{bmatrix} y_1(t) \\ y_2(t) \\ y_3(t) \\ y_4(t) \end{bmatrix} = \begin{bmatrix} \dot{\beta}(t) \\ \dot{r}(t)/r(t) \\ \beta(t) \\ 1/r(t) \end{bmatrix} \tag{B1}$$

where

$$r(t) = \|r(t)\| = \sqrt{r_x^2(t) + r_y^2(t)} \tag{B2a}$$

and

$$\beta(t) = \tan^{-1}[r_x(t)/r_y(t)] \tag{B2b}$$

represent the *relative* target range and bearing angle, respectively. In this coordinate system, the differential equations for arbitrary vehicle motion take the form

$$\begin{bmatrix} \dot{y}_1(t) \\ \dot{y}_2(t) \\ \dot{y}_3(t) \\ \dot{y}_4(t) \end{bmatrix} = \begin{bmatrix} -2y_1(t)y_2(t) + y_4(t)[a_x(t)\cos y_3(t) - a_y(t)\sin y_3(t)] \\ y_1^2(t) - y_2^2(t) + y_4(t)[a_x(t)\sin y_3(t) + a_y(t)\cos y_3(t)] \\ y_1(t) \\ -y_2(t)y_4(t) \end{bmatrix} \tag{B3}$$

where $a_x(t)$ and $a_y(t)$ are the Cartesian components of *relative* acceleration defined in Appendix A by (A4).

A comparison of (A3) to (B3) reveals that the MP equations of motion are considerably more complicated than their Cartesian counterparts and not readily amenable to direct integration. Despite this apparent difficulty, the *exact* general solutions of these differential equations can be obtained by straightforward algebraic manipulation. To this end, observe from Fig. 1 that

$$r_x(t) = r(t)\sin\beta(t) \tag{B4a}$$

$$r_y(t) = r(t)\cos\beta(t). \tag{B4b}$$

Differentiating these expressions with respect to time yields the familiar relations

$$v_x(t) = \dot{r}(t)\sin\beta(t) + r(t)\dot{\beta}(t)\cos\beta(t) \tag{B5a}$$

$$v_y(t) = \dot{r}(t)\cos\beta(t) - r(t)\dot{\beta}(t)\sin\beta(t). \tag{B5b}$$

A one-to-one transformation which maps the MP state vector into its Cartesian counterpart can now be deduced by combining (A1) with (B1), (B4), and (B5). The result is

$$x(t) = f_x[y(t)] = \frac{1}{y_4(t)} \begin{bmatrix} y_2(t)\sin y_3(t) + y_1(t)\cos y_3(t) \\ y_2(t)\cos y_3(t) - y_1(t)\sin y_3(t) \\ \sin y_3(t) \\ \cos y_3(t) \end{bmatrix}. \tag{B6}$$

Observe that (B6) is valid for all values of t such that $y_4(t) \neq 0$. Accordingly, letting $t = t_o$, this transformation may be applied to the right-hand side of (A5) to eliminate $x(t_o)$. Performing the required algebraic operations eventually yields

$$\begin{bmatrix} x_1(t) \\ x_2(t) \\ x_3(t) \\ x_4(t) \end{bmatrix} = \frac{1}{y_4(t_o)} \begin{bmatrix} S_1(t,t_o)\cos y_3(t_o) + S_2(t,t_o)\sin y_3(t_o) \\ S_2(t,t_o)\cos y_3(t_o) - S_1(t,t_o)\sin y_3(t_o) \\ S_3(t,t_o)\cos y_3(t_o) + S_4(t,t_o)\sin y_3(t_o) \\ S_4(t,t_o)\cos y_3(t_o) - S_3(t,t_o)\sin y_3(t_o) \end{bmatrix} \tag{B7}$$

where

$$S_1(t,t_o) = y_1(t_o) + y_4(t_o)[w_1(t,t_o)\cos y_3(t_o) - w_2(t,t_o)\sin y_3(t_o)] \tag{B8a}$$

$$S_2(t,t_o)=y_2(t_o)+y_4(t_o)[w_1(t,t_0)\sin y_3(t_o) + w_2(t,t_o)\cos y_3(t_o)] \quad \text{(B8b)}$$

$$S_3(t,t_o)=(t-t_o)y_1(t_o) + y_4(t_o)[w_3(t,t_o)\cos y_3(t_o) - w_4(t,t_o)\sin y_3(t_o)] \quad \text{(B8c)}$$

$$S_4(t,t_o)=1+(t-t_o)y_2(t_o) + y_4(t_o)[w_3(t,t_o)\sin y_3(t_o) + w_4(t,t_0)\cos y_3(t_o)] \quad \text{(B8d)}$$

and $[w_i(t,t_o);\ i=1,2,3,4]$ are defined in Appendix A by (A7).

The inverse transformation which maps Cartesian states into MP states may also be derived in a straightforward manner by differentiating (B2) with respect to time and then combining the resulting expressions with (A1), (A3), (B1), and (B2). These manipulations lead to

$$y(t)=f_y[x(t)] = \begin{bmatrix} [x_1(t)x_4(t)-x_2(t)x_3(t)]/[x_3^2(t)+x_4^2(t)] \\ [x_1(t)x_3(t)+x_2(t)x_4(t)]/[x_3^2(t)+x_4^2(t)] \\ \tan^{-1}[x_3(t)/x_4(t)] \\ 1/\sqrt{x_3^2(t)+x_4^2(t)} \end{bmatrix}. \quad \text{(B9)}$$

Substituting (B7) into (B9) finally yields, after some elementary algebra, the exact general solution to (B3), viz.

$$y(t)=f[y(t_o);t,t_o]=\begin{bmatrix} f_1[y(t_o);t,t_o] \\ f_2[y(t_o);t,t_o] \\ f_3[y(t_o);t,t_o] \\ f_4[y(t_o);t,t_o] \end{bmatrix} = \begin{bmatrix} [S_1(t,t_o)S_4(t,t_o)-S_2(t,t_o)S_3(t,t_o)]/[S_3^2(t,t_o)+S_4^2(t,t_o)] \\ [S_1(t,t_o)S_3(t,t_o)+S_2(t,t_o)S_4(t,t_o)]/[S_3^2(t,t_o)+S_4^2(t,t_o)] \\ y_3(t_o)+\tan^{-1}[S_3(t,t_o)/S_4(t,t_o)] \\ y_4(t_o)/\sqrt{S_3^2(t,t_o)+S_4^2(t,t_o)} \end{bmatrix}. \quad \text{(B10)}$$

Observe that (B10) is valid for *arbitrary* vehicle motion. Consequently, to obtain MP state equations for bearings-only TMA, components of the vector $w(t,t_o)$ appearing in (B8) must be replaced with those of $-w(t,t_o)$ before $[S_i(t,t_o);\ i=1,2,3,4]$ are substituted into the solution. This simple modification will incorporate the required motion constraint (i.e., constant target velocity) into the generalized expressions.

Derivation of the Measurement Equation

Since target bearing is a component of the MP state vector, the measurement equation for bearing-only TMA may be expressed in the simple linear form

$$\tilde{\beta}(t)=h_y[y(t)]+\eta(t) \quad \text{(B11)}$$

where

$$h_y[y(t)]=[0,0,1,0]\,y(t) \quad \text{(B12)}$$

and $\tilde{\beta}(t),\eta(t)$ are defined in Appendix A.

Although (B11) follows directly from geometric considerations (see Fig. 1), it can also be rigorously derived by combining (A1), (A11), and (A12) with (B6). Utilization of this latter procedure leads to the important functional relation

$$h_x[f_x[y(t)]]=h_y[y(t)]. \quad \text{(B13)}$$

Acknowledgment

The authors gratefully acknowledge K. F. Gong, A. F. Bessacini, and G. M. Hill of the Naval Underwater Systems Center for their technical contributions, and B. A. Clark for preparing the manuscript.

References

[1] W. H. Foy, "Position-location solutions by Taylor series estimation," *IEEE Trans. Aerosp. Electron. Syst.*, vol. AES-12, pp. 187–194, Mar. 1976.
[2] J. L. Poirot and G. V. McWilliams, "Navigation by back triangulation," *IEEE Trans. Aerosp. Electron. Syst.*, vol. AES-12, pp. 270–274, Mar. 1976.
[3] R. R. Tenney, R. S. Hebbert, and N. R. Sandell, Jr., "A tracking filter for maneuvering sources," *IEEE Trans. Automat. Contr.*, vol. AC-22, pp. 246–251, Apr. 1977.
[4] R. C. Kolb and F. H. Hollister, "Bearings-only target estimation," in *Proc. 1st Asilomar Conf. Circuits and Syst.*, 1967, pp. 935–946.
[5] D. J. Murphy, "Noisy bearings-only target motion analysis," Ph.D. dissertation Dep. Elec. Eng., Northeastern Univ., Boston, MA, 1970.
[6] R. W. Bass *et al.* "ASW target motion and measurement models," Computer Software Analysis, Inc., Tech. Rep. TR-72-024-01, Sept. 1972 (this document contains an extensive TMA bibliography).
[7] A. G. Lindgren and K. F. Gong, "Position and velocity estimation via bearing observations," *IEEE Trans. Aerosp. Electron. Syst.*, vol. AES-14, pp. 564–577, July 1978.
[8] V. Petridis, "A method for bearings-only velocity and position estimation," *IEEE Trans. Automat. Contr.*, vol. AC-26, pp. 488–493, Apr. 1981.
[9] D. W. Witcombe, "Pseudo-state measurements applied to recursive nonlinear filtering," in *Proc. 3rd Symp. Nonlinear Estimation Theory and Its Application*, 1972, pp. 278–281.
[10] V. J. Aidala, "Kalman filter behavior in bearings-only tracking applications," *IEEE Trans. Aerosp. Electron. Syst.*, vol. AES-15, pp. 29–39, Jan. 1979.
[11] A. G. Lindgren and K. F. Gong, "Properties of a bearings-only motion analysis estimator: An interesting case study in system observability," in *Proc. 12th Asilomar Conf. Circuits, Syst., and Comput.*, Nov. 1978, pp. 50–58.
[12] V. J. Aidala and S. C. Nardone, "Biased estimation properties of the pseudolinear tracking filter," *IEEE Trans. Aerosp. Electron. Syst.*, vol. AES-18, July 1982.
[13] S. C. Nardone and V. J. Aidala, "Observability criteria for bearings-only target motion analysis," *IEEE Trans. Aerosp. Electron. Syst.*, vol. AES-17, pp. 161–166, Mar. 1981.

[14] A. H. Jazwinski, *Stochastic Processes and Filtering Theory*. New York: Academic, 1970, pp. 272–281.
[15] R. K. Mehra, "A comparison of several nonlinear filters for reentry vehicle tracking," *IEEE Trans. Automat. Contr.*, vol. AC-16, pp. 307–319, Aug. 1971.
[16] G. W. Johnson, "Choice of coordinates and computational difficulty," *IEEE Trans. Automat. Contr.*, vol. AC-19, pp. 77–80, Feb. 1974.
[17] S. I. Chou, "Some drawbacks of extended Kalman filters in ASW passive angle tracking," presented at the ONR Conf. Advances in Passive Target Tracking, Naval Postgraduate School, Monterey, CA, May 1977, Rep. MPS-62 TS 77071, pp. 76–113.
[18] A. D. Cohen and G. W. Johnson, "A new approach to bearings-only ranging," presented at the ONR Conf. Advances in Passive Target Tracking, Naval Postgraduate School, Monterey, CA, May 1977, Rep. NPS-63TS 77-71, pp. 114–133.
[19] H. Weiss and J. B. Moore, "Improved extended Kalman filter design for passive tracking," *IEEE Trans. Automat. Contr.*, vol. AC-25, pp. 807–811, Aug. 1980.
[20] H. D. Hoelzer, G. W. Johnson, and A. O. Cohen, "Modified polar coordinates—The key to well behaved bearings-only ranging," IBM Shipboard and Defense Systems, Manassas, VA, IBM Rep. 78-M19-0001A, Aug. 1978.
[21] G. W. Johnson, H. D. Hoelzer, A. O. Cohen, and E. F. Harrold, "Improved coordinates for target tracking from time delay information," in *Proc. of the Time Delay Estimation and Applications Conf.*, Naval Postgraduate School, Monterey, CA, May 1979, vol. 2, pp. M1–M32.
[22] J. H. Taylor, "The Cramer–Rao estimation error lower bound computation for deterministic nonlinear systems," *IEEE Trans. Automat. Contr.*, vol. AC-24, pp. 343–344, Apr. 1979.
[23] T. O. Lewis and P. L. Odell, *Estimation in Linear Models*. Englewood Cliffs, NJ: Prentice-Hall, 1971, p. 14.
[24] D. E. Ohlms, G. W. Johnson, and H. D. Hoelzer, "KAST performance comparisons on RANGEX sea data," IBM Shipboard and Defense Systems, Manassas, VA, IBM Rep. 80-M19-0099, Mar. 1980.
[25] G. C. Street, A. O. Cohen, and G. W. Johnson, "Evaluation of an integrated target tracking algorithm," IBM Shipboard and Defense Systems, Manassas, VA, IBM Rep. 80-M19-0102, Sept. 1980.

Estimation and Prediction for Maneuvering Target Trajectories

RUSSELL F. BERG, MEMBER, IEEE

Abstract—**The Kalman filter is well suited for application to the problem of anti-aircraft gun fire control. In this paper we make use of the Kalman filter theory to develop an accurate, numerically efficient scheme for estimating and predicting the present and future position of maneuvering fixed-wing aircraft. This scheme was implemented in a radar tracker gun fire control system and tested against a variety of fixed-wing aircraft targets. Actual field test results are presented to demonstrate the high accuracy pointing which can be achieved by this approach.**

Manuscript received February 24, 1982; revised August 17, 1982.
The author is with the General Dynamics Corporation, Pomona, CA 91766.

I. Introduction

THE classical problem of anti-aircraft gun fire control is the accurate prediction of the future position of a given target at the time of projectile intercept. Having obtained this information, the correct gun-pointing angles can be ascertained. Current approaches to the solution of this problem typically employ the use of modern estimation techniques (Kalman filtering) to estimate target velocity and acceleration on the basis of target position

Reprinted from *IEEE Trans. Automat. Contr.*, vol. AC-28, pp. 294–304, Mar. 1983.

measurements [1]–[3]. Once the target's velocity and acceleration estimates have been obtained, the prediction of the target's future position (P_{fx}, P_{fy}, P_{fz}) is generally accomplished via the equations

$$\hat{P}_{fx} = \hat{P}_x + \hat{V}_x t_f + \frac{1}{2}\hat{A}_x t_f^2$$
$$\hat{P}_{fy} = \hat{P}_y + \hat{V}_y t_f + \frac{1}{2}\hat{A}_y t_f^2$$
$$\hat{P}_{fz} = \hat{P}_z + \hat{V}_z t_f + \frac{1}{2}\hat{A}_z t_f^2 \quad (1)$$

where t_f is the projectile time-of-flight and $(\hat{P}_x, \hat{P}_y, \hat{P}_z)$, $(\hat{V}_x, \hat{V}_y, \hat{V}_z)$, $(\hat{A}_x, \hat{A}_y, \hat{A}_z)$ are the current Cartesian coordinate estimates of the target's position, velocity, and acceleration, respectively. The obvious assumption here is that the target's linear acceleration is constant in amplitude and direction over a projectile time-of-flight interval. In this paper a new numerically efficient scheme is described for estimating and predicting the future position of maneuvering fixed-wing aircraft. The scheme was implemented in the fire control computer of a radar tracking gun system which was successfully tested and evaluated against a variety of fixed-wing aircraft. The estimation and prediction algorithms were based on the use of Kalman filter theory and on the estimated aircraft related parameters thrust T (along the aircraft longitudinal axis), lift acceleration L (normal to the aircraft wing plane), and roll rate P. In particular, the following system modeling assumptions (referred to as a coordinated turn) were employed:

1) mean target lift acceleration L is constant

2) mean target thrust T (i.e., thrust minus drag) is constant

3) mean target roll rate P is zero.

This set of assumptions defines a planar maneuver (in the mean sense) which is consistent with the bank-to-turn flight characteristics of fixed-wing aircraft. As such, this modeling approach provides improved estimation and prediction accuracy against maneuvering high "g" targets.

In the interest of numerical efficiency, a novel adaptive Kalman gain calculation was devised to minimize computational burden without compromising target state estimation performance. This approach was developed to replace the on-line implementation of Kalman's covariance propagation equations (or the matrix factorization equations described in [9]). This was required in order to meet the imposed fire control computer memory and execution time constraints. It was observed that pseudosteady-state Kalman filter gains could be computed as simple functions of the ratio Q/R [10].

This technique for computing the adaptive Kalman gains avoids the numerical stability problems associated with propagating Kalman's covariance equations on line, requires less computer memory to implement, and most importantly, it requires an order of magnitude less computer execution time. These results were achieved without any apparent loss in filter estimation accuracy.

The prediction of future target position (one projectile time-of-flight into the future) in terms of the estimated target-related parameters T, L, and P required the solution of a set of coupled nonlinear differential equations. An approximate closed form solution for these equations was derived which extends (1) to the fifth power of t_f. The series coefficients are computed as a function of the estimated target velocity and acceleration.

Field test data are presented to illustrate the target acceleration estimation accuracy which was achieved against high "g" maneuvering targets using a radar tracker and the fire control system software described earlier. The field test data are compared to detailed simulation results to demonstrate the validation of the simulation. Finally, validated simulation results are presented to illustrate the pointing performance improvement which can be obtained by using the Kalman filter coordinated turn estimation and prediction approach.

II. Fire Control System Description

A signal flow diagram of the automated gun fire control system is shown in Fig. 1. A monopulse radar is used to provide measurements of target position (target range and target angle tracking errors) to the fire control computer. The primary functions of the fire control computer are to

1) maintain accurate target tracking by aligning the antenna boresight with the target line-of-sight and

2) compute accurate estimates of the future target position (at the time the projectile arrives at the target) in order to determine the correct gun-pointing angles.

Fig. 2 illustrates the various algorithms which are resident in the fire control computer. In this paper the discussion is restricted to the target state estimator and predictor algorithms in order to illustrate the application of Kalman filter theory to the solution of the gun fire control problem. A complete discussion of the gun fire problem may be found in [8].

III. Target State Estimator Equations

The function of the target state estimator (TSE) is to develop estimates of the target's current position, velocity, and acceleration (target state) by processing the noisy target position measurements from the tracking radar. The monopulse tracking radar provides a measurement of target range (R) and two orthogonal components of target tracking error (ϵ_y, ϵ_z) relative to the track antenna boresight. Hence, if the antenna Cartesian coordinate system x, y, z is defined such that the x axis is oriented along the antenna boresight, then the y and z axes are oriented such that ϵ_y is the measured target tracking error about the antenna y axis and ϵ_z is the measured target tracking error about the antenna z-axis. The measured components of current target position referenced to antenna coordinates are thus given by

$$\begin{bmatrix} P_{x_m} \\ P_{y_m} \\ P_{z_m} \end{bmatrix} = \begin{bmatrix} R \\ R\epsilon_z \\ -R\epsilon_y \end{bmatrix}. \quad (2)$$

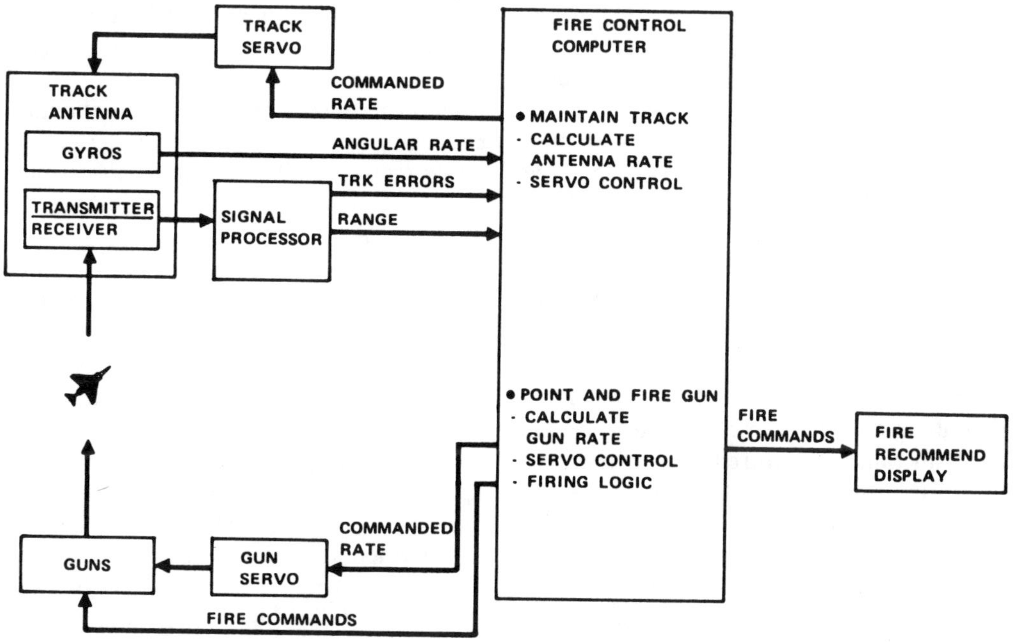

Fig. 1. Fire control system signal flow.

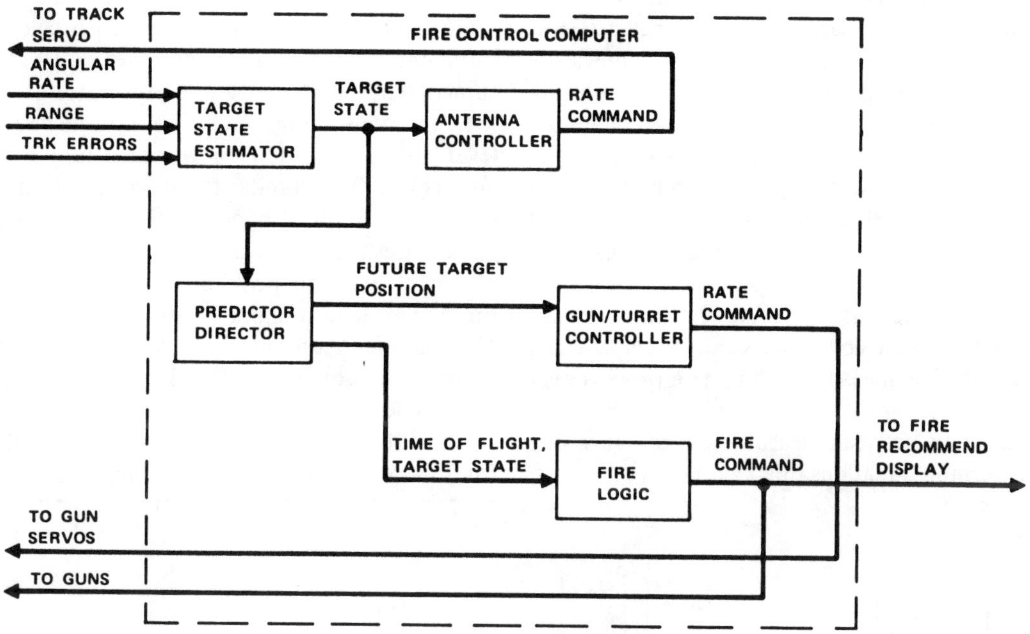

Fig. 2. Fire control software algorithms.

Rewriting (2) such that the noise terms are separated yields

$$\begin{bmatrix} P_{x_m} \\ P_{y_m} \\ P_{z_m} \end{bmatrix} = \begin{bmatrix} P_x \\ P_y \\ P_z \end{bmatrix} + \begin{bmatrix} \eta_R \\ R\eta_{\epsilon y} \\ R\eta_{\epsilon z} \end{bmatrix} \tag{3}$$

where the noise terms $\eta_R, \eta_{\epsilon y}, \eta_{\epsilon z}$ represent, respectively, the target range and tracking error measurement uncertainties. The y and z channel range noise effects are negligible. Equation (3) thus defines the system measurement equations.

The basis of the target dynamics model is the target acceleration model. Following the approach given in [7], target acceleration is modeled as a zero-mean time-correlated random variable with correlation time τ. However, an additional term $\bar{\dot{A}}$ has been added to the Singer model to enhance maneuvering target estimation performance. The target acceleration model takes the form

$$\dot{A} = \frac{-A + W}{\tau} + \bar{\dot{A}}. \tag{4}$$

The term $\bar{\dot{A}}$ is computed as a function of the most recent estimates of target velocity and acceleration. It represents an adaptive estimate of the mean target jerk in terms of the estimated target related parameters of lift, thrust, and roll rate (L, T, and P are computed from the estimates of target velocity and acceleration; see Appendix A). In particular, it is assumed that target lift rate $\dot{L}$, thrust rate $\dot{T}$, and roll rate P are zero mean. These assumptions define the coordinated turn maneuver described in Section I.

In [4] Scheder defined a Kalman filter estimation algorithm based on the above assumptions regarding target maneuver characteristics. However, because of his choice of coordinate basis, a singularity problem was encountered for targets flying straight and level trajectories. By defining an appropriate change in coordinate basis for this work, the singularity problem was eliminated. The derivation of the $\dot{A}$ term is contained in Appendix A.

The target acceleration model defined in (4) may be thought of as describing a trajectory whose mean path is a coordinated turn with random perturbations to this path supplied by the noise term W. Equation (4) is used to model the target acceleration along each of the antenna coordinate axes with the inherent assumption that the antenna frame is instantaneously fixed (nonrotating with respect to inertial space). Obviously, it is not possible to instantaneously stop the track antenna from rotating during target tracking, but it is possible to process the filtering equations mathematically as though the antenna were moving incrementally (at the measurement update rate 0.036 s) from one antenna orientation to the next as it follows the target [5]. That is, the target position measurements are processed by the Kalman filter algorithm as though the antenna (x, y, z) frame were fixed (nonrotating) in the position it held when the target position measurements were last obtained. The TSE computes target velocity and acceleration estimates relative to this assumed instantaneously fixed coordinate frame. To process a new set of measurements, it is necessary to rotate the current estimates and estimation error covariances into the new track antenna orientation. The measurement update process may then be repeated. To compute the change in track antenna coordinate orientation between measurement updates, one can use the quaternion equations [6]. That is,

$$\begin{bmatrix} \dot{A} \\ \dot{B} \\ \dot{C} \\ \dot{D} \end{bmatrix} = \frac{1}{2} \begin{bmatrix} D & -C & B \\ C & D & -A \\ -B & A & D \\ -A & -B & -C \end{bmatrix} \cdot \begin{bmatrix} \omega_{AX} \\ \omega_{AY} \\ \omega_{AZ} \end{bmatrix}$$

$$T_{S_2S_1} = \begin{bmatrix} (A^2 - B^2 - C^2 + D^2) & 2(AB + CD) & 2(AC - BD) \\ 2(AB - CD) & (-A^2 + B^2 - C^2 + D^2) & 2(BC + AD) \\ 2(AC + BD) & 2(BC - AD) & (-A^2 - B^2 + C^2 + D^2) \end{bmatrix} \tag{5}$$

where

A, B, C, D = quaternion parameters

$\omega_{AX}, \omega_{AY}, \omega_{AZ}$ = track antenna measured spatial rate components (x, y, z)

$T_{S_2S_1}$ = transformation from the antenna coordinates at time 1 to the antenna coordinates at time 2.

Several key points are worth noting here. This approach linearizes the target dynamics and measurement equations and decouples a nine-state filter into three three-state filters. This greatly reduces computational complexity, while enhancing performance. The quaternion equations may be processed in parallel with the Kalman filter equations.

IV. Estimator Gain Calculations

For this application the estimator Kalman gains must be calculated on line since the model noise covariance "Q" [variance of W in (4)] and the measurement noise covariance "R" [variance of the noise terms for (3)] are time-varying functions which are not known *a priori*. The "Q" and "R" terms are computed adaptively as a function of the most recent estimates of target state. For this application "Q" was computed on line as a function of estimated target orientation and expected maneuver level [4]. The target orientation is described by the matrix transformation "T_{TA}" derived in Appendix A. Target frame acceleration level statistics are estimated *a priori* on the basis of anticipated target maneuver capabilities. "R" was computed on line as a function of target aspect and expected radar glint noise level. Target aspect is calculated from the estimated target position and velocity vectors. It is assumed that the target's velocity vector is closely aligned with its longitudinal axis. The expected radar glint noise level [12] is computed *a priori* as a function of target size and aspect. The standard Kalman filter equations may be computed on line to derive the necessary Kalman filter gains. However, the Kalman filter covariance propagation algorithm has the problems of numerical instability (on finite word length digital computers) and time inefficiency. More efficient computational algorithms have been devised such as the square root and UDU^T matrix factorization approaches [9].

These approaches were found to be too time consuming for this application. An alternate method for calculating the adaptive Kalman gains was developed. The approach was straightforward. Simulation results demonstrated the fact that for the expected target trajectories and engagement ranges, the incremental transformation of the estimation error covariance matrices (described in Section III) between measurement updates played a negligible role in the determination of the filter gains. Since "Q" and "R"

were slowly varying functions of time (relative to the measurement update rate), this meant that the steady-state filter gains could be closely approximated by steady-state gain values computed strictly as a function of "Q" and "R" for each channel.

Each three-state filter operates on a scalar measurement of target position. Namely, the x-channel filter operates on the measurement of target range. The y- and z-channel filters operate on the track antenna off-boresight measurements of target position. For each channel the target model noise covariance "Q" is a scalar representing the statistics of the expected target acceleration level.

Fitzgerald [10] has shown that for a continuous measurement update rate and [see (4)] large τ, the steady-state filter gains are

$$\begin{aligned} &\text{(position gain)} && K_p = 2\,(Q/R)^{1/6} \\ &\text{(velocity gain)} && K_v = 2\,(Q/R)^{1/3} \\ &\text{(acceleration gain)} && K_a = (Q/R)^{1/2}. \end{aligned} \tag{6}$$

Hence, one would expect that for each channel of the filter, the discrete steady-state filter gains may be fitted to an equation set of the form (for a given τ and measurement update rate)

$$\begin{aligned} \log(K_p) &= A_1 + B_1 * \log(Q/R) \\ \log(K_v) &= A_2 + B_2 * \log(Q/R) \\ \log(K_a) &= A_3 + B_3 * \log(Q/R). \end{aligned} \tag{7}$$

For each channel, steady-state Kalman gains were computed and tabulated (using Kalman's equations) as a function of the target model noise (Q) and the measurement noise (R). Using these data, the "A" and "B" coefficient values in (7) were selected so as to best match the steady-state gains obtained by using Kalman's equations. Simulation results are illustrated for the pop-turn-and-dive target trajectory shown in Fig. 3, with

$$\begin{aligned} A1 &= -1.1 & B1 &= 0.135 \\ A2 &= -1.12 & B2 &= 0.309 \\ A3 &= -1.42 & B3 &= 0.473. \end{aligned}$$

The y-channel Kalman gains computed via Kalman's covariance propagation equations are compared to the results obtained by using (7) in Figs. 4–6. A very close match is obtained when the filter reaches steady state (time > 3 s). During the first 3 s, a set of preprogrammed gains are used (not shown) to match the Kalman covariance propagation equation results. The gains are based on nominal initial expected values of "Q" and "R" and on the initial error covariance values. These preprogrammed initialization gains closely match the results obtained by using Kalman's equations. Thus, a minimum TSE settling time is maintained with much less computational complexity. The computed (on-line) time-varying values of Qy and Ry used to compute the gains are plotted in Figs. 7 and 8, respectively. The Kalman gain peaks which occur at 20.5 s are due to the relatively small value of R_y which occurs at that instant in time. This corresponds to the momentary period of time when the target displays a "nose-on" aspect and the radar glint tracking noise level is minimal.

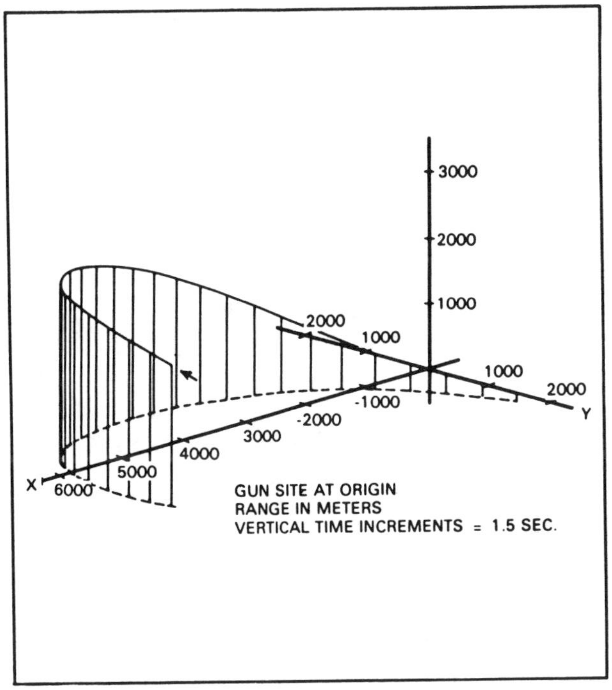

Fig. 3. Pop-turn-and-dive aircraft trajectory.

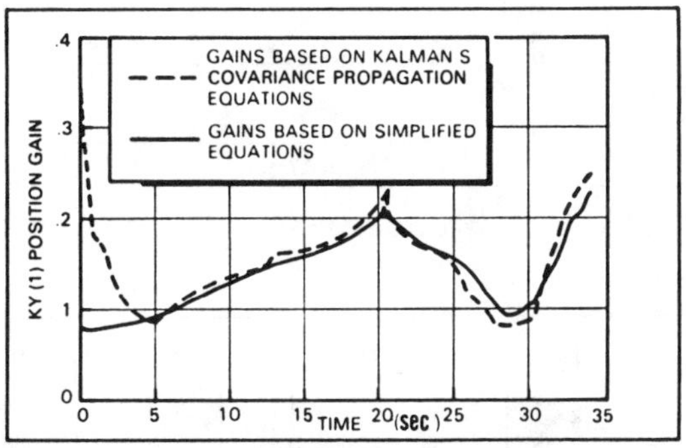

Fig. 4. Y-channel position gains of the Kalman filter.

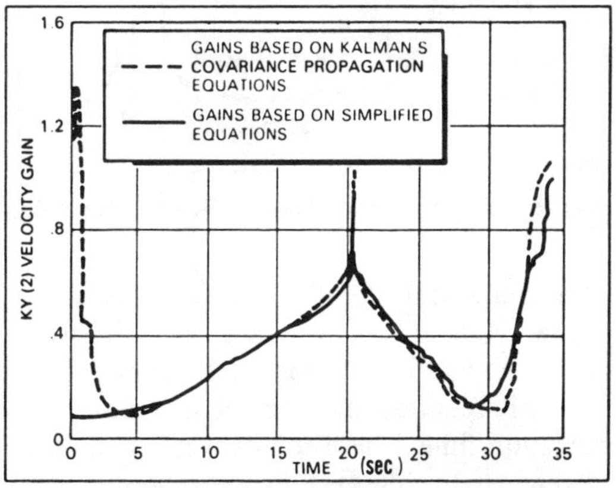

Fig. 5. Y-channel velocity gains of the Kalman filter.

This approach was implemented in the fire control computer with simple one-dimensional table lookup routines. The technique for computing the adaptive Kalman gains

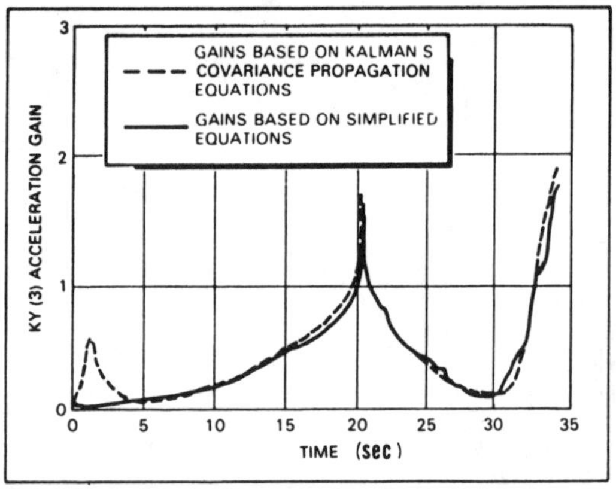

Fig. 6. Y-channel acceleration gains of the Kalman filter.

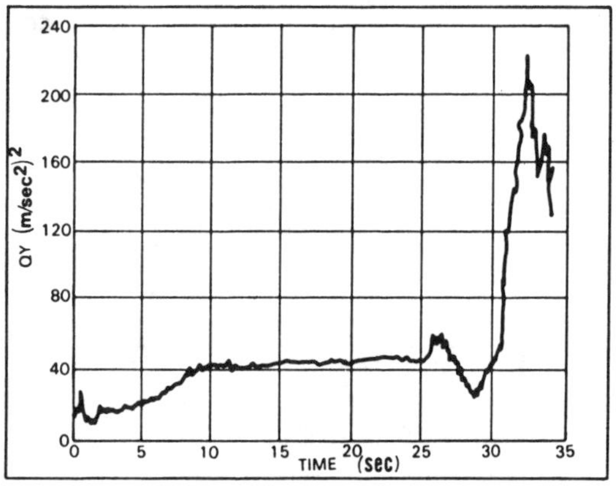

Fig. 7. Y-channel model noise variance (Q_Y).

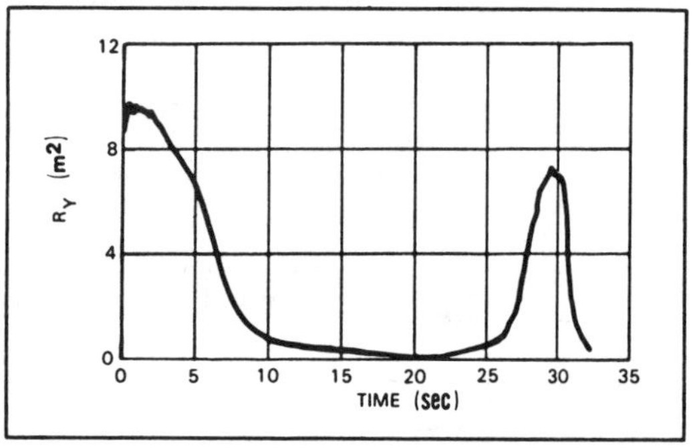

Fig. 8. Y-channel measurement noise variance (R_Y).

avoids the numerical stability problems associated with propagating Kalman's covariance equations on line, requires less computer memory to implement, and most importantly, it requires an order of magnitude less computer execution time. These results were achieved without any apparent loss in filter estimation accuracy.

V. Predictor Equations

Kalman's prediction theory states that the most recent estimates of system state should be used together with the system dynamics model to compute the best estimate of future system state. This is the approach that is used in this

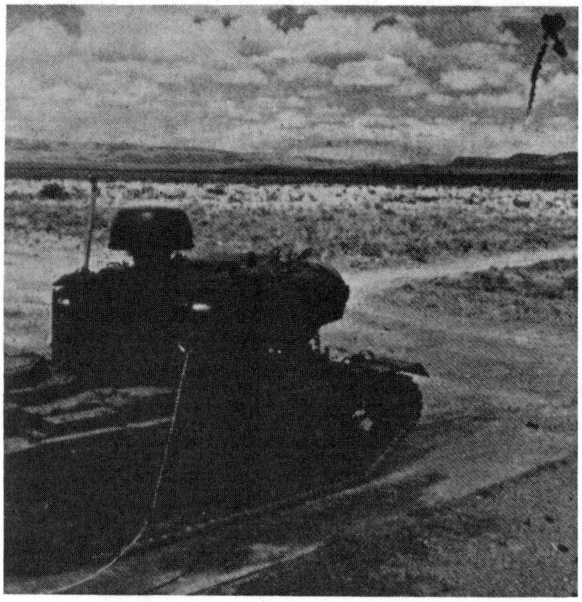

Fig. 9. Prototype radar tracker gun system.

paper. In this case it is necessary to compute future target position at the time of bullet intercept in order to accurately point the gun. The TSE provides a current estimate of target velocity and acceleration in antenna coordinates. Assuming that the bullet time-of-flight has been computed [8], and that the target acceleration is constant in magnitude and direction over the bullet time-of-flight interval, then (1) could be used to predict the future target position. However, a new prediction algorithm has been developed for the fixed-wing aircraft target which more realistically accounts for target maneuvers. It is, of course, based on the coordinated turn (consistent with the estimation algorithm term $\dot{A}$) assumptions that target thrust rate $\dot{T}$, lift rate $\dot{L}$, and roll rate P are zero over the prediction interval. This prediction technique more realistically represents fixed-wing aircraft maneuver characteristics since it is described in target related parameters. The derivation of the coordinated turn predictor equations is extremely tedious since it requires the solution of a set of coupled nonlinear differential equations [see (A12)]. An approximate closed form solution of these equations was derived which extends (1) to the fifth power of t_f. The reader is referred to Appendix B for a derivation of these equations. The predictor equation summary is also given in Appendix B, and a performance comparison of the coordinated turn predictor versus the predictor of (1) is presented in Section VI.

VI. Field Test Results

The estimation and prediction equations described in the previous sections were implemented in the radar tracker automated gun system pictured in Fig. 9. A CDC 469F digital computer was used to process the fire control equations. The system was tested against a variety of fixed-wing aircraft to determine its potential gun-pointing accuracy. An independent radar tracking system was used to provide

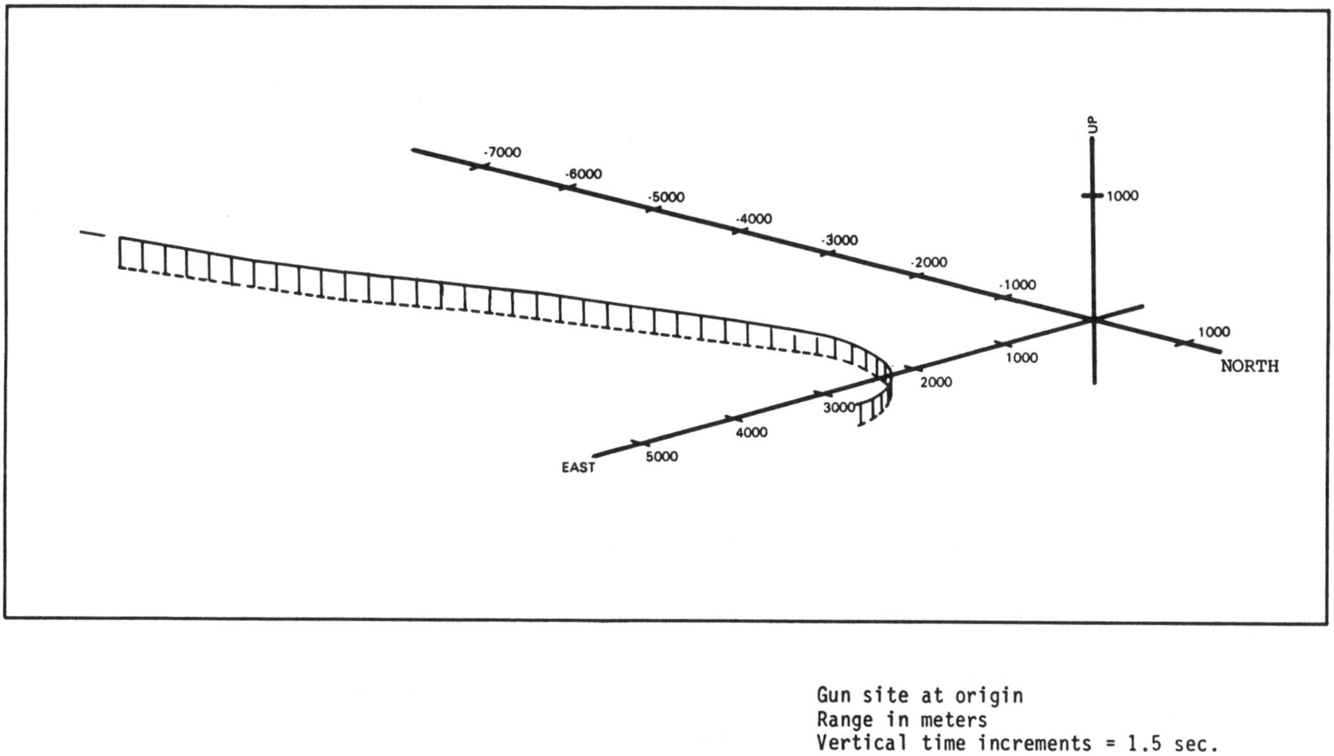

Fig. 10. 3 "g" turn trajectory.

a second source of measured target position information. The second source target position data were used in a double sweep smoother postprocessing algorithm [11] to obtain high accuracy calculations of the target's position, velocity, and acceleration, which were used for subsequent simulation validation and performance studies. These post-processing calculations of acceleration and velocity closely represent the actual target acceleration and velocity. The 3-g turn trajectory of a Jet Drone Test Target is shown in Fig. 10. This plot was made from the postprocessing results from the smoother. Fig. 11 illustrates the recorded target acceleration estimates which were generated by the radar tracker gun system fire control algorithms (discussed in Sections III and IV) against the 3-g turn target. The accuracy of these acceleration estimates was measured by comparing the recorded results to the independent post-processing target acceleration data generated by the smoother. These data are also plotted in Fig. 11 for comparison purposes. The higher disturbance levels in the elevation channel during the first 10 s of flight were due to the effects of radar multipathing [12]. The target was acquired at a relatively long range and low altitude.

A digital computer simulation of the gun system was constructed for the purpose of making system performance predictions. Using the postprocessing data to represent the target trajectory, it was possible to exercise the simulation and directly compare the field test results to the simulation results. Exercising the simulation in a Monte Carlo fashion, one is able to obtain the 5 and 95 percent performance bounds of any system parameters. That is, if the simulation is a good model of the actual system and its environment,

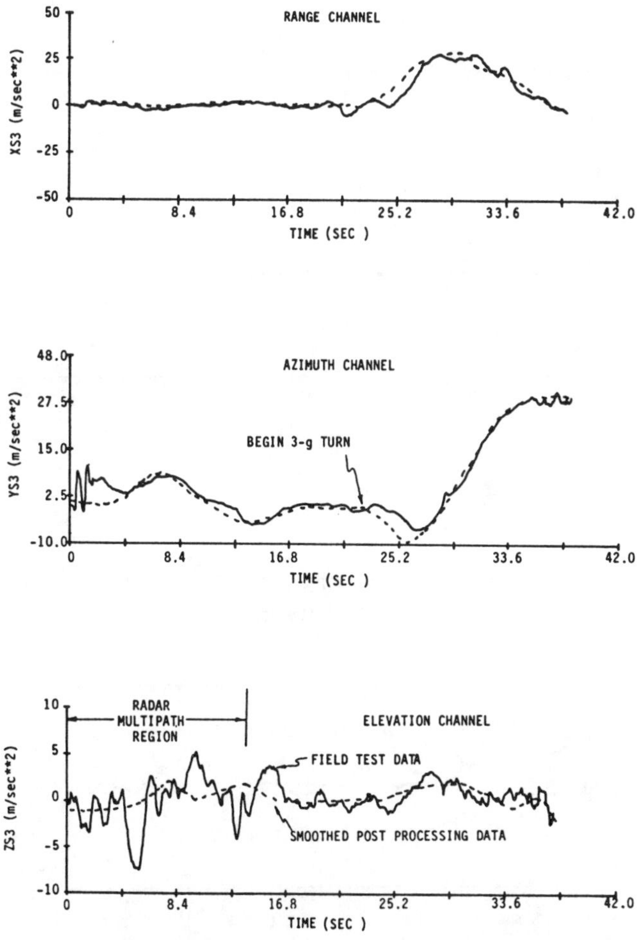

Fig. 11. Field test acceleration estimation data.

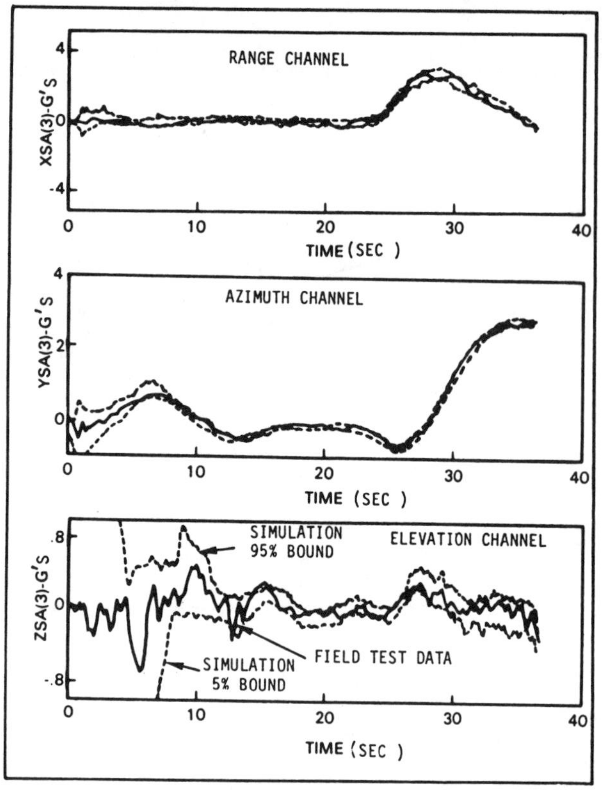

Fig. 12. Simulation validation data (data are measured in g's).

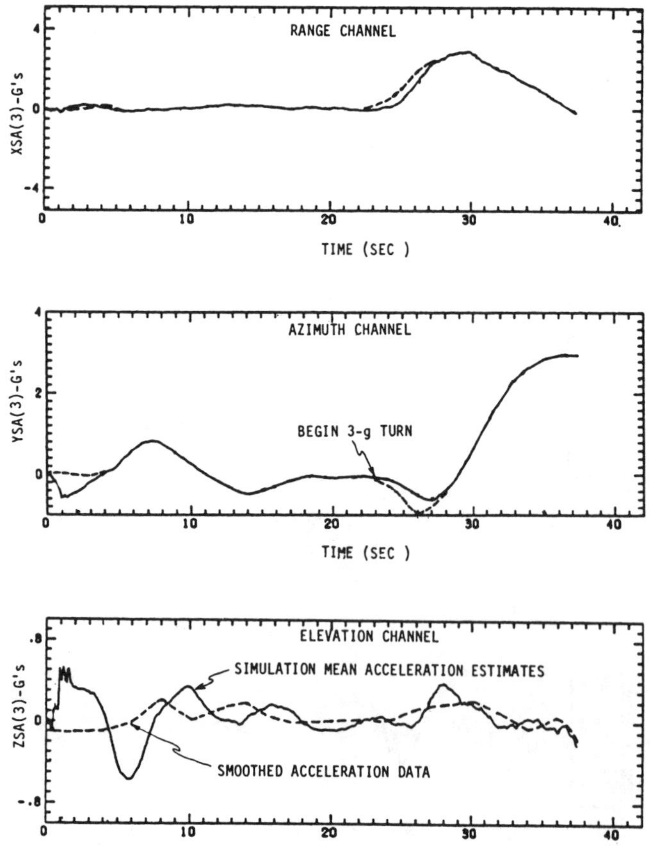

Fig. 13. Validated simulation mean acceleration estimates (all data in g's).

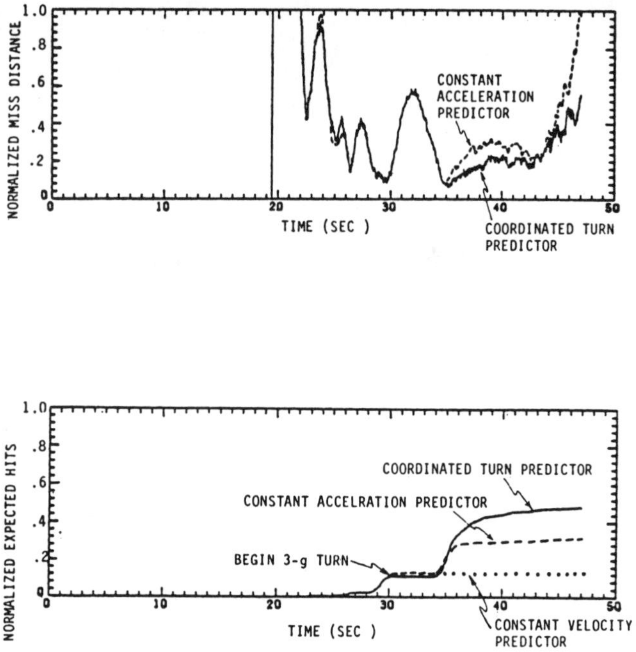

Fig. 14. Predictor performance comparisons (normalized data).

one would expect the field test data to be well contained within these performance bounds 90 percent of the time. This concept is demonstrated in Fig. 12 for the acceleration estimation parameters. Key system parameters were studied in this fashion and the simulation was validated.

Using the validated simulation, various performance studies were conducted. Fig. 13 illustrates the simulation derived Monte Carlo mean acceleration estimates generated by the Kalman filter coordinated turn estimation algorithms. This is compared directly to the smoothed target acceleration data which represent the actual target acceleration.

Fig. 14 illustrates the normalized expected hits and normalized mean miss distance data achieved against the 3-g turn target. The data are based on Monte Carlo simulation results. The coordinated turn predictor achieves nearly twice the performance of the constant acceleration predictor during the interval of time when the target is maneuvering. In all cases, the target state estimates are provided by the coordinated turn TSE so that only the prediction algorithms are being compared directly.

VII. Conclusions

The approach presented in this paper is a state-of-the-art representation of the application of Kalman filtering theory to the solution of the gun fire control problem. It presents the system designer with an approach to dealing with maneuvering targets which was previously unavailable with classical constant velocity prediction gun fire control. The coordinated turn estimator and predictor target model serve to further enhance the capability of the fire control system against fixed-wing maneuvering targets. The numerically efficient estimator gain calculation method of Section IV greatly reduced the computational complexity associated with computing Kalman's covariance propagation equations on line. The performance of this approach has been proven in recent field test demonstrations.

Appendix A
Derivation of the Term $\bar{A}$

Define the target coordinate system $\vec{X}_T, \vec{Y}_T, \vec{Z}_T$ with $\vec{X}_T$ oriented along the target velocity vector, $\vec{Y}_T$ normal to the wing plane, and $\vec{Z}_T$ along the right wing. The forces acting on the target are

$\vec{T}$ = thrust minus drag along $\vec{X}_T$

$\vec{L}$ = lift along $\vec{Y}_T$

$\vec{g}$ = gravity acting along the earth normal axis.

A transformation from the antenna to the target frame T_{TA} is given by

$$T_{TA} = \begin{bmatrix} \vec{X}_T \cdot \vec{X} & \vec{X}_T \cdot \vec{Y} & \vec{X}_T \cdot \vec{Z} \\ \vec{Y}_T \cdot \vec{X} & \vec{Y}_T \cdot \vec{Y} & \vec{Y}_T \cdot \vec{Z} \\ \vec{Z}_T \cdot \vec{X} & \vec{Z}_T \cdot \vec{Y} & \vec{Z}_T \cdot \vec{Z} \end{bmatrix} \tag{A1}$$

where $\vec{X}, \vec{Y}, \vec{Z}$ define the antenna coordinate system. Since the TSE estimates $(\hat{V}_X, \hat{V}_Y, \hat{V}_Z)(\hat{A}_X, \hat{A}_Y, \hat{A}_Z)$ are computed relative to antenna coordinates,

$$\vec{X}_T = \frac{\hat{V}_x}{|V|}\vec{X} + \frac{\hat{V}_y}{|V|}\vec{Y} + \frac{\hat{V}_z}{|V|}\vec{Z}. \tag{A2}$$

Hence,

$$\vec{X}_T = \begin{bmatrix} T_{TA}^{11} \\ T_{TA}^{12} \\ T_{TA}^{13} \end{bmatrix} = \frac{1}{|V|} \cdot \begin{bmatrix} \hat{V}_x \\ \hat{V}_y \\ \hat{V}_z \end{bmatrix}. \tag{A3}$$

The components of target lift acceleration in antenna coordinates (normal to the wing plane) are

$$\begin{bmatrix} L_x \\ L_y \\ L_z \end{bmatrix} = \begin{bmatrix} \hat{A}_x - g_x \\ \hat{A}_y - g_y \\ \hat{A}_z - g_z \end{bmatrix} - \left(\dot{V} - g_{TAN}\right) \cdot \begin{bmatrix} T_{TA}^{11} \\ T_{TA}^{12} \\ T_{TA}^{13} \end{bmatrix} \tag{A4}$$

where

$$\dot{V} = \begin{bmatrix} \hat{A}_x & \hat{A}_y & \hat{A}_z \end{bmatrix} \cdot \begin{bmatrix} T_{TA}^{11} \\ T_{TA}^{12} \\ T_{TA}^{13} \end{bmatrix}$$

$$g_{TAN} = \begin{bmatrix} g_x & g_y & g_z \end{bmatrix} \cdot \begin{bmatrix} T_{TA}^{11} \\ T_{TA}^{12} \\ T_{TA}^{13} \end{bmatrix}$$

and g_x, g_y, g_z are the measured components of gravity in antenna coordinates. Then

$$\vec{Y}_T = \frac{1}{|L|} \cdot \begin{bmatrix} L_x \\ L_y \\ L_z \end{bmatrix} = \begin{bmatrix} T_{TA}^{21} \\ T_{TA}^{22} \\ T_{TA}^{33} \end{bmatrix} \tag{A5}$$

and

$$\begin{bmatrix} T_{TA}^{31} \\ T_{TA}^{32} \\ T_{TA}^{33} \end{bmatrix} = \vec{Z}_T = \vec{X}_T \; X \; \vec{Y}_T. \tag{A6}$$

The T_{TA} matrix may now be used to derive $\bar{A}$. The components of target acceleration in target coordinates are

$$\vec{A} = \begin{bmatrix} \vec{V} \\ 0 \\ 0 \end{bmatrix} + \begin{bmatrix} 0 & -r & q \\ r & 0 & -p \\ -q & p & 0 \end{bmatrix} \cdot \begin{bmatrix} V \\ 0 \\ 0 \end{bmatrix} = \begin{bmatrix} \dot{V} \\ Vr \\ -Vq \end{bmatrix} \tag{A7}$$

where p, q, r are the components of target inertial angular velocity referenced to the target frame. The components of force per unit mass in target coordinates are

$$\frac{\vec{F}}{m} = \begin{bmatrix} T \\ L \\ 0 \end{bmatrix} + \begin{bmatrix} g_{x_T} \\ g_{y_T} \\ g_{z_T} \end{bmatrix} \tag{A8}$$

where

$$\begin{bmatrix} g_{x_T} \\ g_{y_T} \\ g_{z_t} \end{bmatrix} = T_{TA} \cdot \begin{bmatrix} g_x \\ g_y \\ g_z \end{bmatrix}.$$

Since

$$\vec{A} = \frac{\vec{F}}{m}, \tag{A9}$$

then

$$T = \dot{V} - g_{x_T}$$

$$r = \frac{L + g_{y_T}}{V}$$

$$q = \frac{-g_{zt}}{V}. \tag{A10}$$

Finally, the equations for target jerk in antenna coordinates are (since $\dot{\vec{g}} = 0$)

$$\begin{bmatrix} \dot{A}_x \\ \dot{A}_y \\ \dot{A}_z \end{bmatrix} = T_{TA}^T \cdot \left[\begin{bmatrix} \dot{T} \\ \dot{L} \\ 0 \end{bmatrix} + \begin{bmatrix} 0 & -r & q \\ r & 0 & -p \\ -q & p & 0 \end{bmatrix} \cdot \begin{bmatrix} T \\ L \\ 0 \end{bmatrix} \right]. \tag{A11}$$

Using the coordinated turn assumption (i.e., $\dot{\bar{T}}, \dot{\bar{L}}, \bar{p}$ are zero) yields

$$\begin{bmatrix} \bar{A}_x \\ \bar{A}_y \\ \bar{A}_z \end{bmatrix} = T_{TA}^T \cdot \begin{bmatrix} -rL \\ Tr \\ -Tq \end{bmatrix}. \tag{A12}$$

Appendix B
Coordinated Turn Predictor Equations

The motion of a target having constant thrust, lift, and roll angle is confined to a plane. The orientation of this plane is defined by the target's tangential and normal acceleration vectors. This plane will be referred to as the maneuver plane. Using the outputs of the TSE algorithm, one can compute the target's normal acceleration vector, namely,

$$\vec{A}_n = (L + g_{YT})\vec{Y}_T + (g_{ZT})\vec{Z}_T \tag{B1}$$

where the terms L, g_{YT}, g_{ZT} are computed as shown in Appendix A. The transformation from target coordinates to maneuver plane coordinates is

$$T_{MT} = \begin{bmatrix} 1 & 0 & 0 \\ 0 & \cos\Phi & \sin\Phi \\ 0 & \sin\Phi & \cos\Phi \end{bmatrix} \tag{B2}$$

where

$$\cos\Phi = (L + g_{YT})/|\vec{A}n|$$
$$\sin\Phi = (g_{zt})/|\vec{A}n|. \tag{B3}$$

The components of gravity in maneuver coordinates are

$$\begin{bmatrix} g_{Xm} \\ g_{Ym} \\ g_{Zm} \end{bmatrix} = T_{MT} \begin{bmatrix} g_{XT} \\ g_{YT} \\ g_{ZT} \end{bmatrix}. \tag{B4}$$

Hence, the differential equations describing the motion of a target in a coordinated turn (in maneuver plane coordinates) are

$$\dot{V} = T + g_{Xm}\cos\Phi + g_{Ym}\sin\Phi \qquad \text{(tangential)}$$
$$V\dot{\Phi} = L\cos\Phi - g_{Xm}\sin\Phi + g_{Ym}\cos\Phi \qquad \text{(normal)}$$
$$\dot{X}_1 = V\cos\Phi \qquad \Phi_0 = 0$$
$$\dot{X}_2 = V\sin\Phi \tag{B5}$$

where X_1, X_2 are the components of target position in the maneuver plane and Φ is the angular orientation of the velocity vector of the target (in the maneuver plane). The prediction problem is solved if one can efficiently solve (B5). An approximate closed form solution was obtained by using the Taylor series approach [13]. Repeated differentiation of (B5) at zero bullet flight time determined the coefficients of the series expansion. Simulation results indicated that a fifth-order series approximation was adequate for the anticipated maximum bullet flight times. The coefficients of the Taylor series representation of future target position in maneuver coordinates are

$$\ddot{X}_1 = T + g_{xT}$$
$$\ddot{X}_2 = b \qquad b = \sqrt{(L + g_{yT})^2 + g_{ZT}^2}$$
$$\dddot{X}_1 = -WT_2 \qquad w = b/v$$
$$\dddot{X}_2 = WT \qquad T_2 = CL$$
$$C = \begin{cases} 1 & \text{if } b = 0 \\ (L + g_{yT})/b & \text{if } b \neq 0 \end{cases}$$
$$X_1^{(4)} = -W^2T - \dot{W}T_2 \qquad \dot{W} = -(T + 2g_{xT})W/V$$
$$X_2^{(4)} = -W^2T_2 + \dot{W}T \qquad S = \begin{cases} 0 & \text{if } b = 0 \\ g_{zT/b} & \text{if } b \neq 0 \end{cases}$$
$$X_1^{(5)} = -a_1T - a_2T_2 \qquad a_1 = 3W\dot{W}$$
$$X_2^{(5)} = -a_1T_2 - a_2T \qquad a_2 = \ddot{W} - W^3. \tag{B6}$$

Hence, future target position in maneuver coordinates is

$$\begin{bmatrix} X_1 \\ X_2 \end{bmatrix} = \begin{bmatrix} \dot{X}_1 & \ddot{X}_1 & \dddot{X}_1 & X_1^{(4)} & X_1^{(5)} \\ \dot{X}_2 & \ddot{X}_2 & \dddot{X}_2 & X_2^{(4)} & X_2^{(5)} \end{bmatrix} \cdot \begin{bmatrix} t_F \\ t_F^2/2 \\ t_F^3/3! \\ t_F^4/4! \\ t_F^5/5! \end{bmatrix}. \tag{B7}$$

The future target position in antenna coordinates is given by

$$\begin{bmatrix} \hat{P}_{fx} \\ \hat{P}_{fy} \\ \hat{P}_{fz} \end{bmatrix} = \begin{bmatrix} \hat{P}_x \\ \hat{P}_y \\ \hat{P}_z \end{bmatrix} + T_{TA}^T \cdot \begin{bmatrix} X_1 \\ CX_2 \\ SX_2 \end{bmatrix}. \tag{B8}$$

Acknowledgment

The author wishes to thank C. T. Brown for his work on the simulation results of Section VI, D. S. Zinn for his work on the simplified estimator gain equations of Section IV, and W. C. Hall for the solution of the coordinated turn predictor equations.

References

[1] B. L. Clark, "The development of an adaptive Kalman target tracking filter," in *Proc. AIAA Guidance Contr. Conf.*, Aug. 1976, pp. 365–382.

[2] F. W. Nesline and P. Zarchan, "Comparison of Kalman and finite memory filtering for gun fire control applications," in *Proc. 1978 IEEE Conf. Decision Contr.*, Jan. 1979, pp. 95–96.

[3] R. L. Moose and N. H. Gholson, "Adaptive tracking of abruptly maneuvering targets," in *Proc. 1976 IEEE Conf. Decision Contr.*, Dec. 1976, pp. 804–808.

[4] R. A. Scheder, "A self-adapting target state estimator," Army Material Syst. Anal. Activity, Tech. Rep. 166, AD-B016-2286, Dec. 1976.

[5] J. M. Fitts, "Aided tracking as applied to high accuracy pointing systems," *IEEE Trans. Aerosp. Electron. Syst.*, vol. AES-9, May 1973.

[6] J. L. Farrell, *Integrated Aircraft Navigation*. New York: Academic, 1976.

[7] R. A. Singer, "Estimating optimal tracking filter performance for

manned maneuvering targets," *IEEE Trans. Aerosp. Electron. Syst.*, vol. AES-6, July 1970.

[8] W. Wrigley and J. Hovorka, *Fire Control Principles.* New York: McGraw-Hill, 1959.

[9] G. J. Bierman, *Factorization Methods for Discrete Sequential Estimation.* New York: Academic, 1977.

[10] R. J. Fitzgerald, "Simple tracking filters: Closed-form solutions," *IEEE Trans. Aerosp. Electron. Syst.*, vol. AES-17, Nov. 1981.

[11] R. A. Scheder, "A computer program for double sweep optimal smoothing," AMSAA Rep. 246, Jan. 1979.

[12] M. I. Skolnik, *Introduction to Radar Systems.* New York: McGraw-Hill, 1980.

[13] C. M. Bender and S. A. Orzog, *Advanced Mathematical Methods for Scientists and Engineers.* New York: McGraw-Hill, 1978.

Multiconfiguration Kalman Filter Design for High-Performance GPS Navigation

MIN H. KAO AND DONALD H. ELLER

Abstract—This paper describes the design, implementation, and performance of a real-time multiconfiguration Kalman filter for high-performance Navstar global positioning system (GPS) navigation. The design provides extreme flexibility in order to operate with a wide variety of host sensors. It configures automatically (four filter configurations) based upon the host vehicle requirements and sensor availability, in order to process GPS measurements and provide the best estimate of the navigation states. Two new techniques, namely an unaided dead-reckoning Kalman filter implementation and an automatic inertial platform tilt estimation control scheme, are developed to improve the navigation accuracy, especially for high-dynamics applications. Performance results are presented to demonstrate the advantages of these techniques.

I. Introduction

GPS System Operation

THE Navstar global positioning system (GPS) consists of three major segments: space segment, control segment, and user segment. The space segment will eventually deploy 18 half-synchronous satellites in circular 10 898 nautical mile orbits. Each satellite transmits a composite signal at two L-band frequencies modulated by a 10.23 Mbit/s precision (P) code and a 1.023 Mbit/s coarse acquisition (C/A) code. The navigation message, which contains satellite ephemerides and satellite clock drift information, is modulated on the carrier, as an additional lower rate 50 bit/s code.

The message is based on data collected by widely spaced monitor stations, which are actually GPS receivers that measure ranges to the satellites just as a user set would. Since the monitor station positions are precisely known, however, the accumulated ranging information can be processed at a master control station to solve the navigation problem in reverse for satellite orbit determination and systematic error elimination. An upload station then transmits these computed satellite ephemerides and clock drifts to the satellites as required.

The primary objective of a GPS user set is to acquire and recover GPS satellite data, make pseudorange and delta pseudorange measurements, and process these measurements in real time to provide the best estimate of the user position, velocity, and system time. The user receiver maintains a time reference used to generate a replica of the code transmitted by the satellite. The amount of time that the receiver must apply to correlate the replica with the satellite clock referenced code received from the satellite provides a measure of the signal propagation time between the satellite and the receiver. This time of propagation is

Manuscript received March 17, 1982; revised August 17, 1982. This work was supported by a contract from the U.S. Air Force System Command's Space Division.

The authors are with the Magnavox Advanced Products and Systems Company, Torrance, CA 90503.

Reprinted from *IEEE Trans. Automat. Contr.*, vol. AC-28, pp. 304–314, Mar. 1983.

called the "pseudorange" measurement since it is in error by the amount of time synchronization error between the satellite and receiver clocks. The receiver also measures the Doppler shift of the carrier signal from the satellite. By measuring the accumulated phase difference in this Doppler signal over a fixed interval, the receiver can infer the range change increment. This measurement is called the "delta pseudorange" measurement and is in error by an amount proportional to the relative frequency error between the transmitter and receiver clocks. Since the carrier wavelength is short, the delta pseudorange is a highly accurate measurement (2 cm error in Costas track). Measurements from four satellites, after compensation for ionosphere, troposphere, antenna lever arm, satellite clock error, and user clock phase and frequency errors, provide the GPS set with sufficient information to solve for three components of user position and velocity and user clock errors [1]–[4].

Background

GPS is being developed to provide worldwide navigation coverage to a variety of users [5]–[9]. As part of the phase II program, the high-performance user set, which may operate as both an aided and unaided GPS navigation system, is being designed to provide a continuous navigation solution for a wide variety of host vehicles, varying from high-dynamics fighter aircrafts to slow moving ships [10]–[11]. It is also required to integrate with the multitude of sensors (e.g., INS, IMU, Doppler, and speed/heading) that currently exist on the phase II host vehicles.

To meet this broad spectrum of sensor aiding and vehicle dynamics requirements, a Kalman filter, which can automatically configure itself to process GPS measurements and provide the best estimate of the corresponding vehicle navigation states, is implemented. The filter selects among four navigation methods, namely acceleration-coupled (AC), velocity-coupled (VC), high-dynamics unaided, and low-dynamics unaided. The details of the navigation modes and corresponding filter configurations, which are selected based upon sensor availability, vehicle dynamics, and other specific requirements, are presented in Section II. The most significant feature in this multiconfiguration filter is the design commonality/flexibility, which minimizes the interface complexity as well as computer throughput and memory requirements.

The basic GPS measurements provided to the Kalman filter are pseudorange and delta pseudorange to the satellites being tracked. While the Kalman filter has been widely used to process these highly accurate measurements, previous designs have overlooked the more subtle error characteristics of the GPS measurement process. When these characteristics are not properly accounted for in the filter model, unacceptable performance can often result.

Specifically, the GPS delta pseudorange measurement is rather unique since it represents the Doppler relative to the satellite, integrated over a finite time interval. In high-dynamics unaided navigation, the measurement noise and process noise are highly correlated. Previous designs have ignored this effect and, as a result, Kalman filter algorithms were implemented without accounting for the cross-correlation term [12]–[18]. Essentially, the delta pseudorange measurement was merely processed as a pure velocity measurement. While for some applications this approach may be adequate, it is imperative that, to meet the phase II high-dynamics high-performance requirements, the "dead-reckoning" capability of the highly accurate delta pseudorange measurements must be utilized.

Toward this goal, a new technique, which implements a special form of the process noise matrix, is developed and demonstrates a significant performance improvement. No additional effort is needed to implement this process noise matrix, and a great deal of computer throughput is saved by processing the pseudorange measurements at a much slower rate than would otherwise be necessary. The details of this dead-reckoning Kalman filter implementation are presented in Section III of this paper.

Another important feature of the filter design reported herein is the automatic control of the platform tilt estimation in the acceleration-coupled mode. Due to practical limitations, the only IMU error sources which are included as Kalman filter states are platform tilts. It is therefore critical to prevent any unmodeled (as states) error sources from corrupting the existing state estimates. For example, during vehicle dynamics, errors in the accelerometer scale factor can create a velocity error which would result (due to velocity/tilt correlation) in an incorrect tilt estimate. In order to eliminate such effects, these maneuver-dependent error sources (such as accelerometer scale factor and misalignment and gyro scale factor, misalignment, and mass unbalance) are modeled in the filter process noise with coefficients which are tuned, through the use of a detailed system simulator, to provide optimal performance. The details of the tuning process for platform tilt estimation are presented in Section IV.

The remaining sections of this paper are organized as follows: Section II contains a brief description of the phase II Magnavox high-performance user set architecture. It also presents a detailed description of the various navigation modes and corresponding filter configurations. Sections III and IV describe the detailed filter design for acceleration-coupled and high-dynamics unaided navigation. Special emphasis is placed on the description of the performance improvement and computation reduction techniques. Section V presents the simulation results which demonstrate the advantages of these newly developed schemes. Section VI concludes the development of the phase II Kalman filter design for high-performance applications.

II. System Description

User Set Architecture

Fig. 1 shows the major functional elements of the phase II high-performance user sets. The set hardware contains a

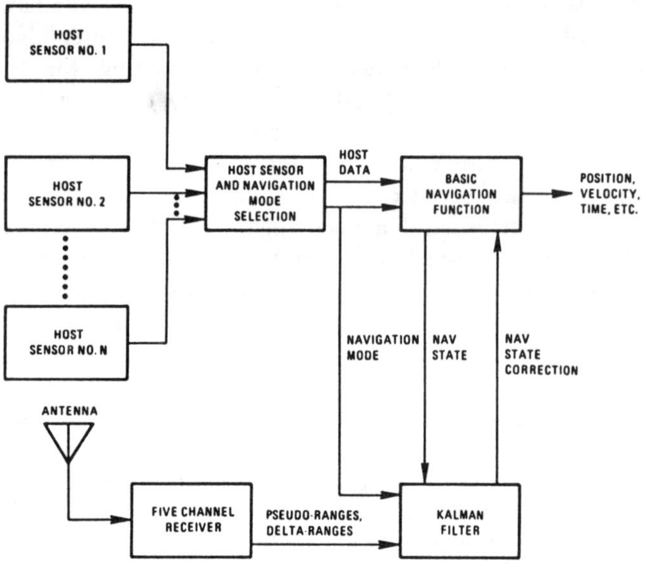

Fig. 1. Phase II high-performance user set architecture.

five-channel receiver which tracks four satellites simultaneously to provide complete observability with each update. Four of the five identical channels are used to continuously track the carrier frequency for the constellation of four satellites and provide delta pseudorange measurements. The fifth channel is positioned sequentially to update the codes for each of the carrier tracking channels and provide pseudorange measurements.

The set software implements a $U-D$ formulated, vector subscripted, extended Kalman filter [19]. It processes four contiguous delta pseudorange measurements every second and four pseudorange measurements every other second. The measurement errors are modeled as uncorrelated which permits their efficient sequential incorporation. The processing of each measurement is totally independent of the presence or absence of the others and the order of the processing is selected to maximize the effectiveness of the process noise matrix design. The set also contains a basic navigation function (the plant) which propagates the state variables (position, velocity, clock bias, clock drift, and acceleration/platform tilts) ahead in time at a 10 Hz rate.

Navigation Modes and Filter Configurations

In order to accommodate the multitude of host sensors and host vehicles available in phase II, three distinct navigation modes have been developed depending on the type of sensor data available and the requirements of the individual host vehicle. The essence of the three modes is determined by the extent to which host sensor data are used in the state propagator (plant). While the host sensor integrated systems provide a more accurate navigation solution, the inclusion of unaided operation, which may be chosen to avoid undesirable GPS/host solution correlation or used as a backup mode in the absence or failure of host sensors, is imperative.

1) Acceleration-Coupled (AC): The GPS set is fully integrated with an INS such that the IMU provided acceleration data are used directly to propagate the navigation states.

2) Velocity-Coupled (VC): The set is integrated with an INS, Doppler, or speed-and-heading sensor, and the host sensor provided velocity data are used directly to propagate the navigation states. (When an INS is available, it would be desirable to acceleration couple; however, since acceleration data are not available from certain host vehicles, velocity coupling is employed.)

3) Unaided (High-Dynamics): Host data are not used for state propagation, and the filter estimated velocity and acceleration are used to propagate the navigation states in a self-contained manner.

4) Unaided (Low-Dynamics): Host data are not used for state propagation and the filter estimated velocity is used to propagate the navigation states in a self-contained manner.

The GPS set selects from the above modes automatically based upon the host vehicle type and sensor availability. As sensors fail and recover, the system is automatically reconfigured to take maximum advantage of the available host data.

In order to mechanize these distinct state propagation schemes, a Kalman filter has been designed which can easily transition among the three commonly used configurations, namely 8-state, 11-state (acceleration), and 11-state (tilt), and estimate vehicle position, velocity, acceleration, tilt, and clock bias and drift [12]–[17], [20]. The states included in each filter configuration are shown in Table I. Also included in the table is a 5-state configuration (the stationary mode reduced filter). This configuration is intended to be used to improve the position and clock solutions by not estimating any dynamic states when the vehicle is stationary. In discretizing the transition matrices for the individual filter configurations, maximum advantage was taken of the commonality and the short (1 s) filter update interval in order to minimize both the throughput and RAM/ROM memory requirements. The corresponding filter configurations to be selected in each of the navigation modes are listed in Table II.

One unique feature in this multiconfiguration filter design is the use of the 8-state filter for all kinds of velocity-cou-

TABLE I
KALMAN FILTER STATES

STATE \ CONFIGURATION	5-STATE (STATIONARY)	8-STATE	11-STATE (ACCELERATION)	11-STATE (TILT)
POSITION (3)	X	X	X	X
CLOCK (2)	X	X	X	X
VELOCITY (3)		X	X	X
ACCELERATION (3)			X	
TILT (3)				X

TABLE II
NAVIGATION MODE/FILTER CONFIGURATION CROSS MATRIX

FILTER CONFIGURATION \ NAVIGATION MODE	ACCELERATION-COUPLED (AC)	VELOCITY-COUPLED (VC)	UNAIDED (HIGH-DYNAMICS)	UNAIDED (LOW-DYNAMICS)
11-STATE (TILT)	X			
11-STATE (ACCELERATION)			X	
8-STATE		X		X

pled sensors. While velocity errors are the proper filter states to be estimated in an INS velocity-coupled system, speed scale factor, and heading sensor bias typically characterize the Doppler type sensors. Due to the limitations on computer memory size, the approach has been to design a minimum number of filter configurations from which the Kalman filter can be operated in conjunction with all possible navigation modes and host sensors. As a result, instead of implementing another filter configuration including heading bias and speed scale factor as states, the same 8-state filter is employed. The distinct error characteristics of this type of sensor are maintained by incorporating the error estimates in a different way. Rather than correcting velocity error estimates, δV_E and δV_N directly, the heading bias and speed scale factor error estimates are computed as[1]

$$\delta HDG = (V_N \delta V_E - V_E \delta V_N)/V^2$$
$$\delta SF = (V_E \delta V_E + V_N \delta V_N)/V^2.$$

These values are then accumulated and used to correct the host sensor speed and heading outputs just as the filter generated velocity outputs would normally be utilized.

The filter reconfiguration upon navigation mode change is accomplished by deleting unused states and/or adding new states in the filter state vector and error covariance matrix. When new states are added, the new portion of the error covariance matrix is initialized as a diagonal submatrix with zero cross-correlation terms.

III. Dead-Reckoning Kalman Filter Implementation for High-Dynamics Unaided Navigation

Dynamics Model

In high-dynamics unaided navigation, the navigation states are propagated with the filter estimated velocity and acceleration. Jerk and higher order terms are thus the driving forces of the error equations. In a single spatial dimension, the position-velocity-acceleration dynamics model can be written as

$$\frac{d}{dt}\begin{bmatrix}\delta X\\ \delta V\\ \delta A\end{bmatrix} = \begin{bmatrix}0 & 1 & 0\\ 0 & 0 & 1\\ 0 & 0 & 1/\tau\end{bmatrix}\begin{bmatrix}\delta X\\ \delta V\\ \delta A\end{bmatrix} + \begin{bmatrix}0\\ 0\\ \delta J\end{bmatrix}$$

[1] V, V_E, and V_N are the averaged ground speed, east velocity, and north velocity, respectively, over the filter update interval.

with

$$E[\delta J(\gamma) \quad \delta J(s)] = \sigma_{\delta J}^2$$
$$E[\delta J(\gamma) \quad \delta x(t_{k-1})] = 0 \qquad s, \gamma \in (t_{k-1}, t_k)$$
$$E[\delta J(\gamma)] = 0$$

where, since dynamics is the main concern in this design, and the interval between filter updates is rather short, a zero-mean constant random jerk, rather than a random white noise jerk, is assumed over the interval.

Measurement Model

The GPS pseudorange from a satellite is defined as

$$PR = \left[(X_S - X_R)^2 + (Y_S - Y_R)^2 + (Z_S - Z_R)^2\right]^{1/2} + b$$

where

(X_S, Y_S, Z_S): satellite position at time of transmission
(X_R, Y_R, Z_R): receiver position at time of reception
b: receiver clock bias.

The linearized observation equation implemented in the extended Kalman filter is

$$\delta PR = H_{PR} X + V_{PR}$$

where X is the state vector with its states defined in Table I; V_{PR} is the additive measurement noise; and H_{PR}, the pseudorange observation matrix, is obtained by linearizing the pseudorange equation with respect to the filter states (Jacobian matrix)

$$H_{PR} = \left.\frac{\partial PR}{\partial X}\right|_{X=\bar{X}} = \left[-U_x, -U_y, -U_z, 1,0,0,0,0,0,0,0,\right]^T$$

where (U_x, U_y, U_z) is the user-to-satellite line-of-sight vector.

The GPS delta pseudorange is defined as the difference between two pseudoranges separated in time

$$DR = PR(t_2) - PR(t_1) \qquad t_2 > t_1.$$

Notice that, since the delta pseudorange represents the Doppler integrated over a finite time interval, theoretically any point within the integration interval can be chosen as the measurement validity time (in practical applications, either the interval start or stop time is selected to reconcile it with the pseudorange validity time).

If the interval stop time is chosen as the reference, the linearized measurement model can be written as

$$\delta DR = H_{DR} X + V_{DR}$$

where the measurement noise V_{DR}, not only accounts for the very small additive tracking error in the highly accurate carrier loop, but also includes the integrated dynamics effects representing unmodeled jerk and higher order terms over the integration interval; and

$$H_{DR} = \left.\frac{\partial DR}{\partial X}\right|_{X = \bar{x}_{\text{stop}}}$$

$$= \Big[-\Delta U_x, -\Delta U_y, -\Delta U_z, 0, \Delta t,$$

$$-\Delta t U_{x_1}, -\Delta t U_{y_1}, -\Delta t U_{z_1},$$

$$\times \frac{\Delta t^2}{2} U_{x_1}, \frac{\Delta t^2}{2} U_{y_1}, \frac{\Delta t^2}{2} U_{z_1}\Big]^T$$

with

Δt:	delta pseudorange integration interval
U_{x1}, U_{y1}, U_{z1}:	user-to-satellite line-of-sight vector at delta pseudorange start time
$\Delta U_x, \Delta U_y, \Delta U_z$:	line-of-sight vector change over delta pseudorange integration interval.

Correlation Between Process and Measurement Noise

The state propagation and delta pseudorange measurement equations can be written in vector form as

$$x_{k+1} = \phi(k+1, k) x_k + W_{k+1}$$
$$z_k = H_{DR}(k) x_k + V_k.$$

Since unmodeled jerk is the main driving force for both process noise W_{k+1} and measurement noise (V_k), the covariance matrix given by

$$E\left\{\begin{bmatrix} w_k \\ v_k \end{bmatrix}\left[w_j^T, V_j^T\right]\right\} = \begin{bmatrix} Q_k & C_k \\ C_k & R_k \end{bmatrix} \delta k_j$$

contains a large cross-correlation term (C_k). In order to account for this, the conventional Kalman filter algorithm must be modified [21]–[23] and batch processing must be used. Although recent simulations have shown that this improves the navigation accuracy, it is computationally expensive and sensitive to the *a priori* knowledge of the correlation noise matrix.

Alternatively, the following technique is developed to compensate for this correlation effect through the use of a special form of the process noise matrix.

"Transparent" Process Noise Design

Theoretically, the process noise matrix, which should account for the effects of all driving forces not being modeled as filter states, is given by

$$Q_k = \int_{t_{k-1}}^{t_k} \int_{t_{k-1}}^{t_k} \phi(t_k, s) E\left[w(v) w^T(s)\right] \phi^T(t_k, v)\, ds\, dv.$$

With the assumed dynamics model, it can be shown that, for each spatial dimension

$$Q = \sigma_{\delta j}^2 \begin{bmatrix} \Delta t^6/36 & \Delta t^5/12 & \Delta t^4/6 \\ \Delta t^5/12 & \Delta t^4/4 & \Delta t^3/2 \\ \Delta t^4/6 & \Delta t^3/2 & \Delta t^2 \end{bmatrix}.$$

Through analysis and simulation it has been determined that, by deleting the position-velocity and position-acceleration correlation terms in the above process noise matrix, the processing of contiguous delta pseudorange measurements, which have a large negative autocorrelation, results in excellent performance without the inclusion of the cross-correlation term described above. This special form of the process noise matrix generates a high Kalman gain for position estimation (from the delta pseudorange measurement) which would otherwise be provided by including the cross-correlation C-matrix in the Kalman filter algorithm. This has the practical interpretation that the filter is dead-reckoning with contiguous delta pseudorange measurements, which totally removes the vehicle dynamics and greatly improves the navigation performance, especially for high-dynamics applications. The set performs like a stationary set refining its solution with pseudorange measurements.

Contrary to all previous designs, in which pseudorange and delta pseudorange measurements are used to estimate position and velocity/acceleration, respectively, this new design concept utilizes the highly accurate delta pseudorange measurements as the main source for both position and velocity/acceleration estimation. Pseudorange measurements provide significant position estimation only when position errors are large or delta pseudorange measurements are not available. This yields the following significant advantages.

1) Minimum sensitivity to vehicle dynamics. The performance during the most violent aircraft maneuvers is almost as accurate as that during benign cruises. This greatly alleviates the filter tuning efforts.

2) Allows a much lower processing rate for pseudorange measurements. This significantly reduces the throughput requirements.

3) Enables the implementation of a very effective divergence control method, in which all delta pseudorange measurements are processed first, and the divergence test is simply based on the pseudorange measurement residual ratios.

Intuitively, since this process noise matrix decouples the processing of pseudorange and delta pseudorange measurements to a great extent, it seems feasible, and beneficial, to tune the position and velocity/acceleration process noise terms independently in order to optimize the filter performance. Simulations have proved this assertion and lead to the development of a class of process noise matrices with the special form

$$Q = \begin{bmatrix} \sigma_p^2 & 0 & 0 \\ 0 & \Delta t^4 \sigma_J^2/4 & \Delta t^3 \sigma_J^2/2 \\ 0 & \Delta t^3 \sigma_J^2/2 & \Delta t^2 \sigma_J^2 \end{bmatrix}$$

where the parameters have the practical significance that σ_J^2 accounts for the effects of the unmodeled jerk on velocity and acceleration, and σ_p^2 represents the remaining position errors after delta pseudorange measurement processing has occurred. Both represent the "tuning parameters" which must be determined by engineering judgment and computer simulation. This process noise matrix is referred to as a "transparent process noise" because its addition has no effect on the Kalman gain and predicted residual variance at delta pseudorange stop time. More analytical properties of the transparent process noise matrix are presented in the Appendix.

IV. Automatic Control of Tilt Estimation for Acceleration-Coupled (AC) Navigation

Process Noise Design

The acceleration-coupled dynamics model used in the filter design contains only the most essential error states (three positions, three velocities, three platform tilts, and two clock terms). Error sources due to gyros and accelerometers, which include dynamics-dependent time-varying components, are not explicitly modeled as states. In order to prevent these effects from corrupting the state estimation or causing filter divergence, the following process noise matrix has been developed.

Considering only the effect of the accelerometer errors, the position-velocity error model for acceleration-coupled navigation can be written as

$$\frac{d}{dt}\begin{bmatrix}\delta X\\ \delta V\end{bmatrix} = \begin{bmatrix}0 & 1\\ 0 & 0\end{bmatrix}\begin{bmatrix}\delta X\\ \delta V\end{bmatrix} + \begin{bmatrix}0\\ \delta a\end{bmatrix}$$

where the accelerometer error is modeled as

$$\delta a = SF\cdot f_1 + M_1\cdot f_2 + M_2\cdot f_3 + b + n$$

where f_1 is the input-axis specific force; f_2 and f_3 are the cross-axis specific forces; SF is the scale factor error; M_1 and M_2 are the misalignment errors; and b and n are bias and random noise, respectively.

The accelerometer scale factor errors and misalignment uncertainties are typically of the same order of magnitude. Assuming uncorrelated error sources and constant specific forces during the covariance update interval (t_{k-1}, t_k), the process noise covariance can be obtained by integrating the equation in the previous section and applying conservative simplifications

$$Q_A = \sigma_{NA}^2\begin{bmatrix}\Delta t^3/3 & \Delta t^2/2\\ \Delta t^2/2 & \Delta t\end{bmatrix} + \sigma_{SMA}^2\Delta V^2\begin{bmatrix}\Delta t^2/4 & \Delta t/2\\ \Delta t/2 & 1\end{bmatrix}$$

where σ_{NA}^2 is the noise variance; σ_{SMA}^2 is the scale factor/misalignment error variance; and $\Delta V^2 = \Delta V_X^2 + \Delta V_Y^2 + \Delta V_Z^2$ is the squared sum of the accelerometer outputs over the filter update interval.

The platform drift error is modeled as

$$\delta\dot{\phi} = SFW_1 + M_1W_2 + M_2W_3 + MU_1f_1 + MU_2f_2 + MU_3f_3 + b + n$$

where W_1 is the input-axis angular rate, W_2 and W_3 are the cross-axis angular rates, SF is the scale factor error, M_1 and M_2 are the misalignment errors, MU_1, MU_2, and MU_3 are the mass-unbalance coefficients, and b and n are bias and random noise, respectively.

Similarly, the platform tilt process noise term is given by

$$Q_G = \sigma_{BG}^2\Delta T^2 + \sigma_{NG}^2\Delta T + \sigma_{SMG}^2\Delta\theta^2 + \sigma_{MUG}^2\Delta V^2$$

where σ_{NG}^2 is the noise variance, σ_{BG}^2 is the gyro drift error variance, σ_{SMG}^2 is the scale factor/misalignment error variance, σ_{MUG}^2 is the mass-unbalance error variance, and $\Delta\theta^2 = \Delta\theta_x^2 + \Delta\theta_y^2 + \Delta\theta_z^2$ is the squared sum of the platform torquing angles over the filter update interval.

Automatic Control of Tilt Estimation

From the above equations, the complete process noise matrix in a single spatial dimension can be written as

$$Q = \begin{array}{c} \begin{array}{ccc}\delta X & \delta V & \delta\theta\end{array} \\ \left[\begin{array}{cc|c} \multicolumn{2}{c|}{Q_A} & 0 \\ \hline \multicolumn{2}{c|}{0} & Q_G \end{array}\right] \end{array} \begin{array}{l}\delta X\\ \delta V.\\ \delta\theta\end{array}$$

This process noise matrix, which is generated in real time using the IMU data $\Delta\theta$ and ΔV, has significant advantages. It reflects the actual maneuver experienced and, as a result, can dynamically control the estimation of the filter states. The actual effect of these terms in the process noise matrix is twofold. For errors which affect the tilt states directly (such as gyro drift and mass unbalance) they merely decrease the filter's confidence in the tilt states. For errors (such as accelerometer scale factor) which affect states that are directly observable and correlated with tilt, they decrease the correlation coefficient as it exists within the filter.

Among all the parameters to be tuned through simulations, σ_{SMA}^2, the accelerometer scale factor/misalignment error variance, has the most significance in that it directly controls the correlation between tilt and velocity states. When properly tuned, it reduces the velocity/tilt correlation in the filter covariance matrix during maneuvers (ΔV^2 is large) and prevents the corruption of the tilt estimates from the processing of GPS measurements. On the other hand, it provides a strong tilt/velocity correlation during benign flights (ΔV^2 is small) and facilitates the effective estimation of tilts from GPS measurements. Details of this filter tuning effect are presented in Section V.

V. Performance

During phase I, a sophisticated GPS simulator was developed at Magnavox to support the system design, implementation, and test efforts. Subsequent to the availability of actual satellite data, the simulator was refined to account for any newly observed effects and provide a precise representation of the GPS space and control segments.

To demonstrate the performance of the GPS phase II Kalman filter design, this simulator (GPSSIM) has been updated to include the phase II navigation routines and host sensor models. The simulator (Fig. 2) consists of two programs, namely, environmental simulator (ESIM) and UE set simulator (UESIM). For specified initial conditions and a set of inputs involving acceleration, pitch rate, and

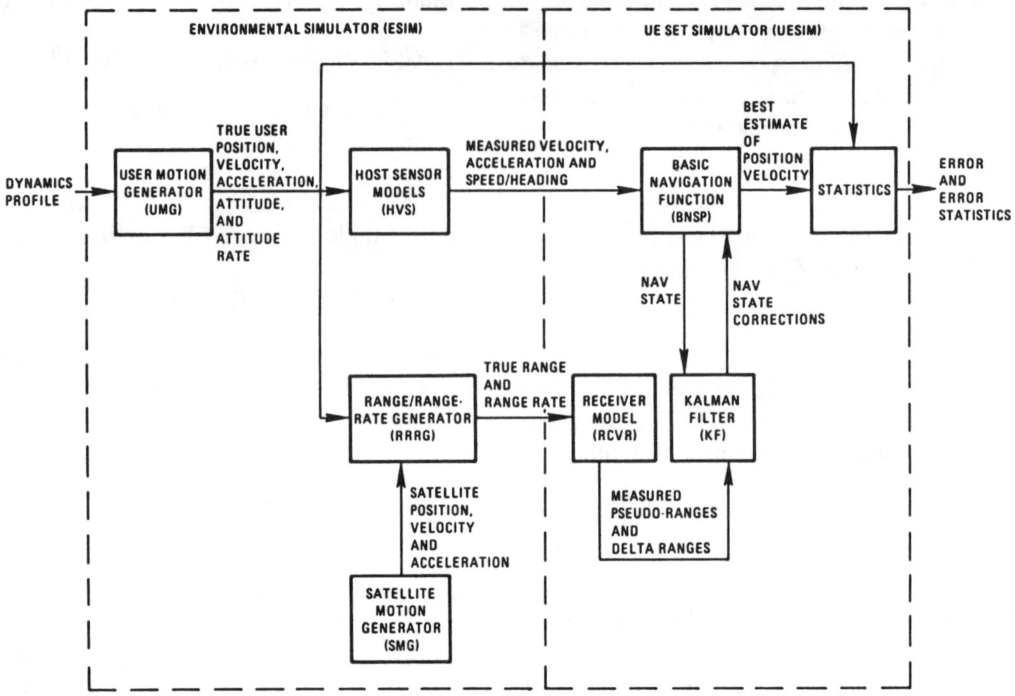

Fig. 2. GPS simulator (GPSSIM) block diagram.

heading rate, the user motion generator (UMG) numerically integrates the host vehicle's equations of motion with a fourth-order Runge–Kutta algorithm. Outputs from UMG at specified time intervals, including vehicle position, velocity, and acceleration, expressed in WGS-72 coordinates, are used to drive the range and range-rate generator (RRRG) and the host sensor models (HVS). The HVS includes all of the existing host vehicle sensors (INS, Doppler, TAS, etc.) and provides simulated position, velocity, acceleration, and speed/heading type measurements. The RRRG computes the true range and range rate to all visible satellites, using the vehicle states from UMG and the satellite states from the satellite motion generator (SMG). These outputs are stored on tape and used to drive the user set simulator (UESIM), which contains the following routines: a receiver model routine (RCVR), which models the receiver tracking loops and provides simulated pseudorange and delta pseudorange measurements; a basic navigation routine (BNSP), which contains the phase II state propagation equations and provides vehicle position, velocity, and acceleration; and a Kalman filter routine (KF), which implements the phase II Kalman filter design and provides the state corrections. The outputs from the simulator are navigation error and error statistics.

The goal of the phase II high-performance user set design is to provide the best estimate of the vehicle states in high-dynamics applications, which are quantitatively specified by upper bounds placed on the host vehicle velocity, acceleration, and jerk [10]. The test trajectory used most extensively is a figure-eight octagon flight which consists of a 45° turn at each corner and a straight flight between turns. The magnitude of the vehicle acceleration reaches a maximum of 9 g, while the magnitude of jerk reaches a maximum of 10 g/s. The GPS phase III satellite orbit constellation is used.

TABLE III
GPS ERROR CHARACTERISTICS

ERROR SOURCE	RESULTING RANGE MEASUREMENT ERROR (1σ)
BIAS	
SATELLITE CLOCK AND NAVIGATION STABILITY	2.7 METERS
SATELLITE PERTURBATIONS	1.0 METERS
CONTROL SEGMENT EPHEMERIS PREDICTION AND MODEL IMPLEMENTATION	2.5 METERS
IONOSPHERIC DELAY COMPENSATION	2.3 METERS
TROPOSPHERIC DELAY COMPENSATION	2.0 METERS
MULTIPATH	1.2 METERS
RANDOM	
PSEUDO-RANGE MEASUREMENT*	1.5 METERS
DELTA PSEUDO-RANGE MEASUREMENT*	0.02 METERS
RECEIVER CLOCK PHASE	1° RMS/SEC
RECEIVER CLOCK FREQUENCY	1E-10 (EXPONENTIALLY CORRELATED WITH 2 HR TIME CONSTANT)

*ASSUMES A (41) dB JAMMING TO SIGNAL RATIO AT THE ANTENNA

The performance results to be presented are of two basic types: comparative and total system. The comparative results demonstrate the performance improvements which are achievable using the design developed herein. The total system results represent the performance capability of the GPS phase II Magnavox high-dynamics set.

The GPS related error sources which affect system performance are presented in Table III [24]. These errors are divided into two classes, bias-like and random. The ran-

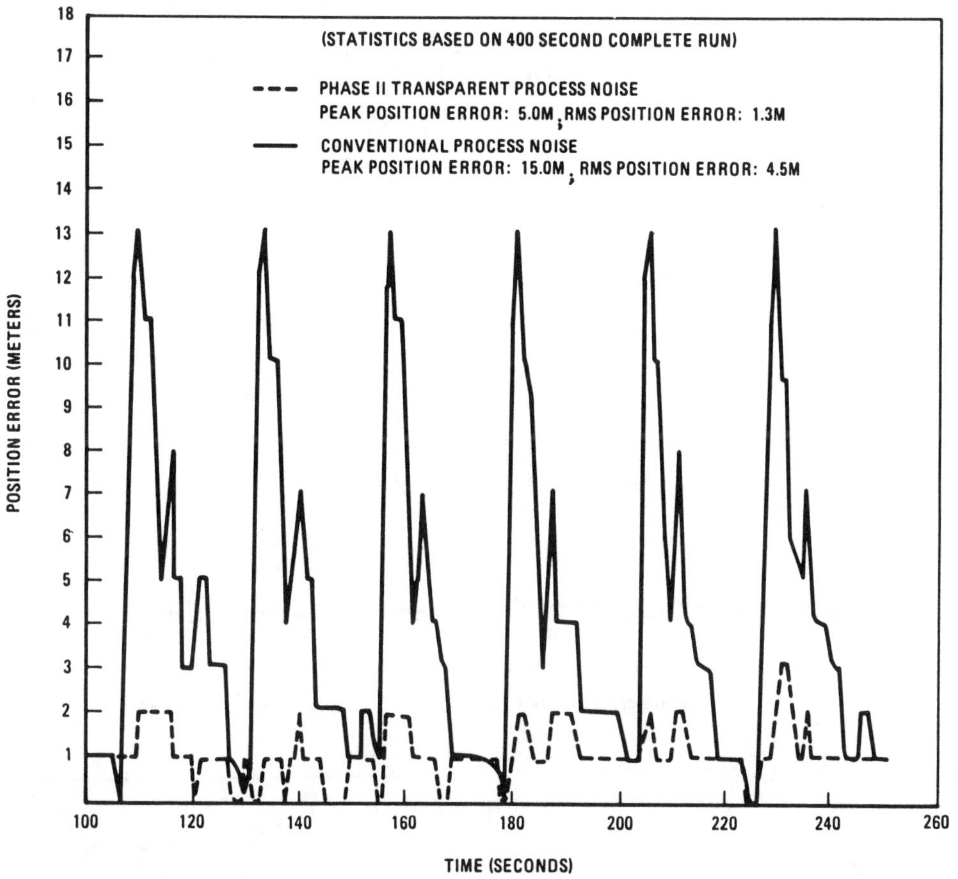

Fig. 3. Position error versus time (unaided, 9 g octagon).

dom errors are a result of the actual phase II receiver hardware performance evaluations and their effect on system performance is determined by the Kalman filter design. The bias errors result in a position error which depends only on the satellite geometry. Since all of the designs being considered react identically to bias type GPS errors, they will be excluded from GPSSIM when generating the comparative results. The total system performance results include all of the GPS error sources.

Comparative System Performance

To compare the performance of the phase II transparent process noise matrix with that of a conventional random jerk process noise matrix [12]–[17], [25] in an unaided high-dynamics environment, numerous simulations have been performed on the 9 g figure-eight trajectory. Fig. 3 shows the resultant position error occurring on a sample run in which both process noise matrices were tuned to optimize their performance. As can be seen in the figure, when the conventional process noise matrix was used, position error spikes of up to 15 occurred as a result of the unmodeled jerk during the turns. The implementation of the transparent process noise matrix, however, has significantly reduced the magnitude of the error spikes.

To further demonstrate the dead-reckoning capability of the transparent process noise matrix, a data-gathering scenario was simulated. (During the data-gathering period, the fifth channel is devoted to collect alternate satellite ephemeris and, as a result, for a period of up to 60 s, no pseudorange measurements are available for position determination.) Fig. 4 presents the position error history for the 9 g octagon trajectory. The advantages of using the transparent process noise matrix are clearly demonstrated.

Fig. 5 presents the effects of the automatic control of tilt estimation during a typical take-off flight. The acceleration during the vehicle take-off (toward north) is 2 g and the cruise speed is 1200 m/s [10]. The accelerometer scale factor error used in the simulation was 200 ppm. As a result, a constant north acceleration error of 400 μg occurred during the entire take-off period. As was explained in the previous section, since the accelerometer scale factor errors are not modeled as filter states, the estimate of the east tilt is corrupted by the scale-factor induced north velocity error. Fig. 5 shows the effects of the process noise compensation technique for two different values of the tuning parameter σ_{SMA}^2, the accelerometer scale factor/misalignment error variance. As shown in the figure, when the parameter was properly tuned, the tilt corruption due to unmodeled scale factor error was minimized. The corresponding velocity history is shown in Fig. 6.

The fact that the process noise parameter σ_{SMA}^2 must be adjusted (tuned) via the simulator to improve performance when the exact value is known from the IMU model is caused by the inherent assumption that all effects modeled as process noise are random, whereas over the short term,

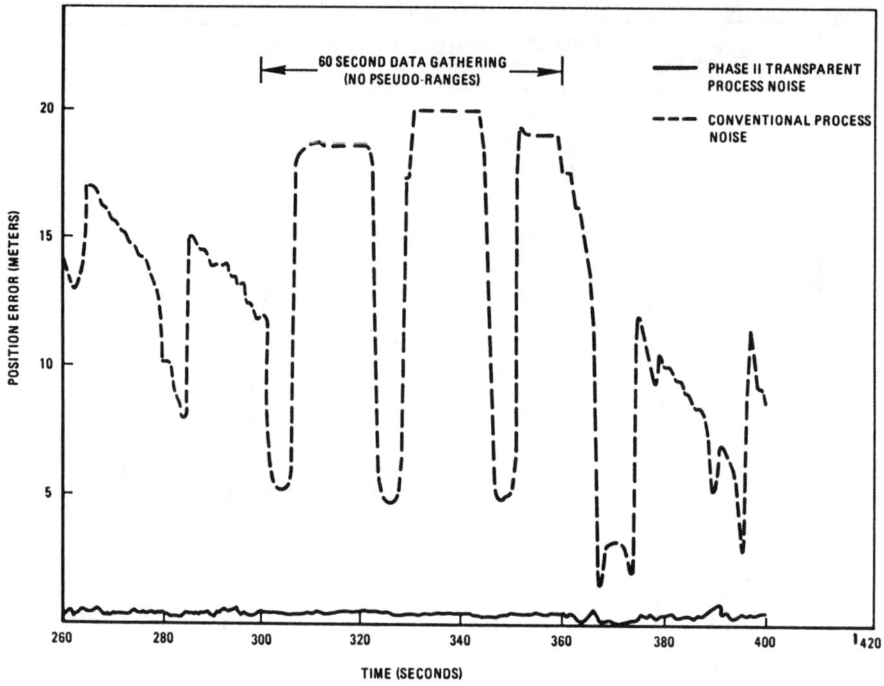

Fig. 4. Position error versus time (unaided, data gathering, 9 g octagon).

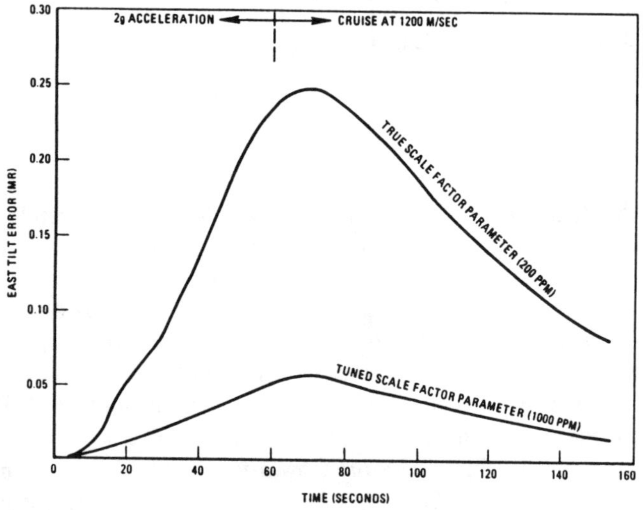

Fig. 5. East platform tilt angle versus time (acceleration-coupled, take-off flight).

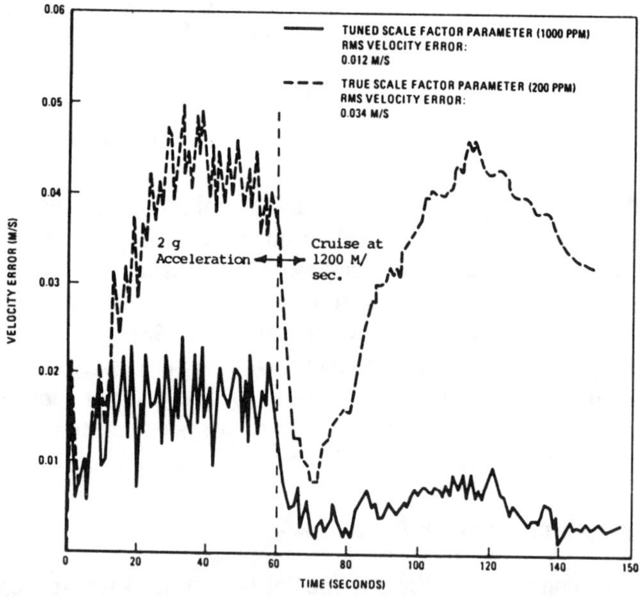

Fig. 6. Velocity error versus time (acceleration-coupled, take-off flight).

the IMU error effects appear much more bias-like. This resulted in a tuned value (1000 ppm) which was much larger than would be expected by examining the actual IMU error source (200 ppm).

To provide further insight into the process noise tuning effect, Fig. 7 depicts the velocity error history for a 9 g figure-eight simulation. It demonstrates that, with the properly tuned process noise compensation, the resultant velocity error due to unmodeled accelerometer errors can be greatly reduced.

Total System Performance

In order to demonstrate the navigation accuracy expected from GPS phase II Magnavox high-dynamics sets, the 9 g figure-eight unaided flight trajectory was simulated with GPSSIM including all of the Table III errors. The resulting position error history is shown in Fig. 8. The decreased accuracy compared with Fig. 3 of the comparative results is caused by the bias errors whose specific effect can be determined from the relative geometry between the user and the satellite constellation being utilized. These geometrical relationships are commonly represented by a solution matrix[2] whose elements provide the position ($3D$) and time errors that result from an uncorrelated set of pseudorange measurement errors with unit variance. The square root of the trace of this matrix, GDOP (geometrical

[2]Simply, the covariance matrix of the least square solution to a single set (4) of pseudorange measurements.

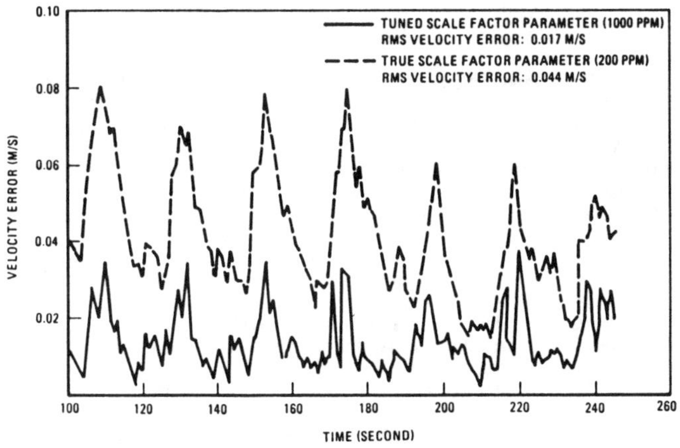

Fig. 7. Velocity error versus time (acceleration-coupled, 9 g octagon).

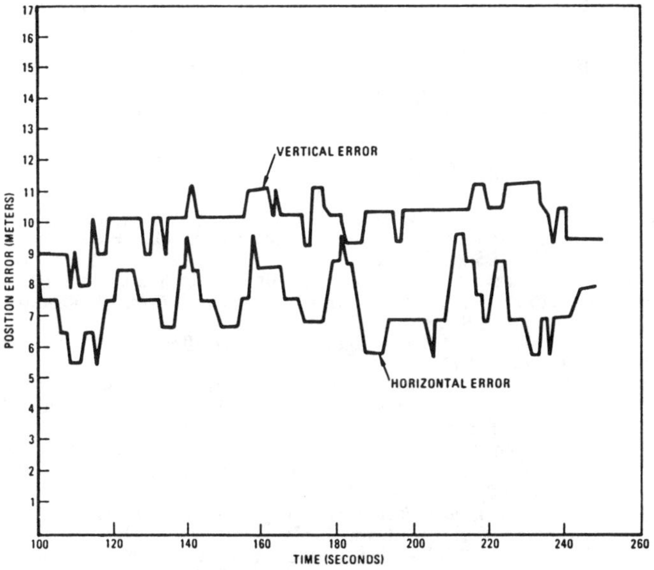

Fig. 8. Total system performance (unaided, 9 g octagon).

dilution of precision), represents the amplification factor of pseudorange measurement errors into a composite user position and time error [6]. All of the results were generated using a typical phase III constellation which has a GDOP of 3. The corresponding horizontal (HDOP) and vertical (VDOP) position components were 1.5 and 2.0, respectively.

VI. Conclusions

The navigation filter for the GPS phase II Magnavox high-performance user set has been developed. Through filter reconfigurations, it can operate with a multitude of host sensors and host vehicles. Several new design concepts are implemented; the most innovative of which is the utilization of the dead-reckoning capability of the highly accurate delta pseudorange measurement. This implementation, not only significantly improves the navigation performance in unaided operation, but also makes it possible to save substantial computer throughput by processing pseudorange measurements at a much slower rate.

Another important design concept is the compensation of the unmodeled accelerometer/gyro error sources through a real-time generated process noise matrix. Excellent performance has been obtained by proper tuning of the single-most critical parameter.

Appendix

The "dead-reckoning" property of the "transparent" process noise matrix can be best explained by the following derivation for a single spatial dimension, in which only three states, namely position (x), velocity (v), and acceleration (a), are considered.

Choosing the delta pseudorange start point as the measurement validity time, the observation matrices for pseudorange and delta pseudorange measurements are

$$H_{PR} = [1,0,0]^T$$
$$H_{DR} = [0, -\Delta t, -\Delta t^2/2]^T.$$

The Kalman gains, as computed by $K = PH^T(HPH^T + R)^{-1}$, are

$$K_{PR} = [P_{xx}/(P_{xx}+R_{PR}), P_{xv}/(P_{xx}+R_{PR}), P_{xa}/(P_{xx}+R_{PR})]^T$$

$$K_{DR} = [-(P_{xv}\Delta t + P_{xa}\Delta t^2/2)/\Delta, -(P_{vv}\Delta t + P_{va}\Delta t^2/2)/\Delta, -(P_{va}\Delta t + P_{aa}\Delta t^2/2)/\Delta]^T$$

where $\Delta = P_{vv}\Delta t^2 + P_{av}\Delta t^3 + P_{aa}\Delta t^4/4 + R_{DR}$, and R_{PR} and R_{DR} are the pseudorange and delta pseudorange measurement noise variances, respectively.

For small position-velocity and position-acceleration correlations, the above equations become

$$K_{PR} \doteq [P_{xx}/(P_{xx}+R_{PR}),0,0]'$$
$$K_{DR} \doteq [0, -(P_{vv}\Delta t + P_{va}\Delta t^2/2)/\Delta - (P_{va}\Delta t + P_{aa}\Delta t^2/2)/\Delta]^T$$

and the resultant navigation state error estimates are[3]

$$\widehat{\delta X}_1 = (P_{xx}/(P_{xx}+R_{PR}))\,\delta PR$$
$$\delta V_1 = -((P_{vv}\Delta t + P_{va}\Delta t^2/2)/\Delta)\,\delta DR$$
$$\widehat{\delta A}_1 = -((P_{va}\Delta t + P_{aa}\Delta t^2/2)/\Delta)\,\delta DR.$$

This indicates the independent estimate of position and velocity/acceleration errors from pseudorange and delta pseudorange measurements, respectively.

The corrected position at delta pseudorange stop time is thus[4]

$$\hat{X}_2(+) = \hat{X}_1(+) + \hat{V}_1(+)\Delta t + \hat{A}_1(+)\Delta t^2/2$$
$$= \hat{X}_1(-) + (P_{xx}/P_{xx} + R_{PR})\,\delta PR + \Delta\tilde{R}$$

[3] δPR and δDR are the pseudorange and delta pseudorange measurement residuals, respectively.

[4] $x(-)$ and $x(+)$ represent the estimated navigation state before and after the incorporation of the measurement processing, respectively.

which is a combination of the delta pseudorange measurement ($\Delta\tilde{R}$) and a correction term based on the pseudorange measurement.

The above assumption of small position-velocity and position-acceleration correlations is a direct result of processing both the rather accurate pseudorange and delta pseudorange measurements and adding the phase II process noise matrix which has zero corresponding correlation terms. In cases, like the code-track mode, in which only pseudorange measurements are processed, the lack of delta pseudorange measurement processing will result in the buildup of the correlations between position and velocity/acceleration, and automatically enable the estimation of velocity and acceleration errors from pseudorange measurements.

If, instead, the delta pseudorange stop point is chosen as the measurement validity time, the measurement observation matrix will become

$$H_{DR} = [0, -\Delta t, \Delta t^2/2]^T.$$

In general the system performance is dependent upon the selection of the measurement validity time and a tradeoff has been necessary to perform the selection. It can be shown that the process noise matrix described above, not only provides a dead-reckoning capability, but also provides the same results in each case. The lemma presented below shows the relationship between the two validity times.

Definition: A process noise matrix Q is called transparent if $H_{2,DR}Q = 0$ where $H_{2,DR}$ is the observation matrix at delta pseudorange stop time.

Lemma: If Q is a transparent process noise matrix, then both the delta pseudorange start and stop time models give the same results and their error covariance (P) matrices and Kalman gains (K) are related by

$$P^k_{k+1,2} = \Phi(k+1,k)P^k_{k,1}\Phi^T(k+1,k) + Q(k)$$
$$K_{k+1,2} = \Phi(k+1,k)K_{k,1}$$

where subscript 1 and 2 designate the use of start and stop point as reference time, respectively.

The proof of the lemma will not be given in this paper.

References

[1] R. P. Denaro, "Navstar: The all-purpose satellite," *IEEE Spectrum*, May, 1981.

[2] L. Jacobson and L. Huffman, "Satellite navigation with GPS," *Satellite Commun.*, July 1979.

[3] J. J. Spilker, Jr., "GPS signal structure and performance characteristics," *J. Inst. Navigation*, vol. 25, no. 2, Summer 1978.

[4] A. J. van Dierendonck *et al.* "The GPS navigation message," *J. Inst. Navigation*, vol. 25, no. 2, Summer. 1978.

[5] "System segment specification for the user system segment of the Navstar global positioning system, Phase I," SS-US-101B, Sept. 1974.

[6] "Final user field test report for the Navstar GPS phase 1; Major field test objective report on navigation accuracy," General Dynamics Electronics Division, June 25, 1979.

[7] D. W. Henderson and J. A. Strada, "Navstar field test results," presented at the Inst. Navigation Nat. Aerospace Symp., Mar. 1979.

[8] R. Denaro *et al.* "GPS phase I user equipment field test," *J. Inst. Navigation*, vol. 25, no. 2, Summer 1978.

[9] M. J. Borel *et al.* "Texas Instruments phase I user equipment," *J. Inst. Navigation*, vol. 25, no. 2, Summer 1978.

[10] "System segment specification for the user system segment—Navstar global positioning system, phase II," SS-US-200, July 1979.

[11] L. J. Jacobson and V. Calbi, "Engineering development of Navstar GPS user equipment," presented at the Inst. Navigation Nat. Aerospace Meet., Trevose, PA, Apr. 8–10, 1981.

[12] *Computer Program Development Specifications for the GPS X User Set (Unaided)*, CP-US-301, July 1979.

[13] T. M. Upadhyay and J. M. Damoulakis, "Sequential piecewise recursive filter for GPS low dynamics navigation," *IEEE Trans. Aerosp. Electron. Syst.*, vol. AES-16, July 1980.

[14] D. W. Klein *et al.* "Navigation software design for the user segment of the Navstar GPS," presented at the AIAA Guidance and Control Conf., San Diego, CA. Aug. 1976.

[15] L. R. Kruczynski, "Aircraft navigation with the limited operational phase of the Navstar global positioning system," *J. Inst. Navigation*, vol. 25, no. 2, Summer 1978.

[16] —, "GPS navigational algorithms," Tech. Rep., Univ. Texas at Austin, May 1977.

[17] T. P. Bower, "Navigation processing design for a low-cost GPS navigation system," presented at the IEEE Position Location and Navigation Symp., Dec. 8–11, 1980.

[18] R. T. Uyeminami, "Navigation filter mechanization for a spaceborne GPS user," presented at the IEEE Position Location and Navigation Symp., Nov. 1978.

[19] G. J. Bierman, *Factorization Methods for Discrete Sequential Estimation*. New York: Academic, 1977.

[20] G. Matchett, "GPS-aided shuttle navigation," presented at the IEEE Nat. Aerosp. and Electron. Conf., 1978.

[21] R. E. Kalman, "New methods in Wiener filtering theory," in *Proc. Symp. Eng. Appl., Random Function Theory and Probability*, J. L. Bogdanoff and F. Kozin, Eds. New York: Wiley, 1963.

[22] A. H. Jazwinski, *Stochastic Process and Filter Theory*. New York: Academic 1970.

[23] J. S. Meditch, *Stochastic Optimal Linear Estimation and Control*. New York: McGraw-Hill, 1969.

[24] "System Specification for the Navstar Global Positioning System," SS-GPS-300B, Mar. 1980.

[25] R. A. Singer, "Estimating optimal tracking filter performance for manned maneuvering targets," *IEEE Trans. Aerosp. Electron. Syst.*, vol. AES-6, July 1970.

Nonlinear Kalman Filtering Techniques for Terrain-Aided Navigation

LARRY D. HOSTETLER, MEMBER, IEEE, AND RONALD D. ANDREAS

Abstract—**The application of nonlinear Kalman filtering techniques to the continuous updating of an inertial navigation system using individual radar terrain-clearance measurements has been investigated. During this investigation, three different approaches for handling the highly nonlinear terrain measurement function were developed and their performance was established. These were 1) a simple first-order extended Kalman filter using local derivatives of the terrain surface, 2) a modified stochastic linearization technique which adaptively fits a least squares plane to the terrain surface and treats the associated fit error as an additional noise source, 3) a parallel Kalman filter technique utilizing a bank of reduced-order filters that was especially important in applications with large initial position uncertainties. Theoretical and simulation results are presented.**

I. Introduction

SINCE the fundamental work of Kalman and Bucy [1], [2] in linear filtering theory, the application of Kalman filters for "optimally" updating navigation systems utilizing auxiliary-sensor data has received considerable attention. Among the navigation systems and aides that have been considered are star trackers, satellite positional fixes, visual positional fixes, landmark trackers, Doppler radars, Omega fixes, VOR/DME data, MICRAD fixes, and air dead-reckoning data [3]–[11].

Investigations are currently being conducted on the updating of inertial navigation systems (INS's) using positional fixes derived from optical, radiometric, or radar obtained terrain data. One of the more successful concepts to date has been the terrain contour matching (TERCOM) system being developed for cruise missile applications [12]–[14]. It basically provides a positional fix by correlating a radar-altimeter-derived terrain profile with a stored topographical map, taking the location of the best match to be the position of the navigator. A sequence of such positional fixes are then utilized as measurements for Kalman updating the position, velocity, and other modeled states in an INS. Although a Kalman filter is eventually utilized, TERCOM, as well as various other approaches to terrain guidance, is still essentially a correlation guidance scheme, i.e., the fundamental measurement utilized as input to the Kalman filter is a positional fix derived from a correlation/matching algorithm on the raw terrain-clearance data.

Manuscript received March 11, 1982; revised September 27, 1982. This work was supported by the U.S. Department of Energy.

The authors are with Sandia National Laboratories, Albuquerque, NM 87115.

A significantly different approach to terrain-aided navigation is the subject of this paper. In contrast to Kalman processing of derived positional fixes, this approach recursively treats each radar terrain-clearance measurement as the measurement to be Kalman processed. By viewing each radar terrain-clearance measurement separately, a new class of "continuously" terrain-aided navigation schemes becomes available that in the past could not easily be developed using correlation type methods. This new class has several significant system features. First, the ability to model significant INS and measurement errors allows the designer to approach optimum accuracy and to desensitize the system to error sources that might otherwise compromise the updating process. Second, updating can occur continuously to a target because the recursive algorithms are not complicated by vehicle maneuvers. Third, the standard practices of covariance analysis inherent in the Kalman technique offer, with appropriate interpretation, an approach to predicting system performance and performing tradeoff studies during preliminary design. And finally, systems in which the error sources do not permit simple correlations algorithms to be derived, e.g., highly inaccurate INS's containing large velocity, attitude, accelerometer and gyro errors, air dead-reckoning and Doppler navigation systems, etc., can now be optimally aided using terrain data.

A program of concept development and flight test evaluation of terrain-aided navigation has been pursued at Sandia National Laboratories since 1974 [15], [23]. This program concentrated on an application with a radar altimeter operating at low altitude, an inertial navigation system, and updating occurring over distances of 40 km or less. Other applications and navigational systems have also been studied, but the important theoretical as well as practical problems are embodied in this setting.

Due to the undulating nature of terrain, the radar terrain-clearance measurement is a nonlinear function of vehicle position. Thus, the basic theory required for optimally processing these measurements is that of nonlinear estimation theory and its approximations [24], [25]. In addition, since the terrain beneath the vehicle is constantly changing, the nonlinearities are functions of time. Properly accounting for these nonlinearities is essential in obtaining satisfactory performance.

The following sections first give the general framework for optimal terrain-aided navigation systems. They then

Reprinted from *IEEE Trans. Automat. Contr.*, vol. AC-28, pp. 315–323, Mar. 1983.

address several specific system configurations that involve different methods for treating measurement nonlinearities. Included are performance comparisons and discussions of the limitations of each method.

II. Nonlinear Kalman Filter Framework

The basic configuration for optimal terrain-aided navigation is shown in Fig. 1.

This structure is typical of Kalman filtering in which nonlinear auxiliary measurements are iteratively processed to estimate and compensate for the errors in a navigation system. At each measurement update time the current state estimate, in conjunction with stored topographical data, is used to obtain a prediction of what the radar ground clearance measurement should be. The actual radar measurement is then compared with this predicted measurement, and their difference is processed by the Kalman filter to generate estimates of the navigation system's error states. The measurement matrix in this case is related to the downrange and crossrange terrain slopes calculated from the stored data. The error estimates are then fed back to compensate the navigation system and thus provide an improved estimate of the actual state (position, velocity, etc.) of the system. This process is iterated many times, e.g., every 30–50 m of distance traveled, as the system maneuvers along its trajectory, thus providing essentially continuous updating to the navigation system.

In order to apply approximate nonlinear filtering algorithms, a set of differential equations describing how the states of the navigation system propagate is required. It is sufficient here to assume the following general form of system model [24], [25]:

$$\dot{x}(t) = f(x(t), t) + w(t) \tag{1}$$

where $x(t)$ is the vector of states in the navigation system and $w(t)$ is the Gaussian system noise vector with $E\{w(t)\} = \mathbf{0}$ and

$$E\{w(t)w^T(t+\tau)\} = Q(t)\delta(\tau).$$

The state estimate propagation equation[1] between terrain-clearance measurements is

$$\dot{\hat{x}}(t) = f(\hat{x}(t), t) \tag{2}$$

and the error covariance propagation equation[1] is

$$\dot{P}(t) = F(\hat{x}(t), t)P(t) + P(t)F^T(\hat{x}(t), t) + Q(t) \tag{3}$$

where the ^ notation refers to an estimate and F is the partial derivative matrix of f with respect to x. The emphasis here is on the use of nonlinear *measurements*. The state dynamics are assumed to be a very good linear approximation. Actually, in practice the position and other necessary errors are included in the state vector and constantly updated in the navigational system to maintain the adequacy of such a representation.

[1]These are the first-order appropriations [25] to the more general nonlinear expressions for propagating the conditional mean and its associated covariance matrix

$$P(t) = E\{[\hat{x}(t) - x(t)][\hat{x}(t) - x(t)]^T\}.$$

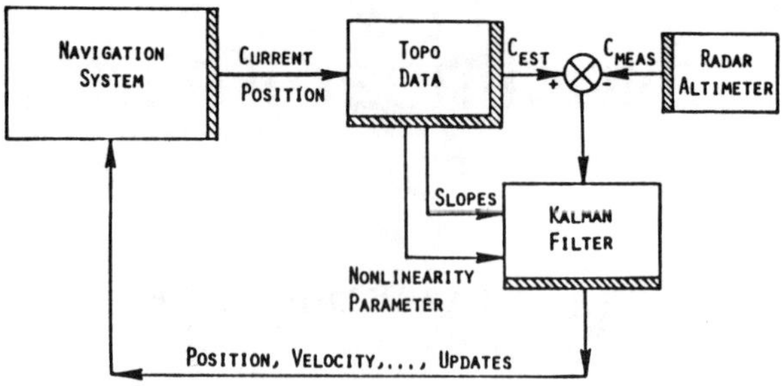

Fig. 1. Optimal terrain-aided navigation.

The terrain-clearance measurement equation is in general a nonlinear function of the antenna pattern, the cross-range/downrange position (CR, DR), altitude (z), and attitude (ψ) of the vehicle. In most instances,[2] a reasonable approximation to the kth terrain-clearance measurement is

$$\begin{aligned} c_k &= c_k(x_k) + v_k \\ &= z_k - h(CR_k, DR_k) + v_k \end{aligned} \tag{4}$$

where $h(CR, DR)$ is the height of the terrain at the position (CR, DR), and v_k is the measurement error with $E\{v_k\} = 0$ and $E\{v_k v_j\} = r_k \delta_{kj}$. The measurement error consists of both radar altimeter error and reference map errors. If these errors are correlated processes it may be necessary to model them as such.

Following the derivation in Gelb [25, ch. 6], when the measurement c_k is taken, the measurement update equations are given by

$$\hat{x}_k(+) = \hat{x}_k(-) + K_k[c_k - \hat{c}_k(x_k)] \tag{5}$$

where $\hat{x}_k(+)$ is the estimate after updating, $\hat{x}(-)$ is the estimate prior to the update, and $\hat{c}_k(x_k) \triangleq E\{c_k(x_k)\}$ is the predicted clearance measurement, where we will now approximate the expectation by assuming x_k is Gaussian with mean $\hat{x}_k(-)$ and covariance matrix $P_k(-)$. The optimal gain matrix is given by

$$\begin{aligned} K_k = E\{&[x_k - \hat{x}_k(-)][c_k(x_k) - \hat{c}_k(x_k)]^T\} \\ &\cdot \left[E\{[c_k(x_k) - \hat{c}_k(x_k)][c_k(x_k) - \hat{c}_k(x_k)]^T\} + r_k\right]^{-1}. \end{aligned} \tag{6}$$

And, finally, the error covariance measurement update equation is given by

$$\begin{aligned} P_k(+) = P_k(-) - K_k E\{&[c_k(x_k) - \hat{c}_k(x_k)] \\ &\cdot [x_k - \hat{x}_k(-)]^T\}. \end{aligned} \tag{7}$$

The techniques utilized in the implementation and evaluation of these equations and expectations are funda-

[2]For cases where the radar altimeter tracks a "closest point" that is not near nadir, the measurement function needs to account for the nonlinear variation with altitude z_k as well as the usual variation with downrange and crossrange positions.

mental in obtaining satisfactory terrain-aided performance, i.e., avoiding divergent filter behavior while maintaining highly accurate state position estimates.

In particular, satisfactory performance depends upon the adequacy of the basic approach used to linearize the measurement function in the calculation of (6) and (7), i.e., to linearize the terrain. Since the expectations must be taken with respect to the current state estimate uncertainties, whatever linearization is used must be valid over a region whose size is consistent with current position estimation errors.

If the position errors are sufficiently small relative to the terrain correlation length, it can be assumed that the terrain is linear and simple first-order spatial derivatives will yield satisfactory performance. In cases where position errors are comparable to or larger than the terrain correlation length, more sophisticated linearization techniques are required.

Three different linearization approaches for treating terrain nonlinearities are discussed in the following sections. The first two, local slope linearization and stochastic linearization, attempt to maintain satisfactory performance within a single Kalman filter framework. The third approach utilizes parallel Kalman filters, each linearized over a different region to handle the case of very large (relative to the terrain correlation length) initial position errors.

III. Local Slope Linearization

The standard extended Kalman approach to linearization is to expand a nonlinear measurement function in a Taylor series about the current state estimates, retain the linear terms, and neglect higher order ones. With the definition of $c_k(x_k)$ from (4), the predicted clearance measurement approximation becomes

$$\hat{c}_k(x_k) = \hat{z}_k(-) - h(\widehat{CR}_k(-), \widehat{DR}_k(-)) = c_k(\hat{x}_k) \quad (8)$$

and the linearization becomes

$$c_k(x_k) \triangleq c_k(\hat{x}_k) + H_k[x_k - \hat{x}_k(-)] \quad (9)$$

where the measurement matrix H_k is defined by

$$H_k = [-\text{crossrange slope}, -\text{downrange slope}, 1, 0, 0, \cdots] \quad (10)$$

with the first three states being taken to be crossrange (CR), downrange (DR), and altitude (z), respectively.

Substituting (9) and (10) into (5), (6), and (7) yields the following standard extended Kalman filter update equations [25]:

$$\hat{x}_k(+) = \hat{x}_k(-) + K_k[c_k - c_k(\hat{x}_k)], \quad (11)$$

$$P_k(+) = [I - K_k H_k] P_k(-), \quad (12)$$

and

$$K_k = P_k(-) H_k^T [H_k P_k(-) H_k^T + r_k]^{-1}. \quad (13)$$

To demonstrate the performance of the standard extended Kalman filter equation, a Monte Carlo simulation of the above equations was performed. A short-range standoff mission scenario was utilized with the system navigating from an initialization point (IP) to a target located approximately 6 km away. The flight time and speed were approximately 40 s and 150 m/s, respectively. A low altitude flight profile was assumed with radar altimeter measurements being processed every 0.4 s. The measurement noise was assumed to be 6 m. Finally, the inertial system errors were modeled for this short mission as being adequately represented by three position and three velocity error states. The initial velocity error standard deviations were taken to be approximately 3 m/s in the east and north states and 0.3 m/s in the vertical state. The initial position error standard deviations were 15 m in all three position error states. The terrain consisted of rolling hills having a terrain height standard deviation (σ_T) of approximately 12 m, a correlation length near 500 m, and a slope standard deviation near 5 percent. Fig. 2 shows the histogram of radial misses at the target for 100 Monte Carlo runs.

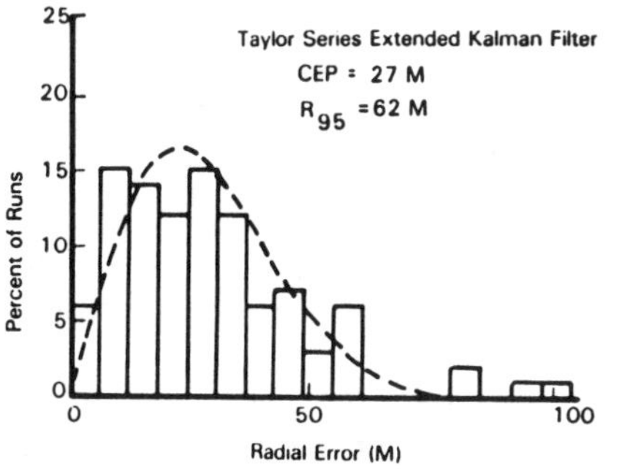

Fig. 2. Standard extended Kalman filter performance.

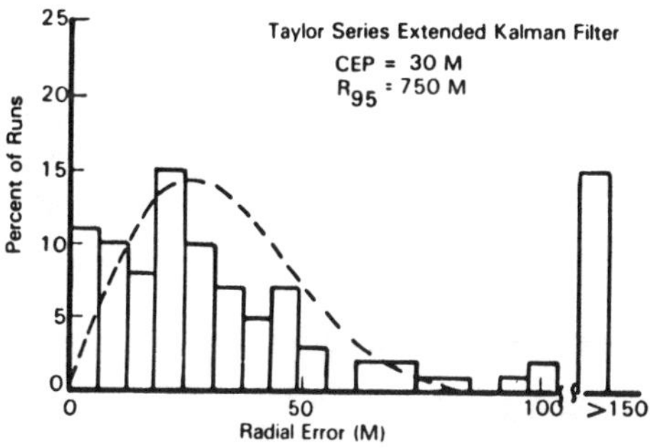

Fig. 3. Standard extended Kalman filter divergence problem.

As seen, the filter achieved a 27 m circular error probable (CEP) at the target. This is a significant improvement over the unaided inertial system which would have had a 120 m CEP due to its large velocity errors. The dashed curve is a theoretical Rayleigh distribution also having a 27 m CEP.

Unfortunately, as mentioned earlier, the highly nonlinear nature of terrain surfaces can lead to filter divergence, especially when the linearization error is comparable to the measurement error. In these cases the standard extended Kalman filter may yield unsatisfactory performance, and divergence can occur in which the actual estimation errors become orders of magnitude larger than the filter's own computation of their covariance [26]. Fig. 3 demonstrates this phenomenon for a simulation test case in which the initial position error standard deviations were 75 m and all other conditions were the same as in the prior simulation. Note that the error distribution is quite non-Rayleigh and has an unexpectedly long tail (more than 15 percent of the Monte Carlo runs had target errors greater than 150 m). This divergence problem is severe, and when it occurs the errors become catastrophic as indicated by a 95th percentile of 750 m for the radial misses. A solution to this problem is the subject of the following section.

IV. Modified Stochastic Linearization

Techniques for obtaining approximate nonlinear filters that attempt to account for measurement nonlinearities and prevent filter divergence have been investigated [24], [25]. This section describes an extension of these techniques and the subsequent application of them to our terrain-aided navigation problem.

Define the stochastic linearization of $c_k(x_k)$ by

$$c_k(x_k) \triangleq \hat{c}_k(x_k) + H_k[x_k - \hat{x}_k(-)] + \epsilon_{\text{fit}} \tag{14}$$

where $\hat{c}_k(x_k) \triangleq E\{c_k(x_k)\}$ is as previously defined, ϵ_{fit} is the linearization fit error, and the measurement matrix H_k is redefined by

$$H_k \triangleq E\{[c_k(x_k) - \hat{c}_k(x_k)][x_k - \hat{x}_k(-)]^T\} P_k^{-1}(-). \tag{15}$$

From the definition of $\hat{c}_k(x_k)$ and H_k, and since we are assuming x_k to be Gaussian with mean $\hat{x}_k(-)$ and covariance $P_k(-)$, one sees that they are simply the least square linear approximation of $c_k(x_k)$ about $\hat{x}_k(-)$ using a Gaussian weighing function with covariance $P_k(-)$.

Making the simplifying approximation that the fit error is uncorrelated with $c_k(x_k)$ and $H_k[x - \hat{x}_k(-)]$ and substituting (14) and (15) into (5), (6), and (7) yields the following measurement update equations:

$$\hat{x}_k(+) = \hat{x}_k(-) + K_k[c_k - \hat{c}_k(x_k)] \tag{16}$$

$$P_k(+) = [I - K_k H_k] P_k(-) \tag{17}$$

and

$$K_k = P_k(-) H_k^T [H_k P_k(-) H_k^T + r_k + r\text{fit}_k]^{-1} \tag{18}$$

where $r\text{fit}_k \triangleq E\{\epsilon_{\text{fit}} \epsilon_{\text{fit}}^T\}$ is the variance of the fit error due to the nonlinearity of the terrain. Note that the usual stochastic linearization equations [25] have been modified by the inclusion of the fit error variance. This additional term is very important in that, similar to second-order techniques, [26], the increased variance keeps the Kalman filter from utilizing data from regions of high nonlinearity. Since this term varies with time, as the terrain nonlinearities vary, the filter will emphasize those measurements with less linearization fit error. For typical terrain, the fit error can be quite large at times relative to the usual measurement noise. In addition, the fit errors are often large in

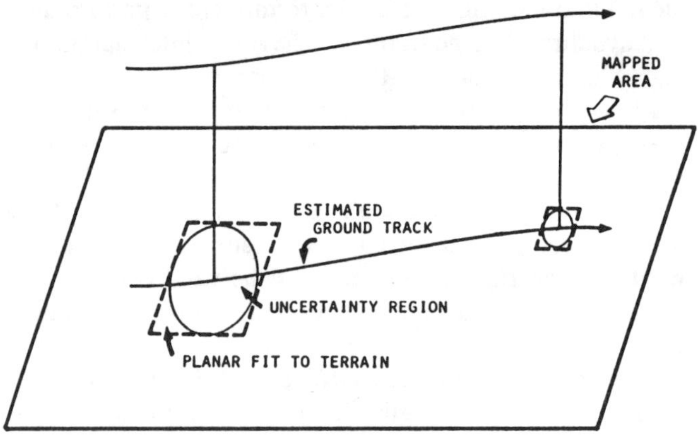

Fig. 4. Modified terrain linearization process.

regions of terrain having steep slopes. Thus, without the additional fit error being considered, the usual stochastically linearized filter would place the most weight on measurements which violated the linearity assumption the worst, obviously not a desirable situation.

Since the linearization is being done using the current covariance estimate $P_k(-)$, the size of the region of linearization is proportional to the position uncertainty at that point. Thus, the technique is adaptive and, assuming large initial position uncertainty, tends to initially smooth out the terrain and its slope variations. However, as the CEP improves and the position uncertainty reduces in size, the smoothing is lessened, the fit error decreases, and the filter is able to take advantage of the higher frequency content in the terrain to further decrease its CEP. This concept is illustrated in Fig. 4.

With the definition of $c_k(x_k)$ from (4),

$$\hat{c}_k(x_k) = \hat{z}_k(-) - \hat{h}(CR_k, DR_k), \tag{19}$$

and treating the terrain height as a function of position only, H_k becomes

$$H_k = (-H_{CR_k}, -H_{DR_k}, 1, 0, \cdots) \tag{20}$$

where the first three states are taken to be crossrange (CR), downrange (DR), and altitude (z), respectively. $\hat{h}(CR_k, DR_k)$, H_{CR_k}, and H_{DR_k} are the parameters of the least square linear approximation of the terrain $h(CR, DR)$ about $\widehat{CR}_k(-)$ and $\widehat{DR}_k(-)$ using the Gaussian covariance matrix corresponding to the CR and DR part of $P_k(-)$.

Fig. 5 illustrates how certain problems are avoided by this modified stochastic linearization technique. For position uncertainties that are large relative to the nonlinearities in the terrain, the simple Taylor series expansions used in both the usual extended Kalman and second-order Gaussian filters can lead to large measurement and linearization slope errors. These errors can quickly lead to filter divergence. Also shown is the fit error due to the nonlinear terrain. It is easy to see that the variance of this error ($r$fit) will vary with time and, in general, will decrease as the position error decreases. In addition, as the position error

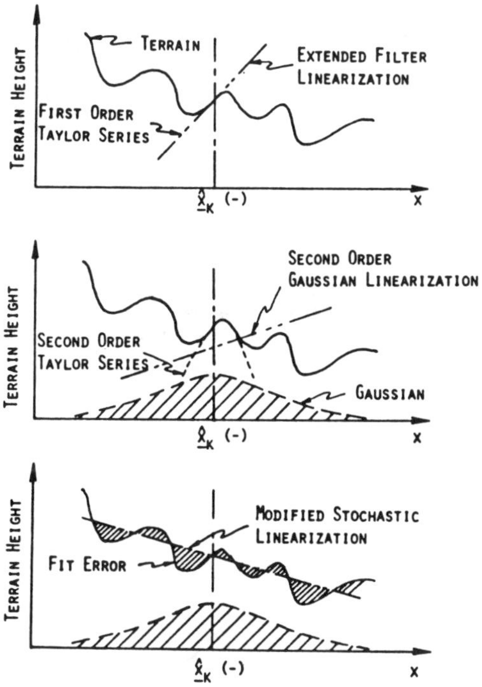

Fig. 5. Linearization techniques.

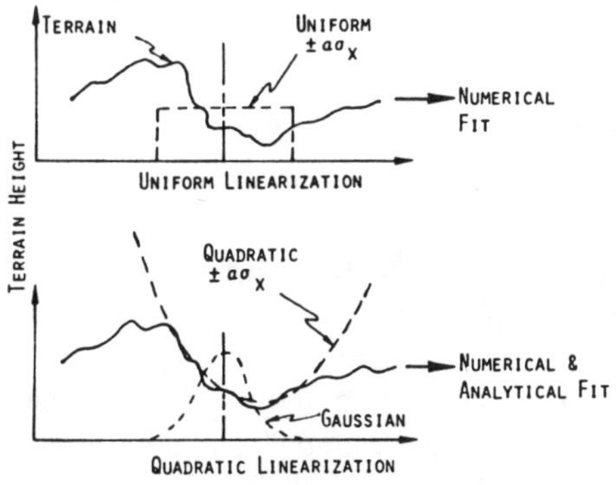

Fig. 6. Modified stochastic linearization.

decreases, the linearization slopes increase. These two effects work together to allow the filter to effectively use the low frequencies in the terrain at first, but to converge and take advantage of the higher frequencies as the system updating proceeds. In the limit as the position errors decrease, the terrain becomes locally linear and the stochastic linearization becomes equivalent to the extended and second-order filter linearizations.

The actual stochastic linearization utilizes the stored topographical data. A straightforward least squares fitting of a plane to the gridded data can be carried out using, as the weighting function, a Gaussian density centered at the current state estimate and having as its covariance matrix the current position error covariance. The variance of the weighted residuals is $r\text{fit}_k$. Two other techniques for approximating this calculation are shown in Fig. 6.

The uniform stochastic linearization essentially approximates the Gaussian weighting function by a uniform density over a region whose size and shape is proportional to the current position uncertainty. Ordinary least squares equations for the plane and residual error result. This is the technique used for the simulations in the following section.

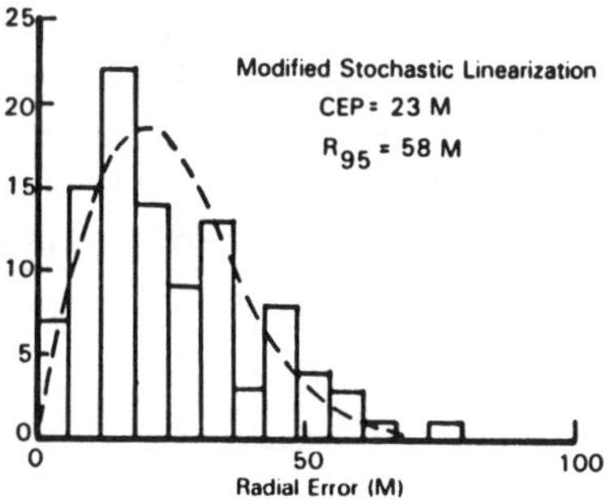

Fig. 7. Performance improvement with modified stochastic linearization.

The quadratic stochastic linearization essentially approximates the terrain as a quadratic surface over a region proportional to the current position uncertainty and then analytically linearizes this quadratic surface using a Gaussian weighting function. The fit variance $(r\text{fit}_k)$ in this case is the sum of the numerically calculated quadratic residual error plus an additional amount due to the subsequent analytical linearization of the quadratic surface.

It should be noted that if there is significant noise in the stored reference data, lower bounds should be placed on the size of the fit regions. This guarantees that the reference errors will not adversely affect the slope calculations.

The above basically describes the modified stochastic linearization technique used for optimally processing the terrain-clearance data. The authors feel that this modified technique should find applications in other nonlinear filtering problems and should provide superior performance over both the usual second-order Gaussian filter as well as the standard stochastically linearized filter when the measurement equation is highly nonlinear. Finally, the consideration of errors in ones knowledge of the actual measurement function and the techniques for handling these errors contribute to preventing filter divergence.

V. Modified Stochastic Linearization Performance

As shown earlier in Fig. 3, the local slope linearization approach can lead to filter divergence under certain conditions. To demonstrate the improved performance of the modified stochastic linearization technique, the same case was simulated using the new approach. Fig. 7 demonstrates the improved performance, where the divergence problem disappeared, and accuracy improved as well. Again, the dashed curve is the Rayleigh density function having the same CEP as the histogram.

System accuracy in general is a function of INS errors, measurement errors, and terrain signature levels. Considerable insight into filter performance and into those parameters influencing system accuracy can be gained through closed-form analytical approximations of predicted performance. Utilizing standard estimation theory techniques, subject to some reasonable limitations on terrain roughness and update interval size, general expressions for filter performance have been derived. For example, if the inertial system is modeled as having only position errors in all three axes, the estimation error covariance matrix $\boldsymbol{P}$ for downrange and crossrange positions is given by

$$\boldsymbol{P}=\left[\begin{array}{c|c} \frac{1}{\sigma_{\mathrm{sdr}}^2} & \frac{\rho}{\sigma_{\mathrm{sdr}}\sigma_{\mathrm{scr}}} \\ \hline \frac{\rho}{\sigma_{\mathrm{sdr}}\sigma_{\mathrm{scr}}} & \frac{1}{\sigma_{\mathrm{scr}}^2} \end{array}\right]\cdot\frac{\sigma_n^2}{\sqrt{N\cdot(1-\rho^2)}} \tag{21}$$

where

$\sigma_{\mathrm{sdr}}, \sigma_{\mathrm{scr}}$ = downrange and crossrange terrain slope standard deviations computed from the terrain slope history obtained by terrain linearization.

ρ = correlation between downrange and crossrange slopes from the terrain slope history.

σ_n = measurement noise standard deviation.

N = number of independent measurements.

Note that this covariance matrix has the property of including the terrain roughness variations and correlations between the crossrange and downrange directions and that the error covariances are proportional to the variance of the measurement errors. Performance characterization can be further simplified as this equation describes the covariance of a bivariate Gaussian density function. If the terrain has uncorrelated slopes in the downrange and crossrange directions with equal slope standard deviations, CEP can be expressed as

$$\mathrm{CEP}=\frac{1.17\sigma_n}{\sqrt{N}\,\sigma_s} \tag{22}$$

where $\sigma_s=\sigma_{\mathrm{sdr}}=\sigma_{\mathrm{scr}}$. Similar position error and velocity error predictions are obtainable for the case when velocity errors are also present in the inertial system and must be estimated. These are

$$\mathrm{CEP}_p=\frac{2.35}{\sqrt{N}}\frac{\sigma_n}{\sigma_s}, \tag{23}$$

$$\mathrm{CEP}_v=\frac{4.08}{\sqrt{N}}\frac{\sigma_n}{T\sigma_s} \tag{24}$$

where

CEP_p = position CEP,

CEP_v = velocity CEP,

T = total flight time.

As indicated, the terrain roughness parameter characterizing performance is the standard deviation of the terrain slopes evaluated along the reference trajectory. Assuming a total flight time of 40 s incorporating 100 independent updates, Fig. 8 shows predicted performance (23) as a function of terrain roughness for several values of measurement noise. For reference, a σ_s of 0.02 is typical of terrain with a standard deviation of 5 m. The simulation results shown in Figs. 2 and 7 basically agree with these curves.

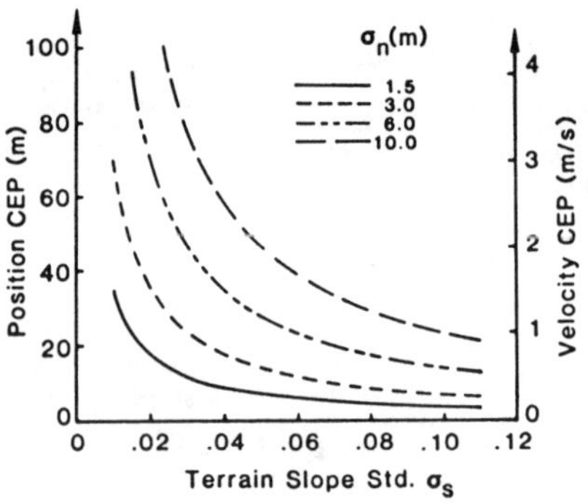

Fig. 8. System performance.

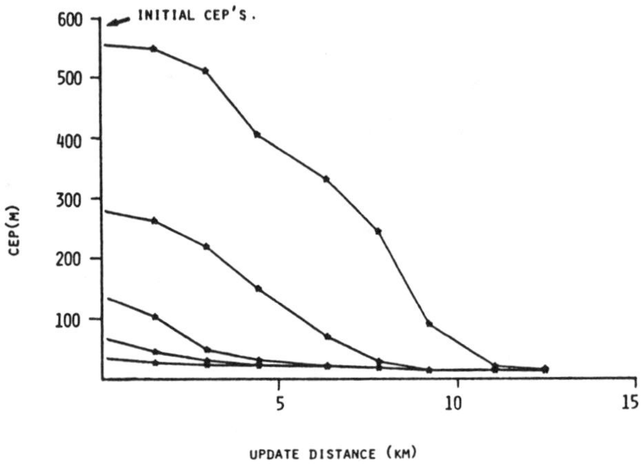

Fig. 9. Terrain smoothing effect on performance.

Because of the time varying nature of both the terrain slopes and the corresponding measurement errors, additional insight and performance and error sensitivity analyses can be obtained through covariance propagation techniques [25]. The general approach is to utilize a terrain slope history (downrange and crossrange) derived from actual topographical data and process it in a covariance analysis computer program to predict performance for any candidate filter design. This procedure was used to generate Fig. 9, which plots CEP histories for various initial position uncertainties. All other conditions were as before and again 6 m of measurement noise was assumed here. These data were calculated by propagating covariances using stochastic linearization along a particular trajectory over the prior mentioned terrain. Note that although the system eventually achieves equivalent accuracy independent of initial CEP, the required flight path is considerably longer for the larger starting CEP's. Thus, the basic effect of stochastic linearization is to smooth or low-pass filter the reference topography so that the calculated terrain slopes are the average slopes over the current position uncertainty region. When the initial position uncertainties are large compared to the terrain correlation length, the resulting terrain smoothing can become excessive. Hence, this method of terrain linearization is clearly restrictive in applications where initial CEP is large and the update distance is limited.

It should be noted here that if batch processing were allowed, this restriction could possibly be overcome by a globally iterated filter-smoother [24] iteratively sweeping through the data, utilizng the state estimate and improved CEP of each previous sweep as the starting point until no further improvement was warranted. This procedure would prevent divergence due to large initialization errors while at the same time allowing for eventual utilization of the full frequency content of the terrain and the correspondingly smaller CEP. This approach not only requires storage of INS and altimeter measurements, but is nonreal-time and is not appropriate for some applications.

Fortunately, a parallel filter structure approach has been developed for handling this delayed convergence problem and is the subject of the next section.

VI. Parallel Filter Structure

The approach developed to handle the case of large initial errors is embodied in the processing approach shown in Fig. 10 where a bank of parallel filters[3] and a set of position displaced INS trajectories are employed [21]–[23]. Each of the parallel filters linearizes around and generates error estimates for its trajectory. Therefore, the operation of each filter can be viewed as altering its trajectory within the degrees of freedom allowed by the number of modeled states to achieve the best match of the measuremens with the terrain in the vicinity of its trajectory. A selection algorithm can then be used that examines the residuals of each filter to select which filter has the best match, i.e., is using the proper terrain and whose error estimates have converged. Note that for a given initial CEP, the number of filters, and hence filter spacing can be adjusted so that the starting uncertainty for each filter is as small as desired. Note also that after an appropriate distance has been flown, the parallel structure would no longer be needed and one filter would be sufficient for continued system operation.

The selection of the convergent filter can be done quite easily by examining the residuals Δ_i for each filter. A selection algorithm based upon the assumed whiteness property of the filter residuals that worked well in practice is to choose the filter with the smallest value of

$$\mathrm{AWRS}_{j\text{th filter}} = \frac{1}{N}\left[\sum_{i=1}^{N} \frac{\Delta_i}{H_i P_i H_i^T + R_i}\right]_{j\text{th filter}} \tag{25}$$

[3]The parallel filter structure discussed here is part of a very broad class of algorithms which has come to be referred to as multiple model estimation algorithms [27]–[29].

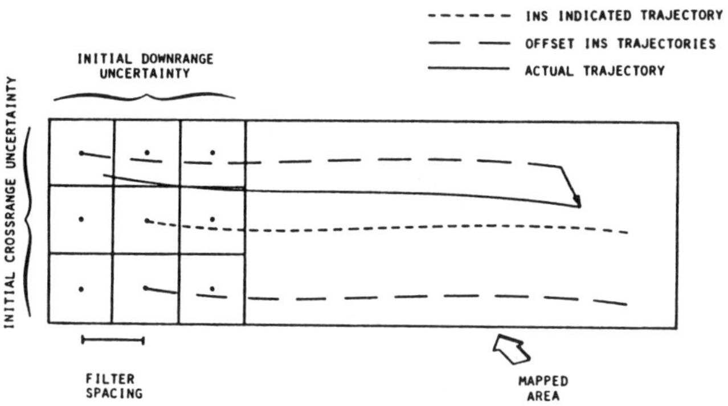

Fig. 10. Parallel filter configuration.

TABLE I
Typical AWRS Values and Final Radial Estimation Errors

Minimum AWRS Selects Filter
With Minimum Estimation Error

5.0	3.3	3.1	4.4	4.8	2.0
4.7	5.1	2.5	4.7	2.7	2.0
7.4	3.9	(1.1)	1.6	2.2	1.8
2.3	3.2	2.4	2.2	2.4	2.9
2.1	2.6	3.9	2.5	2.6	2.6
2.4	3.4	4.3	2.8	4.1	1.9

AWRS

2111	2169	1516	1156	1108	1869
2493	1882	585	1339	395	1554
1556	1045	(13)	145	182	1483
412	893	122	175	1475	1504
1427	607	1073	1485	1486	1504
1758	464	764	2131	1097	2169

Final Radial
Estimation Errors (M)

where H_i is the measurement vector containing the terrain slopes at the ith time interval, P_i is the covariance, and N is the number of measurements processed. This AWRS value is the average weighted residual squared between the predicted ground clearance for each filter and the ground clearance measured by the radar altimeter, for each time t_i. The weighting factor is inherently calculated by each Kalman filter and is simply the expected variance of Δ_i at each measurement. By examining the AWRS values for each filter after a sufficiently large number of measurements have been processed, the correct filter and its associated state error estimates can be chosen.

As an example, Table I shows AWRS values and final radial estimation errors for 36 parallel filters that were used to process a simulated test run. The test conditions were the same as before except very large initial error standard deviations of approximately 500 m were used. Each of the parallel filters utilized modified stochastic linearization and its covariance matrix was initialized with a position error standard deviation of 100 m. Note that the filter whose AWRS value is minimum also has the smallest estimation error, as desired.

It must be pointed out here, however, that if the terrain within the map is not unique, the selection algorithm might choose the wrong filter, which is the classical false fix problem. In this case, of course, the fact that there are competitive AWRS values can be detected and appropriate action taken.

Fig. 11 shows the histogram of radial misses at the target for this 36 parallel filter test case. One sees that for this

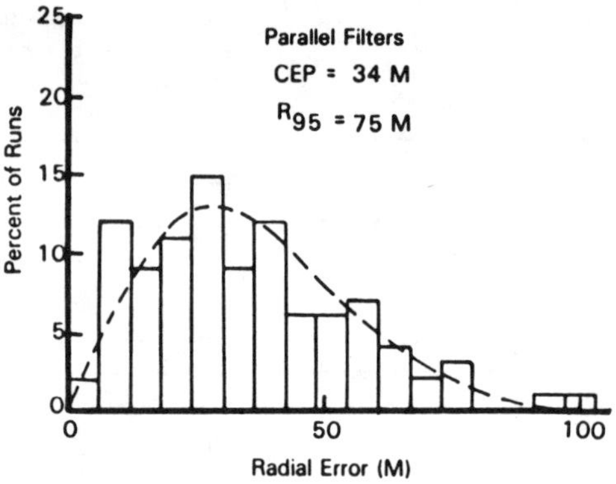

Fig. 11. Parallel filter performance.

short flight distance, the parallel filter formulation performs over on order of magnitude better than the modified stochastic linearization approach (see Fig. 9). The smoothing effect of the single-filter modified approach does not disappear until distances near 12 km, whereas each parallel filter, having been initialized with a position error standard deviation of only 100 m, can draw down much faster. This ability to converge much faster is important in some applications.

An interesting observation can be made regarding this parallel Kalman filter structure, as it can be viewed as a very general processing methodology that includes more well-known techniques as special cases. For example, at one end of the spectrum, postulate many low order filters spaced very close together. If each filter modeled only a vertical channel or measurement error bias, no terrain linearization would be required (!), and the parallel filter structure would reduce to a pure search or correlation technique such as TERCOM with a mean-square difference algorithm. On the other hand, the special case of using only one filter is the modified stochastic linearization described in Section IV. As the number of filters, associated filter spacing, and number of estimated state is varied, a spectrum of different system configuration is represented.

The final topic associated with the parallel filter approach that needs to be dealt with is the potential practical difficulty of the computational burden associated with running many Kalman filters in real time. To this end, the techniques of near optimal modeling have been used to reduce the number of modeled states, and hence simplify filter implementation [22], [23]. In that the dominant INS errors for the final application considered were position errors, three state "near optimal" filters were the basis for design. The time variation of these dominant position errors were modeled in the filters as random walks. Thus, three-state filters with driving process noise were implemented by increasing the diagonals of the position error filter covariances at each update. This technique has, of course, been used by other investigators in the design of suboptimal filters [25]. In our application, the use of three states plus process noise resulted in a 40 percent degradation in performance.

IX. Conclusions

The application of Kalman filtering techniques to the development of a terrain-aided navigation system based upon a radar altimeter sensor and a topographic reference map has been described. The fundamental difficulty in this application was the nonlinear characteristics of the terrain elevation measurement function.

Several different approaches for treating these nonlinearities were developed and limitations of each were identified. These approaches were: 1) an extended Kalman filter using local terrain slopes, 2) a quasi-linear Kalman filter based upon stochastic terrain linearization, and 3) a bank of parallel Kalman filters. This last approach led to a quite general system framework in which some commonly used terrain-aided navigation methods, e.g., TERCOM, can be interpreted as special cases.

References

[1] R. E. Kalman, "A new approach to linear filtering and prediction problems," *Trans. ASME*, vol. 82, pp. 35–45, Mar. 1960.
[2] R. E. Kalman and R. S. Bucy, "New results in linear filtering and prediction theory," *Trans. ASME*, vol. 83, pp. 95–108, Mar. 1961.
[3] R. G. Brown and D. T. Friest, "Optimization of hybrid inertial solar-tracker navigation system," in *IEEE Int. Conv. Rec.*, no. 7, 1964, pp. 121–135.
[4] S. G. Wilson, "Nonlinear filter evaluation for estimating vehicle position and velocity using satellites," *IEEE Trans. Aerosp. Electron. Syst.*, vol. AES-9, pp. 65–75, Jan. 1973.
[5] I. Y. Bar-Itzhack, "Optimal updating of INS using landmarks," in *AIAA Guidance Contr. Conf. Proc.*, Aug. 1977, pp. 492–502.
[6] W. Zimmerman, "Optimum integration of aircraft navigation systems," *IEEE Trans. Aerosp. Electron. Syst.*, vol. AES-5, pp. 737–747, Sept. 1969.
[7] J. A. D'Appolito and J. F. Kasper, Jr., "Predicted performance of an integrated OMEGA/inertial navigation system," in *Nat. Aerosp. Electron. Conf. Proc.*, May 1971, pp. 121–128.
[8] J. C. Bobick and A. E. Bryson, Jr., "Updating inertial navigation systems with VOR/DME information," in *AIAA Guidance Contr. Conf. Proc.*, Aug. 1972, paper 72-846.
[9] R. P. Moore, C. A. Hawthorne, M. C. Hoover, and E. S. Gravlin, "Position updating with microwave radiometric sensors," in *Nat. Aerosp. Electron. Conf. Proc.*, May 1976, pp. 14–19.
[10] P. G. Savage and G. L. Hartmann, "Optimum aiding of inertial navigation systems using air data," in *AIAA Guidance Contr. Conf. Proc.*, May 1976, pp. 14–19.
[11] A. J. Verderese, "Radar navigation via map matching," in *Proc. Electro-Opt. Syst. Des. Conf.*, Sept. 1972, pp. 56–62.
[12] P. J. Klass, "New guidance technique being tested," *Aviat. Week Space Technol.*, pp. 48–51, Feb. 25, 1974.
[13] W. R. Baker and R. W. Clem, "Terrain contour matching (TERCOM) primer," Tech. Rep. ASP-TR-7-61, Aeronaut. Syst. Div., Wright-Patterson AFB, OH, Aug. 1977.
[14] M. D. Mobley, "Air launched cruise missile (ALCM) navigation system development integration and test," in *Nat. Aerosp. Electron. Conf. Proc.*, May 1978, pp.1248–1254.
[15] L. D. Hostetler, "An analysis of a terrain-aided inertial navigation system," Tech. Rep. SAND75-0299, Sandia Lab., Albuquerque, NM, Sept. 1975.
[16] —, "A Kalman approach to continuous aiding of inertial navigation systems using terrain signatures," in *IEEE Milwaukee Symp. Automat. Comput. Contr. Proc.*, Apr. 1976, pp. 305–309.
[17] L. D. Hostetler and R. C. Beckmann, "The Sandia inertial terrain-aided navigation system," Tech. Rep. SAND77-7521, Sandia Lab., Albuquerque, NM, Sept. 1977.
[18] R. D. Andreas, L. D. Hostetler, and R. C. Beckmann, "Continuous Kalman updating of an inertial navigation system using terrain measurements," in *Nat. Aerosp. Electron. Conf. Proc.*, May 1978, pp. 1263–1270.
[19] L. D. Hostetler, "Optimal terrain-aided navigation systems," presented at AIAA Guidance Contr. Conf., Palo Alto, CA, Aug. 1978, and Tech. Rep. SAND78-0874, Sandia Lab., Albuquerque, NM, June 1978.
[20] L. D. Hostetler and R. C. Beckmann, "Expanding the region of convergence for SITAN through improved modeling of terrain

nonlinearities," in *Nat. Aerosp. Electron. Conf. Proc.*, May 1979, pp. 1023–1030.

[21] T. C. Sheives and R. D. Andreas, "An alternate approach for terrain-aided navigation using parallel extended Kalman filters," Tech. Rep. SAND79-2198, Sandia Nat. Lab., Albuquerque, NM, Dec. 1979.

[22] T. C. Sheives, "Reduced-order state estimation for terrain-aided navigation," Tech. Rep. SAND79-2199, Sandia Nat. Lab., Albuquerque, NM, Dec. 1979.

[23] T. C. Sheives, "State estimation using parallel extended Kalman filters on nonlinear measurements," Ph.D. dissertation, Univ. New Mexico, Albuquerque, NM, Dec. 1979, and Tech. Rep. SAND80-0013, Sandia Nat. Lab., Albuquerque, NM, Jan. 1980.

[24] A. H. Jazwinski, *Stochastic Processes and Filtering Theory*. New York: Academic, 1970.

[25] A. Gelb *et al.*, *Applied Optimal Estimation*. Cambridge, MA: M.I.T. Press, 1975.

[26] W. S. Widnall, "Enlarging the region of convergence of Kalman filters that encounter nonlinear elongation of measured range," in *AIAA Guidance Contr. Conf. Proc.*, Aug. 1972, paper 72-879.

[27] D. T. Magill, "Optimal adaptive estimation of sampled stochastic processes," *IEEE Trans. Automat. Contr.*, vol. AC-10, pp. 434–439, Oct. 1965.

[28] D. G. Lainiotis, "Optimal adaptive estimation: Structure and parameter adaptation," *IEEE Trans. Automat. Contr.*, vol. AC-15, pp. 160–170, Apr. 1971.

[29] M. Athans and C. B. Chang, "Adaptive estimation and parameter identification using multiple model estimation algorithm," Lincoln Lab. Tech. Note 1976-28, June 1976.

Application of Multiple Model Estimation to a Recursive Terrain Height Correlation System

GREGORY L. MEALY, MEMBER, IEEE, AND WANG TANG, MEMBER, IEEE

Abstract—**This paper describes the results of an investigation of the performance capabilities of an extended Kalman filter (EKF)-based recursive terrain correlation system proposed for low-altitude helicopter navigation. The major disadvantage of this concept is its sensitivity to initial position error. One method for reducing this sensitivity involves the use of multiple model estimation techniques. In the multiple model approach, a bank of identical EKF's, each of which is initialized at a different point in the *a priori* uncertainty basket, is employed to ensure that one filter is initialized near the true aircraft position. In this manner, the probability of filter convergence is increased substantially, leading to improved navigation performance.**

Manuscript received March 3, 1982; revised September 27, 1982. This work was supported by the U.S. Army Avionics Research and Development Activity (AVRADA) under Contract DAAK80-79-C-0268.

The authors are with The Analytic Sciences Corporation, Reading, MA 01867.

Reprinted from *IEEE Trans. Automat. Contr.*, vol. AC-28, pp. 323–331, Mar. 1983.

I. Introduction

By comparing measured terrain height (obtained as the difference between baro- and radar-altimeter measurements) with stored terrain height data, it is possible to estimate aircraft position. The well-known terrain contour matching (TERCOM) system is one implementation of this concept. Recently, investigators have concentrated on a recursive implementation of such a terrain correlation algorithm (based upon an extended Kalman filter) to continuously update vehicle position [1]–[3]. This paper describes an assessment of the performance capabilities of such a system under conditions typical of helicopter nap-of-the-earth flight. The performance of this algorithm depends upon the local validity of the terrain height linearization. If the estimated initial position is sufficiently far from the true position, the filter may diverge. The possibility of filter divergence is a major obstacle to the use of this approach.

This paper explores the use of multiple model estimation algorithms (MMEA) to reduce the likelihood of filter divergence arising from large initial position uncertainty. In the MMEA approach adopted herein, a bank of identical extended Kalman filters, each of which is initialized at a different point in the *a priori* uncertainty basket, is employed to ensure that one filter is initialized near the true aircraft position. In this manner, the probability of filter convergence is substantially increased, leading to improved navigation performance.

II. Terrain Correlation with an Extended Kalman Filter

The basic TERCOM concept considers ground clearance as the fundamental system measurement from which vehicle position is to be determined. Measured terrain clearance is compared to the value predicted by the on-board navigation system and the measurement model to obtain an error signal proportional to navigation errors. Using a system model which describes how these errors propagate, the filter then implements the equations to calculate an estimate of reference state errors. (The filter may be any linearized filter such as the extended Kalman filter or the Gaussian second-order filter; see, for example, [2], [3]. The examples used here will be for the extended Kalman filter.) The error estimate is subtracted from the reference to give a corrected estimate of the vehicle state.

The update procedure is illustrated in Fig. 1 for a simplified example in which only downrange and altitude errors are considered. The terrain is assumed to be approximately planar in the vicinity of the predicted vehicle location. Based on this approximation, a terrain clearance measurement value is predicted and compared to an actual measurement taken at the true vehicle position. A correction factor Δx can then be computed using the linear slope information to approximate the error associated with the initial x position estimate. A more detailed description of this procedure may be found in [2].

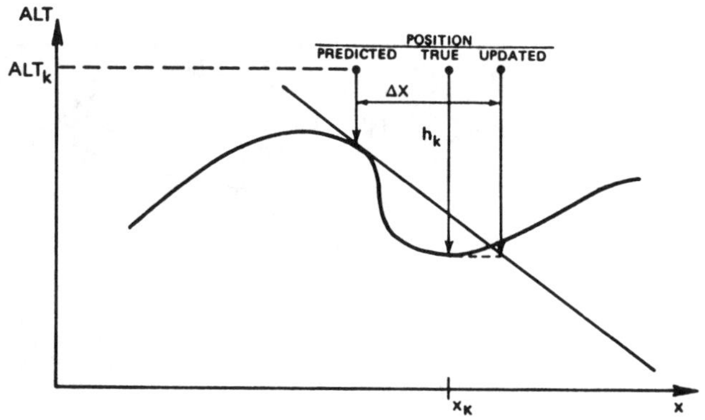

Fig. 1. Single position update.

III. Evaluation Methodology

The terrain correlation implementation considered herein is a straightforward extended Kalman filter [3] involving linear system dynamics and a nonlinear measurement equation.[1] Since the extended Kalman filter (EKF) formulation of the recursive terrain correlation concept requires linearization about current filter state estimates $\hat{x}$, the performance of the system is dependent upon the state-estimate time history $\hat{x}(t)$, i.e., the estimation accuracy is trajectory dependent. The approach to EKF filter evaluation described below is via direct Monte Carlo simulation of vehicle travel over an ensemble of flight paths which provide variation in terrain height profiles. System performance is then obtained by averaging over the resultant state estimation error (estimate-true) time histories. Once the true system performance is obtained, it can be compared to the filter-generated covariance matrix to determine the extent to which the filter model predicts real-world behavior. A detailed Monte Carlo simulation program was developed to allow an exact determination of filter performance under the assumed flight conditions.

The simulation program requires that three pieces of information be supplied: a vehicle flight path, a model characterizing the on-board navigation system, and a detailed specification of the filter to be evaluated. Each of these items is discussed briefly in the following paragraphs.

Flight Path

A typical nap-of-the-earth flight path was used to evaluate filter performance. The trajectory groundtrack is plotted in Fig. 2(a). For this trajectory, a nominal altitude of 100 m (above ground level) and a speed of 10 m/s have been assumed. For later use, the groundtrack is marked with flight time corresponding to a speed of 10 m/s. A

[1] In the formulation considered here, the state vector consists of system errors which may be adequately modeled by a linear dynamical system.

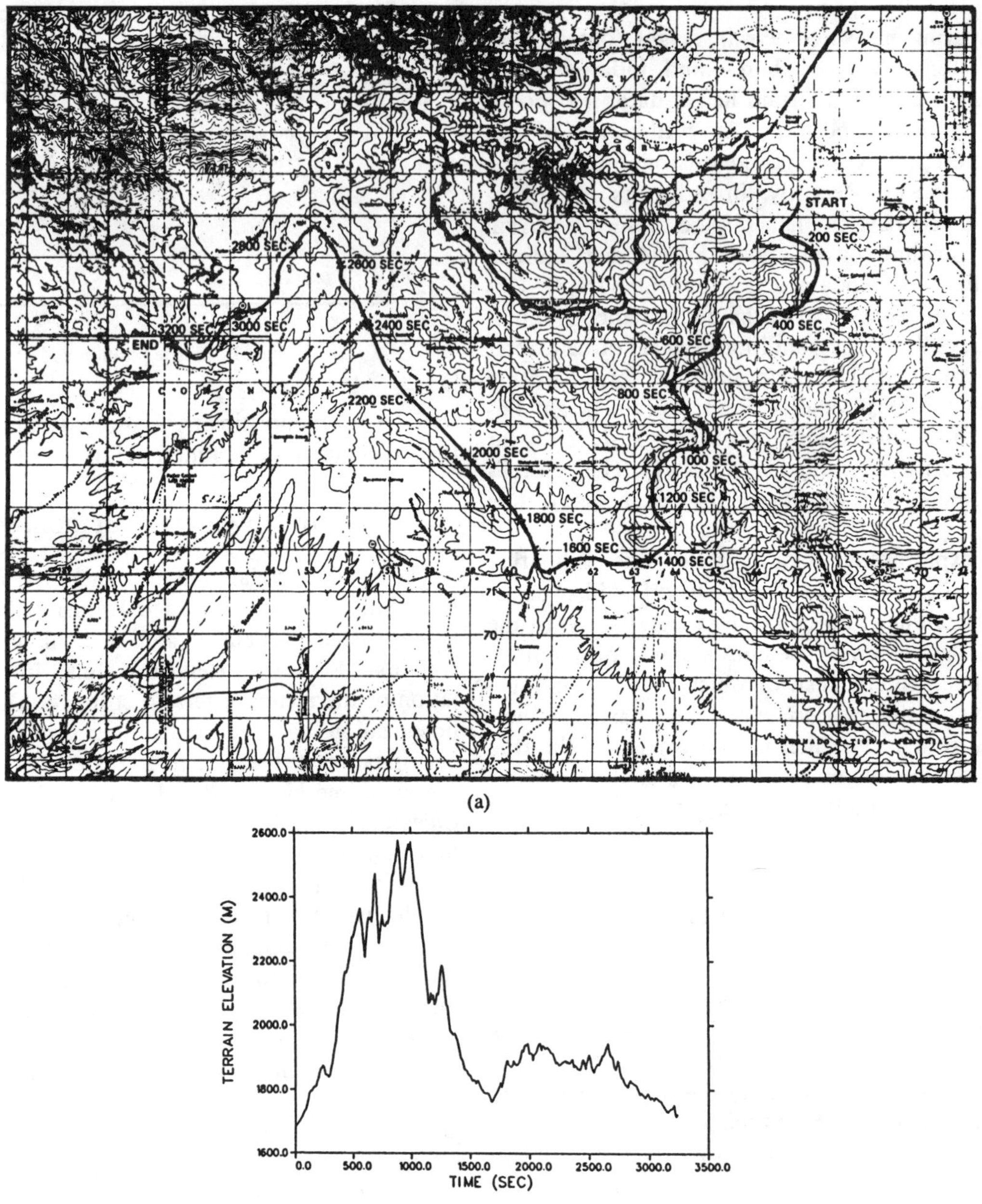

(a)

(b)

Fig. 2. Nap-of-the-earth flight path. (a) Groundtrack. (b) Terrain height profile.

relatively complete characterization of this trajectory is provided by the terrain height profile given in Fig. 2(b). The height profile is constructed from samples taken at 100 m intervals along the flight path.

Real-World Model

The real-world model is used in the Monte Carlo filter evaluation program to simulate the behavior of the navigation system during vehicle flight. The helicopter navigation system consists of a gyromagnetic heading reference, Doppler radar, vertical gyro, and barometric altimeter. Reference altitude is supplied by the baro-altimeter. The heading reference and vertical gyro supply instantaneous aircraft heading, pitch, and roll which are used to resolve the Doppler indicated downtrack, crosstrack, and vertical velocities into navigation coordinates. In addition, a radar altimeter provides the terrain clearance measurements input to the filter.

The instrument *indicated quantities* (vectors consisting of vehicle position P^I, velocity V^I, and attitude θ^I) are obtained by adding the *instrument errors* $(\epsilon_P, \epsilon_V, \psi)$ to the true quantities. The filter attempts to estimate several of these errors such that a correction term can be applied to the appropriate indicated quantity. Elements of the instru-

TABLE I
REAL-WORLD INSTRUMENT ERROR STATE VECTOR

STATE NUMBER	SYMBOL	DESCRIPTION (UNITS)
1	ϵ_{PN}	North Position Error (m)
2	ϵ_{PE}	East Position Error (m)
3	ϵ_{PD}	Down Position Error (m)
4	ϵ_{VZ}	Vertical Velocity Error (mps)
5	Ψ_Z	Heading Error (deg)
6	ϵ_b	Baro-altimeter Error (m)
7	ϵ_{VN}	North Velocity Error (mps)
8	ϵ_{VE}	East Velocity Error (mps)
9	Ψ_r	Roll Error (deg)
10	Ψ_p	Pitch Error (deg)
11	ϵ_r	Roll Gyro Bias Drift (deg/sec)
12	ϵ_p	Pitch Gyro Bias Drift (deg/sec)
13	c_0	Flux Valve Index Error (deg)
14	c_1	Random Gyro Drift (deg)
15	c_2	Single-Cycle Error (deg)
16	c_3	Two-Cycle Error (deg)
17	c_4	Rate-of-Turn Error (deg)
18	m_0	Markov Heading Error (deg)
19	m_1	Markov Heading Error (deg)
20	m_2	Markov Heading Error (deg)

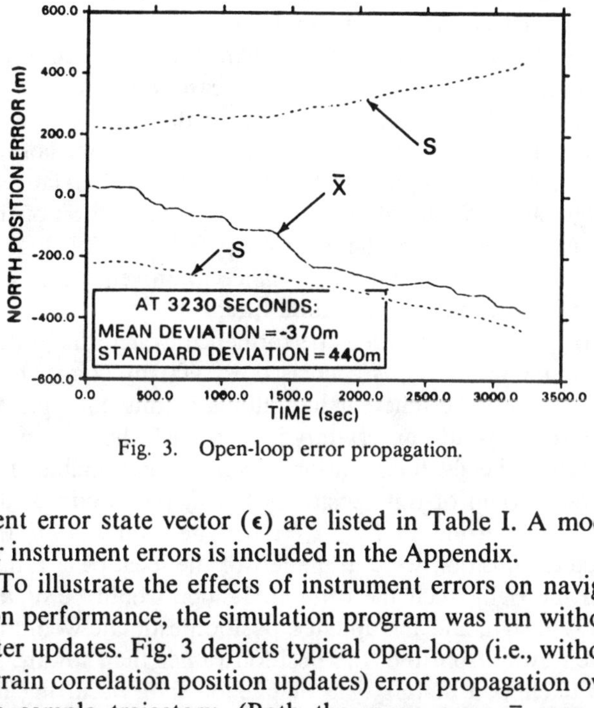

Fig. 3. Open-loop error propagation.

ment error state vector (ϵ) are listed in Table I. A model for instrument errors is included in the Appendix.

To illustrate the effects of instrument errors on navigation performance, the simulation program was run without filter updates. Fig. 3 depicts typical open-loop (i.e., without terrain correlation position updates) error propagation over the sample trajectory. (Both the mean error $\bar{x}$ over 50 Monte Carlo trials and the sample standard deviation s are indicated in the figure.) North position error is due to the integration of Doppler velocity error, and thus exhibits unbounded growth. In the remainder of this paper, north position error will be used to illustrate system navigation performances. East position error behavior is very similar, and provides no new insights into system performance.

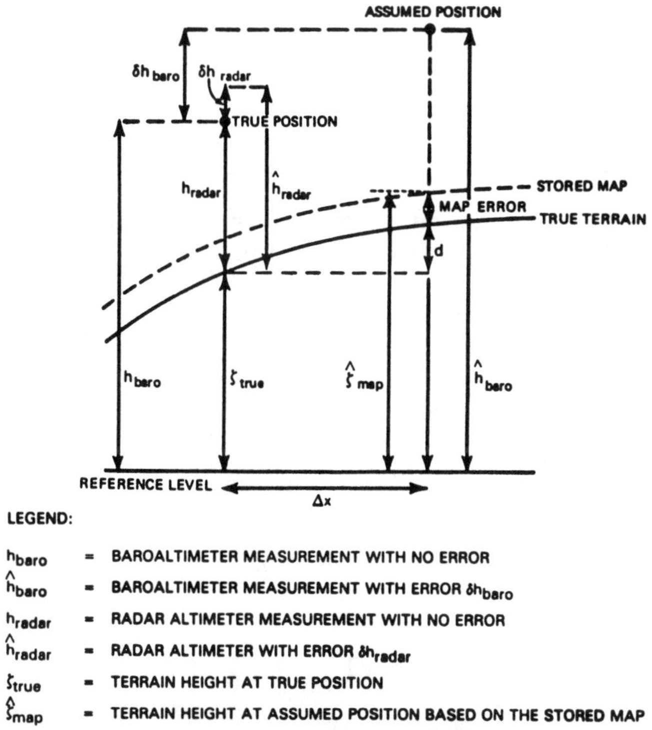

Fig. 4. Measurement geometry in one dimension.

Filter Model

The filter model analyzed in this paper is a low-order approximation to the real world. Since on-board computation capability is limited, a more detailed model would be impractical. The filter state consists of error quantities (e.g., north position error) rather than the navigation variables themselves. With this formulation, the filter estimate is used to correct the navigation system-indicated position and heading at each update time, and the filter is then reset to zero for the next update cycle. The filter state variables are identical to the first six states listed in Table I.

Two distinct measurements (z_1 and z_2) are processed in this formulation of the filter. The first pertains to the mixing of vertical position information. This measurement is simply the difference between the vertical position obtained by integrating Doppler-indicated vertical velocity and the barometric altimeter-indicated altitude. The second measurement (z_2) is the one used for terrain correlation.

Fig. 4 depicts the measurement geometry (in one dimension) associated with the terrain correlation process. The measurement processed by the filter for terrain correlation is of the form

$$\begin{aligned} z_2 &= \hat{h}_{\text{baro}} - \left[\hat{h}_{\text{radar}} + \hat{\zeta}_{\text{map}}\right] \\ &= (h_{\text{baro}} + \delta h_{\text{baro}}) - (h_{\text{radar}} + \delta h_{\text{radar}}) \\ &\quad - (\zeta_{\text{true}} + \text{map error} + d) \end{aligned} \tag{1}$$

where $d = (\partial\zeta/\partial x)\,\Delta x + (\partial\zeta/\partial y)\,\Delta y + \text{higher order terms.}$

Since

$$h_{\text{baro}} - (h_{\text{radar}} + \zeta_{\text{true}}) = 0 \tag{2}$$

(1) can be simplified to

$$z_2 = \delta h_{\text{baro}} - \delta h_{\text{radar}} - \text{map error} - d. \tag{3}$$

Map error is the difference between the stored terrain height map and the actual ground profile, and is a position-dependent function of amplitude quantization and interpolation error. Three error sources Δx, Δy (i.e., north and east position errors) and δh_{baro} correspond to filter states. The remaining errors—δh_{radar}, map error, and the neglected higher order terms—are modeled as additive noise.

Let $z = (z_1, z_2)^T$; then the linearized filter measurement equation becomes

$$z = H_f x_f + v_f \tag{4}$$

with

$$H_f = \begin{bmatrix} 0 & 0 & 1 & 0 & 0 & -1 \\ -\frac{\partial \zeta}{\partial N} & -\frac{\partial \zeta}{\partial E} & 0 & 0 & 0 & 1 \end{bmatrix} \tag{5}$$

where

$\partial\zeta/\partial N$, $\partial\zeta/\partial E$ are the terrain slopes (at the estimated vehicle position) in the north and east directions, respectively, and

v_f is uncorrelated measurement noise with $E[v_f] = 0$, $E[v_f(t_j)v_f^T(t_k)] = R_f\delta_{jk}$ where δ_{jk} is the Kronecker delta.

The last item associated with the filter is the computation of the terrain slopes. The on-board terrain height reference map consists of an array of heights obtained by digitizing over a fixed grid. Since it is unlikely that the estimated vehicle position will coincide with a grid point, some means of interpolation is required. The method used in this paper is bilinear interpolation which requires that the elevation at four different points, generally the four corners of a square, be known. A two-variable polynomial model with four unknown coefficients is assumed, and the unknown coefficients are evaluated by substituting the four known points into the polynomial. The elevation of any point inside a cell can be determined from the polynomial model.

IV. Filter Performance Evaluation

Filter performance was evaluated by running the Monte Carlo simulation program with the terrain correlation filter providing updates every 10 s. North position error behavior is shown in Fig. 5. In this plot (and all graphs of filter performance), *the solid curve is a graph of the average estimation error (over 50 trials)* $\tilde{x}$. Estimation error is defined as the difference between the actual error and the filter estimate of the error.[2] The dashed curves are the standard deviation boundary ($\pm s$) of the estimation error, i.e., at each point in time, s is the standard deviation of the estimation error over the 50 trials.

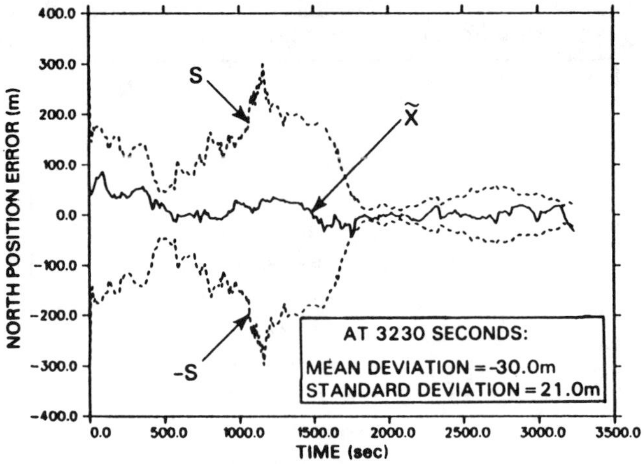

Fig. 5. Filter performance with 200 m (1 σ) initial position uncertainty.

The initial transient period in north position error behavior is associated with the complex interaction in the filter between the initial filter covariance matrix, the linearized measurement matrix, and the simulated altimeter measurements. The nature of this process is such that individual trials in the ensemble may exhibit temporary, large excursions from the true system state before converging to the true value. This is the case with the results depicted in Fig. 5. It is also possible that the filter estimates will not recover. This leads to the phenomenon of filter divergence. Beyond 1800 s, the flight path is relatively benign, allowing the filter to settle out. Comparison of the results in Fig. 5 to the open-loop behavior of Fig. 3 illustrates the dramatic performance improvement attainable with the terrain correlation filter.

Fig. 6 illustrates filter performance when the initial position uncertainty is increased to 400 m (1σ). These results reveal the inherent limitations in this concept. As the initial position uncertainty is increased, the validity of the terrain height linearization becomes questionable. The physical origin of this sensitivity lies in the nature of the terrain. Referring to Fig. 1, it can be seen that if the predicted position were to the left of the local peak rather than the right, then the computed slope would be of the opposite sign, and the updated position estimate would be driven away from the true position rather than toward it. The likelihood of this type of phenomenon occurring increases with the uncertainty in the initial location of the vehicle.

This problem cannot be alleviated within the confines of the existing filter configuration. Since any single terrain

[2] For an unbiased estimator, the average estimation error is zero. The degree to which the simulation results meet this criterion is indicative of one aspect of filter performance.

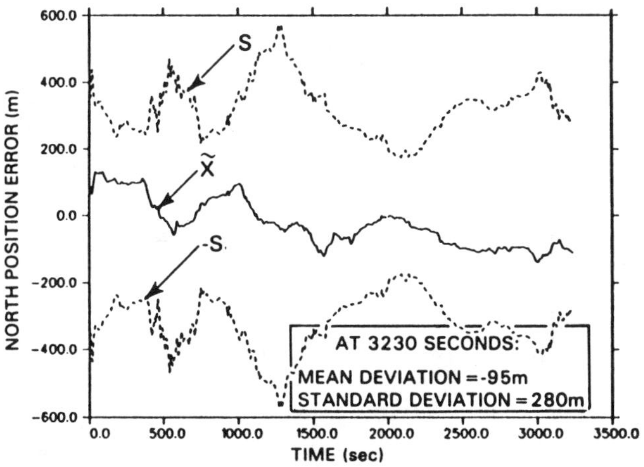

Fig. 6. Filter performance with 400 m (1 σ) initial position uncertainty.

correlation filter has associated with it a (terrain dependent) region within which it will converge with high probability, the only way to ensure convergence is to initialize the filter within this region.[3] If the uncertainty in vehicle position is larger than this convergence region, alternate approaches to the problem must be examined. When the nonlinear terrain height variation is large relative to the system measurement noise, it is possible to retain higher order terms in the Taylor series expansion of the measurement. The resulting Gaussian second-order filter [2]–[4] offers improved performance at the expense of an increased computational burden. In [2], statistical approximation techniques are also discussed as an alternative to retaining higher order terms in the series expansion.

Another technique, termed *multiple model estimation*, is investigated in this paper.[4] The basic idea in this application is to employ multiple filters initialized at various points within the uncertainty region such that one or more filters are sufficiently close to the true vehicle position to ensure convergence. The advantage of this algorithm over the approaches listed above lies in its flexibility; the number of filters as well as their separation can be selected based on *a priori* knowledge of terrain roughness and initial position uncertainty. In addition, it is readily implemented since the multiple model algorithm preserves the structure of the individual extended Kalman filters. The increased computation burden associated with implementing multiple filters may be mitigated by reducing the number of filters as the individual estimates convergence to the true vehicle position.

Multiple model estimation algorithms were first proposed for use in the broad areas of adaptive filtering and parameter identification [5], [6]. The general concept involves the systematic determination of the one filter out of a bank of N filters (each of which is formulated with a different system model) that provides the minimum mean-square error estimate of the observed process. A very readable derivation of the multiple model estimation equations for a linear system may be found in [7]. The major difference between the general formulation and the version used in this paper is in the range of allowable values of the convergence probabilities $\Pr_i$.[5]

The recursive expression for $\Pr_i(k+1)$ developed in [7] is

$$\Pr_i(k+1) = \frac{\beta^i_{k+1}\exp\left[-\frac{1}{2}w^i_{k+1}\right]}{\sum_{j=1}^{N}\beta^j_{k+1}\exp\left[-\frac{1}{2}w^j_{k+1}\right]\Pr_j(k)}\Pr_i(k) \tag{6}$$

where

$$\beta^i_{k+1} = (2\pi)^{-l/2}\left[\det S^i_{k+1}\right]^{-1/2} \tag{7}$$

$$w^i_{k+1} = \left(r^i_{k+1}\right)^T\left(S^i_{k+1}\right)^{-1}r^i_{k+1} \tag{8}$$

and l is the dimension of the measurement vector z. In (7) and (8)

$$r^i_{k+1} = z_{k+1} - H^i_{k+1}\hat{x}^i_{k+1}(-) \tag{9}$$

is the residual associated with the ith filter and

$$S^i_{k+1} = H^i_{k+1}P^i_{k+1}(-)\left(H^i_{k+1}\right)^T + R_{k+1} \tag{10}$$

is the filter indicated residual covariance.

Small values of β^i or large values of w^i (both of which are nonnegative) relative to the other β^j and $w^j, j \neq i$ result in a decrease in $\Pr_i$ relative to the $\Pr_j$, $j \neq i$. It is thus possible for one (or more) of the convergence probabilities to become arbitrarily small. (Since the sum of the probabilities is normalized to unity, there exists at least one filter with a probability greater than or equal to $1/N$.) To prevent $\Pr_i$ from becoming extremely small, which effectively "turns off" the corresponding filter, the $\Pr_i$ were limited to a minimum value of 0.01. This modification does not affect the multiple model algorithm, but it does ensure that none of the filters is eliminated due to a (possibly) temporary increase in the measurement residual. If the residuals are consistently bad, then the filter probability remains at the lower limit.

V. Application of Multiple Model Estimation Techniques

In applying the multiple model technique to the terrain correlation problem, N independent extended Kalman filters are used, each of which is initialized at a different

[3]Since the performance of the EKF is sample path-dependent, this convergence region cannot be determined analytically for arbitrary terrain.

[4]An independent application of this approach to the terrain correlation problem is documented in detail in [8] and discussed briefly in [2].

[5]More precisely, $\Pr_i(k)$ is the *a posteriori* probability that the model for the ith filter best represents the system generating the observation sequence $\{z_1, \cdots, z_k\}$.

position within the *a priori* uncertainty region. A set of five filters was selected as a compromise between thorough coverage of the position uncertainty region and a reasonable computation burden. Assume that the best estimate of vehicle initial position is $(\hat{P}_N, \hat{P}_E)$. Then, the five filters are initialized in a symmetric pattern at $(\hat{P}_N, \hat{P}_E)$, $(\hat{P}_N + \rho, \hat{P}_E)$, $(\hat{P}_N - \rho, \hat{P}_E)$, $(\hat{P}_N, \hat{P}_E + \rho)$, and $(\hat{P}_N, \hat{P}_E - \rho)$, with each filter assigned an *a priori* probability of 0.2. The filter spacing parameter ρ is dependent upon the convergence region associated with the terrain correlation filters. The convergence region is itself dependent upon the underlying terrain.

Since each filter is initialized at a different location, the partial derivatives $(\partial\zeta/\partial N)|_{\hat{P}}$ and $(\partial\zeta/\partial E)|_{\hat{P}}$ in the measurement matrix will differ, resulting in independent position estimates by the filters even though the same measurement sequence is processed. If a filter is initialized near the true vehicle position, the terrain height linearization will closely approximate the real-world environment and the filter will have a relatively high probability of converging to the true position. Of the five filters, it is very likely that one or more will converge to the true location. For the purposes of this study, the filter with the highest convergence (i.e., *a posteriori*) probability was used to estimate vehicle position.

After the initial transients settle out, it is reasonable to assume that the filter with the highest convergence probability will track subsequent vehicle motion. Rather than continuing to process the measurement data through the bank of filters, an algorithm has been included to determine if the multiple filter mode can be terminated. This decision is based on a comparison of the measurement residuals with the filter predicted residual covariance matrix. By switching from five filters to one filter, excess computation is eliminated. The multiple model to single filter transition occurs if the following two conditions hold:

1) one filter has the highest convergence probability for T successive measurements ($T = 30$ was used in this study), and

2) the rms value of the filter residuals over the corresponding T measurements is less than or equal to the filter predicted value for this quantity.

If the first condition is never satisfied, the multiple model formulation is used for the duration of the flight. When the first condition holds but the second does not, a reinitialization mode is entered. In this situation, there is the possibility that none of the filters are tracking vehicle motion properly. An attempt is made to correct this condition in the following manner.

1) The position estimate of the highest probability filter is used as the best estimate of vehicle location.

2) The filters are all reinitialized in a symmetric pattern centered at the best estimate. The filter spacing in this mode is $\rho = 200$ m.

3) The variances associated with the position error states are set to $(200\text{ m})^2$ and the convergence probabilities are reset to 0.2.

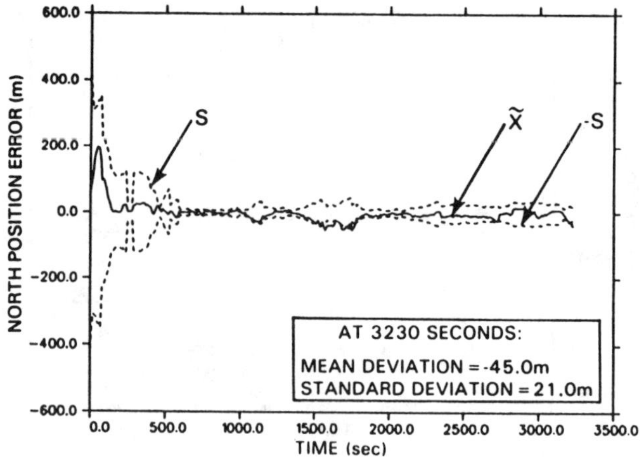

Fig. 7. Multiple model filter performance with 400 m (1 σ) initial position uncertainty.

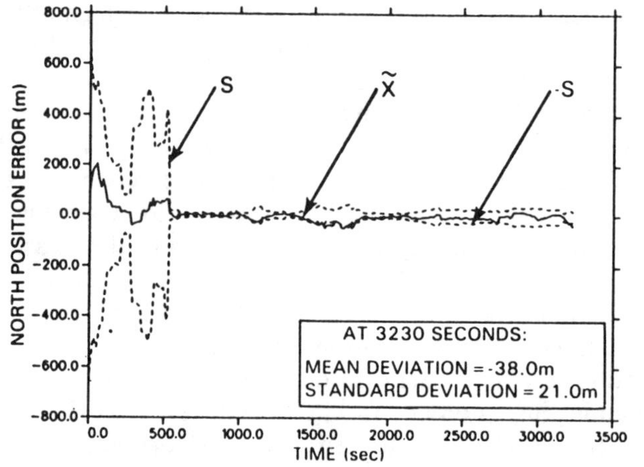

Fig. 8. Multiple model performance with 600 m (1 σ) initial position uncertainty.

4) The multiple model algorithm is then restarted along with the test for switching from the multiple filters to a single filter.

Fig. 7 depicts the performance of the multiple model filter with an initial position uncertainty of 400 m (1σ). Comparing these results to those in Fig. 6 illustrates the substantial improvement that is attainable with this formulation. It is evident that the multiple model formulation is a viable technique for reducing the sensitivity of the terrain correlation filter to initial position uncertainty.

The last example to be presented is for an initial position uncertainty of 600 m (1 σ). Although this value is somewhat unrealistic in the nap-of-the-earth flight scenario, it is included to illustrate the robustness of the multiple model technique. To accommodate the larger uncertainty region, the number of filters was increased from five to nine. Filter performance for this configuration is shown in Fig. 8 for a filter spacing (ρ) of 800 m. These results are actually better than those obtained under any previous condition. Although the position error residuals at the end of the flight are approximately the same as most other cases, the magnitude of the residuals during the flight is much smaller.

VI. SUMMARY

This paper has described the performance evaluation of a terrain correlation algorithm proposed for helicopter navigation in nap-of-the-earth flight regimes. It was shown that the proposed filter configuration affords improved navigation capability if the filter is initialized near the true vehicle location. For larger initial position uncertainties, the proposed formulation exhibits erratic behavior which may include divergent estimates of vehicle position. A modification to the existing filter was proposed to reduce its inherent sensitivity to initial position uncertainty. This formulation, based upon multiple model estimation techniques, was evaluated and shown to afford a significant performance improvement.

APPENDIX

A set of models describing the error characteristics of the navigation system components (e.g., heading reference, Doppler radar, etc.) was developed in [9]. These models may be written in the form

$$\dot{\epsilon} = A\epsilon + w_\epsilon \tag{11}$$

where ϵ is the instrument error state vector of Table I and w_ϵ is the white process noise vector. For the Doppler-aided navigation system evaluated in this paper, the error dynamics matrix A is partitioned as

$$A = \begin{bmatrix} A_1 & A_{12} & A_{13} \\ 0 & A_2 & 0 \\ 0 & 0 & A_3 \end{bmatrix} \tag{12}$$

with

$$A_1 = \begin{bmatrix} 0 & 0 & 0 & 0 & -V_E & 0 \\ 0 & 0 & 0 & 0 & V_N & 0 \\ 0 & 0 & 0 & 1 & 0 & 0 \\ 0 & 0 & 0 & -\beta_1 & 0 & 0 \\ 0 & 0 & 0 & 0 & 0 & 0 \\ 0 & 0 & 0 & 0 & 0 & -\beta_2 \end{bmatrix} \tag{13}$$

$$A_2 = \begin{bmatrix} -\beta_3 & 0 & 0 & 0 & 0 & 0 \\ 0 & -\beta_4 & 0 & 0 & 0 & 0 \\ 0 & 0 & 0 & 0 & 1 & 0 \\ 0 & 0 & 0 & 0 & 0 & 1 \\ 0 & 0 & 0 & 0 & 0 & 0 \\ 0 & 0 & 0 & 0 & 0 & 0 \end{bmatrix} \tag{14}$$

$$A_3 = \left[\begin{array}{c|ccc} 0 & & 0 & \\ \hline & -\beta_5 & 0 & 0 \\ 0 & 0 & -\beta_6 & 0 \\ & 0 & 0 & -\beta_7 \end{array}\right] \tag{15}$$

$$A_{12} = \left[\begin{array}{cccc|c} \cos\theta_{ZT} & -\sin\theta_{ZT} & V_Z\sin\theta_{ZT} & V_Z\cos\theta_{ZT} & \\ \sin\theta_{ZT} & \cos\theta_{ZT} & -V_Z\cos\theta_{ZT} & V_Z\sin\theta_{ZT} & 0 \\ 0 & 0 & (V_E\cos\theta_{ZT} - V_N\sin\theta_{ZT}) & (-V_E\sin\theta_{ZT} - V_N\cos\theta_{ZT}) & \\ \hline & & 0 & & \end{array}\right] \tag{16}$$

$$A_{13} = \left[\begin{array}{ccccc|c|ccc} -V_E\dot{\theta}_{ZT} & -V_E\cos(\theta_{ZT}+\phi_1) & -V_E\cos(2\theta_{ZT}+\phi_2) & 0 & 0 & -V_E & -V_E & -V_E \\ V_N\dot{\theta}_{ZT} & V_N\cos(\theta_{ZT}+\phi_1) & V_N\cos(2\theta_{ZT}+\phi_2) & 0 & 0 & V_N & V_N & V_N \\ \hline & & & & & 0 & & \\ & & & & & 0 & & \\ & & 0 & & & 1 & 0 & \\ & & & & & 0 & & \end{array}\right] \tag{17}$$

where

- V_N, V_E, V_Z are the north, east, and vertical vehicle velocities, respectively
- θ_{ZT} is the true vehicle heading
- $\dot{\theta}_{ZT}$ is true rate-of-turn
- ϕ_1, ϕ_2 are angles associated with installation of the heading reference
- β_1 is the vertical velocity error inverse time constant ($=1/s$)
- β_2 is the baro-altimeter error inverse time constant ($=1/3600$ s)
- β_3 is the north velocity error inverse time constant ($=1/s$)
- β_4 is the east velocity error inverse time constant ($=1/s$)
- β_5 is a Markov error inverse time constant ($=1/s$)
- β_6 is a Markov error inverse time constant ($=1/25$ s)
- β_7 is a Markov error inverse time constant ($=1/400$ s).

ACKNOWLEDGMENT

The authors wish to express their appreciation to Dr. N. Shupe of AVRADA for his support throughout the course of this work.

References

[1] L. D. Hostetler, "A Kalman approach to continuous aiding of inertial navigation systems using terrain measurements," in *Proc. Milwaukee Symp. Automat. Computation Contr.*, Apr. 1976, pp. 305–309.

[2] L. D. Hostetler and R. D. Andreas, "Nonlinear Kalman filtering techniques for terrain-aided navigation," this issue, pp. 315–323.

[3] A. Gelb, Ed., *Applied Optimal Estimation*. Cambridge, MA: M.I.T. Press, 1974.

[4] A. H. Jazwinski, *Stochastic Processes and Filtering Theory*. New York: Academic, 1970.

[5] D. T. Magill, "Optimal adaptive estimation of sampled stochastic processes," *IEEE Trans. Automat. Contr.*, vol. AC-10, pp. 434–439, Oct. 1965.

[6] D. G. Lainiotis, "Optimal adaptive estimation: Structure and parameter adaptation," *IEEE Trans. Automat. Contr.* vol. AC-15, pp. 160–170, Apr. 1971.

[7] M. Athans and C. B. Chang, "Adaptive estimation and parameter identification using multiple model estimation algorithm," M.I.T. Lincoln Lab., Tech. Note 1976-28, June 1976.

[8] T. C. Sheives, "State estimation using parallel extended Kalman filters on nonlinear measurements," Sandia Nat. Lab., Rep. SAND80-0013, Jan. 1980.

[9] W. Tang and G. L. Mealy, "Nap-of-the-earth navigation analysis and multiple model filter design," The Analytic Sciences Corp., Rep. TR-1593-2, Mar. 1981 (DTIC Accession AD-B058 674L).

Design and Analysis of a Dynamic Positioning System Based on Kalman Filtering and Optimal Control

STEINAR SAELID, MEMBER, IEEE, NILS A. JENSSEN, MEMBER, IEEE, AND JENS G. BALCHEN, MEMBER, IEEE

I. Introduction

DYNAMIC positioning systems primarily for drilling vessels and platforms and for support vessels in the offshore oil industry have been manufactured since the early 1960's. In the beginning, these systems were designed using conventional control principles [1]. Since 1974 new systems have been developed based upon "modern control theory," in which state and parameter estimation was employed in the form of extended Kalman filters and multivariable control was designed using LQG control theory [2]–[5]. A Norwegian company has marketed such systems under the name ALBATROSS since 1976.

These recent systems are based upon extensive mathematical modeling of the dynamic behavior of the vessel subject to the forces exerted by the controllable thrusters (propellers) as well as the environmental forces from wind, currents, and waves.

A good system for the dynamic positioning should keep the vessel within a specified position and heading limits with a minimum of fuel consumption and minimum of

Manuscript received March 5, 1982; revised September 27, 1982.
S. Saelid and J. G. Balchen are with the Division of Engineering Cybernetics, Norwegian Institute of Technology, University of Trondheim, Trondheim, Norway.
N. A. Jenssen is with the Maritime Division, A/S Kongsberg Våpenfabrikk, Kongsberg, Norway.

Reprinted from *IEEE Trans. Automat. Contr.*, vol. AC-28, pp. 331–338, Mar. 1983.

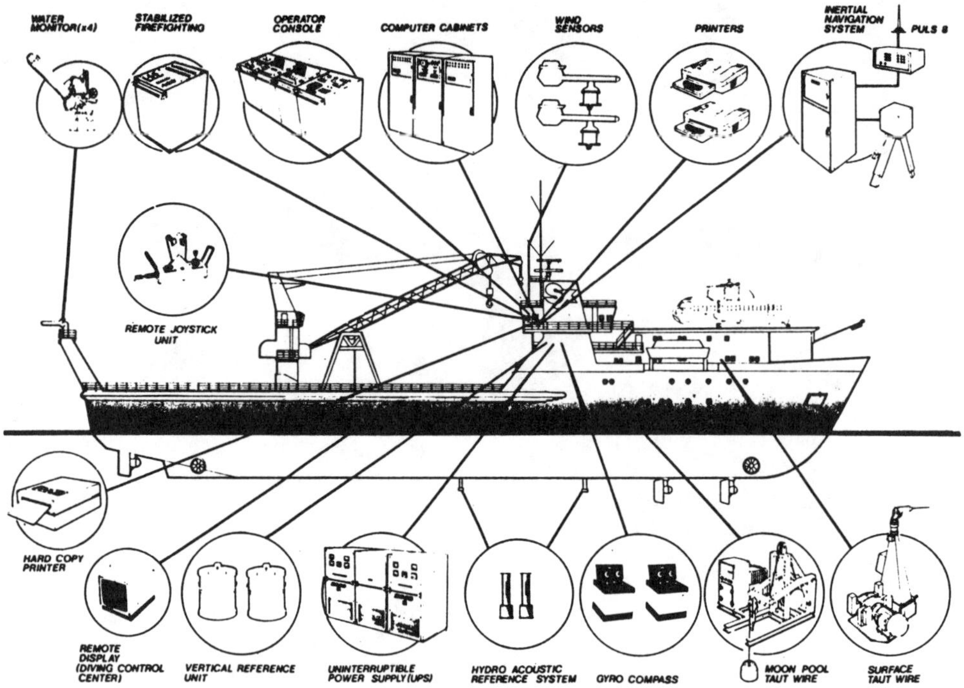

Fig. 1. Seabex One, probably one of the world's most powerful support vessels involved in the offshore industry, is equipped with an ALBATROSS dynamic positioning system.

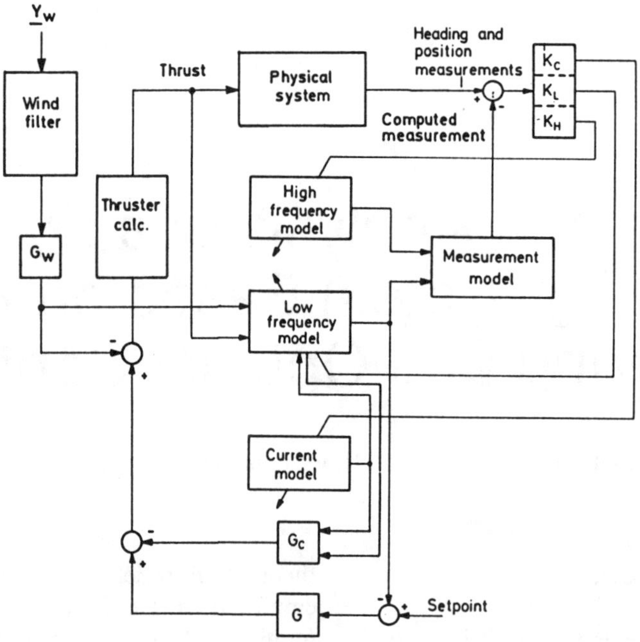

Fig. 2. Control system structure.

wear on the propulsion equipment. Furthermore, the system should tolerate transient failures in the measurement and propulsion systems.

Fig. 1 shows a typical vessel with dynamic positioning and in Fig. 2 is shown a block diagram indicating the main components of such a system.

The behavior of the vessel is modeled as the combination of a "low-frequency drift" and a "high-frequency oscillation." The slow motion is caused by the thrusters and the forces from the wind and water currents. Since the wind can be measured fairly precisely, only the current forces need to be estimated. It is assumed that the current is relatively constant both in magnitude and direction, leading to a model containing only two integrators. Since it is not possible to counteract the oscillatory movement due to waves, the control action of the thrusters is derived from the state of the "low-frequency" model, the state of the current model, and a feed forward from the wind measurement. It is well known that the success of a control system employing a mathematical model can be critically dependent upon the choice of a proper model structure. The original "low-frequency" model [4] had a somewhat inconvenient form and has later been modified. Thereby, it has been possible to explain and remove a tendency toward very low-frequency oscillation. This problem will be dealt with in the next section.

In the early versions of this system [2]–[5] the "high-frequency" model consisted of three harmonic oscillators without damping (two state variables in each) in which the frequencies were updated using a simplified extended Kalman filter. Later this structure had been improved, introducing a damping term in the oscillators and employing a more robust parameter estimation algorithm. These improvements will be dealt with in Section III.

For the details on mathematical modeling of vessel behavior the reader is referred to [4] or [5]. Observations from the improved systems (1982) are presented in Section IV of this paper.

II. Damping of Slow Oscillations

During the years since the first ALBATROSS installations, a considerable amount of operational experience has been obtained. The first ALBATROSS systems have shown that they behave very well and typical operational results are reported in [4].

However, one problem has repeatedly come up during heavy weather conditions. The vessel would suddenly get into slow oscillations showing an amplitude of 2–3 m and a period of 2–3 min. This is of course very unsatisfactory, even if the occurrence of the phenomenon is rather rare.

Examination of the vessel motion during this kind of behavior showed that the oscillations are not only restricted to surge, but a small and correlated oscillation in yaw is also always present. The amplitude of this oscillation could typically be 2–4 degrees.

The ALBATROSS system is based on a simplified Kalman filtering technique and optimal quadratic control, where in most cases controller and filter gains are computed using models linearized at zero current conditions. It is clear that the oscillations occur when the water current is particularly high. This indicates that the problem could be solved by running a complete extended Kalman filter and a scheduled gain optimal controller. This requires running a filter algorithm having more than 20 state variables, and hence requires the installation of increased computer capacity.

It was decided to try to solve the problem by keeping the simplified Kalman filter structure, isolating the reasons for the instability, and seeing if the oscillations could be coped with by minor modifications.

A New LF-Model

In order to simplify the analysis, the low-frequency vessel model given in [4] is reformulated. In [4], the coupling between surge, sway, and yaw motions are taken care of by using drag coefficients which are functions of the angle between vessel heading and the current direction. An alternative model of these interactions is introduced where the drag coefficients are constants and the interactions are taken care of by analytical cross-coupling terms. This approach yields approximately the same dynamic response as the model in [4], but is easier to analyze. The new model is based on Norrbin's work [8], and is described in the following:

$$\dot{x}_L = u_L \tag{1}$$

$$\dot{u}_L = \frac{1}{m_x}\big[m_y(v_L - v_c)r_L + m_x v_c r_L - d_x|u_L - u_c|(u_L - u_c) + F_{wx} + F_{Tx}\big] + \eta_{Lx} \tag{2}$$

$$\dot{y}_L = v_L \tag{3}$$

$$\dot{v}_L = \frac{1}{m_y}\big[m_y u_c r_L - m_x(u_L - u_c)r_L - d_y|v_L - v_c|(v_L - v_c) + F_{wy} + F_{Ty}\big] + \eta_{Ly} \tag{4}$$

$$\dot{\Psi}_L = r_L \tag{5}$$

$$\dot{r}_L = \frac{1}{m_\Psi}\big[-(m_y - m_x)(u_L - u_c)(v_L - v_c) - d_\Psi|r_L|r_L + M_c + M_w + M_T\big] + \eta_{L\Psi} \tag{6}$$

where

x_L, u_L	position and velocity in surge direction (LF-part),
y_L, v_L	position and velocity in sway direction (LF-part),
Ψ_L, r_L	heading and heading rate (LF-part of yaw),
u_c, v_c	current velocities in surge and sway direction,
F_{wx}, F_{wy}	wind forces in surge and sway direction,
F_{Tx}, F_{Ty}	thrust forces in surge and sway direction,
M_w, M_T	moments from wind and thrust, respectively,
M_c	residual current moment.

$\eta_{Lx}, \eta_{Ly}, \eta_{L\Psi}$ are assumed to be zero mean Gaussian white noise processes. d_x, d_y, and d_Ψ are viscous drag coefficients. m_x, m_y, and m_Ψ are inertial coefficients which are assumed to be constants. The residual current generated moment is included to account for errors in the r_L-equation (6).

Analysis of Total System

Now we assume that $v_c \neq 0$ and $u_c = 0$. That is, a water current is flowing in the sway direction. We also assume, for simplicity, that $m_y = 2m_x$ and we linearize (2) and (6) around $v_L = r_L = u_L = u_c = 0$ and $v_c = v_{c0}$. We assume that the system damping is represented by the parameter d and that we disregard the external forces. In that case the linearized versions of (2) and (6) become (note that this is an approximation)

$$\dot{u}_L = -v_{c0} r_L \tag{7}$$

$$\dot{r}_L = \frac{m_x}{m_\Psi} v_{c0} u_L - d r_L. \tag{8}$$

This is clearly an oscillator with an angular frequency

$$\omega_c = v_{c0}\sqrt{\frac{m_x}{m_\Psi}}. \tag{9}$$

Values for a typical vessel indicate a period of, e.g., $T_c \approx 200$ s ($\omega_c \approx 0.03$) when $v_{c0} \approx 1$ m/s. As is seen from (9), ω_c decreases linearly with v_{c0}. This type of oscillatory phenomenon is quite easy to observe in nature. The wobbling motion of a falling leaf is due to exactly the same hydrodynamic mechanism. These oscillatory modes are not accounted for in the feedback control law and in the filter gain computations. Probably, this is the reason for the observed behavior in the surge/yaw coordinates.

This conclusion is confirmed by computer simulation of a typical vessel where

$$m_x = 5 \cdot 10^6 \text{ kg}, \quad m_y = 10^7 \text{ kg},$$

$$m_\Psi = 8 \cdot 10^9 \text{ kg} \cdot \text{m}^2,$$

$$d_x = 8 \cdot 10^4 \text{ N} \cdot \text{s}^2/\text{m}^2, \quad d_y = 4 \cdot 10^5 \text{ N} \cdot \text{s}^2/\text{m}^2,$$

$$d_\Psi = 5 \cdot 10^{10} \text{ N} \cdot \text{s}^2/\text{rad}^2.$$

Simulation results are shown in Section IV.

It is not easy to see immediately what should be done in order to prevent the oscillations. If we disregard the wave frequency adaptation, which will be discussed later, each of the surge, sway, and yaw subsystems has five filter gain parameters K_i, $i = 1, \cdots 5$, and two feedback gain parameters G_i, $i = 1,2$.

In order to analyze the influence of each parameter, we look at the Kalman filter and the feedback law together as one single controller. Again we look at the surge loop, but we include an equation for yaw rate. If we assume that $u_c = 0$, the linearized equations become

$$\dot{\hat{x}}_L = \hat{u}_L + K_1 \epsilon_x \tag{10}$$

$$\dot{\hat{u}}_L = -\hat{v}_c \hat{r}_L + K_2 \epsilon_x + F_{Tx} \tag{11}$$

$$\dot{\hat{x}}_H = \hat{u}_H + K_3 \epsilon_x \tag{12}$$

$$\dot{\hat{u}}_H = \theta \hat{x}_H - 2c\hat{u}_H + K_4 \epsilon_x \tag{13}$$

$$\dot{\hat{r}}_L = \hat{v}_c \frac{m_x}{m_\Psi} \hat{u}_L - d\hat{r}_L + K_\Psi \epsilon_\Psi + M_T \tag{14}$$

$$F_{Tx} = G_1(\hat{x}_L - x_L^{\text{ref}}) + G_2 \hat{u}_L + \hat{v}_c \hat{r}_L \tag{15}$$

$$\epsilon_x = y_s - \hat{x}_L - \hat{x}_H \tag{16}$$

where $\hat{x}_L$, $\hat{u}_L$, and $\hat{r}_L$ are estimates of x_L, u_L, and r_L, respectively. $\hat{x}_H$, $\hat{u}_H$ are estimates of HF-surge position and HF-surge rate, F_{Tx} and M_T are the thrust in the surge and yaw directions, and $\hat{v}_c$ is the water current estimate in the sway direction. y_s is the surge position measurement and ϵ_x and ϵ_Ψ are innovations in surge and yaw. G_1 and G_2 are feedback gains and K_1, K_2, K_3, K_4, and K_Ψ are filter gain constants. x_L^{ref} is the surge position set point. d is a damping parameter of the yaw rate equation. θ is the squared angular frequency and c is a damping parameter both of the HF-model in the surge direction.

Note that the thrust feedback tries to counteract the nonlinear terms due to hydrodynamic effects.

If we now disregard the influence of ϵ_Ψ, the following surge loop transfer function results from the linearized process model equations and (10)–(16)

$$H(s) = \frac{[(G_1K_1 + G_2K_2)s + G_1K_2][s^2 + ds + \omega_c^2] - \omega_c^2 K_2 [G_1 + s(s + G_2)]}{s^2[K_2 + K_1(s + G_2) + (G_1 + s(s + G_2))F(s)][s^2 + ds + \omega_c^2]} \tag{17}$$

where

$$F(s) = 1 + \frac{K_4 + 2cK_3 + sK_3}{s^2 + 2cs + \theta}. \tag{18}$$

$F(s)$ represents the influence of the HF-model.

Tuning of Filter and Control Parameters

Based on (17) and (18) we shall discuss the choice of filter and controller gain parameters. Let us first discuss the HF-filter gain parameters.

The HF-model is given by

$$\dot{x}_H = u_H + C_1 \eta \tag{19}$$

$$\dot{u}_H = -2cu_H - \theta x_H + C_2 \eta \tag{20}$$

where η is white process noise and C_1, C_2 are constants. We assume that $\text{cov}(\eta) = \delta(t)$. C_1 and C_2 should be chosen so that the power spectrum of x_H, $\Phi_{xx}^M(j\omega)$ approximates the observed power spectrum $\Phi_{xx}(j\omega)$ of the real HF motion.

This is done by minimization of

$$J = \int_0^\infty (\Phi_{xx}^M(j\omega) - \Phi_{xx}(j\omega))^2 \, d\omega \tag{21}$$

where $\Phi_{xx}(j\omega)$ is estimated from measurement data or theoretical predictions, whereas $\Phi_{xx}^M(j\omega)$ is given by

$$\Phi_{xx}^M(j\omega) = |H_{\text{HF}}(j\omega)|^2$$

where $H_{\text{HF}}(s)$ is the HF-model transfer function given by

$$H_{\text{HF}}(s) = \frac{sC_1 + (C_2 + cC_1)}{s^2 + 2cs + \theta}. \tag{22}$$

Minimization of J for $\Phi_{xx}(j\omega)$ as given in Fig. 3 yields

$$c = 0.1$$

$$C_1 = 0.4$$

$$C_2 = -0.08$$

$$\theta = 0.2.$$

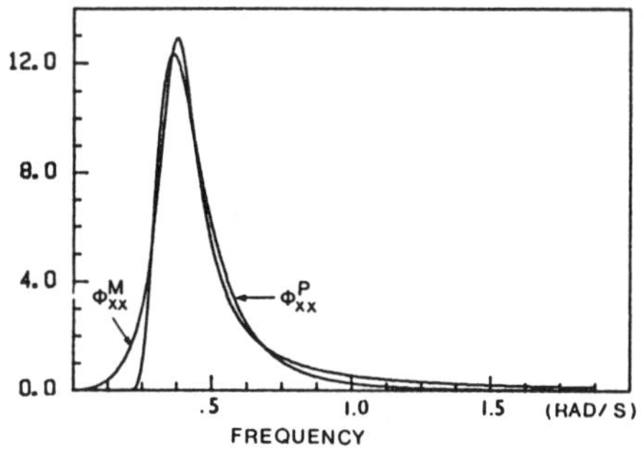

Fig. 3. Power spectrum of actual wave induced motion ϕ_{xx} and of model output ϕ_{xx}^M.

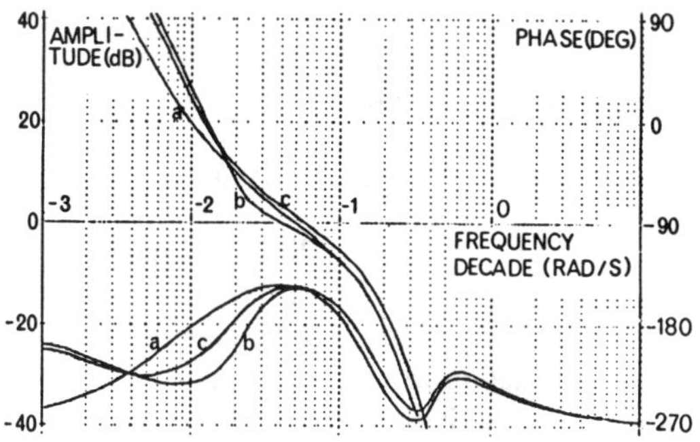

Fig. 4. Phase and gain plots of surge loop.

The resulting $\Phi_{xx}^M(j\omega)$ is shown in Fig. 3. It turns out that we will always have

$$C_2 = -2cC_1. \tag{23}$$

This is natural because in that case $\Phi_{xx}^M(0) = 0$, i.e., no LF components are present in the HF-spectrum.

Now (23) implies a special structure of the process noise covariance matrix of the surge model. It turns out that we will always have

$$K_4 = -2cK_3 \tag{24}$$

when (23) is satisfied. If we now turn to (17) and (18) we observe that $F(s)$ may introduce a decrease in the loop gain $H(s)$. This is due to the HF-model. However, if (24) is satisfied, $F(s) = 1$ for $s = 0$ and the HF-model introduces very small damping of the loop gain in the LF part of the spectrum. In the earlier ALBATROSS systems, (23) and (24) where not satisfied. When these equations are satisfied, the slow oscillation tendency decreases.

In order to further improve the stability of the system, we shall look at (17). The numerator polynomial in (17) becomes

$$N(s) = [(G_1K_1 + G_2K_2)s + G_1K_2]P(s) - \omega_c^2K_2Q(s) \tag{25}$$

where

$$P(s) = s^2 + ds + \omega_c^2 \tag{26}$$

and

$$Q(s) = G_1 + s(s + G_2). \tag{27}$$

The observed slow oscillations can now be explained by a decrease in loop gain when $|v_c|$ increases. This is due to the negative term in (25). Hence, the stability of the system should be improved by increasing $|N(j\omega)|$. From (26) we see that the best solution is to increase K_2 because this gain parameter does influence the poles to a very small degree.

Practical experience shows that if K_2 is doubled during heavy weather/current conditions, the slow oscillations disappear. This is illustrated in the Bode plots of Fig. 4. Curve *a* shows the loop transfer function using nominal parameters and no water current. Curve *b* is computed for nominal parameters and $v_c = 1$ m/s. Curve *c* is computed for $v_c = 1$ m/s and $K_2 = 2 \cdot K_2$ nominal. We observe that by increasing K_2 the phase and gain margins are strongly improved. Hence, in the new ALBATROSS system, K_2 is increased as the current increases. Both simulations and practical experience show that the slow oscillations disappear in that case.

III. The Wave Frequency Adaptation

The wave filtering performance depends strongly on the value of θ, which is the square of the frequency. The dominating wave frequency parameter may change with a factor of three when the weather conditions vary from light breeze to fresh gale (Beaufort numbers 2–8). The earlier ALBATROSS systems have been using steady-state sensitivity analysis in order to compute the gain factors for updating of the frequency parameter [4].

In the most recent system installations, a more refined parameter estimation algorithm is introduced. This algorithm is based on the solution of the complete sensitivity equations. The algorithm is presented in the following, together with a convergence analysis.

The parameter estimation algorithm is a prediction error algorithm of the following form [12]:

$$\hat{\theta}_{k+1} = \hat{\theta}_k + P_k\psi_k\epsilon_k \tag{28}$$

$$P_{k+1} = P_k\left[1 - \frac{\psi_k^2P_k}{(\psi_k^2P_k + \beta_k)}\right]\frac{1}{\beta_k} \tag{29}$$

$$\beta_k = \mu_{k-1}(1 - \mu_k)/\mu_k. \tag{30}$$

Here θ_k is an estimate of the frequency for the actual HF-model, ϵ_k is the prediction error of the measurement, ψ_k is an approximation to $\partial\epsilon_k/\partial\hat{\theta}_k$, P_k is an estimate of $\mathrm{cov}(\psi_k)$, and μ_k is a positive scalar sequence tending to zero such that $k \cdot \mu_k \to \gamma$, $0 < \gamma < \infty$, when the time index $k \to \infty$.

The sensitivity $\psi_k = \partial\epsilon_k/\partial\theta_k$ is computed by derivation of the linearized equations for $\Delta x_k^F = x_k^F - \bar{x}_k^F$ where x_k^F is the full state vector and $\bar{x}_k^F$ is the *a priori* estimate of the full state vector. In order to find ψ_k, the vessel equations are linearized around $u_L = v_L = u_c = v_c = r_L = 0$. This results in a decoupling of the surge, sway, and yaw models. In the following we look at the surge model. The states influencing the surge measurements, when the model is linearized, are $x_k = [x_{L,k}u_{L,k}x_{H,k}u_{H,k}]^T$ where $x_{L,k}$ and $x_{H,k}$ are the LF- and HF-parts of the surge positions, respectively, and $u_{L,k}$ and $u_{H,k}$ are the corresponding velocities. k is the time index of the discretized model. The linearized model of Δx_k becomes

$$\Delta x_k = \hat{\phi}(I - KD)\Delta x_k - \Delta\phi x_k + n_k - \hat{\phi}Kw_k \quad (31)$$

where $z_k \triangleq \partial\Delta x_k/\partial\hat{\theta}$. Now the last term in the above equation is small because $\Delta\phi$ can be moderate or small and because $\partial x_k/\partial\hat{\theta}$ is assumed to be relatively small. If x_k in the second to last term is approximated by $\hat{x}_k$, the above equation may be written

$$z_{k+1} = \hat{\phi}(I - KD)z_k - \frac{\partial\hat{\phi}}{\partial\hat{\theta}}x_k - \Delta\phi\frac{\partial x_k}{\partial\hat{\theta}} \quad (32)$$

where $z_k \triangleq \partial\Delta x_k/\partial\hat{\theta}$. Now the last term in the above equation is small because $\Delta\phi$ can be moderate or small and because $\partial x_k/\partial\hat{\theta}$ is assumed to be relatively small. If x_k in the second to last term is approximated by $\hat{x}_k$, the above equation may be written

$$z_{k+1} = \hat{\phi}(I - KD)z_k - \frac{\partial\hat{\phi}}{\partial\hat{\theta}}\hat{x}_k. \quad (33)$$

Now $\psi_k = Dz_k$ and (33), (28), (29), and (30) constitute the parameter estimation algorithm for the HF-frequency in surge. In the analysis above, we have introduced an approximation by linearizing the equations around zero vessel and current velocities. However, it turns out that the error introduced is very small. In the earlier version of the ALBATROSS system the LF-model was totally neglected in the sensitivity equations [4]. This is also an acceptable solution although inferior. The LF-components in (33) normally account for ~10–15 percent of the ψ_k value.

Convergence Analysis

Ljung [7] has shown that the convergence properties of an algorithm like that given by (28)–(30) can be analyzed by investigating the stability properties of the following set of ordinary differential equations:

$$\dot{\theta} = \frac{1}{R}f(\theta) \quad (34)$$

$$\dot{R} = -R + G(\theta) \quad (35)$$

where $f(\theta) = E(\bar{\psi}_k\bar{\epsilon}_k)$, $G(\theta) = E(\bar{\psi}_k^2)$, and the bar indicates that ψ_k and ϵ_k are computed for a given fixed value $\bar{\theta}$ of $\hat{\theta}$ for all k. The theory says that the recursive algorithm converges if (34) and (35) are stable. We therefore have to find expressions for $\bar{\epsilon}_k$ and $\bar{\psi}_k$ in order to investigate the stability of the parameter estimation algorithm. From the linearized surge filter equations given by

$$\bar{x}_{k+1} = \bar{\phi}\bar{x}_k + \Gamma u_k + \bar{\phi}K\epsilon_k \quad (36)$$

$$y_k = D\bar{x}_k + \epsilon_k \quad (37)$$

we obtain by z-transformation and elimination of $\bar{x}_k$ that (36) and (37) can be expressed as

$$\bar{A}(z)y_k = B(z)u_k + (\bar{A}(z) + \bar{C}(z))\epsilon_k \quad (38)$$

where $\bar{A}(z)$ and $\bar{C}(z)$ are polynomials in z given by $\bar{A}(z) = H_1^2\bar{H}_\theta$ and

$$\bar{C}(z) = \bar{H}_\theta[K_2 + H_1(K_1 + K_2 + K_3 + K_4)] + H_1[(1 - (2c + \bar{\theta}))K_4H_2 - \bar{\theta}K_3H_3]$$

where

$$H_1 = [z - 1];$$
$$H_2 = [z^2 - z - 1];$$
$$H_3 = [z^2 - z + 1];$$
$$\bar{H}_\theta = [z^2 - (2(1 - c) - \bar{\theta})z + (1 - c)].$$

K_1, K_2, K_3, and K_4 are discrete time Kalman filter gains, $\bar{\theta}$ is a fixed parameter value and the bars in $\bar{A}$, $\bar{C}$, etc., indicate that $\bar{\theta}$ is inserted. The B-polynomial does not contain θ and is not important in our discussion of convergence. Now $\bar{\epsilon}_k$ can be written as

$$\bar{\epsilon}_k = \left(\frac{B}{\bar{A} + \bar{C}} - \frac{B}{A_0 + C_0}\right)u_k + \left(\frac{\bar{A}}{\bar{A} + \bar{C}} - \frac{A_0}{A_0 + C_0}\right)y_k + e_k \quad (39)$$

where A_0 and C_0 are the correct polynomials with the actual value $\theta = \theta_0$ inserted. e_k is the innovation signal which would appear for the given u_k and y_k if $\bar{\theta} = \theta_0$. Now, derivation of (38) with respect to θ yields, after some manipulation and using (39),

$$\bar{\epsilon}_k = e_k + \frac{\bar{A} + \bar{C}}{A_0 + C_0}\cdot\frac{\partial\bar{\epsilon}_k}{\partial\bar{\theta}}. \quad (40)$$

From Ljung's work [7] it is then clear that the estimation algorithm is stable if the transfer function

$$F = \frac{\bar{A} + \bar{C}}{A_0 + C_0} - \frac{1}{2} = \frac{1}{2} + \frac{\Delta A + \Delta C}{A_0 + C_0} \quad (41)$$

is positive real, where $\Delta A = \bar{A} - A_0$ and $\Delta C = \bar{C} - C_0$. ΔA and ΔC are given by

$$\Delta A = H_1^2 z\Delta\theta \quad (42)$$

$$\Delta C = -H_1(H_2K_4 + H_3K_3)\Delta\theta + \Delta C_K + 0(\Delta^2) \quad (43)$$

where ΔC_K is a contribution due to incorrect choice of the filter gain matrix and $\Delta\theta = \bar{\theta} - \theta_o$. ΔC_K is given by

$$\Delta C_K = H_\theta \cdot H_1\Delta K_1 + H_\theta z\Delta K_2 + H_1(H_\theta - \bar{\theta}H_3)\Delta K_3 + H_1[H_\theta + (1 - 2c - \bar{\theta})H_2]\Delta K_4. \quad (44)$$

ΔK_i are the differences between the actual and the optimal filter gains. We observe that F is linear in $\Delta\theta$ and ΔK_i except for small second order terms of the form $\Delta K_i\Delta\theta$ represented by $0(\Delta^2)$ in (43). Now F is positive real if $F_R = (F + F^*) > 0$ for all $z = e^{j\omega}$, $\omega \in [-\pi, \pi]$.

Computations of F_R as a function of ω are shown in Fig. 5. The nominal parameter and gain values are $\theta = 0.16$, $K_1 = 0.08$, $K_2 = 0.0028$, $K_3 = 0.4$, and $K_4 = -0.08$. F_R is

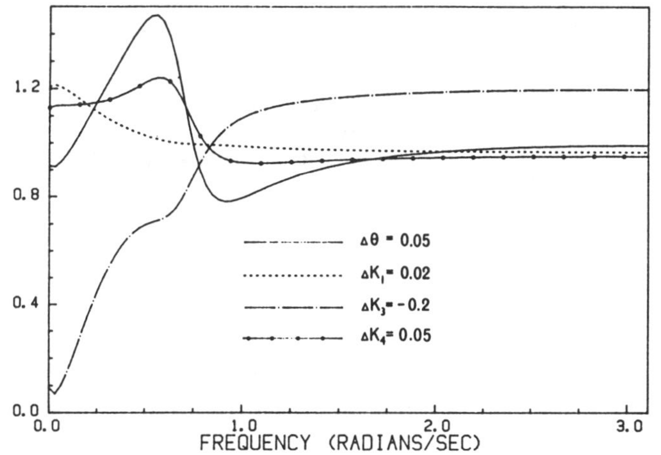

Fig. 5. Plots of $F_R = F + F^*$ as a function of frequency for different values of θ, K_1, K_3, and K_4.

computed for the following perturbations (chosen to give reasonable amplitudes in Fig. 5).

$\Delta\theta = 0.05$ (~ 30 percent), $\Delta K_1 = 0.02$ (50 percent), $\Delta K_3 = -0.2$ (-50 percent), $\Delta K_4 = 0.05$ (~ 60 percent). It turns out that perturbations in K_2 influence the positive realness (and hence, the stability properties) very little. This is quite satisfactory because K_2 is the parameter we have chosen to manipulate in order to increase the system stability during strong current conditions.

We have already seen that $F_R - 1$ is approximately linear in $\Delta\theta$ and ΔK_i. This implies that the stability condition for a general perturbation of $\Delta\theta$ and ΔK_i can be found from Fig. 5. For example, if $\Delta K_i = 0$, the stability domain of the algorithm is given by the constraint $0.05 \lesssim \theta \lesssim 0.4$. Generally, the curves in Fig. 5 indicate that the parameter estimation algorithm is stable for a broad range of parameters.

IV. Simulation Results and Operational Experience

A considerable body of experience has been obtained from operation of ALBATROSS systems during the last few years. However, most of the operational records are not yet released from the vessel owners' because of their company policy. One of the few exceptions is reported in [4].

This section is therefore mainly based on simulation results. These results are qualitatively compared to observed vessel behavior. The simulator of the vessel and its environment used in the following consists of quite complex models which are not identical to the models used in the Kalman filter.

Several Sensors

ALBATROSS systems have been produced having as many as five position reference systems. A semi-submersible emergency support vessel (ESV) operated by the BP tanker company is an example of this kind of system. The BP–ESV has two hydroacoustic position reference systems, a taut wire position reference system, a surface microwave range-bearing position reference system, and a short range radio navigation system. In addition a Doppler speed log is interfaced to the ALBATROSS system. These systems have given extreme safety of operation and very good positioning accuracy.

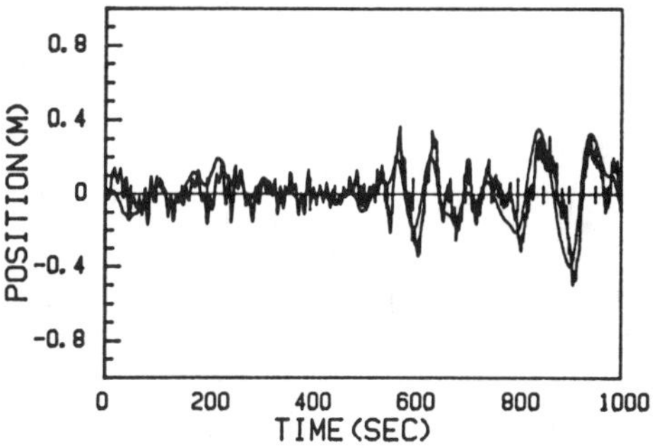

Fig. 6. Simulated and estimated LF position in surge when the Doppler system fails at $t = 500$ s.

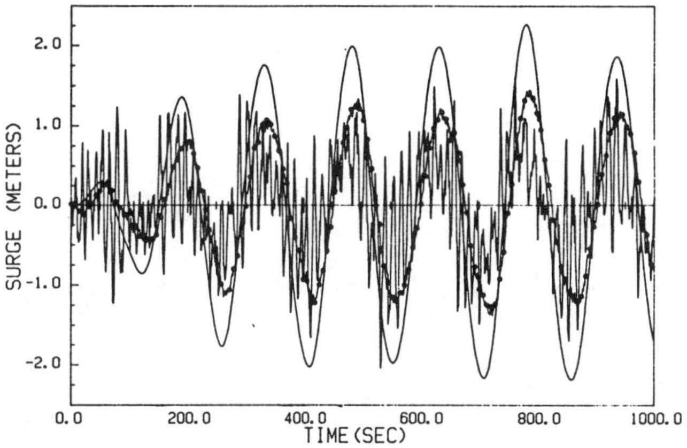

Fig. 7. A slow oscillation builds up when a current of $v_c = -1$ m/s enters in sway direction. The smooth curve is simulated LF-motion in surge, the curve marked by circles is estimated LF-motion and the rapidly oscillating curve is estimated HF-motion.

The effect of using a position reference system of the hydroacoustic type together with a high accuracy Doppler log is shown in Fig. 6. At $t = 500$ the Doppler log has a simulated failure. It is seen that the positioning error doubles or triples thereafter, which certainly is to be expected.

Slow Oscillations

The slow oscillation behavior described and analyzed in Section II is easily demonstrated by simulation. A typical case is shown in Fig. 7 using a simulated system having physical parameters and gains indicated in Sections II and III. Fig. 7 shows simulated "actual" LF-motion in surge together with filtered estimates of LF- and HF-motion in surge. The corresponding simulated "actual" yaw is shown in Fig. 8. It is observed that the HF-estimates are very unsatisfactory because they contain a considerable part of the low-frequency surge motion. The curves in Figs. 7 and 8 correspond very closely to situations observed in real systems.

By modifying the HF-model, increasing the K_2 parameter as indicated in Section III and modifying K_4 to obey (23), the simulated "actual" and estimated LF-motion in

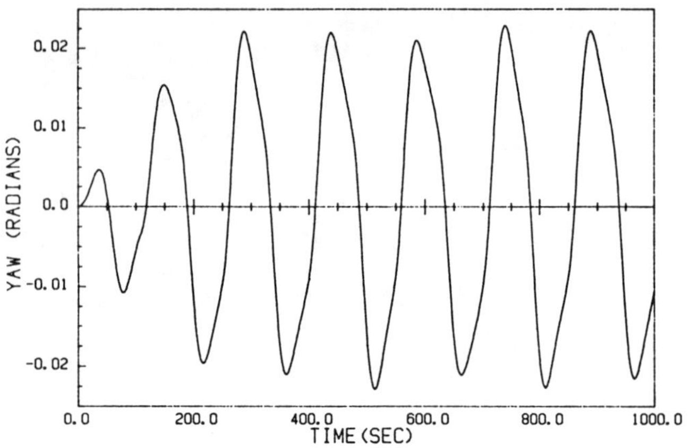

Fig. 8. Simulated LF yaw motion.

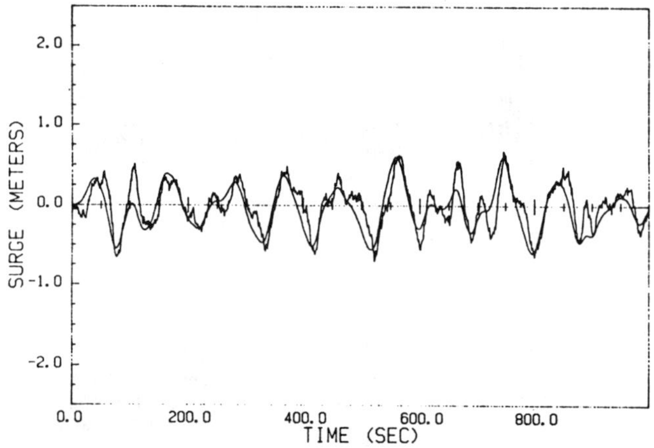

Fig. 9. Simulated and estimated LF-motion in surge using improved Kalman filter.

surge become as shown in Fig. 9. The simulated environmental conditions are identical to those used in producing Figs. 7 and 8. As is seen, the slow oscillations are strongly attenuated. This is in close agreement with observations on real vessels.

Frequency Adaptation Algorithm

The frequency adaptation algorithm works very well. In addition, both operational experience and simulation experience show that the theoretically computed domain of stability for $\hat{\theta}$ is correct. This is important because this information is used to prevent $\hat{\theta}$ entering unstable regions during operation.

Fig. 10 shows estimation of θ as a result of two similar simulation runs. The only difference is that in the second run the HF-motion spectrum is shifted towards the low-frequency end by multiplying the original frequencies by 0.85. This should result in approximately 30 percent decrease in the θ estimate. This is also seen to be the case in Fig. 10.

V. Conclusion

This paper has been concerned with the Kalman filtering algorithms in the ALBATROSS dynamic positioning system.

The early ALBATROSS systems could, under certain rare conditions, turn the vessel into a slow oscillating

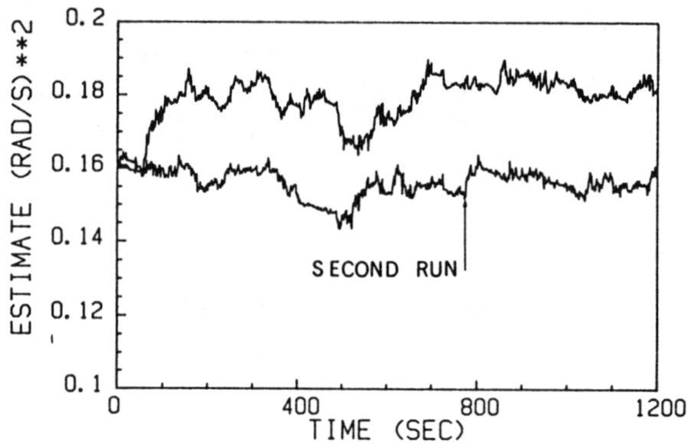

Fig. 10. Frequency estimate in surge. Two different wave motion spectra.

motion. This paper has analyzed the reasons for this instability and devised a simple solution to the problem, based on model adjustments and parameter tuning rules. An improved wave frequency adaptation algorithm is developed and analyzed as well.

The paper illustrates the important fact that a Kalman filter operating in a real life situation very often is a suboptimal filter. This results from a compromise among technical performance, cost of implementation, cost of system modeling, and cost of tuning effort. The tuning of this kind of suboptimal Kalman filter is often a difficult task and should be combined with extensive analysis of the kind presented in this paper.

Acknowledgment

The authors wish to express their appreciation to Kongsberg Våpenfabrikk A/S for permission to publish this material.

References

[1] J. M. Morgan, *Dynamic Positioning of Offshore Vessels*. Tulsa, OK: Petroleum, 1978.
[2] J. G. Balchen, N. A. Jenssen, and S. Saelid, "Dynamic positioning using Kalman filtering and optimal control theory," in *IFAC/IFIP Symposium on Automation in Offshore Oil Field Operation*, Amsterdam, The Netherlands: North-Holland, 1976.
[3] S. Saelid and N. A. Jenssen, "Albatross dynamic positioning system. Estimation and control module," NTH, Trondheim, Norway, 1977, SINTEF Rep. STF48 F77040.
[4] J. G. Balchen, N. A. Jenssen, E. Mathisen, and S. Saelid, "A dynamic positioning system based on Kalman filtering and optimal control, "*Model., Ident. and Contr.*," vol. 1, no. 3, pp. 135–163, 1980.
[5] J. G. Balchen, N. A. Jenssen, and S. Saelid, "Dynamic positioning of floating vessels based on Kalman filtering and optimal control," in *Proc. 19th IEEE Conf. Decision and Contr.*, Albuquerque, NM, 1980, pp. 852–864.
[6] L. Ljung, "Analysis of recursive stochastic algorithms," *IEEE Trans. Automat. Contr.*, vol. AC-22, pp. 551–575, Aug. 1977.
[7] —, "On positive real transfer functions and the convergence of some recursive schemes," *IEEE Trans. Automat. Contr.*, vol. AC-22, pp. 539–551, Aug. 1977.
[8] N. H. Norrbin, "Theory and observations on the use of a mathematical model for ship maneuvering in deep and confined waters," in *Proc. 8th Symp. on Naval Hydrodynamics*, Pasadena, CA, 1970.
[9] H. R. Sørheim, "Dynamic positioning in single-point mooring," The Norwegian Inst. of Technol., Trondheim, Norway, 1981, Dr.ing. thesis, NTH Rep. 81-105-W.
[10] N. A. Jenssen, "Estimation and control in dynamic positioning of vessels," The Norwegian Inst. of Technol., Trondheim, Norway, 1980, Dr.ing. thesis, NTH Rep. 80-90-W.
[11] A. H. Jazwinski, *Stochastic Process and Filtering Theory*. New York: Academic, 1970.
[12] L. Ljung, "Analysis of a general recursive prediction error identification algorithm," *Automatica*, vol. 17, no. 1, pp. 89–99.

Dynamic Ship Positioning Using a Self-Tuning Kalman Filter

PATRICK TZE-KWAI FUNG AND MIKE J. GRIMBLE, SENIOR MEMBER, IEEE

Abstract—**A novel adaptive filtering technique is described for a class of systems with unknown disturbances. The estimator includes both a self-tuning filter and a Kalman filter. The state estimates are employed in a closed-loop feedback control scheme which is designed via the usual linear quadratic approach. The approach was developed for application to the dynamic ship positioning control problem and has the advantage that existing nonadaptive Kalman filtering systems may be easily modified to include the self-tuning feature.**

Manuscript received April 21, 1982; revised September 27, 1982. This work was supported by GEC Electrical Projects Ltd., and the United Kingdom Science and Engineering Research Council.

P. T.-K. Fung was with the Department of Electrical Engineering, University of Strathclyde, Glasgow, Scotland. He is now with the Remote Manipulator Systems Division, Spar Aerospace Ltd., Weston, Ont., Canada.

M. J. Grimble is with the Department of Electrical Engineering, University of Strathclyde, Glasgow, Scotland.

I. Introduction

A DYNAMIC positioning (DP) system is used to maintain a floating vessel on a specified position and at a desired heading. The system involves a position/heading measurement system, a thruster control algorithm, and a set of thrusters (including the main propulsion units in some cases). This type of vessel is used for several applications in the survey and development of offshore mineral and oil resources. The number of countries involved in offshore exploration is increasing rapidly. For example, Saudi Arabia and the Sudan are preparing to dredge rich deposits of zinc, copper, and silver from the Red Sea mud. Manganese nodules of the highest grade have been found

Reprinted from *IEEE Trans. Automat. Contr.*, vol. AC-28, pp. 339–349, Mar. 1983.

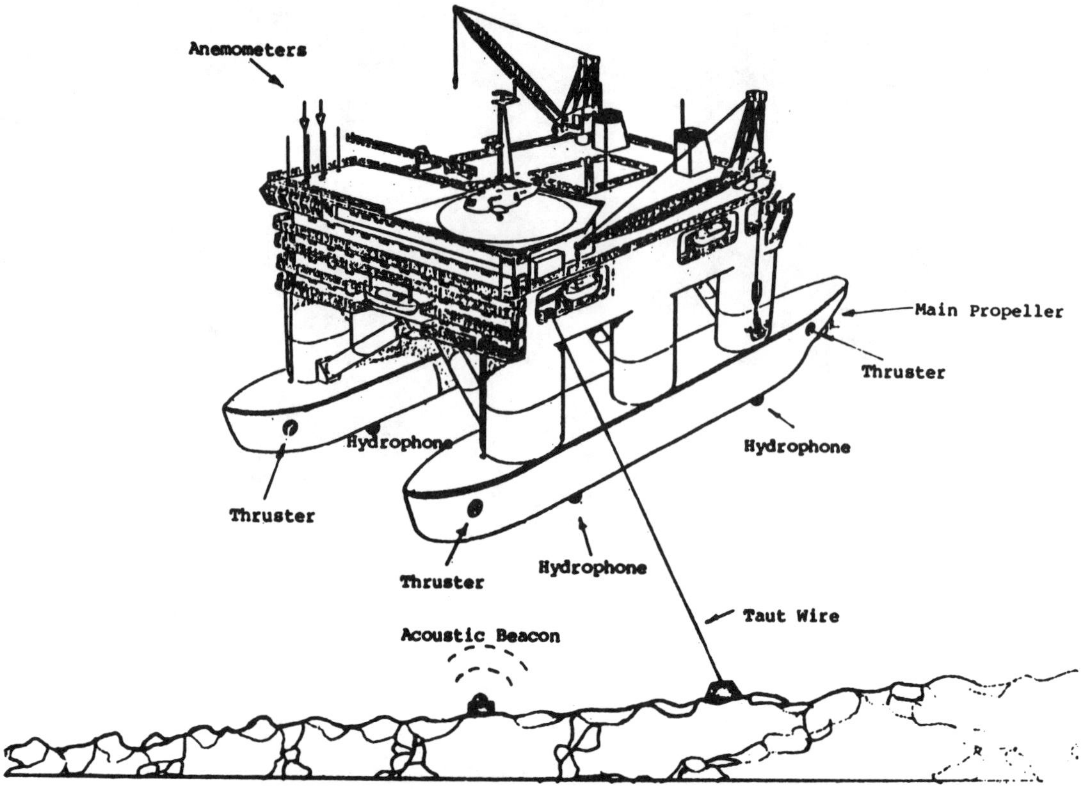

Fig. 1. Basic components in a dynamic positioning system.

in the international waters between Hawaii and California, and hence have become the subject of a prolonged debate at the United Nations Conference on the Law of the Sea.

The basic components in a DP system are illustrated in Fig. 1. Several types of position measurement systems can be used including taut wire [1], short range radio reference, and sonar systems. These measurements can be pooled and this gives rise to a combination of measurement problems. The heading measurement is given by a gyrocompass. Communication satellites are increasingly being used to provide a position fix and this enables vessels to be moved from a reference position in just a few minutes. A maximum allowable radial position error is normally specified, for example, 3 percent of water depth (under 100 m) [2].

The control loops for dynamically positioned vessels include filters to remove the wave motion signals. This is necessary because the thrust devices are not intended and are not rated to suppress the wave induced motions (greater than 0.3 rad/s). The position control system must only respond to the low frequency forces on the vessel. The filtering problem is one of estimating the low frequency motions so that control can be applied. Notice that even though the position measurement includes a noise component, this does not cause the filtering problem. If the total position of the vessel was known exactly there would still be a need to estimate the low frequency motions.

The extended Kalman filtering technique was first applied to dynamic ship positioning systems by Balchen, Jenssen, and Saelid [3]. A simpler, but nonadaptive, constant gain Kalman filtering solution was also proposed by Grimble, Patton, and Wise [4]. In both cases a linearized model was used for the estimation of the low frequency motions and optimal control feedback was employed from these estimates [5]. Balchen assumed in this and subsequent schemes [6] that the high frequency motions were purely oscillatory and could be modeled by a second order sinusoidal oscillator with variable center frequency.

Grimble *et al.* used a fourth order wave model in the specification of the high frequency motions. However, the dominant wave frequency varies with weather conditions and the corresponding Kalman filter gain must therefore be switched for different operating conditions. The extended Kalman filter of Balchen automatically adapted to these varying environmental conditions. The computational load resulting from the gain matrix calculation was reduced by making suitable approximations. An alternative extended Kalman filtering scheme proposed by Grimble, Patton, and Wise [7], [8], employed the higher order wave model, but suggested the use of fixed low frequency filter gains to achieve the necessary computational savings. The self-tuning filter described here is based upon a similar decomposition property. This approach was first proposed by Fung and Grimble [18] using a scalar example and without the theoretical justification given in the following.

The advantages and disadvantages of the self-tuning approach in comparison with the usual extended Kalman filtering schemes can be listed as follows.

Advantages

1) The varying disturbance is represented by single-input single-output channels, and thus the adaptive filter is not multivariable in nature.

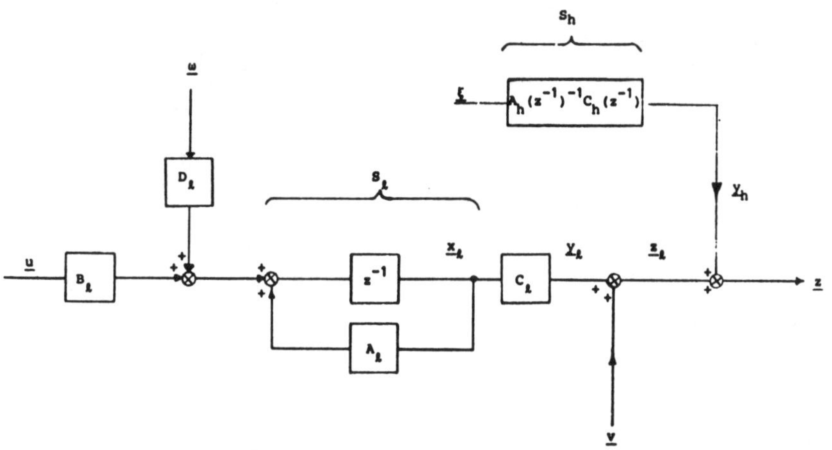

Fig. 2. Low and high frequency subsystems for a ship model.

2) The high frequency adaptive filter forms a separate subsystem to the low frequency Kalman filter, and thus the gain calculations are simplified and the system may be commissioned more easily.

3) The filter gains for the low frequency estimator are fixed and can be computed off-line, whereas all of the gains in an EKF must be computed on-line unless approximations are made [6].

4) Existing constant gain linear Kalman filtering DP systems [4] may easily be modified to include the self-tuning features described here.

5) There is no need to specify the process noise covariance or the form of the high frequency model. Only the total order of the model is assumed known.

6) The high frequency model states which are not needed for control purposes are not estimated in the self-tuning approach.

7) The scheme presented here is relatively insensitive to the presence of nonlinear ship dynamics and thruster nonlinearities [23].

Disadvantages

1) The full EKF in which all of the gains are computed on-line can be classified as being locally optimal (if the linearizations are correct), whereas the self-tuning scheme is suboptimal unless the low frequency estimator gains are calculated on-line using knowledge of the changing high frequency model.

The analysis begins with the system and problem description in Section II. The fixed gain Kalman filter is then considered in Section III and the self-tuning filter is described in Section IV. The errors which are introduced using the self-tuning structure are discussed in Section V and the total estimation algorithm is presented in Section VI. The controller design is considered in Section VII and the simulation and results are described in Sections VIII and IX, respectively.

II. The System Description

The environmental forces acting on a vessel induce motions in six degrees of freedom. In dynamic positioning only vessel motions in the horizontal plane (surge, sway, and yaw) are controlled. To simplify the problem, the motions of the vessel in the sway and yaw directions only are considered. This is possible because the linearized ship equations for the surge motion are normally decoupled from those for the sway and yaw motions [9]. The assumption is also made that the low and high frequency motions can be determined separately and that the total motion is the sum of each of them. Marine engineers often make this assumption since the analysis is simplified and the low frequency motions can also be predicted with more accuracy than the high frequency motions.

The canonical structure of the system under consideration is shown in Fig. 2. The model for a vessel can be separated into low l and high h frequency subsystems. The low frequency motions (subsystem S_l) are controllable via thruster action and the high frequency motions (subsystem S_h) are due to the first order wave forces and are oscillatory in nature. The ship positioning problem is to control the low frequency motions (output of S_l) given that the measured position of the vessel (z) includes both y_l and y_h. The object in the following is to design a state estimator to provide estimates of the low frequency motions x_l. The estimator must be capable of adapting to variations in the high frequency subsystem S_h which occur due to variations in the weather conditions.

The plant S_l can be assumed to be completely controllable and observable and to be represented by the following discrete time-invariant state equations:

$$S_l: \quad x_l(t+1) = A_l x_l(t) + B_l u(t) + D_l \omega(t) \tag{1}$$

$$y_l(t) = C_l x_l(t)$$

$$z_l(t) = y_l(t) + v(t) \tag{2}$$

where

$$E\{\omega(t)\} = 0, \qquad E\{\omega(k)\omega^T(m)\} = Q\delta_{km} \tag{3}$$

$$E\{v(t)\} = 0, \qquad E\{v(k)v^T(m)\} = R\delta_{km} \tag{4}$$

and δ_{km} is the Kronecker delta function, $x_l(t) \in R^n$, $u(t) \in R^m$, $\omega(t) \in R^q$, and $y_l(t) \in R^r$. The process noise $\omega(t)$ is used to simulate the wind disturbance and $v(t)$ represents a white measurement noise signal. The plant matrices A_l, B_l, C_l, and D_l are assumed constant and known. The observed plant output includes the colored noise (wave disturbance) signal $y_h(t)$ and is given as

$$z(t)=z_l(t)+y_h(t). \tag{5}$$

The high frequency disturbance can be represented by the following multivariable autoregressive moving-average model:

$$S_h:\quad A_h(z^{-1})y_h(t)=C_h(z^{-1})\xi(t) \tag{6}$$

which is assumed to be asymptotically stable and $y_h(t)\in R^r$ and $\xi(t)\in R^r$. Here $\xi(t)$ represents an independent zero mean random vector which is uncorrelated with $\omega(t)$ and $v(t)$, and has a diagonal covariance matrix Σ_ξ. The polynomial matrices $A_h(z^{-1})$ and $C_h(z^{-1})$ are assumed to be square and of the form

$$A_h(z^{-1})=I_r+A_1z^{-1}+A_2z^{-2}+\cdots+A_{n_a}z^{-n_a} \tag{7}$$

$$C_h(z^{-1})=C_1z^{-1}+C_2z^{-2}+\cdots+C_{n_c}z^{-n_c} \tag{8}$$

where z^{-1} is the backward shift operator. The matrix $A_h(z^{-1})$ is assumed to be regular (that is, A_{n_a} is nonsingular). The zeros of $\det(A_h(x))$ and $\det(C_h(x))$ are assumed to be strictly outside the unit circle. The order of the polynomial matrices is known, but the coefficient matrices $\{A_i\}$ and $\{C_j\}$, $i=1,\cdots,n_a$, $j=1,\cdots,n_c$ are treated as unknowns, since in practice the wave disturbance spectrum varies slowly with weather conditions. It is also assumed [1] that the disturbances in each observed channel are uncorrelated so that the matrices $\{A_i\}$ and $\{C_j\}$ have diagonal form.

III. The Low Frequency Motion Estimator

Assume for the moment that the colored noise signal y_h can be measured, and hence z_l can be calculated. The plant states x_l can be estimated using a Kalman filter with input z_l assuming the ship equations and noise covariances are known. It is reasonable to assume that a good time-invariant model for the low frequency motions is known and that the noise sources are stationary. This subsystem is stabilizable and detectable and under these conditions the Kalman gain matrix is constant and may therefore be computed off-line. Thus, the solution to this part of the estimation problem is particularly simple.

The Kalman filter algorithm becomes

$$\hat{x}_l(t|t-1)=A_l\hat{x}_l(t-1|t-1)+B_lu(t-1) \tag{9}$$

predictor: $$\hat{y}_l(t|t-1)=C_l\hat{x}_l(t|t-1) \tag{10}$$

$$P(t|t-1)=A_lP(t-1|t-1)A_l^T+D_lQD_l^T \tag{11}$$

$$\hat{x}_l(t|t)=\hat{x}_l(t|t-1)+K_l(t)\epsilon_l(t) \tag{12}$$

corrector: $$\hat{y}_l(t|t)=C_l\hat{x}_l(t|t) \tag{13}$$

$$P(t|t)=P(t|t-1)-K(t)C_lP(t|t-1) \tag{14}$$

$$K(t)=P(t|t-1)\,C_l^T[C_lP(t|t-1)C_l^T+R]^{-1} \tag{15}$$

where

$$\epsilon_l(t)=z(t)-\hat{y}_l(t|t-1)-y_h(t) \tag{16}$$

$$=z_l(t)-\hat{y}_l(t|t-1) \tag{17}$$

and $K(t)$ is the Kalman gain matrix, $P(\cdot)$ is the error covariance matrix. Unfortunately, $y_h(t)$ cannot be separated from $z(t)$ by measurement, and the signal $\epsilon_l(t)$ cannot be calculated. The way in which it is approximated will be discussed in Section V.

IV. High Frequency Motion Estimator

The wave spectrum is represented by the colored noise model (6) and in this section the high frequency motion estimator is constructed based upon this model. The assumption is made that the low frequency motions can be estimated via the technique of Section III. For the present, the problem of generating $\hat{y}_l(t|t-1)$ when $y_h(t)$ is unmeasurable will be ignored.

Define the new variable $m_h(t)$ as

$$m_h(t)=z(t)-\hat{y}_l(t|t-1) \tag{18}$$

and from (16)

$$m_h(t)=\epsilon_l(t)+y_h(t). \tag{19}$$

The innovations signal ϵ_l is white noise and m_h can be treated as the measured output of a plant S_h with measurement noise ϵ_l. The covariance matrix for ϵ_l is denoted by Σ_{ϵ_l}. The innovations signal model becomes

$$A_h(z^{-1})m_h(t)=D_h(z^{-1})\epsilon(t) \tag{20}$$

where $\{\epsilon(t)\}$ is an independent random sequence with covariance matrix Σ_ϵ.

The matrix polynomial $D_h(z^{-1})$ has the form

$$D_h(z^{-1})=I_r+D_1z^{-1}+\cdots+D_{n_d}z^{-n_d} \tag{21}$$

where the zeros of $\det(D_h(x))$ lie strictly outside the unit circle. The parameters of $D_h(z^{-1})$ are determined by the following spectral factorization:

$$D_h(z^{-1})\Sigma_\epsilon D_h^T(z)=C_h(z^{-1})\Sigma_\xi C_h^T(z)+A_h(z^{-1})\Sigma_{\epsilon_l}A_h^T(z). \tag{22}$$

Note that $n_d=n_a$ (since normally $n_a>n_c$) and that by multiplying both sides of (22) by z^{n_d} and taking the limit as $z\to 0$

$$D_{n_d}\Sigma_\epsilon=A_{n_a}\Sigma_{\epsilon_l}. \tag{23}$$

Since $A_h(z^{-1})$ is regular, that is, A_{n_a} is nonsingular, the following identity holds:

$$A_{n_a}^{-1}D_{n_d}=\Sigma_{\epsilon_l}\Sigma_\epsilon^{-1}. \tag{24}$$

Hagander and Wittenmark [10] (for the scalar case) and Moir and Grimble [11] (for the multivariable case) have shown that the optimal estimate of $y_h(t)$ can be calculated using

$$\hat{y}_h(t|t) = m_h(t) - \Sigma_{\epsilon_l}\Sigma_\epsilon^{-1}\epsilon(t) \tag{25}$$

where

$$\epsilon(t) = m_h(t) - \hat{y}_h(t|t-1). \tag{26}$$

Using the identity in (24), $\hat{y}_h(t|t)$ becomes

$$\hat{y}_h(t|t) = m_h(t) - A_{n_a}^{-1}D_{n_d}\epsilon(t). \tag{27}$$

The estimate of $y_h(t)$ is not needed for control purposes, but is required for updating $\hat{x}_l(t|t)$. The wave frequency model changes with environmental conditions and these variations are accounted for in (27) by on-line estimation of A_{n_a}, D_{n_d}, and the innovations $\epsilon(t)$ (Section VI).

V. Modified Estimation Equations

The signal $y_h(t)$ is not measurable and must be replaced in the low frequency Kalman filter by $\hat{y}_h(t|t)$. This substitution causes a difference in the state estimates [denoted $\hat{\bar{x}}_l(t|t)$] and in the calculated innovations

$$\begin{aligned}\bar{\epsilon}(t) &\triangleq z(t) - \hat{y}_l(t|t-1) - \hat{y}_h(t|t)\\ &= \epsilon_l(t) + n_h(t)\end{aligned} \tag{28}$$

where $n_h(t) \triangleq y_h(t) - \hat{y}_h(t|t)$. The signal $n_h(t)$ for the high frequency motion estimator has a zero mean value if the errors in calculating $\hat{y}_h(t|t)$ are neglected. Notice from (16) and (26) that the innovations $\bar{\epsilon}(t)$ are identical to the signal $\epsilon_h(t)$ where

$$\epsilon_h(t) \triangleq m_h(t) - \hat{y}_h(t|t).$$

If the above substitution is made the new low frequency filter has the form

$$\hat{\bar{x}}_l(t|t) = A_l\hat{\bar{x}}_l(t-1|t-1) + B_l u(t-1) + K_l(t)\bar{\epsilon}(t), \tag{29}$$

but this equation may be decomposed into the following two parts:

$$\hat{x}_l(t|t) = A_l\hat{x}_l(t-1|t-1) + B_l u(t-1) + K_l(t)\epsilon_l(t) \tag{30}$$

$$\hat{\tilde{x}}_l(t|t) = A_l\hat{\tilde{x}}_l(t-1|t-1) + K_l(t)n_h(t) \tag{31}$$

where

$$\hat{\bar{x}}(t|t) = \hat{x}_l(t|t) + \hat{\tilde{x}}_l(t|t) \tag{32}$$

and $\hat{\tilde{x}}_l(t|t)$ represents the change brought about by replacing $y_h(t)$ by $\hat{y}_h(t|t)$ in (27). The change in the predicted output

$$\tilde{y}_l(t|t-1) \triangleq \hat{\bar{y}}_l(t|t-1) - \hat{y}_l(t|t-1) \tag{33}$$

where

$$\hat{\bar{y}}_l(t|t-1) = C_l\hat{\bar{x}}_l(t|t-1) \tag{34}$$

but from (9) and (32)

$$\begin{aligned}\tilde{y}_l(t|t-1) &= C_l\big(\hat{\bar{x}}_l(t|t-1) - \hat{x}_l(t|t-1)\big)\\ &= C_lA_l\hat{\tilde{x}}(t-1|t-1).\end{aligned} \tag{35}$$

For later reference note that $\tilde{y}_l(t|t-1)$ is generated from the output of the low frequency subsystem [see (31)] driven by the zero mean signal n_h. The resulting position variations are relatively slow in comparison with the high frequency motions.

The high frequency motion estimator is also modified because the signal $m_h(t)$ in (18) cannot be calculated, but instead $\bar{m}_h(t)$ can be found where

$$\bar{m}_h(t) \triangleq z(t) - \hat{\bar{y}}_l(t|t-1). \tag{36}$$

The basis of the parameter estimation equation (Section VI) follows from (19) and (33) as

$$\begin{aligned}\bar{m}_h(t) &= m_h(t) - \tilde{y}_l(t|t-1)\\ &= A_h(z^{-1})^{-1}D_h(z^{-1})\epsilon(t) - \tilde{y}_l(t|t-1).\end{aligned} \tag{37}$$

Assuming that ϵ and $\tilde{y}_l$ can be calculated the estimate of $y_h(t)$ can be generated using (27) and (37)

$$\hat{y}_h(t|t) = \bar{m}_h(t) - A_{n_a}^{-1}D_{n_d}\epsilon(t) + \tilde{y}_l(t|t-1). \tag{38}$$

The signal $\bar{\epsilon}$ must be calculated to obtain the desired state estimates $\hat{\bar{x}}_l(t|t)$ and this can be found using (18), (27), and (28)

$$\bar{\epsilon}(t) = A_{n_a}^{-1}D_{n_d}\epsilon(t). \tag{39}$$

Recall that the gain $K_l(t)$ is calculated based upon the low frequency subsystem rather than the total system model [7]. This has the advantage that the gain is fixed and independent of variations in the high frequency subsystem. The optimal low frequency position estimate should therefore be calculated from (30), but this is not possible since ϵ_l cannot be computed directly. The state estimates are therefore obtained via (29), but are corrected using the estimated $\tilde{y}_l(t|t-1)$. This can be achieved in the ship positioning problem because the position states are identical to the outputs of the system. Thus, let the corrected estimate

$$\begin{aligned}\hat{y}_l(t|t) &= \hat{\bar{y}}_l(t|t) - \tilde{y}_l(t|t-1)\\ &\equiv \{\text{position states in } \hat{x}_l(t|t)\}.\end{aligned} \tag{40}$$

In the application of Kalman filters it is unavoidable that errors will arise from incorrect models for the plant and noise signals. The signal $\tilde{y}_l(t|t-1)$ will include such errors, but in the following section it is shown how this quantity can be estimated and may be used to correct the low frequency state estimates.

VI. Kalman and Self-Tuning Filter Algorithms

The Kalman and self-tuning filter algorithms are combined below to produce the desired low frequency motion estimator. The Kalman filter to estimate $\hat{\bar{x}}_l(t|t)$ becomes

Algorithm 6.1:
predictor:

$$\hat{x}_l(t|t-1) = A_l\hat{x}_l(t-1|t-1) + B_l u(t-1) \quad (41)$$

$$\hat{y}_l(t|t-1) = C\hat{x}_l(t|t-1) \quad (42)$$

corrector:

$$\hat{\hat{x}}_l(t|t) = \hat{x}_l(t|t-1) + K_l(t)\bar{\epsilon}(t) \quad (43)$$

$$\hat{\hat{y}}_l(t|t) = C_l\hat{\hat{x}}_l(t|t). \quad (44)$$

The signal $\bar{\epsilon}$ is required in the above algorithm, but this can be computed from (39) given the innovations signal ϵ and the matrices A_{n_a} and D_{n_d}. These matrices may be estimated as described in the following. Note that at time $t-1$ the predicted output $\hat{y}_l(t|t-1)$ is known [from (41), (42)] so that $\bar{m}_h(t)$ can be computed from (36). From (38)

$$A_h(z^{-1})\bar{m}(t) = D_h(z^{-1})\epsilon(t) - A_h(z^{-1})\tilde{y}_l(t|t-1). \quad (45)$$

The quantity $\tilde{y}_l$ is a slowly varying signal (from Section V) and can be treated as a constant over a short time interval. Let $s(t) \triangleq A_h(z^{-1})\tilde{y}_l(t|t-1)$ (where using the final value theorem z may be replaced by unity) then (45) becomes

$$A_h(z^{-1})\bar{m}_h(t) = D_h(z^{-1})\epsilon(t) - s(t). \quad (46)$$

The innovations signal model can be represented in the usual form for parameter estimation

$$\bar{m}_h(t) = \psi(t)\theta + \epsilon(t) \quad (47)$$

and the algorithm due to Panuska [12] can be employed to estimate the unknown parameters.

In the ship positioning problem the high frequency disturbances can be assumed to be decoupled, so that $A_h(z^{-1})^{-1}D_h(z^{-1})$ is a diagonal matrix and the parameters for each channel can be estimated separately. Hence, standard extended recursive least squares or maximum likelihood parameter identification algorithms may be used. For the ith channel

$$\bar{m}_{h_i}(t) = \psi_i(t)\theta_i + \epsilon_i(t) \quad (48)$$

where

$$\psi_i(t) = [-\bar{m}_{h_i}(t-1), \cdots, -\bar{m}_{h_i}(t-n_a);$$
$$\epsilon_i(t-1), \cdots, \epsilon_i(t-n_d);\ 1] \quad (49)$$

$$\theta_i^T = [a_{i1}, \cdots, a_{in_a};\ d_{i1}, \cdots, d_{in_d};\ s_i]. \quad (50)$$

Past values of the innovations signal are approximated by

$$\hat{\epsilon}_i(t) = \bar{m}_{h_i}(t) - \hat{\psi}_i(t)\hat{\theta}_i \quad (51)$$

where $\hat{\psi}_i(t)$ is given by (49) with $\epsilon_i(t-j)$ replaced by $\hat{\epsilon}_i(t-j)$ $j = 1, 2, \cdots, n_d$ and $\bar{\theta}_i$ represents the estimated parameter vector.

The recursive Kalman/self-tuning filter algorithm now becomes

Algorithm 6.2:

1) Initialize θ_i, initial parameter covariance for each channel and assign the forgetting factor β. Initialize state estimates.

2) Generate the Kalman filter estimates $\hat{x}_l(t|t-1)$ and $\hat{y}_l(t|t-1)$ using (41) and (42).

3) Calculate $\bar{m}_{h_i}(t)$ using (36) and form $\hat{\psi}_i(t)$.

4) Parameter update:

$$\hat{\theta}_i(t) = \hat{\theta}_i(t-1) + K_i^p(t)\left(\bar{m}_{h_i}(t) - \hat{\psi}_i(t)\hat{\theta}_i(t-1)\right). \quad (52)$$

5) Covariance and gain update

$$P_i^p(t) = \{P_i^p(t-1) - K_i^p(t)$$
$$\cdot(\beta + \psi_i(t)P_i^p(t-1)\psi_i^T(t))K_i^p(t)^T\}/\beta$$

$$K_i^p(t) = P_i^p(t-1)\psi_i(t)$$
$$\cdot(\beta + \psi_i(t)P_i^p(t-1)\psi_i^T(t))^{-1} \quad (53)$$

where $0.95 \leqslant \beta \leqslant 1$.

6) Innovations update:

$$\hat{\epsilon}_i(t) = \bar{m}_{h_i}(t) - \hat{\psi}_i(t)\hat{\theta}_i(t). \quad (54)$$

7) Calculate $\bar{\epsilon}_{li}(t)$ for channel i using (39)

$$\bar{\epsilon}_{l_i}(t) = \hat{a}_{n_a}^{-1}\hat{d}_{n_d}\hat{\epsilon}_i(t). \quad (55)$$

8) If $i <$ number of channels (r) go to step 3).

9) Generate the state $\hat{\hat{x}}_l(t|t)$ [using (43) and (44)].

10) Calculate the estimated $\tilde{y}_l(t|t-1)$ as

$$\hat{\tilde{y}}_{l_i}(t|t-1) = \hat{s}_i(t)/\hat{A}_{h_i}(1)$$
$$\hat{\tilde{y}}_{s_i}(t) = \alpha\hat{\tilde{y}}_{s_i}(t-1) + (1-\alpha)\hat{\tilde{y}}_{l_i}(t|t-1),$$
$$0 < \alpha < 1. \quad (56)$$

11) Correct the position estimates using (40). Return to step 2).

The signal $\hat{\tilde{y}}_{l_i}(t|t-1)$ in step 10) may be processed to produce the smoothed estimate $\hat{\tilde{y}}_{s_i}(t)$ before it is used to correct the state estimates. The algorithm described in the Appendix can predict the velocity as well as smooth the estimation of $\tilde{y}_{l_i}(t)$.

The structure of the self-tuning/Kalman filtering scheme for the dynamic positioning system is shown in Fig. 3. The surge motions are decoupled from the sway and yaw motions, and thus these are normally estimated by separate filters.

VII. Controller Design

The controller design is based on the separation principle of stochastic optimal control theory [16]. The controller with input z and output u is chosen to minimize the

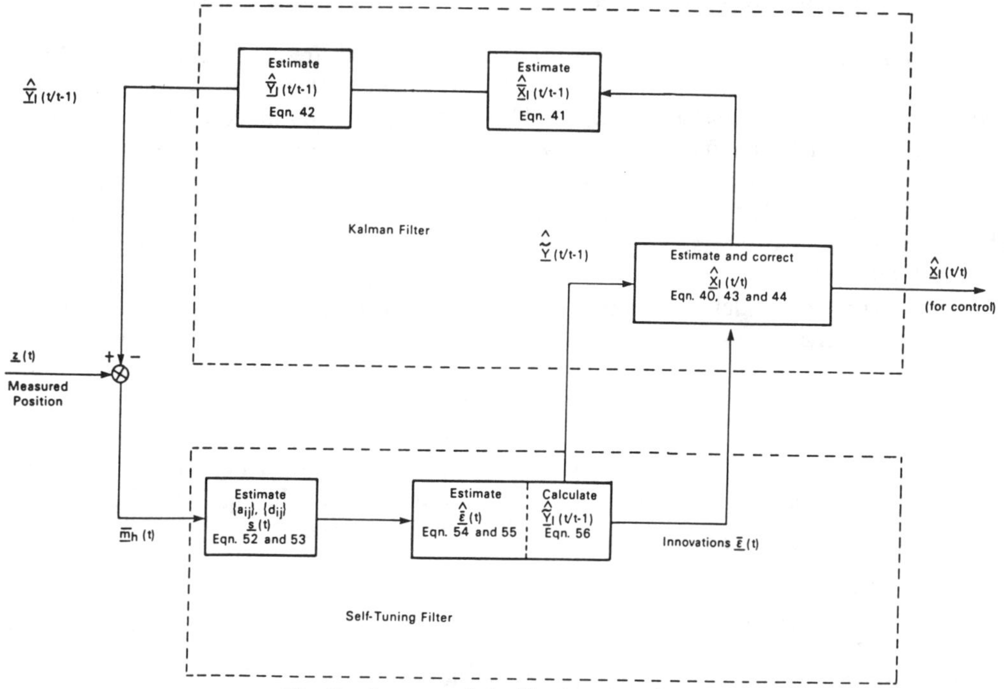

Fig. 3. Structure of the filtering scheme.

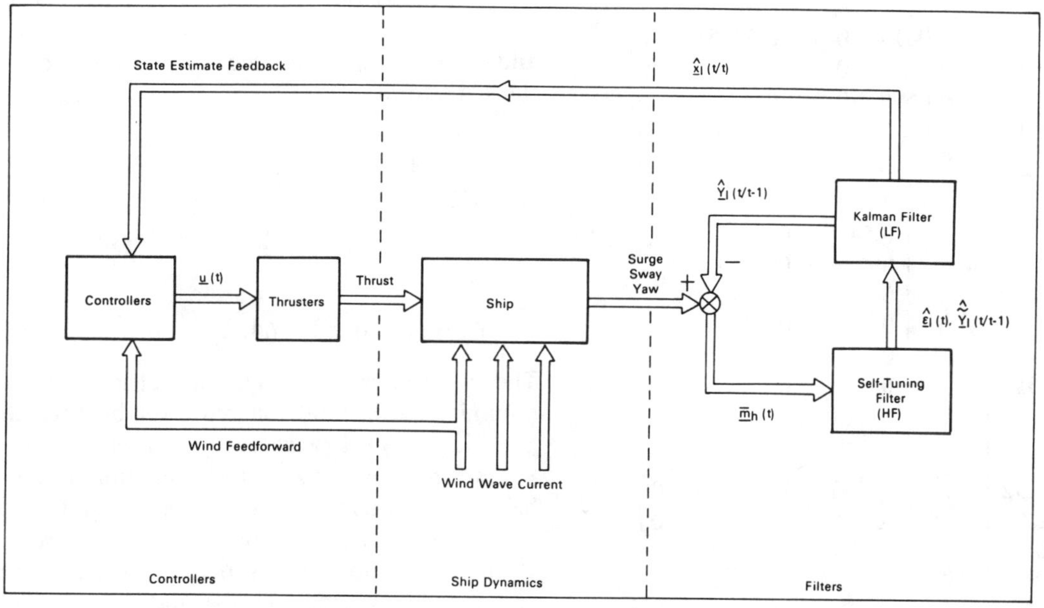

Fig. 4. Kalman and self-tuning filter state estimate feedback scheme.

performance criterion

$$J=\lim_{T\to\infty}\frac{1}{2T}E\left\{\int_{-T}^{T}(x_l-r_l)^TQ_1(x_l-r_l)+u^TR_1u\,dt\right\} \tag{57}$$

where Q_1 and R_1 are positive definite weighting matrices. The optimal control signal is generated from a Kalman filter cascaded with a control gain matrix K_c

$$u(t)=-K_c\hat{x}_l(t). \tag{58}$$

The control gain matrix may be calculated from the steady-state Riccati equation in the usual way. The closed loop control system is shown in Fig. 4.

The optimal control weighting matrices were chosen to penalize the position error corresponding to the low frequency motions (states 2 and 4) and to give an appropriate step response [17]. These were found as

$$Q_c=\text{diag}\{5,60,5,60,1,1\}$$
$$R_c=\text{diag}\{400,400\}.$$

The saturation limits on the control signals were set at ± 0.002 per unit. These represented the actual saturation which can occur when the thrusters are at full load. The selection of the optimal control weighting matrices in the ship positioning problem can be based upon results from asymptotic root loci [17] since the system is uniform rank.

VIII. Simulation and Ship Equations

A. Low Frequency Ship Motions

The low frequency motions of a vessel are determined by nonlinear equations [13] which are linearized for system analysis. The forces which produce the low frequency motions can be listed as follows: 1) forces generated by the thrusters and propellers; 2) wind forces; 3) wave induced forces; 4) hydrodynamic forces.

The linearized low frequency model of the vessel can be represented by (1) and (2) where the state vector is defined as

$$x_l(t)=\begin{bmatrix} x_1(t) \\ x_2(t) \\ x_3(t) \\ x_4(t) \\ x_5(t) \\ x_6(t) \end{bmatrix} \begin{matrix} \text{\} sway velocity} \\ \text{\} sway position} \\ \text{\} yaw angular velocity} \\ \text{\} yaw angle} \\ \text{\} thruster one} \\ \text{\} thruster two.} \end{matrix} \tag{59}$$

The system matrices for Wimpey Sealab [4] corresponding to the zero current condition and the continuous time state equations become

$$A_l=\begin{bmatrix} -0.056 & 0 & 0.0016 & 0 & 0.5435 & 0 \\ 1.0 & 0 & 0 & 0 & 0 & 0 \\ 0.573 & 0 & -0.0695 & 0 & 0 & 9.785 \\ 0 & 0 & 1.0 & 0 & 0 & 0 \\ 0 & 0 & 0 & 0 & -1.55 & 0 \\ 0 & 0 & 0 & 0 & 0 & -1.55 \end{bmatrix}$$

$$B_l=\begin{bmatrix} 0 & 0 \\ 0 & 0 \\ 0 & 0 \\ 0 & 0 \\ 1.55 & 0 \\ 0 & 1.55 \end{bmatrix} \qquad D_l=\begin{bmatrix} 0.5435 & 0 \\ 0 & 0 \\ 0 & 9.785 \\ 0 & 0 \\ 0 & 0 \\ 0 & 0 \end{bmatrix}$$

$$E_l=\begin{bmatrix} 0.384 & 0 \\ 0 & 0 \\ 0 & 6.92 \\ 0 & 0 \\ 0 & 0 \\ 0 & 0 \end{bmatrix} \qquad C_l=\begin{bmatrix} 0 & 1 & 0 & 0 & 0 & 0 \\ 0 & 0 & 0 & 1 & 0 & 0 \end{bmatrix} \tag{60}$$

where E_l is the input matrix corresponding to the wind force disturbances. The above linearized equations are in per-unit form and have been time scaled (real time = 3.104 × simulated time). The following simulation results are also in terms of per-unit quantities and scaled time.

The covariance of the process noise is dependent upon the wind force level and can be defined as

$$Q=\operatorname{diag}\{4\times10^{-6},9\times10^{-8}\}.$$

The standard deviation of the measurement noise (sonar position measurement device) is assumed to be 1/3 and 0.2 degrees, giving the normalized sway and yaw covariances

$$R=\operatorname{diag}\{10^{-5},1.22\times10^{-5}\}. \tag{61}$$

B. High Frequency Motions

The high frequency motions of the vessel are due to the first order wave forces. The worst case high frequency motion is determined by the sea wave spectrum alone and can be represented by the input-output vector difference (6). The order of the polynomial matrices $A_h(z^{-1})$ and $C_h(z^{-1})$ can be assumed to be second and first order, respectively. The parameters of these matrices vary with sea state.

It is usual to test the DP designs for real applications using simulated rather than measured sea wave data. This is partly due to the difficulty in collecting representative sea wave data, but also reflects the fact that tests over a range of different conditions must be made.

The high frequency motions were simulated using two fourth order coloring filters driven by white noise. In state space notation

$$\dot{x}_h=A_hx_h+D_h\xi \tag{62}$$

$$y_h=C_hx_h \tag{63}$$

where

$$A_h\begin{bmatrix} A_h^s & 0 \\ 0 & A_h^y \end{bmatrix} \text{ and } D_h=\begin{bmatrix} D_h^s & 0 \\ 0 & D_h^y \end{bmatrix}$$

and the submatrices for the sway and yaw directions have the same form

$$A_h^s=\begin{bmatrix} 0 & 1 & 0 & 0 \\ 0 & 0 & 1 & 0 \\ 0 & 0 & 0 & 1 \\ -a_4^s & -a_3^s & -a_2^s & -a_1^s \end{bmatrix} \quad D_h^s=\begin{bmatrix} 0 \\ 0 \\ 0 \\ k^s \end{bmatrix} \tag{64}$$

$$C_h^s=[0 \quad 0 \quad 1 \quad 0]. \tag{65}$$

The parameters of the system matrices are calculated to minimize the integral squared error between the modeled and Pierson Moskowitz sea spectra [8].

Tests on the Filters: The simulation results presented below were obtained using the above high frequency model to generate the wave motions. The tests were based on weather conditions corresponding to Beaufort numbers 8 and 5 (wind speeds 19 m/s and 9.3 m/s, respectively) which are typical examples of rough and calm seas, respectively. The first set of "filtering" results (Figs. 5–8) are for Beaufort 8, without closed loop control.

The total sway motion is shown in Fig. 5 and the estimated and modeled low frequency sway motions are shown in Fig. 6. The estimate of the low frequency motion is required for control purposes and it is clear that the estimate is good throughout the time interval (even after initial startup). The high frequency sway motion estimates are not needed for feedback control and are not shown. It is important that the LF motion estimates are relatively smooth to reduce the consequential variations in the control action. The major role of the combined estimator is indeed to separate the HF and LF motion estimates.

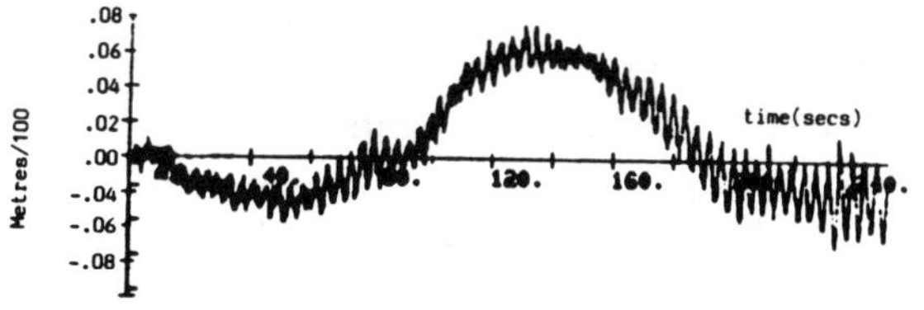

Fig. 5. Observed total sway motion (Beaufort 8).

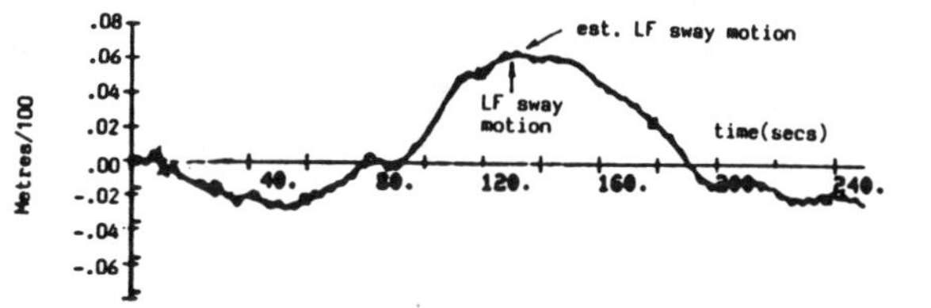

Fig. 6. Estimated and modeled low frequency sway motion (Beaufort 8)

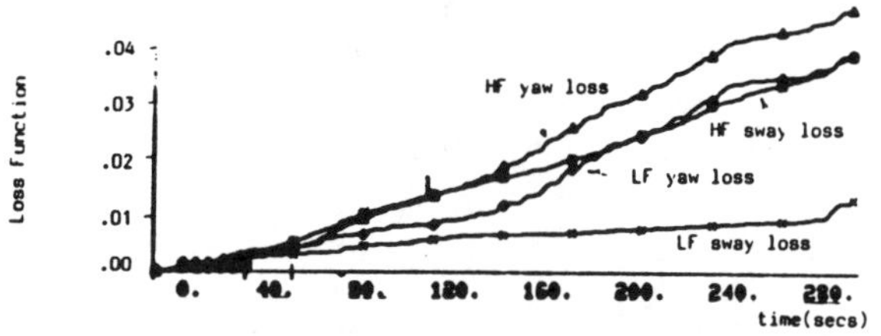

Fig. 7. Sway and yaw loss functions (Beaufort 8).

Because the LF Kalman filter does not have z_l as an input, but rather

$$z(t)-\hat{y}_h(t|t)=z_l(t)+n_h(t)$$

the predicted measurement noise covariance should be increased if the LF estimates contain an HF component. Since the HF wave conditions are slowly varying the amount by which R should be increased is not known exactly, but the system is not oversensitive to such an adjustment (factors of 5 on sway and 10 on yaw were used for the results shown here).

The accumulative loss functions for the position estimation errors in sway and yaw (both HF and LF) are shown in Fig. 7. The LF loss function for sway is defined as

$$J=\sum_{t=1}^{N}\left(y_l^s(t)-\hat{y}_l^s(t|t)\right)^2.$$

If the measurement noise were not artificially increased, when calculating the Kalman filter gain, the HF and LF loss functions for yaw would be found to be similar. This is an indication of optimal performance which has been sacrificed to some extent to obtain smoother position estimates. The parameter estimates for the high frequency model are shown in Fig. 8 where

$$A_h(z^{-1})=I_2+\begin{bmatrix} a_1^s & 0 \\ 0 & a_1^y \end{bmatrix} z^{-1}+\begin{bmatrix} a_2^s & 0 \\ 0 & a_2^y \end{bmatrix} z^{-2} \quad (68)$$

$$D_h(z^{-1})=I_2+\begin{bmatrix} d_1^s & 0 \\ 0 & d_1^y \end{bmatrix} z^{-1}+\begin{bmatrix} d_2^s & 0 \\ 0 & d_2^y \end{bmatrix} z^{-2}. \quad (69)$$

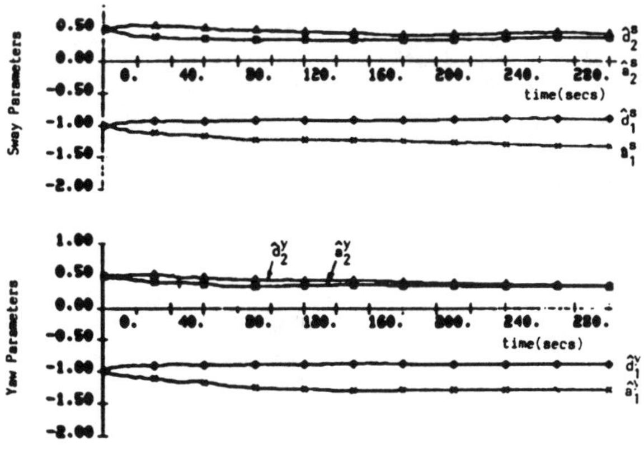

Fig. 8. Sway and yaw estimated parameters (Beaufort 8).

Note that even before the estimated parameters have converged the position estimates are still accurate (see Fig. 8). The initial parameter estimates for the matrices A_h and D_h can be based upon the knowledge that these have stable inverses. The polynomials are all of the form $a=1+a_1z^{-1}+a_2z^{-2}=(m_1z^{-1}+1)(m_2z^{-1}+1)$ and since $|m_1|<1, |m_2|<1$ then $-2<m_1+m_2<2, -1<m_1m_2<1$. Assuming $m_1, m_2<0$ implies that good initial estimates are $a_2=0.5$ and $a_1=-1$. It was found that the initial error covariance for $\hat{s}(t)$ should be small (e.g., 0.1 in this test), but the initial covariance for the other parameters should be high (e.g., 100). The estimate of $s(t)$ may contain a high frequency component and thus this is smoothed by use of a simple first order lag filter.

The filtering results for a calm sea (Beaufort 5) are not shown since the parameter estimates are much better for this case. This is consistent with the theory of Section V that shows that when the modeling errors are negligible, the term $\tilde{y}_l(t|t-1)$ is caused by the estimation error of the high frequency motion [see (35)] which is reduced in a calm sea.

Closed Loop Control: The first set of results are again for the rough sea (Beaufort 8) condition. To allow the parameter estimates to converge (as will be possible in practice) the step response of the system is measured over the time interval 240–360 s. A step reference of 0.06 per unit is input to the system at $t=240$ s. The sway and yaw responses are shown in Figs. 9–12. The low frequency variations, due to wind disturbances, are much reduced under closed loop control, but the high frequency motions are, as required, almost unchanged. The rise time for the step response can be reduced if larger control signal variations are allowed. These are shown in Figs. 13 and 14 and it is clear the sway control enters the saturation limit for a few seconds when the step demand is entered. This is not a problem since in practice position reference changes are not made in steps. One of the main design objectives is to reduce "thruster modulation," that is, variation of the thrusters in sympathy with the wave motions. That this objective has been achieved is clear from the control signals in Figs. 13 and 14.

The equivalent results for the calm sea (Beaufort 5) conditions are not shown. The parameter estimates are

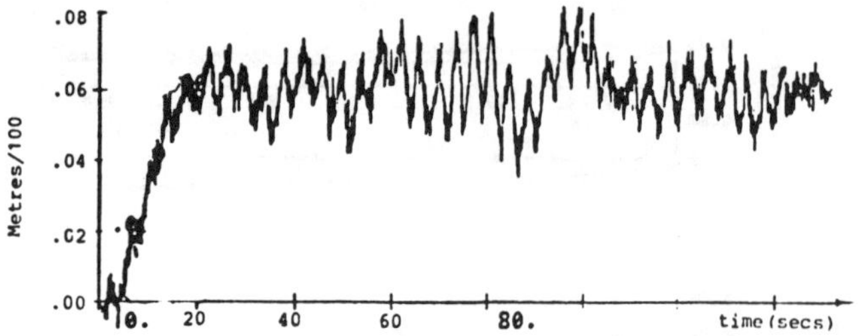

Fig. 9. Controlled total sway motion (sway reference = 0.08, Beaufort 8).

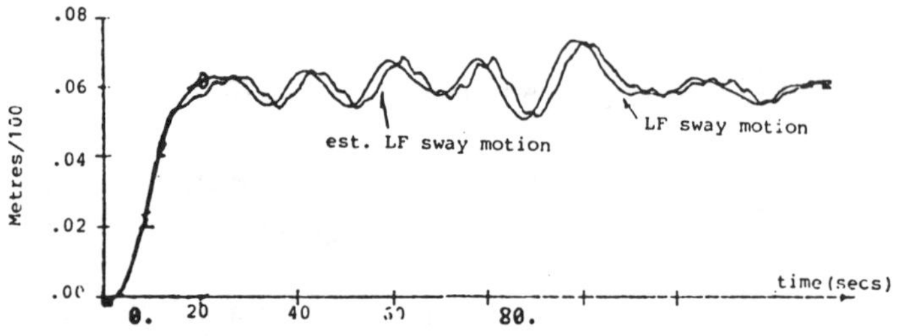

Fig. 10. Controlled LF sway and estimated sway motion (Beaufort 8).

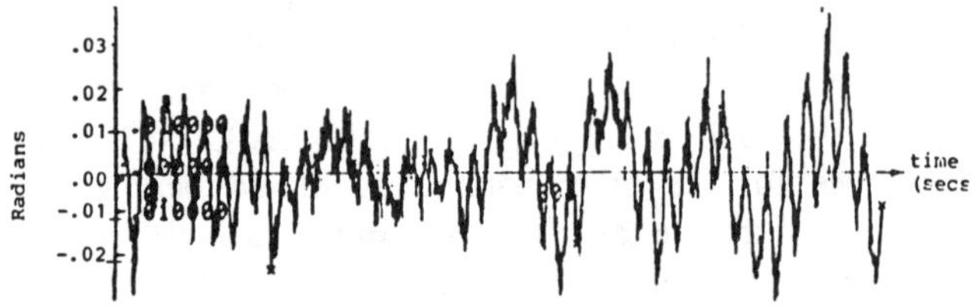

Fig. 11. Controlled total yaw motion (Beaufort 8).

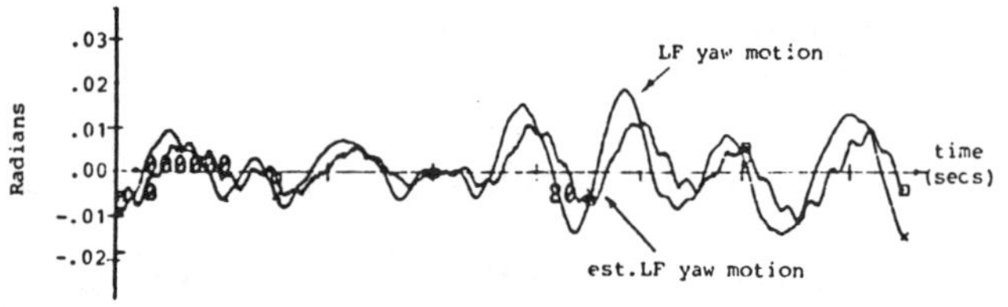

Fig. 12. Controlled LF yaw and estimated yaw motion (Beaufort 8).

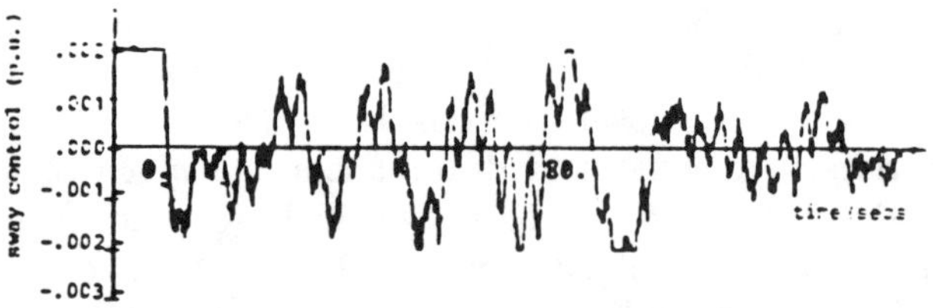

Fig. 13. Sway control signal (Beaufort 8).

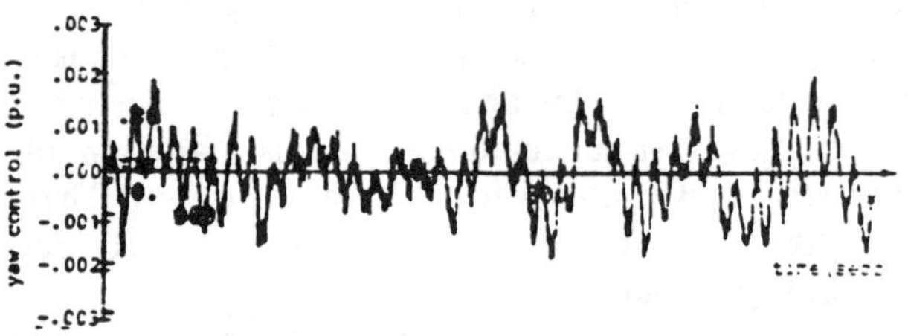

Fig. 14. Yaw control signal (Beaufort 8).

improved and the control signal variations are reduced in this case, as would be expected. Note that in comparing the high frequency motions the magnitude of the HF motion is reduced in the calm sea, but the frequency of the wave motion is higher. The sway motion is less than the allowed limit of the ± 3 for both sea states.

Rapid Weather Changes: The sea state will, relative to the system time constants, take a long time to change. It might therefore be expected that the self-tuning filter could easily track such variations and this has been demonstrated in [18]. If the weather direction changes or if the heading is changed the direction of the disturbances acting on the vessel will also vary. The magnitude of the wind and second-order wave forces will change according to the sine of the angle of incidence of these forces on the vessel and also according to the shape of the superstructure and hull exposed to these forces. The change in the angle of the current forces will be reflected in a change to the low frequency dynamics of the vessel, and hence to the linearized low frequency model [23]. These changes necessitate a variation in the drift estimator or integral action term, and the optimal control gain of this loop must be carefully chosen by posing an appropriate cost function [22]. This design problem is of course common to other Kalman filtering dynamic ship positioning schemes.

Comparison: A comparison between characteristic locus and optimal designs for dynamic ship positioning systems has recently shown [24] that the performance achievable is roughly the same in both cases. The differences lie more in the engineering implications and the relative ease of use of the different design procedures. Similar conclusions may be drawn when comparing the usual and self-tuning Kalman filtering solutions to this problem. The sway step response and control signal variations, shown in [24], for the usual fixed Kalman filtering solution, are very similar to those in Figs. 10 and 13, respectively. If the Kalman filter is matched to the sea state model (by using the same dynamics to the filter as in the wave model [8]) the Kalman filter gives a slightly lower mean square estimation error of about 10 percent. However, whenever the sea state model is significantly mismatched with the Kalman filter the self-tuning filter gives the best results. This is the situation in practice since the HF dynamical model structure is a poor representation of the nonlinear sea spectrum generator. Extended Kalman filtering schemes can also of course adapt the dominant wave frequency parameter, but these usually have a more restrictive structure than the self-tuning wave filter. A full EKF also involves a considerably larger computational burden.

IX. Conclusions

The self-tuning filter replaces the usual fixed high frequency estimator in Kalman filtering DP systems. Thus, systems which do not currently have automatic adaption to varying environmental conditions can be provided with such a feature. The approach has the advantage of simplicity over extended Kalman filtering DP systems. In addition:

1) there is no need to specify the process and measurement noise covariances for the high frequency model, 2) high frequency model states which are not needed for control are not estimated, 3) the structure of the multivariable estimator which involves separate adaptive and nonadaptive subsystems simplifies both implementation and fault finding, and 4) recent simulation results using nonlinear ship models and thruster nonlinearities have demonstrated that the scheme is robust in the presence of such nonlinearities [23].

Appendix

The algorithm for tracking the error $\tilde{y}_l(t)$ based on the estimated position error $\hat{\tilde{y}}_l(t|t-1)$, for the ith channel [19], [20], becomes

$$\tilde{y}_{l_i}^p(t) = \tilde{y}_{l_i}^*(t-1) + T\dot{\tilde{y}}_{l_i}^*(t-1) \tag{70}$$

$$\tilde{y}_{l_i}^*(t) = \tilde{y}_{l_i}^p(t) + k_{1_i}\left[\hat{\tilde{y}}_{l_i}(t|t-1) - \tilde{y}_{l_i}^p(t)\right] \tag{71}$$

$$\dot{\tilde{y}}_{l_i}^*(t) = \dot{\tilde{y}}_{l_i}^*(t-1) + \frac{k_{2i}}{T}\left[\hat{\tilde{y}}_{l_i}(t|t-1) - \tilde{y}_{l_i}^p(t)\right] \tag{72}$$

where

T	sampling interval
k_{1_i}	constant less than unity
k_{2_i}	constant less than unity
$\tilde{y}_{l_i}^p(t)$	predicted position error
$\tilde{y}_{l_i}^*(t)$	updated position error
$\dot{\tilde{y}}_{l_i}^*(t)$	updated velocity error
$\hat{\tilde{y}}_{l_i}(t\|t-1)$	estimated position error from the self-tuning filter.

Acknowledgment

The authors are grateful for the assistance and guidance of D. Wise of GEC Electrical Projects Ltd.

References

[1] A. E. Ball and J. M. Blumberg, "Development of a dynamic ship-positioning system," *GEC J. Sci. Technol.* vol. 42, no. 1, pp. 29–36, 1975.

[2] A. W. Brink, J. B. Van Den Brug, C. Ton, R. Wahab, and W. R. Van Wijk, "Automatic position and heading control of a drilling vessel," Inst. TNO for Mechanical Constructions, The Netherlands, Sept. 1972.

[3] J. G. Balchen, N. A. Jenssen, and S. Saelid, "Dynamic positioning using Kalman filtering and optimal control theory," in *Automation in Offshore Oil Field Operation*, 1976, pp. 183–188.

[4] M. J. Grimble, R. J. Patton, and D. A. Wise, "The use of Kalman filtering techniques in dynamic ship positioning systems," presented at the Oceanology Int. Conf., Brighton, England, Mar. 1978.

[5] M. J. Grimble, "The application of Kalman filters to dynamic ship positioning control," GEC Eng. Memorandum EM188, Feb. 1976.

[6] J. G. Balchen, N. A. Jenssen, E. Mathisen, and S. Saelid, "A dynamic positioning system based on Kalman filtering and optimal control," *Modeling, Ident. Contr.*, vol. 1, no. 3, pp. 135–163, 1980.

[7] M. J. Grimble, R. J. Patton, and D. A. Wise, "The design of dynamic ship positioning control systems using extended Kalman filtering techniques," in *Proc. IEEE Oceans '79 Conf.*, San Diego, CA, Sept. 1979, pp. 488–498.

[8] —, "The design of dynamic ship positioning control systems using stochastic optimal control theory," *Opt. Contr. Appl. Methods*, pp. 167–202, June 1980.

[9] D. A. Wise and J. W. English, "Tank and wind tunnel tests for a drill-ship with dynamic position control," presented at the Offshore Technol. Conf., Dallas, TX, 1975, paper OTC 2345.

[10] P. Hagander and B. Wittenmark, "A self-tuning filter for fixed-lag smoothing," *IEEE Trans. Inform. Theory*, vol. IT-23, pp. 377–384, May 1977.

[11] T. J. Moir and M. J. Grimble, "Optimal self-tuning filtering prediction and smoothing for discrete multivariable processes," *IEEE Trans. Automat. Contr.*, to be published.

[12] V. Panuska, "A new form of the extended Kalman filter for parameter estimation in linear systems with correlated noise," *IEEE Trans. Automat. Contr.*, vol. AC-25, pp. 229–235, Apr. 1980.

[13] J. W. English and D. A. Wise, "Hydrodynamic aspects of dynamic positioning," *Trans. North East Coast Inst. Eng. Shipbuilders*, vol. 92, no. 3, pp. 53–72.

[14] W. G. Price and R. E. D. Bishop, *Probabilistic Theory of Ship Dynamics*. London: Chapman and Hall, 1974, p. 159.

[15] W. J. Pierson and W. Marks, "The power spectrum analysis of ocean wave records," *Trans. Amer. Geophys. Union*, vol. 33, pp. 834–844, Dec. 1952.

[16] H. Kwakernaak and R. Sivan, *Linear Optimal Control Systems*. New York: Wiley Interscience, 1972.

[17] M. J. Grimble, "Design of optimal output regulators using multivariable root loci," *Proc. IEE*, Part D, vol. 128, pp. 41–49, Mar. 1981.

[18] P. T. -K. Fung and M. J. Grimble, "Self tuning control of ship positioning systems," in IEE Workshop on Theory & Application of Adaptive & Self-Tuning Control, Oxford University, Mar. 1981; also in C. J. Harris and S. A. Billings, Eds. London: Peregrinus, 1981, p. 322.

[19] J. A. Cadzow, *Discrete Time Systems; An Introduction with Interdisciplinary Applications*. Englewood Cliffs, NJ: Prentice-Hall, 1973.

[20] S. M. Bozic, *Digital and Kalman Filtering*. London: Arnold, 1979, p. 8.

[21] M. J. Morgan, *Dynamic Positioning of Offshore Vessels*. Tulsa, OK: Petroleum, 1978.

[22] M. J. Grimble, "Design of optimal stochastic, regulating systems including integral action," *Proc. IEE*, vol. 126, pp. 841–848, Sept. 1979.

[23] P. T. -K. Fung, Y. L. Chen, and M. J. Grimble, "Dynamic ship positioning control systems design including nonlinear thrusters and dynamics," presented at the NATO Advanced Study Inst. on Nonlinear Stochastic Problems, Algarve, Portugal, May 1982, paper 16818.

[24] J. Fotakis, M. J. Grimble, and B. Kouvaritakis, "A comparison of characteristic locus and optimal designs for dynamic ship positioning systems," *IEEE Trans. Automat. Contr.* vol. AC-27, Dec. 1982.

On the Feasibility of Real-Time Prediction of Aircraft Carrier Motion at Sea

MENAHEM M. SIDAR, MEMBER, IEEE, AND BRIAN F. DOOLIN, SENIOR MEMBER, IEEE

Abstract —**Landing aircraft on board carriers is a most delicate phase of flight operations at sea. The ability to predict the aircraft carrier's motion over an interval of several seconds within reasonable error bounds may allow an improvement in touchdown dispersion and reduce the value of the ramp clearance due to a smoother aircraft trajectory. Also, improved information to the landing signal officer should decrease the number of waveoffs substantially.**

This paper indicates and shows quantitatively that, based upon the power density spectrum data for pitch and heave measured for various ships and sea conditions, the motion can be predicted well, for up to 15 s. Moreover, the zero crossover times for both pitch and heave motions can be predicted with impressive accuracy.

The predictor was designed on the basis of Kalman's optimum filtering theory (the discrete time case), being compatible with real-time digital computer operation.

I. Introduction

THE landing phase of an aircraft aboard an aircraft carrier represents a complex operation and a demanding task. The last 10–15 s before aircraft touchdown are critical and involve terminal guidance and difficult control problems because not only is the aircraft disturbed by several kinds of stochastic (wind) disturbances, but also the touchdown point (on the ship) is being moved randomly. Despite the wind disturbances and the final point (target) random motion, the landing accuracy specified for carrier operations is very high, i.e., a few tens of feet longitudinal landing dispersion. Such a terminal point problem is made tractable in a most natural way by assuming that the ship's position can be predicted for several seconds ahead so that the airplane is guided toward the future position of the touchdown point. The scope of this study was to establish to what extent a stochastic process, like the ship's motion, is predictable over moderate periods of time.

Quantitative results obtained throughout this predictor's feasibility study concerning the relationship between the prediction error versus the prediction time, the influence of measurement noise, and the "narrowness" effect of the ship motion power density spectrum are presented. Digital simulations show that the prediction accuracy does not degrade prohibitively even for quite large measurement noise.

A variety of possibilities with respect to the incorporation of the prediction algorithm in the aircraft carrier landing system (ACLS) can be investigated, but some of those topics are out of the scope of this paper. Their common denominator consists of the ability to predict the ship's motion in heave and pitch for periods of 10–15 s. Feasibility of predicting the ship's motion within acceptable bounds of error can also lead to improvement of the landing signal officer (LSO) decision policy for waveoffs. Another potential application is the incorporation of the predictor algorithm into a control loop used to stabilize the carriers Fresnel lens system so as to provide improved glide path information to the pilot, for pilot controlled landings.

Manuscript received February 22, 1982; revised August 17, 1982.
M. M. Sidar is with the Ames Research Center, NASA, Moffett Field, CA 94035.
B. F. Doolin was with the Ames Research Center, NASA, Moffett Field, CA 94035. He is now with the Applied Technical Division, Computer Sciences Corporation, Mountain View, CA 94043.

Reprinted from *IEEE Trans. Automat. Contr.*, vol. AC-28, pp. 350–355, Mar. 1983.

The need for prediction for carrier landing operations was pointed out several years ago by Durand [1], Durand and Wasicko [2], Kaplan [3], and Siewert and A'Harrah [4]. Loeb [5] also indicated the need for predicting the ship's motion.

A tremendous amount of theoretical modeling, experimental results, and the collection of a large amount of data over the years are discussed by Powell and Theoclitus [6], Kaplan [3], Johnson [7], and many others.

A study of prediction techniques for aircraft carrier motions at sea was done by Kaplan [3], who considered a deterministic technique based on a convolution integral representation with wave height measurements at the bow serving as input. He derived, from the ship response time-history functions, a Kernel-type weighting function which operated on the measurements in order to provide the predicted motion history. Model test data indicated that this technique yielded ship's pitch prediction for up to 6 s, but the method suffers from severe limitations and practical implementation difficulties. A hybrid prediction technique, based on modern control theory, was suggested in [3] as a possible future approach.

Our approach was to make a rather direct use of the ship's motion characteristics as measured by the measuring instrumentation existing on board the ship, in order to get the predicted motion. Using this information, a predictor based on Kalman's theory of optimum estimation was designed.

Several circumstances contributed to the success of this approach. Because of the size and the mass of the ship, the carrier is significantly filtering out the motion of the sea. A complete landing operation is short enough in time such that the stochastic processes are reasonably assumed to be stationary. Finally, the prediction interval constitute only a small fraction of the time needed for each aircraft to land.

This paper is divided into three parts: 1) the derivation of the mathematical model of the ship's motion, 2) the rationale for the predictor implementation, including the Kalman filter and predictor equations, and 3) discussion of some of the results obtained. Since we are interested only in the most critical aspects of the landing operation, namely the characteristics of the longitudinal channel, we merely investigate in the sequel the predictability of the pitch and heave motion of the ship. The first report on these results appeared in [18].

II. Modeling the Carrier Motion

As we pointed out before, quite a large amount of data exist describing the motion of aircraft carriers at sea. Extensive experiments and simulations have been carried out, both at sea and in water tanks, establishing frequency response curves and power spectral density functions (psdf) as a means of representing ship pitch and heave motion characteristics for utilization in a systems analysis of the whole carrier-aircraft landing system.

Power spectral density functions describing, globally, the statistical behavior of the pitch and the heave motions have been established and measured for several types of carriers at different winds and sea conditions [6], [7]. An analysis and a close comparison of these data reveal that the psdf $\Phi(\omega)$, where ω is the carrier motion frequency in radians per second, is not affected too sensibly either by the type of the carrier or by the sea conditions (see [7]). Moreover, the function peaks quite sharply around a center frequency of about $\omega_o = 0.60\ \mathrm{rad \cdot s^{-1}}$ (Fig. 1), the sea state changing only the value of the peak at $\omega = \omega_o$ (see [7]).

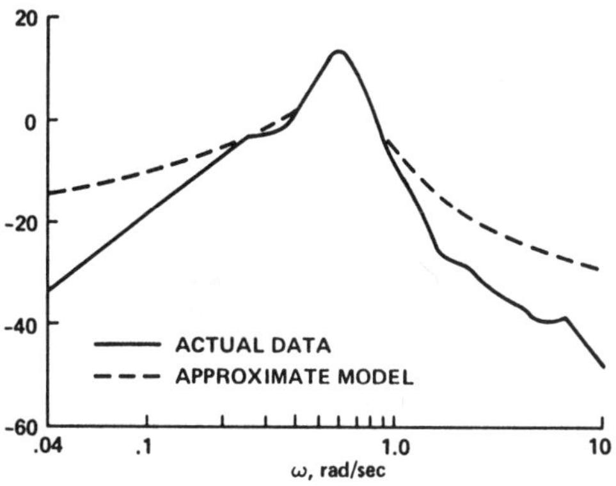

Fig. 1. Ship motion heave spectrum.

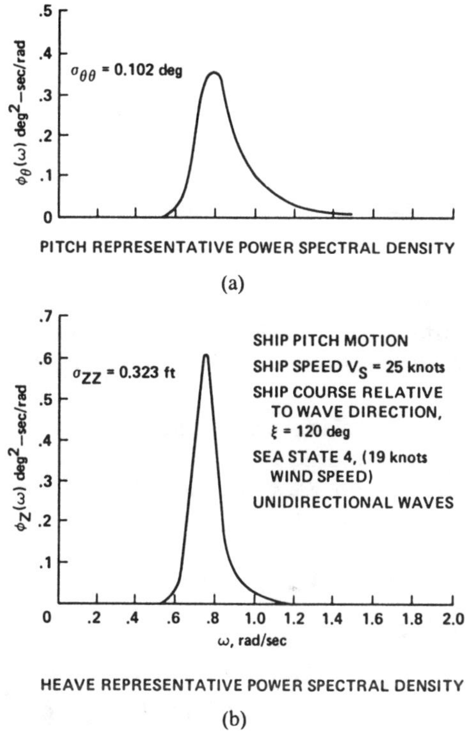

Fig. 2. Pitch and heave representative spectrum.

Data obtained from basin model experimentation confirm (Fig. 2) the measured power density spectrum, showing the same narrow-band aspect, but centered at $\omega_o \cong 0.75\ \mathrm{rad \cdot s^{-1}}$.

It is obvious, also, that for the prediction periods of interest, we can make the plausible assumption that the ship heave motion $z(t)$ and pitch motion $\theta(t)$ are stationary, narrow-band, stochastic processes. Both processes are actually continuously measured aboard the ship as a part of the SPN-42 ACLS, the measurements being contaminated by random noises $v(t)$ and $w(t)$, respectively (see [8]).

In order to obtain the mathematical model of $\Phi_z(\omega)$ and $\Phi_\theta(\omega)$ for the ship's heave and pitch, we take the innovation process point of view; namely, we assume that $z(t)$ and $\theta(t)$ are stochastic processes generated by a white, Gaussian, random process passing through a causal and invertible lumped transfer function $G_z(s)$ or $G_\theta(s)$ (see [9]).

The first step of the procedure is to approximate the (experimentally obtained) density functions $\phi_z(\omega)$ and $\phi_\theta(\omega)$ by analytic expressions that accurately represent the important part of the spectrum. As a result, one obtains a single equivalent transfer function $G_z(s) \cong G_\theta(s) \triangleq G_s(s)$ of the filter acting on the Gaussian noise, which is essentially the same for both the pitch and heave motions and is given in (1)

$$G_s(s) = gs/(s^2 + as + d) \tag{1}$$

with the following nominal values for g, a, and d: $g = 0.6$, $a = 0.06$, $d = 0.36$. This approximation is "optimized" with respect to the prediction problem, as is shown by the comparison with $\Phi(\omega)$ in Fig. 1. The reason for this specific approach is twofold: 1) to avoid running into high-dimensional systems unnecessarily, and 2) to obtain conservative values for the maximum achievable prediction time.

From (1), by making use of the usual "half power" definition of the bandwidth for a narrow-band process, the equivalent bandwidth (BW) of the ship's motion process transfer function $G_s(\omega)$ is obtained

$$\mathrm{BW} = \omega_o\left[\sqrt{1+2\xi} - \sqrt{1-2\xi}\right]\mathrm{rad}\cdot\mathrm{s}^{-1} \quad (\text{centered at } \omega_o). \tag{2}$$

For $\omega_o = 0.6$ rad·s^{-1} and $\xi = 0.05$, $\mathrm{BW} \cong 0.1$, $\omega_o = 0.06$ rad·s^{-1}. Idealized narrow-band random processes, also called ideal bandpass stochastic processes, have been studied by Rice [10], who obtained valuable theoretical results. In particular, the autocorrelation formula and the probability distribution of zero crossing for the ideal bandpass process are given. In this paper we will validate those results, comparing them for our particular problem of ship motion predictability.

Once $G_s(s)$ is obtained, it is necessary to express it equivalently in the state space form. The equations take the form of a pair of linear differential equations driven by a Gaussian white noise process. The s in the numerator of (1) normally implies that the derivative of the input function $u(t)$ is to be used as a forcing term in the state space representation. To avoid differentiating the random process, an adequate transformation was performed [11] with the following set of differential equations being obtained, where $x(t)$ represents either pitch or heave, $u(t)$ is the scalar random input, $v(t)$ is scalar noise in the measurements, and the dot indicates the time derivative

$$\dot{\boldsymbol{x}} = A\boldsymbol{x} + \boldsymbol{b}u \tag{3a}$$

$$y = \boldsymbol{c}^T\boldsymbol{x} + v. \tag{3b}$$

In (3), the following is obtained from (1):

$$A = \begin{bmatrix} 0 & 1 \\ -d & -a \end{bmatrix} \tag{4a}$$

$$\boldsymbol{b}^T = [g, -ga] \tag{4b}$$

$$\boldsymbol{c}^T = [1, 0]. \tag{4c}$$

The following assumptions are made with respect to the noise and the input:

$$E[u] = E[v] = E[vx_i] = 0, \quad \forall i \tag{5a}$$

$$E[u(t)u(s)] = Q\cdot\delta(t-s) \tag{5b}$$

$$E[v(t)v(s)] = R\cdot\delta(t-s). \tag{5c}$$

The system of equations (3) is controllable and observable. In these equations, $x_1(t)$ in $\boldsymbol{x}^T = (x_1, x_2)$ represents position, either heave or pitch angle. Equations (5a)–(5c) express the assumption that the ship motion and the measurement noise are uncorrelated.

III. The Predictor Equations

The following several approaches and algorithms are available for obtaining $\boldsymbol{x}(t+\tau)$, the prediction of $\boldsymbol{x}(t)$ at $\tau > 0$ s from now:

1) the Wiener approach [12], [13], which assumes stationarity, a plausible assumption over the short periods of time required for landing;

2) the Ragazzini–Zadeh approach [13], [14], which is useful for the finite-time measurement case, but is otherwise complex for implementation;

3) the time series analysis and prediction algorithm [15];

4) the Kalman predictor approach [16], [17].

The last approach is adopted here because, by avoiding cumbersome computations, it makes real-time digital computation possible.

The best estimate of the ship's motion at time t, $t \in [o, t_f]$ is denoted by $\hat{\boldsymbol{x}}(t)$. It is the conditional expectation of $\boldsymbol{x}(t)$ based on all prior measurements of $y(t)$. Then the linear optimal least-squares prediction theory of Kalman [16], [17] gives for the best predicted motion $\hat{\boldsymbol{x}}(t+\tau)$ the expression

$$\hat{\boldsymbol{x}}(t+\tau) = \phi(t+\tau, t)\hat{\boldsymbol{x}}(t) \tag{6}$$

where $\phi(t, \sigma)$ is the transition matrix for (3).

The computation process for $\hat{\boldsymbol{x}}(t+\tau)$, therefore, divides into the following two steps: 1) calculate $\hat{\boldsymbol{x}}(t)$; then 2) use (6) to obtain $\hat{\boldsymbol{x}}(t+\tau)$ for the desired prediction time τ (see Fig. 3).

Since the Kalman formulation was adopted to enable real-time digital computation, (3)–(6) should be replaced by their discrete form. The discrete-time representation of the ship motion is given by

$$\boldsymbol{x}_k = \phi_{k,k-1}\boldsymbol{x}_{k-1} + \Gamma_{k-1}u_{k-1} \tag{7a}$$

$$y_k = \boldsymbol{c}^T\boldsymbol{x}_k + v_k \tag{7b}$$

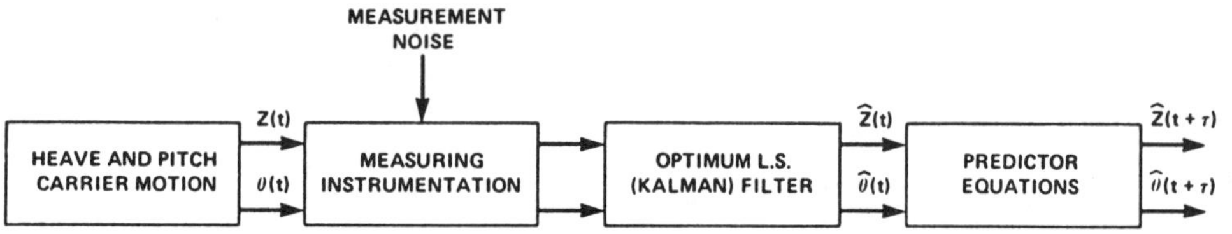

Fig. 3. Ship's motion predictor block diagram.

where

$$\phi_{k,k-1}=\begin{bmatrix}1 & T_s\\ -dT_s & 1-aT_s\end{bmatrix},\qquad \Gamma_{k-1}=\begin{bmatrix}gT_s\\ -agT_s\end{bmatrix}$$

v_k = a sequence of random uncorrelated measurement noise $E(v_k v_j)=R_k\cdot\delta_{kj}$,

u_k = a sequence of random uncorrelated inputs $E(u_k u_j)=Q_k\cdot\delta_{kj}$,

T_s = the sampling time.

The discrete-time version of the optimum filter is obtained from the following set of equations:

$$\hat{x}_k=\phi_{k,k-1}\hat{x}_{k-1}+\boldsymbol{K}_k\left[y_k-c^T\phi_{k,k-1}\hat{x}_{k-1}\right] \tag{8a}$$

$$K_k=P_k'C^T\left[CP_k'C^T+R_k\right]^{-1}. \tag{8b}$$

Note that the measurement vector $\boldsymbol{c}$ has been replaced for convenience by a matrix C where $C\triangleq\begin{bmatrix}1&0\\0&0\end{bmatrix}$ and $\boldsymbol{K}_k$ is the first column of K_k. The *a posteriori* covariance matrix is obtained from

$$P_k'=\phi_{k,k-1}P_{k-1}\phi_{k,k-1}^T+Q_{k-1} \tag{8c}$$

and the *a priori* covariance matrix is given by

$$P_k=P_k'-K_kCP_k'. \tag{8d}$$

P_k, P_k', and K_k (the Kalman filter gain) are square matrices. By time k, as a result of (8a)–(8d), the best estimate $\hat{x}_k$: $[\hat{x}_k|$"past" values of $y_k]$, $\forall k\in[0,N]$ is generated.

The initial value of the error convariance matrix P_k is given by the matrix

$$P_0=\overline{\hat{x}_0\hat{x}_0^T} \tag{8e}$$

which is used as a startup value for the recursive scheme. In this specific case we have to compute only two optimal gains, namely K_{11_k} and K_{21_k}, given by

$$K_{11_k}=P'_{11_k}/\left(P'_{11_k}+R_{11k}\right);\qquad K_{21_k}=P'_{21_k}/\left(P'_{11_k}+R_{11k}\right). \tag{9}$$

The predicted vector $\hat{x}(t+\tau)$ is obtained from (10) which is the discrete equivalent of (6)

$$\hat{x}_{k+m}=\phi(k+m,k)\hat{x}_k \tag{10}$$

where $m\triangleq\tau/T_s$.

The transfer matrix $\phi(k+m,k)$ is given by

$$\phi(k+m,k)\triangleq\phi(\tau)=e^{(-a/2)\tau}\cdot\begin{bmatrix}\cos\beta\tau+(a/2\beta)\sin\beta\tau & (1/\beta)\sin\beta\tau\\ (-d/\beta)\sin\beta\tau & \cos\beta\tau-(a/2\beta)\sin\beta\tau\end{bmatrix} \tag{11}$$

and $\beta^2\triangleq d-a^2/4$.

Finally, since we are interested only in $\hat{x}_1(t+\tau)$, the ship's position optimal predicted value for τ seconds ahead, one obtains the following result:

$$\hat{x}_1(t+\tau)=e^{(-a/2)\tau}\left[(\cos\beta\tau+(a/2\beta)\sin\beta\tau)\hat{x}_1(t)+(1/\beta)\sin\beta\tau\cdot\hat{x}_2(t)\right]. \tag{12}$$

For a narrow-band process, which is the present case, the parameter a in (12) is very small. Thus, for moderate prediction times, (12) can be approximated by

$$\hat{x}_1(t+\tau)\approx\hat{x}_1(t)\cos\beta\tau+\hat{x}_2(t)(\sin\beta\tau)/\beta. \tag{13}$$

Equation (13), which will be used later in calculating the autocorrelation function, suggests that under the stated modeling, the extrapolation is equivalent to the prediction of the state of a harmonic oscillator on the basis of an estimate of its present state. That this approximation is appropriate for the carrier landing problem is easy to see from the quality of the prediction achieved, results of which will be discussed next.

IV. Results and Discussion

The data shown in Figs. 4 and 5 summarize the results of a large number of cases investigated by a digital simulation of the ship's motion and its prediction. The figures show the effect of the achievable prediction time τ in terms of two performance criteria of major significance in the carrier landing problem.

The first performance criterion J_1 measures performance in terms of the error in predicting position. The error in prediction ϵ_p is defined as

$$\epsilon_p(t'+\tau)\triangleq x_1(t'+\tau)-\hat{x}_1(t'+\tau),\qquad \forall t'\in[0,t]$$

or

$$\epsilon_p(kT_s+\tau)\triangleq x_1(kT_s+\tau)-\hat{x}_1(kT_s+\tau),\qquad \forall k\in[0,N]. \tag{14}$$

The criterion J_1 is defined as the following scalar func-

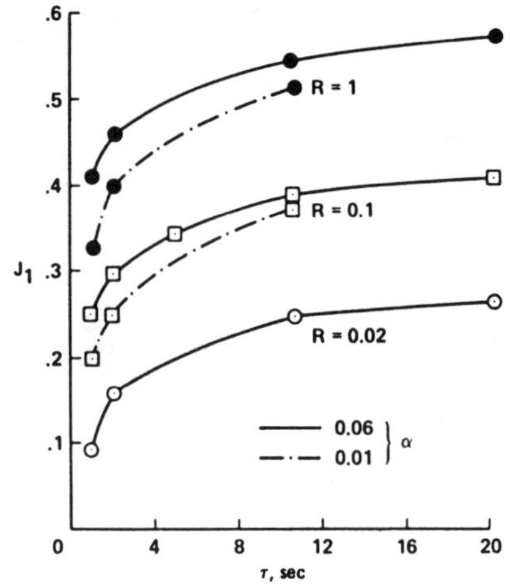

Fig. 4. Value of J_1 versus prediction time τ.

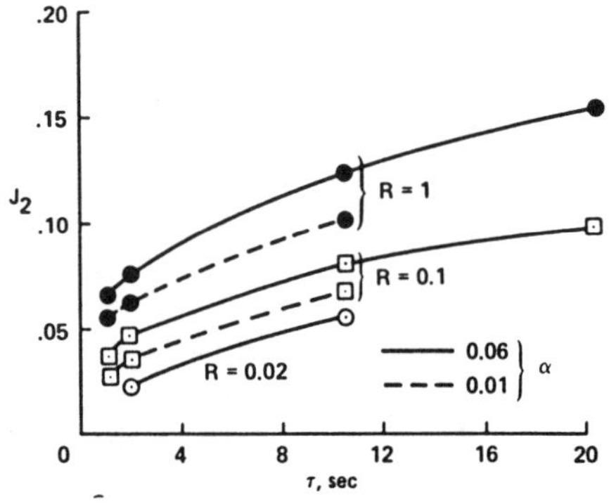

Fig. 5. Value of J_2 versus prediction time τ.

tional

$$J_1 \triangleq (\sigma_{x_1})^{-1}\sqrt{(1/N+1)\sum_{k=0}^{N}\epsilon_p^2(kT_s+\tau)} \tag{15}$$

reflecting the practical need to measure and minimize the touchdown errors. The scalar J_1 depends on the spectral width a and the measurement noise covariance R, in addition to the prediction time τ, as Fig. 4 shows. There it can be seen that if the measurement noise is not prohibitive, prediction times as long as 10–15 s are attainable with reasonable accuracy. For example, a 5 percent measurement noise ($R = 0.02\sigma_{x_1}^2$), with $\sigma_{x_1} = 1$ m (3.1 ft) and $\tau = 10$ s, gives $\sigma_{pr} = \sigma_{x_1}J_1 \cong 0.24$ m (0.8 ft), which can be considered acceptable.

The results obtained during the study suggested the introduction of a second criterion with respect to the predictor quality, namely J_2. This criterion covers for another important practical use of the predictor, e.g., the prediction of the crossover times. This second criterion is defined as follows:

$$J_2 \triangleq (\omega_o/2\pi)\sum_{i=1}^{M}|\Delta T_{CRO}(t_i)| \tag{16}$$

where M is the number of crossover points counted in the fixed finite interval of an experiment, and

$$\Delta T_{CRO}(t_i) \triangleq [T_i\colon x(t_i+\tau)=0] - [T_i\colon \hat{x}(t_i+\Delta t_i+\tau)=0] \tag{17}$$

where

$$0 \leqslant \Delta t_i < t_{i+1} - t_i.$$

Here $[T_i\colon x(t_i+\tau)]$ means the ith time when $\hat{x}$ crosses zero. Thus ΔT_{CRO} is defined as the difference between the crossover time of $x_1(t+\tau)$ and the nearest successive crossover time of $\hat{x}_1(t+\tau)$. It tells how long the actual ship heave, for example, differs in sign from that predicted. Fig. 5 shows that this error is small, even for prediction times of 10–15 s. For example, under the same conditions just given, namely a 5 percent noise-to-signal ratio and a 10 s prediction time, and for $T_o = 2\pi/\omega_o = 10$ s, the total time that the predicted motion differs in sign from the actual motion is 0.5 s over a measurement interval of 60 s. The practical implication of the results shown in Figs. 4 and 5 is clear and important; both the motion of aircraft carriers at sea and the crossover times of their motion can be predicted accurately over periods of time that are long enough and operationally useful.

These results are in good agreement with Rice's analysis [10] of ideal narrow bandpass processes. However, the comparison has to be made with care, since in our study the noise enters into the measurements as well as at the input. Furthermore, our results are derived from simulations of finite duration. Nevertheless, the fact that our results are consistent with his analysis of narrow-band noise strengthens our confidence in the predictability of the motion.

Evidence that the predictability of the motion of the carrier is due to its effective narrow-band character is demonstrated by the high correlation of our results with that to be expected of a narrow-band process. From (3.2-5) in [10] one gets the following expression for the autocorrelation function $R(\tau)$ of such a process:

$$R(\tau) = \sigma^2[\sin(\pi \mathrm{BW})\tau/(\pi\cdot \mathrm{BW})\tau]\cos\omega_o\tau. \tag{18}$$

Setting τ equal to an integral number of periods $\tau = \zeta T_o$ where $f_o = 1/T_o$ is the center frequency of the bandpass (or the peak frequency of the power spectrum of the carrier's motion), one obtains

$$R(\tau)/\sigma^2 = R(\zeta T_o)/\sigma^2 = \sin 0.2\pi\zeta/0.2\pi\zeta. \tag{19}$$

This gives $R(T_o) = 0.935\sigma^2$ and $R(2T_o) = 0.757\sigma^2$.

These results compare well with ours; the correlation between the predicted values of $x_1(t)$, namely $\hat{x}_1(t+\tau)$, and their actual values is very nearly unity over times of the order of 10–15 s.

Furthermore, the influence of the process narrowness BW can also be obtained from (18) and compared with the results in Fig. 4. The autocorrelation function $R(\tau)$ is a sinc function [see (18)] of the bandwidth, a result which is in agreement with the results obtained in our study.

Rice also made an analysis of the expected zero-crossings for a narrow-band process, and again, interpreting his formulas in our terms gives values consonant with the results shown in Fig. 5.

Equation (3.3-12) of [10] gives the expected number of zero crossings of $x_1(t)$ per second as

$$N_z = 2\left[1/3\left(f_h^3 - f_l^3\right)/\left(f_h - f_l\right)\right]^{1/2} \tag{20}$$

where f_h and f_l are the upper and lower frequency limits for the ideal band limited process, and $\mathrm{BW} = f_h - f_l$. When f_h approaches f_l, as in the narrow-band case,

$$N_z \approx f_h + f_l \cong 2f_o \tag{21}$$

which is the number of zero crossings for a sinusoidal heave (or pitch) motion with frequency $f_o = \omega_o/2\pi$. This result agrees with ours (Fig. 5), even for prediction times as large as twice the period T_o.

Also from [10], the probability that a second zero crossing of $x_1(t)$ lies within t and $t + \Delta t$, at a particular time t, say $t = K\theta = K(T_o/2)$, is given (approximately) by (22) for the ideal narrow-band process

$$pr(n_{z_{\Delta t}}) = (1/2)\left(\gamma/\left[1+\gamma^2(t-k\theta)^2\right]^{3/2}\right)\Big|_{\substack{t=k(T_o/2)+\Delta t\\ k=1,2,\cdots}} \tag{22}$$

where

$$\gamma \triangleq \sqrt{3}\left[(f_h+f_l)^2/\mathrm{BW}\right]; \qquad \theta \triangleq 1/(f_h+f_l).$$

Equation (22) shows that the probability density is a symmetrical and decreasing function around $t = k(T_o/2)$, its peak value and dispersion depending on the bandwidth BW. For example, in the specific case of the ship's motion prediction, with $\mathrm{BW} = 0.1\ f_o$ and $T_o = 10.45$ s, $pr(n_{z_o}) = 3.46$, whereas $pr(n_{z_1}) \cong 0.01$. This means that the probability of a zero crossing at $(T_o + 1.0$ s) is very small (0.01). Here again, our results show a nice regularity of the zero-crossing points which are (almost) identifiable with the zero-crossing points of a harmonic ship's motion with frequency f_o. This is, basically, the reason why it is possible to obtain a good prediction of the zero-crossing events for $x_1(t+\tau)$, even for relatively high values of τ.

V. Conclusions

The feasibility of predicting aircraft carrier motion at sea by measuring the actual ship's position was investigated. The ship's motion mathematical model based on statistical data, such as power density spectrum representation, was established. Subsequently, a discrete-time Kalman filter-predictor adapted for real-time computation on a digital computer was investigated. The results obtained show that a maximum achievable prediction time of up to 15 s can be reached within reasonable acceptable errors.

Being able to predict accurately the ship's motion can lead to an improvement of aircraft landing accuracy. This can be accomplished, for instance, by generating new terminal guidance (landing) laws making use of the future ship's position. Moreover, the possibility of prediction can eventually improve the LSO information and policy for landing acceptance or waveoff.

An additional improvement capability consists of using the predictor signal for stabilizing the Fresnel lens system for pilot controlled landings.

The computational possibility of processing the actually measured carrier motion, in order to obtain in (nearly) real time the power density spectrum for the ship's motion by fast Fourier transform algorithms (FFT) is of paramount importance. This possibility may lead toward an adaptive (tuning) predictor scheme, tuning for f_o in real time. It is strongly recommended, therefore, for the future ACLS implementation, to investigate the possibility to incorporate the tuning predictor approach.

Acknowledgment

The authors would like to thank Dr. G. A. Smith for his collaboration and stimulating discussions, and M. Nordstrom for aid in computational work. Thanks are also due to the anonymous reviewers for their constructive and interesting remarks and suggestions which were helpful in improving the paper.

References

[1] T. S. Durand, "Theory and simulation of piloted longitudinal control in carrier approach," Systems Technology Inc., Tech. Rep. 130-1, 1965.

[2] T. S. Durand and R. J. Wasicko, "An analysis of carrier landing," AIAA paper 65-791, 1965.

[3] P. Kaplan, "A study of prediction techniques for aircraft carrier motion at sea," *J. Hydronautics*, vol. 3, no. 3, pp. 121–131, 1969.

[4] R. F. Siewert and R. C. A'Harrah, "Study of terminal flight path control in carrier landings," North American Aviation, Rep. NA66H-289, 1967.

[5] J. L. Loeb, "Automatic landing systems are here," in *Proc. AGARD Conf.*, 1970, paper 14.

[6] F. D. Powell and T. Theoclitus, "Study of an automatic carrier landing environment with the AN/SPN-10 landing control central," Bell Aerosystems Co., Rep. 6026, 1965.

[7] W. A. Johnson, "Analysis of aircraft carrier motions in a high sea state," Systems Technology Inc., Tech. Rep. 137-3, 1969.

[8] AN/SPN-42 Automatic Carrier Landing System, Bell Aerospace Co. publication.

[9] T. Kailath, "An innovations approach to least-squares estimations, Part I: Linear filtering in additive white-noise," *IEEE Trans. Automat. Contr.*, vol. AC-13, pp. 646–655. Dec. 1968.

[10] S. O. Rice, "Mathematical analysis of random noise," in *Selected Papers on Noise and Stochastic Processes*, N. Wax, Ed. New York: Dover, 1954.

[11] K. Ogata, *State Space Analysis of Control Systems*. Englewood Cliffs, NJ: Prentice-Hall, 1968.

[12] N. Wiener, *Extrapolation, Interpolation and Smoothing of Stationary Time Series*. New York: Wiley, 1949.

[13] J. H. Lanning and R. H. Battin, *Random Processes in Automatic Control*. New York: McGraw-Hill, 1961.

[14] J. R. Ragazzini and L. A. Zadeh, "An extension of Wiener's theory of prediction," *J. Appl. Phys.*, vol. 21, pp. 645–655, 1950.

[15] D. M. Jenkins and G. Box, *Time Series Analysis, Forecasting and Control*. San Francisco: Holden-Day, 1970.

[16] R. E. Kalman and R. S. Bucy, "New results in linear filtering and prediction theory," *Trans. ASME*, vol. 831, pp. 95–108, 1961.

[17] H. W. Sorenson, "Kalman filtering techniques," in *Advances in Control Systems*, vol. 3, ch. 5, pp. 219–292, 1966.

[18] M. Sidar and B. F. Doolin, "On the feasibility of real time prediction of aircraft motion at sea," NASA TM X 62454, June 1975.

An Integrated Multisensor Aircraft Track Recovery System for Remote Sensing

WITOLD S. GESING, MEMBER, IEEE, AND D. BLAKE REID, MEMBER, IEEE

Abstract —**This paper describes an application of the Kalman filter in a track recovery system (TRS) for postflight processing of aircraft navigation sensor data. The track recovery system has been successfully used as a key component of the Canadian aerial hydrography pilot project for mapping of shallow coastal waters.**

Recorded data from an inertial navigation system (INS) is combined with data obtained from a number of auxiliary sensors to construct a set of error measurements. The measurements are prefiltered to compress the data and are then processed using a $U-D$ factorized Kalman filter and a modified Bryson–Frazier smoother to produce estimates of the time-correlated sensor errors. The flight profile is obtained by subtracting the computed error estimates from the recorded INS data. The residual errors observed in processing real data collected in a number of field tests are less than 1 m in position and less than 0.03 degrees in attitude.

I. Introduction

An aircraft track recovery system (TRS) has been developed for use in airborne remote sensing applications. A number of these applications require accurate estimates of the aircraft's flight profile parameters, including position, velocity and attitude for the postflight reduction and analysis of remote sensor data. The TRS provides the required information by combining inertial navigation system data with data acquired by other navigation and positioning aids carried on the aircraft. The software package employed for processing of the navigation sensor data implements optimal filtering and smoothing algorithms. The use of optimal estimation methods was motivated primarily by the strict performance requirements of the Canadian aerial hydrography project in which a system was developed for the mapping of shallow coastal waters.

The aerial hydrography system employs an aerial survey camera, a high-powered pulsed laser mounted in adjacent bays in the aircraft, and various navigation sensors to measure their orientation in flight. Ground-based computing and plotting facilities process the data obtained in the air. (Fig. 1.)

The laser measures water depth directly by transmitting pulses of green light, ten times per second, which are reflected from the water surface and from the bottom. The time difference in reception of the surface and bottom

Manuscript received April 14, 1982; revised August 20, 1982.
W. S. Gesing is with the Department of Electrical Engineering, University of Toronto, Toronto, Ont., Canada.
D. B. Reid is with Huntec (70) Limited, Scarborough, Ont., Canada.

Reprinted from *IEEE Trans. Automat. Contr.*, vol. AC-28, pp. 356–363, Mar. 1983.

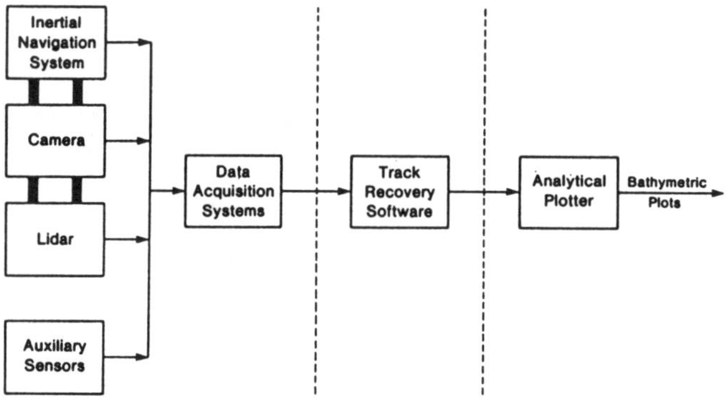

Fig. 1. Aerial hydrography system structure.

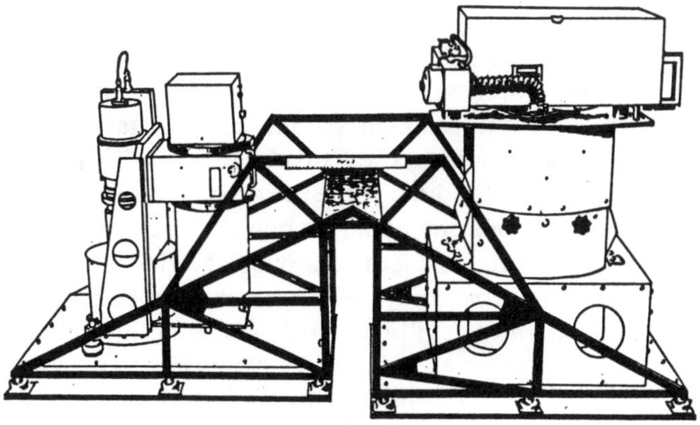

Fig. 2. Laser/camera/INS assembly.

reflections at the aircraft provides an accurate measure of water depth.

At the same time the camera is taking overlapping stereo photographs of the survey area. After the mission has been completed, an analytical plotter forms stereo images of the bottom and near-shore topography. Combined, they produce the shallow water bathymetric plots, which are then merged with the plots of deeper waters, acquired by conventional echo sounding, for compiling nautical charts.

Conventional methods for aerial photo mapping rely on ground control and ground detail points to establish the orientation of the aerial photographs with respect to the earth. In coastal mapping applications the problem of photo orientation is more difficult in that over-water photography may contain little or no land for bridging or photo control. One approach to overcoming this problem is to obtain the position and attitude of the camera at the time each photograph is taken, with the absolute precision sufficient to satisfy requirements for charting of the spot depths on nautical charts, and the relative precision sufficient to be able to form the stereomodels in the analytical plotter from the pairs of adjacent photographs.

The data collected by the individual navigation sensors are not sufficiently accurate to meet these demands and are integrated using the Kalman filter to estimate the sensor errors. The error estimates are then combined with the original data to yield high accuracy position and attitude information.

This paper describes the track recovery system stressing the role played by the Kalman filter. System configuration and the sensors which may be used are described in Section II. The features of implementation which result in high accuracy or a computational savings are described in Section III. Section IV gives a summary of results obtained in field trials of the track recovery system.

II. System Description

The track recovery system consists of an airborne component for acquisition of navigation sensor data and a ground component which processes the data to obtain estimates of flight profile parameters.

The airborne component consists of a research aircraft equipped with a gimbaled inertial navigation system (INS) hardmounted to the top of an aerial survey camera and a number of auxiliary sensors which may include a laser bathymeter (lidar), a microwave ranging system, barometric and radar altimeters, VLF/OMEGA navigation system, and a Doppler radar. The lidar is rigidly connected to the survey camera (Fig. 2) to ensure that the position and the attitude of the entire unit is measured directly by the INS.

The INS is the principal sensor of the hybrid navigation system since it provides dynamically accurate position, velocity, and attitude data. However, over the long term, errors in sensed accelerations and angular rates induce divergent and oscillatory errors in the INS computed positions and velocities [1]. These errors are estimated and calibrated out during postmission processing using the redundant data provided by the auxiliary sensors.

The survey camera is employed as the primary auxiliary sensor for applications requiring accurate position and attitude data. In these applications, aerial photographs containing three or more surveyed ground control points are processed using photogrammetric resection techniques to obtain precise fixes of camera position and attitude [2]. These fixes are utilized to update the inertially derived orientation parameters.

Alternatively, photographs containing landmarks or ground features having known geographic coordinates can be used to obtain camera position fixes without resorting to resection methods. In this mode, the photo coordinates and the corresponding geographic coordinates of the landmarks appearing in each photograph are first measured from selected photographs and topographic maps, respectively, and then differenced with the inertial data [3]. Although this technique does not provide as accurate a position update as photo resections, it is more convenient and less costly to use in the majority of remote sensing applications since it does not require controlled ground points nor access to precise photogrammetric instruments. Moreover, if the coordinates of three or more landmarks appearing in a single photograph are processed using this technique, the resulting fix generated by the update routine will provide a complete solution for all six camera orientation parameters (three position and three attitude) corresponding, in effect, to a low-grade resection.

Auxiliary position data may be provided by a microwave ranging system or a VLF/OMEGA navigation system. The microwave ranging system provides line of sight distance information from a master station mounted on the aircraft to one or more remote stations positioned in known locations. This is accomplished by measuring the roundtrip time of signals transmitted by the master station and returned by the remotes. The VLF/OMEGA radio navigation system operates through the use of very low frequency radio transmissions originating from eight omega and nine communication stations located around the globe.

The laser bathymeter, in addition to its primary function of obtaining water depth measurements along the aircraft flight path, provides an accurate indication of height above water, which, in conjunction with water level and attitude information, is used as an altitude reference. Additional altitude measurements may be provided by the barometric and radar altimeters.

III. Track Recovery Software

The data provided by the above sensors are recorded in flight and they constitute the input to the track recovery software package. The output of the package is a file containing the position, attitude, and velocity time-histories of the survey camera carried by the aircraft at user specified intervals, and a file of estimates of the rms magnitudes of the errors in these parameters.

The basic structure of the TRS package is illustrated in Fig. 3. The INS and barometric altitude data sets are preprocessed by the converter program to obtain the initial position and attitude data required for resection processing as well as to screen the data and reject bad data samples. The prefilter/filter program uses a high order Kalman filter to compute optimum forward-time estimates of the inertial system errors. These estimates are passed to the smoother program which processes the filter data in backward-time to generate improved estimates of the system error state. Finally, the smoothed error estimates are combined with the raw INS navigation parameters in the profile generator routine to compute the optimized position, velocity, and attitude time-histories output by the package.

The prefilter component of the prefilter/filter program performs the following two functions.

1) The prefilter implements the INS vertical channel, which computes altitude and vertical velocity from inertially indicated vertical acceleration. The vertical channel is stabilized by closed-loop feedback from the Kalman filter. That is, the altitude, vertical velocity, and acceleration error estimates computed by the filter are used to reset vertical channel parameters after each filter update.

2) The prefilter constructs measurements of the errors in the INS data by transforming and differencing complementary inertial and auxiliary sensor data. Measurements are constructed at a nominal rate of 1 Hz and are averaged over the user-defined filter update interval to form a smoothed measurement data set for processing by the Kalman filter. Typically, the filter update interval was set to 3–10 s, and reset to update at all camera firing times.

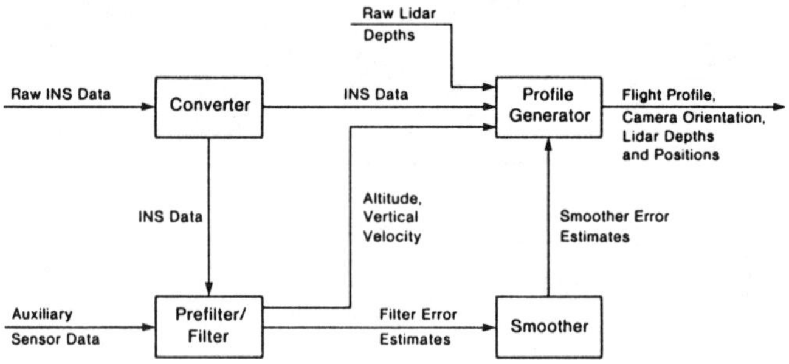

Fig. 3. Track recovery software structure.

Measurement averaging is employed to reduce the computation and storage costs of the filtering and smoothing processes and has been implemented so as to minimize information loss over each filter update interval. The prefilter algorithm is similar to the data compression algorithm given in [10]. Table I contains brief descriptions of error measurements (for more details, see [3]).

A comprehensive error model of the multisensor system is mechanized by the Kalman filter for processing of prefilter measurement data. From this model, any subset of error states or any combination of sensor measurements can be selected to suit specific applications. The model is of the form

$$\dot{x}(t)=F(t)x(t)+u(t) \tag{1}$$

$$z(t_k)=A(t_k)x(t_k)+v(t_k) \tag{2}$$

where x is the n dimensional error state vector, z is the m dimensional vector of measurements, u is the white process noise vector with spectral density Q, v is a white measurement noise vector with covariance R_k, and F and A are the time variable error dynamics and the measurement matrices, respectively.

The error state x is partitioned into four subvectors according to the dynamic characteristics of the component errors

$$x^T=\left(x_I^T, x_E^T, x_W^T, x_C^T\right) \tag{3}$$

where, as shown in Table II, x_i are the state modeling errors of the inertial navigation system, x_E are exponentially correlated error states, x_W are random walk error states, and x_C are random constant error states resulting in a partitioned system model of a form

$$\begin{bmatrix} x_I \\ x_E \\ x_W \\ x_C \end{bmatrix} = \begin{bmatrix} F_I & F_{IE} & F_{IW} & F_{IC} \\ 0 & F_E & 0 & 0 \\ 0 & 0 & 0 & 0 \\ 0 & 0 & 0 & 0 \end{bmatrix} \begin{bmatrix} x_I \\ x_E \\ x_W \\ x_C \end{bmatrix} + \begin{bmatrix} u_I \\ u_E \\ u_W \\ 0 \end{bmatrix}. \tag{4}$$

The equivalent discrete time state propagation model is

$$\begin{bmatrix} x_I \\ x_E \\ x_W \\ x_C \end{bmatrix}_{k+1} = \begin{bmatrix} \Phi_I & \Phi_{IE} & \Phi_{IW} & \Phi_{IC} \\ 0 & \Phi_E & 0 & 0 \\ 0 & 0 & I & 0 \\ 0 & 0 & 0 & I \end{bmatrix} \begin{bmatrix} x_I \\ x_E \\ x_W \\ x_C \end{bmatrix}_k + \begin{bmatrix} \omega_I \\ \omega_E \\ \omega_W \\ 0 \end{bmatrix}_k \tag{5}$$

TABLE I
ERROR MEASUREMENTS COMPUTED IN PREFILTER

INS/Trisponder Range Error Measurements, z_ρ

Slant ranges to two remote stations are processed:

$$z_{\rho j} = \rho_{cj}(\hat{L},\hat{\lambda},\hat{h}) - \rho_{mj}\ ; \qquad j=1,2$$

where ρ_{cj} and ρ_{mj} are computed and measured slant range to the jth remote respectively, and $\hat{L}$, $\hat{\lambda}$, $\hat{h}$ are inertially-derived, latitude, longitude and altitude corrected using filter position error estimates.

INS/Baro, Lidar and Radar Altitutde Error Measurements, $z_{alt}=(z_b,z_\ell,z_r)^T$

$$z_b = h_c - h_b$$

$$z_\ell = h_c - h_\ell - \Delta h$$

$$z_r = h_c - h_r - \Delta h$$

where h_c, h_b are computed and baro altitude respectively, h_ℓ is lidar-indicated height above the water surface computed by transforming the lidar slant range into the local vertical through measured roll and pitch, h_r is radar altitude, and Δh is the height of the water surface above the reference ellipsoid (determined at time of survey).

INS/Resection Position Error Measurements, z_p

$$z_p = R_c^e - R_r^e$$

where R_c^e and R_r^e are the INS-computed and resection-measured geocentric position vectors of the camera perspective center respectively, and superscript e indicates that these vectors are resolved in earth-fixed (geodetic cartesian) coordinates.

INS/Resection Attitude Error Measurements, $z_a=(z_{ax},z_{ay},z_{az})^T$

Inertial position and attitude data are combined with resection attitude measurements to obtain observations of the (small) angles defining the misalignments of the INS and camera reference axes:

$$\Delta\Phi = C_e^c[\hat{C}_e^c]^T \simeq \begin{bmatrix} 1 & \overline{\Delta\phi}_z & -\overline{\Delta\phi}_y \\ -\overline{\Delta\phi}_z & 1 & \overline{\Delta\phi}_x \\ \overline{\Delta\phi}_y & -\overline{\Delta\phi}_x & 1 \end{bmatrix}$$

$$z_{ax} = (\Delta\Phi[2,3]-\Delta\Phi[3,2])/2$$

$$z_{ay} = (\Delta\Phi[3,1]-\Delta\Phi[1,3])/2$$

$$z_{az} = (\Delta\Phi[1,2]-\Delta\Phi[2,1])/2$$

where C_e^c is the direction cosine matrix obtained from resection attitude parameters which relates the earth-fixed and camera coordinate systems, and $\hat{C}_e^c$ is an estimate of C_e^c computed from inertially-indicated latitude, longitude, roll, pitch and azimuth. $\overline{\Delta\phi}_x$, $\overline{\Delta\phi}_y$, and $\overline{\Delta\phi}_z$ are the observed INS to camera misalignment angles taken about the x, y, z camera reference axes.

INS/VLF-OMEGA Measurements, $z_\Omega=(z_{\Omega L},z_{\Omega\lambda})^T$

Inertial and VLF-OMEGA indicated latitude and longitude are differenced and the residuals are transformed into equivalent arc lengths:

$$z_{\Omega L} = (R_N+\hat{h})(\hat{L}-L_\Omega)$$

$$z_{\Omega\lambda} = (R_E+\hat{h})(\hat{\lambda}-\lambda_\Omega)\cos\hat{L}$$

where $\hat{L}$, L_Ω are inertial and VLF-OMEGA indicated latitude, $\hat{\lambda}$, λ_Ω are inertial and VLF-OMEGA indicated longitude, $\hat{h}$ is inertially computed altitude and R_E is the prime radius of curvature of reference ellipsoid R_N is the meridional radius of curvature of reference ellipsoid.

INS/Photo Fix Measurements, $z_{ph}=(z_N,z_E)^T$

Photo coordinates and the corresponding geographic coordinates of landmarks appearing in each photograph are used to obtain observations of north and east components of the position error:

$$z_N = \Delta R_N - \Delta P_N$$

$$z_E = \Delta R_E - \Delta P_E$$

where ΔR_N and ΔR_E are the north and east components of the range vector from the inertially computed camera position to the position of the targeted landmark as measured from a topographic map and ΔP_N and ΔP_E are the corresponding distances computed from measurements taken directly from photographs.

TABLE II
TRS STATE ERROR VECTOR

Type of Error State	Description*
x_I	x,y,z position errors (3)
	x,y,z velocity errors (3)
	x,y,z computer to platform frame misalignment angles (3)
x_E	north, east, down gravity errors due to vertical deflections and gravity anomalies (3)
	baro-altimeter exponentially correlated error (1)
	Trisponder exponentially correlated range errors (line-of-sight to 2 remote stations) (2)
x_W	x,y,z acclerometer biases (3)
	x,y,z gyro drift rates (3)
	north, east VLF/OMEGA position errors (2)
x_C	x,y,z acclerometer scale factor errors (3)
	baro altimeter bias (1)
	baro altimeter scale factor error (1)
	lidar altitude bias (1)
	lidar scale factor error (1)
	radar altimeter bias (1)
	radar altimeter scale factor error (1)
	x,y,z INS to camera misalignment angles in camera coordinates (3)
	Trisponder range biases (line-of-sight to 2 remote stations) (2)

*All error states are resolved in wander azimuth coordinates unless otherwise noted. The number of states is indicated in brackets.

where, if we assume that elements of $F(t)$ remain constant over Δt,

$$\Phi_I(t,\Delta t) \doteq e^{F_I(t)\Delta t} \tag{6}$$

$$\Phi_E(t,\Delta t) \doteq e^{F_E(t)\Delta t} \tag{7}$$

$$\Phi_{IE}(t,\Delta t) \doteq \int_0^{\Delta t}\Phi_I(\Delta t - s)F_{IE}(t)\Phi_E(s)\,ds \tag{8}$$

$$\Phi_{IW}(t,\Delta t) \doteq \int_0^{\Delta t}\Phi_I(\Delta t - s)F_{IW}(t)\,ds \tag{9}$$

$$\Phi_{IC}(t,\Delta t) \doteq \int_0^{\Delta t}\Phi_I(\Delta t - s)F_{IC}(t)\,ds \tag{10}$$

and ω_k is a white noise sequence with covariance Q_k.

Computation of the transition matrix Φ is one of the most time consuming operations of the TRS package and care is taken to evaluate (6)–(10) efficiently. Only the nonzero elements of $F(t)$ are computed and stored and (6)–(10) are evaluated either directly if the closed form of $e^{F(t)\Delta t}$ is known, or by Taylor type series expansions using sparse matrix multiplication. If not all of the error states are selected, matrix Φ of reduced order is computed and all of the subsequent computations are carried out on the system of the reduced order.

The particular estimation algorithm implemented in the prefilter/filter program is Bierman's $U-D$ factorized form of the Kalman filter [7]. In this algorithm, the covariance matrix P is never explicitly computed. Instead, P is propagated in terms of its matrix factors U and D; $P = UDU^T$ where U is a unit upper triangular matrix and D is a diagonal matrix. The $U-D$ algorithm is efficient and provides significant advantages in numerical stability and precision over more conventional filter implementations. The block structure of the transition matrix Φ is utilized in time propagation of the state x and the $U-D$ factors of the covariance matrix [9].

Bierman's algorithm uses scalar rather than batch measurement processing; that is, at each filter update time the measurements aggregated by the prefilter are processed one at a time. This is computationally efficient and allows erroneous measurements to be easily identified and rejected.

The modified Bryson–Frazier (mBF) adjoint-variable smoothing algorithm is employed for processing of filter data. The mBF smoother was selected for its ease of implementation as well as attractive numerical properties [7]. The estimate and covariance recursions in the smoother are independent, so that the covariance computations may be deactivated to obtain a significant saving in computation time. Since the smoother component of the program is relatively inexpensive computationally, no attempt was made to optimize its performance, and in particular, the system structure and sparse matrix techniques were not used for smoother computations. All the data processing was carried out on a DEC 10 computer using double precision arithmetic to avoid possible numerical problems. No numerical problems were observed in the processing of real data using the $U-D$ factorized Kalman filter algorithm and modified Bryson–Frazier smoothing algorithm. Typical computation time using a 33 state filter was 0.9 central processing unit (CPU) min per 1 min of real data with the smoother covariance calculations compared to 0.6 CPU min per 1 min of real data without.

IV. Flight Trial Results

The TRS software package has been tested in three different flight trials. Two of the flight trials were in connection with the aerial hydrography program and used high accuracy photogrammetric resections as the primary position and attitude reference. In the third trial, flown over land, low accuracy photo fixes were taken, and a microwave system was used as the position reference.

The missions to demonstrate the aerial hydrography system were flown in the summer of 1979 over a 20 km reach of the St. Lawrence River, and in the summer of 1980 over a 60 km stretch of the Bruce Peninsula in Lake Huron. In the St. Lawrence trial the aircraft flew a grid pattern comprising a total of nine 20 km lines at a spacing of approximately 600 m. Aircraft ground speed was approximately 75 m/s and altitude was 1500 m above the ground level. The total mission duration was 90 min, including a period of 10 min in which data acquisition tapes were changed, resulting in a data gap. A total of 51 fixes of camera position and attitude was obtained by photogrammetric resection following the mission. The root-mean-square (rms) accuracy of these fixes was estimated to be on the order of 0.5 m per axis in position and 30 arc s per axis in attitude. The navigation sensors used in addition to the INS were the barometric and laser altimeters, and a microwave ranging system with two remote transponders. The Lake Huron flight trial attempted to duplicate the results obtained in the St. Lawrence trial over a different body of water without the use of the microwave ranging system.

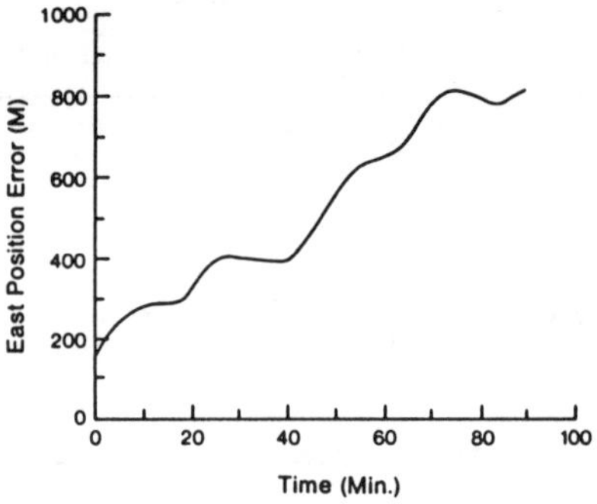

Fig. 4. East position error of inertial system.

Fig. 4 shows the uncorrected east position error of the inertial system over the full duration of the St. Lawrence mission. This error is characterized by an apparent ramp-like trend increasing from about 200 m at the start of the mission to approximately 800 m at the mission's end. A sinusoidal perturbation of about 50 m amplitude caused by acceleration-sensitive error sources, which are excited during aircraft turns, is superimposed over this trend.

Several runs of the TRS package were performed on the data acquired during the St. Lawrence mission to evaluate the sensitivity of TRS performances to sensor configuration and spacing between resection fixes. Only selected fixes were used in filter updates, with the remaining fixes used to provide a check on processing accuracy. As an additional check on system accuracy, the positions of precisely surveyed targets were compared with the positions of these targets computed from the aerial photographs.

The operation of the Kalman filter and smoother is illustrated for a run in which only two resection fixes were processed on each flight line. The fixes lying closest to the start and to the the end of each line were selected. Figs. 6 and 7 show the estimate of the rms time-histories of the east position error after filtering and smoothing, respectively. The actual error estimates computed by the filter and the smoother are not shown as they differ very little from Fig. 4 at this scale. Note that smoothing dramatically reduces the magnitude of the indicated rms error, with peak errors of about 4 m occurring midway between fixes,

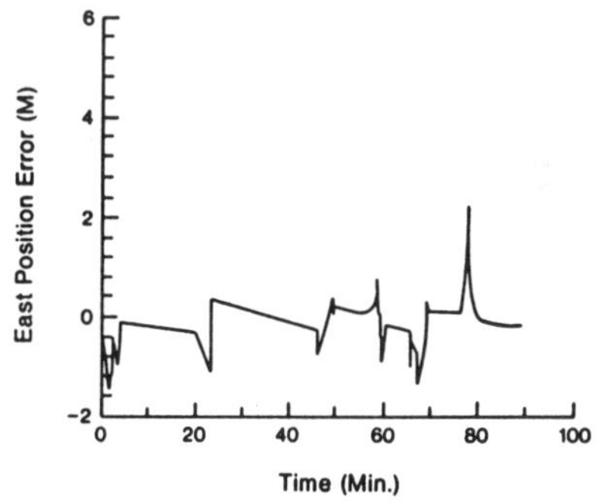

Fig. 5. East position error after filtering and smoothing.

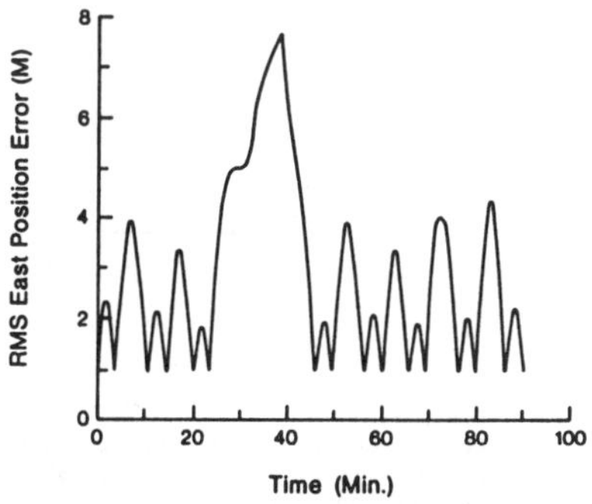

Fig. 7. rms error after smoothing.

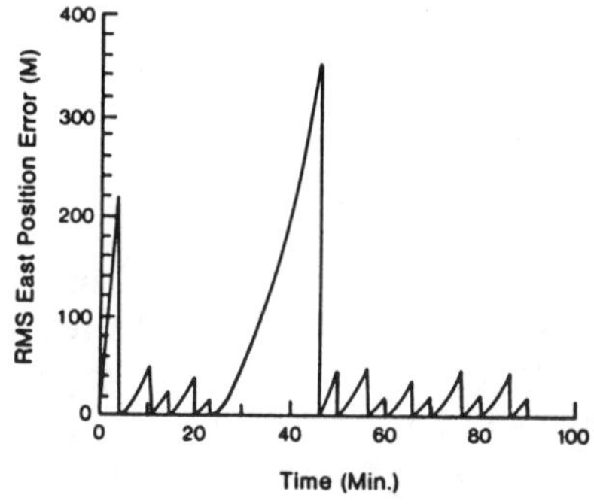

Fig. 6. rms error after filtering.

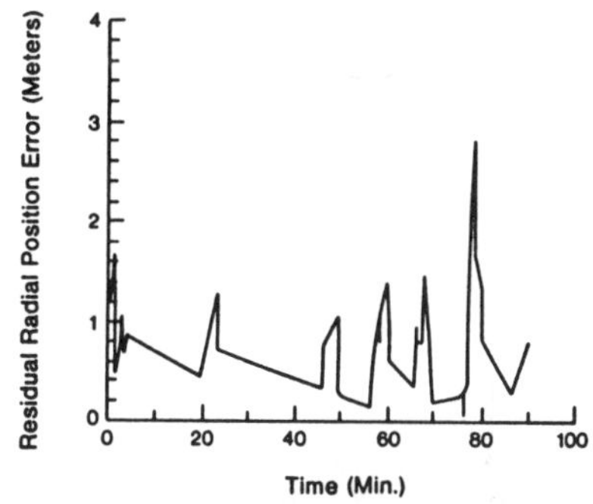

Fig. 8. Residual radial position error.

as opposed to the 40 m peak errors of the filter which occur just prior to fix times. The differences in the character of filter and smoother plots result from the fact that the filter uses only past and present measurement data to compute its current estimate, while the smoother employs all measurement data obtained in the mission to compute each estimate.

Fig. 5 shows the residual error in the east position after the errors estimated by the filter and smoother have been removed from the recorded data. A comparison of this plot with Fig. 4 shows that the error in the east position has been reduced by more than two orders of magnitude, leaving an unbiased residual error which falls on the whole between ± 1 m with peak value of 2.2 m. Figs. 8 and 9 show the residual radial position and attitude error time-histories. Table III shows the sensitivity of the residual attitude and position errors for the St. Lawrence mission as a function of the average spacing between resection updates.

The results of the Lake Huron trial and the runs of the St. Lawrence trial in which the microwave ranging system measurements were not processed demonstrated that the position accuracy is highly dependent on both the spacing between resection fixes and the flight profile. If high accuracy is desired and a microwave ranging system is not used, at least two resections per line should be processed [4].

The third flight trial flown in the vicinity of Ottawa, Ont., in December 1978 over land was used to test the accuracy achievable with lower accuracy photo fix updates described previously. The uncorrected north and east position errors of the inertial navigation system ranged from 250 to 3000 m over the 105 min duration of the mission.

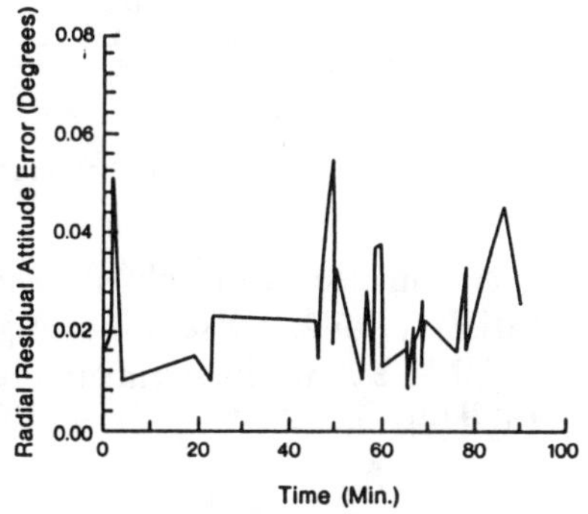

Fig. 9. Residual radial attitude error.

TABLE III
TRS POSITION AND ATTITUDE ERRORS

Average spacing between resection updates (km)	14	25	100	300
RMS Radial Position Error (meters)	.67	.96	4.1	6.4
RMS Radial Altitude Error (degrees)	.025	.026	.028	.029

Fig. 10 shows the residual radial position error using the photo fixes and the barometric altimeter only as the auxiliary navigation sensors. The residual error has been reduced to approximately 40 m (rms). In this trial the accuracy of the various sensor configurations was estimated relative to position computed using all sensors, including the microwave ranging system.

TABLE IV
SAMPLE PERFORMANCE FIGURES

Sensor Configuration	RMS Radial Residual Errors position (m)	attitude (deg)
INS, microwave ranging system, laser and baro-altimeters, 2 photogrametric resection per flight line at an average spacing of 15 km (St. Lawrence flight trial)	0.96	0.026
INS, laser and baro-altimeters, at most 3 photogrammetric resections per flight line at an average spacing of 33 km (Lake Huron flight trial)	1.58	0.050
INS, baro-altimeter, 2 photo fixes per flight line at an average spacing of 25 km (Ottawa flight trial)	42.0* (radial level)	*

*In the Ottawa flight trial no photogrammetric resections were taken, therefore altitude and attitude reference data were not available for computation of error statistics.

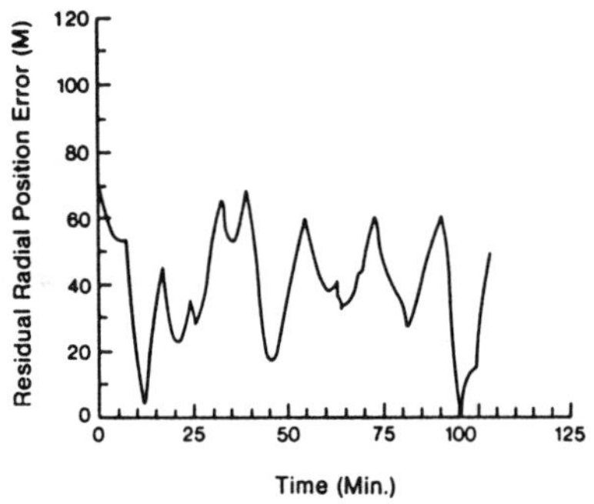

Fig. 10. Residual radial position error (Ottawa flight trial).

Sample rms performance figures obtained from each of the three flight trials for different sensor configurations are given in Table IV. More complete summaries of results can be found in [3] and [4].

V. CONCLUSIONS

A track recovery system based on the Kalman filter and the optimal smoother was described in this paper. The system was used in the Canadian aerial hydrography program and its operation was assessed using real data collected in three flight trials. The residual errors observed in these trials depended on the accuracy and type of measurements processed by the filter and smoother and, for some sensor combinations, were as small as 1 m in position and 0.03 degrees in attitude (total rms).

The performance of the track recovery system met the goals set for both absolute and relative position and attitude accuracy in all flight trials. A statistical analysis of the photo depths indicated that a depth accuracy of 0.65 m (rms) relative to ground truth was obtained in the St. Lawrence mission.

Analysis of the Bruce Peninsula photo-depth readings showed that photo bathymetry can be unreliable in areas having poor bottom contrast (e.g., smooth bottom, turbid water, poor sun angle). As a result, the Canadian program is now concentrating on the development of a scanning laser bathymeter, which will enable rapid and direct sounding of coastal waters. This will result in a relaxation of the relative accuracy requirements and it is expected that the medium-accuracy Litton LTN-51 inertial navigator will be replaced with a lower performance inertial attitude and heading reference system (AHRS). The AHRS would be integrated in real-time with a microwave ranging system and lidar and baro altimeters using a Kalman filter similar to that described here. Smoothing of the position data would be performed postmission in a ground-based minicomputer.

ACKNOWLEDGMENT

The track recovery software package was developed by Philip A. Lapp Limited, Toronto, Canada, under contract to the Canadian Federal Government for use by the Canadian Hydrographic Service (CHS) and the Canada Centre for Remote Sensing (CCRS). The aircraft, sensors, and computer used for acquisition and reduction of flight trial data were provided by CCRS. The cooperation and support of many individuals at the CHS and CCRS, and in particular S. B. MacPhee (CHS), A. J. Dow (CHS), J. R. Gibson (CCRS), and the late J. E. Smyth (CCRS), are gratefully acknowledged. D. Walsh and B. N. McWilliam, both of Philip A. Lapp Limited, made important contributions to software development and to the acquisition and analysis of sensor data.

REFERENCES

[1] K. R. Britting, *Inertial Navigation System Analysis.* New York: Wiley, 1971.
[2] M. M. Thompson, Ed., *Manual of Photogrammetry*, 3rd ed. Falls Church, VA: Amer. Soc. Photogrammetry, 1966.
[3] D. B. Reid, W. S. Gesing, and B. N. McWilliam, "Inertially-controlled aircraft track recovery and updating system," P. A. Lapp Ltd., Tech. Rep. 3/F/80 (prepared for CCRS, Ottawa, Canada), May 1980.
[4] D. B. Reid *et al.*, "Aerial hydrography pilot project," P. A. Lapp Ltd., Tech. Rep. 5/F/80 (prepared for the Canadian Hydrographic Service, Ottawa, Canada), July 1980.

[5] D. B. Reid, W. S. Gesing, B. N. McWilliam, and J. E. Smyth, "Integration of multi-sensor navigation data using optimal estimation techniques," in *Proc. 19th IEEE Conf. Dec. Contr.*, Albuquerque, NM, Dec. 1980, pp. 584–585.
[6] D. B. Reid *et al.*, "An inertially-aided aircraft track recovery system for coastal mapping," presented at the Position, Location, Navigat. Symp., Atlantic City, NJ, Dec. 1980.
[7] G. J. Bierman, *Factorization Methods for Discrete Sequential Estimation*. New York: Academic, 1977.
[8] G. J. Bierman and M. W. Nead, "A parameter estimation subroutine package," JPL Pub. 77-26, Rev. 2, Jet Propulsion Lab., Calif. Instit. Technol., Pasadena, CA.
[9] G. J. Bierman, "Efficient time propagation of $U-D$ covariance factors," *IEEE Trans. Automat. Contr.*, vol. AC-26, pp. 890–894, Aug. 1981.
[10] A. N. Joglehar, and J. D. Powell, "Data compression in recursive estimation with applications to navigation systems," presented at the AIAA Guidance Contr. Conf., Key Biscayne, FL, Aug. 1973.

Bathymetric and Oceanographic Applications of Kalman Filtering Techniques

ROBERT F. BRAMMER, MEMBER, IEEE, RALPH P. PASS, AND JAMES V. WHITE, MEMBER, IEEE

***Abstract*—Ocean currents and seamounts (underwater mountains) can be mapped by analyzing data from satellite radar altimeters. This paper describes the application of Kalman filtering techniques to the analysis of such data acquired during the SEASAT mission. The altimeter data are modeled as samples from autoregressive random processes. Based on these models, matched filters are used to detect the characteristic nonstationarities in the altimeter data caused by seamounts and ocean currents such as the gulf stream. The geostrophic velocities of detected ocean currents are then estimated using Kalman smoothers. A useful formula is derived, which expresses the error power spectrum of the optimal fix lag smoother as a function of the lag and the error spectra of the optimal filter and the optimal infinite-lag smoother.**

Manuscript received April 23, 1982; revised August 26, 1982.
The authors are with The Analytic Sciences Corporation, Reading, MA 01867.

I. Introduction

DURING the past decade satellite microwave radar altimeters have been applied to geodesy, oceanography, and meteorology. After a limited demonstration during Skylab, extensive altimeter data were collected by GEOS-3 and SEASAT [1]–[3]. Further altimeter programs are being planned by the Navy, NASA, and foreign national programs. This paper deals with the application of Kalman filtering techniques to the analysis of radar altimeter data for mapping ocean currents and for detecting seamounts (i.e., underwater mountains, primarily of volcanic origin).

Reprinted from *IEEE Trans. Automat. Contr.*, vol. AC-28, pp. 363–371, Mar. 1983.

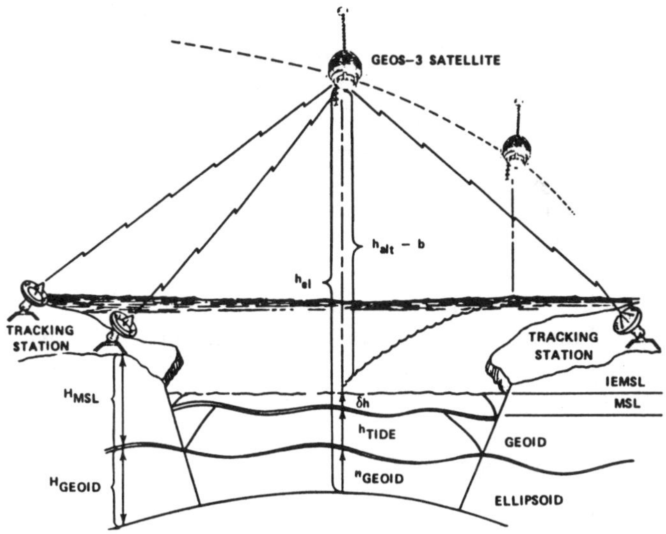

Fig. 1. Satellite radar altimeter measurements.

The utility of satellite altimeters in these applications is a result of the precision with which the altimeter can measure the height of the satellite above the ocean surface. The measurement concept is illustrated in Fig. 1. The satellite is tracked by ground-based tracking stations that determine its orbit and height relative to a reference ellipsoid. After applying various corrections for atmospheric refraction and instrumental bias, the height of the sea surface with respect to the reference ellipsoid can be estimated accurately. Under various assumptions about the ocean geoid, tides, and ocean circulation patterns, these height data can be used to infer gravity anomalies, to estimate ocean tides, and to detect major ocean currents and seamounts.

A. Ocean Current Mapping

The application of satellite altimeter data to ocean current mapping [15] relies on the fact that constant-velocity (geostrophic) currents produce sea-surface height anomalies. The relation between the average velocity v and the height anomaly Δh is given by

$$v = \frac{g}{2\omega \sin\phi} \frac{\Delta h}{\Delta L} \tag{1}$$

where g is the acceleration of gravity, ω is the earth's rotational velocity, ϕ is the latitude, and ΔL is the width of the current. In (1) Δh is measured along the satellite ground track, and hence v is the component of current velocity that is normal to the satellite ground track. Geostrophic currents arise from an approximate balance between the horizontal pressure gradient and the Coriolis force. For major currents like the gulf stream, the height anomaly can be as much as 1 m. For other current systems, such as the ring currents that spin off the gulf stream, the height anomaly can be 30–90 cm. Since the SEASAT radar altimeter had an rms instrumental precision of better than 10 cm, these anomalies are readily measurable.

B. Detection of Seamounts

The location of seamounts is of interest for locating hazards to navigation and for geological and geophysical inferences related to the earth's crust. A seamount is indicated by a change in local gravitational potential due to the density contrast with the surrounding ocean water. Since the seamount is composed of denser material, it will raise the local gravitational potential, which increases the sea-surface height over the seamount. For large seamounts these height anomalies can be as large as 2 m or more. Section III-D describes the detection of seamounts in the gulf of Alaska using SEASAT altimeter data. These detections have been confirmed by conventional bathymetric data [14].

C. Organization

This paper is organized in two technical sections. Section II defines the technical approach to modeling altimeter data as autoregressive stochastic processes. Based on the autoregressive (AR) models, the detection problem is described and matched filters are developed to detect nonstationarities in the mean of the altimeter data caused by seamounts and ocean currents. Following this discussion, the application of Kalman smoothing techniques to the analysis of altimeter data for mapping oceanographic circulation is described. Section III presents results of the analysis of SEASAT altimeter data for detecting both ocean currents and seamounts.

II. Technical Approach

A. Summary of Technical Approach

The approach to the analysis of altimeter data involves three steps. First, a practical technique is presented for developing AR models from individual tracks of data. These models have the form given as

$$z(k) = C_1 z(k-1) + \cdots + C_p z(k-p) + w_p(k). \tag{2}$$

Second, the models are used with matched filters to detect signatures of interest and to estimate the time-varying mean of the data. These filters are matched to parameterized deterministic model signatures for seamounts and ocean currents. Third, a Kalman smoother based on the geostrophic equation is applied to the altimeter data to estimate geostrophic velocities. The interpretation of the results is based on new formulas for the error spectra of the optimal smoother. These error spectra provide insights into the accuracy of the estimation procedure at various spatial wavelengths.

B. Autoregressive Stochastic Modeling

1) Statement of the Modeling Problem: An AR model of order p for a time series $\{z(t); t = 1, 2, \cdots, n\}$ is a difference equation driven by zero-mean white noise $w_p(t)$

$$z(t)=C_1 z(t-1)+C_2 z(t-2)+\cdots+C_p z(t-p)+w_p(t) \qquad t=p+1,p+2,\cdots,n. \quad (3)$$

The AR modeling problem is as follows: given data $\{z(t)\}$ and the maximum order *pmax*, select the order p, the coefficients $C_1 \cdots C_p$, and the white-noise variance $\sigma^2(p)$.

For each order $(p=0,1,2,\cdots,pmax)$ in turn, the AR coefficients $C_1, C_2, \cdots, C_p (C_0=1)$ are selected to minimize the sample mean-square value of the noise

$$\sigma^2(p)=\frac{1}{n-pmax}\sum_{t=pmax+1}^{n} w_p^2(t) \qquad p=0,1,\cdots,pmax. \quad (4)$$

To select the order p that is best supported by the data, (4) is used to compute the Akaike information criterion (AIC in [4], [5]) for each p

$$\mathrm{AIC}(p)=n\ln\left[\sigma^2(p)\right]+2p. \quad (5)$$

The particular order p that minimizes AIC is identified; the corresponding AR parameters $C_1, C_2, \cdots, C_p, \sigma^2(p)$ then define the particular AR model (for the underlying stochastic process) that is best supported by the available data.

2) Kalman-Filter Algorithm: The least-squares problem of minimizing $\sigma^2(p)$ can be solved by processing the altimeter data with a family of Kalman filters [6], one for each order $p=1,\cdots,pmax$. The pth filter estimates the state vector of AR coefficients

$$x(t)=\{C_1 C_2 \cdots C_p\}^T \quad (6)$$

given the measurement equation

$$z(t)=H(t)x(t)+w_p(t) \quad (7)$$

$$H(t)=\{z(t-1)\ z(t-2)\ \cdots\ z(t-p)\}. \quad (8)$$

The variance of $w_p(t)$ is initially set to unity, and the state vector satisfies the noiseless dynamics equation

$$x(t+1)=x(t). \quad (9)$$

After the data in a time series are filtered (with a large initial error covariance), the estimated state provides the desired estimate of the AR coefficients, and these are used in (7) to compute $\{w_p(t)\}$, from which $\sigma^2(p)$ may be computed via (4).

Although this approach is rigorous, it requires an unnecessarily large amount of calculation because each covariance matrix is being updated *n-pmax* times, whereas these matrices are not required for the present application. Moreover, the special structure of the AR model is not being fully exploited. A more efficient algorithm is described in the following.

3) Covariance Algorithm: The covariance algorithm [7], as implemented in the COVAR program [8], solves the least-squares AR modeling problem efficiently by batch processing the data. All the variances $\sigma^2(p)$, $p=0,1,\cdots,$ *pmax* and the coefficients of the highest order model are computed via Cholesky decomposition applied to the least-squares normal equations. The calculation is recursive with respect to model order and is much faster than an equivalent bank of Kalman filters.

An augmented version of COVAR (which computes the AIC for each order model) is used to process the altimeter data in two passes. The first pass is used to select the model order that minimizes AIC. The last pass, with *pmax* set equal to the selected model order, is used to compute the AR coefficients.

C. Matched Filtering

1) Statement of the Detection Problem: The problem of detecting ocean-current or seamount signatures in altimeter data is formalized in the following.

Given $D(t)$ = time series of altimeter data,
$m(t)$ = model signature time series,
$N(t)$ = stationary Gaussian noise model for residual altimeter data that are free of signature $m(t)$,
T = specified time (location) in the data $D(t)$,
A_s = unknown signature amplitude scale factor,
$H(T)$ = hypothesis that $D(t)=N(t)+A_s m(t-T)$, with $A_s \neq 0$
H_o = null hypothesis that $D(t)=N(t)$.

Find An optimal decision rule for correctly choosing between hypotheses H_o and $H(T)$; and an optimal estimate of the amplitude factor A_s when $H(T)$ is chosen.

Optimality Maximize the probability of correct detection for a specified probability of false alarm.

Solution Compute the likelihood ratio

$$LR=\frac{\text{likelihood of } D(t) \text{ under } H(T)}{\text{likelihood of } D(t) \text{ under } H_o}$$

select $H(T)$ when $LR >$ threshold value.
select H_o when $LR \leq$ threshold value.

The optimal decision rule can be efficiently implemented for all values of T (except for values near the ends of the data set) by processing the altimeter data with a time-invariant matched filter [9] followed by a threshold detector as indicated in Fig. 2. The impulse response $h(t)$ of this filter and its Fourier transform $H(f)$ are related by the equations

$$H(f)=\sum_{t=-\infty}^{\infty} h(t)e^{-i2\pi ft} \quad (10)$$

f = normalized frequency (cycles/sample)

$$h(t)=\int_{-1/2}^{1/2} H(f)e^{i2\pi ft}\,df. \quad (11)$$

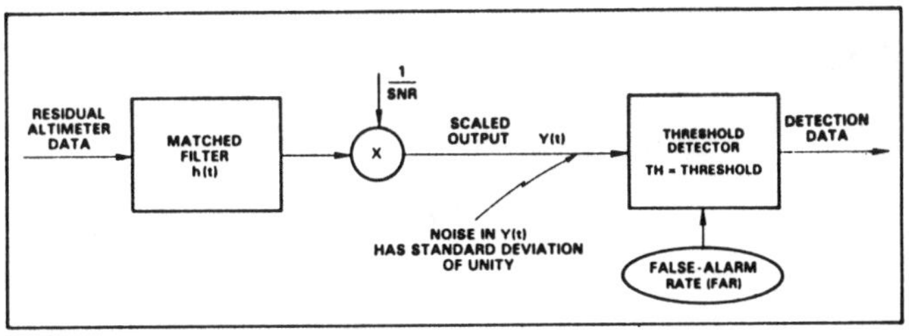

Fig. 2. Matched-filter detector.

The frequency response $H(f)$ of the optimal matched filter can be expressed [9] in terms of the Fourier transform $M(f)$ of the model signature $m(t)$ and the power spectrum $S_{NN}(f)$ of the data noise process $N(t)$

$$H(f)=\frac{M(-f)}{S_{NN}(f)} \tag{12}$$

where the power spectrum of $N(t)$ is the transform of the autocorrelation sequence $R_{NN}(t)$

$$R_{NN}(t)=E\{N(t)N(0)\} \tag{13}$$

$$S_{NN}(f)=\sum_{t=-\infty}^{\infty} R_{NN}(t)e^{-i2\pi ft}. \tag{14}$$

Because the noise $N(t)$ is modeled as a realization of an AR process, the optimal matched filter can be implemented as a finite-impulse-response (FIR) filter. This desirable property is verified by noting that the AR power spectrum of $N(t)$ can be computed from (3) [with $z(t)=N(t)$]

$$S_{NN}(f)=\frac{\sigma^2(p)}{G(f)G(-f)} \tag{15}$$

$$G(f)=1-\sum_{k=1}^{p} C_k \exp(-i2\pi fk). \tag{16}$$

From (12) and (15) it follows that the optimal frequency response is

$$H(f)=\sigma^{-2}(p)G(f)G(-f)M(-f). \tag{17}$$

The corresponding impulse response is

$$h(t)=\sigma^{-2}(p)g(t)*g(-t)*m(-t) \tag{18}$$

where the asterisks denote convolutions. Since $g(t)$ and $m(t)$ have finite supports, it follows that $h(t)$ has finite support.

2) Matched-Filter Detection Algorithm: The processing of altimeter data with a matched filter is depicted in Fig. 2. The output of the matched filter is scaled by the theoretical rms signal-to-noise ratio (SNR), which is defined as

$$\text{SNR}=\frac{\text{peak filter output due to signature } m(t-T)}{\text{rms filter output due to noise } N(t)}. \tag{19}$$

The SNR of an optimized filter satisfies the following equations:

$$\text{SNR}=\sqrt{\sum_{k=-\infty}^{\infty} h(k)m(-k)} \tag{20}$$

$$\text{rms noise in filter output} = \text{SNR} \tag{21}$$

$$\text{peak filter output due to signature } m(t-T)=\text{SNR}^2. \tag{22}$$

The scaled output $Y(t)$ of the matched filter in Fig. 2 is a test statistic that contains a random component having a standard deviation of unity. The threshold detector compares $Y(t)$ against a threshold value TH, which is chosen to produce a specified average false-alarm rate (FAR)

$$TH=\sqrt{2\ln(F_m/\text{FAR})}. \tag{23}$$

In (23) F_m is the mean frequency of the filter output (half the average rate of zero crossings of noise in the filter output). For Gaussian noise and an optimized filter, F_m satisfies the equation

$$F_m=\frac{1}{2\pi}\cos^{-1}\{X/\text{SNR}^2\} \quad \text{(cycles/sample)} \tag{24}$$

$$X=\sum_{k=-\infty}^{\infty} h(k)m(1-k). \tag{25}$$

A false alarm occurs if the sufficient statistic $Y(t)$ crosses the detection threshold TH when the data consist of noise alone. The probability of a false alarm is

$$P_f=Q(TH)=\int_{TH}^{\infty}\frac{1}{\sqrt{2\pi}}\exp(-x^2/2)\,dx. \tag{26}$$

The maximum-likelihood estimate of the location of a detected signature is the value of t for which $Y(t)$ achieves its local maximum value above the threshold TH. Let t_0 denote this estimate of signature location. For SNR $\gg 1$, the Cramer–Rao lower bound (CRLB) on the rms error of this estimate is

$$\text{CRLB}=\left[2\pi F_m Y(t_0)\right]^{-1}. \tag{27}$$

The maximum-likelihood estimate of the signature amplitude scale factor A_s is

$$\hat{A}_s=\frac{Y(t_0)}{\text{SNR}} \tag{28}$$

and the rms error in this estimate is 1/SNR. The scaled estimate of the detected signature is then $\hat{A}_s m(t-t_0)$. The estimated geostrophic velocity corresponding to this signature is

$$v(t)=\frac{g}{2\omega\sin\phi}\hat{A}_s\frac{dm}{dt}(t-t_0)/v_T$$

$$v_T = \text{altimeter ground-track velocity.} \tag{29}$$

D. Kalman Smoothing

1) Statement of the Smoothing Problem: Matched filtering determines the best model signatures for large-scale features such as the gulf stream (~100 km). The model signature is an estimate of the time-varying mean of the sea-surface height (SSH) measurements and does not reflect smaller features of the SSH. The residual differences between the SSH measurements and the model signature reflect both measurement noise and smaller-scale oceanographic signals not present in the model signature. The measurement noise is modeled as uncorrelated, while the oceanographic signal is assumed to be the correlated part of the residuals. The purpose of filtering or smoothing the residuals is to estimate the geostrophic velocity correction $\delta v(t)$ to the geostrophic velocity $v(t)$ computed from the model signature (27).

The geostrophic velocity is estimated from the slope of the sea surface determined from estimates of the correlated component of the residuals between sea-surface height and the model signature. The geostrophic velocity estimate is for that component of velocity which is normal to the satellite ground track.

Kalman filtering and Kalman smoothing algorithms are used to estimate the sea-surface slope. These algorithms require a state-space model for the residual differences between the SSH data and the estimated signature of the ocean current. The signature is estimated by using a bank of five matched filters, each matched to a different-width signature. The determination of this state-space model is discussed later.

For this application there is no reason to limit consideration to a causal filtering approach. By allowing delay in the estimates of the states, the estimation accuracy can be improved by smoothing. Both fixed-interval and fixed-lag smoothers are suitable for this application. The rest of this section will address quantitative comparisons of smoothers.

A standard stationary state-space model is considered

$$x_k = Ax_{k-1} + w_k; \qquad Q = \operatorname{cov}(w_k) \tag{30}$$

$$z_k = Hx_k + v_k; \qquad R = \operatorname{cov}(v_k). \tag{31}$$

The new results presented here are formulas for the covariance and error spectra of the fixed-lag smoother for estimating x_k from z_k. Later, these results will be applied to the estimation of current velocity from SSH data.

The power spectra of the smoothing errors are considered first. The corresponding covariance results will then follow by integration. The steady-state smoother is optimal for a doubly infinite measurement sequence $\{z_k\}_{k=-\infty}^{\infty}$. The spectrum of the optimal smoothing errors follows directly from the Wiener–Khinchine approach and is given by

$$S_s(f) = S_x(f) - S_x(f)H^T\left(HS_x(f)H^T + R\right)^{-1}HS_x(f) \tag{32}$$

where $S_x(f)$ is the spectrum of the state process x. For the state-space model given above, the process spectrum is given by

$$S_x(f) = \left(I - \exp(-2\pi if)A\right)^{-1} \cdot Q\left(I - \exp(2\pi if)A^T\right)^{-1}. \tag{33}$$

Equation (32) is the optimal error spectrum for a noncausal smoother operating on an infinite data set. Considering the other limiting case where no future measurements are used to estimate the state, the resulting error spectrum is the filtering error spectrum given by

$$S_f(f) = \left(I - \exp(-2\pi if)F\right)^{-1}\left((I - KH)Q(I - K^TH^T) + KRK^T\right)\left(I - \exp(2\pi if)F^T\right)^{-1} \tag{34}$$

where $F = (I - KH)A$ is the Kalman-filter error transition matrix and K is the Kalman gain [6]. The optimal fixed-lag smoothing error spectrum is bounded by the limiting cases given in (32) and (34).

The fixed-lag error spectrum can be written as

$$S_N(f) = S_s(f) + G^N\left(S_f(f) - S_s(f)\right)\left(G^T\right)^N \tag{35}$$

where the lag of the smoother is N samples and $G = P_f(P_fF^T)^{-1}$ (assuming A is invertible). Since G has eigenvalues between -1 and 1, (35) shows that the spectrum of the fixed-lag smoothing errors are between those of the filter and the optimal smoother. This simple formula permits evaluation of fixed-lag smoothing performance as a function of the lag N. The error covariance of the fixed-lag smoother is

$$P_N = P_s + G^N(P_f - P_s)(G^T)^N \tag{36}$$

where P_s is the optimal smoothing error covariance and P_f is the filtering error covariance.

The proof of (35) is straightforward starting from the equations of a fixed-lag smoother ([6], p. 175). These equations are transformed using the Z transform and $\exp(2\pi if)$ is substituted for Z. The key to the proof is the following relation:

$$S_x(f)H^T\left(HS_x(f)H^T + R\right)^{-1} = -e^{2\pi if}\left(e^{2\pi if}I - G\right)^{-1}QA^TP^{-1}\left(e^{2\pi if}I - F\right)^{-1}K \tag{37}$$

where $G = A + Q(A^T)^{-1}P_f^{-1} = P_f(P_fF^T)^{-1}$. The proof of (37) starts from the steady-state relation

$$P_f = \left(I - P_fH^TR^{-1}H\right)\left(AP_fA^T + Q\right). \tag{38}$$

Equation (35) follows from (37) and the relation for $S_x(f)$ given in (33).

The restriction to A being invertible is a convenience for the derivation of (35) and (36). By partitioning the state vector into three groups of states, the results can be applied with any matrix A. The first group contains those states for which the transition matrix is nonsingular. The second group contains those states which are linear combinations of the states in the first group. Finally, the third group contains those states which are linear combinations of the states in the first group after a delay. Note that a delayed state in a filter is equivalent to a fixed-lag smoother with the lag number equal to the number of time steps delayed. Thus the error spectra and covariances for any state vector may be determined by using (35) and (36).

An example of the use of (35) and (36) is given in Section III-C, where most of the improvement between the optimal filter and the infinite-lag smoother is captured after two lags.

III. Results

A. Western North-Atlantic Data

Four types of data are used to present the results that follow. First is the SEASAT radar altimeter data which measured the radial distance from the satellite to the sea surface at a rate of 10 samples/s with an instrumental rms precision of approximately 10 cm. (These data were low-pass filtered and resampled at 1 sample/s.) This yielded a 6.7 km separation between successive measurements of the ocean surface with an rms white-noise measurement error of 3–5 cm. This sample spacing is consistent with the resolution needed for the oceanographic applications discussed in this paper because the performance of the algorithms is limited by geoid uncertainties and white measurement noise rather than spatial resolution. For certain bathymetric applications, however, a smaller sample spacing would contain additional useful information about small seamount signatures in the altimeter data.

During the calibration period the orbit of SEASAT was such that data were collected along subtracks that nearly repeated with a three-day period. These nearly repeating subtracks (1–3 km spacing) provide a useful data set for examining the repeatability of the data and the variability of the time-dependent oceanography.

The second data set is the Marsh–Chang gravimetric geoid [10]. This is the most accurate (1 m rms) geoid available in the western North Atlantic. This geoid is the reference surface subtracted from the altimeter data to analyze dynamic oceanographic height anomalies. An error model for the geoid is an integral part of the matched-filter algorithms for ocean-current detection.

The third data set is composed of surface oceanographic measurements collected during the SEASAT operating period from international oceanographic projects [11]. These data locate the gulf stream and associated ring currents and provide quantitative measures of their height anomalies with which to confirm the altimeter estimates.

Finally, bathymetric data describing the topography of the sea floor were provided by the U.S. Navy. This data set is based on sonar bathymetric surveys and is used to confirm the seamount detections derived from the altimeter data.

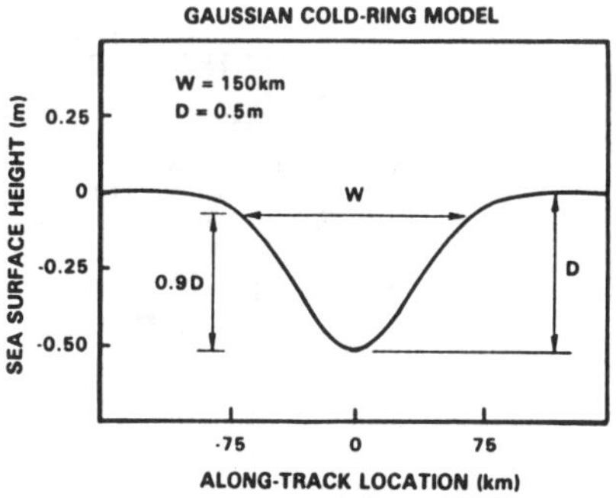

Fig. 3. Cold-ring signature.

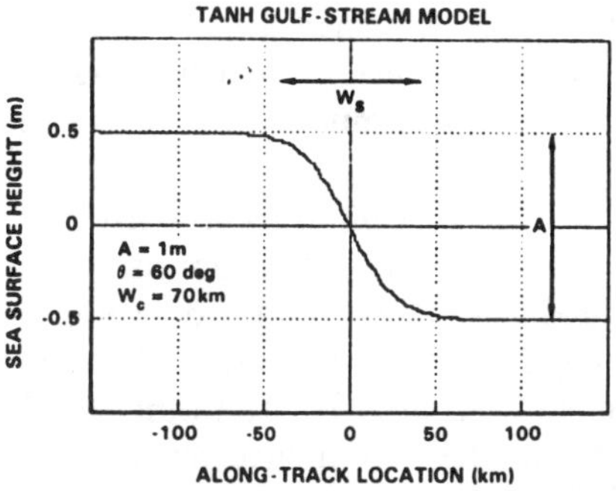

Fig. 4. Gulf-stream signature.

B. Detection of Current Rings and the Gulf Stream

1) Ring Signatures: Cold-ring and warm-ring currents induce depressions and rises, respectively, in the sea-surface topography. These anomalies appear as characteristic signatures in the altimeter data. A generic ring signature is modeled by the following equation for the sea-surface height along the satellite ground track

$$H(x) = -D\exp\left(-9.21(x/W)^2\right)$$
$$x = \text{distance along satellite ground track}$$
$$D = \text{signature depth}$$
$$W = \text{signature width (10 percent height change).} \tag{39}$$

D is positive for cold rings and negative for warm rings. This parametric model has a simple mathematical form and appears to be in reasonable agreement with available data on ring signatures (e.g., [11]–[13]). An example of a model ring signature is shown in Fig. 3, where the central depth is 0.5 m and the width is 150 km.

2) Gulf-Stream Signatures: Boundary currents like the gulf stream cause altimetric signatures that are modeled by the following equation for the sea-surface height along the satellite ground track (tanh = hyperbolic tangent):

$$H(x) = -(A/2)\tanh(3x\sin\theta/W_c)$$
$$A = \text{amplitude of height change}$$
$$\theta = \text{track angle with respect to current velocity}$$
$$W_c = \text{width of current (90 percent height change)}$$
$$W_s = W_c/\sin\theta = \text{signature width.} \tag{40}$$

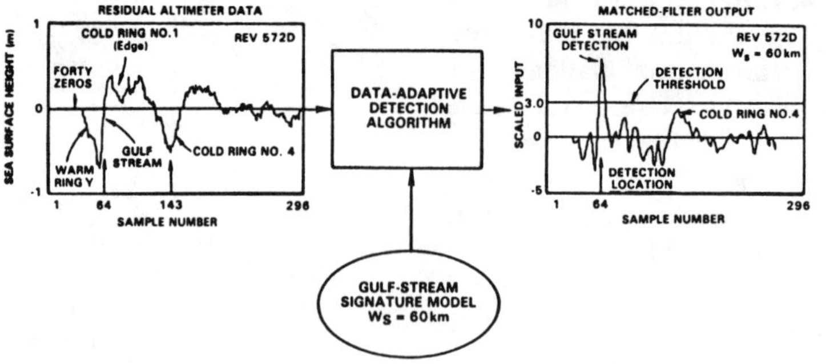

Fig. 5. Demonstration of gulf-stream detection using SEASAT data, Rev. 572D.

The track angle θ is 90° when the satellite ground track is normal to the mean current velocity. Fig. 4 shows the signature for parameter values typical of the gulf stream in the western North Atlantic.

3) Demonstration of Gulf-Stream Detection: Fig. 5 depicts the detection of the gulf stream using SEASAT data with a matched filter optimized for detecting the signature described by (40) (width = 60 km) in the presence of AR noise that models geoid errors and other random fluctuations. The AR model is computed from the last 230 samples of data by using the augmented COVAR algorithm. The matched-filter's scaled output $Y(t)$ crosses the detection threshold and achieves a maximum at sample number 64. The detection threshold is selected in this example to yield the low false-alarm rate of 0.01 false alarm per Mm of track length. As a result, the filter-detector discriminates well against the cold-ring depression at sample number 143. In our experience, the gulf stream is easy to detect so that model signature parameters are not critical; good results are obtained with a variety of filters matched to signature widths ranging from 40–150 km.

4) Demonstration of Cold-Ring Detection: The upper graph in Fig. 6 shows SEASAT altimeter data for Rev. 478A; the depression labeled "cold ring no. 4" is verified by other oceanographic data [11]. A filter was matched to the generic signature described by (39) (width = 300 km) with an AR noise model based on the first 140 data samples. The scaled output $Y(t)$ of this filter for Rev. 478A is shown in the lower part of Fig. 6. The cold ring is correctly detected and located with a detection threshold corresponding to a false-alarm rate (FAR) of 0.03 false alarm per Mm of track length.

C. Estimation of Geostrophic Gulf-Stream Velocities

As was mentioned before, the residual differences between the altimeter SSH data and the estimated mean signature of the gulf stream can be smoothed to refine the geostrophic velocity estimate inferred from a matched-filter analysis. This requires a state-space model for the residuals. Based upon prior statistical analyses of SEASAT altimetry data, the following one-dimensional model is typical for the SSH residuals (z_k = altimeter–geoid–signature)

$$x_k = 0.9x_{k-1} + w_k \tag{41}$$

$$z_k = x_k + v_k \tag{42}$$

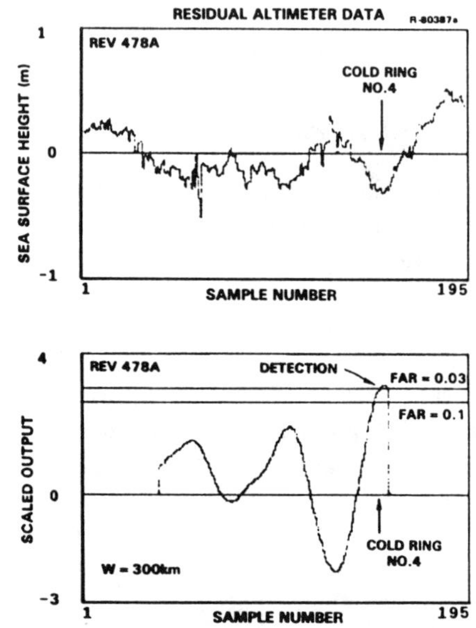

Fig. 6. Demonstration of cold-ring detection using SEASAT data, Rev. 478A.

where $Q = 0.0026\ \text{m}^2$ and $R = 0.000416\ \text{m}^2$. More complicated models for the residuals do not appear to be statistically justifiable with the available data, which are short time series containing 15–20 samples in the viscinity of the gulf stream. For this system, $P_f = 0.000364\ \text{m}^2$ and $P_s = 0.000331\ \text{m}^2$. Thus, smoothing provides a 0.4 dB improvement over filtering. The filtering covariance is a 16 dB reduction in the state covariance. Using (36) all possible smoothing improvement is obtained with a two-lag fixed-lag smoother.

The white measurement noise makes it impractical to estimate instantaneous values of sea-surface *slope* from the SSH data. There is no advantage here to using a 10 sample/s data rate, instead of 1 sample/s because the white measurement noise dominates at the higher frequencies. Therefore, a 5 s (34 km) triangularly weighted running mean of the slope is estimated for computing geostrophic velocity corrections. This quantity was selected to be estimated because it represents a tradeoff between two conflicting needs of the oceanographer. On the one hand, the oceanographer wants to resolve the smallest possible spatial details along the satellite ground track. On the other hand, the oceanographer also wants estimates that are reasonably accurate (not too biased or noisy). The five-

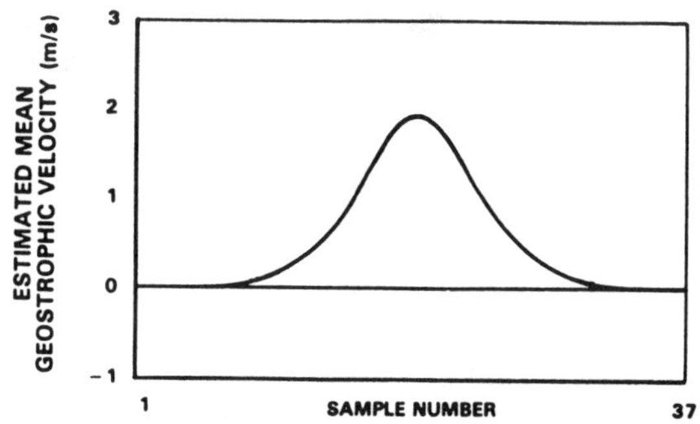

Fig. 7. Geostrophic velocity from mean signature.

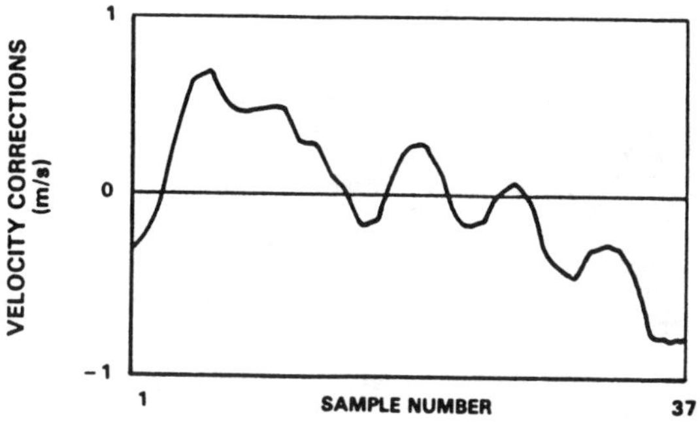

Fig. 8. Geostrophic velocity correction from optimal smoother.

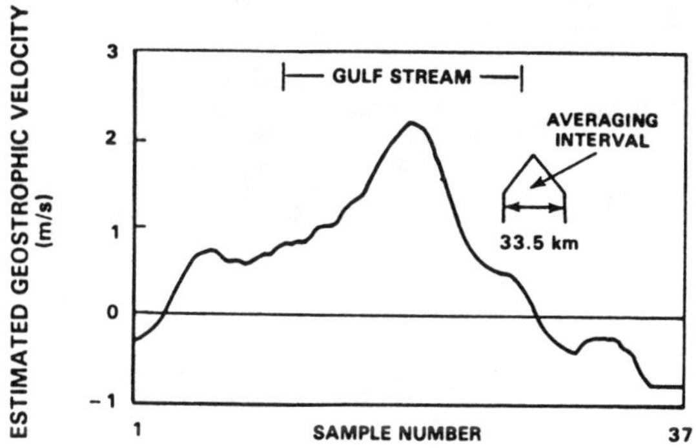

Fig. 9. Corrected geostrophic velocity using mean signature and optimal smoother.

sample mean velocity is judged to be a quantity that strikes a balance between these two goals when the available data are noisy measurements of sea-surface height.

SEASAT Rev. 478A was selected for analysis. This groundtrack crosses the gulf stream near 38° north latitude. Here the geostrophic relationship yields the following equation between changes in state and velocity corrections δv_k

$$\delta v_k = 14.2(x_k - x_{k-1}) \tag{43}$$

where x is in meters and δv is in meters per second.

From the data containing the gulf-stream signature, 37 points were selected, of which 15 were over the gulf stream. SSH data were processed with a bank of five matched filters to determine the most likely mean signature (i.e., the time-varying mean of the SSH data). From these data, the mean geostrophic velocity is estimated using (29). This time-varying mean velocity is shown in Fig. 7. The smoothness of this estimate reflects the simple model used in the matched filtering.

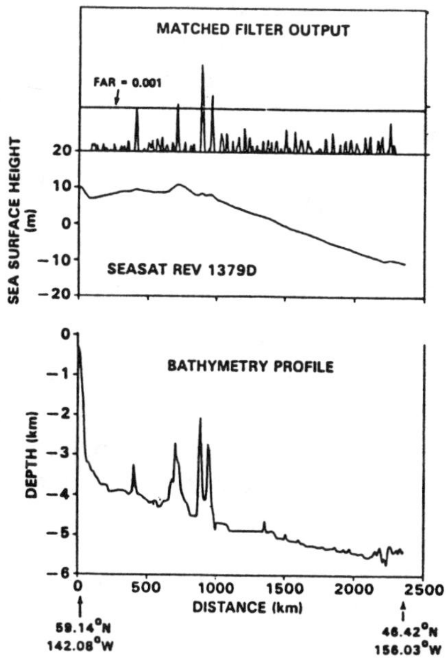

Fig. 10. Example of seamount detections using Rev. 1379D in the Gulf of Alaska.

Fig. 8 shows the geostrophic velocity corrections computed using an optimal smoother on the residual differences between SSH and the model signature. The five-sample triangularly weighted mean of the slope was optimally estimated using a smoother having six states. These velocity corrections have a theoretical rms accuracy of 0.07 m/s.

Fig. 9 shows the corrected geostrophic velocity estimates across the gulf stream obtained by adding the velocity corrections to the time-varying mean velocity.

The geostrophic relation given in (29) holds only for SSH anomalies due to geostrophic currents. The gulf stream can be approximated as a geostrophic current, and hence the corrected velocity in Fig. 9 in the gulf-stream region is consistent with accepted oceanographic modeling theory. Outside the gulf-stream region, the currents are not expected to be geostrophic, and the accuracy will depart from the theoretical rms value of 0.07 m/s. The major effect of the geostrophic velocity correction is to sharpen the peak velocity from the mean-signature estimate.

D. Detection of Seamounts

1) Comparison with Current Detection: Seamount signatures are easier to detect in altimeter data than ocean ring-current signatures. There are three main reasons for this difference. 1) Seamounts cause fixed anomalies in the sea-surface topography, while ocean-current features drift around and change their shapes. 2) The compact altimetric signatures caused by seamounts are easier to distinguish from the background noise as compared with the less-compact ring signatures. 3) Seamount signatures are anomalies in the undulations of the geoid. In contrast, current signatures are superimposed on the geoid undulations. Therefore, an accurate geoid estimate must be subtracted from the altimeter data to detect ring currents reliably.

Despite these differences, the AR modeling and matched-filter detection algorithms are equally applicable to detecting both seamounts and ocean currents. The data-adaptive AR modeling automatically accounts for the different statistics of the data along different tracks, and the proper choice of signature models accounts for the physical differences between the geodetic and oceanographic signatures.

2) Demonstration of Seamount Detection: An example of seamount detection [14] using SEASAT Rev. 1379 in the Gulf of Alaska is shown in Fig. 10. The bottom plot shows ocean depth along the satellite ground track as a function of the along-track distance. Four seamounts are visible. By using a matched filter optimized to detect the signature of a generic conical-shaped seamount, all four seamounts can be detected at a false-alarm rate of 0.001 alarm/Mm. This is demonstrated in the uppermost plot in Fig. 10, which shows the positive excursions of the filter output and the detection threshold. The middle plot shows the SEASAT data that were processed by the filter.

IV. Summary

This paper has described the application of Kalman filtering techniques to bathymetric and oceanographic mapping using data collected by satellite radar altimeters. The principal contributions of this paper are the quantitative analyses of ocean-current and seamount detection and the new results in error spectrum analysis of optimal smoothers. The technique for AR model development should also be widely applicable.

References

[1] H. R. Stanley, Ed., "Scientific results of the GEOS-3 mission," *Spec. Issue J. Geophysical Res.*, vol. 84, July 1979.
[2] G. H. Born, Ed., "Seasat ephemeris analysis," *Spec. Issue J. Astronaut. Sci.*, vol. 28, Oct.–Dec. 1980.
[3] D. E. Weissman, Ed., "Seasat-1 sensors," *Spec. Issue IEEE J. Ocean. Eng.*, vol. OE-5, Apr. 1980.
[4] H. Akaike, "A new look at the statistical model identification," *IEEE Trans. Automat. Contr.*, vol. AC-19, pp. 716–723, Dec. 1974.
[5] —, "Canonical correlation analyses of time series and the use of an information criterion," in *System Identification: Advances and Case Studies*, R. K. Mehra and D. G. Lainiotis, Eds. New York: Academic, 1976, pp. 27–96.
[6] A. Gelb, Ed., *Applied Optimal Estimation*, Cambridge, MA: M.I.T. Press, 1974.
[7] S. M. Kay and S. L. Marple, Jr., "Spectrum analysis—A modern perspective," *Proc. IEEE*, pp. 1380–1419, Nov. 1981.
[8] A. H. Gray and J. D. Markel, "Linear prediction analysis programs (AUTO-COVAR)," in *Programs for Digital Signal Processing*, Digital Signal Processing Committee, IEEE Acoustics, Speech, and Signal Processing Society, Eds. New York: IEEE Press, 1979, pp. 4.1-1–4.1-7.
[9] C. W. Helstrom, *Statistical Theory of Signal Detection*, 2nd ed. London: Pergamon, 1968.
[10] J. G. Marsh and E. S. Chang, "5′ detailed gravimetric geoid in the Northwestern Atlantic Ocean," *Marine Geodesy*, vol. 1, no. 3, pp. 253–261, 1978.
[11] R. E. Cheney, and J. G. Marsh, "SEASAT altimetry observations of dynamic ocean currents in the gulf stream region," *J. Geophys. Res.*, vol. 86, pp. 473–483, Jan. 1981.
[12] P. L. Richardson, R. E. Cheney, and L. V. Worthington, "A census of gulf stream rings, Spring 1975," *J. Geophys. Res.*, vol. 83, pp. 6136–6143, Dec. 1978.
[13] J. G. Marsh and R. E. Cheney, personal communication.
[14] J. V. White, R. V. Sailor, A. R. Lazarewicz, and A. R. LeSchack, "Detection of seamount signatures in SEASAT altimeter data using matched filters," *J. Geophys. Res.*, to be published.
[15] J. V. White, "Development of NOSS algorithms for ocean current mapping," NASA Tech. Rep. CR-156886, vols. 1 and 2, July 1982.

A Kalman Filtering Approach to Natural Gamma Ray Spectroscopy in Well Logging

GUY RUCKEBUSCH, MEMBER, IEEE

Abstract—This paper describes an application of (adaptive) Kalman filtering to a geophysical subsurface estimation problem. The NGT® is a sonde designed to detect the natural gamma rays of various energies emitted from a formation by the radioactive nuclei of potassium (K) and the thorium (Th) and uranium–radium (U) series. Using a minicomputer at the surface, the (Th, U, K) concentrations along the borehole have to be estimated on-line from the detection of the gamma rays in five energy windows. The standard technique in the logging industry has been to compute the elemental concentrations at a given depth using only the observed counting rates at the same depth. The resulting estimates have fairly large statistical errors which have limited the application of the NGT in computer reservoir evaluation. In this paper, it is shown that a Kalman filter based on a dynamical model of the (Th, U, K) vertical variations can produce real-time estimates that are readily usable on a quantitative basis. The paper focuses on the usual critical issues in applying Kalman filtering to real data, namely modeling, adaptivity, and computational aspects.

I. Introduction

The Schlumberger natural gamma ray spectroscopy tool (NGT) was commercially introduced as a logging device several years ago. It provides a direct means to determine the thorium (Th), uranium (U), and potassium (K) concentrations in a formation. These, in turn, are useful in computer reservoir evaluation [1]–[3].

Let us briefly describe the principle of the NGT. It is designed to make a continuous (while logging) spectral analysis of the gamma rays emitted from a formation. The heart of the tool consists of a thallium-activated sodium iodide crystal (12 in × 2 in), optically coupled to a photomultiplier. A gamma ray incident on the crystal may interact in the crystal, leaving all or some of its energy. The energy lost in the crystal is converted into scintillations which cause electron emission in the photocathode of the photomultiplier. The electrons are multiplied to produce an output voltage pulse whose amplitude is proportional to the energy left in the crystal by the incident gamma ray. High-speed comparators sort out this pulse by its amplitude in five "energy windows" covering the energy spectrum from 0.18 to 3 MeV. The pulses in each window are counted separately and sent digitally to the surface to be processed with a general purpose minicomputer (PDP 11/34).

Manuscript received March 2, 1982; revised August 20, 1982.
The author is with Schlumberger-Doll Research, Ridgefield, CT 06877.
® Mark of Schlumberger.

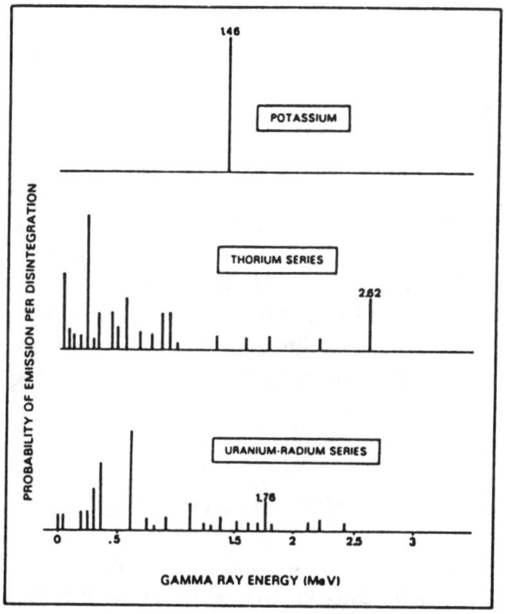

Fig. 1. (Th, U, K) emission spectra.

The gamma rays are produced by the radioactive nuclei of potassium and the thorium and uranium–radium series. The (Th, U, K) emission spectra consist of characteristic discrete energies; see Fig. 1. Between their emission and their absorption in the crystal, gamma rays undergo various interactions with the formation, the borehole, and the tool. Combined with these interactions, the physical processes of partial gamma ray energy loss in the crystal, as well as the intrinsic crystal resolution, result in dramatically broadened energy spectra. These spectra for a standard logging condition (eccentered tool in an 8 in diameter water-filled borehole), and the positions of the five windows are displayed in Fig. 2.

Roughly speaking, the counting rates at a given depth are linear combinations of the (Th, U, K) concentrations at this depth. The problem to be dealt with in this paper is to determine these concentrations along the borehole from a continuous recording of the counts from the five windows.

In the next section, a more mathematical version of the problem will be given. In Section III, a least-squares ap-

Reprinted from *IEEE Trans. Automat. Contr.*, vol. AC-28, pp. 372–380, Mar. 1983.

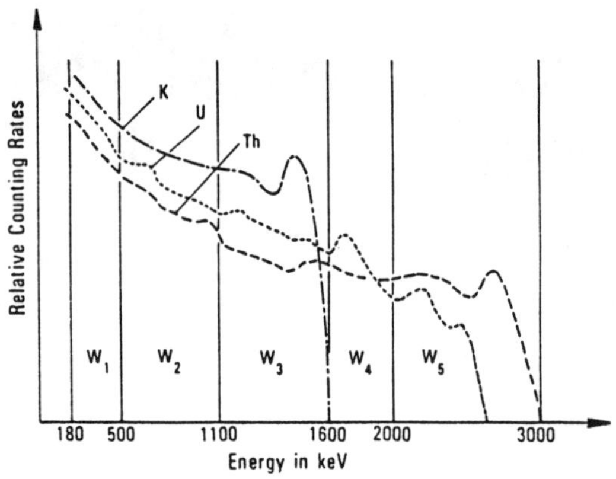

Fig. 2. (Th, U, K) apparent spectra.

proach (one of the standard techniques used in the logging industry) will be described. In the remainder of the paper, we will investigate the application of Kalman filtering theory to this estimation problem. More specifically, Section IV will be devoted to the modeling of the (Th, U, K) vertical variations. The computational aspects will be considered in Section V, while the problems of adaptivity will be discussed in Section VI. Some test results on both stimulated and real examples will be presented in Section VII. Finally, the Conclusion will be devoted to general comments on the applicability of Kalman filtering to well logging.

II. The NGT Estimation Problem

For simplicity, the (Th, U, K) concentrations will be assumed to depend only on the depth variable z. At depth z, the vector of the elemental concentrations (with Th and U in ppm and K in percent) will be denoted by $X(z)$. At time t, if $V(t)$ designates the speed of the tool (the nominal logging speed is 900 ft/h), the corresponding depth reached by the tool is defined by

$$\frac{dz}{dt} = V(t). \tag{1}$$

The counting processes $(N_1(t), \cdots, N_5(t))$ in the five energy windows are five independent inhomogeneous Poisson processes (cf. [4]). The vector of the five counts will be designated by $N(t)$. Let $f_j(E)$ denote the energy spectrum of the jth element ($j = 1,2,3$ for Th, U, and K, respectively) for the standard logging condition. We introduce the 5×3 sensitivity matrix H whose (i, j)th element is given by

$$H_{ij} = \int f_j(E)\, dE \tag{2}$$

(the integration is taken over the ith window). Let t and z be related by (1). The intensity of the Poisson process $N(t)$ is assumed to satisfy

$$\frac{d}{dt} E[N(t)] = HX(z). \tag{3}$$

($E[\cdot]$ designates the mathematical expectation.) It must be pointed out that (3) is but a crude approximation. In fact, it neglects all convolution effects due to the length of the crystal and the scattering of the gamma rays in the formation. Moreover, the logging environment has no reason to be the standard one or even depth invariant. Although Kalman filtering theory could cope with a depth-varying sensitivity matrix (cf. [5]), hereafter we will make the simplifying assumption that H is constant.

In practice, the (Th, U, K) concentrations are to be given every 6 in. Let $\{z_n\}$ denote a set of depths with a regular spacing of 6 in. The problem is then to determine the sequence of elemental concentrations defined by

$$X_n = \frac{1}{z_{n+1} - z_n} \int_{z_n}^{z_{n+1}} X(z)\, dz. \tag{4}$$

Let t_n designate the time when the tool reaches z_n. We define the counting rate sequence $\{Z_n\}$ by

$$Z_n = \frac{1}{t_{n+1} - t_n} [N(t_{n+1}) - N(t_n)]. \tag{5}$$

Clearly, the sequence $\{Z_n\}$ consists of independent five-dimensional vectors whose components are independent. Assuming that the speed of the tool is constant between t_n and t_{n+1}, we have

$$E[Z_n] = HX_n. \tag{6}$$

Equation (5) implies that the covariance matrix R_n of Z_n is given by

$$R_n = \frac{1}{t_{n+1} - t_n} \operatorname{diag}[HX_n] \tag{7}$$

(for any vector Y, diag[Y] means the diagonal matrix whose diagonal elements are equal to the components of Y).

In the sequel, the sum of the components of Z_n will be referred to as the (total) gamma ray, denoted γ_n. It is the practice of the oil industry to calibrate the gamma ray in API units which are proportional to counts per second. From (6) the mathematical expectation of γ_n is given by

$$E[\gamma_n] = a\mathrm{Th}_n + b\mathrm{U}_a + c\mathrm{K}_n \tag{8}$$

where a, b, and c are, respectively, the sum of the components of the first, second, and third columns of H.

The problem to be dealt with in this paper is the "optimal" estimation of the sequence $\{X_n\}$ from the sequence $\{Z_n\}$. Of course, the estimation procedure will depend crucially on the assumptions made on the sequence $\{X_n\}$ and on the chosen optimization criterion. The simplest is to assume that $\{X_n\}$ is a totally unknown, but deterministic, sequence and to take a least-squares criterion. This choice leads to the so-called weighted least-squares filter described in the next section. A different approach will be taken in Section IV where the sequence $\{X_n\}$ will be viewed as the realization of a stochastic process. In this case, the sequence $\{Z_n\}$ will become a doubly stochastic process in the terminology of Snyder [4]. This will lead to the (adaptive) Kalman filter developed in Sections V and VI.

III. The Weighted Least-Squares Filter

In this approach, proposed by Marett *et al.* [1], the estimate of the sequence $\{X_n\}$ is obtained by minimizing the criterion

$$J(\{X_n\}) = \sum_n (Z_n - HX_n)'R^{-1}(Z_n - HX_n) \qquad (9)$$

where R is some *predetermined* diagonal positive definite matrix (in (9) the prime indicates transposition). It should be pointed out that R cannot be taken as the covariance R_n of Z_n because R_n depends on the unknown $\{X_n\}$ through (7). In [1], R has been taken equal to the covariance of the measurements in a *standard shale formation* (Th = 12 ppm, U = 6 ppm, K = 2 percent). Disregarding the positivity constraints on the components of X_n, the minimization of J gives the well-known *weighted least-squares* (WLS) estimate of X_n, namely $(H'R^{-1}H)^{-1}H'R^{-1}Z_n$ (cf. [6]). Note that the WLS estimate is unbiased and depends only on Z_n and not on the other counting rates. To decrease the statistical errors on the sequence of WLS estimates, it is further low-pass filtered with an exponential smoothing (in [1] an "*RC* filter" with a time constant of 4 s is used). The resulting sequence, denoted $\{\hat{X}_n\}$, is defined recursively by

$$\hat{X}_n = \hat{X}_{n-1} + K(Z_n - H\hat{X}_{n-1}) \qquad (10)$$

where $K = (1 - e^{-0.5})(H'R^{-1}H)^{-1}H'R^{-1}$. This algorithm will be referred to as the WLS filter. Figs. 3 and 4 clearly demonstrate that such a filter yields unreliable estimates with high variances and large anticorrelations Th–U and U–K.

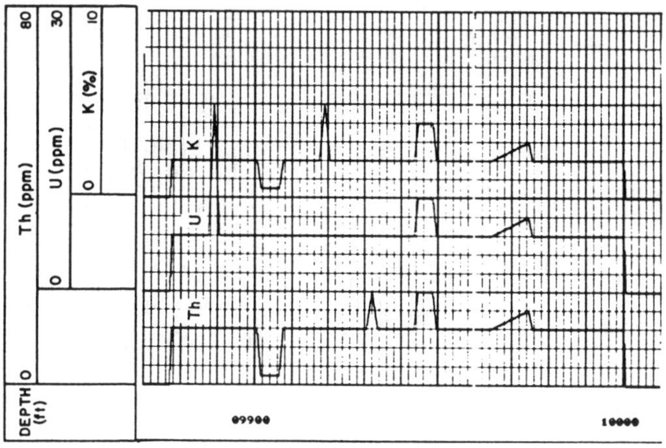

Fig. 3. (Th, U; K) profiles (simulated example).

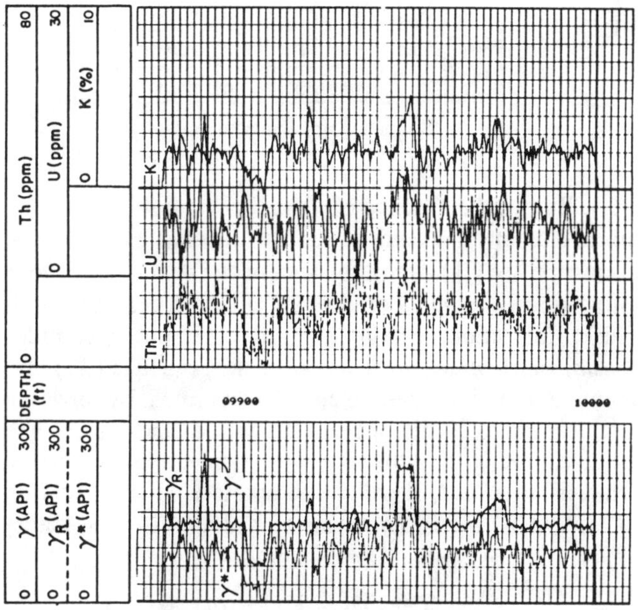

Fig. 4. WLS filter (simulated example).

With some computer power available at the surface (which did not exist when the solution above was elaborated), we may wonder whether better estimates could be constructed. So far, all the estimates which have appeared in the logging literature consist of low-pass filtering (using moving averages or exponential smoothing), a first sequence of estimates of X_n, where each estimate is a function only of the corresponding counting rates (using a WLS approach or a "spectrum stripping" technique as in [3]). Such an approach is doomed to be inadequate for the following reasons.

1) The Cramer–Rao bound (cf. [4]) for the unbiased estimation of X_n from Z_n is equal to $(H'R_n^{-1}H)^{-1}$, where R_n is given by (7). Computation of this bound for typical values of X_n shows that all unbiased estimates of X_n from Z_n alone will be affected by large statistical errors. In particular, this is the case for the maximum likelihood estimate of X_n (cf. [4]), which Monte Carlo simulations have proven to be almost unbiased and to reach the Cramer–Rao bound for all practical values of (Th, U, K). Surprisingly enough, the performances of the WLS estimate (which has a minimum variance in the standard shale) are only slightly degraded with respect to the maximum likelihood estimate. Thus, unless some prior information on X_n is added, there is little improvement to be expected by replacing the WLS estimate by some other (unbiased) estimate.

2) Time-invariant filters cannot satisfy the two conflicting requirements of *high vertical resolution* (for thin beds) and *good noise rejection* (for thick homogeneous beds). For example, it can be seen in Fig. 4 that a time constant of 4 s is already too large for the uranium anomaly at 9889 ft, while it is far too low for the standard shale formation from 10000 to 9975 ft. Thus, time-varying filtering is the only possibility to increase the signal-to-noise ratio while keeping an acceptable vertical resolution.

Clearly, the WLS filter could be made adaptive by replacing the gain K in (10) by some time-varying sequence

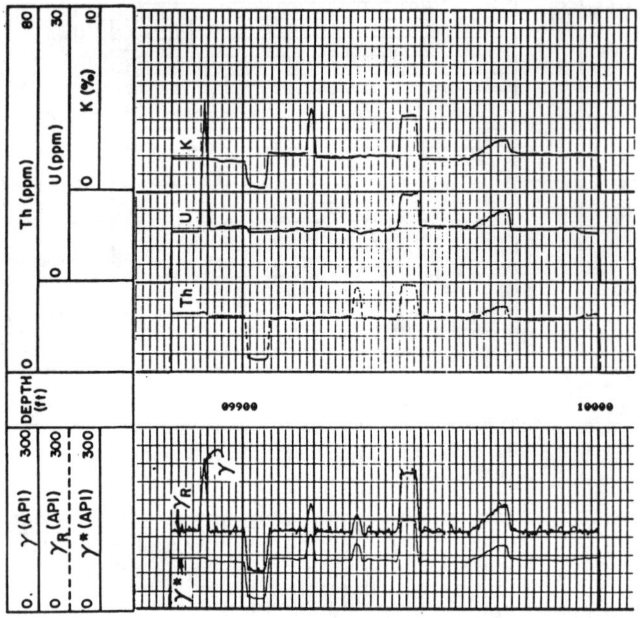

Fig. 5. Kalman filter (simulated example).

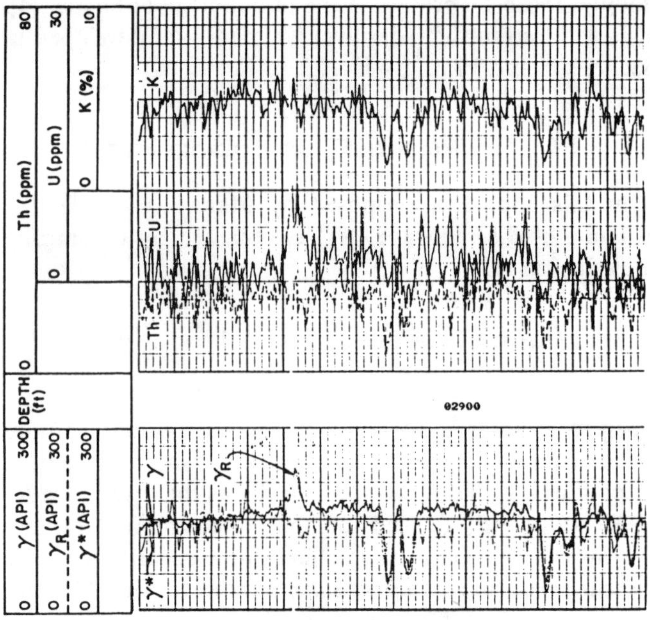

Fig. 6. WLS filter (real example 1).

$\{K_n\}$. But, in the absence of a model of the sequence $\{X_n\}$, the choice of K_n would be hard to justify. (For a related discussion see [7].) As will be seen in the next section, some information does exist on the (Th, U, K) vertical variations. This information will be used to construct a complete *nonstationary probabilistic model* of the sequence $\{X_n\}$. The estimation of $\{X_n\}$ will then be performed by Bayesian rule, leading to a filter of type (10) with a time-varying gain sequence $\{K_n\}$, cf. (20). But now, instead of having K_n arbitrarily defined, it will be completely determined from the (Th, U, K) model by means of Kalman filtering theory.

IV. The (Th, U, K) Modeling

Finding a realistic probabilistic model for the sequence $\{X_n\}$ is a difficult problem. However, the payoff may be important as can be judged from Fig. 5. Petroleum geology and the practice of NGT data interpretation seem to justify the following model:

$$X_{n+1} = X_n + \dot{\gamma}_n A_n + \epsilon_n \tag{11}$$

where $\{A_n\}$ is a slowly varying three-dimensional stochastic process (possibly interrupted by abrupt transitions), $\{\dot{\gamma}_n\}$ is *a priori* an arbitrary deterministic sequence, and $\{\epsilon_n\}$ is a zero-mean Gaussian white noise (of small variance) used to take into account the model inaccuracy. It can be seen from (6) that, without loss of generality, $\dot{\gamma}_n$ can be taken equal to $E[\gamma_{n+1} - \gamma_n]$, thus justifying the notation. Note that (11) apparently contradicts the positivity constraints on the components of $\{X_n\}$. This could be corrected by assuming that the sequence $\{\log X_n\}$, instead of the sequence $\{X_n\}$, satisfies a model of type (11). (Note that $\dot{\gamma}_n$ would then be different from $E[\gamma_{n+1} - \gamma_n]$.) But this would greatly complicate the processing without a significant increase of the performance.

The rationale behind (11) is the following. First of all, homogeneous beds characterized by uniform (Th, U, K) concentrations are well represented by (11) (with $\dot{\gamma}_n = 0$). Radioactive anomalies can be, roughly speaking, classified into "shaly" and "nonshaly." Until the introduction of gamma ray spectroscopy, all anomalies were believed shaly on the basis of the total gamma ray [1]–[3]. In fact, uranium is not really related to shaliness and Serra *et al.* [2] proposed to replace the gamma ray as a shaliness indicator by a uranium-free gamma ray defined by

$$\gamma_n^* = a\mathrm{Th}_n + c\mathrm{K}_n. \tag{12}$$

The shaly anomalies are usually characterized by the fact that the increments of Th, K, and (to a lesser extent) U correlate well. In particular, this implies that they correlate well with $\dot{\gamma}_n$, thus justifying (11). This empirical evidence is strengthened by petroleum geology. In fact, geologists have related the ratio Th/K to the mineralogical composition of the shale [8] and the ratio Th/U to the nature of the sedimentary process [9]. For a homogeneous depositional environment, these ratios should be fairly uniform. But such an assumption would clearly imply that the sequence $\{A_n\}$ must be slowly varying. Examples of (real) shaly anomalies can be seen in Figs. 6 and 7 around 2881 ft or in Figs. 8 and 9 around 7550 ft.

The nonshaly anomalies often consist of streaks of one radioactive element in an otherwise homogeneous medium. The uranium anomalies are especially important as they have been related to high permeability zones [3]. An example of (real) uranium anomaly can be seen in Fig. 7 around 2853 ft. Clearly, the one-element anomalies are well repre-

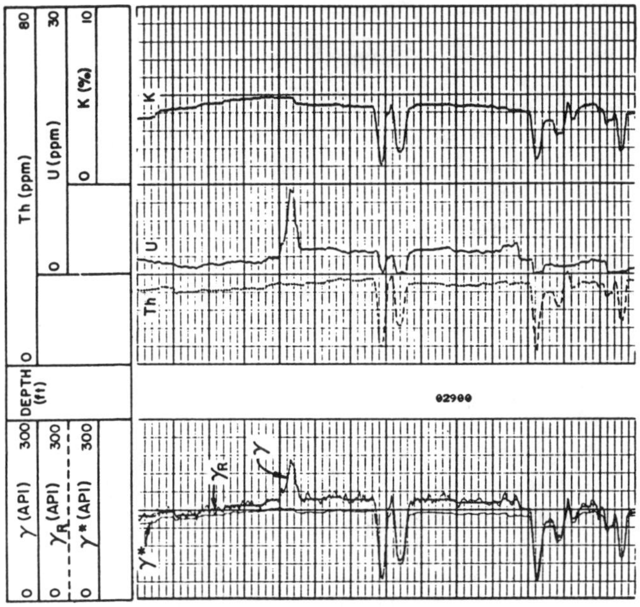

Fig. 7. Kalman filter (real example 1).

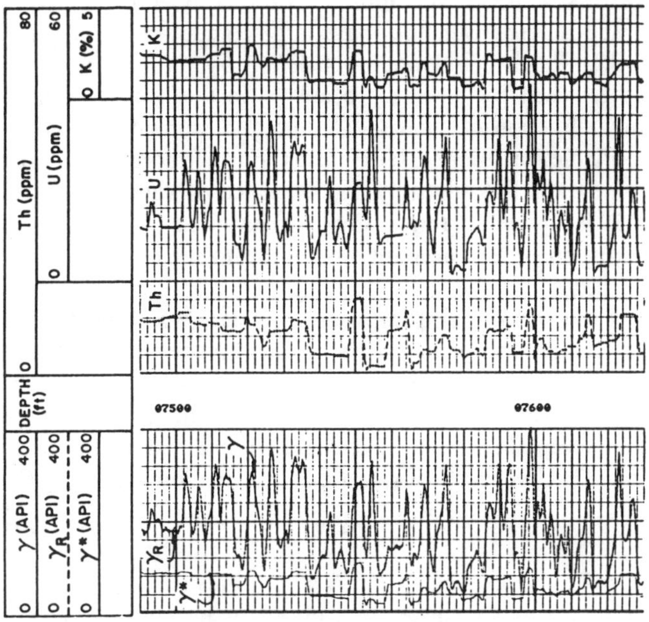

Fig. 9. Kalman filter (real example 2).

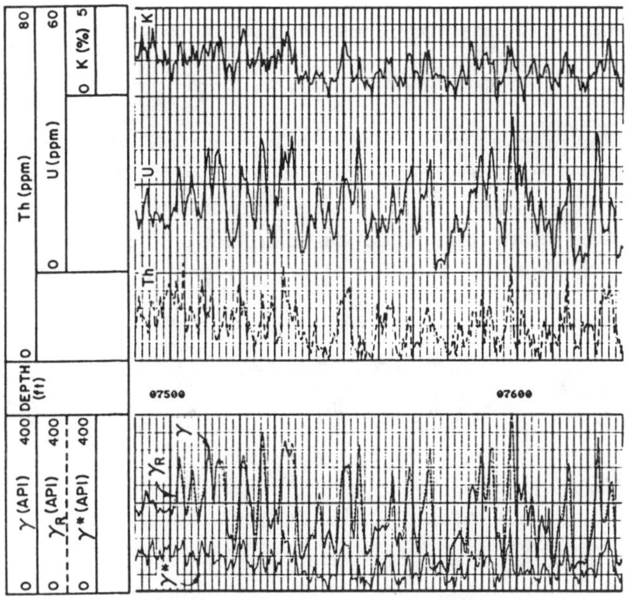

Fig. 8. WLS filter (real example 2).

sented by (11); for a case of a uranium anomaly we have

$$A_n = \left(0, \frac{1}{b}, 0\right)'. \tag{13}$$

The preceding reasoning shows that (11) seems to be a good candidate to model the sequence $\{X_n\}$. To represent the abrupt transitions affecting the sequence $\{A_n\}$ (they correspond to discontinuities in the sedimentary process), the following model could be used:

$$A_{n+1} = A_n + \eta_n + \mu\delta_{n\theta} \tag{14}$$

where $\{\eta_n\}$ is a zero-mean Gaussian white noise of small variance (to track slow variations of $\{A_n\}$), while $\mu\delta_{n\theta}$ indicates a jump of *unknown* amplitude μ at the *unknown* depth θ (δ_{ij} is the Kronecker delta). Clearly, (14) models only one abrupt transition, but it is just a matter of bookkeeping to modify (14) to take into account more (for our model to be meaningful, note that the abrupt transitions have to be assumed not too frequent). That the process A_n may experience abrupt transitions is evidenced by Fig. 9. In fact, the presence of a shaly anomaly around 7550 ft closely surrounded by two uranium anomalies requires two discontinuities for A_n at the depths 7553 and 7548 ft.

The equations (11)–(14) can be written in state-space form as follows:

$$\begin{pmatrix} X_{n+1} \\ A_{n+1} \end{pmatrix} = \begin{pmatrix} I & \dot{\gamma}_n I \\ 0 & I \end{pmatrix} \begin{pmatrix} X_n \\ A_n \end{pmatrix} + \begin{pmatrix} \epsilon_n \\ \eta_n \end{pmatrix} + \begin{pmatrix} 0 \\ \mu \end{pmatrix} \delta_{n\theta} \tag{15}$$

where I denotes the 3×3 identity matrix. To justify the use of Kalman filtering theory, let us rewrite (6) as

$$Z_n = HX_n + V_n. \tag{16}$$

The sequence $\{V_n\}$ is a zero-mean white noise such that the covariance of V_n is R_n. Moreover, the counts in the five windows are generally high enough to justify a Gaussian approximation of the probability distribution of V_n.

The problem is then to estimate the sequence of "state" vectors $\{(X_n', A_n')'\}$ from the observed sequence $\{Z_n\}$. Disregarding the presence of jumps in $\{A_n\}$, the estimation could be performed using standard Kalman filtering theory [10]. In fact, the sequence $\{\dot{\gamma}_n\}$ could be considered as known conditionally to the observations by taking $\dot{\gamma}_n$ as some filtered derivative of the gamma ray at depth z_n. Similarly, R_n could be estimated on-line by (7), where X_n is replaced by its one-step prediction from the Kalman filter.

Taking into account the jumps makes the problem much more involved. Only *suboptimal* techniques, like the generalized likelihood ratio approach of Willsky *et al.* [11], seem practical. However, they have large computational requirements which could not be satisfied by the present computers used for on-line processing (in this respect, only a small fraction of their CPU time and storage capacity is available for filtering NGT data). Furthermore, there exist some (rare, but relatively important) cases where the model (15) is not valid (an example is given in [3, p. 23]) and taking them into account would add to the computational burden.

While work is still in progress to design a sophisticated off-line processing based on a refined model (15), the need of real-time estimates has forced us to look after simpler filtering techniques. Testing on real examples has shown that acceptable estimates can still be obtained by modeling the (Th, U, K) vertical variations as a random walk

$$X_{n+1} = X_n + W_n. \tag{17}$$

In (17), the sequence $\{W_n\}$ is a zero-mean Gaussian white noise which is independent of the sequence $\{V_n\}$. Again note that (17) disregards the positivity constraints on the components of X_n; however, it will enable us to use linear Kalman filtering theory with no real loss of performance. In the sequel, the covariance of W_n will be denoted by Q_n. *A priori*, the sequence $\{Q_n\}$ is totally unknown. However, it will be shown in Section VI how the model (15) can be used to provide a "natural" on-line estimation of the sequence $\{Q_n\}$.

V. Computational Aspects

The equations (16)–(17) have a form amenable to Kalman filtering theory, except that the sequences $\{Q_n\}$ and $\{R_n\}$ are *a priori* totally unknown. To cope with this problem, we will use the following procedure (for a similar approach, see Myers *et al.* [12]).

1) First, design the Kalman filter based on the model (16)–(17), assuming that $\{Q_n\}$ and $\{R_n\}$ are known.

2) Second, replace $\{Q_n\}$ and $\{R_n\}$ in the filtering equations by some real-time estimates.

In this section, we will be concerned with the first issue only. In particular, we will investigate how to propagate the Kalman filter at the minimum computational cost (for a similar problem, see Mendel [13]).

Let us introduce the following notation. For all $n \geqslant 1$, the optimal estimate of X_n given $\{Z_1, Z_2, \cdots, Z_m\}$ will be denoted by $\hat{X}_{n,m}$. The covariance of the error $(X_n - \hat{X}_{n,m})$ will be designated by $\Sigma_{n,m}$. The estimate $\hat{X}_{n,n}$ is obtained recursively with the conventional Kalman filter equations

$$K_n = \Sigma_{n,n-1} H'(H\Sigma_{n,n-1}H' + R_n)^{-1}, \tag{18}$$

$$I_n = Z_n - H\hat{X}_{n,n-1} \tag{19}$$

(I_n is the so-called innovation process),

$$\hat{X}_{n,n} = \hat{X}_{n,n-1} + K_n I_n, \tag{20}$$

$$\Sigma_{n,n} = (I - K_n H)\Sigma_{n,n-1}, \tag{21}$$

$$\hat{X}_{n+1,n} = \hat{X}_{n,n}, \tag{22}$$

$$\Sigma_{n+1,n} = \Sigma_{n,n} + Q_n. \tag{23}$$

The recursion is initialized by $\hat{X}_{1,0}$ and $\Sigma_{1,0}$ which represent the knowledge about X_1 prior to any measurement. We have assumed that $\hat{X}_{1,0}$ is independent of the sequences $\{V_n\}$ and $\{W_n\}$.

When the tool enters a (Th, U, K) homogeneous bed after a sharp transition at depth θ (i.e., Q_n is small compared with Q_θ for $n > \theta$), the Kalman filter estimate $\hat{X}_{n,n}$ will have a relatively large covariance error just after θ. Fig. 7 provides such an example at 2876 ft. To improve on $\hat{X}_{n,n}$, while still operating in real time, we have implemented instead a *fixed-lag smoother* $\hat{X}_{n,n+d}$, where d is some positive delay (the test results of the paper correspond to $d = 20$, but acceptable results can still be obtained with $d = 5$). The estimate $\hat{X}_{n,n+d}$ is conventionally given by

$$\hat{X}_{n,n+d} = \hat{X}_{n,n} + \Sigma_{n,n-1}\left\{ \sum_{k=1}^{d} \left[\prod_{j=0}^{k-1} (I - K_{n+j}H)' \right] E_{n+k} \right\} \tag{24}$$

where E_n is the three-dimensional sequence defined by

$$E_n = H'(H\Sigma_{n,n-1}H' + R_n)^{-1} I_n. \tag{25}$$

The propagation of (18)–(25) represents a large computational burden, especially because of the computation of $H'(H\Sigma_{n,n-1}H' + R_n)^{-1}$. Using the well-known matrix inversion lemma (see Anderson *et al.* [10, p. 138]), it can be shown that the equations (18)–(21) can be propagated more efficiently as follows:

$$\Phi_n = (I + H'R_n^{-1}H\Sigma_{n,n-1})^{-1}, \tag{26}$$

$$\Sigma_{n,n} = \Sigma_{n,n-1}\Phi_n, \tag{27}$$

$$\tilde{E}_n = H'R_n^{-1}(Z_n - H\hat{X}_{n,n-1}), \tag{28}$$

$$\hat{X}_{n,n} = \hat{X}_{n,n-1} + \Sigma_{n,n}\tilde{E}_n. \tag{29}$$

The fixed-lag smoother is then given by

$$\hat{X}_{n,n+d} = \hat{X}_{n,n} + \Sigma_{n,n}\left\{ \sum_{k=1}^{d} \left[\prod_{j=1}^{k} \Phi_{n+j} \right] \tilde{E}_{n+k} \right\}. \tag{30}$$

Clearly, the computation of Φ_n compares favorably with that of $H'(H\Sigma_{n,n-1}H' + R_n)^{-1}$. Moreover, note that the evaluation of $(H'R_n^{-1}H)$ will be needed in the estimation of Q_{n-1}.

VI. Adaptive Estimation of $\{Q_n\}$ and $\{R_n\}$

The filter described in the preceding section is completely specified except for the determination of the sequences $\{Q_n\}$ and $\{R_n\}$. For the sake of clarity, the real-time estimates of Q_n and R_n will be denoted $\hat{Q}_n$ and $\hat{R}_n$. These estimates will be defined recursively, i.e., $\hat{Q}_{n-1}$ and $\hat{R}_n$ will be obtained assuming that the sequence $\{(\hat{Q}_{m-1}, \hat{R}_m), m \leqslant n-1\}$ has been previously determined.

The determination of $\{\hat{R}_n\}$ seems relatively simple, in as much as the following unbiased estimate could be used:

$$\hat{R}_n = \frac{1}{t_{n+1} - t_n} \operatorname{diag}\{Z_n\}. \tag{31}$$

In fact, higher performance was obtained if (31) is replaced by

$$\hat{R}_n = \frac{1}{t_{n+1} - t_n} \frac{\gamma_n}{\hat{\gamma}_{n,n-1}} \operatorname{diag}\{H\hat{X}_{n,n-1}\} \tag{32}$$

where $\hat{\gamma}_{n,n-1}$ denotes the sum of the components of $H\hat{X}_{n,n-1}$ which is defined since it depends only on $\{(\hat{Q}_{m-1}, \hat{R}_m), m \leqslant n-1\}$. To avoid filter divergence (cf. Jazwinski [14, p. 302]), it is necessary to constrain the components of $\hat{R}_n$ to be not less than some positive threshold. We will see later how (32) can be improved upon.

The determination of $\{\hat{Q}_n\}$ is a more involved problem. As it has proven difficult to take into account possible correlations between the (Th, U, K) increments, Q_n will be assumed diagonal everywhere. The estimate of Q_n seems straightforward only in homogeneous (Th, U, K) beds. For these, we simply set

$$\hat{Q}_n = Q_0 \tag{33}$$

where Q_0 is a given diagonal matrix (with small strictly positive diagonal elements) used to track slow variations of $\{X_n\}$. The homogeneous beds can be detected by looking at the innovation process $\{I_n\}$ defined by (19). Under the assumption (33), the covariance matrix of I_{n+1} is exactly $[H(\Sigma_{n+1,n} + Q_0)H' + R_{n+1}]$. It would seem natural to use (33) if, and only if

$$I'_{n+1}\left[H(\Sigma_{n+1,n} + Q_0)H' + \hat{R}_{n+1}\right]^{-1} I_{n+1} < th \tag{34}$$

where $\hat{R}_{n+1}$ is defined by (32) and th is an appropriate threshold. To minimize the computational burden, we have simply replaced (34) by

$$I'_{n+1}\hat{R}^{-1}_{n+1}I_{n+1} < th. \tag{35}$$

To minimize the false alarms rate, (35) has been in fact coupled with some other test based on $(Z_{n+k} - H\hat{X}_{n,n}, 1 \leqslant k \leqslant 6)$. When these two tests are satisfied, note that we can improve on (32) by setting

$$\hat{R}_{n+1} = \frac{1}{t_{n+2} - t_{n+1}} \operatorname{diag}\left[H\hat{X}_{n+1,n}\right]. \tag{36}$$

For nonhomogeneous beds, the problem is more complex. Our estimation procedure below can be justified *heuristically* when (17) is considered as an "approximation" of (11). Clearly we have $W_n = \dot{\gamma}_n A_n$. Assuming that A_n is known, it seems natural to estimate Q_n by the diagonal matrix $\hat{Q}_n$ whose ith diagonal element $[i = 1,2,3]$ is given by

$$\hat{Q}^i_n = \dot{\gamma}^2_n\left[A^i_n\right]^2 \tag{37}$$

where A^i_n denotes the ith component of A_n. When (35) is not satisfied, the increments of the gamma ray are usually high enough to justify the approximation

$$\dot{\gamma}_n = \gamma_{n+1} - \hat{\gamma}_{n,n}. \tag{38}$$

The rationale behind (37) is twofold.

1) On the one hand, $\dot{\gamma}^2_n$ monitors the vertical resolutions of the estimates. Basically, we expect the (Th, U, K) concentrations to be no more irregular than the gamma ray. Note that this is obviously not the case with the WLS filter (compare Fig. 6 to Fig. 7).

2) On the other hand, $[A^i_n]^2$ is used to control the relative bandwidths of the Th, U, and K filters. From Fig. 7, we see that this selectivity is essential to have a good response for both shaly (around 2881 ft) and nonshaly (around 2853 ft) anomalies.

Since A_n is unknown, it will simply be replaced by an estimate $\hat{A}_n$ in (37). To compute $\hat{A}_n$ we proceed as follows. We assume that the model (11) is valid with a constant $\{A_n\}$ in the interval $(n, n+p)$. The choice of p is important. On the one hand, p cannot be too small to have an acceptable accuracy in the determination of $\hat{A}_n$. On the other hand, p cannot be too large to keep the validity of (11). Extensive testing on both simulated and real logs has shown that $p = 6$ seems a good tradeoff (this value has been used for the test results). We then introduce the k-prediction error $\{\tilde{Z}_k, 1 \leqslant k \leqslant p\}$ defined by

$$\tilde{Z}_k = Z_{n+k} - H\hat{X}_{n,n} - HA_n(\gamma_{n+k} - \hat{\gamma}_{n,n}). \tag{39}$$

Neglecting the errors in $H\hat{X}_{n,n}$ with respect to those in Z_{n+k}, it follows from (11) [we have assumed $\epsilon_m = 0$ for all m in $(n, n+p)$] that the sequence $\{\tilde{Z}_k, 1 \leqslant k \leqslant p\}$ is approximately a zero-mean Gaussian white noise of covariance R_{n+k}. Replacing R_{n+k} by the following estimate [cf. (32)]

$$\hat{R}_{n+k} = \frac{1}{t_{n+k+1} - t_{n+k}} \frac{\gamma_{n+k}}{\hat{\gamma}_{n+1,n}} \operatorname{diag}\{H\hat{X}_{n+1,n}\} \tag{40}$$

and assuming a constant speed of the sonde in $(n, n+p)$, the log-likelihood of $\{\tilde{Z}_k, 1 \leqslant k \leqslant p\}$ is then proportional to

$$L\{A_n\} = \sum_{k=1}^{p} \frac{1}{\gamma_{n+k}} \tilde{Z}'_k \hat{R}^{-1}_{n+1} \tilde{Z}_k \tag{41}$$

[the constant additive term has been dropped from (41)]. The maximum likelihood estimate of A_n is then given by

$$\hat{A}_n = \left[\left(H'\hat{R}^{-1}_{n+1}H\right)^{-1}H'\hat{R}^{-1}_{n+1}\right] \cdot \frac{\sum_{k=1}^{p}\left(Z_{n+k} - H\hat{X}_{n,n}\right)\left(\gamma_{n+k} - \hat{\gamma}_{n,n}\right)/\gamma_{n+k}}{\sum_{k=1}^{p}\left(\gamma_{n+k} - \hat{\gamma}_{n,n}\right)^2/\gamma_{n+k}}. \tag{42}$$

It is worth noting that this estimate could be refined by adding constraints on A_n from geological knowledge (for example, the components of the prediction of X_{n+1} from

(11), namely $\hat{X}_{n,n} + A_n(\gamma_{n+1} - \hat{\gamma}_{n,n})$, are nonnegative). Finally, the criterion L can be used to test the validity of the model (11). The idea is to use the procedure above if, and only if, the minimum of L is not too large. Otherwise, the model (11) is not considered valid and we simply set

$$\hat{Q}_n = Q \tag{43}$$

where Q is a predetermined large diagonal covariance matrix (in such a case the (Th, U, K) estimates will look like those given by the WLS filter).

VII. Test Results

The WLS filter and the Kalman filter have been tested on both simulated and real examples. In each case the following results are displayed: the (Th, U, K) estimates, the total gamma ray (denoted γ), the uranium-free gamma ray (denoted γ^*), and the reconstruction of the gamma ray (denoted γ_R) from the (Th, U, K) estimates [cf. (8)].

Figs. 3–5 deal with a simulated example. It consists of several homogeneous standard shale beds, one "ramp" (from 9975 to 9964 ft), three one-element anomalies (around 9932, 9919, and 9889 ft), and two shale anomalies. The first shale anomaly (from 9950 to 9944 ft) shows a high correlation between the increments of (Th, U, K) and those of γ, while the second one (from 9907 to 9901 ft) shows no correlation between the increments of U and those of γ. In each case the estimates from the Kalman filter have much less statistical variability as well as better contrasts than the estimates from the WLS filter.

Figs. 6 and 7 present a real example. Note that the model (15) seems to match the (Th, U, K) vertical variations quite well. As in the simulated example the Kalman filter clearly outperforms the WLS filter. Note that the frequent negative estimates of U and the frequent anticorrelations Th–U exhibited by the WLS filter have completely disappeared with the Kalman filter. Also note that the Kalman filter gives an estimate of the shaliness of the formation (γ^*), which is already usable for quantitative interpretation.

Figs. 8 and 9 display another real example. Note the important activity of the (Th, U, K) vertical concentrations. Model (15) seems a bit less convincing than in the preceding example because of the frequent occurrence of discontinuities for the process A_n. This explains that the performance of the Kalman filter in Fig. 9 is not quite as spectacular as in Fig. 7. However, the Kalman filter still outperforms the WLS filter in statistical stability and vertical resolution. In this respect, note the discrepancies around thin beds between the gamma ray and its reconstruction given by the WLS filter.

VIII. Conclusion

The test results presented in this paper show the kind of improvement that can be obtained with the Kalman filter compared to the WLS filter. In computer reservoir evaluation the Kalman filter has proven to give reliable inputs, unlike the WLS filter, whose results were used mainly on a qualitative basis. However, it must be pointed out that our filtering is far from being optimal with respect to the "system" model (15). In fact, the filter was designed to satisfy the requirements of the computing systems presently used for real-time processing. In the future, these systems will be upgraded and more sophisticated filtering techniques based on (15) will become implementable.

Let us briefly comment on the applications of the Kalman filter to well-logging problems in general. The reader is cautioned that the discussion below will express only the author's viewpoint.

The approach presented in this paper may be extended to *other nuclear tools* which are based on gamma ray or neutron [15] measurements. All these tools involve Poisson statistics like the NGT. A minor complication is that the tool response [the equivalent of (16)] can be nonlinear. As for the NGT, the system model is unknown. Its identification represents a nontrivial problem because geology does not seem to be easily described by linear dynamical systems. Nevertheless, this paper has demonstrated that even simple-minded models like random walks can bring dramatic improvements on present techniques (like the WLS filter).

An other application of Kalman filtering has been in the *segmentation* of a geological formation into "homogeneous" zones using the measurements from a set of well-logging tools [16]. Each homogeneous zone is modeled by a linear depth-invariant dynamical system and the zone boundaries are represented by jump processes. Thanks to its recursivity, Kalman filtering, coupled with a jump detection scheme (cf. [11]), can perform the segmentation at a relatively low computational cost.

References

[1] G. Marett, P. Chevalier, P. Souhaite, and J. Suau, "Shaly sand evaluation using gamma ray spectrometry, Applied to the north sea jurassic," presented at the SPWLA 17th Annu. Logging Symp., June 9–12, 1976.

[2] O. Serra, J. Baldwin, and J. Quirein, "Theory, interpretation and practical applications of natural gamma ray spectroscopy," presented at the SPWLA 21st Annu. Logging Symp., July 8–11, 1980.

[3] W. H. Fertl, "Gamma ray spectral data assists in complex formation evaluation," *The Log Analyst*, vol. 20, pp. 3–37, Sept.–Oct. 1979.

[4] D. L. Snyder, *Random Point Processes*. New York: Wiley, 1975.

[5] G. Ruckebusch, "NGT environmental corrections within a Kalman filtering approach," Int. Rep., Schlumberger-Doll Research, Dec. 3, 1981.

[6] J. M. Mendel, *Discrete Techniques of Parameter Estimation: The Equation Error Formulation*. New York: Marcel Dekker, 1973.

[7] A. Benveniste and G. Ruget, "A measure of the tracking capability of recursive stochastic algorithms with constant gains," *IEEE Trans. Automat. Contr.*, vol. AC-27, pp. 639–649, June 1982.

[8] M. Hassan and A. Hossin, "Contribution á l'étude des comportements du thorium et du potassium dans les roches sédimentaires," *C. R. Acad. Sci., France*, vol. 280, pp. 533–535, 1975.

[9] J. A. S. Adams and C. E. Weaver, "Thorium to uranium ratios as indications of sedimentary processes: Example of concept of geochemical facies," *Bull. Amer. Assoc. Petrol. Geol.*, vol. 42, pp. 387–430, 1958.

[10] B. D. O. Anderson and J. B. Moore, *Optimal Filtering.* Englewood Cliffs, NJ: Prentice-Hall, 1975.

[11] A. S. Willsky and H. L. Jones, "A generalized likelihood ratio approach to state estimation in linear systems subject to abrupt changes," in *Proc. IEEE Conf. Decision Contr.*, Phoenix, AZ, Nov. 1974, pp. 846–853.

[12] K. A. Myers and B. D. Tapley, "Adaptive sequential estimation with unknown noise statistics," *IEEE Trans. Automat. Contr.*, vol. AC-21, pp. 520–523, Aug. 1976.

[13] J. Mendel, "Computational requirements for a discrete Kalman filter," *IEEE Trans. Automat. Contr.*, vol. AC-16, pp. 748–758, Dec. 1971.

[14] A. H. Jazwinski, *Stochastic Processes and Filtering Theory.* New York: Academic, 1970.

[15] G. Ruckebusch, "Contribution to the modeling and filtering of the CNT-X2," Int. Rep., Schlumberger-Doll Research, Nov. 25, 1981.

[16] M. Basseville and A. Benveniste, "Design and comparative study of some sequential jump detection algorithms for digital signals," *IEEE Trans. Acoust., Speech, Signal Processing*, submitted for publication.

Measurement of Instantaneous Flow Rate Through Estimation of Velocity Profiles

MASARU UCHIYAMA, MEMBER, IEEE, AND KYOJIRO HAKOMORI

Abstract—It is very important, but difficult to measure instantaneous flow rate in the unsteady flow regime, since the variation of velocity profile limits the dynamic accuracy of flowmeters. Consideration of the velocity profile variation is crucial to improving their dynamic performance. This paper presents a new method of measuring the instantaneous flow rate of unsteady flow in a circular pipe by estimating the velocity profile of the flow. The estimation is formulated as a discrete Kalman filter problem where the dynamics of the velocity profile, which is governed by a linear partial differential equation, is approximated using a finite-dimensional model. An instantaneous flowmeter consisting of a laser Doppler velocimeter (LDV) and an 8-bit microcomputer which implements the solution of the filter problem has been developed. Successful measurements of a pulsating flow and a stepwise changing flow by this flowmeter are presented.

NOMENCLATURE

a	inside radius of pipe
A	$n \times n$ system matrix defined by (17)
$\boldsymbol{b}$	n-dimensional distribution vector defined by (17)
$\boldsymbol{c}$	m-dimensional measurement vector defined by (34)
C	$m \times n$ measurement matrix defined by (19)
$E[\cdot]$	expectation of $[\cdot]$
$f(\bar{x}, \bar{s})$	unknown function in (8)
$g_i(\bar{r})$	ith eigenfunction defined by (13)
$\boldsymbol{g}(\bar{r})$	n-dimensional vector function with $g_i(\bar{r})$ for ith component
$\bar{G}_p(\bar{r}, \bar{s})$	transfer function of flow in pipe section
$\bar{G}_{pn}(\bar{r}, \bar{s})$	truncated $\bar{G}_p$ with n terms
$h(\bar{r})$	scalar case of $\boldsymbol{h}(\bar{r})$
$h_i(\bar{r})$	ith component of $\boldsymbol{h}(\bar{r})$
$\boldsymbol{h}(\bar{r})$	m-dimensional weighting vector function to be determined by observation process
I	$n \times n$ unit matrix
j	$\sqrt{-1}$
J_0, J_1	Bessel function of first kind of order 0 and 1, respectively
$\boldsymbol{k}_\infty$	n-dimensional steady-state Kalman gain vector for scalar measurement
K_k	$n \times m$ Kalman gain matrix at kth sampling time
K_∞	steady-state value of K_k
M_k	$n \times n$ matrix defined by (31)
p	pressure $p(x, t)$
q_k	flow rate at kth sampling time
Q	z-transform of q_k

Manuscript received March 5, 1982; revised September 13, 1982.
The authors are with the Department of Precision Engineering, Tohoku University, Sendai, Japan.

Reprinted from *IEEE Trans. Automat. Contr.*, vol. AC-28, pp. 380–388, Mar. 1983.

r	coordinate in radial direction
$\bar{r}$	dimensionless r, r/a
$\bar{r}_0$	$\bar{r}$ at measuring point of velocity
R	defined by (6)
Re	Reynolds number
$\bar{s}$	dimensionless Laplace variable
t	time
$\bar{t}$	dimensionless time $\omega_\nu t$
u	flow velocity in axial direction, $u(r, x, t)$
$\bar{u}$	dimensionless u, u/U_0
u_k	u at kth sampling time, $u_k(r, x)$
$\bar{u}_k$	$\bar{u}$ at kth sampling time, $\bar{u}_k(\bar{r}, \bar{x})$
$\bar{U}$	Laplace transform of $\bar{u}$
U_0	representative velocity
w_i	ith component of $\boldsymbol{w}$
w_k	scalar case of $\boldsymbol{w}_k$
$\boldsymbol{w}$	m-dimensional measurement noise vector $\boldsymbol{w}(t)$
$\boldsymbol{w}_k$	$\boldsymbol{w}$ at kth sampling time
W	defined by (6)
W^c	$m \times m$ covariance matrix of $\boldsymbol{w}_k$, $E[\boldsymbol{w}_k \boldsymbol{w}_k']$
x	coordinate in axial direction
$\bar{x}$	dimensionless x, x/a
x_i	ith state variable
$\boldsymbol{x}$	n-dimensional state vector
$\boldsymbol{x}_k$	$\boldsymbol{x}$ at kth sampling time
X^c	$n \times n$ covariance matrix of $\boldsymbol{x}_k$, $E[\boldsymbol{x}_k \boldsymbol{x}_k']$
$\boldsymbol{X}$	z-transform of $\boldsymbol{x}_k$
y_i	ith component of $\boldsymbol{y}$
y_k	scalar case of $\boldsymbol{y}_k$
$\boldsymbol{y}$	m-dimensional measurement vector $\boldsymbol{y}(t)$
$\boldsymbol{y}_k$	$\boldsymbol{y}$ at kth sampling time
Y	z-transform of y_k
Y^c	covariance of y_k, $E[y_k^2]$
z	z-transform variable $e^{\bar{s}\bar{\tau}}$
Z_k^c	$n \times n$ matrix defined by (31)
α_i	ith zero of J_0
γ	dimensionless pressure gradient defined by (2), $\gamma(\bar{x}, \bar{t})$
γ_k	at kth sampling time $\gamma_k^* + \xi_k$
γ_k^*	known part of γ_k
Γ	Laplace transform of γ
δ	Dirac's delta function
δ_{ij}	Kronecker's delta
ϵ_∞	steady-state estimation error
$\boldsymbol{\lambda}$	n-dimensional vector defined by (42)
ν	fluid kinematic viscosity
ξ_k	driving noise = unknown part of γ_k
Ξ^c	covariance of ξ_k, $E[\xi_k^2]$
ρ	fluid density
σ^2	variance of noise
$\bar{\tau}$	dimensionless sampling period
Φ	$n \times n$ transition matrix defined by (25)
$\boldsymbol{\psi}$	n-dimensional distribution vector defined by (25)
ω	angular frequency
$\bar{\omega}$	dimensionless angular frequency ω/ω_ν
ω_ν	viscous characteristic frequency defined by (3)
$(\ )'$	transpose of $(\ )$
$(\hat{\ })$	estimate of $(\ )$

I. Introduction

In recent years, advances in science and technology have made it essential that measurements of instantaneous flow rates be obtained in unsteady pipe flows. However, there are few flowmeters which have excellent dynamic response. Variation of velocity profiles often degrades the dynamic accuracy of flowmeters; for example, such flowmeters as detect local velocity are useless when the shape of the profile changes because the integration performed on the velocity profile becomes invalid.

In order to overcome the above difficulties, the variation in velocity profile must be taken into consideration. Knowing this velocity profile is a key to improving the dynamic characteristics of flowmeters, when we have no practical means of measuring velocity profiles directly.

On the other hand, generally speaking, every flowmeter or velocimeter provides some information on the velocity profile of changing flow even when they do not measure the instantaneous flow rate accurately. Reconstruction of the velocity profile from this information would yield a flowmeter with good dynamic response through the use of an ordinary flowmeter or a velocimeter.

In this paper, we propose a general theory for improving the dynamic characteristics of flowmeters or facilitating the use of velocimeters for flowmeters. The basis of the theory is that the instantaneous flow rate is measured through estimation of the velocity profile and is relevant to laminar flow in a circular pipe.

The estimation is originally a distributed-parameter filter problem because, strictly speaking, the velocity profile variation in the pipe is governed by the Navier–Stokes partial differential equations. In utilizing a digital implementation of the filter, the problem is formulated and solved as a discrete Kalman filter where the dynamics of the velocity profile can be approximated by a finite-dimensional model.

The time-discrete and finite-dimensional solution have been implemented on an 8-bit microcomputer which processes the signal from a laser Doppler velocimeter (LDV) to yield the instantaneous flow rate in the pipe. Experiments for a pulsating flow and a stepwise changing flow have been successfully undertaken using this computerized flowmeter.

The balance of this paper is divided into four parts. The estimation theory and its implementation are developed in Sections II and III, with details of experimental results provided in Section IV. Brief comments and conclusions are presented in Section V.

II. Theory of Profile Estimation

A. Fluid Line Model

In deriving the filter, it is first necessary to formulate a mathematical model which describes the velocity profile variation in a pipe. In this paper, we call this model a fluid line model.

Provided that the flow has fully developed, the velocity profile variation of the laminar flow in a circular pipe is governed by the following linear partial differential equation whether the fluid be compressive or not [1]:

$$\frac{\partial u}{\partial t} = -\frac{1}{\rho}\frac{\partial p}{\partial x} + \nu\left(\frac{\partial^2 u}{\partial r^2} + \frac{1}{r}\frac{\partial u}{\partial r}\right). \tag{1}$$

It should be noted that, in (1), u is assumed to be a function of r, x, and t, while p is assumed to be a function of x and t only. Recent experiments have shown that (1) satisfactorily describes the real flow [2], [3].

According to the theory of a distributed-parameter filter [4], a filter equation can be obtained from (1) directly as a partial differential equation. The equation has to be lumped in some way in order to permit digital implementation using a microcomputer. Here, before derivation of the filter, we consider a finite-dimensional approximation of (1).

First, we introduce the following dimensionless variables:

$$\bar{u} = \frac{u}{U_0}, \qquad \gamma = -\frac{1}{U_0 \rho \omega_\nu}\frac{\partial p}{\partial x},$$
$$\bar{r} = \frac{r}{a}, \qquad \bar{x} = \frac{x}{a}, \qquad \bar{t} = \omega_\nu t \tag{2}$$

where ω_ν is the Nichols' viscous characteristic frequency [5] defined by

$$\omega_\nu = \frac{8\nu}{a^2}. \tag{3}$$

With the above dimensionless variables, we can rewrite (1) as

$$\frac{\partial \bar{u}}{\partial \bar{t}} = \gamma + \frac{1}{8}\left(\frac{\partial^2 \bar{u}}{\partial \bar{r}^2} + \frac{1}{\bar{r}}\frac{\partial \bar{u}}{\partial \bar{r}}\right). \tag{4}$$

Laplace transformation of (4) with respect to $\bar{t}$ yields

$$\bar{s}\bar{U} = \Gamma + \frac{1}{8}\left(\frac{\partial^2 \bar{U}}{\partial \bar{r}^2} + \frac{1}{\bar{r}}\frac{\partial \bar{U}}{\partial \bar{r}}\right). \tag{5}$$

We define new variables W and R as

$$W = \bar{U} - \frac{\Gamma}{\bar{s}}, \qquad R = 2\sqrt{2}\, j\bar{r}\sqrt{\bar{s}}\,. \tag{6}$$

Substituting W and R for $\bar{U}$ and $\bar{r}$ in (5), we obtain

$$\frac{\partial^2 W}{\partial R^2} + \frac{1}{R}\frac{\partial W}{\partial R} + W = 0. \tag{7}$$

This is the Bessel differential equation of order zero and its solution must be finite at $R = 0$. Therefore, we obtain [1]

$$W(R, \bar{x}, \bar{s}) = f(\bar{x}, \bar{s}) J_0(R) \tag{8}$$

where $f(\bar{x}, \bar{s})$ is as yet an unknown function of $\bar{x}$ and $\bar{s}$. Thus

$$\bar{U}(\bar{r}, \bar{x}, \bar{s}) = f(\bar{x}, \bar{s}) J_0(2\sqrt{2}\, j\bar{r}\sqrt{\bar{s}}) + \frac{\Gamma(\bar{x}, \bar{s})}{\bar{s}}. \tag{9}$$

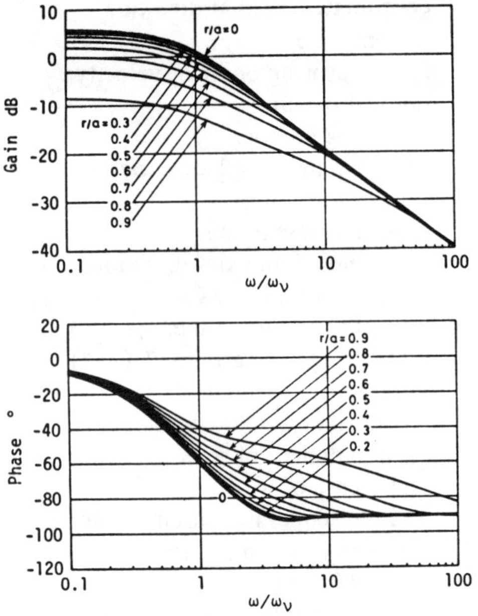

Fig. 1. Bode plot of $\bar{G}_p(\bar{r}, j\bar{\omega})$.

The boundary condition that $\bar{U} = 0$ at $\bar{r} = 1$ determines $f(\bar{x}, \bar{s})$ and yields

$$\bar{U}(\bar{r}, \bar{x}, \bar{s}) = \frac{J_0(2\sqrt{2}\, j\sqrt{\bar{s}}) - J_0(2\sqrt{2}\, j\bar{r}\sqrt{\bar{s}})}{\bar{s} J_0(2\sqrt{2}\, j\sqrt{\bar{s}})}\Gamma(\bar{x}, \bar{s}). \tag{10}$$

The system (10) is an infinite-dimensional one which has an infinite number of poles. The transfer function in (10) is

$$\bar{G}_p(\bar{r}, \bar{s}) = \frac{J_0(2\sqrt{2}\, j\sqrt{\bar{s}}) - J_0(2\sqrt{2}\, j\bar{r}\sqrt{\bar{s}})}{\bar{s} J_0(2\sqrt{2}\, j\sqrt{\bar{s}})}. \tag{11}$$

We show the Bode plot of $G_p(\bar{r}, j\bar{\omega})$ for some equispaced values of $\bar{r}$ in Fig. 1. $\bar{\omega}$ is a dimensionless angular frequency given by $\bar{\omega} = \omega/\omega_\nu$. This dimensionless number is called the Stokes' number.

The partial fraction expansion of (11) yields

$$\bar{G}_p(\bar{r}, \bar{s}) = \sum_{i=1}^{\infty} \frac{\sqrt{2}/\alpha_i}{\bar{s} + \alpha_i^2/8} g_i(\bar{r}) \tag{12}$$

where

$$g_i(\bar{r}) = \frac{\sqrt{2} J_0(\alpha_i \bar{r})}{J_1(\alpha_i)}. \tag{13}$$

The simplest way of approximation is to truncate (12). Let $\bar{G}_{pn}(\bar{r}, \bar{s})$ be the transfer function of the system obtained by the truncation of n terms. The degree of approximation may be assessed with the Bode plot of

$$\bar{G}_{pn}(\bar{r}, j\bar{\omega}) / \bar{G}_p(\bar{r}, j\bar{\omega}). \tag{14}$$

The plot for $n = 8$ is shown in Fig. 2. Generally speaking, the degree of approximation should be determined according to the aim of the filter to be designed. In any case, Fig. 2 shows that the truncation of eight terms gives a good approximation for some purposes.

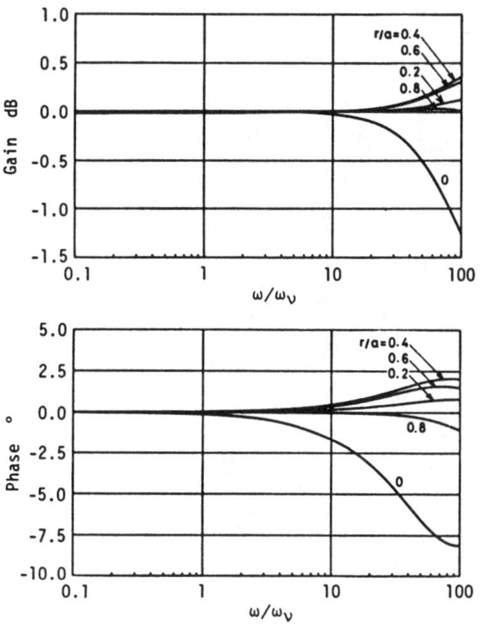

Fig. 2. $\bar{G}_{pn}(\bar{r}, j\bar{\omega})/\bar{G}_p(\bar{r}, j\bar{\omega})$ $(n=8)$.

Assigning the state variables $x_1, x_2, \cdots, x_n$ to each term of $\bar{G}_{pn}$ yields the following state space representation of the fluid line model:

$$\dot{x} = Ax + b\gamma \tag{15}$$

$$\bar{u} = g'x \tag{16}$$

where

$$\begin{aligned} x &= (x_1, x_2, \cdots, x_n)' \\ A &= \mathrm{diag}\left(-\alpha_1^2/8, -\alpha_2^2/8, \cdots, -\alpha_n^2/8\right) \\ b &= \left(\sqrt{2}/\alpha_1, \sqrt{2}/\alpha_2, \cdots, \sqrt{2}/\alpha_n\right)' \\ g &= (g_1, g_2, \cdots, g_n)'. \end{aligned} \tag{17}$$

B. Observation Model

The next stage in designing the filter is the modeling of observation processes. Every flowmeter or velocimeter detects a scalar quantity which conveys a reduced information about velocity profile. For the sake of generality, we integrate the processes to a multiinstrument process modeled as

$$y = Cx + w \tag{18}$$

where

$$\begin{aligned} C &= \int_0^1 h(\bar{r}) g'(\bar{r})\, d\bar{r} \\ y &= (y_1, y_2, \cdots, y_m)' \\ w &= (w_1, w_2, \cdots, w_m)' \end{aligned} \tag{19}$$

where $y_1, y_2, \cdots, y_m$ are measurements taken by the instrument and $w_1, w_2, \cdots, w_m$ are the corresponding measurement noises.

$$h(\bar{r}) = (h_1(\bar{r}), h_2(\bar{r}), \cdots, h_m(\bar{r}))' \tag{20}$$

is a weighting vector function determined by the method of observation. Each component of the vector is determined according to the corresponding instrument.

For example, the weighting function becomes

$$h_i(\bar{r}) = \delta(\bar{r} - \bar{r}_0) \tag{21}$$

in the case of measuring the velocity at the point of $\bar{r} = \bar{r}_0$ by a velocimeter for the measurement of a single point as an LDV. In an ultrasonic flowmeter, the average of velocity profile on a line being measured, the function becomes

$$h_i(\bar{v}) = 1. \tag{22}$$

C. Derivation of the Filter [6]

It is straightforward to derive the filter after the fluid line model and the observation model are set up. We only have to construct a Kalman filter from (15) and (18). The thus constructed filter, however, cannot be implemented with digital systems because it is a continuous-data system. Here, aiming at microcomputer implementation of the filter, we derive the filter after making (15) and (18) discrete in time.

Letting $\bar{\tau}$ be the sampling period and approximating the input of (15) γ by a function like stairs, (15) and (18) become [7]

$$x_{k+1} = \Phi x_k + \psi \gamma_k \tag{23}$$

$$y_k = Cx_k + w_k \tag{24}$$

where

$$\begin{aligned} \Phi &= \exp(A\bar{\tau}) \\ \psi &= (\Phi - I)A^{-1}b. \end{aligned} \tag{25}$$

The suffix k denotes the kth sampling time. The velocity profile is given by

$$\bar{u}_k = g'x_k. \tag{26}$$

The input to the system γ_k may be divided into the known part γ_k^* and the unknown part ξ_k as

$$\gamma_k = \gamma_k^* + \xi_k. \tag{27}$$

For instance, in the case of the measurement of pressure gradient being taken, the measured value is γ_k^* while the measurement noise is $-\xi_k$.

We assume that ξ_k and w_k are white and Gaussian noises with zero means and covariance matrices

$$E\left[\xi_i \xi_j\right] = \Xi^c \delta_{ij}, \quad E\left[w_i w_j'\right] = W^c \delta_{ij}. \tag{28}$$

The optimal filter to estimate velocity profile is obtained as follows:

$$\hat{x}_{k+1} = \Phi \hat{x}_k + \psi \gamma_k^* + K_\infty \{ y_{k+1} - C(\Phi \hat{x}_k + \psi \gamma_k^*) \} \tag{29}$$

$$\hat{\bar{u}}_k = g'\hat{x}_k. \tag{30}$$

The steady-state value of the Kalman gain K_∞ is applied in

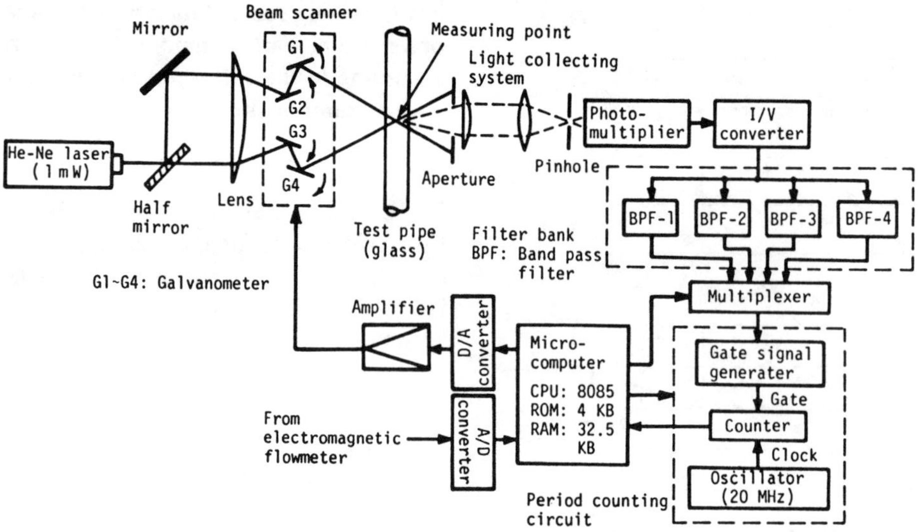

Fig. 3. Construction of experimental equipment.

(29) to make the computation easy. K_∞ is calculated as follows [8]:

$$\begin{aligned} K_\infty &= \lim_{k\to\infty} K_k \\ K_{k+1} &= M_{k+1}C'(CM_{k+1}C'+W^c)^{-1} \\ M_{k+1} &= \Phi Z_k^c\Phi' + \psi\Xi^c\psi' \\ Z_k^c &= (I-K_kC)M_k \\ M_0 &= E[x_0x_0']. \end{aligned} \tag{31}$$

III. Implementation of the Theory

A. Construction of Equipment

The theory was implemented and tested on prototypal equipment consisting of an LDV and an 8-bit microcomputer. Their construction is shown diagrammatically in Fig. 3. The equipment was designed to measure the instantaneous flow rate through estimation of the velocity profile in a pipe. The microcomputer estimates the profile based on the measurements of the velocity at a point in the pipe section taken by the LDV.

The optical system of the LDV is a differential type which was easily handled. The laser is a 1 mW He–Ne laser. The measuring point can be set at any point on a pipe diameter by means of a beam scanner which was originally designed for direct measurement of velocity profiles [3]. In our experiments the point is set at the pipe center because the observation noise is smallest there. The test pipe is a horizontal straight glass pipe with the inside diameter of 10.2 mm. The fluid in the pipe is domestic tap water at a room temperature of 20°C.

Beat signals from the LDV pass a filter bank to enter a period counting circuit. The velocity is inversely proportional to the counted period. The division to obtain the velocity is carried out by the microcomputer. The filter bank consists of four bandpass filters. The most suitable filter for the frequency of signal is selected by the microcomputer.

B. Stochastic Analysis of Measured Data

The optimality of the filter is guaranteed only when both the driving noise ξ_k and the observation noise w_k, which is a scalar process in this case, are white and Gaussian. We have to verify this point before implementation of the filter.

The property of the driving noise cannot be determined unless the circumstances of the flow are given. Any particular circumstances, however, are not specified for our equipment. Then, not expanding this further, we assume *a priori* that the driving noise is white and Gaussian.

The property of measurement noise can be verified through experiments. We examined the property of measurement noise by stochastic analysis of data obtained for the steady laminar flow at various flow rates. The signal-to-noise ratio (SNR) was also examined based on the data obtained. The SNR determines the Kalman gain as described in the following section.

A typical example of observed signals is shown in Fig. 4. We may regard the average as the true value and the deviation from the average as the noise because the data are those for steady flow. The autocorrelation function and probability density function of the noise are also shown in Fig. 4, where σ^2 is the variance of the noise. We can regard the noise as almost white and Gaussian.

We define the SNR to be the ratio of squared average to variance of noise. Experimental values of the SNR against various flow rates are shown in Fig. 5. The SNR is constant while the flow rate changes. The average of the SNR is 2700.

C. Filter Equations

The measuring point of the LDV is set on the pipe axis. Then the weighting function for this case becomes

$$h(\bar{r}) = \delta(\bar{r}). \tag{32}$$

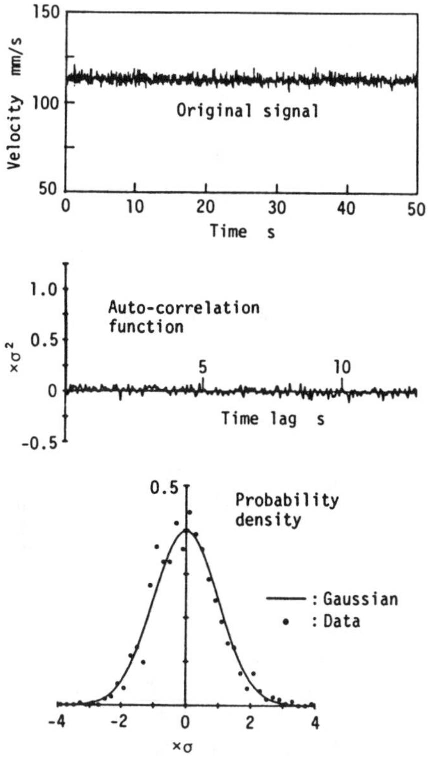

Fig. 4. Properties of measurement noise.

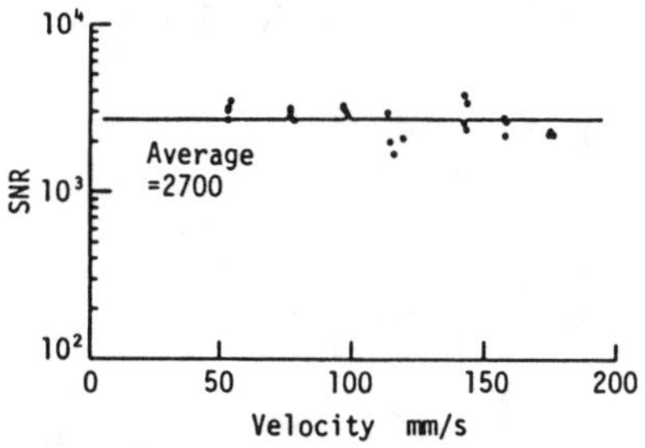

Fig. 5. SNR versus velocity.

Consequently, from (19), (24), and (32), the observation model is obtained as

$$y_k = \boldsymbol{c}'\boldsymbol{x}_k + w_k \tag{33}$$

$$\boldsymbol{c} = \boldsymbol{g}(0) \tag{34}$$

where w_k is white and Gaussian as shown previously in Section III-B.

We let $\gamma_k^* = 0$ because pressure gradient is not measured here and assume that ξ_k is white and Gaussian as mentioned also in Section III-B.

As is clear from (31), the Kalman gain is determined by the ratio of the variances Ξ^c and W^c. The ratio is given by the SNR as follows.

Let

$$X^c = E[\boldsymbol{x}_k\boldsymbol{x}_k'] \tag{35}$$

$$Y^c = E[y_k^2]; \tag{36}$$

we obtain

$$X^c = \Phi X^c \Phi' + \psi \Xi \psi' \tag{37}$$

$$Y^c = \boldsymbol{c}' X^c \boldsymbol{c}. \tag{38}$$

Therefore Y^c is determined when Ξ^c is given. Dividing Y^c by the SNR gives W^c to yield the ratio of the variances.

Thus, the filter equation for our equipment is obtained as

$$\hat{\boldsymbol{x}}_{k+1} = \Phi\hat{\boldsymbol{x}}_k + \boldsymbol{k}_\infty(y_{k+1} - \boldsymbol{c}'\Phi\hat{\boldsymbol{x}}_k) \tag{39}$$

where the Kalman gain is a vector $\boldsymbol{k}_\infty$. The estimated velocity profile is

$$\hat{u}_k(\bar{r}) = \boldsymbol{g}'(\bar{r})\hat{\boldsymbol{x}}_k. \tag{40}$$

The integration of $\hat{u}_k$ leads to the estimated flow rate $\hat{q}_k$:

$$\hat{q}_k = a^2 \int_0^1 2\pi\bar{r}\hat{u}_k(\bar{r})\, d\bar{r} = \boldsymbol{\lambda}'\hat{\boldsymbol{x}}_k \tag{41}$$

where

$$\boldsymbol{\lambda} = \left(\frac{2\sqrt{2}\pi a^2}{\alpha_1}, \frac{2\sqrt{2}\pi a^2}{\alpha_2}, \cdots, \frac{2\sqrt{2}\pi a^2}{\alpha_n}\right)'. \tag{42}$$

It is noted that, in (40) and (41), the velocity and the flow rate are not dimensionless.

D. Assessment

The steady-state accuracy in measurement of flow rate is one of the prime concerns that arises in our application of the Kalman filter. The accuracy can be evaluated by using the final-value theorem of the z-transform. The z-transform of (39) is

$$z\hat{X} = \Phi\hat{X} + \boldsymbol{k}_\infty(zY - \boldsymbol{c}'\Phi\hat{X}). \tag{43}$$

Hence

$$\hat{X} = z(zI - \Phi + \boldsymbol{k}_\infty\boldsymbol{c}'\Phi)^{-1}\boldsymbol{k}_\infty Y. \tag{44}$$

Substituting (44) into the z-transform of (41) gives

$$\hat{Q} = z\boldsymbol{\lambda}'(zI - \Phi + \boldsymbol{k}_\infty\boldsymbol{c}'\Phi)^{-1}\boldsymbol{k}_\infty Y. \tag{45}$$

Provided that y_k is the true velocity of steady laminar flow, the corresponding flow rate q_k is given by

$$q_k = \frac{\pi a^2}{2} y_k \tag{46}$$

because the velocity profile is parabola. Therefore, if the noiseless measurement is available, the steady-state estimation error ϵ_∞ for the unit step input of q_k becomes

$$\begin{aligned}\epsilon_\infty &= \lim_{z\to 1}(z-1)\{z\boldsymbol{\lambda}'(zI - \Phi + \boldsymbol{k}_\infty\boldsymbol{c}'\Phi)^{-1}\boldsymbol{k}_\infty Y - Q\} \\ &= \lim_{z\to 1}\left\{\frac{2z^2}{\pi a^2}\boldsymbol{\lambda}'(zI - \Phi + \boldsymbol{k}_\infty\boldsymbol{c}'\Phi)^{-1}\boldsymbol{k}_\infty - z\right\} \\ &= \frac{2}{\pi a^2}\boldsymbol{\lambda}'(I - \Phi + \boldsymbol{k}_\infty\boldsymbol{c}'\Phi)^{-1}\boldsymbol{k}_\infty - 1.\end{aligned} \tag{47}$$

For our equipment, let $n = 8$ and $\bar{\tau} = 0.01544$; we obtain $\epsilon_\infty = 0.000678$. The cause for the sign of this value to be plus is the model error. The value of $\bar{\tau}$ corresponds to the sampling period of 50 ms.

E. Programming

Programming of (39) and (41) on the microcomputer and their real-time operation brings about an instantaneous flowmeter using a laser Doppler technique. As was noticed in Section III-D, the order of the filter was selected to be eighth. The sampling period is 50 ms. In order to make the computation of the filter equations fast, the coefficients and the variables in the equations were represented by 8- and 16-bit fixed point numbers on the microcomputer, respectively. The program was written in assembly language. Owing to the above elaboration, the execution time of the filter program is about 13 ms which is less than the sampling period. This means that the filter executes in real time.

IV. Experimental Results

A. Steady Flow Measurement and Calibration

A step response of the filter with the initial condition $\boldsymbol{x}_0 = \boldsymbol{0}$ is shown in Fig. 6. The measured flow is a steady flow. The flow rate was measured simultaneously with a measuring glass and a stopwatch. Estimated velocity profiles are shown in Fig. 7. The profile is estimated well after settling of the filter. Fig. 8 shows a result of calibration. The estimated flow rate is the average of estimated values over 5 s after settling. The measurement was taken 40 cm downstream from the inlet of pipe. The flow rate is estimated precisely as long as the measuring point is outside the inlet region, i.e., the flow rate is under about 5 cm^3/s or $Re \leqslant 630$. Another static operation of the filter has been observed by using the input signal from an oscillator instead of the LDV and has shown that the estimation error of the filter is within 0.5 percent when there is no measurement noise. This value is the same order as the quantization error of 8-bit numbers.

B. Transient Flow Measurement

The fluid system to produce changing flow for the dynamic test of the filter is shown in Fig. 9. The flow rate is controlled by a spool valve which is driven by a stepping motor. A microcomputer system controls the stepping motor to produce a variety of changing flows. For comparison, the flow rate is measured simultaneously by an electromagnetic flowmeter which is excited by 50 Hz ac current and can measure changing flow rate of up to several Hz accurately. The positions where the LDV and the electromagnetic flowmeter are installed are separated as shown in Fig. 9. However, we can regard the flow rates at the two points as identical because the fluid in the pipe is incompressible and the two points are connected by a rigid glass pipe.

Results in dynamic operation are shown in Figs. 10 and 11. The initial condition in both cases is $\boldsymbol{x}_0 = \boldsymbol{0}$. The

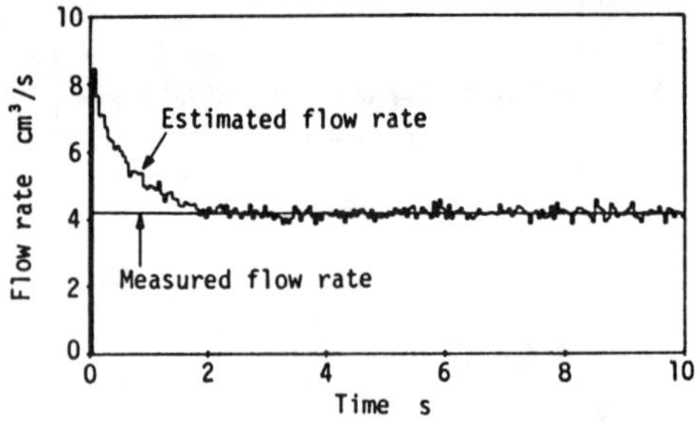

Fig. 6. A step response of the filter.

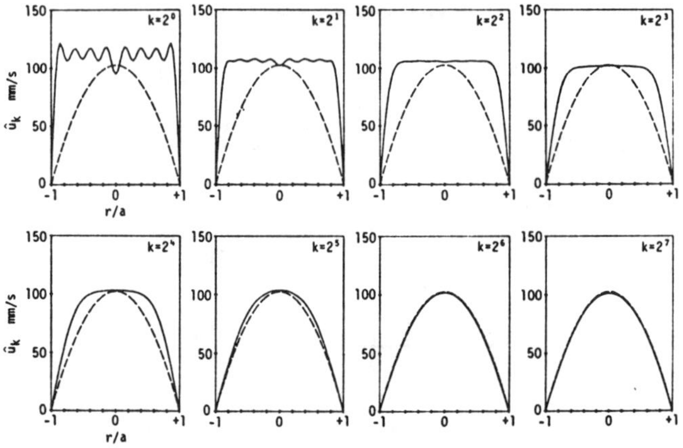

Fig. 7. Estimated velocity profiles in a step response of the filter (- - - -: parabola, ———: estimated profile).

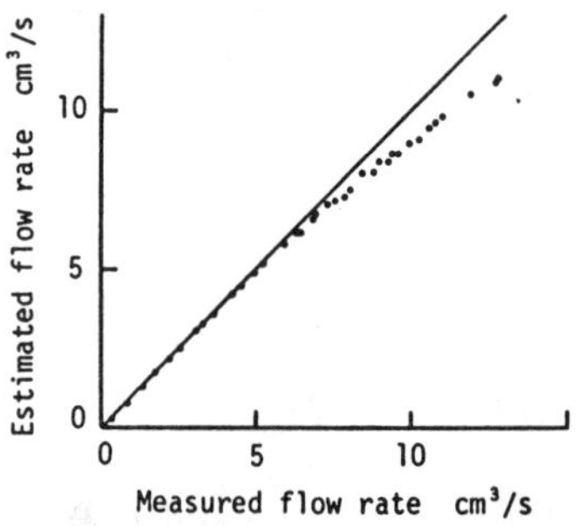

Fig. 8. A result of calibration.

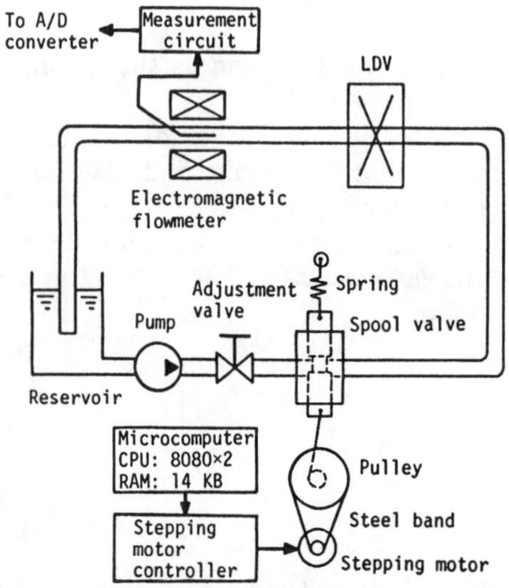

Fig. 9. Fluid system to produce changing flow.

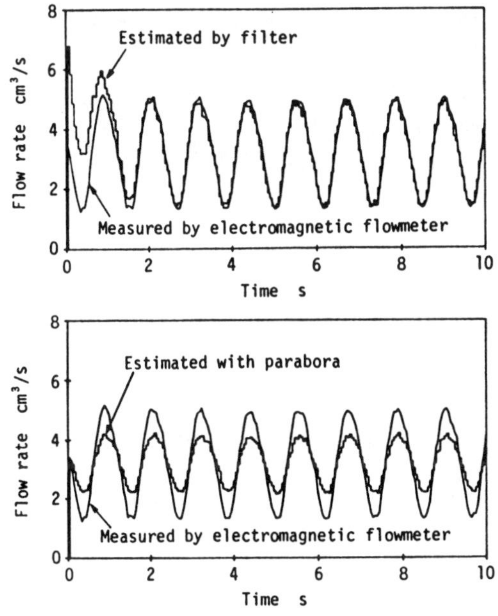

Fig. 10. Experimental results for pulsating flow.

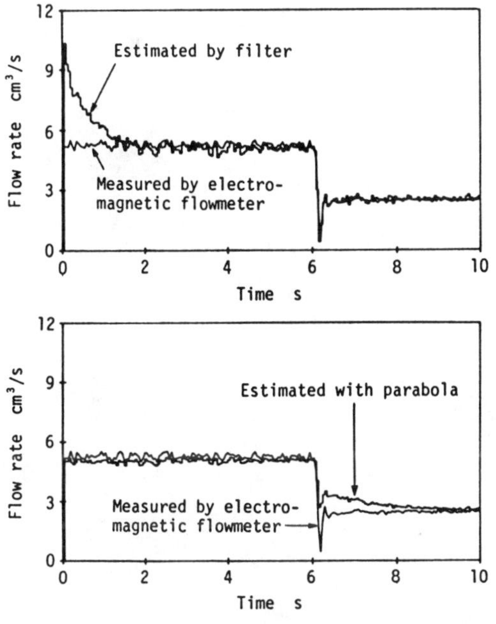

Fig. 11. Experimental results for stepwise changing flow.

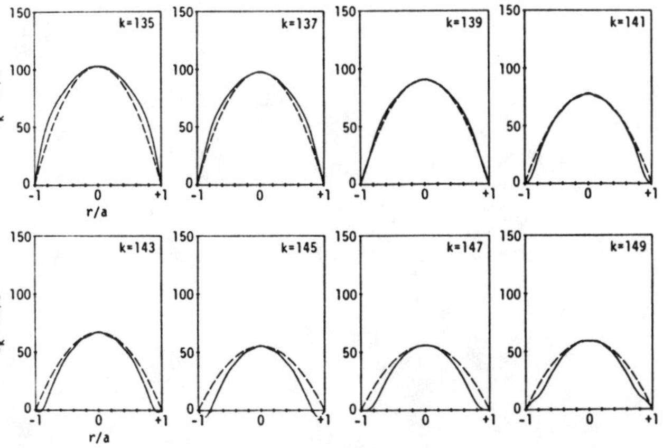

Fig. 12. Estimated velocity profiles of pulsating flow (- - - -: parabola, ——: estimated profile).

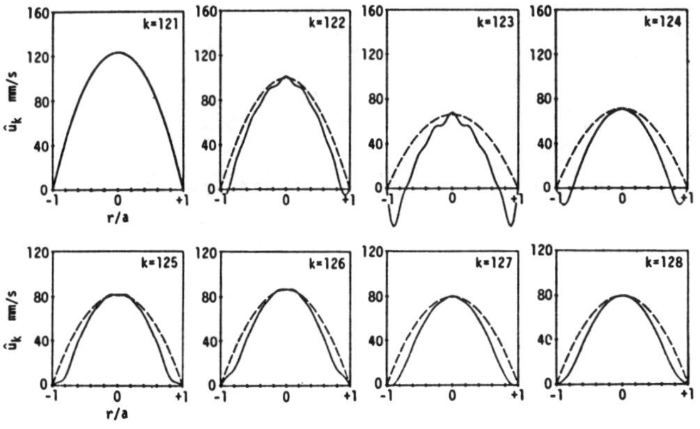

Fig. 13. Estimated velocity profiles of stepwise changing flow (- - - -: parabola, ——: estimated profile).

changing flow rate is estimated precisely after settling of the filter. The flow in Fig. 11 was generated by shutting the valve suddenly. The transient oscillation which is interesting may have been induced owing to the flexibility of the vinyl pipe which joins the test line, reservoir, pump, and valve unit. Flow rates estimated, based on the assumption that the velocity profile is parabolic are also shown in the figures. The estimation based on a parabolic profile leads to inaccurate measurement when the flow changes. This is explained clearly by Figs. 12 and 13 which show velocity profiles estimated by the filter in the respective experiments shown in Figs. 10 and 11. The estimated velocity profiles are significantly different from the parabolas shown.

V. Concluding Remarks

A new method of measuring the instantaneous flow rate through estimation of velocity profiles has been presented. A Kalman filter algorithm for the profile estimation has been derived. In this method, the dynamic characteristics of flow being taken into consideration, the instantaneous flow rate of changing flow, can be measured accurately.

A prototype flowmeter consisting of an LDV and an 8-bit microcomputer which implements the profile estimation algorithm has been developed. Experimental results have proven that this approach to flow measurement is effective.

The proposed method is applicable to a wide range of flow measurement if flow dynamics and observation process can be modeled. Future studies will be directed toward the modeling technique to yield improvement of the dynamic characteristics of various flowmeters and the development of a new type of flowmeter.

Acknowledgment

The authors wish to thank R. Bicker and the reviewers for their useful suggestions which helped to improve this paper.

References

[1] A. F. D'Souza and R. Oldenburger, "Dynamic response of fluid lines," *J. Basic Eng.*, vol. 86, no. 3, pp. 589–598, 1964.

[2] K. Nakano and J. Tanino, "Research on unsteady flow in a tube (the 1st report)" (in Japanese), *J. Japan. Hydrau. Pneu. Soc.*, vol. 7, no. 4, pp. 216–222, 1976.
[3] M. Uchiyama and K. Hakomori, "A beam scanning LDV to measure velocity profile of unsteady flow," *Bull. JSPE*, vol. 16, no. 2, pp. 71–77, 1982.
[4] Y. Sakawa, "Optimal filtering in linear distributed-parameter systems," *Int. J. Contr.*, vol. 16, no. 1, pp. 115–127, 1972.
[5] N. B. Nichols, "The linear properties of pneumatic transmission lines," *ISA Trans.*, vol. 1, no. 1, pp. 5–14, 1962.
[6] R. E. Kalman, "A new approach to linear filtering and prediction problems," *J. Basic Eng.*, vol. 82, no. 1, pp. 35–45, 1960.
[7] D. H. Owens, *Multivariable and Optimal Systems*. New York: Academic, 1981, pp. 52–55.
[8] Y. Takahashi, *Control System Design Manual* (in Japanese), Tokyo: Kyoristu, 1970, pp. 111–113.

Estimation and Prediction of Unmeasurable Variables in the Steel Mill Soaking Pit Control System

VLADIMIR J. LUMELSKY, MEMBER, IEEE

***Abstract*—The objective of the soaking pit operation in the steel-making process is to equalize the temperature throughout the steel ingot masses at some prespecified temperature before the ingots may be rolled at the rolling mill. The ingot temperature, though, is not measurable directly and must be estimated using some indirect measurements (pit wall temperature, fuel flow, etc.). Another piece of information which is important for the whole operation, and specifically for the operation planning purposes, is the predicted moment of time at which the ingots will arrive at the said prespecified temperature. This paper describes a Kalman filter approach and some results of a joint project of General Electric Company–U.S. Steel Company on the estimation and prediction of the ingot temperature in the soaking pit operation.**

Manuscript received March 29, 1982; revised August 20, 1982.
The author is with the Corporate Research and Development Center, General Electric Company, Schenectady, NY 12345.

I. Introduction

In one of the versions of the steel-making process, the steel ingots, before they may be rolled at the rolling mill, typically pass through the soaking pit operation (another version—continuous steel-making process, which does not utilize the soaking pits—is not considered here). The purpose of this operation is to equalize the tempera-

Reprinted from *IEEE Trans. Automat. Contr.*, vol. AC-28, pp. 388–400, Mar. 1983.

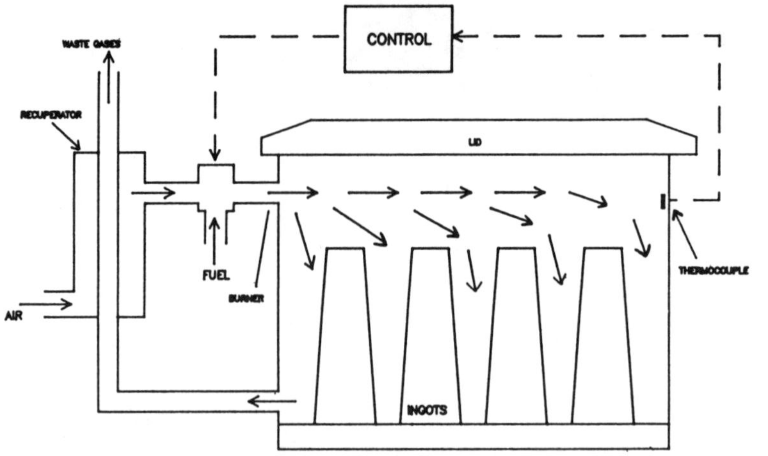

Fig. 1. Soaking pit: schematic cross section.

ture throughout the ingot masses as some prespecified level. Accurate description of the ingot temperature distribution would require the knowledge of analytical space–temperature relationships or of the temperature values at many points along the ingot side, top, and bottom surfaces, as well as throughout its mass. For practical purposes, though, it may be enough to know some average estimates of the ingot surface and center (assuming that both of these terms are defined) temperatures.

Typically, the ingots (5–20 tons each, 5–15 ingots/pit) come from the stripping yard where they are stripped of their molds. At this time, the ingot surface and center temperatures are usually within the ranges of 1300–1900°F and 2000–2800°F, respectively. Sometimes, though, the ingots may come from cold storage, in which case their temperature will be that of the ambient. To do the job, the control system (today it is done by the operator) has to continuously estimate the current ingot temperatures, at least on the surface and at the center, and to stop the operation when these temperatures arrive at the prespecified levels (usually the level is the same for both temperatures). Underheating of ingots results in poor rolling mill performance or in returning the ingots to the pit for additional heat up; overheating (which is often the case) results in a waste of energy. According to the industry estimates, up to 15–20 percent of the fuel savings could be realized in this energy-intensive operation if an efficient ingot temperature estimation system is put in operation.

Unfortunately, direct measurement of the ingot temperature is not feasible because of the harsh conditions inside the pit. Fig. 1 depicts a typical soaking pit. As one can see, the distance between the walls and the ingots prevents contact measurements of the ingot temperatures. Noncontact temperature measurements (using, for example, optical pyrometers) inside the pit do not produce reliable results, according to industry sources. The only direct measurements available are the pit wall temperature, the fuel and air flow, the waste gas temperature, and several other indirect variables. Attempts were undertaken to measure directly the ingot surface temperature in the experimental pits [1] or in small European pits [8] using some contact and noncontact methods. The assumption of measurable ingot surface temperature is, first, questioned by the United States steel industry, and second, still does not solve the problem of measurement of the ingot center temperature.

In theory, knowing some initial conditions (ingot size, time period between the stripping off the mold and starting the soaking, etc.) and assuming some fuel manipulation strategy, one can reconstruct the temperature curves from the beginning until the moment when the temperatures arrive at a prespecified value (this moment is termed in industry *ready-to-roll time*) and, thus, predict the ready-to-roll time. A number of heat transfer models [1]–[6] have been developed utilizing this concept. Unfortunately, these kinds of models cannot take into account any unexpected changes in the pit that may occur within any pit cycle or from cycle to cycle: changes in actual fuel/air flow, unplanned openings/closings of the pit lid, pit aging, accumulation of dust on the pit floor, etc. (We will call a pit cycle the time between charging the ingots into the pit and drawing ready-to-roll ingots. Normally, a cycle takes 5–10 h.) Those changes are evidently reflected in the indirect measurements which could be used for on-line estimation of ingot temperatures. Fig. 2 gives an example of actual temperature and fuel curves for a typical pit cycle. (Pit wall temperature and fuel flow in Fig. 2 were measured directly; ingot surface and center temperatures were estimated using a heat transfer model developed at U.S. Steel Company [6]; under stationary conditions and with accurate values of initial conditions, the model is claimed to have an accuracy of 20–30°F.)

One problem with reconstruction of the temperature curves is that the whole pit cycle is naturally divided into two distinct phases (see Fig. 2): 1) maximum-flow phase where the fuel is set on the maximum level; and 2) in-control phase which starts at the moment when the pit wall temperature reaches some prespecified value; at this moment, a fuel controller (typically this is a proportional-plus-integral (PI) regulator) starts manipulating the fuel flow to maintain constant wall temperature. The presence of two phases complicates the development of a unified pit/ingot model.

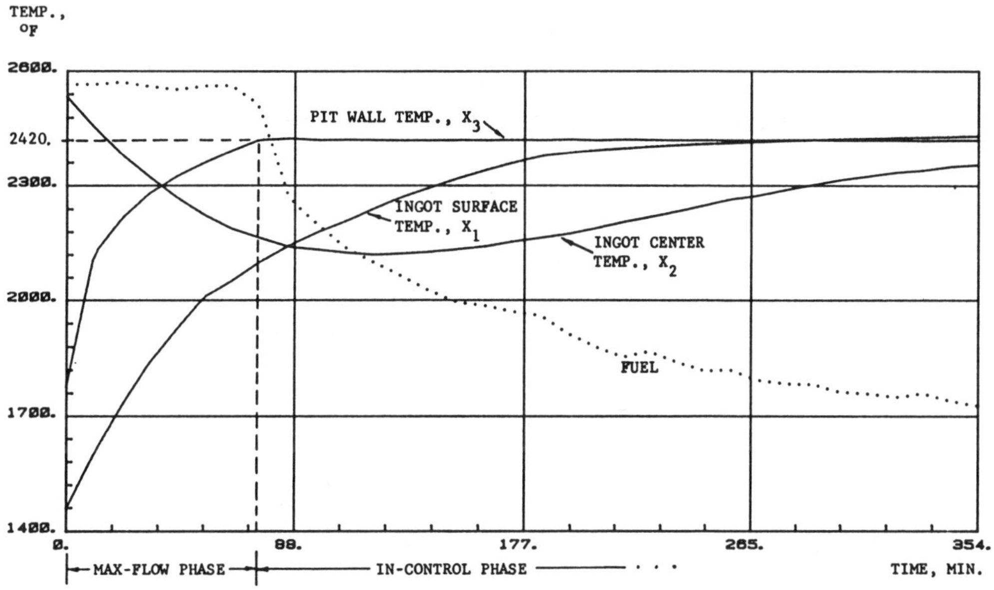

Fig. 2. Example of actual pit data (scale for fuel is not shown).

Another difficulty with direct usage of the pit model for reconstruction of the temperature curves is that it is hard to estimate the initial ingot surface and center temperature with good accuracy. It is not rare for the best estimates of the initial conditions (temperatures) to be off the actual values by 200–400°F.

Therefore, some estimation procedure which would gradually improve the current state estimates by making use of the available on-line measurements would be in place here. This approach was used in [7] and [8] based on use of the Kalman filter procedure; in [7], only the max-flow phase was considered; in [8], both phases are included, but the ingot surface temperature is assumed to be measurable.

Another thing that is desired from the pit control system in addition to the current state estimation is an advance state prediction (in this case, ready-to-roll time prediction). Although the current estimation of the ingot temperature is helpful in determining the ready-to-roll times and thus may result in fuel savings, the real advantage would be realized by a system which could also predict ready-to-roll times far in advance; good predictions would result in smooth scheduling, immediate fuel savings, and increased throughput.

The prediction problem may be approached, again, from two directions.

1) By using the heat transfer (or similar) models and initial conditions as in [1]–[6]; this approach does not allow one to account for intermediate changes in the environment.

2) By using the current state estimates and running dynamic prediction; this way, the on-line indirect measurements allow one to gradually compensate for all kinds of system uncertainties and, thus, continuously improve the state predictions.

The objectives of this work are as follows.

1) Design an estimation procedure based on the Kalman estimation algorithm; assume that both ingot temperatures —surface and center—are not measurable; use a simple lumped pit and ingot linear model (as in [7]); unlike [7], try to describe both max-flow and in-control phases with one model.

2) Design a prediction procedure which would use the estimator output as running "initial conditions" and run the advance state prediction; incorporate in the predictor the model of the controller to simulate the future fuel strategy.

3) Validate the estimator–predictor using actual pit data taken from full size industrial pits under regular production conditions.

Actually, the main role of the Kalman filter here is to compensate for uncertainty in initial conditions. If the Kalman filter at each moment supplies the predictor with reasonable current state estimates, then the predictor using these estimates as "initial conditions" may (assuming a reasonable pit model) run the model up to the moment when the predicted temperatures meet the ready-to-roll requirements; this time moment is used as a ready-to-roll time prediction. Thus, while in the estimation procedure the main emphasis is on the performance of the Kalman filter, in the prediction procedure, the critical part is the model. Since the predictor has to take into account further values of the control variables (fuel, etc.), it is natural to supply the predictor with some model of the controller.

The very nature of the advance prediction problem makes the accuracy and sensitivity objectives of the system "biased" toward the end of the pit cycle. At the beginning of the cycle, we can tolerate larger prediction errors than at its end. On the other hand, because of the physical nature of the object and because of the fact that the control system moves all the temperatures—wall, ingot surface,

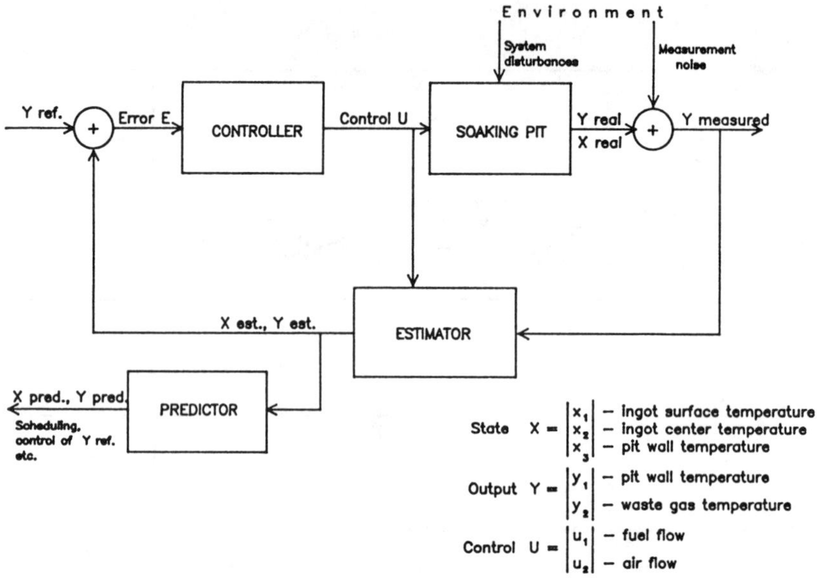

Fig. 3. Control system structure.

ingot center—to the same value, the big differences between some temperature and its estimate in the first half of the cycle will typically correspond to smaller differences between those values at the end of the cycle.

We take advantage of these facts in two ways. First, one model is generated for both phases (maximum flow and in-control; see above), with emphasis on the accuracy of the model in the area of the in-control phase instead of overall model accuracy.

Second, the speed of convergence of the variable Kalman gain is not a consideration as long as the process converges by the second half of the in-control phase. As a matter of fact, the steady-state gain matrix gives satisfactory results, too (see the comparison in Section IV).

The description of the whole system is given in Section II, the pit model considerations in Section III, the estimator structure in Section IV, the predictor structure in Section V, and results of experiments with the system using actual pit data in Section VI.

II. Control System: Simulation System

The structure of the whole system is shown in Fig. 3. The control loop includes the estimator which produces the estimated values of states and outputs for the next moment based on the measured values of the outputs at the current moment and the controls at the previous moment. Besides, the outputs of the estimator are used in the predictor which produces advance predictions of the states/outputs and, as their function, the ready-to-roll time. Apparently, to be able to make valid predictions, the predictor has to simulate the control strategy, and, thus, it has to include some model of the controller as well as the boundary conditions —for example, a limitation on the fuel flow.

For the estimation/prediction purposes, the scheme does not have to be a closed-loop system as shown in Fig. 3. In today's soaking pit control systems, the controller manipulates the control variables (fuel, air) to maintain a constant pit wall temperature. For easier integration of the estimation/prediction scheme in the existing systems, this may be left as is, in which case the estimator/predictor will be used not for direct control, but only for supplying the operator with on-line ingot temperature estimates and with an advance prediction of the ready-to-roll time.

To simulate and study the system shown in Fig. 3, a special software package has been developed. The package allows one

1) to simulate all the elements of Fig. 3
2) to simulate the difference between the "actual" initial conditions and initial conditions "known" to the estimator
3) to simulate the difference between "actual" pit and its model used in the estimator/predictor
4) to apply variable or steady-state Kalman gain matrix or to exclude the Kalman filter from the system
5) to test a number of criteria for the ready-to-roll time in the predictor
6) to substitute different controller models
7) to simulate measurement noise
8) to do various kinds of data/curve graphics presentations, etc.

III. Soaking Pit Model

A simple lumped linear model is assumed. In matrix form, this is

$$X_{k+1} = A \cdot X_k + B \cdot U_k$$
$$Y_{k+1} = C \cdot X_{k+1} \tag{1}$$

where X_k, X_{k+1} = plant state vectors at the moments k and $k+1$, respectively; U_k = control variables vector at the moment k; Y_{k+1} = output vector at the moment $k+1$; and k = time index, $k = 0, 1, 2, \cdots$.

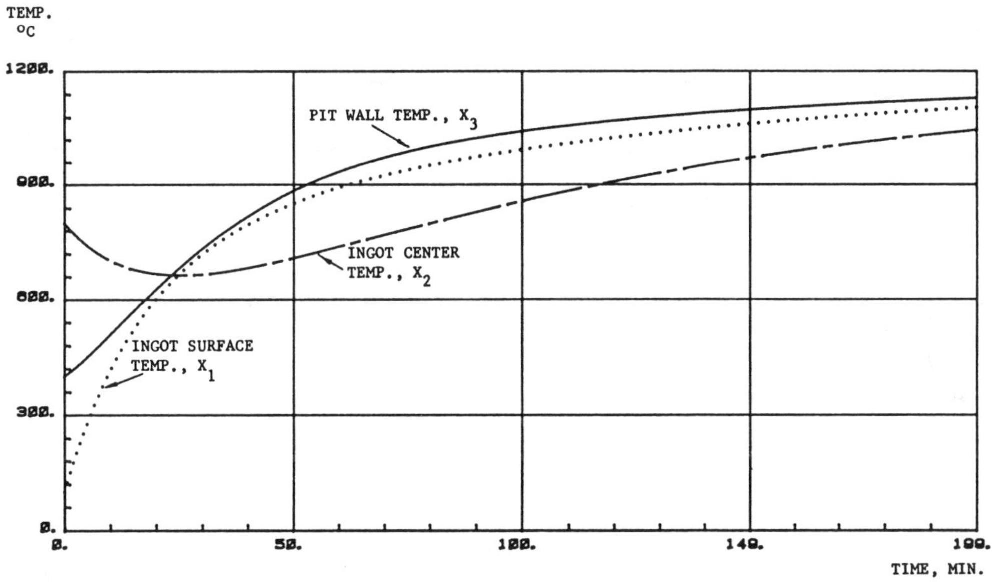

Fig. 4. Temperature curves simulated by linear model (1).

In the component form, the model is

$$\begin{vmatrix} x_1 \\ x_2 \\ x_3 \end{vmatrix}_{k+1} = \begin{vmatrix} a_{11} & a_{12} & a_{13} \\ a_{21} & a_{22} & a_{23} \\ a_{31} & a_{32} & a_{33} \end{vmatrix} \cdot \begin{vmatrix} x_1 \\ x_2 \\ x_3 \end{vmatrix}_k + \begin{vmatrix} b_{11} & b_{12} \\ b_{21} & b_{22} \\ b_{31} & b_{32} \end{vmatrix} \cdot \begin{vmatrix} u_1 \\ u_2 \end{vmatrix}_k$$

$$\begin{vmatrix} y_1 \\ y_2 \end{vmatrix}_{k+1} = \begin{vmatrix} c_{11} & c_{12} & c_{13} \\ c_{21} & c_{22} & c_{23} \end{vmatrix} \cdot \begin{vmatrix} x_1 \\ x_2 \\ x_3 \end{vmatrix}_{k+1}. \tag{2}$$

Here

$$A = \begin{vmatrix} a_{11} & a_{12} & a_{13} \\ a_{21} & a_{22} & a_{23} \\ a_{31} & a_{32} & a_{33} \end{vmatrix}, \quad B = \begin{vmatrix} b_{11} & b_{12} \\ b_{21} & b_{22} \\ b_{31} & b_{32} \end{vmatrix},$$

$$C = \begin{vmatrix} c_{11} & c_{12} & c_{13} \\ c_{21} & c_{22} & c_{23} \end{vmatrix}$$

model coefficient matrices (in this work considered constant),

$X = (x_1, x_2, x_3)$—plant states:
- x_1—ingot surface temperature
- x_2—ingot center temperature
- x_3—pit wall temperature;

$Y = (y_1, y_2)$—plant outputs:
- y_1—pit wall temperature (same as x_3; taken for convenience; see [7])
- y_2—waste gas temperature;

$U = (u_1, u_2)$—control variables:
- u_1—fuel flow
- u_2—air flow

k—time index, $k = 0, 1, 2, \cdots$

A regular least-square procedure was used to identify the model matrices A, B, and C. The data used included actual measurements from two U.S. Steel Company soaking pits (variables x_3, y_2, u_1, u_2) and estimates of the ingot surface and center temperatures received with the U.S. Steel Company heat transfer model mentioned above. A number of sets of model coefficients were identified in an attempt to study the effect of different soaking pits, different ingot sizes, and different total ingot weights.[1]

An example of a set of curves generated by the model for the maximum-flow phase only is shown in Fig. 4.

Fig. 5 demonstrates the level of agreement between the model developed and the data of Fig. 2. Actual initial temperatures were incorporated into the model; both the maximum-flow phase and the in-control phase are present. One can see that, in this example, the accuracy of the model is rather tolerable.

Experiments with the model and the actual data showed that satisfactory estimation/prediction results could be realized with an even more simplified model:

$$X = \begin{vmatrix} x_1 \\ x_2 \\ x_3 \end{vmatrix} \quad Y = y_1 \quad U = u_1$$

and, thus, with simpler coefficient matrices

$$A = \begin{vmatrix} a_{11} & a_{12} & a_{13} \\ a_{21} & a_{22} & a_{23} \\ a_{31} & a_{32} & a_{33} \end{vmatrix}, \quad B = \begin{vmatrix} b_1 \\ b_2 \\ b_3 \end{vmatrix},$$

$$C = \begin{vmatrix} c_1 & c_2 & c_3 \end{vmatrix}.$$

From physical considerations, $a_{23} = a_{32} = b_2 = c_1 = c_2 = 0$, $c_3 = 1$. Thus, only nine unknown parameters are to be estimated by the least-square method. For formal analysis, a_{23}, a_{32}, and b_2 were also included in the list of unknowns.

These variables are used in the material that follows.

[1] The work on the pit/ingot model identification was done by Dr. H. P. Ko.

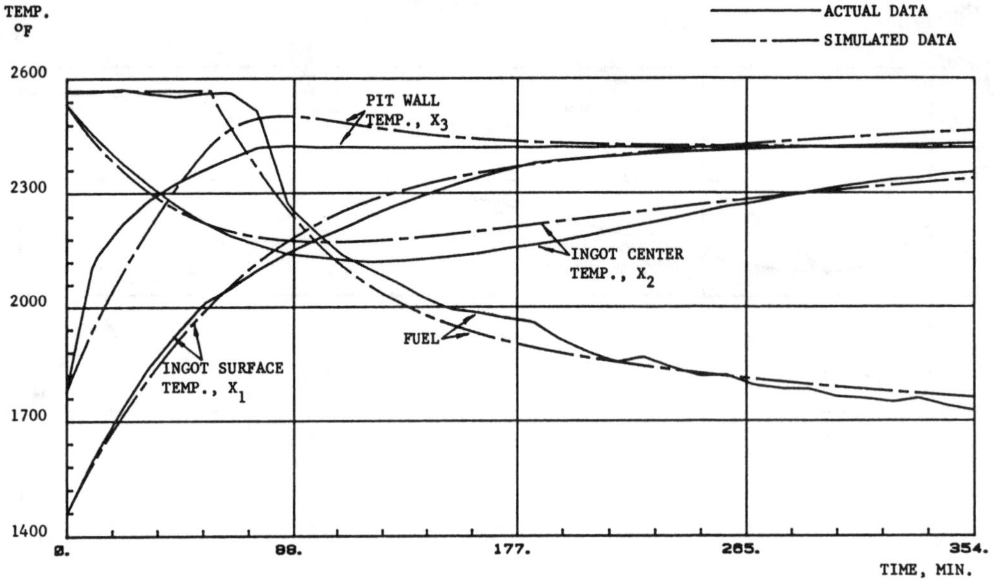

Fig. 5. Actual data versus data simulated by linear model (1) (scale for fuel is not shown).

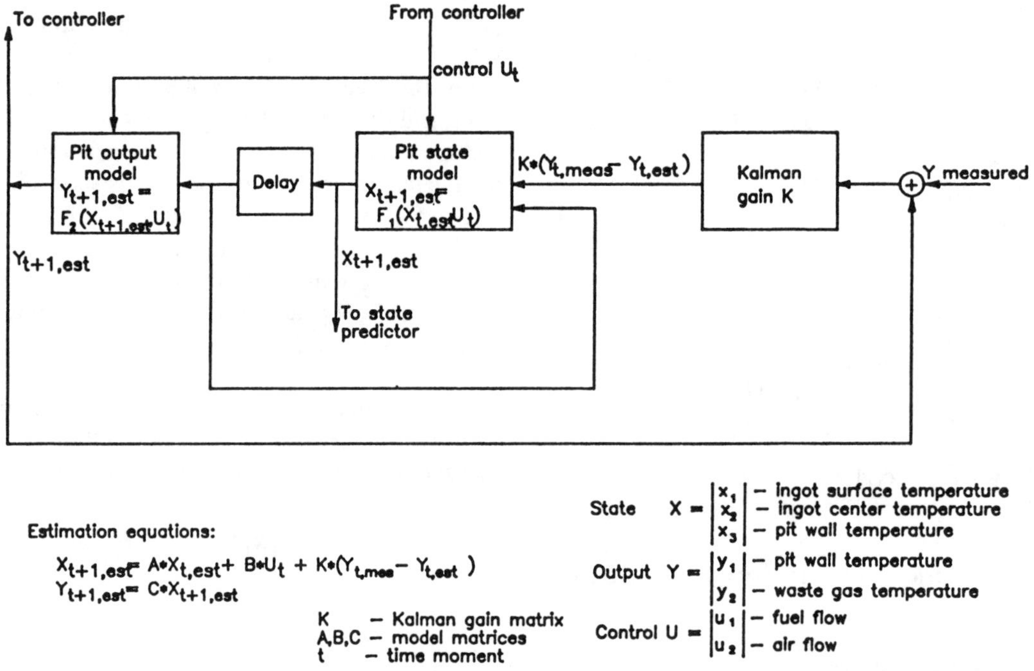

Fig. 6. Estimator (Kalman filter) structure.

IV. Estimator

A regular Kalman filter scheme shown in Fig. 6 was used [9], [10]. F_1 and F_2 in Fig. 6 correspond to (1). The choice of the covariance matrices for system disturbances and measurement noise was rather arbitrary. A number of reasons justify that: the model is simplified, anyway; the errors do not satisfy the statistical assumptions of independence, normality, and zero mean; small errors in the gain are not important because of the filter feedback effect; it is hard to accurately quantify the said covariance matrices. In this work, the covariance matrices needed were assumed to be diagonal; few iterations were made to tune them before they were used in the filter.

Fig. 7 demonstrates the behavior of the elements of the Kalman gain matrix. As one can see, it takes about one-third of a typical pit cycle for the gain elements to settle. This is not a big concern because, as was stressed before, the estimation becomes important at the second half of the cycle. For the same reason, a steady-state gain element can be used instead of the variable Kalman gain. This significantly simplifies the production algorithm and allows one, for example, to use one computer/microprocessor system to control the whole pit battery (a typical steel mill soaking pit operation consists of 20–30 pits). The comparison between the estimator performance with the steady-state and variable gain is given in Figs. 8 and 9 for the ingot center and surface temperature, respectively. One can see

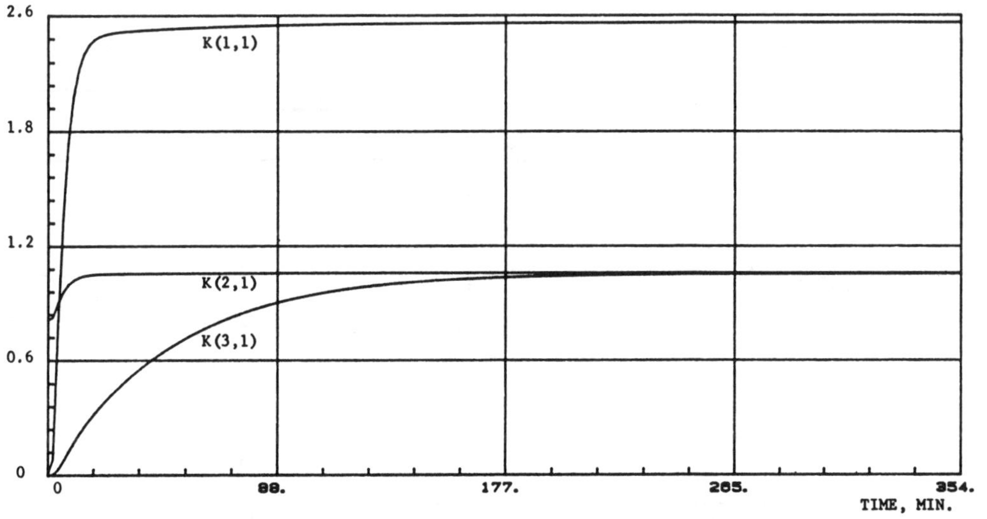

Fig. 7. Kalman gain matrix elements versus time.

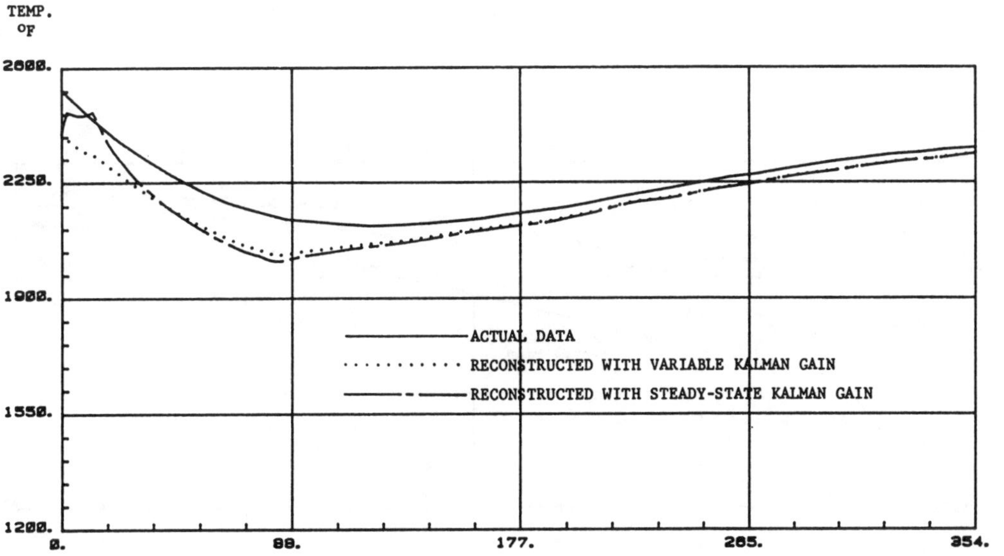

Fig. 8. Ingot center temperature. Comparison between using variable Kalman gain or steady-state Kalman gain. Initial conditions: Actual, $X_2 = 2537°F$; given to the Kalman filter, $X_2 = 2400°F$.

that, although variable Kalman gain produces temperature curves closer to the actual ones, in the second half of the pit cycle it is hard to see any difference in their performance. Notice that initial conditions provided to the estimator (Figs. 8 and 9) differ from the actual ones.

It was stated in the Introduction that the major function of the Kalman filter here is to compensate for uncertainty in the initial conditions. The simulated example shown in Figs. 10 and 11 demonstrates this filter's capability in the estimation of the ingot center and surface temperatures. Only wall temperature is assumed to be measurable. Gaussian noise was added to the wall temperature curve to simulate measurement noise. Initial conditions "known" to the filter are different from "real" ones. Smooth curves correspond to the "real" temperatures and to the temperatures generated by the model, without using the filter. Notice that if the filter is turned on, the ingot temperatures estimated by the filter approach the "real" curves.

Fig. 12 compares the actual data of Fig. 2 to the estimated curves; steady-state gain matrix is used in the estimator.

V. Predictor

Ready-to-roll time may be defined in a number of ways. For example, it is the time moment t_r at which the average ingot temperature (this may be an arithmetic average or an average of weighted components) enters some temperature corridor and stays there. In this work, both ingot surface and center temperatures were handled separately. Thus, t_r is a moment for which

$$\begin{aligned} x_{1,\,\text{set}} - d_1 \leqslant x_1(t_r) \leqslant x_{1,\,\text{set}} + d_1 \\ x_{2,\,\text{set}} - d_2 \leqslant x_2(t_r) \leqslant x_{2,\,\text{set}} + d_2 \end{aligned} \tag{3}$$

assuming that after entering their corridors, both temperatures stay there; here $x_{1,\,\text{set}}$, $x_{2,\,\text{set}}$ are preset temperatures

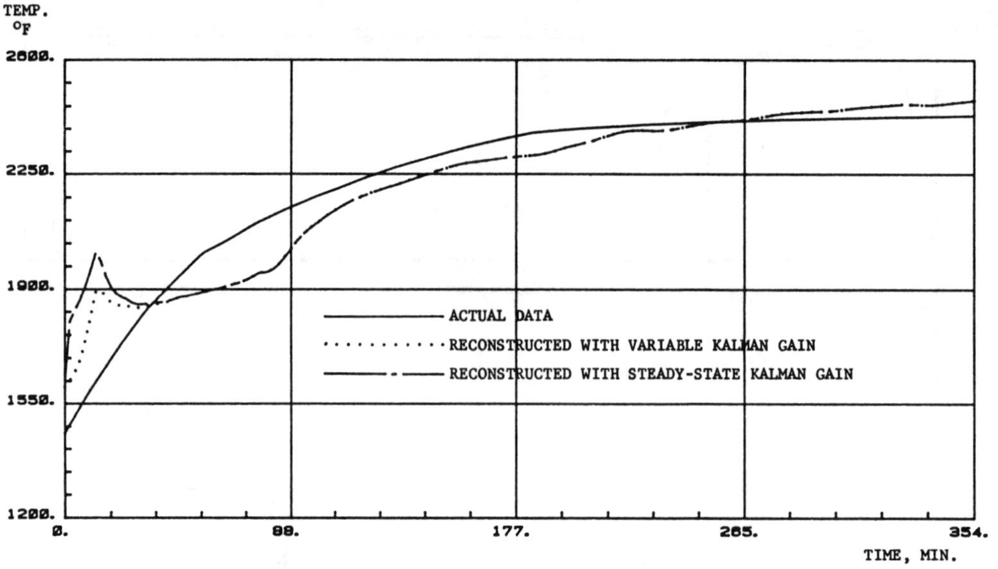

Fig. 9. Ingot surface temperature. Comparison between using variable Kalman gain or steady-state Kalman gain. Initial conditions: Actual, $X_1 = 1452°F$; given to the Kalman filter, $X_1 = 1600°F$.

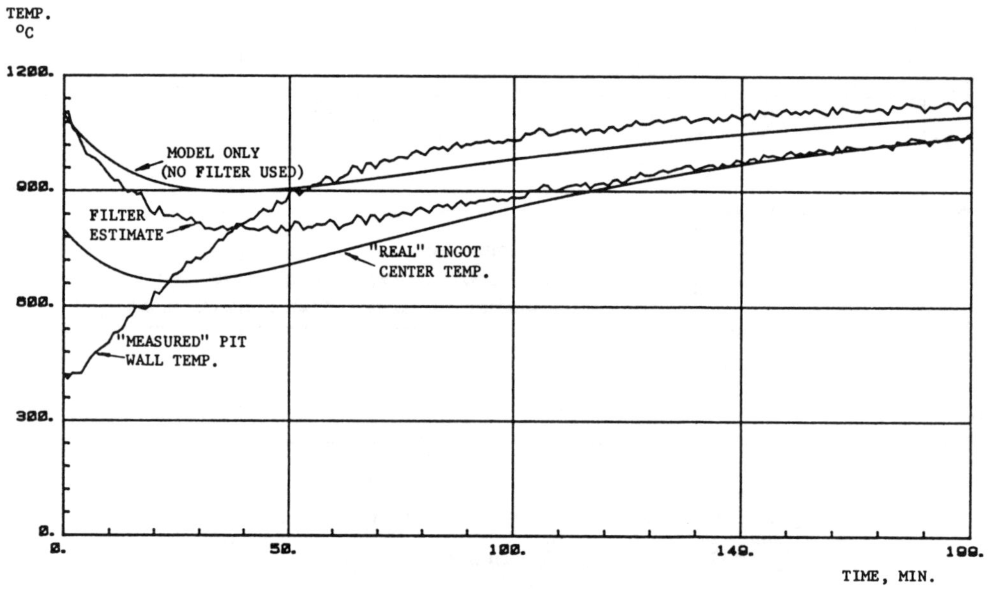

Fig. 10. Ingot center temperature. Effects of the measurement noise and of the uncertainty in initial conditions on the performance of the Kalman filter. Wall temperature is the only "measured" variable. All the curves have been simulated.

and d_1, d_2 define the widths of the corridors. For example, $x_{1,\text{set}} = x_{2,\text{set}} = 2400°F$, $d_1 = 50°F$, $d_2 = 100°F$.

The structure of the predictor is shown in Fig. 13. Once the estimated values of the state for the next moment $(t+1)$ appear on the input of the predictor, the pit/ingot model is run together with the controller model until the state's values meet the requirements for the ingots to be ready. Then the corresponding time value t_r is output as the moment $(t+1)$ prediction of the ready-to-roll time.

One may argue that there is no need to have the model of the real controller to generate the fuel strategy. Instead, one could get the fuel strategy from the state equations (1), assuming that in the second phase (see Fig. 2), the wall temperature must be constant and the fuel is limited by $U_{\max}$. In practice, this approach has a drawback: because of the digitization, at the moment of switch between two phases, instability may occur; some kind of a smoothing process is of help here. Since a PI regulator does just this kind of smoothing, the model of the real controller may be used as well.

Fig. 14 demonstrates the performance of the predictor (actually, of the whole system, Fig. 3) on the actual data, Fig. 2. The horizontal axis in Fig. 14 corresponds to real time and the vertical axis to the ready-to-roll time t_r predictions. For this example, from the operation records it is known that the actual ready-to-roll time corresponded to $t_r = 306$ min ($t = 0$ is the beginning of the pit cycle). Required accuracy of the prediction is ± 15 min; this time corridor is shown in Fig. 14 in broken horizontal lines. Thus, the ready-to-roll time in Fig. 14 is the moment when

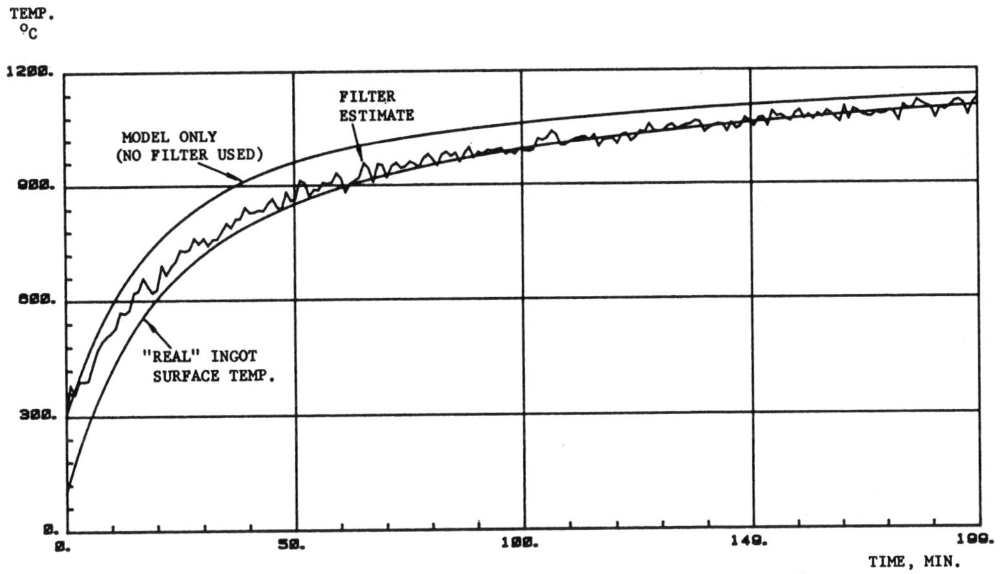

Fig. 11. Ingot surface temperature. Same simulation run (and same story) as in Fig. 10.

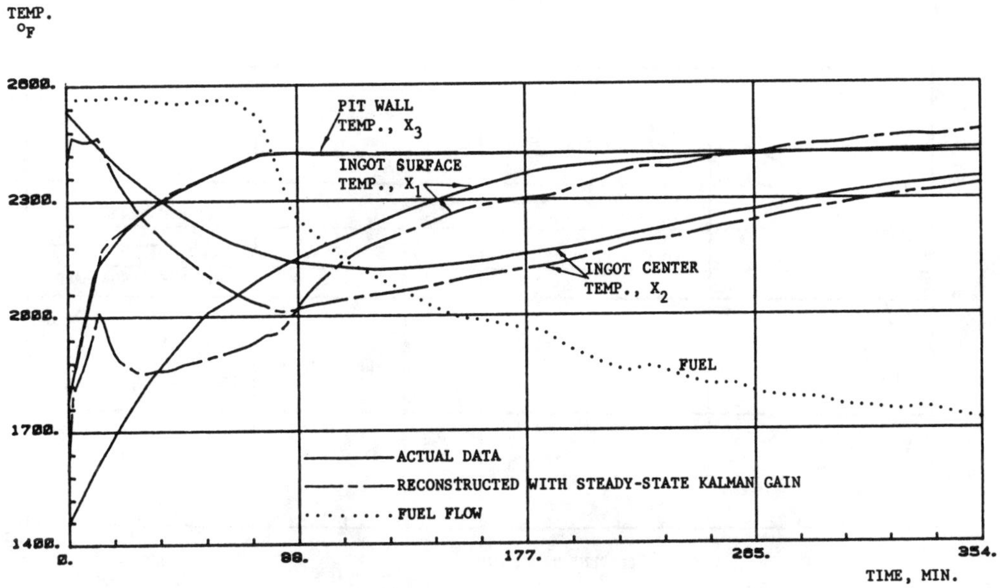

Fig. 12. Example of Kalman filter performance. Actual data. Initial conditions, °F: Actual— $X_1 = 1459$, $X_2 = 2533$, $X_3 = 1776$. Given to the filter— $X_1 = 1600$, $X_2 = 2400$, $X_3 = 1700$.

the prediction curve enters the 306 ± 15 min corridor and stays there. Thus, in our example $t_r = 210$ min; that is, the right prediction was given $(306-210) = 104$ min in advance before the ingots were ready. Initial conditions entered into the estimator in this example differed from the actual ones by amounts that are realistic (see Fig. 14).

VI. Plant Data Analysis

It is relatively easy to tune all the parameters needed for the system (see Fig. 3) to handle one or two cases of possible situations—in terms of ingot sizes, total ingot weight, etc. The main problem for such a system is to be able to cover all or at least the majority of the situations that it may face in real production. This task has been addressed on two levels: analysis of actual multiple pit run data using the system (Fig. 3) and the software package described in Section II, and testing the system in the actual plant environment. Testing and implementation of the system in the plant environment are currently in progress. Some conclusions from the actual data analysis follow.

Since the problem and the approach described in the previous sections do not involve any active control and, thus, do not change the flow of the technological process, the estimator and the predictor, theoretically, could be fully tested on the computer provided that enough actual data from the operating pits are available. In practice, gathering the data in the production environment is a hard problem in itself.

Since each pit cycle takes 5–10 h, not many pit runs can be accumulated within a reasonable time period. Also, there is no choice on what are the specific conditions of a

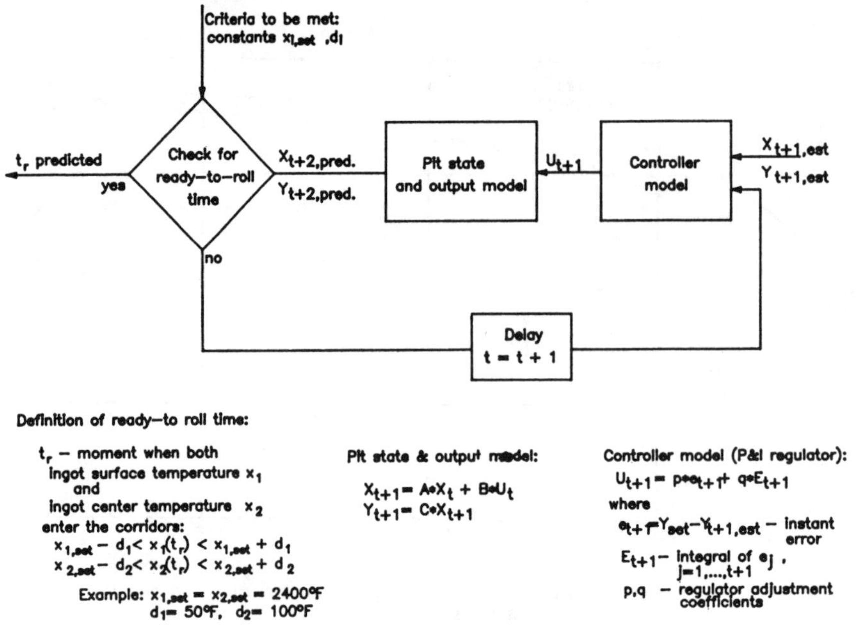

Fig. 13. Ready-to-roll time predictor structure.

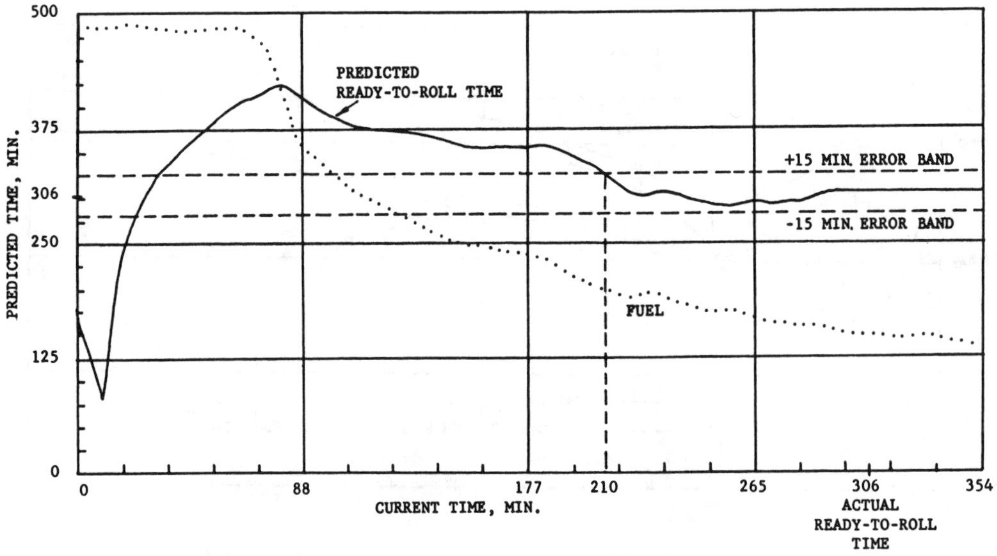

Fig. 14. An example of the predictor performance. Actual data, same pit run as in Figs. 2, 5, 8, 9, and 12. Here the ready-to-roll time is defined as a time moment when the average of the ingot surface and the ingot center temperatures reaches 2370°F and stays there thereafter. For the performance evaluation, 15 min error band around the actual ready-to-roll time (which is 306 min here) shows how far in advance the predictor gives a reasonable prediction (here it happened at a time moment 210 min). Steady-state Kalman gain is used throughout the whole pit cycle, $K = (2.39, 1.29, 0.95)$. Initial conditions—both the actual and what was "known" to the filter—are the same as in Fig. 12.

pit cycle; thus, all the variety of the ingot sizes, weights, initial ingot temperatures, and other pit conditions is reflected in the data gathered.

For this work, the data corresponding to the 44 pit cycle runs or, in other words, 44 cases, were available; the data represent two soaking pits of the same plant. This includes the measurements of the fuel flow, air flow, pit wall temperature, and waste gas temperature. For the 44 runs, the ingot surface and ingot center temperatures were calculated by U.S. Steel Research using their heat transfer model [6]. Those ingot center and ingot temperature values were then considered to be "actual" ingot temperatures and were used for the model identification purposes and for the estimator-predictor performance evaluation. Of course, the immediate question is how actual those "actual" temperatures are; the only satisfaction one can get is from the claim that the model [6] provides an accuracy within 20–30°F.

In the experiments, two standard sets of values for the initial conditions (initial ingot surface temperature x_1, ingot center temperature x_2, pit wall temperature x_3) were used in the estimator—one for hot ingot cases and one for cold ingots:

for hot ingots $X_1 = 1500°\text{F}$, $X_2 = 2300°\text{F}$, $X_3 = 1700°\text{F}$
for cold ingots $X_1 = 0°\text{F}$, $X_2 = 0°\text{F}$, $X_3 = 1700°\text{F}$.

Since the actual initial conditions are unknown, a standard set represents a first guess of what the initial conditions are. The difference between the standard sets and the actual initial conditions in the experiments were up to 500°F, or on the average, about 250°F.

The performance of the estimator has been evaluated by the indexes

$$r_q = \frac{1}{n_q} \sum_{i=1}^{n_q} |x_{\text{est}}(i) - x_{\text{real}}(i)| \tag{4}$$

$$r = \frac{1}{p} \sum_{j=1}^{p} r_j \tag{5}$$

where $x_{\text{est}}(i)$ = the estimated temperature value (for ingot surface, ingot center, or pit wall temperature, respectively) for the ith time point; $x_{\text{real}}(i)$ = the actual value of the temperature; n_q = number of time points in the section of the pit run q being studied; $q = 1, \cdots, 44$; r_q represents the performance of the estimator for the run q, and r = the performance of the estimator for one group or for a number of groups of runs. In other words, the performance of the estimator is measured by the average absolute deviation of the estimated temperature from the actual temperature.

Since, on the one hand, the estimator performs rather poorly at the beginning of the pit cycle (especially if the steady-state Kalman gains are used), and, on the other hand, the importance of the quality of the estimation increases to the end of the cycle, the performance index (4) was evaluated separately for the first half and for the second half of each run.

From the physical considerations of the steel-making process and from a formal analysis, it is clear that one set of pit model coefficient matrices A, B, and C in (1) cannot cover all the variety of possible situations. Thus, according to the ingot sizes, total ingot weight, and "rough" initial conditions (hot/cold ingots), the 44 runs were separated into 12 groups. Each group included a different number of runs according to the data distribution; the distribution of runs among the groups is shown in Table I.

Then, the model matrices A, B, and C were generated using the least-square method for each of the groups. There are two basic choices on how to use the data to generate the models: either some average model is generated for the group of runs using all the runs of the group, or a "representative" run is chosen and used for generation of the model which is then used for the whole group. In practical situations, an average model will probably give better (or, at least, more stable) results. On the other hand, given the small number of runs in the group, using an average model raises a serious study validity question—namely, that the

TABLE I

Group No.	1	2	3	4	5	6	7	8	9	10	11	12
Number of Runs	1	4	4	1	5	1	6	6	3	4	3	6

model generated is being used to analyze the data from which this model had been extracted. In trying to avoid this problem and to make the experiment more meaningful from the standpoint of the pit model validation, a model for only one run per group had been generated by the least-square method.

The "representative" run in each group was chosen on purely subjective basis—if the data curves of the run look "normal," "natural," then the run is considered to be good for the role of a "representative."

Within the approach to the experiment described above, not all the pit runs are of the same value from the standpoint of the performance index (5). Apparently, when the index (4) is evaluated for a group containing only one run (groups 1, 4, and 6 of Table I), the model is being applied to the data from which it has been "extracted," and the index r can be expected to be better (lower) than in average. Similarly, the representative runs in each group will also produce smaller r's. To get a clearer idea about the potential performance of the estimator in the real production, these runs have to be excluded if r is to be averaged over the groups. The excluded runs total 12 runs, one per group, which leaves us with 32 runs and 9 groups to compute the "clean" index r_c.

Examining the data curves of the remaining 32 runs, one is tempted to try to further discriminate among the runs while computing the index r. Out of the 32 runs, two demonstrate unusual fuel behavior (sharp fuel drops within the cycle without corresponding changes in temperatures). For some other 9 runs, the ingot center temperature as computed by the model [6] was constant (at the melting point) throughout the whole cycle. First, it is doubtful that these data reflect the real situation; and second, such cases are not good to handle using the model (1). In total, 11 "bad" cases might be excluded to produce a "superclean" index r_s. Although one may question the "legality" of the index r_s, the unusually high percentage of "bad" cases (11 out of 32) suggests that the r_s index be tried.

Finally, just to demonstrate the reasonable behavior of the estimator in the favorable circumstances, the value of r_r has been computed for the 12 representative cases mentioned above; obviously, this experiment produced the best performance.

The results of the experiments with the Kalman estimator have been compared then with those without the Kalman filter (when only the model is used); the same initial conditions are used in both cases. The results are summarized in Table II.

Studying Table II, a number of observations can be made.

1) The performance of the Kalman estimator is significantly better in the second half of the pit cycle than in the

TABLE II
RESULTS OF THE EVALUATION OF THE KALMAN ESTIMATOR

Performance Index r for: Ingot surface temp. x_1; Ingot center temp. x_2; pit wall temp. x_3; all in °F		First Half of Pit Cycle		Second Half of Pit Cycle	
		Using Model Only	Using Kalman Filter	Using Model Only	Using Kalman Filter
1	2	3	4	5	6
Index r: Index (4) is averaged for all 44 runs	x_1	154	99	144	40
	x_2	276	231	205	131
	x_3	149	6	134	1
Index r_r: (4) is averaged over the "representative" runs (12 runs)	x_1	134	42	115	17
	x_2	183	109	141	52
	x_3	126	5	117	0
Index r_c ("clean"): 32 runs not used for model identification are studied (most practical case)	x_1	161	120	154	49
	x_2	311	277	229	161
	x_3	158	6	140	1
Index r_s ("super-clean"): in addition to 12 representative runs, 11 "bad" runs are excluded. In total, 21=44-12-11 runs are being studied.	x_1	91	111	120	49
	x_2	165	151	116	90
	x_3	120	5	115	1

first half; at the same time, the performance of the system that uses the model only remains roughly the same.

2) Kalman estimator performs better than the system that uses the model only, especially at the second half of the pit cycle when (from the practical standpoint) the estimation becomes really important.

3) Although the simplest linear model (1) has been used, the results of the Kalman estimation, in practical terms, are satisfactory.

4) Not surprisingly, since the wall temperature measurements are available on-line, the estimated wall temperature follows the actual wall temperature almost ideally. When only the model is used, no advantage of the on-line measurements is taken, and equally poor estimations of all three temperatures are produced.

A similar analysis has been made for the performance of the predictor. Because of lack of space, only the summary of this study is presented.

1) At a given moment, when the system is trying to predict the future behavior of the state variables, no on-line measurements can be used. Therefore, the only tool left is the model. This fact makes the model more important for prediction than for estimation. Thus, the quality of the model affects the prediction more than the estimation. The experiment confirms this observation.

2) Besides the model, the accuracy of the prediction depends on the accuracy of the initial conditions. As one can see from Table II, the estimator provides the predictor with better initial conditions than the system that uses the model only. As a result, the Kalman estimator–predictor combination produces better prediction.

3) On the limited data studied so far, the performance of the estimator–predictor included satisfactory and unsatisfactory prediction cases. More data are needed to clarify whether the problem comes from insufficient data used for the model identification or from too simple model (1). In the latter case, more complicated models will be recommended.

VII. CONCLUSIONS

It should be emphasized that, in general, because of the high level of uncertainty in the data accumulation process caused by the variety of the situations involved—different ingot sizes/shapes, different total ingot weight per run, difference in characteristics of individual pits, changing characteristics of a single pit between consecutive pit cleanings, long-run aging of pits, impossibility of direct reliable ingot temperature measurements needed for the model identification, uncertainty introduced by indirect "measurements" using models like [6], etc.—the validation of the system described here becomes a formidable task. The process in question is a genuine poorly defined process, not only in terms of its measurement, but also in terms of measuring the performance of the control strategy.

The analysis of the results of the estimator/predictor performance using the actual plant data available allows one to make the following conclusions.

1) Separate pit/ingot models should be generated for each class of ingot sizes and total ingot weights. A number of similar sizes and total weights may be clustered in one class. Therefore, a number of matrices, A, B, and C [see (1)] will be stored in the control system memory and called upon when the ingot size and the total ingot weight are entered into the system.

2) In spite of using the simplest linear model (1), the estimator part of the system gives satisfactory estimates of the current ingot temperatures. Thus, the estimator seems to be general and powerful enough to handle all the situations in this and similar applications. The estimator allows one to take into account the changes occurring during the pit run (unplanned fuel flow and air flow changes, unplanned pit lid openings, etc.) and, thus, to overcome problems typical for the models which do not use the continuous measurements available during the pit cycle.

3) The predictor part of the system heavily depends on the quality of the pit/ingot model used—which is hardly surprising, taking into account that the only information the predictor uses for an advance prediction is the initial conditions and the model. In this specific application, it may be feasible to get away with the simple model (1) (probably, with some additional correction schemes). Another way is to use more sophisticated (not necessarily linear) models in the predictor; for example, any of the models described in [1]–[6] could be plugged in the predictor. What is important is that the uncertainty of the initial

conditions needed for the predictor would be eliminated by the estimator. In such a case, the estimator and the predictor would be operating using different models.

Acknowledgment

Support from U.S. Steel Company through their actual plant data, active participation in the project, and fruitful discussions, especially with R. O'Shea, is greatly appreciated. Special thanks to H. Chestnut, J. Wheeler, D. Fapiano, and H. P. Ko, all of General Electric Research, for their support and numerous discussions.

References

[1] J. R. Cook, E. J. Longwell, and C. L. Nachtigal, "Algorithms for on-line computer control of steel ingot processing," in *IFAC Proc.*, Helsinki, Finland, 1978, pp. 131–137.

[2] C. Patel, W. H. Ray, and J. Szekely, "Computer simulation and optimal scheduling of a soaking pit-slabbing mill system," *Metallurgical Trans. B*, vol. 7B, pp. 119–130, Mar. 1976.

[3] D. E. Anderson and G. E. Olney, "A soaking pit-slabbing mill productivity simulation," in *AISE Ann. Proc.*, 1976.

[4] R. J. Rohlach, G. T. Miller, P. W. Klatt, and A. Bartel, "A graphical model for soaking pit production planning," *Iron and Steel Eng.*, no. 6, 1981.

[5] K. H. Lee and A. J. Koivor, "Optimal decisions for scheduling of ingot processes," Purdue Lab. Appl. Ind. Contr., Purdue Univ., W. Lafayette, IN, Rep. 80, 1977.

[6] H. W. Schnaible, "Optimization of soaking pit heating times," U.S. Steel Tech. Rep., 1976.

[7] C. N. Cheek, K. Nobbs, and N. Munro, "Estimation and prediction of ingot temperatures on a pilot scale soaking pit," in *IFAC Proc.*, Helsinki, Finland, 1978.

[8] H. J. Wick, "On-line sequential estimation of ingot center temperatures in a soaking pit," presented at the Rolling Mill Conf., Cincinnati, OH, May 1981.

[9] R. E. Kalman and R. S. Bucy, "New results in linear filtering and prediction theory," *Trans. ASME, J. Basic Eng.*, 1961.

[10] K. J. Astrom, *Introduction to Stochastic Control Theory*. New York: Academic, 1970.

Application of Kalman Filters to the Regulation of Dead Time Processes

W. L. BIALKOWSKI

Abstract —**A single-input single-output control algorithm for a process with dead time and dead time uncertainty is described. The process dynamics consist of first-order mixing and pure delay with the pure delay being dominant. The process is disturbed at the upstream end by a disturbance sequence consisting of white noise passed through a first-order shaping filter. The process output is subject to white measurement noise. A discrete Kalman filter is used to produce state estimates for the disturbance mixing and dead time states which are updated from the process output residual error. In order to handle dead time uncertainty of up to *a priori* established limits, the residual error is passed through a dynamic dead band whose magnitude is a function of the dead time states of a separate process dynamic model driven by the process input. The dead band eliminates dead time error components from the residual. Control is achieved by state feedback from the upstream Kalman filter state estimate. The algorithm is in use on paper machine and bleach plant control applications and gives near minimum variance performance when properly tuned.**

Manuscript received February 24, 1982; revised September 27, 1982.
The author is with Central Engineering, Computer Systems Domtar Inc., Islington, Ont., Canada.

Introduction

DEAD time occurs frequently in industrial processes as a result of transport phenomena. In some cases the dead time parameter itself is subject to random variability

Reprinted from *IEEE Trans. Automat. Contr.*, vol. AC-28, pp. 400–405, Mar. 1983.

due to, for instance, hydraulic channeling, with the result that it cannot be determined exactly under given process operating conditions. These combined circumstances present a difficult challenge for the design of a robust stochastic regulator. This paper describes an algorithm which is applicable to the regulation of single input-single output dead time processes with dead time uncertainty. The algorithm uses a discrete steady-state Kalman filter and state feedback regulator. It handles dead time uncertainty within an *a priori* established "window." It is in use in a number of pulp and paper applications and can be tuned to give near minimum variance performance.

Problem Statement

The objective is to achieve close to minimum variance regulation of a single input-single output continuous process with open loop dynamics consisting of a mixing first order lag and dead time, but in which the dead time is long compared to the time constant. The process is disturbed at its upstream end by a correlated random disturbance which can be adequately described by white noise passed through a first order shaping filter. The output of the process is sampled at a fixed sampling rate. The sampling rate is relatively high compared to the dead time. The output is corrupted by white measurement noise. The disturbance shaping filter and process mixing first order lag time constants are known. Normally, the process mixing time constant is short compared to that of the disturbance shaping filter. The process dead time is expected to vary unpredictably between assumed limits under the given steady-state conditions.

Kalman Filter Design

Ignoring the dead time uncertainty for the time being, the stated problem can be described using discrete time state variable notation as follows:

$$x_{t+1} = Ax_t + Bu_t + L\xi_t \tag{1}$$

$$y_t = Cx_t + \theta_t \tag{2}$$

where

x = state vector

u = control input

ξ = white noise disturbance input

y = output

θ = white measurement noise.

Using the following state variable definition:

x_1 = disturbance modeling state

x_2 = first order mixing state

$x_3 \cdots x_n \cdots x_{n+2}$ = dead time states

where

$n = T/D$

= number of states in process dead time

T = process dead time

D = unit delay time (sampling period).

The specific state equations are

$$x_{t+1} = \begin{bmatrix} 1-\frac{D}{\tau_d} & 0 & & & & & \\ \frac{D}{\tau_p} & 1-\frac{D}{\tau_p} & 0 & & & & \\ 0 & 1 & 0 & & & & \\ & 0 & 1 & 0 & & & \\ & & & \cdot & \cdot & & \\ & & & & 0 & 1 & 0 \end{bmatrix} x_t + \begin{bmatrix} 0 \\ k_p\frac{D}{\tau_p} \\ 0 \\ \cdot \\ \cdot \\ 0 \end{bmatrix} u_t + \begin{bmatrix} \frac{D}{\tau_d} \\ 0 \\ 0 \\ \cdot \\ \cdot \\ 0 \end{bmatrix} \xi_t \tag{3}$$

$$y_t = [0 \quad 0 \quad \quad 0 \quad 1] x_t + \theta_t \tag{4}$$

where

τ_d = disturbance shaping filter time constant

τ_p = process first order mixing time constant

k_p = process gain

and the dimension of the A, B, L, C matrices are

$$A = (n+2) \times (n+2)$$
$$B = (n+2) \times 1$$
$$L = (n+2) \times 1$$
$$C = 1 \times (n+2).$$

A series of steady-state Kalman filter design calculations were performed for the above problem description using a range of values for the parameters τ_d, τ_p, n, and the variances of ξ and θ. The objective of these calculations was to establish the values for the Kalman filter update cycle feedback matrix K as given below

$$\hat{x}_{t+1} = \hat{x}_t + K(y_t - \hat{y}_t) \tag{5}$$

where $\hat{x}_t$ and $\hat{y}_t$ are the Kalman filter state vector and output, respectively.

The calculation of K involves the solution of a matrix Riccati equation. The chief results of these calculations are summarized below.

1) The elements of K vary directly with the ratio of disturbance-variance-to-measurement-noise-variance. The greater the noise, the lower the gains.

2) For the case where τ_p is small, the relative values of the elements of K are strongly dependent on τ_d. The elements of K increase monotonically from k_1 through k_{n+2}, with the latter typically having a value in the range 0.3–0.8 depending on the disturbance to noise ratio. For the case where τ_p is equal to D, or the process dynamics

consist of pure dead time, the adjacent elements of K increase in fixed ratio as given in (6).

$$k_{i+1} = \frac{k_i}{(1 - D/\tau_d)} \qquad i = 1 \cdots n+1. \tag{6}$$

Since k_{n+2} has the highest value, the above relationship can be restated as follows:

$$k_{i-1} = (1 - D/\tau_d) k_i \qquad i = n+2, n+1 \cdots 2. \tag{7}$$

Equation (7) can be implemented recursively on-line, thus eliminating the need to store the K matrix. It is interesting to note that (7) is identical to the unforced response of the x_1 state. (x_{t+1} substituted for k_{i-1}.) A preliminary investigation shows that the unforced response of x_2 with a similar substitution will yield the K matrix elements for the case where τ_p has any arbitrary value.

It should be mentioned that the Kalman filter design calculations were done using a version of the linear–quadratic–Gaussian design package developed at the Massachusetts Institute of Technology [1]. The version in question was modified to handle discrete-time problems as well as continuous-time problems. Although no numerical difficulties were experienced during the solution of the above calculations, a subsequent investigation involving multivariable dead time problems resulted in numerical instability of the Riccati solution. The source of the difficulty is the fact that the discrete A matrix is singular in the presence of dead time. The Riccati equation solution algorithm used in this case is based on the work of Kleinman [2]. Reasons as to why it worked for the case described and not for more general cases was not investigated further. However, Pappas, Laub, and Sandell [3] have investigated the general matrix Riccati solution for the singular A matrix.

Regulator Design

Linear–quadratic regulator design calculations were performed using the $L-Q-G$ package for the following problem. For the state equations

$$x_{t+1} = \begin{bmatrix} 1-\dfrac{D}{\tau_p} & 0 & & & \\ 1 & 0 & & & \\ 0 & 1 & 0 & & \\ & & & \ddots & \\ & & & 1 & 0 \end{bmatrix} x_t + \begin{bmatrix} k_p \dfrac{D}{\tau_p} \\ 0 \\ \vdots \\ 0 \end{bmatrix} u_t \tag{8}$$

$$y_t = [0 \quad 0 \quad \cdots \quad 1] x_t \tag{9}$$

calculate the gain matrix G for the state feedback law given in

$$u_t = -G x_t \tag{10}$$

which minimizes the index J below

$$J = \sum_{t=0}^{\infty} (xQx + uRu). \tag{11}$$

Note matrix dimensions are

A	$(n+1)\times(n+1)$
B	$(n+1)\times 1$
C	$1\times(n+1)$
G	$1\times(n+1)$
Q	$(n+1)\times(n+1)$
R	1×1.

The state variable definition for (8)–(11) is identical to that for (3)–(5) except that x_1 of (3), the disturbance state has been eliminated as it is not a controllable state. Thus, x_1 of (8) is identical to x_2 of (3). A series of calculations involving relative weightings of the Q matrix diagonal elements and R matrix values were performed in order to investigate the dominant dynamics of the regulator so that a match could be obtained with the Kalman filter tracking dynamics.

The chief result of these calculations was that in all cases only the g_1 element was nonzero. A practical explanation for this result is that feedback from the most upstream state only is necessary since subsequent dead time states contain older versions of the same information.

From a theoretical point of view, the stochastic regulator should consist of a discrete Kalman filter providing minimum variance state estimates for the states defined in (3). The feedback law should be

$$u_t = -g_2 \hat{x}_{2t}. \tag{12}$$

The value of g_2 should be chosen so that the regulator dynamics match the Kalman filter tracking dynamics.

Dead Time Sensitivity

In the original definition of the problem, it is assumed that the output sampling rate is relatively high compared to the dead time. Thus, if the dead time varies around its nominal value T by up to a maximum of $+/-\Delta T$, and

$$w = \Delta T / D$$

then if the A matrix is expanded by w additional dead time states, then the possible outputs as the dead time varies can be described using an additional set of state variables $\boldsymbol{x}'$ as follows:

$$x'_{t+1} = A' x'_t + B u_t \tag{13}$$

where A' is as in (3) with $a_{11} = a_{21} = 0$ and w additional dead time states.

The earliest, nominal, and latest possible outputs thus are

$$\begin{array}{rlr} & \quad 1 \quad 2 \quad n+2-w \quad n+2 \quad n+2+w \quad \text{(State Index)} & \\ y'_{t-w} = & [0 \quad 0 \quad \cdots 1 \quad \cdots \quad 0 \quad \cdots \quad 0] x'_t & (14) \\ y'_t = & [0 \quad 0 \quad \cdots 0 \quad \cdots \quad 1 \quad \cdots \quad 0] x'_t & (15) \\ y'_{t+w} = & [0 \quad 0 \quad \cdots 0 \quad \cdots \quad 0 \quad \cdots \quad 1] x'_t. & (16) \end{array}$$

In the statement of the problem, the true process output may be any output between y'_{t-w} and y'_{t+w}. The errors in y_t resulting from dead time errors of up to $-\Delta T$ can be bounded by the quantity $|y'_{t-w} - y'_t|$, while errors in y_t resulting from dead time errors of $+\Delta T$ can be bounded by the quantity $|y'_{t+w} - y'_t|$.

One method to desensitize the regulator from the effects of dead time errors of $+/-\Delta T$ is to calculate a time variable dead-band on-line as follows:

$$d = k_d[|y'_{t-w} - y'_t| + |y'_{t+w} - y'_t|]$$
$$d = \text{dead-band half width}$$
$$k_d = \text{tuning parameter} \tag{17}$$

and to modify the Kalman filter feedback equation as follows:

$$e_t = y_t - \hat{y}_t \tag{18}$$
$$e'_t = \text{sgn}(e_t)[|e_t| - d] \qquad \text{for } |e_t| \geqslant d$$
$$= 0 \qquad \text{for } |e_t| < d \tag{19}$$
$$\hat{x}_{t+1} = \hat{x}_t + Ke'_t. \tag{20}$$

On-Line Implementation

On-line implementation differs in some respects from that described above. The Kalman filter update cycle is performed using the following state equations:

$$\hat{x}_{t+1} = \begin{bmatrix} 1 & 0 & & & \\ \frac{D}{\tau_p} & 1 - \frac{D}{\tau_p} & 0 & & \\ 0 & 1 & 0 & & \\ 0 & 0 & 1 & 0 & \\ & & & \cdot & \cdot \\ & & & 1 & 0 \end{bmatrix} \hat{x}_t + \begin{bmatrix} 0 \\ k_p \frac{D}{\tau_p} \\ \cdot \\ \cdot \\ 0 \end{bmatrix} u_t \tag{21}$$

$$\hat{y}_t = [0 \quad 0 \quad \overset{n+2}{1} \quad 0]\hat{x}_t. \tag{22}$$

The same state variable definition is used as that in (3). The A matrix is identical to that in (3) except that the a_{11} element is set to unity (explained later), and a suitable number of additional dead time states has been added to accommodate the maximum possible process dead time.

In the C matrix the nonzero element corresponds to the $n+2$ state. This organization is used to allow the system parameters to be modified on-line with process operating point, based on table lookup. The Kalman filter update cycle is in accordance with (7), (13), (14), (16)–(20). In (13) the same number of additional dead time states are provided as in (21). It is important to note that the reason that the separate state vector x'_t of (13) is generated on-line is because, for the purpose of calculating the time delay uncertainty dead-band, the states must be a function of past values of u_t only. The Kalman filter states $\hat{x}_t$ are unsuitable for this purpose, as they are also functions of uneven Kalman filter updating.

The inclusion of the $\hat{x}_1$ state in the on-line implementation is in itself a modification to normal Kalman filtering practice. The inclusion of x_1 during the design phase being only to augment the state vector in order to allow a mathematical representation of the desired statistical properties of the disturbance. In this case, however, the $\hat{x}_1$ state is retained on-line and represents a state estimate of the disturbance function.

The choice of unity for the a_{11} element has special significance, as it provides the $\hat{x}_1$ state with an integral term to the Kalman filter residual error, as no update cycle contributions to $\hat{x}_1$ are made. Thus,

$$\hat{x}_{1_{t+1}} = \hat{x}_{1_t} + k_1 e'_t. \tag{23}$$

With this modification, $\hat{x}_1$ can track the process disturbance with zero steady-state error regardless of bias. It also ensures that the system becomes steady-state insensitive to calibration errors in u_t and modeling errors in k_p. The sum of $k_p u_t$ plus $\hat{x}_{1t}$ is a particularly useful on-line quantity as it represents a prediction of y_t $n+2$ samples in the future.

The modification does however represent an inconsistency between the assumed disturbance shaping filter implied in the predict and update cycles. The former assumes τ_d to be infinite, while the latter uses τ_d as identified. The theoretical significance of this has not been rigorously investigated.

A final modification is that (12) is not used in the regulator state feedback. Instead, a $P+I$ regulator is used to drive the sum of $k_p u_t$ plus $\hat{x}_{1t}$ to the y_t setpoint. Thus,

$$\text{let } \epsilon_t = (y_t \text{ setpoint}) - (k_p u_t + \hat{x}_{1t}) \tag{24}$$
$$u_{t+1} = K_p \epsilon_t + K_I \sum \epsilon_t. \tag{25}$$

On-Line Tuning

The procedure for tuning the algorithm includes the following steps.

1) Obtain a power spectrum of an open loop steady-state time series of y_t. Estimate τ_d. Calculate $(1 - D/\tau_d)$ for (7).

2) Perform open loop identification tests in order to determine τ_p, k_p, T. Implement τ_p, k_p and T in lookup tables for on-line use. From the observed scatter in the dead time estimates obtained from a series of identification experiments, estimate the likely maximum dead time uncertainty ΔT. Calculate w and store for on-line use. Adjust k_d to a value greater than 1.0 if uncertainty also exists in the values of τ_p and k_p. Care must be taken however not to overestimate either w or k_d, as this results in unnecessarily large dead-band values on the average which in turn results in an unnecessary loss of Kalman filter residual error information. Typical values of k_d are in the range 1.5–2.5.

3) Assuming that τ_p is small and that (7) applies, choose an initial value of k_{n+2}. Adjust k_{n+2} until e_t is close to white noise based on an autocorrelation function test.

If τ_p is greater than $1.5D$ and the sampling period can be adjusted, then D should be increased and all parameters should be adjusted accordingly. Care should be taken not to unduly degrade the dead time uncertainty resolution required by (17) as the sampling period is increased. If the sampling rate cannot be altered, then the Kalman filter design calculations should be repeated to determine the K matrix for the given condition. Equation (7) should be discarded and an alternative method used for implementing the Kalman filter feedback gains on-line.

4) Determine the Kalman filter tracking lock-on time by allowing the filter to respond from an uninitialized state. Tune the regulator K_p and K_I to give a damped step response with a rise time approximately equal to the Kalman filter tracking lock-on time.

5) Check overall performance by calculating an autocorrelation function for y_t and observing that this vanishes within the confidence limits in a time shift of approximately the process dead time plus a small margin.

On-Line Applications

The algorithm is used in a number of process computer applications in the pulp and paper mills of Domtár Inc. There are five applications of paper machine basis weight control, one of paper machine moisture control, and one of bleach plant pulp brightness control. In all cases, the control computers are 16 bit minicomputers operating in 32K or 64K of main memory only. Code is written in assembler language using table driven subroutines. The Kalman filter predict cycle implementation for instance, combines the use of a first order digital filter and time delay push-down stack subroutines. In all cases the problem described in this paper is a small subset of the overall computer functional objectives.

Paper Machine Basis Weight Control—Description

In the paper machine basis weight example, a slurry of pulp in water enters the upstream end of the paper machine. The control computer regulates the volumetric flow of slurry by adjusting a control valve. The slurry is evenly ejected at high velocity onto a wide drainage conveyor to allow the majority of water to drain. The resulting paper sheet is transported continuously through mechanical pressing and steam drying operations to remove the majority of the remaining water. The sheet emerges at the downstream end where the weight per unit area is measured (basis weight). The control problem is to regulate the basis weight to minimum variance if possible by adjusting the upstream pulp slurry control valve. The dead time is a result of both hydraulic and mechanical transport. Dead time uncertainty is caused by channeling in one of the vessels in the upstream end. The relevant process measurements (and sampling rates) are

$$f_t = \text{pulp slurry flow } (1/\text{s})$$
$$c_t = \text{pulp slurry mass percentage } (1/\text{s})$$
$$v_t = \text{paper machine speed } (1/\text{s})$$
$$y_t = \text{basis weight measurement } (5/\text{s}).$$

Although the basis weight measurement is made at a high sampling rate, the sensor mechanically traverses the sheet in order to measure the cross-direction properties of the sheet. One traverse of the sheet takes 40 s.

A tight inner control loop is used to position the pulp slurry control valve through the control of z_t as follows:

$$z_t = f_t c_t k / v_t.$$

z_t is the expected mass per unit area of paper after drainage has occurred. For basis weight regulation

$$u_t = (z_t \text{ setpoint})$$

for the Kalman filter predict cycle

$$u_t = (10 \text{ s average of } z_t).$$

The predict cycle $\hat{x}$ and x' state vectors are updated every 10 s. For one specific machine and paper grade the dynamic parameters are

disturbance time constant τ_d	333 s
process time constant τ_p	27 s
process gain k_p	1.0
dead time T	110 s
dead time uncertainty ΔT	20 s
dead-band gain k_d	2.5
Kalman filter gain k_{n+2}	0.55
$(1-D/\tau_d)$	0.97.

One complication which exists in this application is that Kalman filter update cycle cannot occur at the same frequency as the predict cycle due to the fact that a full traverse of the sheet is required. The two operations thus occur asynchronously, the predict cycle at 10 s and the update cycle at 40 s at which time u_t is adjusted based on $\hat{x}_{1t}$ through a high gain which results in virtually a deadbeat response. This requirement is imposed by other operational needs.

Paper Machine Basis Weight Control—Performance

Fig. 1 shows an autocorrelation function for the Kalman filter residual error e_t. The time series was collected over 4 h at one sample per 40 s. As can be seen, e_t is a highly uncorrelated sequence, but not quite white noise as it should be if the Kalman filter was producing minimum variance state estimated. However, the performance is judged to be good given the use of the uncertainty deadband and the relatively slow update cycle frequency.

Fig. 2 shows an autocorrelation function of paper machine basis weight for the same period and sampling rate as

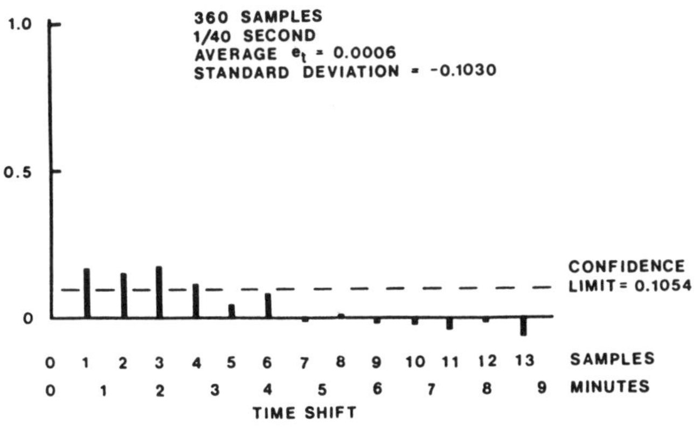

Fig. 1. Autocorrelation function Kalman filter residual error e_t.

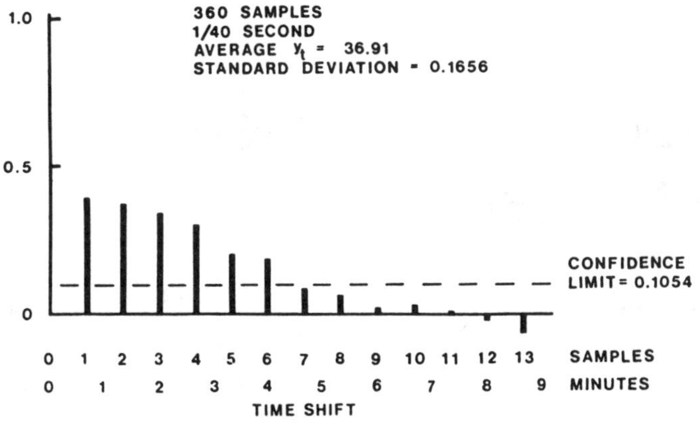

Fig. 2. Autocorrelation function paper machine basis weight y_t.

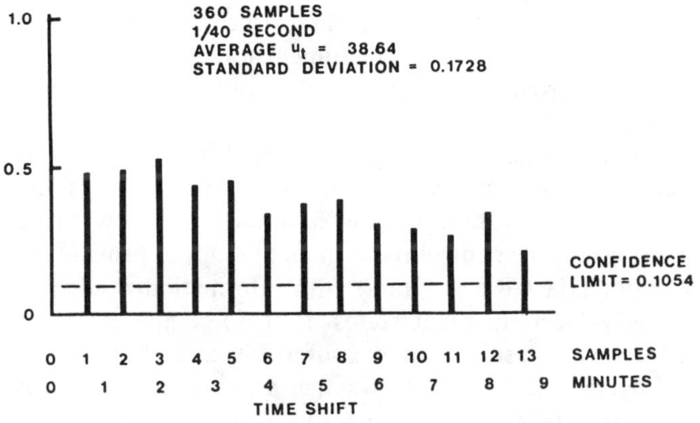

Fig. 3. Autocorrelation function process input u_t.

that for Fig. 1. Since the process dead time is 1.83 min and the sampling period is 40 s, for minimum variance control the autocorrelation function should vanish in 2.5 min. It vanishes in 4.5 min. The standard deviation is 0.45 percent of the y_t average. This performance is judged to be "close to" minimum variance, although some fine tuning would be beneficial for the system in question.

Fig. 3 shows the autocorrelation function of the control input u_t.

Fig. 4 shows the autocorrelation function of the disturbance estimate $\hat{x}_1$. Earlier spectral analysis for this particular grade of paper had shown an initial breakpoint

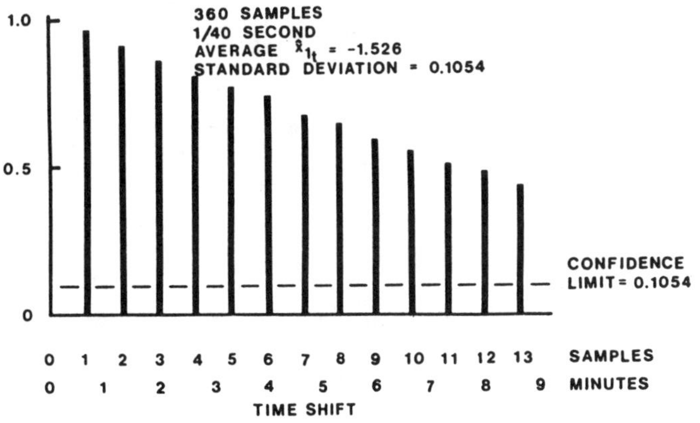

Fig. 4. Autocorrelation function disturbance estimate $\hat{x}_1$.

at 0.003 rad/s corresponding to a first order disturbance shaping filter time constant τ_d of 5.55 min. The time constant of the equivalent shaping filter can be estimated from the autocorrelation function using the expression

$$F_i = (e^{-D/\tau_d})^i \tag{26}$$

where $Fi = i$th autocorrelation function sample.

For the $\hat{x}_1$ autocorrelation the average τ_d calculated from the first 10 points is 12.3 min. The resulting time constant estimate is higher than the true value because the higher frequency components of the disturbance function are attenuated by the Kalman filter and are thus not present in $\hat{x}_1$. The method illustrated, however, provides a useful cross-check of the disturbance model without having to take the control system off control.

Conclusion

A discrete time Kalman filter based regulator has been successfully applied to single input-single output dead time process with dead time uncertainty. It has the capability to achieve near minimum variance control. However, the method of eliminating dead time errors from the Kalman filter residual errors results in a loss of innovative information thus resulting in a net loss of performance. A method of calculating the relative weightings of the Kalman filter gain matrix for a pure delay process has been presented and a method of calculating the general case has been suggested. It has been indicated that state feedback from dead time state estimates is not required. Finally, a set of tuning rules have been evolved and presented.

References

[1] M. Athans, "LQG problem," *IEEE Trans. Automat. Contr.*, vol. AC-16, p. 528, Dec. 1971.
[2] D. L. Kleinman, "On an iterative technique for Riccati equation computations," *IEEE Trans. Automat. Contr.*, vol. AC-13, Feb. 1968.
[3] T. Pappas, A. J. Laub, and N. R. Sandell, "On the numerical solution of the discrete-time matrix Riccati equation," *IEEE Trans. Automat. Contr.*, vol. AC-25, pp. 631–641, Aug. 1980.

On-Line Failure Detection in Nuclear Power Plant Instrumentation

J. LOUIS TYLEE

Abstract—**The functional redundancy approach to detecting instrument failures in nuclear power plant instrumentation is described and evaluated. This real-time method uses a bank of Kalman filters (one for each instrument) to generate optimal estimates of the plant state. By performing consistency checks between the output of each filter, failed instruments can be identified. The technique is used to design an instrument failure detection system for a pressurized water reactor (PWR) pressurizer. Actual pressurizer data are used to demonstrate the capabilities of the functional redundancy methods.**

I. Introduction

In nuclear power plants, it is important that instrument failures be detected and accommodated before significant performance degradation results. If such a detection cannot be made, total shutdown of the plant may be necessary, resulting in lost power production and lost revenues to the operating utility. One technique for detecting failed instruments in a power plant is hardware redundancy, where a two-out-of-three voting logic is used to eliminate faulty measurements. A more common failure detection technique is operator reliance, where the operator uses his knowledge of the plant operating state and dynamics to determine if a certain measurement is unreasonable. With the hundreds of measurements made in a nuclear facility, the cost of redundant sensors for each measurement is high and it is impossible for the plant operator to continually monitor each instrument for possible failures. Clearly, improvement in the detection of failed instruments is needed.

Recent incidents at Three Mile Island and other nuclear facilities have demonstrated this need for better failure detection capabilities. Digital computers are now being brought into the control room in an attempt to provide the operator (via CRT graphics) with information needed to properly ascertain plant status. If these computers could also somehow process the measurements available to them and make a decision as to whether a certain instrument is failed or unfailed, improvements in plant safety and operation would be possible. Several such digital computer-based failure detection schemes have been researched in the past few years [1]. One approach to failure detection, and the one described here, is known as functional (or analytical) redundancy. A functionally redundant instrument failure detection system is shown in Fig. 1.

In Fig. 1, a dynamic system, described by a state vector x, is excited by a vector of known inputs u. These inputs and available measurements y are used to drive a bank of p Kalman filters (one for each of p measurements). Each Kalman filter is designed, via selection of the measurement subvector y^i and other filter characteristics, to be sensitive

Manuscript received February 22, 1982; revised August 17, 1982. This work was supported by the U.S. Department of Energy under DOE Contract DE-AC07-76ID01570.

The author is with EG&G Idaho, Inc., Idaho Falls, ID 83415.

Reprinted from *IEEE Trans. Automat. Contr.*, vol. AC-28, pp. 406–415, Mar. 1983.

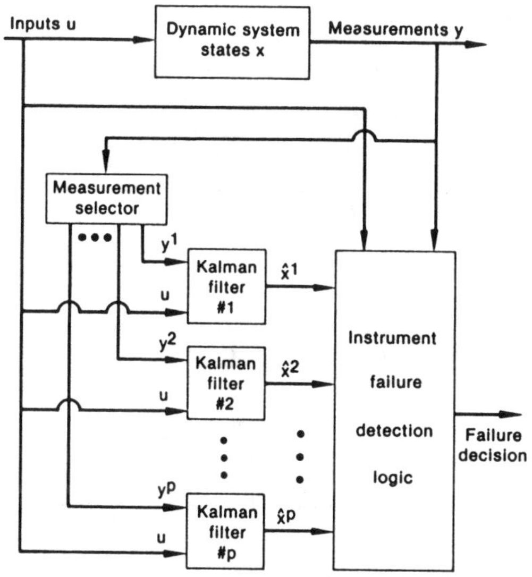

Fig. 1. Instrument failure detection system.

to a failure in just one instrument. Checks are performed on the redundant state estimates $\hat{x}^i$, $i=1,\cdots,p$ in the instrument failure detection logic. If these checks show one estimate to be inconsistent with the other estimates, or inconsistent with the current plant operating point, it is assumed that a failure has occurred in the instrument associated with the filter generating the anomalous estimate.

In this paper, a real-time instrument failure detection system, based on functional redundancy, for a pressurized water reactor (PWR) pressurizer, is developed and evaluated. The pressurizer was selected because it is an essential component in PWR plants and it is imperative that the pressurizer instrumentation provide reliable accurate readings. In the following sections the Kalman filter estimator formulation is developed and a linear discrete time pressurizer model is described. The sensitization of each Kalman filter to instrument failures is addressed and the failure decision process is detailed. Finally, the ability of the system to detect failures is investigated using actual pressurizer data recorded at a small test reactor plant.

II. Kalman Filter Estimators

No effort is made here to derive the Kalman filter equations. Several excellent texts [2]–[4] provide such derivations. Instead, the required form of the plant state and measurement equations and the filter equations are presented with brief descriptions of their use. Also discussed are the properties of the Kalman filter innovations.

A. Plant Dynamics Equations

For the ith Kalman filter in the instrument failure detection system, it is assumed that the plant dynamics can be modeled as a linear shift-invariant discrete time system of the form

$$x^i(k+1)=\Phi x^i(k)+\Theta u(k)+w^i(k) \tag{1}$$

where $x^i(k)$ is the plant state at time k, $u(k)$ a deterministic control input, and $w^i(k)$ a zero-mean white disturbance vector with constant covariance matrix Q^i. Also, Φ is the state transition matrix from k to $k+1$, and Θ is the system input matrix; Φ and Θ are assumed identical for each individual Kalman filter. Note that the work here is based strictly on a linear model. Reasons for this restriction include a simpler formulation, much faster time propagation, and the ability to directly apply optimal control (if desired). The discrete time filter is used instead of a continuous time filter since it is anticipated that the failure detection system will be implemented on a digital, rather than analog, computer. Thus, if the plant dynamics are derived from nonlinear continuous differential equations, a first step in the design of a failure detection scheme is to linearize and discretize the plant model to cast it in the form of (1).

Similarly, a model of plant measurements is required. This must be a linear model of the form

$$y(k)=Cx(k)+v(k) \tag{2}$$

where $v(k)$ is a zero-mean white measurement noise, uncorrelated with plant noise, with covariance matrix R. For the ith Kalman filter, the measurements are defined by

$$y^i(k)=C^ix^i(k)+v^i(k). \tag{3}$$

In (3), C^i is a rowwise partition of C that selects the desired measurement subvector for filter i. The white process $v^i(k)$ is zero-mean with covariance R^i.

Additional information required by each Kalman filter is an initial estimate of the state vector $x^i(0)$ and the covariance of that estimate.

B. Kalman Filter Equations

At each time k, a plant measurement $y(k)$ is taken. From $y(k)$, the various measurement subvectors are formed and used to update the current estimate of the plant state using

$$\hat{x}^i(k|k)=\hat{x}^i(k|k-1)+K^i\left[y^i(k)-C^i\hat{x}^i(k|k-1)\right] \tag{4}$$

where $\hat{x}^i(k|k-1)$ is the ith filter state estimate at k based on measurements up to time $k-1$.

In (4), the matrix K^i is the steady-state Kalman gain for filter i. The use of a steady-state rather than time-varying gain is possible because of the assumption of constant system model matrices (Φ,Θ,C^i) and constant covariance matrices (Q^i, R^i).[1] K^i is computed using

$$K^i=P^i(C^i)^T\left[C^iP^i(C^i)^T+R^i\right]^{-1}. \tag{5}$$

The matrix P^i is the steady-state estimate error covariance which is obtained by solving the algebraic Riccati equation

[1]The use of steady-state Kalman filters for the pressurizer instrument failure detection system is justified because 1) the pressurizer is assumed to have been in steady state for a long period prior to initialization, and 2) the pressurizer dynamics are adequately modeled by fixed system matrices even during large excursions from nominal steady state [5].

[4]

$$P^i = \Phi\left\{ P^i - P^i(C^i)^T\left[C^iP^i(C^i)^T + R^i\right]^{-1}C^iP^i\right\}\Phi^T + Q^i. \tag{6}$$

Prior to taking the next measurement at $k+1$, the state estimate is propagated ahead in time using the state dynamics equation

$$\hat{x}^i(k+1|k) = \Phi\hat{x}^i(k|k) + \Theta u(k). \tag{7}$$

C. Filter Innovations

A concept that will be useful later in defining instrument failure decision functions is that of the Kalman filter innovations, or residuals, sequence. The innovations for filter i, $\delta^i(k)$ are defined as the difference between the plant measurements at k and the measurement predictions at k or

$$\delta^i(k) = y^i(k) - C^i\hat{x}^i(k|k-1). \tag{8}$$

A fundamental property of the Kalman filter is that if the physical system actually evolves according to the state and measurement equations (1) and (3), the filter will generate a $\delta^i(k)$ vector that is zero-mean and white. Furthermore, for a steady-state Kalman filter, the innovations covariance matrix is constant and given by

$$V^i = C^iP^i(C^i)^T + R^i. \tag{9}$$

III. Pressurizer Model

In this section, a discrete time shift-invariant linear model of a PWR pressurizer is developed. First, a nonlinear continuous time third-order pressurizer model is derived from first principles. The required linearization and discretization steps are then described. Finally, the model statistics are obtained. The pressurizer modeled is that of the loss-of-fluid test (LOFT) reactor plant located at the Idaho National Engineering Laboratory. The LOFT plant is a scale model of a PWR and produces 50 MW (thermal) [6]. It is used to study the performance of engineered safety features of a commercial PWR during postulated loss-of-coolant accidents and anticipated operational transients. Due to the LOFT plant similarity to a full-scale plant, the pressurizer model developed here could easily be modified to represent a pressurizer in a commercial reactor facility.

A. Pressurizer Dynamics

In a nuclear power plant, the pressurizer (shown schematically in Fig. 2) is attached at the reactor outlet hot leg and acts as a surge tank in maintaining primary loop pressure. At steady state, the fluid in the pressurizer is approximately two-thirds saturated water and one-third saturated steam at a constant pressure. When the primary coolant temperature increases, the expanding fluid flows into the pressurizer through the surge line, increasing the liquid mass, compressing the vapor region, and increasing pressure. Conversely, a temperature decrease results in an outsurge and corresponding pressure drop. Additional pressure control is provided by a cold water spray and relief valves (to avoid overpressurization) and electric heaters (to increase pressure).

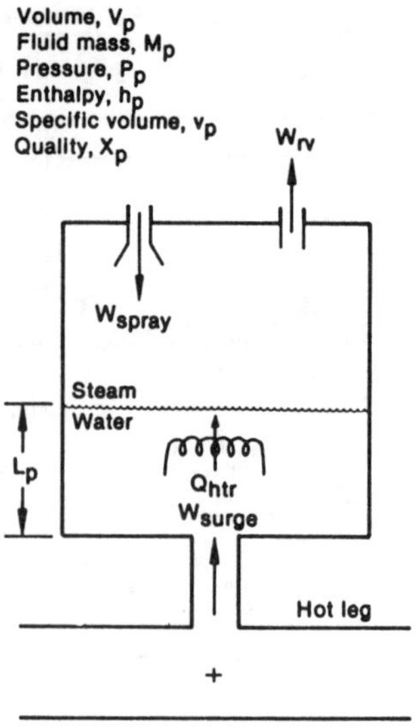

Fig. 2. Pressurizer schematic.

In the model developed here, steam and water in the pressurizer are assumed to be in a homogeneous saturated mixture. Applying mass and energy balances to this mixture results in the following two equations:

$$\frac{dM_p}{dt} = W_{\text{surge}} + W_{\text{spray}} - W_{rv} \tag{10}$$

$$\frac{d(M_ph_p)}{dt} = \frac{V_p}{J}\frac{dP_p}{dt} + Q_{htr} + W_{\text{surge}}h_f + W_{\text{spray}}h_{\text{spray}} - W_{rv}h_g \tag{11}$$

where t is the continuous time variable, J is a units conversion factor, h_f is the saturated water enthalpy, h_g is the saturated steam enthalpy, and h_{spray} is the known spray flow enthalpy. Other variables in (10) and (11) are defined in Fig. 2.

The desired state variables for the pressurizer model are P_p, pressure, and X_p, mixture quality. Equations for these variables are obtained by noting

$$v_p = V_p/M_p = v_f + X_p(v_g - v_f) \tag{12}$$

$$h_p = h_f + X_p(h_g - h_f) \tag{13}$$

where v_f and v_g are the specific volumes of saturated water and steam, respectively. The saturation properties in (12) and (13) can be expressed as direct functions of pressure using standard steam tables.

By substituting (12) and (13) into (10) and (11), one can solve simultaneously for dP_p/dt and dX_p/dt, the desired state equations. This rather lengthy manipulation yields

$$\frac{dP_p}{dt} = \frac{-v_p}{V_p \Lambda}\Bigg\{ v_p \frac{\partial h_p}{\partial X_p}(W_{surge} + W_{spray} - W_{rv}) + \frac{\partial v_p}{\partial X_p}\big[Q_{htr} + W_{surge}(h_f - h_p) + W_{spray}(h_{spray} - h_p) - W_{rv}(h_g - h_p)\big]\Bigg\} \tag{14}$$

$$\frac{dX_p}{dt} = \frac{v_p}{V_p \Lambda}\Bigg\{ v_p(W_{surge} + W_{spray} - W_{rv})\left[\frac{\partial h_p}{\partial P_p} - \frac{v_p}{J}\right] + \frac{\partial v_p}{\partial P_p}\big[Q_{htr} + W_{surge}(h_f - h_p) + W_{spray}(h_{spray} - h_p) - W_{rv}(h_g - h_p)\big]\Bigg\} \tag{15}$$

where

$$\Lambda = \frac{\partial v_p}{\partial P_p}\frac{\partial h_p}{\partial X_p} - \frac{\partial v_p}{\partial X_p}\left[\frac{\partial h_p}{\partial P_p} - \frac{v_p}{J}\right]. \tag{16}$$

An additional state of interest is the measured temperature of the saturated fluid mixture T_{pm}. Assuming first-order dynamics for the temperature sensor, this measurement can be characterized by a single time constant τ_p

$$\frac{dT_{pm}}{dt} = \frac{(T_p - T_{pm})}{\tau_p}. \tag{17}$$

The actual fluid temperature T_p is the saturation temperature based on pressure P_p. Equations (14), (15), and (17) then constitute the nonlinear pressurizer model state equations.

There are four inputs in this pressurizer model. The surge flow is computed using

$$W_{surge} = \begin{cases} K_s(\Delta P_{surge})^{1/2}, & \Delta P_{surge} > 0 \\ -K_s(|\Delta P_{surge}|)^{1/2}, & \Delta P_{surge} < 0 \end{cases} \tag{18}$$

where ΔP_{surge} is the measured pressure difference between the primary loop hot leg and pressurizer, and K_s is an empirically determined constant. Pressurizer spray is an on–off type control established by preset pressure setpoints. Similarly, pressurizer heater output and relief valve flow are on–off controls governed by specific setpoints. Measured pressure is used to establish control status in the pressurizer model. Note this forms a nonlinear feedback path between the pressure state and three of the inputs, which is essentially ignored in the model formulation. Future model changes will eliminate this feedback.

There are six measurements made on the LOFT pressurizer: three redundant level measurements (L_1, L_2, L_3), a pressure measurement, and temperature measurements in the vapor and liquid spaces. With the simple model developed here, the three level measurements cannot be represented individually. Hence, an average pressurizer water level is used

$$L_p = L_1 = L_2 = L_3 = A_p(1 - X_p)\frac{V_p}{v_p}v_f \tag{19}$$

where A_p converts water volume to water level. Recall that pressure is a system state, hence the pressure measurement model is quite simple. Also, since it was assumed that vapor and water in the pressurizer are always at saturation, temperatures in both regions will be equal to the temperature as modeled by (17).

For use in the instrument failure detection system, the nonlinear pressurizer model must be linearized and discretized. In functional form, the nonlinear model is summarized as

$$\dot{x} = f(x, u) \tag{20}$$

$$y = g(x) \tag{21}$$

where the system vectors x, u, and y are

$$x = \begin{bmatrix} X_p \\ P_p \\ T_{pm} \end{bmatrix} \tag{22}$$

$$u = \begin{bmatrix} W_{surge} \\ Q_{htr} \\ W_{spray} \\ W_{rv} \end{bmatrix} \tag{23}$$

$$y = \begin{bmatrix} L_1 \\ L_2 \\ L_3 \\ P_p \\ T_{pm} \end{bmatrix}. \tag{24}$$

Expanding (20) and (21) in a Taylor series (ignoring terms higher than first-order) about a nominal operating point $(\bar{x}, \bar{u})$ will result in three linear differential state equations and five linear measurement equations. Then applying standard discretization methods will yield the required discrete equations [7].

B. Pressurizer Covariance Matrices

The last point to address in developing the LOFT pressurizer model is establishment of the three covariance matrices used in the Kalman filter formulations. These include the initial estimate error covariance

$$P^i(0|0) = E\{[x^i(0) - \hat{x}^i(0)][x^i(0) - \hat{x}^i(0)]^T\} \tag{25}$$

($E\{\cdot\}$ is the statistical expectation operator), the nominal process noise covariance

$$Q = E\{w(k)w^T(k)\}, \tag{26}$$

and the nominal measurement noise covariance

$$R = E\{v(k)v^T(k)\}. \tag{27}$$

The initial estimate error covariance matrix is evaluated using representative uncertainties in initial pressurizer measurements of level, pressure, and temperature [8]. For the pressurizer, this matrix is assumed to be the same for each Kalman filter.

The values for the nominal measurement noise covariance matrix (R) were obtained from analysis of unpublished LOFT steady-state data at various power levels. This matrix is referred to as the nominal measurement noise covariance since in some later filter sensitization work, it will be modified slightly to form each R^i.

The elements of the Q matrix, the nominal process noise covariance, were selected to compensate for modeling errors and inherent noise. By varying the relative magnitudes of Q and R, the Kalman filter is designed to provide what are considered proper weightings to the process model and to the actual plant measurements in generating state estimates. In this work, these filter weightings were selected based on the results of pressurizer model validations [9]. In those validations, it was seen that the pressurizer model provided excellent predictions of water level, yet was somewhat deficient in predicting pressure and temperature. Thus, it was found that better filter performance was obtained with estimates of pressurizer level weighted primarily by the process model and estimates of pressure and temperature more dependent on actual data. Here, each Q^i matrix is assigned the same nominal value.

IV. Kalman Filter Sensitizations

An ideal system configuration for the functional redundancy instrument failure detection system in Fig. 1 would be where each measurement subvector $y^i(k)$ consists of just a single signal, that being the ith element of the measurement vector $y(k)$. That is, Kalman filter i is dedicated to (and sensitive to failures in) instrument i ($i = 1,\cdots,p$). Such a configuration is similar to the dedicated observer approach of Clark [10].

Then, if all instruments are operating properly, the p estimates of the state vector x will be nearly identical. If, however, a failure occurs in instrument 2, the faulty signal $y^2(k)$ will produce a faulty estimate $\hat{x}^2(k)$. The other state estimates will be unaffected since they do not use a signal from instrument 2. Hence, a comparison of the p-state estimates will show that $\hat{x}^2(k)$ is different from the other $p-1$ estimates which identifies instrument 2 as failed.

To implement this ideal configuration of a single sensor driving each Kalman filter, the question of system observability must be addressed. Recall that the instrument failure detection system (as implemented) uses steady-state Kalman gain matrices in each filter. One way to ensure that such steady-state solutions exist is to require the models inherent in the filters to represent completely observable systems [2]. This is what is done here. If observability is not possible using a single output signal, some other way of sensitizing the Kalman filters to instrument failures must be developed. Such a sensitization is described here, as is the choice of failure decision logic.

A. Observability

A system described by (1) and (3) is completely observable when the well-known observability matrix

$$G = \left[(C^i)^T \quad \Phi^T(C^i)^T \quad (\Phi^2)^T(C^i)^T \quad \cdots \quad (\Phi^T)^{n-1}(C^i)^T\right] \tag{28}$$

is of rank n [where n is the dimension of $x^i(k)$]. Computing G for the LOFT pressurizer at nominal conditions showed that complete observability was not possible using only a single instrument, i.e., where C^i is a row vector. Several measurement subvectors that produce a completely observable system were identified. The observable subvector selected for use here is

$$y^i = \begin{bmatrix} L^i \\ P_p \\ T_{pm} \end{bmatrix} \tag{29}$$

where L^i is some indication of water level for the ith Kalman filter.

B. Filter Design

Sensitization of the three filters dedicated to the three level measurements is straightforward. For these filters, the measurement subvectors are

$$y^1 = \begin{bmatrix} L_1 \\ P_p \\ T_{pm} \end{bmatrix} \tag{30}$$

$$y^2 = \begin{bmatrix} L_1 \\ P_p \\ T_{pm} \end{bmatrix} \tag{31}$$

$$y^3 = \begin{bmatrix} L_3 \\ P_p \\ T_{pm} \end{bmatrix}. \tag{32}$$

Examining (30)–(32), it is noted that a failure in one of the level measurements will affect only a single Kalman filter estimate, as is desired. Kalman filters 1–3 each use the nominal process and measurement noise covariance matrices.

The design of Kalman filters 4 and 5, the filters sensitive to pressure and temperature sensor failures, respectively, is more involved. The subvectors used in Kalman filters 4 and 5 are identical:

$$y^4 = y^5 = \begin{bmatrix} \bar{L} \\ P_p \\ T_{pm} \end{bmatrix} \tag{33}$$

where $\bar{L}$ is the average measured water level

$$\bar{L} = (L_1 + L_2 + L_3)/3. \tag{34}$$

The use of $\bar{L}$ rather than a particular level reading provides a desired insensitivity of y^4 and y^5 to failures in a level instrument. Since the same subvectors are used to drive Kalman filters 4 and 5, further sensitization to instrument failures is needed. This additional sensitization is obtained through modification of the individual filter measurement noise covariance matrices.

In the pressurizer failure detection system, the variance of the temperature sensor noise in the R matrix associated with filter 4 is increased by a factor of 100 over the nominal value. This essentially eliminates the temperature measurement in computing $\hat{x}^4(k)$, making this estimate primarily dependent on $\bar{L}$ and P_p. Similarly, to ensure that $\hat{x}^5(k)$, the state estimate sensitive to temperature instrument failures, does not significantly depend on P_p, the variance associated with pressure sensor noise in filter 5 is increased to 100 times nominal. Fortunately, as long as $\bar{L}$ remains, this virtual elimination of pressure and temperature sensors from filters 4 and 5 does not lead to observability problems. That is, steady-state Kalman gains for these two filters do exist.

C. Failure Decision Functions

The design of the five Kalman filters is complete, each filter producing a state estimate sensitive to a particular instrument failure. These state estimates are now used to generate a failure decision function for each filter. The purpose of such decision functions is to provide a clear indication of whether or not an instrument failure exists. The generation of the failure decision functions is detailed in Fig. 3.

In Section II-C, it was seen that the innovations sequence of a Kalman filter is white and zero-mean with known covariance. If an instrument failure occurs, the filter innovations will no longer have these statistics since the process model in the Kalman filter (which was developed assuming no failures) no longer represents the actual dynamics. It is this property that is utilized in developing the failure decision functions. In Fig. 3, the five Kalman filters, driven by the specified measurement subvectors and the input vector, produce the redundant state estimates $\hat{x}^i$, $i = 1,2,3,4,5$. These estimates are then used to generate a prediction of the measurement that each filter has been sensitized to. Subtracting each predicted measurement from the actual measurement forms the scalar innovations γ^i, $i = 1,2,3,4,5$.

A final step is the normalization of each γ^i to the variance in the estimate of γ^i. This variance is obtained

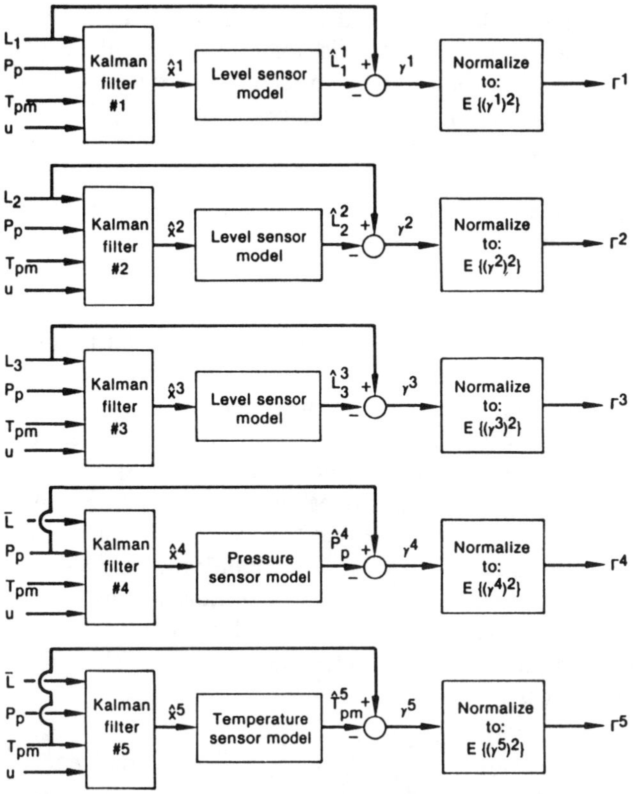

Fig. 3. Pressurizer failure decision functions generation.

from the diagonal elements of the innovations covariance matrix [see (9)] for the corresponding Kalman filter. Hence, the failure decision functions are

$$\Gamma^i = \left[(\gamma^i)^2 / E\{(\gamma^i)^2\}\right]^{1/2}, \qquad i = 1,2,3,4,5 \tag{35}$$

where

$$\gamma^1 = L_1 - \hat{L}_1^1 \tag{36}$$

$$\gamma^2 = L_2 - \hat{L}_2^2 \tag{37}$$

$$\gamma^3 = L_3 - \hat{L}_3^3 \tag{38}$$

$$\gamma^4 = P_p - \hat{P}_p^4 \tag{39}$$

$$\gamma^5 = T_{pm} - \hat{T}_{pm}^5. \tag{40}$$

In (36)–(40), $\hat{L}_1^1$ is the prediction of level 1 by filter 1, $\hat{L}_2^2$ is the prediction of level 2 by filter 2, etc.

D. Failure Decision

Having computed failure decision functions for each Kalman filter, the only question remaining is how to decide if a particular instrument has failed. Several possibilities exist. Detailed statistical significance tests on each Γ^i could be performed to ascertain the probability of failure, comparisons of the decision function signatures could be made to detect possible anomalies, or threshold crossing tests could be made. In this study, the latter test, in addition to simple cross checks between the decision functions, will be used.

Examining (35) shows that a simple interpretation of the failure decision functions is that Γ^i represents (in a Gaussian sense) the number of standard deviations Γ^i is from a zero-mean white sequence produced by Kalman filter i when no failure exists. Hence, the larger Γ^i is, the more likely a failure is to exist. So, a first test for failure is to see if the largest failure decision function (Γ^m) exceeds a predetermined threshold ϵ:

$$\Gamma^m > \epsilon. \tag{41}$$

The satisfaction of the inequality in (41) is not sufficient to declare a failure, however. A second test is required.

Other factors, besides failed instruments, such as sudden changes in process dynamics or inputs, can cause the Kalman filter innovations to exceed the limit specified in (41). These changes should be reflected in each Kalman filter, however. Thus, a second test for the existence of an instrument failure is to see if Γ^m is significantly higher than all other Γ^i values. Such a test would indicate whether the high value is more likely due to an instrument failure or simply a change in the underlying dynamics. This cross-check test is expressed by

$$\Gamma^m > \alpha \Gamma^i, \qquad \text{all } i \neq m. \tag{42}$$

If both (41) and (42) are satisfied, a failure in instrument m is declared.

The selection of values for ϵ and α could be made by considering tradeoffs between the probability of a false alarm and the detection probabilities associated with each instrument. The approach taken here, however, was to simply choose ϵ and α through various simulation studies at a nominal condition. For the LOFT pressurizer, such studies have shown that values of $\epsilon = 6$ and $\alpha = 2$ seem to provide adequate failure detection.

V. Experimental Results

The current version of the pressurizer failure detection system is programmed in Fortran and implemented on a PDP 11/55 minicomputer. Several sets of simulated and actual LOFT plant measurements are stored on magnetic disks and available for retrieval during system operation. The programs run in real-time and, following extensive simulation studies, plans call for the installation of the PDP (with all of the detection system software) at the LOFT facility to test the system performance during actual LOFT operation.

The failure detection system program is divided into two sections: system initialization, and system operation and failure detection. In the initialization portion, input data are read, an initial plant measurement is taken, and the pressurizer operating point ($\bar{x}$, $\bar{u}$) is established. Using this information, the filter initial state estimates [$\hat{x}^i(0|0) = \bar{x}$, for all i] and the required plant model matrices Φ, Θ, and C^i are computed, as are the five filter gain matrices K^i and steady-state innovations covariance matrices V^i. The complete initialization procedure needs about 10 s of computation time on the PDP computer.

In the operating and detection mode, every second a pressurizer measurement is made and the input vector established. As mentioned in the model description, surge flow is computed using an empirical correlation with measured surge line pressure drop and values of other elements of $u(k)$, namely heaters, spray, and relief valve flow, are based on measured pressurizer pressure. Next, the five failure decision functions are calculated and the two failure decision tests applied (if a failure is detected, a declaration is made). The filter state estimates are then updated using (4) and propagated using (7).

One cycle of the failure detection procedure uses 110 ms of PDP processing time. This is adequate since preliminary simulation studies have shown that sampling the plant measurements every second (as is done) provides excellent failure detection results. The remaining time (890 ms) in each cycle of the process allows for future expansion and modification of the failure detection algorithms.

The instrument failure detection system has been evaluated using numerous sets of simulated plant data and some actual LOFT pressurizer data. Three sets of the actual data are examined here. The three transients studied are from the LOFT L6 experiment series [8]. These were experiments run at the LOFT facility to simulate typical PWR operational transients. An actual instrument failure occurs during one of the L6 experiments, hence the detection system is well tested. In the other transients, although no instrument failures occur, the detection system ability to recognize sudden measurement swings as being due to changes in pressurizer dynamics is challenged.

A. L6-1 Loss of Steam Load Transient

The first test, designated L6-1, is a loss of secondary steam load. The test was initiated by closing the main steam control valve (MSCV) that regulates steam flow from the steam generator. Fig. 4 displays pressurizer response during the first 100 s of the L6-1 transient. Closing the MSCV decreases the rate of heat transfer from the primary coolant to the steam generator tubes causing an increase in primary loop temperatures. This results in an increase in ΔP_{surge}. At 21.8 s, an automatic plant trip (reactor scram) occurs due to high primary pressure, causing a rapid drop in primary coolant temperatures and a corresponding decrease in the surge line ΔP. Note both the initial increase in each water level measurement due to primary coolant expansion and the subsequent level drop following plant scram. Also, note the corresponding recorded pressure and water temperature responses. In this and the other transients, water temperature is used to represent the average mixture temperature in the pressurizer model. A check of all the L6-1 curves shows no obvious instrument failures.

Processing these measurements using the detection system yielded the results in Fig. 5. The failure decision

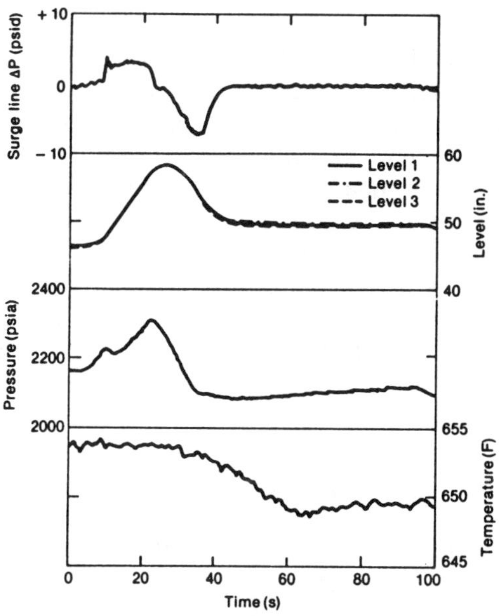

Fig. 4. LOFT L6-1 transient pressurizer measurements.

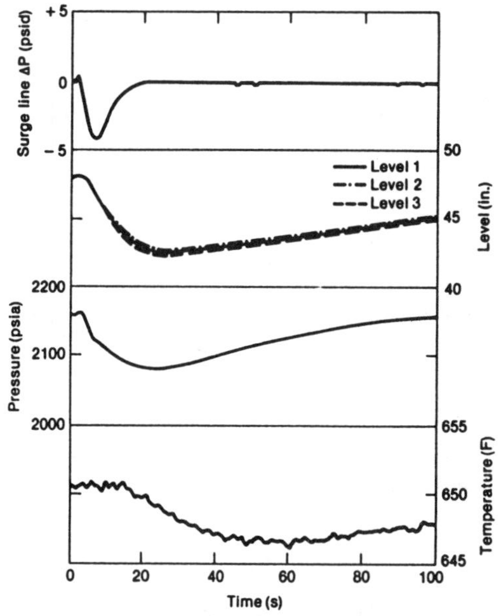

Fig. 6. LOFT L6-2 transient pressurizer measurements.

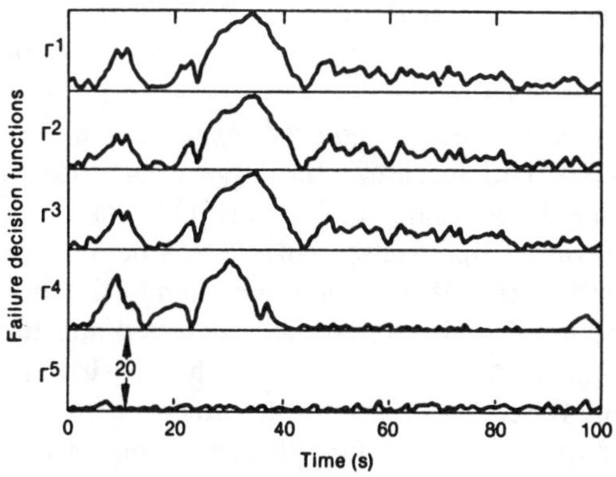

Fig. 5. LOFT L6-1 transient failure decision functions.

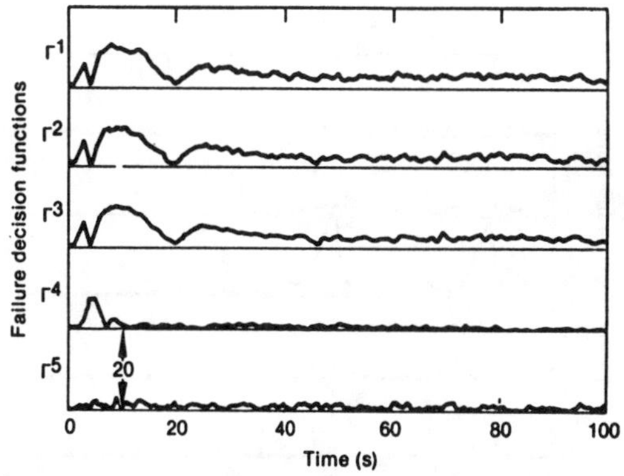

Fig. 7. LOFT L6-2 transient failure decision functions.

functions in this figure suggest the possibility of some type of failure in that several of the Γ^i functions attain high values. As would be expected in such a severe transient, the ϵ threshold is crossed several times by the decision functions. However, the cross-check test (42) between each Γ^i is never successful, correctly indicating that the large Γ's are due to process dynamics changes rather than to instrument failures.

B. L6-2 Loss of Primary Coolant Flow Transient

The LOFT L6-2 experiment was a loss of forced primary coolant flow initiated by tripping power to both primary coolant pumps. Two seconds into the transient, the plant protection system scrammed the reactor on indication of low primary coolant flow. This almost immediate shutdown of the heat generation process resulted in a large negative surge line ΔP and corresponding level, pressure, and temperature measurements in Fig. 6.

As in the L6-1 transient, large values of the failure decision functions in Fig. 7 indicate a possible failure following the plant scram. However, unlike the successful distinguishing of plant dynamics changes from instrument failures in the L6-1 test, the detection system here indicates a pressure sensor failure. Such a failure is not evident in the actual data, so this is most likely a false alarm. At $t = 4$ s, where the failure is declared, the failure decision function values are $\Gamma^1 = 0.11$, $\Gamma^2 = 0.09$, $\Gamma^3 = 0.01$, $\Gamma^4 = 7.46$, and $\Gamma^5 = 0.44$, and hence the ϵ threshold would have to be greater than 7.46 to eliminate this seemingly incorrect detection. Raising ϵ to such a value would drop the failure detection probability, an undesirable result.

A possible reason for the false alarm is a weakness in the surge flow model in (18). Note in Fig. 6 that the false alarm is coincident with reactor scram. Also in Fig. 6, it is seen that upon scram there is a rapid change (almost step-like) in the surge line pressure drop. The static surge model does not account for fluid inertia; hence the sudden change in ΔP_{surge} results in an equally rapid change in calculated surge flow. The linear pressurizer model cannot provide an accurate pressure calculation for sudden surges, resulting in a large innovation (γ^4), producing a false pressure

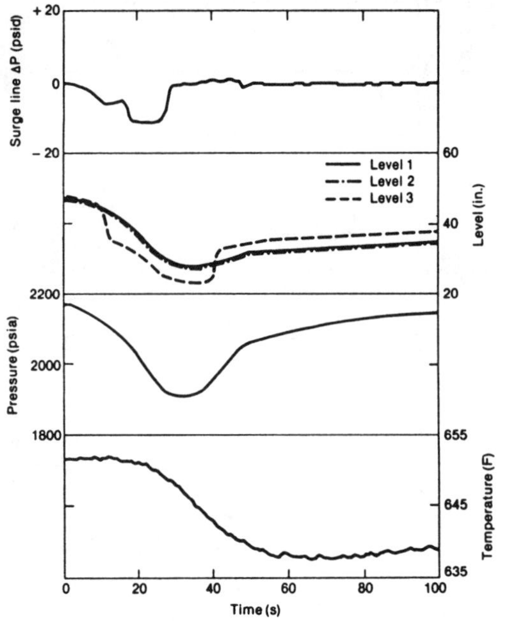

Fig. 8. LOFT L6-3 transient pressurizer measurements.

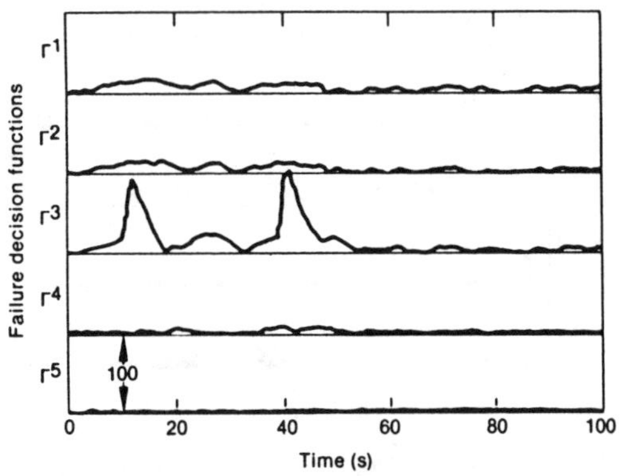

Fig. 9. LOFT L6-3 transient failure decision functions.

sensor failure indication. Adding momentum dynamics to the surge flow model should decrease the false alarm probability of the instrument failure detection system.

C. L6-3 Steam Load Increase Transient

The L6-3 experiment is an interesting test of the functional redundancy failure detection scheme in that an instrument failure is apparent. The L6-3 test simulated an excessive load increase incident in a large PWR. The experiment was initiated by opening the MSCV at its maximum rate. The increased cooling capacity of the steam generator then caused primary coolant temperature and pressure to drop, resulting in a reactor scram on indication of low primary loop pressure at 15.6 s. The surge line ΔP and other pressurizer measurements made during this test are illustrated in Fig. 8. The initial drop in ΔP is due to cooling of the primary loop caused by the MSCV opening. At 15.6 s, ΔP decreases at even a faster rate due to plant scram. At about 26 s, two high pressure injection system (HPIS) pumps were turned on in an effort to increase core cooling. The HPIS pumps were turned off at 50 s.

The apparent instrument failure is shown in the level response. While level 1 and level 2 measurements compare quite closely, the measurement of level 3 exhibits two sudden changes, one at about 10 s and another at 40 s. Using these measurements in the failure detection system resulted in the curves of Fig. 9.

The failure decision functions computed during L6-3 display an obvious indication of the sudden swings in the level 3 measurement. Γ^3 near 10 and 40 s is far larger than any other Γ value. The level 3 sensor is declared to be failed.

VI. Conclusions

In this paper, several capabilities of a real-time functionally redundant instrument failure detection system developed for a PWR pressurizer have been described and evaluated. Tests run using sets of actual LOFT data highlighted the utility of the failure detection system. These LOFT tests, designated L6-1, L6-2, and L6-3, represent fairly severe operational transients that can occur in a commercial reactor plant. In the L6-1 and L6-2 (if the surge model weakness is ignored) experiments, the ability of the system to recognize the large excursions in plant measurements as being due to manipulations of control actuators rather than sensor failures was demonstrated. In the other test (L6-3), two step-like changes in the level 3 sensor occurred and were correctly detected and identified by the system. Many other tests of the instrument failure detection system have been performed using simulated data and other LOFT data with equally promising results [5].

Although tests of the failure detection system using both simulated measurements and LOFT L6 data demonstrated the power and accuracy of functional redundancy methods, they also pointed out a few shortcomings. For example, in cases of pressure sensor failures, incorrect control actions (established based on measured pressure) driving the Kalman filters can cause false alarms. A suggested solution to this problem is to use positive indicators (on–off) of plant control system status. Such indicators would ensure that the control vector in the pressurizer model coincides with actual plant controls and eliminate the existing feedback between the inputs and measure pressure. Also discovered in system tests was a need for an improved surge flow model.

Additional sophistication in the failure decision logic could result in better detection capabilities. The current two-stage logic is very simple and, in some cases, prone to false alarms and missed detections. One possible change would be to use more information from the Kalman filters. For example, in the current configuration, each Kalman filter uses only one component of the measurement estimate at a single time step to generate the failure decision functions. An improvement to this approach would be to use the entire measurement subvector from each filter. These innovations vectors (8) could still be normalized to allow comparisons between the five Γ functions. Using just the innovations at time k, the normalized failure decision function for filter i is

$$\Gamma^i = \{\delta^i(k)\}^T \{V^i\}^{-1} \{\delta^i(k)\}. \tag{43}$$

Another approach is to use the sum of the last N normalized innovations vectors giving Γ^i as

$$\Gamma^i = \sum_{j=k-N+1}^{k} \{\delta^i(j)\}^T \{V^i\}^{-1} \{\delta^i(j)\}. \tag{44}$$

Willsky [1] states that the Γ^i in (44) is a chi-squared random variable with Np degrees of freedom (p is the length of the innovations vector δ^i). Hence, with the aid of chi-squared tables, values of the ϵ threshold could be established based on the desired false alarm probability. This would be a substantial improvement over the current arbitrary assignment of a value to ϵ.

It should be mentioned that the failure detection system developed here is a refinement of a similar concept studied by Clark and Campbell [11]. The improvements made include: presentation of a logical system design procedure, improved Kalman filter sensitizations, normalization of the failure decision functions, development of a failure detection and identification logic, physically realistic covariance data, performance evaluation using actual pressurizer data, and system operation in a real-time environment.

Another approach to detecting failures in the LOFT pressurizer has also been studied [12]. The generalized likelihood ratio (GLR) technique performs statistical tests on the innovations sequence of a single Kalman filter to determine if a failure has occurred. If these failure tests are successful, the GLR method determines in what sensor the failure appears, the failure magnitude, the failure type, and when the failure occurred. The GLR computation also allows on-line validation of the linear process model inherent in the Kalman filter [13]. Specific rules for computing failure thresholds for the GLR test, based on desired false alarm and detection probabilities, are available. The need for only one Kalman filter, the additional failure information provided, and the established theory of the failure detection and identification logic associated with the GLR method make it an attractive choice. A major disadvantage of GLR failure detection, however, is that specific instrument failure modes (e.g., bias, jump, ramp) must be hypothesized *a priori*. The GLR approach has not been evaluated in real time. Only direct comparisons of the real-time performance of the two methods during identical test conditions should be used to select one technique over the other.

The next step in the pressurizer failure detection system development is a further testing of its ability using a hybrid computer model of the LOFT plant to provide simulated plant transient data. Following these simulation studies, the system may be installed at the LOFT facility. Using functional redundancy techniques during actual LOFT operation would help demonstrate feasibility of commercial implementation.

REFERENCES

[1] A. S. Willsky, "A survey of design methods in failure detection in dynamic systems," *Automatica*, vol. 12, Nov. 1976.

[2] A. Gelb, Ed., *Applied Optimal Estimation*. Cambridge, MA: M.I.T. Press, 1974.

[3] A. P. Sage and J. L. Melsa, *Estimation Theory with Applications to Communications and Control*. Huntington, NY: Krieger, 1979.

[4] B. D. O. Anderson and J. B. Moore, *Optimal Filtering*. Englewood Cliffs, NJ: Prentice-Hall, 1979.

[5] J. L. Tylee, "Real-time instrument failure detection in the LOFT pressurizer using functional redundancy," EG&G Idaho, Inc., Rep. EGG-EE-5518, July 1982.

[6] H. L. Coplen and L. J. Ybarrando, "Loss-of-fluid test integral test facility and program," *Nucl. Safety*, vol. 15, Nov.–Dec. 1974.

[7] J. L. Tylee, "Linearization and discretization of nonlinear systems not at equilibrium," submitted to *IEEE Trans. Automat. Contr.*

[8] D. L. Batt and J. M. Carpenter, "Experiment data report for LOFT anticipated transient experiments L6-1, L6-2, and L6-3," EG&G Idaho, Inc., Rep. EGG-2067, Dec. 1980.

[9] J. L. Tylee, "Validation of a simple PWR pressurizer model," in *Proc. 1982 Summer Comput. Simul. Conf.*, Denver, CO, July 1982.

[10] R. N. Clark, "The dedicated observer approach to instrument failure detection," in *Proc. 18th IEEE Conf. Decision Contr.*, Fort Lauderdale, FL, Dec. 1979.

[11] R. N. Clark and B. Campbell, "Instrument fault detection in a pressurized water reactor pressurizer," *Nucl. Technol.*, vol. 56, Jan. 1982.

[12] J. L. Tylee, "A generalized likelihood ratio approach to detecting and identifying failures in pressurizer instrumentation," *Nucl. Technol.*, vol. 56, Mar. 1982.

[13] —, "On-line validation of linear process models using generalized likelihood ratios," in *Proc. 20th IEEE Conf. Decision Contr.*, San Diego, CA, Dec. 1981.

The Application of Kalman Filtering Estimation Techniques in Power Station Control Systems

JOHN N. WALLACE AND RAY CLARKE, MEMBER, IEEE

Abstract—**This paper describes the application of Kalman filter estimation techniques to an operational 500 MW oil-fired boiler/turbine unit. In common with most physical processes, the plant contains nonlinearities, unknown or poorly defined variables, and bias disturbances. These problems are overcome by the use of bias states which are appended to the system state equations.**

List of Symbols

A	system characteristic matrix
B	system input characteristic matrix
$\boldsymbol{b}_k$	bias states
C	system output characteristics matrix
C_1	physical constant in superheater model
C_2	physical constant in superheater model
$D(s)$	superheater transfer function $h_o(s)/h_i(s)$
D_o	superheater steady-state incremental gain
d_o	superheater steady-state of one element of the model
$E(s)$	superheater transfer function $h_o(s)/h_s(s)$
$\boldsymbol{e}_k$	state error vector
$\tilde{\boldsymbol{e}}_k$	modified state error vector
$\boldsymbol{F}_k$	discrete characteristic matrix
$\boldsymbol{F}_k^P$	discrete characteristic matrix (superheater model)
$\tilde{\boldsymbol{F}}_k$	discrete parameter error
$[f_{ij}]$	elements of F_k
$\boldsymbol{G}_k$	discrete input characteristic matrix
$\boldsymbol{G}_k^P$	discrete input characteristic matrix (superheater model)
$\tilde{G}_k$	discrete parameter error (i.e., input matrix)
$[g_{ij}]$	elements of G_k
H_k^P	discrete output matrix (superheater model)
h_s	enthalpy pick-up in superheater or enthalpy of steam at drum outlet in load/pressure model
P_d	drum pressure
P_s	superheat pressure
L_k	bias state input matrix
M_k	bias state characteristic matrix
N_k	bias state output matrix
P_{set}	superheater press set point
P_t	HP turbine inlet pressure
Q	heat input to boiler (or boiler firing rate)
Q_{set}	heat input set point (normalized with respect to Qmax)
Q_{max}	heat input required to produce 500 MW output
Q^f	covariance of process noise in superheater model
R	resistance of governor valve to steam flow
R_s	resistance of superheater to steam flow
R^f	covariance of measurement noise in superheater model
$\boldsymbol{u}$	input vector
$\boldsymbol{v}$	measurement noise vector
$\boldsymbol{w}$	process noise vector
W_{max}	steam flow at 500 MW output
W_s	superheater steam flow
W_f	boiler feedwater flow
$\boldsymbol{x}$	state vector
$\boldsymbol{x}_k^P$	state vector (superheater model)
$\hat{\boldsymbol{x}}_k$	filtered state vector
y_k	output vector
z_k	measurement vector
δt	time step in discrete transition matrix
Δu	variation in control input over successive time steps $u_k - u_{k-1}$
α	gain term in parameter adaption algorithm
α_1, α_2, α_3	physical constants in superheater model
η	$P_t/P_s = R_T/R$—normalized governor valve position.

Manuscript received March 12, 1982; revised August 30, 1982.
The authors are with the Scientific Services Department, Central Electricity Generating Board, South Western Region, Bedminster Down, Bristol, England.

I. Introduction

THIS paper is based upon work which has been carried out at Pembroke Power Station by the southwest region of the Central Electricity Generating Board. Pembroke Power Station operates four × 500 MW oil-fired boiler/turbine units and was first commissioned in 1969/1970. The subsequent increase in the price of fuel oil dictated that the boilers should no longer be operated at base load, but should be flexible and, if necessary, shut down at night and be restarted the following morning (referred to as two-shifting).

Reprinted from *IEEE Trans. Automat. Contr.*, vol. AC-28, pp. 416–427, Mar. 1983.

Operational experience and plant trials indicated that with the existing boiler controls, the new operating regime would lead to increased wear and ultimately to thermal fatigue failure. A high-quality analog control system was already under development at a sister power station at Fawley when the Pembroke scheme was planned in 1976/1977. However, the rapid development in microprocessor technology during the intervening period had led to a large reduction in digital computing costs. At the same time, the development of a real-time high-level operating system—SWEPSPEED—for the display, analysis, and control of plant data was well underway in the southwest region [5]. The decision was therefore taken to reequip Pembroke Power Station with a distributed digital control and monitoring system [11].

II. Theorectical Background

A. Overview

The choice of control strategy for process control need no longer be constrained by computer limitations, and the designer is now faced with a wide choice of competing techniques. Without denigrating any other approach, it was felt that the state-space method had much to commend it for the application under consideration. In particular, it offered the following.

1) State-space offers a generic structured approach which is both flexible and portable.

2) The state-space method can be implemented readily on small digital computers.

3) Many of the apparent complications such as the requirement for low-order plant models, assessment of measurement noise and disturbance levels, and specification of control performance indexes are overcome as part of any major control system exercise.

4) The state-space approach decomposes into the dual problems of state estimation and state feedback.

This paper is concerned with the former problem, although the implications on the control problem will be discussed. A fuller treatment of the control problem is given in a separate paper [4].

B. Model Structure

Implicit in the development of any estimation or optimal control algorithm is the requirement for a mathematical description of the static and dynamic behavior of the plant. The models used in the Pembroke project were developed from a simplified theoretical analysis of the fundamental thermodynamics of the plant. They are formulated in a state-space form

$$x_{k+1} = F_k(x,u)x_k + G_k(x,u)u_k + w_k \tag{2.1}$$

$$z_k = H_k(x,u)x_k + b_k \tag{2.2}$$

where w_k and u_k are independent white noise vectors and all other variables have their usual meaning.

The thermodynamic equations are usually derived in continuous form

$$\dot{x} = A(x,u)x + B(x,u)u \tag{2.3}$$

and the discrete form is computed from the transition matrix [7]

$$\Phi(k) = I + A\delta t + A^2\frac{\delta_t^2}{2!} \tag{2.4}$$

$$F_k = \Phi(k) \tag{2.5}$$

$$G_k = \int_{k\delta t}^{(k+1)\delta t} \Phi[(k+1)\delta t - \tau]Bu(\tau)\,d\tau \tag{2.6}$$

where δt is the sampling interval.

The matrices F, G, and H will normally contain time-varying parameters which are functions of the state vector x and the input vector u. This dependency is usually an intrinsic part of the plant characteristics and cannot be neglected if the model is to be valid over a wide range of operation conditions.

Three principle difficulties now arise if the model is to be incorporated into a Kalman filter operating on plant data.

1) The model only approximates real plant behavior, and so parameters may be ill-defined or time varying.

2) The plant is subject to disturbances of both a measurable and unmeasurable variety.

3) The statistics of both the measurement noise and the process noise are ill-defined and often nonstationary.

The problem of unknown noise statistics has been examined in several papers (e.g., [6]) and is perhaps the least severe of the three problems listed. An initial estimate of the noise variance can usually be derived from physical reasoning, and the variances are then treated as design parameters which can be adjusted to produce an "acceptable" response from the Kalman filter.

Unknown parameters can be dealt with by including them as additional states in an extended Kalman filter. However, a more satisfactory approach is to augment the state equations with additional bias states, which are coupled to the plant states and driven by white noise. This approach is based on earlier work by both Balchen and Jazwinski [1], [7] and can be developed to deal with the problem of unknown parameters or unknown disturbances. In the context of the Pembroke scheme, the most appropriate form of disturbance signal is one which is constant in the short term, but subject to a long term drift. In mathematical terms, this can be represented by the output of an integrator driven by a "white" noise input.

The bias states are thus represented by an equation of the form

$$b_{k+1} = b_k + n_k + L_k u_k \tag{2.7}$$

where n_k is an independent white noise vector and L_k is a known matrix of appropriate dimensions.

The bias states will not in general be functions of the input vector u_k; but the inclusion of the term $L_k u_k$ can be useful in certain instances since it allows the possibility of controllable biases and also enables feedforward information to be incorporated into the bias estimates.

The augmented state equations thus become

$$\begin{bmatrix} x_{k+1} \\ \hline b_{k+1} \end{bmatrix} = \begin{bmatrix} F_k & M_k \\ \hline 0 & I \end{bmatrix} \begin{bmatrix} x_k \\ \hline b_k \end{bmatrix} + \begin{bmatrix} G_k \\ \hline L_k \end{bmatrix} u_k + \begin{bmatrix} w_k \\ \hline n_k \end{bmatrix} \quad (2.8)$$

$$z_{k+1} = [H_k \mid N_k] \begin{bmatrix} x_k \\ \hline b_k \end{bmatrix} + v_k. \quad (2.9)$$

C. Adaptive Kalman Filter Algorithms

Very few of the parameters in the boiler models used at Pembroke Power Station are constants which can be calculated *a priori*. Most parameters can be divided into one of two classes; namely those which can be calculated on-line from known variables, and those which cannot be calculated explicitly, but which must be estimated by some means or other. The former category includes those variables which are known functions of the state variables or plant inputs. Also included are those parameters which are functions of some auxiliary measurement. In some cases an auxiliary variable may be the major source of parameter variations; in these circumstances a virtually constant discrete system model can be obtained by linking the sampling rate to the auxiliary variable. It is a truism, however, that the more measurements a controller requires, the more difficult it is to maintain and the less reliable it becomes.

Even if certain parameters can be measured, it is often better to rely upon good estimates rather than a multiplicity of measurements. Bias states can be used to provide a structured approach to estimating filter parameters.

Consider a plant which can be described exactly by the state equations

$$x_{k+1} = F_k x_k + G_k u_k + w_k \quad (2.10)$$

$$z_k = H_k x_k + v_k. \quad (2.11)$$

If F_k and/or G_k contain unknown or imprecisely known elements, they must be replaced by their estimated values $\hat{F}_k$ and $\hat{G}_k$, respectively. The parameter errors can then be compensated for by the addition of bias states so that the augmented state equations become

$$x_{k+1} = \hat{F}_k x_k + \hat{G}_k u_k + w_k + M_k b_k \quad (2.12)$$

$$b_{k+1} = b_k + n_k \quad (2.13)$$

$$z_k = \hat{H}_k x_k + N_k b_k + u_k. \quad (2.14)$$

The bias states can then be related to the parameter errors by the equations

$$-M_k b_k = \tilde{F}_k x_k + \tilde{G}_k u_k \quad (2.15)$$

$$\tilde{F}_k = \hat{F}_k - F_k \quad (2.16)$$

$$\tilde{G}_k = \hat{G}_k - G_k. \quad (2.17)$$

x_k and b_k can then be estimated by applying the normal Kalman filter to the augmented state equations. The use of bias estimates to update the model parameters was first proposed by Jazwinski [7] who used off-line linear regression on an equation in the form of (2.15); but with b_k replaced by $\hat{b}_k$. This can be developed into an on-line technique by considering the approximate state equation

$$\tilde{x}_{k+1} = \hat{F}_k \tilde{x}_k + \hat{G}_k u_k + w_k \quad (2.18)$$

and the error term

$$e_k = \tilde{x}_k - x_k \quad (2.19)$$

$$e_{k+1} = \hat{F}_k e_k + \tilde{F}_k x_k + \tilde{G}_k u_k. \quad (2.20)$$

If a new error function is defined as

$$\tilde{e}_k = e_{k+1} - \hat{F}_k e_k, \quad (2.21)$$

this leads to a stable parameter adjustment algorithm [3], [9]

$$[f_{ij}]_{k+1} = [f_{ij}]_k - \frac{\alpha}{x_k' x_k} \tilde{e}_k' \frac{\partial \tilde{e}_k}{\partial [f_{ij}]_k}. \quad (2.22)$$

From (2.15), (2.20), and (2.21)

$$-\tilde{e}_k = M_k b_k \approx M_k \hat{b}_k \quad (2.23)$$

and for uncorrelated parameter errors

$$\frac{\partial \tilde{e}_k}{\partial [f_{ij}]_k} = x_k^{ij} \approx \hat{x}_k^{ij} \quad (2.24)$$

$$\frac{\partial e_k}{\partial [g_{ij}]_k} = u_k^{ij} \quad (2.25)$$

where x_k^{ij} is a vector of the same dimension as x_k, but whose elements are all zero except for the ith element, which is equal to the jth element of x_k. u_k^{ij} is derived from u_k in a similar manner. If the true parameters are assumed to be slowly time-varying, the parameter estimates can then be updated by the following algorithm:

$$[f_{ij}]_{k+1} = [f_{ij}]_k + \frac{\alpha}{\hat{x}_k' \hat{x}_k} (M_k \hat{b}_k)' \hat{x}_k^{ij} \quad (2.26)$$

$$[g_{ij}]_{k+1} = [g_{ij}]_k + \frac{\alpha}{\hat{x}_k' \hat{x}_k} (M_k \hat{b}_k)' u_k^{ij}. \quad (2.27)$$

The parameter update algorithm can now be used in conjunction with (2.15) to produce a time update for the bias estimates

$$\hat{b} \Rightarrow \hat{b}_k - (M'M)^{-1} M' (\Delta \hat{F}_k x_k + \Delta \hat{G}_k u_k). \quad (2.28)$$

This leads to an extended Kalman filtering algorithm as follows:

1) time update of state and bias estimates,
2) measurement update of state and bias estimates,
3) recalculation of explicitly dependent model parameters,
4) estimation of nonmeasureable model parameters,
5) model update.

Sections IV and V describe two separate applications of the above Kalman filter on the 500 MW oil-fired boilers at

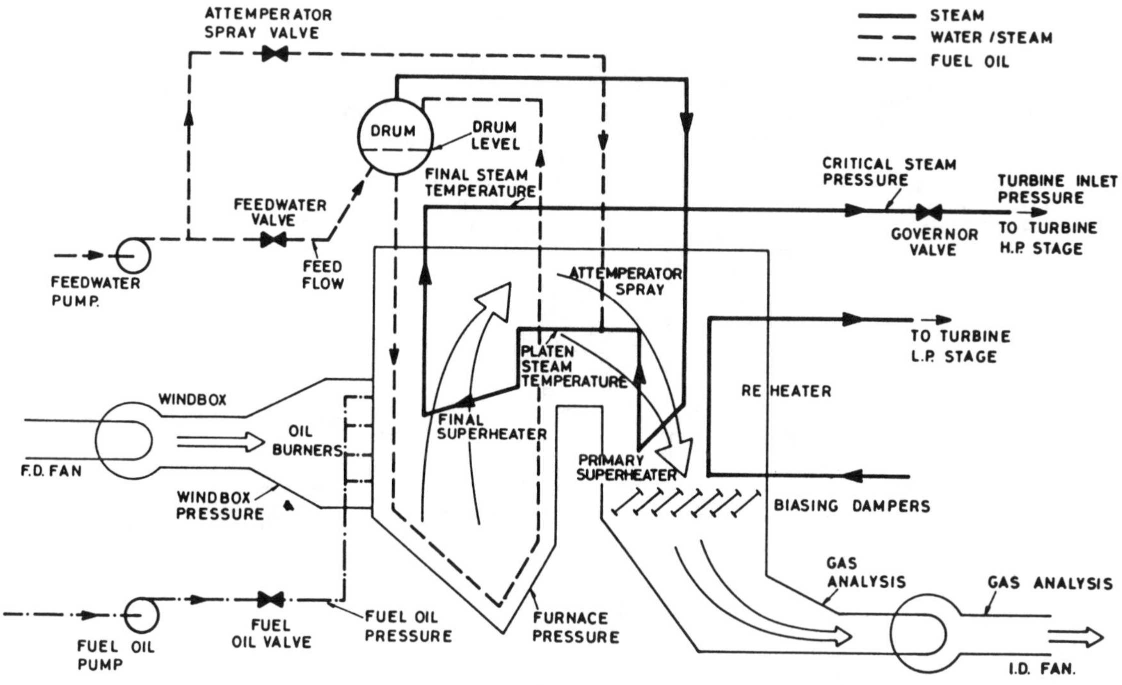

Fig. 1. Simplified schematic diagram of a Pembroke PS boiler.

Pembroke Power Station. These are preceded by a brief description of the plant and equipment.

III. Plant Description

The four main boilers at Pembroke Power Station are 500 MW single reheat oil-fired, with an evaporation capacity of 448 kg/s at 165 bars and 541°C. Each boiler is fitted with a total of 32 burners, mounted in four rows of eight, and supplies steam to a 500 MW single reheat turbo generator on a unit basis.

A schematic diagram of the boiler is shown in Fig. 1. It can be seen from this that although the control problem is a highly interactive one, natural boundaries occur between different plant areas and the problem can be thus subdivided. These natural divisions were used when determining the structure of the new boiler control scheme, which was initiated in 1977 and is now fully operational on all four units. In the new scheme, each boiler is controlled by five LSI 11-based microprocessors with a sixth (feedwater control) to be added at a later date. The processors form a distributed system and are linked by a serial bus which is used for hosting activities and also by point-to-point links for data transmission between processors (Fig. 2). The tasks performed by the processors are allocated on a plant area basis and can be summarized as follows.

Plant Monitor: This processor acts as system host, but also monitors and records metal temperatures and performs creep life calculations.

Combustion Controller: The combustion controller has three primary control loops: the forced draught (FD) fans, the induced draught (ID) fans, and the fuel oil valve. It adjusts these in order to achieve the desired heat input to the boiler and to maintain "optimum" combustion conditions.

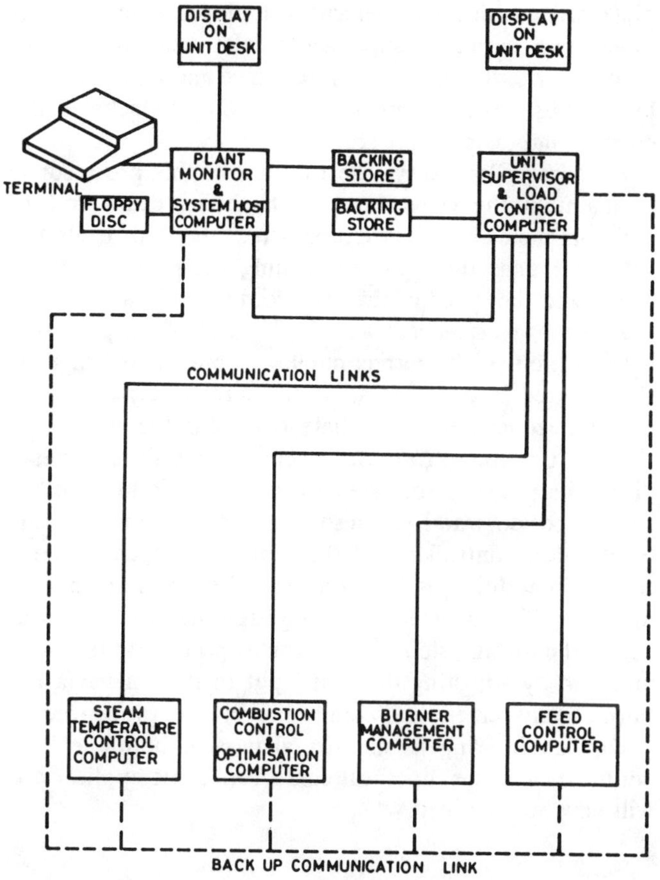

Fig. 2. Distributed computer control system for Pembroke PS.

Superheater Controller: The final superheater is split into four legs, with the final outlet temperature of each leg being maintained at a nominal 541°C (the set point may vary according to operating conditions). It is the function of the superheat controller to control this temperature by adjusting the attemperator spray flow at the superheater

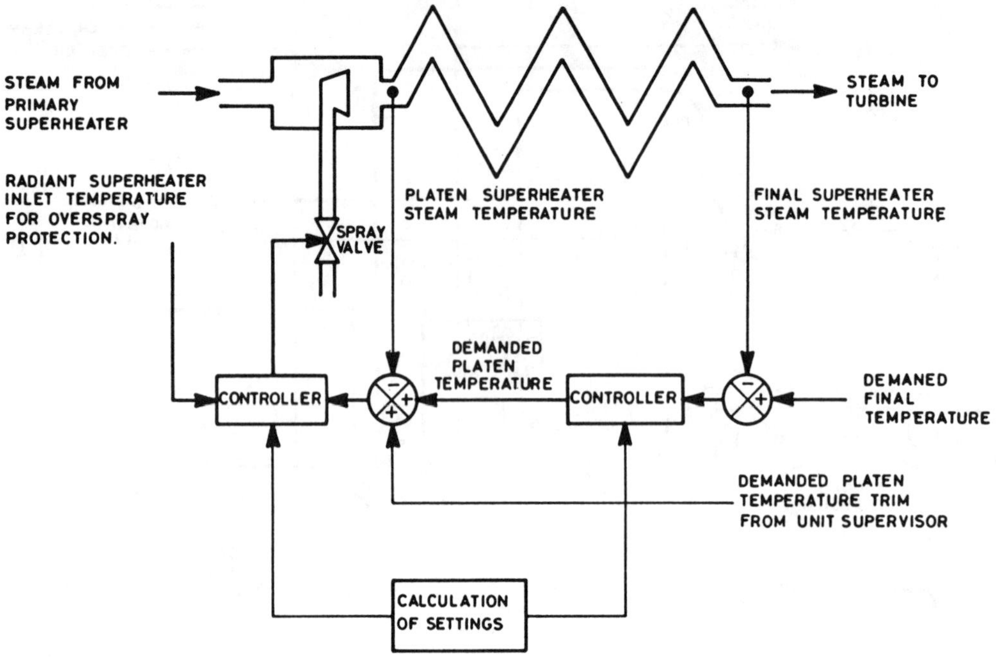

Fig. 3. Schematic diagram of superheater temperature controller.

platens. A cascade arrangement is employed whereby a fast inner loop regulates superheater inlet temperature (attemperator outlet) using the spray valve and a slower outer loop adjusts the attemperator set point to achieve the correct final temperature (Fig. 3).

Feedwater Controller: The function of this processor is to maintain the water level in the boiler drum and to maintain a steady pressure drop across the feed regulation valves. It does this by manipulating the position of the valves and the speed of the main feed water pump.

Burner Management Processor: The task of this processor is to ensure that the correct number of burners is firing for the required load and to select the correct burner pattern to minimize temperature imbalances within the boiler.

Load Controller/Unit Supervisor: The unit supervisor displays all the control data and alarm data from the other slave processors and can also generate set points for the combustion controller and the burner management controller. In addition, it also controls boiler pressure and the unit load. This is achieved by adjusting the pressure ratio across the turbine stop valves using a special interface unit and also by adjusting the heat input to the boiler via the burner management and combustion control processors.

Kalman filtering is an integral part of both the load controller and the superheat controller. These applications will now be described.

IV. The Application of Kalman Filtering Final Superheater Outlet Temperature Control

A. Plant Model Characteristics

As is described in Section III, the superheater control problem is one of adjusting the steam inlet temperature (enthalpy) such that outlet temperature (enthalpy) remains as close as possible to the set point value. In order to visualize the degree of difficulty and the role of Kalman filtering, it is instructive to consider the structure of a simplified mathematical representation of the superheater statistics and dynamics.

TABLE I
Superheater Data Equations

Variable	Description	Units	Function
ΔP	$(P_d + P_s)/2 - 130.0$	bars	
T_{sat}	$331 + 0.6\Delta P - 0.0019\Delta P^2$	°C	
h_{sat}	$2667 - 2.38\,\Delta P - 0.0119\Delta P^2$	kJ/kg	
T_i^{min}	T_{sat} + selectable margin	°C	Minimum inlet temp
T_i^{max}	Constant (eg. 425)	°C	Maximum inlet temp
ΔT	$T_i - T_{sat}$	°C	
Δh	$h_i - h_{sat}$	kJ/kg	
T_i	$\approx T_{sat} + 0.11\Delta h + 2.17\times10^{-4}\,\Delta h^2$	°C	Inlet enthalpy/ temperature conversion
h_i	Inverse of above	kJ/kg	Inlet temperature/ enthalpy conversion
h_o	$2143 + 2.684\,T_o - 1.145\,P_s$	kJ/kg	Outlet temperature enthalpy conversion

The equations are formulated in enthalpy rather than temperature because of the highly nonlinear pressure-dependent relationship between the two quantities. The difference is significant and hence, inlet and outlet temperature measurements and set points are converted to/from enthalpy equivalents using the formulas in Table I.

From a control point of view, the following two relationships are important.

1) The dynamics relating changes in steam enthalpy at the superheater outlet to changes in enthalpy at the inlet (attemperator).

2) The dynamics relating changes in outlet steam enthalpy to changes in gas-side heat flux and steam flow.

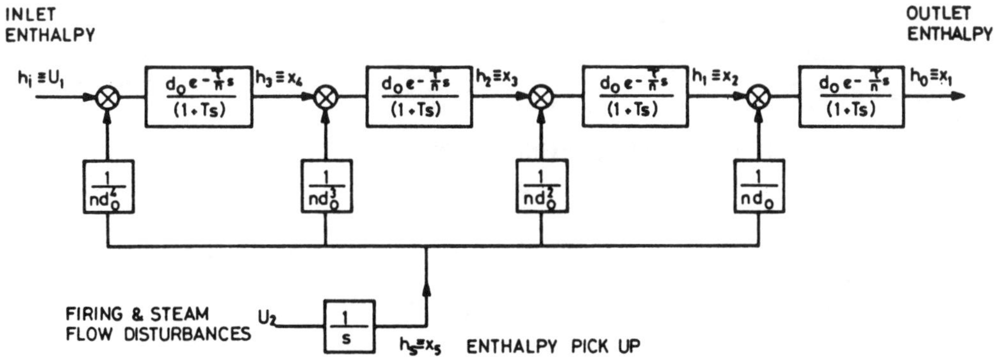

Fig. 4. Simplified superheater model.

Application of linearized analysis produces the small perturbations model

$$\delta h_o = D(s)\delta h_i + \frac{C_1\alpha_1}{(s+\alpha_3)}\left(\frac{1-D(s)}{s}\right)\delta Q$$

$$- \frac{C_2(s+\alpha_2)}{(s+\alpha_3)}\left(\frac{1-D(s)}{s}\right)\delta w_s \quad (4.1)$$

where

$$D(s) = D_o \exp(-\tau(\beta(s)))$$

$$\beta(s) = \frac{s^2 + s(\alpha_1+\alpha_2)}{s+\alpha_2}$$

$C_1, C_2, \alpha_1, \alpha_2, \alpha_3$ are constants derived from physical quantities and

τ is the steam transit time though the superheater.

In the steady state

$$h_s = \frac{Q}{W_s} \quad (4.2)$$

$$h_o = D_o h_i + h_s. \quad (4.3)$$

Substituting these relationships into (4.1) produces the following transfer function representation of the superheater dynamics:

$$h_o(s) = D(s)h_i(s) + E(s)h_s(s) \quad (4.4)$$

where

$$E(s) = E_o(1 - D(s))/s$$

$$E(o) = 1.$$

From a filtering point of view, (4.4) is inconvenient. However, a well proven approximation for $D(s)$ and $E(s)$ takes the form

$$D(s) = (d_o d(s))^n$$

$$E(s) = (d(s) + d^2(s) + \cdots d^n(s))/n \quad (4.5)$$

where

$$d(s) = \frac{e^{-\tau s/n}}{(1+Ts)}$$

$$(d_o)^n = D_o$$

$$T = (\alpha_1 \tau)/n\alpha_2).$$

MODEL :-

$$\underline{x}^P_{k+1} = F^P \underline{x}^P_k + G^P \underline{u}^P_k + \underline{w}_k$$

$$z_{k+1} = H^P \underline{x}^P_{k+1} + v_k$$

WHERE :-

$$F^P = \left[\begin{array}{cccc|c} 1-\lambda & \lambda d_o & 0 & 0 & \lambda/n \\ 0 & 1-\lambda & \lambda d_o & 0 & \lambda/nd_o \\ 0 & 0 & 1-\lambda & \lambda d_o & \lambda/nd_o^2 \\ 0 & 0 & 0 & 1-\lambda & \lambda/nd_o^3 \\ \hline 0 & 0 & 0 & 0 & 1 \end{array}\right]$$

$$G^P = \left[\begin{array}{c|c} 0 & 0 \\ 0 & 0 \\ 0 & 0 \\ \lambda d_o & 0 \\ \hline 0 & 1 \end{array}\right]$$

$$H^P = \left[\begin{array}{cccc|c} 1 & 0 & 0 & 0 & 0 \end{array}\right]$$

Fig. 5. Filter state-space model for superheater.

Plant testing at Pembroke and on other CEGB 500 MW units has demonstrated that the choice of $n = 4$ in (4.5) yields a good approximation to plant input–output response.

With $D(s)$ and $E(s)$ expressed as in (4.5), it is possible to interpret $d(s)$ as the input–output transfer function for the nth section of the superheater. The block diagram structure of Fig. 4 can then be readily derived. It is clear that changes in the heat pick up h_s influence outlet enthalpy h_o more rapidly than the control action via changes in h_i. It is also clear that knowledge of intermediate enthalpies h_1, h_2, and h_3 and of h_s would be useful in achieving optimum superheater control.

The mechanism for obtaining this state information is of course the Kalman filter, which requires the plant model to be expressed in the discrete state-space form of Fig. 5. Of particular interest are the sampling rate δt and time constant factor λ

$$\delta t = t_{k+1} - t_k = \tau/n \quad \text{(by choice)} \quad (4.6)$$

$$\lambda = 1 - \exp\left(-\frac{\tau}{nT}\right). \quad (4.7)$$

The critical item is the choice of sampling interval (δt).

Transport delay τ varies with steam flow and pressure in accordance with the relationship

$$\tau = 10.5\, P_d / w_s. \tag{4.8}$$

If one adapts sampling intervals in accordance with (4.8), then λ remains fixed (since α_1/α_2 is constant). As a consequence of this *sample rate adaption*, the F and G matrices remain time invariant, and hence with constant noise variance, so do the Kalman filter and controller gain matrices. This substantially reduces the on-line computational requirements. State variables $x_1 - x_4$ represent the steam enthalpy at various stages in the superheater, while x_5 represents the imposed firing/steam flow balance.

The noise disturbance vector w is an important part of the Kalman filter. In general, the variance of w is described by

$$E\left[w_k \cdot w_j'\right] = Q^f \quad \text{for } j = k$$
$$= 0 \quad \text{for } j \neq k. \tag{4.9}$$

The elements of w are not necessarily uncorrelated since they represent common uncertainties such as modeling parameter, structural errors, and errors in calculated values of h_s. Thus, noise sources $w_1 \cdots w_4$ were assumed to be identical, and hence

$$w = [w_4, w_4, w_4, w_4, w_5]' \tag{4.10}$$

and

$$Q^f = \Gamma Q'' \Gamma'$$

where

$$Q'' = \begin{bmatrix} q_4^f & 0 \\ 0 & q_5^f \end{bmatrix}$$

$$\Gamma' = \begin{bmatrix} 1 & 1 & 1 & 1 & 0 \\ 0 & 0 & 0 & 0 & 1 \end{bmatrix}.$$

B. Filter Algorithm Design

For simplicity, a constant gain version of the Kalman filter was selected. Hence, the numerical design process consisted of the following.

1) Deterministic selection of measurement (R^f) and disturbance noise variances (Q^f) using approximate estimates of λ and D_o.

2) Computation of the appropriate fixed-filter gain matrix K using data from 1).

3) Filter performance evaluation using plant simulation.

4) Identification of optimum values for λ and D_o using plant data and maximum likelihood criteria.

5) Repetition of step 3).

It was known from observation of plant behavior that neither the measurement nor process noise variances was stationary. R^f was therefore fixed at its average value, which was 1.0 $(\text{kJ/kg})^2$, while Q^f was treated as a design parameter to be selected with regard to transient response and model parameter error sensitivity criteria. Of particular interest was the rate at which the bias estimate $\hat{x}_5$ would track changes in the superheater heat pickup, since the information was to be used in the feedback control algorithm. With noise disturbance variance q_4^f set to zero, q_5^f can be selected to give excellent transient response and low sensitivity to measurement noise. Unfortunately, the performance is somewhat sensitive to modeling errors. Thus, a desensitized design must be employed on a real plant. In this case, the desensitation is achieved by adjusting q_4^f. The corresponding Kalman gain matrices for these two design cases are

$$Q'' = \text{diag}(0.0, 0.014)$$
$$K^f = (0.049, 0.045, 0.037, 0.022, 0.012)'$$
$$Q'' = \text{diag}(0.2, 0.014)$$
$$K^f = (0.140, 0.138, 0.133, 0.107, 0.110)'.$$

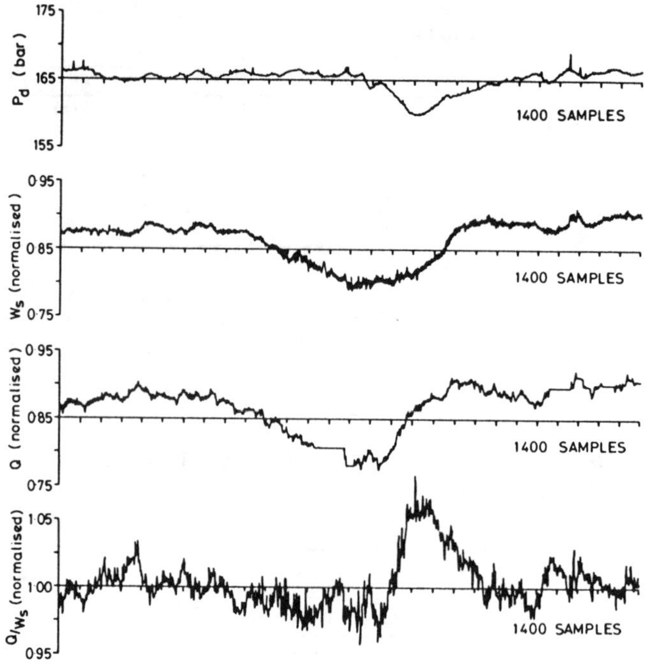

Fig. 6. Plant conditions for superheater filter test run.

In order to optimize filter performance on a real plant, one must identify the correct values of λ and D_o. This was achieved using the maximum-likelihood technique applied to the filter residuals. The application of this (or any other) technique to a real plant is never straightforward because of simplifications in the model and plant disturbance characterization. Nonetheless, good results can be achieved, even with nonideal data records. Figs. 6 and 7 illustrate typical plant test conditions at Pembroke (unit 1). The unit is loaded following in response to grid frequency variations with three of the four superheater legs under computer control. The fourth leg, under test, is on computer manual, with the operator injecting step disturbances onto the (fast) attemperator temperature loop set point (u_1). In attempting the maximum likelihood identification it was found that a more clearly defined optimum was achieved using the sensitive Kalman filter design. This was not surprising since the desensitized design was intended to minimize the

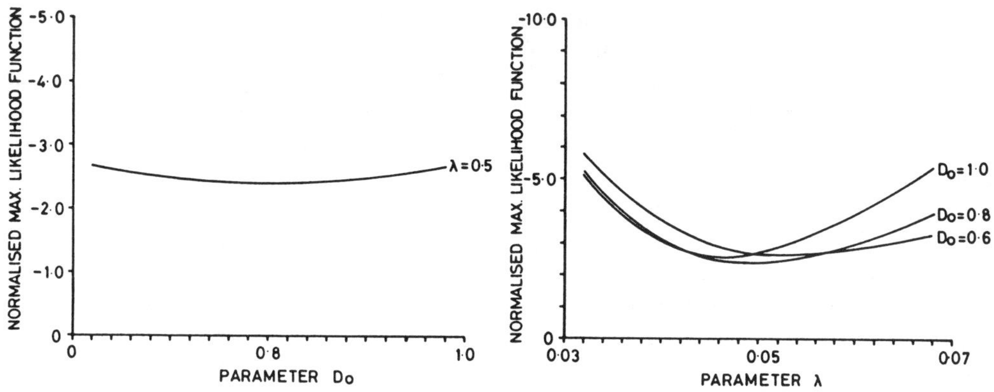

Fig. 7. Parameter optimization using Figs. 6–8 test data.

effects of parameter error. The identification results are presented in Fig. 7. Despite the fact that the prediction residuals were not white, well-defined optimum values for $\lambda(0.05)$ and $D_o(0.8)$ were obtained. These were close to the theoretical values of 0.042 and 0.85, respectively.

C. State Feedback Control Algorithm

The purpose of deriving the state information in this example is feedback control. The control algorithm derived for the superheater decomposes the control input signal into two parts, one aimed at achieving the correct mean value and the other providing regulation about that mean value. Thus,

$$u_1 = u_1^m + u_1^c$$

where

$$u_1^m = (h_o^{\text{set}} - \hat{x}_5)/D_o$$

$$u_1^c = K^c(\hat{x}^c - \hat{\bar{x}}^c)$$

$$x^c = (\hat{x}_1, \hat{x}_2, \hat{x}_3, \hat{x}_4)'$$

$\hat{\bar{x}}^c$ is calculated from h_o^{set}, $\hat{x}_5$, and D_o.

The bias information $\hat{x}_5$ replaces the usual integral term and its response to unexpected disturbances is therefore important. This control structure also allows feedforward information regarding heat balance disturbances to be incorporated into the Kalman filter. It improves the quality of $\hat{x}_5$ and, consequently, control system performance beyond that achievable with conventional techniques.

D. Filter Performance

The quality of the information to be derived from the filter can be judged from the results presented in Fig. 8. As expected, the residuals are not white since the bias estimate $\hat{x}_5$ is continuously adapting to the external disturbances in the superheat heat balance. However, the behavior of $\hat{x}_5$ is excellent, since the filter identifies the trend decrease in x_5 (during the first 800 samples) independently of the inlet perturbations. It also identifies clearly the sudden increase as the unit picks up load in response to grid frequency. If one compares $\hat{x}_5$ with the measured ratio of boiler firing to steam flow (in Fig. 6) the similarity is striking. This is

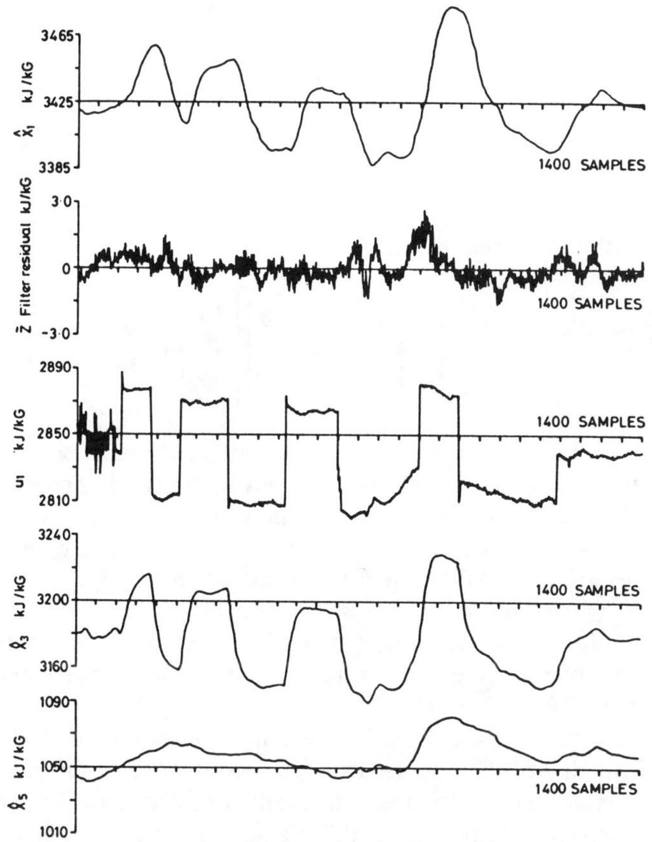

Fig. 8. Filter estimate for superheater.

important in the context of improving the closed-loop control performance.

V. The Application of Kalman Filtering to Load/Pressure Control

Load and pressure are affected by several inputs such as the oil flow per burner, the air flow per burner, the air-to-fuel ratio, the number of burners firing, and the position of governor valve. On the Pembroke Power Station boilers all these variables are controlled by local direct digital control (DDC) loops so that load and pressure can be controlled by two master signals, namely "percentage boiler firing" Q and "percentage governor valve position" η. The latter being defined as the ratio of the "after governor valve" to the "before governor valve" pressure and calculated from $(P_t/P_S)*100$ percent. Thus, the problem has been reduced to a two-input/two-output problem, with the two master signals as the inputs, and boiler pressure (in this case the critical pressure P_S) and steam flow as the output variables. The Pembroke turbines present a constant resistance to steam flow, hence the steam flow and the generated load can be inferred from the measurement of turbine inlet pressure P_t.

The application of basic thermodynamic principles along with the appropriate simplifying assumptions leads to a three-state model of the plant

$$\begin{bmatrix}\dot{P}_d\\ \dot{P}_s\\ \dot{W}_f\end{bmatrix} = \begin{bmatrix} \frac{-(h_s-\beta_1)}{R_sC_d} & \frac{(h_s-\beta_1)}{R_sC_d} & \frac{h_f-\beta_1}{C_d}\\ \frac{1}{R_sC_s} & \left(\frac{-1}{R_sC_s}+\frac{-1}{RC_s}\right) & 0\\ \frac{1}{T_fR_s} & \frac{-1}{T_fR_s} & \frac{-1}{T_f}\end{bmatrix}\begin{bmatrix}P_d\\ P_s\\ W_f\end{bmatrix} + \begin{bmatrix}Q_{\max}\\ 0\\ 0\end{bmatrix}u_1 \tag{5.1}$$

with the measurements

$$z_k = \begin{bmatrix}P_s\\ P_t\end{bmatrix} = \begin{bmatrix}0 & 1 & 0\\ 0 & \frac{R_T}{R} & 0\end{bmatrix}\begin{bmatrix}P_d\\ P_s\\ W_f\end{bmatrix} + \begin{bmatrix}v_1\\ v_2\end{bmatrix}. \tag{5.2}$$

The noise terms v_1 and v_2 both contain a component due to disturbances on the governor valve. This would preclude the use of "one at a time" measurement updates in a filter using these measurements. Therefore, the measurement noise is diagonalized by the transformation $z_k^* = T_k z_k$ where T_k is derived from the noise statistics [1].

Most of the parameters are explicit functions of the state variables or other known quantities and can be calculated as in Table II.

It can be seen from the parameter equations that the enthalpy of the inlet feedwater is unknown and cannot be calculated explicitly from the available data. Secondly, the "percentage governor valve" position enters the model via the parameter matrix and not the input vector. These difficulties can be overcome by replacing h_f and R with their estimated values $\hat{h}_f$ and $\hat{R}$, respectively, and introducing tw new bias states.

Frc. ı the state equations it can be seen that replacing h_f with $\hat{h}_f$ will modify $\dot{P}_d$ by an amount

$$\delta\dot{P}_d = \frac{W_f}{C_d}\left(h_f - \hat{h}_f\right) \equiv \frac{Q_{\max}\delta Q}{C_d} \tag{5.3}$$

where δQ is a bias term corresponding to a deviation in heat input to the boiler.

Similarly replacing R with $\hat{R}$ will modify P_s by an amount

$$\delta\dot{P}_s = \frac{P_s}{C_s}(G-\hat{G}) = \frac{\delta W}{C_s} \tag{5.4}$$

where δW is a bias term corresponding to a deviation in steam flow and $G = 1/R$.

The bias states can now be appended to the original state equations to produce augmented state equations in the form of (2.8) and (2.9) with

$$M_k = \begin{bmatrix}0 & \frac{\delta t Q_{\max}}{C_d}\\ -\frac{\delta t}{C_s} & 0\\ 0 & 0\end{bmatrix} \tag{5.5}$$

$$N_k = \begin{bmatrix}0 & 0\\ \frac{\delta t R_T}{C_s} & 0\end{bmatrix} \tag{5.6}$$

and

$$b_k = \begin{bmatrix}\delta W\\ \delta Q\end{bmatrix}. \tag{5.7}$$

In this case δt is the sampling interval and it is assumed that $F_k = (I + \delta t A)$, although in practice a more precise approximation is normally used.

An extended Kalman filter can now be used to obtain estimates of $\boldsymbol{x}_k$ and $\boldsymbol{b}_k$. The bias estimates can then be used to update the values of $\hat{h}_f$ and $\hat{R}$ to give

$$\delta h_f = \frac{\alpha}{\left(\hat{P}_d^2 + \hat{P}_s^2 + \hat{W}_f^2\right)}\hat{W}_f Q_{\max}\delta\hat{Q} \tag{5.8}$$

$$\delta G = \frac{\alpha}{\left(P_d^2 + P_s^2 + W_f^2\right)}\hat{P}_s\delta\hat{W}. \tag{5.9}$$

The consequential changes in bias estimates $\boldsymbol{b}$ are

$$\delta\hat{w}_k \Rightarrow \delta\hat{w}_k - \frac{\alpha\hat{P}_s^2\delta\hat{w}_k}{\left(\hat{P}_d^2 + \hat{P}_s^2 + \hat{W}_f^2\right)} \tag{5.10}$$

$$\delta\hat{Q}_k \Rightarrow \delta\hat{Q}_k - \frac{\alpha\hat{W}_f^2\delta\hat{Q}}{\left(\hat{P}_d^2 + \hat{P}_s^2 + \hat{W}_f^2\right)}. \tag{5.11}$$

The data shown in Figs. 9–11 were sampled at 5 s intervals from unit 1 at Pembroke Power Station while it was picking up load from 100 to 500 MW. The firing demand was being set by the operator in order to achieve the required load and the governor valve was under computer control maintaining the correct boiler pressure. The curves

TABLE II
LOAD/PRESSURE MODEL PARAMETERS

Variable	Description	Units	Function
h_s	$2667 - 2.38\,(P_D - 130) - .0119\,(P_D - 130)^2$	kJ/kg	Drum outlet steam enthalpy
$\beta 1$	$\dfrac{(h_w \rho w - h_s \rho s)}{\rho w - \rho s}$ (all known quantities)	kJ/kg	Analytic function
h_f	function of feedwater temperature (Unknown)	kJ/kg	Inlet water enthalpy
R_s	$W_s\,(1.5 \times 10^{-4})$	bar/kg/s	Superheater flow resistance
C_d	$(305 - P_D)*1850$	kJ/bar	Drum thermal capacity
C_s	2.0/Rs	kg/bar	C_s*R_s = Superheater time constant
R_T	0.37	bar/kg/s	Turbine steam flow resistance
R	function of governor valve position	bar/kg/s	Governor valve steam flow resistance
T_f	60 s	s	Feedwater flow control time constant
Q_{max}	$1332 * W_{max}$	kJ/s	Max value of heat input to boiler (= 100% firing)
W_{max}	400	kg/s	Max value of steam flow (= 500 MW)

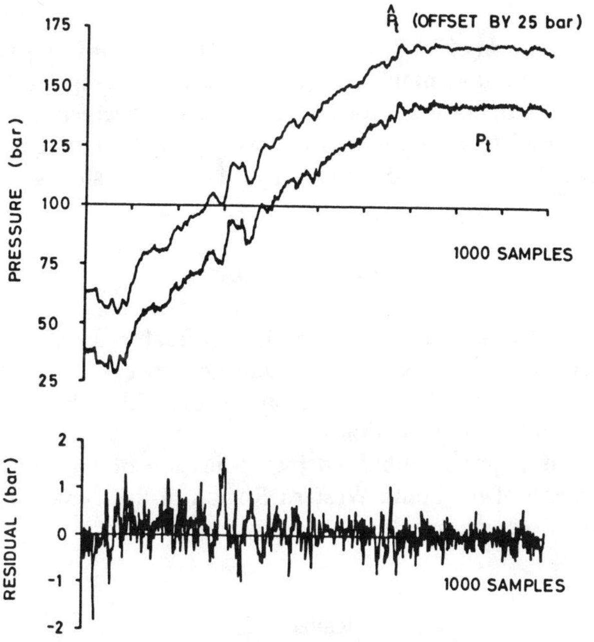

Fig. 9. Smoothing of measured data for pressure/load controller.

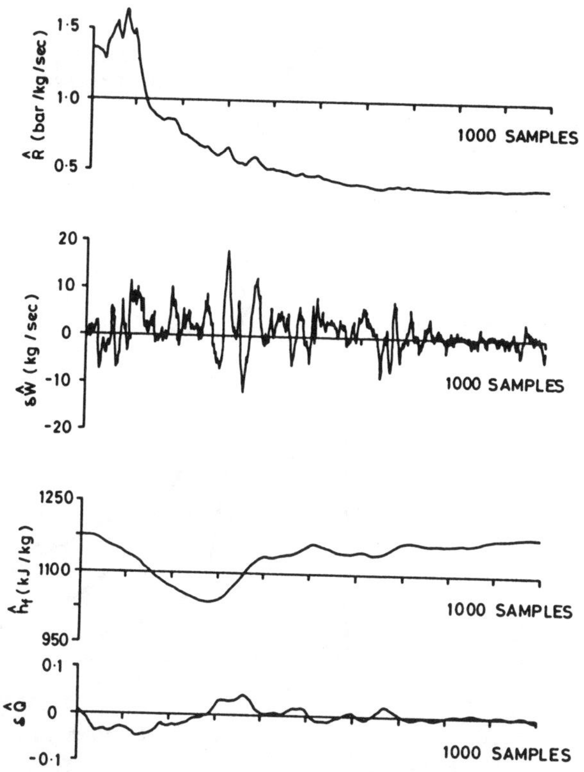

Fig. 10. Parameter adaption for pressure/load controller.

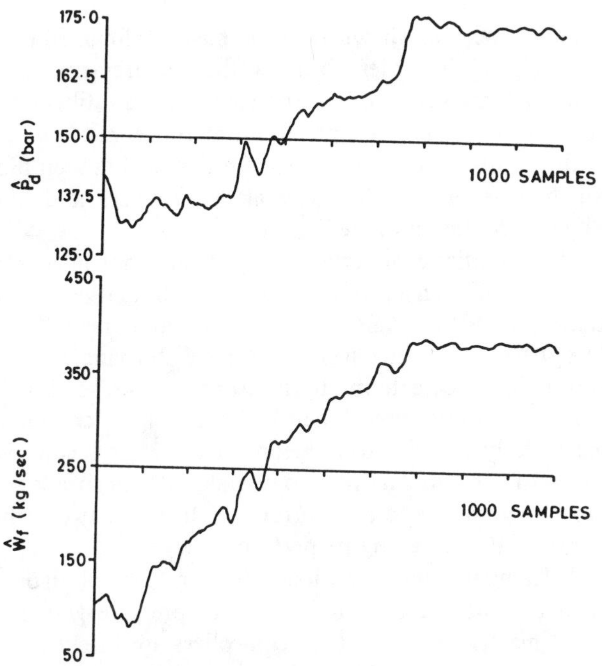

Fig. 11. Estimation of unmeasured variables for pressure/load controller.

illustrate the various applications of the filter, namely the following.

1) Signal-smoothing Fig. 9 shows the filtered value of P_t along with the raw value and the residual. It can be seen that, in this case, the residuals appear to be "white," but it is not true in general that the "best filter" gives the "whitest" residuals.

2) Estimation of plant parameters. In the case of the pressure/load controller the bias states are used to "drive" a parameter estimator. Fig. 10 shows the bias states $\delta\hat{W}$ and $\delta\hat{Q}$ together with the respective parameter estimates $\hat{R}$ and $\hat{h}_f$.

3) Estimation of unmeasured variables. Fig. 11 shows the estimated values of drum pressure $\hat{P}_d$ and feedwater flow $\hat{W}_f$. Both these variables are used in the control algorithms and could be obtained by the use of additional transducers. The Kalman filter is, however, a cheaper and more reliable way of obtaining these data.

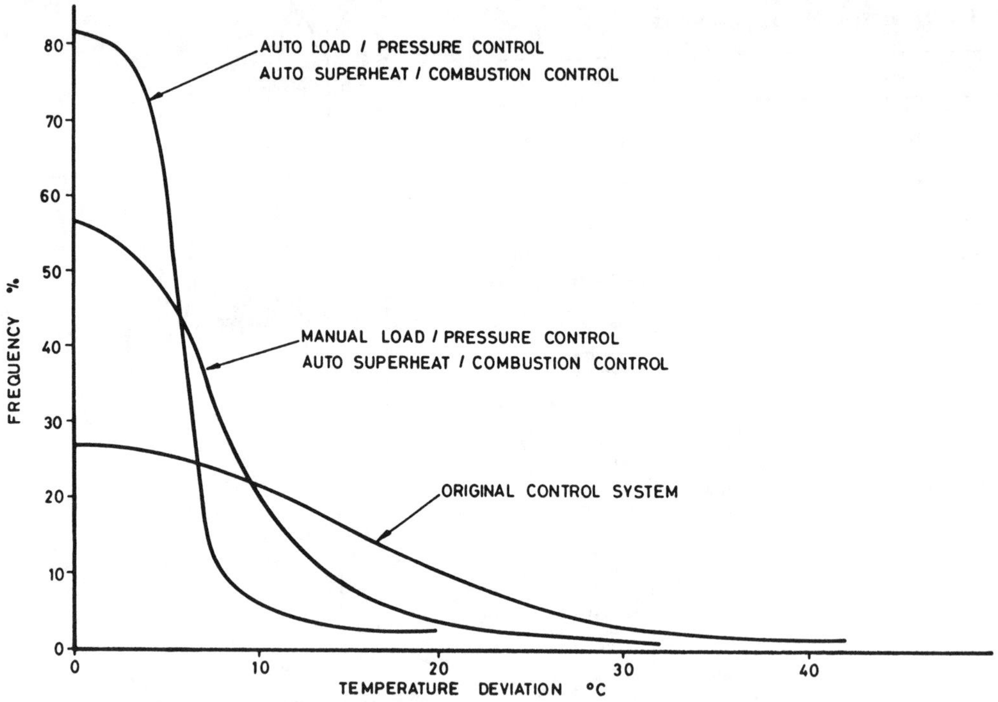

Fig. 12. Superheat temperature deviations during two-shift operation.

VI. Conclusions

This paper has shown how the basic Kalman filter can be modified in order to meet the requirements of an industrial application. In applying Kalman filters to a power generation plant, particular attention has been paid to the plant models and a distinction made between biases which arise due to the external disturbances and biases which arise due to modeling errors.

The operational superheat temperature controller is currently based upon a proportional/integral/derivative (PID) algorithm. Plant trials have shown that a new Kalman filter-based controller has improved disturbance rejection and reduced sensitivity to measurement noise. Furthermore, it is more robust than the PID controller, which is particularly sensitive to modeling errors when feedforward control is added. The filter-based algorithm can allow the model parameters to be in error by a factor of two without significantly degrading its performance.

A Kalman filter-based load/pressure control algorithm is in current operational use. Both "pressure governing" and "master pressure" type controllers are in current use [11] and a multivariable optimal controller is under development [4].

Assessment of the system performance is ongoing, but a simple quantitative measure of its efficacy can be obtained by monitoring the deviation in final steam temperature (i.e., at the boiler outlet) during two shift operations. Fig. 12 shows the distribution of the temperature deviation taken over several two-shift operations. It can be seen that the introduction of the superheat (cascaded PID) and combustion control systems produced a 2.5:1 reduction in the error variance and that this was reduced further by the introduction of load/pressure control (Kalman filter-based "pressure governing" type controller). Reducing the final steam temperature error increases the boiler creep life and reduces fatigue due to thermal cycling. It is just one of the benefits obtained from an improved control system.

Acknowledgment

The authors acknowledge the invaluable assistance of their colleagues both in the Scientific Services Department of the South Western Region of the C.E.G.B. and at Pembroke Power Station.

This paper is published by permission of the Director General of the South Western Region of the Central Electricity Generating Board.

References

[1] J. G. Balchen, T. Endresen, M. Fjeld, and T. Olsen, "Multivariable P.I.D. estimation and control in systems with biased disturbances," *Automatica*, vol. 9, pp. 295–307, 1973.
[2] J. G. Bierman, *Factorisation Methods for Discrete Sequential Estimation*. New York: Academic, 1977.
[3] R. Clarke and J. W. Green, "A method for discrete system identification," in *Proc. 3rd IFAC Symp. Identification Syst. Parameter Estimation*, The Hague, The Netherlands, 1973.
[4] R. Clarke, "The application of multivariable control to a 500 MW oil-fired boiler/turbine unit," presented at the IEEE Conf. Multivariable Contr. Appl., Univ. Hull, England, 1982.
[5] A. G. Entwistle and N. Oldfield, "A high level interactive software system for real time multiprocessor applications," in *IEE Conf. Distributed Comput. Contr. Syst.*, Univ. Aston, Birmingham, England, IEE Conf. Pub. 153, 1977.
[6] M. J. Grimble, "Adaptive Kalman filter for control of systems with unknown disturbances," *Proc IEE.*, vol. 128, part D, no. 6, 1981.
[7] A. K. Jazwinski, "Adaptive sequential estimation with applications," *Automatica*, vol. 10, pp. 203–207, 1974.

[8] B. C. Kuo, *Automatic Control Systems*. Englewood Cliffs, NJ: Prentice-Hall, 1975.

[9] A. J. Udink Ten Cate and H. B. Verbruggen, "A least squares gradient method for discrete process identification," *Int. J. Contr.*, vol. 35, no. 1, 1978.

[10] J. N. Wallace and R. Clarke, "Load control of a 500 MW oil-fired boiler/turbine," presented at the IEE Conf. on Contr. Appl., Warwick, England, 1981.

[11] J. N. Wallace, "Design concepts and experience in the application of distributed computing to the control of large C.E.G.B. power stations," presented at the IAEA Specialist Meeting on Distributed Syst. for Nucl. Power Plants, 1980.

Application of Kalman Filtering to Demographic Models

BRICE G. LEIBUNDGUT, ANDRÉ RAULT, MEMBER, IEEE, AND FRANÇOISE GENDREAU

Abstract—**In this paper demographic systems appear as stochastic and unstable linear systems. Their state representation model is simulated and identified using Kalman filtering. The example of the French beef cattle herd is presented and results are discussed with emphasis on their agroeconomic meaning.**

I. Introduction

SIMULATION models are used in agricultural economics in order to make predictions and to help decision makers by analyzing the effects of alternative policies. These models have been developed rather recently, therefore they can take advantage of the knowledge and methods used in other fields, although the peculiarities of applications generally need an adaptation of these methods. With regard to application areas such as aeronautics or industrial processes, agroeconomic systems are characterized by the following items.

1) Only limited data are generally available and new experiments cannot be performed so that identification is always a passive process.

2) Data are often not observed directly, but through a polling system (sample survey) which introduces large errors in comparison with technological sensors.

3) The simulation time horizon is generally long (10–20 years) and the stationarity assumptions are consequently inappropriate because of structural evolutions.

4) Many uncontrolled factors (man, weather, epidemics,...) can perturb the system so that it is difficult to define the dynamical structure.

Manuscript received March 2, 1982; revised September 13, 1982.
B. G. Leibundgut is with Crédit Lyonnais, Paris, France.
A. Rault and F. Gendreau are with ADERSA, Palaiseau, France.

Reprinted from *IEEE Trans. Automat. Contr.*, vol. AC-28, pp. 427–434, Mar. 1983.

The beef cattle herd is an example of such an agroeconomic system. The problem set by FORMA (Fonds d'Orientation et de Régularisation des Marchés Agricoles) consisted of estimating the number of males and females per age class as accurately as possible using available data from different statistical services.

This problem has been solved by using Kalman filtering which allows a simulation of the system for the past years. Predictions are made through scenarios; they are computed for different assumption sets concerning slaughters (conjunctural[1] scenarios) or parameters (structural scenarios). Decision makers can then decide on the most appropriate policy.

This paper presents the application of this methodology to the beef cattle system. A model structure is first given (Section II), some stochastic aspects of the system are described (Section III), and the identification scheme is detailed (Section IV). In Section V the implementation on a computer is presented with emphasis on the problem of estimation of the covariance matrices. Some results are given and discussed in Section VI.

II. Model Structure

The general structure for demographic models proposed by Leslie and described in [13] has now been widely used [8], [12]. It is based on a distribution of population into age classes and a modeling of life phenomena; the step size of the age classes must be identical to the sampling time interval. A state equation can be derived from this general structure; additionally the available observations are expressed as functions of populations per age classes. Thus, a state representation can be defined and is noted

$$\begin{cases} x(n+1)=Ax(n)-u(n) \\ s(n)=Cx(n). \end{cases}$$

The state equation stands for the demographic evolution of the age pyramid. In the case of the French cattle herd the differentiation of females and males yields the following form:

$$\begin{bmatrix} XF(1,n+1) \\ XF(2,n+1) \\ \vdots \\ \\ XF(N,n+1) \\ \hline XM(1,n+1) \\ \\ \vdots \\ \\ XM(M,n+1) \end{bmatrix} = \left[\begin{array}{ccccc|ccccc} \alpha_1 & \alpha_2 & \cdots & & \alpha_N & & & & & \\ \beta_1 & 0 & \cdots & & 0 & & & & & \\ 0 & \beta_2 & \ddots & & \vdots & & & 0 & & \\ \vdots & \ddots & \ddots & & 0 & & & & & \\ 0 & \cdots & 0 & \beta_{N-1} & \beta_N & & & & & \\ \hline \gamma_1 & \gamma_2 & \cdots & & \gamma_N & 0 & & \cdots & & 0 \\ & & & & & \delta_1 & 0 & & & \\ & & & & & 0 & \delta_2 & \ddots & & \vdots \\ & & 0 & & & & \ddots & \ddots & & \\ & & & & & 0 & & 0 & \delta_{M-1} & \delta_M \end{array}\right] \begin{bmatrix} XF(1,n) \\ XF(2,n) \\ \vdots \\ \\ XF(N,n) \\ \hline XM(1,n) \\ \\ \vdots \\ \\ XM(M,n) \end{bmatrix} + \begin{bmatrix} UF(1,n) \\ UF(2,n) \\ \vdots \\ \\ UF(N,n) \\ \hline UM(1,n) \\ \\ \vdots \\ \\ UM(M,n) \end{bmatrix}$$

where

$XF(i,n)$ [resp. $XM(i,n)$] denotes the number of females (resp. males) of age $(i-1)$ years at the beginning of year (n) (January 1).

$UF(i,n)$ and $UM(i,n)$ represent the balance of slaughters, export and import over class (i) during year (n).

α_i (resp. γ_i) is the average number of alive female (resp. male) calves per female from class (i) per year.

$\beta_i=1-\epsilon_F(i)$ and $\delta_i=1-\epsilon_M(i)$ where $\epsilon_F(i)$ and $\epsilon_M(i)$ are the mortality rates over class (i).

The α_i and γ_i coefficients depend on fertility rates, calf mortality rates, and sex ratio. The above equation results from the combination of

1) a birth process modeled by introducing the α_i and γ_i coefficients;

2) a death process: the β_i and δ_i coefficients take into account the natural mortality while UF and UM represent slaughters, imports, and exports;

3) an aging process which induces the following shift:

$$X(i,n+1)=\beta_{i-1}X(i-1,n)-U(i,n).$$

Basically, the state equation summarizes life phenomena.

The observation equation depends on the number of available statistics. In France information is provided once a year for the male and female populations aged 0–1 year, 1–2 years, and over 2 years. Moreover, there are data

[1] This type of scenario enables the simulation of the herd's evolution for a given economic situation expressed in terms of slaughter policy.

about the number of cows, the milk production, and the natural deaths (published by veterinary services). The observation equation can thus be expressed as follows:

$$
\begin{array}{l}
\text{females 0–1}\\
\text{females 1–2}\\
\text{females +2}\\
\text{males 0–1}\\
\text{males 1–2}\\
\text{males +2}\\
\text{cows}\\
\text{milk prod.}\\
\text{natural deaths}
\end{array}
\left[\begin{array}{c}
S(1,n)\\ S(2,n)\\ S(3,n)\\ \hdashline S(4,n)\\ S(5,n)\\ S(6,n)\\ \hdashline S(7,n)\\ S(8,n)\\ S(9,n)
\end{array}\right]
=
\left[\begin{array}{cccccc|cccccc}
1 & 0 & 0 & & \cdots & & & & & & & \\
0 & 1 & 0 & & \cdots & & & & 0 & & & \\
0 & 0 & 1 & 1 & \cdots & 1 & & & & & & \\
\hline
 & & & & & & 1 & 0 & 0 & & \cdots & \\
 & & 0 & & & & 0 & 1 & 0 & & \cdots & \\
 & & & & & & 0 & 0 & 1 & 1 & \cdots & 1 \\
\hline
\mu_1 & \mu_2 & & & \cdots & \mu_N & & & 0 & & & \\
\xi_1 & \xi_2 & & & \cdots & \xi_N & & & & & & \\
\epsilon_F(1) & \epsilon_F(2) & & & \cdots & \epsilon_F(N) & \epsilon_M(1) & \epsilon_M(2) & & & \cdots & \epsilon_M(M)
\end{array}\right]
\left[\begin{array}{c}
XF(1,n)\\ \vdots \\ XF(N,n)\\ \hdashline XM(1,n)\\ \vdots \\ XM(M,n)
\end{array}\right]
$$

where

μ_i represents the average number of cows per female from class (i).

ξ_i denotes the average milk production of a female from class (i) during a year.

In fact the matrices A and C are not stationary and should rather be noted $A(n)$ and $C(n)$ because α_i, γ_i, and ξ_i are represented as linear functions of time

$$\alpha_i(n)=\alpha_i^0(1-\tau_n) \qquad \gamma_i(n)=\gamma_i^0(1-\tau_n)$$
$$\xi_i(n)=\xi_i^0(1+\rho n).$$

The τ coefficient takes into account the uncontrolled slaughters of calves on the farm (i.e., animals which the farmer slaughters for his own consumption) which have a direct correlation with the increasing use of deep-freezing equipment. The ρ coefficient represents a technical gain on milk production which is induced by artificial feed and cross-breeding.

This formulation takes into account the structural evolution of the system.

III. Stochastic Aspects of the System

In the previous section the system has been considered as deterministic. This statement is true in so far as it is the representation of an aging process; an animal presently i years old will be $(i+1)$ years old next year. But with regard to the birth and death phenomena, the system has to be considered as a stochastic system.

Four kinds of noise must be introduced to improve the previous model structure, and thus render the model more consistent with the actual cattle system.

1) A *process noise* which takes into account the stochastic nature of the system: parameters such as mortality and fertility rates, or sex ratio, or milk production per cow are basically stochastic processes, although they are introduced through average values.

2) A *model noise* which enables us to take into consideration the approximations made in the state representation modeling: linearizations (for instance, the technical gain is modeled as a linear function of time); or integration errors due to the one year sampling period (the actual age of an animal is incremented on its birthday and not on January 1 so that an i years old animal slaughtered during the year might have been either i or $(i+1)$ years old next January.

3) An *input noise* due to the origin of input data: slaughter and export–import data are respectively collected by slaughter houses and custom offices. The animals are first classified according to their number of teeth and are then allocated among age classes. Both collection and allocation introduce errors, which justify the input noise.

4) An *output noise* due to the statistical nature of observations. The agricultural statistical services make an annual sample survey from which they estimate the number of cows, females and males. These data constitute the observations and the associated confidence depends on the sample size.

Consequently, the cattle system is modeled by a stochastic state representation where two noise processes $v(n)$ and $w(n)$ are introduced

$$\begin{cases} x(n+1)=A(n)x(n)-u(n)+w(n)\\ s(n)=C(n)x(n)+v(n). \end{cases}$$

The noise process $w(n)$ represents the combination of the input, process, and model noises which occur in the dynamic evolution and its model (state equation), while $v(n)$ takes into account the output noise and the model noise introduced by modeling observations as linear functions of the state vector. The processes $v(n)$ and $w(n)$ can then be considered as independent.

The basic assumptions concern these noises which are supposed to be white, with zero means, i.e., no bias is allowed on statistical data. It must be kept in mind that the derived results will only be valid if these assumptions are verified. The processes $v(n)$ and $w(n)$ are such that

$$E[v(n)]=0$$
$$E[w(n)]=0$$
$$E[v(n) \quad v^T(n')]=Q\,\delta_{nn'}$$
$$E[w(n) \quad w^T(n')]=R\,\delta_{nn'}$$
$$E[v(n) \quad w^T(n')]=0$$

where $E[\cdot]$ denotes the expectation and $\delta_{nn'}$ the Kronecker delta function.

With these assumptions, and knowledge of the system matrices Kalman filtering can be applied. It yields an unbiased and minimum variance estimator of the state vector $x(n)$.

IV. Identification of the System

Besides the stochastic nature of the system there is another reason which justifies the use of a Kalman filter: the system is unstable; indeed any free population is naturally unstable.

The problem is then to identify a system modeled by a stochastic state representation, given knowledge of noise covariances. A parameter vector θ can be defined; its components are zootechnical parameters such as sex ratio, mortality, and fertility rates, average milk production, technical gain,.... The α_i, β_i, γ_i, δ_i, μ_i, ξ_i, and ϵ_i coefficients introduced in Section II can easily be modeled as functions of θ (see the Appendix), and consequently, the A and C matrices are written as function of θ and n

$$\begin{cases} x(n+1)=A(\theta,n)x(n)-u(n)+w(n) \\ s(n)=C(\theta,n)x(n)+v(n). \end{cases}$$

The identification problem is then to find the optimal parameter vector $\hat{\theta}$. It has been seen in Section III that w and v are much more than input and observation noises. Nevertheless, they appear inside the state representation in the same way as if they were just input and observation noises.

Several solutions, derived from the prediction error identification method proposed by Mehra [9], [10], have been proposed [2], [3], [14]. Two classes of approaches have been made: one assumes process and observation noise [9], [10], while the other assumes a deterministic system with noise on the input and output observations [3]. Actually, they are all based on the conversion of the model into the innovations form [1].

Formulation of the maximum likehood estimation problem yields the following criterion D to be minimized with respect to the parameter vector θ, the input sequence U^*, and the initial state $x(0)$:

$$D(\theta,U^*,x(0))=(x(0)-x_0)^T P_0^{-1}(x(0)-x_0) + \sum_n (s_n-s_n^*)^T Q^{-1}(s_n-s_n^*) + \sum_n (u_n-u_n^*)R^{-1}(u_n-u_n^*)$$

where x, P_0 are the given initial state and its covariance matrix

$$s_n^*=s_n-v_n$$
$$u_n^*=u_n-w_n$$
$$U^*=\{u_0^*,u_1^*,\cdots,u_n^*\}.$$

For a given parameter vector θ, the minimization of $(\theta,U^*,x(0))$ with respect to U^* and $x(0)$ is a problem of optimal control with quadratic cost. This can be classically solved via dynamic programming. Jazwinsky [6] has shown the equivalence between dynamic programming and Kalman filtering for solving such a problem.

So the identification problem is solved in the following two phases.

1) A Kalman filter solves the minimization problem with respect to U^* and $x(0)$ for a given θ

$$\mathcal{D}(\theta)=\min_{U^*,x(0)} D(\theta,U^*,x(0)) = \sum_n v_n^T\left(Q+C_nP_{n/n-1}C_n^T\right)^{-1}v_n$$

with

v_n the innovation process, $v_n=s(n)-C_n\hat{x}_{n/n-1}$

$P_{n/n-1}$ the covariance matrix of the predicted value $\hat{x}_{n/n-1}$.

2) Then a classical nonlinear optimization procedure yields the optimum with respect to θ

$$\mathcal{D}(\hat{\theta})=\min_{\theta}\mathcal{D}(\theta).$$

The flexible polyhedron search [5] has been used to perform the identification of θ. For each step a polyhedron is distorted in a direction opposite to the vertex to which corresponds the highest criterion. If constraints on parameters are to be introduced, penalization methods can be used or variables can be changed, for instance, θ_i is replaced by θ_i' defined by $\theta_i=C_i-e^{-\theta_i'}$ to take into account the constraint $\theta_i<C_i$.

V. Implementation of the Method

The previous methodology has been implemented on a digital minicomputer under the following conditions:

1) simulation time horizon: 1969–1981,
2) 9 dimensional state vector (five female and four male classes),
3) 9 dimensional observation vector,
4) 12 dimensional parameter vector.

The 1969 year has been chosen as initial condition because a complete survey of the French agriculture performed this year gave an estimate x_O of the state vector.

The main problem met when implementing the method was the estimation of covariance matrices P_O, Q, and R. This estimation has been performed as follows.

1) An approximate size of the standard deviation on each class is generally known, thus, diagonal terms can be estimated.

2) Off-diagonal terms are introduced with regard to the definition of variables. When animals are wrongly distrib-

uted into classes, they should belong either to the previous or to the following class. Therefore, $Y(i)$ is only correlated with $Y(i-1)$ and $Y(i+1)$ so that the covariance matrices are mainly tridiagonal. The correlation coefficients are estimated and are generally negative; the absence of an animal in its actual class is balanced by its presence in the preceding or following class.

3) After having estimated the covariance matrices, it must be checked that they actually are positive definite.

Consequently, matrices P_O and R have the following form which also takes into account the fact that male and female calves can be misidentified at the statistical level.

$$P_O, R \rightarrow \left[\begin{array}{ccccc|cccc}
V_{F1} & -C_{F12} & & & & -C_{FM1} & & & \\
-C_{F12} & V_{F2} & -C_{F23} & & & & & & \\
 & -C_{F23} & V_{F3} & -C_{F34} & & & & & \\
 & & -C_{F34} & V_{F4} & -C_{F45} & & & & \\
 & & & -C_{F45} & V_{F5} & & & & \\ \hline
-C_{FM1} & & & & & V_{M1} & -C_{M12} & & \\
 & & & & & -C_{M12} & V_{M2} & -C_{M23} & \\
 & & & & & & -C_{M23} & V_{M3} & -C_{M34} \\
 & & & & & & & -C_{M34} & V_{M4}
\end{array}\right].$$

With regard to the observation covariance matrix Q it must be noted that populations, milk production, and dead animals are independent series. Moreover the correlation between cows and females over two years is positive; it is null between cows and females under two years which are not yet apt for reproduction. (Females are called "cows" when they give birth to a calf.) The form of matrix Q is then

$$Q \rightarrow \left[\begin{array}{ccc|ccc|ccc}
V_{F1} & -C_{F12} & & & & & & & \\
-C_{F12} & V_{F2} & -C_{F23} & & & & & & \\
 & -C_{F23} & V_{F3} & & & & C_{FC} & & \\ \hline
 & & & V_{M1} & -C_{M12} & & & & \\
 & & & -C_{M12} & V_{M2} & -C_{M23} & & & \\
 & & & & -C_{M23} & V_{M3} & & & \\ \hline
 & & C_{FC} & & & & V_{\text{cow}} & & \\
 & & & & & & & V_{\text{milk}} & \\
 & & & & & & & & V_{\text{dead}}
\end{array}\right].$$

Estimation of the respective variance terms (Vx_i) has been based on

1) estimates of input data errors,

2) information concerning the sample size of the survey,

3) estimates of the standard deviations of the stochastic, processes mortality, fertility,..., .

With this estimation of the covariance matrices and the initialization of the state vector, the Kalman filter can be implemented. At each step the inversion of a 9×9 matrix is needed. As the covariance matrices are symmetric positive definite, the Choleski algorithm has been used. The prediction and filtering equations can thus be programmed and are recalled for notational purposes:

$$\text{prediction}\begin{cases} \hat{x}_{n+1/n} = A(n)\hat{x}_{n/n} - u(n) \\ P_{n+1/n} = A(n)P_{n/n}A^T(n) + R \end{cases}$$

$$\text{filtering}\begin{cases} \hat{x}_{n+1/n+1} = \hat{x}_{n+1/n} + K_{n+1}\big(s(n+1) - C(n+1)\hat{x}_{n+1/n}\big) \\ K_{n+1} = P_{n+1/n}C^T(n+1)\big[Q + C(n+1)P_{n+1/n}C^T(n+1)\big]^{-1} \\ P_{n+1/n+1} = \big[I - K_{n+1}C(n+1)\big]P_{n+1/n}. \end{cases}$$

In the identification scheme, Kalman filtering plays the role of simulation algorithm for an unstable system. After some attempts, the parameters have been constrained to preserve their zootechnical meaning. It is indeed important for the structure to remain as comprehensive as possible to ensure the quality of predictions. For instance, some parameters represent mortality rates; negative values cannot be accepted when identification is performed and these parameters have been constrained to be positive. The elements of the parameter vector θ and their relationships with the coefficients of the system matrices are listed in the Appendix.

TABLE I
EXAMPLE OF IDENTIFIED PARAMETERS

Parameter	Identified value	Approximate 95% Confidence Interval
Sex ratio (percentage of females among births)	.4718	[.4654 , .4822]
Mortality rate of over 1 year females	.0166	[.0088 , .0229]
Fertility rate of female 2-3 years	.3482	[.2275 , .4728]

VI. RESULTS AND APPLICATIONS

The first results concern the estimation of the model parameters; they are given in the Appendix. The identification scheme gives estimates of parameters which are unknown in spite of their zootechnical meaning. A confidence interval has also been attached to each parameter; a random search algorithm computes the variations of θ when the optimal criterion $(\hat{\theta})$ is increased as follows:

$$\mathcal{D}(\theta)=\mathcal{D}(\hat{\theta})+\chi^2_{0.05}(q) \qquad \text{where } q=\dim\theta.$$

The assumption that $\mathcal{D}(\theta_0)-\mathcal{D}(\hat{\theta})$ is χ^2 distributed [where θ_0 is the true value of (θ)] is based on the innovation property (independent Gaussian) of a Kalman filter [7]. In fact, it is only approximately true in the case of a model nonlinear in the parameters.

Some examples of these results are given in Table I.

It is obvious that parameters are not equally sensitive. Sex ratio and mortality rates are rather precisely determined while a fertility rate can vary on a rather large interval because it is then balanced by the other fertility rates.

After the parameter identification, the Kalman filter can be run and it provides estimates of the state variables and covariance matrices. The problem is then to interpret these matrices. A rigorous approach would be to diagonalize the covariance matrices and express confidence intervals in the eigenvector basis. But these intervals would have no physical meaning, as a linear combination of age classes does not represent anything. Therefore as diagonal terms are largely greater than the other ones, approximate confidence intervals are computed by neglecting off-diagonal terms. In the following "2σ" intervals are thus built which represent 95 percent confidence intervals under Gaussian assumptions.

Over the time horizon, results can be given concerning either state variables or observations. Fig. 1 presents the number of males 1–2 years which is an observed state variable. Consequently, the observed value can be compared to the filtered estimate.

It appears that the estimation is always near the observation in spite of the large variations of this output. Besides, the computed confidence interval is obviously smaller than the interval given for the observation.

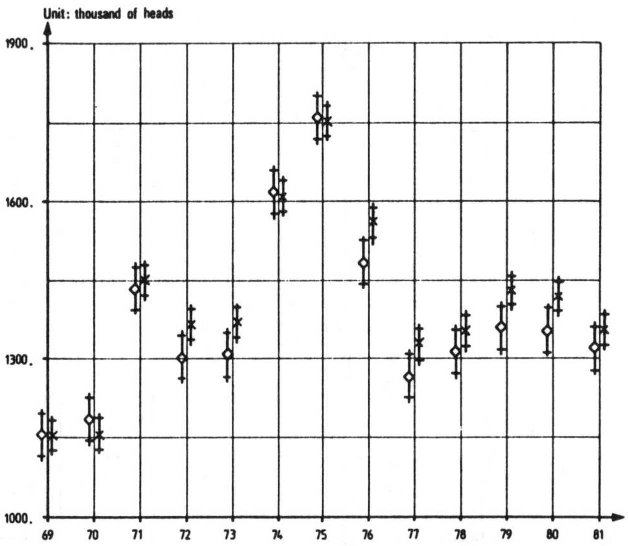

Fig. 1. Number of males 1–2 years (example of an observed state variable). Comparison between the observed value (◇) and the *a posteriori* estimate (×).

TABLE II
NUMBER OF FEMALES 2–3 YEARS (*INITIAL VALUE). COMPARISON BETWEEN PREDICTED AND FILTERED VALUED

Year	Predicted Value	Standard Deviation	Filtered Value	Standard Deviation	Predicted Filtered
1969	2208.9*	33.2*	2206.9	32.7	2.0
1970	2268.8	28.3	2267.3	27.9	1.5
1971	2237.8	24.9	2234.9	24.6	2.9
1972	2292.5	15.5	2296.8	14.5	-4.3
1973	2301.3	14.0	2306.6	13.1	5.3
1974	2341.0	13.4	2349.7	12.4	-8.7
1975	2485.0	13.2	2496.1	12.2	-11.1
1976	2594.3	12.9	2592.2	11.9	2.1
1977	2622.5	12.8	2626.5	11.9	-4.0
1978	2599.0	12.8	2597.6	11.9	1.4
1979	2602.2	12.8	2594.0	11.9	8.2
1980	2548.8	12.8	2594.3	11.9	-0.5
1981	2671.2	12.8	2673.0	11.9	-1.8

The results concerning a state variable which is not directly observed are given in Table II. It permits the comparison between predicted and filtered values for the number of females 2–3 years. The values are not very different, but Kalman filtering prevents the divergence resulting from the addition of these errors.

It is then interesting to look at an observation which is not a state variable. The milk production is given in Fig. 2. It is possible to compute an *a posteriori* estimate from the filtered estimate of the state vector

$$\hat{s}(n)=C(\hat{\theta},n)\hat{x}_{n/n}.$$

The deviation which occurred in 1980 is explained by favorable weather effects on milk production. The other important deviation (1972) is due to an integration error; populations per age are given on January 1, while milk production is summed up on the whole year. Therefore, milk production augmentation due to an increasing num-

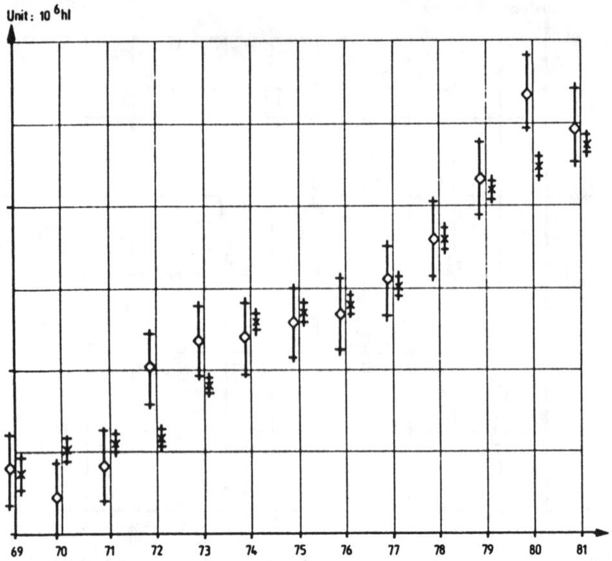

Fig. 2. Milk production—comparison between the observed values (◇) and the *a posteriori* estimate (×).

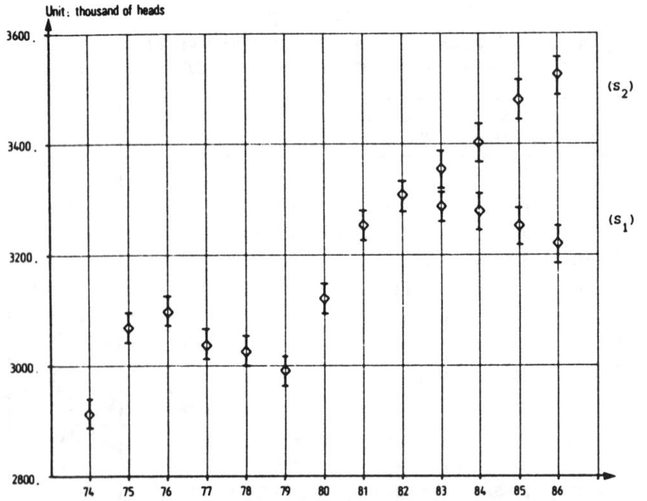

Fig. 3. Number of females 1–2 years—comparison between two slaughter scenarios and the associated predictions.

ber of cows over year 1972 appears on the 1972 observation and on the 1973 *a posteriori* estimate. The model could be improved by summing up milk production between July $(n-1)$ and June (n), but it is not done as predictions have to be given on the whole calendar year.

The three previous examples have shown the simulation aspects of the method on the observed time horizon. Short term and medium term predictions are computed by defining slaughter scenarios and using prediction equations. An example is given in Fig. 3 where the effects of two slaughter scenarios are compared; the slaughter vector is increased (scenario S_1) or decreased (scenario S_2) by 1 percent each year. It enables decision makers to choose the most appropriate slaughter policy for the future.

VII. Conclusion

This paper has presented an application of Kalman filtering in the agroeconomic field. The techniques used for demographic models (i.e., characterization, simulation, and identification methods) have first been used and developed for technical applications. The difficulties met in the agroeconomic applications are mainly related to observations. The classical modeler's problem is to have an observable system, and therefore he generally makes the choice of the observations and their respective sensing devices. In the agroeconomic field he has very few available observations, and thus is limited in the choice of a model because of the observability constraint. Besides, observations are noisy (measured through sample surveys) and only available in the past while in classical Kalman filtering applications, real-time information processing is performed.

A methodology has been developed and applied to the French cattle herd. Kalman filtering enables the simulation of the system as a stochastic one and the identification of an unstable linear system. Implementation of this method has provided results which improve our knowledge of the French cattle herd (identified parameters) as well as the confidence on male and female populations per age classes, over the time horizon. Moreover, predictions are performed through slaughter scenarios in order to help decision makers to control the cattle herd. This implementation has given satisfactory results compared to the simulation and identification of the associated deterministic model.

A similar application is being implemented: a model for pork livestock. The porcine livestock presents some peculiarities concerning both its structure and statistics which require modification and adaptation of the general structure of demographic model which have been presented. Moreover, after implementation of the method, statistical tests of whiteness of the innovations have revealed that some observation data were biased. A method, proposed by Friedland [4], for estimating both the bias and the state vector has been implemented and has brought a solution to this problem.

The present methodology is indeed general and could be used in every field of human sciences in so far as the dynamic evolution of a system can be modeled by a state equation.

Appendix
The Parameter Vector θ

The 12 elements of the identified parameter vector θ are as follows:

$\theta(1)$	sex ratio (percentage of females among births)
$1-\theta(2)$	mortality rate of female calves
$1-\theta(3)$	mortality rate of grown-up females
$1-\theta(4)$	mortality rate of male calves
$1-\theta(5)$	mortality rate of grown-up males
$\theta(6)$	τ (coefficient which takes into account the uncontrolled slaughters)
$\theta(7)$	fertility rate of female 2–3 years
$\theta(8)$	fertility rate of female 3–4 years
$\theta(9)$	fertility rate of female over 4 years
$\theta(10)$	average milk production per cow

TABLE III

i	$\hat{\theta}(i)$	i	$\hat{\theta}(i)$
1	.4718	7	.3482
2	.9477	8	.9850
3	.9834	9	.9198
4	.9268	10	.1567
5	.9898	11	.0023
6	.0013	12	.6837

$\theta(11)$ ρ (technical gain on milk production)

$\theta(12)$ percentage of declared natural deaths.

These elements are related as follows to the coefficients introduced in Section II.

1) The mortality rates are assumed to be constant for grown-up animals, but distinct between males and females. Then

$$\beta_i = \theta(3) \qquad i = 1, \cdots, 5$$
$$\delta_i = \theta(5) \qquad i = 1, \cdots, 4.$$

2) Females less than 2 years are assumed not yet to be cows. Then

$$\alpha_1 = \alpha_2 = 0 \quad \gamma_1 = \gamma_2 = 0$$
$$\xi_1 = \xi_2 = 0 \quad \mu_1 = \mu_2 = 0$$
$$\alpha_i(n) = \theta(i+4)\cdot\theta(1)\cdot\theta(2)(1-\theta(6)n) \qquad i = 3,4,5$$
$$\gamma_i(n) = \theta(i+4)\cdot(1-\theta(1))\cdot\theta(4)(1-\theta(6)n) \qquad i = 3,4,5$$
$$\mu_3 = \theta(7) \quad \mu_4 = \mu_5 = 1$$
$$\xi_i(n) = \mu_i\cdot\theta(10)\cdot(1+\theta(11)n) \qquad i = 3,4,5.$$
$$\epsilon_F(i) = 1-\theta(3)$$
$$\epsilon_M(i) = 1-\theta(5)$$
$$\epsilon_F(i) = \theta(12)[\theta(i+4)\{\theta(1)\cdot(1-\theta(2)) + (1-\theta(1))\cdot(1-\theta(4))\}+1-\theta(2)] \qquad i = 3,4,5.$$

In Table III the estimates obtained from the identification of $\boldsymbol{\theta}$ are given.

The parameters $\theta(5)$, $\theta(8)$, and $\theta(9)$ had been constrained to remain less than 1.

These estimates are generally very close to the values, commonly recognized by zootechnicians, which have been used for initializing the identification algorithm. In some cases, *a priori* ideas were slightly modified; for example the sex ratio was usually assumed to be 48 percent and it was revealed to be closer to 47 percent.

References

[1] B. D. O. Anderson and J. B. Moore, *Optimal Filtering.* Englewood Cliffs, NJ: Prentice-Hall, 1979.

[2] P. E. Caines, "Prediction error identification methods for stationary stochastic processes," *IEEE Trans. Automat. Contr.*, vol. AC-21, p. 500, Aug. 1976.

[3] P. Caspi, "Sur l'identification des systèmes linéaires déterministes instables," Ph.D. dissertation, Université de Grenoble, 1978.

[4] B. Friedland, "Treatment of bias in recursive filtering," *IEEE Trans. Automat. Contr.*, vol. AC-14, p. 359, Aug. 1969.

[5] D. Himmelblau, *Applied Non-Linear Programming.* New York: McGraw-Hill, 1972.

[6] A. H. Jazwinski, *Stochastic Processes and Filtering Theory.* New York: Academic, 1970.

[7] T. Kailath, "An innovations approach to least-squares estimation—Part I: Linear filtering in additive white noise," *IEEE Trans. Automat. Contr.*, vol. AC-13, Dec. 1968.

[8] D. G. Luenberger, *Introduction to Dynamic Systems: Theory, Models and Applications.* New York: Wiley, 1979.

[9] R. K. Mehra, "Identification of stochastic linear dynamic systems using Kalman filter representation," *AIAA J.*, vol. 9, Jan. 1971.

[10] R. K. Mehra, "On-line identification of linear dynamic systems with applications to Kalman filtering," *IEEE Trans. Automat. Control*, vol. AC-16, Feb. 1971.

[11] A. Rault and B. Leibundgut, "Demographic models of French livestocks," in *Proc. 1st Working Conf. on Computer Applications in Food Production and Agricultural Eng.—IFIP*, Cuba, 1981.

[12] R. Rouhani and E. Tse, "Structural design for classes of positive linear systems," *IEEE Trans. Syst., Man, Cybern.*, vol. SMC-11, Feb. 1981.

[13] D. Smith and M. Keyfitz, *Mathematical Demography.* Berlin: Springer-Verlag, 1977.

[14] T. Söderström, "Identification of stochastic linear systems in presence of input noise," *Automatica*, vol. 17, p. 713, 1981.

Author Index

Subject Index

G

H

I

J

K

L

T

U

V

W

Editor's Biography

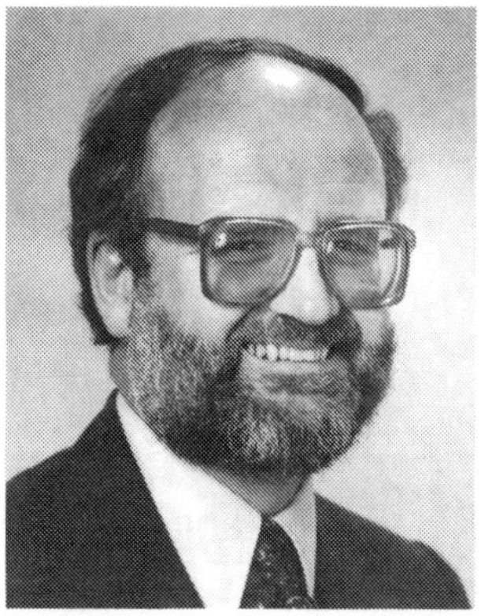

Harold W. Sorenson (M'65–SM'73–F'78) is a native of Omaha, NE. He received the B.S. degree in aeronautical engineering from Iowa State University of Science and Technology, Ames, in 1957, and the M.S. and Ph.D. degrees in engineering from U.C.L.A. in 1963 and 1966, respectively.

He worked at General Dynamics/Astronautics in San Diego, CA, from 1957 to 1962 and at AC Electronics Division of General Motors in El Segundo, CA, from 1963 to 1966. After a year as a Guest Scientist at the Institute for Guidance and Control in Oberpfaffenhofen, West Germany, he joined the Faculty at the University of California, San Diego, in 1968, and is currently Professor of Engineering Sciences in the Department of Applied Mechanics and Engineering Sciences. Sorenson is a co-founder of Orincon Corporation in La Jolla, CA, and served as President from 1973 to 1981. In July 1985, he became Chief Scientist of the United States Air Force in Washington, DC.

Dr. Sorenson has published over seventy papers dealing with Kalman filtering, estimation theory, identification, optimization and control of stochastic systems. He is the author of the book *Parameter Estimation: Principles and Problems*, published by Marcel Dekker Publishing in 1980. He has been an Associate Editor for six journals and was the Guest Editor for the Special Issue of the IEEE TRANSACTIONS ON AUTOMATIC CONTROL entitled ''Applications of Kalman Filtering''. Dr. Sorenson is a Fellow of the IEEE and a Past President of the IEEE Control Systems Society. Currently, he is a member of the Board of Directors of the IEEE, serving as the Director of Division X.